Flora of the Pacific Northwest

Flora of the Pacific Northwest

AN ILLUSTRATED MANUAL

By
C. LEO HITCHCOCK
and
ARTHUR CRONQUIST

Illustrations by Jeanne R. Janish

UNIVERSITY OF WASHINGTON PRESS

SEATTLE AND LONDON

University of Washington Press
PO Box 50096
Seattle, WA 98145–5096, U.S.A.
www.washington.edu/uwpress

Library of Congress Cataloging-in-Publication Data
Hitchcock, Charles Leo.
Flora of the Pacific Northwest.
Condensation of Vascular plants of the Pacific Northwest
by C. L. Hitchcock, and others
1. Botany—Northwest, Pacific.   I. Cronquist, Arthur.   II. Title.
QK144.H452   1973   581.9'795   72-13150
ISBN 0-295-95273-3

The paper used in this publication meets the minimum requirements of
American National Standard for Information Sciences—Permanence of
Paper for Printed Library Materials, ANSI Z39.48-1984. ⊗

Contents

Introduction

THIS manual is essentially a condensation of the five-volume work, *Vascular Plants of the Pacific Northwest*, prepared by C. Leo Hitchcock, Arthur Cronquist, Marion Ownbey, and J. W. Thompson, and published serially from 1955 to 1969 by the University of Washington Press as Volume 17 of the University of Washington Publications in Biology. The sequence of families is the same in the two works except for the transposition of the monocotyledons and dicotyledons; the dicotyledons are in the traditional Englerian sequence, and the monocotyledons are arranged according to the system of Cronquist as presented in *The Evolution and Classification of Flowering Plants* (Boston: Houghton Mifflin, 1968). The genera within each family in this manual are alphabetically arranged, as in the earlier work, but for convenience in presentation it has been necessary to number the species. The sequence of species within a genus has no phylogenetic or systematic implications of its own; it merely reflects the way in which it was convenient to organize the key.

For the purposes of this manual, as in the earlier work, the Pacific Northwest is considered to include all of the state of Washington, the northern half of Oregon (north of approximately the 44th parallel), Idaho north of the Snake River Plains, the mountainous part of Montana, and an indefinite fringe of southern British Columbia. The southern boundary is drawn with the intent of excluding as much as possible of the Klamath and Great Basin elements that become so important to the south of our range. The eastern boundary is intended to exclude most of the Great Plains flora, but the more easterly intermontane valleys of Montana are continuous with the plains, and a considerable number of essentially Great Plains species do enter our range. Relatively few species closely approach our boundary from the north.

In this manual the manuscripts for the vascular cryptogams, the genus *Rosa*, the families Cyperaceae, Salicaceae, and Umbelliferae, and the families in sequence from Polemoniaceae through Compositae are the responsibility of Cronquist, except *Castilleja*, by Noel H. Holmgren. All other groups have been done by Hitchcock, except *Eriogonum*, by James Reveal; *Thlaspi*, by Patricia Kern Holmgren; *Delphinium*, by David Sutherland; and *Panicum*, by Richard W. Spellenberg. The artificial key to the families has been prepared by Hitchcock, and the more synoptical one by Cronquist. The glossary from the earlier work has been pared by Hitchcock in order to save some space. It is a pleasure for Hitchcock to acknowledge the assistance received from Patricia K. Holmgren, who typed the manuscript for his portion of the manual, and from his wife, Evelyn, and from Dr. John Hill, both of whom have shared the joy of proofreading the entire volume.

In the preparation of the manual it has seemed advisable to take account of taxonomic work done since the publication of the various parts of the five-volume flora on which it is based. The correspondence between the two works is therefore not always exact. In order to save space, the synonymy is mostly restricted to names that might be encountered by a student using another flora. More complete synonymy can be found in the larger work.

The concept of species and other taxa in this manual is conservative, both from the standpoint of the number of groups recognized, and from the standpoint of following historical practice in doubtful cases. A fuller exposition of the taxonomic philosophy is given in the introduction to Part 5 of the longer flora.

We trust that the novel format meets the need to present the most useful information in the least space, so that the book can be truly a field manual.

Most of the illustrations have been selected from those in the five volumes of *Vascular Plants of the Pacific Northwest*, which were done by Jeanne R. Janish, except for the family Compositae, done by John H. Rumely. Additional drawings for this manual, consisting mainly of longitudinal sections of flowers, have all been done by Mrs. Janish.

In order to use the limited marginal space available as effectively as possible, most drawings have been reduced to half their size as reproduced in *Vascular Plants of the Pacific Northwest*. This has resulted in an apparent loss of detail. It is therefore suggested that the illustrations be viewed with the aid of a reading glass for at least partial restoration of finer detail. Degree of magnification (or reduction) is not believed to be essential for interpretation of the illustrations, and indication of relative size has been omitted throughout for space economy.

In most instances the drawings are intended to portray certain salient features of particular species, but occasionally they are used to clarify the meaning of technical terms or concepts emphasized in the key. For this reason each drawing, or set of drawings, has been numbered to correspond to leads in the key, rather than to species' numbers.

Although botanists have reached international agreement that there should be one, and only one, valid and correct scientific name for each plant species, that being a Latin binomial, they have taken no formal action concerning vernacular names. Consequently, such nonscientific, so-called "common" names may be lacking for some species but existent, sometimes in multiplicity, for others, often varying from region to region and country to country.

Most manuals such as this one include common names mainly for genera (this sometimes being the generic name itself, e.g., *Aster, Rhododendron, Chrysanthemum, Delphinium*) and for certain well-known, generally widespread, often weedy species. A few list English names for all species, many of which cannot properly be classed as common, often being seldom used, mostly mere English translations of the Latin name. This procedure is followed consistently in at least three works dealing with our area, namely, Abrams, *Illustrated Flora of the Pacific States;* Peck, *Manual of the Higher Plants of Oregon;* and *Northwest Range-Plant Symbols* (USFS Research Paper PNW-40).

Although both authors of this manual dislike the idea of listing coined common names, the senior member feels that he can correctly gauge the nomenclatural desire of the average amateur botanist who (hopefully) will use it, and that space will not be wasted by listing such translations. He is responsible for the insertion of at least one genuine or contrived "common" name for almost every species recognized. However, mere Latin translations, used as common names, are often not included in the Index, especially specific names commemorating persons; e.g., "Hood's phlox" (for *Phlox hoodii* Rich.) appears in the text, but is not included in the Index.

Abbreviations and Signs Used

Provinces of Canada and states of the United States are abbreviated as customary, with the few exceptions shown (Cal, Ore). Unless otherwise indicated, measurements refer to length or height, or to length times width. "Pls 3–6 dm" means that the plants are 3–6 dm tall, "lvs 4–8 cm" that the leaves are 4–8 cm long, and "lvs 4–8 × 1–2 cm" that the leaves are 4–8 cm long and 1–2 cm wide.

adj—adjacent
alp—alpine
alt—alternate(ly)
ann—annual
apet—apetalous
app—approached (by)
   approaching
art—articulate(d)
   articulation
Atl—Atlantic (coast)
aur—auricle
   auriculate
ave—average (ing)

bien—biennial
bl(s)—blade(s)
br(s)—branch(es)
   branched
brlet(s)—branchlet(s)

c—central
ca—about
Cal—California
camp—campanulate
caps—capsule (ar)
Cas—Cascade(s)
cm—centimeter(s)
co(s)—county (ies)
conif—conifer(ous)
Cont Div—Continental
    Divide
CR—Columbia River
CRG—CR Gorge
CRV—CR Valley
cult—cultivated

Dak(s)—Dakota(s)
des—desert
diam—diameter
disart—disarticulate,
   ed, ing
dm—decimeter

e—east (ern, ward)
E Cas—e of the Cas crests
elev—elevation
epig—epigynous
estab—establish(ed)

f.—form(s)
fl(s)—flower(s)
   flowering
   floral
   floriferous
-fld—flowered
flt—floret
for—forest(ed)
fr(s)—fruit(s)
funnelf—funnelform

gamopet—gamopetalous
gamosep—gamosepalous
GB—Great Basin
gen—general(ly)
GL—Great Lakes
glab—glabrous
GNP—Glacier Nat Park
GP reg—Glacier Park region
Gr—Greek
Gr Pl—Great Plains
grassl—grassland(s)

Hem—Hemisphere
hort—horticulture(al)
hypan—hypanthium
hypog—hypogynous

imp—important
infl—inflorescence(s)
inl—inland
intro—introduce(d)
invol(s)—involucre(s)
   involucrate
   involucral
irreg—irregular(ly)
is—islands

L—Latin
lig(s)—ligule(s)
   ligulate
lf—leaf
lft(s)—leaflet(s)
lfy—leafy
lod(s)—lodicule(s)
lowl—lowland(s)
lvs—leaves

m—meter(s)
mainl—mainland
meadowl—meadowland
Mex—Mexico (an)
mm—millimeter(s)
mont—montane
mt(s)—mountain(s)

n—north (ern, ward)
NM—New Mexico
n Mex—northern Mex
no—number

occ—occasional(ly)
OM—Olympic Mts
OP—Olympic Peninsula
opp—opposite(ly)
Ore—Oregon
orn—ornamental

Pac—Pacific (coast)
pan(s)—panicle(s)
   paniculate
ped(s)—pedicel(s)
   pedicellate
per—perennial(s)
peri—perigynium
perig—perigynous
pl(s)—plant(s)
polypet—polypetalous
polysep—polysepalous
prob—probably
PS—Puget Sound (region)
PT—Puget Trough

recep—receptacle
   receptacular
reg—regular(ly)
RM—Rocky Mts
RMS—RM States

s—south (ern, ward)
sagebr—sagebrush
salverf—salverform
scab—scabrous

ix

segm—segment(s)
sp., spp.—species
  (singular and plural)
SR—Snake River
SRC—SR Canyon
SRP—SR Plain
SRV—SR Valley
ssp., sspp.—subspecies
  (singular and plural)
st(s)—stem(s)

stip(s)—stipule(s),
  stipular, stipulate

temp—temperate
tidel—tideland(s)
timberl—timber line
treel—tree line

var., vars.—variety (ies)

VI—Vancouver Island
vic—vicinity

w—west (ern, ward)
wastel—wasteland
W Cas—w of the Cas crests
Wen Mts—Wenatchee Mts
woodl—woodland
W V—Willamette Valley

YNP—Yellowstone National Park

Signs

=   as much as, as many as, equal (to), equalling, equally, equaled (by)

<   less than, fewer than, shorter than, under

>   more than, longer than, greater (than), over

∝   numerous, many

± more or less

⚲—fls perfect

⚲-fld—perfect-fld

♂, ♀—pl dioecious

♂♀—pl monoecious

♂—staminate

♀—pistillate

⚲♂♀—pl polygamo-monoecious

⚲♂, ⚲♀—pl polgyamo-dioecious

×—times (magnification), also (before sp. name) indicates a hybrid

Abbreviations other than the above will be found in the listing of common names and the citation of synonyms (duplicate, nonvalid scientific names) when often only the initial letter is used to indicate generic names, specific epithets, and common names, if the proper antecedents are believed to be apparent.

Glossary

abaxial: On the side away from the axis.

abortive: Hardly or imperfectly developed.

acaulescent: Literally, without a st.

accrescent: Increasing in size with age, especially with reference to calyx.

acerose: Needle-shaped, or with a needle-like tip.

achene, akene: The most generalized type of dry, indehiscent fr.

acicular: Needle-shaped.

acropetal: Proceeding from the proximal to the distal end.

acroscopic: Toward the distal (as opposed to the proximal) end.

actinomorphic: Radially symmetrical, as in a reg fl.

acuminate: Gradually and concavely tapered to a narrow tip or sharp point.

acute: Sharp pointed.

adaxial: On the side toward the axis, or turned toward the axis.

adnate: Grown together, or attached; applied only to unlike parts.

adventitious: Not in the usual place.

adventive: Intro but not naturalized, or only locally estab.

alate: Winged.

allopatric: Occupying different geographical regions.

alternate: Situated singly at each node, as the lvs; situated reg between organs of another kind, as stamens alt with petals.

alveolate: Honeycombed.

ament: A dense, bracteate spike or raceme with a nonfleshy axis bearing ∝ small, naked or apet fls; a catkin.

amplexicaul: Clasping the st.

androecium: The stamens considered collectively.

androgynous: A spike (in *Carex*) with both ♂ and ♀ fls, the ♂ above the ♀.

anemophilous: Pollinated by wind.

angiosperm: One of the group of plants (Magnoliophyta, fl pls) characterized by having ovules enclosed in an ovary.

annual: A pl which germinates, fls, and seeds in a single season.

annulus: A little ring; the partial or complete ring of thick-walled cells encircling the sporangium of typical ferns.

anterior: In front.

anther: The part of the stamen which bears the pollen.

anthesis: The period during which a fl is fully expanded.

anthocyanin: A chemical class of water-soluble pigments, ranging in color from blue or violet through purple to crimson.

antrorse: Directed forward or upward.

apetalous: Without petals.

aphyllopodic: With the lowest lvs reduced to scales, the 1st foliage lvs thus well above the base of the pl, as in some spp. of *Carex.*

aphyllous: Without lvs or without normal foliage lvs.

apiculate: Ending abruptly in a small, gen sharp point.

apocarpous: With carpels free from each other, or with only 1 carpel.

appressed: Pressed flat against another organ.

approximate: Near together.

arachnoid: Provided with a cobwebby, gen sparse pubescence of relatively long, soft, tangled hairs.

arborescent: Treelike.

arcuate: Curved into an arc.

areola, areole: A small, clearly bounded area on a surface.

argenteous: Silvery in color.

aril: A specialized outgrowth from the seed stalk which covers or is attached to the mature seed.

arillate: Provided with an aril.

aristate: Tipped with an awn or bristle.

articulate(d): Jointed; with a predetermined point of natural separation.

ascending: Growing obliquely upward (sts); directed obliquely forward in respect to the organ to which they are attached (parts of pl).

asepalous: Without sepals.

atomate, atomiferous: With scattered sessile or subsessile glands.

atro-: L prefix meaning dark or blackish.

attenuate: Tapering gradually to a very slender tip.

auricle: A small projecting lobe or appendage, gen at base of an organ.

autotrophic: Nutritionally independent.

awn: Slender, gen terminal bristle.

axil: The point of the angle formed by the lf or petiole with the st.

axile placenta: A placenta along the axis of an ovary with 2—more locules.

axillary: Located in or arising from an axil.

baccate: Berrylike.

banner: The upper, gen enlarged petal of a papilionaceous fl.

barbed: Bearing short, firm, retrorse points.

barbellate: Diminutive of barbed.

basifixed: Attached by the base.

basiscopic: Toward the basal or proximal (as opposed to distal) end.

beak: A prolonged, ± slender tip on a thicker organ such as a fr or seed.

berry: A fleshy fr, developed from a single pistil, fleshy throughout.

bi-: L prefix, meaning two.

biennial: A pl which completes its life history and dies in 2 years.

bilabiate: Two-lipped.

bipinnate: Twice pinnate, the pinnae again pinnate.

blade: The expanded part of lf or petal.

bloom: A waxy powder covering a surface.

bract: A specialized lf from the axil of which a fl arises; any ± reduced or modified lf associated with an infl; in conifers, one of the primary appendages of the cone axis, in the axils of which the ovuliferous scales are borne.

bracteole, bractlet: Diminutive of bract.

bud: An undeveloped lfy shoot or fl.

bulb: A short, vertical, underground shoot which has modified lvs or thickened lf bases developed as food storage organs.

bulbil, bulblet: Diminutive of bulb.

bullate: Covered with rounded projections resembling unbroken blisters.

caducous: Falling off very early.

caespitose: Growing in dense, low tufts.

calcarate: Spurred.

calcareous: Limy, or rich in calcium carbonate.

callus: A firm thickening; the firm, thickened base of the lemma in many grasses.

calyculate: Provided with a set of small bracts around the base of an invol.

calyptrate: With a caplike covering.

calyx: All the sepals of a fl, collectively.

companulate: Bell shaped.

canescent: Pale or gray, because of a fine, close, whitish pubescence.

capillary: With the form of a hair.

capitate: Headlike, or in a head.

capsule: A dry, dehiscent fr of > 1 carpel.

carina: A keel.

carinate: Keeled.

carpel: The fertile lf (megasporophyll) of an angiosperm, which bears the ovules; a pistil is composed of 1–more carpels.

carpophore: The part of the recep which in some kinds of fls is prolonged between the carpels as a c axis.

caruncle: An excrescence or appendage near the hilum of certain seeds.

caryopsis: The fr of grasses, typically differing from an achene in having the seed coat adnate to the pericarp.

castaneous: Chestnut-colored; dark reddish-brown.

catkin: See ament.

caudate: With a tail-like terminal appendage.

caudex: A short, ± vertical, often woody, persistent st at or just beneath the ground surface.

caulescent: With an obvious lfy st.

cauline: Of or pertaining to the st.

cell: As used in taxomony, the locule of an ovary or theca of an anther.

chaff: Thin dry scales.

channeled: Marked with 1–more deep longitudinal grooves.

chartaceous: Papery in texture.

chlorophyll: The characteristic green pigment of pls.

ciliate: With a fringe of marginal hairs.

ciliolate: Diminutive of ciliate.

cinereous: Ashy in color, gen because of short hairs.

circinate: Coiled from the tip downward, the apex forming the center.

circumboreal: Occurring all the way around the north pole.

circumscissile: Dehiscing by an encircling transverse line.

cismontane: This side of the mts, gen interpreted to mean w of the Cascade-Sierra crests.

cladophyll: A br st which has the form and function of a lf.

clavate: Shaped like a club or baseball bat.

clavellate: Diminutive of clavate.

claw: The narrow, petiole-like base of some sepals and petals.

cleft: Cut ca half way to the midrib or base, or a little deeper; deeply lobed.

cleistogamous flower: One which sets seeds without opening.

coetaneous: With the fls developing at the same time as the lvs.

collar: The outer side of a grass lf at the juncture of the bl and sheath.

collateral: Side by side.

coma: A tuft of (gen long and soft) hairs, esp on a seed.

commissure: The face by which 2 carpels cohere, gen applied only to schizocarps.

comose: With a coma.

complete flower: One with calyx, corolla, androecium, and gynoecium.

compound leaf: One with 2–more distinct lflets.

compound pistil or *ovary:* One composed of > 1 carpel.

concolored, concolorous: Of uniform color.

cone: A cluster of sporophylls or ovuliferous scales on an axis; a strobilus.

connate: Grown together or attached; applied only to like organs.

connective: The tissue connecting the 2 pollen sacs of an anther.

connivent: Converging or coming together, but not organically united.

convolute: Arranged so that each petal or sepal has one edge exposed and the other edge covered.

cordate: Shaped like a stylized heart, the notch at the base.

cordilleran: Relating to or occurring in the system of mt ranges covering much of w N Am.

coriaceous: Leathery in texture.

corm: A short, vertical, underground st that is thickened as a perennating storage organ.

cormous: Bearing corms.

corniculate: Bearing little horns or crests.

corolla: All of the petals of a fl, collectively.

corona: A set of petal-like structures or appendages between the corolla and androecium.

corymb: A simple, racemose infl that is flat-topped or round-topped because the outer peds are progressively > the inner.

corymbose: In a corymb.

costa: A prominent rib or vein; in ferns, the midrib of a pinna or pinnule.

costate: Longitudinally ridged.

cotyledon: A lf of the embryo of a seed.

crenate: Provided with rounded teeth; scalloped.

crenulate: Diminutive of crenate.

crested: Provided with an elevated, often complex appendage or rib on the summit or back.

crisped: Irreg curled or crooked.

crown: The persistent base of an herbaceous per; the lfy top of a tree.

cruciform: Cross-shaped.

cryptogam: A pl belonging to the nonseed-bearing gen group.

cucullate: Hooded or hood-shaped.

culm: The aerial st of a grass or sedge.

cuneate: Wedge-shaped or triangular, the narrow end at the point of attachment.

cusp: An abrupt, sharp, often rigid point.

cuspidate: Tipped with a cusp.

cuticle: The waxy layer covering the epidermis of a lf or st.

cyme: A broad class of infls characterized by having the terminal fl bloom first.

cymose: With the fls in a cyme.

cymule: Diminutive of cyme.

deciduous: Falling after completion of the normal function.

declined: Curved downward.

decompound: Repeatedly compound.

decumbent: With a prostrate or curved base and an erect or ascending tip.

decurrent: With an adnate wing or margin extending down the st or axis below the point of insertion.

decussate: Arranged opp, with each succeeding pair set at right angles to the previous pair.

deflexed: Bent downward.

dehiscent: Opening at maturity.

deltoid: Shaped ± like an equilateral triangle.

dendritic: Branched in treelike fashion, as the hairs of some Cruciferae.

dentate: With spreading, pointed teeth.

denticulate: Diminutive of dentate.

determinate inflorescence: One in which the terminal fl blooms first.

diadelphous: Connate into 2 groups.

dichasial cyme: A cyme which is repeatedly br in dichasial fashion.

dichasium: A 3-flowered cymule in which the development of the terminal fl is followed by that of the 2 opp or subopp lateral fls.

dichotomous: Forking ± reg into 2 branches of ca = size.

didymous: Developing in pairs.

didynamous: With 4 stamens in 2 unequal pairs.

digitate: Same as palmate.

dimidiate: Halved, as if 1 half were wanting.

dimorphic: Of 2 forms.

dioecious: Producing ♀ and ♂ fls on separate pls.

diplostemonous: With 2 cycles of stamens.

disarticulating: Separating at a pre-existing point.

disciform: With the form of a disk.

discoid: Resembling a disk; in the Compositae, with the fls of a head all tubular and ⚥ (or functionally ♂).

disk: An outgrowth from the recep, surrounding the base of the ovary, often derived by reduction of the innermost set of stamens; in the Compositae, the c part of the head, composed of tubular fls.

dissected: Deeply and often repeatedly divided into ∝ smaller or slender parts.

distal: Toward or at the tip or far end.

distichous: In 2 vertical rows or ranks.

distinct: Not connate with similar organs.

divaricate: Widely spreading from the axis.

divided: Cut into distinct parts, as a lf that is cut to the midrib or the base.

dolabriform: Pick-shaped, attached ± toward the middle.

dorsal: Pertaining to or located on the back.

dorsifixed: Attached by the back.

dorsiventral: Flattened, with the 2 flattened sides unlike.

dorsoventral: From the front to the back, as opposed to from one side to another; also used as a synonym of dorsiventral.

drupe: A fleshy fr with a firm endocarp that permanently encloses the gen solitary seed, or with a portion of the endocarp separately enclosing each of 2–more seeds.

drupelet: Diminutive of drupe.

echinate: Provided with prickles.

echinulate: Diminutive of echinate.

ecotype: The individuals of a sp. that are adapted to a particular environment.

edaphic: Pertaining to soil.

ellipsoid: Elliptic in long-section and circular in cross-section.

elliptic: With approximately the shape of a geometrical ellipse.

emarginate: With a small notch at the tip.

endemic: Confined to a particular geographic area.

endocarp: The inner part of the pericarp.

endosperm: Food storage tissue of a seed derived from the triple fusion nucleus of the embryo sac.

ensiform: Sword-shaped.

entire (margin): Not toothed or otherwise cut.

entomophagy: Insect-consuming.

entomophilous: Pollinated by insects.

epappose: Without pappus.

ephemeral: Lasting for only a short time.

epigynous: With the perianth and stamens attached to the top of the ovary.

epipetalous: Attached to the petals or corolla.

epiphyte: A pl without connection to the soil, growing upon another pl, but not deriving its food or water from it.

equitant: Astride, as if riding.

erose: With an irreg margin, as if gnawed.

excurrent: With a continuing c axis, from which lateral brs arise.

exfoliating: Peeling off in layers.

exocarp: The outer layer or part of the pericarp.

exserted: Projecting beyond an envelope, as stamens from a corolla.

exstipulate: Without stips.

extrorse: Turned toward the outside, or facing outward.

falcate: Sickle-shaped; curved like a hawk's beak.

farinose: Covered with a meal-like powder.

fascicle: A close bundle or cluster.

fastigiate: Crowded close together, ± parallel, and gen erect.

fenestrate: With 1–more windowlike openings.

ferruginous: Rust-colored.

filament: The stalk of a stamen.

filiform: Very slender, threadlike.

fimbriate: Fringed.

fistulose (ous): Hollow.

flabellate: Fan-shaped.

flabelliform: Same as flabellate.

flaccid: Weak and lax, hardly if at all capable of supporting its own weight.

fleshy: Thick and juicy.

floccose: Covered with very long, soft, fine, loosely spreading and ± tangled hairs.

floret: A little fl; an individual fl of a definite cluster.

foliaceous: Lfy in texture.

follicle: A dry fr composed of 1 carpel, which at maturity dehisces along the seed-bearing suture only.

fornix: One of a set of small scales or appendages in the tube or throat of the corolla of some kinds of pls, as many borages.

fovea: A little pit.

free: Not adnate to other organs.

free-central placenta: One consisting of a free-standing column or projection from the base of a compound, 1-celled ovary.

fruit: A ripened ovary, together with any other structures which ripen with it and form a unit with it.

fugacious: Fleeting; lasting only a short time.

fulvous: Tawny, dull yellow.

fuscous: Gray-brown.

fusiform: Spindle-shaped.

galea: The strongly concave or helmet-like upper lip of certain bilabiate corollas.

gametophyte: The generation which has *n* chromosomes and produces gametes as reproductive bodies.

gamopetalous: With the petals connate, at least toward the base; sympetalous.

gamosepalous: With the sepals connate, at least toward the base.

geminate: In pairs.

geniculate: Abruptly bent or twisted.

gibbous: Abruptly swollen on one side, gen near the base.

glabrate: Nearly glabrous, becoming glabrous.

glabrescent: Becoming nearly or quite glabrous.

glabrous: Smooth, without hairs or glands.

gladiate: Sword-shaped.

gland: A spot on the surface of an organ or at the end of a hair which produces a sticky or greasy substance.

glaucescent: Becoming glaucous.

glaucous: Covered with a fine, waxy, removable powder which imparts a whitish or bluish cast to the surface.

globose: ± spherical.

glochidia: Hairs or hairlike outgrowths with retrorse barbs at the tip.

glochidiate: Barbed at the tip.

glomerate: Densely compacted in clusters or heads.

glume: One of the 2 bracts, found at the base of a grass spikelet, which do not subtend fls.

gymnosperm: A member of the group of pls characterized by having ovules that are not enclosed in an ovary.

gynaecandrous spike: One (in *Carex*) with both ♂ and ♀ fls, the ♂ below the ♀.

gynobase: An enlargement or prolongation of the recep of some fls.

gynobasic style: One that is attached directly to the gynobase, as well as to the individual carpels or nutlets.

gynoecium: The carpels of a fl, collectively.

gynophore: A c stalk in some fls, bearing the gynoecium.

habit: The gen appearance or manner of growth of a pl.

habitat: The environmental conditions or kind of place in which a pl grows.

halophyte: A pl adapted to growth in salty soil.

hamate: Hooked at the tip.

haplostemonous: With a single cycle of stamens.

hastate: Shaped like an arrowhead, but with the lobes more divergent.

head: An infl of sessile or subsessile fls crowded closely together at the tip of a peduncle.

helicoid cyme: A sympodial cyme with the apparent main axis curved in ± of a helix.

herb: A pl, either ann, bien, or per, with the sts dying back to the ground at the end of the growing season.

herbaceous: Adjectival form of herb; also, lflike in color or texture, or not woody.

heterosporous: Producing 2 different kinds of spores, gen very unequal in size.

heterostylic: With styles of different (gen 2) lengths in fls of different individuals.

heterotrophic: Parasitic or saprophytic.

hilum: The scar on a seed at its point of attachment.

hirsute: Pubescent with rather coarse or stiff but not pungent hairs.

hirsutulous: Diminutive of hirsute.

hirtellous: Diminutive of hirsute.

hispid: Pubescent with coarse and firm, often pungent hairs.

homosporous: Producing only 1 kind of spore.

hyaline: Thin and transparent or translucent.

hydrophyte: A pl adapted to life in water.

hypanthium: A ring or cup around the ovary, gen formed by the union of the lower parts of the calyx, corolla, and androecium; when the petals and stamens appear to arise from the calyx tube, the hypanthium is that part of the apparent calyx tube which is below the attachment of the petals.

hypogynous: With perianth and stamens attached directly to the recep.

imbricate: Arranged in a tight spiral, so that the outermost member has both edges exposed, and at least the innermost member has both edges covered; a shingled arrangement.

incised: Rather deeply and sharply (and often irreg) cut.

included: Contained within an envelope, not projecting beyond it.

indehiscent: Not dehiscent, remaining closed at maturity.

indeterminate inflorescence: One which blooms from the bottom or the outside toward the top or center, so that it might continue to elongate and produce new fls indefinitely.

indument: The epidermal appendages of a pl considered collectively, as its pubescence.

induplicate: Valvate, with the margins infolded.

indurated: Hardened.

indusium: An epidermal outgrowth or reflexed and modified lf margin which covers the sori of many ferns.

inferior ovary: One with the other fl parts attached to a hypan which is adnate to the ovary and projects beyond it.

inflorescence: A fl cluster of a pl, or the arrangement of the fls on the axis.

internode: The part of a st between 2 adjacent nodes.

intrastaminal: Within (as opposed to outside of) the androecium.

introrse: Turned toward the inside, or facing inward.

involucel: Diminutive of involucre.

involucre: A set of bracts beneath an infl; any set of structures which surrounds the base of another structure.

involute: With edges rolled back.

irregular flower: One in which the petals (less often the sepals) are dissimilar in form or orientation.

keel: A sharp or conspicuous longitudinal ridge; also the 2 partly united lower petals of many Leguminosae.

labellum: Lip.

labiate: Lipped; gen in compounds such as bilabiate.

lacerate: Torn, or with an irreg jagged margin.

laciniate: Cut into narrow and gen unequal segms.

lactiferous: With a milky latex.

laminar: Thin and flat, as in a lf bl.

lanate: Woolly.

lanceolate: Lance-shaped.

latex: A colorless to more often white, yellow, or reddish liquid produced by some pls.

leaflet: An ultimate unit of a compound lf.

legume: The fr of a member of the Leguminosae, composed of a single carpel, typically dry and dehiscing down both sutures; a pl of the family Leguminosae.

lemma: One of the two bracts (lemma and palea) subtending the individual fls in grass spikelets.

lenticel: A slightly raised area in the bark of a st or root.

lenticular: Shaped like a double-convex lens.

lepidote: Scaly; covered with small scales.

ligulate: Having a ligule.

ligule: Term applied to the flattened part of the ray corolla in the Compositae and to the appendage on the inner (upper) side of the lf at the junction of bl and sheath in many Gramineae and some Cyperaceae.

limb: The expanded part of a gamopet corolla above the throat; the expanded part of any petal.

linear: Line-shaped.

livid: Pale lead-colored.

lobe: A projecting segm of an organ.

locule: A seed cavity (chamber) in an ovary or fr.

loculicidal: Dehiscing along the midrib or outer median line of each locule.

lodicule: One of the tiny scales which may represent a vestigial perianth in grasses.

loment: A legume composed of 1-seeded joints.

lunate: Crescent-shaped.

lyrate: Pinnatifid, with the terminal lobe evidently the largest and gen rounded.

malpighiaceous hairs: Dolabriform hairs.

marcescent: Withering and persistent, as petals and stamens in some kinds of fls.

membranaceous, membranous: Thin and flexible, like a membrane.

mericarp: An individual carpel of a schizocarp; 1 of the 2 parts into which a fr of an umbellifer splits at maturity.

-merous: Gr suffix, referring to the parts in each circle of fl organs, gen with a numerical prefix, as 2-merous.

mesic: Moist, neither very wet nor very dry.

micropyle: The opening through the integuments of an ovule to the nucellus.

monadelphous stamens: Stamens with filaments or anthers connate, gen forming a tube.

moniliform: Necklace-like.

monocarpic: Blooming only once and then dying; gen applied to pers.

monoecious: With unisexual fls, both types borne on the same individual.

mucro: A short, sharp, slender point.

mucronate: Tipped with a mucro.

multicipital: Literally, many-headed.

multiple fruit: One derived from several or many fls, as a pineapple or mulberry.

muricate: Beset with small, sharp projections.

mycorrhiza: The symbiotic relationship of a fungus and a pl root.

naked: Lacking various organs or appendages; a naked fl lacks a perianth.

napiform: Turnip-shaped.

naturalized: Well estab, but originally coming from another area.

nectary: Any structure which produces nectar.

nerve: A prominent longitudinal vein of a lf or other organ.

neutral flower: One which has neither stamens nor pistil.

nigrescent: Blackish, or becoming black.

node: A place on a st where a lf is (or has been) attached.

nodose: Knobby, gen used in describing roots.

nodulose: Diminutive of nodose.

nude: Naked.

nut: A hard, dry, indehiscent, gen 1-seeded fr.

nutlet: Diminutive of nut; a very thick-walled achene; 1 of the dry, indehiscent, 1-seeded half-carpels of the Boraginaceae.

ob-: Gr prefix, meaning in a reverse direction.

obcordate: Like cordate, but with the notch at the tip.

obdiplostemonous: With 2 cycles of stamens.

obsolete: So much reduced as to be scarcely detectable or entirely suppressed.

obtuse: Blunt; with the sides coming together at an angle of > 90 degrees.

ochroleucous: Yellowish white.

ochrea, ocrea: A sheath around the st just above the base of the lf, derived from stips.

offset: A short, prostrate shoot, primarily propagative in function.

operculate: Provided with an operculum.

operculum: A little lid; the deciduous cap of a circumscissile fr or other organ.

opposite: Situated directly across from each other at the same node; situated directly in front of (on the same radius as) another organ, as stamens opp the petals.

orbicular: Circular in outline.

orthotropous ovule: One that is straight or unbent.

oval: Broadly elliptic.

ovary: The structure which encloses the ovules of angiosperms; the expanded basal part of a pistil, containing the ovules.

ovate: Egg shaped in outline.

ovoid: Shaped like a hen's egg (applied to solid objects).

ovule: A young or undeveloped seed.

palate: A raised part of the lower lip of a corolla, partly or wholly closing the throat.

palea: One of the pair of bracts (lemma and palea) which gen subtend the individual fls in grass spikelets.

palmate: With 3–more lobes or nerves or lflets or brs arising from a common point.

pandurate: Fiddle-shaped.

panicle: A br, indeterminate infl.

paniculate: Adjectival form of panicle.

papilionaceous flower: One with a banner petal, 2 wing petals, and 2 partly connate keel petals.

papillate, papillose: Covered with papillae (short, rounded, blunt projections).

pappus: The modified calyx crowning the ovary (and achene) of the Compositae.

parietal placenta: A placenta along the walls or on the intruded partial partition of a compound, unilocular ovary.

parted: Deeply cut, gen > half way from the margin to the midvein or base.

pectinate: Comb-like, with a single row of narrow spreading appendages (or hairs) of uniform size, like the teeth of a comb.

pedicel: The stalk of a single fl in an infl.

peduncle: The stalk of an infl or of a solitary fl.

pedunculate: Borne on a peduncle.

peltate: Shield-shaped, attached by the lower surface instead of by the base or margin.

pendulous: Hanging or drooping.

penicillate: With a tuft of short hairs at the end.

perennial: A pl that lives > 2 years.

perfect flower: One with both androecium and gynoecium, whether or not it has a perianth.

perfoliate leaf: One with the basal margins connate around the st, so that the st appears to pass through it.

perianth: The sepals and petals (or tepals) of a fl, collectively.

pericarp: The wall of the fr.

perigynium: A special bract which encloses the achene of *Carex*.

perigynous: With the perianth and stamens united into a basal saucer or cup (hypan) distinct from the ovary; more gen, around the base of the gynoecium, as a perigynous disk.

persistent: Remaining attached after the normal function has been completed.

petal: A member of the 2nd set of fl lvs (just internal to the sepals), gen colored or white and serving to attract pollinators.

petaloid: Petal-like.

petiolate: With a petiole.

petiole: A lf stalk.

petiolule: The stalk of a lflet of a compound lf.

phanerogam: A seed pl.

phyllary: An invol bract of the Compositae.

phylloclade: Same as cladophyll.

phyllode: A ± expanded, bladeless petiole.

phyllopodic: With the lowest lvs well developed instead of reduced to scales.

pilose: With long, straight, rather soft, spreading hairs.

pilosulous: Diminutive of pilose.

pinna: One of the primary divisions of a pinnately compound lf.

pinnate: With 2 rows of lateral brs or appendages, or parts along an axis, like barbs on a feather.

pinnatifid: ± deeply cut in pinnate fashion.

pinnatilobate: With pinnately arranged lobes.

pinnipalmate: Intermediate between pinnate and palmate, as the venation of some lvs in which the 1st pair of lateral veins are much larger than the others.

pinnule: Diminutive of pinna; an ultimate lflet of a lf which is pinnately 2–more times compound.

pistil: The female organ of a fl, composed of 1–more carpels, ordinarily differentiated into ovary, style, and stigma.

pistillate flower: One with 1–more pistils, but no stamens.

placenta: The tissue of the ovary to which the ovules are attached.

plicate: Folded like a fan.

plumose: Feathery.

pod: Any kind of dry, dehiscent fr.

pollinium: A coherent cluster of ∝ pollen grains, transported as a unit during pollination.

polycarpus, polycarpic: With several or ∝ carpels.

polygamo-dioecious: Nearly dioecious, but with some of the fls ♂.

polygamo-monoecious: Nearly monoecious, but with some of the fls ♂.

polygamous: With intermingled ♂ and unisexual fls.

polymorphic: Occurring in several different forms.

polypetalous: With the petals separate from each other.

polysepalous: With the sepals separate from each other.

pome: A fr with a core, like an apple.

poricidal: Opening by pores.

porrect: Directed outward and forward.

posterior: Literally, behind; the posterior side of a fl is the side toward the axis of the infl, rather than the side toward the bract; the upper lip of a bilabiate fl is the posterior lip.

praemorse: Ending abruptly, as if bitten off.

precocious: Developing very early; fls developing before the lvs.

prickle: A sharp outgrowth from the epidermis or bark.

procumbent: Prostrate or trailing, but not rooting at the nodes.

prophyll: One of a pair of bracteoles at the base of a fl, as in some spp. of *Juncus.*

prostrate: Flat on the ground.

proximal: Toward the base or near the end.

pruinose: Strongly glaucous, like a prune.

pseudanthium: A compact infl with small individual fls, the whole simulating a single fl.

puberulent: Minutely pubescent.

pubescent: Bearing hairs (trichomes) of any sort.

pulvinate: Cushion-like.

punctate: Dotted, gen with small pits that may be translucent or glandular.

puncticulate: Diminutive of punctate.

pungent: Firmly sharp-pointed.

pustular, pustulate, pustulose: With little blisters or pustules.

pyrene: The hard or seedlike endocarp of a drupe.

pyriform: Pear-shaped.

pyxis: A circumscissile caps.

raceme: A ± elongate infl with ped fls arising from the bottom upward on an unbr axis.

racemiform: Having the form of a raceme.

racemose: An infl characterized by fl in acropetal sequence.

rachilla: Diminutive of rachis; the axis of the spikelet in the grasses and sedges.

rachis: A main axis, such as that of a compound lf.

radiate: In the Compositae, having the marginal fls of the head lig and the central ones tubular.

ramentum: Thin, elongate scales of the epidermis.

ray: The lig or lig fl in the Compositae; one of the brs of an umbel.

receptacle: The end of the st (ped) to which the other fl parts are attached; in the Compositae, the end of the peduncle, to which the fls of the head are attached.

recurved: Curved backward.

reflexed: Bent backward.

regular flower: A fl in which the members of each circle of parts (or at least the sepals and petals) are similar in size, shape, and orientation.

reniform: Kidney-shaped.

repand: With a shallowly sinuate or slightly wavy margin.

repent: Creeping.

replum: A persistent, septum-like or frame-like placenta which bears ovules on the margins.

resupinate: Upside down; inverted by the twisting of the ped, as in orchid fls.

reticulate: Forming a network, as the veins of a lf.

retrorse: Directed backward or downward.

retuse: With a small terminal notch in an otherwise rounded or blunt tip.

revolute: Rolled back from the margin.

rhizoid: A structure of rootlike form and function, but lacking xylem and phloem.

rhizomatous: Bearing rhizomes.

rhizome: A creeping underground st.

rib: One of the main logitudinal veins of a lf or other organ.

rootstock: Same as rhizome.

rosette: A cluster of lvs or other organs arranged in a circle or disk, often in basal position.

rostellate: Diminutive of rostrate.

rostrate: Beaked.

rotate: Flat and circular in outline; saucer-shaped.

rotund: Round or rounded.

ruderal: Weedy, growing in waste places.

rudimentary: Imperfectly developed, but reflecting an early evolutionary stage rather than evolutionary reduction.

rugose: Wrinkled.

rugulose: Diminutive of rugose.

runcinate: Sharply cleft or pinnatifid, with backward-pointing segms.

runner: A slender stolon.

saccate: Furnished with or in the shape of a sac or pouch.

sagittate: Arrow-shaped.

salverform: With a slender tube and an abruptly spreading limb.

samara: A dry, indehiscent, gen 1-seeded, winged fr.

sanguineous: Blood red.

saprophyte: A pl which lives on dead organic matter, neither parasitic nor making its own food.

scaberulous, scabrellate: Diminutive of scabrous.

scabrid: Roughened.

scabridulous: Minutely roughened.

scabrous: Rough to the touch.

scale: Any small, thin or flat structure.

scape: A lfless (or merely bracteate) peduncle arising from the ground level in acaulescent pls.

scapose: With the fls on a scape.

scarious: Thin, dry, and chaffy in texture, and not green.

schizocarp: A fr which splits into its separate carpels at maturity.

sclerophyll: A firm lf, with a relatively large amount of strengthening tissue.

scorpioid cyme: A sympodial cyme with a zigzag rachis, the successive brs of which make up the rachis arising on different sides.

secund: With the fls or brs all on 1 side of the axis.

sepal: A member of the outermost set of fl lvs, typically green or greenish and ± lfy in texture.

sepaloid: Sepal-like, esp in color and texture.

septate: Provided with partitions.

septicidal: Dehiscing through the septa, between the locules.

septum: A partition; in an ovary, a partition formed by the connate walls of adj carpels.

sericeous: Silky, from the presence of long, slender, soft, ± appressed hairs.

serotinous: Literally late; with fls developing after the lvs are fully expanded.

serrate: Toothed along the margin with sharp, forward-pointing teeth.

serrulate: Diminutive of serrate.

sessile: Attached directly by the base, without a stalk.

seta: A bristle.

setaceous: Bristle-like.

setose: Beset with bristles.

sheath: An organ which partly or wholly surrounds another organ, as the sheath of a grass lf, which surrounds the st.

sigmoid: Doubly curved, like the letter S.

silicle: A fr like a silique, but short, not much longer than wide.

silique: An elongate caps in which the 2 valves are deciduous from the persistent, seed-bearing partition or replum.

simple leaf: One with the bl all in 1 piece, not compound.

simple pistil: One composed of only 1 carpel.

sinuate: With a strongly wavy margin.

sinus: The cleft or recess between 2 lobes or segms of an expanded organ such as a lf.

sordid: Of a dull, dingy, or dirty hue.

sorus: A cluster of sporangia, as in ferns.

spadix: A spike with small, crowded fls on a thickened, fleshy axis.

spathe: A large, gen solitary bract subtending and often enclosing a spadix or other infl.

spatulate: Shaped like a spatula, rounded above and narrowed to the base.

spicate: Arranged in a spike.

spike: A ± elongate infl of the racemose type, with sessile or subsessile fls.

spine: A firm, slender, sharp-pointed structure, representing a modified lf or stip; loosely, any body having the appearance of a true spine.

spinulose: Provided with small spines.

sporangium: A case or container for spores.

spore: A 1-celled reproductive structure other than a gamete or zygote, in vascular pls always representing the 1st cell of the gametophyte generation.

sporocarp: A specialized body (not obviously a lf or a cone scale) within which sporangia are borne.

sporophyll: A lf (often ± modified) which bears or subtends 1–more sporangia.

sporophyte: The generation which has 2n chromosomes and produces spores as reproductive bodies.

spur: A hollow appendage of the corolla or calyx.

squarrose: Abruptly spreading or recurved at some point above the base.

stamen: Loosely, the male organ of a fl, consisting of anther and gen a filament.

staminate flower: One with 1–more stamens, but no pistil.

staminode: A modified stamen which does not produce pollen.

standard: The uppermost petal of a papilionaceous fl; the banner.

stellate: Star-shaped; stellate hairs have several—∞ brs from the base.

sterile: Unproductive or infertile.

stigma: The part of the pistil which is receptive to pollen.

stipe: The stalk of a structure, without regard to its morphological nature.

stipitate: Borne on a stipe.

stipulate: Provided with stipules.

stipule: One of a pair of basal appendages found on many lvs.

stolon: An elongate, creeping st on the surface of the ground.

stoloniferous: Bearing stolons.

stramineous: Straw-colored.

striate: Marked with fine, ± parallel lines.

strict: Very straight and upright.

strigillose: Diminutive of strigose.

strigose: Provided with straight, appressed hairs all pointing in ± the same direction.

strobilus: A cluster of sporophylls on an axis; a cone.

strumose: Covered with cushion-like swellings; bullate.

style: The slender stalk which typically connects the stigma(s) to the ovary.

stylopodium: An enlargement or disklike expansion at the base of the style, as in the Umbelliferae.

sub-: Latin prefix, meaning under, almost, or not quite.

suberose: Corky in texture.

subtend: To be directly below and close to, as a lf subtends its axillary bud.

succulent: Fleshy and juicy.

suffrutescent: Half-shrubby, or somewhat shrubby, or dying back to a persistent, woody base.

sulcate: Marked with longitudinal grooves.

superior ovary: One which is attached to the summit or center of the recep and free of all other parts.

suture: A seam or line of fusion; gen applied to the vertical lines along which a fr may dehisce.

sympatric: Occupying the same geographic region.

sympetalous: With the petals connate, at least toward the base; gamopetalous.

sympodial: With the apparent main axis actually consisting of a series of short axillary brs.

syncarpous: With united carpels.

syngenesious: With connate anthers.

synsepalous: With connate sepals; gamosepalous.

taxon (plural *taxa*): Any taxonomic entity, of whatever rank.

tendril: A slender, coiling or twining organ by which a climbing pl grasps its support.

tepal: A sepal or petal, or member of an undifferentiated perianth.

terete: Cylindrical; round in cross-section.

ternate: In 3's.

tessellate: With a checkered or paving-block pattern.

testa: The seed coat, derived from the integuments of the ovule.

tetradynamous: With 4 long and 2 short stamens, as in the Cruciferae.

thalloid: Resembling or consisting of a thallus.

thallus: A pl body which is not clearly differentiated into roots, sts, and lvs.

theca: A pollen sac or locule of an anther.

thorn: A stiff, woody, modified st with a sharp point.

throat: The orifice of a gamopet corolla or gamosep calyx, or the somewhat expanded part between the proper tube and the limb; in grasses, the upper margins of the sheath.

thyrse: An elongate, narrow, mixed pan.

tomentose: Covered with tangled or matted, woolly hairs.

tomentulose: Diminutive of tomentose.

tomentum: A covering of tangled or matted, woolly hairs.

torus: The recep of a fl or of a head.

trailing: Prostrate but not rooting.

trichome: Any hairlike outgrowth of the epidermis.

trichotomous: Forking in 3's.

trifoliate, trifoliolate: With 3 lvs; with 3 lflets.

trigonous: With 3 angles.

trilocular: With 3 locules.

trimerous: With 3 parts of a kind.

triquetrous: With 3 sharp or projecting angles.

truncate: With the apex (or base) transversely straight or nearly so, as if cut off.

tuber: A thickened part of a rhizome, gen at the end, serving in food storage and often in reproduction.

tubercle: A small swelling or projection.

tuberculate: Bearing tubercles.

tuberous: Thickened like a tuber.

tubular: With the form of a tube.

tunicate: Covered or provided with sheathing lf bases which form concentric circles when viewed in cross-section, as the bulb of an onion.

turbinate: Top-shaped.

turgid: Swollen.

turion: A small, bulblike offset, as in some spp. of *Epilobium.*

umbel: A racemose infl with a very short axis and more elongate peds which seem to arise from a common point.

umbellate: In umbels.

umbellet: One of the ultimate umbellate clusters of a compound umbel.

umbelliform: With the form but not the structure of an umbel.

umbo: A blunt or rounded elevation or protuberance on the end or side of a solid organ, as on the scales of many pine cones.

umbonate: Provided with an umbo.

uncinate: Hooked at the tip.

undulate: Wavy.

unisexual flower: One with an androecium or a gynoecium, but not both.

urceolate: Urn-shaped or pitcherlike, contracted at or just below the mouth.

utricle: A small, thin-walled, 1-seeded, ± inflated fr.

valvate: Arranged with the margins of the petals or sepals adjacent throughout their length, without overlapping; opening by valves.

valve: One of the portions of the ovary wall into which a caps separates at maturity; in anthers opening by pores, the portion of the anther wall covering the pore.

vascular: Pertaining to conduction. Vascular pls are those which have xylem and phloem; a vascular bundle is a strand of xylem and phloem and associated tissues.

vein: A vascular bundle, esp if externally visible, as in a lf.

velutinous: Velvety.

venation: The mode of veining.

ventral: Pertaining to or located on the front or belly side.

ventricose: Inflated or swelling out on 1 side only, or unequally, as the corolla of many species of *Penstemon.*

vernation: The arrangement of lvs in the bud.

verrucose: Warty.

versatile anther: One attached near the middle and therefore capable of swinging freely on the filament.

verticil: A whorl of lvs or fls.

verticillaster: A false whorl.

verticillate: Arranged in verticils.

vespertine: Pertaining to or opening or functional in the evening.

vestigial: Much reduced and hardly or not at all functional.

villose: Same as villous.

villosulous: Diminutive of villous.

villous: Pubescent with long, soft, often bent or curved but not matted hairs.

virgate: Wand-like; slender, straight, and erect.

viscid: Sticky or greasy.

viscidulous: Diminutive of viscid.

whorl: A ring of 3–more similar structures radiating from a node or common point.

wing: A thin, flat extension or projection from the side or tip of a structure; 1 of the 2 lateral petals in a papilionate fl.

xero-: Gr prefix, meaning dry.

xerophyte: A pl adapted to life in dry places.

zygomorphic: Bilaterally symmetrical; most irreg fls are zygomorphic.

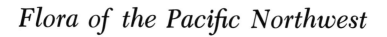

Flora of the Pacific Northwest

Synoptical Key to the Families of Vascular Plants in Our Region

(An artificial, somewhat less technical, illustrated key starts on p. 12.)

1a Pls not producing seeds; spores the typical disseminules

  2a Lvs always small or very narrow, simple, and with a single, unbr mid-vein; sporophylls or sporangiophores often aggregated into a terminal cone, but never forming specialized sporocarps; pls rooted to the substrate (sometimes epiphytic), not free floating

    3a Lvs gen alt or opp (occ subverticillate), distinct; sporangia borne in the axils of vegetative or ± modified lvs that are often aggregated into a terminal cone; st of ordinary type, not as in the following group (Division Lycopodiophyta)

      4a Pls homosporous, the spores $\propto$ in each sporangium, all ca alike; lvs without a ligule (Class Lycopodiopsida)   **Lycopodiaceae** p. 40

      4b Pls heterosporous, the microspores very $\propto$ and tiny, the megaspores much larger and fewer; lvs ligulate, i.e., with a small ventral appendage a little above the base (Class Isoetopsida)

        5a Lvs small, not > ca 5 mm; sts br and ± elongate; megaspores 1–4; pls terrestrial or epiphytic   **Selaginellaceae** p. 41

        5b Lvs elongate, gen well > 10 mm; st very short, unbr; megaspores more $\propto$, gen 50–300; pls ± aquatic or amphibious, occ merely in wet ground   **Isoetaceae** p. 42

    3b Lvs whorled, connate around st, only the tips distinct; sporangia borne on highly modified sporangiophores that are aggregated into a characteristic terminal cone; st longitudinally ribbed and grooved, hollow, jointed, with a persistent meristematic region at base of each internode (Division Equisetophyta, Class Equisetopsida)   **Equisetaceae** p. 42

  2b Lvs in most spp. larger and with a br vein system, often cleft or compound, but if small as in the foregoing group, then the pls aquatic and either free-floating (Salviniaceae) or with the sporangia borne in specialized, woolly sporocarps 2–3 mm thick (*Pilularia*, in Marsileaceae); pls not cone-bearing (Division Polypodiophyta, Class Polypodiopsida)

    6a Pls homosporous, terrestrial (though often of moist places) or epiphytic; sporangia borne on the lvs, not enclosed in sporocarps

      7a Pls eusporangiate, the sporangia thick-walled, 2-valved, without an annulus, borne in a "fertile spike" that projects from upper side of lf, not forming sori, and without indusium; lvs sometimes bent over in bud, but not circinate   **Ophioglossaceae** p. 44

      7b Pls leptosporangiate, the sporangia thin-walled, with definite annulus, gen forming definite sori on lower surface or along margins of lf, each sorus often subtended by an indusium; lvs circinate in bud   **Polypodiaceae** p. 46

    6b Pls heterosporous, aquatic or semiaquatic; sporangia borne in specialized sporocarps

      8a Pls rooted to the bottom; microsporangia and megasporangia in the same sporocarp; lvs (petiole included) well > 1 cm   **Marsileaceae** p. 55

      8b Pls gen free-floating; microsporangia and megasporangia borne in separate sporocarps; lvs in ours well < 1 cm   **Salviniaceae** p. 56

1b Pls producing seeds, which are the typical disseminules

  9a Ovules naked, not enclosed in an ovary, the pollen germinating at or near the micropyle; woody plants without vessels and with simple narrow (sometimes vestigial) lvs that do not have an anastomosing vascular system (Division Pinophyta, Class Pinopsida)

    10a Ovules and seeds borne in definite cones, the cone woody to fleshy, with several or $\propto$ scales, the seeds ultimately shed from the dry and shrunken scales; lvs not at once notably sharp-pointed and with a petiole-like base.

      11a Ovulate cones small to large (often > 3 cm), woody, with $\propto$ spirally arranged scales; lvs needlelike (though often flat), borne singly or in fascicles of 2–5 or pseudoterminally on short lateral brs   **Pinaceae** p. 59

      11b Ovulate cones small (< 3 cm), woody or often fleshy, with 2–12 opp or ternate scales; lvs opp or in whorls of 3 or 4, scalelike to occ needlelike   **Cupressaceae** p. 57

    10b Ovules and seeds borne singly, not in cones, the seed with a fleshy, orange-red covering; lvs needlelike (but flat), sharp-pointed and with an abrupt, short, petiole-like base   **Taxaceae** p. 56

  9b Ovules enclosed in an ovary, the pollen germinating on the stigma; herbs or woody plants, with or less often without vessels, and with various sorts of lvs that gen have an anastomosing vascular system (Division Magnoliophyta)

3

12a Embryo gen with 2 cotyledons; pls gen with intrafascicular cambium; vascular bundles in herbaceous forms gen borne in a ring enclosing a pith; floral parts, when of definite number, most often borne in sets of 5, less often 4, seldom 3 or 2 (carpels often fewer); lvs gen pinnately or palmately veined; herbs or woody pls (Class Magnoliopsida)   the **Dicotyledons**, lead 35a

12b Embryo with 1 cotyledon (or undifferentiated); pls without intrafascicular cambium; vascular bundles often scattered; floral parts, when of definite number, typically borne in sets of 3, seldom 4, never 5 (carpels often fewer); lvs in most spp. parallel-veined; ours all herbs (Class Liliopsida) the **Monocotyledons**, lead 13a

Monocotyledons

13a Fls with separate carpels, or seemingly or actually with only a single carpel; pls aquatic or semiaquatic, sometimes thalloid

14a Perianth differentiated into evident sepals and petals, at least in many of the fls; pls never thalloid

15a Ovules several or ∝ , scattered over inner surface of carpel; lvs linear, not differentiated into bl and petiole   **Butomaceae** p. 557

15b Ovule solitary; lvs with or without an expanded bl   **Alismataceae** p. 557

14b Perianth, when present, not differentiated into sepals and petals, the tepals always relatively small and inconspicuous (fls rarely produced in Lemnaceae, which are free-floating and thalloid)

16a Pls thalloid, free-floating aquatics without vessels or tracheids; fls unisexual, the ♂ of a single stamen, the ♀ of a single carpel   **Lemnaceae** p. 677

16b Pls with roots, stems, and lvs, not thalloid, and with tracheids and sometimes also vessels, seldom free-floating

17a Fls unisexual, grouped into dense, unisexual spikes or heads, both sexes on the same pl; seeds with well developed endosperm

18a Infl of globose heads; vestigial perianth gen present; achenes sessile or nearly so, not wind-distributed   **Sparganiaceae** p. 674

18b Infl of dense, cylindrical spikes; perianth 0; achenes slenderly long-stipitate, with long hairs on the stipe, wind-distributed   **Typhaceae** p. 675

17b Fls ⚥ or unisexual, if unisexual then not grouped into dense, unisexual heads or spikes with both sexes on the same pl; seeds without endosperm

19a Emergent marsh-pls with simple, unbr stem terminating in a raceme or spike (*Lilaea*, of Juncaginaceae, sometimes wholly submerged but acaulescent, with clustered basal lvs); fr dehiscent (except in *Lilaea*); tepals and stamens each 6 (except in *Lilaea*)

20a Ovules 2; lvs cauline as well as basal; peds individually subtended by bracts **Scheuchzeriaceae** p. 560

20b Ovule 1; lvs wholly basal; peds not individually subtended by bracts, except in *Lilaea* **Juncaginaceae** p. 561

19b Pls submersed or with floating lvs and sts, only the infls sometimes emergent; sts gen br and lfy; fr indehiscent; tepals and stamens < 6; fls not individually subtended by bracts

21a Carpels gen 2 or more; stamens 1–4

22a Fls ⚥, borne in spikes or racemes; lvs alt or the upper sometimes opp; stamens 2 or 4

23a Tepals 4; stamens 4; spikes axillary; fr carpels sessile; pls of fresh water **Potamogetonaceae** p. 562

23b Tepals 0; stamens 2; spikes terminal; fr carpels stipitate; pls of brackish water **Ruppiaceae** p. 566

22b Fls unisexual, borne in axillary cymes or solitary in the axils; lvs opp; stamen 1 **Zannichelliaceae** p. 566

21b Carpel 1; stamen 1

24a Fls solitary in the axils; ovule basal, anatropous; pls of fresh or brackish water **Najadaceae** p. 562

24b Fls in a small, unilateral spadix subtended by a tardily rupturing spathe; ovule apical, pendulous, orthotropous; submersed marine pls   **Zosteraceae** p. 567

13b Fls syncarpous; endosperm gen present; pls terrestrial or aquatic

25a Perianth much reduced and chaffy or bristly or fleshy, or wanting

26a Fls very small, ∝ , and ± fleshy, aggregated into a spadix that is subtended by a spathe; lvs broad and net-veined to narrow and parallel-veined   **Araceae** p. 676

26b Fls various but never aggregated into a spadix; lvs narrow and parallel-veined, or ± reduced

27a Ovary with > 1 ovule; infl various but not as in the following group

28a Ovules 1 per locule

29a Locules 2; fls in dense, unisexual heads; fr indehiscent   **Sparganiaceae** p. 674

29b Locules 3 or 6; fls ⚥, in a terminal raceme; fr dehiscent   **Juncaginaceae** p. 561

28b Ovules 3–∝ per locule; locules 1 or 3; fr dehiscent   **Juncaceae** p. 567

27b Ovary with a single ovule and locule; fr indehiscent; fls in characteristic spikes or spikelets

30a Fls spirally or less often distichously arranged on the axis of a spike or spikelet, gen each fl seemingly or actually subtended by only a single bract, without an evident bract between the fl and the axis; seed coat gen free from pericarp; lf sheath gen closed; st gen solid, often triangular; fls often with a perianth of evident bristles; carpels 3 or less often 2; embryo embedded in the endosperm **Cyperaceae p. 576**

    30b Fls distichously arranged on the axis of the spikelet (or only 1 per spikelet), each fl gen subtended by a pair of bracts (lemma and palea), the palea inserted between the fl and the axis; seed coat gen adnate to pericarp; lf sheath gen open; st gen hollow, never triangular; fls without a perianth, unless the lodicules are so interpreted; embryo peripheral to the endosperm **Gramineae p. 602**

25b Perianth well developed, one or both sets of tepals petaloid and ± showy (pet of ♂ fls vestigial in some Hydrocharitaceae)

    31a Pls aquatic, submersed to partly floating, the fls borne at the water surface

        32a Ovary inferior, unilocular but with ± deeply intruded partial partitions, the ovules borne on the surface of the partitions; fls mostly unisexual; perianth of well differentiated sepals and petals, or the petals vestigial in the ♂ fls; endosperm none **Hydrocharitaceae p. 559**

        32b Ovary superior, ours unilocular with 3 parietal placentae; fls ⚲; endosperm well developed; perianth of 6 petaloid tepals **Pontederiaceae p. 678**

    31b Pls terrestrial to semiaquatic

        33a Stamens 3-6; seeds of ordinary number and structure, with well differentiated embryo and well developed endosperm; ovary superior or inferior and perianth reg or irreg

            34a Stamens gen 6 (only 3 or 4 in a few spp. with superior ovary); ovary superior or less often inferior **Liliaceae p. 678**

            34b Stamens 3; ovary inferior **Iridaceae p. 697**

        33b Stamen 1 (2), adnate to the style; seeds very ∝ and tiny, with undifferentiated embryo and very little or no endosperm; ovary inferior and perianth irreg **Orchidaceae p. 698**

Dicotyledons

35a Petals separate from each other (or sometimes some or all of them ± connate distally, above a free base), or wanting

    36a Carpels 2 or more and separate, or the carpel seemingly or actually solitary, with 1 locule, 1 placenta, and 1 stigma (carpel solitary and falsely bilocular in spp. of *Astragalus*) **Group 1, lead 41a**

    36b Carpels 2 or more (as shown by no of styles, stigmas, locules, or placentas) and united (at least toward base) to form a compound pistil (carpels merely loosely connate by margins in Phytolaccaceae)

        37a Fls ± strongly reduced, unisexual or less often ⚲, lacking a perianth, or with very small sepals only, individually inconspicuous, very often some of them borne in catkins or pseudanthia, often appearing before the lvs; pls autotrophic **Group II, lead 55a**

        37b Fls gen more normally developed, ⚲ or less often unisexual, with an evident, often ± showy perianth; some parasitic pls with ± reduced fls are keyed here rather than with the foregoing group

            38a Ovary superior, or less than half inferior

                39a Pls herbaceous or merely suffrutescent at base, or, if occ shrubby, then with simple lvs and a single basal ovule **Group III, lead 68 a**

                39b Pls woody (trees, shrubs, or woody vines), not at once simple-lvd and with a single basal ovule **Group IV, lead 97a**

            38b Ovary inferior (fully so, or at least half) **Group V, lead 105a**

35b Petals united, at least at base (♀ fls without corolla in a few Compositae; calyx lobes inserted at sinuses of corolloid perianth in Cucurbitaceae)

    40a Ovary superior, or if inferior, then stamens more ∝ than corolla lobes, and not arranged as in *Adoxa* **Group VI, lead 120a**

    40b Ovary inferior; stamens alt with corolla lobes or fewer than corolla lobes, except in *Adoxa*, which has 10 unithecal stamens, 2 at each sinus of corolla **Group VII, lead 142a**

Group I

Fls seemingly or actually apocarpous, with 1–∝ carpels; petals distinct or wanting

41a Carpels gen 2–∝ (solitary in some Ranunculaceae, these terrestrial pls with ± ∝ stamens opening by longitudinal slits); herbs, except some Rosaceae

    42a Stamens attached directly to recep, or to lower part of ovary; stips wanting, or seldom present and small

        43a Stamens gen ± ∝, > 10 except in a few spp., these with caducous sepals, or > 5 carpels, or both

            44a Pls aquatic, with large, floating (ours peltate) lf bls *(Brasenia)* **Nymphaeaceae p. 122**

            44b Pls terrestrial to occ aquatic, the aquatic spp. not with large, floating, peltate lf bls **Ranunculaceae p. 124**

43b Stamens few, not > 10; sepals gen persistent; carpels 2–5
    45a Carpels 4 or 5; pls succulent   **Crassulaceae** p. 182
    45b Carpels 2 or occ 3; pls not notably succulent   **Saxifragaceae** p. 184
42b Stamens attached to hypan (often seemingly to calyx tube), or stamens attached to outer base of a lobed, perig disk
    46a Stamens attached to outer base of a lobed, perig disk; lvs exstip, ours ternately dissected with broad, flat ultimate segms   **Paeoniaceae** p. 124
    46b Stamens attached to hypan; lvs various, gen with stips, and only seldom ternately dissected
                                                                      Rosaceae p. 205
41b Carpel solitary, or seemingly so; herbs or woody pls; forms with > 10 stamens are either aquatic or have the anthers opening by terminal, uplifting valves
    47a Pls aquatic; fls apet, solitary in the lf axils
        48a Submerged, rootless aquatics with alt, dissected lvs; stamens 10–16   **Ceratophyllaceae** p. 123
        48b Emergent, rhizomatous aquatics with opp, simple lvs; stamen 1   **Hippuridaceae** p. 313
    47b Pls terrestrial; fls with or without petals
        49a Fls much reduced, mostly or all unisexual, always individually small and inconspicuous, some-times with minute sepals, but never with petals; lvs simple
            50a Trees with milky juice *(Maclura)*   **Moraceae** p. 75
            50b Herbs with watery juice
                51a Fr a 2-valved, 1-seeded caps; prostrate, mat-forming, stellate-hairy ann *(Eremocarpus)*
                                                    Euphorbiaceae p. 284
                51b Fr an achene; sprawling to erect ann or per, hirsute or glab, but not stellate   **Urticaceae** p. 76
        49b Fls ± well developed, ☿ or occ unisexual, gen with evident sepals or petals or both (small and without perianth in *Achlys*, of the Berberidaceae, which has trifoliolate lvs)
            52a Perianth consisting of both sepals and petals, or *(Achlys)* wholly wanting; stamens (5) 6–13
                53a Corolla reg, or wanting; stamens 6–13 (most often 6), distinct; anthers opening by terminal, uplifting valves   **Berberidaceae** p. 142
                53b Corolla irreg, gen papilionaceous (see descr); stamens gen 10 (occ 9 or 5), all or all but 1 connate by their filaments; anthers opening by longitudinal slits   **Leguminosae** p. 228
            52b Perianth consisting only of a ± corolloid calyx; stamens gen 4, 5, or 8
                54a Woody pls with tetramerous fls   **Elaeagnaceae** p. 302
                54b Herbs with pentamerous perianth, the stamens 5 or sometimes only 3 or 4
                                                           Nyctaginaceae p. 102

Group II

Carpels 2 or more and united; fls ± strongly reduced, unisexual or less often ☿, lacking a perianth or with very small sepals only, individually inconspicuous, very often some of them borne in catkins or pseudan-thia; lvs simple, except in Haloragaceae; pls autotrophic

55a Seeds centrospermous (i.e., the embryo elongate, peripheral or nearly so, and gen ± curved around a gen copious perisperm, or less often spiral or merely folded and the perisperm scanty or none); ovary with a single locule and a single basal ovule; fr indehiscent or more often with circumscissile or irreg dehiscence; herbs or des shrubs, seldom aromatic, often of alkaline places, sometimes weedy, very often scurfy-pubescent   **Chenopodiaceae** p. 93
55b Seeds not centrospermous; ovary otherwise, except in Myricaceae, this family markedly aromatic; fr indehiscent, or opening by 2–4 valves; pls not notably of alkaline places, and not scurfy-pubescent
    56a Herbs
        57a Pls aquatic
            58a Lvs pinnately dissected; stamens 4 or 8; styles 4; vestigial perianth present; ovary inferior
                                                  Haloragaceae p. 312
            58b Lvs simple; stamen 1; styles 2; perianth wanting; ovary nude   **Callitrichaceae** p. 285
        57b Pls terrestrial, sometimes of wet places
            59a Ovary with 3 or more ovules; fr dehiscent; fls in pseudanthia; lvs entire or merely toothed
                60a Pseudanthium consisting of ∝ ♂ fls on an axis, collectively subtended by several peta-loid bracts; pls scapose; juice watery; ovules and seeds relatively ∝, ca 18–40 (6–10 per carpel)   **Saururaceae** p. 64
                60b Pseudanthium consisting of an invol (often with marginal petal-like glands) subtending several ♂ fls (each ♂ fl consisting of a single, naked stamen) and a single central ♀ fl (con-sisting of a naked, 3-carpellate, eventually stipitate pistil); pls lfy-std; juice milky; ovules and seeds 3 (1 per carpel)   **Euphorbiaceae** p. 284
            59b Ovary with a single ovule; fr indehiscent; fls not in pseudanthia; lvs coarsely lobed or cleft
                                                   Moraceae p. 75
    56b Woody pls; fls often in catkins
        61a Fr caps, unilocular, 2–4-valved, with ∝ loosely long-hairy seeds   **Salicaceae** p. 64
        61b Fr indehiscent, 1–2-seeded

   62a Lvs opp

     63a Lvs simple; evergreen shrubs; stamens 4   **Garryaceae** p. 339

     63b Lvs pinnately compound; deciduous trees; stamens 2 *(Fraxinus)*   **Oleaceae** p. 356

   62b Lvs alt

     64a Pls strongly aromatic; ovary 1-celled, with a single orthotropous basal ovule; fr drupaceous; styles 2   **Myricaceae** p. 72

     64b Pls not notably aromatic; ovule(s) anatropous, apical or nearly so, or from the apex of a partial partition; fr and styles various

       65a Pls with a milky juice; fr multiple, fleshy; pistil gen with 2 styles and locules but only 1 ovule; endosperm gen present   **Moraceae** p. 75

       65b Pls with a watery juice; fr not multiple; endosperm wanting

         66a Ovary superior, gen unilocular and uniovulate; perianth present, though small; fls ♂ or unisexual, not in catkins; fr a drupe or samara; styles 2   **Ulmaceae** p. 74

         66b Ovary inferior or nude, partly or wholly 2–3-locular; perianth present and minute, or wanting; fls unisexual, at least the ♂ ones borne in catkins; fr a nut or samara

           67a Styles 2; ovary partly bilocular; ♀ fls (except in *Corylus*) in strobiloid catkins; fr a nut (± enclosed by a hull in *Corylus*) or a narrowly winged samara   **Betulaceae** p. 72

           67b Styles and locules 3; ♀ fls solitary or few together, not in catkins; fr a nut, gen subtended by a hull   **Fagaceae** p. 74

Group III

Fls normally developed, polypet to apet; carpels 2 or more and united; ovary superior; herbs

68a Aquatic pls with large, floating lf bls and large, long-pedunculate fls at the water surface; stamens ∝   **Nymphaeaceae** p. 122

68b Terrestrial to aquatic pls, neither with large, floating lf bls, nor with large, long-pedunculate fls at the water surface; stamens few–∝

   69a Pls modified toward entomophagy, the lvs or some of them producing digestive juices and either provided with sticky hairs or modified into pitcherlike structures; fls ♂, reg, hypog

     70a Lvs modified into pitchers; stamens ∝ ; placentation axile   **Sarraceniaceae** p. 182

     70b Lvs beset with glandular, insect-catching hairs, not modified into pitchers; stamens (4) 5; placentation parietal   **Droseraceae** p. 182

   69b Pls not modified toward entomophagy; other characters various

     71a Seeds (except in some Polygonaceae) centrospermous (i.e., the embryo elongate, peripheral or nearly so, and gen ± curved around a gen copious perisperm, or less often spiral or merely folded and the perisperm scanty or none); placentation most often free-central or basal in a unilocular compound ovary, but different in Phytolaccaceae, Aizoaceae, and some Caryophyllaceae

       72a Carpels ca 10, loosely connate into a ring, ripening collectively into a fleshy, berrylike fr; ovules 1 in each carpel, basal   **Phytolaccaceae** p. 103

       72b Carpels 2–5, united to form a compound ovary, the styles gen ± distinct

         73a Ovary several-celled, with axile placentation; our sp. an apet ann weed with whorled lvs   **Aizoaceae** p. 104

         73b Ovary gen 1-celled, with free-central or basal placentation (in Caryophyllaceae occ partitioned at base, rarely throughout, these spp. with evident petals)

           74a Sepals fewer than petals, gen 2 (up to 8 in *Lewisia rediviva*, which has 12–18 petals and ∝ stamens)   **Portulacaceae** p. 104

           74b Sepals > 2, of the same no as the petals when the fl has petals

             75a Lvs opp, rarely alt or whorled; sepals 4 or 5; petals gen of the same number as the sepals, or occ wanting; ovules gen several to ∝ on a free-central (to rarely axile) placenta, in a few genera solitary on a basal placenta; herbs   **Caryophyllaceae** p. 109

             75b Lvs gen alt, seldom opp or whorled; sepals 1–6, sometimes in 2 cycles; petals none, unless the inner set of 2 or 3 sepals in many Polygonaceae be so interpreted; ovule solitary on a basal placenta; herbs or occ shrubs

               76a Fr a utricle, circumscissile or irreg dehiscent, or in a few genera indehiscent; fls with the perianth in only 1 whorl; pls without sheathing stips

                 77a Sepals and bracts ± dry and scarious   **Amaranthaceae** p. 101

                 77b Sepals and bracts greener and more herbaceous   **Chenopodiaceae** p. 93

               76b Fr an achene; most genera (notable exception: *Eriogonum*) with well developed, sheathing stips; most genera (notable exception: *Polygonum*) with the perianth in 2 whorls   **Polygonaceae** p. 78

     71b Seeds not centrospermous; placentation axile to parietal, not free-central or basal in a unilocular ovary

       78a Pls either with ∝ (gen > 15) stamens, or with parietal placentation, or both, always chlorophyllous and photosynthetic; some pls with carpels connate below (there with axile placentation)

and free above (there with parietal placentation), as in some Crassulaceae and Saxifragaceae, are
keyed here rather than with the following group
79a Sepals only 2 or 3
    80a Fls reg; stamens ∝ to less often as few as 3; petals 4–6   **Papaveraceae** p. 143
    80b Fls irreg; stamens 6; petals 4   **Fumariaceae** p. 144
79b Sepals 4 or more
    81a Stamens ∝ , monadelphous by their filaments; carpels (5) 6–∝   **Malvaceae** p. 291
    81b Stamens few to ∝ , but not monadelphous, carpels seldom > 5, except in a few Ericaceae
        82a Styles 2 or more, distinct for most or all of their length, or styles wanting and stigmas 2
        or more, separate and sessile
          83a Lvs opp; stamens ∝ , often basally connate into 3–5 groups   **Hypericaceae** p. 294
          83b Lvs alt; stamens few–∝ , not connate into groups
            84a Fls irreg, hypog; stamens ca 10–∝ , borne internal to a partial disk; carpels 3–4 (6),
            open adaxially near the tip, each with a sessile stigma   **Resedaceae** p. 181
            84b Fls reg or nearly so, hypog to perig; stamens 3–10; disk wanting; ovary closed be-
            fore maturity of the fr, with separate styles or sessile stigmas
              85a Carpels exactly as many as the petals, gen 5 or 4; fls hypog; pls succulent
                        Crassulaceae p. 182
              85b Carpels gen 2, less often 3 (*Lithophragma*) or 4 (*Parnassia),* fewer than the petals
              (or fewer than the sepals, if the petals are reduced or wanting); fls gen perig,
              seldom hypog, sometimes with the other floral parts adnate to lower part of ovary;
              pls not succulent   **Saxifragaceae** p. 184
        82b Style 1, with 1 or more stigmas, or stigma 1 (sometimes compound) and sessile
          86a Fls reg or nearly so; carpels 2; stamens seldom < 6
            87a Ovary gen 1-celled; fls (except in *Polanisia*) with an evident gynophore; stamens (5)
            6 or more, all ca alike; lvs trifoliolate or palmately compound, with well defined lflts,
            not dissected   **Capparidaceae** p. 180
            87b Ovary gen 2-celled, with the ovules attached to the margins of the partition; fls
            lacking a gynophore except in a few genera such as *Stanleya;* stamens typically 6, the
            2 outer shorter than the 4 inner, or occ less than 6; lvs simple and entire or toothed to
            often pinnately dissected, but without distinct lflts   **Cruciferae** p. 146
          86b Fls distinctly irreg; carpels 3–4; stamens 5   **Violaceae** p. 296
78b Pls with few (up to 10 or seldom 15) stamens and with axile placentation, or if with distinctly
    parietal placentation (some Ericaceae) then nongreen, lacking chlorophyll
    88a Styles 2–5, distinct for most or all of their length, or styles wanting and stigmas 2–5, separate,
    and sessile
        89a Lvs compound or deeply cleft
          90a Style gynobasic; ovules 1 per carpel; ann   **Limnanthaceae** p. 287
          90b Style terminal; ovules 1–∝ per carpel; ann or per
            91a Fr a beakless, loculicidal caps with 1–∝ seeds per carpel; lvs trifoliolate
                         Oxalidaceae p. 281
            91b Fr a beaked, gen septicidally dehiscent caps or schizocarp, the mericarps 1-seeded
            and gen again dehiscent; lvs not trifoliolate   **Geraniaceae** p. 279
        89b Lvs simple, entire or merely toothed
          92a Ovules and seeds ∝ in each locule; lvs opp; pls of wet places   **Elatinaceae** p. 295
          92b Ovules and seeds only 1 or 2 in each locule; lvs alt or in 1 sp. opp; pls of ordinary
          habitats   **Linaceae** p. 282
    88b Style 1, with 1 or more stigmas
        93a Lvs compound or dissected; stamens 10–15   **Zygophyllaceae** p. 282
        93b Lvs simple, entire or merely toothed, sometimes much reduced; stamens 2–10
          94a Fls strongly perig, reg; lvs or some of them gen opp or whorled   **Lythraceae** p. 302
          94b Fls hypog, reg or irreg; lvs alt, or less often opp or whorled (or all basal)
            95a Carpels 2; stamens 2; petals none *(Besseya)*   **Scrophulariaceae** p. 413
            95b Carpels 4 or 5; stamens 5–10; petals present or absent
               96a Fls strongly irreg; stamens 5; pls green, not mycotrophic   **Balsaminaceae** p. 289
               96b Fls reg; stamens > 5, gen 8 or 10; pls mycotrophic, often lacking chlorophyll
                         Ericaceae p. 340

Group IV

Fls normally developed, polypet to apet; carpels 2 or more and united; ovary superior; woody pls

97a Tendril-bearing vines; stamens opp the petals   **Vitaceae** p. 291
97b Shrubs or trees, or vines without tendrils; stamens various
    98a Lvs compound or palmately lobed
        99a Carpels 3; fr a berry like drupe; lvs pinnately compound or trifoliolate; ours shrubs or vines
                         Anacardiaceae p. 287

99b Carpels not 3; fr samaroid
  100a Carpels 2, remaining united in fr to form a double samara; lvs merely lobed, or in 1 sp. 3–5-foliolate; shrubs or trees   **Aceraceae** p. 288
  100b Carpels several, gen 5, separating in fr to form a cluster of simple samaras; lvs pinnately compound, ours with > 5 lflts; ours a tree   **Simaroubaceae** p. 283
98b Lvs simple, entire or merely toothed, sometimes much reduced
  101a Lvs scale-like, gen < 2 mm; styles 2–5, separate; seeds ∝   **Tamaricaceae** p. 296
  101b Lvs ± well developed, gen 4 mm or more; style 1
    102a Ovules and seeds ∝ ; stamens 5–12, gen 10   **Ericaceae** p. 340
    102b Ovules and seeds 1–2 per carpel; stamens 3–5
      103a Carpels 6–9; stamens 3; lvs small, in ours not > ca 1 cm   **Empetraceae** p. 286
      103b Carpels 2–5; stamens 4 or 5; lvs larger, often > 1 cm
        104a Stamens alt with petals   **Celastraceae** p. 288
        104b Stamens opp petals   **Rhamnaceae** p. 290

Group V
Fls normally developed, polypet to apet; carpels 2 or more and united; ovary inferior

105a Stamens 6– ∝
  106a Lfless, spiny succulents; sepals, petals, and stamens ∝   **Cactaceae** p. 301
  106b Pls lfy, unarmed or occ thorny, seldom succulent; sepals and petals seldom > 6 each
    107a Woody pls
      108a Stigma and style 1; pollen sacs apically elongate, dehiscing by a terminal pore; creeping shrubs (2 spp. of *Vaccinium*)   **Ericaceae** p. 340
      108b Stigmas (and often also styles) 2–5; pollen sacs otherwise
        109a Lvs opp, exstip; seeds with abundant endosperm   **Hydrangeaceae** p. 204
        109b Lvs alt, gen stip; seeds in most genera without endosperm   **Rosaceae** p. 205
    107b Herbs
      110a Sepals only 2 or 3
        111a Sepals 2; styles 3-9, separate for much of their length; petals present; placentation free-central *(Portulaca)*   **Portulacaceae** p. 104
        111b Sepals 3; style 1, with a lobed stigma; petals wanting or vestigial; placentation axile   **Aristolochiaceae** p. 78
      110b Sepals 4 or 5 (6)
        112a Styles (or sessile stigmas) 2 or 3; stamens 10 (8 in 1 sp. of *Chrysosplenium*)
                                                                                      Saxifragaceae p. 184
        112b Style 1; stamens 2, 4, 8, or ∝ (10 in *Jussiaea*)
          Stamens ∝ ; sepals and petals 5   **Loasaceae** p. 300
          Stamens 2–10 (gen 8); sepals and petals gen 4 (2 in *Circaea*, 5 in *Jussiaea*)   **Onagraceae** p. 303
105b Stamens 2–5
  113a Pls hemiparasitic (but still with chlorophyll); ovary unilocular; style 1, sometimes very short
    114a Root-parasites with normally developed lvs, ours alt   **Santalaceae** p. 78
    114b Stem-parasites, not attached to the ground, lvs opp, often much reduced   **Loranthaceae** p. 77
  113b Pls autotrophic; ovary 1–several-locular, with 1–several styles
    115a Lvs opp; style 1, with capitate to 4-parted stigma
      116a Woody pls; fr fleshy; stamens 4   **Cornaceae** p. 339
      116b Herbs; fr dry; stamens 2 *(Circaea)* or 4 *(Ludwigia)*   **Onagraceae** p. 303
    115b Lvs alt (or all basal); styles (or sessile stigmas) 2–5, except in *Hedera* (Araliaceae)
      117a Ovules and seeds several to ∝ ; fls in various sorts of infls, but not in umbels or dense heads
        118a Shrubs; fr a berry   **Grossulariaceae** p. 199
        118b Herbs; fr caps   **Saxifragaceae** p. 184
      117b Ovules and seeds 1 per carpel; fls in umbels or *(Eryngium)* dense heads, these often grouped into compound umbels (Umbelliferae) or (Araliaceae) into racemes, corymbs, or pans
        119a Pls (except *Aralia*) woody; carpels 2–6; fr in ours a berry   **Araliaceae** p. 313
        119b Pls herbaceous; carpels 2; fr a schizocarp   **Umbelliferae** p. 314

Group VI
Fls sympet; ovary superior, or if inferior then the stamens more ∝ than corolla lobes

120a Ovaries several (gen 5), connate only below the middle, or free nearly throughout; pls succulent; stamens 10 (spp. of *Sedum*)   **Crassulaceae** p. 182
120b Ovary 1, of 2–several carpels, or ovaries 2 but with a common style or stigma; pls not notably succulent; stamens of diverse nos, sometimes 10
  121a Placentation free-central or basal in a unilocular ovary
    122a Corolla reg; stamens at least as many as corolla lobes; pls terrestrial or sometimes aquatic, but not insectivorous

123a Stamens as many as the corolla lobes and alt with them or more often twice as many as the corolla lobes (occ genera and spp. of) **Caryophyllaceae** p. 109
   123b Stamens as many as and opp the corolla lobes
      124a Styles 5; ovule solitary **Plumbaginaceae** p. 355
      124b Style 1, with capitate or slightly lobed stigma; ovules gen several or ∝ , seldom (*Douglasia*) only 1 **Primulaceae** p. 350
122b Corolla irreg; stamens 2; pls ± insectivorous, often aquatic **Lentibulariaceae** p. 445
121b Placentation otherwise
   125a Stamens twice as many as corolla lobes (thus 8 or 10), except in *Rhododendron occidentale*, a large-fld shrub with anthers dehiscent by terminal pores **Ericaceae** p. 340
   125b Stamens as many as corolla lobes and alt with them (thus gen 4 or 5), or fewer than corolla lobes; anthers gen dehiscing longitudinally
      126a Ovules and seeds 4 (2 per carpel), or fewer by abortion, without endosperm; fr of 4 ± distinct or separating 1-seeded nutlets, or the nutlets fewer by abortion; style often gynobasic
         127a Stamens 5; corolla reg or seldom somewhat irreg; lvs gen alt (seldom some or all opp), gen entire; style gen (not always) gynobasic **Boraginaceae** p. 384
         127b Stamens 4 or 2; corolla slightly to strongly irreg; lvs opp, entire to more often toothed or cleft
            128 Ovary barely or scarcely lobed, the style not gynobasic, the nutlets separating only at maturity; pls not notably aromatic **Verbenaceae** p. 398
            128b Ovary ± deeply 4-cleft, the style gen gynobasic; pls gen ± aromatic **Labiatae** p. 399
      126b Ovules and seeds either > 4, or with endosperm, or both; style never gynobasic and fr never of 4 nutlets
         129a Corolla reg or nearly so; stamens, except in Oleaceae and some Plantaginaceae, isomerous with corolla lobes
            130a Corolla scarious, persistent; perianth 4-merous; stamens 4 or 2; carpels 2; lvs gen all basal, in 1 sp. cauline and opp **Plantaginaceae** p. 447
            130b Corolla gen petaloid, not at all scarious; fls and habit diverse
               131a Lvs gen opp; carpels 2
                  132a Carpels united, forming a compound ovary which ripens into a single fr (gen a caps); pls without milky juice
                     133a Woody pls; placentation axile; stamens 2 or 4; corolla lobes 4
                        134a Stamens 4 **Buddlejaceae** p. 356
                        134b Stamens 2 **Oleaceae** p. 356
                     133b Herbs; placentation parietal; stamens and corolla lobes 4 or 5 **Gentianaceae** p. 356
                  132b Carpels largely separate, united only by the stigmas and sometimes also the styles, ripening as separate follicles (or 1 follicle sometimes abortive); pls with milky juice
                     135a Pollen granular; stamens of ordinary type, separate or merely connivent, only slightly or not at all adnate to stigma, and with an ordinary type of connective; fls without a corona **Apocynaceae** p. 362
                     135b Pollen agglutinated into pollinia; stamens of unusual structure, monadelphous and adnate to the stigma, and with an enlarged, ornate connective; fls with a corona **Asclepiadaceae** p. 363
               131b Lvs gen alt (many exceptions among the Polemoniaceae, which gen have 3 carpels, occ exceptions among the Hydrophyllaceae)
                  136a Ovules 2– ∝ per carpel, on parietal or axile placentae, not basal and erect; pls erect to prostrate or occ scrambling, but not twining or trailing
                     137a Carpels 3 (2–4); corolla convolute in bud; placentation axile **Polemoniaceae** p. 366
                     137b Carpels 2, very rarely more; aestivation and placentation various
                        138a Aquatic pls with stip lvs, the bl floating or emergent **Menyanthaceae** p. 361
                        138b Terrestrial pls (sometimes of wet places) with exstip lvs
                          139a Corolla lobes gen imbricate in bud; style shallowly (but evidently) to deeply cleft, except in *Romanzoffia*; placentation gen parietal (except in *Romanzoffia*) **Hydrophyllaceae** p. 377
                          139b Corolla lobes variously folded, contorted, or valvate in bud, but not imbricate; style undivided, with capitate or slightly bilobed stigma; placentation axile **Solanaceae** p. 409
                  136b Ovules 2 per carpel, axile-basal, erect; ours twining or trailing pls
                     140a Pls autotrophic, chlorophyllous, with well developed lvs; embryo (esp the well developed cotyledons) plicate **Convolvulaceae** p. 363
                     140b Pls twining stem-parasites, without chlorophyll, and without well developed lvs; embryo ± spirally curved, not plicate, the cotyledons scarcely developed **Cuscutaceae** p. 364
         129b Corolla ± strongly irreg; stamens gen fewer than corolla lobes, but isomerous with corolla lobes in *Verbascum*, of Scrophulariaceae

141a Pls with chlorophyll and normally developed lvs, though sometimes partly parasitic; placentation axile **Scrophulariaceae** p. 413

141b Root-parasites, lacking chlorophyll, the lvs reduced and scale-like; placentation parietal, but the placentae often intruded **Orobanchaceae** p. 444

Group VII

Fls sympet; ovary inferior; stamens as many as or fewer than corolla lobes, except in *Adoxa*

142a Fls in various sorts of infls, if in invol heads then the ovules pendulous, the fls not blooming in centripetal sequence, and the anthers not connate

143a Tendril-bearing vines; fls unisexual; lvs alt **Cucurbitaceae** p. 457

143b Pls of various habit, sometimes trailing, but without tendrils; fls gen ♀; lvs alt to opp or whorled

144a Stamens free from corolla, or attached only to its very base; lvs alt (sometimes all basal), exstip **Campanulaceae** p. 457

144b Stamens attached to corolla distinctly above its base; lvs opp or whorled (rhizome scales alt in *Adoxa*), with or without stips

145a Stips present, cauline (i.e., interpetiolar), sometimes enlarged and lflike, the lvs then apparently whorled; corolla reg; stamens as many as corolla lobes; endosperm well developed; ours herbs **Rubiaceae** p. 448

145b Stips wanting, except in a few Caprifoliaceae, in which they are adnate to the petiole; other features not combined as above

146a Stamens twice as many as corolla lobes; paired in the sinuses, each with only a single pollen sac; herbs with subcapitate infl and drupaceous fr; endosperm well developed **Adoxaceae** p. 454

146b Stamens as many as corolla lobes, or fewer, with 2 pollen sacs each; other features not combined as above

147a Fls in various sorts of infls, but not in invol heads; ovaries not individually enclosed by a gamophyllous involucel; corolla gen 5-lobed

148a Pls ± woody (barely so in *Linnaea*); stamens gen 5 (4 in *Linnaea*); endosperm well developed; frs diverse, often fleshy **Caprifoliaceae** p. 450

148b Pls herbaceous; stamens 3; endosperm wanting; fr dry, indehiscent, 1-seeded **Valerianaceae** p. 454

147b Fls in dense, invol heads, the ovaries individually enclosed by a gamophyllous, cupulate involucel; fr dry, indehiscent, 1-seeded; corolla 4–5-lobed; stamens 4 or 2; endosperm wanting **Dipsacaceae** p. 456

142b Fls borne in dense, invol, centripetally fl heads (invol wanting in *Psilocarphus*); ovary 1-celled, with a solitary, erect ovule; anthers gen connate into a tube around the style; endosperm wanting **Compositae** p. 461

Artificial Key to the Families of Vascular Plants in Our Region

1a Pls aquatic, either sterile (i.e., without fls, frs, or spores) or with fl parts variable and not in recognizable multiples of 3, 4, or 5 **Group I** p. 13
1b Pls either terrestrial or bearing fls, frs, or spores; fls gen with parts in multiples of 3, 4, or 5
2a Pls not producing seeds, reproducing by means of tiny 1-celled spores borne in small sacs or vessels (sporangia) produced on the back of the lvs, or in the axils of lvs sometimes aggregated into a terminal cone, or in specialized globose to ovoid bodies (sporocarps) borne along st or at base of lvs; per (all ours) with well marked alternation of generations, the sporophyte well developed, differentiated into roots, sts, and lvs, forming the phase of the life cycle that is gen seen and collected, and for which this key is designed; gametophyte much the smaller of the 2 generations, but not remaining organically attached to the sporophyte, sometimes developing largely or wholly within the spore wall (vascular cryptogams)
  3a Lvs with single unbr midvein, simple, always small or very narrow; spore-bearing lvs or stalks often grouped into a terminal cone
    4a Lvs gen alt or opp, distinct; sporangia in axils of vegetative or ± modified lvs often aggregated into terminal cone; st not ribbed or grooved lengthwise, neither hollow nor jointed
      5a Lvs elongate and ± onionlike, with 4 hollow air chambers running lengthwise; st unbr; pl ± aquatic **Isoetaceae** p. 42
      5b Lvs flattened and neither hollow nor at all onionlike; pls terrestrial (occ at edge of bogs, etc.)
        6a Spores all of 1 kind, ∝ in each sporangium; pls relatively coarse, the lvs mostly well > 3 mm; sts gen at least 2 mm thick; spore-bearing lvs often either interspersed among the vegetative lvs or borne in cones, the cones sometimes separated from the vegetative lvs by merely bracteate stalks **Lycopodiaceae** p. 40
        6b Spores of 2 kinds, the smaller ones ∝ in each sporangium, the larger ones only 1–4 per sporangium; lvs gen < 3 mm; sts gen < 2 mm thick; spore-bearing lvs aggregated into cones at the end of the brs, the cones immediately subtended by vegetative lvs **Selaginellaceae** p. 41
    4b Lvs whorled, connate around st, only the tips distinct; sporangia on much-modified stalks or brs aggregated into terminal cones; st ribbed and grooved lengthwise, hollow, jointed **Equisetaceae** p. 42
  3b Lvs mostly with br vein system, often large and cleft or compound; pls not cone-producing
    7a Pl terrestrial or epiphytic, sometimes in moist places, producing only 1 kind of spore; sporangia on lvs, not enclosed in special bodies (sporocarps)
      8a Sporangia thick-walled, 2-valved, without ring of elastic cells (annulus) around them, not grouped in clusters or sori and without special covering or indusium, borne on a fertile portion of the lf, rest of lf sterile and unlike the fertile part; lvs sometimes bent over in bud, but not coiled **Ophioglossaceae** p. 44
      8b Sporangia thin-walled, with well-defined ring of elastic cells (annulus) around them, gen grouped in clusters or sori on lower surface or along margins of lf, each often partially or wholly covered by an epidermal outgrowth of the lf or by the lf margin (indusium); lvs coiled (circinate) in bud **Polypodiaceae** p. 46

12

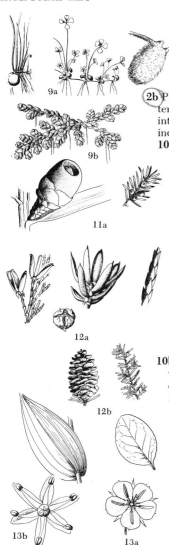

7b Pl aquatic or semi-aquatic, producing both large and small spores; sporangia in specialized bodies (sporocarps) borne along st or from base of lvs

    9a Pls rooted to bottom; lvs, incl petiole, much > 1 cm; sporocarps producing both large and small spores   **Marsileaceae p. 55**

    9b Pls gen free-floating; lvs much < 1 cm; sporocarps producing only large or small spores   **Salviniaceae p. 56**

2b Pls producing seeds either in cones or in the remains of fls; pls with alternation of generations, but the sporophyte generation differentiated into roots, sts, and lvs, the gametophyte never becoming physiologically independent and gen not recognizable as separate from the sporophyte

    10a Ovules (immature seeds) not enclosed within an ovary, gen borne on the surface of a cone scale; pls woody, with narrow lvs lacking a br vein system, gen producing a fleshy to woody cone (the gymnosperms)

        11a Pls ♂, ♀; seeds borne singly, with pulpy and ultimately reddish covering; lvs needlelike, flat, spreading and 2-ranked (brs spraylike), without resin ducts, ours abruptly sharp-pointed and with slender petiole-like base   **Taxaceae p. 56**

        11b Pls ♂♀ or ♂, ♀; seeds borne in woody to fleshy cones of several–∝ scales, ultimately shed as scales dry and shrink; lvs scalelike or needlelike, but often not in flat sprays, or with 1 or more resin ducts as seen in cross-section with 10× magnification

            12a Lvs scalelike or ± needlelike, but opp or whorled; ovulate cones small, woody or often fleshy, scales few (2–12), opp or in 3's   **Cupressaceae p. 57**

            12b Lvs needlelike, borne singly or in small bunches of 2–5 or pseudo-terminally on short lateral brs; ovulate cones woody, with ∝ spirally arranged scales   **Pinaceae p. 59**

    10b Ovules enclosed within the ovary of a fl; pls herbaceous to woody, with various kinds of lvs gen having a br vein system, and producing clearly defined to ± conelike fls (the angiosperms)

        13a Floral parts, when of definite no, gen in sets of 5, 4, or rarely 3, but carpels often fewer; lvs mostly pinnately or palmately veined; embryo gen with 2 cotyledons; vascular bundles of herbaceous sts gen in a ring enclosing a pith; herbs, trees, or shrubs (Magnoliatae, the Dicots)   **Group II p. 16**

        13b Floral parts, when of definite no, gen in sets of 3, occ 4, but never 5, carpels often fewer; lvs gen parallel-veined; embryo with 1 cotyledon; vascular bundles often scattered and a definite pith lacking; pls mainly herbaceous, ours rarely with a woody st above ground (Liliatae, the Monocots)   **Group III p. 36**

Group I

Pls aquatic, sterile or with floral parts not definite in number

1a Lvs ± like those of an onion (quill-like), round or ± square in cross-section, divided lengthwise into 4 hollow compartments, expanded near base and tufted on very short st, ultimately gen bearing a subbasal, adaxial sporangium (*Isoetes*)   **Isoetaceae p. 42**

1b Lvs flattened or if round in cross-section (terete) gen not hollow, or hollow but with only 1 or 2 cavities

    2a Lvs linear, 3–15 cm, semiterete, hollow, borne in basal tufts

        3a Lvs cross-septate; pl rhizomatous; ovary 2-seeded (*Lilaeopsis*)   **Umbelliferae p. 314**

        3b Lvs not cross-septate; pl tufted, not rhizomatous; ovary ∝ -seeded (*Lobelia dortmanna*)   **Campanulaceae p. 457**

    2b Lvs not terete and hollow

        4a Lvs coiled in bud, uncoiling from base upward, either quill-like and without lflets or clover-like and with 4 lflets; pl rhizomatous, gen with small, biconvex, oval, pill-like bodies (sporocarps) along rhizomes   **Marsileaceae p. 55**

        4b Lvs not coiled in bud, sometimes quill-like, but never clover-like

            5a Pls free-floating, small, gen without well-differentiated roots, sts, and lvs, and often resembling mosses or liverworts, or lvs very small and arising from short, fork-br sts

      6a Sts and lvs present, roots arising from sts   **Salviniaceae** p. 56

      6b Sts and lvs lacking, roots (if any) arising from a flattened, ± discoid body   **Lemnaceae** p. 677

   5b Pls not free-floating except when fragmented, always with obvious st and recognizable lvs

      7a Pls strictly marine; lvs linear, alt, 2-ranked, sheathing, rarely < 3 dm; fls sessile on 1 side of a flattened axis, enclosed in a sheath   **Zosteraceae** p. 567

      7b Pls mostly in fresh or brackish water, if marine then lvs much < 3 dm

         8a Lvs in whorls of 3–10 (or more) on an elongate st

            9a Lvs compound to dissected; fls ♂♀ or ♂̣♂̣♀

               10a Lf bls palmately to dichotomously dissected; perianth of 8–15 greenish segms; stamens 10–16; pistil 1-carpellary   **Ceratophyllaceae** p. 123

               10b Lf bls pectinately dissected to almost entire; fls mostly 4-merous   **Haloragaceae** p. 312

            9b Lvs simple, entire to denticulate

               11a Lvs in large part alt and opp as well as in 3's, narrowly linear, 10–45 × scarcely 1.5 mm; main st brd above (*Howellia aquatilis*)   **Campanulaceae** p. 457

               11b Lvs whorled, or opp and whorled, none alt

                  12a Lvs gen at least 5 (4–12) per whorl   **Hippuridaceae** p. 313

                  12b Lvs gen 3–4 per whorl

                     13a Lvs broadly expanded and ± sheathing at base, denticulate to strongly dentate   **Najadaceae** p. 562

                     13b Lvs not expanded and sheathing at base, often entire

                        14a Lvs in part gen ± expanded and rounded near tip, smoothly entire, nearly always broadest in distal half, > 1 cell thick near margin; fls often present, naked and gen ♂♀ or ♂, ♀; pistil 2-carpellary; stamens gen 1   **Callitrichaceae** p. 285

                        14b Lvs mostly tapered to distal end, broadest near or below midlength, often very minutely denticulate, only 1-cell thick near margins; fls recognizably 3-merous so fl pls prob will not be referred here (*Elodea*)   **Hydrocharitaceae** p. 559

         8b Lvs alt, opp, or all in a basal tuft

             15a Lvs with a ± expanded, triangular to rotund bl; venation obscurely to plainly netted   **Subgroup Ia, lead 16a**

             15b Lvs mostly linear, flat to terete, if bl expanded then main veins running the full length   **Subgroup Ib, lead 32a**

        Subgroup Ia

   16a Lvs gen floating; stips membranous, mostly connate around st above base of petiole; pistil 1, stigmas 3 (*Polygonum*)   **Polygonaceae** p. 78

   16b Lvs often submerged; stips not connate and sheathing; pistil various

      17a Pl rooted to bottom; fls with greatly elongate, flaccid petioles and peltate to cordate, subentire, floating bls   **Nymphaeaceae** p. 122

      17b Fls often floating; lvs never as above

         18a Lvs alt, trifoliolate or simple and cordate-ovate, bls or lflets 3–12 cm; petioles elongate, sheathing at base   **Menyanthaceae** p. 361

         18b Lvs often opp or whorled, neither trifoliolate nor cordate-ovate, bls or lf segms mostly < 3 cm; petioles rarely sheathing

            19a Submerged lvs dissected and bearing small, ovoid, insect-trapping bladders (*Utricularia*)   **Lentibulariaceae** p. 445

            19b Submerged lvs not bladder-bearing

               20a Fls naked, axillary, tiny; lvs opp, the submerged ones linear and 1-nerved, the floating or emergent ones broadened and 3–5-nerved   **Callitrichaceae** p. 285

               20b Fls with perianth; lvs not at once as above

                  21a Pistils several; lvs often dissected (*Ranunculus*)   **Ranunculaceae** p. 124

Figure labels: 6a, 6b, 7a, 10a, 10b, 11a, 13a, 14a, 14b, 12a, 16a, 17a, 18a, 19a, 20a, 21a

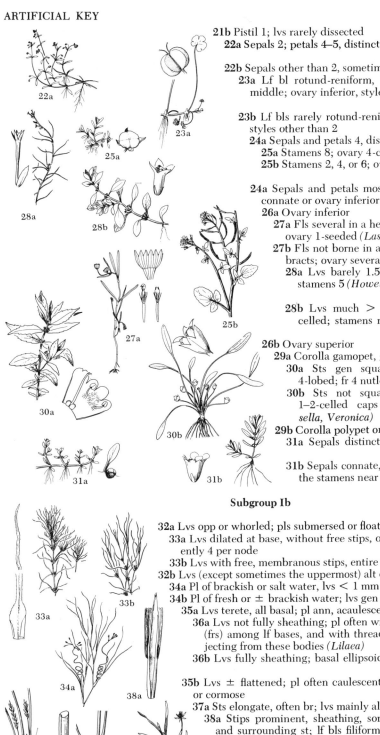

21b Pistil 1; lvs rarely dissected
  22a Sepals 2; petals 4–5, distinct *(Montia)*
                                              Portulacaceae p. 104
  22b Sepals other than 2, sometimes lacking; petals various
    23a Lf bl rotund-reniform, 1–6 cm broad, lobed to ca middle; ovary inferior, styles 2 *(Hydrocotyle)*
                                              Umbelliferae p. 314
    23b Lf bls rarely rotund-reniform, if so ovary superior or styles other than 2
      24a Sepals and petals 4, distinct; ovary superior
        25a Stamens 8; ovary 4-celled   **Elatinaceae** p. 295
        25b Stamens 2, 4, or 6; ovary 1–2-celled
                                              Cruciferae p. 146
    24a Sepals and petals mostly other than 4, if 4 either connate or ovary inferior
      26a Ovary inferior
        27a Fls several in a head, surrounded by ∝ bracts; ovary 1-seeded *(Lasthenia)*   **Compositae** p. 461
        27b Fls not borne in a head, not surrounded by ∝ bracts; ovary several-seeded
          28a Lvs barely 1.5 mm broad; ovary 1-celled; stamens 5 *(Howellia)*
                                              Campanulaceae p. 457
          28b Lvs much > 1.5 mm broad; ovary 4–5-celled; stamens rarely 5 *(Jussiaea, Ludwigia)*
                                              Onagraceae p. 303
      26b Ovary superior
        29a Corolla gamopet, gen irreg
          30a Sts gen square in cross-section; ovary 4-lobed; fr 4 nutlets   **Labiatae** p. 399
          30b Sts not square; ovary not 4-lobed; fr 1–2-celled caps *(Bacopa, Gratiola, Limosella, Veronica)* **Scrophulariaceae** p. 413
        29b Corolla polypet or lacking
          31a Sepals distinct; stamens hypog
                                              Elatinaceae p. 295
          31b Sepals connate, forming a long tube bearing the stamens near the top   **Lythraceae** p. 302

Subgroup Ib

32a Lvs opp or whorled; pls submersed or floating
  33a Lvs dilated at base, without free stips, often either toothed or apparently 4 per node   **Najadaceae** p. 562
  33b Lvs with free, membranous stips, entire   **Zannichelliaceae** p. 566
32b Lvs (except sometimes the uppermost) alt or all basal
  34a Pl of brackish or salt water, lvs < 1 mm broad   **Ruppiaceae** p. 566
  34b Pl of fresh or ± brackish water; lvs gen at least 1 mm broad
    35a Lvs terete, all basal; pl ann, acaulescent
      36a Lvs not fully sheathing; pl often with hardened ellipsoid bodies (frs) among lf bases, and with thread-like structures (styles) projecting from these bodies *(Lilaea)*   **Juncaginaceae** p. 561
      36b Lvs fully sheathing; basal ellipsoid bodies lacking
                                              Juncaceae p. 567
    35b Lvs ± flattened; pl often caulescent, gen per, often rhizomatous or cormose
      37a Sts elongate, often br; lvs mainly along st
        38a Stips prominent, sheathing, sometimes projecting above bl and surrounding st; lf bls filiform to greatly broadened, often toothed   **Potamogetonaceae** p. 562
        38b Stips small, not sheathing; lvs linear, entire, 1–5 mm broad
                                              Pontederiaceae p. 678
      37b Sts produced only at time of fl, unbr; lvs mainly or all basal
        39a Lvs with sheathing base gen several cm, bl linear, scarcely tapered, petiole (other than sheath) lacking
          40a Lvs pith-filled and terete, or strongly cross-septate and often equitant; sheaths gen open   **Juncaceae** p. 567

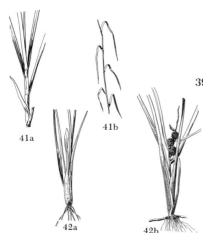

40b Lvs neither pith-filled nor strongly cross-septate, rarely equitant; sheaths open or closed

 41a Lvs in 3 vertical rows; st gen solid or pith-filled, often 3-cornered; sheaths gen closed **Cyperaceae** p. 576

 41b Lvs in 2 vertical rows; st gen hollow, mostly terete or bilateral; sheaths gen open **Gramineae** p. 602

39b Lvs sheathless or with short (rarely completely)-sheathing base, bl often expanded, petioles often elongate

 42a Lvs with very slender, ± terete petioles gen enlarged and sheathing at base and with broadened bl from narrowly elliptic to oval or sagittate **Alismataceae** p. 557

 42b Lvs sessile, slightly sheathing at base, elongate, flattened and ± tapelike, mostly 5–15 mm broad, never expanded into broader bl **Sparganiaceae** p. 674

Group II

Dicots, typically with floral parts in 4's or 5's and with netted-veined lvs

1a Pls br-parasites without roots in the soil, their sts arising above ground

 2a Sts twining, threadlike, pinkish or yellowish; fls gamopet **Cuscutaceae** p. 364

 2b Sts not twining, gen greenish; fls apet **Loranthaceae** p. 77

1b Pls not br-parasites, either rooted (and gen arising) underground, or aquatic and free-floating or anchored to the bottom by roots

 3a Pls saprophytic or parasitic and without visible green color in lvs and sts, gen lfless or with merely bractlike lvs on sts

  4a Corolla bilabiate, gamopet; stamens 4; placentae 2–3; root-parasites **Orobanchaceae** p. 444

  4b Corolla not bilabiate, often polypet; stamens 6–10; placentae 4–8; saprophytes **Ericaceae** p. 340

 3b Pls greenish lvd or std (although sometimes partially parasitic)

  5a Fls without normal parts, but modified into bulbil-like propagules; pls herbaceous

   6a Stips sheathing st; lvs entire (*Polygonum*) **Polygonaceae** p. 78

   6b Stips not sheathing st; lvs toothed (*Lithophragma, Saxifraga*) **Saxifragaceae** p. 184

  5b Fls normal, not modified into bulblets; pls herbaceous to arborescent

   7a Lvs modified to function as traps for insects either by means of tubular, much-enlarged petioles open at the top, or by flat bls covered with long, often purplish, gland-tipped hairs; pls of wet places; corolla polypet

    8a Lvs relatively large, curved near tip and with enlarged, hollow petiole opening under the curve **Sarraceniaceae** p. 182

    8b Lvs small; petioles not hollow, but bls obovate to oblanceolate and covered with long, gen purplish, gland-tipped|hairs **Droseraceae** p. 182

   7b Lvs not modified to function as insect traps as described above, occ insectivorous by means of broad glandular lvs, but then corolla gamopet

    9a Sts greatly enlarged, succulent, flattened and jointed or globose to cylindric and often ribbed or fluted, armed with stiff, needle-like spines; lvs absent or very small and scalelike **Cactaceae** p. 301

    9b Sts gen with green lvs, either not succulent and greatly enlarged or not at all spinose

     10a Sts woody or at least persistent and per above ground; pls mostly trees, shrubs, or woody vines—including many mat-forming spp. and several with herbaceous and greenish but persistent sts (e.g., English ivy), but excluding most pls with ± erect herbaceous sts with only a small persistent woody base **Subgroup IIa** (lead 11a)

     10b Sts herbaceous—including all anns and those pers that gen die back to the crown or to no more than a small persistent base **Subgroup IIb** (lead 25a)

Subgroup IIa

Dicots with woody sts

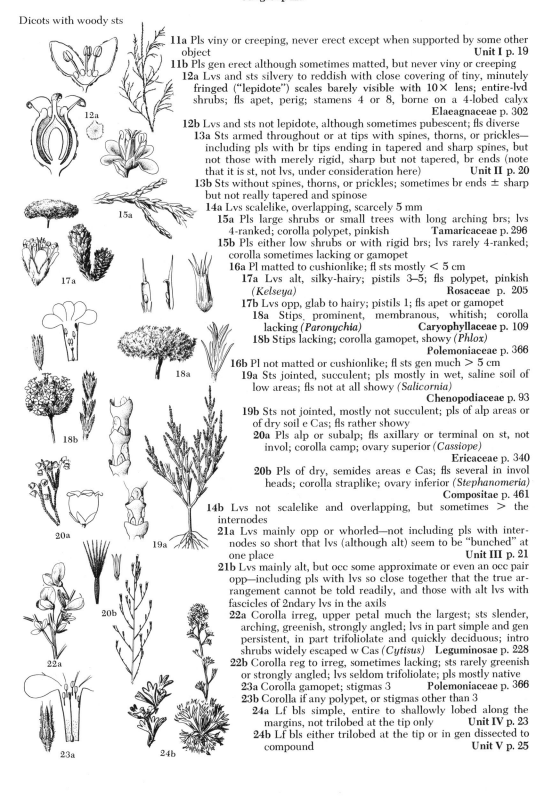

11a Pls viny or creeping, never erect except when supported by some other object   **Unit I p. 19**
11b Pls gen erect although sometimes matted, but never viny or creeping
 12a Lvs and sts silvery to reddish with close covering of tiny, minutely fringed ("lepidote") scales barely visible with 10× lens; entire-lvd shrubs; fls apet, perig; stamens 4 or 8, borne on a 4-lobed calyx **Elaeagnaceae p. 302**
 12b Lvs and sts not lepidote, although sometimes pubescent; fls diverse
  13a Sts armed throughout or at tips with spines, thorns, or prickles—including pls with br tips ending in tapered and sharp spines, but not those with merely rigid, sharp but not tapered, br ends (note that it is st, not lvs, under consideration here) **Unit II p. 20**
  13b Sts without spines, thorns, or prickles; sometimes br ends ± sharp but not really tapered and spinose
   14a Lvs scalelike, overlapping, scarcely 5 mm
    15a Pls large shrubs or small trees with long arching brs; lvs 4-ranked; corolla polypet, pinkish **Tamaricaceae p. 296**
    15b Pls either low shrubs or with rigid brs; lvs rarely 4-ranked; corolla sometimes lacking or gamopet
     16a Pl matted to cushionlike; fl sts mostly < 5 cm
      17a Lvs alt, silky-hairy; pistils 3–5; fls polypet, pinkish (*Kelseya*) **Rosaceae p. 205**
      17b Lvs opp, glab to hairy; pistils 1; fls apet or gamopet
       18a Stips prominent, membranous, whitish; corolla lacking (*Paronychia*) **Caryophyllaceae p. 109**
       18b Stips lacking; corolla gamopet, showy (*Phlox*) **Polemoniaceae p. 366**
     16b Pl not matted or cushionlike; fl sts gen much > 5 cm
      19a Sts jointed, succulent; pls mostly in wet, saline soil of low areas; fls not at all showy (*Salicornia*) **Chenopodiaceae p. 93**
      19b Sts not jointed, mostly not succulent; pls of alp areas or of dry soil e Cas; fls rather showy
       20a Pls alp or subalp; fls axillary or terminal on st, not invol; corolla camp; ovary superior (*Cassiope*) **Ericaceae p. 340**
       20b Pls of dry, semides areas e Cas; fls several in invol heads; corolla straplike; ovary inferior (*Stephanomeria*) **Compositae p. 461**
   14b Lvs not scalelike and overlapping, but sometimes > the internodes
    21a Lvs mainly opp or whorled—not including pls with internodes so short that lvs (although alt) seem to be "bunched" at one place **Unit III p. 21**
    21b Lvs mainly alt, but occ some approximate or even an occ pair opp—including pls with lvs so close together that the true arrangement cannot be told readily, and those with alt lvs with fascicles of 2ndary lvs in the axils
     22a Corolla irreg, upper petal much the largest; sts slender, arching, greenish, strongly angled; lvs in part simple and gen persistent, in part trifoliolate and quickly deciduous; intro shrubs widely escaped w Cas (*Cytisus*) **Leguminosae p. 228**
     22b Corolla reg to irreg, sometimes lacking; sts rarely greenish or strongly angled; lvs seldom trifoliolate; pls mostly native
      23a Corolla gamopet; stigmas 3 **Polemoniaceae p. 366**
      23b Corolla if any polypet, or stigmas other than 3
       24a Lf bls simple, entire to shallowly lobed along the margins, not trilobed at the tip only **Unit IV p. 23**
       24b Lf bls either trilobed at the tip or in gen dissected to compound **Unit V p. 25**

Subgroup IIb

Dicots with herbaceous sts

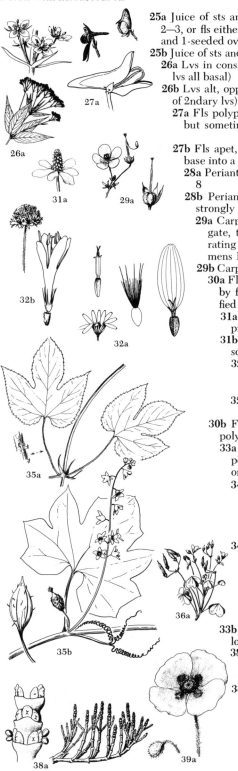

25a Juice of sts and often also of lvs milky; lvs simple and opp, or sepals 2–3, or fls either funnelf or borne in heads and with an inferior, 1-celled, and 1-seeded ovary   **Unit VI p. 26**

25b Juice of sts and lvs nonmilky

    26a Lvs in considerable part 3 or more per node (not including pls with lvs all basal)   **Unit VII p. 26**

    26b Lvs alt, opp, or all basal (sometimes main lvs with axillary fascicles of 2ndary lvs)

        27a Fls polypet and strongly irreg, gen showy; petals gen all distinct but sometimes at least 1 pair (or even 2 pairs) ± connate at base   **Unit VIII p. 27**

        27b Fls apet, or polypet but reg, or gamopet (the petals all joined at base into a common tube) and either reg or irreg

            28a Perianth 2- or 4-merous, apet or polypet; stamens gen 2, 4, 6, or 8   **Unit IX p. 28**

            28b Perianth gen 5-merous (occ lacking), if 2- or 4-merous then strongly gamopet or stamens other than 2, 4, 6, or 8

                29a Carpels 5 (2–8), 1-seeded, weakly connate around the elongate, tapered recep which extends above the calyx; frs separating at maturity, the styles persistent; petals 5, distinct; stamens 10, but sometimes only 5 fertile   **Geraniaceae p. 279**

            29b Carpels not as above; recep not greatly elongate

                30a Fls clustered in heads or thick, crowded spikes surrounded by few–∝ bracts; ovary inferior; calyx often greatly modified or lacking; corolla gamopet or lacking

                    31a Corolla lacking; invol bracts 5–8, showy, white or pinkish; seeds several   **Saururaceae p. 64**

                    31b Corolla gamopet; invol bracts ∝, gen greenish; seeds solitary

                        32a Stamens 5, gen connate around style; calyx modified into scales, bristles, or awns, or sometimes absent   **Compositae p. 461**

                        32b Stamens 2–4, distinct; calyx cupulate and several-lobed   **Dipsacaceae p. 456**

                30b Fls not clustered in heads, or ovary superior, or corolla polypet

                  33a Lf bls pinnately, ternately, or palmately divided to compound, the divisions extending > half way to the midrib or lf base

                    34a Pls strong vines with 3–7-lobed lvs

                      35a Lvs opp; tendrils lacking; fls apet *(Humulus)*   **Moraceae p. 75**

                      35b Lvs alt; tendrils present; fls gamopet   **Cucurbitaceae p. 457**

                    34b Pls not viny or lvs not 3–7-lobed

                      36a Lvs once-ternate; lflets obcordate; juice sour   **Oxalidaceae p. 281**

                      36b Lvs not ternate or lflets not obcordate; juice rarely sour

                        37a Corolla polypet or lacking   **Unit X p. 29**

                        37b Corolla gamopet   **Unit XI p. 30**

                33b Lf bls from entire or toothed to lobed or hastate, the lobing often not > half way to the midrib or base

                  38a Lvs opp and scalelike; sts succulent, jointed; fls minute, borne in fleshy terminal spikes *(Salicornia)*   **Chenopodiaceae p. 93**

                  38b Lvs not opp and scalelike or sts not succulent and jointed; fls often showy

                    39a Sepals 2–3, fewer than petals, shed as fl opens; pistil 1, compound; stamens gen ∝, if < 10 then ovary with parietal placentation   **Papaveraceae p. 143**

                    39b Sepals either > 3 or not shed as fl opens; pistil often simple, sometimes > 1; stamens sometimes few,

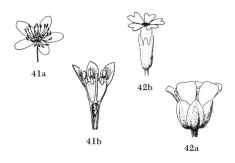

41a

41b

42b

42a

if < 10 then ovary seldom with parietal placentation
40a Perianth parts alike, not consisting of 2 dissimilar
   sets (corolla gen the set lacking), sometimes of 2 or
   more series of similar bractlike members
   41a Stamens hypog, not borne on the perianth
                                                   Unit XII p. 30
   41b Stamens perig or epig, borne on the perianth
      or on top of ovary   **Unit XIII p. 32**
40b Perianth consisting of both calyx and corolla, the
   two gen evidently dissimilar
   42a Corolla polypet   **Unit XIV p. 33**
   42b Corolla gamopet   **Unit XV p. 34**

Unit I

Sts woody, creeping or vining

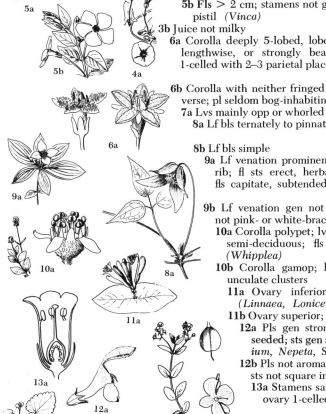

1a Lvs alt, exstip, persistent, pinnately compound; lflets leathery, spiny-
   toothed; sts unarmed *(Berberis)*   **Berberidaceae p. 142**
1b Lvs·and lflets not at once as above
   2a Sts herbaceous, unarmed, stolonous or ± rhizomatous; lvs of each year
      gen a single pair at tip of st, bl entire, cordate to reniform, persistent
                                                    Aristolochiaceae p. 78
   2b Sts and lvs not at once as above
      3a Juice milky; corolla gen twisted in the bud
         4a Lvs alt; corolla funnell, very shallowly lobed **Convolvulaceae p. 363**
         4b Lvs opp; corolla deeply lobed if at all funnell
            5a Fls < 2 cm; stamens grown together and to style and stigma
                                                    Asclepiadaceae p. 363
            5b Fls > 2 cm; stamens not grown together, almost or quite free of
               pistil *(Vinca)*   **Apocynaceae p. 362**
      3b Juice not milky
         6a Corolla deeply 5-lobed, lobes either with fringed scales running
            lengthwise, or strongly bearded with bristlelike scales; ovary
            1-celled with 2–3 parietal placentae; trailing bog pls
                                                    Menyanthaceae p. 361
         6b Corolla with neither fringed scales nor strong bearding; ovary di-
            verse; pl seldom bog-inhabiting
            7a Lvs mainly opp or whorled
               8a Lf bls ternately to pinnately compound *(Clematis)*
                                                    Ranunculaceae p. 124
            8b Lf bls simple
               9a Lf venation prominent, lateral veins ± parallel with mid-
                  rib; fl sts erect, herbaceous, with whorl of lvs near tip;
                  fls capitate, subtended by (gen 4) white or pinkish bracts
                                                    Cornaceae p. 339
               9b Lf venation gen not at all parallel; lvs not whorled; fls
                  not pink- or white-bracteate
                  10a Corolla polypet; lvs ovate-elliptic, short-petiolate, hairy,
                     semi-deciduous; fls several in small, stalked clusters
                     (Whipplea)   **Hydrangeaceae p. 204**
                  10b Corolla gamop; lvs diverse; fls rarely in small ped-
                     unculate clusters
                     11a Ovary inferior; lvs sometimes connate-perfoliate
                        (Linnaea, Lonicera)   **Caprifoliaceae p. 450**
                     11b Ovary superior; lvs never connate-perfoliate
                        12a Pls gen strongly aromatic; ovary 4-lobed and 4-
                           seeded; sts gen square in ✕−section *(Glecoma, Lam-
                           ium, Nepeta, Satureja)* **Labiatae p. 399**
                        12b Pls not aromatic; ovary not 4-lobed, gen ∝ -seeded;
                           sts not square in x-section
                           13a Stamens same no as corolla lobes and opp them;
                              ovary 1-celled, with free-central placentation
                                                    Primulaceae p. 350
                           13b Stamens fewer than corolla lobes, or alt with
                              lobes, or both; ovary gen 2-celled, never with free-
                              central placentation **Scrophulariaceae p. 413**

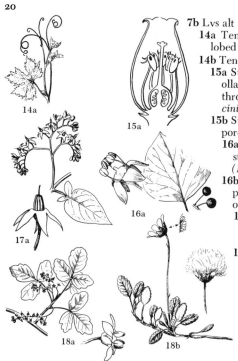

7b Lvs alt
   14a Tendrils present; lvs deciduous, bls simple, gen palmately
      lobed   **Vitaceae** p. 291
   14b Tendrils lacking; lvs gen persistent if palmately lobed
      15a Stamens 8 or 10; anthers opening by terminal pores; cor-
         olla gamopet; fr 4–5-celled caps or berry, often persistent
         through winter (*Arctostaphylos, Cassiope, Gaultheria, Vac-*
         cinium)   **Ericaceae** p. 340
      15b Stamens often other than 8 or 10, not opening by terminal
         pores; corolla polypet or lacking; fr various
         16a Pl evergreen; lvs leathery, simple, entire to 3–5-lobed;
            sts often with aerial roots; stamens 5; ovary inferior
            (*Hedera*)   **Araliaceae** p. 313
         16b Pls mostly deciduous, if evergreen the lvs often com-
            pound and sts without aerial roots; stamens often > 5;
            ovary mostly superior
            17a Lf bls gen in part simple, in part 3–5-lobed to subpin-
               natifid; st tips herbaceous; corolla gamopet, rotate (*So-*
               lanum)   **Solanaceae** p. 409
            17b Lf bls often all compound; sts sometimes woody
               throughout; corolla polypet or lacking
               18a Lvs 3–5-foliolate, deciduous, exstip, smooth and
                  shining; pl erect or clambering **Anacardiaceae** p. 287
               18b Lvs often simple, persistent, or stip, mostly not
                  shining; sts often trailing (*Dryas, Luetkea, Potentilla,*
                  Rubus)   **Rosaceae** p. 205

Unit II

Sts woody, armed with spines, thorns, or prickles

   1a Juice milky or reddish; lvs alt, deciduous, simple; br tips rigid and spiny
      (*Maclura*)   **Moraceae** p. 75
   1b Juice not milky; either lvs opp, evergreen, or compound, or brs not spinose
      at tip
      a1 Pl freely br, rhizomatous shrub 1–1.5 m, with needle-like, axillary spines;
         fls pea-like; Old World sp., estab in Grant Co, Wn (*Alhagi*)
                                     Leguminosae p. 228
      a2 Pl not as above in all respects
         2a Lateral brs spine-tipped, grooved, covered with rigid, sharp, needlelike,
            persistent lvs; fls yellow, irreg, polypet (*Ulex*)   **Leguminosae** p. 228
        2b Lateral brs either not spine-tipped or not bearing sharp needlelike lvs;
         fls not at once yellow, irreg, and polypet
         3a Sts prickly or spiny other than at tips; lvs palmately lobed; pistil 1;
            ovary inferior, 5–∝-seeded
            4a Lf bls 1-3 dm broad, strongly prickly along veins (*Oplopanax*)
                                      Araliaceae p. 313
            4b Lf bls gen < 1 dm broad, not prickly along veins
                                    Grossulariaceae p. 199
        3b Sts various, if prickly then either lvs not palmately lobed, or pistils >
         1, or ovary superior or with < 5 seeds
         5a Lvs opp, either grayish and mealy-pubescent or leathery and
            persistent; brs spine-tipped
            6a Lvs grayish mealy-pubescent; fls sessile in 2–several dense ver-
               ticils at br end; pl of sagebr des (*Salvia*)   **Labiatae** p. 399
            6b Lvs not grayish-mealy; fls umbellate to pan; pl of foothills and
               mts (*Ceanothus*)   **Rhamnaceae** p. 290
        5b Lvs alt or deciduous or brs not spine-tipped
         7a Brs and lvs ± grayish-villous or -tomentose; lvs linear or
            pedately 3–5-parted (*Artemisia, Tetradymia*) **Compositae** p. 461
        7b Brs and lvs not grayish-tomentose or lvs neither linear nor
         pedately parted
         8a Brs glab, grooved lengthwise, spine-tipped, green; lvs alt,
            entire, 3–15 mm; petioles jointed with a fleshy, cushionlike,
            gen purplish base; stips linear, persistent (*Glossopetalon*)
                                   Celastraceae p. 288
        8b Brs not grooved or not spine-tipped, or lvs mostly > 15 mm
         and otherwise not as above

9a Pl pinnate-lvd, arborescent; some stips gen modified into
   spines; fls in drooping racemes *(Robinia)*
                                                   Leguminosae p. 228
9b Pl either not arborescent or not pinnate-lvd; stips rarely
   spinose; fls seldom in drooping racemes
   10a Corolla polypet; stamens > 5 *(Rosa, Rubus)*
                                                   Rosaceae p. 205
   10b Corolla gamopet or lacking; stamens not > 5
       11a Corolla lacking; fls inconspicuous; ovary 1-celled
           and 1-seeded, fr dry; lvs gen fleshy, scurfy or terete
           (Atriplex, Sarcobatus)   **Chenopodiaceae** p. 83
       11b Corolla present; fls ± showy; ovary 2-celled, ∝
           -seeded; fr fleshy; lvs various   **Solanaceae** p. 409

Sts woody; lvs opp or whorled   **Unit III**

1a Lvs whorled (sometimes only on the peduncles or at 1 st node); fls not at
   once gamopet and with inferior ovary, 5 stamens, and 3 stigmas
   2a Sts or lvs gen woolly; lvs mostly entire *(Eriogonum)* **Polygonaceae** p. 78
   2b Neither sts nor lvs woolly; lvs often toothed
       3a Lvs linear or linear-lanceolate, entire
           4a Brs woody, not dying back in winter; lvs heatherlike, longer than
              the internodes, margins revolute, lower surface grooved; ovary
              superior   **Empetraceae** p. 286
           4b Brs semiwoody, partially dying back in winter; lvs not revolute|and
              grooved beneath; ovary inferior *(Galium)* **Rubiaceae** p. 448
       3b Lvs either broader than linear-lanceolate or plainly toothed
           5a Lf bls entire, main veins ± parallel; corolla polypet; stamens 4
                                                   Cornaceae p. 339
           5b Lf bls gen toothed, main veins not parallel; corolla gamopet or sta-
              mens 8–10
               6a Fertile stamens 4; corolla irreg *(Penstemon, Scrophularia)*
                                                   Scrophulariaceae p. 413
               6b Fertile stamens gen 8–10; corolla reg *(Chimaphila)*
                                                   Ericaceae p. 340
1b Lvs opp on most parts of st (sometimes also deeply palmatifid and pseu-
   do-whorled but then corolla gamopet, ovary superior, stamens 5, and
   stigmas 3)
   7a Pl only slightly woody, sts dying back most of length each year
                                                   Subunit IIIa, lead 8a
   7b Pl woody throughout, sts dying back only slightly or not at all each year
                                                   Subunit IIIb, lead 20a

Subunit IIIa

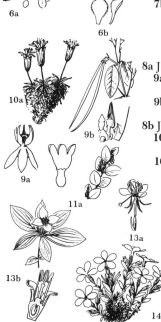

8a Juice milky; seeds gen hairy
   9a Stamens fully connate, surrounding and adnate to the style and stigma
                                                   Asclepiadaceae p. 363
   9b Stamens free or only partially connate, weakly or not at all adnate to
      style and stigma   **Apocynaceae** p. 362
8b Juice not milky; seeds rarely hairy
   10a Lvs linear, connate at base; fls solitary on short peduncles; pl matted;
       corolla polypet *(Silene)*   **Caryophyllaceae** p. 109
   10b Lvs not linear and connate or fls > 1 per fl st; corolla often gamopet
       11a Corolla gamopet; ovary inferior   **Caprifoliaceae** p. 450
       11b Corolla sometimes polypet or lacking, if gamopet then ovary supe-
           rior
           12a Fls 4-merous; ovary inferior, gen 4-celled
               13a Stamens gen 8; free hypan gen evident; fls not in
                   white-bracteate clusters   **Onagraceae** p. 303
               13b Stamens gen 4; free hypan lacking; fls in white (pink) -bracteate
                   clusters   **Cornaceae** p. 339
           12b Fls mostly other than 4-merous; ovary superior
               14a Corolla reg, strongly gamopet; stigmas 3, ovary 3-celled and
                   3-∝ -seeded   **Polemoniaceae** p. 366
               14b Corolla irreg, polypet, or lacking; ovary often only 1-seeded if
                   stigmas 3

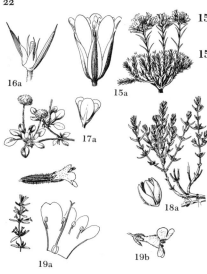

15a Nodes gen swollen; styles or stigmas 2–5; ovary 1-celled, several-seeded, placentation free-central; lvs often pungent (*Arenaria*)   **Caryophyllaceae** p. 109

15b Nodes mostly not swollen; ovary gen either 1-seeded or 2-celled; lvs rarely pungent

   16a Lvs with prominent scarious stips; ovary 1-seeded; stigmas 2 (*Cardionema, Paronychia*)   **Caryophyllaceae** p. 109

   16b Lvs exstip, or ovary > 1-seeded, or stigmas other than 2

      17a Sts or lvs woolly, never mealy or succulent (*Eriogonum*)   **Polygonaceae** p. 78

      17b Neither sts nor lvs woolly, but sometimes mealy or succulent

         18a Lvs gen succulent or farinose (covered with a meal-like powder); corolla lacking; ovary 1-seeded   **Chenopodiaceae** p. 93

         18b Lvs neither succulent nor farinose; corolla present, gamopet; ovary > 1-seeded

            19a Ovary 4-ovuled, gen 4-lobed; fr 4 nutlets; sts gen square in ✕–section (*Agastache, Hedeoma, Monardella*)   **Labiatae** p. 399

            19b Ovary ∝ -ovuled, seldom 4-lobed; fr caps; sts not square in ✕–section (*Penstemon*)   **Scrophulariaceae** p. 413

Subunit IIIb

20a Lvs palmately lobed to pinnately compound

   21a Corolla polypet or lacking; fr a 1- or 2-winged achene (samara); ovary superior

      22a Lvs mostly palmately lobed (in 1 sp. pinnately compound but lflets mostly 3 or rarely 5); samara double   **Aceraceae** p. 288

      22b Lvs pinnately compound, lflets (3) 5–7 (9); samara single   **Oleaceae** p. 356

   21b Corolla gamopet; fr fleshy, 1–several-seeded; ovary inferior   **Caprifoliaceae** p. 450

20b Lvs simple, gen entire or only toothed

   23a Sts and lvs often ± pubescent with appressed, 2-armed hairs; lvs prominently veined, main lateral veins ± parallel with midrib; sts often reddish; fls sometimes in white-bracteate heads   **Cornaceae** p. 339

   23b Sts not pubescent with 2-armed hairs, rarely red; fls never in white-bracteate heads

      24a Stamens 10, hypog, anthers often opening by terminal pores; fls complete, gamopet, reg; ovary 5-celled (*Chimaphila, Kalmia*)   **Ericaceae** p. 340

      24b Stamens either < 10, or epig; fls often incomplete or polypet, gen irreg if gamopet; ovary rarely 5-celled

         25a Fls ♂, ♀, borne in drooping racemes; lvs leathery, subentire, persistent, bl 4–8 cm; petioles slightly connate by a small ridge   **Garryaceae** p. 339

         25b Fls ☿ to ♂ ♀, not borne in drooping racemes; lvs various, often deciduous

            26a Calyx 4-lobed; stamens 8, borne at top of calyx tube; corolla lacking; lvs silvery or rusty beneath (*Shepherdia*)   **Elaeagnaceae** p. 302

            26b Calyx not 4-lobed if stamens 8; corolla gen present; lvs rarely silvery or rusty

               27a Lvs (sometimes also sts) gen covered with meal-like powder, often also ± glandular

                  28a Lvs grayish-mealy, minutely glandular and aromatic; fls gamopet, sessile in 2–several dense verticils at br ends (*Salvia*)   **Labiatae** p. 399

                  28b Lvs gen green or pale, neither glandular nor strongly aromatic; fls apet, not strictly verticillate   **Chenopodiaceae** p. 93

               27b Lvs and sts never mealy, rarely glandular, often hairy

                  29a Lvs coriaceous, persistent, oblong to obovate, 1–2.5 cm, with persistent, thickened stips 1–2 mm; stamens 5, opp petals; fls perig (*Ceanothus*)   **Rhamnaceae** p. 290

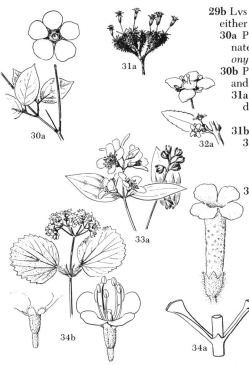

29b Lvs diverse, mostly exstip or inconspicuously stip; stamens either alt with petals (if any) or fls epig

  30a Pl deciduous; lf bls oblong-lanceolate, serrate, acuminate; stips tiny, glandlike; fls perig; brs green-barked (*Euonymus*)   **Celastraceae** p. 288

  30b Pl often evergreen; lvs various, often exstip or leathery and persistent; fls often epig

    31a Lvs linear, connate at base; fls solitary on short peduncles; pl matted, < 10 cm; corolla polypet (*Silene*)   **Caryophyllaceae** p. 109

    31b Lvs, pl, and fls not at once as above

      32a Fls perig, with a prominent disc surrounding the ovary; ovary ultimately protruding above disc (superior); lvs leathery, glab, glossy, persistent, 1–3 cm, finely serrate (*Pachistima*)   **Celastraceae** p. 288

      32b Fls perig to epig; ovary at least half inferior; lvs mostly deciduous, sometimes lobed

        33a Lvs gen with 3 prominent veins running from base, often serrate but never lobed, petiolate; corolla polypet   **Hydrangeaceae** p. 204

        33b Lvs seldom strongly 3-veined, sometimes sessile or entire, if serrate and 3-veined then also lobed; corolla gamopet

          34a Lvs connected at base by a small ridge (interpetiolar stips), tomentose beneath; calyx and corolla 4-lobed   **Buddlejaceae** p. 356

          34b Lvs not connected by interpetiolar stips; calyx and corolla gen 5-lobed   **Caprifoliaceae** p. 450

Unit IV

Sts woody; lvs alt, bls simple, entire to shallowly lobed

1a Pl low, woody-br, subalp to alp, heatherlike; lvs closely crowded, linear, entire, 4–15 (18) mm, evergreen, often revolute-margined

  2a Fls pedunculate, borne singly or in clusters; fr caps, gen persistent through winter (*Cassiope, Phyllodoce*)   **Ericaceae** p. 340

  2b Fls subsessile in lf axils; fr fleshy, rarely remaining on pl through winter   **Empetraceae** p. 286

1b Pl often in large part herbaceous or cushionlike, not heatherlike, if with linear, revolute-margined lvs then gen of des, saline, or lowl areas

  3a Pl gen semishrubby, ann brs tending to die back toward woody base

    4a Lvs exstip, often narrow, gen covered with meal-like powder, but sometimes grayish- or yellow-hairy (sometimes stellate-pubescent) or glab and very succulent; fls apet, inconspicuous, often ♂, ♀, or ♂♀, never borne in umbels; styles 2 (rarely 3–4); ovary 1-celled and 1-seeded   **Chenopodiaceae** p. 93

    4b Lvs and fls various, but not at once as above

      5a Corolla polypet; sepals and petals 4; stamens gen 6 (*Smelowskia*)   **Cruciferae** p. 146

      5b Corolla gamopet or fls with other than 4 sepals and petals and 6 stamens

        6a Fls in heads or umbels, gen invol; lvs often linear or woolly, at least beneath

          7a Fls mostly in umbels; ovary superior   **Polygonaceae** p. 78

          7b Fls in heads; ovary inferior   **Compositae** p. 461

        6b Fls not invol, rarely in heads or umbels; lvs diverse

          8a Corolla polypet, very irreg, upper petal largest; pistil 1, ovary 1-celled and 1-placentary (*Astragalus*)   **Leguminosae** p. 228

          8b Corolla gen gamopet if irreg; ovary gen > 1-celled or > 1-placentary

            9a Lvs gen stip; stamens gen > 10; fls perig or epig, polypet or apet; pistil sometimes > 1, neither 2-celled nor 1-celled with 2–3 parietal placentae   **Rosaceae** p. 205

            9b Lvs often exstip; stamens rarely > 10; fls often hypog, sometimes gamopet; pistil 1, sometimes 2-celled or 1-celled with 2–3 parietal placentae

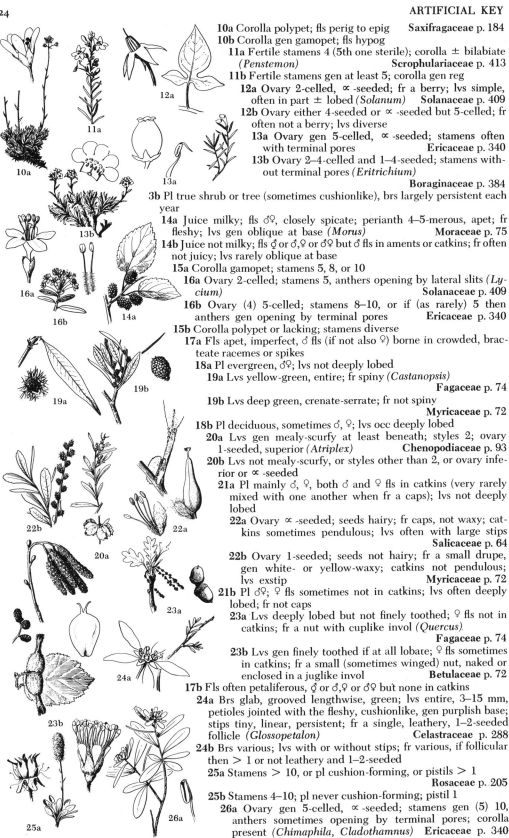

10a Corolla polypet; fls perig to epig   **Saxifragaceae** p. 184

10b Corolla gen gamopet; fls hypog

  11a Fertile stamens 4 (5th one sterile); corolla ± bilabiate (*Penstemon*)   **Scrophulariaceae** p. 413

  11b Fertile stamens gen at least 5; corolla gen reg

    12a Ovary 2-celled, ∝ -seeded; fr a berry; lvs simple, often in part ± lobed (*Solanum*)   **Solanaceae** p. 409

    12b Ovary either 4-seeded or ∝ -seeded but 5-celled; fr often not a berry; lvs diverse

      13a Ovary gen 5-celled, ∝ -seeded; stamens often with terminal pores   **Ericaceae** p. 340

      13b Ovary 2–4-celled and 1–4-seeded; stamens without terminal pores (*Eritrichium*)

                        Boraginaceae p. 384

3b Pl true shrub or tree (sometimes cushionlike), brs largely persistent each year

  14a Juice milky; fls ♂♀, closely spicate; perianth 4–5-merous, apet; fr fleshy; lvs gen oblique at base (*Morus*)   **Moraceae** p. 75

  14b Juice not milky; fls ♂ or ♂,♀ or ♂♀ but ♂ fls in aments or catkins; fr often not juicy; lvs rarely oblique at base

    15a Corolla gamopet; stamens 5, 8, or 10

      16a Ovary 2-celled; stamens 5, anthers opening by lateral slits (*Lycium*)   **Solanaceae** p. 409

      16b Ovary (4) 5-celled; stamens 8–10, or if (as rarely) 5 then anthers gen opening by terminal pores   **Ericaceae** p. 340

    15b Corolla polypet or lacking; stamens diverse

      17a Fls apet, imperfect, ♂ fls (if not also ♀) borne in crowded, bracteate racemes or spikes

        18a Pl evergreen, ♂♀; lvs not deeply lobed

          19a Lvs yellow-green, entire; fr spiny (*Castanopsis*)

                        Fagaceae p. 74

          19b Lvs deep green, crenate-serrate; fr not spiny

                        Myricaceae p. 72

        18b Pl deciduous, sometimes ♂, ♀; lvs occ deeply lobed

          20a Lvs gen mealy-scurfy at least beneath; styles 2; ovary 1-seeded, superior (*Atriplex*)   **Chenopodiaceae** p. 93

          20b Lvs not mealy-scurfy, or styles other than 2, or ovary inferior or ∝ -seeded

            21a Pl mainly ♂, ♀, both ♂ and ♀ fls in catkins (very rarely mixed with one another when fr a caps); lvs not deeply lobed

              22a Ovary ∝ -seeded; seeds hairy; fr caps, not waxy; catkins sometimes pendulous; lvs often with large stips   **Salicaceae** p. 64

              22b Ovary 1-seeded; seeds not hairy; fr a small drupe, gen white- or yellow-waxy; catkins not pendulous; lvs exstip   **Myricaceae** p. 72

            21b Pl ♂♀; ♀ fls sometimes not in catkins; lvs often deeply lobed; fr not caps

              23a Lvs deeply lobed but not finely toothed; ♀ fls not in catkins; fr a nut with cuplike invol (*Quercus*).

                        Fagaceae p. 74

              23b Lvs gen finely toothed if at all lobate; ♀ fls sometimes in catkins; fr a small (sometimes winged) nut, naked or enclosed in a juglike invol   **Betulaceae** p. 72

      17b Fls often petaliferous, ♂ or ♂,♀ or ♂♀ but none in catkins

        24a Brs glab, grooved lengthwise, green; lvs entire, 3–15 mm, petioles jointed with the fleshy, cushionlike, gen purplish base; stips tiny, linear, persistent; fr a single, leathery, 1–2-seeded follicle (*Glossopetalon*)   **Celastraceae** p. 288

        24b Brs various; lvs with or without stips; fr various, if follicular then > 1 or not leathery and 1–2-seeded

          25a Stamens > 10, or pl cushion-forming, or pistils > 1

                        Rosaceae p. 205

          25b Stamens 4–10; pl never cushion-forming; pistil 1

            26a Ovary gen 5-celled, ∝ -seeded; stamens gen (5) 10, anthers sometimes opening by terminal pores; corolla present (*Chimaphila, Cladothamnus*)   **Ericaceae** p. 340

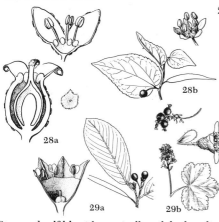

26b Ovary 1–5-celled, often only 1-seeded; stamens 4–6, anthers not opening by terminal pores; corolla sometimes lacking

  27a Ovary 1-celled and 1-seeded; petals lacking; stamens opp sepals

    28a Fls 4-merous, perig; shrubs; lvs equal-sided (± symmetrical) at base   **Elaeagnaceae** p. 302

    28b Fls 4–6-merous, hypog; small tree; lvs unequal-sided (asymmetrical) at base   **Ulmaceae** p. 74

  27b Ovary 2–3-celled or several-seeded; petals often present; stamens sometimes alt with sepals

    29a Stamens opp petals (alt with sepals); lvs never palmately lobed; ovary superior   **Rhamnaceae** p. 290

    29b Stamens alt with petals; lvs gen palmately lobed; ovary inferior   **Grossulariaceae** p. 199

Unit V

Sts woody; lf bls either apically trilobed or dissected to compound

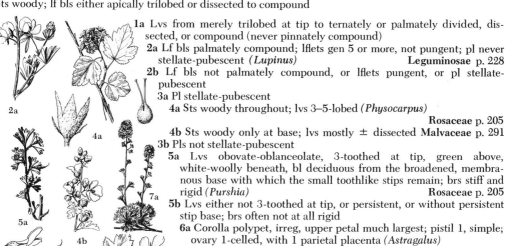

1a Lvs from merely trilobed at tip to ternately or palmately divided, dissected, or compound (never pinnately compound)

  2a Lf bls palmately compound; lflets gen 5 or more, not pungent; pl never stellate-pubescent *(Lupinus)*   **Leguminosae** p. 228

  2b Lf bls not palmately compound, or lflets pungent, or pl stellate-pubescent

    3a Pl stellate-pubescent

      4a Sts woody throughout; lvs 3–5-lobed *(Physocarpus)*   **Rosaceae** p. 205

      4b Sts woody only at base; lvs mostly ± dissected **Malvaceae** p. 291

    3b Pls not stellate-pubescent

      5a Lvs obovate-oblanceolate, 3-toothed at tip, green above, white-woolly beneath, bl deciduous from the broadened, membranous base with which the small toothlike stips remain; brs stiff and rigid *(Purshia)*   **Rosaceae** p. 205

      5b Lvs either not 3-toothed at tip, or persistent, or without persistent stip base; brs often not at all rigid

        6a Corolla polypet, irreg, upper petal much largest; pistil 1, simple; ovary 1-celled, with 1 parietal placenta *(Astragalus)*   **Leguminosae** p. 228

        6b Corolla gamopet, or reg, or pistil compound or >1

          7a Lvs persistent, triternate; pl trailing and ± matted *(Luetkea)*   **Rosaceae** p. 205

          7b Lvs not persistent and triternate or pl not trailing and matted

            8a Lf bls palmately lobed; corolla polypet

              9a Stips prominent, not fused lengthwise with petiole; stamens and pistils ∝ *(Rubus)*   **Rosaceae** p. 205

              9b Stips lacking, or fused their length with petiole; stamens 5, pistil 1   **Grossulariaceae** p. 199

            8b Lf bls not palmately lobed, or corolla not polypet

              10a Fls in invol heads; lvs rarely ternate   **Compositae** p. 461

              10b Fls not in invol heads; lvs ternate, segms shallowly to deeply lobed

                11a Stamens at least 25; pistils ∝ *(Rubus)* **Rosaceae** p. 205

                11b Stamens < 15; pistil 1   **Anacardiaceae** p. 287

1b Lvs pinnately compound

  12a Lflets glossy and shining, pungent-toothed; pl evergreen *(Berberis)*   **Berberidaceae** p. 142

  12b Lflets gen not glossy, not pungent-toothed; pl often deciduous

    13a Corolla irreg, polypet, upper petal largest; pistil 1, simple; ovary 1-celled with 1 parietal placenta *(Astragalus, Robinia)*   **Leguminosae** p. 228

    13b Corolla not irreg if polypet; pistil gen compound or > 1

      14a Lvs stip (stips sometimes deciduous); stamens 15–100 *(Chamaebatiaria, Potentilla, Rubus, Sorbus)*   **Rosaceae** p. 205

      14b Lvs exstip; stamens < 15

        15a Lvs gen > 4 cm; lflets elliptic to ovate or obovate, toothed, never tomentose

16a Pl arborescent; lflets 10–25, mostly > 4 cm, with 1–3 rounded lobes or coarse teeth on each side near base, each lobe with a large, sessile gland; fr a samara   **Simaroubaceae** p. 283
16b Pl shrubby; lflets < 10 or otherwise not as above; fr not a samara   **Anacardiaceae** p. 287
15b Lvs often < 4 cm; lflets gen linear, often woolly (*Artemisia, Eriophyllum, Tanacetum*)   **Compositae** p. 461

Unit VI

Sts herbaceous; juice milky

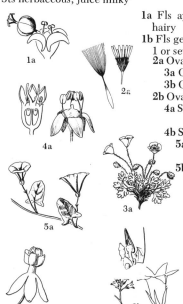

1a Fls apet; ovary 3-carpellary and -celled, each cell 1-seeded; seeds not hairy   **Euphorbiaceae** p. 284
1b Fls gen with ± showy corolla; ovary 2-carpellary; seeds often hairy, either 1 or several
  2a Ovary inferior
    3a Ovary 1-celled and 1-seeded   **Compositae** p. 461
    3b Ovary 3–5-celled and ∝ -seeded   **Campanulaceae** p. 457
  2b Ovary superior
    4a Sepals 2 or 3, shed as the fl opens; corolla polypet   **Papaveraceae** p. 143
    4b Sepals 5 (4), persistent; corolla often gamopet
      5a Pistil 5 (4–6) -carpellary; petals distinct, gen quickly shed   **Araliaceae** p. 313
      5b Pistil 2-carpellary; petals joined at base, persistent
        6a Corolla large and funnelf, very shallowly lobed; seeds 2–4, not hairy   **Convolvulaceae** p. 363
        6b Corolla either not large and funnelf or divided into prominent lobes; seeds ∝ , mostly hairy
          7a Corolla lobed much > half the length, lobes reflexed; stamens completely fused, surrounding and adnate with the style and stigma   **Asclepiadaceae** p. 363
          7b Corolla either lobed < half the length, or with non-reflexed lobes; stamens free or partially connate, weakly if at all adnate to the style and stigma   **Apocynacea** p. 362

Unit VII

Sts herbaceous; lvs whorled

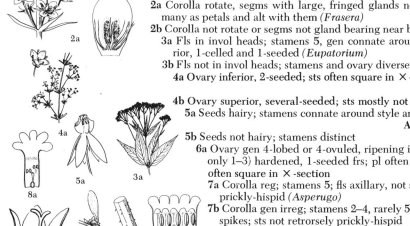

1a Corolla gamopet
  2a Corolla rotate, segms with large, fringed glands near base; stamens as many as petals and alt with them (*Frasera*)   **Gentianaceae** p. 356
  2b Corolla not rotate or segms not gland bearing near base
    3a Fls in invol heads; stamens 5, gen connate around style; ovary inferior, 1-celled and 1-seeded (*Eupatorium*)   **Compositae** p. 461
    3b Fls not in invol heads; stamens and ovary diverse
      4a Ovary inferior, 2-seeded; sts often square in ✕ -section   **Rubiaceae** p. 448
      4b Ovary superior, several-seeded; sts mostly not square in ✕ -section
        5a Seeds hairy; stamens connate around style and stigma   **Asclepiadaceae** p. 363
        5b Seeds not hairy; stamens distinct
          6a Ovary gen 4-lobed or 4-ovuled, ripening into 4 (or by abortion only 1–3) hardened, 1-seeded frs; pl often harsh-pubescent; sts often square in ✕ -section
            7a Corolla reg; stamens 5; fls axillary, not spicate; sts retrorsely prickly-hispid (*Asperugo*)   **Boraginaceae** p. 384
            7b Corolla gen irreg; stamens 2–4, rarely 5; fls often in terminal spikes; sts not retrorsely prickly-hispid
              8a Stamens 4–5; corolla only slightly irreg; style terminal on ovary or nutlets   **Verbenaceae** p. 398
              8b Stamens 2 or 4; corolla gen strongly irreg; style attached toward base of nutlets or lobes of ovary   **Labiatae** p. 399
          6b Ovary neither 4-lobed nor developing into 4 nutlets; pls not harsh-pubescent; sts rarely square in ✕ -section
            9a Stigmas gen 3 (2); stamens alt with corolla lobes; placentation not free-central   **Polemoniaceae** p. 366

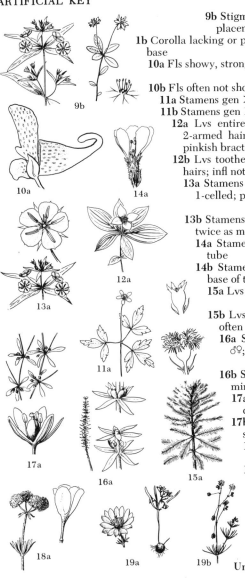

9b Stigmas gen single and capitate; stamens opp corolla lobes; placentation free-central **Primulaceae** p. 350

1b Corolla lacking or polypet, rarely some or all the segms weakly connate at base

10a Fls showy, strongly irreg; corolla spurred or saccate
Balsaminaceae p. 289

10b Fls often not showy, never strongly irreg

11a Stamens gen > 10; pistils several, simple **Ranunculaceae** p. 124

11b Stamens gen 10 or fewer; pistil 1, gen compound

12a Lvs entire, ± parallel-veined, sparsely pubescent with sessile 2-armed hairs; fls clustered, infl surrounded by 4 large white or pinkish bracts **Cornaceae** p. 339

12b Lvs toothed, not parallel-veined, or not pubescent with 2-armed hairs; infl not subtended by large bracts

13a Stamens hypog, as many as corolla lobes and opp them; ovary 1-celled; placentation free-central; style and stigma 1
Primulaceae p. 350

13b Stamens sometimes perig or epig, gen alt with corolla lobes or twice as many as they; ovary and style diverse

14a Stamens and petals (if any) borne toward top of the calyx tube **Lythraceae** p. 302

14b Stamens borne with petals, if any, on the recep or at the base of the calyx tube

15a Lvs pinnately dissected; stamens 4 or 8, epig
Haloragaceae p. 312

15b Lvs not pinnately dissected, or stamens other than 4 or 8, often hypog

16a Stamen 1; fls greenish, tiny, sessile in lf axils, mostly ♂♀; pls of wet places; lvs linear, sessile, 4–10 per node
Hippuridaceae p. 313

16b Stamens > 1; fls often colored to showy, ped or in terminal infl, mostly ♀; pls mostly of dry places; lvs diverse

17a Fls axillary, inconspicuous, greenish, apet; ovary 3 (5)-celled; styles 3 (5) **Aizoaceae** p. 104

17b Fls mostly in terminal infl, often colored or even showy; ovary mostly 1-celled; styles diverse

18a Ovary 1-seeded; stigmas gen 3; perianth segms 3–6, much alike, often greenish **Polygonaceae** p. 78

18b Ovary > 1-seeded; stigmas 3–5; perianth segms gen > 6, in 2 dissimilar series

19a Sepals 2 or petals ca 15; lvs whorled at only a few nodes **Portulacaceae** p. 104

19b Sepals and petals 5; lvs whorled at all nodes (*Spergula*) **Caryophyllaceae** p. 109

Unit VIII

Sts herbaceous; fls polypet and strongly irreg

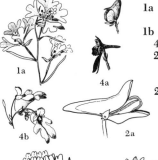

1a Sepals and petals 4; stamens 4 or 8; ovary inferior, 4-celled (*Clarkia*)
Onagraceae p. 303

1b Sepals and petals 5, or stamens 6, or ovary either superior or other than 4-celled

2a Stamens 6, connate into 2 groups of 3 each; sepals 2, tiny, often deciduous; petals 4, one or both of the outer pair spurred or saccate
Fumariaceae p. 144

2b Stamens often 5, 10, or ∝, never connate into 2 equal groups; sepals at least 3; petals spurred or nonspurred

3a Calyx gen petaloid and more showy than the corolla, 1 sepal strongly saccate, hooded, or spurred

4a Pistils 2 or more; stamens ∝, free (*Āconitum, Delphinium*)
Ranunculaceae p. 124

4b Pistil 1; stamens 5, often connate **Balsaminaceae** p. 289

3b Calyx mostly not petaloid, rarely more showy than the corolla, gen greenish, when spurred (as occ) the spur formed from 2 sepals

5a Stamens 8–50, not connate, inserted at edge of a lopsided disc; ovary 3–4 (5) -celled **Resedaceae** p. 181

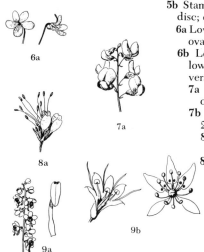

5b Stamens 5–10, sometimes ± connate, not inserted at edge of a disc; ovary 1-, 2-, or 5-celled

  6a Lowest petal of corolla largest; stamens 5, connivent around pistil; ovary 1-celled with 3 parietal placentae   **Violaceae** p. 296

  6b Lowest petal of corolla gen not largest, although sometimes lower pair of petals larger than upper 3; stamens and ovary diverse

    7a Upper petal of fl much the largest; stamens often connate; ovary 1-celled with 1 parietal placenta   **Leguminosae** p. 228

    7b Upper petal gen not the largest; stamens not connate; ovary 2–4-celled or with 2 parietal placentae

      8a Sepals and petals 4; stamens often 6 (4–∝), hypog; lvs gen 3–5-foliolate; ovary long-stipitate **Capparidaceae** p. 180

      8b Sepals and petals gen 5; stamens rarely 6, often perig to epig; lvs rarely 3–5-foliolate; ovary not long-stipitate

        9a Stamens hypog, anthers dehiscent by terminal pores; ovary (4) 5-celled *(Pyrola)*   **Ericaceae** p. 340

        9b Stamens perig or epig, anthers not dehiscent by terminal pores; ovary gen 1–2 (3)-celled *(Saxifraga, Tolmiea)*   **Saxifragaceae** p.184

Unit IX

Sts herbaceous; fls (except pistil) 2- or 4-merous

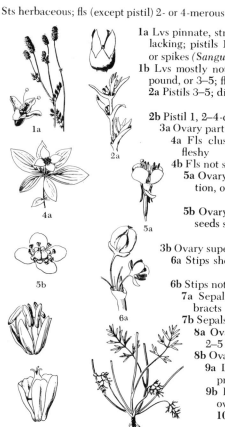

1a Lvs pinnate, strongly stip; pinnae toothed to pectinately dissected; corolla lacking; pistils 1–3, simple, 1-seeded; stamens perig; fls ∝ in dense heads or spikes *(Sanguisorba)*   **Rosaceae** p. 205

1b Lvs mostly not pinnate; corolla often present; pistils either 1 and compound, or 3–5; fls often racemose, pan, or scapose

  2a Pistils 3–5; diminutive ann with opp, entire lvs *(Tilaea)*   **Crassulaceae** p. 182

  2b Pistil 1, 2–4-carpellary; pl diverse

    3a Ovary partially to wholly inferior; stamens, 4, 6, or 8

      4a Fls clustered, subtended by several white or pinkish bracts; fr fleshy   **Cornaceae** p. 339

      4b Fls not subtended by large colored bracts; fr caps

        5a Ovary wholly inferior, either 2–4-celled and with axile placentation, or 1-celled and 1–2-seeded; corolla gen present   **Onagraceae** p. 303

        5b Ovary only partially inferior, 1-celled with 2 parietal placentae; seeds several; corolla lacking *(Chrysosplenium)*   **Saxifragaceae** p. 184

    3b Ovary superior

      6a Stips sheathing st above petiole; stigmas strongly fringed *(Oxyria)*   **Polygonaceae** p. 78

      6b Stips not sheathing st; stigmas entire or lobed

        7a Sepals scarious; fls subtended by sharp-pointed, perianth-like bracts   **Amaranthaceae** p. 101

        7b Sepals not scarious; fls not subtended by sharp-pointed bracts

          8a Ovary deeply 2–5-lobed, segms subglobose, 1-seeded; styles 2–5   **Limnanthaceae** p. 287

          8b Ovary not divided into globose segms; style gen 1

            9a Lvs opp, stip; stamens 8; ovary 4-celled; diminutive, gen prostrate ann *(Elatine)*   **Elatinaceae** p. 295

            9b Lvs rarely if ever both opp and stip; stamens seldom 8; ovary diverse

              10a Stamens 2; corolla absent or vestigial; calyx unequally (3) 4-lobed; ovary 2-celled, placentation axile *(Besseya)*   **Scrophulariaceae** p. 413

              10b Stamens mostly other than 2; corolla gen present; calyx gen equally lobed and sepals distinct; ovary diverse

                11a Sepals 2 or 3, always shed as fl opens   **Papaveraceae** p. 143

                11b Sepals gen 4 (rarely more), gen persistent during fl period

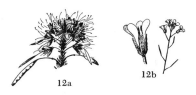

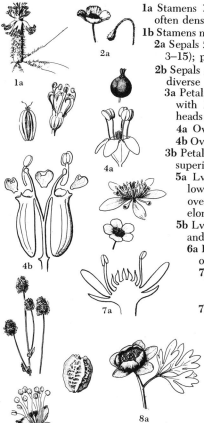

12a Ovary with free-central placentation, 1-celled
Caryophyllaceae p. 109
12b Ovary with parietal placentation, 1- or 2-celled
    13a Fr 1-celled, ∝-seeded, gen strongly stipitate; stamens occ > 6 **Capparidaceae** p. 180
    13b Fr gen 2-celled or occ 1-celled but then seldom at all stipitate and only 1- or 2-seeded; stamens 2-6 **Cruciferae** p. 146

Unit X

Apet or polypet herbs with deeply lobed to divided lvs

1a Stamens > 10, monadelphous, forming a tube surrounding the pistil; pl often densely stellate-pubescent **Malvaceae** p. 291
1b Stamens not > 10 if monadelphous; pl gen not stellate-pubescent
    2a Sepals 2–3 (fewer than petals), shed as fl opens; stamens mostly ∝ (occ 3–15); pistil compound **Papaveraceae** p. 143
    2b Sepals gen as many as petals (if any), rarely lacking; stamens and pistils diverse
        3a Petals, sepals, and stamens gen 5; pistil 1, ovary inferior, 2–6-celled, with 1 seed per cell; fls mostly in flat- to round-topped umbels or heads
            4a Ovary 4–6-celled; styles 4–6 *(Aralia)* **Araliaceae** p. 313
            4b Ovary 2-celled; styles 2 **Umbelliferae** p. 314
        3b Petals, sepals, and stamens often not all 5; pistils mostly > 1 or with superior ovary or with > 1 seed per cell; fls rarely umbellate
            5a Lvs pinnately compound, stip; corolla lacking; calyx 4-merous, lower portion persistent, hardened, and forming a 4-angled covering over the 1–2 simple ovaries; fls closely crowded in short spikes or elongate heads *(Sanguisorba)* **Rosaceae** p. 205
            5b Lvs not pinnate and stip, or calyx not 4-merous, or fls not crowded and spicate
                6a Fls with > 15 stamens or > 1 pistil, or both; pistil simple, with only 1 placenta and sometimes only 1 seed
                    7a Stamens hypog, borne on the recep, a disc lacking; sepals often deciduous, frequently petaloid, not leathery; lvs exstip.
Ranunculaceae p. 124
                    7b Stamens perig to epig, either borne on the calyx or on (or at the edge of) a disc surrounding the pistils; sepals sometimes leathery and persistent; lvs often stip
                        8a Lvs exstip, ternately decompound; stamens borne on and at edge of a prominent lobed disc free of the calyx; fr large, leathery, ∝-seeded follicles; sepals leathery, persistent **Paeoniaceae** p. 124
                        8b Lvs gen strongly stip; stamens gen borne on the calyx or on a disc adnate to the calyx; fr and sepals diverse
Rosaceae p. 205
                6b Fls with only 1 pistil, gen with < 15 stamens (if stamens > 13 then pistil compound)
                  9a Carpels 3–6, partially open on inner margin before maturity; fls irreg; stamens 10–50, inserted on or just within a large, lopsided disc **Resedaceae** p. 181
                  9b Carpels 1–5, completely closed prior to dehiscence; fls reg to irreg, but very seldom with a disc other than adnate to the calyx; stamens few –∝ **Subunit Xa** (lead 10a)

Subunit Xa

10a Ovary gen 1-celled and with 1 parietal placenta, but occ with a false partition formed by intrusion of upper or lower suture, the two apparent placentae then beside each other on the ovary wall; stamens either 5 and distinct or 10 and connate; lvs mostly strongly stip *(Astragalus, Petalostemon)*
Leguminosae p. 228
10b Ovary with other than 1 parietal placenta, or stamens other than 5 (and distinct) or 10 (and connate)
    11a Fls perig or epig
        12a Pistil 3-carpellary, ovary 3-celled or 1-celled with 3 parietal placentae; stamens ∝; pl very rough and scab with finely barbellate hairs; lvs exstip **Loasaceae** p. 300

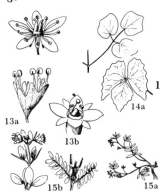

12b Pistil other than 3-carpellary, or stamens 5–10, or lvs stip; pl never scab with barbellate hairs

    13a Stamens mostly 3, 5, or 10; pistil 2–3 (4) -carpellary; calyx not bracteate   **Saxifragaceae** p. 184

    13b Stamens often > 10; pistil 1-carpellary, with only 1 seed or placenta; calyx often with bracts alt with lobes   **Rosaceae** p. 205

11b Fls hypog

  14a Ovary 1-celled; lvs ternate to triternate   **Berberidaceae** p. 142

  14b Ovary 2–5-celled; lvs ternate to pinnatifid or pinnate

    15a Lvs alt, exstip; delicate herbs of shady or moist places; ovary lobed into 2–5 subglobose, 1-seeded segms   **Limnanthaceae** p. 287

    15b Lvs often opp, stip; vigorous herbs mostly of dry areas; ovary 4–5-celled, each cell 2–∝-seeded   **Zygophyllaceae** p. 282

Unit XI

Gamopet herbs with deeply lobed to compound lvs

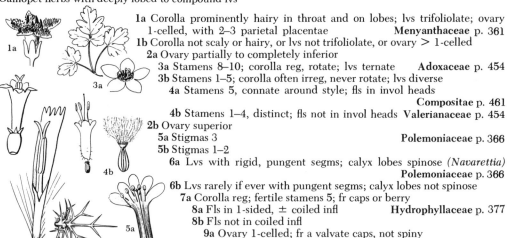

1a Corolla prominently hairy in throat and on lobes; lvs trifoliolate; ovary 1-celled, with 2–3 parietal placentae   **Menyanthaceae** p. 361

1b Corolla not scaly or hairy, or lvs not trifoliolate, or ovary > 1-celled

  2a Ovary partially to completely inferior

    3a Stamens 8–10; corolla reg, rotate; lvs ternate   **Adoxaceae** p. 454

    3b Stamens 1–5; corolla often irreg, never rotate; lvs diverse

      4a Stamens 5, connate around style; fls in invol heads   **Compositae** p. 461

      4b Stamens 1–4, distinct; fls not in invol heads   **Valerianaceae** p. 454

  2b Ovary superior

    5a Stigmas 3   **Polemoniaceae** p. 366

    5b Stigmas 1–2

      6a Lvs with rigid, pungent segms; calyx lobes spinose (*Navarettia*)   **Polemoniaceae** p. 366

      6b Lvs rarely if ever with pungent segms; calyx lobes not spinose

        7a Corolla reg; fertile stamens 5; fr caps or berry

          8a Fls in 1-sided, ± coiled infl   **Hydrophyllaceae** p. 377

          8b Fls not in coiled infl

            9a Ovary 1-celled; fr a valvate caps, not spiny   **Hydrophyllaceae** p. 377

            9b Ovary 2–5-celled; fr a berry or sometimes a spiny or circumscissile caps   **Solanaceae** p. 409

        7b Corolla irreg; fertile stamens often other than 5; fr sometimes 1–4 1-seeded nutlets

          10a Ovary becoming deeply 4-lobed toward maturity, forming 4 1-seeded nutlets; sts gen square in ✕-section; pl often strongly aromatic

            11a Style terminal on ovary or nutlets; corolla only slightly irreg; stamens 4–5; fls in terminal, bracteate spikes   **Verbenaceae** p. 398

            11b Style subbasal on ovary segms or on nutlets; corolla strongly irreg; stamens 2 or 4; infl diverse   **Labiatae** p. 399

          10b Ovary unlobed or bilobed; fr a ∝-seeded caps; sts rarely square in ✕-section; pl rarely aromatic

            12a Sepals 4–5; ovary 2-celled; stamens gen 4–5, distinct   **Scrophulariaceae** p. 413

            12b Sepals 2; ovary 1-celled; stamens often 6, joined into 2 groups   **Fumariaceae** p. 144

Unit XII

Herbs with entire to shallowly lobed lvs; fls hypog, lacking calyx or corolla or both

1a Pistils > 1, simple; stamens often > 10   **Ranunculaceae** p. 124

1b Pistil 1, mostly compound, but sometimes the carpels fused only near base; stamens gen not > 10

  2a Perianth 2-lipped; stamens 2; ovary 2-celled, ∝-seeded (*Besseya*)   **Scrophulariaceae** p. 413

  2b Perianth not 2-lipped; stamens seldom 2; ovary gen 1-seeded, or with 1 seed per cell, or 1-celled with free-central placentation

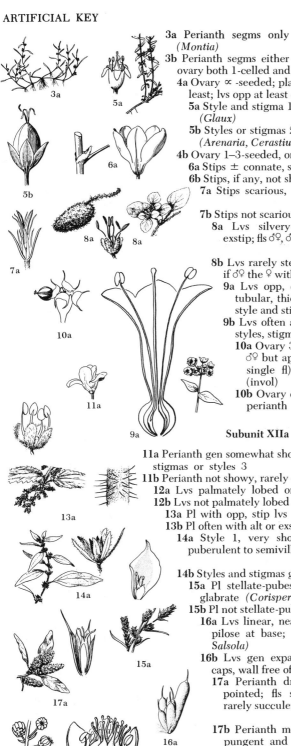

3a Perianth segms only 2; stigmas 3; ovary 1-celled, 2–∝ -seeded (*Montia*)   **Portulacaceae** p. 104

3b Perianth segms either lacking or > 2; stigmas gen other than 3 if ovary both 1-celled and several-seeded

    4a Ovary ∝ -seeded; placentation free-central in upper part of ovary at least; lvs opp at least on lower part of st, nodes often swollen

        5a Style and stigma 1; stamens as many as the sepals, alt with them (*Glaux*)   **Primulaceae** p. 350

        5b Styles or stigmas 2–5; stamens often twice as many as the sepals (*Arenaria, Cerastium, Stellaria*)   **Caryophyllaceae** p. 109

    4b Ovary 1–3-seeded, or placentation basal, or lvs alt

        6a Stips ± connate, sheathing st above nodes   **Polygonaceae** p. 78

        6b Stips, if any, not sheathing

            7a Stips scarious, prominent; lvs opp; stamens 5 (*Paronychia*)   **Caryophyllaceae** p. 109

            7b Stips not scarious or lvs alt; stamens often other than 5

                8a Lvs silvery-stellate-pubescent and hirsute-hispid, opp, exstip; fls ♂♀, ♂ with 5–6 sepals, ♀ naked, style 1 (*Eremocarpus*)   **Euphorbiaceae** p. 284

                8b Lvs rarely stellate-pubescent, often alt or stip; fls mostly ⚥, if ♂♀ the ♀ with perianth or perianth-like invol

                    9a Lvs opp, exstip; perianth connate, camp to funnelf or tubular, thickened at base and closely surrounding ovary; style and stigma 1; ovary 1-seeded   **Nyctaginaceae** p. 102

                    9b Lvs often alt or stip; perianth gen not strongly connate; styles, stigmas, or seeds mostly > 1

                        10a Ovary 3-lobed, 3-celled, and 3-seeded; fls naked and ♂♀ but apparently consisting of several stamens (each a single fl), 1 pistil (separate fl), and a cuplike calyx (invol)   **Euphorbiaceae** p. 284

                        10b Ovary other than 3-celled and 3-seeded; fls gen with perianth   **Subunit XIIa** (lead 11a)

Subunit XIIa

11a Perianth gen somewhat showy, the segms mostly in 2 series; stamens 3–9; stigmas or styles 3   **Polygonaceae** p. 78

11b Perianth not showy, rarely colored

    12a Lvs palmately lobed or dissected (*Cannabis*)   **Moraceae** p. 75

    12b Lvs not palmately lobed

        13a Pl with opp, stip lvs and stinging hairs (*Urtica*)   **Urticaceae** p. 76

        13b Pl often with alt or exstip lvs, never with stinging hairs

            14a Style 1, very short, stigma subsessile, tufted; lvs alt, finely puberulent to semivillous, 3-nerved; fr lenticular (*Parietaria*)   **Urticaceae** p. 76

    14b Styles and stigmas gen 2 or more; lvs various

        15a Pl stellate-pubescent in infl, at least, but sometimes tardily glabrate (*Corispermum, Eurotia*)   **Chenopodiaceae** p. 93

        15b Pl not stellate-pubescent

            16a Lvs linear, nearly terete, rigidly pungent-tipped, sometimes pilose at base; ovary wall ± adherent to seed (*Halogeton, Salsola*)   **Chenopodiaceae** p. 93

            16b Lvs gen expanded, not pilose at base, rarely pungent; fr caps, wall free of seed

                17a Perianth dry and ± membranous, segms often sharp-pointed; fls subtended by scarious, pungent bracts; lvs rarely succulent, or scurfy, or densely soft-hairy   **Amaranthaceae** p. 101

                17b Perianth mostly not dry and membranous, segms rarely pungent and fls seldom pungent-bracteate; lvs often succulent, or strongly scurfy, or densely soft-hairy

                    18a Stamens and styles ca 10; fr fleshy, ca 10-seeded; pl glab   **Phytolaccaceae** p. 103

                    18b Stamens and styles 5 or fewer; fr dry to fleshy, 1-seeded; pl often scurfy or hairy   **Chenopodiaceae** p. 93

Unit XIII

Herbs; lvs entire to ± lobed; fls perig-epig, lacking calyx or corolla

1a Fls in heads surrounded by 1–several series of bracts (invol); ovary inferior, 1-celled and 1-seeded; stamens gen connate around style
        Compositae p. 461
1b Fls rarely in invol heads, if capitate ovary gen superior or > 1-seeded; stamens not connate
    2a Perianth irreg, gen spurred at base; stamens fewer than perianth lobes; lvs opp
        Valerianaceae p. 454
    2b Perianth reg, not spurred; stamens mostly at least as many as perianth lobes; lvs often alt
        3a Stamens ca 12, monadelphous; perianth deeply 3-parted
            Aristolochiaceae p. 78
        3b Stamens distinct, mostly < 12; perianth rarely 3-parted
            4a Styles or stigmas 2–3; ovary 1-celled and 1-seeded
                5a Lvs opp, strongly scarious-stip (*Paronychia*)
                    Caryophyllaceae p. 109
                5b Lvs either alt or without scarious stips
                    6a Perianth dry and scarious; fls surrounded by sharp-pointed bracts; fr caps, circumscissile   **Amaranthaceae** p. 101
                    6b Perianth not dry and scarious; fls not pungent-bracteate; fr not circumscissile
                        7a Lvs opp, linear, semiterete, fleshy, not basally connate; stigmas 2; perianth not at all showy; ovary wall membranous (*Nitrophila*)   **Chenopodiaceae** p. 93
                        7b Lvs either alt or not at once linear, fleshy, and semiterete, often connate at base; perianth sometimes colored; ovary wall often hardened
                            8a Perianth gen colored, base not hardened and enveloping the fr; lvs rarely both opp and connate at base
                                Polygonaceae p. 78
                            8b Perianth greenish, inconspicuous, with hardened base enclosing the fr; lvs opp, connate at base (*Scleranthus*)
                                Caryophyllaceae p. 109
             4b Styles or stigmas either other than 2–3 or ovary not 1-celled and 1-seeded   **Subunit XIIIa** (lead 9a)

Subunit XIIIa

9a Lvs opp, not all basal, exstip
    10a Ovary 1-celled and 1-seeded; fr achene; perianth often fleshy, gen showy, closely surrounding fr   **Nyctaginaceae** p. 102
    10b Ovary 2–4-celled; fr ∝ -seeded caps; perianth rarely either showy or closely investing fr   **Lythraceae** p. 302
9b Lvs alt, all basal, or stip
    11a Fls umbellate to subcapitate, with or without invol; styles 2; ovary 2-celled, each cell 1-seeded   **Umbelliferae** p. 314
    11b Fls not umbellate, or ovary either 1-celled or > 2-seeded
        12a Pistil 1, ovary inferior, 1-celled, ultimately 1-seeded; lvs entire, glab, exstip, ± leathery or fleshy; pl a root-parasite   **Santalaceae** p. 78
        12b Pistils 1–several; ovary either superior or > 1-seeded or lvs not leathery, entire, and exstip; pl not a root-parasite
            13a Pistils 1, 2–3-carpellary; styles or stigmas 2–3
                14a Ovary several-seeded; lvs diverse but gen neither fleshy nor rigidly pungent   **Saxifragaceae** p. 184
                14b Ovary 1-seeded; lvs mostly fleshy, terete, or rigidly pungent at tip
                    15a Fr a circumscissile caps; fls subtended by sharp-pointed bracts; lvs flattened, gen not succulent
                        Amaranthaceae p. 101
                    15b Fr indehiscent, ovary wall gen grown to seed; fls gen not subtended by sharp-pointed bracts; lvs sometimes narrow and often terete or rigidly sharp-pointed   **Chenopodiaceae** p. 93
            13b Pistils sometimes > 1, simple, 1-carpellary; style and stigma 1; ovary 1-celled, with only 1 placenta   **Rosaceae** p. 205

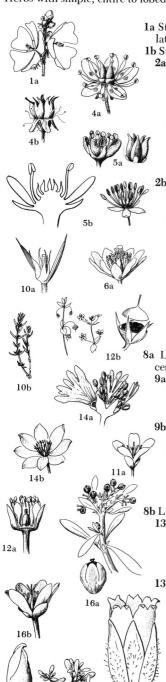

Unit XIV

Herbs with simple, entire to lobed lvs; fls with both calyx and polypet corolla

1a Stamens ∝, monadelphous, connate around ovary and style; pl often stellate-pubescent   **Malvaceae** p. 291
1b Stamens either < 15 or with filaments free or only slightly connate at base
  2a Pistils 2–several, simple
    3a Fls perig; lvs gen stip
      4a Pistils 2–3, slightly fused basally; stamens 5–10
            Saxifragaceae p. 184
      4b Pistils > 3 or stamens > 10   **Rosaceae** p. 205
    3b Fls hypog or lvs not stip
      5a Stamens (4) 5 or 10; pistils (4) 5; fr 4–5 follicles; lvs fleshy
            Crassulaceae p. 182
      5b Stamens > 10 or fr other than 4–5 follicles; lvs rarely fleshy
            Ranunculaceae p. 124
  2b Pistil 1, simple or compound
    6a Stamens 15–100; lvs opp, exstip; pistil 3–5-carpellary, placentation not free-central   **Hypericaceae** p. 294
    6b Stamens < 15 or pistil either not 3–5-carpellary or with free-central placentation
      7a Ovary 1-celled (at least near tip), often with free-central placentation or with 1 seed, never with > 1 parietal placenta
            Subunit XIVa, lead 8a
      7b Ovary > 1-celled or 1-celled but with 2 or more parietal placentae and several seeds   **Subunit XIVb**, lead 18a

Subunit XIVa

8a Lvs gen opp, never all basal; nodes often swollen; placentation free-central or basal
  9a Fr 1-seeded, indehiscent; lvs semiterete; stigmas 2
    10a Lvs with prominent scarious stips (*Cardionema*)
            Caryophyllaceae p. 109
    10b Lvs exstip (*Nitrophila*)   **Chenopodiaceae** p. 93
  9b Fr caps, gen several seeded and dehiscent; lvs gen flat; stigmas various
    11a Sepals 2; petals 5–9   **Portulacaceae** p. 104
    11b Sepals gen the same no as the petals
      12a Styles or stigmas gen > 1; stamens 3–10; caps not circumscissile   **Caryophyllaceae** p. 109
      12b Style 1; stigma 1; stamens the same no as the petals and opp them (*Anagallis*)   **Primulaceae** p. 350
8b Lvs alt or all basal, or nodes not swollen, or placentation not free-central
  13a Stamens > 10
    14a Fls irreg, petals deeply 7–11-lobed; sepals and petals 4
            Resedaceae p. 181
    14b Fls mostly reg, petals never deeply 7–11-lobed; sepals and petals mostly other than 4   **Portulacaceae** p. 104
  13b Stamens 10 or fewer
    15a Fls perig
      16a Lvs alt, strongly scarious-stip; styles 3; ovary 1-seeded (*Corrigiola*)   **Caryophyllaceae** p. 109
      16b Lvs various, rarely if ever scarious-stip; styles and seeds various   **Portulacaceae** p. 104
    15b Fls hypog
      17a Sepals fewer than petals (if any), gen 2–3 or rarely apparently 4 because of a pair of juxtaposed bracts   **Portulacaceae** p. 104
      17b Sepals gen same no as petals   **Caryophyllaceae** p. 109

Subunit XIVb

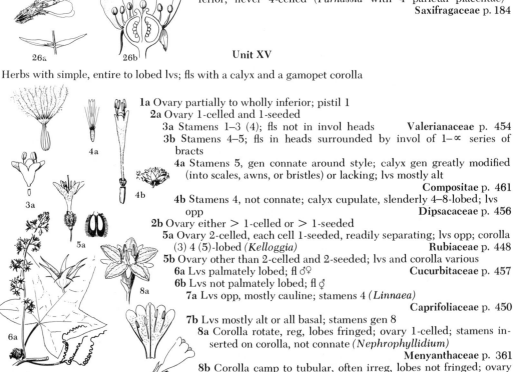

18a Ovary 2-celled, inferior, each cell 1-seeded   **Umbelliferae** p. 314
18b Ovary 1–several-celled, either superior or with > 1 seed per cell
   19a Lvs opp, entire, nodes swollen; styles 3–5; stamens 10, hypog; calyx
      strongly gamosep   **Caryophyllaceae** p. 109
   19b Lvs, styles, stamens, and calyx not at once as above
      20a Fls hypog
         21a Pl tiny, growing in moist places; lvs opp, stip   **Elatinaceae** p. 295
         21b Pl not greatly reduced and in wet places, or lvs alt or not stip
            22a Style 1; ovary 4–5-celled or 1-celled with 2 parietal placentae;
               anthers sometimes opening by terminal pores
               23a Ovary 4–5-celled; anthers gen opening by terminal pores
                                     Ericaceae p. 340
               23b Ovary 1-celled; placentae 2, parietal, sometimes intruded
                  and nearly partitioning ovary; anthers not opening by terminal
                  pores *(Frasera, Lomatogonium)*   **Gentianaceae** p. 356
            22b Styles 2–5; anthers not opening by terminal pores
               24a Lvs opp; calyx gamosep; fertile stamens 10
                                     Caryophyllaceae p. 109
               24b Lvs often alt; calyx not strongly gamosep; fertile stamens not
                  > 5   **Linaceae** p. 282
      20b Fls perig or epig
         25a Ovary inferior, 4–6-celled; lvs alt *(Jussiaea)*   **Onagraceae** p. 303
         25b Ovary superior or < 4-celled; lvs occ opp
             26a Lvs gen opp, exstip; ovary superior, often 4-celled
                                       Lythraceae p. 302
            26b Lvs alt, often stip, sometimes toothed; ovary often partially in-
               ferior, never 4-celled *(Parnassia* with 4 parietal placentae)
                                     Saxifragaceae p. 184

Unit XV

Herbs with simple, entire to lobed lvs; fls with a calyx and a gamopet corolla

1a Ovary partially to wholly inferior; pistil 1
   2a Ovary 1-celled and 1-seeded
      3a Stamens 1–3 (4); fls not in invol heads   **Valerianaceae** p. 454
      3b Stamens 4–5; fls in heads surrounded by invol of 1–∞ series of
      bracts
         4a Stamens 5, gen connate around style; calyx gen greatly modified
         (into scales, awns, or bristles) or lacking; lvs mostly alt
                                      Compositae p. 461
         4b Stamens 4, not connate; calyx cupulate, slenderly 4–8-lobed; lvs
         opp   **Dipsacaceae** p. 456
   2b Ovary either > 1-celled or > 1-seeded
      5a Ovary 2-celled, each cell 1-seeded, readily separating; lvs opp; corolla
      (3) 4 (5)-lobed *(Kelloggia)*   **Rubiaceae** p. 448
      5b Ovary other than 2-celled and 2-seeded; lvs and corolla various
         6a Lvs palmately lobed; fl ♂♀   **Cucurbitaceae** p. 457
         6b Lvs not palmately lobed; fl ☿
             7a Lvs opp, mostly cauline; stamens 4 *(Linnaea)*
                                   Caprifoliaceae p. 450
            7b Lvs mostly alt or all basal; stamens gen 8
               8a Corolla rotate, reg, lobes fringed; ovary 1-celled; stamens in-
                  serted on corolla, not connate *(Nephrophyllidium)*
                                   Menyanthaceae p. 361
               8b Corolla camp to tubular, often irreg, lobes not fringed; ovary
                  often 2–5-celled; stamens often nearly free of corolla, some-
                  times connate   **Campanulaceae** p. 457
1b Ovary superior; pistils very occ > 1
   9a Carpels (4) 5 (6), nearly or quite distinct; fr separate follicles; stamens
   gen 4, 8, or 10   **Crassulaceae** p. 182
   9b Carpels gen 1–3, if 4–5 then completely fused into a compound ovary
   gen with a single style; stamens often fewer than the corolla lobes

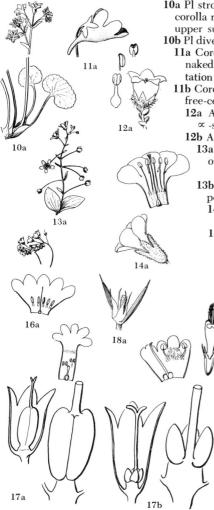

10a Pl strongly rhizomatous; lvs sheathing at base, bl cordate-reniform; corolla rotate, lobes with several fringed scales running lengthwise on upper surface (*Nephrophyllidium*)   **Menyanthaceae** p. 361

10b Pl diverse; corolla lobes not fringed lengthwise

  11a Corolla strongly irreg, with prominent basal spur; fl solitary on naked peduncle; lvs basal; ovary 1-celled with free-central placentation (*Pinguicula*)   **Lentibulariaceae** p. 445

  11b Corolla not spurred, or fls > 1 per st, or ovary not 1-celled with free-central placentation

    12a Anthers opening by terminal pores; ovary mostly 5-celled, ∝ -seeded   **Ericaceae** p.340

    12b Anthers not opening by terminal pores; ovary gen < 5-celled

      13a Sepals 2; corolla reg, petals gen 4–6, connate only at base; ovary 1-celled with free-central placentation   **Portulacaceae** p. 104

      13b Sepals and petals gen same no; corolla mostly strongly gamopet, often irreg; ovary diverse

        14a Style brs or stigmas 3; ovary 3-celled   **Polemoniaceae** p. 366

        14b Style brs or stigmas other than 3; ovary rarely 3-celled

          15a Ovary 2-celled, 4-lobed, gen becoming 4-celled and 4-seeded; fr 4 (or by abortion 1, 2, or 3) hardened, 1-seeded nutlets; pl mostly either opp-lvd and with 4-angled sts (squarish in × -section) or harshly pubescent and with fls in curled infl

            16a Corolla gen reg (except *Dasynotus*), (4) 5-lobed; stamens (4) 5; fls often in curled infl; pubescence often harsh; sts not square in × -section; lvs mostly alt   **Boraginaceae** p. 384

            16b Corolla ± irreg, mostly bilabiate; stamens mostly 2 or 4 (5); infl not curled; pubescence seldom harsh; sts often square; lvs gen opp

              17a Style terminal on nutlets; stamens 4–5; corolla gen only slightly irreg; pl seldom aromatic   **Verbenaceae** p. 398

              17b Style subbasal on nutlets; stamens 2 or 4; corolla gen strongly irreg; pl often strongly aromatic   **Labiatae** p. 399

          15b Ovary 1–more-celled, never developing into 4 nutlets; pl various

            18a Lvs opp, acicular, with prominent scarious stips (*Cardionema*)   **Caryophyllaceae** p. 109

            18b Lvs alt, or not acicular and with scarious stips   **Subunit XVa (lead 19a)**

Subunit XVa

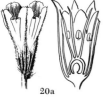

19a Stamens as many as corolla lobes and opp them, or twice as many; ovary 1-celled; corolla reg

  20a Ovary 1-seeded; styles 5; stamens same no as corolla lobes   **Plumbaginaceae** p. 355

  20b Ovary >1-seeded, or styles 1, or stamens twice as many as corolla lobes

    21a Styles and stigmas 1; stamens as many as the petals; caps often circumscissile   **Primulaceae** p. 350

    21b Styles or stigmas 2–5; stamens gen twice as many as the petals; caps not circumscissile   **Caryophyllaceae** p. 109

19b Stamens fewer than corolla lobes or as many and alt with them; ovary not 1-celled and with free central placentation; corolla often irreg

  22a Calyx and corolla 4-lobed, reg

    23a Corolla scarious, persistent; caps gen circumscissile; fls spicate; spikes terminal (pl scapose) or axillary (pl not scapose, but opp-lvd)   **Plantaginaceae** p. 447

    23b Corolla not scarious, gen deciduous; caps not circumscissile; fls rarely spicate; pl seldom scapose

      24a Ovary 1-celled, with 2 parietal placentae; pl gen glab; corolla gen twisted in bud   **Gentianaceae** p. 356

24b Ovary gen 2-celled; pl often pubescent; corolla not twisted in bud (*Synthyris, Veronica*) **Scrophulariaceae** p. 413

22b Calyx and corolla other than 4-lobed, or fls irreg

25a Corolla gen funnelf, twisted in bud; lvs alt, bls often cordate or hastate; ovary mostly (1–3) 4-seeded; seeds never hairy; pl often vining **Convolvulaceae** p. 363

25b Corolla not funnelf and twisted in bud, or lvs opp, or ovary ∞ - seeded, or seeds hairy; pl rarely vining

26a Pl gen glab; lvs opp; corolla either rotate and segms with prominent basal glands or fringed appendages, or lobes ± twisted in bud; ovary 1-celled, with 2 parietal placentae **Gentianaceae** p. 356

26b Pl often pubescent; lvs often alt; corolla lobes neither glandbearing nor twisted in bud; ovary often 2-celled

27a Ovary 1-celled; placentae 2, parietal, sometimes intrusive and nearly contacting, but never joined in the center

28a Stamens as many as corolla lobes (gen 5) **Hydrophyllaceae** p. 377

28b Stamens 4; corolla lobes 5 (*Limosella*) **Scrophulariaceae** p. 413

27b Ovary 2 (rarely 4) -celled, or ovaries apparently 2; placentation often axile

29a Fertile stamens gen 2 or 4, occ 5 but then fl slightly irreg **Scrophulariaceae** p. 413

29b Fertile stamens 5; fls reg

30a Lvs opp, simple, gen entire; seeds (except in *Vinca*) hairy; ovaries apparently 2, union of the 2 carpels mostly above ovaries

31a Corolla lobed much > half length, lobes reflexed; stamens connate around style and stigmas and adnate to them **Asclepiadaceae** p.363

31b Corolla either lobed < half length or lobes not reflexed; stamens distinct or only slightly connate, only slightly if at all adnate with stigma **Apocynaceae** p. 362

30b Lvs gen alt (sometimes in pairs at nodes but on 1 side of st and not truly opp), mostly not entire; seeds never hairy; ovary 1, carpels 2 or 3, completely fused

32a Lvs reniform-orbicular, toothed to lobed; pl often developing basal tubers; fls white, gen with yellow eye, borne in 1-sided and ± coiled infl (*Romanzoffia*) **Hydrophyllaceae** p. 377

32b Lvs rarely at all reniform-orbicular; pl not developing basal tubers; infl not at all coiled

33a Lvs with narrow, gen pungent lobes or teeth; fr caps (*Navarettia*) **Polemoniaceae** p. 366

33b Lvs gen subentire, lobes (if any) neither narrow nor pungent; fr often berry **Solanaceae** p. 409

Group III

Monocots; fls gen 3-merous; lvs gen parallel-veined

1a Fls naked or merely subtended by much-modified, gen chaffy or bristle-like bracts

2a Pl floating, thallus-like, roots lacking or few and unbr; fls consisting of 1 or 2 stamens and a naked pistil, borne in small, marginal pockets of the thallus **Lemnaceae** p. 677

2b Pl terrestrial or aquatic but rooted to bottom and with only part of sts and lvs floating

3a Lvs opp or whorled; pl aquatic, submersed or floating

4a ♀ fls with 1 pistil and 2–4 stigmas; lvs often 4 per node, often toothed, dilated and sheathing at base, without free stips **Najadaceae** p. 562

4b ♀ fls with (1) 2–8 (9) pistils, each with single stigma; lvs 2 per node, entire and with free, membranous stips **Zannichelliaceae** p. 566

3b Lvs alt (except sometimes the uppermost), or pl non-aquatic, or both

5a Pl growing in salt or brackish water, often marine; sts and lvs floating or submersed; pistils sometimes several and produced on a

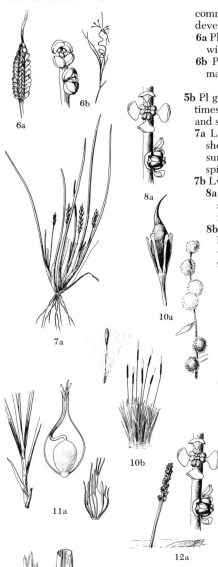

common peduncle and separate peds that elongate greatly as frs develop

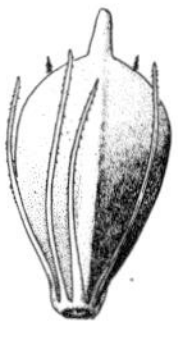

6a Pl strictly marine; lvs 1.5–4 mm broad; frs not long-ped, enclosed within a spathe   **Zosteraceae** p. 567

6b Pl mostly of brackish water along coast or inl, but sometimes marine; lvs < 1 mm broad; frs long-ped, not enclosed in a spathe   **Ruppiaceae** p. 566

5b Pl growing on land or in fresh water, mostly not marine, but sometimes in brackish water; pistils not borne on an elongating peduncle and separate peds

7a Lvs terete, basal; pl acaulescent; ♀ fls of 2 types, some among sheaths of basal lvs and with long styles often reaching to water surface, others in mixture of ♂ fls (single stamens) in pedunculate spikes (*Lilaea*)   **Juncaginaceae** p. 561

7b Lvs gen flattened; pl caulescent; ♀ fls (if any) gen all alike

8a Pistils and stamens 4; connective of anthers broadened and sepal-like; pl with submersed or floating sts and lvs, only the infl emergent   **Potamogetonaceae** p. 562

8b Pistils and stamens gen other than 4; connective of anthers not broadened and sepal-like; pl mostly partially or wholly out of water

9a Fls largely ♀♂, ♂ and ♀ borne separately in heads or in a single elongate, terminal spike 10–20 cm, ♂ fls uppermost; pl not grasslike

10a Fls several in globose-capitate clusters, each subtended by 3–5 chaffy bracts; fr hardened, strongly beaked, plainly visible in head   **Sparganiaceae** p. 674

10b Fls ∝ in a terminal, terete spike, each naked or subtended merely by slender hairs; fr neither hardened nor strongly beaked, concealed in spike by subtending hairs   **Typhaceae** p. 675

9b Fls mostly ☿, if ♀♂ then gen borne in 1–several greatly reduced spikes < 10 cm, ♂ often not uppermost; pl grasses or grasslike

11a Lvs in 3 vertical ranks; st often solid or pithy, not swollen-noded; sheaths closed; each fl gen subtended by 1 (occ 2) bracts and often with several inner subtending scales or bristles; ovary sometimes enclosed in a sac-like covering; fr achene, gen ± beaked; styles often 3   **Cyperaceae** p. 576

11b Lvs in 2 vertical ranks; st gen terete, mostly swollen-noded, gen hollow; sheaths either open (as commonly) or partially to completely closed; each fl gen subtended by 2 bracts but with not > 2 (3) inner subtending scales (scales gen obscure); ovary never enclosed in a sac; fr a grain (ovary wall grown tightly to seed), rarely at all beaked; styles (in ours) always 2   **Gramineae** p. 602

1b Fls with perianth; perianth segs either all alike or differentiated into calyx and corolla

12a Pls submersed or floating aquatics, only infl emersed; fls 4-merous, perianth of 1 sepal-like series   **Potamogetonaceae** p. 562

12b Pls mostly terrestrial, if aquatic rarely either entirely emersed or floating, and with other than 4-merous fls

13a Fls ♂♀, borne separately in capitate-globose clusters, each subtended by 3–5 chaffy bracts; achene hardened and beaked   **Sparganiaceae** p. 674

13b Fls ☿ or if (as rarely) ♂♀, ♂ and ♀ not borne separately and not subtended by chaffy bracts; fr diverse

14a Infl a terminal, fleshy spike subtended by a large whitish-yellow spathe; fls closely crowded and ± grown together; perianth parts and stamens 4   **Araceae** p. 676

14b Infl not as above; perianth parts and stamens mostly other than 4

15a Pl with gen triangular sts and 3-ranked lvs, grasslike; perianth several bristles or scales, not in 2 series of (2) 3 each; fr an achene, lenticular to trigonous   **Cyperaceae** p. 576

15b Pl with gen terete sts and other than 3-ranked lvs, rarely at all grasslike; perianth segs gen in 2 series of (2) 3 each; fr diverse

16a Pistils > 1, quite distinct, or carpels ± connate but separating into 3 or 6 follicles
    17a Perianth inconspicuous, greenish, segms all ca alike; carpels 3–6
        18a Fls in 3–12-fld, bracteate raceme, peds in fr 12–25 mm; follicles strongly compressed, divergent, 1–2-seeded
                            Scheuchzeriaceae p. 560
        18b Fls in > 12-fld, ebracteate spike or raceme, peds gen < 6 mm in fr; follicles neither strongly compressed nor divergent, 1-seeded   **Juncaginaceae** p. 561
    17b Perianth ± showy, sepals green, petals white to pink or purplish; carpels at least 6
        19a Pistils 6, verticillate, fr a follicle; fls long-pedicellate in a simple umbel   **Butomaceae** p. 557
        19b Pistils gen > 6, fr gen not a follicle; fls in racemes, pans, or compound umbels   **Alismataceae** p. 557
16b Pistil 1, compound, 1–3-celled, fr not follicular
    20a Pl grasslike; fls inconspicuous, perianth mostly greenish to brownish- or purplish-green, segms all ca alike
                            Juncaceae p. 567
    20b Pl scarcely or not at all grasslike; perianth gen ± showy, corolla often not like the calyx in color
        21a Ovary superior
            22a Pl aquatic; sepals and petals connate into a basal tube; stamens 3, adnate to perianth tube
                          Pontederiaceae p. 678
            22b Pl not aquatic or sepals and petals distinct and the 4 or 6 stamens borne on the recep   **Liliaceae** p. 678
        21b Ovary inferior
            23a Pl aquatic, gen submersed, mostly ♂, ♀; stamens (1) 3–12   **Hydrocharitaceae** p. 559
            23b Pl terrestrial, ⚥-fld; stamens 1–3
                24a Stamens 3, free of style; fls reg   **Iridaceae** p. 697
                24b Stamens 1 or 2, joined with style; fls very irreg
                            Orchidaceae p. 698

18b

18a

19a

18b

19b

20a

22a

22b

23a

24a

24b

Descriptive Flora

LYCOPODIACEAE   Clubmoss Family

St dichotomously to monopodially br; lvs mostly alt or opp, seldom in pseudowhorls, narrow or very small or both (ours never as much as 1.5 cm), elig, with a single unbr vascular bundle; sporangia axillary (or seemingly so) to the sporophylls or adnate to the base thereof; sporophylls resembling vegetative lvs or often ± modified, often but not always aggregated into a terminal cone; gametophyte tiny, thalloid; pls homosporous, spores ∝ in each sporangium, all ca alike.

Lycopodium L.   Clubmoss

Mostly evergreen herbs with elongate, lfy sts; our spp., except *L. inundatum*, with subterranean, mycorrhizal gametophyte. (Gr *lycos,* wolf, and *pous,* foot, from appearance of br shoot-tips of some spp.).

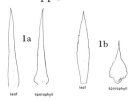

1a Sporophylls elongate, gen several times as long as wide, green and photosynthetic, not very different from the vegetative lvs
  2a Sporophylls borne in zones on the st alt with zones of vegetative lvs; sts per and evergreen, all ± erect, forming a cluster, gen 5–20 (30) cm; lvs crowded, in ca 8 ranks, from subappressed and only 3–5 mm to spreading and up to 11 mm, × 0.6–1.4 mm; exposed cliffs and talus slopes to dense moist woods, always where humid; circumboreal, s to n Ore (Mt Hood), n Ida, nw Mont, and NC; fir c. (*L. lucidulum* f. or var. *occidentale*)
  1 L. selago L.
  2b Sporophylls aggregated into sessile terminal cones; sts ann, elongate, prostrate or arching, with scattered, erect cone-bearing brs; lvs crowded, in 8–10 ranks, mostly 4–8 × < 1 mm; wet places, esp sphagnum bogs; interruptedly circumboreal; our phase, var. *inundatum,* s to nw Cal, n Ida, nw Mont, Minn, and Va; bog c., marsh c.   **2 L. inundatum** L.
1b Sporophylls obviously different from the vegetative lvs, not strongly photosynthetic, seldom > ca twice as long as wide
  3a Lvs in 6–10 ranks, often > 4 mm
    4a Cones sessile, solitary, terminating the densely lfy sts; lvs with a sharp, somewhat spinulose, but not at all hairlike tip
      5a Prostrate sts aerial, lfy; erect sts mostly with few, long brs; lvs gen 5–11 mm; gen in moist conif for, occ on more open, rocky slopes below timberl; circumboreal, s to ne Ore, n Ida, nw Wyo, Minn, and Va; isolated in Colo; interrupted c., stiff c.   **3 L. annotinum** L.
      5b Prostrate sts deeply subterranean; erect sts much br; lvs gen 2.5–5 mm; woodl; boreal Am and Asia, s to Wn (Skagit Co), n Ida, nw Mont, Ind, and Ala; groundpine   **4 L. obscurum** L.
    4b Cones (1) 2–several on sparsely lfy-bracteate peduncles; lvs tapering to a very long, hairlike (often deciduous) tip; prostrate sts aerial and lfy; moist conif woods and swamps; circumboreal, s to nw Cal, n Ida, nw Mont, Mich, and NC; isolated in parts of tropics; elk-moss; ground- or running-pine; stag's horn moss   **5 L. clavatum** L.
  3b Lvs (at least of vegetative brs) in 4–5 ranks, up to ca 3.5 mm (excluding any adnate basal portion)
    6a Vegetative brs appearing lfy, st largely hidden by lvs; free tips of at least some lvs gen 2–3.5 mm; peduncles absent or not well defined
      7a Lvs of ultimate vegetative brs gen (4) 5-ranked, seldom decussately opp, all ca alike, not laterally twisted, the st not appearing flattened; meadows and open, often rocky places; mont, to above timberl, less commonly in conif for; boreal Am and e Asia, s to c and ne Ore, c Ida, and Ravalli Co, Mont; Alas c. (*L. complanatum* var. *s., L. sabinaefolium* var. or spp. *s.*)   **6 L. sitchense** Rupr.
      7b Lvs of ultimate vegetative brs strictly 4-ranked, decussately opp, every other pair decurrent on st as a pair of flanges, the flange continuous with one margin of the lf, so that the st appears somewhat flattened or wing-margined with the lateral rows of lvs twisted; rocky slopes and open conif for at high elev, often above timberl; circumboreal, s to s BC, nw Mont (GNP), and Que; alpine c.   **7 L. alpinum** L.
    6b Vegetative brs appearing as flattened, wing-margined, remotely scaly sts, the free tips of lvs up to ca 2 mm; peduncles well defined; moist conif for, lowl to midmont; circumboreal, s to s Wn, n Ida, and nw Mont; ground cedar   **8 L. complanatum** L.

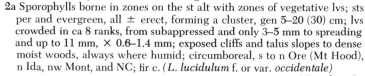
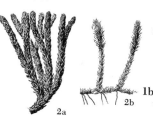
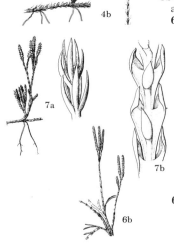

SELAGINELLACEAE Selaginella Family

Sts br, short or ± elongate; lvs lig (i.e., with a small ventral appendage near the base); lvs and sporophylls small, in ours not > ca 6 mm; sporophylls slightly to strongly differentiated from the vegetative lvs, in terminal cones; sporangia axillary; pls heterosporous, microspores ∝ and tiny, megaspores (1–) 4.

Selaginella Beauv.   Selaginella; Lesser-clubmoss

Evergreen herbs with dichotomously to monopodially br, lfy sts; roots originating from slender brs called rhizophores; cones gen ♂ ♀, lower sporophylls each subtending a megasporangium, upper ones each subtending a microsporangium. (Diminutive of *Selago*, ancient name of some sp. of *Lycopodium*)

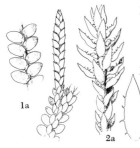

1a Vegetative lvs basically opp, displaced so as to form 4 rows, those of the 2 marginal rows larger and differently shaped than those of the 2 dorsal rows; cliffs, banks, and tree trunks at low elevs in humid regions; CRG, esp on Ore side, w occ near CR to Portland, Ore, and Cowlitz Co, Wn; disjunct along Clearwater R, Ida; Douglas' s. 1 S. douglasii (Hook. & Grev.)·Spring
1b Vegetative lvs spirally arranged, not conspicuously dimorphic
  2a Lvs thin and soft, without a dorsal groove; cone not quadrangular; sporophylls loose, gen 2.5–5 mm; wet places such as mossy banks; circumboreal, s irreg at midmont elev to s BC, ne Nev, w Wyo, Mich, and NS; seldom collected s of Can; lesser club-moss (*Lycopodium s.*)
  2 S. selaginoides (L.) Link
  2b Lvs thick and firm, with an evident dorsal groove; cone quadrangular; sporophylls ± appressed, not > 2.5 mm
    3a Sts rather loosely to very loosely br, sometimes forming open mats, but not forming cushionlike mats
      4a Sts elongate and very loosely br, up to several dm, gen pendent from trees; lvs with an evidently adnate-decurrent base gen 0.4–0.9 mm; spores whitish or ochroleucous; hanging from (chiefly angiospermous) trees or sprawling over moist banks or rocks, lowl (up to ca 700 ft); W Cas, Wn to nw Cal, esp along coast and lower parts of OM and Coast ranges; Ore s.   3 S. oregana D. C. Eat.
      4b Sts shorter (up to 1–2 dm); terrestrial, forming open mats; lvs merely sessile, without an adnate-decurrent base; spores ± orange; exposed rocky sites; lowl to midmont; s BC and sw Alta, to c Cal, c Ida, and w Mont; Wallace's s. (*S. rupestris* f. or var. w., *S. montanensis, S. r.* var. *columbiana, S. w. f. c.*)   4 S. wallacei Hieron.
    3b Sts very compactly br, forming cushion-mats with short, crowded, erect or often curved-ascending brs; lf-base shortly adnate to st
      5a Vegetative lvs slightly dissimilar, those on lower side of st (or on convex side of curved br) somewhat longer than others; terminal seta of lvs (0.3) 0.5–2.0 mm; widespread cordilleran sp. of exposed, often rocky sites; compact s.; 3 vars.   5 S. densa Rydb.
        a1 Terminal seta of lvs lutescent and translucent, carrying their thickness nearly or quite to tip, in profile the lf-tip abruptly rounded off or even subtruncate; margins of sporophyll as in either of other vars.; uncommon, in our range only at or above timberl (*S. s.*)   var. **standleyi** (Maxon) Tryon
        a2 Terminal seta of lvs whitish and opaque, except sometimes toward the base; leaf-profile more gradually beveled toward the tip
          b1 Sporophylls minutely ciliolate-serrulate along margins all the way to the tip; common on n GP, extending w occ into w Mont, sw Alta, and se BC, gen in valleys and foothills (*S. rupestris* var. *d.*)   var. **densa**
          b2 Sporophylls essentially smooth-margined at least in distal third, sometimes throughout; common phase in our range, distinctly more mont than var. *densa*
          var. **scopulorum** (Maxon) Tryon
      5b Vegetative lvs at a given level on the st all alike; setae < 0.5 mm; exposed rocky sites, midmont to above timberl; GB sp., isolated in Wallows Mts, Ore, and in sw Mont; Watson's s.
      6 S. watsonii Underw.

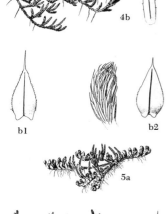

ISOETACEAE   Quillwort Family

Lvs clustered in a close spiral on summit of a 2–3-lobed stock, some spp. wholly aquatic and commonly evergreen, other spp. amphibious or terrestrial with lvs deciduous soon after spore maturity; pls heterosporous, microspores ∝ and tiny, megaspores fewer (up to 300); lvs lig (with a small ventral appendage slightly above the base), elongate, gen much > 1.5 cm; outer lvs often sterile, next inner ones each bearing a megasporangium, next inner ones each bearing a microsporangium, or pls seldom partly or wholly dioecious; roots originating in the furrows of the stock, but pushed progressively farther from the center of the furrow.

Isoetes L.   Quillwort

Stock very short, corm-like; lvs elongate, slender, ± cylindrical and containing 4 large, longitudinal air-cavities (in cross-section a circle containing a cross), broadened to a flattened base; air cavities with frequent transverse partitions; peripheral tissues of lf often containing longitudinal strands of fibers; sporangia borne at base of lvs, partly or wholly covered by the velum, a thin flap of tissue which extends down from above. (Ancient Gr name for some other pl).

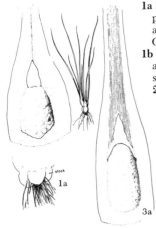

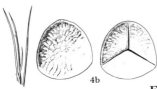

1a Stock 3-lobed; velum covering the whole inner surface of the sporangium; peripheral strands of fibers 3, well developed (1 median abaxial, 2 at the adaxial margins); terrestrial, gen on wet ground at low elevs; W Cas and CRG: Nuttall's q. *(I. suksdorfii)*   1 **I. nuttallii** A. Br.

1b Stock 2-lobed; velum incomplete, not covering the lower part of the sporangium, or rarely obsolete; peripheral strands present or absent; widespread

2a Megaspores seldom > 0.5 mm, with scattered low tubercles (as seen at 40×), some of which are sometimes confluent into short, low ridges or wrinkles; pls gen amphibious or in shallow water

3a Hyaline wing-margins of sporophylls extending > 1 cm above sporangium; peripheral strands well developed; lvs often > 15 cm; lowl and foothills; amphibious, in and around temporary lakes and pools, gen in lowl and foothills; Wn to Cal, e across n Ida and nw Mont, apparently disjunct in n Utah; Howell's q. *(I. underwoodii)*
                                                                2 **I. howellii** Engelm.

3b Hyaline wing-margins of sporophylls not extending > 1 cm above sporangium; peripheral strands obsolete or nearly so; lvs seldom > 15 cm; shallow water of mt pools, mid- to rather high-mont, often near timberl; BC and Mont to Cal, Ariz, and Colo; Bolander's q. *(I. b. var. parryi)*   3 **I. bolanderi** Engelm.

2b Megaspores often > 0.5 mm, and (as seen at 40 ×) ± densely beset with spines or jagged crests or high ridges; pls amphibious (in forms of no 4) or more often wholly submerged, often in deep water

4a Megaspores gen 0.25–0.5 (0.6) mm wide, beset with ∝ sharp to truncate or in part bifid spines; lvs thin and soft, gradually tapering from near the base to a long, very slender tip; variously amphibious or permanently immersed in and around ponds and lakes, sea level to midmont; circumboreal, s to Cal, Utah, Colo, Minn, and NJ; bristle-like q. *(I. maritima; I. echinospora var. m.; I. macounii; I. truncata; I. echinospora var. flettii; I. f.; I. e. var. hesperia; I. muricata var. h.; I. setacea,* misapplied)   4 **I. echinospora** Dur.

4b Megaspores gen 0.5–0.8 mm wide, beset with jagged crests or high ridges; lvs coarse and firm, carrying their width well upwards toward the acute tip; submerged, often in deep water, from sea level to midmont; interruptedly circumboreal, s in w Am to Cal, ne Utah, and Colo; lake q. *(I. occidentalis, I. piperi, I. howellii var. p.)*
                                                                5 **I. lacustris** L.

EQUISETACEAE   Horsetail Family

Sts without 2ndary thickening, becoming hollow except at the nodes, longitudinally ribbed and grooved, jointed, simple or with whorls of brs at the nodes, the epidermis with silicified cell walls and very rough to the touch; lvs whorled, gen connate to form a sheath at each node, and gen projecting above the sheath as separate teeth; sporangia borne on sporangiophores that are closely grouped in successive whorls to form a terminal cone; spores ∝ , minute, alike, spherical, each with 4 spirally wound bands (elaters) with enlarged tip.

Equisetum L.   Horsetail; Scouring-rush

Pls gen < 2 m, rhizomatous; aerial sts ann, or per and evergreen; sporangiophores peltate, each with 5–10 elongate sporangia under a polygonal cap. (L *equus*, horse, and *seta*, bristle, from a fancied resemblance of the br sts of some spp. to a horse's tail). Many of the spp. are garden and field pests, but several of those with per sts are valued as orn.

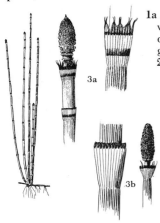

1a Aerial sts per and evergreen (ann in no 2), simple, or at least not with reg whorled brs, never dimorphic; cones apiculate (except often in no 2); teeth of the sheaths sometimes art and deciduous, sometimes persistent; stomates gen sunken below the level of the epidermis

 2a Sts robust, gen 2–15 dm×(2) 3–14 mm, (14) 16–40-ridged; central cavity > half the diameter of the st; teeth art to the sheath, but not always deciduous

  3a Aerial sts per, evergreen; cones distinctly apiculate; sheaths gen with a basal or suprabasal black band as well as an apical one, the teeth deciduous or often ± persistent; streambanks and other moist or wet places, lowl to midmont; circumboreal, s to Cal and Fla; Dutch rush, common s.; ours the var. *affine* (Engelm.) A. A. Eat.; the sp. hybridizes with no 2, producing sterile pls called **E. × *ferrissii*** Clute (**E. intermedium**, *E. h.* var. *i.*, *E. h.* var. *suksdorfii*) **1 E. hyemale** L.

  3b Aerial sts ann; cones blunt or inconspicuously apiculate; sheaths black-banded at the tip only, the teeth promptly deciduous; wet places, lowl to midmont; s BC to Baja Cal, e to O and Tex; smooth s. (*E. funstoni*, *E. kansanum*) **2 E. laevigatum** A. Br.

 2b Sts smaller and more slender, gen 1–3 (5) dm × (0.5) 1–2 (3) mm, 3–12 (14)-ridged; central cavity < ⅓ diameter of st; teeth of sheath persistent, not art

  4a Teeth of most sheaths (3) 4–10 (14), as many as the ridges of the internode; central cavity evident, gen larger than the (3) 4–10 (14) cavities outside the c cavity; sts 1–3 dm × 1–2 (3) mm, ascending to erect; wet places, lowl to rather high mont; circumboreal, s to s Wn, ne Ore, s Utah, Ill, and Pa; northern s., variegated h.; 2 vars.

   3 E. variegatum Schleich.

   a1 Sts relatively robust, up to 5 dm, with gen 10–14 ridges; teeth of sheath wholly or nearly wholly black, with narrow or no hyaline margins, terminal seta poorly developed or wanting; Alas to s BC, barely reaching our area (*E. a.*) var. **alaskanum** A. A. Eat.

   a2 Sts 1–3 dm, with gen (3) 5–10 (12) ridges; teeth of sheath green with black or blackish midstripe and white-hyaline margins, with hairlike, deciduous tip 0.5–1 mm; our common phase var. **variegatum**

  4b Teeth of sheath 3, half as many as ridges of the internode; c cavity wanting, other cavities 3; sts 0.7–2.5 dm × 0.5–1 mm, flexuous; moist places, esp in conif woods; circumboreal, s to ne Wn, nw Mont, Minn, and NY; sedgelike h. or s.; a desirable garden subject

   4 E. scirpoides Michx.

1b Aerial sts ann, often dimorphic, gen at least some of them with reg whorls of brs; cones blunt, not apiculate; teeth of sheaths persistent, not art; stomates not sunken below level of epidermis

 5a Fertile and sterile sts alike; ridges of the st smooth or often minutely cross-wrinkled, but without tubercles or spicules; pls fr in summer

  6a St shallowly 9–25-grooved, c cavity commonly > half diameter of st and much > the small or obsolete peripheral cavities; teeth of sheaths 1.5–3 (3.5) mm, not hyaline-margined or only very narrowly and inconspicuously so; wet places, often in bogs or other standing water; lowl to midmont; circumboreal, s to s Wn, n Ida, nw Wyo, Minn, and Pa, but doubtfully in Ore; water h. (*E. limosum*) **5 E. fluviatile** L.

  6b St deeply 5–10-grooved; c cavity small, < ⅓ diameter of st, scarcely > the well-developed peripheral cavities; teeth 3–7 mm, evidently hyaline-margined; wet places, lowl to midmont; circumboreal, s to s Wn, n Ida, nw Mont, Nev, and Pa; marsh h. **6 E. palustre** L.

 5b Fertile and sterile sts unlike, sterile ones green and br, fertile ones at first simple and not green, sometimes later green and br; ridges of st

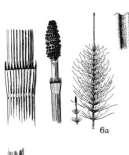

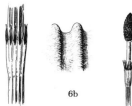

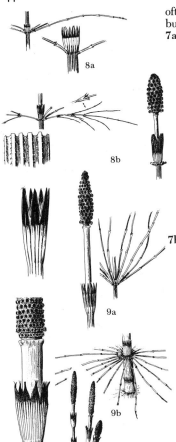

often with tubercles or spicules, or merely papillate or almost smooth, but seldom at all cross-wrinkled; pls fr in the spring

7a Fertile sts ± persistent, becoming green and br; ridges of the st beset with spicules, or high tubercles, or very short, high transverse ridges

    8a Teeth of sheaths free or nearly so, as many as the internodal ridges, black or blackish with pale margins; brs simple, and with the first internode short, scarcely > the associated sheath of the main st; sterile sts smoothish toward base, otherwise beset with spreading, high, blunt siliceous tubercles or very short, high, transverse ridge-crests on the 10–18 ridges; streambanks and moist woods; circumboreal, s to s BC, nw Mont, Ia, and NJ; shady h.
                                    7 E. pratense Ehrb.

    8b Teeth cohering in several broad brown lobes; brs gen again br, and with the 1st internode relatively long, commonly > the associated sheath of the main st; sterile sts beset with 2 rows of small, spreading or recurved spicules on each of the 10–18 ridges (or nearly smooth toward the base of the internode); wet places to moist woods; circumboreal, s to s BC, n Ida, and nw Mont (GNP), SD, Ia, and Ky; wood h.
                                    8 E. sylvaticum L.

7b Fertile sts permanently whitish or brownish, unbr, soon withering; ridges of st with inconspicuous low tubercles or papillae, or practically smooth

    9a Pls less robust; sterile sts 1.5–6 (10) dm × 1.5–5 mm, evidently (4) 10–12-ridged, with small c cavity ca 1/4 diameter of st; sheaths 5–10 mm, greenish, teeth 1–3 mm; fertile sts up to ca 3 dm × 8 mm; sheaths 14–20 mm, with large, partly connate teeth 5–9 mm; cone 0.5–3.5 cm; moist to moderately dry places throughout our range, often weedy; common h., field h. (*E. saxicola*)   **9 E. arvense** L.

    9b Pls more robust; sterile sts 5–30 dm × 0.5–2 cm, gen 20–40-ridged, c cavity > 1/4 diameter of st; sheaths 1–2.5 cm, teeth 3–8 mm; fertile sts 2.5–6 dm × 1–2.5 cm; sheaths 2–5 cm, with teeth connate in groups of 2–4; cone 4–10 cm; moist low places; interruptedly circumboreal; in N Am from Alas to s Cal, wholly W Cas; giant h.; ours gen called var. *braunii* Milde (*E. t.* var. *hillii*)   **10 E. telmateia** Ehrh.

OPHIOGLOSSACEAE   Adder's-tongue Family

St erect, underground; lvs gen only 1 each year, consisting of a fertile and a sterile segm which are united below into a short or elongate common stalk; stalk of fertile segm (fertile spike) continuous with the common stalk, the 2 forming an axis on which the sterile segm (sterile bl) is sessile or petiolate, the common stalk enlarged at base and partly or wholly enclosing the bud; sterile bl erect in bud or often bent over, but not circinate; sporangia ∝, relatively large (commonly ca 1 mm thick), stalked to sessile or even embedded in the axis of the simple or br fertile spike, thick walled, 2-valved, without an annulus; spores all of 1 kind.

1a Sterile segm (apparent lf-bl) gen lobed or compound; veins free, forked; sporangia short-stalked   **Botrychium**

1b Sterile segm entire; veins anastomosing; sporangia sessile and coherent in 2 rows on the axis of the spike   **Ophioglossum**

Botrychium Sw.   Grape-fern; Moonwort

Sterile segm gen pinnately or ternate-pinnately compound to dissected, seldom simple; veins free, forked; fertile segm pinnate to decompound; sporangia ∝, gen short-stalked and free. (Gr *botrys*, bunch of grapes).

1a Sterile bl once-pinnate or ternate-pinnate (simple), or only the basal segms again divided, always relatively small, not > ca 7 cm

    2a Sterile bl attached near ground level (common stalk seldom > 3 cm), evidently petiolate, tending to be somewhat ternate-pinnate rather than strictly pinnate, basal pair of pinnae somewhat larger than the next pair, often stalked and sometimes again cleft; pinnae mostly 2–4 (5) pairs; midmont meadows; circumboreal, s to s Cal, NM, and NJ, in our range chiefly e Cas; little g.   **1 B. simplex** E. Hitchc.

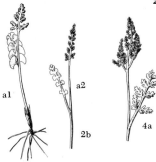

2b Sterile bl attached well above ground level (common stalk well >3 cm in well-developed pls), sessile or petiolate, distinctly pinnate and often with > 4 pairs of pinnae; moist or wet places but seldom in meadows, mid- to high-mont; widespread esp in cooler part of N Hem; moonwort; 2 vars. in our area   2 B. **lunaria** (L.) Swartz

   a1 Sterile bl sessile or on a short stalk gen < 5 mm, typically bright green; pinnae gen crowded and somewhat overlapping, flat, mostly flabellate and wider than long, the proximal margin of each pinna forming an obviously retrorse angle with the axis of the sterile bl; wet places, but gen not in deep shade; chiefly e Cas in our range

var. **lunaria**

   a2 Sterile bl with petiole gen > 5 mm, typically yellowish-green; pinnae tending to be ± remote, often somewhat spoon-shaped rather than flat, not strongly flabellate, not notably wider than long, proximal margin diverging widely from axis of sterile bl; more often in dense for at lower elevs, as in Cas and OM and n Ida (*B. o.*)

var. **onondagense** (Underw.) House

1b Sterile bl gen subbipinnate or bipinnate to dissected, often much > 7 cm

  3a Sterile bl gen longer than wide, not > ca 9 cm; fertile spike erect or nearly so in bud

    4a Sterile bl gen oblong and evidently petiolate, ultimate segms mostly longer than wide; pl gen 1–3 dm; woods, rare in our area, where known only from n Ida, but interruptedly circumboreal; camomile g. (*B. lunaria* var. *m.*)   3 B. **matricariaefolium** (Doell) A. Braun

    4b Sterile bl gen ovate or ovate-oblong and sessile, ultimate segms nearly or fully as long as broad; pl 1–2 dm; moist or wet, ± open places, mont; Alas and BC to Wn, ne Ore, and ne Nev; Eurasia; northern g. (*B. b.* var. *obtusilobum, B. pinnatum*)   4 B. **boreale** Milde

  3b Sterile bl ca as wide, or wider than, long, often much > 9 cm; fertile spike reflexed in bud

    5a Sterile bl evergreen, evidently petiolate (petiole at least 1 cm), attached near ground level; pl 1–5 dm, sparsely hairy when young, later ± glab; moist or wet, open or shaded places, from sea level to fairly high mont; circumboreal, s to Cal, Ida, Wyo, Ia, and NC; leathery g. (*B. californicum, B. coulteri, B. occidentale, B. silaifolium*)

5 B. **multifidum** (Gmel.) Trevis.

    5b Sterile bl deciduous, sessile or with a short petiole up to 5 mm, attached well above ground level

      6a Sterile bl large, gen (5) 7–20 × (6) 10–30 cm; stalk of fertile spike gen 4–17 cm; pl 1.5–5 dm, slightly hairy when young, soon glab; moist woods and thickets, seldom in meadows, valleys to midmont; circumboreal, s irreg to Mex, in our range chiefly in s BC, n Ida, and nw Mont, also in n and w Wn; Va g. (*B. v.* var. *occidentale*)

6 B. **virginianum** (L.) Swartz

      6b Sterile bl smaller, gen 1–6 × 1–9cm; stalk of fertile spike short, gen 1–3 cm; pl 0.5–3.5 dm, glab from the first; moist or wet places, mont to high mont; interruptedly circumboreal, s to s Wn, Utah, Colo, Wis, and Pa; lance-lvd g.   7 B. **lanceolatum** (Gmel.) Angstr.

Ophioglossum L.   Adder's-tongue

Lf solitary (rarely 2–several); sterile bl simple, entire, with reticulate venation; veins anastomosing; fertile spike unbr, with 2 rows of coalescent, sessile or embedded sporangia along 1 side. (Gr *ophis*, snake, and *glossa*, tongue).

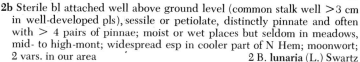

   O. vulgatum L. Pl gen (0.5) 1–3 dm; sterile bl attached well above ground level (common stalk gen 3–15 dm), sessile or nearly so, elliptic to ovate, entire, rounded to obtuse or merely acutish at the tip, gen 2.5–10 × 1–4 cm; fertile stalk gen 3–15 cm, the fertile spike erect, 1–4 cm; meadows and woods; circumboreal (but not high-mont), s irreg to Fla and Mex, known from widely scattered stations in our area.

POLYPODIACEAE   Polypody or Common Fern Family

St consisting of a creeping to short and erect rhizome or a br caudex, bearing scales or hairs; lvs large as compared to the st, coiled in bud, petiolate, simple to gen compound or decompound (gen pinnately or ternate-pinnately so); fertile and sterile lvs alike or unlike; sporangia gen grouped into sori, these marginal or on the lower surface, naked or more often ± covered by an indusium, at least when young; sporangia small, those of a given sorus not all developing at once, gen long-stalked, thin-walled, elastically dehiscent, with a well-developed vertical annulus interrupted by the stalk; spores all alike; gametophytes green, terrestrial or ephiphytic, dorsiventral, monoecious, the sex organs gen on the lower side.

1a Rhizome and lvs with hairs only, lacking scales; sori marginal, confluent, protected by the reflexed lf margin; petiole stout, erect, with several bundles; spores tetrahedral   **Pteridium**
1b Rhizome (and often lvs) scaly (scales sometimes narrow, but flattened), often hairy also; other characters not all as above
  2a Petiole with a single bundle, always slender, very often dark and wiry; indusium formed by the reflexed lf margin, or lacking; spores tetrahedral
    3a Sporangia following along the veins, wholly exposed, scarcely forming definite sori; lf margins very narrowly reflexed, but not forming an indusium; lower surface of lf bls gen covered with tiny, waxy scales, appearing powdery   **Pityrogramma**
    3b Sporangia submarginal, covered, when young, by the reflexed indusial margins of the lf segs (also extending back a short distance from the margin and some of them exposed from the beginning in 1 sp. of *Pellaea*); lvs not powdery-waxy
      4a Indusium formed by the individually reflexed tips of the lobelets of the relatively broad ultimate lf segs, the sori borne on the reflexed indusial flap; lvs deciduous   **Adiantum**
      4b Indusium formed by the continuously reflexed and modified margins of the fertile pinnules; lvs in most spp. evergreen
        5a Fertile lvs with notably narrow and elongate ultimate segs, these 1–3 (5) mm wide; lvs glab, often dimorphic
          6a Lvs strongly dimorphic, the sterile ones well developed but obviously shorter than the fertile ones and with evidently shorter and relatively broader ultimate segs; petiole greenish or greenish-stramineous, at least distally   **Cryptogramma**
          6b Lvs weakly or not at all dimorphic, either all alike and fertile, or with a few sterile ones that are not very different from the fertile ones; petiole dark brown   **Aspidotis**
        5b Fertile lvs with the ultimate segs either relatively or absolutely broader than in the 2 foregoing genera; lvs monomorphic or subdimorphic, glab or often scaly or hairy, esp beneath
          7a Lf bls glab or sparsely hairy, not conspicuously woolly or scaly; ultimate segs of lvs, or many of them, at least 5 mm   **Pellaea**
          7b Lf bls ± woolly beneath, sometimes also beset with scales; ultimate segs of lvs smaller, all well < 5 mm   **Cheilanthes**
  2b Petiole with 2–more bundles (at least toward the base), sometimes slender and dark as in the foregoing group, more often stouter and paler; sori laminar, with or without an indusium, the indusium not formed from the reflexed lf margin (occ seemingly so in *Blechnum*); spores bilateral
    8a Lvs conspicuously dimorphic, pinnate or seldom subbipinnate, the fertile ones with very narrow and elongate pinnae that have the confluent sori protected by a continuous indusium which is so close to the margin as to be difficult to differentiate from it   **Blechnum**
    8b Lvs not conspicuously dimorphic, variously cleft or dissected; sori discrete or sometimes ± confluent, but not forming a continuous submarginal row that is protected by a continuous indusium
      9a Veinlets of lvs br and anastomosing, forming a row of areolae on each side of the midvein of the ultimate lf segs; lvs pinnate-pinnatifid, with elongate sori parallel to the mid-vein

                                  Woodwardia
      9b Veins all free or only slightly anastomosing; sori, indusia, and lf form various

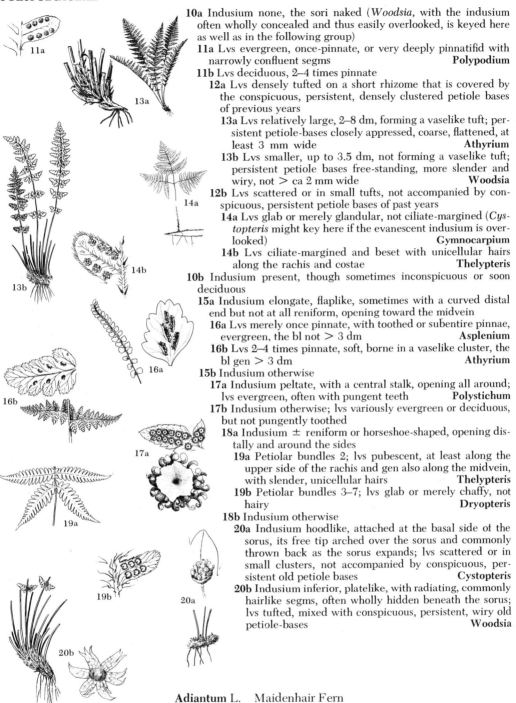

10a Indusium none, the sori naked (*Woodsia*, with the indusium often wholly concealed and thus easily overlooked, is keyed here as well as in the following group)

11a Lvs evergreen, once-pinnate, or very deeply pinnatifid with narrowly confluent segms **Polypodium**

11b Lvs deciduous, 2–4 times pinnate

12a Lvs densely tufted on a short rhizome that is covered by the conspicuous, persistent, densely clustered petiole bases of previous years

13a Lvs relatively large, 2–8 dm, forming a vaselike tuft; persistent petiole-bases closely appressed, coarse, flattened, at least 3 mm wide **Athyrium**

13b Lvs smaller, up to 3.5 dm, not forming a vaselike tuft; persistent petiole bases free-standing, more slender and wiry, not > ca 2 mm wide **Woodsia**

12b Lvs scattered or in small tufts, not accompanied by conspicuous, persistent petiole bases of past years

14a Lvs glab or merely glandular, not ciliate-margined (*Cystopteris* might key here if the evanescent indusium is overlooked) **Gymnocarpium**

14b Lvs ciliate-margined and beset with unicellular hairs along the rachis and costae **Thelypteris**

10b Indusium present, though sometimes inconspicuous or soon deciduous

15a Indusium elongate, flaplike, sometimes with a curved distal end but not at all reniform, opening toward the midvein

16a Lvs merely once pinnate, with toothed or subentire pinnae, evergreen, the bl not > 3 dm **Asplenium**

16b Lvs 2–4 times pinnate, soft, borne in a vaselike cluster, the bl gen > 3 dm **Athyrium**

15b Indusium otherwise

17a Indusium peltate, with a central stalk, opening all around; lvs evergreen, often with pungent teeth **Polystichum**

17b Indusium otherwise; lvs variously evergreen or deciduous, but not pungently toothed

18a Indusium ± reniform or horseshoe-shaped, opening distally and around the sides

19a Petiolar bundles 2; lvs pubescent, at least along the upper side of the rachis and gen also along the midvein, with slender, unicellular hairs **Thelypteris**

19b Petiolar bundles 3–7; lvs glab or merely chaffy, not hairy **Dryopteris**

18b Indusium otherwise

20a Indusium hoodlike, attached at the basal side of the sorus, its free tip arched over the sorus and commonly thrown back as the sorus expands; lvs scattered or in small clusters, not accompanied by conspicuous, persistent old petiole bases **Cystopteris**

20b Indusium inferior, platelike, with radiating, commonly hairlike segms, often wholly hidden beneath the sorus; lvs tufted, mixed with conspicuous, persistent, wiry old petiole-bases **Woodsia**

Adiantum L.   Maidenhair Fern

Medium-sized to small, mesophytic ferns; rhizome long-creeping to short and ascending, beset with brown to blackish scales; lvs shining, resistant to wetting, deciduous (ours); petiole slender and wiry, dark and polished; bl gen glab, 1–several times pinnate or partly dichotomous, ultimate segms petiolulate, often ± fan shaped, veins free; indusium formed by the separately reflexed tips of the lobelets of some lf-segms, the sori borne on the reflexed margins. (Gr *a*, without, and *dianein*, to wet, referring to the rain-shedding of the foliage). Excellent garden subjects for the shaded or moist place.

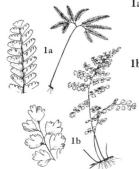

1a Lf bl nearly or fully as wide as long, the 2 subequal primary divisions re-
curved-spreading, the pinnae all on the convex side; larger pinnae with gen
15–35 pairs of pinnules, these 12–22 ⨉ 5–9 mm, obliquely oblong; moist
woods, lowl to midmont; widespread in wooded parts of temp N Am;
northern m. *(A. p.* var. or ssp. *aleuticum)*   1 **A. pedatum** L.

1b Lf bl elongate, gen 2–4 times as long as wide, with a continuous rachis
bearing pinnae on both sides; larger pinnae with up to ca 6 pairs of pinnules, these 8–25 mm and almost or fully as wide; wet places, esp on calcareous or basic rocks; widespread in warmer regions, n irreg to Va, Mo, Colo,
s Utah, and Cal, disjunct at Fairmont Hot Springs, BC, where growing in
warm water; Venus-hair f.   2 **A. capillus-veneris** L.

Aspidotis (Nutt.) Copel.   Aspidotis

Small, mesophytic rock-ferns with a short rhizome that is beset with firm, narrow scales; lvs glab,
evergreen, monomorphic or only slightly dimorphic, 2–4 times pinnate, with free veins, the pinnules often confluent; petioles slender and wiry, gen brown, with a single vascular bundle; margins of fertile pinnules reflexed and abruptly white-scarious (in ours), forming a well-defined
common indusium for the submarginal sori, or the sori separately indusiate with individually reflexed, scarious bits of the lf-margin. (Gr *aspis*, shield).

A. densa (Brackenr.) Lellinger. Podfern. Lvs ∝ , clustered, gen all alike and
fertile, the sterile ones, when present, few, smaller, with smaller, relatively
broader, sharply toothed or incised segms; petioles castaneous, gen 4–8 cm
and > bl; fertile bls 2.5–6 cm, a third to fully as wide, tripinnate; pinnae gen
4–7 offset pairs, lowest ones largest; ultimate pinnules ∝ , crowded, ± linear,
3–12 ⨉ 1–2 mm, with firm, cartilaginous-mucronate tip; cliff crevices and
moist, rocky slopes, foothills to near timberl; s BC to Cal, inl to e Ore, c and n
Ida, and nw Mont, and irreg to Que *(Onychium d., Pellaea d., Allosorus d.,*
Cheilanthes d., Cryptogramma d., Ch. *siliquosa).*

Asplenium L.   Spleenwort

Small to medium-sized, gen evergreen ferns, rhizome scaly, in our spp. short and densely rooting;
lvs in our spp. elongate and narrow, once-pinnate with toothed to subentire pinnae; veins free,
simple or forked, not reaching the margin; petiole slender and wiry, with 2 oval bundles at base,
these united below bl into an x-shaped bundle; sori elongate, each borne on a veinlet; indusium
hyaline, flaplike, attached along the sorus-bearing vein, opening along the side toward the midline
of the pinna. (Gr *a*, without, and *splen*, spleen, referring to supposed medicinal properties). Both
spp. attractive garden subjects.

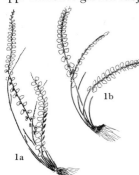

1a Lvs 7–35 cm, glab, evergreen; petiole shining, reddish-brown, the rachis
the same color; pinnae gen 12–35 pairs, 2.5–9 ⨉ 1.5–7 mm, with a few
rounded teeth; cliff crevices and talus slopes, gen where moist; interruptedly circumboreal, s to Ore, Ariz, Tex, and Ga; maidenhair s.
                                                            1 **A. trichomanes** L.

1b Lvs 5–15 cm, glab or with few septate, glandular hairs along rachis, hardly
evergreen; petiole brown or brownish toward base, green or greenish-
yellow distally; rachis greenish; pinnae 7–20 pairs, 3–8 ⨉ 2–6 mm,
coarsely round-toothed; cliff crevices, esp on limestone, often to near or
above timberl; circumboreal, s to c Wn, ne Nev, Colo, and NY; green s.
                                                            2 **A. viride** Huds.

Athyrium Roth   Lady-fern

Medium-sized to large, mesophytic ferns; rhizome scaly, in our spp. short and suberect, covered
by the conspicuous, flattened, densely clustered old petiole-bases; lvs in our spp. tightly bunched,
erect-ascending and out-curved; petiole coarse, with 2 bundles at base, these uniting distally into
a single trough-shaped bundle; blade 2–4 times pinnate, with free venation; sori round to elongate,
each borne on a veinlet which continues to the margin (veinlets obscure in no 2); indusium hyaline, flaplike, attached along the veinlet which bears the sorus, and opening on the side toward the
midvein of the pinnule, or sometimes reduced or obsolete. (Gr *a*, without, and *thyreos*, shield, referring to the non-peltate indusium).

1a

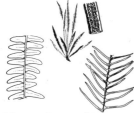

1b

1a Indusium present, though sometimes small and soon deciduous; veinlets obvious on lower surface of pinnule at 10×; lvs 3–20 dm, larger pinnae gen 4–25 × 1–7 cm; sori gen at least 1 mm wide or long; moist woods, meadows, and swamps, very common and often something of a pest in the garden; lowl to mont; circumboreal, and throughout our area in suitable habitats (*A. f.* var. *sitchense, A. s., A. f.* var. *cyclosorum, A. c., A. c.* var. *strictum, A. c.* f. *hillii, A. f.* var. or ssp. *californicum*)   1 **A. filix-femina** (L.) Roth.

1b Indusium wholly wanting; veinlets ± obscure; lvs gen 2–8 dm, larger pinnae 2.5–8 (12) × 1–3.5 (5.5) cm; sori < 1 mm wide or long; rocky slopes and borders of rushing streams; mid- to high-mont, often near timberl; circumboreal, s to Cal, Colo, and Que; alpine 1. (*A. alpestre*, misapplied); ours the var. *americanum* (Butters) Cronq.   2 **A. distentifolium** Tausch

Blechnum L.   Blechnum

Small to large, mesophytic ferns, gen with lvs tufted on a stout, suberect to short-creeping rhizome; petiole with 2 bundles that unite distally; blades gen coriaceous and glab, gen once pinnate, often (including our sp.) strongly dimorphic; sterile lvs with lateral veins of the pinnae simple or few-forked, free; fertile pinnae with a pair of longitudinal veins parallel to the costa, the sori confluent into a coenosorus along each of these veins; indusium a continuous flap along the fertile vein (in ours appearing almost like an inrolled pinna-margin), opening toward the costa. (Gr name for some other pl).

B. spicant (L.) Roth. Deer-fern, hard-f. Rhizome short-creeping; lvs once pinnate with broadly sessile pinnae; sterile lvs 2–8 (10) dm, petiole 3–25 cm, reddish- or purplish-brown, pinnae gen 35–70 pairs, largest ones borne near or above middle of bl, gen 1–5.5 cm × 3–7 (10) mm; fertile lvs gen surpassing the sterile and with petiole up to 50 cm, fertile pinnae ca as many and as long as the sterile, but only 1.5–2 mm wide; moist to wet places, gen in heavy shade, from near sea level to midmont; interruptedly circumboreal, s to Cal, chiefly W Cas in our range, but also in n Ida; an easily grown, choice garden pl.

Matteucia struthiopteris (L.) Todaro, a widespread circumboreal sp. that has been reported to enter our range in s BC, resembles *B. spicant* in having notably dimorphic lvs. Its sterile lvs are relatively large (to 2 m × 1.5–3.5 dm), distinctly > fertile ones, and pinnate-pinnatifid. The fertile lvs have ∝ and crowded pinnae with inrolled margins, but the proper indusia are attached near the costa and open outward. N Am pls are considered to represent a distinct var., var. *pensylvanica* (Willd.) Morton.

Cheilanthes Sw.   Lip-fern

Rather small, ± xeromorphic ferns; rhizome short and much-br to longer and simple, beset with slender, brown to blackish scales; lvs firm, evergreen, 2–4 times pinnate, ± woolly or scaly or both, ultimate segms small, with clavate-tipped free veins; petioles slender and wiry, brown to blackish-purple, with a single bundle; sori borne on vein ends just within margins of pinnules, often confluent, point of attachment gen covered by the reflexed or merely inrolled margins which form a continuous, unmodified (our spp.) or partly scarious indusium, but mature sporangia often conspicuously exserted. (Gr *cheilos*, margin, and *anthos*, fl, referring to the submarginal sori).

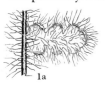

1a

2a

1a Lf bls merely villous-hirsute beneath (surface readily visible between the hairs), 5–20 × 1.5–5 cm, ca 3 times pinnate, larger ones with gen 12–20 pairs of pinnae, larger pinnae with gen 7–12 pairs of pinnules; rock crevices; widespread in e US, and disjunct at 1 station in OM, Wn; lanate 1.

      1 **C. lanosa** (Michx.) D. C. Eat.

1b Lf bls tomentose (and sometimes also scaly) beneath, the surface hidden by the hairs, often smaller

2a Lf bl tomentose beneath, but without scales, 3–13 × 1.5–4 cm, 3–4 times pinnate, with gen 6–12 opp or offset pairs of pinnae, the ultimate pinnules gen 1–1.5 mm, broadly rounded; cliff crevices, esp on limestone; foothills to midmont; sw US and n Mex, n irreg to Wis, s Alta, and s BC, E Cas; Fee's 1. (*C. gracilis*)   2 **C. feei** Moore

2b

2b Lf bl with evident scales on rachis and costae beneath, as well as tomentose on lower surface, gen 4–11 × 1–2 cm, bipinnate or partly tripinnate, pinnae gen 9–20 offset pairs; ultimate pinnules 1–3 mm, gen longer than wide; cliff crevices, often on igneous rock; low mont to near timberl; s BC to s Cal, esp along Cas-Sierran axis, inl to ne Ore, c and n Ida, and nw Mont; lace 1.   3 C. gracillima D. C. Eat.

Cryptogramma R. Br.   Rock-brake

Small rock-ferns with short to elongate rhizome beset with tenuous to firm and elongate scales; lvs glab, gen 2–3 times pinnate with free veins, the fertile ones sharply differentiated from the sterile, with relatively long and narrow ultimate pinnules; petioles slender and often wiry, gen green or stramineous at least distally, with a single vascular bundle, those of fertile bls often > those of sterile, or even > sterile blade; margins of fertile pinnules reflexed to form a continuous, slightly to strongly modified indusium, sori borne on thickened vein-ends just short of the margins on the reflexed side. (Gr *kryptos*, hidden, and *gramma*, line, referring to the lines of sori covered by the reflexed lf margins). Attractive and more amenable to cult than most rock ferns.

1a

1b

1a Lvs densely tufted on a short, much-br rhizome beset with slender scales and old petiole-bases; petioles greenish or straw-colored; fertile bl 4–14 × 1.5–5 cm, ultimate pinnules 4–12 × 1–2 mm, with indusial margins broadly reflexed and often meeting in a median line beneath but not much modified; cliff crevices and talus; midmont to timberl, but down to near sea level in CRG; interruptedly circumboreal, and widespread in cordilleran N Am; rock-brake, parsley-fern (*Blechnum c., C. acrostichoides*); ours the var. *acrostichoides* (R. Br.) Clarke   1 C. crispa (L.) R. Br.

1b Lvs scattered on a more elongate, slender rhizome bearing inconspicuous hyaline-reticulate, small and relatively broad scales; petioles brown or dark purple at least toward base, gen greenish or straw-colored distally; ultimate pinnules of fertile bl up to 2 cm × 3–5 mm, with narrow, evidently hyaline-scarious indusial margins that seldom meet in the center; moist, shaded cliffs and ledges, mid- to upper-mont; circumboreal, s interruptedly to n Wn, ne Nev, Colo, Ia, and NJ; Steller's r. (*Pellaea s.*)   2 C. stelleri (S. G. Gmel.) Prantl

Cystopteris Bernh.   Bladder-fern

Small to medium-sized ferns with short to elongate rhizomes covered with ovate to lanceolate scales; lvs (1) 2–4 times pinnate or ternate-pinnate, with free venation; petiole slender, with 2 bundles, scaly only at base; sori each borne on a veinlet which continues to the margin; indusium hyaline, basiscopic (i.e., attached under the proximal margin of the sorus on the entering veinlet), ± hoodlike, its free tip arched over the sorus and gen thrown back as the sorus expands, sometimes tiny and caducous. (Gr *kystis*, bladder, and *pteris*, fern, referring to the hoodlike indusium).

1a

1b

1a Lf bl 3.5–25 × 0.5–10 cm, at least twice as long as wide, gen > petiole, (1) 2-3 times pinnate, the lowest pair of pinnae no > the pair above; moist to moderately dry, often rocky places, lowl to above timberl; widespread in N Hem and common throughout our range; brittle b.   1 C. fragilis (L.) Bernh.

1b Lf bl 4–15 cm, nearly or fully as wide, gen much < the petiole, 3–4 times ternate-pinnate, the lowest pair of pinnae obviously the largest and longest; moist or wet woods, mont; circumboreal, s to s BC, nw Mont (GNP), n shore of Lake Superior, and Que, disjunct in Colo; mt b.
                                2 C. montana (Lam.) Desv.

Dryopteris Adans.   Shield-fern; Wood-fern

Medium-sized to large ferns with short, stout, horizontal to erect rhizomes beset with entire or toothed, chaffy scales; lvs clustered at end of rhizome, petiole with 3–7 free bundles and few–∝ chaffy scales; bl subbipinnatifid to tripinnate, sometimes beset with flattened or hairlike, multicellular scales, otherwise glab or merely glandular, the minor axes decurrent on the major ones, the ridges on the upper side of the rachis not raised above the level of insertion of the costae; veins free; sori laminar, roundish, each borne on a vein; indusium well developed and persistent, reniform or horseshoe-shaped, glab or sometimes glandular. (Gr *drys*, oak, and *pteris*, fern). Excellent garden pls.

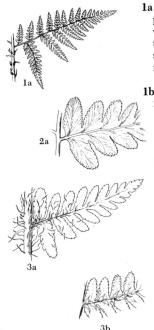

1a Lvs essentially tripinnate, deciduous, bl (1) 2–5 dm, gen 1/2–2/3 as wide; pinnae ∝, the lower and gen also the middle ones again twice pinnatifid with their 2nd-order segms softly spinulose-toothed or sometimes incised, the upper pinnae progressively less dissected; moist or wet woods and streambanks; circumboreal, s to Ore (W Cas), n Ida, nw Wyo, Ia, and NC; mt wood-fern, spreading w. (D. spinulosa, D. dilatata)

1 D. austriaca (Jacq.) Woynar

1b Lvs merely bipinnate or subbipinnate, bl gen 1/4–1/2 as wide as long

2a Lvs subbipinnate, the primary pinnae pinnatisect, with a broad foliaceous midstrip gen 2–5 mm wide; largest pinnae near or a little below the middle of the bl; sterile lvs evergreen, smaller and more spreading than the deciduous fertile ones; moist woods, circumboreal, s to n Ida, nw Mont, Ark, and NC; crested s., c. Buckler-fern   2 D. cristata (L.) Gray

2b Lvs bipinnate, essentially alike, the primary pinnae with midstrip not > 1.5 mm wide

3a Ramentum of lf bl chiefly or entirely hairlike, except sometimes along the rachis; lvs deciduous, not glandular; largest pinnae near or a little below the middle, lowest ones smaller and gen a little more distant; moist woods and stream banks, circumboreal, s to s BC, Cas of Wn, ne Ore, ne Ida, nw Mont, and irreg to Ariz and NM and e US; male fern

3 D. filix-mas (L.) Schott

3b Ramentum chiefly or entirely of evident scales; lvs evergreen, often glandular along rachis and midveins; pinnae gen all or nearly all approximate, lower ones as large as or only slightly smaller than those above; wooded places often in drier and more open sites than no 3; s BC to s Cal, e to Ariz, chiefly W Cas in our range; coastal s.

4 D. arguta (Kaulf.) Watt.

Gymnocarpium Newm.   Oak-fern

Delicate, mesophytic ferns with slender, rather sparsely scaly rhizomes, the scales fibrous, glab, entire; petiole slender, gen > bl, sparsely scaly toward the dark base, the 2 bundles free to the tip; lf bls thin, glab or glandular, deltoid or pentagonal, with opp pinnae; veins free, reaching the margin, sori rotund or broadly elliptic, seated on veins not far within the margin, indusium none. (Gr gymnos, naked, and karpos, fr, referring to the lack of indusia).

G. dryopteris (L.) Newm. Petiole gen 1–3 dm, = or > bl; bl up to 18 × 25 cm, from a little longer than wide to more often distinctly wider than long, bipinnate-pinnatifid to tripinnate-pinnatifid; pinnae several pairs, lowest pair each nearly as large as rest of bl, distinctly asymmetrical; moist woods, streambanks, and wet cliffs; circumboreal, s at low to moderate altitudes to Ore, Ariz, Ia, and Va (Lastrea d., D. d., Thelypteris d., Phegopteris d. var. disjunctum, D. linnaeana).

Pellaea Link   Cliff-brake

Small, ± xeromorphic ferns, ours all with short, br rhizomes so densely beset with long, narrow, brown scales as to appear brown-woolly; lvs firm, evergreen, 1–4 times pinnate, with free veins, the pinnae and often the pinnules with a minutely to evidently petiolulate base, not broadly sessile; petioles slender and wiry, green to (in ours) reddish-brown or blackish-purple, with a single bundle, often breaking off well above the ground to leave a persistent base; sori borne on vein-ends just within the margins of the lf, the point of attachment covered (except in no 1) by the reflexed margin, which forms a continuous indusium. (Gr pellos, dark, referring to the petioles).

1a Pinnae undivided, with plane or very narrowly reflexed margins which leave many or all the sporangia wholly exposed from the first; bls 4–13 × 1.3–5 cm; pinnae 4–14 pairs, blue-green; open, rocky slopes, midmont; Sierran Cal, disjunct in Wallowa Mts. Ore, and in c Ida; Bridges' c.

1 P. bridgesii Hook.

1b Pinnae, or many of them, gen bifid or ternate or again pinnate, ultimate segms with reflexed margins which form a continuous indusium covering the young sporangia

2a Persistent petiole-bases notably more numerous than the green lvs; middle and lower pinnae gen bifid, often with unequal segms, the lowest pinnae not obviously larger than the others; petiole with a series of

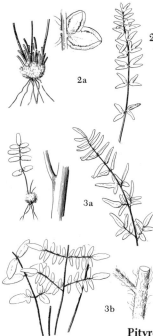

cross-grooves; bls 2.5–12 × 1.5–3 cm; rocky places, esp crevices or talus slopes, foothills to ca timberl; E Cas from c Wn to s Cal, e to sw Mont, Wyo, and Utah; Brewer's c.   2 **P. breweri** D. C. Eat.

2b Persistent petiole-bases relatively few, not notably more ∝ than the green lvs; lvs various, but not as in no 2

   3a Lvs of 1 form, glab or sparsely long-hairy; longest ultimate segms of lvs not > ca 2 cm; midrib of lowest pinnae departing from the rachis at a narrow angle, with a buttress-like base shortly decurrent on the rachis; cliff crevices and ledges, gen on limestone, lowl to midmont; NH to Tenn, w to s BC, Wn, Utah, and Ariz; smooth c.; 2 vars. in our area   3 **P. glabella** Mett.

     a1 Lvs relatively small (up to 15 cm, including petiole), mostly only once pinnate, or with the lower pinnae only bifid or trifid or mitten-shaped, subsessile or with the stalk and rachis up to ca 2 mm; spores 64 per sporangium; mainly Gr Pl and Atl slope of RM (*P. pumila, P. o.*)   var. **occidentalis** (E. Nels.) Butters

     a2 Lvs ave larger, up to 25 cm, basal pinnae often again pinnate with 3–5 pinnules, stalk and rachis together up to 2.5 cm; spores 32 per sporangium; CRG, n to s and se BC; also in GB region *(P. suksdorfiana)*   var. **simplex** Butters

   3b Lvs dimorphic, some with the pinnules of at least the lower pinnae relatively broad; longest ultimate segms of lvs gen 2–5 cm; midrib of lowest pinnae departing at a wide angle, without a decurrent base; petiole and midribs evidently hairy; on limestone; e US to SD, Colo, and Ariz, and with outlying nw stations in Sask and se BC; purple c.   4 **P. atropurpurea** (L.) Link

Pityrogramma Link   Gold-fern; Gold-back Fern

Small to medium-sized ferns with lvs tufted on a short rhizome beset with slender scales; petiole with a single bundle, slender, dark and shining, gen glab above the scaly base; lf bl relatively broad, pinnately or ternate-pinnately dissected, powdery-waxy at least beneath, otherwise glab or merely glandular; veins free; sporangia borne along the veins, hardly forming distinct sori; indusium none. (Gr *pityron*, bran, and *gramme*, line, referring to the branlike indumentum about the lines of sporangia).

P. triangularis (Kaulf.) Maxon. Petioles chestnut-brown to purplish-brown, gen 6–22 cm, notably > the bls; lf bls glab above, yellowish-powdery beneath, gen 4–10 cm and nearly or fully as broad, ternate-pinnately compound; pinnae few, opp, lowest pair much the largest; lf margins very narrowly revolute, but not covering the sporangia; rock crevices and open, sometimes rocky slopes, valleys and foothills; s BC (VI) to Baja Cal, in CRG, and also in se Wn and reputedly w Ida *(Gymnogramma t.)*; ours is var. *t.*

Polypodium L.   Polypody

Small to medium-sized, mesophytic ferns with br, widely creeping rhizomes beset with chaffy scales; lvs jointed to rhizome, evergreen; petiole with 3 bundles near base that unite distally; bl (ours) deeply pinnatifid to once pinnate, the segms entire to toothed, broadly sessile and often confluent; sori laminar, rotund or elliptic, borne subterminally on the vein-forks, without indusium. (Gr *polys*, many, and *pous*, foot, referring to the br rhizomes). Excellent garden subjects.

   1a Lvs very thick and leathery, some of the veins anastomosing; pinnae gen broadly rounded distally, gen with a few brown scales on midvein, largest ones gen 2.5–9 × 1–2.5 cm; sori 2.5–5 mm; rhizome nearly tasteless; tree trunks and exposed cliffs and banks, sometimes within reach of ocean spray; near coast, s BC to s Cal; Scouler's p., leather-lf p.
                                          1 **P. scouleri** Hook. & Grev.

   1b Lvs firm but not notably thick and leathery, veins all free; pinnae seldom > 1 cm wide, and then distinctly pointed; sori 1.5–2.5 mm; rhizome variously licorice-flavored or acrid

     2a Lf bl (10) 15–50 cm, the pinnae acute or acuminate, the largest ones (2.5) 3–12 cm, gen at least 4 times as long as wide; rhizome licorice-flavored; tree trunks and moist banks and rocks, W Cas, Alas to Cal,

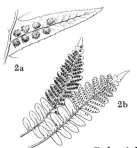

2a

2b

and in CRG; licorice-fern (*P. occidentale, P. vulgare* var. *o., P. falcatum, P. g.* var. *f.*) 2 **P. glycyrrhiza** D. C. Eat.

2b Lf bl gen smaller, 4–15 (20) cm, with rounded or obtuse pinnae, the largest ones (0.6) 1–2.5 (3) cm, rarely > 3 times as long as wide; cordilleran sp. of moist cliffs, ledges, and rock crevices, less often of ordinary soil in woods, from lowl to well up in mts; the diploid cytotype (n = 37), mainly Cas and W Cas, with acrid rhizomes, and tending to differ in other small ways, has been segregated (perhaps on insufficient grounds) as *P. montense* Lang from the nomenclaturally typical tetraploid cytotype, mainly E Cas, with licorice-flavored rhizomes

3 **P. hesperium** Maxon

Polystichum Roth   Holly-fern; Christmas-fern; Sword-fern

Small to large evergreen ferns with lvs tufted on a short, stout, scaly, erect to short-creeping rhizome; petiole < bl, with 4–5 free bundles and with dimorphic scales (some broad and often toothed or cleft, others smaller and almost hairlike); bl scaly esp on rachis and costae, pinnate to subtripinnate, the principal pinnae subsessile on a narrow, very shortly petiolulate base; veins free, mostly once-forked; sori round, borne on the veins, gen in 1–more definite rows on each side of the costa or midvein; indusium peltate. (Gr *polys*, many, and *stichos*, a row, referring to the reg rows of sori). All fine garden subjects.

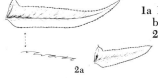

2a

2b

4a

4b

1a Pinnae undivided (though often oblique at base and with a projecting basal lobe on the upper side)

2a Spinulose teeth of pinnae appressed or closely ascending or incurved; lvs relatively large, 2–15 dm overall, petiole 0.5–5 dm; larger pinnae gen 2–15 × (0.6) 0.8–1.5 cm, lowest ones seldom much reduced; 2 vars.

1 **P. munitum** (Kaulf.) Presl

a1 Pls relatively large and coarse, lvs gen 5–15 dm overall, the petiole gen chaffy throughout and 1.5 dm or more, the larger pinnae gen (3) 4–15 cm and gradually tapering to the slender, spinulose tip; gen in moist, conif woods, in open or in deep shade, from near sea level to midmont; Alas to Cal, esp abundant in and w Cas, extending e across n Wn to n Ida and nw Mont, occ in ne Ore; common C. or s. (*P. m.* var. *incisoserratum*)   var. **munitum**

a2 Pls smaller, lvs gen 2–5 (6) dm, petiole 0.5–2 (2.5) dm, less chaffy, the distal half gen ± smooth; pinnae nearly or quite as numerous as in var. *munitum* and often overlapping, firmer, and less obviously veiny, with few or no scales, the larger ones gen 2–4 (5.5) cm and ± abruptly tapering to the spinulose tip; in distinctly drier places, as in rock crevices and dry conif woods; Cas-Sierran axis from c Wn to s Cal, also in OM, Wn, sometimes on serpentine; imbricate s. or C.

var. **imbricans** (D. C. Eat.) Maxon

2b Spinulose teeth of pinnae ± strongly spreading or ascending-spreading; lvs smaller, 1–6 dm, petiole < 1 dm, larger pinnae (gen near or above middle of lf) gen 1–4.5 × 0.3–1.3 cm; lowest several pinnae progressively reduced and often distant, lowest ones subtriangular and up to ca 1.5 cm and more than half as broad; talus slopes and cliff-crevices, occ in conif woods, midmont up to timberl; circumboreal, s to Que, sw Mont, c Ida, and nw Wn, and interruptedly to Colo, and Cal; mt h.

2 **P. lonchitis** (L.) Roth

1b Pinnae, or some of them, conspicuously cleft (> half way to midrib) or again pinnate, at least toward the base

3a Lvs relatively large, 3–10 dm overall, larger pinnae gen (3) 4–10 cm; teeth or segms of lvs spinulose-tipped

4a Lvs more dissected, relatively soft, gen producing 1–more scaly vegetative buds near tip of rachis; middle pinnae with 1st 8–more pairs of sinuses extending > 2/3 way to midvein; mature petiole gen conspicuously chaffy throughout; thickets and deep woods, mont; Alas and s BC to nw Mont, n Ida, and nw Wn (OM and Cas) to n Ore (Mt Hood); Anderson's s. (*P. braunii* spp. *a., P. jenningsi*) 3 **P. andersonii** Hopkins

4b Lvs less dissected, firm, not producing vegetative buds; middle pinnae with only lowest 2–8 pairs of sinuses extending as much as 2/3 way to midvein; mature petiole with few or no scales on distal 1/2–2/3; woods, streambanks, and open rocky places, midmont; Wen Mts and

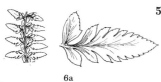

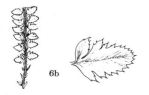

Cas, c Wn, to s Cal, rarely collected in our area; Cal s. *(P. aculeatum var. c.)* **4 P. californicum** (D. C. Eat.) Diels

3b Lvs smaller, 1–4 dm overall, larger pinnae gen 1–3 (4) cm, the teeth with or without spinulose tip

5a Lvs bipinnate or subtripinnate, teeth and segms blunt or rounded to merely acute, soft or slightly callous-thickened, not at all spinulose or mucronate; open rocky slopes, mont but below timberl, esp on serpentine; c Cal and sw Ore, also in Wen Mts, c Wn, in c Ore, and in S Am; Shasta fern *(P. lemmoni, P. m. var. l.)* **5 P. mohrioides** (Bory) Presl

5b Lvs subbipinnate, the teeth or segms with firm, mucronate or spinulose tip

6a Principal pinnae gen (0.8) 1.5–3.2 cm, gen 2–3 times as long as wide (width measured above the large basal lobe), the lobes or teeth gen with ascending or incurved, callous-mucronate, but hardly spinulose tip; common, cliff crevices and open rocky slopes, midmont up to near timberl but lower in CRG; s BC to s Cal, e to w Mont and n Utah, wholly E Cas in our region; also in Que; rock s *(P. aculeatum var. s.)* **6 P. scopulinum** (D. C. Eat.) Maxon

6b Principal pinnae up to 1.5 (1.8) cm, gen 1–2 times as long as wide, the spinulose points of the teeth prominent and tending to be spreading or widely incurved; rare, cliff crevices and talus slopes, midmont upward to near timberl; s BC and nw Wn (OM) irreg to n Cal, c Ida and n Utah; Kruckeberg's s. **7 P. kruckebergii** Wagner

Pteridium Gled.   Bracken; Brake-fern

Medium-sized to large mesophytic fern with deep-seated, elongate, br, hairy (not scaly) rhizome; lvs scattered, deciduous; petiole coarse and firm, ± erect and stemlike, with several distally united bundles; bl firm, pinnately or ternate-pinnately ca thrice compound, ultimate segms ∝, crowded, sessile and often confluent; veins connected by a hidden submarginal strand on which the confluent sori are borne, otherwise free; sori protected by the narrowly inrolled indusial lf-margin and (on the other side) by a delicate, hyaline, concealed inner indusium (Gr *pteris*, a fern).

P. aquilinum (L.) Kuhn. Lvs gen (3) 5–20 (up to 50) dm overall, petiole up to 10–15 dm, gen < bl; bl glab or inconspicuously hairy above, ± densely villous or villous-puberulent beneath and on the indusial flap; basal pair of pinnules often the largest, but seldom so large as to make the bl appear ternate; moist to dry woods or open slopes; cosmopolitan and throughout our range, lowl to mont; ours the var. *pubescens* Underw.

Thelypteris Schmidel   Wood-fern

Medium-sized, mesophytic ferns with gen slender and elongate (occ stout and erect), blackish, sparingly scaly rhizome; petiole bundles 2, united below the bl; lvs thin, light green, deciduous, solitary to tufted, bipinnatifid or subbipinnatifid, sparingly or not at all scaly, pubescent with slender, unicellular hairs, at least on the rachis; rachis grooved, marginal ridges continuous, elevated above the insertion of the midveins, veins free; sori roundish, each borne on a vein which continues to the margin; indusium·when present small and inconspicuous, hyaline, gen ciliate and glandular, reniform to horseshoe-shaped, becoming engulfed by the mature sorus, or in some spp. wholly wanting. (Gr *thelys*, female, and *pteris*, fern).

1a Rhizome slender and elongate; pubescence extending to the costae and often also the foliar surface, often on both surfaces

2a Lf bl (4) 10–20 (25) cm, 1–2 × as long as wide, = or < the petiole, the pinnae all ± approximate, the lowest several pairs the largest, indusium lacking; cliff crevices and moist banks in wooded regions; circumboreal, s to n Ore, se BC, Ia, and NC; beechfern, northern b., long b. *(Phegopteris p., P. vulgaris, Dryopteris p.)* **1 T. phegopteris** (L.) Slosson

2b Lf bl gen 25–60 cm and 3–5 × as long as wide, > the petiole, the lowest several pairs of pinnae reduced and distant; indusium horse-shoe shaped, inconspicuous; moist woods and springy places, foothills to midmont; Cas-Sierran region (chiefly on w side), Mt Rainier, Wn, to nw and c Cal; Sierra w. (*Dryopteris n., D. oregana, T. o.*)

2 **T. nevadensis** (Baker) Clute

1b Rhizome short and stout; pubescence chiefly confined to upper side of rachis; bl 3–7 dm, 2.5–5 times as long as wide, bipinnate with entire, often revolute-margined segms; lowest 2–several pairs of pinnae gen ± reduced and distant; indusium reniform or horseshoe-shaped, evanescent; mt fern (*Dryopteris montana, Phegopteris oreopteris, D. o., T. o., T. o. var. hesperia*)

3 **T. limbosperma** (All.) Fuchs

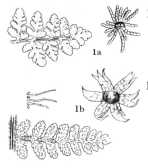

Woodsia R. Br.   Woodsia

Small to medium-sized ferns with tufted, pinnate or bipinnate to subtripinnate lvs and persistent petiole-bases on a short rhizome covered with yellowish to dark brown scales; petiole with 2 bundles at base, these united distally; sori provided with an inconspicuous, inferior indusium. (For Joseph Woods, 1776–1864, English architect and botanist).

1a Bl glab or merely glandular, 4–15 × 1–4.5 cm, subbipinnate to subtripinnate; indusium platelike, with radiating, hairlike, often somewhat beaded segms, gen largely hidden beneath the mature sorus; rock crevices, ledges, and talus slopes, less arid parts of lowl and in mts; widespread cordilleran sp., e irreg to Que, gen E Cas in our range; w. (*W. o. f. glandulosa*)

1 **W. oregana** D. C. Eat.

1b Bl glandular and with glandless, septate hairs, at least on the lower side, up to 22 × 7 cm; indusium much as in no 1; rock crevices to talus slopes, mont and less arid lowl; widespread cordilleran sp., e irreg to Que, gen E Cas in our region; RM w. (*W. obtusa* var. *lyallii*) 2 **W. scopulina** D. C. Eat.

Woodwardia J. E. Smith   Chain-fern

Large mesophytic ferns with coarse and woody br rhizome beset with chaffy scales; petiole gen with 2 bundles; lvs evergreen (ours); bls pinnatifid to pinnate-pinnatifid; veins partly anastomosing, forming one or more series of areoles along the midvein of the pinna or pinnule; sori often set in pits, oblong to linear, spanning 2–more areoles, arranged in chainlike rows on each side of the costa, with a conspicuous, flaplike indusium opening toward the costa. (For Thomas J. Woodward, 1745–1820, English botanist). Excellent garden pl.

W. fimbriata J. E. Smith. Bls large and firm, nearly erect, 4–15 dm or more, pinnate-pinnatifid; pinnae alt along rachis, 1–3 (4) dm, the foliar-winged costa 3–8 mm wide, the segms (pinnules) gen 10–20 opp or subopp pairs, serrulate, 2–5 (10) × up to 1 cm; largest pinnae near or below middle of bl; lowermost ones somewhat reduced and distant; sori (2) 3–6 mm; streambanks and wet places in gen, sometimes maritime, lowl to midmont; s BC to s Cal, chiefly near the coast, irreg inl to Ariz and Nev (*W. chamissoi, W. radicans* var. *americana*).

MARSILEACEAE   Pepperwort Family

St a creeping, superficial rhizome; lvs alt, in 2 rows, erect, filiform or with a long petiole and 2–4 approximate lflets; sporocarps compressed-ovate to globose, slender-ped or sessile, the stalk arising directly from the rhizome or adnate to the petiole; pls heterosporous; sori enclosed in a sporocarp representing a modified, folded pinna with connate margins, each sorus bearing both mega- and microsporangia; sori 2–∝ in each sporocarp; microspores ∝ and minute, megaspores solitary in each megasporangium; sporangia thin-walled, with vestigial or no annulus.

1a Lvs with a slender petiole and 4-foliolate bl   **Marsilea**
1b Lvs filiform, without an expanded bl   **Pilularia**

Marsilea L.   Pepperwort; Clover-fern

Lvs long-petiolate, bl floating or emergent (submerged), with 4 approximate lflets suggesting a 4-lvd clover; sporocarps dehiscent by 2 valves, somewhat compressed, ± ovat 'o elliptic. (For Giovanni Marsigli, Italian botanist of 18th century).

M. vestita Hook. & Grev. Petiole gen 2–15 cm; lflets gen 5–15 × 3–17 mm, pubescent with long, loose to subappressed hairs, or sometimes ± glab; sporocarps 4–8 mm, purple-punctate, densely pubescent to glabrate; shallow water and mud; foothills, valleys, and lowl, s BC to Sask and Minn, s to Cal, n Mex, and Ark (*M. longipes, M. oligospora*).

Pilularia L.   Pillwort

Lvs bladeless, with unbr midvein; sporocarps eventually dehiscent at top by 2–4 valves, globose, short-pedunculate, axillary. (Diminutive of L *pila*, a ball, referring to the globose sporocarps).

P. americana A. Br. Rhizome filiform, long-creeping; lvs glab, filiform, 1.5–5 cm × ca 0.5 mm; sporocarps solitary on short, deflexed stalks from the nodes, 2–3 mm thick, appressed-woolly; shallow vernal pools; irreg from c Ore to s Cal, disjunct in Ark, Ga, and Chile; seldom collected.

SALVINIACEAE   Water-Fern Family

St filiform, br, free-floating or seldom on wet soil; lvs small, crowded, alt or whorled, simple or bilobed or dissected, sessile or nearly so; pls heterosporous; microsporangia and megasporangia borne in separate sori, each sorus enclosed by its basally attached indusium, which forms a sort of sporocarp; sporocarps soft and thin-walled, each containing a single sorus; male sorus with ± ∝ microsporangia; female sorus with a single megasporangium, and a single megaspore; sporangia thin-walled, with vestigial or no annulus.

Azolla Lam.   Water-fern; Mosquito-fern

Delicate, mosslike pls, subdichotomously br above every 3rd lf; roots inconspicuous; lvs minute, alt, sessile, unequally bilobed, 1 lobe submersed, serving as a float, the other smaller and emersed; sporocarps when present borne on the 1st lf or a br, paired at the tip of a short, bipartite 2ndary lobe, appearing almost axillary; microspores 32 or 64, aggregated into 4 or more groups, each group with its own thin common wall which bears ∝ slender appendages (glochidia) with terminal, retrorse hooks; megaspore with a hemispheric or broadly bell-shaped basal part and a sharply marked, complex, conical terminal part that eventually shows 3 basally spreading, apically confluent, longitudinal valves. (Name of uncertain derivation).

1a Glochidia with scattered septa; basal portion of the megaspore pitted; pl compactly br and forming mats gen < 2 cm across; upper lobe of lf gen slightly < 1 mm, often reddish-purple; ponds and backwaters; s BC to Wis, s to Mex and S Am; Mex w.; our more common sp.
                                                                                    1 **A. mexicana** Presl
1b Glochidia without septa or rarely septate just beneath the tip; basal portion of the megaspore tessellate; pls forming mats (1) 2–5 cm broad; upper lobe of lf gen ca 1 mm; ponds and backwater, or even occ on wet soil; much of S and C Am and Mex, n in coastal states of w US, occ to Alas; seldom collected in our range; duckweed fern   2 **A. filiculoides** Lam.

TAXACEAE   Yew Family

Stamens (1) 3–12 in tiny globose clusters in lf axils, subtended by several bracts, each ± peltate; pollen sacs 2–9, pendent; typical ♀ cones lacking, ovules solitary (paired), terminal on a short scaly stalk, ripening in 1 season into a hardened seed surrounded by and almost immersed in a disc that tends to become juicy and highly colored; cotyledons 2; trees or shrubs with spirally-arranged, persistent lvs.

Taxus L.   Yew

Ovule solitary in lf axils, immediately subtended by 2 pairs of bracts, at maturity surrounded by a fleshy, gen reddish disc (aril), the whole often loosely termed a berry; brs mostly drooping or spreading, gen imperfectly or not at all whorled. (L name for European yew; Gr name for yew, *taxos*, was believed to have been derived from *taxon*, a bow, for manufacture of which the wood has been used from antiquity).

T. brevifolia Nutt. Western or Pac y. Tree mostly 5–10 (25) m, with erect and straight to commonly contorted trunk; bark very thin, with outer purplish scales covering newly formed reddish or reddish-purple inner bark; lvs (12) 14–18 × 1–2 mm, yellow-green, paler beneath, abruptly mucronate, persistent 5–6 years; ovules borne on lower side of brs; seeds 5–6 mm, aril fleshy, reddish, ovoid; ± moist for, esp along streams; Alas to Sierran Cal, e to e BC, Ida, and Mont; not only excellent for bows, but wood durable, hard, and attractive (*T. baccata* var. *b.*).

CUPRESSACEAE   Cypress Family

Pls ♂♀ or ♂, ♀; ♂ cones small, mostly terminal on brlets, stamens broadly peltate, pollen sacs 2–6, suspended; ♀ cones terminal, comparatively small, with 2–12 scales, 1–2.5 cm, dry to greenish or purplish and ± fleshy, scales opp or in 3's, flattened and peltate or fleshy and grown together, with 1–several erect ovules near base; seeds often winged; cotyledons mostly 2 (5–6); shrubs or trees with opp or whorled, gen scale-like lvs, but juvenile lvs often strongly acicular and sometimes all lvs acicular or ± needle-like.

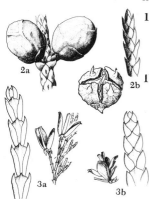

1a ♀ cones subglobose, scales peltate or ± fleshy and fused; lvs opp or in 3's
  2a Cones fleshy, greenish to purplish; scales fused, fertile ones each with
    1–2 seeds; lvs opp or in 3's, scale-like or sometimes needle-like **Juniperus**
  2b Cones dry; scales peltate, not fused at maturity, fertile ones with 2–5
    seeds; lvs all opp, scale-like   **Chamaecyparis**
1b ♀ cones elongate, scales flattened, opp, distinct, dry; lvs opp or apparently
  in 4's
  3a Tips of dorsal and ventral lvs freed at ca same point as the lateral pair,
    hence lvs apparently in 4's; cone scales reg 6, only the middle pair
    bearing seeds   **Calocedrus**
  3b Tips of dorsal and ventral lvs freed at different level than those of lateral
    pair, hence lvs more obviously opp (in 2's); cone scales gen 8–12, 2–3
    pairs bearing seeds   **Thuja**

Calocedrus Kurz   Incense-cedar

Pl ♂♀ (ours); ♂ cones terminal on lateral brlets, stamens gen 10–18, opp, ± peltate, with 4 (3–5) pollen sacs near the base; ♀ cones single and terminal on short brlets, ± pendent, maturing in 1 season, scales 6, opp, flattened, middle 2 fertile and bearing 2 seeds each, basal pair sterile, sharply reflexed, terminal pair sterile, dorsoventrally flattened, largely connate, = middle pair; seeds unequally 2-winged; cotyledons 2; large trees with scale-like persistent lvs borne in 4 rows, lateral pairs strongly keeled, dorsoventral ones more flattened (hence brs flattened), all strongly decurrent, their tips of ca equal length, brs largely in ± erect, flat sprays. (Gr *calo*, beautiful, and *kidros*, cedar).

C. decurrens (Torr.) Florin. Pyramidal trees 24–40 m, with rather thick yet smooth, reddish- to purplish-brown bark and gen with at least upper brs and leader ± erect; lvs yellowish-green, mostly 4–6 (20) mm, tips gen incurved, those of all 4 rows freed at ca same level; ♂ cones 5–7 mm; ♀ cones ca 2.5 cm, deep reddish-brown, tardily deciduous; seeds ca 8 mm; mixed conif for; lower se slope of Mt Hood, Ore, s in Cas and through Cal to Baja Cal, e to c Ore and w Nev (*Thuja d.*, *Libocedrus d.*, *T. craigana*). Wood resistant to fungal attack and often used in contact with soil; tree itself a fine orn.

Chamaecyparis Spach   Cedar; Cypress

Stamens 6–10 pairs, each bearing 2–4 globose, often reddish pollen sacs; ♀ cones erect on short brlets, subglobose, scales 3–6 pairs, spreading, strongly peltate, slightly elevated in center; seeds prominently winged, maturing 1st year; cotyledons 2; ♂♀ trees (ours) with opp, scale-like lvs in 4 rows, lateral rows not greatly different from the dorsoventral, but ± keeled and brlets therefore ± flattened. (Gr *chamai*, dwarf, and *kuparissos*, cypress, genus somewhat resembling *Cupressus*, "true" cypress).

C. nootkatensis (D. Don) Spach, Alas, Nootka, Sitka, or yellow c. Tree 20–50 m, with narrow crown and slender, drooping leader, brs mostly tending to droop strongly; bark thin, grayish-brown, cinnamon-brown inside, tending to come off in long narrow strips; wood sulfur-yellow; lvs malodorous when crushed, bluish-green, pungent, glandular on back; mature ♀ cone scarcely 10 mm, globose, brownish and glaucous, scales gen with short point in center of the peltate tip; seeds ca 3 mm; Alas s at increasing altitudes to n Cal, in OM and Cas in Wn, e to Blue Mts, Ore, mostly at 2000–7000 ft elev (*Cupressus n.*).

Juniperus L.   Juniper

♂ cones spherical to ovoid, 3–5 mm, stamens eccentrically peltate, with mostly 3–6 subglobose pollen sacs; ♀ cones small, subglobose, scales mostly 3–8, opp or in 3's, coalescent, becoming fleshy, when ripe the cone often loosely termed a berry, 1–5 (10) -seeded, gen with only the scale tips (sometimes also ovule tips) projecting, glaucous-green or -brown; seeds wingless; cotyledons 2–6; ♀, ♂ or occ ♀♂ shrubs or trees with scale-like to needle-like lvs arranged in 2's or 3's, at least the basal portion of lf decurrent on st. (L name for juniper). (*Sabina*). The several spp. in our area are valuable orn, esp the prostrate ones; wood durable, used chiefly for fence posts and pencils.

1a Lvs in 3's, all linear-lanceolate and ± needle- or awl-like, 5–19 mm, pungent, not glandular on dorsal surface, jointed at base of free portion; pl mostly decumbent, ours seldom > 2 m; lowl woods to mt valleys and open rocky alp slopes; Alas to Cal, e to Newf and Greenl, Neb, Mich, O, and Ga, Eurasia; common or mt j.; several vars.   1 **J. communis** L.
  a1 Pl erect rounded tree or shrub gen > 3 m; lvs spreading, straight, 12–19 × scarcely 1 mm; mostly n and e of our area, to be expected along Can border in Ida and Mont (*J. c.* var. *erecta*)   var. **communis**
  a2 Pl rarely > 1 m; brs spreading to decumbent; lvs gen ascending, 7–15 × 1–2 mm

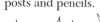

    b1 Main st partially erect, 5–20 dm; brs spreading, becoming decumbent; occ in Ida and Mont, more common e   var. **depressa** Pursh
    b2 Main st, as well as brs, trailing
      c1 Sts freely br, pl well-matted; range of sp.; common phase in our area, almost to exclusion of other vars. (*J. nana, J. sibirica*)
          var. **montana** Ait.
      c2 Sts sparingly br, long and trailing, pls not densely matted; serpentine areas in sw Ore, reported as ranging n to Wn, but not seen from our area   var. **jackii** Rehder
1b Lvs in 2's or 3's, scale-like (or juvenile lvs ± awl-like but only 5–7 mm), often glandular on dorsal surface, not jointed at base of free portion; pls often erect and > 2 m
  2a Lvs in large part in 3's, their margins minutely ciliolate-denticulate under 20× magnification
    3a Lvs with prominent gland on dorsal surface; ♀ cone bluish-black, 7–8 mm, 1–3-seeded; freely br tree 4–10 (15) m; des foothills and lower mts; Kittitas to Klickitat Co and e to Whitman and Asotin cos, Wn, s and abundant e Cas, to s Cal, e to Blue Mts, Ore, and sw Ida and Nev; western or Sierra j.   2 **J. occidentalis** Hook.
    3b Lvs not glandular; ♀ cones bluish-glaucous, aging brown, mostly 7–10 mm, 1 (2) -seeded; rounded, shrubby sp. of the GB and RM area, app but not known from our area; Utah j.   **J. osteosperma** (Torr.) Little
  2b Lvs almost all in 2's, entire-margined
    4a Sts decumbent, pl nearly or quite prostrate, rarely as much as 3 dm; mature (non-juvenile) lvs strongly apiculate; foothills to dry mont ridges; Alas and c BC to Mont, esp e RM, s to Wyo and Colo, e to Minn, Ia, and Me; creeping j. (*J. prostrata, J. sabina* var. *p., J. virginiana* var. *p., J. sabina* var. *procumbens*)   3 **J. horizontalis** Moench
    4b Sts ascending to erect, mature pl up to 10 and rarely < 1 m; mature (non-juvenile) lvs obtuse or acute to slightly apiculate, but needle-like juvenile lvs often long-persistent, esp in coastal form; coastal is to inl valleys and lower mts, VI e, in BC and Wn, to Mont, Daks, Neb, Colo, NM, and Neb; local in e Wn near CR and in e Ore; RM j. (*J. virginiana* var. *montana, J. v.* var. *s., J. excelsa, J. s.* var. *patens*)
          4 **J. scopulorum** Sarg.

Thuja L. Arborvitae; Cedar

♂ cones tiny, tipping short brlets, stamens (3) 4 pairs, laterally peltate, pollen sacs (3) 4, subglobose; ♀ cones solitary at end of short brlets, reflexed, globose-ovoid, scales 5 (4–6) pairs, mucronate near tip, lowermost pair much shortest, sterile, uppermost pair also sterile, flattened, and partially fused laterally, intervening pairs flattened and gen each with 2 basal ovules on ventral surface; seeds with narrow lateral wings, maturing 1st year; cotyledons 2; ♂♀ trees with scale-like, opp lvs in 4 rows, lateral lvs keeled and ± spreading, brlets thus flattened and brs ± spray-like. (Gr name *thuia*, for an evergreen tree, perhaps some juniper).

T. plicata Donn. Red, western red, or Pacific red c.; giant c. Trees up to 70 m, becoming strongly fluted and buttressed at base, often with long clear bole; bark thin, reddish-brown, ridged and fissured lengthwise, readily pulled off in long fibrous strips; ♂ cones subglobose, ca 2 mm; juvenile ♀ cones bluish, becoming brown and 8–12 mm at maturity, ovoid-ellipsoid; seeds ca 6–7 mm; mainly where moist or swampy; Alas to Humboldt Co, Cal, from coast inl to se BC and n and w Mont; wood notably resistant to rot; pl valuable orn.

PINACEAE   Pine Family

♂ cones small, stamens ∝, spirally arranged, ours each with 2 elongate pollen sacs on the lower (dorsal) surface; ♀ cones mostly large and woody, scales ∝, spirally arranged, each with 2 inverted ovules on upper surface and subtended by a variously modified, free to completely adnate bract, persistent, or individually deciduous in *Abies;* seeds winged; cotyledons 2–15; ♂♀ trees (ours) with mostly whorled brs; lvs needle-like, deciduous or persistent, spirally arranged, borne singly or in fascicles of 2–5 on main brs, or closely aggregated and falsely whorled on short lateral spur shoots.

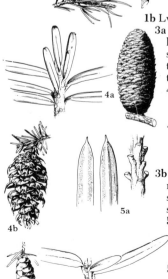

1a Lvs borne in clusters of 2–5 along main brs, or ∝ in apparent whorls on short lateral spur brs
  2a Lvs deciduous, ∝ and apparently whorled on short lateral shoots   **Larix**
  2b Lvs persistent, spirally arranged along brs in small clusters of 2, 3, or 5
          Pinus
1b Lvs borne singly and spirally arranged along brs
  3a Needles not leaving a persistent base when shed, the naked brs smooth, lf scar itself often in a slight crater; cones either stiffly erect and shed scale by scale, or ± pendent and with prominent 3-lobed bract protruding past each ovuliferous scale; needles mostly flat and blunt, retained on cut brs indefinitely
    4a Cones erect, shed scale by scale, never remaining intact; lf scar suborbicular; dormant terminal buds ± rounded and blunt; brlets uniformly opp and paired, only occ unpaired   **Abies**
    4b Cones ± pendent, shed as a unit (gen to be found under the tree); lf scar transversely elliptic-oval; dormant terminal buds sharp-pointed; brlets in part alt or imperfectly paired as well as opp (branching± irreg)   **Pseudotsuga**
  3b Needles leaving a small persistent protuberance when shed, naked brs roughened, lf scar on top of the persistent lf base; cones not stiffly erect, shed as a unit, without prominent bracts free of scales; needles often square in section and sharp-pointed, soon deciduous from cut brs
    5a Lvs sharp, gen square or only slightly flattened in section, narrowed slightly and rather gradually at base, not slenderly petiolate, lower surface not much whiter than upper; main shoot erect; naked young brs very rough   **Picea**
    5b Lvs blunt, strongly flattened, abruptly narrowed to short, slender petiole, lower surface often much whiter than upper; main shoot often drooping; naked young brs slightly roughened   **Tsuga**

Abies Mill. Fir

Cones borne at tips of previous year's brs, ♂ single, pendent from lower side of brs mostly near middle of the tree or above, 7–20 mm; stamens with expanded knoblike tips; ♀ cones gen near top of tree, borne singly and stiffly erect, globose to cylindric, maturing in 1 season, scales ∞, much-flattened, longer or shorter than their thin, non-adnate, subtending bracts, shed singly with the bracts in early winter, axis of cone peglike, persistent for sometimes several years; cotyledons 4–10; tall evergreen trees with neatly whorled brs, primary brs repeatedly brd in 1 plane to form large flat sprays; bark at first thin and often bulged by resin vesicles, old bark often thick, gen finely roughened, often grayish; lvs gen twisted to lie in 1 plane or turned upward, flat, blunt or ± pointed but not pungent, white on at least the lower (dorsal) surface, narrowed to a short, stout, petiole-like base, wholly deciduous, leaving a slightly depressed, nearly circular lf scar; winter buds and young ♀ cones gen resin-covered. (L name of the fir). All spp. have stomata on the lower (dorsal) surface, visible with 10× magnification as tiny white dots running lengthwise in several rows on either side of the midvein; collectively they appear as 2 white stripes. Some spp. have such white stripes on the upper surface also.

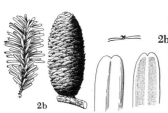

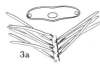

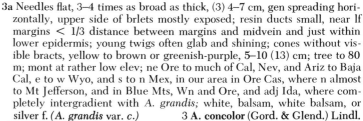

1a Lvs gen blunt or retuse, white-stomatous on lower surface only, green on upper surface; lvs either all horizontally spreading and exposing the br above, or in part horizontally spreading and in part closely appressed and pointing forward on upper side of brs, hiding the br itself; resin canals < 1/4 as broad as midvein and located near lower epidermis of lf < half distance between margins and midvein

  2a Lvs (2) 3–4 (5) cm, horizontally spreading; upper side of twigs bare except for twisted lf bases; ♀ cones light green, 6–11 cm, bracts much < scales and concealed; large trees up to 90 m; from near sea level to ca 3000 (5000) ft elev in Cas and to 7000 ft in RM; VI and mainl BC to Cal, e to se BC, Ida, Mont, and ne Ore, where intergradient with *A. concolor;* grand, lowl, lowl white, silver, or balsam f. *(A. excelsior)*

1 A. grandis (Dougl.) Forbes

  2b Lvs mostly < 2.5 (to 3) cm, longer ones spreading horizontally, others (gen shorter) strongly appressed, pointing forward, and ± completely hiding upper side of twigs; ♀ cones deep purple (to green), 8–15 cm, bracts < scales and concealed; tall straight trees to 70 m; mostly at 1000–4000 ft elev and gen slightly above *A. grandis,* but down to near sea level on OP and is of PS, gen replaced upward by *A. procera* and *A. lasiocarpa,* but occ to 6000 ft in Cas, when often with *Tsuga mertensiana;* s Alas to n Cal, mostly w Cas in Wn and Ore; amabilis, Pac silver, Cas, lovely, red, or silver f. *(A. grandis* var. *densifolia)*

2 A. amabilis (Dougl.) Forbes

1b Lvs often acute to pungent, white on both sides with lines of stomata; lvs (except in *A. concolor*) tending to turn upward or to spread more nearly in all directions rather than mainly horizontally; resin canals sometimes (*A. lasiocarpa*) larger, up to half as broad as midvein, and then medianly located between upper and lower epidermis and between lf margins and midvein

  3a Needles flat, 3–4 times as broad as thick, (3) 4–7 cm, gen spreading horizontally, upper side of brlets mostly exposed; resin ducts small, near lf margins < 1/3 distance between margins and midvein and just within lower epidermis; young twigs often glab and shining; cones without visible bracts, yellow to brown or greenish-purple, 5–10 (13) cm; tree to 80 m; mont at rather low elev; ne Ore to much of Cal, Nev, and Ariz to Baja Cal, e to w Wyo, and s to n Mex, in our area in Ore Cas, where is almost to Mt Jefferson, and in Blue Mts, Wn and Ore, and adj Ida, where completely intergradient with *A. grandis;* white, balsam, white balsam, or silver f. *(A. grandis* var. *c.)*   3 **A. concolor** (Gord. & Glend.) Lindl.

  3b Needles much the thickest in center, mostly < 3 × as broad as thick, gen < 3 (but rarely up to 4) cm, tending to turn upward more than to spread horizontally; resin ducts located ca halfway between lf margins and midvein, either small and just within lower epidermis or much larger and midway between the 2 lf-surfaces; young twigs finely pubescent; cones sometimes with prominent bracts

    4a Resin ducts 1/3–1/2 as broad as midvein, midway between upper and lower epidermis; rows of stomata on upper surface in broad, median band scarcely interrupted down center; mature ♀ cones 6–11 cm,

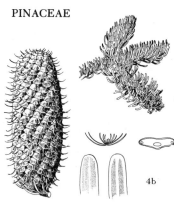

mostly deep purple (rarely green), without visible bracts; tree to 30 m, but often dwarfed and shrublike; subalp to alp slopes, from perhaps 2500 ft upward to timberl; Alas and Yuk to s Ore, in both Cas and OM, e to c Ida and Mont and s to NM and Ariz; alp, subalp, balsam, or white balsam f. (*A. balsamea* spp. *1., A. subalpina* var. *fallax*)

4 **A. lasiocarpa** (Hook.) Nutt.

4b Resin ducts < 1/3 as broad as midvein, just within lower epidermis; rows of stomata on upper surface often forming 2 bands, a very narrow area above the midvein greenish and without stomata; mature ♀ cones (10) 11–18 cm, with prominent straw- to olive-colored bracts exserted past the reddish brown scales; symmetrical tree to 70 m; deep for of Cas mostly, Wn to n Cal, in Wn barely n of Stevens Pass, Chelan Co, w to Mt St. Helens, and in Willapa Hills, Wahkiakum Co, Wn, also on Saddle Mt, Clatsop Co. Ore; noble, red, or white f. (*A. nobilis* of auth.)

5 **A. procera** Rehder

Larix Adans.   Larch

Cones terminal on short lateral spur brs, ♂ globose to cylindric, mostly 7–12 (15) mm, yellowish, ♀ ultimately woody, maturing 1st season, gen red until dry, scales persistent, subtended and exceeded by thin, long-caudate, acute to retuse bracts; seeds with thin, cuneate wing; cotyledons 5–7; lvs consisting in part of thin, scale-like, ± tomentose, persistent bracts at the tips of short, spur-like side brs and at the base of the cones, but mostly greenish needles that are deciduous each fall, these either spirally arranged and strongly decurrent as on seedling sts and lead shoots of older pls, or crowded in false whorls of 15–40 at tips of short, slow-growing, stubby spur-shoots. (L name for the larch). Orn trees with fresh green summer foliage turning deep golden in fall before dropping.

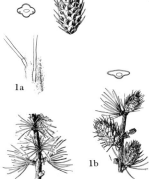

1a Young brs strongly tomentose with long, tangled hairs; needles ca as thick as broad, gen showing 2 small resin ducts in cross-section under ca 25 × magnification; ♀ cones (3) 3.5–4.5 cm; small, subalp or alp trees 10–15 (25) m; gen near timberl, often on n facing slopes; s BC to c Cas and Wen Mts, Wn, e to sw Alta, n Ida, and w Mont; alp, subalp, or Lyall 1.; tamarack

1 **L. lyallii** Parl.

1b Young brs glab to ± pubescent, hairs short and not tangled; needles much broader than thick, nearly plane on upper surface but strongly ridged beneath and broadly triangular in section, without visible resin ducts; ♀ cones 2.5–3 (3.5) cm; large trees up to 80 m; from foothills to midmont, rarely subalp, often where swampy; s BC s, e Cas, to Deschutes Co, Ore, e to n Ida, nw Mont, and ne Ore; w, Mont, or mt 1.; hackmatack; tamarack, w t.

2 **L. occidentalis** Nutt.

Picea A. Dietr.   Spruce

Cones borne on year-old twigs, ♂ ± pendent, yellow to purple, 1–2 cm; ♀ cones ovoid to cylindric, gen pendent, maturing in 1 season, deciduous as a whole, scales thin but ± woody, non-prickly, much exceeding and concealing the narrow, thin bracts; seeds 2 per scale, strongly winged; cotyledons(4) 5–15; evergreen trees with rather thin, scaly bark and short, rigid needles tending to spread in all directions around the twigs. (Old L name for some conifer, from *pix, picis*, pitch). Besides the following spp., *P. mariana* (Mill.) B.S.P. occurs in BC, and is sometimes reported for our area. It has cones 1.5–3.5 cm, smaller than ave for any of our spp.

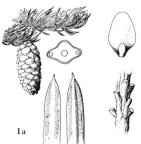

1a Young brs gen pubescent; needles ca as thick as broad, 4-angled; stripes of stomata nearly equally broad on upper and lower surfaces; ♀ cones 4–5 (6) cm, scales somewhat pointed and ± rhomoidal in shape; spirelike trees up to 50 m, but trunk rarely > 1 m thick near base; mont, gen in swampy places, mostly above 3000 ft elev; BC s, on e side Cas, to Cal, e to sw Alta, Mont, and other RMS to NM and Ariz; our common sp. e Cas; Engelmann, Columbian, white, mt, or silver s. (*P. glauca* spp. *e.*) 1 **P. engelmannii** Parry

1b Young brs glab; needles often semiflat on upper surface or without stomata on lower surface; ♀ cones sometimes > 6 cm, scales sometimes rounded or truncate at tip; both sides Cas

2a Needles almost flat on upper surface, nearly twice as broad as thick, lower surface either green and without stomata or the 2 stripes of sto-

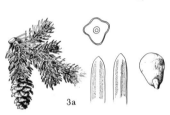

2a

3a

mata much narrower than those on upper surface; massive tree up to 70 m × 4–5 m near base; Alas to n Cal, from coast to w side Cas, where up to ca 2000 ft; Sitka s. (*P. falcata*)   2 **P. sitchensis** (Bong.) Carr.

2b Needles 4-angled, 4-sided, ca equally stomatiferous on 4 sides; e Cas (*P. engelmannii* reported from OP)

   3a ♀ cones mostly 2.5–3.5 (6) cm, scales rounded to blunt at tip; needles mostly 12–20 (25) mm; stunted and deformed to erect trees up to 25 m; lowl swamps to mont slopes; Alas to s BC, e to Me and Newf, s e RM to e Mont and Wyo, reported for our region in Mont; white, Can, Black Hills, cat, or skunk s. (*P. albertiana, P. canadensis* of auth., *P. c.* var. *g.*)   3 **P. glauca** (Moench) Voss

   3b ♀ cones gen at least 4 cm, scales ± pointed and roughly rhomboidal in shape; needles often > 2.5 cm

      4a Cones 4–5 (6) cm; needles neither particularly rigid nor pungent; e Ore and Wn to much of Mont (see lead 1a); var. *glabra* Goodm. (*P. columbiana*), a not uncommon variant of   1 **P. engelmannii** Parry

      4b Cones gen at least 6 cm; needles rigid and strongly pungent; app our area in Wyo and se Ida, common in s RM; blue s.   4 **P. pungens** Engelm.

Pinus L.   Pine

♂ cones clustered at base of shoots of current season; ♀ cones single to clustered toward tip of brs of current season, maturing in 2 years (ours), becoming large and woody, deciduous or retained on trees for sometimes many years; scales flattened and ± broadened upward to a gen wider and more thickened exposed area (apophysis) having a terminal or subterminal scar-like protuberance (umbo) that sometimes extends into a sharp spine; bracts small, adnate to scales; seeds almost wingless to broadly winged terminally; cotyledons 3–15 or more; lvs of 2 kinds, some needle-like and green and borne (ours) in clusters (fascicles) of 2, 3, or 5, on small, ultimately deciduous spur brs, others scale-like, membranous and non-green, forming bundle sheath around the needle clusters. (Old L name for the pine).

2a

2b

4a

1a Needles 2–3 (very occ some 4 or 5) per fascicle; tip (umbo) of ♀ cone scales ending in a sharp, hard prickle

   2a Needles largely in 2's, 3–6 cm; ♀ cones 4–6 cm, ± lopsided, some gen long-persistent on tree; tree rounded to pyramidal, 10–30 (35) m; coastal dunes and bogs to dry mt slopes, where occ to timberl, often in pure stands in areas long since burned over; Alas and Yuk to Baja Cal, e to Alta, Sask, Daks, and through RMS to Colo; lodgepole, black, scrub, shore, coast, or tamarack p.; 2 vars.   1 **P. contorta** Dougl.

      a1 Trees rarely > 15 m, gen with rounded crown; bark dark brown to grayish-black, becoming 2–2.5 cm thick; needles deep green; coastal from Alas to n Cal (*P. c.* var. *hendersoni*)   var. **contorta**

      a2 Trees up to 35 m, gen columnar when in close stands; bark reddish-brown, rarely > 1 cm thick; needles yellow-green; range of sp. except not coastal (*P. murrayana, P. c.* var. *m., P. tenuis*)   var. **latifolia** Engelm.

   2b Needles mostly in 3's, (10) 12–25 cm, yellow-green; ♀ cones 8–14 cm, symmetrical, all deciduous when mature; large tree up to 70 m, with bark at first dark brown or blackish but becoming very thick and changing to cinnamon red in old trees and divided into large plates scaling off freely; lower levels chiefly inl, mostly where rather dry, often in open, pure stands; BC s to Baja Cal, mostly e Cas, occ w Cas in sw Wn and (more abundantly) in Ore, as in WV, e to se BC, Mont, and Daks, s in RMS to w Tex and n Mex; w yellow, yellow, pondosa, ponderosa, blackjack, or bull p. (*P. p.* var. *scopulorum*)   2 **P. ponderosa** Dougl.

1b Needles 5 per fascicle; tip (umbo) of ♀ cone scales not at all prickly

   3a Cones 3.5–25 cm, sessile or with very short stalk scarcely 1 cm; cone scales thick, woody, sometimes remaining closed; lvs 3–7 cm; alp or subalp, gen dwarfed and contorted trees 4–15 (20) m; wing < seed and often adherent to cone scale

      4a Cones ± slenderly ovoid, (5) 8–25 cm, light brown to greenish-brown, rather quickly opening at maturity and soon deciduous, scales thick but thinner toward tip; bark grayish when young, becoming brown to blackish; at or near timberl in mostly semi-arid ranges from se BC to

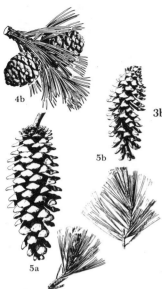

sw Alta and Daks, s in Ida and Mont to NM, Cal, and Nev, also in ne Ore; limberpine   3 **P. flexilis** James

4b Cones ovoid, (3.5) 5–8 cm, deep red to purple, tending to remain closed and to shed seeds slowly, rarely falling from trees intact, scales much thickened distally and narrowed and thinner only on the short, ± upturned point; bark thin, divided into whitish scales brownish beneath; at or near timberl, s BC through Cas to Cal, e to ne Ore, Ida, wc Mont, and n Nev; white bark p. (*P. flexilis* var. *a.*)
    4 P. albicaulis Engelm.

3b Cones (10) 15–45 cm, on stalk 1–7 cm; cone scales thin, opening to shed seeds; lvs 5–10 cm; stately for trees up to 80 m, producing excellent lumber, never alp; wing considerably > seed, not adherent to cone scale but remaining attached to seed

5a Cones 25–45 cm; needles strongly acute; seed wing rounded at tip; tree up to 80 m; mixed conif for, primarily in Cal, s to n Baja Cal, n to sw Ore and very occ to headwaters of Santiam R, e Linn Co; sugar p.
    5 P. lambertiana Dougl.

5b Cones 15–25 cm; needles gen obtuse; seed wing acute; moist valleys to somewhat dry slopes, from near sea level up to 6000 ft elev in RM; s BC s, in both OM and Cas, to Cal and w Nev, e to Ida and w Mont; w white p.   **6 P. monticola** Dougl.

Pseudotsuga Carr.

Cones borne near tip of previous year's growth, ♀ subapical, subsessile, cylindric-ovoid, maturing in 1 year, with exserted 3-lobed bracts subtending and exceeding the scales; seeds broadly winged; cotyledons 4–12; lvs flattened, needle-like, short-petiolate, wholly deciduous and leaving an indented scar somewhat broader than long; buds conical, sharp, non-resinous, glab; large for trees. (Gr *pseudo*, false, and *Tsuga*, Japanese name for the hemlock).

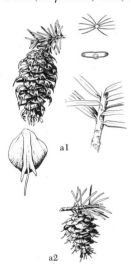

P. menziesii (Mirbel) Franco. Douglas fir, Oregon pine (to the lumberman). Up to 90 m, with very thick, rough, dark brown bark, brs spreading to drooping, leader stiffly erect; needles (1.5) 2–3 (3.5) cm, retained 7–10 yrs, mostly uniformly spreading or upswept on brs, greenish on upper surface but white-banded with stomata beneath; cones 4–10 cm, pendent, yellowish-green to purplish before maturity, soon deciduous, scales pubescent, bracts 4–7 mm broad across tip, middle lobe much the longest; seeds 5–6 mm, wing up to twice as long; cotyledons 4–10; moist to dry areas, sea level to near timberl in RM; sw BC to Cal, e to sw Alta, Mont, and other RMS to n and c Mex; 2 vars.

a1 Cones 6–10 cm; bracts straight and appressed toward cone tip; lvs deep yellowish-green; twigs pubescent; trees to 90 m; coastal, e to e Wn and Ore, and to w Nev, not at high elev; coast D. f. (*Abies douglasii*, *P. caesia*, *P. menziesii* var. *viridis*, *P. mucronata*, *P. taxifolia*, *P. t.* var. *elongata*, *P. vancouveriensis*)   var. **menziesii**

a2 Cones mostly 4–7 cm; bracts appressed to spreading or reflexed; lvs more bluish-green; twigs glab or pubescent; rarely > 40 (50) m; primarily RM, w to e Wn, Ore, Nev, and Ariz; RM D. f. (*P. lindleyana*, *P. douglasii* var. *glauca*, *P. taxifolia* var. *g.*)   var. **glauca** (Beissn.) Franco

Tsuga Carr.   Hemlock

Cones borne on previous year's brs mostly in upper part of tree, ♀ single at br tips, greenish to purplish, ± pendent, maturing in 1 year and shed as a unit, scales thin, concave, and rounded; seeds small, with large, broadly oblong wing; cotyledons 3–6; graceful trees gen with slender and drooping leader; needles short, flat, whitish-stomatose on lower or on both surfaces, mostly of unequal length, deciduous in 5–6 yrs (or almost immediately on cut brs). (Japanese name).

1a Needles nearly flat, whitish beneath, but without stomata on upper surface and hence greenish, tending to spread and form flat, spraylike brs; leader always drooping; ♀ cones green to brownish, 1.5–2.5 cm; common for tree, esp where moist, Alas to n Cal, e to se BC, n Ida, and nw Mont, from lowl up to perhaps 5000 ft elev, not subalp or alp; w, Pac, or lowl h.
    1 T. heterophylla (Raf.) Sarg.

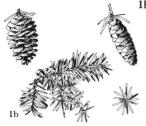

1b Needles ± thickened in middle and ± 4-sided, equally stomatiferous on upper and lower surfaces, tending not to form flat sprays but to surround twig or mostly to turn upward; leader of young trees gen erect; ♀ cones green to deep purple or brownish-purple, slender, 2.5–7 cm; subalp to alp trees, mostly at (3500) 4000–7000 ft elev, often to timberl, where greatly dwarfed; s Alas and BC to n Cal, in both Cas and OM, e to n Ida, w Mont, and ne Ore; black or mt h. *(T. hookeriana)* 2 **T. mertensiana** (Bong.) Carr.

SAURURACEAE   Lizard-tail Family

Fls (ours) ⚥, naked, infl spikelike, surrounded by several large white to reddish invol bracts; stamens 6–8, ± adnate to the pistil; pistil 3–4-carpellary, carpels distinct to connate; ovary superior to inferior, simple or compound with parietal placentation; fr fleshy, follicular or caps; per herbs, ours with long-petioled lvs and semiscapose fl sts.

Anemopsis Hook.   Yerba Mansa

Fls bracteate in fleshy conical spikes; invol bracts 5–8, leathery, corolloid; rhizomatous per of marshy places. (Gr *anemone*, and *opsis*, like, from resemblance of fl).

A. californica Hook. Basal lvs with petioles 5–20 cm, bl oblong to elliptic, 5–20 cm, rounded to cordate at base; sts 1–5 dm, naked except for a sheathing lf ca midlength, often with 1–3 reduced, petiolate lvs in the axil; invol bracts white to red or purplish, 1.5–4 cm; spike 2–4 cm; marshy, alkaline areas; Tex to Cal, intro and said to be well estab at Hermiston, Umatilla Co, Ore.

SALICACEAE   Willow Family

♂, ♀ shrubs or trees; fls in catkins (aments), naked, each subtended by a small, scalelike bract (hereafter referred to as a scale) and with either 1 or 2 enlarged basal glands (in *Salix*), or with an obliquely cup-like disc (in *Populus*); ♂ fls of (1) 2–∝ stamens; ♀ fls of 1 pistil with 2–4 carpels and as many stigmas, stigmas sometimes bifid or cleft, with or without a common style; placentation parietal or (sometimes) basal; fr caps, 2–4-valved; seeds ∞, small, covered with long white hairs and very buoyant; lvs alt, simple, gen stip (but stips often deciduous).

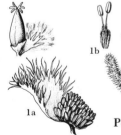

1a Fls each subtended by an obliquely cup-shaped disc, but without obvious glands; lf buds covered by several scales; stamens 6-∝ ; scales laciniate or fimbriate, or seldom *(P. alba)* subentire; trees with pendulous catkins
                                                                                                              Populus
1b Fls lacking a disc but provided with 1 or 2 enlarged basal glands; lf buds covered by 1 scale; stamens gen 2, but sometimes 5–8 or only 1; scales entire or occ somewhat toothed; trees or gen shrubs, with erect or sometimes pendent catkins   **Salix**

Populus L.   Poplar; Cottonwood; Quaking-aspen; Aspen

Catkins drooping, appearing before the lvs; scales laciniate to rarely entire; fls each on a cupulate, oblique disc; stamens 6-∝ ; caps 2–4-valved; stigmas as many as valves of caps, sometimes deeply cleft; trees with ovate to cordate or reniform or occ rather narrow lvs; bud scales several, gen resinous. (Classical L name).

Several commonly cult spp. are often found as escapes in our area, but prob none is truly estab. *P. alba* L., white or silver p., has tomentose buds and lvs conspicuously tomentose beneath; a form of this with erect brs is called var. *pyramidalis* Bunge, or var. *bolleana* Lauche, or *P. bolleana* Lauche. *P. nigra* L., European black p., is represented with us mainly as a ♂ clone with suberect brs and a long, narrow crown; this phase, the Lombardy poplar, has often been called var. *italica* Duroi. The lvs ± resemble those of *P. deltoides,* but are more cuneate and finely toothed, the teeth slightly or not at all incurved, those of *P. deltoides* being prominently toothed, the teeth callous-incurved at the point. The original Carolina poplar was *P. deltoides,* but for many years most of the nursery stock sold under that name has been a hybrid cultigen prob from the cross

P. deltoides × *P. nigra;* the oldest name for such hybrids is prob *P. canadensis* Moench, the later name *P. eugenei* Simon-Louis applying to the particular ♂ clone gen distributed as Carolina p.; it is intermediate in growth form and lvs between *P. deltoides* and the Lombardy p., and it has also been called *P. canadensis* var. *eugenei* (Simon-Louis) Schelle.

1a Smooth-barked tree; buds shiny but not resinous; bracts ± persistent, deeply few-lobed and strongly long-white-hairy; stamens 6–14; caps narrow and small, lanceolate or lance-ovate, 4–6 mm, 2-carpellate; stigmas slenderly lobed; petioles strongly flattened laterally; lvs not distinctly cartilaginous-margined; widespread in mts and mostly E Cas, down to near sea level w Cas, not confined to streambanks and lakeshores; Alas to Lab, s to Cal, n Mex, Tenn, and NJ; asp, q. aspen, trembling a. *(P. vancouveriana, P. t.* var. *v.)* 1 **P. tremuloides** Michx.

1b Rough-barked trees; buds resinous; bracts deciduous at anthesis, laciniate-fringed, otherwise glab or weakly short-hairy; stamens 12–60 or more; caps gen larger and relatively broader, 2–4-carpellate; stigmas broadly dilated, with irreg margins, petioles and lvs various; pls largely confined to streambanks and lakeshores

2a Ovary and young fr gen evidently hairy, seldom glab; lvs very strongly resinous, esp when young, distinctly paler beneath than above; petioles terete or subterete; stamens gen 30–60; carpels 3; lvs not notably cartilaginous-margined; along streams and lakeshores, or in moist regions occ in better-drained soil away from water; Alas to Baja Cal, both sides Cas, to sw Alta, w Mont, Wyo, and Utah; black c. *(P. balsamifera* var. *californica, P. hastata);* closely related to the more e and n *P. balsamifera* L., which differs in its narrower, glab, 2-carpellate ovaries and fewer (ca 20) stamens. Around Flathead Lake, Mont, apparently *P. t.* hybridizes with *P. deltoides* var. *occidentalis;* such hybrids have passed in herbaria, prob erroneously, as *P. besseyana* Dode. 2 **P. trichocarpa** T. & G.

2b Ovary and fr glab; lvs not very strongly resinous, only slightly if at all paler beneath; other characters various but not combined as above

3a Lf bls relatively broad, gen 0.8–1.3 times as long as wide, with truncate or subtruncate to reniform-cordate base, evidently cartilaginous-margined; petioles laterally flattened, elongate, gen > 3/5 as long as bl; stamens gen 30–60; carpels 3–4; streambanks at low elev; mostly e RM, sw Alta to Colo and Tex, e to Que and Fla, w in Mont to Flathead and Sanders cos; ours is var. *occidentalis* Rydb.; necklace p., Gr Pl c. *(P. o., P. monilifera* var. *o., P. sargentii)* 3 **P. deltoides** Marsh.

3b Lf bls relatively narrow, 1.4–5 times as long as wide, with narrowly to fairly broadly cuneate or rounded-ovate base, obscurely if at all cartilaginous-margined; petioles shorter, seldom > 3/5 as long as bl, dorsoventrally compressed to subterete; stamens 12–30; carpels 2–3

4a Lf bls mostly 1.4–2 times as long as wide, broadly cuneate to rounded at base, with petioles 1/3–3/5 as long; carpels 2–3; frs gen developing normally, but ♂ fls apparently unknown; perhaps a hybrid between *P. angustifolia* and *P. deltoides* var. *occidentalis* or (farther s) *P. fremontii;* stream banks, Alta and Mont (Flathead Lake) to e Wyo, sw NM, Utah, and se Ariz, e to SD; Rydberg's c. 4 **P. acuminata** Rydb.

4b Lf bls mostly 2–5 times as long as wide, cuneate (rounded on large lvs of vigorous young shoots) at base, with petioles up to 1/3 as long; carpels 2; streambanks, high plains to low mts; sw Alta, w Mont, c and s Ida, se Ore, e Cal, and Ariz, e to SD, w Neb, NM, and n Mex; black c., mt c., narrow-lvd c. *(P. fortissima, P. balsamifera* var. *a., P. canadensis* var. *a.)* 5 **P. angustifolia** James

Salix L. Willow

Aments gen erect or spreading, developing before the lvs (precocious), with them (coetaneous), or after them (serotinous); scales entire to erose or slightly toothed; fls with 1 or 2 short and broad to slender and elongate protuberances (glands); disc none; stamens (1) 2 (–8); caps 2-valved, stigmas 2, entire to gen bifid; shrubs or occ trees, sometimes prostrate, gen in wet or moist places, esp along streams and around bodies of water; winter buds with a single nonresinous scale. (The classical L name).

Two Old World spp. are often cult and occ apparently estab; both are trees with long, narrow, serrulate lvs, coetaneous aments on lfy-bracteate peduncles, deciduous yellow scales, and 2 (3 or 4) stamens with filaments hairy toward the base. *S. babylonica* L., weeping w., not only has a weeping habit, but also unusually short caps (1–2.5 mm) with minute, subsessile stigmas. *S. alba* L., white w., has more-spreading, fragile twigs, caps 3–5 mm, and short styles 0.2–0.4 mm; our pls mostly have yellow twigs and are called var. *vitellina* (L.) Stokes; golden w.

1a

1a Stamens 3–8; arborescent shrubs or small trees 2–12 m; lvs narrow, > 3 times as long as wide, finely serrulate, gradually tapered to the long narrow tip; filaments strongly hairy toward base; scales yellowish, deciduous; aments coetaneous on lfy-bracteate peduncles   **Group I, lead 5a**

1b Stamens 2, or sometimes only 1; other characters various, but not combined as in Group I

2a

  2a Pls depressed, seldom > 1 (2) dm, forming dense, lawnlike or cushionlike mats high in the mts, gen near or above timberl   **Group II, lead 6a**

  2b Pls ascending to erect, rarely < 2 dm, not mat-forming, occurring at various elevs

    3a Scales pale, gen yellowish, soon deciduous; filaments strongly hairy toward base; lvs relatively narrow, (2.5) 3–20 times as long as wide, tapered (rounded) to a subsessile or shortly subpetiolar base, the petiole proper up to ca 5 mm; shrubs or small trees (1) 2–8 m, often colonial by running roots   **Group III, lead 9a**

3a

    3b Scales brown to blackish in most spp., pale and yellowish or anthocyanic in a few, persistent; filaments glab or sometimes weakly hairy toward base; lvs various, often broader or with longer petioles, or both; pls gen (always?) without running roots

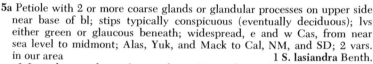

4b

      4a Ovaries and caps glab, or in a few spp. sometimes sparsely villous, esp toward the tip   **Group IV, lead 11a**

      4b Ovaries and caps ± densely short-hairy, or becoming rather sparsely so in age   **Group V, lead 23a**

4a

4b

Group I

5a Petiole with 2 or more coarse glands or glandular processes on upper side near base of bl; stips typically conspicuous (eventually deciduous); lvs either green or glaucous beneath; widespread, e and w Cas, from near sea level to midmont; Alas, Yuk, and Mack to Cal, NM, and SD; 2 vars. in our area   **1 S. lasiandra** Benth.

  a1 Lvs glaucous beneath; mainly w Cas, and in n Ida and nw Mont in our area; Pac w., red w. (*S. speciosa*, *S. lyallii*)   var. **lasiandra**

  a2 Lvs green on both sides; widespread E Cas; whiplash w., caudate w. (*S. c.*, *S. pentandra* var. *c.*, *S. c.* var. *parvifolia*)

                                       var. **caudata** (Nutt.) Sudw.

5a

5b Petiole gen without obvious glands; stips seldom well developed, gen minute and quickly deciduous, or obsolete; lvs pale and glaucous beneath; streambanks in plains and foothills, widespread E Cas, e to Atl; peach-lf w.   **2 S. amygdaloides** Anderss.

Group II

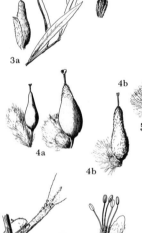

5b

6a Aments serotinous, terminating the short lfy shoots of the season; scales pale, inconspicuously short-hairy within, otherwise glab; filaments weakly short-hairy toward base; style very short, it and the stigmas together gen < 1 mm; lf bls up to 3 (3.5) × 2 cm; s BC and Alta to Cal, Utah, and NM; 2 doubtful vars.   **3 S. nivalis** Hook.

  a1 Lvs mostly 7–15 mm; aments relatively small, ♀ ones up to 1 cm at maturity, with up to 12 caps; range of sp., but in gen more n; snow w. (*S. solheimii*)   var. **nivalis**

  a2 Lvs larger, the larger ones mostly 15–30 (35) mm; aments as in var. *nivalis*, or sometimes larger, the ♀ ones up to 2 cm at maturity, with up to 25 or more caps; range of sp., but in gen more s; RM w. (*S. s.*)   var. **saximontana** (Rydb.) Schneid.

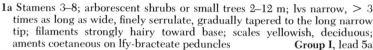

6a

6b Aments coetaneous, terminating short, lfy, lateral brs, but not the principal vegetative shoots of the season; scales dark, gen conspicuously longhairy (at least on margins), hairs much > body of scale; filaments glab; style often longer, it and the stigmas together often well > 1 mm

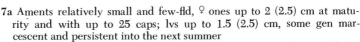

7a Aments relatively small and few-fld, ♀ ones up to 2 (2.5) cm at maturity and with up to 25 caps; lvs up to 1.5 (2.5) cm, some gen marcescent and persistent into the next summer

8a Caps and gen also the ovaries glab; aments very small, 1–7-fld, the scales merely fringed-ciliate; style and stigmas together < 1 mm; lf bls 3–7 mm; sw Mont and w Wyo; rarely collected; Dodge w.

4 **S. dodgeana** Rydb.

8a

8b Caps and ovaries villous-tomentulose, rarely glab; aments larger, the ♀ mostly 12–25-fld, the scales gen hairy on the body as well as on the margins; style and stigmas together 1–1.5 (2) mm; larger lf bls gen 10–15 (25) mm; sw BC s, in Cas, to Mt Rainier, Wn, occ e to Mont and Wyo and s to Colo and Utah, reported for ne Ore; Cas w. *(S. tenera, S. brownii var. t.)* 5 **S. cascadensis** Cockerell

7b Aments larger and with more fls, ♀ ones mostly (1) 2–4 (6) cm at maturity and with 25–50 (75) caps; larger lvs mostly (1.2) 1.5–5 cm, few if any marcescent and persistent into the next summer; circumboreal, s in w cordillera to Cal and NM, but only in n Cas (Okanogan Co) in Wn, and in Ore only in the Wallowa Mts; arctic w. *(S. brownii var. p., S. petrophila);* our pls are var. *petraea* Anderss., except the robust forms of nc and nw Wn, these not yet clearly placed as to the var.

6 **S. arctica** Pall.

8b

Group III

9a Stigma lobes relatively long and slender, gen borne on a definite common style (seldom 2 styles), the whole 0.8–1.5 mm; W Cas and e through CRG to Deschutes R

10a Lvs copiously and persistently villous or villous-puberulent with loose hairs; lf bls (2.5) 3–7 (10) times as long as wide, larger ones mostly 3–10 × 1–3.5 cm; streambanks w Cas, and inl in CRG to e end; soft-lvd w. *(S. macrostachya, S. longifolia var. s., S. s. var. villosa)*

7 **S. sessilifolia** Nutt.

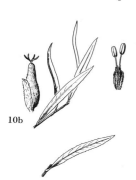

7b

10b Lvs silvery-strigose or subsericeous with appressed hairs when young, soon ± glabrate and greener; lf bls 3–15 (20) times as long as wide, mostly 5–15 cm × 4–15 mm; banks of CR, from mouth of Deschutes R to mouth and lower few miles of the Willamette R; CR w.

8 **S. fluviatilis** Nutt.

9b Stigma lobes relatively short and stout, each of the 2 stigmas lobed nearly to the base and essentially sessile on the ovary (seldom with a very short common style), the whole 0.2–0.5 (0.6) mm; lf bls mostly 5–15 cm × 4–20 mm, (4) 5–15 (20) times as long as wide; plains and foothills to midmont; transcontinental, Alas to Cal and e to Atl, in our area wholly E Cas; several sspp. and vars. in our area 9 **S. exigua** Nutt.

10a

a1 E ssp.; caps 5–8 mm, glab, distinctly ped, mature ♀ aments loose and up to 8 cm; lvs gen more obviously veiny than in other sspp., mostly toothed as in ssp. *melanopsis*, but often more persistently hairy; scales variable, tending to be pointed; Atl coast to e base RM, encroaching into our area in Mont; sandbar w., river-bank w. *(S. i., S. longifolia* Muhl., *S. l. var. i.)* ssp. **interior** (Rowlee) Cronq.

a2 Cordilleran sspp.; caps 3–5 (6) mm, typically sessile or subsessile, ♀ aments mostly dense and short, typically 3–5 (10) cm

10b

b1 Lvs ± evidently serrulate with callous or glandular teeth, often relatively wider than in ssp. *exigua*, though sometimes very narrow; pls less hairy, ovary and caps glab, lvs early glabrate, gen essentially smooth when expanded to full size; scales mostly broad and blunt; E Cas, from Banff, Alta, to Wyo, n Colo, n Utah, n Nev, and Cal, gen at higher elev than ssp. *exigua;* dusky w. *(S. tenerrima, S. longifolia var. t., S. fluviatilis var. t., S. melanopsis var. t.);* ours is var. *m.*

ssp. **melanopsis** (Nutt.) Cronq.

b2 Lvs entire or merely with a few scattered, inconspicuous, small teeth; pls more hairy, lvs more persistently hairy (often eventually glabrate, esp above), ovaries and caps often hairy; scales mostly narrow and pointed; s Alta and BC (e Cas), s through RM and GB to n Mex; in our area mostly in foothills and plains; 2 vars. in our area

ssp. **exigua**

9b

c1 Ovaries and caps ± hairy, the main phase in our area; coyote w., slender w. *(S. argophylla, S. longifolia var. a., S. l. var. e., S. flu-*

viatilis var. *e.*)   var. **exigua**

c2 Ovaries and caps gen glab or subglab from the 1st; chiefly s of our range, but n occ to lower part of Clearwater R, Ida; narrow-lf w. (*S. s.*)   var. **stenophylla** (Rydb.) Schneid.

Group IV

11a Lvs ± evidently hairy at least on 1 side when first fully expanded, but sometimes eventually glabrate

  12a Style elongate, gen 1.5–3 mm; aments almost or quite sessile on twigs of previous year, some apparently terminal; lvs more densely and persistently hairy above than beneath, but not glaucous; mont shrub up to 3 m, with precocious to coetaneous aments; moist to boggy areas, moderate to fairly high elev, prob not above timberl; s BC and Okanogan Co, Wn, to w Mont, Ida, and n Wyo, rare except in Big Horn Mts, Wyo; Tweedy's w. (*S. rotundifolia* Nutt., *S. barrattiana* var. *t.*)   10 **S. tweedyi** (Bebb) Ball

  12b Style shorter, up to ca 1.8 mm; other characters various but not combined as in no 10

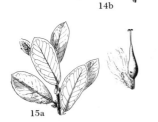

    13a Aments precocious; coarse shrubs or small trees (1) 2–6 (12) m; foothills and lowl

      14a Lvs 1.4–3 times as long as wide, villous-tomentose with long hairs beneath; caps glab to villous, 6–8 mm, ped < 1 mm, inconspicuous; style 0.8–1.6 mm; maritime, along seacoast and around PS, from s BC to n Cal, often on stabilized dunes; Hooker w. (*S. h.* var. *tomentosa*)   11 **S. hookeriana** Barratt

      14b Lvs (2) 3–7 times as long as wide, strigose-puberulent with short hairs beneath, or glabrate; caps glab, 3–5 mm, ped (1) 2 mm; style 0.3–0.8 mm; E Cas, s BC to Baja Cal, e to Ida, Utah, Tex, and n Mex; arroyo w. (*S. boiseana, S. sandbergii, S. suksdorfii*)   12 **S. lasiolepis** Benth.

    13b Aments coetaneous or serotinous; mont spp. of various sizes

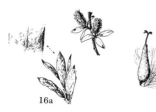

      15a Lvs more densely and persistently hairy above than beneath, the lower side ± glaucous and soon glab, upper side green but persistently hairy or only tardily glabrate; style (0.5) 0.8–1.8 mm; midmont to subalp; Alas and Yuk s to OM and Cas as far as Mt Adams, Wn, n Ida, and Alta; Barclay's w. (*S. conjuncta*)   13 **S. barclayi** Anderrs.

      15b Lvs more densely and persistently hairy beneath than above, or ca =hairy on both sides, not at all glaucous, but sometimes pale beneath because of the pubescence; style 0.5–1.0 (1.2) mm

        16a Lvs closely and permanently hairy, entire, small and relatively narrow, up to 4.5 × 1.5 cm, mostly 3–4.5 times as long as wide; ♀ aments 1.5–3 (4) cm at maturity; pl 0.6–2 m; midmont to subalp; ne Ore and n Nev to sw Mont and Colo, not common; Wolf's w.; 2 vars.   14 **S. wolfii** Bebb

          a1 Caps glab or occ subglab; chiefly s, but n to extreme sw Mont   var. **wolfii**

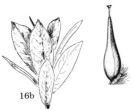

          a2 Caps more hairy; pl in gen larger in most respects; more n and w (*S. i.*)   var. **idahoensis** Ball

        16b Lvs loosely long-woolly-villous when young, less so and sometimes even glab in age, entire or ± toothed, larger and relatively wider, larger ones (4) 5–8 × (1.3) 1.7–3.5 cm, 1.5–3 × as long as wide; ♀ aments mostly 3–6 cm; pl (0.5) 1–3 m; midmont to barely above timberl; s Alas and Yuk s, in both Cas and OM, to n Cal, e occ to w Mont, Ida, w Wyo, and n Utah; undergreen w., variable w. (*S. c.* vars. *denudata, puberula,* and *sericea*)   15 **S. commutata** Bebb

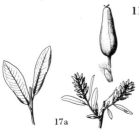

11b Lvs glab by the time they are fully expanded

  17a Scales pale or partly anthocyanic; style 0.1–0.3 mm, or obsolete; bog shrub 4–12 dm, sparingly br; lvs very pale and glaucous beneath, entire, gen (2) 3.5–6 (8) × 0.5–2 (3) cm; Yuk and BC to Que and NJ, s to Wn and sw Ore, in our range chiefly in and w Cas; bog w. (*S. myrtilloides* var. *p.*, *S. dieckiana, S. p.* var. *hypoglauca*)   16 **S. pedicellaris** Pursh

  17b Scales brown to blackish; style (except sometimes in no. 18) > 0.3 mm; pls either taller, or more brd, or both; lvs various

    18a Lvs glaucous beneath; style often > 0.7 mm

      19a Pl low, mont, < 1 m; lvs entire or subentire, up to 5(7) × 2(3) cm; aments lfy-pedunculate; ♀ aments 1.5–3 cm at maturity; style

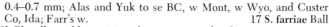

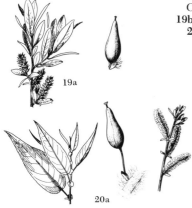

19a

20a

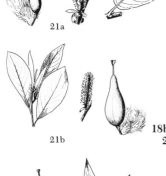

21a

21b

22a

23a

25a

0.4–0.7 mm; Alas and Yuk to se BC, w Mont, w Wyo, and Custer Co, Ida; Farr's w.   17 S. farriae Ball
19b Pls taller, seldom < 1 m; lvs, aments, and styles various
    20a Style 0.2–0.7 mm; lvs (2) 3–5 times as long as wide, larger ones 5–10 (15) × 1–3 (6) cm; ♀ aments 3–6 (9) cm at maturity; foothills and lowl into lower intermont valleys; widespread, transcontinental sp.; 3 vars. in our area   18 S. rigida Muhl.
        a1 Young twigs conspicuously spreading-hairy, pubescence often extending to petioles and along midrib on upper surface; otherwise much like var. mackenzieana; w Cas, PS to WV, (S. mackenzieana var. m.)   var. macrogemma (Ball) Cronq.
        a2 Young twigs glab or inconspicuously hairy and soon glabrate; widespread except within range of var. macrogemma
            b1 Twigs mostly reddish-brown; lvs rarely < 3 times as long as wide and rarely entire; caps peds mostly 2–4 mm; scales gen glab or subglab; nw, from Yuk and Mack to s BC, Cas of Wn and Ore, e to w Mont and Wyo; Mackenzie w. (S. monochroma, S. mackenzieana, S. cordata var. m.)
                var. mackenzieana (Hook). Cronq.
            b2 Twigs ± distinctly yellowish; lvs often < 3 × as long as wide, often ± entire; peds 0.5–2 (3) mm; scales glab or more often hairy like axis of the ament; more s and e, from ne Ore and SRP, Ida, to Cal, Ariz, and NM, and to Gr Pl, and in forms transitional to var. mackenzieana to Ida, Ore, and nc Wn; Watson w. (S. lutea, S. cordata var. l., S. flava, S. watsonii, S. cordata var. w.)   var. watsonii (Bebb) Cronq.
    20b Style 0.7–1.8 mm; lvs often < 3 times as long as wide
        21a Lvs very shiny above, rather coarsely and remotely toothed to entire; ♂ catkins stout, ca 1.5 cm thick or more; W Cas, s BC to Cal, up CRG to Bingen, Wn, lowl to midmont; Piper's w. (S. hookeriana var. laurifolia)   19 S. piperi Bebb
        21b Lvs not very shiny above, closely and finely toothed to subentire; ♂ aments more slender, much < 1.5 cm thick; E Cas, midmont to alp; RM, Alas to c Ida, NM, and e Utah, e to Lab and Que; mt w. (S. padifolia, S. padophylla, S. pseudomonticola var. p., S. barclayi var. p., S. cordata var. m.)
                20 S. monticola Bebb
18b Lvs green beneath, not glaucous; style short, 0.2–0.7 mm
    22a Pls midmont to subalp; ped of caps short, up to ca 1 mm; Alas and BC to Newf and Que, s irreg to Cal, Wyo, s Utah, and possibly Colo; blueberry w., bilberry w. (S. novae-angliae var. m.)
                21 S. myrtillifolia Anderss.
    22b Pls of foothills and lowl to midmont; ped elongate, 2–4 mm; var. mackenzieana (see lead 20a)   18 S. rigida Muhl.

Group V

23a Aments serotinous, evidently naked-pedunculate, terminating some of the main shoots of the season; high-mont, near or above timberl, seldom > 1 m; lvs thick and firm, 2–6 × 1–4 cm; RM region, s Alta and BC to n Cas, Wn, and ne Ore, and c Mont; rock w.; ours is var. erecta Anderss. (S. fernaldii)   22 S. vestita Pursh
23b Aments precocious to coetaneous, sessile or ± lfy-bracteate-pedunculate, not terminating principal lfy shoots of the season; habit and habitat various
    24a Stigmas elongate, gen 0.5–1.0 mm; twigs not glaucous
        25a Style gen 0.7–1.7 mm; lvs gen glab when fully expanded; pls 0.2–4 m; moist or wet places to open woods; circumboreal s to Cal, NM, and N Eng; tea-lvd w.; our 3 vars. are all referrable to ssp. planifolia (Pursh) Hiitonen   23 S. phylicifolia L.
            a1 Tall shrubs, mostly (1) 2–4 m; larger lvs gen 4.5–8 (13) × 1–3 (5) cm; filaments glab; high mont, often near timberl; wc Mont and c Ida to Nev and Sierran Cal, e to Utah and NM (S. nelsoni, S. chlorophylla var. n., S. planifolia var. n., S. p., S. phylicifolia ssp. p.)
                var. planifolia
        a2 Smaller shrubs, up to 2 m
            b1 Larger lvs gen 2.5–3.5 (5) × 0.8–1.5 (2.3) cm; low shrubs 0.2–1 (2) m; filaments glab; high mont, often near timberl; wc Mont and

25b

26a

27a

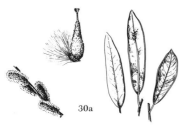

30a

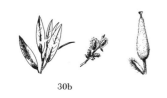

30b

29b

32a

c Ida to Sierran Cal, Utah, and NM (*S. m.*, *S. phylicifolia* var. *m.*)
var. **monica** (Bebb) Jeps.

b2 Larger lvs mostly 4.5–6.5 × 2–3 cm; shrub gen 1–2 m; filaments glab or occ inconspicuously hairy toward base; catkins elongate, the ♀ ones up to 8 cm; Cas, n Wn to Mt Hood, Ore (*S. p.*)
var. **pennata** (Ball) Cronq.

25b Style mostly < 0.6 (0.8) mm; lvs typically reddish-strigillose beneath, varying to more conspicuously hairy or glab; pls robust, (1) 2–12 m; lowl to midmont; Alas and Yuk to Cal, Ariz, and NM, throughout our range and both e and w Cas; Scouler w. (*S. flavescens*, *S. s.* var. *f.*, *S. nuttallii*, *S. stagnalis*, *S. s.* vars. *coetana* and *thompsoni*)
24 **S. scouleriana** Barratt

24b Stigmas shorter, 0.2–0.5 mm (up to 0.6 mm in no 26, which has glaucous twigs)

26a ♂ catkins very stout, (1.5) 2–2.5 cm thick; maritime or submaritime, rarely > 5 mi from salt water (see lead 14a) 11 **S. hookeriana** Barratt

26b ♂ catkins more slender, rarely > ca 1.5 cm thick; sometimes near the ocean, but no sp. exclusively maritime

27a Twigs and lower surface of lvs tomentose with long, slender, tangled, appressed hairs; lvs slender, 3.5–10 times as long as wide, larger ones 4.5–8 (15) × 0.7–1.5 (2.3) cm; pl (0.2) 0.5–1.2 (1.5) m; bogs and swamps, Alas to Lab, s to s BC, SC, Ia, and NJ, and in the mts to Ida and Colo; seldom collected in our area; hoary w.
25 **S. candida** Fluegge

27b Twigs and lvs glab or variously hairy, lvs often ± sericeous or spreading-hairy beneath, but not tomentose; lvs and habit various

28a Lvs conspicuously discolored with pubescence, densely and persistently gray- or white-hairy beneath, green and only slightly short-hairy or glab above; ped of caps from nearly obsolete up to ca 1.5 mm

29a Twigs soon strongly glaucous, otherwise essentially glab except when very young; stamens 2; lvs gen soon glab above; mainly e Cas

30a ♂ catkins 2–3 cm; ♀ catkins 2–6 cm at maturity; style (0.4) 0.6–1.3 mm; shrub (1) 2–3 (4) m; moist places to open slopes, foothills to midmont or subalp, onto the Palouse prairie in e Wn; BC and Alta, s to Sierran Cal, Nev, Utah, and NM, only in extreme e Wn, e across s Can and n US; Drummond w. (*S. bella*, *S. d.* var. *b.*, *S. subcoerulea*, *S. d.* var. *b.*)
26 **S. drummondiana** Barratt

30b ♂ catkins 0.7–1.5 cm; ♀ catkins 1–2.5 cm at maturity; style up to ca 0.4 mm (see lead 32a); forms of
28 **S. geyeriana** Anderss.

29b Twigs not glaucous, but ± densely and persistently velvety; stamen 1 (unique among our spp. in this regard); lvs only tardily if ever glabrate above; lowl to midmont; s Alas to Cal, mainly in and w Cas, but also in ne Ore, e Wn, n Ida, and se BC; Sitka w. (*S. s.* vars. *congesta* and *denudata*)
27 **S. sitchensis** Sanson

28b Lvs not conspicuously discolored with pubescence, or if somewhat so then caps with ped at least 2 mm

31a Style very short, up to ca 0.4 mm; scales often yellowish or light brown, but sometimes dark brown or blackish

32a Lvs relatively small and narrow, up to 1.5 cm wide, gen (3) 4–6 (7) times as long as wide; twigs often strongly glaucous; fr peds 1–2.5 mm; ♂ catkins 7–12 (15) mm; ♀ catkins 1–2 (2.5) cm; scales yellowish to brown or blackish; foothills to high mont, but down to near sea level w Cas; s BC to Cal, e to n Mont, Utah, and Colo; Geyer w.; 2 vars.
28 **S. geyeriana** Anderss.

a1 Twigs strongly glaucous; pubescence not rusty in color; mainly e Cas (*S. macrocarpa* var. *argentea*, S. g. var. *a.*)
var. **geyeriana**

a2 Twigs slightly or not at all glaucous; pubescence somewhat rusty in color; w Cas (*S. macrocarpa*, *S. meleiana*)
var. **meleiana** Henry

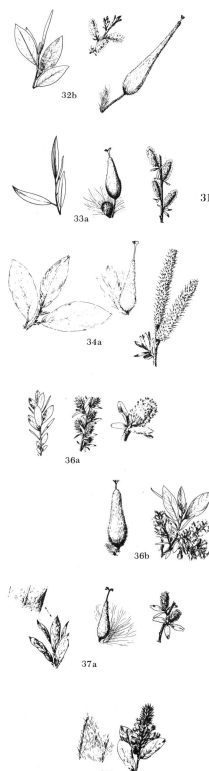

32b

33a

34a

36a

36b

37a

37b

32b Lvs mostly larger and relatively wider, larger ones seldom < 1.5 cm wide, mostly 2–3 (3.5) times as long as wide; twigs not at all glaucous; peds 2–4 mm; ♂ catkins (10) 12–20 (25) mm; ♀ catkins (2.5) 3–6 cm; scales yellowish, becoming light brown; lowl to midmont; transcontinental in Can and n US, in most of our range, but not W Cas; Bebb w.; 2 vars.   **29 S. bebbiana** Sarg.

a1 Lower surface of lvs thinly appressed-hairy, soon glabrate, seldom very strongly raised-reticulate-veiny; common in most of the cordilleran region of US *(S. p., S. rostrata* var. *p.)*   var. **perrostrata** (Rydb.) Schneid.

a2 Lower surface of lvs ± densely and loosely villous or villous-puberulent and evidently raised-reticulate-veiny; e and n in range, extending w occ to Wyo, Mont, n Ida, and adj Wn *(S. rostrata, S. cinerascens* var. *occidentalis)*

var. **bebbiana**

31b Style longer, (0.4) 0.5–2.0 mm; scales light to dark brown or blackish

33a Lvs quickly becoming glab, even before they are fully expanded; twigs glaucous; shrubs 1.5–4.5 m; foothills to midmont; Hood R Co, Ore, along e side Cas to Sierran Cal e to Owyhee Co, Ida, and Nev; Lemmon's w.

30 S. lemmonii Bebb

33b Lvs ± persistently hairy; twigs not glaucous; shrubs seldom > 2 m

34a Catkins essentially sessile, not lfy-bracteate, some of them terminating twigs of the previous year; ♀ catkins (3) 4–9 cm; style (0.7) 1.0–2.0 mm; mont, up to timberl; Alas and Yuk s to sw Alta, se BC and GNP, Mont; Barratt w. *(S. albertana)*   **31 S. barrattiana** Hook.

34b Catkins borne on ± lfy-bracteate peduncles, none of them terminating twigs of the previous year; more widespread spp.

35a Lower side of lvs distinctly glaucous as well as gen hairy

36a Petioles very short, gen 1–3 (4) mm, often not exceeding the bud; catkins very short and compact, the ♀ ones mostly 1.5–2 cm at maturity and with the caps sessile or on very short peds up to 0.5 mm; anthers < 0.5 mm; in widely variant habitats, foothills to high mont, often where salty or alkaline; Alas to Que, s through w Can and irreg to Kittitas Co, Wn, ne Ore, Ida, Mont, and Colo; short-fruited w. *(S. stricta);* ours the var. *b.*

32 S. brachycarpa Nutt.

36b Petioles longer, gen 3–10 mm; catkins somewhat longer and looser, ♀ ones mostly 2–5 cm at maturity and the caps borne on peds 0.5–2 mm; anthers > 0.5 mm; moist places to open slopes, midmont to above timberl; circumboreal, s to s BC, Alta, Man, and Que, and in the RM region to NM, in Ida only from near Henry's Lake; not known from Wn or Ore; glaucous w. *(S. desertorum* var. *elata, S. villosa,* S.g. var. *v.)*

33 S. glauca L.

35b Lower side of lvs merely hairy, not glaucous

37a Lvs closely and permanently hairy, entire, relatively narrow, gen 3–4.5 times as long as wide, seldom as much as 2 cm wide; ♀ aments 1.5–3 (4) cm at maturity; var. *idahoensis* (see lead 16a)

14 S. wolfii Bebb

37b Lvs loosely long-woolly-villous when young, less so in age, or eventually glab, entire or somewhat toothed, wider, gen 1.5–3 times as long as wide, the larger ones (1.3) 1.7–3.5 cm wide; ♀ aments gen 3–6 cm at maturity; (see lead 16b) forms of

15 S. commutata Bebb

MYRICACEAE   Sweet Gale Family

Fls hypog, naked, ♀, ♂ or ♀♂, the ♀ and ♂ separate (rarely mixed) in short, crowded spikes; ♂ fls with 1 bract, stamens 3–12, either with filaments free and without inner bracteoles or ± connate and gen with 1–3 2ndary bracteoles, anthers 2-celled; ♀ fls 1-bracteate and 2ndarily bracteolate; pistil 1, 2-carpellary; styles 2; ovary 1-celled, 1-seeded; fr drupaceous; deciduous to evergreen shrubs or trees.

Myrica L.   Sweet Gale; Wax-myrtle

Mostly aromatic shrubs or small trees, characteristically of bogs or swamps, often glandular or waxy, esp on frs; lvs alt, exstip (ours), simple, entire to toothed. (Old Gr name, *myrike*, possibly applied first to another pl). Rather nice orn, esp *M. californica*.

1a Pl shrubs, deciduous, ♀, ♂, gen < 2 m; lvs oblanceolate to oblong-obovate, mostly obtuse, 3–6 cm, entire to few-toothed on the upper half, brightly yellow-dotted, pubescent on both surfaces, or glabrate above; lowl to mont swamps and bogs, Alas to Ore, in our area w Cas and near coast, e to Newf, Wisc, and Va, s to NC; Eurasia; sweet gale   1 **M. gale** L.

1b Pl evergreen shrubs or small trees, ♀♂, gen > 2 m; lvs elliptic to elliptic-oblanceolate, ± acute, 5–8 (10) cm, entire to remotely serrate full length, blackish (black and pale yellow) -dotted; near coast, Grays Harbor Co, Wn, to Cal; Pac w., Cal w.   2 **M. californica** Cham.

BETULACEAE   Birch Family

Fls ♀♂, the ♂ always in pendent catkins, the main bracts subtending 2 smaller bracts and 3–6 fls of 1–4 (8) stamens each; ♀ fls subcapitate or in catkins, 2–3 per bract, naked or with 3–4-merous perianth; ovary inferior (or naked), 2-carpellary, 1–2-celled, 2–4-ovuled; styles 2; fr a nut or a winged or wingless nutlet; deciduous shrubs or trees with alt, simple, petiolate, serrate lvs.

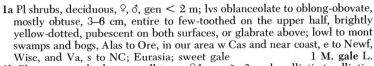

1aFr a rounded nut in a juglike invol; ♀ fls few, clustered, not in catkins; lf bls cordate   **Corylus**

1bFr a flattened, ± winged nutlet, not invol; ♀ fls ∝ in elongate or conelike catkins; lf bls seldom cordate

2a ♀ catkins hardened and conelike, scales persistent after shed of nutlets **Alnus**

2b♀ catkins not hardened or conelike, scales deciduous with nutlets   **Betula**

Alnus Hill   Alder

♂ catkins pendent, main bracts ± peltate, subtending 2 (1–4) bracteoles and 3–6 fls, each fl with (3) 4-parted perianth and (1) 3–4 stamens; ♀ catkins conelike, scales hardened and persistent, each with (1) 2 naked fls and 2–3 bracteoles; trees or shrubs with simple, dentate to lobed lvs. (Old L name for the pl).

1a Catkins developing and fl with the lvs, on twigs of current year; frs with thin, membranous wings at least half as wide as nutlet; lf bls ovate, 3–10 cm, not revolute, only slightly paler beneath than above, finely 1–2-serrate and ± sinuate; winter buds sharply acute; shrubs, 2–4 (8) m; moist places, lowl to subalp; Alas s, in Cas and OM, to n Cal, e through Wn and Ore to much of Ida and Mont; Sitka a., wavy-lvd a.   1 **A. sinuata** (Regel) Rydb.

1b Catkins developing and fl before the lvs, on twigs of previous year; either frs wingless or lf bls slightly revolute and gen much paler beneath than above, margins various; winter buds blunt to acute

2a Fr with wing 1/5–1/2 as wide as nutlet; lf bls broadly elliptic to ovate-elliptic, 5–15 cm, sinuate, margins slightly revolute, much paler (rusty-gray) beneath than above; new twigs not puberulent, gen glab; trees up to 25 m; moist lowl; Alas s, W Cas and on OP, to Cal, also in Ida; red a., Ore a. (*A. oregana*)   2 **A. rubra** Bong.

2b Fr wingless; lf bls sinuate or not, neither revolute nor rusty-gray beneath; new twigs and petioles strongly puberulent; shrubs to small trees

3a Lf bls sinuate or lobed, serrate-denticulate, elliptic or ovate-oblong, 3–7 (11) cm; stamens mostly 4, filaments scarcely 1/2 as long as the

anthers; shrubs, 2–5 (12) m; moist to wet places, low to high mont; Alas s, E Cas, to Cal, e to NS, s in RM to NM; mt a. *(A. occidentalis, A. rugosa, A. tenuifolia)*; ours the var. *occidentalis* (Dippel) Hitchc.

3 **A. incana** (L.) Moench

3b Lf bls 1–2-serrate, elliptic or oblong-rhombic, 4–8 cm; stamens 1–3, filaments mostly subequal to anthers; trees, 5–20 m; BC s, E Cas, to Baja Cal, e to w Ida; white a.   4 **A. rhombifolia** Nutt.

Betula L.   Birch

♂ catkins pendent, fls 3 per cluster, each with 2–3 perianth segms, mostly 2 stamens, and 1 primary and 2 2ndary subtending bracts; filaments short, bifid, each br with a single anther sac; ♀ catkins mostly erect, the bracts 3-lobed, subtending (1) 2–3 naked fls; styles spreading, stigmatose, fr a winged nutlet, deciduous with bracts of catkin; trees or shrubs with gen serrate lvs; bark gen ± separable into thin layers strongly marked by elongate lenticels. (Old L name for the genus).

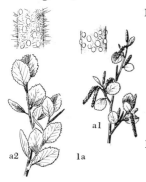

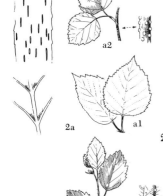

1a Shrubs up to 3 (6) m; wing of fr not so broad as nutlet; young twigs densely puberulent and warty-glandular; lf bls 1–2 (4) cm, crenate-serrate or serrate, rather leathery; wet places, often in swamps; Alas s, in Cas and OM, to Cal, e to Newf and ne US, and occ in Ida and Mont and in RM to Colo; bog b., scrub b.; 2 vars.   1 **B. glandulosa** Michx.

    a1 Young twigs finely puberulent; lf bls mostly oval, suborbicular. or obovate; pl 1–2 m; range of sp. except not w of Cas in Wn, and only occ so in n Ore *(B. crenata)*   var. **glandulosa**

    a2 Young twigs ± short-pubescent as well as puberulent; lf bls ± elliptic; pl (1) 2–5 (6) m; bogs in lowl of PT, Wn, to nw Ore, CRG, and s in Cas to Klamath Lake, Ore; Hall's b. *(B. h.)*   var. **hallii** (Howell) Hitchc.

1b Shrubs or trees 4–20 m; wing of fr gen as broad as nutlet; young twigs glab to crisp-puberulent, often not warty; lf bls often > 4 cm, slightly lobed to biserrate, scarcely leathery; mostly not in swamps

    2a Lf bls ovate, (2) 3.5–7 (11) cm, ± acuminate, often slightly lobed, gen biserrate and with tufts of hairs in vein axils beneath; young twigs gen finely puberulent or thickly crisp-pubescent (or both), mostly with few (if any) crystalline-warty glands; bark often gray to pale coppery, ± readily peeling; moist, open to dense woods of lowl or lower mts; Alas s, mostly E Cas, except in nw Wn, to ne Ore, e to Atl coast and s to n Ida and to Mont, Colo, Ia, Ill, and NC; paper b.; 2 vars.

2 **B. papyrifera** Marsh.

        a1 Lf bls 4–8 (11) cm, acuminate; twigs not crystalline- glandular; range of sp. in w US, except for that of var. *subcordata*; w p. b. *(B. andrewsii, B. montanensis, B. occidentalis)*   var. **commutata** (Regel) Fern.

        a2 Lf bls 3.5–5 cm, acute to slightly acuminate; twigs crystalline-glandular; se Wn, ne Ore, and w Ida; nw p. b. *(B. s.)*

var. **subcordata** (Rydb.) Sarg.

    2b Lf bls often elliptic, suborbicular, or obovate, (1) 2–4.5 (7) cm, rounded to acute, not lobed, mostly once-serrate, rarely hairy beneath in vein-axils; young twigs crisp-puberulent to glab, mostly strongly crystalline-warty; bark coppery-red to purplish-brown, not readily peeling

        3a Lvs gen elliptic, acute at base, acute or acuminate at tip, (2) 4–6 cm, with or without tufted hairs in vein axils beneath; st ± crystalline-glandular, gen puberulent or pubescent (or both) on young twigs; occ hybrid of *B. papyrifera × B. occidentalis,* known from se Wn and vic, and from RM *(B. × utahensis)*   3 **B. × piperi** (Britt.) Hitchc.

        3b Lvs ± ovate, base rounded to subcordate, rounded to acute at tip, (1) 2–4.5 (6) cm; sts gen strongly crystalline-glandular, but sometimes otherwise glab; along streams or in moister for; Alas s, E Cas (always?) to Cal, e to Sask, Dak, Wyo, Colo, Utah, and Ariz; water b., spring b., red b.; 2 vars.   4 **B. occidentalis** Hook.

            a1 Young twigs strongly pubescent and often puberulent; lf bls gen very hairy and with tufts of hairs in vein axils beneath; s BC and sc Ida to Utah and n Cal   var. **inopina** (Jeps.) Hitchc.

            a2 Young twigs glab to moderately pubescent; lf bls rarely tufted-hairy beneath; Alas s, mostly e of Cas in Wn and Ore, to Cal, common in Ida and Mont, and to Sask, Dak, and NM *(B. elrodiana, B. fontinalis, B. guthriei, B. obovata)*   var. **occidentalis**

Corylus L.   Hazelnut; Filbert

♀ fls in small clusters at br ends, borne in pairs at base of large bract, each fl at anthesis with a 2ndary, erose bractlet; ovary inferior, 1-celled; styles purplish-red, elongate; nut hard-shelled, ours scarcely 15 mm; ♂ fls in pendent catkins at end of brs of previous year, each fl with primary bract and 2 2ndary bracts, stamens (6) 8; filaments short, anthers with 1 (2) pollen sacs; shrubs or small trees with caducous-stip, toothed lvs. (L name for hazelnut).

C. cornuta Marsh. Glabrate shrub 1–3 (5) m; young growth hirsute and glandular-pubescent; lf bls ± cordate, acute to acuminate, biserrate; invol hispid-hirsute, 1.5–2.5 cm; BC to Cal, e to Newf and Ga; ours the var. *californica* (DC.) Sharp; widespread at lower elev on well-drained soil, BC to Cal, on both sides Cas, e to Ida *(C. c.)*.

FAGACEAE   Beech Family

Fls ♀♂, apet; ♂ fls in bracteate spikes or catkins, perianth 4–7-lobed, stamens 4–20; ♀ fls single or in 2's or 3's, perianth adnate to ovary, 4–6-lobed; pistil 1; ovary inferior, 3 (6)-celled, each cell 2-ovuled; styles 3 (6); fr gen a 1-seeded, invol nut, borne singly in a cuplike scaly invol or 1–3 in a spiny bur; evergreen to deciduous trees or shrubs with simple lvs.

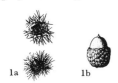

1a Invol spiny, burlike, enclosing 1–3 nuts; lvs entire; ♂ catkins spreading to erect   **Castanopsis**
1b Invol cuplike, bearing a single nut (acorn); lvs lobed to pinnatifid (ours); ♂ catkins pendent   **Quercus**

1a   1b

Castanopsis Spach   Chinquapin

♂ catkins stiffly erect or spreading; ♀ fls borne at base of ♂ catkins or in short spikes; invol becoming spiny, 4-valved; shrubs or trees. (Gr *kastanos*, name for chestnut, and *opsis*, like). *(Chrysolepis)*.

C. chrysophylla (Dougl.) DC. Golden c., giant c. Evergreen shrub or small tree (3) 5–30 m, with thick, heavily furrowed bark; lf bls lanceolate to oblong-elliptic, thick and leathery, yellow-green beneath, (3) 5–15 cm; bur 1.5–2 cm broad; nuts 10 (15) mm; dry, open sites to fairly thick woodl; OP, Mason Co, and Skamania Co, Wn; CRG and nw and nc Ore, to Cal *(Castanea c.)*.

Quercus L.   Oak

♂ fls 1 per node, in loose, drooping catkns; ♀ fls surrounded by scaly invol that becomes cap- or cup-like; ovary 3-celled; styles 3; nuts (acorns) largely projecting from invol; trees or shrubs, deciduous (ours) to evergreen. (L name for the oak).

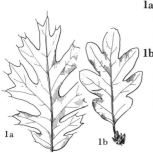

1a Lf lobes with acuminate and bristle-tipped teeth; petioles mostly > 2 cm; acorns oblong, 2–3 cm, maturing in 1 year, shell tomentose inside; tree (7) 10–20 (35) m; foothills and lower mts; Lane Co, Ore, to Cal; Kellogg's o., Cal black o. *(Q. californica)*   1 **Q. kelloggii** Newberry
1b Lf lobes entire to rounded or abruptly apiculate; petioles mostly < 2 cm; acorns ovoid to subglobose, 2–3 cm, maturing in 2 years, shell glab inside; tree (5) 10–20 (30) m; prairies and foothills; VI to Cal, mostly w Cas, but in CRG and n along e base of Cas to Yakima Co, Wn; Garry o., Ore white o. *(Q. gilberti, Q. jacobi)*   2 **Q. garryana** Dougl.

1a   1b

ULMACEAE   Elm Family

Fls 1–several in lf axils, small, apet, ♂ or ♀♂; perianth greenish, (3) 4–6 (8)-lobed; stamens as many as the tepals and opp them, or sometimes more, erect in bud; pistil 1, 2-carpellary, gen 1-celled and 1-ovuled; styles 2; fr drupe or samara; trees or shrubs with alt, simple, often oblique-based, mostly deciduous-stip lvs.

1a Fls ♂; fr broadly winged samara; various cult spp. sometimes persistent
   around abandoned homes and mistakable for native   **Ulmus**
1b Fls in part ♀♂; fr rather hard-walled drupe; native   **Celtis**

Celtis L.   Hackberry

Fls developing with lvs, lower ones mostly ♂, the upper ones ♂ or ♀; tepals 4–6, nearly distinct; styles spreading; deciduous shrubs or small trees. (Derivation obscure, name used by both Gr and L authors).

C. reticulata Torr. Shrub or tree 4–10 m; lf bls 3–10 cm, obliquely ovate, ± acuminate, serrate, scab on both surfaces, strongly reticulate-veiny beneath, 3-nerved from near the base; peds 3–20 mm; perianth ca 3 mm; drupe reddish-brown to purple, ellipsoid-globose, 6–8 mm, the pulp thin but sweet; open, often rocky areas, esp along rivers such as SR; e Wn to Cal, e to Ida, Utah, and Ariz *(C. douglasii).*

MORACEAE   Mulberry Family

Fls ♀♂ or ♀, ♂, small, apet, mostly crowded and cymose to capitate; ♀ fls naked or with 4 (2–6) ± connate tepals; ♂ fls mostly with stamens as many as tepals and opp them (or only 1–2); pistil 1, 2-carpellary; stigmas 2 (1); ovary 1 (2)-celled, superior or with adnate perianth; fr drupe or achene, often multiple; ours vines or trees, often with milky juice; lvs alt (opp), simple to palmately parted, stip.

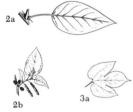

1a Pls shrubs or trees; juice ± milky; lvs alt
   2a Lvs entire; sts ± thorny   **Maclura**
   2b Lvs serrate to lobed; sts not thorny   **Morus**
1b Pls herbaceous, st ± fibrous, often vining; juice not milky; lvs in part opp
   3a Lvs coarsely 3–7-lobed; stout vines   **Humulus**
   3b Lvs palmately 5–9-parted; sts not vining; widely cult and often escaped,
      but apparently not persistent in our area; hemp   **Cannabis sativa** L.

Humulus L.   Hop

Strongly twining, ♀, ♂ vines. (L name for the hop).

H. lupulus L. Lvs opp, cordate-based, deeply 3–5(7)-lobed, very scab, 4–10 cm; fls axillary, ♀, ♂, the ♂ pan, with 5 distinct perianth segms; ♀ fls in 2's in foliaceous-bracteate spikes, calyx (or bract?) covering the ovary, entire; fr an achene, enclosed by perianth and hidden by the enlarged bracts, the "hop" thus formed 3–5 cm, all parts yellow-glandular; Eurasian, widely cult for the bitter resinous material in the ♀ infl, and often persistent or escaped in old fields.

Maclura Nutt.   Osage Orange

Thorny, brownish-barked trees 5–20 m. (For William Maclure, 1763–1840, Am geologist).

M. pomifera (Raf.) Schneider. Thorns 1–2 (5) cm; lf bls ovate-lanceolate, 5–15 cm; fls ♀♂, the ♂ in subglobose, pedunculate racemes, slender-pedicellate, perianth deeply 4-lobed; ♀ fls in dense, globose, subsessile heads, perianth 4-lobed, closely covering the 1-celled ovary; style 1, filiform, in fr the calyx and recep enlarged and corky, the whole infl becoming roughened and orange-like, 7–12 (15) cm thick; native in much of sc US; intro in many places, in our area estab only in SRC, s of Asotin, Wn.

Morus L.   Mulberry

Fls ♀♂ or ♀, ♂, the ♂ loosely spicate, with 4 tepals and 4 stamens; ♀ fls in short, crowded spikes, perianth 4-parted; ovary at first 2-celled, but becoming 1-celled; styles 2; fr an achene covered by fleshy perianth, the ♀ infl becoming succulent; trees or shrubs with toothed to lobed, deciduous lvs. (L name for mulberry).

M. alba L. White m. Shrub or small tree to 15 m; lf bls ovate, 3–12 cm, coarsely serrate to several-lobed, acute to acuminate; ♂ spikes drooping; ♀ spikes 10–15 mm; fr white to red or black; Asian, widely intro in warm dry areas of US, in our region only in the SRC of se Wn and adj Ida.

URTICACEAE   Nettle Family

Fls ♀, ♂ or ♀♂ to partly ⚥, hypog, mostly small, greenish, apet, single or cymose to pan; perianth 4-parted or absent; stamens 4, opp tepals; pistil 1, 1-carpellary, style and ovule 1; stigma tufted; fr achene; ann or per herbs (ours), often with stinging hairs; juice watery; lvs simple, opp or alt, mostly stip.

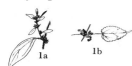

1a Lvs entire, alt, not stip; hairs not stinging   **Parietaria**
1b Lvs dentate, stip, opp; hairs stinging   **Urtica**

1a   1b

Parietaria L.   Pellitory

Fls in axillary clusters surrounded by linear bracts, ♀ and ♂ (and often a few ⚥) mixed. (Old L name for the pl, from *paries*, wall, in reference to habitat).

P. pensylvanica Muhl. Ann, finely pubescent, prostrate to erect, sts up to 4.5 dm; lf bl oval to narrowly elliptic, 2–4 cm; tepals 1–2 mm; achene ca 1 mm; woods to shaded banks or shelter of large rocks; BC s, e Cas, to Cal, e to Atl and s to Ala and Mex (*P. occidentalis*).

Urtica L.   Nettle

Fls ♀, ♂ or ♀♂, in axillary pans or spikes; ann or (mostly) per herbs. (L name for nettle, from *uro*, to burn).

1a Pl ann, 1–5 dm; lf bl elliptic to ovate or obovate, (1) 1.5–4 cm; stips 2–3 mm; European; not common in our area, but known from Wn and Ore; dog n., dwarf n.   **1 U. urens** L.
1b Pl per, rhizomatous, 1–3 m; lf bl lanceolate to ovate or cordate, (5) 7–15 cm; stips 5–15 mm; moist areas from sagebr des to deep woods; much of US and s Can; Eurasia; stinging n.; several vars.   **2 U. dioica** L.
    a1 Pl ± completely ♀, ♂; Eurasian, intro in N Am and perhaps in our area   ssp. **dioica**
    a2 Pl ♀♂; gen in N Am   ssp. **gracilis** (Ait.) Seland.
       b1 Lf bl lanceolate, ave at least 3 times as long as broad; petioles short, scarcely 1/3 as long as bl; infl crowded, upper fl bracts reduced and ± exceeded by some of the infl
          c1 Pl strongly pubescent-cinereous; sw US, but occ in e Wn and Ore and in Ida; hoary n. (*U. h.*)   var. **holosericea** (Nutt.) Hitchc.
          c2 Pl not cinereous-pubescent, sometimes only bristly
            d1 St bristly near base, otherwise not hairy; lvs lightly pubescent; chiefly e and c US and s Can, w to RM (*U. p., U. strigosissima*)   var. **procera** (Muhl.) Wedd.
            d2 St hairy as well as bristly; lvs strongly hairy; Mont to NM (*U. serra*)   var. **angustifolia** Schlecht.
       b2 Lf bl ovate-lanceolate, < 3 times as long as broad; petioles mostly at least 1/3 as long as bl; infl not crowded; upper (fl) lvs mostly > infl
          e1 St hirsute, grayish throughout; lvs coarsely bristly and strongly pubescent beneath; chiefly Cal, but occ e Cas, to Okanogan Co, Wn, and to e and c Ida; Cal n., coast n. (*U. c.*)   var. **californica** (Greene) Hitchc.
          e2 St bristly but often not otherwise hairy near base; lvs greenish, glab to sparsely pubescent
            f1 Lvs broadly ovate, sparsely hairy and bristly, length rarely > twice width; petioles slender, ca half length of bl; st subglab; Alas to Cal and w Mont; common var. w Cas; Lyall n. (*U. l.*)   var. **lyallii** (Wats.) Hitchc.
            f2 Lvs lanceolate to ovate, often fairly hairy, 2–3 times as long as broad; petioles 1/3–1/4 as long as bl; st often hairy; BC to Atl, common in Ida and w Mont, occ in Wn and Ore; slim n. (*U. cardiophylla, U. g., U. viridis*)   var. **gracilis**

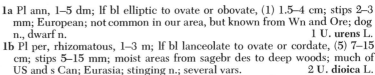

LORANTHACEAE   Mistletoe Family

Fls ♀, ♂ (ours) or ⚥, spicate to pan; perianth of 1 or 2 series of 2–3 similar, greenish and inconspicuous (ours), free or ± connate segms; stamens as many as tepals and inserted on them; ovary inferior, 1-locular; fr berry or drupe, 1 (2–3)-seeded, often viscid; shrubs (ours) with mostly opp lvs, swollen nodes, and simple, exstip, scalelike to broad lvs, parasitic on brs of other shrubs or trees.

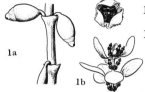

1a Lvs scalelike; ♀ fls mostly with 2 tepals; fr ± flattened, borne on recurved
  peds; anthers 1-celled   **Arceuthobium**
1b Lvs scalelike to well developed (broad); ♀ fls mostly with 3 tepals; fr subglobose, sessile; anthers 2-celled   **Phoradendron**

Arceuthobium Bieb.   Dwarf Mistletoe

♂ fls with (2) 3–4 (5) yellowish to green, ± fleshy tepals; fr ovoid, greenish to bluish, ± sticky; greenish to yellowish, glab shrubs, parasitic on conifs; sts jointed; lvs scalelike, opp-connate. (Gr *arkeuthos*, juniper, and *bios*, life, the pl often parasitic on juniper). (*Razoumofskya*).

1a Fls sessile or ♀ with slender peds up to 1 mm, gen 2 at each node; ♂ fl buds
  lenticular, lateral tepals keeled; st often > 2 mm thick and seldom < 3 (4)
  cm long; widespread, Alas to Cal and Mex, to e of RM, on many conif
  hosts; w d. m.; several growth forms   1 **A. campylopodum** Engelm.
  a1 Parasitic on *Pinus*
    b1 St greenish-brown, mostly 1–3 cm; mainly on *P. flexilis*, but also on *P.
      albicaulis* in our area; Ore, Ida, and Mont, s
                                                   f. **cyanocarpum** (Nels.) Gill
    b2 St mostly > 2–3 cm, often yellow or brownish
      c1 Twigs greenish; on *P. monticola*   f. **blumeri** (Nels.) Gill
      c2 Twigs mostly orange or yellow (brown); on *P. contorta* and *P. pon-
        derosa*   f. **campylopodum**
  a2 Parasitic on genera other than *Pinus*: on *Picea engelmannii*—f. *micro-
    carpum* (Engelm.) Gill; on *Abies*—f. *abietinum* (Engelm.) Gill; on *Tsuga*
    —f. *tsugensis* (Rosend.) Gill; on *Larix*—f. *laricis* (Piper) Gill
1b Fls terminal on short lateral brs, brs bearing ♂ fls often more than 2 per
  node; ♂ fl buds subglobose, tepals not keeled; st scarcely as much as 2 mm
  thick, often < 3 cm long
  2a 2ndary brs of twigs lying in same plane, thus together flat and fanlike; st
    mostly 0.5–3 cm, internodes scarcely 1 mm thick and only 3–6 times as
    long; on *Pseudotsuga menziesii*; Douglas d. m.; BC to Cal, e to Alta,
    Mont, and NM   2 **A. douglasii** Engelm.
  2b 2ndary brs of twigs distinctly whorled and not at all fanlike; st 2–6 cm,
    internodes mostly at least 1 mm thick and 7–15 times as long; on *Pinus*
    spp.; in our area chiefly on *P. contorta*, but reported on *P. albicaulis*; Am
    d. m.; s BC to Cal, e to Mont and s to NM   3 **A. americanum** Nutt.

Phoradendron Nutt.   Mistletoe

Fls ± sunken in short, jointed spikes, with 3 (2–5)-merous perianth; ♂ fls with triangular-rounded segms; fr viscid; fleshy shrubs, parasitic on both conifs and dicots, with gen jointed sts and scalelike to well-developed lvs. (Gr *phor*, thief, and *dendron*, tree, in reference to the parasitic habit).

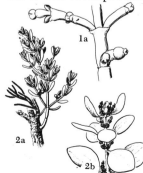

1a Lvs thin and scalelike, not jointed with st; c Ore to Baja Cal, e to Utah,
  NM, and Mex; on *Juniperus* in c Ore, but also on *Calocedrus* and *Cupressus* in other areas; juniper m. (*P. libocedri, P. ligatum*); ours the var.
  ligatum (Trel.) Fosb.   1 **P. juniperinum** Engelm.
1b Lvs well developed, jointed with st
  2a Pl glab or granular-puberulent; lvs 1–2 (2.5) cm, oblanceolate to nar-
    rowly obovate, sessile; se Ore to Baja Cal, e to Tex and NM; on *Juni-
    perus* and *Cupressus*, but also reported on *Abies* and *Arbutus*, prob
    reaching our area on *J. occidentalis* in e Ore; dense m.; ours the var.
    densum (Torr.) Fosb. (*P. d.*)   2 **P. bolleanum** (Seeman) Eichl.
  2b Pl finely velvety-pubescent on new growth; lvs (1) 1.5–3 (4) cm, ovate to
    obovate, petiolate; Ore to Cal and to c and e US, mostly on *Quercus*, in
    our area in w Ore, on *Q. garryana*; Am m.; ours the var. *villosum* (Nutt.)
    Engelm. (*P. v.*)   3 **P. flavescens** (Pursh) Nutt.

SANTALACEAE   Sandalwood Family

Fls 1–several and axillary, to ∝ in terminal or subterminal cymes, reg, apet; perianth greenish, white, or purple, connate, (3) 4–5-lobed, adnate to ovary; stamens as many as tepals and opp them, gen on a free hypan; pistil 3–5-carpellary but 1-celled; ovules 1–3, pendulous from an elongate basal placenta; fr drupelike (ours); ♀, ♂ or ♀♂ to ⚥ per herbs (ours), often parasitic, with simple, entire lvs.

Comandra Nutt.   Comandra

Fls in axillary or terminal clusters, ⚥ or ⚥♀♂; hypan disc-lined, saucerlike to turbinate; filaments tufted-hairy at base; rhizomatous root parasites with alt, ± fleshy or leathery lvs. (Gr *kome*, hair, and *aner* or *andros*, man, the stamens hairy at base).

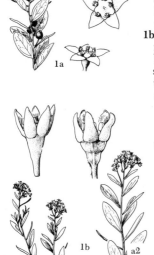

1a Fls gen 3 (1–4) in upper lf axils; hypan flared, tepals ca 1.5 mm and as broad, disc nearly flat; lvs narrowly ovate to obovate-oblanceolate, (1) 1.5–4 cm; fr red, juicy; bogs and open woods to lower alp slopes; Alas to nc and ne Wn and to n Ida; n c. *(Geocaulon l.)*   1 C. **livida** Richards.
1b Fls in subterminal or terminal cymes; tepals lanceolate to ovate, 1.5–4 mm, not so broad; hypan camp to turbinate, disc-lined; fr blue to purplish or brownish; dry to moist, well-drained soil, lowl to subalp, esp common in sagebr; much of US and s Can; parasitic on ∝ conif and angiosperms; bastard toad-flax; 2 or 3 vars.   2 C. **umbellata** (L.) Nutt.
  a1 Tepals narrowly lanceolate, mostly 3–4 mm; fr 6–9 mm; lvs thick, glaucous, midnerve (but not lateral veins) sometimes visible; e Cas, Wn and Ore, to Ida and Mont, e and s; pale b. *(C. linearis, C. p.)*

var. **pallida** (DC.) Jones
  a2 Tepals ovate to ovate-lanceolate, 1.5–3 mm; fr 4–6 mm; lateral veins of lvs often visible
    b1 Lvs ± fleshy, very glaucous, margins rarely revolute, lateral nerves rarely visible in pressed pl; w Wn and Ore to Cal, also e Cas from Kittitas Co, Wn, to CRG *(C. c.)*   var. **californica** (Eastw.) Hitchc.
    b2 Lvs rather thin, not strongly glaucous, margins ± revolute, lateral nerves mostly plainly visible in pressed pl; common in c and e US and s Can, prob not reaching our area, but near it in BC *(C. u.)*

var. **umbellata**

ARISTOLOCHIACEAE   Birthwort Family

Fls solitary (ours) in the axils, ⚥, reg (ours); perianth adnate to ovary and often with a connate tube above it, 3-lobed (ours); corolla lacking; stamens 6–∝, filaments short and ± connate; pistil (4) 6-carpellary; style 1, short and thick; stigma (4) 6-lobed; ovary 6-celled (ours), placentation axile; fr caps; herbaceous to woody vines with simple, alt, nonstip lvs.

Asarum L.   Wild Ginger

Trailing herbs with long-petioled, cordate to reniform (ours), entire, sometimes mottled lvs. (Gr *asaron*, the ancient name for one of the spp.).

A. **caudatum** Lindl. Rootstock extensive, pl matted; lvs 2 per node, persistent, bl 4–10 × up to 15 cm; fls pedunculate, peduncles 1–3 cm; perianth camp, brownish-purple to yellowish or greenish, segms flared, 2.5–8 cm; bracts (petals?) alt with sepals, linear, 1–3 mm, erect to recurved; stamens 12, connectives broad, apiculate; moist shady woods, lowl to mid-mont; BC to Ore, e and w Cas, e to n Ida and w Mont *(A. hookeri, A. c. f. chloroleucum)*.

POLYGONACEAE   Buckwheat Family

Fls mostly ⚥ (♀♂), gen ∝ in large infl, often invol; perianth deeply 3–6-parted, green to highly colored and corolloid, often in 2 series of 3 segms each, sometimes strongly accrescent; stamens mostly 3–9, distinct, inserted at or above base of perianth; pistil gen 3 (2–4)-carpellary; ovule 1; styles 3 (1–4); fr an achene, often winged; ours ann or per herbs to small shrubs, sometimes vining; lvs alt (opp), often with papery, sheathing stips; bls simple and gen almost or quite entire.

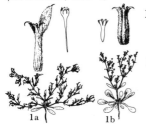

1a Lvs with well-developed, gen membranous stips sheathing st above each node; fls not borne within an invol
 2a Lf bl reniform; tepals 4; stamens 6; pistil 2-carpellary, styles 2, stigmas fimbriate; ovary strongly compressed; fr lenticular   **Oxyria**
 2b Lf bl not reniform; tepals often 5 (6); pistil gen 3-carpellary
  3a Tepals 6 (rarely 4), outer ones not enlarging in fr, inner ones (valves) gen accrescent, 1 (or more) often with c grainlike callosity on outer surface; lvs not jointed to stip base   **Rumex**
  3b Tepals gen 5 (rarely 4 or 6), remaining subsimilar in size, inner series not accrescent and never with c callosity; lvs often jointed to stip base   **Polygonum**
1b Lvs without stips; fls 1–several in camp to tubular, 3–10-toothed invols   **Eriogonum**
 4a Segms of invol 3–10, not spine-tipped
 4b Segms of invol 3–6, spine-tipped
  5a Invol turbinate, 4–5-lobed, 2–several-fld; perianth 6-parted to base   **Oxytheca**
  5b Invols narrowly cylindric, (3) 5–6-toothed, 1-fld; perianth lobed < half length   **Chorizanthe**

Chorizanthe R. Br.   Chorizanthe; Spine-flower

Invols with slender, recurved, spine-tipped teeth, in capitate to spreading, bracteate cymes; perianth ± turbinate, ca = invol, mostly 6-lobed; stamens mostly 3 or 9 (ours); filaments glab; stigmas 3, capitate; achene 3-angled; ours di- or trichotomously br anns with entire, gen alt lvs. (Gr *chorizo*, divide, and *anthos*, flower, referring to the deeply divided perianth).

1a Teeth of invol 5, 1 much the largest (up to 15 mm); perianth 3–4 mm, yellow, lobes sparsely pubescent; stamens 9, anthers and filaments subequal; sagebr des; c Wn and e Ore to c Ida, s to Cal, Utah, and Ariz; Watson's s.   **1 C. watsonii** T. & G.
1b Teeth of invol 6, 4–5 mm, 3 ca twice as long as others; perianth glab, (2) 3 mm; stamens 3, filaments 4–5 times as long as anthers; sagebr des; se Ore to se Ida, on both sides of SR, s to Cal; brittle c. or s.; ours the var. *spathulata* (Small) Hitchc.   **2 C. brevicornu** Torr.

Eriogonum Michx.   Eriogonum; Buckwheat; Wild Buckwheat; Umbrella-plant
(Contributed by James Reveal)

Invols camp to turbinate, 3–10-lobed or -toothed, 2–several-fld, solitary or in clusters, terminal, capitate, umbellate, or cymose; peds stipelike in some; perianth white to yellow, rose, or purple, glab to pubescent, 6-parted at base; stamens 9; pistils 3-carpellary; stigmas mostly capitate; ann to per herbs or subshrubs with alt to whorled, entire, gen ± tomentose, exstip lvs. (Gr *erion*, wool, and *gonu*, knee, referring to the woolly nodes and sts). Dwarf spp. excellent rock garden subjects e Cas; grazing value slight.

1a Pl per
 2a Perianth abruptly stipelike at the attenuated base; bracts on fl stem foliaceous, indefinite in no (2–several)   **Group I**
 2b Perianth not stipelike or attenuated at the base; bracts gen scalelike, mostly 3-parted   **Group II**, lead 13a
1b Pl ann or bien   **Group III**, lead 24a

Group I

3a Invol lobes at least half as long as tube and gen reflexed or spreading
 4a Perianth pubescent externally
  5a Fl st without subtending bracts below the solitary terminal invol; sagebr flats to open, rocky mt ridges; se Ore to Cal, e to c Ida, Mont, Wyo, and Colo; mat b. (*E. andinum*)   **1 E. caespitosum** Nutt.
  5b Fl st with whorled subtending bracts at base of umbel or near middle of st

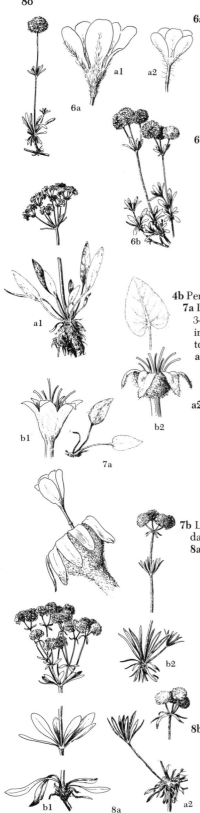

6a Invols solitary, infl capitate, not immediately subtended by lfy bracts, these restricted to a whorl near middle of fl st; pl matted, rarely > 1 dm; lvs ± tomentose on both surfaces, gen not > 3 mm wide; perianth whitish to lemon or pink, never truly yellow; sagebr des and hills to juniper and ponderosa pine for; c Wn s, E Cas, to ne Cal, e to w Ida and w Nev; Douglas' b.; 2 vars. 2 **E. douglasii** Benth.
  a1 Perianth strongly hairy, esp on stipe; c Wn, e Ore, and w Ida
    var. **douglasii**
  a2 Perianth sparsely hairy to nearly glab or weakly glandular; Yakima Co, Wn, s to Wasco Co, Ore   var. **tenue** (Small) Hitchc.
6b Invols gen > 1, often umbellate, immediately subtended by several lfy bracts; fl st lfless or with 1 or more alt (rarely whorled) bracts near middle in some; pl ascending to erect, (0.5) 1.5–3 dm; lvs gen more tomentose below than above, gen > 3 mm wide; perianth white or pink to yellow; sagebr des or juniper flats and scablands to ponderosa pine for; Chelan Co, Wn, s to Cal, E Cas, e to w Ida and Nev; round-headed e., rock b.; 2 vars. 3 **E. sphaerocephalum** Dougl.
  a1 Perianth yellow; common phase in Wn, also in e Ore and w Ida, s to n Cal and Nev (*E. fasciculifolium, E. geniculatum*)
    var. **sphaerocephalum**
  a2 Perianth off-white to pink, or pale to greenish-yellow; c Ore to wc Ida and s, rarely if ever in Wn (*E. fruticulosum*)
    var. **halimioides** (Gandg.) Stokes
4b Perianth glab externally
  7a Lvs broadly lanceolate to triangular, often truncate to cordate at base, 3–20 × (1) 1.5–5 cm (including petiole); fl st 1.5–5 dm, lfless below infl; rocky places, foothills into mts; Chelan Co, Wn, s in Cas to Cal, e to wc Ida and ne Ore; northern b.; 3 vars.   4 **E. compositum** Dougl.
    a1 Lf bl lanceolate, mostly ∝ times as long as broad, cuneate to ± rounded at base, not truncate or cordate; Wen Mts, Chelan and Kittitas cos, Wn, sporadic and rare elsewhere
      var. **lancifolium** St. John & Warren
    a2 Lf bl mostly ovate to deltoid, often ± truncate to cordate at base and < 3 times as long as broad
      b1 Invols glab or mostly very weakly glandular-puberulent, not lanate; st gen glab; E Cas, Chelan Co, Wn, to Mt Rainier where common, occ in e to wc Ida and s to Baker Co, Ore
        var. **leianthum** Hook.
      b2 Invols sparsely to densely woolly; st ± floccose-lanate to glab; common in most of Ore and n Cal, n to Mt Rainier and across sc and se Wn to Asotin Co (*E. johnstonii, E. pilicaule*)
        var. **compositum**
  7b Lvs < 1 cm wide or in other ways not as above, not truncate or cordate at base; fl st mostly < 2 dm or whorled-bracteate near midlength
    8a Fl st gen whorled-bracteate near midlength, bracts reduced or lacking in some; lvs gen linear to linear-lanceolate or oblanceolate, grayish-lanate on both surfaces or only sparsely tomentose and much less grayish above, bl at least 3 (mostly > 4) times as long as broad; rocky places from sagebr des to ponderosa pine for and mont ridges up to 6000 ft elev; BC s, E Cas, to ne Cal, e to Mont, Wyo, and n Utah; Wyeth b., parsnip-fld e.; 3 vars.
      5 **E. heracleoides** Nutt.
      a1 Fl st gen bracted ca midlength
        b1 Lvs gen (6) 8–15 mm wide; chiefly se Wn, ne Ore, and wc Ida   var. **heracleoides**
        b2 Lvs gen < 6 mm wide; range of sp. but only occ within ranges of other vars. (*E. a.*)   var. **angustifolium** (Nutt.) T. & G.
      a2 Fl st mostly bractless; lvs linear; Chelan, Kittitas, and Douglas cos e to Spokane, Wn, sporadic elsewhere   var. **minus** Benth.
    8b Fl st lfless near midlength; lvs gen elliptic, rarely as much as 3 times as long as broad; sagebr des to alp rocky ridges, mostly E Cas, s BC to Cal, e to Mont, Wyo, Colo, and Ariz; sulfurfl, sulfur b.; highly variable, with several vars.   6 **E. umbellatum** Torr.
      a1 Primary rays of umbels simple, not br or bracteate in the middle
        b1 Pl prostrate, dwarf, fl st rarely up to 1 dm; fls cream to yellow; umbels subcapitate; lvs oblong-elliptic, gen 1–1.5 cm, ca half as broad, white below, gray above; alp or subalp in Cas, s Wn

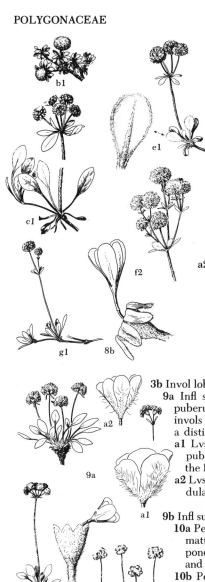

to c Ore *(E. montanum)*   var. **hausknechtii** (Dammer) Jones

b2 Pl gen at least 1 dm; umbels ± open; lvs various

   c1 Perianth white to cream or pale yellow; similar to var. *umbellatum* except for fl color, often mistaken for *E. heracleoides* which has much narrower lvs; foothills upward to 9000 ft; s BC to Ore, e to Mont and Colo *(E. s.)*

           var. **subalpinum** (Greene) Jones

   c2 Perianth mostly bright yellow

     d1 Lvs essentially glab on both surfaces

       e1 Lvs gen slightly tomentose on margins and midribs, mostly obovate; primarily Wen Mts, Wn

            var. **hypoleium** (Piper) Hitchc.

       e2 Lvs glab on both surfaces, mostly elliptic; s Ore and Ida to Wyo, s to Nev and Utah *(E. neglectum, E. umbelliferum)*   var. **intectum** Nels.

     d2 Lvs tomentose below, sparsely tomentose to glab above; range of sp. except largely replaced by var. *subalpinum* in s Can and Mont *(E. tolmieanum, E. rydbergii)*

                   var. **umbellatum**

a2 Primary rays of umbels br or at least bracteate near middle

   f1 Lvs glab on both surfaces

     g1 Rays of infl bracteate but not compoundly brd; sc Ore *(E. torreyanum, E. u.* var. *t.* of Ore authors)

          var. **glaberrimum** (Gandg.) Reveal

     g2 Rays of infl compoundly brd; s Wn s to sc Ore

              var. **chrysanthum** Gandg.

   f2 Lvs tomentose at least below; infl compoundly brd; c and sw Ida to ne Ore and se Wn, s to Cal *(E. croceum, E. ellipticum)*

             var. **stellatum** (Benth.) Jones

3b Invol lobes erect, often < half as long as the tube

  9a Infl subtended by 2 linear bracts 5–20 mm; perianth ± glandular-puberulent outside and gen crisp-pilose to lanate, hairs often rufous; invols turbinate-camp, 4–6 mm; pl alp or subalp, thick-crowned from a distinct tap-root, herbaceous; 2 vars.   **7 E. pyrolifolium** Hook.

   a1 Lvs lanate below; perianth mostly strongly pilose-lanate, glandular puberulence concealed; Cas, n Wn s to Cal, and Valley Co, Ida, to the Bitterroot Mts, Ida-Mont; alpine b.   var. **coryphaeum** T. & G.

   a2 Lvs not lanate (gen glab) below; perianth sparsely if at all lanate, glandular puberulence conspicuous; e Ida, Wen Mts, Wn, and Cal; oarlf b.

               var. **pyrolifolium**

  9b Infl subtended by > 2 bracts; perianth glab or densely pubescent

   10a Perianth glab externally; pls forming large low mats, but not densely matted; lvs rhombic to elliptic, 2–4 cm; gravelly flats in lodgepole and ponderosa pine for to alp talus and ridges; Cas, Linn Co, Ore, to Cal and nw Nev; mt b.   **8 E. marifolium** T. & G.

   10b Perianth pubescent externally; pl suffrutescent or matforming, but mats dense; lvs various

     11a Fl st with whorl of linear bracts near middle; lvs tightly revolute, narrowly linear; pl ♀, ♂; sagebr flats to low mt ridges; Chelan Co, Wn, E Cas to n Ore, e to sw Ida; thyme b., thyme-lvd e.

               9 E. thymoides Benth.

     11b Fl st bractless; lvs not tightly revolute

      12a Lvs rarely as much as 2 cm × 1–3 mm; ± revolute; perianth not truly stipitate, base attenuate but much broader than ped; st gen < 1 dm; open, rocky places, mont to subalp areas in RM; e BC to sw Alta, s to n Mont; androsace b., rockjasmine b.

          10 E. androsaceum Benth.

      12b Lvs mostly at least 3 cm, gen much > 3 mm wide; perianth with stipelike base 0.5–1.5 mm; st often > 1 dm; open knolls of grassl to alp ridges and scree, ec Alas, BC, s through extreme e Wn to ne Ore, e to Alta, and s to Ida, Mont, and Wyo to Colo and Ariz; yellow b.; 3 vars.   **11 E. flavum** Nutt.

      a1 Stipe of perianth mostly thicker than ped, gen < 1 mm; pl grayish-tomentose

        b1 Fl st 4–20 cm; lvs mostly 10–15 mm wide, thick; low elev, mostly e RM, but in our area in c Mont *(E. crassifolium, E. sericeum)*   var. **flavum**

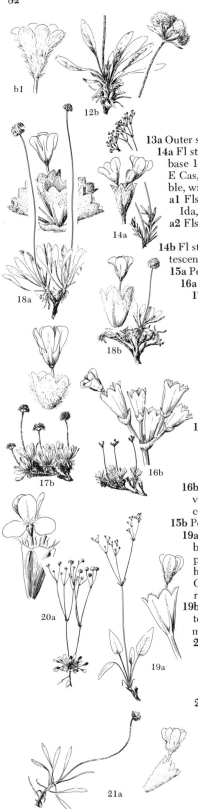

b2 Fl st 1–4 cm; lvs mostly 5–8 mm wide; alp, RM, nw Wyo
and sw Mont   var. **polyphyllum** (Small) Jones
a2 Stipe of perianth slender, often no thicker than ped, 1–1.5 mm;
pl moderately pubescent; lvs ± greenish above, not thick, variable in width; chiefly w RM, s BC to Alta, s to Ore, e to Wyo;
Piper's b. (*E. p., E. f.* var. *linguifolium*)
var. **piperi** (Greene) Jones

Group II

13a Outer segms of perianth ± similar to inner ones
14a Fl st lfy 1/3–4/5 length; pl gen not forming mat, gen with suffrutescent
base 1–4 dm; invols solitary in broad, compound, cymose infl; nc Wn s,
E Cas, to Cal, e to Mont, s to NM and Ariz; slenderbush b.; highly variable, with 2 vars. in our area   **12 E. microthecum** Nutt.
a1 Fls yellow; chiefly John Day Valley, Ore, sporadic elsewhere in sw
Ida, n Nev, and ne Cal (*E. idahoense*)   var. **microthecum**
a2 Fls white or pink; n part of sp. range, more common
var. **laxiflorum** Hook.
14b Fl st lfless or lfy < 1/3 length; pl often forming large mat, rarely suffrutescent; invols mostly in capitate clusters (solitary)
15a Perianth glab or glandular, not pubescent
16a Infl tightly capitate with clusters of several invols
17a Invols rigid; perianth pale yellow to bright yellow or
cream-colored
18a Invols 6–8-lobed, narrowly turbinate, (3) 3.5–5 mm; perianth
pale yellow to cream-colored, 2–3 mm; barren ridges, often in
pale soil of volcanic origin, mostly with sagebr at low elev;
barely within our range in s Baker and Malheur cos, Ore, e to
Twin Falls Co, Ida, s to n Nev; ochre-fld e., ochre b.; ours the
var. *calcareum* (Stokes) Peck   **13 E. ochrocephalum** Wats.
18b Invols 5-lobed, mostly camp; perianth yellow, 2.5–3 mm; e
Ore to Bitterroot Mts, Ida-Mont; golden b.
14 E. chrysops Rydb.
17b Invols membranaceous, not rigid, 5-lobed, 2.5–3 mm; perianth
cream and strongly pink-tinged, glab; sagebr flats and grassy hillsides; Granite and Powell cos, Mont, to Lemhi, Clark, and Custer
cos, Ida, often on calcareous soils; imperfect b.
15 E. mancum Rydb.
16b Infl of loose cymose-umbellate heads of solitary, short-rayed invols; perianth yellow; barely in our area in Harney and Deschutes
cos, Ore; Cusick's b.   **16 E. cusickii** Jones
15b Perianth pubescent externally
19a Pubescence of perianth short and appressed; pl 3–8 (10) dm; lf
bl lanceolate to ovate, (5) 7–15 (25) cm, abruptly narrowed to
petioles ca as long, often truncate or sagittate at base, only slightly
hairy below, hairs not appressed; sagebr des to open mont ridges;
Okanogan Co, Wn, s E Cas to Cal and n Nev, e to Ida; tall b.,
rush b.   **17 E. elatum** Dougl.
19b Pubescence of perianth mostly longer and tangled, often tomentose; lf bl never as above, if as much as 7 cm, then gen strongly tomentose below; pl mostly < 3 dm
20a Fls mixed with semiplumose, linear bractlets that protrude from
invols; infl diffusely cymose; tepals ± clawed at base; sandy or
rocky places from lowl to subalp areas; Cas, from Lewis Co, Wn,
s to Cal, toward coast in Ore; barestem b. (*E. oblongifolium*)
18 E. nudum Dougl.
20b Fls not plumose-bracteolate, or if so bractlets not protruding
from invols; infl mostly loosely to tightly capitate, umbellate, or
cymose; tepals not clawed
21a Fl st lfy 1/6–1/3 length; lvs mostly at least 2 cm; invol lobes
0.5–0.8 mm, acute; perianth 2–2.5 mm; badlands and rocky
ridges, prairies to e foothills of RM; Mont to Wyo, e to Sask,
Daks, and Neb, apparently in n Ida; few-fld b.; 2 vars.
19 E. pauciflorum Pursh
a1 Perianth white; infl capitate to umbellate; range of sp. (*E.
multiceps*)   var. **pauciflorum**

a2 Perianth yellow; infl cymose; Treasure Co, Mont

var. **canum** (Stokes) Reveal

21b Fl st lfless above base; lvs rarely as much as 2 cm; invol teeth mostly < 0.5 mm (see lead 12a)   10 **E. androsaceum** Benth.

13b Outer segms of perianth up to twice as wide as inner ones

22a Infl br or umbellate, not capitate

23a Infl lfy-bracteate at lower forks, gen (5) 10–20 cm, brs strongly ascending, mostly trichotomous at lowest nodes but dichotomous above; invols all solitary, strongly tomentose, gen 3 (4)-toothed; perianth rarely yellow; sagebr des to ponderosa pine for; E Cas, BC to Ore and wc Ida; snow b. (*E. album, E. decumbens*)   20 **E. niveum** Dougl.

23b Infl gen not lfy-bracteate, mostly < 10 cm, brs often 2–3 per node, tending to spread; invols sometimes clustered, sometimes glab, mostly 5-toothed; perianth often yellow; sagebr des to ponderosa pine for; E Cas, Okanogan Co, Wn, to n Cal, e to Ida, Mont, and Nev; strict b.; several infraspecific elements   21 **E. strictum** Benth.

a1 Invols glab (except margin of teeth), solitary; infl open, freely dichotomous or trichotomous, glab, brs nearly filiform; lf bl elliptic or ovate-elliptic, mostly 1–2 cm, greenish above, on slender petioles (1) 2–4 times as long; perianth white or ochroleucous to pink; Blue and Wallowa mts, Wn and Ore, and adj Ida; Blue Mts e.   ssp. **strictum**

a2 Invols mostly tomentose or if glab then often clustered; infl congested, primary brs mostly 4 or more, stout; lvs = grayish on both surfaces; 3 vars.   ssp. **proliferum** (T. & G.) Stokes

b1 Perianth yellow; mostly w of CR in Wn, from e base Cas, Chelan Co, to sc and se Ore and ne Cal and adj Nev (*E. a., E. flavissimum, E. s.* var.*f.*)   var. **anserinum** (Greene) Davis

b2 Perianth white or cream

c1 Invols and infl almost or quite glab; local (sporadic) in Yakima and w Grant and Douglas cos, Wn   var. **glabrum** Hitchc.

c2 Invols and infl floccose to tomentose; range of sp., but much less common within range of other vars., often transitional to *E. niveum* and *E. ovalifolium* (*E. bellum, E. fulvum*) var. **proliferum**

22b Infl tightly capitate, not br nor with elongated rays; sagebr des and juniper and ponderosa pine for to alp ridges above timberl; BC s, in Cas and RM, to Cal, e to Alta, Mont, and NM; oval-lvd e., cushion b.; several geographic races   22 **E. ovalifolium** Nutt.

a1 Pl alp or subalp, greatly dwarfed; lvs mostly < 1.5 cm (including petiole); scapes mostly 1–6 (10) cm; invols mostly ± camp, ca 3 mm, teeth ca 1/2 (1/3) as long as connate portion; fl mostly in July–Aug

b1 Lvs white or silvery on both surfaces (or ± rusty), thickish, bl nearly rotund, seldom as much as 1.5 times as long as wide; s BC s, in Cas and OM, to Cal, e to Wallowa and Steens mts, Ore, e across Nev to extreme w Utah; freely intergradient with var. *depressum* (*E. roseiflorum*)   var. **nivale** (Canby) Jones

b2 Lvs paler, often greenish at least on upper surface, bl elliptic to oblong or spatulate, often much > 1.5 times as long as wide; chiefly in Mont and Ida, n to Can RM, w to Wallowa and Steens mts, Ore, s to extreme n Utah (*E. rubidum*)   var. **depressum** Blank.

a2 Pl of the plains and sagebr des to mont slopes; lvs mostly > 1.5 cm; fl st often > 10 cm; invols commonly 4–5 mm, narrowly turbinate; fl mainly in May–June

c1 Lf bl mainly oval to rhombic, mostly (0.5) 1–1.5 cm, mostly < twice as long as wide; Alta to NM, on both sides RM, much less common w to Ida, e Ore, and s to Cal, Nev, and Utah; fl mostly in June (*E. purpureum*)   var. **ovalifolium**

c2 Lf bl oblanceolate or spatulate to oblong or obovate, mostly either > 1.5 cm or much > twice as long as wide

d1 Lvs mostly 1.5–3 (4) cm, bl gen at least twice as long as wide; fls mostly white or cream to pink; RM, Alta and se BC to Colo, occ w to Ida, most abundant in c Mont; mostly fl July–Aug (*E. ochroleucum*)   var. **macropodum** (Gandg.) Reveal

d2 Lvs mostly 3–6 cm, bl gen < twice as long as wide; fls gen yellow; mainly in sagebr des to ponderosa pine for; e Ore to Cal, e along SR to w Wyo, s to n Utah and Colo, sporadically in n Nev and ne Cal, n to Chelan Co, Wn, and to sw Mont; fl May–June (early July) (*E. orthocaulon*)   var. **celsum** Nels.

Group III

24a Invols erect and strongly appressed to st, 1.5–3 mm, often strongly ribbed; perianth 2–2.5 mm, lobes oval to obovate

25a Invols cylindric to narrowly conic-cylindric, mostly 2–3 mm, strongly ribbed, lobes acute, < 1/4 as long as tube; common e Cas, c Wn to nw Cal, e to Ida; broom b.; 2 vars.   23 **E. vimineum** Dougl.

a1 Sts mostly glab above; invols glab; common from c Wn to nw Cal var. **vimineum**

a2 Sts and invols floccose to tomentose; sc to e Ida *(E. s., E. commixtum, E. v.* var. *c.)*   var. **shoshonense** (Nels.) Stokes

25b Invols campanulate, not heavily ribbed, 1.5–2 mm, lobes rounded, ca 1/3 as long as the tube; perianth (1) 1.5–2 mm, lobes ovate-lanceolate, gen > 1/2 as long as tube; extreme sc Wn, e Ore, and adj Ida to s Cal and Ariz; Bailey's b. *(E. vimineum* var. *b.)*   24 **E. baileyi** Wats.

24b Invols erect or deflexed, peduncled, not closely appressed to st, neither strongly angled nor ribbed

26a St tomentose or glandular, not glab; invols erect

27a Lvs and st tomentose or floccose; lvs not linear

28a Perianth white, glab externally, tepals subsimilar; st up to 1 m, lfy nearly the entire length; invols tomentose; grassl and foothills; mostly e RM, Mont and Daks, s to Mex; annual b.   25 **E. annuum** Nutt.

28b Perianth yellowish with purplish tinge or spots, tepals dissimilar, glandular; pl up to 3 dm; invols glandular; sandy to heavy clay soil chiefly in sagebr des; Yakima Co, Wn, and se Ore to Cal, Nev, and Utah; spotted b. *(E. angulosum,* misapplied*)*   26 **E. maculatum** Heller

27b Lvs sparsely hirtellous, linear to linear-oblanceolate; st glandular-puberulent; invols glab; sandy to rocky soil, streambanks to hillsides; c Wn, e Ore, and wc Ida to Nev and e Cal; spurry b.; ours the var. *reddingianum* (Jones) Howell *(Oxytheca r.)*   27 **E. spergulinum** Gray

26b St glab; invols deflexed

29a Perianth 1.5–2 mm, outer segms ± quadrangular or fiddle-shaped, crisp-margined, tips truncate or retuse; invols broadly turbinate, 1.5–2 mm; sandy valleys and hills, se Ore to Cal, e to Mont and s to Nev and NM; bordering our area in se Ore, s and e Ida, and e Mont; nodding b.   28 **E. cernuum** Nutt.

29b Perianth 2–2.5 mm, segms oblong to obovate, not crisp-margined, tips rounded; invols narrowly turbinate, 2–3 mm; des plains to lower mts, se Ore and sw Ida s to wc Nev; barely within our area in n Malheur Co, Ore; Watson's b. *(E. deflexum* var. *multipedunculatum)*   29 **E. watsonii** T. & G.

Oxyria Hill   Mountain Sorrel

Fls ♂; perianth 4-parted almost to base; fr broadly winged; per herbs with mostly basal lvs with ± reniform bl and sheathing stips. (Gr *oxys*, sharp, in reference to acid juice of the pls).

O. digyna (L.) Hill. Glab, often reddish-tinged; lf bl 1–5 cm broad; fl st 1–several, 1–4 (6) dm; perianth ca 1.5 mm, 2 segms narrow and ± strongly keeled, other 2 erect and not keeled; fr 4–6 mm broad; moist ground, alp or subalp, circumboreal; Alas s through Cas and OM to Cal, e to Lab, NH, and NM.

Oxytheca Nutt.   Oxytheca

Fls ♂, 2–several in each turbinate invol; stigmas capitate; fr ovoid-lenticular; dichotomously (trichotomously) br ann with mostly basal, simple, exstip lvs; invol single and pedunculate to subsessile at each dichotomy, lobes spinulose-tipped. (Gr *oxys*, sharp, and *theke*, cover, in reference to invol).

O. dendroides Nutt. Lvs ∝, rosulate, linear to linear-oblanceolate, 1–2.5 (3) cm; st 5–30 cm; invol 1–1.5 mm, lobes with slender, unequal awns 0.5–3 mm; perianth camp, pubescent, ca 1 mm, white to deep pink; sandy areas, sagebr des; c Wn and e Ore to Cal, e to s Ida; S. Am.

Polygonum L.   Doorweed; Knotweed; Smartweed

Fls ♂ to ♂♀♂ or ♀♂, solitary or clustered, gen in spikelike to pan racemes; peds jointed at tip; perianth mostly 5 (4–6)-parted at least half length, greenish, white, or pink, outer 2–3 tepals often keeled and largest; stamens gen 8; pistil 3 (2–4)-carpellary; stigmas gen capitate; fr lenticular to 3 (4)-angled; ann or per herbs or subshrubs, often climbing; stips gen sheathing, sometimes becoming lacerate; lvs alt, mostly entire and petiolate, often jointed with stip base. (Gr *poly*, many, and *gonu*, knee, because of swollen nodes of some spp.). (*Aconoganum, Bilderdykia, Bistorta, Persicaria, Tiniaria*). Pls of little or no grazing value; many noxious weeds.

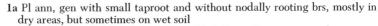

1a Pl ann, gen with small taproot and without nodally rooting brs, mostly in dry areas, but sometimes on wet soil

  2a St twining; lf bl sagittate-ovate, (2) 3–6 cm; European weed, common on waste or cult land; dullseed; cornbind; black, climbing, knot, or ivy bindweed   **1 P. convolvulus** L.

  2b St not twining; lvs mostly linear to oblong, not sagittate

    3a Stips strongly sheathing, oblique to truncate (not lobed) at top, not lacerate; pl mostly 3–10 dm   **Group I,** lead 16a

    3b Stips sheathing at first, but gen strongly bilobed above and becoming lacerate and often recurved, sometimes nearly to base; pl sometimes < 3 dm

      4a Fls mostly solitary (2–3) in lf axils, sessile; lvs linear-lanceolate, subulate, not narrowed to petioles and not jointed at base, 3-nerved, lateral nerves submarginal; stips strongly lacerate into filiform segms either rigid and straight (semisetose) or flexible and mostly recurved

        5a Segms of stips rigid, straight, stiffly erect, almost setose; fls often > 1 per axil; stamens 5–8; perianth ca 2.5 (3) mm

          6a Outer tepals mostly < half as long as inner; fls 2–3 (4) per axil; pl mostly < 5 cm; stamens 5–6, only 3 fertile; dry, open flats (often on wastel) to ponderosa pine for; sw Ida and e Ore, from ne Crook Co to Malheur, Harney, and Lake cos; dwarf des k.   **2 P. heterosepalum** Peck & Ownbey

          6b Outer tepals ca = the inner; fls mostly 1 (2) per axil; pl (6) 10–20 cm; fertile stamens mostly 8; dry, gravelly to heavy soil; Klickitat Co, Wn, and WV, Ore, to n Cal, mostly w Cas; Cal k. (*P. greenei*)   **3 P. californicum** Meisn.

        5b Segms of stip slender, not rigid, gen ± crisped or curled; fls 1 per axil; perianth gen reddish, ca 2 (2.5) mm; stamens gen 8, only 3 fertile; pl 2–8 cm; dry areas or vernal pools, in gravelly to heavy soil; Klickitat Co, Wn, s (E Cas) to Cal; Parry's k.   **4 P. parryi** Greene

      4b Fls gen 2–several per axil, ped; lvs various, mostly 1-nerved, petiolate, or jointed near base; stips often entire or only slightly lacerate, segms never stiff and semisetose

        7a Ped becoming sharply recurved, older fls reflexed; pl spreading to erect; racemes elongate; fls not crowded, axillary to much-reduced bracts which are gradually transitional to lvs below   **Group II,** lead 19a

        7b Ped erect or spreading, not sharply reflexed; pl sometimes prostrate; fls often much-crowded in short terminal racemes, often much < the bracts

          8a Fls in terminal spikelike racemes, crowded and mostly overlapping, often much < the subtending bracts; lvs linear to linear-lanceolate or linear-oblanceolate, rarely > 2 mm broad, lateral veins indistinct; st mostly ± sharply angled   **Group III,** lead 23a

          8b Fls mainly axillary to lvs, seldom greatly crowded, upper ones often axillary to and > the reduced bracts; lvs various, often elliptic or oblong to obovate, commonly > 2 mm broad, lateral veins often prominent; st mostly terete and only striate

            9a Lvs elliptic to obovate, pointed at each end, 5–15 (20) mm, often at least half as broad as long, only slightly reduced upward; achenes greenish-black (almost black), shining; pl mostly < 1 (to 2.5) dm; open, semi-barren soil of subalp or alp flats and slopes; BC s, in Cas and higher coastal mts, to

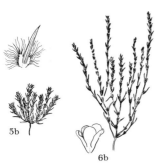

Cal, e to RM, Mont to Colo; broadlf k., lfy dwarf k.
5 **P. minimum** Wats.

9b Lvs various, often rounded, or > 20 mm; achenes often greenish-brown or coal-black; pl often > 1 dm

  10a Achenes black, smooth and shining; perianth connate for only 1/4–1/3 length, segms not yellow-margined
Group IV, lead 29a

  10b Achenes mostly yellow-green to brown, often slightly roughened and not shining, if black then perianth connate for >1/3 length or segms yellow-margined

    11a Tepals yellow-margined, outer 3 strongly cucullate and longer and broader than inner 2

      12a Lvs linear to narrowly oblong, (1) 2–6 cm, at least 5 times as long as broad, gradually reduced upward; floral lvs (bracts) scarcely exceeding fls; pl mostly 1.5–7 (10) dm; moist to dry soil, often on wastel; e Wn and Ore to Cal, e through Ida and Mont to Can and Atl coast; bushy k., yellow-fld k. *(P. latum)*
6 **E. ramosissimum** Michx.

      12b Lvs oblong or elliptic to ovate, 2–4 times as long as broad, only slightly reduced upward; floral lvs greatly exceeding fls; pl mostly 2–4 dm; dry, gen ± wastel; mostly c and e US and Can, but occ in our area, as at White Salmon, Klickitat Co, Wn; erect k. *(P. aviculare* var. *e.)*   7 **P. erectum** L.

    11b Tepals green, white- or pink-margined, outer 3 mostly neither cucullate nor much broader and longer than inner 2   **Group V**, lead 32a

1b Pl per, often rhizomatous, gen growing in or near water, sts often freely rooting

  13a Fls in a terminal, pedunculate head or spicate raceme on a lfless to bracteate or only few-lvd st; pl with short, thick rhizomes; lvs chiefly basal

    14a Racemes mostly < 1.5 cm broad when pressed, gen much > 3 times as long as broad; lower fls (at least) gen replaced by bulblets; pl 1–3 dm; shady woods, meadows, and streambanks to alp slopes; Alas s to n Cas, Wn, Wallowa Mts, ne Ore, and much of Ida and Mont, in the RM to NM, and e to Minn, Me, and Newf; Eurasia; European bistort, alpine b., viviparous b.   8 **P. viviparum** L.

    14b Racemes mostly > 1.5 cm broad when pressed, rarely > 2.5 times as long as broad; bulblets lacking; pl 1–7.5 dm; streambanks and wet meadows to alp slopes in most of our higher mts; BC to Cal, e to Alta, Mont, and NM; Am bistort, w b., snakeweed *(P. glastifolium, P. linearifolium, P. vulcanicum)*   9 **P. bistortoides** Pursh

  13b Fls in axillary and terminal racemes on lfy sts; pl various

    15a Pl erect (escape from cult), strongly rhizomatous, never alp or subalp; ann sts often > 1 m, ± wandlike, freely br; lvs cordate to broadly lanceolate, some much > 6 cm broad   **Group VI**, lead 38a

    15b Pl gen ± decumbent, native, often alp or subalp; ann sts gen much < 1 m; lvs various, gen < 6 cm broad   **Group VII**, lead 41a

Group I

16a Perianth or peduncles, or both, strongly glandular-punctate, glands sessile, much broader than high; lvs ovate to narrowly lanceolate, (3) 5–12 cm

  17a Achenes deep brown or black, smooth and shining; perianth 3–3.5 mm, segms 5; stamens 8; common in much of N and S Am in moist places; throughout our area, but collected mostly w Cas; water smartweed, dotted s. *(P. acre)*   10 **P. punctatum** Ell.

  17b Achenes light brown, granular, dull; perianth (2.5) 3–4 (4.5) mm, segms mostly 4 (3 or 5); stamens gen 4 or 6; European weed, fairly common in much of N Am; in our area in wet places at lower elev; smartweed, marshpepper s.   11 **P. hydropiper** L.

16b Perianth and peduncles not glandular-punctate, although sometimes glandular-puberulent or papillate-glandular, glands mostly few and as high as broad; lvs 3–20 cm

10a 10b

12b

12a

14b

14a

15a

17a

17b

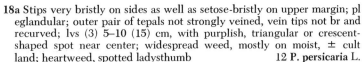

18a Stips very bristly on sides as well as setose-bristly on upper margin; pl eglandular; outer pair of tepals not strongly veined, vein tips not br and recurved; lvs (3) 5–10 (15) cm, with purplish, triangular or crescent-shaped spot near center; widespread weed, mostly on moist, ± cult land; heartweed, spotted ladysthumb   **12 P. persicaria** L.

18b Stips glab to pubescent (not bristly) on sides, gen ciliate on upper margin; pl sometimes ± glandular-puberulent, at least on perianth; outer pair of tepals strongly nerved, vein tips br and recurved; lvs rarely purple-spotted; fairly widely intro European weed; in our area not common, mostly in wet places; willow weed, curltop ladysthumb *(P. incanum)*   **13 P. lapathifolium** L.

Group II

19a Achenes 2–2.5 mm, gen < twice as long as broad, often > the (1.5–2.5 mm) perianth; pl gen 1–2 dm

20a Basal lvs ovate to broadly elliptic, 5–15 mm, gen at least 1/3 as broad as long; pl mostly 5–10 (20) cm; dry to moist flats from sagebr plains into lower mts, often in ponderosa pine for; e Ore to Cal, e to sc Mont and Wyo; Austin's k.   **14 P. austiniae** Greene

20b Basal lvs mostly linear-lanceolate to linear-oblanceolate, < 1/3 as broad as long

21a Pl with ∝ basal, nearly erect brs, commonly 10–20 cm; lvs 10–20 mm, gradually but strongly reduced upward, upper bracts < the fls; dry to moist, well-drained soil; sagebr des to lower mts; chiefly Wyo and Colo, but occ to Alta and w to Ida; Engelmann's k.   **15 P. engelmannii** Greene

21b Pl gen with 1 main st, 5–12 (15) cm, with 1–several divergent basal brs; lvs 5–20 mm, scarcely reduced upward, upper bracts mostly much > the fls; dry, gen rocky slopes, often on serpentine; Cas from Lane to Klamath cos, in the Calapooya Range, and in Grant Co, Ore; Cascadian k.   **16 P. cascadense** Baker

19b Achenes mostly > 2.5 mm and at least twice as long as broad; perianth rarely < 3 mm, gen > the achene; pl 1–4 (5) dm

22a Perianth mostly (3.5) 4–5 mm; lvs linear to oblong, 2–5 (6) cm, scarcely 1/6 as broad; dry, light to heavy soil, sagebr des to ponderosa pine for or mont slopes; E Cas, n Wn to Cal, e to Ida and w Mont; wiry k., Palouse k.   **17 P. majus** (Meisn.) Piper

22b Perianth gen (2.5) 3–3.5 (4) mm; lvs various, but sometimes 1/6 as broad as long; widespread, on dry to moist land, esp in w US; 2 weakly defined vars.   **18 P. douglasii** Greene

a1 Lvs linear or narrowly oblong, rarely > 4 mm broad; range of sp. but not common at higher elev; Douglas' k.   var. **douglasii**

a2 Lvs narrowly to ± broadly oblong, often > 4 mm broad; mont mostly, esp in Cas and OM, s to Cal, not common in Ida and Mont; mt k.   var. **latifolium** (Engelm.) Greene

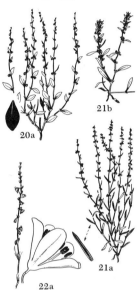

Group III

23a Achenes black, smooth and shining

24a Perianth 2–2.5 mm; achenes ovoid, mostly < 2.5 mm; styles distinct

25a Sts gen 5–15 cm, fl almost full length, upper lvs and bracts strongly overlapping; fls gen 4–5 per axil (see also lead 21b)   **16 P. cascadense** Baker

25b Sts gen 10–35 cm, fl chiefly above middle, lvs rarely overlapping; fls gen 1–3 per axil; dry prairies and open knolls in the lower mts w Cas, BC to ne Ore; Nuttall's k.   **19 P. nuttallii** Small

24b Perianth gen 3–4 mm; achenes ± lanceolate in outline, gen at least 3 mm; styles gen connate (sometimes to midlength); sts up to 5 dm; gravelly to heavy, gen moist soil; BC s, mostly w Cas in Wn, to Cal, e to se Wash and across Ida to Colo; fall k., spurry k.   **20 P. spergulariaeforme** Meisn.

23b Achenes yellow to dark brown, often minutely striate, rarely shining

26a Fl bracts expanded, much < the lower lvs (but broader), broadly white-margined; fertile stamens 8; infl mostly short, often as broad as long; achenes gen dark brown and plainly striate, rarely at all shining; pl mostly 6–20 (25) cm; meadowl and vernal pools to mont ridges and flats;

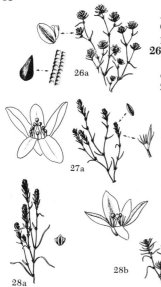

Adams and Spokane to Asotin cos, Wn, Blue and Wallowa mts to Harney Co, Ore, e through Ida to Powell Co, Mont, and possibly to nw Wyo; pokeweed fleecefl, white-margined k.   21 **P. polygaloides** Meisn.

26b Fl bracts rarely both expanded and white-margined; fertile stamens often only 3; infl often narrow, up to 2–3 times as long as broad; achenes often yellow, smooth, and shining

  27a Fertile stamens gen 8; bracts of infl gen not white-margined, mostly 2–3 times as long as fls; pl 2–8 cm; vernal pools and meadows to open flats, mostly in lowl; e Cas, Grant Co, Wn, to Cal, e to Alta, Mont, Wyo, and NM; water k. *(P. esotericum)*   22 **P. watsonii** Small

  27b Fertile stamens often only 3; bracts of infl often white-margined, sometimes 4–6 times as long as fls

    28a Fl bracts gen plainly white-margined, upper ones often no > the fls; pl often > 7 cm; meadows and vernal pools to dry open ground; e Cas, Kittitas Co, Wn, to Cal, e to Ida, c Mont, and possibly nw Wyo, gen, foothills to midmont, except in CRG; closefld k.

                                23 **P. confertiflorum** Nutt.

    28b Fl bracts only slightly if at all white-margined, upper ones gen > the fls; pl mostly < 7 (to 9) cm; meadows and vernal pools to dry subalp slopes; BC s, mostly E Cas, to Cal, e to Mont, Wyo, Colo, and Ariz; Kellogg's k. *(P. minutissimum, P. unifolium)*

                                24 **P. kelloggii** Greene

Group IV

29a Lvs mostly linear or broadest ones sometimes narrowly oblong-lanceolate, gen < 4 mm broad; infl gen closely crowded toward st tips; pl 5–50 cm

  30a Perianth 2–2.5 mm; styles distinct (see also lead 25b)

                                19 **P. nuttallii** Small

  30b Perianth 3–4.5 mm; styles ± connate, sometimes to midlength (see also lead 24b)   20 **P. spergulariaeforme** Meisn.

29b Lvs gen elliptic-oblong to elliptic-oblanceolate, larger ones nearly always > 4 mm broad; infl loose, only slightly if at all congested toward st tips; pls 5–25 cm

  31a Pl gen erect; lvs seldom > twice as long as broad, only slightly reduced upward, concealing fls; alp and subalp, rarely below 3600 ft elev, even in w Wn (see also lead 9a)   5 **P. minimum** Wats.

  31b Pl gen spreading at the base; lvs mostly much > twice as long as broad, much reduced upward, mostly not concealing upper fls; foothills to lower elev in mts, dry to moist areas; nc Cas, Wn, to Cal, e to Ida, Mont, Daks, and Colo; sawatch k.   25 **P. sawatchense** Small

Group V

32a Mature achenes 1–3 mm > the perianth, yellowish-brown, smooth and shining (not striate), slenderly lanceolate-ovoid

  33a Pl erect; perianth 6-lobed, 3 outer segms much the longest and broadest (not cucullate); long-fr k.; reported from our area, but not seen, apparently always e of RM, perhaps only a phase of *P. aviculare*

                                26 **P. exsertum** Small

  33b Pl tending to br basally and to spread or become prostrate; perianth mostly 5-lobed, segms subequal

    34a Perianth ca 3 mm; pls coastal, mostly of strand, but also around salt marshes, often more like per than ann; Alas to Cal, ne N Am; ne Asia; Fowler's k.   27 **P. fowleri** Robins.

    34b Perianth ca 2.5 (2–3) mm; pl weedy, rarely maritime; widespread weed, often on very poor soil; doorweed, prostrate k.

                                28 **P. aviculare** L.

32b Mature achenes no > the perianth, yellow to brown, often ± striate, ovoid

  35a Pl erect, 3–6 dm; brs many, ascending; lvs linear; perianth connate scarcely 1/3 length (see also lead 12a)   6 **P. ramosissimum** Michx.

  35b Pl mostly spreading to prostrate, often < 3 dm, if erect either lvs not linear or perianth connate at least half length

    36a Lvs linear, rarely > 10 mm, obscurely jointed at base; brs strongly angled and mostly papillate-scaberulous, not prominently striate; per-

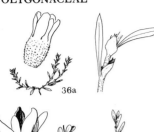

36a

37a

37b

38a

40a

41a

43a

44a

ianth connate ca 1/2 length, gen strongly papillate; stips mostly > the internodes on upper part of st; European weed, known in our area only from near Pine, Elmore Co, Ida
                              29 **P. polycnemoídes** Jaub. & Spach

36b Lvs often > 10 mm or plainly jointed near base; brs often prominently striate, not strongly angled, never papillate; perianth connate 1/3–1/2 length, never papillate; stips gen much < the internodes
    37a Outer 3 tepals cucullate-keeled, longer and broader than inner 2; lvs oval to obovate, rounded (obtuse); achenes smooth, mostly included in perianth; dry ± waste ground, Mont, Ida, e Ore, e and s
                              30 **P. achoreum** Blake
    37b Outer tepals mostly not cucullate-keeled, gen only slightly if at all longer or broader than inner 2; lvs linear to lanceolate or oblanceolate, often ± acute; achenes mostly faintly striate, often exserted; ± universal weed; doorweed, knotweed (*P. littorale, P. neglectum, P. rubescens*) (see also lead 34b)   28 **P. aviculare** L.

Group VI

38a Lvs strongly cordate, without basal lobes; fls ♀, tepals keeled on back, not strongly accrescent; sts up to 4 m; mostly w Cas; an Asiatic, attractive but overly aggressive pl; sachaline, giant knotweed
                              31 **P. sachalinense** Schmidt
38b Lvs mostly truncate to cuneate at base, or cordate but with 2 basal lobes; fls often ♀♂; tepals often not keeled; sts rarely > 2 m
    39a Perianth pink to red; pl strongly pubescent, mostly 7–11 dm, escaping and prob estab w Cas, esp in Seattle, Wn   32 **P. campanulatum** Hort.
    39b Perianth greenish to white, or only slightly pinkish; pl glab, mostly at least 10 dm
        40a Fls ± ♀♂; tepals ca 2 mm at anthesis, becoming strongly accrescent and very prominently winged in ♀ fls; Asiatic, frequently estab w Cas in our area; Japanese k.   33 **P. cuspidatum** Sieb. & Zucc.
        40b Fls ♀; tepals 3–4 mm at anthesis, not accrescent and not winged in fr; estab in Polk Co, Ore, and prob elsewhere
                              34 **P. polystachyum** Wall.

Group VII

41a Sts per, woody-based, covered with hyaline, lacerate stips; along immediate shoreline, mostly on sand dunes, VI to Cal; black k., nailwort k. (*P. confertiflorum*)   35 **P. paronychia** Cham. & Schlecht.
41b Sts mostly ann, herbaceous, not covered with lacerate, hyaline stips; gen, sometimes even subalp or alp
    42a Pl subalp to alp, with enlarged, mostly fleshy crowns or rootstocks; lvs ∝, all cauline and alike, scarcely reduced upward; sts mostly freely br, gen erect
        43a Fls in bracteate to lfy, loose pans; lvs lanceolate, gen in part > 8 cm; sts erect, (5) 8–20 dm; meadows, talus slopes, and mt ridges; Alas s, in Cas of Wn and Ore, to Cal, e Ore, Ida, and w Mont, s to Nev; alpine knotweed, pokeweed, fleecefl (*P. polymorphum*)
                          36 **P. phytolaccaefolium** Meisn.
        43b Fls in small clusters in lf axils; lvs mostly ovate to oblong, rarely as much as 8 cm; sts ascending to erect, < 5 dm
            44a Lvs short-petiolate, glab or soft-pubescent; slopes and ridges in the Cas and OM, Wn, s to Cal, e to Ida; Newberry's fleecefl; 2 vars.
                          37 **P. newberryi** Small
            a1 Lvs ± finely pubescent; Cas of Wn and Ore to Cal; Valley Co, Ida   var. **newberryi**
            a2 Lvs glab or subglab; OM and Wen Mts, Wn var. **glabrum** Jones
        44b Lvs sessile or subsessile, glab to ± scab-puberulent; s Ore to Cal, app but not quite within our area?   38 **P. davisiae** Brewer
    42b Pl of various habitats, if subalp or alp then gen aquatic; sts mostly unbr, often floating or decumbent; lvs various
        45a Fls rose-colored, gen in 1 or 2 terminal or subterminal spikelike pans; achenes lenticular; styles 2–4 mm; perianth not glandular
            46a Peduncles gen glandular-pubescent; infl gen at least 4 cm, cylin-

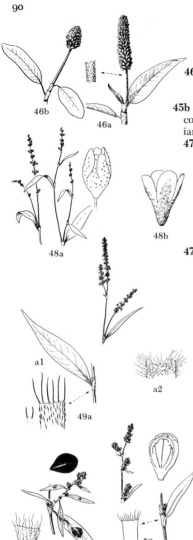

dric; wet places, often partially submerged, amost throughout N Am; common both sides Cas, lowl into mts; water smartweed (*P. amphibium* var. *c.*, *P. oreganum*, *P. cusickii*, *P. chelanicum*)
39 **P. coccineum** Muhl.

46b Peduncles gen glab; infl gen oblong-ovoid or ovoid-conic, mostly < 4 cm; aquatic or semi-aquatic, cosmopolitan; water smartweed or w. ladysthumb (*P. natans*, *P. villosulum*)   40 **P. amphibium** L.

45b Fls mostly white, greenish-white, or pink, borne in several to many contracted pans; achenes mostly triquetrous; styles gen < 2 mm; perianth sometimes glandular

47a Perianth strongly glandular-punctate, glands sessile

48a Achenes brownish-black to black, smooth and shining; perianth gen 5-lobed; stamens gen 8; common in N and S Am, in wet places, in our area mostly w Cas, but not rare elsewhere (see lead 17a)   10 **P. punctatum** Ell.

48b Achenes brown, glandular, dull; perianth mostly 4-lobed; stamens gen 6; European weed, common in much of N Am; wet places throughout our area at mostly lower elev (see lead 17b)
11 **P. hydropiper** L.

47b Perianth not glandular-punctate on the exposed surface (rarely glandular-puberulent)

49a Infl ending in spicate, interrupted, gen paired racemes mostly > 3 cm; pl gen an obvious per; nerves of tepals not br and recurved; mostly in mud, lowl to lower mt valleys; widespread, BC to Mex and Que, rather gen; 2 vars.   41 **P. hydropiperoides** Michx.

a1 Stip hairs appressed, partially adnate; pls commonly only slightly hairy or subglab; common; waterpepper
var. **hydropiperoides**

a2 Stip hairs spreading from pustular base, not adnate; pls mostly very pubescent; rare with us, collected in Okanogan Co, Wn
var. **setaceum** (Baldwin) Gleason

49b Infl gen with several to many short, thick, mostly continuous racemes gen < 3 cm; pl gen ann, sometimes simulating per; nerves of outer tepals sometimes br and recurved

50a Nerves of outer pair of tepals prominent, veins br and recurved at tip; stips mostly not bristly on sides; lf bl not spotted; wet places, not common (see lead 18b)
13 **P. lapathifolium** L.

50b Nerves of tepals not prominent, not br and recurved at tip; stips mostly very bristly on sides; lf bl gen purplish-spotted near center; moist to dry wastel or cult areas; common weed, esp w Cas (see lead 18a)   12 **P. persicaria** L.

Rumex L.   Dock; Sorrel

Infl gen large pan, often lfy-bracteate; fls gen jointed with ped, mostly ♂ (♀, ♂); styles 3, stigmas expanded, peltate, stellate-fimbriate; achenes strongly 3-angled, mostly smooth, brownish to nearly black; ann, bien, or (mostly) per herbs, often reddish-tinged; lvs gen petioled, bls large. (Ancient L name for the pl). Of slight grazing value; many spp. weedy. (*Acetosa*, *Lapathum*).

1a Fls mostly ♀, ♂ (♀♂); lvs either sagittate or hastate or pl mont per with ∝ elliptic to oblanceolate lvs

2a Lf bl tapered at both ends, neither sagittate nor hastate, (4) 6–10 (13) cm; pl alp to subalp, with thick taproot, 1.5–7 dm; outer tepals spreading to erect in fr; meadows and open slopes; BC s, in Cas of Wn and Ore, to Cal, e to Alta, Ida, and Mont to Colo; alp or mt s.
1 **R. paucifolius** Nutt.

2b Lf bl ± hastate to sagittate; ± weedy spp., spreading by extensive shallow roots; outer tepals sometimes sharply reflexed in fr

3a Ped jointed about midlength; lf bl up to 10 cm, sagittate, lobes pointing downward; pl (3) 4–10 dm, not spreading from shallow roots; valves (inner tepals) with basal callosity, much larger than the achene and not closely investing it; outer tepals sharply reflexed in fr; native to Eurasia; used as a salad green and known from nc Mont, but prob elsewhere in our area; garden, kitchen, green, tall, or meadow s.
2 **R. acetosa** L.

3b Ped jointed close to fl; lf bl 1–5 cm, hastate, lobes spreading to divaricate; pl 1.5–4 dm, spreading widely from long slender roots; valves without callosities, not much larger than achene, but closely investing it; outer tepals erect; European weed, common almost everywhere in semiwaste areas and gardens, into the mts; sheep, red, cow, mt, field, or horse s.; sour weed   **3 R. acetosella** L.

1b Fls mostly ♂; lvs varied in shape but never hastate and only rarely sagittate; pl mainly of lowl

4a Pl strongly rhizomatous, forming large patches; lvs gen leathery

5a Valves (inner tepals) much > 1 cm in fr, without callosities; lf bl 6–15 cm; pls 1.5–5 dm, mainly on sand dunes or riverbanks, grassl and sagebr des to lower mt valleys; BC s, e Cas in Wn and Ore (except CRG), to Cal, e to Sask, Daks, Neb, Colo, and NM; veiny d., winged d., wild begonia   **4 R. venosus** Pursh

5b Valves < 1 cm, sometimes with callosities

6a St mostly > 5 dm, unbr below infl; bl of basal lvs much > 10 cm; valves thin, without callosities; Wyo-Colo, tentatively reported for Ida, but not known in our area?   **R. densiflorus** Osterhout

6b St 2–4 dm tall, freely br; bl of basal lvs 3–8 (10) cm; valves thick, with fairly large callosities; S Am, intro here and there, collected several times near Portland, Ore; wedgelvd d.
                                        5 R. cuneifolius Campd.

4b Pl gen not rhizomatous, not forming large patches, gen with strong taproot; lvs various

7a Inner tepals becoming ± strongly toothed or dissected (teeth at least 1 mm), 1 or more with prominent callosity   **Group I**

7b Inner tepals entire to slightly erose or denticulate, never toothed or dissected (teeth < 1 mm), sometimes none with callosity
                                        Group II, lead 12a

Group I

8a Pl ann, 2–7 dm; all valves with a callosity; ped strongly thickened upward to fl, jointed well below midlength; lf bl 2–12 (15) cm

9a Callosities at least 0.5 mm broad, rounded on back; pl coastal

10a Valves mostly 3–4 mm, teeth gen > the width of the non-toothed portion of valve; callosity mostly 0.5–0.7 mm wide and nearly as long as valve; lf bl 4–12 cm; occ in Wn and Ore, Que to Mass; yellow d., seashore d.   **6 R. persicarioides** L.

10b Valves (4) 5–6 mm, teeth < the width of the non-toothed portion of the valve; callosity ca 1 mm wide, scarcely 1/2 as long as valve; lf bl 2–6 cm; European weed, occ on ballast near Portland, Ore; toothed d.
                                        7 R. dentatus L.

9b Callosities rarely > 0.4 mm broad, but gen somewhat thicker, acute at tip, rarely half as broad as the non-toothed portion of valve; lf bl 4–12 (15) cm; gen in wet places; Alas s, at lower elev on both sides of Cas, to Cal, e to Que and NJ; S Am; Eurasia; seaside d., golden d. *(R. feuginus)*
                                        8 R. maritimus L.

8b Pl per, 4–12 dm; some valves without callosity; ped not strongly thickened upward, jointed near midlength; lf bl 5–20 (35) cm

11a Ped thick, mostly no > the fr valves, prominently jointed ca midlength; mature callosities coarsely warty; lf bl 5–15 cm; pan brs ± spreading; moist to dry, ± waste areas; common in se and sw US, known in our area in WV, Ore; fiddle d.   **9 R. pulcher** L.

11b Ped slender, mostly > the fr valves, jointed somewhat below midlength; mature callosities smooth or very slightly wrinkled; lf bl 10–20 (35) cm; pan brs mostly ascending to erect; cosmopolitan weed of moist roadsides and wastel; common w Cas; bitterd., butterd., broad-lvd d.
                                        10 R. obtusifolius L.

Group II

12a Fr valves all without callosity

13a Sts br from lower nodes, 2–6 dm; valves seldom > 3 mm; basal and lower st lf bl 5–10 (15) cm, mostly at least 5 times as long as broad, acute (rounded) at base, margins plane to undulate but not crisped (see lead 14b)   **16 R. salicifolius** Weinm.

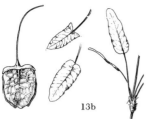

13b Sts simple below infl, (5) 10–20 dm; valves 3–10 mm; basal and lower
st lf bl 10–30 cm, < 5 times as long as broad, some gen truncate to cordate at base, ± crisp-margined; moist to swampy areas, lowl to mont;
Alas to Cal, e to Que and Daks, and in RM to NM; western d.; 2 vars.
11 R. occidentalis Wats.

  a1 Mature valves (6) 7–10 mm; in our area mostly in Wn and Ore, both
sides Cas, to Ida, rare in RM *(R. confinis, R. fenestratus, R. p.)*
var. **procerus** (Greene) Howell

  a2 Mature valves (3) 4–5 (6) mm; mainly RM, occ w to Wn and Ore
var. **occidentalis**

12b Fr valves all or in part with prominent callosity

  14a Sts unbr below infl; petioles mostly ± papillate- or scabrid-puberulent;
callosities often reticulate-pitted

   15a Valves gen cordate, (5) 6–7 (9) mm, gen at least twice as long and 6
times as broad as callosity; basal and lower st lf bls ovate to oblong-lanceolate, ± truncate or cordate at base, up to 30 cm; pl (6)
10–20 dm; occ European weed, chiefly in se Wn; patience d.
12 R. patientia L.

   15b Valves gen triangular-ovate or oblong, 2.5–5 (6) mm, < twice as
long and 6 times as broad as callosity; basal and lower st lf bls gen
rounded to acute at base

    16a Pan open, verticils not contiguous; valves oblong-ovate (ovate),
often much < twice as long and < 3 times as broad as callosity; lvs
not strongly crisp-margined; pl gen 6–15 dm

     17a All valves gen tubercled; callosity oblong, nearly twice as long
as broad; brs of infl lfy to above midlength; ped stout, ca = calyx;
European weed, known from w Cas, BC to Cal; clustered d.
13 R. conglomeratus Murr.

     17b Only 1 valve gen tubercled; callosity semiglobose, nearly as
broad as long; brs of infl lfy only below midlength; ped slender,
much > calyx; weed of e US, known in w Wn and Ore; red-
veined d.   **14 R. sanguineus** L.

    16b Pan dense, the verticils ± contiguous, at least toward br ends;
valves ovate-cordate to triangular, ca twice as long and 3–5 times as
broad as callosity; lvs strongly crisp-margined; pl 5–10 dm; European weed common in most of N AM, in our area from lowl to lower
mts; curly, sour, or yellow d.   **15 R. crispus** L.

  14b Sts with axillary, often much reduced brs at some or all nodes below
the infl; petioles rarely scabrid- or papillate-puberulent; callosities not
reticulate-pitted; coastal dunes to mt meadows and rocky slopes; Alas to
Baja Cal, Que to Me, NY, Tex; willow d., narrow-lvd d.; several poorly
defined vars.   **16 R. salicifolius** Weinm.

   a1 Valves thin in texture, at least 1 gen almost covered by large callosity
(> 2/3 as long and broad as valve); mostly w Cas in our area; variable in size of valves and of callosities, often treated as several vars.
or spp. *(R. crassus, R. transitorius)*   ssp. **salicifolius**

   a2 Valves thickened, gen triangular, mostly ± denticulate, all 3 (or 0)
gen with callosity little > half as long as valve and < 1/2 as broad;
chiefly in and on e side Cas in our area; 4 vars.
ssp. **triangulivalvis** Danser

    b1 Callosities well developed

     c1 Valves ca 4 mm; se Ore to Mex *(R. m.)*
var. **mexicanus** (Meisn.) Hitchc.

     c2 Valves ca 3 (3.5) mm; widespread, Cas and e *(R. t.)*
var. **triangulivalvis**

    b2 Callosities not well developed, often lacking

     d1 Callosities lacking; valves broadly triangular, as broad as long;
Yuk to se BC and sw Alta, Mont, Wyo, and Colo, w to Ida, Utah,
ne Ore, and Cal *(R. hesperius, R. utahensis)*
var. **montigenitus** Jeps.

     d2 Callosities present but < 1 mm broad; valves narrowly triangu-
lar, ca 2/3 as broad as long; Cas and e to c Wn
var. **angustivalvis** Danser

CHENOPODIACEAE Goosefoot Family

Fls small, glomerate (1–∞) in lf axils or in bracteate or ebracteate spikes, pans, or cymes, ⚥ to ⚥♂ or ♀, ♂, apet; perianth mostly greenish, persistent, 5 (2–6)-lobed or -parted (or a single scale) or rarely lacking in the ♀ fls; stamens gen = the tepals and opp them (fewer), free or ± connate; pistil 2–3 (5)-carpellary, ovary 1-celled and 1-ovuled, superior or rarely with perianth adnate; fr indehiscent or irreg rupturing (utricle), sometimes several cohering by fleshy perianths; ann or per herbs to fairly large shrubs, often ± fleshy, glab to pubescent, mostly with mealy excrescence on lvs and younger brs; lvs simple, alt (opp), sometimes scalelike; sts sometimes jointed.

1a St jointed; lvs small alt or opp scales; fls sessile, borne in depressions of joints of fleshy spikes or along st
  2a Lvs and brs opp; embryo folded; endosperm lacking   **Salicornia**
  2b Lvs and brs alt; embryo ± annular, surrounding an abundant endosperm; reported for Ore and Ida but not known from our area
                                          Allenrolfea occidentalis (Wats.) Kuntze
1b St not jointed; lvs flattened to terete but not scalelike, sometimes linear
  3a Bracts and lvs strongly spinulose at tips; fls ⚥; fr perianth broadly winged crosswise   **Salsola**
  3b Bracts and lvs not spinulose, spines (if any) on st and cauline in origin; fr perianth gen not cross-winged, but if so fls at least partially ⚥♂
    4a Lvs opp; tepals strongly overlapping; rhizomatous, low, per herbs with linear, entire, fleshy lvs   **Nitrophila**
    4b Lvs mostly alt; tepals rarely overlapping; pl seldom rhizomatous
      5a Pls spinose-br shrubs with linear, semiterete lvs; embryo spirally coiled; endosperm lacking; fls spicate, ⚥♂, the ♂ uppermost, naked, with 2–3 stamens nearly covered by a long-stalked, peltate scale; ♀ fls 1–2 in axils of lflike bracts, pistil surrounded by cuplike, shallowly lobed to nearly entire perianth which becomes enlarged and top-shaped in fr, its upper portion flaring to form a winged border   **Sarcobatus**
      5b Pls herbs or shrubs, if shrubby and at all spinose then lvs flattened, embryo annular, and endosperm present
        6a Those lvs present at anthesis linear to narrowly lanceolate, entire, densely hairy, sometimes stellate; pl grayish to rufous; embryo annular, surrounding endosperm   **Group I**
        6b Those lvs remaining at anthesis not at once linear or linear-lanceolate, entire, and grayish-hairy; embryo sometimes spiral
                                                      Group II, lead 9a

Group I

7a Pl ann; tepals all or in part with dorsal tubercle or (in fr) a straight to hooked spine; fls ⚥ and ♀, 1–∞ in lf axils and in lateral spikes; perianth depressed-globose, enclosing fr; fr much-flattened horizontally   **Bassia**
7b Pl per; tepals neither tubercled nor spinose, but sometimes cross-winged
  8a Pubescence mostly stellate; fls ⚥♂, the ♀ enclosed in 2 ± connate bractlets, but otherwise naked; seed erect; ♂ fls 4-merous; shrubs with alt, entire, narrow lvs   **Eurotia**
  8b Pubescence simple; fls mostly ⚥, all 5-merous, tepals finally ± carinate and cross-winged; seed horizontal   **Kochia**

Group II

9a Embryo spirally coiled; endosperm scant or lacking; lvs linear and ± terete; fls all axillary, borne singly, in clusters of 2–5 along main st, or in lateral spikes; fr tepals cross-winged, cross-corrugate, or ± horned
  10a Lvs with a short, apical, conical, mostly bristle-tipped tubercle; fr tepals strongly cross-winged near tip; infl somewhat pilose; alt-lvd ann herb   **Halogeton**
  10b Lvs without a terminal bristle-tipped tubercle; fr tepals cross-corrugated or dorsally horned; infl not pilose; ann or per   **Suaeda**
9b Embryo annular; endosperm gen abundant; fls often in crowded spikes or pans; perianth often not winged, cross-corrugate, nor horned (sometimes lacking)
  11a Perianth lacking, or consisting of a single (2–3) bractlike tepal < (and

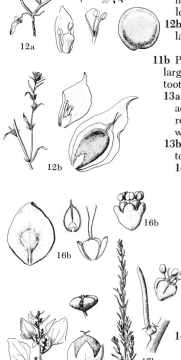

not enclosing) fr; seed vertical; ann, lvs entire to hastately lobed

12a Pl prostrate to ascending, gen < 1.5 dm, ± farinose to glabrate, never stellate-pubescent; fls 1–several per axil; lvs sometimes hastately lobed; fr not broadly wing-margined   **Monolepis**

12b Pl erect to strongly axcending, gen > 1.5 dm, not farinose, gen stellate-pubescent, at least in infl; fls 1 per axil; fr broadly wing-margined   **Corispermum**

11b Perianth 3–5-lobed; fr at least partially enclosed by perianth or by large subtending bracts; seed often horizontal; ann or per; lvs often toothed

13a Root mostly enlarged and fleshy; perianth hardened at base and adherent to ovary and often also to other fls; stamens perig; glab, often reddish herb with ovate to oblong lvs; cult beet, often persistent on waste ground   **Beta vulgaris** L.

13b Root rarely enlarged and fleshy; perianth not hardened and adherent to ovary or adj fls; stamens hypog; pl various

14a Fls ♀♂, the ♀ naked or with greatly reduced perianth but each subtended and enclosed by 2 accrescent, sepaloid bracteoles; ♂ fls ebracteate but with 3–5-lobed perianth

15a Stigmas 4–5; glab herbs; lf bls triangular-ovate, 5–12 cm; common spinach, sometimes persistent in old garden areas   **Spinacia oleracea** L.

15b Stigmas 2–3; herbs or shrubs; lf bls various

16a Pl ann; lvs orbicular to broadly ovate, dentate; bractlets of ♀ fls laterally compressed, narrowly crenulate-winged dorsally, strongly bidentate at tip; e Mont, prob not quite reaching our area   **Suckleya suckleyana** (Torr.) Rydb.

16b Pl ann or per, often shrubby, if ann the lvs rarely ovate and bractlets of ♀ fls dorso-ventrally compressed and not dorsally winged   **Atriplex**

14b Fls mostly ⚥ and with reg, (3–4) 5-lobed perianth

17a Pl ann; perianth not winged in fr   **Chenopodium**

17b Pl per; perianth becoming carinate and cross-winged in fr   **Kochia**

Atriplex L.   Orache; Greasewood; Shadscale; Saltbush; Silverscale

Fls ♀♂ or ♀, ♂, single to glomerate in the axils or in simple to pan spikes; ♂ fls ebracteate, perianth (3–4) 5-merous; stamens ± adnate to tepals, sometimes ± connate below, bracts of ♀ fls distinct to connate, often laciniate or toothed or appendaged on back; stigmas 2–3; embryo annular, surrounding endosperm, radicle pointing away from base of ovary (superior), toward it (inferior), or intermediately (lateral or ascending); herbs or shrubs, often ± spiny, glab to strongly farinose. (L name for the pl). *(Grayia)*.

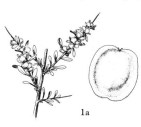

1a Pl stellate-scurfy on new growth; bractlets of ♀ fls connate and closely investing fr, much compressed laterally, wing-margined, nearly orbicular; grayish-barked shrubs up to 1.5 m, with entire, oblanceolate lvs mostly 1–2.5 (3.5) cm; sagebr des (in our area) e Cas; e Wn to s Cal, e to sw Mont, Wyo, and Colo, often in alkaline soil; spiny hopsage *(Chenopodium s., Grayia s.)*   1 **A. spinosa** (Hook.) Collotzi

1b Pl not stellate-scurfy; bractlets of ♀ fls mostly not connate, sometimes flattened on back; herbs or shrubs, if shrubby then lvs mostly not oblanceolate and sometimes toothed

2a Pl per shrubs, ♀, ♂; main brs woody, ± rigid, sometimes spine-tipped; lvs entire

3a Lateral brs rigid, gen sharply spinose; main lvs mostly petiolate, bl elliptic to broadly ovate or obovate, 1–2 cm, at least half as broad; ♀ bracts connate < half their length, neither winged nor appendiculate on back; freely br shrubs 4–8 (10) dm; ± alkaline des plains and foothills; nc Ore to Cal, e through s and c Ida and s Mont to Daks, Colo, and n Mex; sheepfat, sh., spiny sa. *(A. spinosa, A. subconferta)*   2 **A. confertifolia** (Torr. & Frem.) Wats.

3b Lateral brs mostly not very rigid, rarely spinose; main lvs sessile or subsessile, bl mostly linear to narrowly oblong or narrowly obovate, mostly > twice as long as broad; ♀ bracts either strongly winged or appendiculate on back, connate > half length

4a Pl woody at base only, 1—5 dm; lvs linear-spatulate to narrowly obovate, 2–5 cm × 2–10 mm; fr bractlets not prominently winged; mostly where alkaline, e Wn to nw Cal, e through s Ida and Mont to Sask, Daks, NM, and Ariz; saltsage, moundscale; 2 vars.
                                                             3 A. nuttallii Wats.
  a1 Lvs linear-spatulate, 2–6 mm broad; bracts fusiform, with beak up to 2 mm, often short-stalked; se Wn to Cal, e to Mont, Utah, and Nev (A. f.)   var. **falcata** Jones
  a2 Lvs oblong to narrowly obovate, mostly 5–10 mm broad; bracts ovate to subrotund, often sessile, beak lacking or broad and much < 2 mm; e Ida to Sask, Mont, Wyo, Colo, and Utah (A. oblanceolata)   var. **nuttallii**
4b Pl woody almost throughout, (2) 4–20 dm; lvs elliptic to spatulate or linear, (1) 2–5 cm × 2–8 (12) mm; fr bractlets ± prominently winged; saline areas, e Wn and Ore to Baja Cal, e to Alta, Daks, Kans, Tex, and Mex; wingscale, fourwing sa.; hoary sa.; shadscale; 2 vars.   **4 A. canescens** (Pursh) Nutt.
  a1 Fr bracts mostly 4–6 (8) mm; pl low and spreading, 1–4 dm, mostly e of RM, but in our area in ec Mont
                                                       var. **aptera** (Nels.) Hitchc.
  a2 Fr bracts mostly 6–20 mm; pl ± erect, gen much > 4 dm; common phase almost throughout range of sp.   var. **canescens**
2b Pl herbaceous ann, ♀♂; lvs often toothed
  5a Lvs opp below, gen alt above, often glabrate, mostly not grayish-mealy on either surface, meal (if any) covering < half the surface, therefore lvs greenish   **Group I**
  5b Lvs mostly or all alt, grayish-mealy on lower surface, at least, the meal gen covering most of surface   **Group II**, lead 10a

Group I

6a Fr bracts ovate to ± orbicular, (3) 5–12 mm broad, entire-margined and without tubercles on back
  7a Fr bracts 6–12 mm broad, 1-veined at base; lf bls ovate-triangular to broadly lanceolate, 5–20 × 2–10 cm, often dentate; Old World cult pl, occ escape reported from various parts of our area; orache, garden o., sea purslane   **5 A. hortensis** L.
  7b Fr bracts mostly 3–5 mm broad, 5-veined at base; lf bls ± triangular-hastate, cauline ones mostly ave 6–7 cm and as broad, entire to dentate; Eurasian, intro and weedy in many places and reported from our area in Okanogan Valley, Wn, and from Mont   **6 A. heterosperma** Bunge
6b Fr bracts in gen elongate, never > 5 mm broad, often denticulate-margined or tubercled on back

  8a ♂ fls mostly clustered in terminal or axillary spikes; tepals with small fleshy crest on back; bracts of ♀ fls connate to the tip, compressed, ovate, ca 2 mm, fls with tiny, membranous, 3–4-merous perianth; pl 1–3 dm; lvs 1–3 cm × 3–10 mm, entire; alkaline and badland regions; mostly e RM, in our area in sc Mont; rillscale   **7 A. dioica** (Nutt.) Macbr.
  8b ♂ fls mostly mixed with ♀; tepals not crested on back; bracts of ♀ fls connate much < half their length, 3–20 mm in fr, fl often naked; pl 1–10 dm; lvs 2–10 cm, entire to dentate
9a Radicle pointing upward in relation to base of ovary; fr bracts deltoid to lanceolate and with rounded basal lobes; ♀ fls sometimes with small membranous perianth; se Ore to Cal, Nev, and Utah, prob not quite extending into our area   **A. phyllostegia** (Torr.) Wats.
9b Radicle pointing downward; fr bracts either hastately or not at all lobed at base; ♀ fls naked; pls (1) 2–10 dm; lvs 2–10 cm, entire to dentate; common, saline or alkaline soil, along coast and inl, almost throughout temp and arctic N Am as well as in the Old World; several vars.   **8 A. patula** L.
  a1 Lvs linear, rarely > 4 mm broad, entire; ♀ bracts narrowly lanceolate to linear
    b1 Bracts mostly denticulate and tubercled on back, 3–4 mm; very occ intro in our area; shore o. (A. l.)   var. **littoralis** (L.) Gray
    b2 Bracts entire, not tuberculate, 8–12 mm; coastal from s BC to Alas, rarely reaching our area (A. z.)
                                              var. **zosteraefolia** (Hook.) Hitchc.

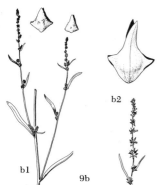

d1

c1

9b

d2

a2 Lvs mostly oblong to lanceolate or ovate, often hastate or > 4 mm broad, sometimes denticulate; ♀ bracts lanceolate to deltoid, often denticulate or tuberculate on back
  c1 Lf bls at least in part triangular-hastate; ♀ bracts gen denticulate, and tuberculate on back; throughout much of US but not typically along the coast; fat-hen *(A. carnosa, A. h., A. lapathifolia)* var. **hastata** (L.) Gray
  c2 Lf bls seldom hastate; ♀ bracts entire to denticulate-margined, gen smooth on back
    d1 Bracts of ♀ fls entire-margined, (4) 5–12 mm, gen smooth on back; lvs lanceolate to oblong, not lobed; coastal from Alas to Cal *(A. gmelini)* var. **obtusa** (Cham.) Hitchc.
    d2 Bracts mostly denticulate, 3–6 mm, often tuberculate on back; lvs various, often denticulate or slightly divaricate-lobed at base; range of sp.; common or spear o. var. **patula**

Group II

10a Lvs 3–12 mm, sessile, entire, ovate to elliptic, all but lowest pair alt, overlapping along st; pl mostly 5–15 (25) cm; fls single or paired in upper lf axils; ♀ bracts connate nearly to tip, ovate (broadest near base), 1–2 mm, entire, smooth; alkaline flats; se Ore to ne Cal and n Nev, app but not presently known in our area **A. pusilla** (Torr.) Wats.
10b Lvs mostly either > 12 mm, or petiolate, or toothed, sometimes more than the lowest pair opp; pls gen > 25 cm; fls gen ± glomerate; ♀ bracts gen either broadest above midlength, or > 2 mm and tuberculate or denticulate
  11a Fr bracts 2–3 mm, connate to tip, cuneate, truncate or ± rounded and slightly 2–3 (5)-toothed across tip, mostly smooth on back; pl 2–12 (20) dm; lvs sessile to short-petiolate, truncate to subcordate or ± hastate at base, entire to undulate, but not toothed, 2–5 (7) cm; gen on alkaline soil; BC s through e Wn and Ore to Cal, e to Mont, Colo, and NM; wedgescale o. **9 A. truncata** (Torr.) Gray

11a

13a

13b

  11b Fr bracts mostly > 3 mm, not at once cuneate and with subtruncate tip, often toothed or tuberculate on back; lvs often either cuneate at base, or petiolate, or with toothed margins
    12a Mature bracts of ♀ fls 3–4 mm, broadly spatulate or oblong, ending in an entire greenish lobe; lvs prominently 3-veined, entire, lower ones slender-petiolate; Alta and Daks through e Mont to Utah, NM, and Ariz, just possibly reaching our area in e Mont **10 A. powellii** Wats.
    12b Mature ♀ bracts 4–10 mm, varied in shape; lvs gen not prominently 3-veined but when so with additional lateral nerves gen evident and margins mostly ± toothed
      13a Pl rounded, mostly 2–5 (8) dm, not weedy; lf bls entire or slightly toothed, lanceolate to deltoid-ovate, 2–5 (6) cm; ♀ bracts connate > half length, often stipelike at base, broadest above midlength, with greenish, foliaceous, gen strongly laciniate margins and strongly tuberculate-appendaged and greatly hardened sides; radicle superior; mostly on alkaline soil; se Ore to Cal, e to Sask, Daks, and NM, in our area in Ida and Mont; silverscale **11 A. argentea** Nutt.
      13b Pl erect, mostly not rounded, 2–20 dm, ± weedy; lf bls dentate, lanceolate to ovate or subrotund, 2–4 (7) cm; ♀ bracts connate < half their length, broadest at or below midlength, greenish margin mostly dentate, sides not greatly hardened and mostly with short tubercles; radicle intermediate between superior and inferior; intro from Old World; weedy, esp along roadsides and wastel, mostly e Cas; BC to Cal, Ida, and Mont; red o. *(A. spatiosa)* **12 A. rosea** L.

Bassia Allioni   Bassia

Fls ♂ and ♀, 1–∞ in lf axils and in lateral spikes; perianth depressed-globose, enclosing fr, segms all or in part with dorsal tubercle or (in fr) with a straight to hooked spine; ann herbs with alt, entire, linear or narrowly lanceolate, sessile lvs, mostly ± pilose or tomentose. (For Ferdinando Bassi, 1710–1774, Italian botanist).

B. hyssopifolia (Pall.) Kuntze. Pl erect, up to 2 m, sericeous and ± lanate, esp in infl; lvs 1–4 cm × 1–2 (4) mm, each tepal with a rather stout, curved to hooked spine; Eurasian weed, well estab in irrigated areas e Cas; BC to Cal, Ida, Mont, and Nev. *B. hirsuta* (L.) Asch. may occur in our area also, but has not been seen; it has shorter, straight spines on some (but not all) tepals.

Chenopodium L.   Lamb's Quarters; Goosefoot; Pigweed

Fls ⚥, greenish, mostly ± glomerate in terminal and often lateral crowded infl; ours ann, rather freely br, glab to farinose or glandular-hairy herbs; lvs alt, ± fleshy, entire to hastate or toothed, often becoming reddish- or purplish-tinged. (Gr *chen*, goose, and *podos*, foot, in reference to lf shape of some spp.). *(Botrys)*.

1a Pl ± resinous with sessile to stalked glands, strongly aromatic

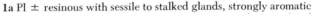

  2a Perianth strongly reticulate, tubular, ± turbinate, shortly 4–5-lobed, eglandular and gen not pubescent; seed erect; fr laterally flattened; widely distributed weed from S Am, reported on ballast here and there, but apparently not estab   **C. multifida** (L.) Moq.

  2b Perianth not strongly reticulate, mostly camp and lobed at least to mid-length, either glandular or hairy; seed often horizontal, fr then flattened on top

    3a Perianth glab or lightly puberulent, gen not glandular; seeds mostly horizontal; fr flattened on top; fls sessile and ± glomerate in large pans of short spikes; pl up to 1 m; lf bls rhombic-ovate to lanceolate, up to 10 cm, sinuate-pinnatifid to undulate-dentate; occ weed in Wn and Ore, widely distributed; Mexican tea wormseed *(Atriplex anthelmintica)*   1 **C. ambrosioides** L.

    3b Perianth strongly glandular, gen also pubescent; seeds either erect and fr flattened from the sides, or fls short-pedicellate or cymose; pl mostly <6 dm; lvs various

      4a Seed erect, fr laterally flattened; fls in axillary clusters; fl lvs gradually reduced; Australian weed known from se Ore; to be expected in our area   **C. pumilio** R. Br.

      4b Seed horizontal, fr flattened on top; fls in ∝ small, axillary, dichotomously br cymes; pl sweetly aromatic, 1.5–4 (6) dm; leaf bls ovate to oblong, 2–4 (5) cm, shallowly to deeply sinuate-lobed; fl lvs greatly reduced or lacking; common weed from Eurasia, mostly along roadways, waste areas, or gravelly streambanks; Jerusalem-oak, Jerusalem-o. g., feather geranium   2 **C. botrys** L.

1b Pl eglandular (except infl of *C. hybridum*), often glab, never strongly aromatic

  5a Seed gen erect, fr laterally flattened; lvs large, triangular, often hastate, greenish on both sides; fls all glomerate in axils and in single terminal spike, or in both axillary and terminal, simple to compound spikes; perianth often red at maturity   **Group I**

  5b Seed gen horizontal, fr flattened at top, if seed erect (as sometimes) then lvs either farinose or not hastate; fls not all in axillary glomerules or axillary and terminal spikes; perianth not reddish at maturity

                                                                    Group II, lead 9a

Group I

6a Glomerules < 4 mm broad, aggregated into ∝ axillary, simple to compound spikes; perianth not reddish or fleshy; pl 1–5 (10) dm; lf bls deltoid to rhombic-ovate, often ± hastate, 1.5–6 cm; widely distributed weed in our area, often on moist saline soil; red g. *(C. humile)*   3 **C. rubrum** L.

6b Glomerules often > 4 mm broad at maturity of pl, mostly in single terminal spike, but sometimes in compound spike, fls terminal on main brs; perianth often reddish and fleshy; pl 1.5–4 (6) dm; lf bls ± triangular, hastate, entire to sinuate-dentate or lobed, up to 10 cm

  7a Perianth becoming reddish and fleshy in fr; all seeds erect

    8a Spikes lfless or with greatly reduced bractlets above; seeds oblong in outline, ca 0.9 × 0.6 mm, margins acute; styles 0.4–0.6 mm; semi-weedy sp., widespread in our area; Alas to Cal, e to Que, Minn, NJ, and NM; strawberry blite *(Blitum c., B. hastatum)*

                                                                    4 **C. capitatum** (L.) Asch.

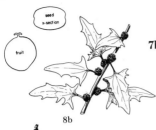

8b

8b Spikes lfy-bracteate to near tip; seeds ± rotund in outline, 1–1.2 mm; Eurasian weed known in our area from Wn, Ore, and Ida; lfy g. (*Blitum virgatum*) **5 C. foliosum** (Moench) Asch.

7b Perianth greenish to ± pinkish, but not fleshy, in fr; some of seeds horizontal in terminal and subterminal glomerules; pl spreading to erect, sts 3–6 dm; lf bls triangular to broadly lanceolate, gen ± hastate, often sinuately lobed, (2) 4–15 cm; sagebr des to mt slopes, often roadside weed; BC to Cal, e to Sask, Nev, and Colo; Eurasia; red g. (*C. rubrum*, misapplied; *Blitum c.*) **6 C. chenopodioides** (L.) Aellen

Group II

7b

9a Pl prostrate or low and spreading (erect), st 1–4 dm; lf bls lanceolate to oblong-ovate, green above, grayish-farinose beneath, 1–3 (5) cm, sinuate-dentate; seeds erect; widespread weed, esp on saline or alkaline soil; in our area chiefly in Mont, s Ida, lower CR, and in se Ore; glaucous g., oaklf g. (*C. salinum*) **7 C. glaucum** L.

9b Pl gen erect; lvs various but if grayish-farinose beneath then seeds all horizontal

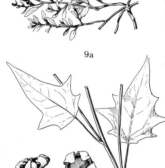

9a

12a

    10a Lvs large, mostly 3–20 cm, ovate to deltoid, only slightly if at all farinose, gen green on both surfaces; tepals in fr not strongly keeled along midvein; pericarp tightly adherent to the lightly sculptured seed

      11a Fr rounded on margins; seeds apparently smooth but reticulate-lined; tall, nearly glab weed; occ in much of N Am, reported from Ore **C. urbicum** L.

      11b Fr sharply angled on margins; seeds finely pitted

        12a Lf bls acuminate, sinuately 2–4-toothed or shallowly lobed on each side, (5) 7–20 cm, rounded to cordate at base; fls ± glomerate, in slender, open, pan spikes or cymes, slightly or not at all farinose; seeds (1) 1.5–2.5 mm broad; pl 2–15 dm; widespread weed in much of our area; sowbane, maple-lvd g. (*C. gigantospermum*) **8 C. hybridum** L.

        12b Lf bls acute to rounded, coarsely several-toothed (more than 4 teeth per side), 2–6 (8) cm, cuneate to semitruncate at base; fls single to glomerate in rather compact pans, gen thickly farinose; seeds 1–1.5 mm broad; pl spreading to erect, 2–5 dm; common weed, esp along roadsides; sowbane, nettle-lvd g. **9 C. murale** L.

12b

b2

a1

b1

15a

    10b Lvs various, often densely farinose, sometimes entire; tepals in fr often strongly carinate on midvein; pericarp sometimes loose on seed; seed often smooth and shining

      13a Lvs greenish (not farinose) on both surfaces, entire to undulate, ovate to lanceolate; seeds ± undulately ridged, margin rounded; fr tepals not strongly keeled on midvein; weedy sp. reported from Ore **C. polyspermum** L.

      13b Lvs mostly either toothed or grayish-farinose beneath; fr tepals gen strongly keeled on midvein; seeds often smooth

        14a Lf bls entire to hastately 1–2-lobed at base, linear or oblong to triangular-ovate, margins mostly not toothed; pericarp not tightly adherent to seed

          15a Lvs linear or narrowly lanceolate, only lightly if at all hastately lobed, mostly at least 4 times as long as broad, (1) 2–4 (5) cm × (1) 2–4 (6) mm; des e of Cas; BC to Baja Cal, e to Atl as weed; slimlf goosefoot; several vars. **10 C. leptophyllum** (Moq.) Wats.

            a1 Lvs greenish, lightly farinose, mostly 1-nerved; occ, e Wn and Ore to the Daks (*C. s.*) **var. subglabrum** Wats.

            a2 Lvs mostly white-farinose beneath, sometimes 3-nerved

              b1 Lvs 1-nerved, mostly < 4 mm broad; range of sp. (*Botrys l.*) **var. leptophyllum**

              b2 Lvs in part 3-nerved, often > 4 mm broad; with var. *leptophyllum*, but more common to s and e of our range (*C. inamoenum, C. dessicatum, C. pratericola, C. o.*) **var. oblongifolium** Wats.

          15b Lvs broadly lanceolate or ovate to triangular, often with large hastate lobes, 1–4 (6) cm, 1/3–3/4 as broad; sagebr des to lower mont for; BC s, e Cas, to Baja Cal, e to Daks and Tex; Fremont's g.; 3 vars. **11 C. fremontii** Wats.

            a1 Lvs lanceolate, gen not hastate, mostly thin, often only sparsely farinose and greenish beneath; throughout our area

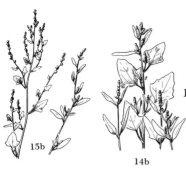

15b

14b

(*C. a., C. wolfii*) var. **atrovirens** (Rydb.) Fosberg
a2 Lvs mostly triangular and hastate, commonly grayish-farinose beneath
  b1 Pl low and spreading, 1–4 dm, densely farinose throughout; prob always to e of our area *(C. i.)* var. **incanum** Wats.
  b2 Pl mostly erect, up to 8–10 dm; upper surface of lvs gen greenish; most common phase with us var. **fremontii**
14b Lf bls mostly toothed on margin and often hastate (teeth sometimes inconspicuous), sometimes lanceolate but gen ovate to cuneate-rhombic, (2) 3–10 cm; pericarp gen tightly adherent to seed; ubiquitous weed of gardens and wastel; lambsquarter, white g. or pigweed *(C. berlandieri, C. covillei, C. lanceolatum, C. paganum)*
  12 **C. album** L.

Corispermum L. Bugseed

Fls ⚥, conspicuously bracteate; perianth a single (2–3) papery, erose tepals between fl and axis of infl; stamens (1–2) 3–5; styles 2; fr strongly flattened, wing-margined; ann, gen freely br, glab to stellate-hirsute; lvs alt, linear, entire, sessile. (Gr *koris*, bedbug, and *sperma*, seed, because of appearance of seed).

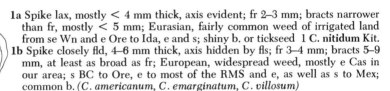

1a Spike lax, mostly < 4 mm thick, axis evident; fr 2–3 mm; bracts narrower than fr, mostly < 5 mm; Eurasian, fairly common weed of irrigated land from se Wn and e Ore to Ida, e and s; shiny b. or tickseed 1 **C. nitidum** Kit.
1b Spike closely fld, 4–6 mm thick, axis hidden by fls; fr 3–4 mm; bracts 5–9 mm, at least as broad as fr; European, widespread weed, mostly e Cas in our area; s BC to Ore, e to most of the RMS and e, as well as s to Mex; common b. *(C. americanum, C. emarginatum, C. villosum)*
  2 **C. hyssopifolium** L.

1a 1b

Eurotia Adans. Winterfat; White Sage; Winter Sage

Fls ⚥♂, the ♂ ebracteolate, perianth 4-merous, ♀ naked but enclosed in 2 villous, partially connate, slightly keeled bracteoles with divergent and hornlike tips; stellate-tomentose, alt-lvd shrublets. (Gr *euros*, mold, because of grayish, moldlike vesture).

E. lanata (Pursh) Moq. Shrubby base up to 2 dm; ann herbaceous sts up to 5 dm, stellate and villous-tomentose, grayish but becoming rufous; lvs linear to narrowly lanceolate, ± fascicled; plains and foothills, often in saline or alkaline areas; e Wn and Ore to Cal, e to Ida and Mont, and farther n, e, and s.

Halogeton C. A. Meyer Halogeton

Fls ♂⚥♀, bracteolate; perianth deeply 5-parted, persistent and often cross-winged near tip; ann with slender bracts and alt, terete to narrowly oblong, fleshy, setiferous, pilose lvs. (Gr *halos*, salt, and *geiton*, neighbor, in reference to habitat).

H. glomeratus Meyer. Brs mostly basal, up to 5 dm; pl tomentose in lf axils, but otherwise glab and ± glaucous; lvs linear, fleshy, ± terete; poisonous pl intro from Asia, rapidly spreading in grassl and sagebr des of w US, mostly on ± waste or overgrazed land.

Kochia Roth

Fls sessile in axils of foliose bracts, mostly ⚥ but often in part ♀; perianth 5-merous, persistent and enclosing fr, becoming ± prominently carinate and cross-winged; ann to shrubby per with alt (opp), linear to narrowly lanceolate, often fleshy and terete, entire lvs. (For W. D. J. Koch, 1771–1849, German botanist).

1a Pl per, shrubby-based, 1–5 dm; lvs subterete, not strongly veined; perianth mostly white-tomentose; plains and foothills, mostly where alkaline; in our area in se Ore, s Ida, and s Mont, to Cal, Colo, and NM; red sage (*K. vestita*) **1 K. americana** Wats.

1b Pl ann, erect, reddish-std, 5–15 dm; lvs narrowly lanceolate, acute or acuminate, in part strongly 3–5-nerved; perianth and often infl rusty-villous or villous-lanate; Eurasian, freely estab and rapidly spreading weed of roadside and wastel in irrigated areas e Cas; summer or mock cypress, burning-bush, red belvedere **2 K. scoparia** (L.) Schrad.

Monolepsis Schrad.   Monolepsis; Povertyweed

Fls 1-several in axillary glomerules, ☿♀♂; perianth mostly a single (2–3, or 0) bractlike, greenish tepal; stamens 1 (2); fr ± flattened; seed erect, smooth, lenticular; ± farinose ann with alt, entire to hastate, fleshy lvs. (Gr *monos*, solitary, and *lepis*, scale, referring to the gen solitary tepal).

1a Pl freely br dichotomously; lvs inconspicuous, 3–13 mm, bl entire, oblong to ovate, narrowed abruptly to very short petiole; fls 1–5 per axil; tepals 1–3, linear; des valleys, mostly on alkaline soil; se Ore to Cal, e across s Ida to Wyo and Colo; dwarf m. **1 M. pusilla** Torr.

1b Pl not dichotomously br; lvs mostly > 13 mm, often either spatulate or obovate or hastate; fls gen > 5 per axil; tepals mostly 1, ± spatulate or oblong

2a Lvs hastately lobed; tepal acute, 1.5–2.5 mm; seed ca 1 mm broad; dry to moist, often saline or alkaline soil, des plains to lower mts; s BC s, e Cas, to Cal, e to Alta, the RMS, and e; patata (*M. chenopodioides*) **2 M. nuttalliana** (Schultes) Greene

2b Lvs entire; tepal rounded, < 1 mm; seed ca 0.5 mm broad; des regions, often where alkaline or saline; barely reaching our area in se Ore and s Ida, s to Baja Cal and Nev; prostrate m. **3 M. spathulata** Gray

Nitrophila Wats.

Fls ☿, 1 or 3 (2) per axil, each with 1–2 bracteoles ca = the deeply 5 (6–7)-parted perianth; stamens slightly connate at base; seed erect, lenticular, embryo annular; low per herbs with mostly opp, entire, fleshy lvs. (Gr *nitron*, soda, and *philos*, loving, in reference to the habitat).

N. occidentalis (Moq.) Wats. Borax weed. Rhizomatous per; st freely brd, 1–3 dm; lvs (5) 10–25 mm, semiterete; perianth segms erect, 1-nerved, enclosing fr; ± alkaline soil, e Wn (?) and Ore, to Nev, Cal, and Mex.

Salicornia L.   Samphire; Glasswort; Saltwort

Fls mostly ☿ (some ♀♂), sessile in groups of 3 (1–7), sunken in depressions of joints of fleshy spikes just above axil of opp, scalelike, connate bracts; perianth ± pyramidal in outline, connate and saccate below, nearly closed above, but with shallowly lobe-margined, puckered, and slitlike opening through which the 2 stamens protrude; fr strongly compressed laterally; embryo folded; glab succulent ann or per herbs or subshrubs with jointed sts and opp, scalelike lvs. (L *salsus*, salt, and *cornu*, horn, in reference to habitat and to hornlike appearance of brs).

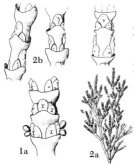

1a Pl per, rhizomatous, gen matted; c fl only a little > lateral fls in each cluster; salt marshes and beaches along coast; Alas to Baja Cal; pickleweed, woody g. (*S. ambigua, S. pacifica, S. perennis*) **1 S. virginica** L.

1b Pl ann, mostly erect; c fl much > lateral fls

2a Joints of spike ca 2 (2.5) mm and ca as thick; upper margin of c fl extending almost to node above; saline or alkaline soil; s BC s, through e Wn and Ore to Nev, e through Ida and Mont to Sask, Kans, and NM; occ as waif w Cas, as in freight yards in Seattle, Wn; red g. **2 S. rubra** Nels.

2b Joints of spike (2) 2.5–4 × ca 2 (3)mm; upper margin of c fl 0.5–1 mm below node above; Eurasia; coastal in our area, in salt marshes; European g. **3 S. europaea** L.

Salsola L.

Fls ♂, solitary or clustered in axils of spiny bracts, each with 2 smaller bracteoles; perianth distinct almost to base, gen becoming carinate crosswise and often prominently winged; ann (ours) or per herbs with alt, linear, entire, succulent, ± spinulose lvs. (L *salsus*, salt, in reference to habitat and to taste of pl).

Salsola kali L. Russian thistle, tumbleweed, wind witch. Glab to pubescent, freely brd; st striate, often purplish, up to 1 m, becoming much hardened; bracts becoming spinescent; ubiquitous Eurasian weed of dry, often alkaline regions, mostly e Cas (*S. pestifer*). *S. collina* Pall., a sparingly brd, otherwise rather similar sp. is spreading in c US, and will possibly be found in our area.

Sarcobatus Nees   Greasewood

Fls ♀♂, borne in ∞ axillary, ∞ -fld spikes, ♂ uppermost, naked, of 2–3 stamens almost covered by long-stalked, peltate, rhombic scale; ♀ fls 1–2 in axils of scarcely reduced, lflike bracts; pistil closely surrounded by a cuplike, shallowly lobed to nearly entire perianth which becomes enlarged in fr and adherent to lower part of ovary, the upper portion flaring and winglike; spiny, freely brd shrubs with fleshy, alt, linear lvs. (Gr *sarco*, flesh, and *batos*, bramble, referring to the succulent lvs and spiny brs).

S. vermiculatus (Hook.) Torr. Black g. Whitish-barked, up to 2.5 m; lvs deciduous, (1) 1.5–4 cm; saline or alkaline soil of dry regions; e Cas, Wn and Ore to se Cal, e to Alta, Daks, Wyo, Colo, and NM.

Suaeda Forsk.   Seablite

Fls ♂ to ♀♂, solitary to glomerate and spicate in axils of small bracts, each gen with 2 small membranous bractlets; perianth fleshy, lobed ca half length or more, segms erect to spreading, ± cucullate, rounded to keeled or cross-corrugate or sometimes ± horned; ann or per, glab to pubescent herbs or subshrubs, mostly glaucous, with alt, linear, terete or ± flattened lvs; pl gen of saline or alkaline soil, often where marshy. (Arabic name *suwayd*, for one of the spp.). *(Dondia).*

1a Lvs abruptly contracted to a narrow petiole, linear to narrowly lanceolate, subterete, 5–25 mm; seed erect; tepals alike, not horned on back; pl up to 7 dm; se Ore to Cal, Wyo, and NM, possibly s of our range; bushy s. (*S. diffusa*)   1 **S. nigra** (Raf.) Macbr.
1b Lvs sessile; seed horizontal; tepals often unequal and ± horned on back
  2a Pl per, up to 6 dm; tepals erect, =, not horned; lvs linear, subterete, 10–20 mm; e Wn and Ore to Cal, e to Alta and in the RMS to n Mex; tall s.   2 **S. intermedia** Wats.
  2b Pl ann, mostly < 5 dm; tepals often unequal and horned on back
    3a Tepals = or subequal, not horned; seed 1.3–2 mm broad; lvs 10–20 (50) mm, more flattened than terete; coastal salt marshes; Wn to Alas, widespread on other continents; herbaceous s.
                                                                3 **S. maritima** (L.) Dumort.
    3b Tepals unequal, or (more often) horned; seed mostly < 1.3 mm broad
      4a Fl bracts broadly lanceolate to ovate, 2–3 mm, mostly broader than the semiterete, 10–40 mm lower lvs; e Cas, BC to Cal, e to Sask, Minn, and Tex; pahute weed (*S. erecta*)   4 **S. depressa** (Pursh) Wats.
      4b Fl lvs linear, 2–5 mm, gen scarcely broader than the linear, subterete, 10–30 mm lower lvs; e Wn and Ore to Nev, e through s Ida to Wyo and Colo; slender s.   5 **S. occidentalis** Wats.

AMARANTHACEAE   Amaranth Family

Fls small, apet, glomerate in the axils or in simple or pan spikes, hypog, ♂ to ♀♂ or ♀, ♂; tepals 3–5, distinct or ± connate, gen dry and scarious; stamens gen as many as tepals and opp them, sometimes fewer; pistil 2–3 (5)-carpellary, ovary 1-celled; fr a 1-seeded, circumscissile caps (ours) or a utricle, drupe, or berry; endosperm abundant; embryo curved; ann or per herbs (± shrubby); lvs alt or opp, exstip, simple.

Amaranthus L.   Amaranth; Pigweed

Fls ♀♂ or ♀, ♂, subtended and mostly exceeded by gen several scarious, pungent bracts; tepals (1–4) 5, distinct, scarious, midvein mostly ± excurrent; ann with alt, petiolate, entire to dentate lvs. (Old Gr name, from *a*, not, and *marainein*, to wither, because of the papery, nonwithering fls and bracts). (*Galliaria, Mengea, Pyxidium*).

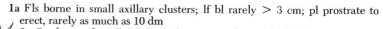

1a Fls borne in small axillary clusters; lf bl rarely > 3 cm; pl prostrate to erect, rarely as much as 10 dm
  2a Seeds mostly 0.7–0.8 (0.6–1) mm; tepals mostly 3 (1); stamens 1–3
    3a Pl prostrate; tepals of ♀ fls unequal, 1 well developed, 2 much reduced or lacking; lf bl 5–15 mm, rhombic-elliptic to obovate; moist, often alkaline flats or lake shores and vernal pools; e Wn and Ore to Cal, e to Alta, Mont, and Nev; Cal a. (*A. carneus*)
                                    1 A. californicus (Moq.) Wats.
    3b Pl erect, up to 10 dm; sepals of ♀ fls ca = and like those of the ♂ fls; lf bl 10–30 mm, ovate-rhombic to obovate; widespread weed of cult or wastel, in most of our area; tumbleweed, white p.   **2 A. albus** L.
  2b Seeds 1.3–1.7 mm; tepals 4–5; stamens 3–4; pl prostrate to ascending, up to 10 dm broad; lf bl obovate to ovate-rhombic, 5–25 (30) mm; chiefly in sagebr des; e Wn to Cal, e to Daks, Kans, NM, and Tex, and as weed elsewhere on cult or waste ground, even w Cas; prostrate p., tumbleweed, tumbleweed a. (*A. blitoides*)   **3 A. graecizans** L.
1b Fls borne in terminal and axillary, simple to compound spikes; pl erect, 5–20 dm; lf bl mostly much > 3 cm
  4a Pl scurfy-villous below infl, hairs flattened, crisped, multicellular; lvs mostly hairy beneath along veins; ♀ fls with perianth 2.5–4 mm, ca 1.5–2 times as long as ovary and gen > the mature fr; tepals broadly rounded to retuse, membranous and whitish but with greenish midnerve; stamens 5 (4); weed of wastel in much of N Am and elsewhere; rough p., pigweed a., redroot a., green a.   **4 A. retroflexus** L.
  4b Pl glab to puberulent (never villous) below infl; ♀ fls with perianth 2–3 (3.5) mm, rarely > mature fr; tepals acute to rounded, ± bristle-tipped, midnerve not very green; stamens 3 (4); common, very widespread weed; Powell's a.   **5 A. powellii** Wats.

NYCTAGINACEAE   Four-o'clock Family

Fls reg, hypog, ours ⚥, cymose to racemose or umbellate, bracteate, bracts distinct or connate, often forming greenish and calyxlike or highly colored invol; corolla absent; calyx tubular to camp, petaloid, the segms gen 5, connate > half length; stamens 1–∞, sometimes apparently adnate to the perianth, unequal in length; pistil 1-carpellary; ovary 1-seeded, superior, but sometimes apparently with the perianth adnate; fr an achene, often enclosed in thickened base of perianth; embryo straight to curved; ann or per herbs (ours), with gen opp, simple, extip lvs.

1a Fls sessile in heads; invol bracts distinct; fr gen winged or strongly keeled
                                              Abronia
1b Fls mostly ped; invol bracts connate; fr often ribbed, but not winged
                                             Mirabilis

Abronia Juss.   Sandverbena; Abronia

Heads pedunculate, the several–∞ fls subtended by 4–6 distinct, herbaceous to scarious and colored bracts; perianth showy, white, yellow, or pink to purple, narrowly salverf, constricted and finally deciduous above the ovary; stamens 4–5, included; achene ellipsoid-cylindric, included in the accrescent perianth-base; ann or per herbs, often glandular, prostrate to ascending; lvs opp, petiolate. (Gr *abros*, delicate or graceful). (*Tripterocalyx*).

1a Pl per, prostrate-creeping, glandular-puberulent; coastal beaches and dunes
  2a Perianth yellow, tube 8–10 mm; invol bracts ovate-lanceolate to broadly elliptic, 6–8 mm; body of fr broader than the thick, keel-like wings; lf bls suborbicular or ± reniform to oblong or ± deltoid, 1.5–4 (6) cm; s VI to s Cal; yellow s.   **1 A. latifolia** Eschsch.

2b Perianth pink to purple, tube 6–8 mm; invol bracts narrowly lanceolate, 5–6 mm; body of fr no broader than the thin wings; lf bls elliptic to oblong, 2–6 cm; BC to Baja Cal; pink s.; ours the var. *acutalata* (Standl.) Hitchc.   **2 A. umbellata** Lam.

1b Pl sometimes ann, gen decumbent to ascending, sometimes not glandular; CRG and e Cas

3a Pl ann; perianth pinkish; fr not > 6 mm; se Ore to Cal and Nev; prob always to the s of our area   **A. turbinata** Torr.

3b Pl per or perianth white to greenish

4a Pl ann; perianth greenish-white; fr prominently winged all around; lf bls elliptic-lanceolate to ovate-rhombic, 2–6 cm; Beaverhead Co, Mont, e to Daks, s to Nev and NM; smallfld a.   **3 A. micrantha** Torr.

4b Pl per; perianth white, sometimes pinkish tinged; fr winged only on sides if at all; lf bls various

5a Invol bracts 1.5–4 mm broad; wings of fr firm and thin, mostly wider than fr body toward top; fls scarcely fragrant, perianth white, tube 12–20 mm; pl finely puberulent and glandular, but not hirsute; CRG and e Wn and Ore, e along the SRP in Ida; white s., white a. (*A. lanceolata, A. suksdorfii*)   **4 A. mellifera** Dougl.

5b Invol bracts (2) 4–12 mm broad; wings of fr thick and hollow, seldom wider than body of fr; fls fragrant, perianth white or pinkish tinged, tube 16–20 mm; pls often ± hirsute above as well as glandular-puberulent; sw Mont and n border of SRP of Ida, s and e to Daks, Wyo, and Tex; fragrant white a. (*A. ammophila, A. cheradophila, A. elliptica, A. nudata*)   **5 A. fragrans** Nutt.

Mirabilis L.   Four-o'clock; Umbrellawort

Fls 1–several in 5-lobed, calyxlike, camp to saucer-shaped, mostly accrescent invols borne singly or in pans or cymes; perianth pink to purple (ours), funnelf to ± camp, limb rotate, 5-lobed; stamens gen exserted; fr ellipsoid to obovoid, smooth to rugose, often ribbed, hardened and nutletlike; per herbs (ours) with gen several sts from a thick root; lvs opp, entire, sessile to petiolate. (L, meaning wonderful, in reference to fls). (*Allionia*).

1a Lvs linear to narrowly lanceolate, 4–9 × mostly < 1 cm; perianth pink to purple, ca 10 mm; mostly e RM, Mont to Mex and Ariz; narrowlvd f. (*M. angustifolium*)   **1 M. linearis** (Pursh) Heimerl

1b Lvs mostly broadly lanceolate or broader, some > 1 cm broad; perianth various, sometimes much > 10 mm

2a Pl densely and evenly retrorse-puberulent above; invol 1–fld; Malheur Co, Ore, to Cal and Ariz, just s of our area
            M. bigelovii var. **retrorsa** (Heller) Munz

2b Pl gen sparsely puberulent in lines above, sometimes glab; invol 3–7-fld

3a Fl invol 15–25 mm; perianth rose-purple, 15–25 mm; fr lightly 10-ribbed; rocky slopes, SRC and lower Salmon R, Ida, to adj Ore; Macfarlane's f.   **2 M. macfarlanei** Constance & Rollins

3b Fl invol 5–7 mm; perianth pink (white), ca 10 mm; fr strongly 5-ribbed; Mont to Wisc, s to Ala and Mex; intro in our area in Bonner Co, Ida; heartlvd u.   **3 M. nyctaginea** (Michx.) MacM.

PHYTOLACCACEAE   Pokeweed Family

Fls racemose, ♂, hypog to slightly perig, reg; sepals (4) 5, distinct; petals lacking; stamens 5–30; carpels 5–12, connate basally but styles and stigmas free; fr a berry; ours alt-lvd herbs.

Phytolacca L.

Racemes bracteate, apparently lateral and opposed to the alt, petiolate, entire lvs; sepals petaloid, rounded; ovary deeply lobed, becoming fleshy in fr; seeds 1 per cell (carpel) of the ovary, erect; embryo annular; pl herbaceous to woody. (Gr *phyton*, plant, and L *lacca*, crimson-lake, referring to the color of the juice of the berry).

P. americana L. Pokeberry, pokeweed, pigeonberry. Glab herbaceous per up to 3 m, ± malodorous; lvs petiolate, bl elliptic to ovate-lanceolate, up to 2.5 dm; racemes peduncled, ∝ –fld, up to 2 dm; fls saucer-shaped, greenish-white to pink, 5–7 mm broad; stamens and carpels mostly 10; berry deep purple, ca 1 mm thick; native to e US, often grown for the edible young lvs, and sometimes long-persistent in gardens or waste areas, as reliably reported from The Dalles, Ore *(P. decandra)*.

AIZOACEAE   Carpetweed Family

Fls reg, ♀, apet, hypog to epig, solitary in lf axils or in axillary to terminal clusters; sepals 5–8, distinct (ours) to ± connate, often adnate to ovary; stamens (2) 3–∝, gen twice as many as sepals, sometimes more ∝ and mostly sterile and petaloid; pistil 3–5 (20)-celled; fr caps or leathery and either dehiscent or indehiscent; ann or per herbs, often succulent, with alt to whorled, gen simple and exstip lvs.

Mollugo L.   Carpetweed

Fls axillary, inconspicuous; sepals 5, distinct, persistent; stamens (2) 3–5 (10), hypog; pistil 3 (to 5) -carpellary, styles 3 (to 5); fr caps, loculicidal; seeds ∝; ann herbs with whorled (ours) or alt lvs. (L name, from *mollis*, soft, the significance obscure).

M. verticillata L. Sts glab or slightly glandular-pubescent, prostrate, dichotomously br, up to 3 dm; lvs (3) 4–6 per whorl, linear-oblanceolate to obovate, 5–30 × 1–3 (6) mm; moist, ± waste ground; BC s, e Cas in Wn, to Baja Cal, e and throughout most of US.

PORTULACACEAE   Purslane Family

Fls ♀, mostly reg, often showy; sepals gen 2 (up to 9 in *Lewisia*), persistent (except in *Talinum*); petals gen 5 (up to 15), sometimes reduced, rarely lacking, distinct or ± connate; stamens gen opp petals and often adnate to them, but sometimes fewer or more ∝ (up to 50 in *Lewisia*); pistil 2–3-carpellary; ovary superior to partially inferior, 1-celled, with basal placentation; style brs and stigmas 2–8; fr caps, circumscissile or 2–3-valve from tip; seeds 1–∝, smooth and shining; ann to per, ± succulent herbs with alt or opp, simple, gen entire, stip or exstip lvs.

1a Petioles persistent as slender spines on caudex or brs; lvs semiterete; sepals 2, deciduous after anthesis; per with lfless, pedunculate, ∝ –fld cymes
                                                                                       Talinum
1b Petioles not persistent and spinose; lvs gen flattened; sepals persistent after anthesis, sometimes > 2; ann or per; fls seldom pedunculate in lfless cymes
   2a Ovary ± inferior, upper part of calyx circumscissile with ovary; succulent, prostrate, ann weed with small yellow fls   **Portulaca**
   2b Ovary superior, often valvate, if circumscissile the calyx persistent in its entirety and pl per with other than yellow fls
      3a Fls tightly clustered in scorpioid umbels or pans; sepals 2, ± scarious; petals 2–4, smaller than the sepals; stigmas 2; stamens 1–3
         4a Sepals scarious only on the margins, 2–4 mm; petals 2; stamen 1; style < 1 mm   **Calyptridium**
         4b Sepals largely scarious; petals 4; stamens 3; styles > 1 mm
                                                                                       Spraguea
      3b Fls solitary or in ∓ open racemes or cymes; sepals occ > 2, mostly not scarious; petals mostly 5 or more and > the sepals; often stigmas > 2 and stamens > 3
         5a Fls in lfy-bracteate racemes, lower ones in axils of scarcely reduced lvs; ann with alt lvs; caps 3-valved, 15–40-seeded   **Calandrinia**
         5b Fls gen solitary or in minutely bracteate or ebracteate racemes or cymes; pl various; caps either circumscissile or < 15-seeded
            6a Pl ann or per, sometimes slenderly rhizomatous or stoloniferous, but never with fleshy corms or thick, fleshy rootstocks; petals gen white (pinkish); sepals 2; stamens 2–5   **Montia**

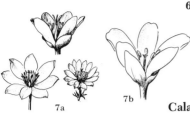

6b Pl per, with fleshy corms or thick fleshy taproots; petals gen pink to red; sepals sometimes > 2 and stamens often > 5
   7a Caps circumscissile near base, the lid sometimes splitting upward; sepals > 2, or petals or stamens > 5, or stigmas > 3   **Lewisia**
   7b Caps dehiscent downward by 3 valves; sepals 2; petals and stamens mostly 5; stigmas 3   **Claytonia**

7a   7b

Calandrinia H.B.K.   Calandrinia

Sepals 2; petals 5 (3–7), crimson or rarely white in ours, quickly withering; stamens 5–12, free or slightly connate; stigmas 3; succulent ann (ours) or per herbs with alt, fleshy lvs. (For J. L. Calandrini, 1703–1758, Swiss botanist).

C. ciliata (R. & P.) DC. Red maids, desert rockpurslane. Lvs linear to oblanceolate, 1.5–7 cm × 2–10 mm, coarsely ciliate; sepals and petals 3–8 mm; mostly where moist in early spring; occ w Cas in Wn, and e and w Cas in Ore, to s Cal; S Am (*C. caulescens, C. c.* var. *menziesii, C. tenella*).

Calyptridium Nutt.   Calyptridium

Petals withering-persistent, shorter than the sepals; caps valvate; ann with fleshy, mostly basal lvs. (Gr *kalyptra*, veil, referring to the persistent sepals).

C. roseum Wats. Rosy c. Lvs spatulate-oblanceolate, basal ones 5–30 mm; sepals 1–3 mm; petals white, scarcely 1 mm; stamen 1; sagebr des to mont for; c Ore to Cal, e to Ida and Nev, barely in our area in sc Ida.

Claytonia L.   Springbeauty; Claytonia

Fls racemose, rather showy; petals free or basally connate; ovary mostly 6-ovuled and 2–6-seeded; glab per with fleshy roots or corms, and 1–several fl sts mostly with a pair of opp lvs below raceme. (For John Clayton, 1685–1773, botanist who collected mostly in Virginia). °*C. embellata* Wats., another cormose sp., now known from Wasco Co., Ore (but otherwise s of our area), differs from *C. lanceolata* in its much broader, blunt to rounded, petiolate st lvs.

1a Pl from a semiglobose corm; basal lvs 1–few, petioles largely subterranean; petals 5–12 (15) mm; sagebr foothills to alp slopes; BC to Cal, both sides Cas, e to Alta and NM; Western s., lanceleaf s.; several vars.
                      1 **C. lanceolata** Pursh

a1 Fls pale (to deep) yellow or orange
   b1 St lvs elliptic or ovate, (1) 1.5–2.5 (4) cm, mostly > 1/3 as broad; pl mostly < 10 cm; corolla pale yellow; Mt. Baker and Glacier Peak, Cas, Wn   var. **chrysantha** (Greene) Hitchc.
   b2 St lvs lanceolate or narrowly oblong, several × as long as broad; pl mostly > 10 cm; petals deep yellowish-orange; Henry's Lake, Fremont Co, Ida   var. **flava** (Nels.) Hitchc.
a2 Fls white to deep pink
   c1 St lvs narrowly lanceolate, at least 5 × as long as broad, rarely > 1 cm broad; sts many, corms 1–4 cm broad; YNP and vic
                     var. **multiscapa** (Rydb.) Hitchc.
   c2 St lvs broader, elliptic-lanceolate to ovate-lanceolate, mostly < 5 × as long as broad; sts 1–several, corms mostly < 2 cm broad; range of sp.
                     var. **lanceolata**
1b Pl from fleshy, thickened taproot; basal lvs ∝, in large rosettes, petioles mostly above ground; gravelly soil, rock crevices, and talus slopes; mont; Wen Mts, Wn, and Cas, Ore, e and s to Alta, Mont, Utah, Colo, and NM; alpine s.; several vars.   2 **C. megarhiza** (Gray) Parry
   a1 Petals deep pink, 11–15 mm; sepals thin, 7–9 mm, gen acute; infl ca = lvs; paired fl bracts linear, 10–25 × 1–1.5 mm, gen 5–10 mm below 1st bract of infl; Wen Mts, Wn; Wen s.   var. **nivalis** (English) Hitchc.
   a2 Petals mostly pale pink and < 11 mm; sepals mostly thickened and 4–7 mm, often rounded to obtuse; infl mostly < lvs; paired fl bracts often up to 5 mm broad, scarcely 5 mm below 1st bract of infl
      b1 Petioles narrowly winged, mostly with only main vein evident; cauline lvs linear, 1–1.5 mm broad; bls of basal lvs not > 1.5 cm broad; Wallowa, Blue, and c Cas mts, Ore, to Nev
                   var. **bellidifolia** (Rydb.) Hitchc.

b2 Petioles more broadly winged and with 2 (4) lateral veins (as well as midvein) evident; cauline lvs 1–5 mm broad; bls of basal lvs often > 1.5 cm broad; RM, Alta to NM, w to c Ida and Utah   var. **megarhiza**

Lewisia Pursh   Lewisia

Fls either single and pedunculate or several– ∝ and corymbose, pan, or cymose; sepals 2 or 4–9; petals 5–18, white to salmon, rose, or magenta, soon withering; stamens (4) 5–50; style 1, stigmas 3–8; caps circumscissile at base, but sometimes also dehiscent upward by 3–5 valves; seeds (1) 3–50; ± succulent per herbs with fleshy or cormose roots and mostly basal lvs, scapes 1 or more, with alt, opp, or whorled bracts. (For Meriwether Lewis, 1774–1809, of Lewis and Clark Expedition). *(Oreobroma). L. cotyledon* (Wats.) Robins. has been transplanted into one area of Goat Rocks, Cas, Wn, where it may persist. It has ∝ fls per st, with 2 sepals, pink petals 12–20 mm, and gen 6–9 stamens.

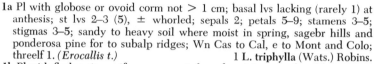

1a Pl with globose or ovoid corm not > 1 cm; basal lvs lacking (rarely 1) at anthesis; st lvs 2–3 (5), ± whorled; sepals 2; petals 5–9; stamens 3–5; stigmas 3–5; sandy to heavy soil where moist in spring, sagebr hills and ponderosa pine for to subalp ridges; Wn Cas to Cal, e to Mont and Colo; threelf l. *(Erocallis t.)*   1 L. **triphylla** (Wats.) Robins.
1b Pl with fleshy, ± napiform or carrot-shaped root > 1 cm; basal lvs present at anthesis (often withered and drying in *L. rediviva*); st lvs commonly reduced and bractlike, gen alt
   2a Sepals (4) 5–9; fls 1 per scape, peduncle readily disart above the whorl of linear, scarious bracts borne ca midlength of scape; petals 12–18, white to rose, 18–35 mm; gravelly to heavy, mostly dry soil, sagebr plains to lower mts; BC s, E Cas in Wn and Ore, to Cal, e to Mont, Colo, and Ariz; bitterroot   2 L. **rediviva** Pursh
   2b Sepals 2, or apparently 4 because of 2 closely subtending bracts; fls often 2– ∝ , ped not disart, bracts not whorled
      3a Fls white, borne singly on scapes 5–20 mm, immediately subtended by 2 sepaloid bracts; petals (5) 6–7 (11), 10–15 mm; stamens 8–15; gravelly soil, where moist in early spring, esp near snowbanks; mts of c Ida, and Sierran Cal; Kellogg's l.   3 L. **kelloggii** Brandg.
      3b Fls either pinkish or white but gen > 1 per scape, never immediately subtended by sepaloid bracts
         4a Fl st (5) 10–20 cm, several-fld, bracts alt or lacking; petals 6–11
            5a Petals 5–13 mm; stamens 5–6; lvs linear-oblanceolate to narrowly spatulate, 2–10 cm × 3–8 mm; exposed gravelly and rocky slopes or rock crevices; s BC s, in Cas and OM, to Cal, also in ne Ore and adj Ida; Columbia l.; 3 vars.
                                         4 L. **columbiana** (Howell) Robins.
               a1 Lowest bracts of st (below infl) mostly entire or dentate but not glandular; lvs obtuse to acute; fls gen rather deep pink
                  b1 Basal lvs thick,∝, mostly 3–10 cm × up to 8 mm; pl gen 1.5–3 dm; sts with entire bracts to above midlength; petals gen 7–10 (11) mm; e side Cas, Okanogan Co, Wn, to CRG; prob n into s BC   var. **columbiana**
                  b2 Basal lvs fewer and thinner, mostly < 4 cm; pl 0.5–1.5 (2.5) dm; sts with only lowest bracts entire; petals mostly 5–8 mm; Wallowa Mts, Ore, and Seven Devils Mts, Idaho Co, Ida.   var. **wallowensis** Hitchc.
               a2 Lowest bracts of sts serrulate, gen glandular; lvs gen rounded; fls light to deep pink or rose; Mt. Rainier and OM, Wn, s to Saddle Mt, Clatsop Co, Ore   var. **rupicola** (English) Hitchc.
            5b Petals 25–40 mm; stamens 12–25; lvs ovate or oblong-lanceolate to obovate, 10–20 cm × mostly > 10 mm; rock crevices, talus slopes, and rocky banks, chiefly in ponderosa pine area; Wen Mts, s Chelan and n Kittitas cos, Wn; Tweedy's l.
                                         5 L. **tweedyi** (Gray) Robins.
         4b Fl st gen < 10 cm, 1-fld, with pair of opp bracts at midlength or below; petals ca 7 (5–9); open, moist to dry places in mts to above timberl; Cas and OM, Wn, to Cal, e to Mont, NM, and Ariz; 2 vars.
                                         6 L. **pygmaea** (Gray) Robins.

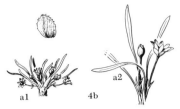

a1 Sepals mostly rounded, heavily veined, glandular-serrulate, 2–5 mm; lvs gen linear, < 8 cm; range of sp.; least l., dwarf l., alpine l. (*L. aridorum, L. exarticulata, L. minima*)   var. **pygmaea**
a2 Sepals ± acute, entire, 5–12 mm; lvs often narrowly oblanceolate, up to 15 cm; with var. *p.*, but more w, rare in Ida or Mont (Bitterroot Mts); Nev l.   var. **nevadensis** (Gray) Fosberg

Montia L.   Montia

Fls mostly racemose, bracteate or ebracteate; sepals 2, persistent; petals 5 (2–6), ± unequal, distinct or rarely ± connate, white to red or purplish; stamens (2) 3–5, opp and often adnate to petals; stigmas 3, linear; caps 3-valved; seeds gen 1–3, black and shining, smooth (to muricate), with small strophiole; ann or per, ± succulent herbs, without fleshy underground sts or taproot; lvs linear to rhombic-obovate, the cauline ones from several and alt to only 2 and opp. (For Giuseppi Monti, 1682–1760, Italian botanist). (*Limnia, Montiastrum, Naiocrene*).

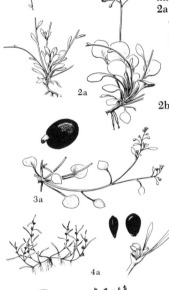

1a St lvs mostly several and alt, petiolate or with narrow base gen slightly enlarged and scarious at juncture with st; per, or ann with petals scarcely 5 mm and rarely > sepals
  2a Pl per, gen slenderly rhizomatous; petals 7–15 mm; basal lvs 1.5–3 (6) cm, bl obovate to oblanceolate, 2–20 mm broad; Alas to Cal, from Cas to coast, also in RM to Mont, Ida, and Utah; littlelf m., small-lvd m., streambank springbeauty; 2 vars.   **1 M. parvifolia** (Moc.) Greene
    a1 Basal lvs mostly < 5 mm broad; petals 7–12 mm; seeds 1–1.2 mm; common, esp in RM (*Claytonia filicaulis, C. rupicola*)   var. **parvifolia**
    a2 Basal lvs mostly much > 5 mm broad; petals 10–15 mm; seeds up to 1.5 mm; chiefly coastal (*M. sweetseri*)   var. **flagellaris** (Bong.) Hitchc.
  2b Pl ann, nonrhizomatous (sometimes stoloniferous); petals scarcely 5 mm; lvs various
    3a Lvs lanceolate to suborbicular, rarely < 3 mm broad; seeds shallowly papillate, strophiole 0.3–0.5 mm; stamens 5; moist places mostly W Cas, BC to Cal; branching m. (*Claytonia d.*)
           2 M. diffusa (Nutt.) Greene
    3b Lvs linear to spatulate, mostly < 3 mm broad; seeds ± smooth, but minutely reticulate, strophiole not > 0.1 mm; stamens gen 3
      4a Sepals ca 1 mm; lvs linear-spatulate to linear-oblanceolate, 5–20 × 0.5–2 mm; fls clustered at nodes, opp and much < the lvs, subtending bract of each cluster scarious, often connate with lf base and resembling sheathing stips; moist lowl areas W Cas, BC to nw Cal; Howell's m.   **3 M. howellii** Wats.
      4b Sepals mostly > 2 mm; fls racemose, exceeding subtending lvs; racemes terminal or axillary and terminal
        5a Sepals ca 2 mm; seeds ca 1 mm; pl 2–8 cm; lvs linear, 6–15 (20) × ca 0.5 mm; moist lowl areas; w Wn to Cal, e to Ida and w Mont; dwarf m. (*Claytonia d., C. howellii*)
              4 M. dichotoma (Nutt.) Howell
        5b Sepals 2.5–6 mm; seeds 1.5–2 mm; pl mostly 5–20 cm; lvs linear, 10–50 × ca 1 (3) mm; lowl to foothills, where dry or moist; BC to Cal, from coast e to Mont and Utah; narrowlvd m., line-lf Indianlettuce (*Claytonia l.*)   **5 M. linearis** (Dougl.) Greene
1b St lvs gen opp (often sessile), rarely alt and then pl ann with petals at least twice as long as sepals
  6a Pl per with several pairs of opp obovate st lvs, rhizomatous and stoloniferous, gen with offsets along stolons and often with some of fls changed to bulbils; wet places, mont at middle to lower elevations; Alas to Cal, from Cas e to Man, Minn, Ia, and NM; Chamisso's m., water m. (*Claytonia c., C. aquatica*)   **6 M. chamissoi** (Ledeb.) Robins. & Fern.
  6b Pl ann or if per with only 1 pair of st lvs (other than bracts), rarely with offsets or bulbils
    7a St lvs several pairs, opp; pl ann, flaccid-std, gen rooting at nodes; petals connate ca half length, < 2 mm; stamens 3; seeds mostly muricate; wet places, Alas to Cal, from coast e to Ida and through Can to Me; Eurasia; water chickweed; 2 vars.   **7 M. fontana** L.
      a1 Seeds 0.7–0.9 mm, dull black, strongly muricate; common phase, both sides Cas, to Ida   var. **tenerrima** (Gray) Fern. & Wieg.
      a2 Seeds 1–1.2 mm, brownish-black, shining, rounded-papillate;

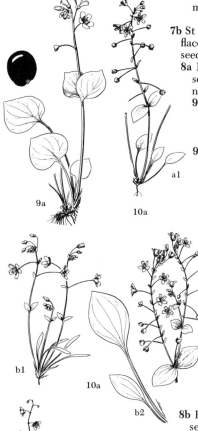

mostly n or s of our area, but known in s Ida

var. **lamprosperma** (Cham.) Fenzl

7b St lvs 1 pair, if more then gen not truly opp; pl ann or per; sts rarely flaccid or nodally rooting; petals either distinct or much > 2 mm; seeds never muricate

8a Petals mostly 6–13 mm, when > 6 mm at least twice as long as sepals; pl sometimes per; racemes often bracteate throughout; st lvs not connate

9a Racemes ebracteate except for lowest fl; st lvs 1–5 cm broad; petals white; wet places, often mont; BC to Cal, both sides Cas, e to w Mont and Utah; broadlvd m. (*Claytonia c.*)

8 M. cordifolia (Wats.) Pax & Hoffm.

9b Racemes bracteate above lowest fl, sometimes throughout; st lvs sometimes < 1 cm broad; petals often pink

10a St lvs 0.5–5 cm broad, obovate-rhombic to broadly lanceolate or oblanceolate; moist shaded places, lowl to midmont; Alas to Cal, both sides Cas, e to Mont and Utah; w springbeauty, Siberian m., candyfl. (*Claytonia s.*); 3 vars.

9 M. sibirica (L.) Howell

a1 True bulbils developed in axils of some basal lvs; mostly s of our area, but approached by occ pls as far n as nw Wn

var. **bulbifera** (Gray) Robins.

a2 True bulbils lacking, although some lvs often reduced to a fleshy petiole base

b1 Basal lvs narrowly lanceolate, mostly < 1 cm wide; st lvs lanceolate to oblanceolate, often short-petiolate; occ with var. *sibirica* and mergent with it; CRG s, w of Cas, to Cal

var. **heterophylla** (T. & G.) Robins.

b2 Basal and cauline lvs mostly much > 1 cm wide; Alas s, on both sides Cas, to Cal, e to Mont and Utah (*C. asarifolia*)

var. **sibirica**

10b St lvs < 0.5 cm broad, linear to narrowly spatulate; moist to dry places; Lincoln Co, Wn, to Wallowa Co, Ore, and adj Ida; sand m. (*Claytonia a.*)   **10 M. arenicola** (Hend.) Howell

8b Petals mostly < 5 mm, rarely longer, but not > twice as long as sepals; pl ann; racemes ebracteate above lowest fl; st lvs often connate

11a St lvs linear-lanceolate to lanceolate-ovate, free or connate on one side for < half their length; basal lvs linear to linear-spatulate; pl 1–8 cm; moist to dry soil; BC s, mostly w of Cas in Wn, to CRG, and on both sides Cas in Ore, to Cal; pale m., common m. (*Claytonia s., C. exigua*)

11 M. spathulata (Dougl.) Howell

11b St lvs gen connate much > half their length, rarely ovate and not connate; basal lvs diverse; pl often > 8 cm; gen where moist at least in spring, valleys to lower mts; BC s, both sides Cas, to Baja Cal, e to Daks, Wyo, Utah, and Ariz; miner's lettuce (*Claytonia p., C. depressa, C. parviflora*)

12 M. perfoliata (Donn) Howell

Portulaca L.

Fls solitary in axils or clustered at br tips, opening in sunlight; sepals 2, adnate to base of ovary; petals 5 (4–6), white to yellow, red, or purple; stamens 6–∞; styles 3–9; ovary ± inferior; ours ann with exstip, mostly alt lvs. (L name, used by Pliny; derivation obscure).

P. oleracea L. Common purslane. Glab and fleshy, mostly prostrate; lvs spatulate to obovate, 1–3 cm × 3–10 mm; sepals and petals ca 3–4 mm; corolla yellow; stamens 6–10; styles 4–6; European weed, widespread in gardens and newly waste ground.

Spraguea Torr.   Pussypaws

Fls capitate in ± scorpioid umbels or pans on lfless or bracteate peduncles; sepals scarious-margined, white to pink, accrescent and much larger than the white to pink petals; stigma bilobed; caps flattened, 2-valved; seeds several, black and shining, ca 1 mm; ann to per, glab, fleshy herbs with tufted basal lvs. (For Isaac Sprague, 1811–1895, botanical artist).

S. umbellata Torr. Ann or per, forming mats up to 1.5 dm broad; lvs obovate-spatulate to oblanceolate; sepals 4–10 mm broad; pine woods to subalp ridges; BC s, mostly E Cas, to Baja Cal, e to Mont, Wyo, and Utah; 2 vars.
a1 Pl per; lvs mostly < 15 mm, often rounded; mostly e side Cas, BC, Wn, and Ore, to Cal; Mt Hood p.   var. **caudicifera** Gray
a2 Pl often ann; lvs gen > 15 mm, often obtuse to acute; occ in Cas, but common e to Mont and s; umbellate p. *(Calyptridium u.)* var. **umbellata**

Talinum Adans.   Fameflower; Talinum

Fls showy, diffusely cymose on lfless penducles; sepals ± orbicular but short-pointed; petals 5 (6–7), white to red, quickly withering; stamens 15–30, hypog; style 3-cleft; caps dehiscent by 3 valves; seeds lenticular, gen shining; glab per with ∞ linear, nearly terete basal lvs mostly with the midribs persistent and often spinescent. (Believed to be a colloquial African name of uncertain meaning).

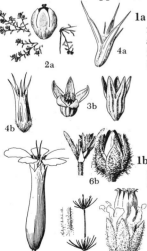

1a Petals pale to deep red, 7–10 mm; lvs 15–25 mm, midribs persistent, straight, hardened, 5–15 mm; mostly on basaltic outcrops and scablands in sagebr des of c Wn to nc Ore; spinescent f. or t.   **1 T. spinescens** Torr.
1b Petals white (very pale pink), 6–8 mm; lvs 5–12 mm, midribs not always persistent, more bristlelike than spinescent, gen curved, mostly < 3 mm; exposed rocky slopes in lower mts of Okanogan Co, Wn, n to s BC; Okanogan f. or t.   **2 T. okanoganense** English

1b

1a

CARYOPHYLLACEAE   Pink Family

Fls complete or sometimes apet or ♀♂ or ♀, ♂, mostly cymose and ped; sepals distinct to connate, gen 5 (4); petals 5 (4, rarely 0), often deeply bilobed to laciniate or fringed, sometimes with well defined claw and bl and mostly with appendages on adaxial side at juncture of the two, claw sometimes prolonged upward into aurs; stamens mostly = or 2× sepals, sometimes fewer, inserted under ovary and often ± fused with petals at base, or sometimes inserted at the edge of a small perig disc; pistil 2–5-carpellary; ovary mostly 1-celled, with free-central or basal placentation, but sometimes imperfectly 3–5-celled, often borne on a short stalk; styles 1–5, mostly distinct; fr gen caps but sometimes 1-seeded and indehiscent; ann to per herbs (a few suffrutescent), mostly with swollen nodes and opp (alt), simple, entire, exstip (scarious-stip) lvs.

1a Fr 1-seeded, indehiscent, gen closely invested by the hardened calyx; stamens borne at the edge of a perig disc surrounding the ovary and often adnate to the calyx
  2a Styles 3; lvs alt; calyx ca 1 mm   **Corrigiola**
  2b Styles 1 or 2; lvs opp; calyx much > 1 mm
    3a Calyx lobes much > the tube, ± hooded and spinose at the tip; style 1, cleft above
      4a Sepals very unequal, 3 outer ones largest and nearly enclosing the inner 2, the spine nearly = the body   **Cardionema**
      4b Sepals more nearly =, the spine < the body   **Paronychia**
    3b Calyx lobes gen ca = tube, neither hooded nor spinose at the tip; styles 2   **Scleranthus**
1b Fr several-seeded, dehiscent, not closely invested by the hardened calyx; stamens mostly hypog
  5a Nodes each with 2 fairly prominent scarious stips
    6a Styles and caps valves 5; lvs apparently whorled but really in 2 clusters at each node   **Spergula**
    6b Styles and caps valves 3; lvs opp or fascicled   **Spergularia**
  5b Nodes without scarious stips
    7a Sepals almost or quite distinct   **Group I**
    7b Sepals ± connate, gen for > half their length   **Group II**, lead 15a

Group I

8a Pl per, glab, fleshy, occurring only along seacoast; fls single in lf axils; disc prominent; seeds smooth, ca 3 mm **Honkenya**
8b Pl of varied habit, often ann, seldom along seacoast; fls mostly cymose, rarely with disc; seeds ± rugose or papillate, < 3 mm
  9a Sepals very unequal, setaceous, gen recurved; pl ann; fls apet; c Harney Co, Ore, and s, prob not in our area **Loeflingia squarrosa** Nutt.
  9b Sepals = or subequal, seldom setaceous, rarely recurved; pl often per
    10a Styles mostly 5; caps narrowly cylindric, delicately membranous, (1) 2–3 times as long as the calyx and curved near the tip, dehiscent by 10 gen revolute-margined teeth **Cerastium**
    10b Styles gen 3 or 4, when 5 the caps neither cylindric nor curved near the tip, and dehiscent by < 10 teeth or valves
      11a Styles normally 5 (4), alt with the sepals; valves of caps opp the sepals, entire **Sagina**
      11b Styles normally 3, if 4 or 5 then opp the sepals and the valves of the caps alt with the sepals
        12a Caps cylindric, delicately membranous; petals ± toothed or jagged but not lobed; fls ± umbellate **Holosteum**
        12b Caps mostly ovoid, sometimes hardened, occ delicately membranous, but then the petals bilobed; fls mostly solitary or cymose
          13a Petals ± deeply bilobed, sometimes reduced or wanting; caps dehiscent by twice as many teeth as styles; stamens and petals inserted under the ovary
            14a Caps delicately membranous, cylindric, curved to one side; ann (ours) **Cerastium**
            14b Caps ovoid to ovoid-cylindric, not curved to one side; ann or per **Stellaria**
          13b Petals gen rounded (rarely ± shallowly retuse); caps often dehiscent by the same no of valves as styles; valves entire or 2-toothed; stamens and petals often inserted at the edge of a disc **Arenaria**

Group II

15a Styles commonly 2 (rarely 3)
  16a Fls closely subtended by 1–several pairs of approximate, often ± connate bracts
    17a Calyx 20–40- nerved, ca 15 mm **Dianthus**
    17b Calyx 5-nerved, 4–6 mm; very occ, perhaps not estab, but collected at Coeur d'Alene, Ida **Tunica saxifraga** (L.) Scop.
  16b Fls not closely subtended by bracts
    18a Calyx much < 1 cm, strongly membranous below the sinuses, 5-nerved **Gypsophila**
    18b Calyx at least 1 cm, not particularly membranous below the sinuses, 10–20-nerved
      19a Petals appendaged at juncture of claw and bl **Saponaria**
      19b Petals neither appendaged nor clearly divided into claw and bl **Vaccaria**
15b Styles gen 3–5
  20a Petals unappendaged; calyx teeth > the considerably hardened calyx tube; ovary not stalked **Agrostemma**
  20b Petals mostly appendaged; calyx teeth < the tube; ovary gen stalked within the calyx
    21a Styles gen 3, if (as occ) 4 or 5 then fls gen ♂ **Silene**
    21b Styles gen 5, if (as occ) 4 then fls gen ♂, ♀ **Lychnis**

Agrostemma L.

Fls several in an open dichasially br cyme, but apparently terminal on the main and lateral sts, ♂, showy; calyx teeth > the strongly 10-ribbed tube; pl ann. (Gr *agros*, field, and *stemma*, wreath; the fls are showy).

A. githago L. Common corncockle, corn campion. Finely white-hirsute, up to 10 dm; lvs linear-lanceolate, 5–12 cm; calyx tube 12–16 mm, the lobes 20–40 mm; petals red; stamens 10, inserted with the petals around the base of the sessile ovary; styles 5; ovary 1-celled; caps 5-valved; seeds black, ca 3 mm, tuberculate; European, a weed of wastel and roadsides, fairly common in Wn and Ore, occ in Ida and Mont.

Arenaria L.   Sandwort

Fls mostly several to ∝ in diffuse to capitate cymes, but occ single and terminal, complete or sometimes apet or functionally ♂ ♀; sepals 5, free or slightly connate; petals 5, < to 2–3 × > the calyx, entire to emarginate, not bilobed; stamens mostly 10, gen inserted at the edge of a small perig disc; styles 3 (2–5); caps 1-celled; ann to per herbs. (L *arena*, sand, the pls often growing in sandy places). (*Alsinopsis, Moehringia*).

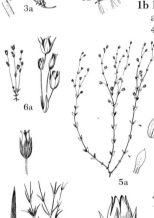

1a Lvs narrowly oblong to ovate or obovate, not sharp-pointed, mostly much > 2 mm broad; pl per
   2a St flaccid, 1–2 mm thick, glab; lvs ± linear, 2–5 cm × 2–5 mm; swampy places, mostly along coast, from Tacoma "prairies" and sw Wn to Cal, rarely collected; swamp s. (*A. palustris*)   1 **A. paludicola** Robins.
   2b St firm, < 1 mm thick, scabrid-puberulent; lvs various
      3a Sepals acute or acuminate, mostly 3–6 mm; petals < to 1.5 times > sepals; lvs linear-elliptic to lanceolate, mostly 2–5 (7) cm × 2–15 mm, acute; moist to dry, shaded to open places, meadowl to mont slopes; BC s, on both sides Cas, through Ore to Cal; e to Lab, s in RM to NM; biglf s.   2 **A. macrophylla** Hook.
      3b Sepals oblong, rounded, 2.5–3 mm; petals ca twice as long as sepals; lvs narrowly elliptic-oblong to oblanceolate, 1–3 (4) cm × 2–12 mm, gen rounded to obtuse; moist woods to dry or moist meadows and mont slopes; e Cas, through Wn and Ore to Cal, and in RM to NM, circumboreal; bluntlf s.   3 **A. lateriflora** L.
1b Lvs mostly narrowly linear and sharp-pointed, if > 1.5 mm broad then pl always ann
   4a Pl ann or short-lived per of lowl, rarely at all matted
      5a Lvs ovate to lanceolate, 3–7 × 1.5–4 mm, 3 (5)-nerved; pl retrorsely puberulent, often ± glandular above; European, fairly common weed in our area in dry to moist, barren to wooded places; thyme-lf s.
         4 **A. serpyllifolia** L.
      5b Lvs linear, < 1.5 mm broad, 1-nerved
         6a Pl 2–5 cm, glab and ± glaucous; lvs 2–4 mm, 2ndary lvs not fasciculate in axils of the primary; sepals 2–3 mm; dry areas in sagebr and ponderosa pine for; Klickitat Co, Wn, to Cal, e to se Wn, ne Ore, and Ida; dwarf s. (*Alsine p.*)   5 **A. pusilla** Wats.
         6b Pl 10–30 cm, ± glandular-puberulent above; lvs 3–20 mm, 2ndary lvs gen fasciculate in axils of the primary; sepals 2.5–3.5 mm; lowl, esp along coast; s BC to n Ore, mostly w Cas; slender s. (*A. tenella*); ours the var. *puberulenta* (Peck) Hitchc.   6 **A. stricta** Michx.
   4b Pl per, either mont or matted
      7a Caps dehiscing by 3 entire valves; pl gen matted and < 10 (up to 15) cm; primary lvs mostly < 10 mm, often subtending fascicles of 2ndary lvs; fls 1–∝ in open cymes; seeds plump, 0.5–1.5 mm   **Group I**
      7b Cap dehiscing by 6 valves; pls often much > 10 cm; fls often congested; lvs often > 10 mm, rarely subtending fascicles of 2ndary lvs; seeds flattened and ± wing-margined, (1.25) 1.5–3 mm
         Group II, lead 11a

Group I

8a Pl glab; sepals ca 2.5 (3–3.5) mm; fl st 1–3 cm; caps globose-ovoid, < the sepals; lvs linear, 4–6 mm, 1-nerved; tufted alp or subalp pl of gravelly soil; Ross s.; 2 vars.   7 **A. rossii** R. Br.
   a1 Petals gen = or > calyx; pl with trailing, slender sts; Alas to Cas and OM of Wn, and in RM to Colo and c Ida, possibly in ne Ore; Siberia and Greenl   var. **rossii**
   a2 Petals rudimentary or lacking; pl densely tufted, without trailing sts; common phase in Ida, Mont, and Wyo   var. **apetala** Maguire

8b Pl mostly glandular-pubescent; sepals often > 3 mm; fl st often much > 3 cm; caps gen narrowly ovoid and much > the sepals; lvs often 3-nerved

  9a Fl st brittle, gen shattering at the nodes; lvs subulate to linear-lanceolate, 3–10 mm, mostly pungent, 3-nerved; caps < the acuminate to pungent sepals; seeds ca 1.5 mm, prominently elongate-papillate in concentric rows; sagebr hills to alp slopes, esp on gravelly benches or talus; Nuttall's s.; 2 vars.   8 **A. nuttallii** Pax

    a1 Lvs ascending or strict, not particularly yellowish; pl glandular-pubescent to pilose; BC s, in Cas, to Ore, e to Alta, Wyo, and Utah (*A. pungens*)   var. **nuttallii**

    a2 Lvs arcuate or squarrose, yellowish-green; pl glandular-pubescent with shorter, thicker hairs; Grant and Malheur cos, Ore, to Nev and Cal   var. **fragilis** (Mag. & Holmg.) Hitchc.

  9b Fl st not brittle and shattering; lvs mostly obtuse or only slightly mucronate, 1- or 3-nerved; caps often > the obtuse or slightly acute sepals; seeds rarely > 1 mm, very lightly reticulate to tessellate-tuberculate

    10a Sepals 4–5 mm, the tip mostly obtuse and ± erose, slightly incurved and hooded, purplish; coarse, matted per, 1–4 dm broad, from thick taproot; fl st 1–6 (8) cm, mostly 1 (2–3)-fld; seeds 0.8–1 mm; subalp to alp ridges and talus slopes; Alas to Ore, in both Cas and OM, e to RM as far s as NM, and in Ida and ne Ore; Gaspé, Greenl; arctic s. (*A. obtusa*, misapplied; *Al. o.*)   9 **A. obtusiloba** (Rydb.) Fern.

    10b Sepals 3–4 mm, obtuse to acute, not hooded; delicate per with slender taproot, forming cushions barely 1 dm broad; fl st 1–10 (15) cm, (1) 2–5 (7)-fld; seeds mostly < 0.8 mm; alp or subalp meadows to ridges, Alas to Greenl, s through Cas and OM to Cal and Nev, and in RM to Colo and NM, through Ida to ne Ore; Eurasia; reddish s., boreal s. (*A. aequicaulis*, *A. propinqua*, *A. quadrivalvis*)   10 **A. rubella** (Wahlenb.) J. E. Smith

Group II

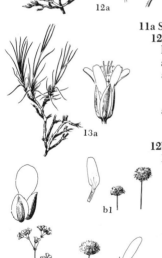

11a Sepals acute or acuminate or infl considerably congested to capitate

  12a Infl openly to diffusely cymose, not congested; pl gen pulverulent-pubescent (not glandular) on lower lvs and st (glab); sagebr hills to alp ridges; King's s.; 2 vars.   11 **A. kingii** (Wats.) Jones

    a1 Sepals ca 4 (3.5–4.5) mm; seeds 1.5–2 mm; pl mostly finely pulverulent-puberulent; se Ore to Cal, e to c Ida and s Utah, mostly to ca midmont (*A. g.*)   var. **glabrescens** (Wats.) Maguire

    a2 Sepals ca 5–6 mm; seeds 2–2.5 mm; pl mostly glab below but glandular-pubescent above; se Ida and w Wyo to Utah (*A. u.*)   var. **uintahensis** (Nels.) Hitchc.

  12b Infl ± congested to glomerate; pl gen glandular-pubescent

    13a Sepals 3–6 mm; petals ca half again > the sepals; lvs linear, 0.5–8 cm, ± pungent; fl st (1) 1.5–5 dm, nearly lfless; sagebr des to alp slopes; capitate s., ballhead s.; several vars.   12 **A. congesta** Nutt.

    a1 Sepals obtuse; pl not glandular-puberulent, gen glab

      b1 Fls capitate, peds not evident; c Wn to Cal, Mont, Utah, and Colo (*A. burkei*, *A. c.*)   var. **congesta**

      b2 Fls ± loosely arranged, peds evident, often > the sepals; c Ida to Mont, Wyo, Colo, and Utah, esp at higher elev   var. **lithophila** Rydb.

    a2 Sepals acute or with short, acute to acuminate tip projecting past the membranous margins; pl sometimes glandular-puberulent

      c1 Fls capitate, peds not evident; c Wn to w and c Ida, s to s Ore (*A. c.*)   var. **cephaloidea** (Rydb.) Maguire

      c2 Fls gen evidently cymose, peds visible

        d1 Pl glandular-puberulent above; Douglas Co, Wn, and Valley and Washington cos, Ida   var. **glandulifera** Maguire

        d2 Pl glab (rarely slightly glandular); common along e side Cas from Okanogan Co, Wn, to n Ore   var. **prolifera** Maguire

    13b Sepals 5–12 mm; petals only slightly or no > the sepals; lvs various; fl st mostly 3–10 (15) cm, gen lfy

14a Fl st glab, brittle and easily shattered, very lfy; petals rarely > the sepals; lvs linear-acicular, (6) 10–20 × ca 1 mm; sand dunes to sagebr slopes; 2 vars.   13 **A. franklinii** Dougl.

    a1 Sepals 5–8 mm, barely > the petals; cymes not tightly congested, peds 1–4 mm; along CR, c Ore; Thompson's s.   var. **thompsonii** Peck

    a2 Sepals 8–12 mm, often 2–3 mm > the petals; cymes congested, peds rarely > 1 mm; c Wn to s Ore and Nev, e to Butte Co, Ida; Franklin's s.   var. **franklinii**

14b Fl st pubescent, not brittle or easily shattered, gen with only 1–4 pairs of lvs; petals slightly > the sepals; lvs acicular, 6–10 × ca 1 mm; sagebr plains to lower mont slopes, mostly where sandy or rocky; c Mont to Tex and Neb, gen e of RM, but in ne Utah; Hooker's s.   14 **A. hookeri** Nutt.

11b Sepals obtuse or only shortly acute; infl diffusely cymose

    15a Lvs 1–2 (3) cm, rigid and pungent, often ± squarrose, mostly basal; fl st with 1–3 pairs of much-reduced lvs scarcely 5 mm and always < half as long as the basal lvs; seeds 2.5–3 mm; ± gravelly sagebr hills to rocky alp slopes; Beaverhead Co, Mont, to se Ore and Utah, Nev, and ne Cal; prickly s., needlelf s. *(A. salmonensis)*   15 **A. aculeata** Wats.

    15b Lvs often > 2 cm, rather soft, more acuminate than rigidly pungent, often erect or curved but not squarrose; st lvs 2–5 pairs and at least half as long as the basal lvs; seeds 2–2.5 mm; sagebr plains to subalp rocky ridges; Alas to n Ore, in Cas and OM, e to Alta, Mont, and Nev; Eurasia; mt s., thread-lvd s. *(A. nardifolia*, misapplied); ours the var. *americana* (Mag.) Davis   16 **A. capillaris** Poir.

Cardionema DC.

Fls complete; sepals and petals 5; stamens 3–5; low per herbs with linear, pungent lvs and prominent, scarious stips. (Gr *kardio*, heart, and *nema*, thread, the anthers ± heart-shaped on slender filaments).

C. ramosissima (Weinm.) Nels. and Macbr. Sandmat. Pl matted; sts up to 3 dm, finely pubescent to ± tomentose, gen covered at base with marcescent lvs and stips; lvs crowded, 5–10 mm; fls sessile in clusters in the upper lf axils, inconspicuous; calyx turbinate, sepals unequal, distinct except at the base, 4–5 mm, the spines as long, divergent; petals scalelike, connate at base with the stamens, forming a short tube surrounding the ovary but free of the calyx; sandy beaches along the coast; se Wn to Mex and Chile *(Pentacaena r.)*.

Cerastium L.   Chickweed; Cerastium

Fls hypog, ☿, mostly cymose, or single and terminal; sepals 5, distinct, mostly scarious-margined; petals 5 (3); styles opp sepals; caps 1-celled; ann to per herbs. (Gr *keras*, horn in reference to the slender curved caps).

1a Pl per, native; petals at least 1.5 times > the sepals

    2a Fl bracts rarely at all scarious; lvs narrowly oblong, oblong-lanceolate, or spatulate, mostly 10–15 (30) × 2–5 (8) mm, without axillary fascicles of smaller lvs or sterile brs; calyx glandular-hairy, hairs multicellular and up to 0.5 mm; cymes 1–5 (7)-fld; alp cirques and talus, BC and Alta, Mont, and Ida, to Colo; also in Cal, Ariz, and rarely in Wn Cas; alpine ce., alpine ch. *(C. alpinum, C. earlei)*   1 **C. berringianum** Cham. & Schlecht.

    2b Fl bracts scarious-margined; lvs linear to narrowly lanceolate, 10–30 (40) × 2–4 (8) mm, those of the st gen with axillary whorls of reduced lvs or sterile brs; calyx mostly with hairs much < 0.5 mm; cymes (3) 5–9-fld; gen, coastal cliffs to subalp (rarely alp) slopes; field ch., starry ce., field mouse-ear *(C. angustatum, C. campestre, C. elongatum, C. graminifolium, C. leibergii, C. scopulorum, C. thermale, C. variabile)*   2 **C. arvense** L.

1b Pl ann or weedy per; petals < 1.5 times > the sepals

    3a Styles 3; valves of caps 6; European, estab weed in Clark Co, Wn; doubtful ch.   3 **C. dubium** L.

    3b Styles 5; valves of caps 10

        4a Fls ± glomerate; ped < the sepals, gen < 5 mm; petals mostly

subequal to sepals; pl ann; lvs spatulate to obovate, mostly 8–20 × up to 8 mm

5a Seeds ca 1 mm; sepals 7–9 mm; European, rare weed with us, but known from Pullman, Wn, and from Union Co, and from near La Grande, Ore; dry ch. 4 **C. siculum** Guss.

5b Seeds ca 0.5 mm; sepals 4–5 (6) mm; European, widespread weed at lower elev, mostly w Cas; sticky ch. or ce. 5 **C. viscosum** L.

4b Fls not glomerate; ped up to twice as long as the sepals; petals and lvs various

6a Sepals nearly 1 mm > the shallowly retuse petals, broadly scarious-margined; ann; known from Vancouver, BC, and possibly in our area 6 **C. semidecandrum** L.

6b Sepals only slightly if any > the ± bilobed petals, not broadly scarious-margined; sometimes bien or per; lvs mostly spatulate to oblanceolate, 7–40 × 2–5 (10) mm

7a Pl ann, ascending to erect, not matted; ped and calyx gen glandular-pilose; petals = or > sepals; widespread, occ in dry to moist areas at low elev; nodding ch. or ce. 7 **C. nutans** Raf.

7b Pl bien or per, ± matted; ped and calyx gen hirsute, mostly nonglandular; petals subequal to the sepals; widespread, bad lawnweed w Cas; common or big ch., mouse-ear ch. 8 **C. vulgatum** L.

Corrigiola L.

Fls ∝ in axillary and terminal clusters, tiny, complete; calyx gamosep, 5-lobed; petals 5, entire; stamens 5; ann to per herbs with alt, scarious-stip lvs. (L *corrigia,* shoestring, because of the slender sts).

C. litoralis L. Strapwort. Glab and glaucous ann or bien; sts ∝, decumbent, freely br, up to 2.5 dm; lvs oblanceolate, 0.2–2 (3) cm; stips white, ca 1 mm, denticulate; ped gen < the fl, jointed just below the calyx; calyx ca 1 mm, red or greenish, the lobes white-margined, ca = tube; petals subequal to the calyx, white or pinkish; stamens included, anthers purple; fr ca 1 mm; European, intro and apparently estab near Portland, Ore.

Dianthus L.   Pink

Fls solitary to cymose, mostly showy, complete (ours) to ♂♀; calyx gamosep, narrowly tubular, closely bracteate at base; petals white to red or purplish, clawed, but without aurs or appendages; stamens 10, fused with petals around ovary stalk; styles 2; caps 1-celled, dehiscent by 4 valves; ann to per herbs mostly with stiff, erect sts. (Gr *dios,* of the gods, and *anthos,* flower).

1a Pl glab; basal lvs 10–20 mm broad; European, escaped from gardens in w Wn and Ore; sweet william 1 **D. barbatus** L.

1b Pl gen pubescent; basal lvs rarely > 5 mm broad

2a Fls single or in loose 2–3 (5)-fld cymes; ovary stalk ca 3 mm; pl scabrid-puberulent; European per, occ estab in w Wn and w Mont, at least; maiden p. 2 **D. deltoides** L.

2b Fls several in much-congested cyme; ovary stalk ca 1 mm; pl gen ± crisp-puberulent; garden ann from Europe, frequently escaping; known from Wn, Ida, Mont, and Ore; grass p., Deptford p. 3 **D. armeria** L.

Gypsophila L.

Fls complete, diffusely cymose; calyx gamosep; petals white; stamens 10, hypog; styles 2; caps 1-celled, 4–6-valved, 2–5-seeded; per (ours). (Gr *gypsos,* gypsum, and *philos,* friendship, the preference of some spp.).

G. paniculata L. Baby's breath. Pl glab and glaucous, freely br, 4–8 (12) dm; lvs linear to lanceolate, 2–5 (10) × up to 1 cm; ped 2–4 × > fls; calyx turbinate, ca 2 mm, membranous in the sinuses, the lobes purplish with white margins; petals oblanceolate, slightly > the sepals, distinct, lacking well-defined claws or appendages; Asian, escaped and estab in several areas, esp in Wn and Ida.

Holosteum L.

Fls complete; sepals 5, distinct; petals 5 (4), whitish; stamens 5–10; styles 3; caps 1-celled, ovoid-cylindric, dehiscent by 6 slightly recurving valves, ∝ -seeded; ann (ours). (Gr *holosteon*, whole bone, possibly a humorous reference to the delicate nature of the pl).

H. umbellatum L. Jagged chickweed. Sts 1–several, ± glandular-pubescent above, up to 3 dm; lvs mainly in basal rosette, 1–2 cm, oblong to lanceolate; fls (2) 4–16 in a terminal umbel-like cyme with a tiny bracteate invol; ped slender, 2–3 cm; sepals ca 3 (5) mm; petals gen slightly > the sepals, erose-fimbriate; caps gen twice the length of the calyx, valve tips recurved to revolute; Eurasian weed, common, but mostly along CR and SR.

Honkenya Ehrb.

Fls complete or a few ♀ ♂; stamens 10(8), inserted with the petals at the edge of a prominent disc; styles 3 (2–6), short; caps 3–5-celled until maturity and then 1-celled; fleshy per herbs. (For G. A. Honckeny, a German botanist, 1724–1805). (*Ammodenia*).

H. peploides (L.) Ehrb. Sea purslane, seabeach sandwort. Pl rhizomatous, fleshy, ± trailing; lvs 1–3 (4) cm; sepals 5 (4), distinct, 4–7 mm; petals 5 (4), whitish, entire, slightly clawed, = sepals or only half as long; caps 5–8 mm, globose-ovoid, tardily dehiscent by 3–5 valves; seeds ca 3 mm; coastal beaches, Alas to n Ore; e N Am, Greenl, Iceland, n Eurasia.

Lychnis L.   Campion

Fls hypog, few to several and cymose, or sometimes single and terminal, complete or sometimes ♀ ♂; calyx gamosep, tubular; corolla white to red, petals clawed and often with aurs, bl gen retuse to 2–4-lobed and with 2(4) appendages; stamens 10, connate at base and fused with the petals, forming a short tube around the ovary stalk; styles 5 (4); caps gen 1-celled but sometimes ± completely 5 (4)-celled, dehiscent by (4) 5 or (8) 10 teeth, ∝ -seeded; ann to per herbs. (Gr *lychnos*, lamp, in reference to the bright colors of the fls).

1a Pls grayish-tomentose, eglandular; calyx lobes twisted; petals red, bl broadly obcordate; European orn, often persisting in waste areas, chiefly w Cas; rose c.   1 L. coronaria (L.) Desr.
1b Pls glandular above, not grayish-tomentose; calyx lobes not twisted; petals white to red
  2a Pl ♀, ♂; petal bl > 5 mm
    3a Fl white, vespertine, fragrant; European weedy per (bien); widely distributed in waste or undisturbed, mostly rather dry areas; white c., evening c.   2 L. alba Mill.
    3b Fl red, diurnal, odorless; European, escaped and estab in several areas w Cas, BC to Ore; red c. (*L. rosea, L. rubra*)   3 L. dioica L.
  2b Pl mostly ♀; petal bl < 5 mm
    4a Fr calyx inflated and camp; fls 1–3; petals pink to mauve, bl (2) 3–4 mm; buds nodding; seeds 1–2 mm, wing-margined; pl rarely > 2.5 dm; circumpolar, s in high RM of Mont and Ida to Colo; apet c. (*Wahlbergella a.*); ours the var. *montana* (Wats.) Hitchc.
                                  4 L. apetala L.
    4b Fr calyx tubular, not inflated; fls 1–9; petals white or pink, bl 1–3 mm; seeds ca 0.7 mm, not wing-margined; pl 2–5 dm; sagebr slopes to alp ridges, BC to Sask and s to Colo and Ariz; in our area in Ida and Mont; Drummond c. (*Wahlbergella d., W. striata*)
                              5 L. drummondii (Hook.) Wats.

Paronychia Adans.

Fls ♀, single and terminal or in small cymes; calyx 5-lobed; stamens 5, opp the sepals; per with subulate, crowded lvs and prominent scarious stips. (Gr name for a finger infection, whitlow, for which the pl was believed to be a cure).

P. sessiliflora Nutt. Whitlow wort. Puberulent to glab, caespitose and matted, yellow-green per up to 2 dm broad; lvs linear, ca 5 mm, greatly crowded; fl sts 1–3 (14) cm; fls inconspicuous, apet; calyx gamosep, brownish-yellow, ca 3 mm, lobes ± hooded and short-awned; stamens borne on a short, perig disc lining the calyx, sometimes alt with tiny linear staminodia or vestiges of petals; dry sterile soil of grassl to ponderosa pine woodl; Alta and Daks to Tex, mostly e RM, in our area in Mont.

Sagina L.   Pearlwort

Fls small, hypog, complete to sometimes apet, slenderly ped, axillary to cymose; sepals 4–5, distinct; petals 5 (4), rarely lacking, white, entire to retuse; stamens gen 5 but sometimes fewer or up to 10; styles gen as many as the sepals and alt with them; caps valves as many as the sepals and opp them; seeds ∝ ; ann or per, mostly matted herbs with opp, linear, basally connate lvs mostly (2) 5–20 × ca 1 mm. (L, meaning nourishing or fattening).

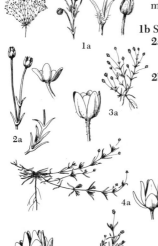

1a Sepals 4; petals 0; pl finely glandular-puberulent above; lvs 2–4 (5) mm; filiform-std, ann European weed; w Wn to Cal; common p. *(S. ciliata)*
                                                                    1 **S. apetala** L.
1b Sepals 5, or petals present, or pl not glandular-pubescent
   2a Pl of moist mont areas, per, matted, glab; lvs seldom fascicled; sepals erect in fr; caps often ± reflexed; common in most of mts of w N Am; Eurasia; alpine or arctic p.   2 **S. saginoides** (L.) Britt.
   2b Pl rarely mont, sometimes ann, or glandular-pubescent, or not matted; lvs sometimes fascicled; sepals sometimes spreading in fr; caps mostly erect
     3a Ped and calyx ± weakly glandular-puberulent; pl ann, gen with lvs neither fasciculate nor in basal rosette; fls 5-merous; low elev where moist, mostly w Cas, reported e to Ida; western p.
                                                                    3 **S. occidentalis** Wats.
     3b Ped and calyx gen glab; pl often per with basal rosette of fascicled lvs; fls sometimes 4-merous
       4a Sepals 4 or 5, often spreading in fr, mostly < 2 (2.5) mm; petals often lacking; caps 4–5-valved; seeds ca 0.3 mm, dull; European, intro weed, neither fleshy not strictly coastal; w Cas, BC to n Cal in our area; procumbent p.   4 **S. procumbens** L.
       4b Sepals 5, erect in fr, (2.5) 3–3.5 mm; petals present; caps 5-valved; seeds ca 0.5 mm, lustrous; pl native, fleshy, strictly coastal, mostly on moist sand or rocks, Alas to Cal; stick-std p.
                                                                    5 **S. crassicaulis** Wats.

Saponaria L.

Fls cymose, complete, hypog; calyx gamosep, tubular; petals clawed, appendaged at top of the claw; stamens 10, filaments coherent with the petals at base and surrounding the short stalk of the ovary; styles 2 (3); caps 1-celled; per herbs. (L *sapo*, soap, the juice of 1 sp. lathering with water).

S. officinalis L. Bouncing bett, soapwort. Pl rhizomatous, glab or subglab, forming clumps 4–9 dm; lvs lanceolate to oblanceolate, 4–12 × 1.5–4 cm; fls fragrant, showy, white to pink; calyx 15–20 mm at anthesis, up to 25 mm in fr, 20-nerved, narrow, often deeply cleft in 1–more places; corolla showy, white to pink, bl 10–15 mm, obovate-cuneate, shallowly retuse, appendages linear; seeds black, ca 1.7 mm; European orn, occ escaped and estab, mostly in Wn and Ore.

Scleranthus L.

Fls tiny, greenish, apet, ⚥, perig, subsessile in dichotomously br, bracteate cymes; calyx gamosep, 5 (4)-merous, becoming hardened in fr; stamens mostly 5–10; ann (ours) with small, entire, opp and slightly connate lvs. (Gr *skleros*, hard, and *anthos*, fl, the calyx tube very hard).

S. annuus L. Ann knawel. Low and spreading, glab to strongly crisp-puberulent; sts up to 15 cm; lvs linear, subulate-pungent, 5–15 (20) mm, slightly scarious near the base; calyx 3–4 mm, the lobes narrowly lanceolate, = or slightly > the camp tube, strongly 10-nerved; filaments short, broadened at base into a membranous, disciform ring at the top of the calyx tube; European weed of garden or waste-place, occ throughout much of our area.

Silene L.   Campion; Catchfly; Wild Pink; Silene

Fls cymose, ♂ to ♀, ♂, often showy, hypog; calyx gamosep, 5-lobed, often becoming inflated; petals white, pink, red, or purple, clawed and gen with aurs and appendages; stamens 10, the filaments fused at base with the petals and ovary stalk; styles mostly 3 (4 or even 5); ovary 1-celled or sometimes 3 (4–5)-celled; caps dehiscent by 6 (8–10) teeth, ∞ -seeded; ann to per herbs with opp or whorled lvs. (Gr *sialon*, saliva, in reference to the sticky sts, or from *seilenos*, a woodland deity—*Silene* was a companion of *Bacchus*). (*Cucubalus*).

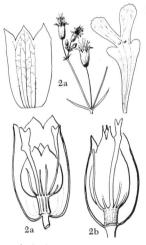

1a Calyx glab (except cilia on lobes), often umbilicate-camp, lightly 15–20-nerved, becoming membranous; petals white, bl 4–6 mm, bilobed, appendages lacking or merely tiny callosities; caps 3-celled; lvs ovate-lanceolate to oblanceolate, 3–8 × 0.5–3 cm; pl mostly 3–10 dm; bladder campion
   2a Calyx strongly inflated, up to 2 cm in fr, umbilicate; caps included in calyx, stalk 2–3 mm, glab; widely distributed weed from Europe, gen in waste places; bladder campion *(S. behen)*   **1 S. cucubalus** Wibel
   2b Calyx slightly inflated, scarcely at all umbilicate, rarely as much as 15 mm in fr; caps gen protruding from calyx, stalk < 2 mm, minutely puberulent; European and closely related to *S. cucubalus*, but much less common, although well estab in Mont   **2 S. cserei** Baumg.
1b Calyx hairy, or pl not otherwise as above
   3a Pl ann, ± weedy, mostly of waste places, all intro from Europe, but sometimes apparently native   **Group I**
   3b Pl per, native, mostly mont or on undisturbed soil   **Group II**, lead 10a

Group I

4a Calyx 25–30-nerved, obconic-ovoid, umbilicate
   5a Petal bl mostly shallowly bilobed, white to reddish, 3–6 mm; appendages 1–2 mm; lvs linear-lanceolate to oblanceolate, 2–5 cm × 2–7 mm; calyx up to 16 mm; seeds scarcely 0.8 mm; occ along coast in our area; striated cat.   **3 S. conica** L.
   5b Petal bl rounded to shallowly retuse, white to purplish, 8–12 mm; appendages 2–5 mm; lvs oblanceolate to lanceolate, 3–12 cm × 5–12 mm; calyx becoming 20–30 mm; seeds 1.2–1.5 mm; occ in our area along coast and inl to Latah Co, Ida; conoid cat.   **4 S. conoidea** L.
4b Calyx 10-nerved, mostly tubular and not umbilicate
   6a Pl glab or rarely sparsely puberulent; ovary stalk 7–8 mm; petals pinkish-lavender, bl rounded to shallowly lobed, nearly = claw; European orn, fairly common escape from gardens; sweet william cat.
                                                              5 S. armeria L.
   6b Pl strongly hairy, at least at base; ovary stalk < 5 mm; petals white to pink, red, or purple
      7a Fls subsessile in 1-sided, raceme-like infl; calyx strongly nerved and stiffly hirsute
         8a Petal bl 2–5 mm, entire to shallowly retuse; appendages ca 1 mm; lvs oblanceolate to spatulate, 1.5–4 cm × 2–8 (15) mm; common weed from Wn to Cal, esp w Cas; windmill pink, French s. (*S. anglica*)   **6 S. gallica** L.
         8b Petal bl 5–9 mm, bilobed at least half length; appendages ca 0.2 mm; lvs lanceolate to oblanceolate, 3–8 cm × 5–35 mm; widespread weed but not common in our area; forked cat.   **7 S. dichotoma** Ehrh.
      7b Fls in open to contracted cymes; calyx sometimes lightly nerved and non-hirsute
         9a Pl glab above, often glandular in bands just below the nodes; petal bl 2–4 mm, obcordate; appendages barely 0.4 mm; calyx 4–10 mm, glab; common weed throughout our area, often well up in mts; sleepy cat.   **8 S. antirrhina** L.

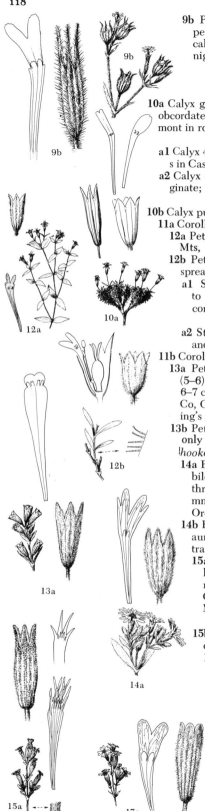

9b Pl pubescent throughout, not glandular in bands below nodes; petal bl 7–10 mm, bilobed ca 1/3 length; appendages 0.5–1.5 mm; calyx 15–30 mm, hirsute; widespread weed of cult or waste areas; nightfl s., sticky cockle   **9 S. noctiflora** L.

Group II

10a Calyx glab; matted per, rarely as much as 1 dm; petals pink (white), bl obcordate, < 5 mm; appendages scarcely 1 mm; ovary stalk 1–2 mm; high mont in rock crevices or on talus; moss cam., moss pink, cushion p.; 2 vars.
                                                                      10 S. acaulis L.
   a1 Calyx 4–6 (7) mm; petal bl mostly obovate and emarginate; circumpolar, s in Cas and OM to Ore, and to Ida and Mont   var. **exscapa** (Allioni) DC.
   a2 Calyx (6) 7–11 mm; petal bl oblong-oblanceolate, rounded to ± emarginate; RM, s Mont to Ariz and NM
                                         var. **subacaulescens** (Williams) Fern. & St. John
10b Calyx pubescent; pl gen > 1 dm
   11a Corolla gen < 10 (12) mm; fls in open leafy cymes
      12a Petals purplish; fls ♂; lvs mostly 1–2 cm; rare, cliffs and talus, Wen Mts, Wn; Seely's s.   **11 S. seelyi** Morton & Thompson
      12b Petals white; fls functionally ♀, ♂; lvs mostly 2–6 (10) cm; widespread in w N Am; Menzies' s.; 2 vars.   **12 S. menziesii** Hook.
         a1 St glandular-pubescent below; BC s, in Cas (but not OM), to Cal, e to Mont and Wyo, not in Ariz, Utah, and NM, and less common in Wyo, s Ida, and c Ore than the var. *menziesii*
                                         var. **viscosa** (Greene) Hitchc. & Mag.
         a2 St nonglandular-pubescent below; widespread, BC to Cal, in Cas and OM, e to Alta, NM, and Ariz   var. **menziesii**
   11b Corolla gen > 12 mm; fls mostly in reduced-bracteate infl
      13a Petal bl 1–2 mm, entire or shallowly emarginate; appendages 4 (5–6); ovary stalk glab; seeds ca 2 mm; lvs lanceolate to oblanceolate, 6–7 cm × 5–15 mm; rare, sagebr plains to ponderosa pine for; Wallowa Co, Ore, to Spokane Co, Wn, adj Ida, and Flathead Co, Mont; Spalding's s.   **13 S. spaldingii** Wats.
      13b Petal bl mostly > 2 mm and deeply lobed; appendages sometimes only 2; ovary stalk mostly hairy; seeds rarely > 1.5 mm (except S. *hookeri*)
         14a Bl of petals 7–22 mm, ± equally 4-lobed > half length or rarely bilobed and with lateral teeth; aurs lacking; pl eglandular throughout; sts prostrate and ± rhizomatous at base; seeds ca 2 mm; valleys and open to wooded hillsides w of Cas from Polk Co, Ore, to nw Cal; Hooker's s. *(S. ingrami)*   **14 S. hookeri** Nutt.
         14b Bl of petals mostly < 7 mm, or bilobed but without lateral teeth; aurs often present; pl, or at least calyx, often glandular; sts rarely trailing or rhizomatous at base; seeds mostly < 1.5 mm
            15a Appendages 4 or 6, linear; petals equally 4-lobed or middle lobes deeply divided; aurs gen lacking; sagebr slopes to ponderosa pine for or subalp ridges, e slopes Cas in Wn to ne Cal, w in CRG to Skamania Co, Wn, e through se Wn and e Ore to w Mont, nw Wyo, and ne Nev; Ore s. *(S. filisecta, S. gormani)*
                                                               15 S. oregana Wats.
            15b Appendages mostly 2, or if 4 often not linear, or petals bilobed or unequally 4-lobed or with aurs
               16a Ovary stalk 3–7 mm long; petal bl 3–8 mm, bilobed up to 1/3 length but otherwise entire, or rarely with small lateral tooth on each margin
                  17a Caps 3–4-celled to near top; calyx purplish, strongly hirsute and ± glandular, tubular-clavate, 10–15 mm; pl of alp to subalp slopes, mostly on scree; occ from c Ida to sw Mont and nw Wyo, Alas, Yuk; Eurasia; creeping s.; ours the var. *australe* Hitchc. & Mag.   **16 S. repens** Pers.
                  17b Caps 1-celled; calyx rarely purplish, hirsute but eglandular to glandular-pubescent; pl rarely if ever in subalp or alp habitats
                     18a Pl copiously glandular-pubescent at least in infl and on calyx; appendages mostly oblong and ± erose (see lead 19a)   **18 S. scouleri** Hook.

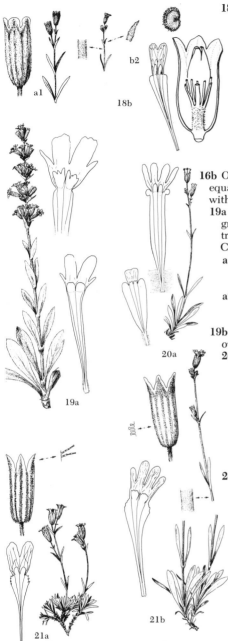

18b Pl eglandular or only weakly glandular-puberulent above; appendages linear or narrowly oblong but not erose; sagebr plains to mont ridges; BC to Cal, in both Cas and OM, e to w Mont, n Nev, and Utah; Douglas' s.; 3 vars.   **17 S. douglasii** Hook.

  a1 Lvs fleshy, those of fl st only very slightly if at all reduced; grassy slopes at Cascade Head, Tillamook Co, Ore *(S. o.)*   var. **oraria** (Peck) Hitchc. & Mag.

  a2 Lvs not fleshy, those of fl st much reduced upward

    b1 Pubescence sparse, sometimes glandular, short; lower CRG *(S. m.)*   var. **monantha** (Wats.) Robins.

    b2 Pubescence abundant, nonglandular; range of sp. except for var. oraria *(S. columbiana, S. dilatata, S. lyallii, S. macrocalyx, S. multicaule)*   var. **douglasii**

16b Ovary stalk often < 3 mm, if that long; petal bl either ± equally 4-lobed at least 1/3 its length or deeply bilobed and with prominent tooth on each side

  19a Lvs 6–25 cm × 4–30 mm, those of st ave at least 3 pairs, gradually reduced upward; ovary stalk 3–6 mm; pl never truly mont, mostly of prairies and open timberland; BC to c Cal and NM; Scouler's s.; 2 vars.   **18 S. scouleri** Hook.

    a1 Lvs < 15 mm broad, the cauline ave 3–6 pairs; petals with mostly nonciliate claw; widespread except along coast   var. **scouleri**

    a2 Lvs sometimes > 15 mm broad, fleshy, the cauline ave 3–11 pairs; claw of petals often ciliate; strictly coastal, BC to c Cal *(S. grandis)*   var. **pacifica** (Eastw.) Hitchc.

  19b Lvs often < 6 cm × 4 mm, or those of st only 1–2 pairs; ovary stalk often > 3 mm; pl often subalp to alp

    20a Ovary stalk 1.5–2.5 mm; styles mostly 4; aurs gen present; petal bl merely obcordate to 4-lobed; appendages often 4; pls of sagebr and pinyon-juniper areas of Ore, Ida, and n Nev, never high-mont, rather rare; scapose s.; 2 vars.   **19 S. scaposa** Robins.

      a1 Petal bl subentire to merely cordate, not 4-lobed; Wheeler and Gilliam cos to the Blue Mts, Ore

                               var. **scaposa**

      a2 Petal bl ± equally 4-lobed; se Ore, Lost River Mts, Ida, and n Nev   var. **lobata** Hitchc. & Mag.

    20b Ovary stalk (2) 2.5–3.5 mm; styles gen 3; aurs lacking; petal bl mostly bilobed and with 2 prominent lateral teeth, rarely with > 2 appendages; mont, to above timberl

      21a Basal lvs mostly 1–2 cm × 1.5–2 (4) mm; pl rarely > 1.5 dm; calyx hairs purple-septate; alp and subalp slopes, often on talus; Cas, from Mt Baker, Wn, to Douglas Co, Ore, but on only a few peaks; Suksdorf's s.   **20 S. suksdorfii** Robins.

      21b Basal lvs 3–8 cm × 4–10 mm; pl gen > 1.5 dm; calyx hairs often not purple-septate; mont to subalp; BC s to Cas and OM, Wn, e to RM in BC and s to c Ida and w Wyo; Parry's s. *(Wahlbergella p., S. macounii, S. tetonensis)*   **21 S. parryi** (Wats.) Hitchc. & Mag.

Spergula L.

Fls small, diffusely cymose, complete, hypog; sepals 5, distinct; petals 5, white, entire; stamens gen 10 (5); ovary 1-celled; caps dehiscent by 5 valves opp sepals, several-seeded; ann herbs with falsely whorled, linear lvs and scarious stips. (L *spargo*, to scatter, the seeds scattering when ripe).

S. arvensis L. Stickwort, starwort, cornspurry. Pl ± glandular-pubescent (subglab), light to dark green; sts mostly several, up to 6 dm, ascending to erect; lvs 1.5–4 cm × < 1 mm; ped slender, up to 4 cm, often reflexed in fr; sepals ovate, 2–3.5 mm; petals ca = or < the sepals, white; caps ovoid, slightly > calyx; seeds almost black, 1.2–1.5 mm, narrowly acute-margined, from almost smooth to thickly white-papillate; European weed, common in much of N Am, mostly w Cas in our area.

Spergularia (Pers.) J. & C. Presl   Sandspurry

Fls small, ⚲, hypog, laxly cymose; sepals 5, distinct, often scarious-margined; petals 5 (sometimes lacking), white to rose, entire, mostly < sepals; stamens 2–10; ovary 1-celled; caps dehiscent by 3 valves; seeds ± flattened, often delicately winged; ann or per, rather fleshy herbs with linear lvs and prominent scarious stips, mostly of coastal saline or inl alkaline soil. (Derived from *Spergula*). *(Tissa)*.

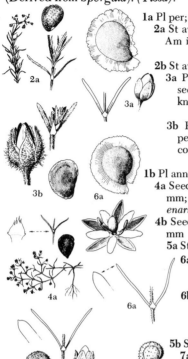

1a Pl per; stamens 7–10

  2a St ave < 1 mm thick; lvs mostly 0.3–1 mm broad; seeds 0.4–0.6 mm; S Am in origin, in our area only near Portland, Ore; hairy s.

                                                     1 **S. villosa** (Pers.) Camb.

  2b St ave > 1 mm thick; lvs mostly much > 1 mm broad; seeds 0.6–1 mm

    3a Pl glab to sparsely pubescent; sepals 3–6 mm, glab; petals white; seeds gen with wing 0.1–0.4 mm broad; European, widely intro, but known in our area only from a few collections in nw Ore; stout s.

                                              2 **S. media** (L.) Presl

    3b Pl ± glandular-pubescent; sepals 5–9 mm, glandular-pubescent; petals pinkish; seeds wingless or with wing scarcely 0.1 mm broad; common along coast, often in saltmarshes; BC to Baja Cal; beach s.

                                  3 **S. macrotheca** (Hornem.) Heynh.

1b Pl ann; stamens often < 7

  4a Seeds blackish, not winged; lvs mostly < 1 mm broad; stips mostly < 2 mm; European, intro along SR and CR, Wn, Ore, and Ida; alkali s. *(Arenaria d., A. salsuginea, S. s.)*   4 **S. diandra** (Guss.) Boiss.

  4b Seeds brown, often winged; lvs often > 1 mm broad; stips mostly > 2 mm

    5a Stamens 2–5

      6a Lvs not mucronate; seeds 0.8–1.1 mm, gen wing-margined; coastal, often on tidel, common from Alas to n Cal; Can s.

                                6 **S. canadensis** (Pers.) G. Don

      6b Lvs ± mucronate; seeds 0.5–0.9 mm, mostly not winged; European, widely intro and in our area along the coast and inl to RMS, but nowhere common; saltmarsh s. *(S. salina, S. sparsiflora)*

                                   7 **S. marina** (L.) Griseb.

    5b Stamens mostly 10, rarely as few as 6 or 7

      7a Sepals glab; seeds mostly broadly winged (see also lead 3a)

                                     2 **S. media** (L.) Presl

      7b Sepals glandular-pubescent; seeds not winged

        8a Lvs rarely > 15 mm, fascicled, stips thus several at each node, mostly 3.5–6 mm; common European weed, abundant in gardens and waste areas; red s.   8 **S. rubra** (L.) Presl

        8b Lvs often > 15 (to 40) mm, rarely fascicled, stips thus gen only 2 per node, mostly < 2.5 (1.5–4) mm; European, known in our area only from old collections near Portland, Ore; Bocconi's s.

                                 9 **S. bocconii** (Scheele) Fouc.

Stellaria L.   Starwort

Fls rather small, ⚲, axillary to cymose; sepals 5 (4), distinct; petals 5 (4), white, deeply to shallowly bilobed, but sometimes greatly reduced or lacking; stamens hypog, mostly 10 (5 or fewer); styles gen 3 (4 or 5); caps ± ovoid, 1-celled, tardily dehiscent by twice as many teeth as styles, ∞ - seeded; ann or per, gen ± rhizomatous herbs, often with 4-angled st and sessile or short-petiolate, linear to ovate lvs. (L *stella*, star, referring to fl shape). *(Alsine, Stellularia)*.

1a Sts, petioles, and peds pubescent in longitudinal lines; lvs ovate to ovate-elliptic, lower ones distinctly petiolate, ovate to ovate-elliptic, (5) 10–30 mm; European; widespread, common, noxious weed; chickweed

                                      1 **S. media** (L.) Cyrill.

1b Sts, petioles, and peds glab or ± uniformly pubescent; lvs various

  2a Pl glandular-pubescent above; petals much > the sepals

    3a Styles gen 5; sepals 5–9 mm; lvs petiolate; European weed, reported for BC, not seen in our area   **S. aquatica** (L.) Scop.

    3b Styles gen 3 (4); sepals rarely > 5 mm; lower lvs sessile

      4a St mostly > 1.5 dm; lvs (2) 3–15 cm × (1.5) 3–10 (20) mm, ave >

4 times as long as broad; cymes diffusely fld; dry or moist woods to subalp meadows; Cas, Chelan Co, Wn, s to Sierra Nevada, Cal, e to Ida, Utah, Wyo, and NM, but not in Mont; sticky s., sticky chickweed (*A. curtisii, A. j.*)   2 **S. jamesiana** Torr.

4b St mostly < 1.5 dm; lvs 1–3 cm × up to 15 mm, ave < 4 times as long as broad; cymes few-fld; higher mts, on talus mostly, sc Mont to sw Alta; Am s.   3 **S. americana** (Porter) Standl.

2b Pl glab to nonglandular-pubescent; petals often < the sepals, or lacking

    5a Basal lvs (at least) oblong, ovate, or ovate-lanceolate and ± petiolate; fls often axillary and single; petals mostly much< the sepals, often lacking   **Group I**

    5b Basal lvs similar to the cauline, linear to lanceolate, all gen sessile; fls rarely single, mostly cymose or umbellate; petals often > the sepals, but sometimes < or even lacking   **Group II, lead 11a**

Group I

6a Pl ann, delicate, filiform-std, not rooting at nodes; st lfy chiefly near base; sepals plainly 3-nerved; gravelly meadows to grassy hillsides, often where moist, common; BC to Baja Cal, e to Ida, w Mont, and Utah; shining chickweed *(S. praecox)*   4 **S. nitens** Nutt.

6b Pl either per or freely rooting at nodes; st ± equably lfy; sepals often 1-nerved

    7a St pubescent above, often matted; petals much < sepals, deeply bilobed (lacking); mont; Cas and OM, Wn, to c Cal, e to Ida, Mont, and nw Wyo; Simcoe Mt s. (*A. s., S. borealis* var. *s., S. washingtoniana*)
                                          5 **S. simcoei** (Howell) Hitchc.

    7b St glab; petals various, often lacking

        8a Fls single in lf axils

            9a Lvs gen glab, minutely crisp-margined; sepals scarious-margined, acute, strongly 3-nerved, 2.5–4 mm; moist places, lowl to foothills or lower mts; Alas to Cal, Cas to coast, e to Alta, Mont, and Wyo; crisped s.   6 **S. crispa** Cham. & Schlecht.

            9b Lvs ciliate at base, not crisp-margined; sepals scarcely scarious-margined, ± obtuse, obscurely nerved, 2–2.5 mm; mont meadows and streambanks; BC, Cas and OM, Wn, to Cal, e to Alta, se Wn, ne Ore, Ida, Mont, and s to Colo; bluntsepaled s. (*S. viridula*)   7 **S. obtusa** Engelm.

        8b Fls in axillary or terminal cymes

            10a Cymes lfy-bracteate; lvs narrowly lanceolate, margins smooth; seeds very slightly roughened; native and mostly mont (see lead 15a)   13 **S. calycantha** (Ledeb.) Bong.

            10b Cymes scarious-bracteate; lvs mostly oblong to oblanceolate, margins often undulate-crisped; seeds strongly papillate-rugose; European weed, chiefly in lawns w Cas; bog stitchwort (*S. uliginosa*)   8 **S. alsine** Grimm

Group II

11a Petals lacking or < half as long as sepals; sepals 2–3 mm; st very slender; fls ± ∝ in scarious-bracteate, umbellate cymes; ped filiform, up to 3 cm; moist mont meadowl and for; Cas and Blue mts, Ore, to Cal, e to Mont and Colo; Siberia; umbellate s. (*A. baicalensis, S. gonomischa*)
                                          9 **S. umbellata** Turcz.

11b Petals gen > half as long as sepals; sepals often > 3 mm; st not notably slender; fls single to cymose, often leaf-bracteate; ped often not filiform

    12a Fls axillary, not cymose; petals at least = sepals

        13a Lvs gen linear or linear-lanceolate, stiff, sharply acute; sepals ca 4 mm at anthesis, evidently 3-nerved; moist streambanks to rocky slopes; Alas s, and in OM and Cas, to Cal, e to Newf, NY, Minn, Ariz, and NM; Eurasia; longstalk s.; 2 vars.   10 **S. longipes** Goldie

            a1 Pls low and compact, gen < 1.5 dm; fls gen only 1 or 2; common in our range (*S. monantha, S. m.* var. *a.*)
                                      var. **altocaulis** (Hultén) Hitchc.

            a2 Pls gen > 1.5 dm; fls in scarious-bracteate cymes; with above var. and at least as common (*S. laeta, S. stricta, S. subvestita*)
                                            var. **longipes**

        13b Lvs linear-lanceolate to oblong, acute to obtuse, not stiff; sepals often < 4 mm at anthesis, sometimes not evidently 3-nerved

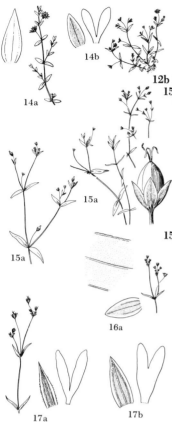

14a Sepals 3.5–5 mm at anthesis, ultimately > caps; lvs fleshy; pls circumpolar; in our area along the coast, gen in salt marshes, at a few places in Wn and Ore; low or spreading s.   11 **S. humifusa** Rottb.
14b Sepals 2–3 mm at anthesis, ultimately < caps; lvs thin; pls of moist areas from Alas to Newf, s in RM to Colo; in our area in Mont and Ida; thicklvd s.   12 **S. crassifolia** Ehrb.
12b Fls few, to ∝ and cymose; petals often < the sepals
  15a Cymes lfy-bracteate; petals < the sepals, or wanting; circumpolar, widespread in mts of our area; northern s.; 3 poorly defined vars.
                                13 **S. calycantha** (Ledeb.) Bong.
    a1 Calyx 2–2.7 mm at anthesis; caps 3–5 mm; lvs ovate to lanceolate, gen not > 2.5 cm; n, possibly not quite within our range (*S. borealis, S. lanceolata, S. oxyphylla*)   var. **calycantha**
    a2 Calyx 2.7–4.5 mm at anthesis; caps 4.5–7.5 mm; lvs narrowly to broadly lanceolate, mostly > 2.5 cm
      b1 Cymes bracteate, terminal, several-fld; upper lvs much-reduced; throughout our area (*S. brachipetala*) var. **sitchana** (Steud.) Fern
      b2 Cymes leafy, few-fld, or fls mostly axillary; with var. *sitchana*
                                var. **bongardiana** Fern.
  15b Cymes membranous-bracteate, or petals > sepals
    16a Lf margins finely tuberculate-scaberulous under 30 × magnification; sepals glab, 3–4 mm; mont streambanks and meadows; Alas to Newf, s through Wn and Ore to Cal, and in Ida and Mont to NM and SC; Eurasia; longlvd s.   14 **S. longifolia** Muhl.
    16b Lf margins smooth; sepals often ± pubescent, 3.5–5.5 mm
      17a Sepals strongly ciliolate, 3.5–5.5 mm, ca = caps; seeds prominently scalloped-tuberculate concentrically; European weed, mostly in lawns w Cas, but e to Ida; lesser starwort or stitchwort
                                 15 **S. graminea** L.
      17b Sepals gen glab, ca 4 mm, mostly slightly < caps; seeds lightly reticulate; streambanks, meadowl, and moist mont slopes; Alas to Cal, e to Newf, NY, and NM; Eurasia (see lead 13a)
                              10 **S. longipes** Goldie

Vaccaria Medic.

Fls cymose, complete, hypog; calyx gamosep, 10-nerved, winged along main nerves to the lobes; petals pink or red, ± clawed but without appendages or aurs; styles 2 (3); caps 1-celled; ann herbs. (L *vacca*, cow, application not clear).

V. segetalis (Neck.) Garcke. Cowcockle, cowherb. Pl glab and ± glaucous, taprooted, up to 8 dm; lvs mostly cauline, lanceolate to oblong-lanceolate, mostly sessile and ± aur or connate, 4–9 × 0.5–3 (4) cm; fls ∝ in open, gen flat-topped, ± lfy-bracteate cymes; calyx tubular at anthesis, 11–14 mm, often purplish, glab, much inflated in fr; petals pink, bl 5–8 mm, erose and slightly retuse; seeds reddish-black, globose, ca 2 mm; common European weed, mostly in waste areas and along roadsides and railroads (*Saponaria vaccaria, V. v., S. s.*).

NYMPHAEACEAE Water-lily Family

Fls solitary on long peduncles, showy, ⚥; sepals 3–∝, greenish to yellow or purplish, blending with the 3–∞ white, yellow, reddish, or bluish petals; stamens 12–∞ (ours) or sometimes only 3–6; pistils 3–∝ and 1-carpellary, or only 1 but multicarpellary and multiloculed, gen superior, but sometimes partially to almost completely inferior; styles short or lacking, stigmas linear to discoid; fr a follicle or leathery caps; herbaceous, aquatic per (ours) with thick rhizomes and floating to rarely emersed, peltate to cordate, entire (ours) to dissected lvs.

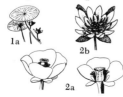

1a Sepals and petals 3 (4) each, purplish; lvs peltate; pistils several   **Brasenia**
1b Sepals and petals 4–∝, greenish, yellow, or white (purple-tinged); lvs cordate, not peltate; pistil 1
  2a Sepals 5–12, larger and more showy than the petals, greenish to bright yellow; stigma 1, discoid; ovary superior   **Nuphar**
  2b Sepals 4, ca as large as the petals but less showy; petals white or pinkish; stigmas > 1, spreading; ovary partially inferior   **Nymphaea**

Brasenia Schreb.   Water-shield; Water-target

Fls small, hypog; stamens 12–30; pistils mostly 9–15, 1-carpellary, 1–2 (3)-ovuled; fr leathery, 1–2-seeded; brs and lvs floating. (Origin of the name unknown).

B. schreberi Gmel. Petioles 5–40 cm; lf bls broadly elliptic, 5–12 cm; peduncles 1-fld, 5–20 cm; perianth segms distinct, 10–17 mm, narrowly oblong, purplish; stamens purplish, = or subequal to the petals; stigma ca = ovary; fr narrowly ovoid, 6–8 mm; shallow ponds and sluggish streams; BC to Cal, e over most of Can and US; widespread in other parts of the world except Europe.

Nuphar J. E. Smith   Cow-lily; Yellow Water-lily

Fls showy, hypog, outer sepals smaller and greenish, the inner ones yellow or with some purplish tinge; petals 10–20, not much larger than the stamens and merging with them; pistil several-loculed, stigma subsessile, flattened and with a scalloped margin; fr leathery and ± fleshy; per with fleshy rhizomes from which arise the long peduncles and long-petioled, gen cordate, commonly floating lvs. (Arabic name, *naufar*, for some water lily).

1a Sepals gen 6, 2.5–3.5 cm; stamens yellow; fr 3–5 cm; seeds ca 5 mm; lf bls 1–2.5 (3) dm; BC to e Can, s to ne Ida and n Mont in our area *(Nymphaea v.)*   1 **N. variegatum** Engelm.

1b Sepals gen 9, (3) 3.5–6 cm; stamens ± reddish or purplish; fr 5–9 cm; seeds ca 4 mm; lf bls 1–4.5 dm; Alas s, both sides of Cas, to Cal, e to Alta, SD, and Colo; common in most of our area from lowl into the mts; spatterdock, Indian pond lily, Wakas *(Nymphaea p.)*   2 **N. polysepalum** Engelm.

Nymphaea L.   Water-lily

Fls showy; petals ∝, ours white or pinkish, inserted with the ∝ stamens on the ovary; pistil ∝-celled; stigmas 10–30, broad, spreading; ovary partially inferior; caps tardily dehiscent, leathery. (Gr *nymphe*, a water nymph, referring to the habitat).

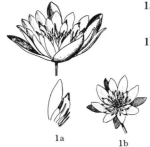

1a Fls fragrant; sepals 3–6.5 cm; petals 20–30; stamens 50–100; lf bls suborbicular, up to 25 cm; intro from e N Am, and common in many lowl lakes on both sides Cas; fragrant or Am w.   1. **N. odorata** Ait.

1b Fls not fragrant; sepals 2–3 cm; petals 7–15; stamens 30–45; lf bls elliptic-oval, 5–12 cm; common in e N Am but rare in our area, where known from Ida and from Whatcom Co, Wn; pygmy w. *(Castalia leibergii)*   2 **N. tetragona** Georgi

CERATOPHYLLACEAE   Hornwort Family

Fls minute, ♂♀, apet, hypog, solitary in the lf axils; perianth (or an invol?) greenish, of 8–15 tepals; stamens 10–16; pistil 1, simple; fr hardened, 1-seeded, gen with spiny tip or margin; submersed, aquatic per with slender sts and sessile, palmately to dichotomously dissected, whorled lvs.

Ceratophyllum L.   Hornwort; Coontail

Characters of the family; the only genus. (Gr *keratos*, horn, and *phyllon*, lf, because of the hornlike segms of the lvs).

C. demersum L. Sts delicate, up to 4 m, freely br; lvs 5–12 per whorl, 1–2 (3) times palmately and dichotomously dissected into linear or filiform, gen ± finely toothed segms, often polymorphic on different parts of the pl; achenes flattened, the body 4–6 mm, with a terminal hooked spine up to 12 mm and 2 divergent basal spines 1–6 mm; standing to slow-moving water; cosmopolitan, but in our area most common in Wn and Ore.

PAEONIACEAE   Peony Family

Fls reg, ± showy; sepals 5, leathery, persistent; petals 5, purplish to red or white; stamens ∝, developing centrifugally (from the center outward) on and at the edge of an irreg, lobed, prominent, fleshy disc; pistils 2–8; fr a leathery, ∝-seeded follicle; herbaceous (ours) to suffrutescent per with large, ± compound, exstip lvs.

Paeonia L.   Peony

Characters of the family; the only genus. (Gr name used by Theophrastus, believed to have been meant to commemorate *Paeon*, physician of the gods, who was supposed to have used the pl medicinally).

P. brownii Dougl. Brown's p., w p. Pl glab and glaucous, gen clumped, 2–5 dm; lvs large, fleshy and ± leathery, from ternate and with the primary lfts again deeply ternately parted (the segms less deeply 1–2 times dissected) to biternate and again deeply 1–2 times lobed into ultimate oblong segms 3–10 mm broad; fls on short terminal peduncles, tending not to open fully; sepals unequal, leathery, greenish to purplish, 1–2 cm, persistent; petals 5, brownish-reddish-purple, ca = the sepals, deciduous; follicles 4–5 cm; sagebr des to ponderosa pine for; e Cas, Chelan Co, Wn, to Cal, e to Ida, w Wyo, and Utah.

RANUNCULACEAE   Buttercup Family

Fls hypog, reg to irreg, inconspicuous to showy, complete to apet or ♀, ♂, single to pan; sepals 3–20, distinct; petals mostly 5–10 (3–20), gen gland-bearing (gen near base); stamens mostly ∝ and spirally arranged, but sometimes only 5 or 10; pistils 2–∝, rarely only 1, simple (except in *Nigella*), 1–∝-ovuled; fr mostly follicles or achenes, rarely a berry or caps; ann or per herbs or woody vines with alt to opp or whorled, simple to compound, exstip lvs.

1a Fls strongly bilaterally symmetrical, showy
  2a Upper sepal spurred but not hooded; petals 4, not concealed by the sepals **Delphinium**
  2b Upper sepal hooded but not spurred; petals gen 2, concealed by the sepals **Aconitum**
1b Fls nearly or quite reg, often not showy
  3a Petals prominently spurred, showy; fr several follicles; pl per **Aquilegia**
  3b Petals either not spurred or not showy, sometimes lacking; fr often not follicular; pl sometimes ann
    4a Pl scapose, linear-lvd, ann; fls inconspicuous; sepals spurred; frs ∝ achenes on a slender, elongate recep **Myosurus**
    4b Pl either not scapose, or not linear-lvd, mostly per; fls mostly showy; fr various, but rarely achenes borne on an elongate recep
      5a Pistil 1; fr a several-seeded berry; lvs large, several times dissected **Actaea**
      5b Pistils 2 or more; fr achene or follicle; lvs various
        6a Ovules gen 1 per pistil; fr a 1-seeded achene
          7a Corolla present, gen more showy than the calyx
            8a Petals without nectary; terrestrial ann with dissected lvs and white to red or sometimes bluish-tinged petals **Adonis**
            8b Petals with nectary near base; mostly per, sometimes aquatic, if ann either lvs not dissected, or petals yellow, or both **Ranunculus**
          7b Corolla lacking or much less showy than the calyx
            9a Sepals greenish or white, much less evident than the stamens; st lvs alt or lacking; fls ∝; pl herbaceous
              10a Lvs simple, palmately lobed; fls ♂ **Trautvetteria**
              10b Lvs ternately decompound; fls often ♀, ♂ **Thalictrum**
            9b Sepals mostly white or pink to blue, much more evident than the stamens; st lvs opp or whorled; fls often solitary; pl sometimes a woody vine
              11a St lvs opp; pl often a woody vine; styles long and plumose; sepals gen 4 **Clematis**

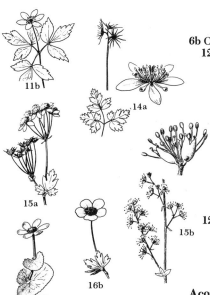

11b St lvs gen in a whorl of 3; pl never woody or viny; styles often short or not plumose; sepals mostly at least 5
Anemone
6b Ovules 2–∝ per pistil; fr a 2–∝ -seeded follicle
12a Follicles distinct or only weakly connate at base; fls not closely subtended by dissected invol lvs
13a Lvs 1–3 times ternately divided to compound
14a Lvs evergreen, leathery; petals (or sterile stamens?) gland-bearing; low scapose per with slender yellow rhizomes
Coptis
14b Lvs herbaceous; petals and stamens not gland-bearing; pl without slender yellow rhizomes
15a Fls solitary or 3–10 and corymbose or umbellate
Isopyrum
15b Fls many > 10, in compound raceme
Cimicifuga
13b Lvs simple to palmately compound, not ternately dissected
16a Lvs simple, merely crenate or dentate
Caltha
16b Lvs deeply 5-lobed to semi-compound
Trollius
12b Follicles connate to near middle, upper free half beaklike and not seed-bearing; fls subtended by finely dissected invol lvs; sepals petaloid, bluish; occ escape from gardens; love-in-a-mist
Nigella damascena L.

Aconitum L.   Monkshood

Fls in terminal, simple to bracteate racemes, complete; sepals 5, petaloid, white to yellow or purple, upper one largest, arched and hoodlike; 2 petals concealed within hood, narrowly clawed at base and with expanded bl and ± coiled spur; vestiges of 3 other petals sometimes present; stamens ∝ ; pistils 3–5; follicles 3–5, ∝ -seeded; herbaceous per with alt, palmately lobed lvs. (Ancient Gr name used by both Dioscorides and Theophrastus).

A. columbianum Nutt. Columbian m. Sts 5–20 dm, ± fistulose; hood 15–30 mm; seeds strongly winged lengthwise and transversely sinuous-lamellate; moist woods to subalp meadows, often along streams; Alas s, e Cas, to Cal, e to sw Alta, w Mont, Colo, NM, and SD; with 3 highly intergradient vars.
a1 Sepals greenish-yellow, sometimes purplish-tinged; sw Wn and sc Ida to sc Mont and to NM, chiefly RM **var. ochroleucum** A. Nels.
a2 Sepals light to deep bluish-purple, or occ white (albino forms)
b1 Bulblets present in some of lf axils or along brs of infl; sts often ± reclining; occ from Mt Hood, Ore, s in Cas to Cal
var. howellii (Nels. & Macbr.) Hitchc.
b2 Bulblets lacking above ground; sts gen erect; range of sp. as a whole
var. columbianum

Actaea L.   Baneberry

Fls inconspicuous, mostly complete, racemose; sepals 3–5, petaloid, quickly deciduous; petals mostly 5–10 (0), narrowly spatulate; stamens ∝ ; per herbs. (Gr name, *aktea*, for elder tree, prob with reference to lf similarity).

A. rubra (Ait.) Willd. W red b. Sts 4–10 dm; lvs all cauline, 2–3 times ternate and ± pinnatifid, segms mostly 3–9 cm, sharply toothed and lobed; sepals white to purplish-tinged, 2–3 mm; petals white, subequal to the sepals; stamens longer than the petals; berry gen red, much less commonly white (the f. *neglecta* (Gillman) Robins.); moist woods, Alas to Cal, both sides Cas, e through Can and n US to Atl and s in RM to NM (*A. arguta, A. eburnea, A. neglecta*).

Adonis L.   Pheasant-eye

Fls complete, ± showy, solitary at br tips; sepals 5, < the corolla; petals 5–10 (20), white to red or bluish-based; stamens and pistils ∝ ; fr achenes; ann (ours) or per herbs. (Gr name, from mythology, for *Adonis*, from whose blood the pl was alleged to have sprung).

A. aestivalis L. Glab ann 2–6 dm; sepals greenish, 5–10 mm; achenes dorsally short-toothed at base and with short, straight, terminal beak; intro from Europe, occ escaping and persistent in parts of Ida and Mont. *A. annua* L. may also occ be found as an escape; scarcely recognizable from *A. aestivalis*, but petals perhaps more spreading and mature achenes lacking a basal tooth.

Anemone L.   Windflower; Anemone

Fls mostly showy, solitary and terminal on naked or invol peduncles, apet, ⚥; sepals mostly 5–6 (to 9); stamens and pistils ∝ ; achenes glab to silky or lanate, style short to elongate, glab to plumose; per herbs with palmately to ternately lobed or dissected basal lvs and fl sts naked except for a whorl of gen 3 sessile to short-petiolate, simple to dissected invol lvs; peduncles 1–several, often with a 2ndary invol about midlength. (From Gr *anemos*, wind, application obscure). *(Pulsatilla)*.

1a Styles plumose, (1.5) 2–3.5 cm in fr; sepals 2–4 cm; lvs dissected into linear segms
   2a Sepals blue to purplish (rarely albino); basal lvs mostly 2 (3) times dissected, the ultimate segms (2) 2.5–4 mm broad; sts mostly 0.5–2.5 (3.5) dm; prairies to mt meadows, where soil well-drained; Alas to Wen Mts, Wn, e to Alta, s through Mont to Tex, and to Ill; pasqueflower, wild crocus, lionsbeard (*A. hirsutissima, A. ludoviciana, A. patens* var. *multifida*)
                                                                 1 **A. nuttalliana** DC.
   2b Sepals white or purplish-tinged; basal lvs 3–4 times dissected, the ultimate segms mostly < 2 mm broad; sts mostly (1) 2–6 dm; midmont to alp; BC s through OM and Cas, Wn, to Cal, e to Alta, Mont, ne Ida, and ne Ore; w pasquefl, mt p.   2 **A. occidentalis** Wats.
1b Styles rarely plumose, < 1.5 cm at maturity; sepals often < 2 cm; lvs often not dissected into linear segms
   3a Invol lvs simple, bl crenate to deeply divided, never dissected
      4a Invol lvs mostly crenate-serrate, sometimes lobed to near midlength; basal lvs gen solitary; scapes 1-fld; sepals white, 1.5–2.5 cm; achenes 2.5–4 mm, hirsute on lower part, somewhat > the glab style; woods, often with shrubs, dry to moist areas; King Co, Wn, s on w side Cas, but not on OP, to CRG and through Ore to n Cal; threefl a., western white a., Columbia w.   3 **A. deltoidea** Hook.
      4b Invol lvs mostly lobed at least half their length; basal lvs gen several; scapes often > 1-fld; sepals white or bluish-tinged, mostly 1–1.5 (2) cm; achenes ± strigose to woolly
         5a Pl mostly > 2 dm; lvs 6–12 cm broad; peduncles 1–3; achenes 3–5 mm, strigose; Alta to NM e of the RM, prob not in our range
                                                                 A. canadensis L.
         5b Pl mostly < 2 (3.5) dm; lvs 1–3 cm broad; peduncles solitary; achenes 2–2.5 mm, densely woolly; streambanks and mt meadows; Alas to Cas of n Wn, e to Atl and s in RM to Colo; c Ida and ne Ore; Asia; northern a., small-fld a.   4 **A. parviflora** Michx.
   3b Invol lvs either ternately compound (lfts short-petiolulate) or 2 or more times dissected into narrow segms
      6a Pl caespitose, nonrhizomatous; basal lvs several, tufted; achenes woolly
         7a Recep (with attached achenes) cylindric, 2–4 cm × scarcely 1 cm; basal lvs ternate (or quinate), main divisions ternately cleft, ultimate segms mostly much > 5 mm broad; prairies and lower mts; BC s in RM to Mont, NM, and Ariz, e to SD, Mo, and NJ, not in Ida? thimbleweed, long-headed a.   5 **A. cylindrica** Gray
         7b Recep with attached achenes subglobose, mostly 1–2 cm and nearly as thick; basal lvs gen dissected into linear or oblong segms mostly < 5 mm broad
            8a Styles straight, slender, (1.5) 2–4 mm, gen yellow; lvs 3–4 times ternate, ultimate segms often < 2 mm broad; peduncles mostly solitary; sepals blue (white) on dorsal surface; alp and subalp; Alas to Cal, in both OM and Cas, e to Alta and s Mont; Drummond's a.; 2 vars.   6 **A. drummondii** Wats.
               a1 Lf segms mostly 1.5–2.5 mm broad; styles (1.5) 2–2.5 (3) mm; Alas to Alta, s to c and s Mont *(A. l.)*
                                                          var. **lithophila** (Rydb.) Hitchc.

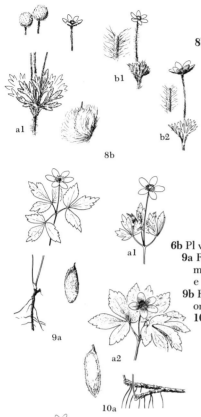

a2 Lf segms mostly 1–1.5 (2) mm broad; styles 2–4 mm; s BC to Sierra Nevada, Cal; in OM and Cas, also in c Ida
var. **drummondii**
8b Styles thickened at base, 1–1.5 mm, gen pink or red; lvs rarely quite triternate, ultimate segms gen at least 2 mm broad; peduncles often 2–3; sepals ochroleucous to yellowish and tinged with red, blue, or purple on dorsal surface; foothills to arctic-alp, Alas to Cal, e to NY and s in RM to NM; S Am; Pac a., cliff a.; 3 vars.
7 **A. multifida** Poir.
a1 Sts 2–6 dm, mostly with 2–3 peduncles; ultimate segms of lvs mostly > 3 mm broad; range of sp., esp at lower elev (*A. globosa*)
var. **multifida**
a2 Sts seldom > 2 dm, mostly with only 1 peduncle; ultimate segms of lvs rarely > 3 mm broad
b1 Pl grayish, copiously villous-hirsute, hairs 2–3 mm; mainly in OM, Wn, less common in Cas from Mt Rainier to s BC, rarely in RM
var. **hirsuta** Hitchc.
b2 Pl greenish, sparsely villous-hirsute, hairs < 2 mm; peaks of c Ida and s Mont to NM, and Charleston Mts, Nev (*A. t.*)
var. **tetonensis** (Porter) Hitchc.
6b Pl widely rhizomatous; basal lvs solitary or few; achenes not woolly
9a Rhizomes dark brown, erect or ascending, often br near tip; sepals mostly white or pinkish, gen > 8 mm; stamens 35–60; shady woods, e Wn and ne Ore to c Ida and w Mont; Piper's a. 8 **A. piperi** Britt.
9b Rhizomes gen pale brown, horizontal; sepals sometimes either blue or < 8 mm; stamens sometimes < 35
10a Stamens (30) 35–100; sepals (10) 12–20 mm, bluish-purple to reddish-purple, blue, pale pink, or white; moist woods to open hillsides; mostly e Cas from Chelan Co, Wn, to CRG and n Cas and coastal mts of Lincoln Co, Ore, also in Grays Harbor Co, Wn, and reported for Blue Mts, Ore; Ore a.; 2 vars. 9 **A. oregana** Gray
a1 Sepals white or pink; stamens mostly > 60; marshes near coast, Grays Harbor Co, Wn, and Lincoln Co, Ore (*A. f.*)
var. **felix** (Peck) Hitchc.
a2 Sepals gen bluish; stamens mostly < 60; range of sp. (*A. cyanea*)
var. **oregana**
10b Stamens gen < 35; sepals mostly < 10 mm, gen white (bluish or ± rose); open prairies to subalp ridges; sw BC to n Cal, from near coast into Cas; Lyall's a.
10 **A. lyallii** Britt.

Aquilegia L. Columbine

Fls reg, showy, white to blue, yellow, or red, complete; sepals 5, petaloid, spreading, short-clawed; petals 5, spurred at base, bl rounded to emarginate, gen erect; stamens ∝; pistils 5 (to 10); follicles erect to spreading, ∝-seeded; per herbs with bi- or triternately compound, mostly basal lvs and cuneate, ± deeply lobed lfts. (L *aquila*, eagle, petals suggestive of an eagle's claws).

1a Sepals blue or purple to white
2a Pl scapose, scarcely 1.5 dm; sts 1 (2)-fld; lf bls leathery, scarcely > 1 cm broad, lfts greatly crowded, strongly pubescent; spurs 5–10 (15) mm, nearly straight; sw Alta to nw Wyo, mostly along Cont Div or immediately e; gen on subalp limestone scree or in rock crevices; limestone c. (*A. j.* var. *elatior*)
1 **A. jonesii** Parry
2b Pl lfy-std, gen > 1.5 dm; sts gen 2–more-fld; lf bls not leathery, rarely < 3 cm broad, lfts not greatly crowded, glab to slightly pubescent; spurs either hooked or much > 15 mm
3a Spur < petal bl, hooked; Alas and n BC to SD, prob not in our area
A. brevistyla Hook.
3b Spur 2–4 cm, ca twice as long as petal bl, nearly or quite straight; woods and slopes; RM, sw Mont and c Ida to Colo, NM, and Utah; Colo c.; 2 vars.
2 **A. coerulea** James
a1 Sepals pale to deep blue; chiefly s, rarely n to s Ida and sw Mont (*A. leptocera*)
var. **coerulea**
a2 Sepals white, with little or no tinge of blue; chiefly n Utah to w Wyo and c Ida, into sw Mont and rarely to Colo (*A. c.* var. *albiflora*)
var. **ochroleuca** Hook.

1b Sepals yellow to pinkish or red
    4a Sepals yellow (rarely ± pinkish); spurs ± incurved, petal bl (4.5) 6–13
       mm; moist mt meadows to alp slopes; s BC s in Cas to Kittitas Co, Wn, e
       to Alta, Ida, Mont, Colo, and Utah, also in ne Ore; yellow c. (*A. f.*
       var. *miniana, A. formosa* var. *f.*)   3 **A. flavescens** Wats.
    4b Sepals pale to deep red; spurs nearly straight, petal bl 2–6 mm; woods
       and moist meadows to midmont; s Alas to Baja Cal, e to w Alta, Mont,
       and Utah; in OM and Cas; red c., Sitka c. (*A. columbiana, A. truncata, A.*
       wawawensis)   4 **A. formosa** Fisch.

Caltha L.   Marshmarigold

Fls apet, white or yellow, showy, reg, solitary or 2–3 per st on long peduncles; sepals 5–12, petaloid; stamens ∝; follicles 5–10, several-seeded; glab, fleshy, herbaceous per with reniform to cordate-oblong, simple, crenate or dentate lf bls. (Early Gr name for some yellow-fld pl).

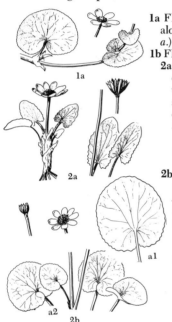

1a Fl sts several-lvd, decumbent to creeping; sepals yellow; mostly in bogs
    along coast, Alas to Ore, not e Cas; yellow m., western m. (*C. palustris* var.
    a.)   1 **C. asarifolia** DC.
1b Fl sts lfless or with only 1 (2) lvs, erect; sepals mostly white
    2a Lf bls always longer than broad, up to twice as long, often ± sagittate
       or aur at base, margins more dentate than crenate, teeth obtuse to acu-
       tish, often callous at tip; fls mostly 1 (2–3) per st; wet places, alp to
       subalp; mainly RM, Alas to Colo, Utah, c Ida, and ne Nev and se Ore;
       elkslip, e. m.; 2 vars.   2 **C. leptosepala** DC.
       a1 Sepals canary-yellow; local in alp areas of Lost River Mts, Custer Co,
         Ida   var. **sulfurea** Hitchc.
       a2 Sepals white or greenish or ± bluish-tinged (*C. auriculata, C. cheli-*
         donii, C. uniflora)   var. **leptosepala**
    2b Lf bls almost or quite as broad as long, reniform to cordate-based, mar-
       gins subentire to ± crenate; fls mostly 2 per st; wet, alp or subalp places;
       Alas to Cal, e to Ida, Utah, and Colo; white m., twinfld m., broadlvd m.;
       2 vars.   3 **C. biflora** DC.
       a1 Main lf bls gen > 5 cm; scapes (with peduncles) mostly > 3 dm in fr;
         e Ore through Ida, se to Colo, very occ in Cas of BC and Wn (*C. r.*)
         var. **rotundifolia** (Huth) Hitchc.
       a2 Main lf bls mostly < 5 cm; scapes (with peduncles) gen < 3 dm; Alas
         to Cal, in our area from the Cas w (*C. leptosepala* var. *b., C. howellii,*
         C. malvaceus, C. macounii)   var. **biflora**

Cimicifuga L.   Bugbane

Fls apet, rather small, in large compound racemes or pan; sepals (4) 5, cream to pinkish, shed at
anthesis; stamens ∝, cream-colored, outer 5 often broadened and petaloid, clawed at base and
bilobed above; follicles 1–5, several-seeded; herbaceous per with large, bi- to triternate lvs and
shallowly to laciniately lobed and toothed lf segms. (L *cimex*, bug, and *fugere*, to repel; one sp.
sometimes used as a bedbug repellent).

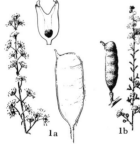

1a Infl loose, pubescent but not glandular; staminodia gen present; follicles
    mostly 2–5, with stipe 3–6 mm; lfts coarsely and laciniately cleft; sts
    0.8–1.5 m; moist woods at base of Mt Hood, Ore, and in Skamania Co, Wn;
    Mt Hood b., cut-lvd b.   1 **C. laciniata** Wats.
1b Infl closely fld, puberulent and glandular; staminodia lacking; follicles 1 or
    2 (3), subsessile, stipe < 1 mm; lfts shallowly lobed and ± finely ser-
    rate-dentate; sts 1–2 m; moist woods at lower elev; s BC to nw Ore, w Cas;
    tall b.   2 **C. elata** Nutt.

Clematis L.   Virgins-bower; Clematis

Fls rather showy, apet, ☿ to ♀, ♂, single to pan on axillary or terminal peduncles; sepals mostly 4
(5), spreading to erect and sometimes ± connivent, white to reddish, bluish, or purplish (ours);
stamens ∝, outer ones sometimes sterile and ± petaloid; achenes ∝, style long, plumose, persistent; herbaceous per to woody vines with opp (ours) or whorled, 1–several times ternate or pinnate lvs. (Gr name, application obscure). (*Atragene, Viorna*).

1a Sepals white or yellowish, 6–15 mm; fls pan; pl woody vines; lvs pinnately compound, lfts (3) 5–7

  2a Fls ♂; lfts gen entire or upper one 3-lobed; intro, known from Puget Sound region, Wn; travelers-joy   1 **C. vitalba** L.

  2b Fls ♂, ♀; lfts all coarsely toothed; sagebr des to ponderosa pine for; BC s to s Cal, e of Cas in Wn, but in CRG and on both sides of Cas in Ore, e to Dak and NM; western c. or v. (*C. brevifolia, C. suksdorfii*)
                                                     2 **C. ligusticifolia** Nutt.

1b Sepals bluish to reddish or brownish-purple, at least 20 mm; fls solitary; pl sometimes herbaceous and not vining; lvs gen ternately compound or dissected into ∝ narrow segms

  3a Lvs 2–4 times pinnately dissected into narrow segms; pl herbaceous, not vining; sepals erect, leathery, connate at base, tips free and gen recurved; grassl and sagebr des to ponderosa pine for; sc BC to e Wn and Ore, e to Mont and Wyo; Douglas' c., leatherfl., vasefl., sugarbowls (*C. douglasii, C. wyethii*)   3 **C. hirsutissima** Pursh

  3b Lvs ternate to triternate, segms ± ovate; pl ± woody, gen vining; sepals ± spreading, not connate at base

    4a Lvs ternate, lfts entire to sharply toothed or occ deeply lobed; pl creeping to climbing; foothills to lower mts e Cas; BC to n Ore, e to Alta, Mont, and Wyo; Columbia c., Columbia v., rock c.; 2 vars.
                                           4 **C. columbiana** (Nutt.) T. & G.

      a1 Lfts entire or merely toothed; sepals blue; pl mostly climbing; range of sp. except not in Wen Mts (*A. grosseserrata*)
                                          var. **columbiana**

      a2 Lfts often deeply lobed; sepals blue to reddish-purple; pl more creeping than climbing; Wen Mts, Chelan and Kittitas cos, Wn
                                         var. **dissecta** Hitchc.

    4b Lvs more nearly biternate to triternate

      5a Lvs ca biternate, 2ndary divisions gen coarsely toothed to rounded-lobate; pl vining; mont, e slopes RM, Mont to Colo, NM, ne Ariz, and e Utah; climbing purple v., RM c. (*A. repens*)
                                    5 **C. pseudoalpina** (Kuntze) Nels.

      5b Lvs ca triternate; pl forming small mats, but not vining; calcareous ridges and slopes in foothills and lower mts; e base RM, Mont and ne Wyo to SD; matted purple v. (*C. pseudoalpina* var. *t.*)
                                    6 **C. tenuiloba** (Gray) Hitchc.

Coptis Salisb.   Goldthread

Fls reg, not showy, mostly (1) 2–several on naked scapes, ♂, to ♂ ♀, ♂ ♂; sepals 5–8, greenish and ± petaloid, deciduous; petals 5–7, clawed, bl gen narrow, lig, and gland-bearing near base, but sometimes hooded or tubular and glanduliferous near the tip; stamens 12–25; follicles 5–10, several-seeded, stipe conspicuous, flattened; scapose, glab to occ ± puberulent per with extensive, slender, yellowish rhizomes and long-petiolate, evergreen, gen shiny, ternately divided to biternately or ternate-pinnately compound lf bls. (Gr *kopto*, to cut, referring to the dissected lvs).

1a Lfts 3, < 2 cm, not appreciably lobed; scapes 1-fld; sepals whitish, narrowly lanceolate to spatulate, 5–10 mm; petals ca half as long as the sepals, fleshy, hollowed and nectariferous at the tip, without a flattened bl; deep woods, often where swampy; Alas to VI and s BC; Japan and Siberia; threelft g.   1 **C. trifolia** (L.) Salisb.

1b Lfts > 3, or > 2 cm, or lobed at least 1/3–1/2 the length; scapes often > 1-fld; sepals narrowly lanceolate, 5–10 mm; petals mostly much > half as long as sepals, slenderly clawed below the narrowly lanceolate, basally glanduliferous bl

  2a Lvs ± biternately or ternately-pinnately compound; lfts at least 5; scapes mostly > the lvs at anthesis; moist woods and bogs, Alas to VI and Snohomish Co, Wn; spleenwort-lvd g.   2 **C. asplenifolia** Salisb.

  2b Lvs ternate, lfts 3, from shallowly to deeply lobed or parted; scapes gen < the lvs at anthesis

    3a Lfts deeply lobed or parted almost to base, 3–4 (5) cm; scapes 1–2 (3)-fld, 5–12 cm; moist coastal woods; Grays Harbor Co, Wn, and lower CRG to nw Cal; cutlf g., Ore g.   3 **C. laciniata** Gray

    3b Lfts mostly lobed ca half length, 3–6 (7) cm; scapes 2–5-fld, 10–20 cm; moist woods, ne Wn and adj BC to nw Mont; western g., Ida g.
                                      4 **C. occidentalis** (Nutt.) T. & G.

Delphinium L.   Larkspur; Delphinium

Fls in simple to compound racemes, showy, complete, strongly irreg; sepals 5, blue to green-ish-yellow or ochroleucous (ours) or sometimes red, the upper one prominently spurred; petals gen 4 and paired, smaller than the sepals, upper pair spurred inside main spur of upper sepals; stamens ∝; follicles gen 3 (1–5), ∝-seeded; ann or (ours) per herbs with large, fibrous to tuberous and ± fleshy roots; sts often tall and fistulose; lvs alt, bl gen palmately parted or divided into 3 (5–7) main segms ± ternately 1–3 times lobed, ultimate segms entire to toothed. (L form of Gr name, *delphinion*, for the larkspur, used by Dioscorides). (*Delphinastrum*).

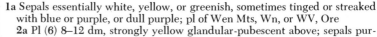

1a Sepals essentially white, yellow, or greenish, sometimes tinged or streaked with blue or purple, or dull purple; pl of Wen Mts, Wn, or WV, Ore
 2a Pl (6) 8–12 dm, strongly yellow glandular-pubescent above; sepals pur-plish but strongly streaked with yellow or green, spur 7–10 mm; follicles 6–8 mm; in or at the edge of meadows in the Wen Mts, Wn, s Chelan and n Kittitas cos, in areas ± moist at least early in season; Wen 1., greenish 1. **1 D. viridescens** Leiberg
 2b Pl often < 5 dm or not glandular-puberulent; sepals white or cream, rarely bluish-streaked, spur gen > 10 mm; follicles (10) 12–20 mm
  3a Roots large and fibrous, br; follicles mostly 15–22 mm; pl of dry grassy hillsides and ponderosa pine for; Wen Mts, Wn, s Okanogan Co to s Chelan and w Douglas cos; yellow-white 1.
   2 D. xantholeucum Piper
  3b Roots small, semiglobose, tuberlike; follicles gen < 15 (to 16) mm; pl of prairies or small openings in woodl, gen where moist early in season; WV, Ore
   4a Lateral sepals 9–14 mm; spur 10–14 mm; fls not glandular; follicles 8–12 mm, erect, nonglandular-puberulent; dry bluffs and open ground, now restricted almost entirely to fencerows and protected woodl; Multnomah, Clackamas, and Yamhill cos; pale 1. (*D. nut-tallii* var. l., *D. willametense*) **3 D. leucophaeum** Greene
   4b Lateral sepals 12–18 mm; spur 14–20 mm; petals ± glandul-ar-pubescent; follicles 10–16 mm, spreading, often glandul-ar-pubescent; roadsides and open fields to dry hillsides; mostly in Benton Co, but also in Polk and Marion cos; peacock 1.
    4 D. pavonaceum Ewan
1b Sepals gen bluish or purplish, if (as rarely) whitish pl from neither Wen Mts nor WV
 5a St strongly fistulose, (6) 10–30 dm; roots extensive, tough and fibrous; lvs with 5–7 main lobes variously toothed but not dissected; fls not par-ticularly showy, sepals rarely > 12 mm, cupped forward; pl of wet places, gen along streams or mt meadows
  6a Follicles gen evenly pubescent (rarely glab); racemes compact, ∝-fld, peds often not > calyx spurs; fls often whitish-streaked or pale blue; ne Ore to Mont and s in RM to Colo; western 1., duncecap 1. (*D. cu-cullatum, D. reticulatum, D. o.* var. *griseum*) **5 D. occidentale** Wats.
  6b Follicles glab or pubescent only along suture; racemes ± open, peds gen > calyx spurs; calyx deep purple; seldom in the range of *D. occi-dentale*; c Cas, OM, and extreme ne Wn, s in Ore Cas to Cal; pale 1. (*D. splendens, D. scopulorum*) **6 D. glaucum** Wats.
 5b St not fistulose, or pl not otherwise as above; lvs often dissected; sepals often flared or > 12 mm; pl often of dry areas or with small, fleshy roots
  7a Pl primarily of wet habitats (often drying by midseason), with exten-sive, br, fibrous root systems and rather tall, often ± fistulose sts rarely < 4 dm
   8a Lvs rather evenly distributed, those of st much like the basal, mostly sharply toothed; infl loose, lower peds much > calyx spurs; follicles spreading, glab; sepals flared, acuminate, (15) 18–25 mm; pl (3) 7–15 dm; both sides CRG, s on w side Cas to nw Cal; poison 1., Columbia l. **7 D. trolliifolium** Gray
   8b Lvs often much more ∝ near base of pl, those of st greatly modified upward, none really sharply toothed; infl more spikelike, peds mostly < calyx spur; follicles erect, mostly pubescent; sepals often cupped forward, mostly 7–11 mm, blunt to acute; pl often < 7 dm; e of Cas
    9a Pl ashy with abundant, fine, crisp puberulence at least above,

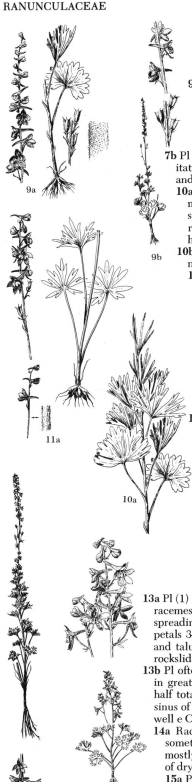

9a

9b

11a

10a

15a   13a

mostly 4–7 (10) dm; lvs strongly dimorphic; sts often single; swales and wet meadows, sagebr des to ponderosa pine for, where ground often dry by midsummer; BC to nc and e Ore, e to Ida and n Wyo; Burke's l., meadow l. (*D. simplex,* misapplied; *D. s.* var. *distichiflorum*)   8 **D. burkei** Greene

9b Pl not ashy, but gen yellow glandular-villous and ± clammy, 4.5–10 dm; lvs not strongly dimorphic; sts gen several; along rocky, gen intermittent streams or springs in sagebr hills to ponderosa pine woodl; Wen Mts, s Chelan Co to n Yakima Co, Wn, but only known from a comparatively few populations; Kittitas l. (*D. cyanoreios* f. *m.*)   9 **D. multiplex** (Ewan) Hitchc.

7b Pl gen in dry habitats, such as sagebr des or open for, if from wet habitats then root system small and more fleshy or tuberous than tough and fibrous

10a Lvs strongly dimorphic, those of midstem closely overlapping and much shorter-petiolate, more finely divided, and more closely spaced than lower ones; pls with small clusters of somewhat fleshy roots, finely crisp-puberulent, sometimes with glandular spreading hairs above as well (see also lead 9*a*)   8 **D. burkei** Greene

10b Lvs not noticeably dimorphic, mostly rather uniformly spaced and not overlapping; pls various

11a Pl of fairly moist habitats, primarily of meadowl that dries by midsummer, with small cluster of somewhat fleshy roots, gen glandular-pubescent (but occ eglandular and crisp-puberulent or even glab above); lvs few, mainly basal or near-basal, the 3 or 5 main divisions gen rather shallowly 1–2-lobed into broad, oblong to ovate, ± rounded segms, upper portion of st often with only very much reduced bracts; fls deep blue, mostly in tight raceme, ped gen not > sepals, if >, then strongly ascending; sts gen 1 (2), (1) 3–6 (10) dm; sagebr valleys to subalp meadowl; se Wn and c Ore to w Mont, s to ne Cal and c Nev; slim l., dwarf l. (*D. cyanoreios, D. c.* f. *idahoense* and *wallowense, D. diversicolor, D. diversifolium, D. d.* ssp. *harneyense*)   10 **D. depauperatum** Nutt.

11b Pls more commonly of dry habitats, such as gravelly slopes, talus slides, and well-drained meadows, often with thickened, tough, br root system; lvs more evenly spaced and often dissected into linear or narrowly oblong, often toothed segms; fls often light blue or borne in open raceme, ped often > sepals

12a Roots extensive, fibrous, freely br, firmly attached to st; pl either from e Cas or from subalp to alp areas in Cas and OM
                                                          Group I

12b Roots 1–several, globose to fusiform-thickened, not extensively br or fibrous, often readily detached from st; pl of des or grassl or from lower levels in mts, at least in Cas
                                            Group II, lead 18a

Group I

13a Pl (1) 2–3 (4) dm, with large clump of fleshy basal lvs; sts mostly fistulose; racemes open, freely br, gen > half total height of pl; lower peds spreading-ascending, mostly several times as long as sepals; sinus of lateral petals 3–4 mm deep; follicles mostly 11–14 (18) mm; alp or subalp ridges and talus slopes; Ore Cas and Cas (Chelan to Yakima cos) and OM, Wn; rockslide l. (*D. caprorum, D. bicolor* var. *g.*)   11 **D. glareosum** Greene

13b Pl often > 4 dm, esp if sts fistulose, lvs not particularly fleshy and often in greater part cauline; racemes often ± compact, or simple, or much < half total height of pl; lower peds often spreading, or scarcely > sepals; sinus of lateral petals mostly < 3 mm deep; follicles often > 15 mm; mostly well e Cas, or occ along e base Cas at lower elev

14a Racemes compact, fl late June–Aug; peds < sepals, or lower ones sometimes slightly > but strongly ascending; sepals cupped forward; pl mostly crisp-puberulent; sts rather strongly fistulose, (2) 3–15 dm; pl of dry mts or valleys, not of open sagebr plains

15a Pl finely crisp-puberulent throughout, never glandular, (6) 8–15 dm; sts abundantly lfy; lvs only gradually reduced upwards; peds mostly ascending to semi-erect; dry sagebr slopes to ponderosa or lodgepole pine for, c Ore to ne Cal and sw Ida; hedgenettle l., tall mt l. (*D. umatillense*)   12 **D. stachydeum** (Gray) Nels. & Macbr.

15b Pl glab throughout, or pubescent above where often ± glandular, (2) 3–10 dm; sts gen lfiest below; lvs often abruptly reduced upwards; peds more spreading than ascending; sagebr or bunchgrass-covered slopes to open mont for; Custer Co, Ida, to YNP and Madison Co, Mont; palish 1. (*D. g.* var. *multicaule*)   13 **D. glaucescens** Rydb.

14b Racemes open, fl mostly Apr–June; peds (lower ones at least) up to 2–4 times as long as sepals, gen widely spreading (ascending); sepals widely flared; pl mostly not crisp-puberulent; sts 1–5 (7) dm, mostly not fistulose; pl of various habitats, including open sagebr plains

16a Sts 1–several, ± fistulose, mostly glab up to infl (where often crisp-puberulent but not glandular); lower peds ascending to erect; basal lvs gen ± withered by anthesis; pl of open des or juniper woodl into the lower des mts; c and se Ore to Cal, e to sw Mont, Utah, and Nev; Anderson's 1., des 1. (*D. megacarpum, D. zylorrhizum*)
                                                                         14 **D. andersonii** Gray

16b Sts mostly single, not fistulose, often strongly pubescent or glandular, or both; lower peds spreading; basal lvs rarely withered by anthesis; pl more commonly of grassl to ponderosa pine for or subalp areas, mostly n of above range, at least in sagebr des, where occ

17a Lower petals deep blue, bl gen suborbicular and shallowly notched, sinus rarely > 1/5 length of bl; sepals unequal, lower pair much the longest; grassl and ponderosa pine for to subalp scree; nc Ida through most of Mont to nw Wyo, w SD and n to Alta and Sask; Mont 1., little 1.   15 **D. bicolor** Nutt.

17b Lower petals variously colored, often white or pale blue, but then very heavily blue-pencilled, gen elliptic, deeply notched, sinus gen 1/4–1/2 length of bl; sepals more nearly =; sagebr des to (commonly) mt valleys and slopes, esp in ponderosa pine belt; sw BC s, along e foothills of Cas, to n Cal, e to Alta, Mont, Wyo, and Neb, s to Colo and Ariz; upland 1.; 3 well marked vars.
                                                                   16 **D. nuttallianum** Pritz.

a1 Lower petals brownish or yellowish-purple; pl mostly crisp-puberulent and eglandular (glandular); e Wn, n Ida, nw Mont, and Alta, occ in Wallowa Mts, Ore   var. **fulvum** Hitchc.

a2 Lower petals white to deep purple, never brown or yellowish-purple; pl various

b1 Sepals pale blue to lavender; petals white, very prominently lined with purplish-black, bl small, < claw, not covering stamens; foothills and valleys of Wen Mts, s Chelan and n Kittitas cos, Wn *(D. l.)*   var. **lineapetalum** (Ewan) Hitchc.

b2 Sepals gen deep blue, if light in color (as pl from wc Ida esp) then petal bl > claw and not prominently lined purplish-black; range of sp. except rarely with above vars. (*D. bicolor* var. *mccallae, D. b.* f. *helleri, D. b.* var. *nelsonii, D. n., D. helleri, D. lineapetalum* f. *viscidum, D. n.* vars. *levicaule* and *pilosum*)
                                                                        var. **nuttallianum**

Group II

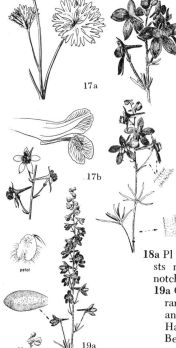

18a Pl gen eglandular and ± shaggy (at least calyx) with crisp or wavy hairs; sts rather readily detached from root system; lower petals shallowly notched, sinus < 1/5 length of bl; w of Cas, except in and near CRG

19a Calyx cupulate, sepals only slightly spreading, mostly < 12 mm, spur rarely > 12 mm; racemes ∝ - fld, compact, peds often < sepals; prairies and basaltic cliffs, Pierce, Clark, Cowlitz, Thurston, Mason, and Grays Harbor cos, Wn, to Clackamas Co, Ore, and e along CRG as far as Benton Co, Wn; Nuttall's 1. (*D. columbianum, D. oreganum*)
                                                                         17 **D. nuttallii** Gray

19b Calyx not cupulate, sepals widely flared, mostly > 12 mm; spur 13–15 mm; racemes mostly few-fld, loose, peds commonly much > sepals; coastal bluffs and prairies to lower mt meadows; w Cas from BC to nw Cal; Menzies' 1.; 2 vars.   18 **D. menziesii** DC.

a1 Pl coarsely pubescent almost throughout with shaggy hairs, not glandular; BC to PS area, Wn, and sporadically s along coast to Cal (*D. chilliwacense, D. pauperculum*)   var. **menziesii**

a2 Pl partially glab, or finely pubescent, or ± glandular; follicles often glab; WV and Cas of Ore to Cal   var. **pyramidale** (Ewan) Hitchc.

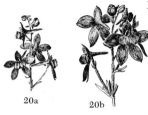

18b Pl often glandular, esp above; lower petals often white or pale blue; wholly e of Cas except in CRG
    20a Blades of lower petals gen elliptic and notched > 1/5 their length, mostly blue, but sometimes white or pale blue (see lead 17b)
                                      16 D. nuttallianum Pritz.
    20b Blades of lower petals gen orbicular and notched < 1/5 bl length, deep blue (see lead 17a)
                                        15 D. bicolor Nutt.

20a   20b

Isopyrum L.   Isopyrum; Rue-anemone

Fls solitary to few and corymbose-umbellate, apet (ours), ♂, reg, ± showy; sepals 5 (6–9), petaloid, white to rose; stamens 10–∝; follicles 2–20, few-seeded; style prominent, persistent; glab to pubescent, ± glaucous per herbs with alt, (1) 2–3 times ternate lvs. (Gr name, from *isos*, equal, and *pyros*, wheat, referring to the grainlike frs).

1a Fls scarcely 10 mm broad, solitary, white (pinkish); peduncle barely > lvs; stamens mostly 10, filaments enlarged near midlength; pl 0.3–1.2 dm; shady places, Yamhill Co, Ore, to nw Cal, w Cas; dwarf i., w i.
                                      1 I. stipitatum Gray
1b Fls 12–20 mm broad, in 3–10-fld corymbs or umbels > the lvs, white to rose; stamens ∝, filaments enlarged near tip; pl 3–10 dm; moist woods and streambanks; Lewis and Thurston cos, Wn, to CRG and WV, Ore, as far s as Marion Co; Hall's i., Willamette i.   **2 I. hallii** Gray

Myosurus L.   Mouse-tail

Fls solitary on lfless scapes, inconspicuous, complete or sometimes apet, reg; sepals 5 (6–7), deciduous, greenish, spurred at base and with a 1–3-nerved bl; petals mostly 5 or fewer, linear, > the sepals; stamens 5 or 10; achenes (10) 25–200, in spirals on a much-elongate, slender recep, sessile, ± flattened, with dorsal keel and prominent stylar beak; glab, acaulescent ann with linear lvs and 1–several scapes, gen in moist areas, esp in vernal pools. (Gr *mus*, mouse, and *oura*, tail, referring to the appearance of the spike).

1a Scapes < 1 cm at maturity (fl subsessile); sepals ca 2.5 mm; recep 1–2.5 cm; beak of achene ca 1 mm; alkali flats, Umatilla Co, Ore, and in c Cal; shortstd m.   **1 M. sessilis** Wats.
1b Scapes 1–15 cm at maturity; sepals 1.5–4 mm; recep 0.5–5 cm; beak of achene often < 1 mm
    2a Beak of achene (0.5) 0.8–1.5 mm; recep (mature) 5–10 mm, with 20–50 (90) achenes; sepals mostly 1-nerved, 1.5–2.5 mm; wet places in sagebr and grassl areas, or occ subalp; BC to Cal, e Cas, e to Mont, Wyo, and Utah; sedge m., bristly m.   **2 M. aristatus** Benth.
    2b Beak of achene rarely > 0.5 (0.7) mm; recep mostly (5) 15–50 mm when mature, often with > 100 achenes; sepals 3 (5)-nerved, 1.5–3 (4) mm; BC s, on both sides Cas, to s Cal, e to Ont, Va, NC, and Tex; Europe; least m., tiny m., common m.   **3 M. minimus** L.

Ranunculus L.   Buttercup; Crowfoot; Water-buttercup

Fls complete, reg, ± showy; sepals 5 (3–6), spreading to reflexed, often ± petaloid; petals mostly 5 (6–10), or sometimes ∝ (rarely lacking), commonly yellow, but sometimes white, or reddish-tinged, with small, gen scale-covered gland on upper side near base; stamens (5) 10–∝, occ some sterile and simulating petals; pistils 5–∝; achenes in an ovoid or globose to cylindric cluster, each compressed to inflated, often ribbed or striate, glab to hairy or echinate, often keeled on the back and with a short to prominent, straight to hooked, gen persistent style; stylar beak glab or pubescent; aquatic to terrestrial, gen per (ann) herbs with erect to creeping or floating sts and simple to compound, ± stip, alt lvs. (L name, used by Pliny, from *rana*, frog, in reference to the aquatic habitat). (*Arcteranthis, Batrachium, Beckwithia, Cyrtorhyncha, Flammula, Halerpestes, Hecatonica, Kumlienia, Oxygraphis*).

1a Lvs entire to serrate, neither dissected nor compound, sometimes lobed at tip and pl then stoloniferous; scale of nectary forming a true pocket, lateral margins adnate to petal for 1/2 to full length; achenes glab or very rarely sparsely pubescent, cluster gen < 1 cm broad   **Group I,** lead 9a

1b Lvs at least in part deeply lobed to compound; nectary glands various; achenes sometimes hairy or spiny (if lvs simple then either achenes finely pubescent or cluster of achenes ca 1 cm broad)

  2a Fl sts 1 (rarely 2)-fld, lfless or with a single reduced bract

    3a Pl ann, ± finely tomentose throughout; fls inconspicuous; sepals 4–6 mm, persistent long after anthesis; petals yellow, drying white and usually pinkish tinged, 5–8 mm; stamens 10–15; European weed, rapidly spreading mostly in sagebr des e Cas; e Wn and Ore to Nev, Ida, and Colo; hornseed b.   1 **R. testiculatus** Crantz

    3b Pl per, glab; fls more showy, sepals and petals both gen > 7 mm; stamens at least 20

      4a Petals 5, reddish, 10–20 mm; achenes obovate, 6–12 mm, strongly compressed, glab and smooth, stylar beak ca 0.5 mm; sagebr valleys and ponderosa pine for, always where well drained; e Ore to c Ida, s to Cal and Nev; pink b., Anderson's b.   2 **R. andersonii** Gray

      4b Petals 7–12 (15), yellowish, 7–10 mm; achenes ellipsoid, ca 2.5 mm, prominently nerved lengthwise, stylar beak ca 1 mm; damp slopes and rock crevices; Alas to Wn, where presently known only from Mt Colonel Bob, OM, and from Del Campo Peak, Snohomish Co; Cooley's b.   3 **R. cooleyae** Vasey & Rose

  2b Fl sts mostly 2–several-fld, lfy

    5a Mature achenes bristly, spiny, or papillate, not cross-corrugated; scales of nectary fairly prominent, margins free at least half their length; petals yellow   **Group II,** lead 15a

    5b Mature achenes not bristly, spiny, or papillate, although sometimes cross-corrugated and sometimes hairy (nectary scale then often with margins adnate most of length and forming a pocket); petals sometimes white

      6a Achenes prominently cross-corrugated; petals white or with white bl and yellowish claw; pl aquatic; nectary scale very short to almost obsolete, broader than long, forming a minute pocket or a mere thickening   **Group III,** lead 18a

      6b Achenes not prominently cross-corrugated; petals gen yellow; pl mostly terrestrial; nectary scale gen rather prominent, often with free margins

        7a Pl aquatic or semi-aquatic, often with some finely dissected submerged lvs; sts gen creeping and often nodally rooting; nectary scale conspicuous, ± adnate to petal, margins and tip gen free, bearing gland on exposed ventral surface rather than concealed beneath it; achenes glab, either beakless or with beak much < 0.5 mm, or with corky thickenings on margins **Group IV,** lead 21a

        7b Pl not aquatic and with dissected lvs; sts mostly not nodally rooting; achenes either hairy or with conspicuous beak, not corky-thickened on margin

          8a Achenes only moderately compressed (thickness at least 1/3 the dorsiventral width), often finely pubescent; nectary scale gen short, the margins adnate almost to their tips, forming a small pocket; receptacle gen elongating considerably in fr and sometimes becoming as much as 10 times as long as at anthesis   **Group V,** lead 25a

          8b Achenes strongly compressed, often discoid (thickness < 1/3 the dorsiventral width), never pubescent; nectary scale elongate, the margins free above for at least 1/2–2/3 of the length; receptacle in fr not > 2–3 times as long as at anthesis   **Group VI,** lead 37a

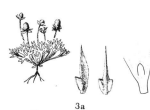

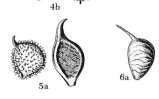

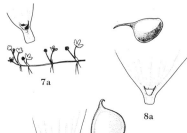

Group I

9a Sepals 3 (4); roots tuberous; achenes beakless; petals 8–12, narrow, gen 10–18 mm; lvs cordate, entire to sinuate; European sp., sometimes cult and occ estab, as in w Wn; lesser Celandine *(Ficaria f.)*   **R. ficaria** L.

9b Sepals gen 5; pl otherwise gen not at once as above

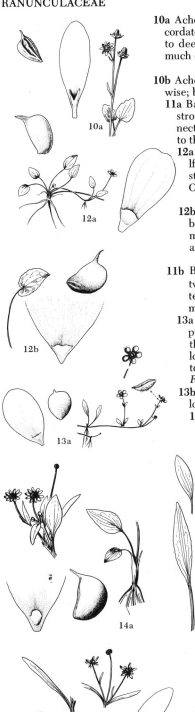

10a Achenes 50–200, ribbed lengthwise, borne in a columnar mass; lf bl cordate-rotund to cordate-lanceolate, lanceolate, or rhombic, shallowly to deeply crenate-toothed to ± 3-lobed at tip; wet places, throughout much of N and S Am; Eurasia; shore or seaside b. *(R. c. f. hebecaulis)*
                                                          4 R. cymbalaria Pursh
10b Achenes either < 50 or borne in a globular mass, not ribbed lengthwise; bl various, often entire
  11a Basal lvs ovate to cordate, not twice so long as broad; roots rather strongly enlarged and fleshy toward base, whitish and carrot-shaped; nectary scale at least as long as broad, margins adnate almost or quite to the tip
    12a Sts reclining or prostrate to ascending, rooting at the nodes; basal lf bls 1–3 cm, gen entire; petals mostly ca 4 (to 6) mm; mudflats and streambanks to subalp meadows; Cas from Deschutes and Lane cos, Ore, to Cal; Gorman's b. *(R. reptans var. g., R. terrestris)*
                                                          5 R. gormanii Greene
    12b Sts erect to sometimes prostrate but never rooting at the nodes; basal lf bls 2–5 cm, mostly ± crenate-serrulate; petals 4–7 mm; wet mont areas; Cas from c Ore to Cal, to Blue and Wallowa mts, Ore, and through Ida to c Mont; mt b. *(R. cusickii, R. alismellus var. p.)*
                                                          6 R. populago Greene
  11b Basal lvs linear to elliptic or lanceolate (rarely ovate), mostly > twice as long as broad; roots scarcely enlarged near base, not whitened; nectary scale sometimes much broader than long or with margins adnate to not much beyond midpoint if at all
    13a Sts decumbent or prostrate, nodally rooting, often appressed-pubescent; sepals gen hairy; nectary scales mostly much broader than long; lvs entire; petals (2) 4–5 (7) mm; mudflats mostly, lowl to mt valleys, often where brackish; Alas to Cal, e to Atl and to RMS; creeping b., lesser spearwort *(R. filiformis, R. intermedius, R. reptans, R. samolifolius, R. unalaschensis)* **7 R. flammula** L.
    13b Sts erect, not nodally rooting, gen glab; nectary scale at least as long as broad; lvs sometimes ± serrate; petals sometimes > 7 mm
      14a Roots glab; lf bls mostly at least 1 cm broad; petals rarely < 5 mm; basal lvs slenderly petiolate; stamens 25–90; mudflats to alp meadows in mts of most of w US and sw Can; water-plantain b.; 5 vars.   **8 R. alismaefolius** Geyer
      a1 Petals gen 10, narrow, mostly only 2–3 mm broad; mt meadows, Valley Co, Ida, to c Nev, e to s Wyo and to Colo
                                                          var. **montanus** Wats.
      a2 Petals gen 5, mostly 3–7 mm broad
        b1 Pl mostly 3–6 dm, glab; basal lf bls broadly lanceolate, often ± serrulate, petioles 2–6 mm broad; petals 8–10 mm; occ from sw BC s in mts of w Wn to WV, Ore, and to ne Cal, e to c Ida and w Mont; plantainlvd b.   var. **alismaefolius**
        b2 Pl mostly < 3 (1–4) dm, sometimes pilose; basal lvs entire, variable in shape, but gen slender-petiolate; petals mostly < 8 mm
          c1 Basal lf bl gen tapered gradually to petiole; petals ave > 6 (to 8) mm; roots thickened only slightly toward base; mont, e side Cas, Wn to Cal, e to Ida and Wyo
                                                          var. **hartwegii** (Greene) Jeps.
          c2 Basal lf bl narrowed abruptly to petiole; petals 4–8 mm; roots ± thickened toward base; mont to subalp meadows
            d1 Pl mostly pubescent on peds or petioles; e side Cas, Ore, to Ida, Mont, and Wyo   var. **davisii** Benson
            d2 Pl glab; e side Cas, Kittitas Co, Wn, to Cal, e to Ida and w Mont; dwarf plantainlvd b.   var. **alismellus** Gray
      14b Roots finely tomentose; lf bl mostly < 1 cm broad, broadly petiolate; petals mostly < 5 mm; stamens ca 15; wet places in foothills and lower mt meadows; Union Co, Ore, and Washington, Adams, Valley, and Elmore cos, Ida; Blue Mt b.
                                                          9 R. oresterus Benson

Group II

15a Achenes ca 2 mm, covered with basally papillate, hooked bristles; pl ann; petals < 3 mm, often lacking; nectary gen above midlength of petal; moist to dryish hills and woodl; e Wn and adj Ida to Ariz and Baja Cal; downy b.   **10 R. hebecarpus** H. & A.
15b Achenes mostly much > 2 mm (if not, pl per), with bristles, papillae, or spines, but these never hooked; petals 5–9 mm; nectary basal or nearly so
   16a Achenes 2–3 mm, papillose or bristly but never spiny; pl per, hirsute; stamens 25–50; European weed, widely distributed in US: known in our area only from near Portland, Ore; hairy b.   **11 R. sardous** Crantz
   16b Achenes 4–6 mm, spiny; pl often glab, ann (or occ per); stamens mostly 10 (15)
      17a Lvs cordate to reniform, shallowly lobed into 3 deeply crenate to lobate segms, not divided into narrow lfts; pl glab or slightly hirsute above; nectary scale about half as broad as contiguous petal; European weed in much of US; in our area mostly w Cas, Wn and Ore to Cal; spiny-fr b.   **12 R. muricatus** L.
      17b Lvs oblanceolate to obovate, gen divided into linear to oblanceolate or obovate segms; pl ± pubescent all over; nectary scale nearly as broad as contiguous petal; widely intro European weed, mostly on waste ground or in dry areas in the lowl; known in our area from e Wn, Ore, and Ida; field b., corn c., hungerweed   **13 R. arvensis** L.

Group III

18a Pl glab; floating lvs ternately deeply divided, ultimate segms elliptic-obovate, entire or with 2–3 rounded teeth; recep glab; petals 4–6 mm; achenes 2–7, merely apiculate, not beaked; lowl vernal pools along coast, where occ in Wasco and Benton cos, Ore, and in Cal; Lobb's w-b. (*R. aquatilis* var. *l.*)   **14 R. lobbii** (Hiern) Gray
18b Pl rarely completely glab; floating lvs often finely dissected; recep hairy; achenes 10–50 or more; petals 5–10 (14) mm; achenes sometimes with prominent persistent stylar beak
   19a Mature achenes with persistent stylar beak ca 1 mm; lvs all finely dissected, bl sessile on stip base (without petiole), mostly holding form when pl withdrawn from water; marshes and slow streams, often where brackish; RMS, Mont to NM, e to Que, Ala, and Tex, w to se Ida, Utah, and Nev; longbeaked w-b.   **15 R. longirostris** Godr.
   19b Mature achenes beakless, or remnant of style scarcely 0.5 mm; lvs sometimes petiolate
      20a Lf bl sessile on stip base, not collapsing when pl withdrawn from water; achenes 30–80; ponds and sluggish streams; BC to Cal and Mex, e to Que and Mass, s to Minn; stiff-lvd w-b. (*R. aquatilis* var. *s.*)   **16 R. subrigidus** Drew
      20b Lf bl mostly with slender petiole 1–2 cm, gen collapsing when pl withdrawn from water; achenes mostly 10–25; sluggish streams and ponds, in much of N Am and Europe; white w-b., water c.; 3 vars.
                                                                                17 R. aquatilis L.
         a1 Floating lvs simple, very much unlike the submerged (filiformly dissected) lvs; Alas to Cal, s and w of Cas, e to Mont, Wyo, and Utah; Europe   var. **hispidulus** Drew
         a2 Floating lvs, as submerged ones, filiformly dissected
            b1 Lvs ca 3 times dissected, ultimate segms 0.5–1.5 mm broad; se Ida to n Utah, also near Spokane, Wn   var. **porteri** Benson
            b2 Lvs 4–7 times dissected into filiform segms < 0.5 mm broad; gen in range of sp. as whole, but not in Europe (*R. c.*, *R. trichophyllus*)   var. **capillaceus** (Thuill.) DC.

Group IV

21a Pl ann, erect, not nodally rooting, (1.5) 2–5 dm; achenes (50) 100–200, in a cylindrical cluster, stylar beak scarcely 0.2 mm; petals 2–5 mm; semi-aquatic (often brackish) areas to moist meadows; most of N Am, n of Mex; Eurasia; celerylvd c. or b., blister b.; 2 vars.   **18 R. sceleratus** L.
   a1 Achenes cross-corrugate or -reticulate on central area of faces; basal lvs deeply divided into 3 main divisions, sinuses between segms narrow, 2ndary lobing shallow; mostly European, but widely distributed in N

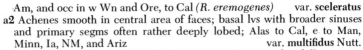

·Am, and occ in w Wn and Ore, to Cal *(R. eremogenes)* var. **sceleratus**

a2 Achenes smooth in central area of faces; basal lvs with broader sinuses and primary segms often rather deeply lobed; Alas to Cal, e to Man, Minn, Ia, NM, and Ariz var. **multifidus** Nutt.

21b Pl per, gen < 15 cm; sts mostly floating or reclining and nodally rooting; stylar beak gen at least 0.2 mm; petals often > 5 mm

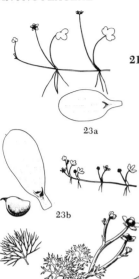

23a

22a Lf bls scarcely 15 mm, 3–5-lobed, lobes entire or only shallowly notched; petals 2–4.5 mm; stylar beak < 0.3 mm; nectary scale adnate to petal on sides as well as at base

23a Lvs cordate, sometimes > 10 mm; sts and petioles often ± hairy; fr recep 2–4 mm; ponds and muddy shores; Alta s in the RM to Colo and c Ida; floating w-b.; ours the var. *intertextus* (Greene) Benson (var. *n.* in Eurasia) 19 **R. natans** Meyer

23b

23b Lvs mostly rounded to acute basally, scarcely 10 mm; sts and petioles glab; fr recep mostly < 2 mm; mud flats and shallow ponds; circumpolar and s in N Am to Mont, Wyo, and Newf; farnorthern b., arctic b. 20 **R. hyperboreus** Rottb.

22b Lf bls either > 15 mm or ± finely dissected; petals 3–10 (15) mm; stylar beak mostly > 0.5 mm; nectary scale adnate to petal only at base, upper margin (as well as sides) free

24a

24a Beak of achene (1) 1.5–2 mm; lf bls 2–8 cm, 3–5 times ternately dissected into segms scarcely 2 mm broad; petals (7) 9–15 mm; achenes 1.8–2.5 mm, puffy-margined for ca 3/4 of the circumference and strongly keeled on distal dorsal margin; ponds and mudflats at lower elev; BC s, e Cas in Wn, to CRG, e Ore, and Cal, Nev, and Utah; e to Ont, Que, Atl coast, and Miss Valley, not in RM in our area; yellow w-b. *(R. fluviatilis)* 21 **R. flabellaris** Raf.

24b Beak of achene scarcely 1 mm; lf bls 1–2 (3) cm, shallowly lobed to parted, but the divisions gen > 2 mm broad; petals 3–7 (10) mm; achenes 1–1.5 (2) mm, ± puffed at base for 1/2–2/3 of the circumference, upper distal region only slightly keeled; shallow ponds to mudflats, over much of N Am and Asia; Gmelin's b., small yellow w-b.; 2 vars. in our area (var. *g.* in Asia) 22 **R. gmelinii** DC.

a1 Pl hirsute; lf bls ca 10–15 mm, mostly shallowly lobed into short, rounded segms; sepals 2.5–4 mm; petals mostly 4–5 mm var. **limosus** (Nutt.) Hara

24b

a2 Pl glab to hirsute; lf bls 10–30 mm, mostly rather shallowly 2–3 times dissected into linear segms; sepals 4–6 mm; petals 4–10 mm var. **hookeri** (G. Don) Benson

Group V

25a Nectary scale (1) 1.5–3 mm, ± oblong, adnate ca whole length along the margins; achenes in a semiglobose cluster (7) 10–20 mm broad; petals (6) 8–15 mm; pl glab to sparsely hirsute above, but long-hirsute on basal stips and ± tomentose toward the tips of main roots; sts rarely > 15 cm, ascending to ± decumbent; e Cas, sagebr or grassl to ponderosa pine for, gen on well drained soil

26a

26a Lvs triternately dissected into linear segms; nectary scale rarely ciliate on upper margin; sagebr slopes between Wasco Co, Ore, and Goldendale, Klickitat Co, Wn; obscure b. *(R. triternatus)* 23 **R. reconditus** Nels. & Macbr.

26b Lvs entire to broadly lobed; nectary scale gen ciliate on upper margin; BC s, e Cas, to Cal, e to Daks, Neb, and NM; sagebr b.; 2 vars. 24 **R. glaberrimus** Hook.

a1 Basal lf bls mostly entire, elliptic to oblanceolate; st lvs entire to 3-lobed, middle lobe much largest; mostly mont, often in juniper or ponderosa pine or on mont sagebr slopes; BC to Cal, e Cas, e to Daks and NM *(R. e., R. waldronii)* var. **ellipticus** Greene

a2 Basal lf bls mostly ovate to obovate, sometimes broader than long, more often shallowly lobed than entire; st lvs often entire; mainly lowl sagebr and grassl areas; with var. *e.* through much of its range, but not quite so far e and s var. **glaberrimus**

a2 a1

25b Nectary scale often < 1 mm, or broader than long, margins often free part of length; achenes mostly in more elongate or narrower cluster; petals often < 6 mm; pl often of wet places or truly mont, sometimes in or w of Cas, often pilose, but roots never tomentose; sts gen erect, often > 15 cm

26b

27a Achenes finely pubescent, but sometimes glabrate with age

28a Roots enlarged toward tip (± clavate), abruptly rounded, tuberlike, 1–2.5 (3) dm, 2–5 mm thick; sts mostly < 10 cm; basal lvs deeply divided into 3–5 oblanceolate lobes; nectary scale narrowly fan-shaped, lateral margins adnate full length and forming pocket; sagebr slopes to lower for; se Ida to Nev, YNP, sw Wyo, and Utah, fl soon after appearance from under snow; Jove's b. *(R. digitatus)*   25 **R. jovis** A. Nels.
28b Roots tapered gradually to tip; sts mostly > 10 cm
  29a Petals 2–6 (8) mm (or lacking); basal lf bls cuneate, some merely crenate; mt meadows and moist slopes; Alta to Colo in RM, w to e BC and Wn, much of Ida, Utah, Nev, and Ariz; unlovely b. *(R. alpeophilus, R. utahensis)*   26 **R. inamoenus** Greene
  29b Petals either > 8 mm or basal lvs deeply lobed or divided, often cordate
    30a Pl glab or ± yellowish- or brownish-pilose on calyx and upper lvs and bracts, otherwise glab; basal lf bls reniform or oval to broadly obovate, 1–3 cm, from shallowly 3-lobed to 2–3 times dissected; stylar beak straight to slightly curved, 0.9–1.5 mm; mt meadows and talus slopes; Alas to Cal, e to Alta, RMS, and Ariz; subalp b.; 4 vars.   27 **R. eschscholtzii** Schlecht.
      a1 Middle division of basal lvs only once lobed; styles ca 1 mm; achenes glab; pl gen glab
        b1 Lf segs mostly round-tipped, middle lobe of basal lvs entire to shallowly 3-lobed; mainly w, from Alas to Cal, but occ elsewhere, mostly boreal *(R. helleri)*   var. **eschscholtzii**
        b2 Lf segs mostly acute, middle lobe of basal lvs gen deeply lobed to entire; mostly in w and s part of sp. range; Wn to Alta and s *(R. eximius, R. saxicola)*   var. **suksdorfii** (Gray) Benson
      a2 Middle division of basal lvs gen twice lobed into narrow or linear segs; styles mostly > 1 mm; achenes sometimes pubescent; pl gen brownish-pilose above
        c1 Achenes mostly pubescent; ultimate segs of basal lvs mostly > 2 mm broad; calyx sparsely pilose or glab; e Ore to Ida   var. **trisectus** (Eastw.) Benson
        c2 Achenes glab; ultimate segs of basal lvs gen < 2 mm broad; calyx often strongly brownish-pilose; w Ida to sw Wyo, Utah, and Colo   var. **alpinus** (Wats.) Hitchc.
    30b Pl ± pilose throughout
      31a Basal lf bls cordate, 1–3 cm, pedately to irreg parted into 5–7 simple or lobed divisions; stylar beak recurved, ca 1 mm; petals obovate, 8–10 mm; nectary scale glab, deeply retuse on upper margin; circumpolar, s through Can, sporadic in RM, in Flathead Co, Mont, and in Wyo, Colo, NM, and Ariz; birdfoot b., n b.   28 **R. pedatifidus** J. E. Smith
      31b Basal lf bls cordate, (1.5) 2–6 cm, deeply crenate to shallowly lobed and narrowly toothed; stylar beak scarcely curved, 0.6–1 mm; petals 8–15 mm; nectary scale ciliate on upper margin, not retuse; mt meadows, BC to Alta and Sask, sporadically s to ne Wn, and in Wyo, Utah, NM, and Ariz; heartlvd b.   29 **R. cardiophyllus** Hook.
27b Achenes and young ovaries glab
  32a Petals mostly > 8 mm, if < then basal lf bls all lobed at least half length; pl of mt meadows and talus, mostly 5–20 cm (see lead 30a)   27 **R. eschscholtzii** Schlecht.
  32b Petals mostly < 5 mm, but if (as rarely) >, then basal lf bls in part merely deeply crenate or sts much > 20 cm; pls sometimes of lowl
    33a Sts erect, 1–5 dm; basal lf bls lobed < half length, if at all
      34a Achenes 20–40 (50) per fl, 1–1.5 mm; stylar beak 0.1–0.2 mm; petals up to 3.5 mm; moist woodl to meadows, lowl to subalp; Alas to ne Wn, e to Lab, Ida, Tex, and Fla; Cuba; smallfld b., kidney-lvd b.   30 **R. abortivus** L.
      34b Achenes (40) 50–100, 1.5–2 mm; stylar beak 0.7–0.9 mm; petals (if any) 2–8 mm; (see lead 29a)   26 **R. inamoenus** Greene
    33b Sts often spreading, seldom up to 1.5 dm; basal lf bls lobed at least half length
      35a Petals 1.5–3 (3.5?) mm; pl gen < 6 cm; basal lf bls < 1 cm; stylar beak mostly straight; stamens 10–20; Alas to Greenl, s in

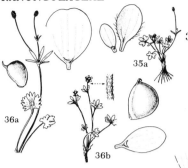

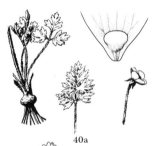

high alp meadows of RM to Mont, Wyo, and Colo; dwarf b.

31 **R. pygmaeus** Wahlenb.

35b Petals 3–5 mm; pl gen > 6 cm; basal lf bls > 1 cm; stylar beak recurved; stamens 20–50

36a Sts 10–20 cm; achenes 1–1.5 mm, bulged slightly on dorsal margin near base; Alas s, mostly on talus slopes or in alp meadows, to Cas-OM, Wn, and to ne Ore and Alta, Mont, and Ida; modest b.

32 **R. verecundus** Robins.

36b Sts 3–9 cm; achenes 2–2.5 mm, rounded rather uniformly to base and without bulge on dorsal margin; Alas and n Can, s on talus slopes and in alp meadows, to BC, Alta, Mont, and Colo; arctic b. (*R. grayi, R. hookeri, R. ramulosus*)

33 **R. gelidus** Kar. & Kir.

Group VI

37a Achenes with straight stylar beak 2.5–4 mm; basal lf bls pinnately 5–7-lobed; petals 9–18 mm; nectary scale 0.5–1 mm, adnate along margins for not > 1/3 length; moist fields to mont meadows; Alas to Cal, e to Mont, Wyo, and Utah; straightbeak b.; 2 vars.

34 **R. orthorhynchus** Hook.

a1 Petals yellow but reddish-tinged, mostly > twice as long as broad; stylar beak 3.5–4 mm; s Alas and BC s, w side Cas, to s Ore and Cal

var. **orthorhynchus**

a2 Petals yellow, mostly < twice as long as broad; style gen < 3.5 mm; BC s, e side Cas, to Cal, e to Mont, Wyo, and Utah (*R. macranthus*)

var. **platyphyllus** Gray

37b Achenes with stylar beak either not straight or < 2.5 mm; basal lf bls rarely pinnately lobed; petals mostly < 10 mm; nectary scale often adnate > 1/3 length

38a Petals much > sepals, 5–18 (ave > 7) mm; basal lf bls mostly simple, when compound sts either nodally rooting and weedy or enlarged and cormose at base

39a Sts tending to root at lower nodes, gen prostrate or decumbent; fls often double; basal lvs compound; stylar beak stout, 0.7–1.2 mm, recurved; pl hirsute, weedy, widely estab in US, intro from Europe; creeping b.; represented in our area by 2 vars.

35 **R. repens** L.

a1 Petals > 10 (fls double); occ in Wn, Ore, and Ida

var. **pleniflorus** Fern.

a2 Petals gen 5 (up to 10); common phase

var. **repens**

39b Sts not rooting at lower nodes; fls never double; basal lvs various

40a Sts thickened and cormose at base; basal lf bls compound; intro from Europe; known from near Bingen, Wn, and from WV, Ore; bulbous b.

36 **R. bulbosus** L.

40b Sts not enlarged or cormose at base; basal lf bls sometimes deeply divided but never quite compound

41a Petals gen 8–16, oblong, ca 3 times as long as broad; sts spreading to decumbent; beak of achene stout, 0.5–1 mm, recurved; grassy coastal bluffs, nw Wn, Clatsop Co, Ore, to Cal; Cal b.

37 **R. californicus** Benth.

41b Petals gen 5, when (as sometimes) 8 or more then scarcely twice so long as broad, or achenes with slender beak at least 1 mm; sts gen erect

42a Achene body 2–2.5 mm; beak 0.3–0.6 mm; basal lf bls pentagonal, cut or incised (3) 4–5 times into short acute teeth; sepals spreading; European, widely estab on moist to well-drained soil; Alas to Cal, e to Ida and Mont; meadow b., tall b.

38 **R. acris** L.

42b Achene body 2.5–3.5 mm; beak 1–2 mm; basal lf bls neither pentagonal nor > 2–3 times cut, segms blunt to acute; sepals mostly reflexed

43a Bls of basal lvs parted almost to base, 3 primary segms cuneate, shallowly toothed or lobed; beak of achene not greatly flattened, 0.7–2 mm, scarcely recurved; mostly coastal, Alas through w Wn and Ore to Cal, e to ne Ore; western b.; 4 vars.

39 **R. occidentalis** Nutt.

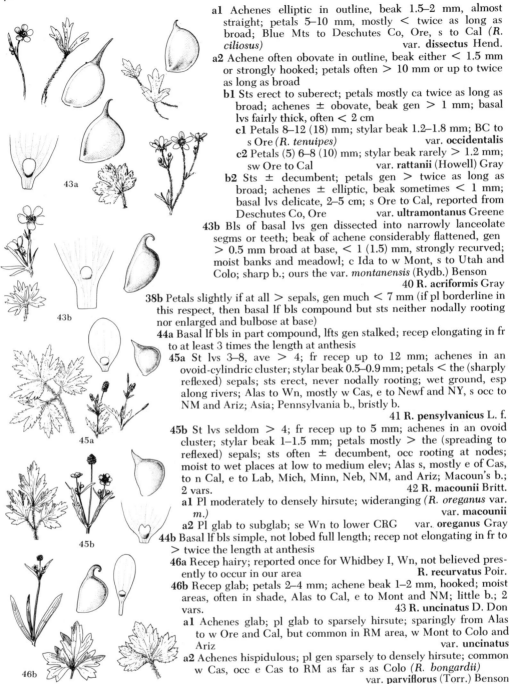

al Achenes elliptic in outline, beak 1.5–2 mm, almost straight; petals 5–10 mm, mostly < twice as long as broad; Blue Mts to Deschutes Co, Ore, s to Cal (*R. ciliosus*)   var. **dissectus** Hend.

a2 Achene often obovate in outline, beak either < 1.5 mm or strongly hooked; petals often > 10 mm or up to twice as long as broad

   b1 Sts erect to suberect; petals mostly ca twice as long as broad; achenes ± obovate, beak gen > 1 mm; basal lvs fairly thick, often < 2 cm

      c1 Petals 8–12 (18) mm; stylar beak 1.2–1.8 mm; BC to s Ore (*R. tenuipes*)   var. **occidentalis**

      c2 Petals (5) 6–8 (10) mm; stylar beak rarely > 1.2 mm; sw Ore to Cal   var. **rattanii** (Howell) Gray

   b2 Sts ± decumbent; petals gen > twice as long as broad; achenes ± elliptic, beak sometimes < 1 mm; basal lvs delicate, 2–5 cm; s Ore to Cal, reported from Deschutes Co, Ore   var. **ultramontanus** Greene

43b Bls of basal lvs gen dissected into narrowly lanceolate segms or teeth; beak of achene considerably flattened, gen > 0.5 mm broad at base, < 1 (1.5) mm, strongly recurved; moist banks and meadowl; c Ida to w Mont, s to Utah and Colo; sharp b.; ours the var. *montanensis* (Rydb.) Benson

                                  40 **R. acriformis** Gray

38b Petals slightly if at all > sepals, gen much < 7 mm (if pl borderline in this respect, then basal lf bls compound but sts neither nodally rooting nor enlarged and bulbose at base)

44a Basal lf bls in part compound, lfts gen stalked; recep elongating in fr to at least 3 times the length at anthesis

   45a St lvs 3–8, ave > 4; fr recep up to 12 mm; achenes in an ovoid-cylindric cluster; stylar beak 0.5–0.9 mm; petals < the (sharply reflexed) sepals; sts erect, never nodally rooting; wet ground, esp along rivers; Alas to Wn, mostly w Cas, e to Newf and NY, s occ to NM and Ariz; Asia; Pennsylvania b., bristly b.

                                 41 **R. pensylvanicus** L. f.

   45b St lvs seldom > 4; fr recep up to 5 mm; achenes in an ovoid cluster; stylar beak 1–1.5 mm; petals mostly > the (spreading to reflexed) sepals; sts often ± decumbent, occ rooting at nodes; moist to wet places at low to medium elev; Alas s, mostly e of Cas, to n Cal, e to Lab, Mich, Minn, Neb, NM, and Ariz; Macoun's b.; 2 vars.   42 **R. macounii** Britt.

      a1 Pl moderately to densely hirsute; wideranging (*R. oreganus* var. *m.*)   var. **macounii**

      a2 Pl glab to subglab; se Wn to lower CRG   var. **oreganus** Gray

44b Basal lf bls simple, not lobed full length; recep not elongating in fr to > twice the length at anthesis

   46a Recep hairy; reported once for Whidbey I, Wn, not believed presently to occur in our area   **R. recurvatus** Poir.

   46b Recep glab; petals 2–4 mm; achene beak 1–2 mm, hooked; moist areas, often in shade, Alas to Cal, e to Mont and NM; little b.; 2 vars.   43 **R. uncinatus** D. Don

      a1 Achenes glab; pl glab to sparsely hirsute; sparingly from Alas to w Ore and Cal, but common in RM area, w Mont to Colo and Ariz   var. **uncinatus**

      a2 Achenes hispidulous; pl gen sparsely to densely hirsute; common w Cas, occ e Cas to RM as far s as Colo (*R. bongardii*)   var. **parviflorus** (Torr.) Benson

Thalictrum L.   Meadowrue

Fls apet, scarcely showy, ♀ to ♀, ♂, racemose to pan; sepals 4–5, greenish-white or green to purplish, gen caducous; stamens 8–∝, filaments mostly white (ours), filiform to spatulate; pistils (2) 4–9 (15), style short, stigma unilateral; achenes sessile to short-stipitate, gen strongly ribbed lengthwise; rhizomatous, glab and often glaucous, mostly malodorous per herbs with alt, bi- to tri-ternately compound lvs. (Gr *thaliktron*, Dioscorides' name for some unknown pl).

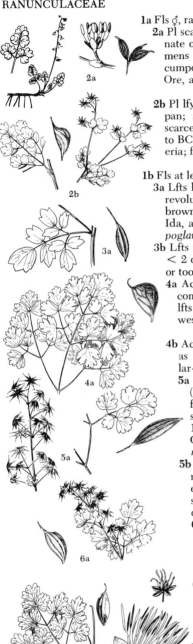

1a Fls ♂, racemose or pan; pls either subscapose or anthers not > 1 mm
    2a Pl scapose or with single lf near base, mostly 0.3–1.8 (2.8) dm; lvs biternate or ternate-pinnate; fls loosely racemose; sepals grayish-purple; stamens 8–15, anthers ca 1.5 mm, subequal to the filiform filaments; circumpolar, in N Am s in RM from Mont to NM, in parts of Ida and ne Ore, also in Cal and Nev; dwarf or alp m.; ours the var. *hebetum* Boiv.
                                                         1 T. alpinum L.
    2b Pl lfy-std, 3–8 dm; lvs 3–4 times ternate; fls in lfy-bracteate racemes or pan; sepals white or greenish; stamens 12–20, anthers ca 0.7 mm, scarcely 1/3 as long as the (slightly clavate) filaments; damp woods, Alas to BC and Alta, and in RM to Colo, w to Utah, Ida, e Ore, and Cal; Siberia; fewfld m.; ours the var. *saximontanum* Boiv.
                                               2 T. sparsiflorum Turcz.
1b Fls at least in part ♀, ♂, pan; pls lfy-std; anthers at least 1 mm
    3a Lfts leathery to membranous, heavily veined and gen hairy beneath, ± revolute, ca 2–3 cm, acutely 3-lobed, lobes mostly entire; achenes gen brownish-hairy; moist woods and meadows; e BC to Ont (rare in ne Wn, Ida, and Mont), to NM and Ariz, and to La and Tex; purple m. (*T. hypoglaucum*)
                                3 T. dasycarpum Fisch. & Avé-Lall.
    3b Lfts mostly membranous, glab or sparsely glandular-puberulent, often < 2 or > 3 cm, if 3-lobed then lobes commonly again less-deeply lobed or toothed; achenes gen glandular-puberulent or glab
        4a Achenes glab, body 4–6 (8) mm, gen > half as broad, inflated but compressed, lateral nerves evidently br, sinuous, and anastomosing; lfts gen glab; moist woods, WV, Ore, to Baja Cal and w Nev; tall western m. (*T. campylopodum, T. campylocarpum*)
                                     4 T. polycarpum (Torr.) Wats.
        4b Achenes often glandular-puberulent, or > 6 mm, or < half as broad as long, or with prominent, unbr lateral nerves; lfts often glandular-puberulent
            5a Achenes spreading to reflexed, elliptic-fusiform in outline, (4) 5–8 (10) mm, gen < half as broad, membranous-walled, moderately flattened; stigma commonly purplish, gen 3–4.5 mm, including the style; ♀ infl open, rather lfy, peds slender, = or > the achenes; lfts 1.5–3 cm, not prominently veined; BC s, and common, both sides Cas, to n Cal, e to Alta, Mont, Wyo, Colo, and Utah; western m. (*T. megacarpum, T. propinquum, T. rainierense*)   **5 T. occidentale** Gray
            5b Achenes often erect or ascending, 3–6 mm, either subterete and narrowly ellipsoid, or strongly compressed and obliquely obovate-elliptic in outline; stigma rarely purple, 1.5–3.5 mm, including style; ♀ infl often congested, inconspicuously bracteate, some peds often < the achenes; lfts gen 1–2 cm, often prominently veined
                6a Achenes obliquely obovate-elliptic in outline, strongly compressed, 4–6 mm; veins of lower surface of lfts not prominently raised; damp thickets and stream edges; WV to Baja Cal, e to ne Ore, s Ida, Wyo, Colo, and NM; Fendler's m.
                                    6 T. fendleri Engelm.
                6b Achenes obliquely elliptic in outline, terete or subterete, 3–4 (6) mm; veins of lower surface of lfts much-raised; moist to fairly dry woods, BC s, e Cas, to se Ore, e to Alta and to Que, and through Ida and Mont to Wyo, Daks, and Minn; veiny m. (*T. columbianum, T. fissum*)   **7 T. venulosum** Trel.

Trautvetteria F. & M.   False Bugbane

Fls ♂, gen apet, reg, loosely corymbose; stamens ∝, whitish; pistils several; rhizomatous per herbs with mostly basal, palmately-lobed to -divided lvs. (For Ernest Rudolf van Trautvetter, 1809–1889, Russian botanist).

T. caroliniensis (Walt.) Vail. Pl mostly 5–8 dm; lf bls 1–2.5 (3) dm broad; sepals gen 4 (3–7), white or greenish, caducous; filaments flattened and ± clavate, 3–5 mm; achenes 3–4 mm, the style short, hooked, persistent; moist woods and stream borders; widely distributed in N Am and Japan; pls of w N Am, the var. *occidentalis* (Gray) Hitchc., differ only slightly from var. *caroliniensis* of c and e US.

Trollius L.   Globeflower

Fls showy, reg, ♂, apet, solitary on terminal peduncles; sepals 5–12; stamens ∝ , outer 5–15 stami-nodial (perhaps petals), smaller, oblong-tubular, and glanduliferous; pistils distinct; glab per herbs with alt, palmately lobed to compound lvs. (L form of German name, *Trollblume*, for the pl).

T. laxus Salisb. Am g. Sts 1–4 (5) dm; basal lvs long-petiolate, the bl pal-mately cleft into gen 5 obovate main segms; st lvs mostly 2–3 (1–5); sepals 5–9, petaloid, greenish white or white, ovate to obovate, ca 15 (10–20) mm; staminodia oblong, 2–5 mm; follicles 8–11 mm, with stout, often recurved styles, thin-walled, several-seeded; swamps to alp meadows, gen blossoming near snow; BC to OM and Cas, Wn, and Wallowa Mts, Ore, e to Conn, s in the RM to Colo; ours the var. *albiflorus* Gray; var. *laxus*, of e and c US, has more-yellowish sepals.

BERBERIDACEAE   Barberry Family

Fls ♂, reg, hypog; perianth mostly 4 or 5 whorls of 3 (2–4) distinct members each, the outer 1–2 series often deciduous and bractlike, sepals and petals apparently 6 (4) each (rarely lacking), alike or dissimilar, the inner series of petals sometimes glandular at base and often regarded as scales or nectaries; stamens gen as many as the petals and opp them, or fewer or more ∝ (up to 15); an-thers gen opening by 2 uplifting valves; fr a 1-carpellary, 1-celled, 1–several-seeded follicle or berry; shrubs, or per, often scapose herbs with alt, simple to compound, evergreen to deciduous lvs.

1a Lvs pinnate, spinulose-margined, persistent; shrubs   **Berberis**
1b Lvs ternately compound, not spinulose, not persistent through the winter; herbaceous per
  2a Fls spicate, naked; lvs once-ternate   **Achlys**
  2b Fls pan, with perianth; lvs bi- or triternate   **Vancouveria**

Achlys DC.

Stamens ca 10; fr ± fleshy but dry when ripe, indehiscent, 1-seeded; herbs. (For *Achlys*, Gr god-dess of night or obscurity).

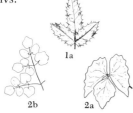

A. triphylla (Smith) DC. Vanillaleaf, deerfoot. Pl widely rhizomatous, scapose, glab throughout; petioles 1–3 dm; lf bls 5–20 cm broad, 3–foliolate, segms fan-shaped, coarsely sinuate-dentate; scapes 2–4 dm; fls inconspicuous, in tight spikes 2.5–5 cm; fr reddish-purple, 3–4 mm, puberulent, the ventral side incurved and with a prominent cartilaginous ridge; deep woods to open parks, often near streams; w BC to nw Cal, from e base Cas to coast (at lower elev) in Wn and nw Ore, but only w Cas from s WV to Cal.

Berberis L.   Barberry; Oregongrape; Mahonia

Fls racemose (ours), yellow; perianth whorls 5, each 3-merous, the outer series bractlike, the inner 2 series gen slightly smaller than the sepals, bilobed, glandular at base; anthers with 2 uplifting valves; berry 1–several-seeded, glaucous-blue; low to tall shrubs with yellowish inner bark and wood and coriaceous lvs. (Arabic name, *berberys*, for the pl). *(Mahonia, Odostemon).*

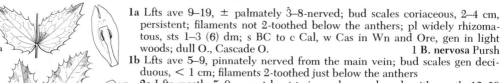

1a Lfts ave 9–19, ± palmately 3–8-nerved; bud scales coriaceous, 2–4 cm, persistent; filaments not 2-toothed below the anthers; pl widely rhizoma-tous, sts 1–3 (6) dm; s BC to c Cal, w Cas in Wn and Ore, gen in light woods; dull O., Cascade O.   1 **B. nervosa** Pursh
1b Lfts ave 5–9, pinnately nerved from the main vein; bud scales gen deci-duous, < 1 cm; filaments 2-toothed just below the anthers
  2a Lfts mostly 5–9, ave at least twice as long as broad, with mostly 12–29 fairly prominent spinulose teeth, mostly shiny on upper surface and shiny to dull (but not papillose) beneath; pls 1.5–20 (45) dm, from stoloniferous to stiffly erect; thin woods to sagebr covered hills; s BC to CRG and s WV, Ore, from e base Cas to coast, e in BC to ne Ida; shining O., tall O. *(B. nutkana)*   2 **B. aquifolium** Pursh

2b Lfts mostly 5–7, ave < twice as long as broad, with mostly 15–43 rather inconspicuous spinulose teeth, glossy to dull on upper surface, always dull and ± glaucous beneath where covered with minute papillae visible with 15× magnification; pls 1.5–8 (10) dm; foothills to lower mont for; e Wn s, e Cas, to ne Cal, e to Alta, SD, Tex, NM, Utah, and Nev; creeping O., low O. (*B. nana*)   3 **B. repens** Lindl.

Vancouveria Morr. & Dec.   Inside-out-flower

Fls nodding in open, pedunculate pan; perianth segms 18–21, outer 6–9 sepaloid but deciduous by anthesis, inner 6 longer, yellow or white, sharply reflexed; petals 6, < the sepals, reflexed, clawed at base and with ± hooded bl; stamens 6; per herbs. (For Capt. George Vancouver, 1757–1798, early British explorer in the Pacific Northwest).

V. hexandra (Hook.) Morr. & Dec. White i. Scapose, rhizomatous, herbaceous per; lvs 1–3 (4) dm, long-petiolate, biternate or incompletely triternate, the lfts cordate-ovate, 3-lobed, sparsely pubescent; petioles and base of scapes gen brownish-pilose; fls 10–30, white; peds slender, recurved, 15–40 mm; perianth segms 5–8 mm; sepals minutely glandular-puberulent; filaments flattened, erect; follicles several-seeded; moist shady woods; s PT to ne Cal, w Cas (*V. brevicula, V. picta*).

PAPAVERACEAE   Poppy Family

Fls solitary to several, reg, ⚥, gen hypog but sometimes perig, mostly showy, white, yellow, red, or bluish; sepals 2–3, shed as the fl opens; petals 4 (6–12); stamens mostly ∞, to few (3); pistil 1 (ours), 2–several-carpellary; fr caps, elongate to obovoid, dehiscent by valves or subterminal pores; ann to per herbs (ours), often with colored or milky juice and alt or basal, entire to palmately or pinnately decompound lvs.

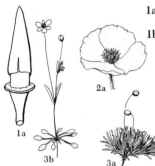

1a Sepals coherent, deciduous as a unit; recep hollowed and prolonged above base of the ovary, ringlike at the top   **Eschscholzia**
1b Sepals distinct and shed separately as the fl opens; recep not hollowed
  2a Stamens ∞; fls showy, petals mostly > 1 cm; caps falsely several-celled, dehiscent by subterminal pores; lvs toothed to lobed   **Papaver**
  2b Stamens mostly 3–15; fls not showy, petals not > 5 mm; caps valvate; lvs often entire
    3a Lvs linear, all basal and tufted; petals yellow; caps ovoid, 2.5–3 mm   **Canbya**
    3b Lvs broader, not tufted, some cauline and opp; petals white; caps linear, 10–18 mm   **Meconella**

Canbya Parry   Canbya

Petals 6 (7); stamens 6–15; pistil 3 (4)-carpellary; caps 1-celled; ann with small, tufted lvs. (For William W. Canby, 1831–1904, Am botanist).

C. aurea Wats. Golden c. Minute glab ann with ∞ short brs and fleshy, glaucous, linear lvs < 1 cm × scarcely 0.5 mm, forming tiny cushions 0.5–2 × 0.5–4 cm; peduncles several, subfiliform, 1–6 cm; sepals 3, shed separately, ca 2 mm; petals yellow, 3–4 mm; stamens (6) 9–15; caps ovoid, 2.5–3 mm; dry sandy areas in sagebr des; Deschutes and Crook to Harney cos, Ore, to Nev.

Eschscholzia Cham.

Fls solitary to several on ± scapose peduncles; sepals 2; petals gen 4; stamens ∞; caps 1-celled, with 2 parietal placentae; per herbs (ours) with ternately decompound lvs. (For Johann Friedrich Eschscholtz, 1793–1831, Russian scientist with Capt. Kotzebue; he visited Cal in 1816).

E. californica Cham. Cal or gold poppy. Glab and glaucous per, or in our area often blossoming as an ann and not surviving the winter, 1–several-std, 1–5 dm; peduncles 6–20 cm; recep with a prominent spreading outer rim 1–2 mm wide; sepals connate into a conicle (dunce-cap like) unit, shed as the fl opens; petals 8–40 mm, pale yellow to orange (occ white); caps linear; CRG to s Cal, intro elsewhere, esp w Cas (*E. columbiana, E. douglasii, E. c.* var. *d.*).

Meconella Nutt.   Meconella

Fls tiny, white to yellow, 1–few on slender peduncles ± lfy at the base; petals (4–5) 6; stamens 3–6 (12); pistil 3 (2)-carpellary, 1-celled; small, opp-lvd ann. (Diminutive of Gr *mekon*, poppy). (*Platystigma*).

M. oregana Nutt. White m. Glab and glaucous ann; st simple to ± br, 2–10 cm; lvs entire, the basal spatulate-obovate, 6–18 mm, the cauline opp, sessile; fls solitary on axillary and terminal peduncles 2–4 cm; sepals 3, shed separately, ca 1.5 mm; petals white, 2.5–4 mm; stamens (3) 6, anthers oval, much < filaments; caps linear, 10–18 mm; open ground, gen where wet in the spring; VI and PT to WV, Ore, and to Cal; rare in Wn (*Platystemon o.*).

Papaver L.   Poppy

Fls showy, pedunculate, white, yellow, or red to violet; petals 4 (6); caps globose to obovoid; ann or per herbs with colored or milky juice and alt or basal, lobed to pinnately dissected lvs. (L name for the poppy).

1a Pl per, scapose; petals orange-yellow, ca 1 cm; caps bristly, 10–15 mm; higher mts, on talus; nw Mont to se Alta; alpine p.   1 **P. pygmaeum** Rydb.
1b Pl ann; sts lfy; petals not yellow; intro spp., rarely escaped
  2a Lvs simple, those of the st cordate-clasping, mostly glab; petals white to purple; caps glab, subglobose, 2.5–5 cm; opium p., rarely persistent > 1 or 2 seasons   2 **P. somniferum** L.
  2b Lvs pinnatifid to compound, not cordate-clasping, often hairy; petals scarlet, gen black at base
    3a Caps subglobose to broadly obovoid, glab; petals 4–7 cm; occ in waste places; corn p., field p.   3 **P. rhoeas** L.
    3b Caps slenderly oblong-obovoid, bristly; petals 1.5–2 cm; occ in waste places, esp e Cas; long prickly-headed p.   4 **P. argemone** L.

FUMARIACEAE   Fumitory Family

Fls mostly showy, ⚥, hypog, racemose or pan, irreg; sepals 2, small, gen soon deciduous; petals 4, in 2 pairs, at least 1 of the outer pair spurred or saccate at base, often slightly connate and sometimes ± crested at the tip, inner pair gen both connate and crested at the tip; stamens 6, gen ± connate into 2 groups of 3 each; pistil 2-carpellary; ovary 1-celled with 2 parietal placentae; fr a 1-seeded nut or several-seeded caps; ann to per herbs with watery juice and much-compounded, glaucous and glab to pubescent lvs.

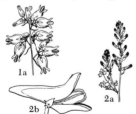

1a Outer petals alike, both spurred or saccate at base; pl scapose   **Dicentra**
1b Outer petals dissimilar, only 1 spurred or saccate; pl lfy-std
  2a Fls reddish-purple; fr subglobose, 1-seeded, indehiscent; pl ann   **Fumaria**
  2b Fls yellow or white to pinkish; fr caps, elongate, several-seeded, dehiscent; pl per, bien, or (rarely) ann   **Corydalis**

Corydalis Medic.   Corydalis

Outer pair of petals gen ± hooded above, but only 1 spurred at base; caps linear to ovoid or obovoid, dehiscent (often explosively); pl with fleshy roots or rhizomes and large, decompound lvs. (Gr name for the crested lark, which was possibly suggested by the crested hood or by the spur of the upper petal). (*Capnoides*).

1a Pl either yellow fld or ann to bien; caps linear, not explosively dehiscent
  2a Fls yellow
    3a Pl per with large fleshy root; lvs triternate, ultimate segms ± elliptic-obovate; spur of corolla ca 2–3 mm; caps 8–11 mm; European sp., often cult and very rarely escaped, as at Elk Rock, Multnomah Co, Ore; yellow c.   1 **C. lutea** DC.

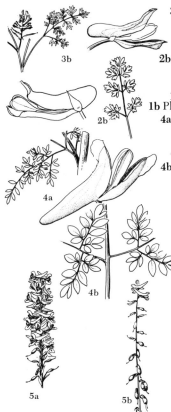

3b Pl ann or bien, taprooted; lvs bi- to quadripinnate, ultimate segms ± linear; spur (2) 3–7 mm; caps 20–25 mm; moist to dry soil, common over much of N Am; in our area mostly e Cas, but very occ w Cas (OP); golden c., g. smoke   **2 C. aurea** Willd.

2b Fls pinkish, yellow-tipped; pl ann or bien, erect, 2–8 dm; lf bl 3–5 × pinnate; caps (25) 30–45 mm; mostly in dry, open to shaded, often disturbed soil; Alas and BC to Newf and Ga, in our area only in GNP; pink c.   **3 C. sempervirens** (L.) Pers.

1b Pl strong per, never yellow-fld; caps ± obovoid, explosively dehiscent

4a Lower st lvs 4–6× pinnate, ultimate segms 5–15 × 2–5 (6) mm; clawed petal erose-crested; corolla 10–20 mm, pale to deep pinkish or rose-lavender; wet places in Cas, Clackamas and Multnomah cos, Ore; Clackamas c.   **4 C. aquae-gelidae** Peck & Wilson

4b Lower st lvs 2–4 × pinnate, segms mostly > 15 × > 5 mm; clawed petal often not erose-crested

5a Main racemes mostly with > 50 fls; corolla white to pink, deep pinkish or purple at tip, 17–25 mm; spurred petal not strongly crested but with free margins; st lvs 3–6; wet places, mostly along streams; occ from ne Ore and Ida to Colo and Cal; Case's c.; 2 vars.

5 C. caseana Gray

a1 Racemes simple or few-br, compact; spurred petal gen with margin reflexed > 1.5 mm; Baker and Union cos, Ore, and Boise, Camas, Elmore, and Valley cos, Ida   var. **cusickii** (Wats.) Hitchc.

a2 Racemes openly br, loose; spurred petal with margin reflexed much < 1.5 mm; Shoshone, Clearwater, and Ida cos, Ida

var. **hastata** (Rydb.) Hitchc.

5b Main racemes 15–35-fld; corolla pink, not much more deeply tinged at tip; spurred petal crested but without free margins; st lvs mostly 2–3; moist, shaded soil, BC to Ore, e and w slopes Cas to coast; Scouler's c. (*C. allenii, C. macrophylla*)   **6 C. scouleri** Hook.

Dicentra Bernh.   Bleedingheart

Fls solitary to pan, white to purplish (yellow); caps ± elongate; scapose per with fleshy, fascicled roots or rhizomes and ± ternately decompound lvs. (Gr *dis*, twice, and *kentron*, spur, in reference to the spur of outer petals). (*Bicuculla*).

1a Fls solitary, terminal; scape 4–8 (10) cm; roots fusiform, fascicled; corolla white to pinkish, cordate at base, outer petals widely spreading and slightly recurved; inner petals gen purplish-tipped, 12–15 mm; well-drained soil, foothills to subalp slopes; Cas, Wn, to Cal, e to Ida, Wyo, and Utah; steer's head   **1 D. uniflora** Kell.

1b Fls several; scape gen > 10 cm; roots not fusiform

2a Fls pan, mostly deep pink; outer petals rounded-saccate at base, sac 2–4 mm deep; inner petals 10–15 mm; pl rhizomatous, not tuber-bearing; moist woods, lowl to lower mts; BC to Cal, Cas to coast; Pac b.

2 D. formosa (Andr.) Walp.

2b Fls racemose, white or pale pink; outer petals spurred-saccate, sac 7–11 mm; inner petals ca 10 mm; pl with erect rootstocks covered with small, ± globose tubers; moist woods to gravelly banks, along CR in Wn, Ore, Ida, and in c and e N Am; dutchman's breeches (*D. occidentalis*)

3 D. cucullaria (L.) Bernh.

Fumaria L.   Fumitory

One of the outer petals spurred-saccate at base; fr nutlike, tiny; caulescent ann with decompound lvs. (L *fumus*, smoke, prob in reference to the smell or to the pale color of the fls).

F. officinalis L. Common f. Glab and glaucous ann, simple and erect to freely br and diffuse, the sts 2–4 (8) dm; fls in axillary, pedunculate racemes, minutely bracteate, reddish-purple but more deeply purple toward the tip, 7–12 mm, the spur ca 3–4 mm; fr ca 2–2.5 mm, warty, truncate or depressed at the tip, subglobose, ca 2.5 mm broad; European, intro, and occ weedy, in Ore, Wn, and Mont.

CRUCIFERAE   Mustard Family

Fls ⚥, reg to ± irreg, hypog, solitary or racemose, often showy; sepals 4, sometimes ± saccate at base; petals 4 (rarely lacking), white to yellow, red, or blue-purple, often clawed; stamens mostly 6 (outer 2 < inner 4), or sometimes only 4 or even 2 °; fr sessile or sometimes stalked (stipitate) above point of insertion of perianth, gen with 2 completely deciduous valves, seeds attached marginally (parietally), septum gen complete, less commonly fr indehiscent or tardily breaking crosswise between seeds, sometimes inflated or terete, but often flattened either parallel with septum (compressed) or contrary to it (obcompressed), from several times as long as broad and in gen slender (silique) to < 3 times as long as broad (silicle) and elliptic or oblong to rhombic in outline; seeds 1–∝ per cell, mostly in a single row in each locule (uniseriate) or in 2 rather evident rows (biseriate); cotyledons folded lengthwise (conduplicate) or not folded upon themselves, but rather folded back parallel with radicle, and with the back of one against the radicle (incumbent), with one margin of each against the radicle (accumbent), or in an intermediate position (oblique)°°; ann to herbaceous or suffrutescent per, often with characteristic, simple to br pubescence; lvs simple to compound, mostly alt, exstip.

1a Fr stipitate (stipe at least 1 mm)
  2a Fr a silicle, elliptic to oval; petals blue to purple; st lvs petiolate, cordate-based, gen ± hirsute   **Lunaria**
  2b Fr a silique, slender; petals rarely blue or purple; st lvs often sessile or not cordate-based, rarely hirsute
    3a Petals yellow; stipe mostly > 7 mm   **Stanleya**
    3b Petals white, greenish, or purplish; stipe mostly < 7 mm
      4a Basal lvs gen hastate; petals white, dentate on claw; pl glab   **Chlorocrambe**
      4b Basal lvs not hastate; petals often purplish, not dentate on claw; pl sometimes pubescent   **Thelypodium**
1b Fr sessile or subsessile (stipe < 1 mm)
  5a Pod divided crosswise into 2 rather dissimilar segms, basal segm gen seed-bearing and with evident valves (often indehiscent), but sometimes sterile, upper segm often not valvular, mostly not seed-bearing, sometimes merely the slightly narrowed, often flattened, beaklike style   **Group I, lead 15a**
  5b Pod uniform, not divided crosswise into 2 segms; style rarely beaklike, often prominent but always thinner than body of fr
    6a Fr a silicle, cordate to oval, ± oblong, elliptic, or obcordate, sometimes deeply 2-lobed or only 1-seeded
      7a Silicles strongly compressed to ± inflated, never globose, terete, or deeply 2-lobed
        8a Seeds 1 per cell   **Group II, lead 24a**
        8b Seeds 2 or more per cell   **Group III, lead 32a**
      7b Silicles slightly to strongly obcompressed, or greatly inflated and subterete to terete, sometimes deeply 2-lobed
        9a Fr samara-like, 1-celled and 1-seeded, indehiscent; pl bien or per, glab and ± glaucous; st lvs aur   **Isatis**
        9b Fr not samara-like, gen dehiscent, 2-celled and gen several-seeded; pl sometimes ann, or pubescent, or with non-aur lvs
          10a Pl aquatic, scapose; lvs linear, ± terete, acicular **Subularia**
          10b Pl either non-aquatic or with lfy sts and non-acicular lvs
            11a Pls fleshy maritime herbs; basal lvs with slender petiole much > the reniform to oval, simple bl; seeds ∝, biseriate   **Cochlearia**
            11b Pls seldom maritime and fleshy; basal lvs various; seeds mostly few, often uniseriate   **Group IV, lead 39a**
    6b Fr a silique, elliptic to linear, > 1-seeded
      12a Petals yellow (sometimes very pale) to orange   **Group V, lead 51a**
      12b Petals white or pinkish to purple, never yellow
        13a Pl caespitose per from br caudex, semiscapose (fl sts merely bracteate); basal lvs 3–15 cm, entire, grayish with br hairs; petals 11–15 mm, pinkish to purple; sagebr plains and ponderosa pine belt; e Cas   **Phoenicaulis**
        13b Pl with well-developed fl sts or with lvs and petals not as above

°To be considered as 6 in number if not mentioned in generic description.
°°Distinctions utilized only in a few genera where necessary.

14a Seeds biseriate   Group VI, lead 68a
14b Seeds uniseriate   Group VII, lead 74a

Group I

15a Silique segms much alike, or lower one much the smaller, each 1 (2)-seeded (or lower one seedless), indehiscent, the 2 finally breaking apart; style lacking; pl coastal

  16a Petals light yellow (cream); basal lvs elliptic-ovate to oval, 20–40 cm; pl maritime   **Crambe**

  16b Petals white to purplish; basal lvs oblong-ovate to oblong-obovate, 2–7 cm   **Cakile**

15b Silique segms very dissimilar, lower one gen the larger and (1–) several-seeded, upper one gen seedless and beaklike, but several-seeded in *Raphanus*

  17a Pl stipitate-glandular; petals purple; fr indehiscent but breaking crosswise into 1-seeded segms; style pointed and stigmatic for ca 1 mm, ultimately beaklike and 7–20 mm   **Chorispora**

  17b Pl not stipitate-glandular; petals mostly white, yellow, or purple; fr often dehiscent; style sometimes not beaklike

    18a Style beaklike, gen much unlike body of silique, often strongly flattened; lower silique segm seed-bearing

      19a Beak stout, terete to greatly flattened, not styliform, gen much > 5 mm; cotyledons folded

        20a Seeds uniseriate   **Brassica**

        20b Seeds biseriate

          21a Siliques 3–5 mm thick; beak often at least half as long as valves   **Eruca**

          21b Siliques < 3 mm thick; beak < half as long as valves   **Diplotaxis**

      19b Beak mostly not flattened, ± styliform, < 5 mm; cotyledons often not folded

        22a Petals yellow; lvs pinnatifid; cotyledons folded   **Erucastrum**

        22b Petals white, sometimes purplish-tinged; lvs entire to sinuate-dentate; cotyledons not folded

          23a Fr an indehiscent (or tardily dehiscent) 2-seeded silicle; pl pubescent with br hairs   **Euclidium**

          23b Fr a dehiscent, several-seeded, slender silique; pl glab   **Streptanthella**

    18b Style not beaklike, scarcely differentiated from upper portion of the terete, beaked valves; silique with very short, sterile, basal (stipelike) segm distinct from the valvular, fertile segm   **Raphanus**

Group II

24a Pl per, densely stellate with sessile, appressed, ∝-rayed hairs, yellow fld

  25a Silicles 7–20 mm, often deeply bilobed or cordate at base   **Physaria**

  25b Silicles mostly < 7 mm, never deeply lobed or cordate at base   **Lesquerella**

24b Pl either ann (bien), nonstellate, or other than yellow-fld

  26a Silicles broadly winged, oblong, 1-seeded, ± samaroid; pl glab, bien or per; st lvs aur   **Isatis**

  26b Silicles more nearly oval or rotund, often 2-seeded; pl either hairy or ann; cauline lvs often not aur

    27a Petals yellow; silicles inflated, broadly biconvex in cross section, strongly reticulate-pitted; upper st lvs aur; pl stellate-pubescent **Neslia**

    27b Petals white or cream, or purplish-tinged; silicles gen much-compressed, never strongly reticulate-pitted; st lvs often not aur

      28a Silicles 2-celled, gen 2- or more-seeded

        29a Pl scapose; pubescence neither stellate nor 2-pronged   **Draba**

        29b Pl lfy-std; pubescence either stellate or 2-pronged

          30a Silicle valves nerveless; pubescence stellate   **Alyssum**

          30b Silicle valves 1-nerved; pubescence 2-pronged   **Lobularia**

      28b Silicles 1-celled, 1-seeded

        31a Silicles < 5 mm, pubescent with hooked hairs   **Athysanus**

        31b Silicles gen > 5 mm, glab or pubescent with non-hooked hairs   **Thysanocarpus**

Group III

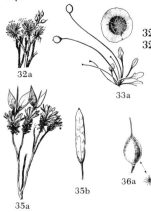

32a Pl scapose, per, white-fld, 1–3 cm; silicles ovoid, leathery   **Draba**
32b Pl often not scapose, sometimes ann or not white-fld, mostly > 3 cm; silicles seldom leathery
  33a Fls solitary on naked scapes; seeds broadly wing-margined; pl small glab ann   **Idahoa**
  33b Fls few to ∝ , often on lfy sts; seeds not winged; pl often per or hairy
    34a Silicles oblong to elliptic, gen > twice as long as broad
      35a Seeds densely cellular-papillate, silvery; silicles 2–4 cm × 6–10 mm; petals white to purplish   **Anelsonia**
      35b Seeds not silvery and cellular-papillate; silicles < 2 cm × < 6 mm; petals white or yellow   **Draba**
    34b Silicles oval to obovate, not > twice so long as broad
      36a Petals yellow (sometimes purple-tinged); pl per, densely appressed stellate-pubescent   **Lesquerella**
      36b Petals white or pl not appressed-stellate per
        37a Style not > 1 mm; silicles oval in outline, 3–4 mm; petals not bilobed   **Alyssum**
        37b Style 0.5–3 mm; silicles oval-elliptic to obovate, gen > 4 mm; petals sometimes bilobed
          38a Petals white, deeply bilobed; seeds slightly winged; silicles compressed, at least twice as broad as thick   **Berteroa**
          38b Petals cream, not bilobed; seeds not winged; silicles more turgid, < twice as broad as thick   **Camelina**

Group IV

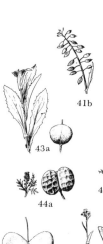

39a Silicles inflated
  40a Petals yellow (rarely purplish); pubescence often stellate; seeds gen > 2 per silicle
    41a Pubescence stellate; pl mostly in dry areas
      42a Silicles 7–20 mm, often deeply bilobed, or with an apical notch at least 2–4 mm, or cordate at base   **Physaria**
      42b Silicles mostly < 7 mm, never deeply bilobed or with an apical notch as much as 2 mm deep and never cordate at base **Lesquerella**
    41b Pubescence unbr or lacking; pl gen in wet places   **Rorippa**
  40b Petals white or purple; pubescence (if any) non-stellate; seeds (1) 2 per silicle
    43a Pl per, rhizomatous; lvs simple, those of st aur or sagittate   **Cardaria**
    43b Pl ann; lvs sometimes pinnatifid, neither aur nor sagittate
      44a Lvs pinnatifid; silicles glab, cordate and deeply bilobed, strongly rugose; style inconspicuous   **Coronopus**
      44b Lvs entire to dentate; silicles pubescent, neither cordate nor deeply bilobed, not rugose; style prominent, beaklike   **Euclidium**
39b Silicles ± obcompressed, but not inflated
  45a Silicles obcordate to triangular, 4–8 mm, > 2-seeded; st lvs aur; pls ann   **Capsella**
  45b Silicles rarely obcordate or triangular, often only 2-seeded, or much > 8 mm; st lvs often not aur; pls often per
    46a Pubescence stellate
      47a Silicles (5) 7–20 mm, deeply bilobed or with apical notch at least 2 mm deep   **Physaria**
      47b Silicles mostly < 7 mm, never deeply bilobed or notched > 1 mm deep   **Lesquerella**
    46b Pubescence simple or sparingly br, or lacking
      48a Seeds 2 per silicle   **Lepidium**
      48b Seeds > 2 per silicle, often ∝
        49a St lvs ± aur; silicles gen > 4 mm   **Thlaspi**
        49b St lvs, if any, not aur; silicles not > 4 mm
          50a Pl semiscapose; filaments with small, scalelike, basal appendages; seeds 2 per cell   **Teesdalia**
          50b Pl lfy-std; filaments not appendaged; seeds > 2 per cell   **Hutchinsia**

Group V

51a Pl strongly pubescent with 2-rayed or sometimes 3–4-rayed, sessile, appressed hairs oriented with axis of lf or st   **Erysimum**

51b Pl glab to pubescent with other than 2-rayed or sessile 3–4-rayed hairs

  52a Style (or apparent style) beaklike, nearly 1/2 as broad as valves

    53a Siliques 1-celled, indehiscent, gen torulose   **Raphanus**

    53b Siliques 2-celled, tardily dihiscent, not torulose

      54a St lvs ± aur; basal lvs lyrate-pinnatifid   **Barbarea**

      54b St lvs not aur; basal lvs not lyrate-pinnatifid

        55a Seeds uniseriate; racemes lfy-bracteate   **Erucastrum**

        55b Seeds biseriate; racemes ebracteate   **Diplotaxis**

  52b Style not beaklike, if prominent always much narrower than valves

    56a Seeds biseriate

      57a Pubescence unbr or lacking

        58a Lvs simple, entire to deeply lobed; pl often scapose; siliques strongly compressed

          59a Siliques 10–15 times as long as broad; styles 2–3 mm, 1/3–1/2 as broad as valves; basal lvs deeply toothed to subpinnatifid   **Diplotaxis**

          59b Siliques rarely as much as 8 times as long as broad; styles mostly very slender if > 1 mm; basal lvs entire to remotely toothed   **Draba**

        58b Lvs often pinnate; pl never scapose; siliques terete to slightly compressed   **Rorippa**

      57b Pubescence br

        60a Pl scapose per; lvs simple, entire

          61a Seeds with loose silvery covering; siliques 20–40 × 6–10 mm; petals white or purplish   **Anelsonia**

          61b Seeds without a loose silvery covering; siliques < 20 mm long or < 6 mm broad; petals white to yellow   **Draba**

        60b Pl ann or bien, lfy-std; lvs 1–3 times pinnatifid   **Descurainia**

    56b Seeds uniseriate

      62a Pubescence stellate; lvs 1–3 times pinnatifid   **Descurainia**

      62b Pubescence simple; lvs simple to pinnatifid

        63a Style at least 0.3 mm

          64a St lvs aur or cordate-clasping

            65a Lvs at most deeply dentate; pl ann, glab, glaucous; siliques 8–13 cm   **Conringia**

            65b Lvs lyrate to pinnate; pl bien or per, not glaucous; siliques mostly < 8 cm   **Barbarea**

          64b St lvs neither aur nor cordate-clasping

            66a Siliques quadrangular, valves strongly keeled, 2–4.5 cm; cotyledons conduplicate; petals 4–7 mm   **Erucastrum**

            66b Siliques terete to compressed, valves not keeled; cotyledons accumbent; petals < 4 mm or siliques < 2 cm **Rorippa**

        63b Style < 0.3 mm

          67a Pl per, rhizomatous; lvs linear, often entire   **Schoenocrambe**

          67b Pl ann or bien, not rhizomatous; lvs gen pinnatifid **Sisymbrium**

Group VI

68a Pl scapose per; siliques compressed

  69a Seeds silvery, loosely cellular-papillose; siliques 2–4 cm × 6–10 mm **Anelsonia**

  69b Seeds not silvery or loosely cellular-papillose; siliques < 2 cm long or < 6 mm broad   **Draba**

68b Pl lfy-std, sometimes ann; siliques sometimes not compressed

  70a Lvs gen pinnate but sometimes simple and then some at least 2 dm, and pl ± aquatic   **Rorippa**

  70b Lvs no more than pinnatifid, < 2 dm; pl not aquatic

    71a Siliques gen at least 8 times as long as broad

      72a Pl glab and glaucous, 5–15 cm; st lvs aur   **Arabidopsis**

      72b Pl often hairy or much > 15 cm; st lvs often not aur

        73a Siliques either compressed, or subquadrangular and pl mostly with unbr pubescence   **Arabis**

        73b Siliques subterete; pl gen with ± br pubescence   **Halimilobos**

    71b Siliques < 8 times as long as broad   **Draba**

Group VII

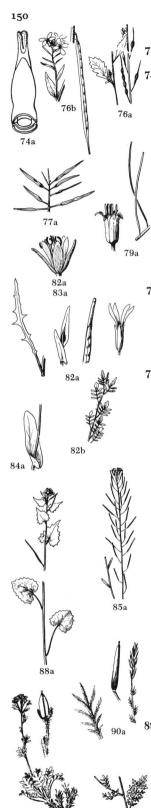

74a Stigma bilobed, lobes ca 1 mm, but erect and ± connivent and gen not evident; petals 18–25 mm; seeds 3–4 mm; siliques ± torulose   **Hesperis**
74b Stigma entire or bilobed but lobes < 1 mm and spreading; petals rarely as much as 18 mm; seeds often < 3 mm; siliques often not torulose
    75a Siliques not fully dehiscent, at least tips of valves remaining intact; lf bls never reniform or deltoid
        76a Fr torulose at maturity, completely indehiscent   **Raphanus**
        76b Fr not torulose at maturity, only the tip 3–5 mm portion remaining indehiscent and beaklike   **Streptanthella**
    75b Siliques fully dehiscent (sometimes ± tardily); lf bls occ reniform or deltoid
        77a Pl subscapose, per; lvs entire; siliques broadly linear to oblong, strongly compressed; petals pink to purple, 11–15 mm   **Phoenicaulis**
        77b Pl lfy-std, occ ann; lvs seldom entire; siliques not strongly compressed; petals various, sometimes < 11 mm
            78a Pubescence simple or lacking   **Subgroup VIIa**
            78b Pubescence in part or entirely br   **Subgroup VIIb, lead 89a**

Subgroup VIIa

79a Calyx urn-shaped at anthesis; petal bl crispate or strongly grooved, or both; siliques (4) 5–15 cm
    80a Siliques strongly flattened; seeds strongly flattened and slightly winged; lvs (ours) merely dentate, the cauline clasping at base   **Streptanthus**
    80b Siliques terete or only slightly flattened; seeds slightly flattened but not winged; lvs (ours) lyrate-pinnatifid, none clasping   **Caulanthus**
79b Calyx not urn-shaped; petal bl neither crispate or strongly grooved; siliques < to > 5 cm
    81a St lvs in part pinnate or pinnatifid
        82a Pl ann or bien; upper st lvs not pinnatifid   **Thelypodium**
        82b Pl either per or with most of the st lvs pinnatifid   **Cardamine**
    81b St lvs toothed to entire
        83a Anthers tending to coil after dehiscing   **Thelypodium**
        83b Anthers not coiling after dehiscing
            84a Pl ann, glab, glaucous, 3–7 dm; st lvs cordate-clasping; siliques 8–13 cm   **Conringia**
            84b Pl per or ann but not otherwise as above; lvs often not clasping, silicles mostly < 8 mm
                85a Pl ann, scarcely 4 dm; siliques terete, 10–16 × < 1 mm   **Arabidopsis**
                85b Pl bien or per, often > 4 dm; siliques often > 16 mm long or > 1 mm broad
                    86a Lvs all or in part compound   **Cardamine**
                    86b Lvs at most lyrate-pinnatifid
                        87a Pl rhizomatous, tuberiferous, or glab and not > 10 cm; style 1–2 mm   **Cardamine**
                        87b Pl neither rhizomatous nor tuberiferous, gen ± hairy and mostly > 10 cm; style sometimes < 1 mm
                            88a Lvs slender-petiolate, bl reniform to deltoid; pl garlic-odored when crushed   **Alliaria**
                            88b Lvs sessile or petiolate but bl neither reniform nor deltoid; pl not garlic-odored   **Arabis**

Subgroup VIIb

89a Lvs pinnatifid to pinnate
    90a Lvs rigid, segms linear, acicular; pl of sagebr des and ponderosa pine for   **Polyctenium**
    90b Lvs not rigid, segms not acicular, often oblong to obovate; pl often alp or subalp
        91a Pl per, native, pulvinate, alp or subalp, white-fld   **Smelowskia**
        91b Pl ann or bien, not at all pulvinate, never alp and rarely subalp except as a weed
            92a St lvs gen > once-pinnatifid, if once-pinnatifid then segms toothed   **Descurainea**

92b St lvs shallowly pinnatifid, the segms mostly entire   **Halimilobos**
89b St lvs simple
  93a Lower fls bracteate, or axillary to scarcely reduced lvs; sepals persistent
    until siliques well developed; pl ann, 1–4 dm; lvs remotely dentate; si-
    liques slightly torulose; style ca 1–1.5 mm, acute, stigmatic full length;
    petals pink   **Malcolmia**
  93b Lower fls gen not bracteate or axillary to lvs; sepals gen quickly deci-
    duous; pl mostly bien or per, if ann either > 4 dm or with other than
    pink petals; sometimes lvs deeply toothed or style < 0.5 mm
    94a Pl ann, 0.5–4 dm; style < 0.5 mm   **Arabidopsis**
    94b Pl rarely ann, often either > 4 dm or with styles at least 0.5 mm
      95a Siliques gen with stipe 0.5–1 mm   **Thelypodium**
      95b Siliques sessile or with stipe < 0.5 mm
        96a Pl bien, glab and glaucous   **Thelypodium**
        96b Pl either not bien or not glab and glaucous
          97a St lvs sinuately pinnatifid, segms spreading, entire to toothed;
            petals cream; siliques terete, reflexed   **Thelypodium**
          97b St lvs mostly not sinuately pinnatifid; petals sometimes not
            cream; siliques often not compressed or reflexed
            98a Siliques mostly glab, but if pubescent then > 2.5 cm or >
              1.5 mm broad   **Arabis**
            98b Siliques pubescent, not > 2.5 cm or > 1.5 mm broad
                                                                         Halimilobos

Alliaria Adans.

Racemes simple to compound, bractless or lower fls lfy-bracteate; siliques linear, terete but ±
4-angled, valves 3-nerved; stigma capitate. (L *allium*, onion or garlic, because of garlic smell of
pl).

A. officinalis Andrz. Garlic mustard, hedge garlic. Taprooted bien (sometimes
fl 1st season), glab or sparsely hirsute, garlic-scented; sts erect, 3–10 dm; lvs
all slender-petioled, basal bls reniform, st lvs cordate-deltoid, coarsely sinuate-dentate, (1.5) 3–6 cm broad; petals white, spatulate, short-clawed, ca 6
mm and ca twice length of sepals; siliques spreading but ± arched upward,
4–6 cm, gen ± torulose; seeds ca 3 mm; native to Old World, occ intro in
parts of US; estab in our area in Portland, Ore.

Alyssum L.   Alyssum

Racemes bractless; petals pale yellow, gen fading to white; silicles rotund to ovate-elliptic,
strongly compressed, mostly with 1–2 seeds per cell; stellate-pubescent ann or per herbs with
simple lvs. (Gr *a*, without, and *lyssa*, madness; pl supposed to sooth madness and anger).

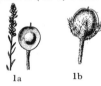

1a Silicles glab; style 0.5–1 mm; sepals soon deciduous after anthesis; Old
  World weed; in our area occ in c Wn and Ore, and in Ida and Mont; desert
  a.   **1 A. desertorum** Stapf
1b Silicles stellate-hairy; style < 0.5 mm; sepals persistent until fr enlarges;
  European weed, widespread on dry, waste soil; pale a.   **2 A. alyssoides** L.

Anelsonia Macbr. & Pays.   Daggerpod

Racemes bractless, subsessile among lvs; silicles erect, ovate-elliptic, strongly compressed; style
1–2 mm; seeds ∞, biseriate; scapose, caespitose per. (For Aven Nelson, 1859–1952, authority on
flora of RM).

A. eurycarpa (Gray) Macbr. & Pays. Lvs entire, oblanceolate, 1–2 cm, densely
tufted, marcescent, grayish-pubescent with mixture of simple, forked, and
dendritic hairs; racemes scarcely exceeding lvs, 1–10-fld; petals white or ±
brownish-purple, 3–6 mm; silicles erect, elliptic-lanceolate, ± acuminate,
glab, 2–4 cm × 6–10 mm; style 1–2 mm; seeds ca 2.5 mm, silvery with loose
covering of cellular-papillae; subalp to alp slopes and ridges, often on talus; c
Ida to Nev and Sierran Cal (*Draba e., Parrya e., Phoenicaulis e., Phoenicaulis
huddelliana*).

Arabidopsis (DC.) Schur

Racemes ebracteate; fls small (2–3 mm), white or pinkish; siliques subterete or slightly compressed, 10–16 × 1–1.5 mm; stigma shallowly lobed; seeds ∝, uniseriate or biseriate, not mucilaginous when wet; ours ann with erect sts and simple lvs. (Meaning "like *Arabis*"). (*Stenophragma*).

1a Cauline lvs aur; pl glab and glaucous, 5–15 cm; seeds at least partially biseriate; Man, s to Colo, RM onto plains; Asia; salt water cress (*Sisymbrium s., S. glauca, Thellungiella g., A. g.*)   1 **A. salsuginea** (Pall.) Busch
1b Cauline lvs not aur; pl gen ± pubescent, 10–40 cm; seeds uniseriate; European weed now common from BC to Ore, e to Ida, esp in gardens; Thale cress, common wall cress, mouse-ear (*Arabis t., Sisymbrium t.*)
                                                        2 **A. thaliana** (L.) Schur

Arabis L.   Rockcress

Racemes ebracteate; petals white to red or purple, obovate-spatulate; siliques linear, mostly 1-nerved, quadrangular to strongly compressed, sessile or subsessile; stigma entire to slightly bilobed; seeds ∝, uniseriate to biseriate; bien (ann) to per and sometimes suffrutescent herbs with simple to strongly dendritic pubescence, entire to lyrate-pinnatifid basal lvs, and mostly ± aur st lvs. (Named for Arabia). (*Turritis*).

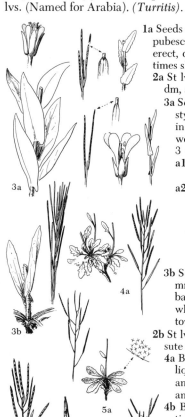

1a Seeds wingless or with wing barely 0.3 mm broad; basal lvs glab, ciliate, or pubescent on lower surface with coarse, forked to 4-rayed hairs; siliques erect, often tightly appressed, not > 2 mm broad; fls gen white, but sometimes slightly pinkish or lavender-tinged
  2a St lvs gen aur; pl bien (ann?) or very short-lived per, mostly much > 3 dm, stellate to strongly hirsute at base
    3a Seeds uniseriate; siliques 3–8 cm × 1–2 mm, moderately compressed; style 0.5–1 mm; pl gen br at base, with 2–several fl sts; circumpolar, s in N Am to Cal, Ariz, and NM, from sea level to mont, in meadows, woods, or open slopes, but sometimes semiweedy; hairy r. (*A. ovata*); 3 vars.   1 **A. hirsuta** (L.) Scop.
      a1 Petals 3–5 mm, cream; common in Ida and Mont and e (*A. p.*)
                                                        var. **pycnocarpa** (Hopkins) Rollins
      a2 Petals mostly > 5 mm, white to pinkish
        b1 Siliques 1–1.3 mm broad; stigma nearly entire; st lvs mostly entire; mts, BC to Cal (*A. rupestris, T. spathulata*)
                                                        var. **glabrata** T. & G.
        b2 Siliques 1.3–2 mm broad; stigma lobed; lower st lvs toothed; Alas s, along coast and in lower coastal mts, to s Ore (*A. e.*)
                                                        var. **eschscholtziana** (Andrz.) Rollins
    3b Seeds biseriate, at least in lower part of fr; siliques 6–10 cm × 1–1.5 mm, very lightly compressed; style scarcely 0.5 mm; pl rarely br at base, fl st gen 1; creek banks to light woods, mostly in clearings where often weedy; BC to n Cal, e to RMS and beyond; Europe; towermustard   2 **A. glabra** (L.) Bernh.
  2b St lvs not aur; pl per, mostly < 3 (4) dm, often glab or only slightly hirsute at base
    4a Basal lvs all or in part lyrate-pinnatifid; pl glab to sparsely hirsute; siliques 2–4 cm × 1–1.5 mm, subterete; style 0.2–1 mm; damp woods and stream and lake margins, subalp; Alas to Mt Baker, Wn, e to Alta and Sask; lyrelvd r.; ours the var. *kamchatica*   3 **A. lyrata** L.
    4b Basal lvs entire to serrate-dentate, never lyrate-pinnatifid; pl sometimes cruciform-hairy; siliques various; style ca 1 mm
      5a Basal lvs cruciform-hairy on lower surface, mostly remotely dentate-serrate, not ciliate; sts often br; siliques 20–25 × ca 1 mm; wet banks and conif woods; SRC, Wn, to wc Ida; cross-haired r.
                                                        4 **A. crucisetosa** Const. & Rollins
      5b Basal lvs glab, or hirsute with simple or forked hairs, mostly entire and strongly ciliate; sts simple
        6a Siliques 2.5–4 cm; seeds with terminal wing ca 0.3 mm, not winged on sides; st lvs few, much narrower than basal lvs, ± aur (see lead 9a)   8 **A. furcata** Wats.
        6b Siliques 1.2–2 cm; seeds not winged; st lvs several, not much narrower than basal lvs, not aur; moist flats, often sheltered by shrubs, foothills to midmont; e Wn to Mont and Alta, s to Nev,

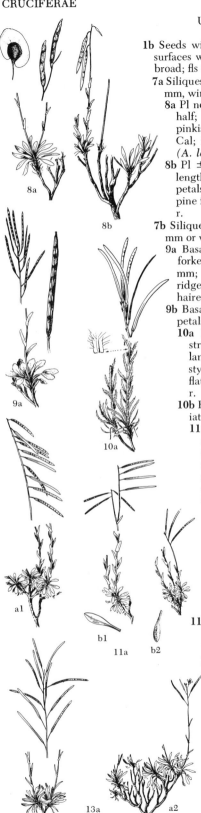

Utah, and Wyo; Nuttall's r. (*A. bridgeri, A. macella*)

5 A. nuttallii Robins.

1b Seeds with wing > 0.3 mm broad, or basal lvs densely hairy on both surfaces with many-br hairs, or siliques spreading or reflexed or > 2 mm broad; fls often pink, red, or purple

  7a Siliques (3) 4–6 mm broad; pl sometimes woody-based; seeds mostly 3–5 mm, wing gen at least 1 mm broad

    8a Pl not shrubby at base; siliques erect, valves 1-nerved only on lower half; st lvs 1–5, remote; seed wing 2–3 mm broad; petals 5–7 mm, pinkish-purple to nearly white; dry ridges, Mt Hood, Ore, s in Cas to Cal; flatseed r., broad-seeded r.; ours the var. *howellii* (Wats.) Jeps. (*A. leibergii*)   **6 A. platysperma** Gray

    8b Pl ± shrubby at base; siliques pendent, valves 1-nerved almost full length; st lvs gen > 5, often overlapping; seed wing 1–2 mm broad; petals 5–8 mm, pink to purplish; rock slides and dry ridges, often in pine for or in sagebr; Yakima Co, Wn, to Cal, e in Ore to c Ida; woody r.   **7 A. suffrutescens** Wats.

  7b Siliques < 3 mm broad, or pl not woody-based, or seeds either < 3 mm or with wing < 1 mm broad

    9a Basal lvs rosulate, oblanceolate to obovate, ciliate with simple or forked hairs; petals white, 6–9 mm; siliques erect, 2.5–4 cm × 1.5–2 mm; sts glab or sparsely hirsute at base; alp and subalp slopes and ridges; Chelan Co, Wn, to c Ore, in OM and Cas, also in CRG; forkhaired r., Cas r. (*A. olympica, A. suksdorfii*)   **8 A. furcata** Wats.

    9b Basal lvs not at once oblanceolate or obovate, rosulate, and ciliate; petals often pink or purple; siliques various; sts often hairy above

      10a Basal and lower st lvs linear-oblanceolate, 1.5–3 mm broad, strongly hirsute and ciliate; st lvs many, upper ones linear-lanceolate; siliques ascending-spreading, 4–7 cm × 2–3 (3.5) mm; style lacking; petals 6–10 mm, deep pink to nearly white; sagebr flats to ponderosa pine woodl; ec Wn to nc Ore, e to c Ida; Cusick's r.   **9 A. cusickii** Wats.

      10b Basal and lower st lvs either > 3 mm broad or not hirsute and ciliate; siliques often spreading to pendent or with noticeable style

        11a Basal lvs oblanceolate to oblanceolate-obovate, rounded or only slightly acutish, gen pannose with tiny, grayish, freely br hairs (except var. *paddoensis*); pl 0.5–2 (4) dm, arising from br caudex; racemes 3–10-fld, secund; ped 2–6 mm; siliques ascending to spreading or slightly reflexed; petals rose-purple, 4.5–7 mm long; alp talus slopes or meadows and ridges; BC s, in OM and Cas, to Cal, e to RM; Lemmon's r.; 3 vars.   **10 A. lemmonii** Wats.

        a1 Siliques > 2 mm broad; sts few, mostly > 2 dm; Alta to Wyo (*A. d.*)   var. **drepanoloba** (Greene) Rollins

        a2 Siliques gen not > 2 mm broad; sts gen several and not > 2 dm

          b1 Pl subglab; basal lvs not at all pannose; e side Mt Adams, Wn; perhaps also in Wen Mts, Wn   var. **paddoensis** Rollins

          b2 Pl ± grayish-hairy; basal lvs gen pannose; range of sp. as a whole (*A. bracteolata, A. latifolia, A. oreocallis, A. polyclada*)   var. **lemmonii**

        11b Basal lvs either not pannose or pannose but acute, or pl much > 2 dm, or racemes > 10-fld and not secund; siliques and petals various

          12a Pl mostly < 2 (3–7) dm, when > 3 dm sts scarcely 1 mm thick at midlength; caudex gen br; basal lvs elliptic-oblanceolate to narrowly oblanceolate, 1.5–3 cm × 1.5–4 (5) mm, acute; siliques ascending-erect to arcuate-spreading but not reflexed; seeds uniseriate; lvs often glab and fleshy

            13a Sts scarcely 1 mm thick at midlength; siliques 1.1–1.8 (very rarely to 2) mm broad; basal lvs gen toothed, glab or (commonly) ± grayish-pubescent with several-br hairs; petals 5–8 mm, pink to purple; low mont to subalp; BC to nc and e Ore, e to Mont and Wyo; littlelf r.; 2 vars.   **11 A. microphylla** Nutt.

              a1 Caudex simple or few-br; siliques 6–15, spreading to ascending; c Ida to Beartooth Mts, Mont, Big Horn Mts, Wyo, and Utah   var. **saximontana** Rollins

14a

13b

15a

16a

15a

16b

17a

18a

a2 Caudex gen freely br; siliques often fewer than 6 and recurved or pendent; BC s E Cas to Ore, e to Mont and
Wyo *(A. densicaulis, A. macounii, A. tenuicula)*
var. **microphylla**

13b Sts gen much > 1 mm thick at midlength; siliques at least
2 (3) mm broad; basal lvs entire, greenish to fleshy, mostly
glab (finely hairy); petals 6–10 mm, purple; subalp to alp
ridges and meadows; BC through OM and Cas to Cal, e to
Alta, Mont, Wyo, and Utah; Lyall's r. *(A. densa, A. multiceps, A. nubigena, A. oreophila)* 12 **A. lyallii** Wats.

12b Pl mostly > 3 dm, if shorter gen with simple caudex or with
reflexed siliques or with lvs > 4 mm broad; sts mostly > 1 mm
thick at midlength; seeds occ biseriate; lvs gen pubescent

14a Lvs, other than those of upper st, densely pubescent and
often grayish with small, freely br (dendritic) hairs, never ciliate with simple or forked pubescence; siliques spread-
ing-arcuate to pendulous, but peds curved downward rather
than abruptly reflexed; st lvs aur **Group I**

14b Lvs greenish to grayish, pubescence 2–3-rayed to stellate
or sometimes more freely br (dendritic), but then st lvs either
not aur or pubescent with long simple or forked cilia, or siliques pendent on sharply reflexed peds **Group II**, lead 17a

Group I

15a Pl strong per, gen with br caudex; sts mostly br above; basal lvs linear-
oblanceolate, 1–3 mm broad, entire; st lvs few, remote, 1–2 (3) mm broad;
petals 4–6 mm, white to deep pink; seed wing ca 0.5 mm broad at widest
point; mostly sagebr areas on des mts from c Ore e to sc Ida, n Nev, and
Wyo; Cobre r. *(A. canesens)* 13 **A. cobrensis** Jones

15b Pl short-lived per, often with unbr caudex; sts often simple; basal lvs 2–6
mm broad, sometimes remotely toothed; st lvs often ∝ and sometimes overlapping, 2–10 mm broad; petals 6–9 mm, mostly pink, rose, or purple;
wing of seed < 0.5 mm wide

16a Siliques densely pubescent (glabrate), 2.2–3 mm broad, 1-nerved for <
half length; basal lvs grayish-hairy; petals 8–10 mm, rose-pink to almost
white; foothills to midmont; Yakima Co, Wn, e Cas to Cal, e to c Ida and
Nev; hoary r. *(A. lignipes var. impar, A. subpinnatifida var. i.)*
14 **A. puberula** Nutt.

16b Siliques gen glab (slightly puberulent), 1–1.5 (rarely 2) mm broad,
1-nerved > half length; basal lvs more greenish than gray; petals 6–9
mm, rose-pink to purplish; sagebr des; sc and e Ida to Wyo, Colo, and
Ariz; woody-br r. 15 **A. lignifera** Nels.

Group II

17a Seeds biseriate, wing almost 1 mm broad at widest point; siliques strictly
erect, 2–3 mm broad; pubescence (if any) of lower part of st simple to
forked, often appressed; basal lvs mostly with 2-rayed pubescence; petals
7–12 mm, white or pale pink; Alas and Yuk s, in OM and Cas, to Cal, e to
RM from Alta to Colo, and through GL region to Atl; mont to subalp in our
region; Drummond's r. *(A. albertina, A. connexa)* 16 **A. drummondii** Gray

17b Seeds uniseriate (partially biseriate in *A. divaricarpa*), wing not > 0.5
mm broad; siliques often not erect, 1–2.5 mm broad; pubescence of lower
part of st often stellate or dendritic; basal lvs mostly with 3-rayed to stellate
or dendritic hairs

18a Pubescence of lower st and lvs appressed, mainly 3-rayed; siliques
mostly straight, stiffly erect to spreading, 3–8 cm × 1.5–2.5 mm; petals
6–9 mm, pink to purplish-red; mont, Alas s, in OM and Cas, to Cal, e to
RM from Alta to Colo, and through GL region to Atl; spreadingpod r. *(A.
brachycarpa, A. stokesiae)* 17 **A. divaricarpa** Nels.

18b Pubescence more-freely br; siliques gen arcuate or reflexed; petals various

19a Siliques ascending to spreading-drooping, 4–12 cm × 1.5–2 mm,
mostly strongly arcuate; peds not geniculate and abruptly curved
downward at base; st lvs mostly aur; petals 6–14 mm, white to deep
purple; mostly in sagebr or ponderosa pine for; BC s, e Cas, to Cal, e

to Alta, Mont, Wyo, and Utah; elegant r., sicklepod r.; 4 vars.

18 **A. sparsiflora** Nutt.

a1 Petals deep purple; peds stout, ascending, often glab; basal lvs dentate; e base Cas, Chelan to Klickitat cos, Wn (*A. atriflora, A. atrorubens*)   var. **atrorubens** (Greene) Rollins

a2 Petals white to light purple; peds various, sometimes strongly hirsute; basal lvs entire to dentate

  b1 Petals white, 6–8 mm; peds hirsute; mainly BC and Mont (*A. c.*)
                  var. **columbiana** (Macoun) Rollins

  b2 Petals pink to purplish, mostly > 8 mm; peds sometimes glab

    c1 Peds sparsely hairy to glab; basal lvs entire; Ida to Utah; ne Cal (*A. arcoidea, A. peramoena*)   var. **sparsiflora**

    c2 Peds mostly strongly hirsute; basal lvs toothed; e side Cas, Wn to Cal, e to Mont and Wyo (*A. elegans, A. perelegans, A. subserrata*)   var. **subvillosa** (Wats.) Rollins

19b Siliques pendulous, 3–7 cm × 1–2 (2.5) mm, nearly or quite straight; peds gen geniculate and abruptly reflexed at base; st lvs often not aur; petals 5–10 mm, white to pink or (more commonly) pinkish-purple; sagebr plain and ponderosa pine for to subalp ridges; BC s, in Cas, to Cal, e to Alta, Mont, Mich, Neb, Colo, and Que; Greenl; Holboell's r.; 5 vars.   19 **A. holboellii** Hornem.

a1 St lvs not aur; mont, BC s, in OM and Cas, to Cal, e to Mont and Colo (*A. p.*)   var. **pendulocarpa** (Nels.) Rollins

a2 St lvs aur, often ± clasping

  b1 Peds geniculate and gen abruptly reflexed at base, rather than curved; siliques gen straight, pendulous-secund to pendulous-appressed; basal lvs finely pubescent

    c1 Lower portion of st coarsely hairy with simple or forked to stellate hairs; petals mostly < 7 mm, white to pale pinkish; valleys and lower mts; Alta and Mont to Wyo, e to Que (*A. c.*)
                  var. **collinsii** (Fern.) Rollins

    c2 Lower portion of st finely and uniformly pubescent with small, dendritic, often appressed hairs; petals mostly > 7 mm, white to deep pink or purple

      d1 Siliques 2–2.5 mm broad; st lvs not revolute, upper ones mostly glab, basal gen greenish and not pannose; mostly mont to subalp; BC to OM and Cas, Wn, e in Can to Que and Greenl   var. **holboellii**

      d2 Siliques mostly 1.5–2 mm broad; cauline lvs gen revolute, upper ones mostly pubescent, basal ones often pannose or at least grayish; ponderosa pine for to midmont ridges; Yuk and BC s, mostly e Cas, to Cal, e to Alta and Colo and occ to Que (*A. caduca, A. consanguinea, A. exilis, A. lignipes, A. retrofracta, A. rhodantha*)   var. **retrofracta** (Grah.) Rydb.

  b2 Peds more uniformly recurved; siliques pendent-drooping, often incurved at tip; basal lvs coarsely pubescent; s Cas, Wn, to Cal, e to se BC, Alta, Mont, and s to Colo (*A. p.*)
                  var. **pinetorum** (Tidest.) Rollins

Athysanus Greene   Sandweed; Athysanus

Racemes slender, unilateral, ebracteate; sepals quickly deciduous; petals (sometimes wanting) white, inconspicuous; silicles suborbicular, much-compressed, apically notched, indehiscent; pl ann. (Gr *a*, without, and *thysanos*, fringe or tassel, referring to the lack of a wing on the silicle).

A. pusillus (Hook.) Greene. Pl 0.4–3.5 dm, gen br from lfy base, hirsute with simple, forked, and 4-rayed hairs; lvs oblanceolate to obovate, 6–30 × 2–10 mm, entire to ± toothed, short-petioled to sessile, but not aur; petals spatulate, 1–2 mm; racemes ∞ -fld, much-elongate and open; peds 1–4 mm, gently recurved; silicles ca 2.5 mm; style 0.2–0.3 mm; dry, ± grassy areas, often in sagebr; VI s, e and w Cas, to s Cal, e to e Wn, ne Ore, and Ida.

Barbarea R. Br.   Wintercress

Racemes bractless; petals yellow, rather showy; siliques linear, almost terete-quadrangular, valves 1-nerved, tardily dehiscent; style ± beaklike; seeds uniseriate, nonmucilaginous when wet; glab to sparsely hirsute bien or per herbs with ± lyrate-pinnatifid basal lvs and mostly ± aur st lvs. (For St. Barbara, 4th century A. D.). *(Campe)*.

1a Style 2–3 mm, beaklike; petals 6–8 mm; siliques 1–3 cm; upper st lvs mostly lobed; European sp., in wet places; w Wn and Ore; bitter w., yellow rocket *(C. stricta)*   **1 B. vulgaris** R. Br.
1b Style 0.5–1.5 (2) mm, not beaklike; petals sometimes < 6 mm; siliques 1.5–8 cm; upper lvs often pinnatifid
   2a Basal lvs with (8) 10–20 lateral lobes; petals 6–8 mm; siliques 4.5–8 cm; European sp., sometimes grown as salad vegetable and occ escaping to wastel; Wn to Cal; Belle Isle cress, early w., landcress
                                                                                                2 B. verna (Mill.) Asch.
   2b Basal lvs with 2–8 (10) lateral lobes; petals 2–5 mm; siliques 1.5–5 cm; moist areas from lowl into mts; Alas s, both sides Cas, to Cal, e to RMS and N Eng; Eurasia; Am w. *(B. americana, C. barbarea)*
                                                                                                3 B. orthoceras Ledeb.

Berteroa DC.   Berteroa

Racemes ebracteate; petals white, bilobed; silicles compressed but also ± inflated, 2-locular, tardily dehiscent; stellate-pubescent ann with simple, entire lvs. (For Giuseppe Bertero, 1789–1831, Italian botanist).

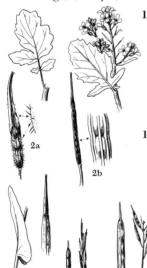

B. incana (L.) DC. St. 3–11 dm, gen br both at base and above; basal lvs oblanceolate, 3–5 cm, slender-petioled, st lvs several, sessile, reduced upward; racemes ∝ -fld; peds 5–10 mm; petals 4–6 mm; silicles oblong-oval, 5–7 × 3–3.5 mm, ca 1.5 mm thick, stellate; seeds 3–7 per cell; style slender, 2–3 mm; European weed of dry or waste places in our area; BC, Ida, and Mont.

Brassica L.   Mustard

Racemes ebracteate; petals pale to bright yellow in ours, rather showy; siliques linear, subterete or ± quadrangular, gen torulose, tardily dehiscent, with prominent, compressed to subterete, 1–3-nerved beak and 1–3-nerved valves; seeds ∝, subglobose, uniseriate, not mucilaginous when wet; ann (ours) to per, glab to hirsute-hispid herbs, often with lyrate-pinnatifid basal lvs. (L name for cabbage). *(Sinapis)*.

1a Beak subterete to strongly flattened, 3-nerved on each side, sometimes seed-bearing in base; valves 3-nerved
   2a Siliques bristly-hairy, 3.5–4.5 mm, strongly flattened, seedless, from ca 1/3 as long to nearly twice as long as valves; seeds 3–8, ca 3 mm; European, often cult, sometimes escaped on waste land, gen; white m. *(B. alba)*   **1 B. hirta** Moench
   2b Siliques subterete, gen glab, 2–3 mm broad; beak 1/3–2/3 as long as valves, often bearing a single seed; seeds 7–12, ca 2 mm; European, intro in N Am, and often escaped in waste places; charlock, wild m. *(B. arvensis)*   **2 B. kaber** (DC.) Wheeler
1b Beak ± terete, gen 1-nerved on each side, seedless
   3a St lvs sessile, aur-clasping, upper ones subentire; siliques spreading to ascending, 3–7 cm × 2.5–3.5 mm; seeds 1.5 mm; pl glab or subglab, weedy; in much of our area, European; field m., common m., bird rape
                                                                                                3 B. campestris L
   3b St lvs sometimes petiolate, neither aur nor clasping, gen all toothed to lobed; siliques various; seeds ca 2 mm
      4a Siliques rather strictly erect, 1–2.5 cm × 1–1.5 mm, midnerve as prominent as sutures; pl gen hirsute-hispid near base; European, weedy in many areas; black m.   **4 B. nigra** (L.) Koch
      4b Siliques spreading-ascending, 2–4 cm × 2–3.5 mm; midnerve much less prominent than sutures; pl mostly glab or subglab; Asian, occ in waste places in much of our area; brown m., leaf m., Chinese m., Indian m.   **5 B. juncea** (L.) Coss.

Cakile Mill.   Searocket

Racemes ± bracteate or ebracteate; petals white to purplish, long-clawed; silique fleshy but hardened in fr, 2-jointed, lower segm ± cylindric and much thinner than the ± ovoid, distinctly beaked upper segm, each portion 1-celled and 1 (2)-seeded, or lower one seedless, segms separating after maturity, but neither one dehiscent; seeds not mucilaginous when wet; fleshy, mostly semiglab ann with alt, simple to pinnatifid lvs. (Old Arabic name for this plant). *(Bunias)*.

1a Lvs deeply crenate to sinuate-dentate, 2–7 cm; petals 6–8 mm; siliques 15–25 mm, lower joint ± terete, without hornlike projections; along coast, Alas to Cal (more common in Wn and Ore), GL region and Atl; American s. *(C. californica, C. e.* var. *c.)*   1 C. edentula (Bigel.) Hook.
1b Lvs pinnatifid, 4–8 cm; petals 8–10 mm; siliques 12–18 mm, lower joint flared into 2 (sometimes 4) prominent hornlike lobes at tip; coastal, occ from BC to Cal (more common s, esp in Cal); Europe; European s.   2 C. maritima Scop.

Camelina Crantz   Falseflax

Racemes ebracteate; petals pale yellow, spatulate; silicles ± compressed, wing-margined, obovate in outline, slightly stipitate; styles 2–2.5 mm; seeds several per cell, mucilaginous when wet; hirsute to stellate ann with alt, simple, entire to repand and sometimes clasping lvs. (Gr *chamai,* on the ground, and *linon,* flax, because of prevalence in flax fields).

1a Sts strongly hirsute-stellate near base; silicles mostly 5–7 × 3–4 (5) mm, obtuse; peds seldom > 15 mm; Eurasian weed, widespread in dry places; littlepod f., hairy f.   1 C. microcarpa Andrz.
1b Sts sparsely hairy to glab near base; silicles mostly 7–9 × 5–6 (7) mm, subtruncate; peds often > 15 mm; European weed, much less common than *C. microcarpa,* but in similar places; falseflax, gold-of-pleasure   2 C. sativa (L.) Crantz

Capsella Medic.   Shepherd's-purse

Racemes ebracteate; petals white; silicles strongly obcompressed, cuneate-obcordate; seeds ∞, mucilaginous when wet; pl ann. (L *capsella,* little box, referring to the silicle). *(Bursa)*.

C. bursa-pastoris (L.) Medic. Pl hirsute and stellate, 1–5 dm, sts simple to br; basal lvs ± lyrate-pinnatifid, petiolate; upper st lvs sessile and clasping; racemes ∞ -fld; peds slender, spreading; petals 1.5–4 mm; silicles 4–8 × 3–5 mm, truncate to retuse, valves strongly keeled; style ca 1 mm; ubiquitous weed of European origin, common in waste places and in gardens and along roadsides.

Cardamine L.   Bittercress; Toothwort

Racemes bracteate or ebracteate; petals white to pink or rose, gen ± clawed; siliques linear, slightly compressed, readily dehiscent, valves 1-nerved near base, mostly not nerved above; style (region above valves) 0.5–8 mm; seeds ∞, uniseriate, not mucilaginous when wet; glab to hirsute ann or per herbs, often with elongate to thickened and tuberlike rhizomes, and with simple to pinnate lvs. (Ancient Gr name, *kardamon,* used by Dioscorides for some pl of this family). *(Dentaria)*.

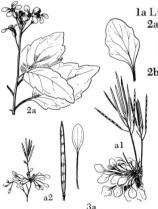

1a Lvs mainly simple
2a Petals pink, 2–2.5 cm; lvs ± oblong-ovate, borne mostly near st tips, 5–13 cm, undulate to shallowly lobed; slenderly rhizomatous per; moist woods along Selway and Clearwater rivers, Idaho Co, Ida; Constance's b.   1 C. constancei Detl.
2b Petals white, 3–12 mm; lvs various, not clustered near st tips; pl sometimes taprooted
3a Pl 2–10 cm, taprooted; lvs chiefly basal and rosulate, slenderly petiolate, bl ± rhombic-elliptic to ovate, 5–30 mm, entire to shallowly 2–4-toothed; st lvs only 1–3, reduced; petals 3–5 mm; circumpolar, alp or subalp; s to Cal, Alta, and NH; alpine b.; 2 vars. 2 C. bellidifolia L.
a1 Lvs thick, mostly 10–30 mm, often with shallow lobes; scapes 4–10 cm; Ore Cas to Cal   var. pachyphylla Cov. & Leib.
a2 Lvs thin, mostly 5–15 mm, gen entire; scapes mostly 2–7 cm; Cas of Wn n and e   var. bellidifolia
3b Pl mostly 20–60 cm, rhizomatous; lvs chiefly cauline, crenate to lobate, reniform to cordate-deltoid, (2) 3–10 cm; petals 7–12 mm; mt

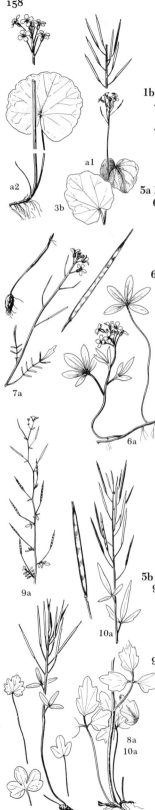

streambanks to alp meadows; BC s through Cas (but not OM) to Cal, e to RM from Wyo to NM; large mt b.; 2 vars.   3 **C. cordifolia** Gray

al Petals 7–9 mm; BC to Cal, e to w Ida and Nev (*C. l.*)

var. **lyallii** (Wats.) Nels. & Macbr.

a2 Petals 10–12 mm; c Ida to Wyo, s to NM and Ariz (*C. c.* var. *pilosa*)

var. **cordifolia**

1b Lvs all or in part compound

4a Petals 6–14 mm, often pink; style (1.5) 3–8 mm; pls (except possibly *C. pattersonii*) per, rhizomatous or tuber-bearing   **Group I**

4b Petals 2–7 mm, white; styles 0.5–2 mm; pl sometimes taprooted ann or bien   **Group II, lead 12a**

Group I

5a Petals white

6a Basal and lower st lvs palmately or pinni-palmately (3) 5 (7)-foliolate or very deeply palmatifid; lflets oval to oblanceolate, mucronulate, entire, 10–30 × 5–15 mm; pl slenderly and extensively rhizomatous, mostly 10–15 cm; siliques 3–4 cm; style 5–6 mm; petals 8–13 mm; known only from Swan, Mission, and Flathead mt ranges, Mont; gen on talus or cliffs; cliff t. (*C. californica* var. *r.*)   4 **C. rupicola** (Rydb.) Hitchc.

6b Basal and lower st lvs various, often pinnate or with toothed or narrow lflets; pls from areas w of Mont

7a St lvs gen pinnately (5) 7–11-foliolate; lflets entire or shallowly toothed, the lateral ones linear-lanceolate, 5–12 mm, terminal lflet gen oblanceolate, 2–3.5 cm, and saliently few-toothed; pl rhizomatous, swollen and tuberous at some nodes, producing decumbent and nodally rooting fl sts at others; petals 8–11 mm; siliques spreading to erect, 2–4 cm × ca 1.5 mm; style 4–6 mm; wet places, often in standing water; WV, Ore; WV b. (*C. rariflora*)   5 **C. penduliflora** Schulz

7b St lvs mostly ternate or semipalmate, segms all about same size; pl gen rhizomatous, the rhizomes often thickened and tuberous, if pls with decumbent and nodally rooting sts and growing in wet places then lflets 3 on all lvs

8a Rhizomes widely spreading, scarcely 1.5 (2) mm thick; sts 2–8 dm, hirsute at base; lvs all much alike, 3 (rarely a few 5)-foliolate; lflets ovate to ovate-lanceolate, 1.5–7 cm, 3–5 (7) -lobed or -deeply toothed; siliques 2–4 cm × ca 2 mm; petals 8–13 mm, sometimes pinkish; style 1.5–4 mm; wet ground at lower elev, often in deep shade; Alas s to n Cal, w Cas and on OP, Wn; angled b., seaside b. (*D. grandiflora*)   6 **C. angulata** Hook.

8b Rhizomes not widely spreading, at least in part thickened and fleshy and > 2 mm thick; pls mostly of drier places, rarely in wet places; petals gen pink (see leads 11a-11b for distinction between)

8 **C. integrifolia** (Nutt.) Greene

9 **C. pulcherrima** Greene

5b Petals pink

9a Petals 6–9 mm; racemes bracteate throughout, peds very slender, 1–3.5 cm, upper ones gen bracteolate; siliques 2.5–3.5 cm × ca 1.5 mm; style 2–3 mm; lvs pinnate, lflets 3–5, ovate to obovate, 3–15 mm, entire to apically 3-lobed; pl delicate, 1–2 dm, said to be ann or bien, but very possibly a very slenderly rhizomatous per; known only from Onion Peak and Saddle Mt, Clatsop Co, Ore; Saddle Mt b.   7 **C. pattersonii** Hend.

9b Petals 7–14 mm; racemes ebracteate, peds ebracteolate; rhizomes evident, sometimes greatly elongate, sometimes shortened and tuberlike

10a Basal and st lvs gen ternate; rhizomes extensive, slender, scarcely 1.5 (2) mm thick (see lead 8a)   6 **C. angulata** Hook.

10b Basal and st lvs gen dissimilar, some either simple or with > 3 lflets; rhizomes thickened and fleshy, 3–6 mm thick

11a Rhizomes short and tuberlike, fleshy, ovoid or subglobose, mostly 6–10 × 3–8 mm; lvs dimorphic, gen purplish beneath, those from rhizomes gen simple (3-foliolate), cordate-ovate to cordate-orbicular, 2–7 cm; st lvs 3–5-foliolate, lflets lanceolate, entire to few-toothed; petals 7–12 mm; siliques 2–5 cm × 1–2 mm; style 4–7 mm; moist, shady woods and slopes; w side coastal mts, Tillamook Co, Ore, to Cal; milk maids, toothwort; ours the var. *sinuata* (Greene) Hitchc. (*C. s.*)   8 **C. integrifolia** (Nutt.) Greene

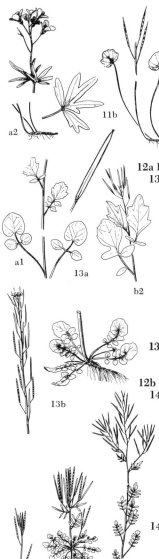

11b Rhizomes ultimately 2–5 cm × 2–3 (4) mm; pl otherwise similar to, and scarcely distinguishable from, *C. integrifolia;* BC to Cal, from w side Cas westward, also in CRG; slender t.; 2 vars.

9 **C. pulcherrima** Greene

  a1 Basal lvs gen simple, if deeply lobed the lobes entire or subentire; BC to n Cal *(C. nuttallii)*   var. **tenella** (Pursh) Hitchc.

  a2 Basal lvs in part gen palmately or pinnipalmately compound, lflets 3–7, gen lobed to dissected; OP and Mt Rainier, Wn, to sw Ore, also in and adj to CRG *(D. macrocarpa* Nutt., *C. quercetorum, C. tenella* var. *dissecta)*   var. **pulcherrima**

Group II

12a Pl rhizomatous, per; petals 3–7 mm

  13a Basal lvs not all pinnate, at least some simple; rhizomes slender and elongate; petals 3–7 mm; wet places, Alas and BC to much of w US; Brewer's b.; 3 vars.   10 **C. breweri** Wats.

    a1 Terminal lflet of lower st lvs cordate-based, lateral lobes ± oval; sepals 1.5–2 mm; petals 3–6 mm; Alas to Cal, chiefly w Cas in Wn and Ore   var. **orbicularis** (Greene) Detl.

    a2 Terminal lflet of st lvs rounded to cuneate at base; sepals mostly 2–2.5 mm; petals 5–7 mm

      b1 Lflets, esp terminal one, sharply 7–11-lobed; nc Ida to nw Wyo *(C. foliacea)*   var. **leibergii** (Holz.) Hitchc.

      b2 Lflets mostly shallowly sinuate, lobes gen 3–4; BC to Cal, mostly e Cas in our region; Wn and Ore, e to Nev and (s of our area) to Wyo *(C. callosicrenata, C. hederaefolia, C. oregana, C. vallicola)*   var. **breweri**

  13b Basal lvs all pinnate; rhizomes somewhat enlarged near base of sts; petals 3.5–5 mm; wet places, esp along streams; Alas to s Ore, w Cas to the coast; w b.   11 **C. occidentalis** (Wats.) Howell

12b Pl taprooted ann or bien, often weedy; petals 2–4 mm

  14a Siliques (1) 1.3–1.5 mm broad, 15–22 (24)-seeded; seeds 1.5–2 mm; lateral lflets of st lvs more nearly ovate or obovate than lanceolate or narrowly oblanceolate, mostly ± petiolulate; widespread, mostly in wet places; Alas to Cal, e to RM; Siberia; little western b., few-seeded b.; 2 vars.   12 **C. oligosperma** Nutt.

    a1 Racemes subumbellate, rachis mostly 1–2 cm; alp or subalp cliffs and talus slopes; Alas to w Ore; Siberia *(C. k.)*   var. **kamtschatica** (Regel) Detl.

    a2 Racemes elongate, rachis gen > 3 cm; lowl and lower mts; BC to Cal, mostly w Cas, where a pernicious garden weed, occ e to Mont *(C. umbellata, C. unijuga)*   var. **oligosperma**

  14b Siliques 0.7–1 (1.5) mm broad, (20) 24–40-seeded; seeds 1–1.5 mm; lateral lflets of st lvs more nearly linear, lanceolate, or narrowly oblanceolate, mostly sessile; moist places, BC to n Cal, w and (esp) e Cas, e to RMS and Atl; Pa b.   13 **C. pensylvanica** Muhl.

Cardaria Desv.   Whitetop; Hoarycress

Racemes corymbose, ebracteate; sepals quickly shed; petals white, 3–4 mm; silicles inflated but ± obcompressed, ovoid to cordate or reniform-cordate, tardily or not dehiscent; styles slender, (1) 1.5–2.5 mm; seeds 1 (2) per cell, mucilaginous when wet; rhizomatous per herbs with simple, dentate, ± sagittate lvs, strongly pubescent with unbr hairs. (Gr *kardia,* heart, with reference to silicle shape). *(Hymenophysa)*.

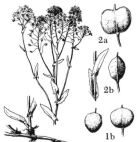

1a Silicles glab, often cordate-based, deltoid-ovoid to subreniform or cordate-ovoid, 3–5 × 4–6 mm

  2a Silicles cordate at base, rounded to retuse at tip; widespread weed in lowl of much of temp N Am; heart-podded h., hoary pepperwort *(Lepidium d.)*   1 **C. draba** (L.) Hand.

  2b Silicles acutish to rounded at base, gen acute at tip; with *C. draba* and intergradient with it: Chalapa h.   2 **C. chalapensis** (L.) Hand.

1b Silicles finely pubescent, more acute than cordate at base, obovoid-subglobose, 4–6 × 4–6 mm; Eurasian weed, closely resembling *C. draba* in appearance and habitat, but less common with us; globepodded h.   3 **C. pubescens** (Meyer) Jarm.

Caulanthus Wats.   Wild Cabbage; Caulanthus

Racemes elongate, ebracteate; calyx ± urceolate, outer sepals often ± saccate at base; petals white to purplish (ours), mostly clawed and with ± crispate bl; anthers large, often coiling slightly after dehiscence; siliques linear, terete or slightly compressed, valves 1-nerved and fully dehiscent; style minute or lacking; seeds ∞, uniseriate, not mucilaginous when wet; ann to short-lived per, glab to sparsely hirsute, gen glaucous herbs with sometimes inflated sts and ± lyrate-pinnatifid (entire) basal lvs; st lvs sometimes sessile and aur. (Gr *kaulos*, st, and *anthos*, fl, with reference to type sp., in which fls seem to arise directly on the thick st).

1a Pl per (bien?), glab to the infl; sts gen inflated; sepals 8–12 mm, greenish-purple; petals 10–14 mm, dull purple; siliques ascending to suberect, 10–15 cm × ca 1.5 mm; des plains and lower mts; se Ore to se Cal, e to s Ida, Utah, and Nev; thickstd w. c. *(Streptanthus c.)*
                                                                                     1 **C. crassicaulis** (Torr.) Wats.

1b Pl ann or bien, pilose to hirsute below the infl; sts not inflated; sepals 6–9 mm, mostly greenish-yellow; petals 8–12 mm, white with purple veins to purplish-rose but white-bordered; siliques spreading-ascending, 7–10 cm × ca 1 mm; des flats and lower mts; Baker and Deschutes cos, Ore, and Washington Co, Ida, to se Cal and w Utah; hairy w. c. *(Streptanthus p.)*
                                                                                     2 **C. pilosus** Wats.

Chlorocrambe Rydb.   Chlorocrambe

Racemes ebracteate, elongate; petals laterally toothed or incised near base, greenish-white or yellowish; anthers apiculate, tending to coil; glab per herb. (Gr *chloros*, light green, and *Crambe*, another member of the family).

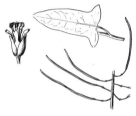

C. hastata (Wats.) Rydb. Spearhead. Sts 1–several, erect, simple or br above, 6–15 dm; petioles slender, 1–10 cm; lf bls 5–12 cm, deltoid to lanceolate, ± hastate, entire to coarsely sinuate-lobed, lower ones sometimes ± lyrate; sepals greenish-white to yellowish, 4–6 mm, ± spreading, not gibbous at base; petals 6–8 mm; anthers 2–3 mm; siliques 4–9 cm, linear, terete but slightly compressed, valves 1-nerved full length, completely dehiscent; seeds ∞, uniseriate, not mucilaginous when wet; stipe (1) 2–7 mm; mt slopes and canyons, Wallowa Mts, Ore, to Utah.

Chorispora R. Br.   Chorispora; Blue Mustard

Racemes ebracteate except at base; petals purple; silique terete, constricted between seeds, eventually rupturing crosswise, not dehiscent, upper part sterile and forming a long, tapered, sharp beak; seeds several, uniseriate, not mucilaginous when wet; ann or per herbs with simple, sinuate-dentate lvs. (Gr *choris*, separate, and *spora*, seed, referring to widely spaced seeds in the silique).

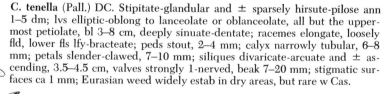

C. tenella (Pall.) DC. Stipitate-glandular and ± sparsely hirsute-pilose ann 1–5 dm; lvs elliptic-oblong to lanceolate or oblanceolate, all but the uppermost petiolate, bl 3–8 cm, deeply sinuate-dentate; racemes elongate, loosely fld, lower fls lfy-bracteate; peds stout, 2–4 mm; calyx narrowly tubular, 6–8 mm; petals slender-clawed, 7–10 mm; siliques divaricate-arcuate and ± ascending, 3.5–4.5 cm, valves strongly 1-nerved, beak 7–20 mm; stigmatic surfaces ca 1 mm; Eurasian weed widely estab in dry areas, but rare w Cas.

Cochlearia L.   Scurvygrass; Spoonwort

Racemes ebracteate; petals white (ours); silicles inflated but ± obcompressed, ovoid to ellipsoid (ours), dehiscent, valves strongly nerved; seeds ∞, biseriate, not mucilaginous when wet; glab ann to per herbs. (Gr *cochlear*, spoon, referring to the ± spoon-shaped basal lvs).

C. officinalis L. (Ann?) bien or per, br from base; sts ± decumbent, mostly 1–3.5 dm; basal lvs rosulate, slenderly petiolate, bls reniform to cordate-oblong or ovate, entire to sinuate, 5–20 × 5–15 mm; st lvs with ± larger bls; petals 3–5 mm; peds 5–15 mm; silicles (3) 4–7 mm; style 0.2–0.5 mm; circumboreal, s from Alas to Grays Harbor Co, Wn, and reported from extreme nw Ore; rarely collected in our area.

Conringia Adans.   Hare's-ear

Racemes ebracteate; petals cream; siliques linear, terete-quadrangular, valves 1–3-nerved, fully dehiscent; seeds ∝, uniseriate, not mucilaginous when wet; glab and gen glaucous ann herbs (ours) with aur -clasping, simple lvs. (For Herbert Conring, 1606–1661, German professor).

C. orientalis (L.) Dumort. H. mustard, h. cabbage, treacle mustard. Taprooted ann; sts 3–7 dm, erect; basal lvs oblong-lanceolate to oblanceolate, 5–9 cm, entire or subentire, narrowed gradually to base; st lvs oblong-lanceolate to -elliptic, mostly 4–12 × 2–5 cm, entire; siliques 8–13 × 1.5–2 mm, ± toru-lose, slender-tipped and with thickish style ca 1 mm; European weed, common in dry areas, mainly e Cas, esp abundant in Mont.

Coronopus Boehm.   Wartcress

Racemes ebracteate; sepals spreading; petals white (ours), tiny; stamens 2 or 4; silicles inflated but ± obcompressed, cordate and deeply lobed, indehiscent, valves hardened, each almost com-pletely enclosing its single seed; seeds not mucilaginous when wet; ann (ours), ± foetid herbs with alt lvs. (Gr *korone*, crow, and *pous*, foot, referring to lf shape).

C. didymus (L.) J. E. Smith. Decumbent-based, glab to scurfy or ± hirsute, freely br ann 2–5 dm; lvs ∝, ovate-oblong, 1.5–3 cm, deeply pinnatifid, segms narrow, entire to toothed; racemes ∝, ∝ -fld, 1–4 cm; peds 1.5–2.5 mm; sepals ca 0.5 mm, deciduous; petals linear, 2–2.5 mm; silicles ca 1.5 × 2 mm, didymous, strongly rugose; intro from Europe; occ weed of waste places, gardens, and roadsides; BC to Cal and e to Atl.

Crambe L.   Crambe

Racemes ± br, ebracteate except lowest fls; petals short-clawed, obovate, cream; siliques inde-hiscent, 2-jointed, lower segm ± stipelike, seedless, upper segm ± globose, 1-celled and 1-seeded; thick-std herb with cabbage-like lvs. (Gr name, *Krambe*, for cabbage).

C. maritima L. Sea-kale. Pl glab and glaucous, st stout, freely br, 2–5 dm; lvs fleshy, elliptic-ovate to oval, 1–4 dm, short-petiolate, sinuate-dentate to sub-pinnatifid and coarsely dentate; peds stout, 15–30 mm; sepals cream, spreading; petals 7–12 mm; lower segm of fr 2–3 mm, upper segm 7–10 × 7–10 mm; European, said to be estab at Yaquina Head, near Newport, Lin-coln Co, Ore, otherwise not known in our area.

Descurainia Webb & Berth.   Tansymustard

Racemes ebracteate; sepals spreading; petals yellow to cream, clawed; siliques linear to clavate, terete to ± quadrangular, valves prominently nerved; seeds ∝, uniseriate to ± biseriate, mucila-ginous when wet; ann or bien, stellate-pubescent (subglab) and sometimes stipitate-glandular herbs. (For Francois Descurain, 1658–1740, French botanist). *(Sophia).*

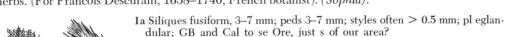

1a Siliques fusiform, 3–7 mm; peds 3–7 mm; styles often > 0.5 mm; pl eglandular; GB and Cal to se Ore, just s of our area?
                                                           1 **D. californica** (Gray) Schulz
1b Siliques linear to clavate or (if ± fusiform) pl not otherwise as above
   2a Lvs (at least lower ones) bi- or tripinnate; siliques narrowly linear,
      mostly > 20-seeded; seeds uniseriate; septum of silique 2–3-nerved; pl
      finely stellate-pubescent and often grayish, 3–10 dm; common European
      weed, esp in dry areas; Alas to Cal, e to Atl; flixweed *(Sisymbrium s.)*
                                                           2 **D. sophia** (L.) Webb
   2b Lvs only once-pinnate, or siliques clavate, or seeds either < 20 or par-
      tially biseriate; pod septum 1-nerved or nerveless
      3a Siliques ± clavate, rounded above, only slightly or not at all torulose;
         septum gen not nerved; weedy ann, gen finely stellate-pubescent
         above; in most of US and s Can; w t.; 6 vars.
                                                           3 **D. pinnata** (Walt.) Britt.
         a1 Pl glandular, esp infl, often grayish-pubescent; siliques 2–10 (12)
            mm

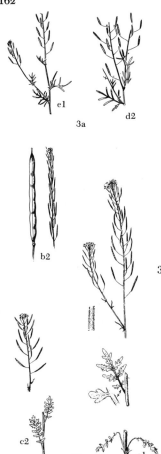

bl Siliques narrowly elliptic, 2–4 mm; peds 3–6 mm; seeds biseriate; pl grayish; petals 1–1.5 mm, pale yellow; c Ore to Nev and
Cal *(D. p.)*   var. **paradisa** (Nels. & Kenn.) Peck
b2 Siliques linear or linear-obovate, 5–10 (12) mm; seeds only partially biseriate; petals off-white to dark yellow, often > 1.5 (3)
mm
   cl Pl canescent; petals white to pale yellow, gen not > 2 mm; se
Ore to Wyo and Mex, not presently known in our area *(S. an-
drenarum, S. h.)*   var. **halictorum** (Cockerell) Peck
   c2 Pl greenish; petals yellow, 2–3 mm; e base RM, Can to Tex
and to Atl *(D. b., S. magna)* var. **brachycarpa** (Richards.) Fern.
a2 Pl eglandular, or if (as rarely) ± glandular then siliques 10–20 mm
d1 Siliques 4–12 mm, from slightly < to > the peds
   el Petals ca 1.5 mm; peds mostly 4–6 mm; e Wn to Nev, e to
Mont and Colo *(S. n.)*   var. **nelsonii** (Rydb.) Peck
   e2 Petals 2–3 mm; peds mostly 6–12 mm; RM from e BC to Colo;
var. **intermedia** (Rydb.) Hitchc.
d2 Siliques 10–20 mm, mostly < the peds; BC s, in e Wn, to e Cal
and Nev, e to Mont and Colo *(D. canescens, D. incisum var. f., D.
longipedicellata, S. f., S. glandifera, S. gracilis)*
var. **filipes** (Gray) Peck
3b Siliques linear or at least not clavate, pointed above, strongly torulose; pod septum 1-nerved; widespread in w N Am, mostly at lower to
middle elev in the mts; mt t.; 4 vars.   4 **D. richardsonii** (Sweet) Schulz
a1 Peds and siliques erect and closely appressed
b1 Pl grayish-stellate; Yuk s, mostly e RM, to Colo, e to GL, occ in
Mont and Ida   var. **richardsonii**
b2 Pl greenish, moderately stellate to subglab; s Mont to NM, occ in
Ida *(S. brevipes, S. hartwegianum, S. procera)*
var. **macrosperma** Schulz
a2 Peds and siliques ascending to ± widely spreading
c1 Pl stipitate-glandular; common from BC s, e Cas, to Cal, e to RM
in Alta and s to NM *(D. rydbergii, S. californica, S. v.)*
var. **viscosa** (Rydb.) Peck
c2 Pl eglandular; c Cal to Baja Cal, e to RM, n to Mont, very occ in
Wn, Ore, and Ida *(S. incisa, S. leptophylla, S. purpurascens, S. s.)*
var. **sonnei** (Robins.) Hitchc.

Diplotaxis DC.

Racemes ebracteate; petals clawed, pale yellow (ours); siliques linear, almost terete but ± compressed, style stout, beaklike; seeds ∞, biseriate; ann or bien, gen hirsute herbs. (Gr *diplos*, double, and *taxis*, arrangement, referring to biseriate seeds).

D. muralis (L.) DC. Wall rocket, w. mustard, stinkweed. Glab or ± hirsute-hispid ann (bien) 2–5 dm; lvs all petiolate, mostly rosulate, oblong to
oblanceolate, toothed to broadly subpinnatifid, mostly 5–10 × 1–2.5 cm; fl sts lfy
only at base, becoming much elongate and > half total height of pl; peds
1–2 (3) cm; sepals ± spreading, 4–5 mm; petals 4–7 mm; siliques subterete,
2–3.5 cm × ca 3 mm, beak 2–3 mm; occ near coast, as around Portland, Ore;
much more common in c and e US; Europe.

Draba L.   Whitlow-grass; Whitlow-wort; Draba

Racemes ebracteate or bracteate below; sepals erect to slightly spreading; petals white or yellow
(ours), clawed, rounded to rarely bifid; silicles mostly strongly compressed to ± inflated, sometimes elongate, narrow and ± silique-like, quickly to tardily dehiscent; style prominent to obsolete; seeds gen ∞ and biseriate, but sometimes only 1 or 2, mostly ± flattened, occ strongly
winged, not mucilaginous when wet; ann to per, scapose to lfy-std, glab to pubescent, with entire
to dentate, rather small lvs and very distinctive br pubescence. (From *drabe*, name first used by
Dioscorides for some member of the Cruciferae).

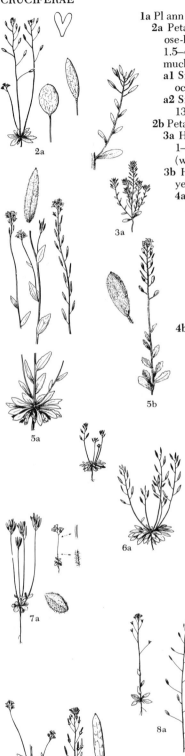

1a Pl ann

  2a Petals deeply bilobed, ca 2.5 mm; fl sts lfless, 5–25 cm, glab, or pilose-hirsute with br hairs near base; silicles elliptic to obovate, 3–10 × 1.5–4 mm; open grassy plains to sagebr des and lower mts; common in much of US and Can, Eurasia; spring w.; 2 vars.   **1 D. verna** L.

    a1 Silicles elliptic or elliptic-oblanceolate, ca 7 × 2 mm; ovules 30–60; occ, Wn to Cal, e to Mont, more common in e US   var. **verna**

    a2 Silicles elliptic-obovate to obovate, ca 4–5 × 3.5 mm; ovules mostly 13–32; common, Wn to Cal and in e US   var. **boerhaavii** Van Hall

  2b Petals not bilobed, often yellow; fl sts often lfy

    3a Hairs small, sessile, 2–4-rayed; silicles ca 3 (2–5) mm; petals white, 1–3 mm; sts lfy; native in c and e US, known from Mont and WV, Ore (where adventive?); shortpod w.   **2 D. brachycarpa** Nutt.

    3b Hairs simple or br and stalked; silicles gen > 5 mm; petals often yellow

      4a Upper portion of st (and gen peds also) hairy; petals white

        5a Silicles narrowly lanceolate, 8–14 × 1.5–2.5 mm, soft-pubescent with mixture of simple and forked hairs; pl bien or short-lived per, sometimes fl 1st year and mistaken for ann; mont woods to subalp ridges; BC to Cas of Wn and Ore, e to Alta, Wyo, and c Nev; tall d. (*D. cascadensis, D. columbiana*) 3 **D. praealta** Greene

        5b Silicles oblong-obovate, 6–10 × 2.5–4 mm, glab to hispidulous with simple hairs; sagebr plains to des washes and hillsides; in our area only along SR and lower CR, in Ida, Ore, and Wn; wedgelvd d.; common in sw US; ours the var. *platycarpa* (T. & G.) Wats. (*D. viperensis*)   **4 D. cuneifolia** Nutt.

      4b Upper portion of st (and also peds) glab; petals often yellow

        6a Sts rarely with > 1 or 2 lvs, if any; petals 2–3 mm, yellow, fading to white; lvs basal and rosulate, 10–25 × 2–4 mm, ciliate with unbr hairs, upper surfaces mostly glab, lower surfaces with few to many simple, forked, or stellate hairs; silicles narrowly elliptic or lanceolate-elliptic, 5–12 × 2–3 mm, mostly glab; style scarcely 0.15 mm; alp or subalp meadows, talus, and ridges; BC s, in OM and Cas, to Cal and Ariz, e to ne Ore, Ida, and Mont, and s to Colo; thicklvd d. (*D. parryi*)   **5 D. crassifolia** R. Grah.

        6b Sts gen with several lvs; petals sometimes white, often > 3 mm; lvs various, often pubescent on upper surfaces; silicles various; pl often of lowl

          7a Petals white, mostly 2–5 mm (wanting); lvs gen entire; silicles linear, 5–20 × 1–1.75 (2) mm, glab to finely hispidulous; des plains and foothills in much of US; Carolina w.; 2 vars.   **6 D. reptans** (Lam.) Fern.

            a1 Upper surface of st lvs and both surfaces of many of basal lvs with mostly simple hairs; sporadic in our area, but the common phase in s RM and e (*D. caroliniana, D. coloradensis, D. micrantha*)   var. **reptans**

            a2 Upper surface of all lvs with mostly br hairs; common, c Wn to Cal and Ariz, e to Mont and Utah   var. **stellifera** (Schulz) Hitchc.

          7b Petals yellowish (sometimes fading to off-white); lvs in part gen denticulate to dentate; silicles often > 2 mm broad; pl often mont

            8a Ped 1–5 (rarely < 1.5) times as long as fr; silicles elliptic to elliptic-oblanceolate, 4–11 × 2–3 mm, glab to hirsutulous; Eurasia; common in valleys and lower mts of w N Am, gen where rather dry; fl Mar-June; woods d.   **7 D. nemorosa** L.

            8b Ped rarely as much as 1.5 times as long as fr; silicles linear to narrowly oblong, 8–12 (22) × 1.5–2.3 mm, gen glab; moist banks or meadows or dry slopes, foothills to near timberl, almost throughout mts of w N Am; Alas to Cal, e to Alta, Mont, Wyo, and Colo; fl May-Aug; our pls with simple or forked hairs, the var. *nana* (Schulz) Hitchc.; var. *stenoloba*, with forked to 4-rayed hairs, mostly to n of our range; Alas w., slender d.   **8 D. stenoloba** Ledeb.

1b Pl bien or per

  9a Petals white; pl scapose, per, 2–10 cm; lvs oblanceolate, 5–12 × 1–2 mm, strongly ciliate with simple hairs and often also pubescent with

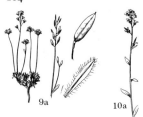

1–2-forked hairs, midnerve prominent; silicles oblong-ovate, 3–6 (9) ×
1.5–2 mm, apiculate (style 0.1–0.3 mm), 10–20-seeded; alp, on highest
peaks, Park Co, Mont, to Colo and se Utah, more common from BC to
Alas; Eurasia; Austrian w. 9 **D. fladnizensis** Wulfen
9b Petals yellow or lvs, pubescence, or silicles not as above
  10a Fl sts lfy; pl mostly not matted **Group I**
  10b Fl sts lfless (or with 1–2 near-basal lvs); pl gen matted
                                            Group II, lead 19a

Group I

11a Style scarcely 0.2 mm, often lacking; pls bien or short-lived per, sometimes fl in 1 year
  12a Lvs densely pubescent on both surfaces with both simple and forked to
stellate hairs; petals white or cream; pl 1–3 dm; silicles narrowly lanceo-
late, 8–14 ×|1.5–2.5 mm, soft-pubescent with simple and forked hairs;
moist mont woods to subalp ridges (see lead 5a) 3 **D. praealta** Greene
  12b Lvs gen sparsely pubescent, upper surface often glab; petals mostly
yellow (fading with age); silicles various
    13a Peds gen < the silicles; silicles narrowly elliptic or narrowly lanceo-
late-elliptic, 5–12 × 2–3 mm; st lvs only 1 or 2; alp or subalp (see lead
6a) 5 **D. crassifolia** R. Grah.
    13b Peds mostly = or > the silicles; silicles linear to narrowly oblong,
8–22 × 1.5–2.3 mm; st lvs gen several; moist banks to dryish meadows
or slopes, foothills to ± subalp, gen in most of w N Am (see lead 8b)
                                           8 **D. stenoloba** Ledeb.
11b Style evident, at least 0.2 mm; pls mostly strong per
  14a Petals white
    15a St lvs only 1 or 2, reduced; pl 1–12 cm, scapose; petals not emargin-
ate, 2.5–5 mm; silicles linear to elliptic or narrowly oblong-
oblanceolate, (3) 7–12 (20) × 1–2 (3.5) mm, glab to stellate; style
0.2–0.5 mm; lvs densely stellate-pannose, midribs gen long-persistent;
alp (subalp) talus slopes and rock crevices; Alas to ne Ore, in OM and
Cas in Wn, e to Ida, Mont, Wyo, and Colo; lancefr d.; 3 vars.
                                        10 **D. lonchocarpa** Rydb.
      a1 Silicles 7–10 × 1.7–3.5 mm, elliptic or ± oblanceolate; Cas, Chelan
and Whatcom cos, Wn var. **thompsonii** (Hitchc.) Rollins
      a2 Silicles gen narrowly elliptic to linear and not > 2 mm broad
        b1 Silicles elliptic, 4–7 (10) cm × 1.5–2 mm; hairs of lvs more nearly
4-rayed than stellate; highest peaks, Beartooth Mts, n Wyo (and
prob adj Mont) to Colo var. **exigua** Schulz
        b2 Silicles mostly > 10 (7–20) mm, more nearly linear than elliptic;
hairs of lvs mostly stellate, often pannose; Alas to OM and Cas,
Wn, Wallowa Mts, Ore, and e to Mont and Wyo var. **lonchocarpa**
    15b St lvs several; pl 5–40 cm, not scapose; petals gen emarginate; sili-
cles various
      16a Hairs of lvs soft-stellate to ∝ -br, but not pectinately br; silicles
soft-pubescent; alp and subalp, on open knolls or dry meadows to
rock crevices; Alas and Yuk s in RM to Mont, Ida, and Colo, w to
Utah and Nev, e to Atl; Eurasia; lance lvd d. (*D. cana*)
                                         11 **D. lanceolata** Royle
      16b Hairs of lvs appressed, pectinately br from central axis paralleling
midvein of lf; silicles mostly glab; reported from Beartooth Mts,
Wyo-Mont border, otherwise known only from Alas and Yuk;
smoothish d. 12 **D. glabella** Pursh
14b Petals yellow
  17a St lvs 2–6; basal lvs fleshy, sometimes ciliate but otherwise mostly
glab; silicles glab, elliptic-lanceolate, crispate-undulate to contorted,
10–16 × 3–5 mm; style 0.75 mm; petals 4–8 mm; mont ridges and talus
slopes; RM, sc Mont to Colo; thicklvd d. 13 **D. crassa** Rydb.
  17b St lvs gen > 6; basal lvs not fleshy, gen densely pubescent on both
surfaces; silicles commonly hairy, 7–20 × 2–6 mm; style 0.3–1.5 mm;
petals 4.5–6 mm
    18a Mature silicles ovate-oblong to oblong-elliptic, 11–15 (20) × 3–6
mm; st lvs ∝, closely crowded at base of st; pubescence mostly
simple to br but not cruciate or stellate; style 1–1.5 mm; volcanic
peaks, mostly above timberl; Mt Rainier, Wn, to Mt Lassen, Cal;

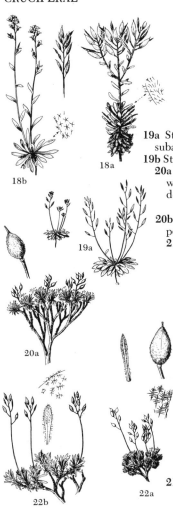

alpine d., Mt Lassen d.   14 **D. aureola** Wats.
18b Mature silicles narrowly to broadly lanceolate, 7–20 × 2–4 mm;
   st lvs 3–30, not closely crowded at base of st; pubescence gen in
   part cruciform or semi-stellate; style 0.3–1.5 mm; mt for to alp
   ridges or meadows; Alas s in RM to NM and Ariz; in our area in
   Wn, Ida, and Mont; golden d. (*D. aureiformis, D. decumbens, D.*
   luteola, D. mccallae, D. surculifera)   15 **D. aurea** Vahl

Group II

19a Style lacking (rarely up to 0.15 mm); pl short-lived per, not matted;
   subalp to alp (see lead 6a)   5 **D. crassifolia** R. Grah.
19b Style mostly much > 0.15 mm; pl commonly matted and caespitose
   20a Silicles inflated, ovoid, 1–2-seeded, 3–7 mm; seeds ca 2 mm; petals
      white, 4–5 mm; styles 0.5–1.8 mm; scapes 1–3 cm; sagebr des and low
      des mts; sc Wn to se Ore and c Nev; fl Apr–June; Douglas' d.
                                                    16 **D. douglasii** Gray
   20b Silicles mostly not inflated, always > 2-seeded; seeds gen < 2 mm;
      petals often yellow; scapes mostly > 3 cm
      21a Pubescence of lvs (esp on lower surfaces) wholly or in large part of
         sessile or short-stalked, 2-rayed, doubly pectinate hairs, many lying
         parallel to midvein; petals yellow, 3–5 mm; lvs linear to somewhat
         spatulate, 3–13 (15) × 0.75–3.5 mm
         22a Lvs ave < 1.5 mm broad, closely appressed-pubescent; scapes
            1–10 cm; silicles ovate to elliptic or oblong-obovate in outline,
            2.5–8 × 2–4 mm, plane or ± inflated; seeds 2–10; style 0.1–1 mm;
            widespread from plains to mont slopes in much of w Can and US;
            few-seeded d.; 2 vars.   17 **D. oligosperma** Hook.
            a1 Pubescence of silicle valves fine, at least in part doubly pectinate;
               nw Park Co, Wyo, and in Utah; to be expected in Mont
                                          var. **pectinipila** (Rollins) Hitchc.
            a2 Pubescence of silicle valves coarse, simple and forked, mostly
               retrorsely appressed; common, BC s, in Cas, to Ore and Cal, e to
               Alta, Mont, Wyo, and Colo (*D. andina*)   var. **oligosperma**
         22b Lvs ave ca 2 mm broad, not closely appressed-pubescent; scapes
            1–20 cm; silicles ovate or ovate-lanceolate to elliptic, 4–12 × 1.5–3
            mm, scarcely at all inflated; seeds 8–14; style 0.4–1 mm; subalp and
            alp; BC to OM and Cas, Wn, e to Alta, Ida, Mont, and Wyo; Yel-
            lowstone d.   18 **D. incerta** Pays.
      21b Pubescence of lvs simple to stellate, never doubly pectinate; petals
         sometimes white; lvs various
         23a Lvs subglab except for prominent, straight, simple cilia along
            margins, lower surface sometimes rather sparsely pubescent with
            forked to stellate hairs   Group IIa
         23b Lvs pubescent (often cinereous) on both surfaces
                                                        Group IIb, lead 27a

Group IIa

24a Petals white, (3) 4–6 mm; lvs obovate-oblanceolate, 3–12 × 1–2 mm;
   scapes 3–7 cm; silicles glab, narrowly elliptic-oblong, 7–10 mm × ca 2
   mm; style 0.5–1 mm; lower canyons to subalp rock crevices, always on
   limestone? GNP and Lost River Mts, Butte Co, Ida; limestone d.
                                           19 **D. oreibata** Macbr. & Pays.
24b Petals yellow; lvs various, often > 2 mm broad; scapes often > 7 cm; silicles often pubescent; style sometimes < 0.5 mm
   25a Pl glab except for stiff cilia on lvs; lvs fleshy, lanceolate to obovate or
      oblanceolate, 3–8 (10) × 1–2.5 mm; scapes 0.5–4.5 cm; silicles ovate to
      oblong-ovate or oblong-elliptic in outline, 3–8 × 2–4 mm, glab; style
      0.2–0.5 mm; alp meadows and talus slopes; Bitterroot Mts, Ida-Mont, ne
      Park Co and Grand Tetons, Wyo, and Uinta and Wasatch mts, Utah;
      pointed d.; 2 vars.   20 **D. apiculata** Hitchc.
      a1 Lvs 3–6 × 1–2 mm; cilia few, mostly < 0.3 mm; scapes 0.5–3 cm;
         racemes 2–5-fld; peds mostly < silicles; range of sp. except for Bitter-
         root Mts; to be expected in Mont along Wyo border   var. **apiculata**
      a2 Lvs mostly 4–8 × 1.5–2.5 mm; cilia ∝, many > 0.4 mm; scapes

26a

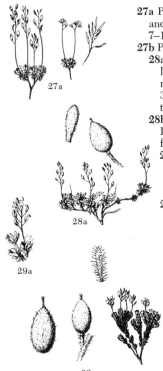

26b

27a

28a

29a

30a

31a

31b

2–4.5 cm; racemes 5–12-fld; peds mostly at least = silicles; Bitterroot Mts, Ida-Mont   var. **daviesiae** Hitchc.

25b Pl gen pubescent on sts or lower surface of lvs; style often > 0.5 mm; scapes and silicles various

  26a Lvs fleshy, obovate to oblanceolate, 2–9 (ave at least 3) mm broad, midnerve not prominent; silicles 4–11 × 2.5–5 mm, 4–16-seeded; seeds 1–1.5 mm; style 0.2–0.75 mm; petals 4–6 mm; alp ridges, talus, and rock crevices; Wallowa Mts, Ore, in our area, also Sierra Nevada, Cal; Lemmon's d.; ours the var. *cyclomorpha* (Pays.) Schulz
                                                  21 **D. lemmonii** Wats.

  26b Lvs not fleshy, linear to narrowly oblanceolate, 0.5–3 mm broad, midnerve prominent; silicles 2–7 × 2–3.5 mm, 2–12-seeded; seeds ca 2 mm; style 0.5–1 mm; petals 2–6 mm; midmont to subalp or alp; s BC s, e Cas, to Sierran Cal, e to Blue and Wallowa mts, Ore, and adj Wn, Ida, Mont, w Wyo, and Utah; Nuttall's d. (*D. caeruleomontana, D. nelsonii, D. sphaerula*)   22 **D. densifolia** Nutt.

Group IIb

27a Petals white; style 0.2–0.5 mm; lvs finely cruciate- or stellate-pubescent and pannose; silicles linear to elliptic or narrowly oblong-oblanceolate, (3) 7–12 (20) × 1–2 (3.5) mm (see lead 15a)   10 **D. lonchocarpa** Rydb.

27b Petals yellowish; style often > 0.5 mm; pubescence and silicles various

  28a Silicles ovoid to lance-ovoid, ± inflated, finely stellate, 2–5 mm, at least half as broad; style ca 1 mm; lvs narrowly obovate, 3–7 × 1–2.5 mm, cinereous with tiny, tangled, multibr hairs; scapes 2–5 cm; petals 3–4 mm; gravel bars along mont streams, and subalp to alp ridges; Sawtooth Mts, c Ida; globefruited d.   23 **D. sphaerocarpa** Macbr. & Pays.

  28b Silicles various, seldom inflated, but often > 5 mm; styles sometimes < 1 mm; lvs often with tangled pubescence on lower surface, upper surface often with simple or merely forked hairs

    29a Lvs fleshy, obovate to oblanceolate, 5–30 × 2–9 mm, ± hirsute with simple and forked hairs; silicles oval or elliptic, 2.5–5 mm broad; alp rocky ridges and talus; ne Ore and Sierra Nevada, Cal (see lead 26a)
                                              21 **D. lemmonii** Wats.

    29b Lvs various, not fleshy; silicles various

      30a Lvs 0.75–1.5 mm broad, strongly ciliate with long, simple and forked hairs, lower surface with prominent midrib and long, tangled, br hairs, upper surface with simple and forked hairs; silicles ovate to ovate-lanceolate in outline, ± inflated, 3–8 × 2.5–4.5 mm, mostly hispidulous with simple and br hairs; petals 2–4.5 mm; style 0.5–1 mm; subalp and alp BC s, in OM and Cas, to Sierra Nevada, Cal, e to RM from Alta to Wyo, and in Utah and Nev; Payson's d.; 2 vars.   24 **D. paysonii** Macbr.

        a1 Silicles 5–8 × 3–4.5 mm; style ca 1 mm; Little Belt Mts, c Mont, to nw Wyo, occ n to GNP   var. **paysonii**

        a2 Silicles 3–5 × 2.5–3.75 mm; style 0.5–0.75 mm; range of sp. except for var. *paysonii* (*D. novolympica*)
                                      var. **treleasii** (Schulz) Hitchc.

      30b Lvs gen at least 2 mm broad, pubescent with cruciate to stellate hairs, midnerve not prominent; silicles, petals, and style various

        31a Silicles 5–8 × 3.5–5 mm, oval to ovate, densely hairy; style 0.6–1 mm; scapes 2–4 cm; lvs 5–12 × 2–4 mm, densely and coarsely pubescent with simple, forked, and cruciate to stellate hairs; rock crevices and slopes at or above timberl; BC, Wn, Wyo, and Utah; Wind R d.; 2 vars.   25 **D. ventosa** Gray

          a1 Pubescence mixed, gen including simple, cruciate, and stellate hairs; known only from Alas, Glacier Peak, Wn, and Mt Waddington, BC   var. **ruaxes** (Pays. & St John) Hitchc.

          a2 Pubescence more uniformly stellate throughout; YNP to Uinta Mts, Utah   var. **ventosa**

        31b Silicles 5–12 × 2–4 mm, elliptic-lanceolate, glab to puberulent; style 1–1.8 mm; scapes 3–9 cm; lvs 5–12 × 2–4 mm, cinereous and ± pannose with stellate or more freely br and appressed hairs, without simple or forked hairs; subalp and alp, gen in rock crevices; Sawtooth and Smoky mts, c Ida; silvery d.
                                        26 **D. argyraea** Rydb.

Eruca   Adans.

Racemes bractless; petals white to yellow, gen purplish-veined; siliques indehiscent or tardily dehiscent, ± inflated and almost terete, tapered to prominent flattened stylar beak, valves 1–3-nerved; seeds several, biseriate, not mucilaginous when wet; ann to per, glab to pubescent herbs with pinnatifid lvs. (L name for the pl).

E. sativa Gars. Garden-rocket, rocket-salad. ± pilose ann 2–10 dm, simple to br below; lvs 5–15 cm, sinuate-pinnatifid, much smaller above; sepals erect, 8–10 mm; petals clawed, 15–20 mm; peds stout, 4–8 mm; siliques erect, lance-ovoid, 15–30 × 3–5 mm, beak flattened, up to half as long as valves; European, sparingly intro in US, rarely in our area, but collected in Klickitat Co, Wn (*Brassica e.*).

Erucastrum   Presl

Fls in bracteate racemes (ours); sepals ± saccate at base; petals light yellow (ours); siliques sessile, terete-quadrangular, tardily dehiscent, style prominent and ± beaklike; seeds several, uniseriate, not mucilaginous when wet; ann or per herbs with ± pinnatifid lvs and simple hairs, if any. (*Eruca*, plus L suffix *-astrum*, like, meaning similar to *Eruca*).

E. gallicum (Willd.) Schulz. Dog mustard. Strigose-pilose, gen freely br ann (bien?) 1.5–8 dm; sts lfy into infl; lvs 3–20 cm, pinnatifid and again sinuately lobed, reduced upward; racemes ± lfy-bracteate, elongate and loosely fld; peds slender, 7–15 mm; sepals ca 4 mm, sparsely pilose; petals 5–7 mm; siliques 2–4.5 cm, ca 1.5 mm thick; style 1.5–3 mm; European weed, occ in Mont, Ida, Wn, and prob in Ore.

Erysimum   L. Wallflower

Racemes ebracteate, gen ∝ - fld; sepals erect, outer ones mostly saccate at base; petals lemon to golden or reddish, long-clawed; siliques linear, compressed to terete-quadrangular, valves strongly nerved; styles prominent, 1–4 mm in ours, often beaklike; seeds ∝ , uniseriate, wingless to wing-margined, not mucilaginous when wet; ann to per herbs with alt, entire to sinuate-dentate lvs. (Gr name, *erusimon*, supposedly derived from *eryo*, to draw, as some spp. were used as is mustard, to cause blistering). (*Cheirinia, Cheiranthus*).

1a Petals gen < 10 (3.5–11) mm; style rarely > 1.5 mm
  2a Peds nearly or fully as thick as fr; siliques ± divaricate, 5–10 cm × ca 1.5 mm, constricted between the nonwinged seeds; petals (4) 6–10 mm, light yellow; ann, (1) 2–5 dm; weedy European sp. of waste places mostly; des plains and lower mts, e Wn and Ore, Ida, and Cal, scattered e to Atl; spreading w.   1 **E. repandum** L.
  2b Peds scarcely half as thick as fr; siliques ascending to erect, gen < 5 cm × 1–1.5 mm, mostly not constricted between the nonwinged seeds; petals various; ann to per
    3a Siliques 1.5–3 cm; petals 3.5–5 mm, pale yellow; pl sparsely pubescent, mostly greenish; European, prob intro in N Am; ± weedy, frequently in moist places; Alas to Cal, e through RM to Atl, Newf to NC; treacle mustard, wormseed m.   2 **E. cheiranthoides** L.
    3b Siliques mostly 2.5–5 cm; petals 7–11 mm, pale yellow; pl gen grayish with abundant pubescence, bien or short-lived per; dry, often alkaline soil; Alas s, e Cas, to se Ore, e to RM, Mont to Colo, and sc Can to c US; smallfld rocket, small w. (*E. parviflorum, E. asperum* var. *p.*)
                                        3 **E. inconspicuum** (Wats.) MacM.
1b Petals > 11 mm or styles > 1.5 mm
  4a Seeds wing-margined all around; silique strongly flattened, not torulose, 7–10 cm × 2.5–3 mm; style 3–4 mm; pl bien, 1.5–4.5 dm; lvs many, linear-oblanceolate, 4–8 cm × 2–4 mm, mostly entire; sagebr des, e Wn and Ore to Ida and Nev; pale w.   4 **E. occidentale** (Wats.) Robins.
  4b Seeds winged only near tip or not at all; silique 1–2.5 mm broad, either quadrangular or strongly flattened, if flattened then pl per or siliques torulose; style (1) 2–4 mm
    5a Pl per (sometimes blossoming as bien), gen greenish; siliques flattened, (1.5) 2–2.5 mm broad, mostly torulose; open ridges, talus slopes,

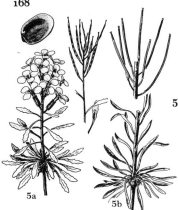

and rock crevices; Cas and OM, Wn to s Ore; sand-dwelling w.; 2 vars.   **5 E. arenicola** Wats.

a1 Pl strong per with freely br caudex; siliques mostly 3–5 cm; stylar beak ca 3 mm; OM, Wn   var. **arenicola**

a2 Pl short-lived per, often with simple caudex; siliques 3–12 cm; stylar beak 1–4 mm; Cas of Wn and Ore, from foothills to alp ridges   var. **torulosum** (Piper) Hitchc.

5b Pl bien (sometimes not fl until 3rd year?), gen grayish-hairy; siliques ± quadrangular, 1–1.5 (2) mm broad, slightly or not at all torulose; s BC s, e Cas in Wn, CRG and WV, Ore, to Cal, e and common from plains into mts, to well e of RMS; rough w., prairie rocket (*Cheirinia argillosa, C. asperrima, C. elata, E. capitatum*)

                                                        6 E. asperum (Nutt.) DC.

Euclidium R. Br.   Euclidium

Racemes spikelike, ebracteate except sometimes at base; petals white, spatulate (ours ca 1 mm); silicles 2-celled and 2-seeded, nearly globose, tardily dehiscent, style stout and beaklike; ann herbs with simple, nonaur lvs. (Gr *eu*, well or good, and *kleidos*, key, referring to the indehiscent, or "locked" fr).

E. syriacum (L.) R. Br. Pl strongly pubescent with forked hairs, 1–4 dm, freely br; lvs oblong-oblanceolate, petiolate, 2–5 cm × 3–8 mm, ± dentate; racemes few-fld; fls subsessile; sepals ca 1 mm; silicles hispid, body 2–3 mm, beak stout, nearly = the body, subterete; Eurasian weed, estab in much of US, as in c Wn and Ida.

Halimolobos Tausch   Halimolobos

Racemes ebracteate; petals white (ours), often pinkish veined; siliques linear, terete to ± compressed, often torulose but readily dehiscent, valves 1-nerved, style fairly prominent (0.2–1.5 mm); seeds ∝, uniseriate to biseriate, gen mucilaginous when wet; bien or per herbs with simple or forked to stellate pubescence and simple but sometimes strongly toothed, petiolate to sessile and sometimes aur lvs. (Gr *halimos*, maritime or sea, and *lobos*, lobe, the application obscure).

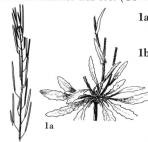

1a St lvs aur; siliques glab, 1.5–4 cm × ca 1 mm; seeds biseriate; style 0.2–0.5 mm; petals ca 4 mm; open prairies to lower mts; Yuk to Alta and Sask, s to e Ida, Wyo, Utah, and Colo; twiggy h. **1 H. virgata** (Nutt.) Schulz

1b St lvs not aur; siliques hairy (or glabrate), 1–2.2 cm × 1–1.3 mm; seeds uniseriate; style 0.3–1.5 mm; petals 3–8 mm

2a Siliques only slightly if at all torulose, stellate at maturity, somewhat compressed; lvs entire to slightly toothed; stigma scarcely lobed; style 0.3 to barely 0.5 mm; pl 2–5 dm; sagebr and des scabland, Chelan, Douglas, and Okanogan cos, Wn; Whited's h.

                                                2 H. whitedii (Piper) Rollins

2b Siliques strongly torulose, sometimes glabrate, subterete; lvs often ± lyrate to subpinnatifid; stigma often evidently lobed; style often > 0.5 mm; rock slides and gravel banks, mostly in ponderosa pine for, Ida; puzzling h.; 2 vars.   **3 H. perplexa** (Hend.) Rollins

a1 Peds 5–10 mm; styles 0.3–0.6 mm; lvs greenish, moderately stellate; Little Salmon R Canyon, Payette, Idaho, and Adams cos, Ida   var. **perplexa**

a2 Peds 10–17 mm; styles ca 1 (1.5) mm; lvs densely pubescent, grayish; Salmon R drainage, Lemhi Co, Ida   var. **lemhiensis** Hitchc.

Hesperis L.

Racemes ebracteate; sepals erect; petals white to purple, showy, clawed; siliques linear, terete to ± quadrangular, tardily dehiscent, valves 1 (3)-nerved; stigma deeply 2-lobed, lobes erect; seeds ∝, uniseriate; bien or per herbs with lanceolate to ovate, entire to lyrate lvs and simple to forked pubescence. (Name used by Theophrastus, from Gr *hesperos*, evening, when the fl fragrance is strongest).

H. **matronalis** L. Damask violet, dame's v., sweet rocket. Per, 5–13 dm; hirsute with simple and forked hairs and also pubescent with shorter, forked hairs; sts simple to sparingly br; lvs lanceolate to ovate-lanceolate, serrate-dentate, 5–20 cm, long-petiolate below to sessile above; fls fragrant, white or rose to purple; sepals hairy; petals 18–25 mm; siliques 4–10 cm, gen torulose; seeds 3–4 mm; European garden fl, sometimes escaping and persisting, gen on waste ground.

Hutchinsia R. Br.   Hutchinsia

Racemes ebracteate; petals white; silicles sessile, strongly obcompressed, dehiscent; seeds several, mucilaginous when wet; ours small glab ann. (For Ellen Hutchins, 1785–1815, Irish botanist).

H. **procumbens** (L.) Desv. Pl 3–15 cm; lvs mostly on lower part of st, 5–20 mm, from obovate and entire and with petiole = bl to lyrate-pinnatifid, reduced upward; peds slender, spreading, 3–10 mm; sepals scarcely 1 mm, ca = the cuneate-obovate petals; silicles elliptic to elliptic-obovate, 3–3.5 mm; style ca 0.2 mm; sagebr plains to coastal sand dunes; BC to Cal, e to Wyo and Colo, also in Lab and Newf; Old World.

Idahoa Nels. & Macbr.   Scalepod

Fls solitary on lfless scapes, inconspicuous; petals white; silicles suborbicular, strongly compressed, 6–12 mm, dehiscent; seeds 6–12, biseriate, strongly flattened and broadly winged, not mucilaginous when wet; glab acaulescent ann. (For state of Idaho, where pl was believed to be most common).

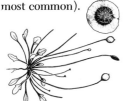

I. **scapigera** (Hook.) Nels. & Macbr. Lvs rosulate, bls ovate, entire to lyrate, 5–15 mm, petioles slender, 1–3 times as long as bls; scapes several, (2) 3–13 cm; sepals spreading, 1.5–2 mm, gen red or purplish, ca = petals; seeds ca 5 mm; foothills and valleys chiefly in sagebr; e Cas, Wn to Ida and Cal (*Platyspermum s.*).

Isatis L.

Fls in bractless racemes; petals yellow (ours); silicles strongly flattened contrary to normal position of septum, but with single median seed, a septum lacking; style lacking; stigma bilobed; ann or per herbs with simple lvs. (Gr name used by Dioscorides for this pl).

I. **tinctoria** L. Dyer's woad. Bien or per 4–12 dm, glab and glaucous; basal lvs up to 18 cm, oblanceolate to elliptic-oblanceolate, ± crenulate, ciliate and pubescent with simple hairs; st lvs sessile, aur, mostly entire, glab; infl much-compounded; petals spatulate, ca 3.5 mm; silicles oblong to oblong-oblanceolate, 12–18 × 5–7 mm; formerly grown as a source of blue dye and now a weed of dry places in much of our area; Europe.

Lepidium L.   Peppergrass; Pepperweed

Racemes ebracteate; sepals quickly deciduous to long-persistent; petals white or yellow, sometimes lacking; stamens 2, 4, or 6; silicles ovate to obovate, strongly obcompressed, sometimes wing-margined, often ± emarginate to bilobed at tip; style lacking to as much as 3 mm; seeds 2, mucilaginous when wet; cotyledons incumbent to accumbent; ann to somewhat suffrutescent per with simple (if any) hairs; lvs entire to bi- or tripinnatifid, sometimes those of st aur or sagittate. (Gr *lepis*, scale, referring to small "scalelike" silicles).

2a

1a St lvs aur to apparently (but not truly) perfoliate
  2a Petals yellow, ca 1.5 mm; silicles rhombic-ovate, ca 4 mm, slightly emarginate (sinus ca 0.2 mm deep); basal lvs bi- or tripinnatifid, upper st lvs cordate-clasping (falsely perfoliate); ann European weed, often abundant on dry, waste or overgrazed land throughout our area; clasping p.
                                   1 **L. perfoliatum** L.
  2b Petals white, 2–2.5 mm; silicles 5–6 mm; basal lvs entire to lyrate, upper st lvs aur-clasping
    3a Anthers purple; pl per, multistd; silicles not white-pustulose; known in BC as far s as Victoria, but not seen from our area
                            2 **L. heterophyllum** (DC.) Benth.
    3b Anthers cream; pl ann (occ bien?), gen single-std; silicles pustulose,

oblong-ovate, margins and tip ± winged and upturned, slightly emarginate; style 0.2–0.6 mm; common European weed in much of US, occ in our area; field p., f. cress, pepperwort   3 **L. campestre** (L.) R. Br.

1b St lvs neither aur nor perfoliate

  4a Silicles ovate-rotund, gen sparsely pilose, ca 2 mm; pl per, mostly 1–2 m, spreading by rootstocks; basal lvs up to 30 × 6–8 cm, long-petioled; st lvs reduced, becoming subsessile, entire to dentate; petals white, ca 1.5 mm; Eurasian weed, occ in Ida and Mont, perhaps elsewhere on irrigated land; pepperwort, broadlvd p.   4 **L. latifolium** L.

  4b Silicles mostly > 2 mm, often glab; pl often ann or without rootstocks; lvs often ± pinnatifid

    5a Pl per; lvs pinnatifid to pinnate; style 0.3–1 mm, exceeding sinus of silicle; petals white, 3–4 mm; silicles ovate to ovate-elliptic, 2.5–3 mm, glab; des areas of much of w US; mont p.; 2 vars. 5 **L. montanum** Nutt.

      a1 Filaments bearded; des of Canyon and Payette cos, Ida, where not collected in recent years, perhaps extinct   var. **papilliferum** Hend.

      a2 Filaments glab; se Ore across s Ida to w Wyo, s to Utah and Nev, perhaps not quite reaching our area   var **montanum**

    5b Pl ann or lvs merely toothed; style mostly < 0.3 mm or included in sinus of silicle; petals often < 3 mm

      6a Style gen evident, 0.1–0.3 mm, mostly almost = sinus of silicle; peds terete to slightly flattened; stamens 6

        7a Lvs pinnatifid; silicles 5–6 mm, margins winged and upturned, sinus ca 0.4 mm deep, > style; cotyledons lobed; erect ann 2.5–7 dm; salad herb, occ escaping in Wn and Ore; garden cress   6 **L. sativum** L.

        7b Lvs entire to toothed; silicles 2.5–4 mm, margins neither winged nor upturned, sinus not > 0.2 mm deep; cotyledons not lobed; per, 3–7 dm; Old World sp. collected near Portland, Ore, on ballast, but prob not estab in our area   7 **L. graminifolium** L.

      6b Style lacking or minute, much < sinus of silicle; peds often strongly flattened; stamens sometimes < 6

        8a Sepals persistent long past anthesis; peds slightly flattened and narrowly wing-margined, not > twice as broad as thick; silicles plainly reticulate, oval to oblong-obovate, 2.5–3.5 mm, glab or sparsely ciliate, sinus open, ca 0.4 mm deep; style lacking; spreading, pubescent ann with bipinnatifid basal lvs; petals vestigial or lacking; stamens 2; rare in our area, but known from Portland, Ore; more common in Cal; upright p. (*L. pubescens*)   8 **L. strictum** (Wats.) Rattan

        8b Sepals gen deciduous during or shortly after anthesis; peds various, if flattened silicles either not prominently reticulate or with winged, ± upturned margins and narrow sinus

          9a Peds strongly flattened, gen at least twice as broad as thick; silicles either strongly reticulate or with winged, upturned margins; basal lvs once-pinnatifid into linear segms   **Group I**

          9b Peds terete or silicles neither strongly reticulate nor with winged, upturned margins; basal lvs various, often more than once-pinnatifid   **Group II**, lead 12a

Group I

10a Silicles not strongly veined, ovate to oval, 3.5–4.5 mm, both surfaces shining, glab, margins slightly upturned, tip rounded and with narrow sinus 0.3–0.4 mm deep; stamens mostly 6 (4); petals white, 1–2 mm (vestigial); dry areas, Klickitat Co, Wn, s in w Ore to much of Cal; shining p.   9 **L. nitidum** Nutt.

10b Silicles strongly reticulate-veined, glab to pubescent, not shining, sometimes < 3.5 mm, tip sometimes with divergent lobes

  11a Peds gen > silicles, ca twice as broad as thick; silicles ovate, 2.5–3.5 mm, deeply emarginate, the 2 lobes scarcely 0.5 mm, divergent; stamens gen 4; mostly in Cal, on saline soil, but collected at least once in Victoria, BC; sharpfruited p.   10 **L. oxycarpum** T. & G.

  11b Peds gen < silicles, > twice as broad as thick; silicles ovate to oblong-ovate, 3.25–4.5 mm, tip winged and deeply emarginate, but sinus narrow, 0.3–1 (2) mm deep, the 2 lobes parallel to ± divergent; stamens 4 or 6; gen on ± saline soil in lowl; e Cas, Wn to Cal, e to Ida and Utah; veiny p.; 2 vars.   11 **L. dictyotum** Gray

12a

14a

14b

c1

c2

bl

a1   13b

a1 Apical lobes of silicle < 1 mm, rounded to acute, mostly not divergent; sc Wn to Cal, e to Ida and Utah   var. **dictyotum**
a2 Apical lobes of silicle > 1 mm, gen ± divergent, acuminate; with var. *dictyotum* in Wn and Ore, but much less common, not in Ida or Utah   var. **acutidens** Gray

Group II

12a Silicles oblong-obovate to obovate, widest above middle, 2–3.5 mm; petals mostly lacking or vestigial; cotyledons incumbent or very slightly oblique; widespread, semiweedy sp. of dry areas in much of US and Can; Eurasia; prairie p., common p.; 4 vars.   12 **L. densiflorum** Schrad.
  a1 Silicles ave ca 2.5 mm, broadest near or just above middle; peds only slightly flattened; st lvs mostly toothed; common weed from RMS e, occ in most w states (*L. neglectum*)   var. **densiflorum**
  a2 Silicles ave ca 3 mm, broadest somewhat above middle; peds rather strongly flattened; st lvs commonly entire or subentire; apparently native in much of our area, but sometimes weedy
    b1 Silicles glab; mostly e of coastal mts, Alas to Cal and Ariz, e to Mont and Colo   var. **macrocarpum** Mulligan
    b2 Silicles pubescent at least on margins
      c1 Silicles pubescent on both surfaces; c Mont to Utah, w to Cal, occ in Wn, Ore, and Ida (*L. p.*)   var. **pubicarpum** (Nels.) Thell.
      c2 Silicles ciliate, but otherwise glab; e of Cas, BC to n Ore, e to Ida (*L. e., L. simile*)   var. **elongatum** (Rydb.) Thell.
12b Silicles elliptic to oval, widest at middle or below; petals sometimes 1–3 mm; cotyledons sometimes oblique or accumbent
  13a Silicles ± elliptic, distinctly longer than broad; cotyledons incumbent; petals lacking or vestigial
    14a Pl foetid; silicles 2–3 mm; racemes terminal, elongate, not greatly compounded; European weed, rather common in e US, but very occ w of RM; known in our area at least from near Portland, Ore; narrowlvd p.   13 **L. ruderale** L.
    14b Pl not foetid; silicles 2.5–3.5 mm; racemes much compounded, axillary ones often much < terminal ones; dry grassl to woodl, often somewhat weedy; Alta to Man, s through Mont and e Ida to NM, occ intro farther w; branched p. (*L. bourgeauanum, L. divergens*)
                        14 **L. ramosissimum** Nels.
  13b Silicles elliptic-rotund to nearly orbicular, almost or fully as broad as long; cotyledons accumbent to oblique (ours); petals gen evident, 1–3 mm, widely distributed in N Am and weedy in much of rest of world; tall p.; 4 vars.   15 **L. virginicum** L.
    a1 Cotyledons accumbent; silicles mostly longer than broad; common in e N Am, but only occ in our area   var. **virginicum**
    a2 Cotyledons oblique to incumbent
      b1 Peds and upper part of st glab; cotyledons oblique to nearly incumbent; e Wn to Cal, e to Ida, Wyo, and Ariz to NM, Tex, and Okla (*L. idahoense, L. intermedium*)   var. **medium** (Greene) Hitchc.
      b2 Peds and upper part of st puberulent to hirsute; cotyledons oblique
        c1 Basal lvs pinnate, pubescent with crisped hairs, pinnae deeply lobed; ann or bien; silicles sometimes ciliate; along coast from PS region, Wn, to BC (*L. m.*)   var. **menziesii** (DC.) Hitchc.
        c2 Basal lvs mostly only toothed; gen ann; pubescence mostly straight; silicles glab; common, Cal to Colo and NM, occ in Wn, Ore, Ida, and Mont, s (*L. hirsutum, L. occidentale*)
                        var. **pubescens** (Greene) Hitchc.

Lesquerella Wats.   Bladderpod

Racemes ebracteate, sometimes compound; peds mostly slender, straight to S-curved; petals yellow to reddish- or purplish-tinged; silicles subrotund to ovate, elliptic, or oblong in outline, inflated but from ± compressed to distinctly obcompressed, acute to slightly retuse; style persistent, gen prominent; seeds 2–10 per cell, biseriate, not mucilaginous when wet, ± flattened and sometimes slightly winged; ann to per (ours) herbs with dense, sessile, rather closely appressed, stellate pubescence and simple, entire to toothed, nonaur lvs. (For Leo Lesquereux, 1805-1889, American bryologist). (*Vesicaria*).

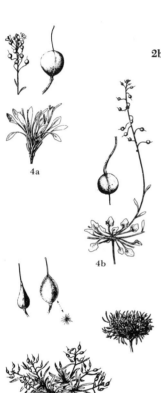

1a Silicles obcompressed, not inflated, 5–8 mm, oblong-elliptic, acute at each end, margins keeled; style 2-3 mm; peds 5–12 mm, spreading and sigmoid in fr; basal lvs 1.5–3.5 cm, bls rhombic to oblong-obovate, 0.5–1.5 cm, entire; petals 8–10 mm, yellow; from sagebr des to above timberl; Lemhi and Lost R ranges, Ida, and Granite Co, Mont; keeled b.   1 **L. carinata** Rollins
1b Silicles compressed or inflated, often semirotund in outline; pl various
  2a Peds gen recurved or arched-recurved in fr, not sigmoid; basal lvs ∝ , mostly narrowly to broadly oblanceolate, 2–10 cm, bl tapered gradually to petiole; silicles subglobose to oblong-rotund, 3–4 mm; styles 2.5–5 mm; petals yellow but often reddish- or purplish-tinged, 6–9 mm; plains region to e edge of RM; Louisiana b., silvery b.; 2 vars.
                          2 **L. ludoviciana** (Nutt.) Wats.
    a1 Petals mostly reddish- or purplish-tinged; sts slender and mostly decumbent, 5–15 cm; racemes often secund in fr; Alta to Man, s to e Mont and Wyo and the Daks (*L. macounii, L. rosea*)
                          var. **arenosa** (Richards.) Wats.
    a2 Petals yellow; sts ± erect, mostly 15–40 cm; racemes not secund; e Mont to Minn and Ill, s to Utah, Colo, and Neb (*L. arenosa, L. argentea*)   var. **ludoviciana**
  2b Peds gen spreading to erect, often sigmoid; basal lvs often broader than oblanceolate or bl narrowed abruptly to petiole; silicles and styles various
    3a Silicles globose to subglobose, not flattened along margins or near tip, rarely > 4 mm
      4a Basal lvs tapered rather gradually to petiole, 3–12 cm, bl oblanceolate to obovate; sts gen ∝ , erect, ave 2–4 dm; style and silicle each ca 3–4 mm; common in sagebr des, esp near or in juniper or ponderosa pine woodl; BC to n Ore, e to Nez Perce Co, Ida; Columbia b.
                          3 **L. douglasii** Wats.
      4b Basal lvs tapered abruptly to petiole, mostly not > 3 cm, bl commonly ovate to obovate; sts gen < 2 dm; style mostly slightly to much < fr; silicle barely 4 mm; fairly common sp. in des ranges, Utah to Cal; ours the var. *sherwoodii* (Peck) Hitchc., known only from Wallowa Mts, ne Ore; King's b.   4 **L. kingii** Wats.
    3b Silicles mostly ovate to elliptic-oblong, gen ± flattened along margins or at tip, often > 4 mm
      5a Basal lvs rarely as much as 4 mm broad, linear to narrowly oblanceolate; sts seldom > 10 cm; silicles ovate, (3) 4–5.5 mm, ± compressed toward tip; petals 5–7 mm; dry benchl to mt ridges, mostly in or e of RM; Alta to Sask, s to c and e Mont, Daks, Colo, and se Ida; alpine b.; 3 vars.   5 **L. alpina** (Nutt.) Wats.
        a1 Pl lower mont to alp; sts (including racemes) up to 16 cm, often ± prostrate; ovules 2–4 (6) per cell; s Alta and Sask to c Custer Co, Ida, n Colo, w Neb, and Daks   var. **alpina**
        a2 Pl of foothills, greatly dwarfed; fls and frs scarcely surpassing the basal lf clusters; ovules gen 2 per cell
          b1 Stellae tightly appressed; Meagher Co, Mont, to Clark and c Custer Co, Ida, s to se Wyo   var. **laevis** (Pays.) Hitchc.
          b2 Stellae not tightly appressed, rays ± spreading; s Madison and Beaverhead cos, Mont, to w Wyo (*L. c., L. parvula*)
                          var. **condensata** (Nels.) Hitchc.
      5b Basal lvs in part > 4 mm broad, bl ovate to oblanceolate, obovate, or ± orbicular; sts 5–20 cm, prostrate to erect; silicles ovate-oblong to elliptic-obovate, 4–6 mm, slightly inflated but compressed and ± flattened along upper margins and below tip; petals 7–10 mm; sagebr valleys to alp slopes; ne Ore to Cal, c Ida, and Utah; western b.; 3 vars.   6 **L. occidentalis** Wats.
        a1 Pl bien; crown gen simple, without persistent old lvs; middle elev in mts of ne Ore, to Nev and Utah (*L. c.*)
                          var. **cusickii** (Jones) Hitchc.
      a2 Pl per; crown often br, gen with ∝ persistent old lvs
          b1 Basal lvs gen toothed to lobed, mostly grayish; sts often ascending to erect, mostly at least 10 cm; ne Ore to Cal
                          var. **occidentalis**
          b2 Basal lvs gen entire, often greenish; sts mostly prostrate, often < 10 cm; se Wn and w Ida to Nev (*L. d.*)
                          var. **diversifolia** (Greene) Hitchc.

Lobularia Desv.

Racemes ebracteate; petals white or bluish-tinged (ours); silicles oval-elliptic, strongly compressed; seeds 1 per cell; ann with entire lvs. (L *lobulus*, little lobe, in reference to the small silicles).

L. maritima (L.) Desv. Sweet alyssum, sweet alison. Pl 1–3 dm, br from base, grayish with appressed, 2-br hairs; lvs linear-oblanceolate, 1–4 cm × 1–4 mm; peds spreading, slender, 5–10 mm; fls fragrant; sepals 1.5–2 mm, quickly deciduous; petals obovate, 3–4 mm; silicles gen purplish, 2.5–3.5 mm, tardily dehiscent, sparsely hairy, the valves prominently 1-nerved; style ca 1 mm; Eurasian orn, often escaping and persistent in waste places.

Lunaria L.

Racemes ebracteate or partially bracteate; petals showy, ca 2 cm; silicles slenderly stipitate, oblong-oval, strongly compressed; seeds flattened and ± wing-margined, 3–5 per cell; ann to per, ± pubescent herbs. (L *luna*, moon, referring to the large, roundish silicles).

L. annua L. Honesty. Sparingly pubescent ann or bien 5–10 dm, freely br; lvs mostly petiolate except on upper st, bl cordate-ovate, deeply dentate, 4–10 cm; outer sepals saccate at base, 8–10 mm; petals bluish-purple, ca 2 cm; silicles 3.5–4.5 × 2–4 cm; stipe 7–12 mm; style 6–8 mm; seeds 7–9 mm broad; European, cult and sometimes escaped from gardens, esp in PS area, Wn.

Malcolmia (L.) R. Br.   Malcolmia

Racemes ± lfy-bracteate; petals pinkish; siliques linear, subterete, readily dehiscent; style tapered to an acute tip, stigmatic full length; seeds uniseriate; ann with simple lvs. (For William Malcolm, 1778–1805, British horticulturist).

M. africana R. Br. Pl 1.5–4 dm, strongly pubescent with small, freely br hairs; lvs simple, petiolate, bls elliptic to oblanceolate, 3–6 × 0.5–2 cm, remotely but deeply dentate, reduced upward; sepals 4–5 mm, erect, persistent until siliques well developed; petals 6–8 mm; siliques subsessile, 4–6 cm × ca 1 mm, terete-quadrangular, stellate-hairy, slightly torulose; style 1–1.5 mm; African weed, well estab in GB area, and now in Ida.

Neslia Desv.

Racemes ebracteate; petals yellow; silicles subrotund; styles slender; seeds gen 1 per cell, not mucilaginous when wet; entire-lvd ann. (For J. A. N. de Nesle, French botanist).

N. paniculata (L.) Desv. Ball mustard. Pl erect, 3–9 dm, pubescent with br and stellate hairs; lvs entire, basal ones oblanceolate, 3–7 cm, petiolate, becoming sessile, aur, and narrowly oblong-lanceolate upward on st; racemes much-elongate; peds very slender, 6–10 mm; sepals 1.5–2 mm; petals ca 2 mm; silicles ca 2 mm and as broad, valves much-hardened and strongly reticulate-alveolate; style slender, 0.5–0.8 mm; European weed, sparingly intro in BC and Wn.

Phoenicaulis Nutt.   Daggerpod

Racemes ebracteate, ∝-fld; sepals ± gibbous-based; petals showy; siliques strongly compressed, oblong-lanceolate to linear; seeds 3–6 per cell, uniseriate, not wing-margined; caespitose per with bracteate fl sts; lvs mostly in basal rosette. (Gr *phainos*. to shine, and *kaulos*, stem, referring to the glab, ± shining fl sts). *(Parrya)*.

P. cheiranthoides Nutt. Lvs rosulate, grayish with 4-rayed to dendritic hairs, bls oblanceolate, 3–15 × 0.5–1.5 cm, entire; fl sts 5–20 cm, mostly glab; bracts ± aur, 5–20 mm; petals 11–15 mm, pink to reddish-purple; siliques 2–8 cm × 2–6 mm, often ± falcate, valves strongly 1-nerved, fully dehiscent; style ca 1 mm; seeds 3–4 mm; sagebr des and ponderosa pine for; c Wn to Cal, e to Ida and Nev *(P. menziesii)*.

Physaria (Nutt.) Gray   Twinpod; Double Bladderpod

Racemes ebracteate; sepals mostly erect, not saccate; petals yellow or sometimes purplish or aging to purple, 5–14 mm, mostly spatulate; silicles gen with deep apical sinus and often cordate base, inflated but sometimes slightly to strongly obcompressed; style prominent, slender, persistent; seeds 1–4 per cell, biseriate, not mucilaginous when wet; stellate, gen several-std per herbs with simple, rosulate basal lvs and nonaur st lvs. (Gr *physa*, bellows, referring to inflated silicles).

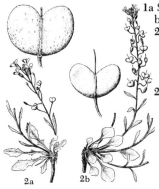

1a Silicles strongly obcompressed, only slightly inflated, obtuse to truncate at base; styles 1–7 mm

  2a Styles 1–2 mm; silicles (8) 10–20 mm; basal lvs 2–5 cm, oblanceolate to obovate, petioles with toothed to lyrate margins; petals light yellow, 8–12 mm; silicle septum elliptic-lanceolate, acute at each end, 2–3 mm broad; dry places in SR and Salmon R canyons, Ida and Ore, and in Blue Mts, Wn and Ore; Ore t.   **1 P. oregana** Wats.

  2b Styles (3) 4–7 mm; silicles 5–7 mm; basal lvs 3–6 cm, ovate to obovate, petioles mostly with entire (toothed) margins; petals yellow, aging to rose, 5–11 mm; silicle septum lanceolate, 1.5–3 mm broad; hillsides and gravelly streambanks; e Wn to Ida and Mont; Geyer's t.; 2 vars.

                                  2 P. geyeri (Hook.) Gray

    a1 Petals purplish in age; ovules 2–3 per cell; Custer and Lemhi cos, Ida

                                    var. **purpurea** Rollins

    a2 Petals yellow in age; ovules seldom > 2 per cell; e Wn to c Mont

                                        var. **geyeri**

2a   2b

1b Silicles not obcompressed, strongly inflated, mostly shallowly cordate at base; styles 5–9 mm

  3a Silicle septum narrowly lanceolate, 8–14 × 2–2.5 mm; silicles (10) 14–18 mm, apical sinus open, 2–3 mm deep; petals yellow, 8–14 mm; basal lvs 4–8 cm, bl obovate to oblanceolate, tapered ± gradually to petiole; mont to subalp rocky ridges and talus slopes; e side Cas, Chelan Co to Mt Adams, Wn; alpine t.   **3 P. alpestris** Suksd.

  3b Silicle septum obovate to oblanceolate, ± obtuse at tip, 3–6 × 2–3 mm; silicles 10–20 mm, apical sinus very narrow, 2–4 mm deep; petals yellow, 9–12 mm; basal lvs 2–8 cm, bl obovate to broadly oblanceolate or ovate, narrowed somewhat abruptly to petiole; sagebr slopes to wooded hillsides; e Wn to Alta and Wyo; common t.; 2 vars.

                            4 P. didymocarpa (Hook.) Gray

    a1 Silicle valves (12) 15–20 mm; lvs mostly with broad, toothed petioles; Salmon R and tributaries near Salmon City, Ida   var. **lyrata** Hitchc.

    a2 Silicle valves 8–12 (15) mm; lvs entire to toothed; ne Wn to e side RM in Mont   var. **didymocarpa**

3a, a2, a1, 3b

Polyctenium Greene   Combleaf

Racemes ebracteate, congested; petals white or purplish; siliques linear, ± obcompressed, fully dehiscent; seeds several, uniseriate, not mucilaginous when wet; per herb with rigid, sessile or subsessile lvs, pubescent with small br hairs. (Gr *poly*, many, and *ctenos*, comb, referring to the deeply pinnatifid lvs).

P. fremontii (Wats.) Greene. Desert c., d. smelowskia. Pl grayish-glaucous, hairy almost throughout; sts 5–20 cm, gen several; lvs 1.5–3 cm, pinnatifid into linear, prominently veined, ± acicular segms scarcely 1 mm wide; peds 2–5 mm; petals 4–6 mm, cuneate-spatulate; silicles ascending to erect, 5–12 × 1.4–1.8 mm; seeds ca 1 mm; style ca 0.5 mm; sagebr des and pinyon and ponderosa pine woodl; c Ore and c Ida *(Smelowskia f.)*.

Raphanus L.

Racemes ebracteate, often compound; sepals erect, often hooded at tip and ± saccate at base; petals clawed, yellow to purple (white); siliques linear to lance-ovoid, terete, 2-segmented, lower segm stipelike, sterile, very short, upper segm 1-celled, indehiscent but eventually breaking crosswise between seeds, tapered to beaklike tip; seeds 2–15, uniseriate, not margined and not mucilaginous when wet; mostly hirsute-hispid ann or bien herbs (ours) with lyrate-pinnatifid, nonaur lvs. (Gr *raphanos*, quick to appear, because of rapid seed germination).

1a Siliques (2) 4–12-seeded, 4.5–6 cm × 3–6 mm; strongly constricted between seeds, deeply grooved lengthwise, basal sterile segm 1–1.5 mm; beak 1–2 cm; petals mostly yellow (white), often purplish-veined or -tinged, 15–20 mm; pl 3–8 dm; occ Eurasian weed in Wn and Ore, gen on waste ground; jointed charlock   1 **R. raphanistrum** L.

1b Siliques 1–3 (5)-seeded, 3–6 cm × 5–10 mm, slightly if at all constricted between seeds, lightly grooved, basal sterile segm 0.5–1.5 mm; sometimes obsolete, beak conical, 1/4–1/2 total length of silique; petals gen purplish, but sometimes white or yellow with purplish venation, 15–20 mm; pl 4–12 dm; European; common weed in waste places in much of our area; wild radish   2 **R. sativus** L.

Rorippa Scop.   Cress; Yellowcress

Racemes ebracteate; sepals deciduous to persistent long after anthesis, scarcely saccate at base; petals white to yellow, sometimes purplish tinged, rather small (1–6 mm); siliques terete to slightly compressed, oval to linear, valves 1-nerved, fully dehiscent; style tiny to fairly prominent; seeds few −∝, biseriate to partially uniseriate, seldom much-flattened, not mucilaginous when wet; ann to per, glab to hirsute herbs with simple to pinnate lvs, gen in moist to marshy places. (Name prob derived from *rorippen*, Saxon name for this pl). *(Radicula)*.

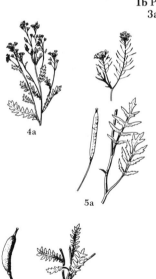

1a Petals white (purple-tinged), 3–6 mm
   2a Pl aquatic; lvs pinnate, 4–12 cm; siliques mostly ca 15 mm; European sp., often used as salad herb, widely distributed in quiet streams and marshy areas; water-cress *(Sisymbrium n.)*
                                    1 **R. nasturtium-aquaticum** (L.) Schinz & Thell.
   2b Pl terrestrial; lvs simple, 15–30 cm; widely cult for peppery root, occ persistent; horseradish *(Armoracia rusticana, Cochlearia a.)*
                                              R. armoracia (L.) Hitchc.
1b Petals yellow, often < 3 mm
   3a Pl per, slenderly rhizomatous; petals 3–5 mm; styles 1–2.5 mm
      4a Siliques ovate to oblong in outline, ± compressed, gen slightly arcuate, 3–7 × 2–2.5 mm, soft-pubescent; style 1–2 mm; pl finely pubescent or papillose throughout, sepals long-persistent; petals ca 4 mm; lvs 3–7 cm, sinuate to pinnatifid; widespread in w N Am, on moist, gen sandy soil; persistentsepal y.; 2 vars.
                                        2 **R. calycina** (Engelm.) Rydb.
         a1 Siliques ovate, 2–5 mm; lvs coarsely toothed or shallowly lobed; mainly Mont   var. **calycina**
         a2 Siliques more nearly oblong, 4–7 mm; lvs ± pinnatifld or lyrate-pinnatifld; Wn to Cal, occ e to Mont, Neb, and NM *(R. c.)*
                                        var. **columbiae** (Suksd.) Rollins
      4b Siliques linear, subterete, 5–15 × ca 1.5 mm, glab; pl mostly glab or glabrate, at least above; sepals gen deciduous shortly after anthesis; petals 3–5 mm; lvs various
         5a Lvs deeply lobed to pinnatifid, 4–10 cm, divisions narrow, sharply toothed (to entire); style rarely > 1 mm; petals 3–4 mm; European, rare in our area, OP, Wn, and WV, Ore; creeping y.
                                        3 **R. sylvestris** (L.) Besser
         5b Lvs mostly deeply lobed, 3–8 cm, divisions entire to shallowly toothed; style (1) 1.5–2.5 mm; petals 4–5 mm; moist woods and wet lowl, often where alkaline; Wn to Cal, e of Cas, e to Sask, Ill, and Tex; spreading y.   4 **R. sinuata** (Nutt.) Hitchc.
   3b Pl ann or bien, nonrhizomatous; petals 1–2 mm; styles 0.3–1 mm
      6a Peds (3) 4–12 mm, mostly at least = silique; siliques ovate to oblong, 3–8 (12) × 2–3 mm; sts mostly erect, 3–10 dm; petals (1) 2 mm; circumboreal, common in much of N Am; marsh y., hispid y.; 4 vars.
                                        5 **R. islandica** (Oed.) Borbás
         a1 St lvs pinnatifid, tender in texture, glab; siliques ave 3–4 mm; circumpolar, mostly n of our area, but occ, perhaps where intro *(R. palustris)*   var. **islandica**
         a2 St lvs mostly merely cleft to shallowly lobed, firm in texture, sometimes pilose; siliques various
            b1 Siliques 6–12 mm, often 3–4-carpellary; pl glab; Alas to w Wn, occ e to Mont *(R. clavata, R. pacifica)*
                                        var. **occidentale** (Wats.) Butters & Abbe

b2 Siliques 3–6 mm, rarely > 2–carpellary; pl sometimes hirsute
    c1 Pl hirsute; occ in much of N Am, but not common *(Radicula h.)*
                          var. **hispida** (Desv.) Butters & Abbe
    c2 Pl glab or subglab; common in most of US and s Can
                          var. **glabrata** (Lun.) Butters & Abbe
6b Peds mostly (1) 2–4 (7) mm and much < silique; siliques linear to oval, 2–15 mm; sts commonly ± spreading to decumbent, rarely > 4 dm
7a Siliques oval to oblong-lanceolate, 2–6 (8) × 2–2.5 mm, seldom arcuate; moist sandy soil; BC to Cal, e to Mich, Mo, and Tex; bluntlvd y. *(R. sphaerocarpa)*   **6 R. obtusa** (Nutt.) Britt.
7b Siliques linear, 6–15 × 1–1.5 mm, gen arcuate; moist soil, lowl to mont; BC to Baja Cal, e to Mont and Colo; western y.; 2 vars.
                       7 R. curvisiliqua (Hook.) Bessey
    a1 Sepals mostly < 1.5 mm, often long-persistent; petals mostly 1–1.5 mm; pl low, diffusely br, rarely > 15 cm; lvs mostly lyrate; with var. *curvisiliqua* in most of range *(R. l.)*
                       var. **lyrata** (Nutt.) Peck
    a2 Sepals mostly 1.5–2 mm, quickly deciduous; petals 1–2 mm; pl ± erect, 1–5 dm; lvs seldom lyrate; range of sp. var. **curvisiliqua**

Schoenocrambe Greene   Plainsmustard

Racemes ebracteate; petals yellow; siliques linear, terete, slightly torulose, valves indistinctly nerved, fully dehiscent; style almost wanting; seeds ∝, uniseriate, not mucilaginous when wet; rhizomatous per herbs with linear lvs. (Gr *schoenus*, a rush, and *Crambe*, another genus of this family).

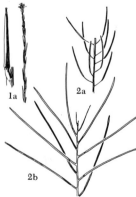

S. linifolia (Nutt.) Greene. Flaxlvd p. Pl glab and ± glaucous; sts several, slender, mostly 2–7 dm; lvs 2–9 cm × 1–4 (6) mm, entire to remotely toothed or pinnatifid, lower ones gen deciduous by anthesis; pedicels slender, 5–10 mm; sepals 4–5 mm, yellowish, ± saccate at base; petals 6–10 mm; siliques ± erect, 2.5–5 cm × ca 1 mm; seeds ca 1.5 mm; sagebr plains and lower des mts; BC to Nev, e of Cas, e to Mont and NM *(S. decumbens, S. pinnata, S. pygmaea)*.

Sisymbrium L.   Tumblemustard

Racemes ebracteate; sepals ± spreading; petals clawed, pale yellow; siliques terete, linear to ± subulate, straight to arcuate, valves mostly 3-nerved, fully dehiscent; style short or none; seeds ∝, uniseriate (ours), not mucilaginous when wet; ann or bien, ± hirsute herbs with petiolate to sessile, nonaur, ± pinnatifid to deeply lobed lvs. (L form of Gr name for some pl in mustard family).

1a Siliques closely and tightly appressed, 8–15 mm, tapered from base to short beaklike stylar tip; petals 3–4 mm; seeds ca 1.3 mm; pl strongly hirsute-hispid, 3–8 dm; European weed, esp common w of Cas in our area; hedge mustard *(Erysimum o.)*   **1 S. officinale** (L.) Scop.
1b Siliques spreading to erect, but not closely appressed, much > 15 mm; petals 5–8 mm; seeds 0.7–1 mm; pl mostly glab above, 3–15 dm
    2a Ped slender, 7–20 mm, much thinner than silique; siliques ascending to erect, 2–3.5 cm; Eurasian, well estab in much of US; common in Mont, also in Ida, Wn, and Ore; Loesel t.   **2 S. loeselii** L.
    2b Ped stout, 4–10 mm, nearly as thick as silique; siliques rigidly spreading, 5–10 cm; European weed, widespread, mostly in waste places; Jim Hill mustard, t. *(Norta a.)*   **3 S. altissimum** L.

Smelowskia C. A. Mey.   Smelowskia

Racemes ebracteate except the lowest 1–2 fls; sepals ± persistent; petals spatulate, white or cream to purplish-tinged, 4–8 mm; siliques ovate to linear, subterete to compressed or obcompressed, valves prominently 1-nerved or keeled, fully dehiscent, glab to pubescent; style rather prominent (0.3–1 mm); seeds several, uniseriate, not mucilaginous when wet; cushion-forming per with persistent old lvs, gen canescent with br and unbr hairs; lvs in ours pinnate to pinnatifid. (For Timotheus Smelowsky, 1770–1815, Russian botanist).

1a Siliques ovate to narrowly oblong, 3–6 × 2–3 mm, septum obtuse at tip; sepals persistent long after anthesis, 2.5–3.5 mm; petals 4–5 mm; basal lvs not stiffly long-ciliate on petiole, 2.5–7 cm; subalp to alp moraines, ridges, and rock crevices; Cas from Chelan Co, Wn, to Cal; shortfr s.
1 **S. ovalis** Jones

1b Siliques linear-oblong, 5–11 × 1.5–2.5 mm, septum acute at tip; sepals deciduous shortly after anthesis, 2–3 mm; petals 4–8 mm; basal lvs stiffly ciliate on petiole, 1–10 cm; widespread in subalp to alp or arctic regions of e Asia and w N Am; ours, from BC to Alta, s to Cas and OM, Wn, and in RM to Colo, is var. *americana* (Regel & Herd.) Drury & Roll.; alpine s., Siberian s.
2 **S. calycina** (Steph.) C. A. Mey.

Stanleya Nutt.   Stanleya; Princesplume

Racemes ebracteate; buds elongate and cylindric; sepals linear, spreading to reflexed; petals yellowish or greenish-yellow (ours); anthers often coiling after dehiscence; siliques long-stipitate, linear, terete to compressed, valves 1-nerved, readily dehiscent; style absent or as much as 2 mm; seeds several to ∝, uniseriate, not mucilaginous when wet; ann to per, glab to pilose-lanate herbs with simple to pinnate lvs. (For Lord Edward Stanley, 1775–1851, British ornithologist).

1a Pl long-hairy on lower part of st; basal lvs runcinate, long-hairy, 7–15 cm; st lvs few, much reduced upward, petiolate; sepals greenish-yellow, 12–16 mm; petals narrow, 13–20 mm; siliques suberect, nearly or quite straight, body 4–7 cm × ca 2 mm, compressed; stipe 1–2 cm; style minute; seeds flattened, ca 2 mm; sagebr des; c Ida to Wyo; woolly s. (*S. runcinata*)
1 **S. tomentosa** Parry

1b Pl semiglab to glab; lvs various, but scarcely runcinate; sepals, petals, and siliques various

2a St lvs petiolate, often pinnate (entire); basal lvs 5–15 cm, lobed to bipinnatifid; sepals yellow, 10–16 mm, glab; petals 10–16 mm, villous-tomentose on inner surface of brownish claw; siliques subterete, body 3–7 cm, ± arcuate, spreading to ascending; stipe 1–2 cm; style lacking; widespread, from plains to lower mts; se Ore to s Cal, e to Daks, Kan, and Tex; bushy s.
2 **S. pinnata** (Purish) Britt.

2b St lvs sessile, aur, mostly entire; petals 12–25 mm, glab; style 0.5–2 mm

3a Pl per, 3–12 dm; basal lvs with evident petioles; sepals greenish-white, 12–20 mm; petals 15–20 mm; siliques arcuate-spreading, subterete, body 4–8 cm, stipe 15–25 mm; plains to lower mts; c Ore to c Mont, s to Wyo, Utah, and Nev; per s.
3 **S. viridiflora** Nutt.

3b Pl ann or bien, 2–8 dm; basal lvs nearly or quite sessile; sepals pale yellow, 6–12 mm; petals 12–25 mm; siliques erect, nearly straight, terete, body 2–5 cm, stipe 7–20 mm; plains and low sand or clay hills; Baker Co to sw Ore; bien s.
4 **S. confertiflora** (Robins.) Howell

Streptanthella Rydb.   Streptanthella

Racemes open, ebracteate; petals white, purplish-veined; anthers not coiling after dehiscence; siliques reflexed, strongly compressed, valves 1-nerved, extending into a beaklike, indehiscent tip; seeds several, uniseriate, not mucilaginous when wet; glab ann with sinuate-dentate to entire, nonaur lvs. (Diminutive of *Streptanthus*, from Gr *streptos*, twisted, and *anthos*, fl, referring to the crispate petals).

S. longirostris (Wats.) Rydb. Beaked s. Pl ± glaucous, 1.5–5 dm, gen freely br; basal and lower st lvs gen deciduous by anthesis, narrowly oblanceolate, 2–6 cm, sinuate-dentate; upper st lvs linear; racemes open and elongate; peds 1–3 mm; sepals 2–3 mm, ± spreading, slightly saccate at base; petals narrowly spatulate, 3–3.5 mm; siliques 3.5–4.5 cm × ca 1.5 mm, tip indehiscent, ca 3.5 mm, the upper limit of the valves not marked in any way; stigma minute; seeds flattened and wing-margined; sagebr des, se Wn to s Cal, e to Wyo and NM (*Streptanthus l.*).

Streptanthus Nutt.   Streptanthus

Racemes open, ebracteate; calyx cylindric-urceolate; sepals often anthocyanic, at least distally, 2 (sometimes all 4) gen saccate at base; petals white or pale yellow to pinkish or purple, rather narrow, gen channeled, the bl gen crisped; stamens 6; silique linear, distinctly compressed, non-stipitate, the valves (ours) 1-nerved; seeds flattened, gen ± winged, uniseriate; ann or per (ours), glab or with unbr hairs. (Gr *streptos*, twisted, and *anthos*, fl, in reference to the crisped petals).

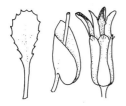

S. cordatus Nutt. Heart-lvd s. Glaucous, short-lived per 3–8 dm, glab except gen with few short stiff hairs on basal lvs and at tip of sepals; basal lvs broadly spatulate-obovate, slenderly petiolate, bl dentate, up to 3 cm; cauline lvs several, broadly oblong, sessile and cordate-clasping at base, rounded at tip, up to 5 cm, gen entire; sepals 5–10 mm; petals ca 1/2 > sepals, reddish-purple, the bl ± recurved; silique spreading to ascending, 4–8 cm × 3–4 mm; seeds slightly wing-margined, ca 2.5 mm; gen on ± sterile, rocky, midmont slopes; se Ore to s Cal, e to RM, Wyo to Colo; in our area in Grant Co, Ore.

Subularia L.   Awlwort

Racemes ebracteate; fls tiny, petals white; silicles slightly obcompressed, fully dehiscent; style lacking; seeds several, ± biseriate, mucilaginous when wet; small, glab, scapose aquatic ann. (L *subula*, awl, referring to the awl-shaped lvs).

S. aquatica L. Lvs basal, linear, subterete, subulate, 1–5 cm; scapes 2–10 cm; racemes loosely 2–8-fld; sepals and petals ca 1 mm; peds 2–5 mm; silicles oval to elliptic in outline, ca 2.5 mm; pl of fresh water ponds and lakes, gen growing submerged; Eurasia and much of N Am; Alas to Cal, on both sides Cas, e to RM and s to Wyo, also farther e to Atl coast.

Teesdalia R. Br.

Racemes ebracteate; petals white; silicles oblong-obovate, strongly obcompressed, valves keeled and ± winged above, fully dehiscent; style lacking; scapose ann herbs with mostly basal, entire to lyrate-pinnatifid lvs. (For Robert Teesdale, English botanist).

T. nudicaulis (L.) R. Br. Shepherd's cress. Pl glab or very sparsely hirsute; scapes 0.5–2.5 dm, lfless or with 1 or more lvs near base; lvs 1.5–5 cm, slender-petiolate, bl oval to obovate or oblanceolate, entire to lyrate-pinnatifid; sepals ca 0.5 mm; petals ca 1 mm; filaments with expanded scalelike base; silicles ovoid, ca 2.5 mm; European, intro in w Wn and Ore, gen in sandy or gravelly soil.

Thelypodium Endl.   Thelypody

Racemes ebracteate; sepals erect to spreading, greenish to white or purplish; petals white to purplish; anthers sagittate-based, mostly coiling after dehiscence; siliques subsessile to long-stipitate, terete or ± compressed, straight to arcuate, often torulose, valves 1-nerved, fully dehiscent; style short; seeds ∝, uniseriate, flattened but not wing-margined, not mucilaginous when wet; ann to per, glab to sparsely hirsute, mostly glaucous herbs with entire to lyrate-pinnatifid basal lvs and petiolate to sessile and aur st lvs. (Gr *thelys*, female, and *podoin*, little foot, referring to the stipitate ovary of many spp.).

1a St lvs mostly sessile and aur
  2a Pl per, taprooted, caudex with dried petioles of previous years; low alkaline flats from Burns, Harney Co, Ore, s and e, prob not quite reaching our area   **T. flexuosum** Robins.
  2b Pl bien, or per with creeping roots, caudex without withered remains of old lvs
    3a Racemes elongate and open at anthesis; basal lvs lyrately lobate, 2–4 cm, often ± hirsute; sepals erect, 4–9 mm; petals greenish-white to purple, narrowly spatulate, 8–20 mm, bl crispate; siliques ascending to erect, 2–4 cm × ca 1 mm, stipe ca 0.5 mm; style 0.5–1.5 mm; moist

3a

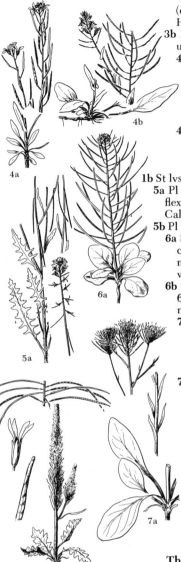

(often alkaline) plains or river valleys; ne Ore to Cal, possibly into Wn; Howell's t.   1 **T. howellii** Wats.

3b Racemes ± corymbiform at anthesis; basal lvs entire or merely sinuate to dentate, gen glab; fls and siliques various

    4a Siliques straight, ascending, sessile or with stipe scarcely 0.5 mm, body (2) 2.5–4 (6) cm × 1–2 mm, torulose; pl bien, without creeping roots; sepals 5–8 mm; petals light to deep violet-purple, 10–16 mm; moist (often alkaline) meadows that dry by summer; des plains to lower mts; se Wn to Cal, e to Mont and Wyo; slender t. (*T. macropetalum*)   2 **T. sagittatum** (Nutt.) Endl.

    4b Siliques arcuate-spreading, stipe 1–5 mm, body 3–4.5 cm × 1–2 mm, torulose; pl per with creeping roots; sepals 4–5 mm; petals lilac-purple, 6–8 (11) mm; lower canyons of Blue Mts, Ore; arrowlf t.   3 **T. eucosmum** Robins.

1b St lvs not aur, often petiolate

    5a Pl ann, hirsute, (1.5) 3–15 dm; siliques terete, very slender, 2–6 cm, reflexed, straight to curved, sessile; near coast in Wn and Ore, s to Baja Cal; cutlf t.   4 **T. lasiophyllum** (H. & A.) Greene

    5b Pl bien, often glab; siliques gen spreading to erect

      6a Sts 1–6 dm; pl glab; sepals and petals purple, 2.5–4 (5) mm; anthers ca 1 mm; stipe ca 0.5 mm; siliques 4–7 cm × ca 1.5 mm; style 0.5–1.5 mm; shale banks along Salmon R and tributaries; Custer Co, Ida; wavylf t.   5 **T. repandum** Rollins

      6b Sts 3–25 dm; pl often hairy; sepals and petals white to purple, petals 6–20 mm, mostly much > the sepals; anthers 1.5–4 mm; stipe 0.7–4 mm; siliques various

        7a St lvs gen entire; siliques 2–3.5 cm, stipe 0.7–2 mm; style 0.5–1 mm; anthers scarcely 2 mm; sagebr plains to lower des mts and e foothills of Cas; Okanogan Co, Wn, to c Ore, and Nev, e to Mont, Neb, and Colo; entirelvd t. (*Pleurophragma i.*)
                                                          6 **T. integrifolium** (Nutt.) Endl.

        7b St lvs mostly toothed to subpinnatifid; siliques 3–14 cm, stipe 1–4 mm; style (0.5) 1–3.5 mm; anthers 1.5–4 mm; widespread in sagebr des of w N Am; thicklvd t.; 3 vars.
                                                          7 **T. laciniatum** (Hook.) Endl.

      a1 Sepals purplish, often ± saccate at base; siliques spreading, 6–14 cm; Grant Co, e in Wn and into n Ore (*T. s.*)
                                                          var. **streptanthoides** (Leiberg) Pays.

      a2 Sepals white, not saccate; siliques various

        b1 Siliques ascending-erect, 3–6 cm; c Wn, chiefly along CR, to Nev and Ida (*T. m.*)   var. **milleflorum** (Nels.) Pays.

        b2 Siliques spreading, 4–11 cm; gen, Wn to Cal, e to Ida and Nev (*T. leptosepalum*)   var. **laciniatum**

Thlaspi L.   Pennycress; Stinkweed

Racemes ebracteate; petals white (ours); silicles sessile, fully dehiscent, strongly obcompressed, often wing-margined, oblong-elliptic to obovate or obcordate, acute to rounded and emarginate; style minute to conspicuous; seeds 2–several per cell, striate-corrugate concentrically, not mucilaginous when wet; ann to per, glab herbs with simple, entire to dentate or lobed lvs, those of the st aur. (Old Gr name for some pl of this family).

1a Pl ann, weedy, 1–5 dm

    2a Silicles oval to oblong-obcordate, 10–17 mm, strongly wing-margined all around, the wing 1–3 mm broad; seeds concentrically ridged; European weed, common in N Am, even into lower mts; field p., fanweed
                                                              1 **T. arvense** L.

    2b Silicles ± obovate, 5–8 mm, wing margin scarcely 1 mm broad, nearly lacking at base; seeds smooth; European weed, rare with us but apparently estab near Pullman, Wn; perfoliate p.   2 **T. perfoliatum** L.

1b Pl per, rarely as much as 3 dm tall; style 0.5–3 mm; silicles mostly > 10 (up to 13) mm

    3a Petals (2) 2.5–3 (4) mm; style ca 0.5 mm; sepals 1–1.5 (2) mm; sagebr foothills, meadows, dry grassy slopes, or limestone cliffs in mts of c Ida to nw Wyo and sw Mont, at middle elev; smallfld p.   3 **T. parviflorum** Nels.

3b Petals 4–11 (13) mm; style 1–3 (4) mm; sepals 1.5–4 (5) mm; Alas to Cal,
e to RM, Mont to NM; Fendler's p., wild candytuft; 2 vars.

4 **T. fendleri** Gray

    a1 Pls strongly caespitose and low growing; basal lvs ∝, mostly oblan-
ceolate, 2–4 (5) mm wide, narrowed very gradually to the petiole;
confined to mts of c Ida *(T. i.)*   var. **idahoense** (Pays.) Hitchc.

    a2 Pls gen not strongly caespitose or low growing; basal lvs oval to ob-
long, 4–9 (16) mm wide, with definite petiole; range of sp., but apparently identical with Eurasian material properly called *T. montanum* L.
(T. glaucum, T. hesperium)   var. **glaucum** (Nels.) Hitchc.

Thysanocarpus Hook.   Lacepod; Fringepod

Racemes ebracteate; fls small, white to purplish-tinged; silicles strongly compressed, 1-celled and
1-seeded, indehiscent, wing-margined; glab to hirsute ann. (Gr *thysanos*, fringe, and *carpos*, fruit, referring to the fringelike wing of the silicle).

T. curvipes Hook. Sand f. Simple to freely br, 1.5–8 dm, hirsute below, glab
above; basal lvs petiolate, oblanceolate, 1.5–5 cm, dentate to shallowly runcinate, st lvs smaller, oblong to narrowly lanceolate, sessile and aur, entire to
dentate; peds subfiliform, recurved, 4–7 mm; sepals and petals ca 1 mm; silicles oval to elliptic, orbicular, or obovate, 5–8 mm, bordered with a broad
radiately-nerved wing, glab to ± tomentose; style 0.1–0.5 mm; open hillsides
or borders of woodl; mostly e Cas, BC to Cal, e to Ida.

CAPPARIDACEAE   **Caper Family**

Fls ± irreg, hypog, mostly ♂, bracteate-racemose; sepals 4 (ours), distinct to ± connate; petals 4
(ours), yellow, white, or pink to purple; stamens 4– ∝ (6 in most of ours), gen much > the petals,
often borne on a thickened disclike area; pistil 2-carpellary, sometimes sessile, but gen with short
to elongate stipe; fr (ours) a dehiscent 1-celled caps with 2 parietal placentae, but sometimes an
indehiscent berry or nut; ann to per herbs (ours) or ± woody, with mostly alt, 3–5-foliolate lvs.

1a Stamens gen > 6; stipe of pistil < 4 mm; petals with white bl and purple
claw   **Polanisia**
1b Stamens gen 6; stipe at least 5 mm; petals with yellow, pink, or purple bl
(except albinos)
    2a Caps much longer than broad, terete to ± compressed, 13–60 mm
                                      Cleome
    2b Caps rhomboidal, little if any longer than broad, < 10 mm   **Cleomella**

Cleome L.   Spiderflower; Cleome

Fls showy, petals yellow or light pink to purple; caps linear to lanceolate or oval in outline, terete
to compressed, several-seeded; stipe slender, 10–20 mm; glab to pilose or glandular ann (ours)
with 3–5 (7)-foliolate lvs. (Name used by Theophrastus for some mustard-like pl). *(Celome, Peri-
toma)*.

1a Petals pink to purplish (occ albino), 8–11 mm; lflets 3, mostly 1.5–7 cm;
caps 3–6 cm, linear-elliptic; sepals connate ca half length; pl glab to
sparsely pilose, mostly 2.5–10 (15) dm; plains to lower mts, often on wastel;
e Wn to Cal, e to Sask and s to NM; RM bee plant, stinkweed

1 **C. serrulata** Pursh

1b Petals yellow
    2a Pl glandular, 1–7 dm; lflets 3, mostly 1–2.5 cm; caps ovate-oblong in
outline, 13–25 mm; sepals distinct; sandy to heavy soils, sagebr des; e
Ore and adj Ida to Nev and Cal; golden s. or c.   2 **C. platycarpa** Torr.
    2b Pl eglandular, 5–15 dm; lflets (3) 4–5 (7), mostly 3–5 cm; caps linear,
nearly terete, 15–35 mm; sepals connate at base; sandy des plains and
lower mont valleys; e Wn to Cal, e to Mont, Neb, and Tex; yellow bee
plant, yellow s.   3 **C. lutea** Hook.

Cleomella DC.   Cleomella

Fls slightly irreg, ours bracteate and racemose; sepals distinct; stamens inserted on small, lopsided disc; caps 2–4 (6)-seeded, rhomboidal, valves laterally distended; ann herbs with 3-foliolate lvs. (Diminutive of *Cleome*).

C. macbrideana Pays. Pl glab, 5–40 cm; lflets mostly 1–2 cm; fls fragrant, petals pale to orange-yellow, (2) 5–7 mm; caps 6–9 × 5–6 mm; style 1–2 mm; stipe 6–15 mm; des valleys and foothills; Malheur Co, Ore, to Lemhi Co, Ida, s to Nev (*C. grandiflora*); two other spp. approach our area—*C. parviflora* Gray, with stipe scarcely 2 mm, and *C. plocasperma* Wats., with caps and stipe > 2 but < 6 mm.

Polanisia Raf.

Fls slightly irreg, bracteate and racemose; stamens (5) 8–16, much > corolla; caps sessile or short-stipitate, linear to oblong, terete to ± compressed, ∞ -seeded; ours ann with 3-foliolate lvs. (Gr *polys*, many or much, and *anisos*, unequal, referring to the unequal stamens).

P. trachysperma T. & G. Clammy-weed. Viscid-pubescent, ill-scented, gen freely br ann 1–8 dm; lflets 15–35 mm, ovate-elliptic to oblanceolate; peds slender, ascending, 10–20 mm; sepals lanceolate, 3–4 mm, purplish, deciduous; petals 7–9 mm, bl white or cream, obovate, notched at the tip, narrowed to a slender purple claw; caps 2.5–4.5 cm, pubescent; stipe 1–3 mm; sandy plains and foothills, often in des washes; se Ore to ne Cal, Ariz, and Mex, e through c Ida and s Mont to Minn, and s to Tex (*Jacksonia t., Cleome t.*).

RESEDACEAE   Mignonette Family

Fls hypog, irreg, gen ⚥; calyx deeply 4–6-lobed; petals 4–6 (ours), yellow to white or greenish-white, gen unequal in size, rounded at base but with a deeply lobed to fimbriate bl; stamens 10–∞, borne inside a 1-sided disc; fr caps, carpels 3–4 (6), open on the inner (adaxial) side near the tip, but otherwise indehiscent; ann to per, glab herbs with simple to pinnatifid lvs.

Reseda L.   Mignonette

Fls spicate-racemose; ours ann or bien. (Ancient L name, used by Pliny, from *resedare*, to calm or soothe; the pl believed to have sedative power).

1a Petals 4, yellowish, upper 1 much the largest, 3–5 mm, unequally 7–11-lobed ca half length; stamens 30–50; lvs narrowly lanceolate to oblanceolate, entire to callus-dentate; caps globose, 3–5 mm; European, occ on ballast or escaped from gardens, formerly used as the source of a yellow dye; dyer's rocket, weld   1 **R. luteola** L.
1b Petals mostly 5–6, sometimes greenish-white or white, mostly subequal; petals 3-lobed or dissected nearly to the base; upper lvs mostly lobed to pinnatifid; caps ± cylindric
  2a Basal lvs mostly entire, those of upper st 3-lobed; petal bl lobed most of length into linear segms; garden mignonette, sometimes escaped, but prob never truly estab   2 **R. odorata** L.
  2b Basal lvs and those of st 3–several-lobed; petals mostly 3-lobed
    3a Lvs 3–several-lobed, mostly on upper half of bl; petals yellow, ca 5 mm; stamens 15–25; seeds pear-shaped, black and shining; occ escape from gardens, persisting in waste places; Europe; yellow m.
                                                          3 **R. lutea** L.
    3b Lvs pinnatifid most of length into 10–16 segms; petals dirty- or greenish-white, 4–6 mm; stamens ca 10; seeds reniform, dull, finely papillose; sparingly intro along coast, mostly on ballast; Europe; upright m.   4 **R. alba** L.

3a   3b

SARRACENIACEAE   Pitcher-plant Family

Fls reg, hypog, solitary on long scapes (ours); sepals and petals 5 in ours, distinct; stamens ∝; pistil 3–5-carpellary, style 1, gen peltate and lobed; fr caps, 3–5-carpellary, placentation axile; seeds ∝, tiny; bog-inhabiting, herbaceous per with large, tubular basal lvs modified for entrapment of insects.

Darlingtonia Torr.   California Pitcher-plant; Cobra-plant

Fls showy, complete; stamens 12–15; pistil 5-carpellary; style 2–3 mm, deeply 5-lobed; caps 5-celled, loculicidal; rhizomatous, glab herb; lvs large, pitcherlike, glandular-hairy inside, lip overarching the opening under the crook, bordered in front by a prominent, 2-lobed appendage. (For Dr. William Darlington, 1782–1863, botanist of Philadelphia).

D. californica Torr. Lvs 1–5 dm, greenish-yellow, but the hood ± purplish, opening directly under transparent area in the crook; appendage yellow to purplish-green; scapes to 1 m; fl pendent; sepals oblong-elliptic, yellowish-green, 4–6 cm; petal dark purple, ca 2/3 as long as sepals; caps turbinate-obovoid, 2–3 cm; bogs along the coast, often along trickling streams and on serpentine rocks; s Lane Co, Ore, to nw Cal. Although not native in Wn, this sp., and at least one sp. of the closely related *Sarracenia* have been transplanted into bogs in w Wn, and tend to persist for a few years, as they do in gardens.

DROSERACEAE   Sundew Family

Fls several and cymose (ours), complete, reg, hypog to ± perig; sepals and petals gen 5 (4); stamens 4–20; pistil 3–5-carpellary; ovary superior or slightly inferior, 1-celled and with 3–5-parietal placentae (ours), or 3–5-celled; styles 3–5, distinct, sometimes bifid; fr caps; seeds small; per herbs (ours) with mostly basal lvs curled in bud, bl covered on one or both surfaces with stalked glands that entrap small insects.

Drosera L.   Sundew

Cymes 1-sided, racemelike, on naked scapes; calyx, corolla, and stamens 5 (4); lvs glandular-viscid, stip. (Gr *droseros*, dewy, referring to the gland-tipped hairs of the lvs).

1a Lvs erect or ascending, bl (2) 3–5 (7) mm broad and 2–several times as long; scapes 6–18 cm; calyx 4–6 mm; petals somewhat > the sepals, white; styles (3) 4–5; seeds with loose, longitudinally striate seed coat that is not flattened at the tips; swamps and bogs, Alas to Cal, e to Ida and Mont and to GL and e Can; Eurasia; great s. (*D. longifolia*) 　　1 **D. anglica** Huds.

1b Lvs mostly spreading, bl (5) 6–12 mm broad and gen little if any longer; scapes to 25 cm; styles gen 3; seed coat loose, longitudinally striate, prolonged and flattened at the tips; otherwise ca like *D. anglica*; swamps and bogs; Alas to Nev and Cal, e to Ida, Mont, Minn, Fla, Va, Lab, and N Eng; Eurasia 　　2 **D. rotundifolia** L.

CRASSULACEAE   Stonecrop Family

Fls ⚥, or ♂, ♀, reg, hypog, often showy, ours cymose, 4–5-merous; calyx ± connate at base; petals white, yellow, pink, or red, distinct to ± connate; stamens gen as many to 2× as many as petals; pistils 3–5 (ours), distinct or basally connate, 1-carpellary; follicles gen ∝-seeded; ann or per, ± strongly succulent herbs (ours) with alt to whorled, exstip lvs.

1a Pl per; fls mostly 5 (occ 4)-merous; stamens twice as many as the petals 　　　**Sedum**

1b Pl ann; fls mostly 4-merous; stamens as many as the petals 　　　**Tillaea**

Sedum L.   Stonecrop

Fls ⚥ or (in *S. roseum*) ♂, ♀, in congested to open cymes; ours per herbs, often rhizomatous or stoloniferous, with alt or opp, terete to strongly flattened, entire to serrulate lvs. (Old L name, said to have been derived from *sedeo*, to sit, in reference to the squatty habit of the pls). *(Clementsia, Gormania, Rhodiola)*.

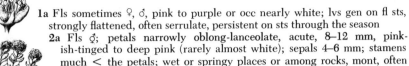

1a Fls sometimes ♀, ♂, pink to purple or occ nearly white; lvs gen on fl sts, strongly flattened, often serrulate, persistent on sts through the season

  2a Fls ♂; petals narrowly oblong-lanceolate, acute, 8–12 mm, pinkish-tinged to deep pink (rarely almost white); sepals 4–6 mm; stamens much < the petals; wet or springy places or among rocks, mont, often above timberl; sc Mont to Utah, Colo, and Ariz; rose crown
                                              1 S. rhodanthum Gray

  2b Fls gen ♀, ♂; petals mostly deep purple, oblong, obtuse to acute, 1–3 mm; sepals 1–2 mm; stamens = or > the petals; cliffs, talus, and ridges, gen where moist in early summer, subalp to alp; BC to Cal, e to Alta, Mont, GL, Me, and Colo; Eurasia and Greenl; king's crown, roseroot, midsummer-men *(Rhodiola integrifolia)*   **2 S. roseum** (L.) Scop.

1b Fls ⚥, yellow (sometimes drying to pinkish); lvs mostly basal, those of fl sts entire, gen deciduous by anthesis

  3a Lvs of fl sts mainly opp, broadly oval to obovate or spatulate, < 1 cm; petals light yellow, oblong-lanceolate, 6–9 mm, slightly apiculate

    4a Mature follicles strongly divergent; petals distinct; dried fl sts 1.5–2 mm thick; subalp to alp, on rocky ridges and talus slopes; Cas and OM, Wn, to Mt Hood, Ore, and s BC; spreading s.
                                          3 S. divergens Wats.

    4b Mature follicles erect; petals connate ca 1 mm at base; dried fl sts ca 1 mm thick; mont ledges and gravel bars to alp ridges and talus slopes; c Ida and adj Mont, se Ore, Nev, Utah, and w Wyo; weakstd s.
                                               4 S. debile Wats.

  3b Lvs of fl sts alt, of various shapes, often > 1 cm

    5a Follicles glandular-pustular or -tuberculate, widely divergent at maturity; sepals 1–1.5 mm; petals distinct, 5–8 mm; lvs spatulate to oval, strongly papillate, 5–25 mm; cymes with spreading to recurved brs; mossy rocks and talus of canyons; Yakima Co, Wn, to se Wn and w Ida, mostly along SR, s to nc Ore; Leiberg's s.   **5 S. leibergii** Britt.

    5b Follicles either not glandular-pustular or not widely divergent at maturity; sepals gen much > 2 mm; petals and lvs various; cymes often with short or erect brs; pls often mont or coastal

      6a Lvs gen broadest below (or only slightly above) midlength, tapering from midlength to the tip or to both ends, either strongly keeled dorsally or ± terete; petals distinct

        7a Lvs strongly keeled, often persistent (sometimes only the midribs) on lower part of old sts, 15–25 mm, linear or linear-lanceolate, strongly acuminate; follicles widely divergent at maturity; upper st lvs gen bearing sterile, bulbil-like, modified fls (propagules); sepals ca 2 mm; follicles widely divergent; sagebr des to ponderosa pine for or subalp ridges or rock slides; BC to Cal, only occ on OM, Wn, and in WV, Ore, e to w Mont; wormlf s. *(S. douglasii)*   **6 S. stenopetalum** Pursh

        7b Lvs not strongly keeled, sometimes completely deciduous from fl sts by anthesis, ± terete; follicles sometimes erect; upper st lvs rarely bearing propagules

          8a Follicles divergent; lvs smooth, ovoid, 3–5 mm; petals ca 5 mm; European cultivar, sometimes escaping and persisting; wall-pepper *(S. elrodi)*   **7 S. acre** L.

          8b Follicles erect; lvs mostly finely papillate, often linear or linear-lanceolate, 5–20 (30) mm; petals 7–9 mm; open dry areas, mostly on rocks or gravel, from sea level to high mont; Alas and Yuk to Cal, e to Alta, SD, Neb, and Colo; lancelvd s.; 3 vars.   **8 S. lanceolatum** Torr.

            a1 Lvs gen smooth, overlapping, even on fl sts, up to 30 mm, not scarious when dried; cymes loose, brs ± divergent and recurved; coastal, and on is of PS, BC to Wn *(S. n.)*
                                    var. nesioticum (Jones) Hitchc.

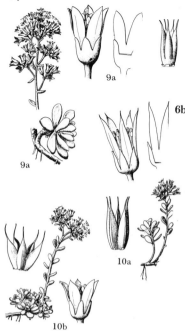

a2 Lvs gen finely papillate, rarely overlapping on fl sts, seldom
   > 20 mm, gen scarious when dried; cymes gen closely fld
  b1 Lvs of fl sts mostly ovoid to ellipsoid, slightly flattened,
     4–9 (20) mm, incurved; chiefly Wen Mts, Wn, in Kittitas
     and Chelan cos (S. r.)   var. **rupicolum** (Jones) Hitchc.
  b2 Lvs of fl sts linear to linear-lanceolate, 7–20 mm, gen not
     strongly incurved; range of sp. (S. subalpinum)
                                                        var. **lanceolatum**
6b Lvs gen broadest above midlength and tapered to base, strongly
   flattened; petals often connate for 1–4 mm
  9a Lvs of sterile shoots mostly opp, spatulate-obovate, (15) 20–30
     (35) × 5–12 (15) mm;_petals connate for ca 3 mm, 7–10 mm,
     acute, gen drying salmon; follicles erect; lava flows and rocky
     slopes; Cas, Ore, from near Mt Hood to near Cal; creamy s.
                                            9 S. oregonense (Wats.) Peck
  9b Lvs all alt; petals sometimes distinct or lanceolate- acuminate;
     follicles sometimes divergent
   10a Petals (8) 10–13 mm, connate for (1.5) 2–3 mm, narrowly lan-
      ceolate and long-acuminate; follicles erect; mont, on cliffs,
      ridges, and talus; BC to Cal, in Cas and OM, to near or along
      coast in Ore; Ore s. (Gormania o.)   **10 S. oreganum** Nutt.
   10b Petals 7–10 mm, nearly or quite distinct, narrowly oblong-
      lanceolate but not long-acuminate; follicles ± divergent;
      coastal cliffs and ledges into drier foothills; s BC to Cal, and in
      the CRG and WV, Ore; spatula-lf s., broadlf s.
                                         11 S. spathulifolium Hook.

Tillaea L.   Pigmy-weed

Petals distinct, whitish; pistils mostly 4 (3–5), distinct; diminutive, glab ann with opp, entire, succulent lvs. (For Michael Angelo Tilli, 1655–1740, Italian botanist).

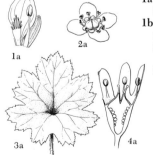

T. aquatica L. Sts weak, glab, 2–6 (10) cm, prostrate to ascending, freely br;
lvs 3–6 mm, linear to linear-lanceolate, connate-sheathing at base; fls single in
the axils, minute, complete, 4-merous; peds up to 6 mm in fr; calyx cup-
shaped, 0.5–1 mm, lobed ca half length; petals ca 1.5 mm, oblong-lanceolate,
membranous, whitish; stamens 4, much < the petals; follicles scarcely 2 mm,
erect, purplish, 6–12-seeded; mud flats and vernal pools; widespread in Eurasia and much of N Am, but infrequent in Wn and Ore, mostly w Cas (Cras-
sula a., Hydrophila a., Tillaeastrum a., Tillaea drummondii, T. a. var. d.).

SAXIFRAGACEAE   Saxifrage Family

Fls (1) few to ∝ , mostly ⚥, reg or occ ± irreg, perig to epig; calyx large, greenish to petaloid, gen ±
adnate to pistil, often with free, connate, flared to tubular hypan, lobes 5 (4–6), equal or· ± unequal
and calyx irreg; petals as many as sepals or sometimes fewer or lacking, gen borne above (or with)
the stamens on a hypan or at edge of a disclike nectary, often less showy than the sepals; stamens
gen (3) 5 or 10, sometimes 5 staminodial; carpels mostly 2–3 (4 or even 5 or 6), from completely
fused to almost distinct, ovary superior to completely inferior, 1–several-celled, cells or placentae
as many as the carpels; fr caps or follicular, with mostly alt (opp), stip or gen exstip, simple to
compound lvs.

1a Ovary 1-celled with 4 parietal placentae; styles almost or quite lacking;
   fertile stamens 5, alt with 5 staminodia   **Parnassia**
1b Ovary 2–3-celled or 1-celled with 2–3 parietal placentae; styles gen prominent; fertile stamens (3) 5–10, staminodia lacking
  2a Petals lacking; sepals 4; stamens 4–8; fls greenish, inconspicuous
                                                           Chrysosplenium
  2b Petals present or sepals 5 and stamens (3) 5 or 10; fls often showy
    3a Fl sts scapose, 7–20 dm, developing before lvs; lf bls mostly > 1 dm
       broad, peltate; petioles 2–15 dm   **Peltiphyllum**
    3b Fl sts lfy or < 7 dm, or developing with lvs; lf bls gen < 1 dm broad;
       petioles gen < 2 dm
      4a Ovary 1-celled, with 2 (3) parietal (to semibasal) placentae **Group I**
      4b Ovary 2 (3–5)-celled, with axile placentation   **Group II**, lead 13a

Group I

5a Fr with 2 very unequal, sterile, dehiscent, upper portions (valves) >
    fertile basal portion; stamens 10; petals entire, linear   **Tiarella**
5b Fr with 2 (3) ± equal sterile upper portions or fertile to the tip; stamens
    3, 5, or 10; petals entire to laciniate
    6a Styles or stigmas 3; petals white or pinkish, laciniate; pl often bulblet-
        bearing in lf axils or underground   **Lithophragma**
    6b Styles or stigmas mostly 2; petals often greenish or not laciniate; pl
        seldom bulbiferous
        7a Petals entire (occ lacking)
            8a Fls pan, spicate, or in a thyrse; seeds finely echinulate in longitu-
                dinal rows; stamens mostly 5; calyx never narrowly turbinate
                Heuchera
            8b Fls racemose; either seeds not echinulate or calyx narrowly turbi-
                nate; stamens sometimes 3 or 10
                9a Stamens gen 3; calyx greenish-purple, 7–10 (15) mm, not cylin-
                    dric but oblique at base and irreg above where split much more
                    deeply on 1 side than elsewhere   **Tolmiea**
                9b Stamens 5 or 10; calyx rarely purplish, gen < 7 mm, often cylin-
                  dric or saucer-shaped, reg
                    10a Seeds finely muriculate in longitudinal rows; racemes loosely
                      5–12-fld, not secund; lf bl reniform, 1–4 cm broad, gen ciliate
                      Conimitella
                    10b Seeds not muriculate; racemes closely 10–46-fld, strongly
                      secund; lf bl cordate, 2–7 cm broad, mostly hirsute on 1 or both
                      surfaces   **Mitella**
        7b Petals trifid to pectinately lobed
            11a Calyx, including adnate portion, 2–4 mm, or up to 6 mm, but then
               petals trilobed, styles < 1 mm, and fl sts lfless   **Mitella**
            11b Calyx, including adnate portion, (6) 7–10 mm; petals mostly with
               at least 5 lobes; styles > 1 mm; fl sts ± lfy
               12a Stamens 10   **Tellima**
               12b Stamens 5   **Elmera**

Group II

13a Stamens 10
    14a Carpels distinct almost to base, adnate to calyx <1/5 length; lvs leath-
        ery, broadly ovate-elliptic to elliptic-obovate, 3–6 cm, crenate but not
        lobed; pl rhizomatous   **Leptarrhena**
    14b Carpels ± fused or adnate to calyx at least 1/5 length or pl not otherwise
        as above
        15a Styles partially connate; petals pink to deep red; calyx camp, gen
            reddish, (5) 6–10 mm; lvs alt, petiolate   **Telesonix**
        15b Styles free above fertile part of ovary; gen either petals white or
            calyx not at once camp, red, and as much as 6 mm; lvs sometimes ses-
            sile and opp   **Saxifraga**
13b Stamens 5
    16a Calyx not adnate to ovary, camp, (10) 12–16 mm, sepals lanceo-
        late-acuminate; st lvs prominently toothed-stip   **Bolandra**
    16b Calyx gen ± adnate to ovary, various in size and shape; st lvs often
        lacking or without toothed stip
        17a Pl bulbiferous at crown, neither stoloniferous nor strongly rhizoma-
            tous, rootstocks filiform or lacking; fl sts rarely >2 dm; upper st lvs
            prominently stip; petals white or violet   **Suksdorfia**
        17b Pl not bulbiferous at crown, gen stoloniferous or evidently rhizoma-
            tous; either fl sts > 3 dm or upper st lvs not prominently stip; petals
            white
            18a Petals 1.5–2.5 mm, withering-persistent; calyx mostly 2.5–3 mm;
               sts gen < 2.5 dm; pl stoloniferous   **Sullivantia**
            18b Petals mostly 4–7 mm, deciduous; calyx rarely < 4 mm; sts (1.5)
               2–8 (10) dm; pl rhizomatous, never stoloniferous   **Boykinia**

Bolandra Gray   Bolandra

Fls complete, bracteate-pan; calyx tubular-camp, 5-lobed; petals and stamens 5, borne at top of hypan; carpels 2, fused 1/4–1/5 length, free of calyx; ovary 2-celled at base; caps loculicidal; per herbs with reniform, palmately veined, stip lvs. (For Dr. Henry N. Bolander, 1831–1897, collector for Cal State Geological Survey).

B. oregana Wats. Pl slightly glandular-pubescent, 2–4 (6) dm, with short, bulbiferous rhizomes; stips small on lower lvs but large on fl bracts; petioles up to 15 cm; bls 2–7 cm broad, shallowly lobed and acutely dentate or serrate-dentate; fl bracts ± clasping, 1–3 cm; calyx accrescent and up to 14–18 mm, lobes narrow, purplish, = or > the camp-tubular hypan; petals purple, linear, = or slightly > sepals; stamens alt with and ca 1/3 as long as petals, filaments purple; caps ca 1 cm, 2-celled basally, free portion of carpels tapered upward and hollow; moist rocks; lower CRG and SR of se Wn, ne Ore, and adj Ida (*B. imnahaensis*).

Boykinia Nutt.   Boykinia

Fls pan; calyx turbinate to camp, adnate to ovary 1/3–1/2 length and short-tubular above, lobes 5, equal, lanceolate; petals white, exceeding sepals, short-clawed; stamens 5, opp sepals; caps 2-celled, ca 1/2 inferior, styles distinct, ± beaklike; seeds finely tuberculate; glandular-pubescent, often brownish-pilose, rhizomatous herbs with alt, cordate to reniform, freely cleft and toothed bls and large and semifoliose to bristle-like stips. (For Dr. Samuel Boykin, 1786–1848, naturalist of se US). (*Therophon*).

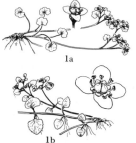

1a Stips merely slight, membranous, bristle-bordered expansions at base of petiole; lf bl 2–6 (8) cm broad; sts slender, 1.5–6 dm; stream banks and moist woods; BC s, w Cas to coast, through Wn and Ore to Sierran Cal; slender b., Santalucia b., coast b. (*B. occidentalis*) 1 **B. elata** (Nutt.) Greene
1b Stips (upper ones) ± lfletlike, often clasping or connate; lf bl up to 20 cm broad; sts stout, (2) 3–10 dm; meadows and streambanks, w Wn to Cal, ne through Ore to Ida and Mont; mt b., greater b., Sierra b.; 2 vars.

2 **B. major** Gray

a1 Petals ± undulate-margined, narrowed abruptly to short claws, bls mostly ovate to suborbicular; pan close-fld, rounded to flattened; w Ore to Cal, e to Ida and Mont   var. **major**
a2 Petals nearly obovate to spatulate, plane-margined, attenuate to claws; pan more open, pyramidal; OP, Wn, to Tillamook Co, Ore (*B.i.*)

var. **intermedia** Piper

Chrysosplenium L.   Golden-carpet; Golden-saxifrage; Water-carpet

Fls perig, solitary or cymose; calyx largely adnate to ovary, broadly camp, lobes spreading; stamens at edge of small disc; filaments no > the anthers; carpels 2, ovary 1-celled, immersed in the disc but styles protruding; caps ca 1/2 inferior; ours glab per herbs with ± stolonous sts and opp or alt, exstip, crenate lvs. (Gr *chrysos*, gold, and *splen*, spleen, name thought to have been derived from fancied medicinal value of pls).

1a Lvs alt, bl oval to reniform, with 3–7 broad crenations, 5–12 mm broad, scarcely as long, gen < the slender petioles; fls ca 3 mm broad; stamens 4; ± circumpolar, Alas s to Okanogan Co, Wn, and in RM to Colo; northern g. (*C. alternifolium* ssp. *t.*)   1 **C. tetrandrum** (Lund) Fries
1b Lvs opp, bl oval or broadly ovate to reniform, with 15–20 shallow crenate-serrations, 3–15 mm broad and up to 2 cm, > petioles; fls ca 4 mm broad; stamens 8; moist to swampy places; BC s, w Cas and mostly along coast, to nw Cal; western g., Pac w. (*C. scouleri*)

2 **C. glechomaefolium** Nutt.

Conimitella Rydb.   Conimitella

Fls complete; calyx turbinate-obconic, adnate to ovary ca 1/2–1/3 length; petals 5, white, entire, clawed, inserted with stamens near base of free hypan; stamens 5, opp sepals; per, scapose, ± rhizomatous herbs with long-petioled, exstip lvs. (L *conus*, cone, and *mitella*, a related genus; i. e., like *Mitella* but with a conelike hypan).

C. williamsii (Eaton) Rydb. Fl sts lfless, 2–4 (6) dm; lf bls reniform, leathery, 1–4 cm broad, shallowly bicrenate, stiffly ciliate, gen purple beneath; fls 5–12, minutely bracteate, racemes blossoming upward; peds 1–5 mm; calyx 4–6 (9) mm, free hypan tubular, ca = the ovate lobes; petals rhombic to oblanceolate, 4–5 mm; caps dehiscent along the stylelike superior portion; seeds black, ca 1 mm; rock crevices, cliffs, and mt slopes; RM and adj ranges, Mont to nw Wyo and e Ida (*Lithophragma w., Tellima nudicaulis*).

Elmera Rydb.   Elmera

Fls racemose on mostly lfy fl sts; calyx cuplike, adnate to ovary at base; free hypan gland-lined, reg to slightly lopsided, ca as long as sepals; petals white, short-clawed, bl mostly 3–7-cleft; stamens opp sepals; ovary adnate for ca 1/4 length, placentae 2; styles short, merging with hollow conical tips of ovary; caps ovoid, ∞-seeded, dehiscent by ventral sutures on upper portion; seeds finely warty in longitudinal rows; per herbs with slender rhizomes and long-petioled, reniform, strongly pubescent-glandular lvs. (For A. D. E. Elmer, 1870–1942, American botanist who collected in much of w N Am).

E. racemosa (Wats.) Rydb. Sts 1–2.5 dm; lf bls gen 3–5 cm broad and not so long; stips brownish; st lvs (0) 1–3; racemes 10–35-fld; calyx greenish-yellow, hypan 3–4 mm; seeds ca 0.6 mm; rock crevices and mt ridges, mont to subalp, Ore and Wn; with 2 vars.

a1 Petioles and lower sts hirsute-glandular to hirsute and puberulent; infl glandular-puberulent and hirsute; Cas, Mt Rainier to Mt Adams, and OM (*Heuchera r., Tellima r.*)   var. **racemosa**

a2 Petioles and lower sts glandular-puberulent; infl rarely other than glandular-puberulent or with few scattered longer hairs; Cas, from Hart's Pass, Wn, to Mt Stuart in Wen Mts; disjunct in nw Klamath Co, Ore
var. **puberulenta** Hitchc.

Heuchera L.   Alumroot; Heuchera

Fls reg to irreg, racemose-thyrsiflorous; calyx yellow-green, green, or reddish, saucer-shaped to tubular-camp, adnate almost to top of ovary but with free hypan above, lobes mostly 5; petals 5 or sometimes fewer or lacking, white to greenish-yellow, gen clawed; stamens mostly 5, but 1 or more sometimes rudimentary, opp sepals; ovary half to almost completely inferior, projecting above into hollow, stylelike, centrally dehiscent beaks; styles short to prominent; caps ∞-seeded, dehiscent on the beaks; seeds spinulose in longitudinal rows; herbaceous per, mostly with br crowns, often rhizomatous; fl sts mostly scapose, ± glandular; lf bls palmately lobed and deeply 1–2-crenate-dentate. (For Johann Heinrich von Heucher, 1677–1747, Prof. of Medicine at Wittenberg).

1a Stamens gen much > sepals; styles slender, 1.5–3 mm; fls in open to diffuse pan

2a Calyx 1.5–3 (3.5) mm, nearly as broad, free hypan rarely > 0.5 mm, considerably < adnate portion; petals white, 2–4 times as long as sepals, slenderly clawed; pan diffuse

3a Seeds narrowly ellipsoid, slightly curved, 3–4 times as long as broad, brown, prominently echinulate; petioles and basal portion of fl st glab or occ glandular-puberulent; stips ciliate; lf bls gen broader than long, acutely lobed 1/3–1/2 length, sinuses narrow; stream banks and moist rocks, from coast to above timerl; Alas to Mt Hood, Ore, in both Cas and OM in Wn, e in Wn to Wen Mts, and in BC to Selkirk Mts, and possibly to RM; smooth a., alpine a.   1 **H. glabra** Willd.

3b Seeds broadly oblong-ellipsoid, not curved, < twice as long as broad, nearly black, finely echinulate to tessellate; basal portion of fl st and petioles gen strongly villous with white or brownish hairs, or if (as rarely) glandular-puberulent or glab then stips long-villous margined; lf bls often longer than broad and with rounded, shallow lobes; stream banks and rock crevices, from near sea level to subalp; BC s, in Cas

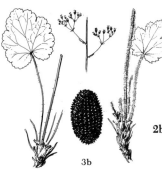

3b

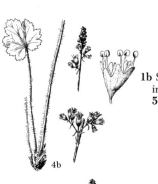

4b

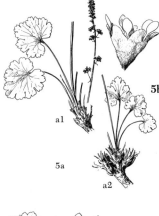

a1

5a

a2

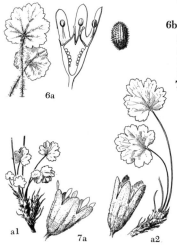

6a

a1   7a   a2

and w, through Wn and Ore to Cal, e in Ore to Blue and Wallowa mts and to adj Ida; smallfl a.; 2 vars.   **2 H. micrantha** Dougl.

a1 Lf bl at least as broad as long, shallowly lobed much < 1/3 length, lobes rounded; petioles and lower sts villous to puberulent or subglab; occ in Wn, but common from CRG to Marion Co, Ore, e in Ore to Ida *(H. glaberrima, H. lloydii, H. nuttallii)*   var. **micrantha**

a2 Lf bl gen less broad, more deeply and acutely lobed up to 1/3 length; petioles and lower sts gen strongly villous; common phase in Wn and BC *(H. barbarossa, H. longipetala, H. d.)*
var. **diversifolia** (Rydb.) R. B. & L.

2b Calyx (3) 3.5–10 mm, gen much less broad, free hypan > 0.5 mm, often = or > adnate portion; pan open to contracted, but not diffuse

4a Calyx (3) 3.5–5 mm, pinkish; hypan 0.5–1.5 mm; se Ore to s Cal, e to sw Ida, Utah, Colo, and Ariz; not known from quite within our area
3 H. rubescens Torr.

4b Calyx 5–10 mm, greenish; hypan strongly lopsided, 1–5 mm; lf bl 2–6 cm broad, mostly glab above and ± whitish-hirsute beneath; moist banks to rock crevices and sandy prairie; n Can Rockies s, mostly e RM, to Colo, e to Sask and Man, Minn, Wisc, and Ind; Richardson's a. *(H. ciliata)*   **4 H. richardsonii** R. Br.

1b Stamens gen < sepals; styles short and thick, rarely > 1 mm; fls mostly in congested pan

5a Calyx reg, mostly 2–3 mm at anthesis, adnate portion turbinate, free hypan flared and ± saucer-shaped, subequal to spreading lobes, 1–1.5 mm, lined with a thin glandular disc ± covering the nearly completely inferior ovary; petals broadly elliptic to ovate, ca 1.5 times as long as sepals; seeds ovoid-ellipsoid, ca 1 mm, finely echinulate in longitudinal rows; mont, Alta to NM; common a., small-lvd a.; 2 vars.
5 H. parvifolia Nutt.

a1 Fl sts gen > 2 dm; lvs gen 3–6 cm broad, deeply cordate with narrow sinus, lobed 1/5–1/3 length; Wyo to Colo and Utah, w to c Ida and Nev *(H. u.)*   var. **utahensis** (Rydb.) Garrett

a2 Fl sts gen < 2 dm; lf bls 1–4 cm broad, deeply cordate, and with narrow sinus, to shallowly cordate, mostly lobed 1/3–1/2 length; common phase in Mont, Alta, and c Ida *(H. flabellifolia)*
var. **dissecta** Jones

5b Calyx gen much > 3 mm at anthesis, often oblique, mostly camp at base and with cup-shaped hypan, sepals gen erect, ca 1.5 mm; hypan either not gland-lined or disc not covering top of ovary, ovary mostly not > half inferior; petals often lanceolate to oblanceolate or wanting

6a Petioles and lower part of sts densely brownish-villous (when dried), mostly with eglandular hairs 2–5 mm; filaments often > twice as long as the dehisced anthers; seeds dark brown, 0.7–0.8 mm, slenderly echinulate, with only slightly tapered spines in longitudinal rows; gravelly prairies to thickets in lower mts; w Cas, BC to Douglas Co, Ore, up CRG to e base of Cas and s along Cas in Ore to Klamath Co; meadow a., narrowfld a., green-fld h.
6 H. chlorantha Piper

6b Petioles and lower portion of sts glab to glandular-pubescent or ± glandular-villous, hairs < 2 mm or whitish when dried and gen glandular; filaments mostly < twice as long as the dehisced anthers; seeds 0.6–0.9 mm, echinulate in longitudinal rows with conical spines; mostly e of Cas crests

7a Calyx 4–6 (6.5) mm at anthesis; petals 5 on nearly all fls, half as long to nearly twice as long as sepals, bl oblanceolate to spatulate; stamens not > 1.5 mm; lf bls cordate-orbicular to subreniform, 1–7 cm broad and somewhat shorter; hillsides and rocky canyons to talus and alp ridges, CRG and ne Ore, c Ida, and sw Mont; gooseberrylvd a.; 2 vars.   **7 H. grossulariifolia** Rydb.

a1 Lvs gen 1–2.5 (3.5) cm broad; petals = to half again as long as sepals; calyx 4–5 (6) mm; scapes rarely > 4.5 dm; sw Mont and Lemhi, Custer, Blaine, Elmore, Boise, and Valley cos, Ida, and SRC in Payette Co, Ida, and Wallowa and Baker cos, Ore *(H. cusickii, H. gracilis)*   var. **grossulariifolia**

a2 Lvs mostly (2.5) 3–7 cm broad; petals gen < sepals; calyx 5–6.5 mm; scapes up to 8 dm; pl tending to be larger in most respects than the var. *grossulariifolia*; Payette and Salmon R

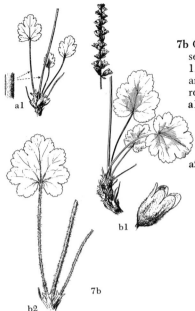

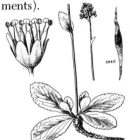

drainages in n Elmore and Boise cos, and in Valley and Idaho cos, Ida; CRG of Wasco and Hood R cos, Ore, and adj Wn (*H. t.*)   var. **tenuifolia** (Wheelock) Hitchc.

7b Calyx (4.5) 6–8 mm at anthesis; petals gen not > half as long as sepals, mostly linear and fewer than 5, or absent; stamens gen > 1.5 mm; lvs gen narrower; rocky hillsides, cliffs, and talus; BC in and e of Cas, to ne Cal, e to Alta, Mont, nw Wyo, and n Nev; roundlf a., lava a.; 3 vars.   **8 H. cylindrica** Dougl.

    a1 Lvs and st base finely glandular-pubescent (to hirsute), bls thick, 1–2.5 cm broad, slightly longer, mostly rounded or ± acute to somewhat cordate at base; Wen Mts, s Chelan Co, to Grant Co, Wn, s to ne Cal, e in Ore to sw Ida and n Nev (*H. a., H. ovalifolia*)   var. **alpina** Wats.

    a2 Lvs and st base variously pubescent; bls often > 2.5 cm broad, gen cordate and with definite sinus

        b1 Base of st and petioles glab or sparsely glandular-pubescent; with var. *cylindrica* in much of range, but scarce or lacking in most of BC and Alta; the most common phase in most of Mont and much of Ida and Wash; infrequent in Ore (*H. g.*)   var. **glabella** (T. & G.) Wheelock

        b2 Base of st and petioles glandular-pubescent to hirsute; bracts of sts large, esp in pls of n Cas and occ in those of RM; common phase in n Cas and BC to Mont s to w Wyo, also in Ida and s in Wn to Ore where ± intermediate to other vars. (*H. columbiana, H. saxicola, H. suksdorfii*)   var. **cylindrica**

❧

Leptarrhena R. Br.   False Saxifrage; Leatherleaf Saxifrage

Fls ∝ in a tight thyrse, perig; calyx shallowly camp, 5-lobed; petals white, small, persistent; stamens 10; carpels 2, connate only at the slightly inferior base, tapered into hollow, ventrally dehiscent, beaklike tips; follicles 2, almost distinct; rhizomatous per herbs with leathery, persistent lvs and sparsely lfy fl sts. (Gr *leptos*, fine or slender, and *arrhen*, male, referring to the slender filaments).

L. pyrolifolia (D. Don) R. Br. Lvs narrowly obovate to elliptic or ovate-oblong, 3–15 cm, narrowed to short broad petiole < 1/2 as long as bl, crenate-serrate, bright green on upper surface, pale green beneath; infl much-congested at anthesis, becoming much more open; calyx 2–3 mm broad, scarcely as long, basal portion adnate to ovary, lobes ca 1–1.5 mm; petals spatulate, 1–2 mm, gen unequal; seeds 2–3.5 mm, light brown, ± lunate, with testa attenuate at each end into a tubular tail; stream banks and meadows to moist slopes; Alas s, in OM and Cas of Wn, to Cas of Ore, e in BC and n Wn to n Ida, w Mont, and sw Alta (*L. amplexifolia, Saxifraga p.*).

Lithophragma Nutt.   Fringecup; Prairiestar; Woodlandstar; Lithophragma

Fls racemose, often ± showy; calyx narrowly obconic to camp or cupshaped, shallowly 5- lobed, partially adnate to ovary but with free hypan; petals white to pink or purplish-tinged, often ± unequal, narrowly clawed and with digitately cleft or divided bl; stamens 10; ovary 1/5–1/6 inferior; caps 3-beaked; per, mostly glandular-pubescent herbs with simple, lfy fl sts and orbicular-reniform to reniform, palmately parted or cleft to laciniately lobed lvs. (Gr *lithos*, stone, and *phragma*, wall, presumably referring to pl habitat).

1a Seeds muricate; basal lvs glab or sparsely pubescent; fls mostly 2–5 (7), ± corymbose at anthesis but becoming racemose, lower peds gen elongate and 1.5–3 times as long as calyx; st lvs often bulbiferous; petals mostly 5-lobed; calyx camp to cup-shaped, 3–4 mm at anthesis, 4–6 mm in fr; seeds minutely muricate, ca 0.5 mm

    2a St lvs bulbiferous; pl mostly purplish above; grassy hillsides and sagebr des to ponderosa pine and Douglas fir for e Cas; s BC to Cal, e to Alta, the Daks, and Colo; bulbiferous f., p., or w., rocketstar (*Tellima b., L. tenella* var. *ramulosa*)   **1 L. bulbifera** Rydb.

    2b St lvs not bulbiferous; pl greenish to purplish above; grassl and sagebr plains to ponderosa pine or oak woodl; e base Cas, Chelan Co, Wn, to

CRG, e to ne Ore, se Wn, and across n Ida to nw Mont; smooth f., p., or w. (*Tellima g.*)   2 **L. glabra** Nutt.

1b Seeds smooth or slightly wrinkled to verrucose or reticulate, never muricate; basal lvs ± strongly pubescent on lower surface at least; fls (4–5) 6–11, racemose; lower peds from slightly < to half again > calyx; st lvs not bulbiferous; petals often only 3-lobed

3a Calyx obconic to vase-shaped, (3) 4–6 mm at anthesis, 6–10 mm in fr; sepals 1–2 mm; ovary at least 2/3 inferior; petals gen 3-lobed; sagebr des and grassl to lower mont for; BC s, on both sides of Cas, to n Cal, e to Alta, Daks, and Colo; smallfl f., p., or w. (*Tellima p.*)
3 **L. parviflora** (Hook.) Nutt.

3b Calyx ± camp, 2–3 (3.5) mm at anthesis, 3.5–5 mm in fr; sepals scarcely 1 mm; ovary ca half inferior; petals mostly 5- or 7-lobed; sagebr des to ponderosa pine for; e side Cas in Wn and Ore to s Ida and Mont, Wyo, Colo, Ariz, and Nev; slender f., p., or w.; 2 vars.   4 **L. tenella** Nutt.

a1 Peds rarely > 1/4 > fr calyx; basal lf bls gen divided or lobed almost or quite the full length; sc Ore to Cal and Nev, e to s Mont and Wyo (*Tellima t.*)   var. **tenella**

a2 Peds slender, lower ones often 1/2 > fr calyx; basal lf bls in part lobed no > half their length; e side Cas; Okanogan and Grant to Yakima Co, Wn (*L. t.*)   var. **thompsonii** (Hoover) Hitchc.

Mitella L.   Mitrewort; Bishops-cap

Fls in simple racemes; calyx saucer-shaped to turbinate-camp, adnate to ovary, free portion deeply 5-lobed but also with thinly gland-lined and gen ± flared hypan; petals greenish, white, or pinkish- to purplish-tinged, slenderly clawed, bl filiformly dissected to trilobed; stamens 5 or 10; ovary 1-celled, from < half to nearly completely inferior, bilobed and with short hollow or solid stylar beaks; caps ± circumscissile; seeds blackish and shining; rhizomatous, often glandular-puberulent per with lfless or 1–3-foliate fl sts. (Diminutive of L *mitra*, cap, referring to fr shape). (*Mitellastra, Ozomelis, Pectiantia*).

1a Stamens 10; ovary < 1/2 inferior; calyx lobed at least half length; petals greenish-yellow, ca 4 mm, filiformly dissected into gen 8 lateral divisions; scapes lfless, finely glandular-puberulent, 3-20 cm; stream banks, moist woods, and bogs; Alas to Cas of nw Wn, e to Lab and Newf and s to c Mont, Minn, and Pa; e Asia; bare-std m., stoloniferous m.   1 **M. nuda** L.

1b Stamens 5; ovary at least 1/2 inferior; calyx and petals various; scapes sometimes lfy

2a Stamens alt with sepals (opp petals); anthers cordate-reniform, much broader than long; petals greenish, 2–3 mm, pectinately dissected into (4) 8 (10) filiform lateral segms; ovary almost completely inferior; scapes mostly 2–3 (1–4) dm, lfless or with 1–2 reduced near-basal lvs; moist woods, often along streams; Alas s along coast to ne Cal, e to Alta and Colo; common in much of our area; alpine m., five-stamened m. (*P. latiflora*)   2 **M. pentandra** Hook.

2b Stamens opp sepals (alt with petals); anthers gen at least as long as broad; petals often white, variously lobed

3a Fls blossoming downward in infl; fl sts with 1–3 lvs, mostly 2–4 dm, glandular-pubescent; calyx 5-6 mm broad, adnate to ovary for < 1/3 length, free hypan ca = sepals; styles ca 1 mm; lf bls cordate, mostly 5-lobed 1/4–1/3 length and crenate-dentate, sparsely hirsute on upper surface, (2) 3-7 cm broad; deep woods to swampy ground, seacoast to midmont; BC to nw Cal, e to n and wc Ida and nw Mont; lfy m., star-shaped m.   3 **M. caulescens** Nutt.

3b Fls blossoming upward in infl; fl sts mostly lfless, sometimes < 2 dm; calyx often < 5 mm broad; styles often < 1 mm; lf bls various

4a Calyx shallowly saucer-shaped, much broader than long; sepals triangular, spreading-recurved; anthers cordate; stigmas bilobed; petals gen greenish-yellow

5a Lf bls mostly 4–8 cm broad, cordate to reniform, always broader than long, slightly if at all white-hirsute; peds 1–2 (in fr to 5) mm; petals 1–2 mm, pinnatisect into 5–9 filiform, mostly paired segms; moist mt valleys and wooded slopes, upward to timberl; BC s, in both Cas and OM, to Sierran Cal, e to Alta, nw Mont, and n Ida; Brewer's m.   4 **M. breweri** Gray

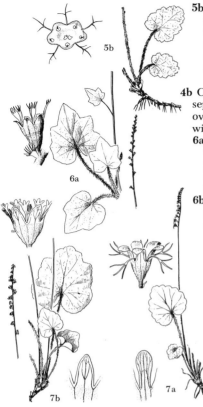

5b Lf bls mostly 1.5–3.5 cm broad, cordate-ovate to cordate-oblong, always longer than broad, upper surface gen strongly and coarsely white-hirsute; peds mostly 0.5–1.5 mm; petals ca 1.5 mm, pinnatifid into 4–7 linear, unpaired segms; deep, moist woods, creek bottoms and wet banks; BC s, on w side Cas to coast, to c Cal; oval-lvd m., coastal m. *(M. hallii)*
                                                              5 M. ovalis Greene
   4b Calyx cup-shaped to camp, gen considerably longer than broad, sepals ovate to oblong, often erect, or spreading only at tips; anthers ovate to oblong; stigmas not bilobed; petals gen whitish or tinged with pink to purple
      6a Lf bls ± cordate-ovate to cordate-triangular, ca as long as broad, with 5 (7) distinctly angular lobes, terminal lobe often acute; petals 3–5 lobed at tip; scapes often with 1 or 2 lvs near base; moist woods and along streams; Cas, from Mt Adams, Wn, to nw Cal; angle-lvd m., varied-lvd m.   **6 M. diversifolia** Greene
      6b Lf bls reniform to cordate-ovate, broader than long, shallowly rounded-lobed, terminal segm not acute; petals not > 3-lobed; scapes rarely with foliage lvs; gen in our area
         7a Racemes strongly secund, 10–48-fld; calyx (3) 4–6 mm, sepals with simple c vein and br lateral veins; petals mostly (1.5) 2–4 mm, bl trilobed into divaricately spreading or ascending, mostly filiform segms; open to dense, moist woods; extreme e Wn and ne Ore to RM from Mont to Colo; side-fld m., cross-shaped m. *(M. stenopetala)*   **7 M. stauropetala** Piper
         7b Racemes weakly if at all secund, (4) 10–20-fld; calyx 1.5–3.5 mm, sepals with br central vein and gen simple lateral veins; petals (1) 1.5–2.5 (3.5) mm, digitately cleft into 3 ascending to erect, narrow segms (rarely entire); deep for to mont slopes; BC s, in OM and Cas, to n Cal, e in Can and Wn to Alta and s Mont and to ne Ore, not known from Ida; three-tooth m., Pac m. *(M. micrantha, M. violacea, O. pacifica)*   **8 M. trifida** Grah.

Parnassia L.   Grass-of-Parnassus

Fls solitary on naked or bracteate peduncles, reg, often ± showy; calyx broadly obconic, ± adnate to ovary, deeply 5-lobed; petals 5, white, rounded to clawed at base, entire to pectinate-fimbriate proximally; ovary superior to ca 1/4 inferior; caps dehiscent apically; seeds with loose, ± inflated testa; scapose, glab per herbs with short rhizomes and entire, petiolate lvs. (For Mt. Parnassus, Greece).

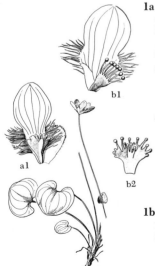

   1a Petals fimbriate-pectinate on proximal half, 8–12 mm; staminodia thickened and scalelike, upwardly flared and gen with large c lobe and several lateral lobes or with 5–∝ slender segms bulbous at tip; lvs mostly reniform, 1.5–5 cm broad; bogs, stream banks, and wet meadows, mont to alp; Alas to Cal, on both sides Cas, e to w Alta, Mont, Wyo, Colo, and NM; fringed g., RM g.; 3 vars.   **1 P. fimbriata** Konig.
      a1 Staminodia short and thick, marginal segms short and rounded, not at all filamentlike; range of sp. as whole except for that of the other vars.
                                                              var. **fimbriata**
      a2 Staminodia ending in longer, more slender, filamentlike, gen capitate segms
         b1 Segms of staminodia mostly < 10, slender, strongly capitate, all marginal, = the narrow basal scale; n Cas of Ore (Washington, Clackamas, and Hood River cos)   var. **hoodiana** Hitchc.
         b2 Segms of staminodia mostly at least 12, 1 or 2 median and subterminal, all < the strongly flared basal scale; sc Ida to n Nev *(P. i.)*
                                                    var. **intermedia** (Rydb.) Hitchc.
   1b Petals not fimbriate-pectinate on lower half; staminodia various; lvs often not reniform
      2a Scapes bractless or with near-basal bract; petals 1–3-nerved, ca same length as sepals; staminodia with (1) 3–5 segms, gen ca half as long as filaments; anthers < 1 mm; lf bls gen ovate to elliptic, 5–15 (20) mm, truncate to tapered at base; arctic regions, s to Cas of Okanogan Co, Wn, and to Lab and Gaspé; ours the var. *pumila* Hitchc. & Ownbey, with

staminodia reduced to linear, entire to lacerately 2–3-toothed scales; known only from Okanogan Co, Wn; Kotzebue's g. 2 **P. kotzebuei** Cham.

2b Scapes gen with bract above level of basal lvs; petals 5–13-nerved, gen much > sepals; staminodia with 5–∝ segms, gen much > half as long as filaments; anthers at least 1 mm

3a Lf bls elliptic to elliptic-ovate, neither truncate nor cordate at base; bract never clasping; petals mostly 5-veined, 4–7 mm, rarely > 1.5 times as long as sepals; bogs, stream banks, and wet meadows; BC to Que, s to n Ida, c Mont, SD, and Minn; small-fld g. 3 **P. parviflora** DC.

3b Lf bls lanceolate to broadly ovate, often ± cordate; bract often clasping; petals 7–13-veined, (6) 7–12 mm, mostly > 1.5 times as long as sepals; shaded areas in mts to arctic-alp moist areas; Alas to Que, s to BC, Minn, and in the RM to Colo, Utah, and w to s Nev and Sierran Cal; northern g., wideworld g.; ours the var. *montanensis* (Fern. & Rydb.) Hitchc. 4 **P. palustris** L.

Peltiphyllum Engl.   Shieldleaf; Umbrella-plant; Indian Rhubarb

Fls in large pan-corymbose cymes, reg, showy; calyx adnate to base of ovary, deeply 5-lobed, without free hypan; petals 5, white to deep pink, entire; stamens 10; carpels gen 2, free above; fr follicular; per herbs with thick rhizomes. (Gr *pelte*, shield, and *phyllon*, lf, referring to the peltate lvs). Nice orn for wet places.

P. peltatum (Torr.) Engl. Lvs basal, membranous-stip, petiole 1–15 dm, rough-hirsute, bl suborbicular, peltate, cupped in center, 5–40 cm broad, shallowly 7–15-lobed and again lobed and serrate-dentate; scapes stout, 1–20 dm, hirsute and glandular; sepals 2.5–3.5 mm; petals oblong-elliptic to -obovate, 4.5–7 mm; filaments flattened, tapered from a broad base upward; styles thick, tapered upward to discoid-capitate stigmas; follicles purplish, 6–10 mm, fused with calyx for 1–2 mm, dehiscent full length; seeds ca 1 mm; cold mt streams, where firmly anchored; Benton Co, Ore, to sw Ore and nw and Sierran Cal (*Saxifraga p.*).

Saxifraga L.   Saxifrage

Fls reg to irreg, complete or sometimes apet, perig to almost completely epig, solitary, cymose, pseudoracemose, or pan; calyx saucer-shaped to conic or camp, often with short, free hypan, ± adnate to ovary at base; petals white to greenish or purple, often flecked with yellow or red, alike or dissimilar, often with slender claw; stamens 10, inserted with petals at top of hypan or on calyx surrounding ovary, filaments slender to expanded and ± petaloid; carpels 2 (3–4), from distinct except at extreme base to connate to above placental area, gen tapered into first hollow then solid stylar portions; fr caps or follicles; seeds smooth to wrinkled or muricate, sometimes with loose and inflated testa; ours per herbs, scapose to lfy-std, with alt or opp, entire to toothed, lobed, or pinnatifid, sessile to petiolate lvs. (L *saxum*, rock, and *frangere*, to break). (*Antiphylla, Cascadia, Hexaphoma, Heterisia, Hirculus, Hydatica, Leptasea, Lobaria, Micranthes, Muscaria, Ponista, Spatularia, Steiranisia*). The ann, European **S. tridactylites** L. is well estab in Vancouver, B.C.

1a Fls purple (except albino pls), borne singly on short lfy sts; lvs opp, decussate, entire, marcescent, 3–5 mm, coarsely ciliate; pls pulvinate, 2–4 cm; arctic tundra to alp mt slopes; circumboreal; s in N Am to OM and Cas, Wn, Wallowa Mts, Ore, and RM of Mont, Ida, and n Wyo; purple s., twinfld s. 1 **S. oppositifolia** L.

1b Fls white or yellow (rarely pinkish- or purplish-tinged), gen > 1 per st; st lvs lacking or alt and often toothed or lobed

2a Pl rhizomatous, delicate, with slender, succulent, trailing to erect, glab to sparsely glandular-puberulent sts; lvs mainly cauline, bl entire to 3 (5) -toothed, elliptic to obovate or flabellate, lower ones 4–19 mm, upper ones sometimes larger, all with slender petioles; petals oblong-lanceolate, 2–4 times as long as sepals; calyx camp-turbinate, 2.5–3.5 mm, slightly keeled below sinuses of acute sepals; seeds ca 0.6 mm, crescent-shaped, with ca 8 longitudinal rows of prominent, comb-like spines; streams and wet banks; CRG and lower WV, Saddle Mt, and s along Ore coast to sw Ore, also from near Montesano, Grays Harbor Co, Wn; Nuttall's s. 2 **S. nuttallii** Small

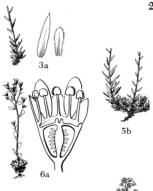

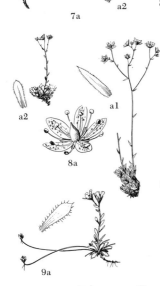

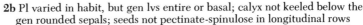

2b Pl varied in habit, but gen lvs entire or basal; calyx not keeled below the gen rounded sepals; seeds not pectinate-spinulose in longitudinal rows

   3a Lvs entire, mostly linear to lanceolate, gen 1–more borne on fl st below infl   **Group I**, lead 7a

   3b Lvs toothed to lobed, or entire but all basal

      4a Lvs in part with 3 apical cuspidate teeth, otherwise entire; petals white, gen reddish- to purplish-spotted; ovary not > 1/5 inferior

         5a Most of lvs 3-toothed and > 10 mm; circumboreal, Alas to Lab, s to BC and reported (but not seen) from our area

                               3 S. tricuspidata Rottb.

         5b Most of lvs entire, mostly < 10 mm (see under lead 8a)

                               5 S. bronchialis L.

      4b Lvs various, but not with 3 apical cuspidate teeth; petals various; ovary often > 1/5 inferior

         6a Fls rarely > 10; ovary gen at least 1/2 inferior at anthesis, if < half inferior then most of fls replaced by bulbils; lf bls gen not > 2 cm long or broad; st lvs gen 2 or 3   **Group II**, lead 12a

         6b Fls gen > 10; ovary often < 1/2 inferior at anthesis; lf bls mostly > 2 cm long or broad; st lvs gen lacking (except in *S. mertensiana*) or reduced to mere bracts   **Group III**, lead 15a

Group I

7a Filaments clavate and ± petaloid; lvs fleshy, only (0) 1–3 on fl st, rounded at tip, 3–12 mm, pyriform to spatulate, glab or with few basal cilia; petals white, 3–4 times as long as sepals; seeds with loose, plainly cellular testa; mt meadows and stream banks, gen alp; Alas to Cal, Ida, and Mont; alpine s., Tolmie's s.; 2 vars.   **4 S. tolmiei** T. & G.

   a1 Lvs with few cilia near base; fl sts glandular-pubescent; Alas s to s Ore, in OM as well as Cas   var. **tolmiei**

   a2 Lvs (and pl) glab; known in our area only from Bitterroot Mts, Ravalli Co, Mont, and Hazard Lakes, Idaho Co, Ida, common in Sierran Cal (*S. l.*)   var. **ledifolia** (Greene) Engl. & Irmsch.

7b Filaments not clavate; lvs often > 3 on fl st, sometimes ciliate full length, mostly acute; petals often yellow; testa of seeds not loose and inflated

   8a Petals white, gen purplish-spotted; lvs mostly rigidly pungent, marcescent for many years, (3) 5–18 × 1–2.5 (3) mm, entire or very rarely with 2 tiny, subapical, lateral lobes, coarsely and sometimes harshly ciliate; sea level to arctic-alp; circumboreal, in N Am from Alas to Greenl, s to n Ore, Ida, and in the RM to Colo; matted s., spotted s.; 2 vars.

                               5 S. bronchialis L.

    a1 Lvs lanceolate to linear, entire, acute, mostly > 4 times as long as broad, cilia coarse, gen < 0.3 mm; BC to most of mont Wn, s, in RM esp, to much of Ida, w Mont, ne Ore, and in RM to Colo and NM (*S. a., S. cognata*)   var. **austromontana** (Wieg.) Jones

    a2 Lvs oblong to spatulate, occ slightly 3-lobed near tip, mostly not > 4 times as long as broad, rounded to obtuse, cilia ∝, 0.3–0.5 mm; CRG and adj Cas of Wn and Ore (n to Mt Rainier), Saddle Mt, Clatsop Co, Ore, and on Mt Baldy, OM, Wn (*S. v.*)

                         var. **vespertina** (Small) Rosend.

8b Petals yellow; lvs not rigidly pungent

   9a Pl with ∝ lfless, filiform, axillary stolons; ovary half inferior at anthesis; hairs of infl purplish, glandular; lvs oblanceolate to linear-oblanceolate, 10–20 × 2–4 mm, glandular-ciliate, acute-acicular at tip; moist rocks and alp scree; circumboreal, s in N Am to BC and in RM from s Mont to NM and Ariz; stoloniferous s. (*S. setigera*)

                         6 S. flagellaris Willd.

   9b Pl not stoloniferous; ovary gen much < half inferior at anthesis; hairs of infl often not purplish; lvs various

      10a Petals ovate to obovate, 5–7 mm, 5–9-nerved, finely cross-rugulose with orange below middle, abruptly narrowed to short but distinct claw; lvs obtuse to rounded, 3–7 (10) mm, glab; sepals reflexed; alp rocky slopes and moraines; RM from Beartooth Mts, n Park Co, Wyo, to NM, to be expected in adj Mont; golden s.

                           7 S. chrysantha Gray

10b Petals not abruptly narrowed to claw, or lvs acute or strongly ciliate; sepals slightly spreading to erect

   11a Basal lvs gen 1.5–5 cm, slender-petiolate; fls 1 per st; calyx and
peduncle rusty-pilose, eglandular; circumboreal; s from Alas to n
BC, also in RM of Colo and to be expected in Mont and Ida
                                          S. hirculus L.

   11b Basal lvs gen ca 1.5 cm × 1–2 mm, not slender-petiolate; fls
often > 1 per st; calyx and peduncle pubescent or puberulent
and gen glandular, never rusty-pilose; RM of Can, s to s BC; evergreen s. *(S. van-bruntiae)*   **8 S. aizoides** L.

Group II

12a Lvs gen narrowly oblong-cuneate to oblanceolate-obovate, apically 3
(5–7)-toothed or -lobed into sometimes linear segms; lvs (at least basal
ones) entire, petiole lacking or broad, no longer than (and not clearly distinguishable from) bl; bulbils lacking

   13a Pl strongly tufted or matted; lvs gen lobed but not toothed, the cauline
mostly entire or at least less deeply lobed than basal ones; sts simple;
petals gradually narrowed to broad base and only slightly or not at all
clawed; filaments > sepals; cliffs, moist rocks, and rocky slopes, from
near sea level to arctic-alp; circumboreal, s in w N Am to most of w US;
tufted s.; 3 vars.   **9 S. caespitosa** L.

      a1 Pubescence of lvs and sts scant, not pilose; petals more nearly oblan-
ceolate than obovate, rarely > 2.5 times as long as the triangular to
ovate sepals; Can RM, s through Mont and c Ida to Colo, NM, and
Ariz *(S. delicatula, S. monticola, Mu. micropetala)* var. **minima** Blank.

      a2 Pubescence abundant, that of lvs often pilose-arachnoid; petals more
obovate than oblanceolate, gen > 2.5 times as long as the oblong-
oblanceolate sepals

         b1 Lvs with short to long cilia, seldom pilose, closely imbricate on gen
tightly compacted sterile shoots; subalp in OM and Cas from c BC
to Mt Rainier *(S. e.)*   var. **emarginata** (Small) Rosend.

         b2 Lvs long-pilose, often ± arachnoid, not closely imbricate; pl
tending to sprawl and br diffusely, lower st lvs often with buds in
axils; CRG, Saddle Mt, and coastal Ore to Lincoln Co, n along coast
of Wn to Island Co, and in OM
                var. **subgemmifera** (Engl. & Irmsch.) Hitchc.

   13b Pl not strongly caespitose, gen 1-std from basal rosette of (more
toothed than lobed) lvs, the cauline commonly more prominently toothed
than basal ones; sts often brd; petals abruptly narrowed to short but distinct claw; filaments < sepals; glacial moraines and alp gravelly
meadows; n RM, s in BC to n Cas, Wn, and to Utah and Colo, w to c Ida
and Wallowa Mts, Ore; Europe; wedge-lf s. *(S. incompta)*; ours the var.
oregonensis (Raf.) Breit.   **10 S. adscendens** L.

12b Lvs at least in part reniform and evenly lobed, basal ones with slender
petiole at least as long as bl; bulbils gen present in infl or in axils of basal
lvs

   14a Bulbils developed in axils of upper st lvs and replacing lower fls in infl;
fl sts gen > 10 cm; ovary scarcely 1/4 inferior; petals 2–5 times as long as
sepals, retuse; stream banks, moist rocks, and glacial detritis, alp, circumboreal; Alas s to Cas, n Wn, and in RM to NM, w to c Ida and Nev;
nodding s. *(S. simulata)*   **11 S. cernua** L.

   14b Bulbils lacking in axils of upper st lvs and in infl, but sometimes present in axils of basal lvs; fl sts gen < 10 cm; ovary ca 1/2 inferior; petals
2–3 times as long as sepals, rounded at tip; damp cliffs, rock crevices,
and talus, alp; BC to Cas-Om, Wn, Blue and Wallowa mts, Ore, Sierran
Cal, and Ariz, e in BC to RM, and s through Mont to e Utah and Colo,
prob in Ida; pygmy s., weak s.   **12 S. debilis** Engelm.

Group III

15a Lvs ± orbicular, sometimes reniform or cordate at base, gen broader
than long; filaments clavate

   16a Stips thin, membranous, mostly sheathing; petioles and fl sts gen pilose; lf bls mostly 3–10 cm broad and nearly as long, commonly shallowly lobed, primary segms 3-toothed; fl sts gen with 1–more lvs or

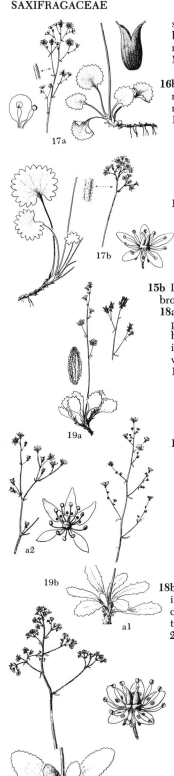

scales below infl; fls often partially replaced by bulbils; basal lvs gen with bulblets in axils; caps reflexed; wet banks and along streams, lowl to high mont; Alas s, in our area from Cas to coast, to nw Cal, e in BC to nw Mont, c Ida, and ne Ore; wood s., Mertens' s. *(S. heterantha)*

13 S. mertensiana Bong.

16b Stips narrow, nonsheathing; petioles and sts glab to pubescent but rarely pilose; lf bls 1.5–8 cm broad, mostly only once-toothed; fl sts gen naked below infl; bulbils lacking in infl; caps erect

17a Infl glandular-pubescent to -puberulent; hairs mostly 1–3-celled, often reddish-purple, gland-tipped; petals dissimilar, larger ones with bl oblong to oval and not > 1.5 times as long as wide, often truncate to slightly cordate near base and narrowed abruptly to slender claw; mont to alp, in moist to wet places; Alas s, through Cas and OM, to s Cal, e to Alta and s in RM to NM and Ariz; common in Mont, Ida, and ne Ore; brook s. *(S. odontophylla, S. odontoloma)* **14 S. arguta** D. Don

17b Infl gen ± pilose with often wavy, several-celled, mostly eglandular hairs; petals alike, bl ± oblong-elliptic, gen much > 1.5 times as long as broad, rounded to cuneate at base and narrowed gradually to rather wide claw; mont streams and wet areas; Alas s, in both OM and Cas, to Ore, e in BC to Alta; Eurasia; dotted s. *(S. aestivalis, S. nelsoniana, S. paddoensis);* ours the var. *cascadensis* (Calder & Savile) Hitchc.

15 S. punctata L.

15b Lvs mostly flabellate to lanceolate or obovate, gen much longer than broad, never reniform; filaments often not clavate

18a Seeds with several longitudinal rows of closely set, flattened, ribbonlike papillae; petals white, abruptly narrowed to short slender claw; lvs flabellate to cuneate-oblanceolate, with 3–11 (17) prominent teeth on apical portion and tapered rather uniformly to from slender to broadly winged petioles

19a Filaments clavate; pl sparsely pubescent with uniseriate hairs; lvs flabellate, (5) 10–25 (50) mm and nearly as broad, narrowed to eventually slender petioles, none of fls replaced by bulbils; mont to alp, along streams or in moist to wet places; Alas to n Cas of Wn, e in BC to w Alta, and s in RM to n Ida and w Mont; red-std s., Lyall s.

16 S. lyallii Engl.

19b Filaments slender, not clavate; pl gen strongly pubescent, larger hairs multiseriate (composed of > 1 row or series of cells); lvs ± cuneate-oblanceolate, narrowed gradually to broadly winged petioles; fls gen partially replaced by bulbils; moist cliffs and banks to stream borders; Alas s in Coastal and Cas mts, through Wn and Ore to nw Cal, e in BC to Ida and w Mont; rusty s., Alas s.; 2 vars.

17 S. ferruginea Grah.

a1 Fls partially replaced by bulbils; more s in range, gradually replaced n by var. *ferruginea (S. nootkana, S. nutkama)*

var. **macounii** Engl. & Irmsch.

a2 Fls mostly if not all normal; Alas s along coast to s VI, occ further s

var. **ferruginea**

18b Seeds variously wrinkled or reticulate but not ridged with rows of papillae; petals white to greenish, sometimes lacking, gen rounded to broad, clawed or clawless bases; lvs ovate to elliptic, bls entire to shallowly toothed most of length, petioles gen not broadly winged

20a Ovary < half inferior at anthesis; stamens inserted at edge of narrow bandlike gland or nectary surrounding but not covering top of ovary; pl mostly 1–2.5 (3) dm; lvs reg crenate-serrate; filaments often clavate

21a Filaments always strongly clavate and petaloid; lvs abruptly rounded to subcordate at base; sepals strongly reflexed; ovuliferous portion of ovary almost completely superior; infl open and diffuse; petals yellow-spotted; gland or nectary visible on maturing or ripe fr; rock crevices and moist to wet banks; Lane Co, Ore, s, mostly near coast, to ne Cal, intermediate farther n to *S. occidentalis;* Marshall's s. *(S. hallii, S. laevicaulis, S. petiolata)*

18 S. marshallii Greene

21b Filaments from strongly clavate to subulate; lvs mostly tapered at base; sepals spreading to reflexed; ovuliferous portion of ovary 1/5–1/3 inferior; infl diffuse to narrow and congested; petals rarely spotted; gland or nectary not visible on maturing fr; wet banks to moist meadows or cliffs; BC s, in OM and Cas, to nw Ore, e to Alta,

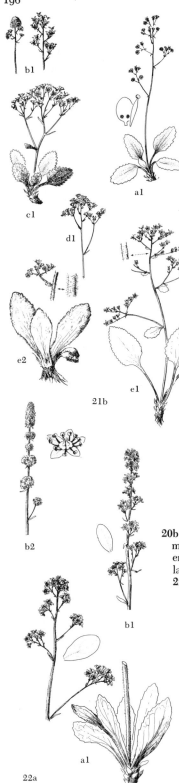

Ida, Mont, nw Wyo, and n Nev; western s., redwool s.; 6 vars.

19 S. occidentalis Wats.

a1 Filaments petaloid, very strongly clavate, 3–4 times as broad above middle as at base; petals sometimes with 2 yellow spots near base; foothills of Wen Mts, chiefly in and near Tumwater Canyon, Chelan and Kittitas cos, Wn, and in lower tributaries of Clearwater and adj SRC, Ida and Ore *(S. i., S. marshallii* var. *divaricata)*   var. **idahoensis** (Piper) Hitchc.

a2 Filaments not petaloid although sometimes clavate, commonly ± subulate; petals not yellow spotted

b1 Infl rather small and ± compact, rounded to pyramidal, mostly < 5 cm at anthesis and < 10 cm in fr, brs ascending to erect; filaments clavate; BC s in Cas to Whatcom and Chelan cos, Wn, e in BC to Alta and s in RM to Wyo, w to Ida, ne Ore, and se Wn *(S. microcarpa, S. o.* var. *wallowensis, S. saximontana)*   var. **occidentalis**

b2 Infl open, often flat-topped, brs erect to spreading or divaricate, often > 5 cm at anthesis and > 10 cm in fr; filaments clavate to subulate

c1 Infl mostly flat-topped; bracts and calyces often reddish pilose-lanate; filaments slightly or not at all clavate; ovary 1/3–1/4 inferior at anthesis; scapes, sepals, ovaries, filaments (and sometimes petals) strongly purple-tinged; mostly w of Cas and near coast; BC to nw Ore and in CRG to Wasco Co, *(S. klickitatensis, S. r.)*   var. **rufidula** (Small) Hitchc.

c2 Infl gen pyramidal; bracts and calyces seldom reddish- pilose; ovary mostly < 1/3 inferior at anthesis; pl often not purple-tinged

d1 Filaments clavate; infl not diffuse, brs mostly ascending; s BC s along e side Cas, at fairly high elev, to Mt Rainier and s side Mt Adams, into CRG *(S. a., S. lata)*   var. **allenii** (Small) Hitchc.

d2 Filaments only slightly or not at all clavate; infl diffuse, brs often stiffly divaricate; pl mostly of lowl valleys or foothills

e1 Lvs mostly narrowed to distinct petiole almost or quite as long as bl; hairs of infl mostly short, with purplish- or reddish-glandular tips; lower CRG to Tillamook Co, Ore *(S. gormani)* var. **dentata** (Engl. & Irmsch.) Hitchc.

e2 Lvs gradually narrowed to broad petiole much < bl; infl densely pubescent with wavy, yellow-glandular hairs; Onion Peak and Saddle Mt, Clatsop Co, Ore   var. **latipetiolata** Hitchc.

20b Ovary either > half inferior at antheses or plant ave > 3 dm; stamens at edge of flattened, lobed disc ± covering greater part of ovuliferous portion of ovary; lvs gen entire to sinuate or remotely denticulate; filaments not clavate

22a Fl sts 3–12 dm; infl elongate and narrow to openly corymbose, rarely < 1 dm in late anthesis; lvs gen much > 5 cm, slenderly ovate to oblanceolate, narrowed gradually to sessile base or to broadly winged petiole; neither lvs nor bracts ever reddish pilose-lanate; bogs to wet meadows and stream banks, w slope Cas from Snohomish Co, Wn, to Cas and WV, Ore, and to Sierran Cal, e in Ore to mont Ida and Mont and s to Wyo and Colo; bog s., Ore s.; 3 vars.   **20 S. oregana** Howell

a1 Petals white, mostly 3–5 mm and half as broad; infl broadly pyramidal at early anthesis, becoming open and with remote, long, often spreading lower brs; ovary gen not > 1/3 inferior at anthesis, in fr almost superior; scapes often strongly long-hirsute; range of sp. in w Wn and Ore, e to e Ore and occ to w Mont   var. **oregana**

a2 Petals white to greenish-white or greenish, mostly 2–3 (4) mm, < half to > half as broad, occ some or all lacking; infl narrow and elongate, brs nearly erect; ovary gen at least 1/3 (1/2) inferior at anthesis, in fr ca 1/3 inferior; scapes sparsely hirsute to glandular-pubescent

23a

b1 Filaments, frs, and sepals gen greenish; petals gen > 2 mm (not rarely lacking); Ida and Mont to Colo, w to e Ore (*S. arnoglossa, S. m.*)   var. **montanensis** (Small) Hitchc.

b2 Filaments purplish-tinged to deeply reddish-purple; fr gen and sepals often also reddish or purplish; petals lacking or barely 1 mm; c Mont to Wyo, sometimes with var. *montanensis* (*S. s.*)   var. **subapetala** (E. Nels.) Hitchc.

22b Fl sts gen < 3 dm; infl often congested and < 1 dm, even in fr; lvs various, frequently with narrow petioles, sometimes reddish-pilose, as also the bracts

23a Lvs with deltoid to rhombic-deltoid, crenate-serrate to crenate-dentate bls and short, broadly winged petioles, glab on upper surface but gen rusty-arachnoid-pilose beneath; infl closely congested and glomerate or interrupted-glomerate; fl sts 1–2 (0.5–3) dm; strongly glandular-pubescent above; moist areas in sagebr des to subalp meadows of RM; s BC and Alta to Colo, w to Utah and c Ida; diamondlf s. (*S. greenei, S. rydbergii*)
                          21 S. rhomboidea Greene

23b Lvs gen entire or only denticulate, pubescent or glab; infl mostly more open, sometimes diffusely cymose-pan; fl sts 1–3 (4) dm; prairies and wet banks to subalp meadows; BC to Cal, e to c Ida and Nev; swamp s.; several vars.   **22 S. integrifolia** Hook.

a1 Petals white, 1.5–3 (4.5) mm, ovate to obovate, gen at least half as broad as long

b1 Lf bls narrowly rhombic to ovate-lanceolate or rhombic-ovate and narrowed to broad petioles rarely as long, mostly strongly ciliate-pilose and slightly to densely coarse-hirsute as well as rusty-arachnoid beneath; w Cas, BC to Lincoln Co, Ore, CRG to Wasco Co (*S. aphanostyla, S. bidens, S. laevicarpa*)   var. **integrifolia**

b2 Lf bls deltoid to ovate or lanceolate, narrowed abruptly to slender petioles often as long or longer, glab except for weak ciliation or sometimes loose rusty floccosity on lower surface; e Cas, Okanogan Co, Wn, to s Ore, CRG and WV, Ore, e to se Wn, w Ida, and ne Ore (*S. fragosa*)
                          var. **claytoniaefolia** (Canby) Rosend.

a2 Petals gen yellowish or greenish-white, often purple or pink tinged (lacking), 1–2.5 mm, mostly spatulate or oblanceolate, (2) 2.5–3.5 times as long as broad

c1 Lvs rhombic-lanceolate, narrowed gradually to broad, often strongly ciliate-pilose petioles; sts gen strongly hirsute at base and glandular-pubescent above; infl sometimes glomerate

d1 Infl densely congested, ± capitate, rarely with any lower brs showing; petals gen lacking; anthers yellow; occ along e side Cas, Okanogan Co to Yakima Co, Wn (*S. a.*)
                          var. **apetala** (Piper) Jones

d2 Infl open, lower brs ± distant (1–2 cm apart); petals gen present; anthers mostly orange; s BC s, e side Cas, to n Ore, down CRG, e to e Wn, w Ida, and perhaps Mont (*S. plantaginea*)   var. **leptopetala** (Suksd.) Engl. & Irmsch.

c2 Lvs mostly rhombic-ovate, narrowed to rather slender, weakly ciliate to glab petioles; sts sparsely pilose to nearly glab at base and gen moderately glandular-pubescent above; infl not glomerate; Blue and Wallowa mts, Ore, to adj Wn and c and sw Ida, s to n Nev and ne Cal (*S. c.*)
                          var. **columbiana** (Piper) Hitchc.

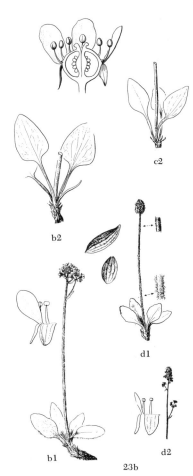

c2

b2

b1

d1

d2

23b

Suksdorfia Gray   Suksdorfia

Fls few–∞ in an elongate to ± flat-topped cymose infl, borne opp rather than axillary to bracts, if any; sepals erect to spreading, as long as adnate portion and free portion of hypan combined; petals white or purplish-violet, entire or ± bilobed; stamens 5, opp sepals; ovary 2-celled, from ca 1/2 to almost completely inferior, prolonged into hollow, tapered, stylelike beaks; stigmas nearly sessile, capitate; seeds ± prismatic, lightly to strongly warty; herbaceous pers with short, sparsely to copiously bulbiferous rhizomes, lfy fl sts, and crenate to deeply divided, cordate to reniform basal lvs and strongly stip st lvs. (For W. N. Suksdorf, 1850–1932, outstanding collector and student of flora of northwest). *(Hemieva)*.

1a Petals violet (white), erect, 6–9 mm; fls 1–10 in an open, loose, racemelike or pan cyme; calyx narrowly turbinate-camp; hypan not disc-lined; mossy banks, moist rock crevices, and shaded cliffs where wet at least early in spring; BC s at scattered intervals on e side Cas, to CRG, e to n Ida and nw Mont; violet s.   **1 S. violacea** Gray
1b Petals white, spreading, 2.5–4 mm; fls ∝ in a ± flat-topped pan cyme; calyx shallowly and broadly camp; hypan lined with thick disc partially covering ovary; foothills to subalp slopes, on wet, mossy rocks, often where dry by midsummer; BC s, mostly on e side Cas, to n Cal, e to Alta, Mont, c Ida, and ne Ore; buttercuplvd s. *(Saxifraga r., Boykinia r.)*
                                                                        2 S. ranunculifolia (Hook.) Engl.

Sullivantia T. & G.   Sullivantia

Fls mostly ∝ in paniclelike or racemelike cymes, perig; calyx broadly turbinate, 5-lobed; petals 5, white; stamens 5, opp sepals; ovary 2-celled; caps dehiscent along ventral suture of beaks; seeds several; per, stoloniferous herbs with lfy fl sts and toothed lvs. (For W. S. Sullivant, 1803–1873, a moss specialist).

S. oregana Wats. Pl yellowish-green, glandular above; lvs membranous-stip, bl cordate-reniform, 1–10 cm broad, incised-lobed 1/3–1/2 length into cuneate, sharply toothed segms; st lvs 1–3, reduced upward to ± lfy bracts; fl sts 5–25 cm; calyx 2.5–3.5 mm, lobes triangular, ca = the adnate and free portions of hypan combined; petals slightly > sepals; stamens < sepals; fertile portion of ovary ca 2/3 inferior, sterile upper part tapered and beaklike, true styles lacking; caps ca 4 mm long; seeds ± slenderly fusiform, slightly wing-margined, lightly reticulate-pitted; moist cliffs, CRG and lower WV, Ore.

Telesonix Raf.

Fls complete, reg, showy, bracteate-pan; calyx turbinate-camp, adnate to lower part of ovary, forming a free hypan above and with 5 ovate-lanceolate sepals; petals 5; ovary 2-celled, ca 1/2 inferior, styles free or partially connate; per herbs. (Derivation obscure).

T. jamesii (Torr.) Raf. James' saxifrage. Sts 5–20 cm, glandular-pubescent, often reddish or purplish above, 1–2-lvd; lvs with broad stip base, mostly from the thick rootstocks, bls reniform, (1) 2–6 cm broad, doubly crenate to shallowly lobed and doubly crenate-dentate; fls 5–25, often secund, reddish-purple; calyx 9–13 mm, lobes ovate-lanceolate, ca 2/5 total length of calyx; petals obovate to spatulate, reddish-purple, 2–3 mm; seeds brown, shining, 1–2 mm; moist rock crevices and talus slopes, often on limestone; Alta to SD, s in high mts of Mont and e Ida to Colo, Utah, and s Nev *(Saxifraga j., Boykinia j., Therofon heucheraeforme)*; ours the var. *heucheriformis* (Rydb.) Bacigalupi.

Tellima R. Br.   Fringecup

Fls complete, in long, minutely bracteate racemes; calyx camp-tubular, 5-lobed; petals 5, pinnately divided; ovary 2-carpellary, 1-celled; per with sparsely lfy fl sts and membranous-stip lvs. (Anagram of *Mitella*).

T. grandiflora (Pursh) Dougl. Pl to 8 dm, rhizomatous; basal lf bls 3–10 cm broad, cordate-ovate to ± reniform, shallowly (3) 5–7-lobed and once or twice crenate-dentate; st lvs 1–3, reduced upward; petioles 5–20 cm; calyx greenish, 5–8 mm at anthesis and up to 11 mm in fr, free hypan well developed, scarcely flared, much > adnate base of calyx; sepals erect, triangular-ovate; petals greenish-white to deep reddish, short-clawed; ovary tapered into hollow, beaklike, eventually dehiscent portions nearly as long as fertile portion; moist woods, stream banks, and lower mt slopes; s Alas to San Francisco Bay region, mostly w of Cas in Wn and Ore, but up CRG and e in BC to se BC, n Ida, and ne Wn *(Mitella g., Tiarella alternifolia, T. breviflora, T. odorata)*.

Tiarella L.   Coolwort; Foamflower; False Mitrewort; Laceflower

Fls complete, pan; calyx irreg, upper sepal gen largest, hypan 1/2–1/4 as long as sepals, camp, almost or quite free of ovary; petals white, resembling the filaments; ovary superior, 1-celled, divided > half length into 2 erect, sterile, unequal, hornlike extensions ending in filiform styles; caps few-seeded, dehiscent along sterile valves above fertile portion; rhizomatous per herbs with cordate and ± palmately lobed lvs. (L diminutive of Gr *tiara*, ancient Persian headdress, which fr resembles). *(Blondia, Petalosteira)*.

T. trifoliata L. Pl 1.5–4 dm, hirsute and often glandular; calyx finely glandular-puberulent, 1.5–2.5 mm, lobed ca half length; petals linear; styles 1.5–3 mm; valves of caps unequal, one ca 3–5 and other 7–10 mm; damp woods; Alas s, along coast and inl, to Cal, e to Alta, Ida, and Mont; 3 vars. often treated as separate spp.

a1 Lf bls simple or (very rarely) some of upper st lvs 3-foliolate, shallowly to rather deeply 3–5-lobed; basal bls up to 12 cm broad; Alas to Cal, in both OM and Cas, mostly above 2000 ft, e to sw Alta, w Mont, n Ida, and ne Ore; coolwort f.   var. **unifoliata** (Hook.) Kurtz.

a2 Lf bls 3-foliolate, lflets petiolulate; basal bls rarely > 9 cm broad

b1 Lflets gen lobed no > half length, not laciniately cleft; Alas to n Ore, from coast up to 3500 ft elev in s Cas, e in BC to RM and s to n Mont and w Ida; not on e side Cas in Wn and Ore; trefoil f. *(T. stenopetala)*
                                                                        var. **trifoliata**

b2 Lflets cleft or divided full length and ± laciniate into ultimate narrow, oblong segms; VI and adj PS is., Wn; cutlf f.
                                                              var. **laciniata** (Hook.) Wheel.

Tolmiea T. & G.   Youth-On-Age; Pig-a-Back- Plant; Thousand Mothers

Fls complete, racemose; calyx tubular-camp, cleft almost to base between the smaller sepals; petals gen 4, linear-subulate; stamens 3, opp upper sepals; ovary 1-celled, 2-carpellary, superior; fr caps; seeds ovoid, finely spinulose; per herbs with cordate, palmately veined and shallowly lobate, membranous-stip, long-petiolate lf bls. (For Dr. William F. Tolmie, 1812–1886, surgeon for Hudson's Bay Co., Ft. Vancouver). *(Leptaxis)*. Reproducing vegetatively by buds at base of lf bls; a good house pl.

T. menziesii (Pursh) T. & G. Pl hirsute and ± glandular, rhizomatous, up to 8 dm; lf bls 3–10 cm broad, not quite so long, shallowly 5–7-lobed and once or twice crenate-dentate, reduced upward; racemes loosely ∝ -fld, 1–3 dm; calyx greenish-purple to chocolate colored, oblique at base, free hypan 5–9 mm; sepals 3–5 mm, unequal; petals chocolate, up to twice as long as sepals; stamens unequal; caps slender, 9–14 mm, prolonged above fertile portion into 2 divergent, obconic, hollow, ventrally dehiscent, unequal beaks ending in slender styles; seeds ca 0.5 mm; moist woods, esp near streams; Alas s, from coast to w side Cas, through Wn and Ore to Cal *(Heuchera m., Tiarella m.)*.

GROSSULARIACEAE   Currant or Gooseberry Family

Fls racemose, mostly 2–several, ours complete; calyx adnate to ovary and with tubular to saucer-shaped free hypan; sepals 5 (4–6), gen more showy than petals; petals and stamens 5 (4–6), borne near top of free hypan; ovary inferior, 1-celled with 2 parietal placentae; styles 2, distinct to fully connate; fr a several-seeded berry; ours spinose or unarmed, deciduous shrubs with simple, alt, palmately lobed lvs; stips absent or adnate to the petiole.

Ribes L.   Currant; Gooseberry

Fls greenish, white, yellow, pink, red, or purplish, sometimes showy; peds often jointed immediately below ovary; sepals erect or spreading to sharply reflexed; petals always smaller than the sepals, oblong to cuneate or reniform, mostly ± clawed; berry globose or subglobose, ∝ -seeded, yellow, red, blue, or black, often glaucous, glab to pubescent and often glandular or ± spiny, palatable to ill-tasting; glab to pubescent and often glandular shrubs, unarmed, or with 1–several nodal spines and often also with internodal bristles. (From *ribas*, Arabic name for these pls). *(Calobotrya, Cerophyllum, Coreosma, Grossularia, Limnobotrya)*.

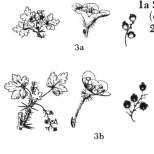

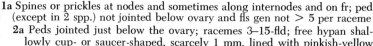

1a Spines or prickles at nodes and sometimes along internodes and on fr; ped (except in 2 spp.) not jointed below ovary and fls gen not > 5 per raceme
  2a Peds jointed just below the ovary; racemes 3–15-fld; free hypan shallowly cup- or saucer-shaped, scarcely 1 mm, lined with pinkish-yellow disc; petals pinkish to purple; berry setose-glandular, not prickly

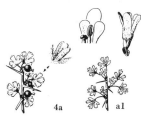

  3a Lvs strongly pubescent and ± glandular, 1–2.5 (4) cm broad; berries reddish; peds stout, gen < twice as long as bracts; pl 2–5 (7) dm; subalp to alp talus slopes and rocky bluffs; s BC s, on e slope Cas, to Sierran and s Cal, e to RM, Mont to NM; alpine prickly c., mt g. *(R. lentum, R. molle, R. nubigenum)*   **1 R. montigenum** McClatchie
  3b Lvs glab to sparsely pubescent, rarely glandular, gen 1–5 (7) cm broad; berries dark purple; peds slender, often > twice as long as bracts; pl 10–15 (20) dm; moist woods and stream banks to for slopes and subalp ridges; Alas s, in both OM and Cas, to Cal, e to Newf, Daks, Mich, Pa, and Colo; swamp g., swamp black g., prickly c. *(R. echinatum, R. parvulum)*   **2 R. lacustre** (Pers.) Poir.

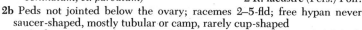

2b Peds not jointed below the ovary; racemes 2–5-fld; free hypan never saucer-shaped, mostly tubular or camp, rarely cup-shaped

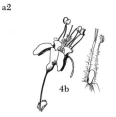

  4a Styles no > petals, connate full length; stigma 1, enlarged, bilobed, oblong; fls < 8 mm (fresh or boiled), strongly crisp-pubescent outside; lf bls 3 (5)-cleft for > half their length, mostly < 15 mm broad; petals white or yellowish, oblong-obovate, ca half as long as sepals; hypan 1.5–2 (2.5) mm, ca as broad; des hillsides to ponderosa pine for; se Wn and e Ore to s Cal, e to sw Ida, Utah, Nev, and Ariz; 2 vars.
                                                                 3 R. velutinum Greene
    a1 Lvs mostly glab, cleft > half length into 3 (5) slender, cuneate, entire to 3-toothed segms; ovary glab or sometimes sparsely crisp-hairy, never glandular; SRC and tributaries, se Wn and e Ore to Lemhi Co, Ida; Goodding's g. *(R. g.)*
                                                          var. **gooddingii** (Peck) Hitchc.

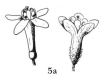

    a2 Lvs densely pubescent and often glandular, cleft mostly < half length into broader, 2–3-toothed segms; ovary pubescent to glandular (glab); Crook Co, Ore, to Cal and Utah; des or plateau g. *(R. glanduliferum)*   var. **velutinum**
  4b Styles mostly > petals, gen distinct for at least 1/4 length; stigmas 2, capitate-discoid; fls gen > 8 mm, or smaller but never crisp-pubescent (sometimes sparsely pilose) outside; lf bls seldom cleft > half length, often much > 15 mm broad   **Group I,** lead 6a

1b Spines and prickles lacking; ped jointed below ovary; fls gen > 5 (rarely only 2) per raceme
  5a Free hypan camp to cylindric, never saucer-shaped or shallowly cup-shaped, at least as long as broad; ovary without sessile, yellow, crystalline glands; fls sometimes bright yellow   **Group II,** lead 18a
  5b Free hypan saucer-shaped or very shallowly cup-shaped, broader than long; ovary sometimes with sessile, yellow, crystalline glands; fls never bright yellow   **Group III,** lead 24a

Group I

6a Styles hairy on lower half, at least; either sepals not crimson or berry neither prickly nor glandular
  7a Stamens mostly > twice as long as petals, sometimes > sepals and much-exserted; sepals white or greenish to purple
    8a Sepals white or slightly greenish, rarely pinkish-tinged, 5–8 mm, mostly at least twice as long as the 2–3 mm hypan; stamens gen 1–3 mm > sepals; anthers hairy; lvs pubescent at least on lower surface; styles gen connate much > half length; thickets along streams and watercourses or on open rocky hillsides; nc Ore to n Nev, n to se Wn and e to w Ida; snow g.   **4 R. niveum** Lindl.
    8b Sepals greenish to pink or purple, sometimes < twice as long as hypan; stamens mostly < sepals; anthers glab; lvs sometimes glab except for marginal cilia; styles seldom connate > half length

      9a Sepals strongly purplish-tinged, 5–7 mm; stamens 1–2 mm > sepals; open woods to moist hillsides, w of Cas in our area; BC to

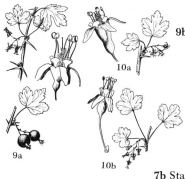

s Cal, from coast to lower mts, up CRG to Klickitat Co, Wn; straggly g., coast black g. *(R. suksdorfii)*   **5 R. divaricatum** Dougl.

9b Sepals greenish or only slightly purplish-tinged, 3–4 (5) mm; stamens only slightly if any > sepals; e of Cas in our area

    10a Lvs mostly glab on at least one surface; calyx seldom hairy on outside; hypan (2) 2.5–3.5 mm, but not so broad, nearly or fully = sepals; stamens mostly no > sepals; thickets, wooded slopes, and open ridges; BC s, e Cas, to Sierran Cal, e to Mont, Wyo, Colo, and NM; whitestem g. *(R. purpusii)*   **6 R. inerme** Rydb.

    10b Lvs mostly thickly pubescent on both surfaces; calyx gen hairy on outside; hypan 1.2–2 mm and as broad, rarely > half as long as sepals; stamens gen slightly > sepals; edge of streams and marshes from Jefferson Co, Ore, s (e Cas) to n Cal; Klamath g.
        7 R. klamathense (Cov.) Fedde

7b Stamens rarely as much as twice as long as petals, never = sepals, not strongly exserted even with sepals reflexed; sepals white or pale greenish, sometimes reddish tinged

    11a Hypan flared, ± camp, 2–3.5 (4) mm, ca as broad (at top) as long; fls gen < 10 (11) mm with sepals extended; anthers barely 1 mm

        12a Fls 8–11 mm with sepals extended; sepals at least half again as long as hypan; styles often connate to above midlength; pls mostly erect, 0.5–2 m, brs slender, rarely at all bristly when young; moist to dry canyons and wooded hillsides; se BC to se Wn and adj Ore. e to w Mont, not e of Con Div; Ida g. *(R. leucoderme)*
            8 R. irriguum Dougl.

        12b Fls rarely > 8 mm with sepals extended; sepals scarcely > hypan; styles connate to no more than midlength; pl often low and intricately brd, bristly when young

            13a Styles from < hypan to ca = petals, connate for < 1 mm; petals 1–1.5 mm, ca half as long as sepals; anthers = filaments; young brs sparsely or not bristly; nodal spines crowded; pl gnarled and much-brd; limestone cliffs and talus; Lost River and adj mts of Custer and Lemhi cos, Ida, and Anaconda Range, Deerlodge Co, Mont; Henderson's g. *(R. neglectum)*
                9 R. hendersonii Hitchc.

            13b Styles > petals, often ca = sepals, connate for ca half length; petals 2–2.5 mm, nearly = sepals; anthers ca half as long as filaments; pl gen armed more with internodal bristles than nodal spines, but sometimes with only nodal spines; prairies and lower mts; e BC to Newf, s to NC and Mich, perhaps not s of Can border in our area; northern g.   **10 R. oxyacanthoides** L.

    11b Hypan tubular, only slightly if at all flared, (4) 5–6 mm, not so broad; fls mostly > 10 mm with sepals extended; anthers gen > 1 mm

        14a Calyx glab outside; fls 10–13 mm; hypan 4–5 mm; pl spreading, 0.5–1 (3) m; along valley watercourses and lower hillsides; Alta to Assiniboia, s to c and e Ida, Mont, Wyo, Neb, and Mich; Missouri g. *(R. camporum, R. saximontanum)*   **11 R. setosum** Lindl.

        14b Calyx ± finely pilose outside; fls 11–16 mm; hypan 5–6 mm; pl erect to clambering, up to 3.5 m; dry to moist watercourses and lower hillsides; se Wn and adj Ore to nc Ida; Umatilla g.
            12 R. cognatum Greene

6b Styles glab; sepals often crimson and berry prickly or glandular

    15a Anthers lanceolate, at least 2 mm, broadest at base and tapered to indehiscent mucronulate tips, smooth on back; sepals crimson; berry prickly

        16a Pl semiglab, lvs never glandular; ped < the subtending, (3) 4–5 mm bracts; anthers reddish; brs minutely puberulent but not bristly; bark reddish-brown; petals 3.5–5 mm, deeply erose; hypan (4.5) 5–6 mm; sepals 8–11 mm; gravel bars, riverbanks, and lowl canyons to subalp ridges; Lane Co, Ore, to c Cal, w Cas; shinylf g.
            13 R. cruentum Greene

        16b Pl strongly pubescent, lvs and calyx glandular; ped 2–3 times as long as the 2–3 mm bracts; anthers white; brs bristly, becoming grayish or straw-colored; petals 3–4 mm, shallowly and coarsely erose; hypan 2.5–3.5 mm; sepals 5–10 mm; thickets and open slopes; Lane Co, Ore, to c Cal, w Cas; prickly g., Menzies' g.   **14 R. menziesii** Pursh

    15b Anthers ± oval, gen < 2 mm, broadest near midlength, not mucronu-

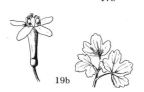

17a

17b

late, dehiscent to tip, smooth or warty to capitate-papillate on back; sepals often greenish; berry stipitate-glandular to prickly

17a Sepals greenish or only slightly reddish tinged, (5.5) 6–9 mm, nearly 2 × as long as petals and stamens; anthers white, smooth on back; berry strongly bristly with spines 2–4 mm; canyons and ridges; both sides Cas and in Wen Mts, Chelan Co, Wn, to e side Mt Hood, Ore; spiny g., Watson g.   **15 R. watsonianum** Koehne

17b Sepals crimson, 10-13 mm, ca twice as long as petals and ca = stamens; anthers reddish or purple, warty or capitate-papillate on back; berry stipitate-glandular but not spiny; lowl valleys and streamcourses to for or open mont slopes; BC s to Cal, mostly in Cas but to coast, more abundant on e side Cas in Wn; gummy g., Lobb's g.

16 R. lobbii Gray

Group II

19b

18a Fls bright yellow, glab (golden currants)

  19a Hypan > 10 mm, ca twice length of sepals; e RM, prob not native in our area, but possibly an escape; golden c. (*R. longiflorum* Nutt.)

17 R. odoratum Wendl.

  19b Hypan < 10 mm, gen < twice as long as sepals; stream banks and washes in grassl or sagebr des to ponderosa pine for; e slope Cas, nc Wn to Cal, e to e side RM, Sask and SD to NM; golden c.

18 R. aureum Pursh

18b Fls not bright yellow, mostly glandular or pubescent

  20a Pl (except ovaries) with scattered sessile, yellowish, crystalline glands; fls greenish-white, hypan 3–4 mm, = or slightly > sepals; petals white, 2/3–4/5 as long as sepals; berry smooth, not palatable; swamps to moist canyons; Alta to NM, mostly e RM, but in our area in c Mont, where almost w to Missoula, e to NS and Va; black c. (*R. floridum*)

19 R. americanum Mill.

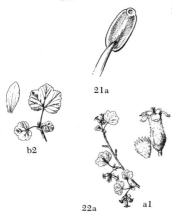

20a

21a

b2

22a   a1

20b Pl eglandular or with glands other than sessile and crystalline; fls diverse

  21a Anthers tipped with small, cuplike gland visible under 10X magnification; fls white or pink to green, rarely reddish-tinged; free hypan = to nearly twice as long as sepals

    22a Hypan twice as long as sepals, greenish-white or white to pinkish-tinged, (5) 6–9 mm; sepals 1.5–3 mm; lf bls 1.5–2.5 (0.5–4) cm broad; petals 1–2 mm; berry dull to bright red, unpalatable; BC s, on e side Cas, to s Cal, e to Mont, Neb, Colo, and NM; squaw c., wax c.; 3 vars.   **20 R. cereum** Dougl.

      a1 Fl bracts ± fan-shaped, blunt to broadly rounded and several-lobed to prominently toothed; lvs glab to pubescent and ± glandular on both surfaces; e side Cas, BC to Cal, e to c Mont, Ida, and w Nev (*R. pumilum, R. reniforme, R. spathianum*)

var. cereum

      a2 Fl bracts ovate to obovate, gen pointed, entire to sharply denticulate or with 2–3 shallow lobes

        b1 Pl mostly strongly pubescent; calyx pubescent and glandular; lvs mostly < 15 (20) mm broad; c Ida to c Mont, e and s to Neb, NM, Utah, and e Nev (*R. i.*)

var. inebrians (Lindl.) Hitchc.

        b2 Pl mostly semiglab on lvs and calyx; lvs 15–30 mm broad; SR and tributaries, wc Ida and adj Wn and Ore

var. colubrinum Hitchc.

    22b Hypan ca = sepals, whitish to greenish, often pinkish tinged, (5) 6–8 mm; sepals subequal to hypan; lf bls (2) 3–10 cm broad; petals 2.5–4 mm; berry deep bluish-black, unpalatable; creek banks to moist or dry, open to heavily timberl slopes, up to timberl; BC s, mostly e Cas through Wn and Ore to Sierran Cal, e to Mont, Wyo, Colo, and Ariz; sticky c.; 2 vars.   **21 R. viscosissimum** Pursh

      a1 Ovary glab or very sparsely stipitate-glandular; berry glaucous; Cas, occ from Mt Rainier s, becoming common in Cal (*R. hallii*)

var. hallii Jancz.

      a2 Ovary rather strongly glandular, gen also soft-pubescent; berry slightly if at all glaucous; BC to Mt Rainier, Wn, e through Wn

22b

23a

23b

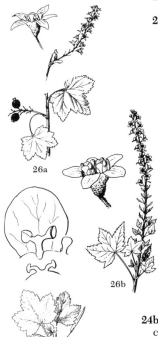

26a

26b

28a

28b

and ne Ore to Mont and Colo, also in Utah, Nev, Ariz, and s Sierran Cal   var. **viscosissimum**

21b Anthers not gland-tipped; fls gen pale to deep rose (at least when dried); hypan < sepals

    23a Hypan 3–5 mm, tubular-camp; sepals only slightly or no > hypan; fls pale to deep pink when fresh or dried; petals 2.5–3.5 mm; berry glaucous black, unpalatable; open to wooded, moist to dry valleys and lower mts; BC to below San Francisco, Cal, from coast to e slope Cas in Wn and n Ore; red or blood c., redfl c.

                                          22 R. sanguineum Pursh

    23b Hypan 1–2 (2.5) mm, saucer- or bowl-shaped; sepals slightly to nearly 2 × > hypan; fls greenish-white to yellowish when fresh, sometimes drying pinkish; petals 1–2 mm long; berry black; mts of Utah and Colo to Ariz, known in our area from Blue Mts, se Wn, and Seven Devils Mts, Ida; Wolf's c. (*R. mogollonicum*)

                                          23 R. wolfii Rothr.

Group III

24a Ovary and gen young lvs with scattered sessile, yellow, crystalline glands; lvs mostly 5–12 (20) cm broad

    25a Racemes drooping; hypan ± cup-shaped; sepals gen purplish-tinged; cult black currant, occ escaped   **24 R. nigrum** L.

    25b Racemes erect or spreading; hypan saucer-shaped, widely flared, 1–1.5 mm; petals white, ca 1.5 mm; sepals often whitish or green; pl with sweetish, rather disagreeable odor; berry bluish-black, not palatable

        26a Hypan and sepals white; racemes mostly 4–10 (17) cm; bracts narrowly linear-lanceolate, 1–3 mm, all < peds, membranous, not greenish; lvs 3-lobed < half length, lower lobes again much less deeply but very unequally 2-lobed; stream banks, moist woods, and margins of meadows in mts; Alas to n Cal, E Cas in our area, e to Ont, Minn, and Utah; stinking c., n black c.; 2 vars.

                                  25 R. hudsonianum Richards.

          a1 Pl rather hairy; lf bls pubescent on lower surface and gen also on upper; ovary often glandless; Alas to s BC and occ to Okanogan Co, Wn, e to Hudson's Bay, n Ida, Mont, and Minn; Hudsonbay c.

                                      var. **hudsonianum**

          a2 Pl glab except for glands, to lightly pubescent on fls, twigs, petioles, and along veins of lower surface of lvs; ovary glandular; s BC s in Cas to sw Ore and n Cal, e to Ida, Mont, Utah, and n Wyo (*R. p.*)

                                var. **petiolare** (Dougl.) Jancz.

        26b Hypan and sepals greenish, gen strongly tinged with purplish-brown; racemes mostly 15–30 cm, lower bracts often ± lflike and > peds, upper ones reduced but rarely < 4 mm, greenish-tipped; lvs primarily 5-lobed much > half their length, lower segms gen again less deeply divided into unequal to nearly = lobes (bls then rather = 7-lobed); Alas to nw Cal, mostly w Cas, but very occ e Cas, as in Okanogan Co, Wn; stink c., blue c.   **26 R. bracteosum** Dougl.

24b Ovary either glab or with stipitate glands, never with sessile, yellow, crystalline glands; herbage rarely with crystalline glands, but if so ovary alwys stipitate-glandular; lvs mostly 4–8 (12) cm broad

    27a Ovary glab and smooth; ripe fr red; sepals mostly at least as broad as long

        28a Calyx greenish-yellow; petals yellowish-red; pl gen erect, 10–15 dm; anthers dumb-bell shaped, pollen sacs separated by almost the width of the filament; cult red currant, sometimes escaped or bird-disseminated near habitations; known from supposedly wild collections in BC, Wn, and Ore   **27 R. sativum** (Reichb.) Syme

        28b Calyx deep purplish or purplish-tinged; petals reddish-purple; pl often decumbent and nodally rooting; anthers broadly cordate, retuse, sacs almost contiguous; frs sour; moist woods to mt rock slides; Alas to Newf, s in Cas to CRG, and to Blue Mts, Wn and Ore, and to SD and Va, but in our area only in BC, Ore, and Wn; Asia; wild or Am red currant (*R. ciliosum, R. migratorium*)   **28 R. triste** Pall.

    27b Ovary pubescent or stipitate-glandular or both; ripe fr mostly dark bluish to black; sepals longer than broad

        29a Racemes drooping to pendent; filaments much widened near base,

29a

32a

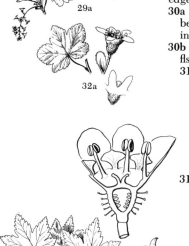

33a

borne at edge of raised-rimmed disc; sepals spreading; petals reddish, obovate-cuneate, 1–1.5 mm, spreading; bracts 3–5 mm, gen = peds; mont to alp stream banks, thickets, and rock slides; s BC to n Ore, in both Cas and OM, also in n Ida; maplelf c. (*R. acerifolium*)
                                                                                      29 **R. howellii** Greene
    29b Racemes spreading to erect; filaments neither flattened nor borne at edge of raised-rimmed disc; petals and bracts various
        30a Hypan ca 2 mm, nearly = sepals, bowl-shaped or shallowly bell-shaped; fls red or deep pink; s Ore to Nev and Cal, not known in our area, but reported for Mont   **R. nevadense** Kell.
        30b Hypan < 2 mm, considerably < sepals, shallowly saucer-shaped; fls various
            31a Bracts of racemes oblong to obovate, 3–4 mm, at least half as long as peds
                32a Fls greenish-white or yellowish-green; lvs gen lobed < half their length, lobes triangular; berry black (see lead 23b)
                                                                                      23 **R. wolfii** Rothr.
                32b Fls yellowish or pinkish; lvs gen lobed > half length, lobes oblong-ovate or oblong-obovate, rounded; berry red; local in Cas of sw Ore where not quite reaching our area
                                                                                      R. erythrocarpum Cov. & Leib.
            31b Bracts of racemes narrowly lanceolate to ovate, mostly 1–2 mm, much < half as long as peds
                33a Petals with reniform or crescent-shaped bl and short claw, at least as broad as long, red to purplish; disc brownish to red, almost flat; ovary with stalked glands < 0.5 mm; coastal woods to mont slopes; Alas to Cal, w Cas in Ore, but e in BC and Wn to sw Alta and n Ida; western or trailing black c. (*R. affine*)
                                                                                      30 **R. laxiflorum** Pursh
                33b Petals cuneate to flabellate, longer than broad, whitish to pink; disc pale pink to greenish-white, saucer-shaped; ovary with stalked glands up to 1.2 mm; e Alas to Lab, s to n BC, Minn, Me, and Appalachian Mts; s limits in BC and Alta uncertain, but not believed to enter our area
                                                                                      R. glandulosum Grauer

HYDRANGEACEAE   Hydrangea Family

Fls often showy, mostly cymose (racemose or solitary), ours complete; calyx ± adnate to ovary, gen without free hypan; lobes 4–6 (ours); petals 4–6, white (ours), distinct; stamens mostly ∝ or at least twice as many as petals; pistil (3) 4–5-carpellary (ours), ovary half to completely inferior; fr a 3–5-celled, ± woody caps with 1–several seeds per cell; trees or erect to trailing shrubs with deciduous to persistent, opp, simple, exstip lvs.

1a

1b

    1a Stamens 25–50; petals at least 5 mm; caps ∝ -seeded   **Philadelphus**
    1b Stamens 8–12; petals < 5 mm; caps with 1 seed per cell   **Whipplea**

Philadelphus L.   Mockorange; Syringa

Fls showy, mostly 3–11 per false raceme; calyx fully adnate to ovary, 4 (5)-lobed; petals 4 (5); stamens ∝ ; pistil 4 (3–5)-carpellary, ovary gen 4-celled; styles connate ca half length; caps woody, loculicidal; erect to spreading deciduous (ours) shrubs. (Gr *philos,* love, and *delphos,* brother, in commemoration of Ptolemy Philadelphus, king of Egypt).

P. lewisii Pursh. Pl 1.5–3 m; lf bl ovate to broadly elliptic, entire to serrate-dentate, glab to strigose, 2–9 cm; petals oblong to ± ovate or obovate, mostly 10–20 (25) mm; along watercourses and on cliffs, talus slopes, and rocky hillsides of sagebr des to ponderosa pine or Douglas fir for; BC to n Cal, from coast inl to Mont and n and c Ida, up to 7000 ft elev e Cas (*P. angustifolius, P. columbianus, P. confusus, P. gordonianus, P. helleri, P. intermedius, P. oreganus, P. platyphyllus, P. zelleri);* a valued orn.

Whipplea Torr.   Yerba de Selva; Whipplevine

Fls inconspicuous, compactly cymose-pan; calyx partially fused with ovary, without free hypan, lobes (4) 5–6; petals (4) 5–6, ca twice as long as sepals; stamens ca twice as many as petals; pistil (3) 4–5-carpellary, ovary ca 1/2 inferior; styles free or connate at base; fr caps; matted shrub with ± marcescent lvs. (For A. W. Whipple, 1818–1863, commander of US RR Expedition in 1853–54).

W. modesta Torr. Sts trailing, freely rooting, up to 1 m, with erect fl shoots 0.5–2 dm, rather coarsely pubescent throughout; lvs semideciduous, subsessile to short-petiolate, ovate to ovate-elliptic, 1–2.5 cm, remotely crenate-serrate; peduncles 2–5 cm; calyx 2.5–3.5 mm, lobes oblong-lanceolate, 1.5–2 mm; petals ca twice as long as sepals, spreading; stamens ca as long as petals, filaments flattened, anthers dumb-bell shaped; caps depressed-globose, separating into 3–5 leathery, ventrally dehiscent, 1-seeded segms; dry, often rocky, open to wooded areas w Cas, from OP, Wn, to Monterey Co, Cal.

ROSACEAE   Rose Family

Fls solitary to ∝, complete or sometimes apetalous or ♀, ♂, gen reg, perig to epig; calyx 5 (4–10)-merous, ± gamosep, often bracteolate between lobes, either adnate to pistil or free and saucerlike to camp or tubular, gen disc-lined; petals 5 (4–10) or sometimes lacking; stamens mostly ∝ but rarely reduced to few or only 1, borne with petals on calyx; pistils 1–several and simple and free of hypan, or sometimes only 1 but 2–5-carpellary, ovary then partially to completely inferior; fr an achene, follicle, drupe, or pome, sometimes an aggregation of achenes or drupelets with recep then dry to fleshy; trees, shrubs, or ann to per herbs, sometimes armed, with alt, basal, or rarely opp, simple to compound, deciduous or evergreen, mostly stip lvs.

1a Pl prickly-std; lvs pinnate; carpels several, borne within a globose to urn-shaped, ultimately thickened and fleshy hypan   **Rosa**
1b Pl not at once prickly-std, pinnate-lvd, and with a thickened and fleshy hypan enclosing several carpels
  2a Calyx adnate to ovary; pistil 1, ovary 2–5-carpellary; fr fleshy; deciduous trees or shrubs with ♂ fls and inferior ovaries   **Group I**
  2b Calyx free of ovary, but sometimes enclosing it; pistils 1–∝, simple, ovary superior; fr dry to fleshy; pl herbaceous to arborescent, occ ♀, ♂
                                                  Group II, lead 7a

Group I

3a Lvs pinnate, lflets 5–17   **Sorbus**
3b Lvs simple
  4a Pl with strong sharp thorns; carpels with hardened, shell-like covering, each 1-seeded   **Crataegus**
  4b Pl unarmed; carpels with papery to cartilaginous walls, 2-seeded
    5a Lvs oblanceolate, 1–4 cm, entire to finely serrulate, narrowed to slender base, but without clearly defined bl and petiole
                                       Peraphyllum
    5b Lvs mostly ovate or oblong to obovate, serrate to strongly toothed or lobed, with clearly defined bl and petiole
      6a Fls racemose; carpels partitioned into 2 cells each, each cell 1-seeded; lvs unlobed   **Amelanchier**
      6b Fls corymbose; carpels not partitioned, each 1-celled and 2-seeded; lvs sometimes lobed   **Pyrus**

Group II

7a Pl shrub or small tree; lvs simple, entire to serrate, never lobed, deciduous; fls corymbose, umbellate, or racemose; calyx turbinate-camp, not completely enclosing fr; pistils 1–5; fr 1–5 drupes; fls sometimes ♂, ♀
  8a Fls ♀, ♂; pistils gen 5; lvs entire   **Oemleria**
  8b Fls ♂; pistil 1; lvs crenulate to serrate   **Prunus**
7b Pl herb or shrub; lvs gen either persistent or lobed to compound (or both); pistils 1–∝; frs various, but always > 5 if drupaceous; fls gen ♂; calyx diverse, often enclosing fr

9a Petals lacking; pistils 1–2 (3); ovary enclosed in calyx
 10a Pl shrubby; lvs entire to toothed, persistent   **Cercocarpus**
 10b Pl herbaceous; lvs pinnate or palmately-lobed or -dissected
  11a Lvs odd-pinnate; style terminal; fls spicate **Sanguisorba**
  11b Lvs palmately-lobed or -dissected; style basal; fls other than spicate   **Alchemilla**
9b Petals gen present; pistils sometimes > 3; ovary sometimes not enclosed
 in calyx
 12a Hypan hooked-bristly at top, enclosing 1–2 achenes; pl herbaceous,
  pinnate-lvd   **Agrimonia**
 12b Hypan not hooked-bristly at top, often not enclosing fr; pl various
  13a Lvs deciduous, simple, cuneate, deeply 3-toothed at tip, otherwise
   entire, revolute-margined, greenish above, tomentose beneath; pistils 1 (2); pl shrub, erect, rigid-brd   **Purshia**
  13b Lvs often persistent or not cuneate and 3-toothed; pistils often >
   2; pl often herbaceous
   14a Lvs persistent, bi- or triternately dissected into linear segms
    15a Pl herbaceous, erect, taprooted; fls cymose; fr 5–20 achenes
                Chamaerhodos
    15b Pl trailing, ± shrubby; fls racemose; fr 4–6 follicles **Luetkea**
   14b Lvs deciduous if either bi- or triternately dissected
    16a Lvs pinnate-pinnatifid into ultimate segms 0.5–1.5 × ca 0.5
     mm; low, glandular, stellate-pubescent shrubs **Chamaebatiaria**
    16b Lvs rarely pinnate-pinnatifid into tiny ultimate segms, if so pl
     not a glandular-stellate pubescent shrub
     17a Pl a prostrate, matted or cushion-forming, woody shrub
      with persistent, ± marcescent, entire to coarsely crenate lvs;
      fls single at br ends or in spikelike racemes
      18a Fls ∝ in pedunculate racemes; perianth 5-merous,
       small; petals 1.5–2.5 mm, white; lvs entire **Petrophytum**
      18b Fls solitary at br ends or on scapes; perianth often
       8–10-merous; petals 2–12 mm, sometimes pink or yellow;
       lvs often crenate
       19a Perianth mostly 8–10-merous; petals 8–12 mm, white
        or yellow; styles plumose in fr; fls pedunculate; lvs
        10–40 mm   **Dryas**
       19b Perianth 5-merous; petals 2–3 mm, light pink; styles
        not plumose; fls subsessile; lvs 2.5–4 mm **Kelseya**
     17b Pl either herbaceous or not trailing; lvs often lobed to
      compound; fls often neither tightly racemose nor solitary at
      tips of br or scapes
      20a Pl shrubby, erect, unarmed; lvs toothed or lobed but
       never compound; pistils (1) 2–7   **Subgroup** IIa
      20b Pl often herbaceous or prickly; lvs often compound; pis-
       tils often > 7   **Subgroup IIb**, lead 23a

Subgroup IIa

21a Lvs palmately 3–5-lobed; calyx gen stellate; fr follicular, ± inflated,
 several-seeded, dehiscent on 2 sutures   **Physocarpus**
21b Lvs pinnately ∝ -lobed to -toothed; calyx not stellate; fr either an achene
 or dehiscent on 1 suture only
 22a Fr a follicle, 2–several-seeded; stamens 25–50; lvs gen finely toothed,
  but sometimes shallowly lobed; petals often pink, red, or purple **Spiraea**
 22b Fr an achene, gen 1 (2)-seeded; stamens gen 20; lvs gen shallowly
  lobed as well as toothed; petals white   **Holodiscus**

Subgroup IIb

23a Pl ♀, ♂, rhizomatous, herbaceous, 1–2.5 m; lvs ternate-pinnatisect to triternate-pinnatisect; pan 1–5 dm; petals white, scarcely 1 mm **Aruncus**
23b Pl mostly ⚥, often not rhizomatous, sometimes shrubby or < 1 m; lvs
 rarely ternate; petals mostly > 1 mm
 24a Lvs pinnate or pinnately divided, terminal segm much largest, broadly
  cordate, palmately (3) 5–7-cleft and doubly serrate, 8–20 cm broad; pl
  herbaceous, not prickly, strongly rhizomatous, 1–2 m; petals white, ca 6
  mm; achenes much-flattened, the stipe 2–3 mm   **Filipendula**

24b Lvs not as above; pl gen woody, prickly, < 1 m, or with other than white petals

25a Hypan narrowly obconic, enclosing the ovaries; fr 2–6 canescent achenes, ultimately largely exserted from the hypan; filaments persistent, stiffly erect in fr; pl herbaceous, rhizomatous; lvs cordate-orbicular, shallowly lobed to toothed **Waldsteinia**

25b Hypan saucer-shaped to camp; pistils gen ∝, frs rarely enclosed by hypan; filaments not persistent and stiffly erect; pl and lvs various

26a Stamens 5; pistils gen 2–15

27a Lvs ternate; style laterally attached to ovary **Sibbaldia**

27b Lvs pinnately dissected; style subterminal on ovary **Ivesia**

26b Stamens at least 10; pistils mostly ∝

28a Stamens 10; pl per; lflets (5) 7–19, gen dissected **Horkelia**

28b Stamens > 10, or pl ann or bien; lflets often < 7

29a Calyx ebracteolate; pl rhizomatous, stoloniferous, or trailing to erect shrubs, sometimes armed; frs drupelets, weakly adherent in a cluster **Rubus**

29b Calyx bracteolate; pl various, often herbaceous, never armed; frs achenes, often partially embedded in fleshy recep

30a Recep enlarged, hemispheric, fleshy or spongy in fr; lvs mostly ternate; sts stolonous, freely rooting at nodes

31a Petals yellow; calyx bracteoles obovate, 3–5-lobed, gen somewhat larger than sepals **Duchesnea**

31b Petals white or pinkish; calyx bracteoles entire to bilobed, gen smaller than sepals **Fragaria**

30b Recep gen smaller than sepals, not enlarged and spongy or fleshy in fr; lvs various; sts gen not stolonous

32a Style or achene, or both, ± pilose; pl never shrubby; style slender, apical, straight to bent or geniculate, persistent in fr although sometimes jointed near midlength, the upper segm deciduous; lvs mostly lyrate-pinnatifid **Geum**

32b Style glab; achene glab except in *P. fruticosa*, a shrub; style apical to lateral or sub-basal, straight, jointed to achene and gen wholly deciduous or readily broken off; lvs rarely lyrate-pinnatifid **Potentilla**

Agrimonia L.   Agrimony

Fls in bracteate, spikelike racemes, small, complete; calyx with obconic free hypan, becoming 10-grooved and much hardened, hooked-bristly near top; sepals persistent; petals 5, yellow, borne with the 5–15 stamens near top of hypan; pistils 2 (3), enclosed by hypan, but ovary superior; fr gen single (2) achene; rhizomatous per herbs with large pinnate lvs and foliaceous stips. (Said to be a variant of *Argemone*).

A. **striata** Michx. Sts 5–10 (15) dm, hirsute above; lflets 5–13, very unequal, upper ones up to 6 cm; stips to 2 cm; racemes 5–20 cm; hypan ca 3 mm; dry to moist, well-drained soil; e BC to NS, w on e side RM to NM, Ia, and NY; barely in our area in c Mont.

Alchemilla L.   Lady's-mantle

Fls small, greenish, perig, in axillary clusters or in cymes or pans; calyx with tubular to camp hypan completely enclosing (but free of) pistil, almost closed by disc at throat, lobes 4 (5), gen alt with smaller bracteoles; stamens 1–4 (5), inserted at edge of disc; pistils 1 (2), 1-carpellary; ovule 1; fr an achene; ann or per herbs. (Said to be from *Alkemelych*, Arabic name for the pl, or from *alchemy*, in reference to use).

1a Fls clustered along st in axils of connate stips; stamens 1 (2); pl ann, ± villous-hirsute, 5–10 (20) cm; lf bls cuneate-obovate or flabelliform, ± biternately lobed, 4–8 mm; open fields to woods, weedy; s BC to Cal and e Wn, e Ore, and possibly Ida and Mont, e US: western 1. (*A. cuneifolia, Aphanes macrosepala*) 1 **A. occidentalis** Nutt.

1b Fls in terminal cymes or pans; stamens 4; pl per, rhizomatous, 15–50 cm; lf bls cordate-reniform, shallowly round-lobed, 3–10 cm broad; occ weed from Europe; known from c Mont; common 1. 2 **A. vulgaris** L.

Amelanchier Medic.   Serviceberry; Shadbush

Fls complete, racemose, often ± showy; calyx ± camp, adnate basally to ovary and with free, ± flared hypan, lobes 5, lanceolate, persistent; petals 5, white (pink); stamens mostly 12–20, inserted with petals at top of hypan; pistil 2–5-carpellary; ovary 2–5-celled, inferior; fr ± fleshy, reddish to purple, pomaceous; seeds 2 per carpel, separated by false partitions; unarmed trees or shrubs with alt, deciduous, simple lvs and linear, quickly deciduous stips. (Derivation obscure, perhaps from French name for a European sp.). Fairly choice orn shrubs.

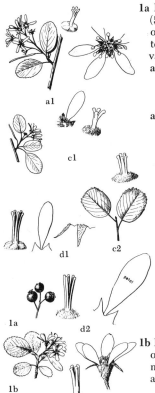

1a Lvs glabrate or sparsely sericeous on lower surface; petals mostly 10–20 (5–25) mm; fr glab; styles mostly 5 (4); pl spreading to erect, 0.5–10 m; open woods to canyons and hillsides, from near sea level to subalp; s Alas to Cal, e to Alta, Daks, Neb, NM, and Ariz; western s.; 5 poorly defined vars.   1 A. alnifolia Nutt.

  a1 Top of ovary glab or with a few hairs; petals mostly < 15 mm; lvs rather thick and leathery, sparsely pubescent to glab; se Wn to w Mont, s to se Ore, ne Cal, Utah, and Colo (A. basalticola, A. cuneata, A. glabra)
                                    var. pumila (Nutt.) Nels.

  a2 Top of ovary mostly hairy; petals often much > 15 mm; lvs mostly thin, often strongly hairy

    b1 Petals gen < 12 mm; top of ovary strongly pubescent

      c1 Styles gen 4; lvs elliptic-oblong, gen 2–3 cm × little > half as broad, subentire or with few teeth much above midlength; w Cas, s BC to sw Wn   var. humptulipensis (Jones) Hitchc.

      c2 Styles mostly 5; lvs various but gen strongly toothed most of upper half; s Alas to s Ore, mostly e Cas, e to Alta, Daks, Neb, and Colo
                                    var. alnifolia

    b2 Petals gen much > 12 mm; top of ovary occ sparsely if at all hairy

      d1 Ovary densely grayish tomentose on top; petals mostly < 16 mm × < 4 mm; calyx lobes mostly < 3 (3.5) mm; pl often treelike; common var. in and w of Cas, s Alas to Cal, occ e Cas in s BC, n Wn, n Ida, and Mont (A. ephemerotricha, A. florida, A. gormani, A. oxyodon, A. parvifolia)   var. semiintegrifolia (Hook.) Hitchc.

      d2 Ovary glab to tomentose, but not grayish, on top; petals mostly > 16 (to 25) × up to 8.5 mm; calyx lobes 3–5 mm; freely intergradient with other vars. e Cas; BC to Ore, e to RM, Mont to Wyo and Utah
                                  var. cusickii (Fern.) Hitchc.

1b Lvs ± permanently lanate, at least on lower surfaces; petals 5–10 mm; fr often pubescent; styles mostly 2–4 (5); valleys and hillsides, sagebr des to mont open for; se Ore to se Cal, e to c Ida, s Mont, Wyo, Colo, and Tex, also in Yakima Co, Wn; Utah s. (A. oreophila)   2 A. utahensis Koehne

Aruncus L.   Goatsbeard

Fls pan-racemose, ♂, ♀, white, rather small, perig; sepals and petals 5; stamens ∞, borne with petals at top of hypan; pistils 3–5, distinct; fr follicular, (1) 2–4-seeded; tall per herbs with alt, compound, exstip lvs. (Gr aryngos, goat's beard, referring to appearance of infl). Both foliage and fls attractive; good garden per.

A. sylvester Kostel. Sylvan g. Pl rhizomatous; sts several, erect, 1–1.5 m; lvs mostly cauline, long-petiolate, reduced upward, ± triternate-pinnatisect below to ternate-pinnatisect above, lflets ovate to oblong-lanceolate, gen acuminate, up to 15 × 8 cm, sharply biserrate; pans 1–8 dm; calyx broadly and shallowly camp to saucer-shaped, hypan free of ovaries, 1.5–2 mm broad, lobes spreading, ca = hypan; petals ca 1 mm, or slightly smaller in ♀ fls; stamens 15–20; follicles ca 3 mm, erect except for divergent, persistent, 0.5 mm style; seeds 2–2.5 mm; moist woods, esp along small streams; Alas to Cal, from coast to Cas in Ore and Wn, e in BC to Selkirk Mts; Eurasia.

Cercocarpus H.B.K.   Mountain-mahogany

Fls ♂, apet, 1–several in axillary clusters on short lateral shoots; calyx turbinate, 3–8 mm, often accrescent in fr, hypan persistent around (but free of) ovary, bearing near summit 15 or more stamens, lobes 5; pistil 1, 1-carpellary; style exserted, elongate and plumose in fr; achene hardened, terete; shrubs or small trees with alt, persistent, simple, entire to toothed lvs with small stips adnate to petiole. (Gr *kerkos*, tail, and *carpos*, fr, referring to long styles persistent on fr). Attractive, slow growing shrubs for the garden.

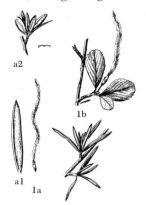

1a Lvs entire, narrowly elliptic to elliptic-lanceolate, 1–3 (3.5) cm × up to 10 mm, acute at both ends, margins gen revolute; stamens 20–30; stylar tail of achene 5–8 cm; des foothills and mts; se Wn to Cal and Ariz, e to Mont and Colo; curl-lf m.; 2 vars.   1 C. ledifolius Nutt.

   a1 Lvs revolute-margined, linear to linear-lanceolate, mostly < 3 mm broad; se Wn and adj Ore across mont Ida to Mont and n Wyo, occ s to Colo and Ariz (*C. hypoleucus*)   var. **intercedens** Schneid.

   a2 Lvs scarcely revolute, narrowly elliptic to elliptic-lanceolate, 4–10 mm broad; occ from se Wn to Nev and Cal, where common   var. **ledifolius**

1b Lvs toothed, lanceolate to deltoid-rotund, 1–4 cm × (5) 10–25 mm, mostly obtuse to rounded, dark green above, pale beneath, margins scarcely revolute; stamens 25–40; stylar tail mostly 3–6 cm; coastal chaparral and pineland to des mts; Lane and Deschutes cos, Ore, to Baja Cal, e to RM, Wyo to c Mex; birchlf m. (*C. betuloides*); ours the var. *glaber* (Wats.) Martin
                                                                        2 C. montanus Raf.

Chamaebatiaria (Porter) Maxim.   Fern-bush; Desert-sweet

Fls white, pan, ± showy, complete, perig; sepals and petals 5; stamens ∝, borne with petals near top of hypan; pistils 5 (4–6), ovaries free, superior; follicles several-seeded; shrubs with alt, deciduous, compound, stip lvs. (Resembling *Chamaebatia*).

C. millifolium (Torr.) Maxim. Spreading, aromatic shrub 1–2 m, glandular and stellate-pubescent; lvs 2.5–5 cm, finely divided into 16–30 narrow pinnae, pinnae pinnatifid into ultimate lobes 0.5–1 (1.5) × 0.5 mm; pans 5–20 cm, lfy-bracteate; calyx 4–6 mm, turbinate-camp, lobes erect in fl, ca = hypan; petals slightly > the sepals, broadly obovate-cuneate; stamens ca 50; follicles 5–6 mm, finely pubescent, coriaceous, freely dehiscent on ventral suture and very slightly on the dorsal; des canyons and mt sides, esp in lava; Deschutes Co, Ore, to se Cal, e Cas, e across Ida to Utah (*Spiraea m.*, *C. glutinosa*).

Chamaerhodos Bunge   Chamaerhodos

Fls small, white, complete, perig, bracteate-cymose; sepals and petals 5; stamens 5, opp petals and borne at top of hypan; pistils 5–10 (20); ovary 1-seeded; fr achene; small bien or per herbs with bi- or triternately divided lvs. (Gr *chama*, on the ground, and *rhodon*, rose).

C. erecta (L.) Bunge. Strongly taprooted, glandular-pubescent and hirsute, short-lived per 1–3 dm, often reddish- or purplish-tinged; st gen 1, freely br above; lvs ∝, basally rosulate and marcescent, 1.5–3 (4) cm, narrowly petiolate, ternately 2–3 times dissected into linear segms; fls ∝ in large, ± flat-topped pan cymes; calyx 4–5 mm, turbinate, hypan ca = lobes, ± hirsute-hispid; petals obovate-cuneate, ca = sepals; filaments subulate; styles ca = stamens; achenes ca 1.5 mm, membranous, ovoid-pyriform, brownish; plains and foothills; RM and e, Yuk to Colo, in our area in sw Mont (*Sibbaldia e.*, *C. nuttallii*); ours the var. *parviflora* (Nutt.) Hitchc.

Crataegus L.   Haw; Hawthorn; Thornapple

Fls corymbose, complete, showy, epig; calyx with short, disc-lined, free hypan above ovary, lobes 5; petals 5, white or pink; stamens mostly 10–25; pistil (1) 2–5-carpellary; pome yellow to red or black, with (1) 2–5 1-seeded segms (stones) surrounded by ± fleshy floral tube; shrubs or small trees with thorny brs, alt, deciduous, toothed to lobed lvs, and small deciduous stips. (Gr *kratos*, strength, referring to notably strong wood). Our native spp. not valued orns.

1a Lvs deeply 3–7-lobed for ca half width; styles 1–2 (3); fr red or yellow, mostly with only 1 or 2 stones; orn shrubs or small trees escaped from cult through bird dissemination; chiefly w Cas
   2a Style 1; fr 1-stoned; lvs 3–7-lobed   C. monogyna Jacq.
   2b Styles 2–3; fr mostly 2-stoned; lvs 3–5-lobed   C. oxyacantha L.
1b Lvs subentire to shallowly lobed < 1/2 width; styles 2–5; fr gen > 2-stoned; native, both sides Cas

3a

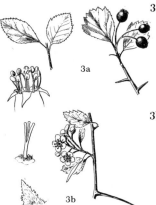

3a Styles gen 5; stamens 10–20; fr black; thorns 1–2 (3) cm; Alas to Cal, from coast inl to Alta, Daks, Wyo, and Ont; black h.; 2 vars.

1 **C. douglasii** Lindl.

a1 Stamens gen 10; ovary often ± hairy; lvs gen broadest above middle, often ± lobed; mostly e Cas (*C. brevispina, C. rivularis*)

var. **douglasii**

a2 Stamens gen 20; ovary gen glab; lvs mostly broadest near middle, serrate or biserrate; w Cas, BC to Cal, and in Fraser R Canyon, BC, and CRG

var. **suksdorfii** Sarg.

3b Styles gen 2–4 (5); stamens 10; fr dark red; thorns (2) 4–7 cm; e Cas, s BC to n Ore, e to Mont and Ida; Columbia h.; 2 vars.

2 **C. columbiana** Howell

a1 Infl hairy to glabrate; calyx and ovary glab to pubescent, but not lanate; fr mostly glab; range of sp. except for that of var. *piperi* (*C. williamsii*)

var. **columbiana**

a2 Infl grayish-hairy; calyx and fr ± lanate; se Wn and adj Ida and ne Ore (*C. p.*)

var. **piperi** (Britt.) Eggleston

Dryas L.   Mountain-avens; Dryad; Dryas

Fls solitary and scapose, complete, rather showy, perig; calyx glandular-stipitate and tomentose, hypan saucerlike, lobes 8–10; petals mostly 8–10, white or yellow; stamens ∝, borne with petals at edge of disc of hypan; pistils ∝, styles persistent, elongating and becoming plumose as achenes ripen; prostrate, nodally-rooting shrubs with crenate-serrate to entire, gen revolute, persistent, stip lvs. (L *dryas*, a wood nymph). Choice rock garden pls.

1a Sepals ovate, 4–6 mm; petals yellow, ascending, 8–12 mm; filaments hairy near base; lf bl cuneate-based, mostly 1.5–3 × (0.5) 1–2 cm; peduncles with 1–4 tiny bracts; high mont, often above timberl, but down to foothills along streams; Alas and RM s to ne Wn, Wallowa Mts, Ore, and Mont, e to GL region and St. Lawrence R; yellow m. (*D. tomentosa*)

1 **D. drummondii** Richards.

1b Sepals narrowly oblong-lanceolate; petals white or cream, spreading; filaments glab; lf bl mostly cordate or rounded at base; peduncles bractless or 1-bracteate

2a Lvs entire or crenate on lower half of bl, bl 1–2 (2.5) cm, 1/5–1/3 as broad, midrib nonglandular on lower surface; alp, Alas to Greenl, s in RM to near Mont border; entire-lvd white m.   2 **D. integrifolia** Vahl

2b Lvs crenate full length of bl, bl 1–3 cm, 1/5–1/2 as broad, midrib often glandular on lower surface; midmont to high alp; Alas to Cas, n Wn, e to Lab, and in RM s to Ida, Mont, Colo, and w to ne Ore; Asia; white d.; 2 vars.

3 **D. octopetala** L.

a1 Lvs ave at least 5 mm broad, glandular on lower surface; wideranging except in range of var. *angustifolia*   var. **hookeriana** (Juz.) Breit.

a2 Lvs ave < 5 mm broad, often not glandular; c Ida and c Mont

var. **angustifolia** Hitchc.

Duchesnea Smith

Fls solitary and pedunculate, complete, perig; calyx with saucer-shaped hypan < the 5 lobes, lobes alt with longer, 3 (5)-lobed bracts; petals 5, yellow; stamens 20–25, inserted with petals at edge of hypan; pistils ∝, style slender, attached near midlength of ovary; achenes reddish, recep spongy and ± fleshy but not palatable; herbaceous per with ternately compound, stip lvs in basal rosettes and alt along trailing sts. (For A. N. Duchesne, 1747–1827, French botanist).

D. **indica** (Andr.) Focke. Indian strawberry. Stips toothed or lobed; lflets elliptic or ovate-elliptic, 2–4 cm, crenate-serrate; petals 3–5 mm > the stamens; fr up to 1 cm broad, strawberrylike, but of poor flavor; native of India, sometimes grown as orn and escaped w Cas, from BC to Ore.

Filipendula Mill.

Fls ∝, cymose-pan, ⚥, perig; calyx mostly 5-lobed, hypan flat, much< the sepals; petals 5 (4–7), borne with 20–40 stamens at edge of hypan; pistils 5–15, distinct, erect, style short, stigma ± reniform, ovules 1–2; achenes flattened, 1-seeded; per, rhizomatous herbs with palmately lobed to pinnatisect, stip lvs. (L *filum*, thread, and *pendulus*, hanging, referring to hanging tubers found on one sp.). Attractive garden per.

F. occidentalis (Wats.) Howell. Queen-of-the-forest. Sts simple, 1–2 m; lvs pinnately divided into 2–9 remote lflets mostly 5–15 mm, except terminal lflet 8–20 cm, and as broad, palmately (3) 5–7-cleft into biserrate lobes; infl flat-topped; sepals 3–4 mm; petals white, ca 6 mm; stamens white, ca = petals; achenes ca 4 mm; rock crevices near high water line; Clatsop Co, and Trask, Wilson, and Tillamook rivers, Tillamook Co, Ore *(Spiraea o.).*

Fragaria L.   Strawberry

Fls complete, rather showy, single and scapose to bracteate-cymose, perig; calyx saucerlike, hypan narrow, lobes 5, ovate to lanceolate, alt with sepaloid bracteoles; petals 5, white or pinkish; stamens 20–25, sometimes some or all sterile; pistils ∝, ovaries sometimes sunken in recep; achenes small, pyriform, borne on enlarged, fleshy and juicy recep; per herbs with stolonous sts; lvs basal, stip, trifoliolate, coarsely crenate-serrate. (L *fraga*, name for strawberry). Not unattractive ground covers in garden).

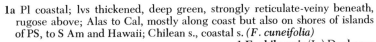

1a Pl coastal; lvs thickened, deep green, strongly reticulate-veiny beneath, rugose above; Alas to Cal, mostly along coast but also on shores of islands of PS, to S Am and Hawaii; Chilean s., coastal s. *(F. cuneifolia)*

1 F. chiloensis (L.) Duchesne

1b Pl not coastal; lvs either thin or not reticulate-veiny, gen bright yellow-green or bluish-green

  2a Lvs yellow-green, not glaucous above, relatively thin, ± veiny, upper surface slightly bulged between main lateral veins, gen ± pilose-silky; terminal tooth of lflets well developed, gen > the uppermost lateral teeth; fls mostly = or > the lvs; meadows and stream banks to light woods; N Am and Europe; woods s.; 3 vars.   **2 F. vesca** L.

  a1 Achenes superficial on recep; not known from our area   var. **vesca**
  a2 Achenes partially sunken in shallow pits in recep

    b1 Peduncles gen > lvs in fr; lflets thin, lower surface not > half-covered by pubescence; BC to Cal, mostly in and e of Cas in Wn and Ore, occ in PT and WV, e to Alta, Mont, Wyo, and NM; *(F. b., F. helleri, F. retrorsa)*   var. **bracteata** (Heller) Davis

    b2 Peduncles mostly < lvs in fr; lflets rather thick, often evidently hairy, lower surface nearly concealed by silky pubescence; BC to Cal, chiefly in valleys w Cas, but in Cas and OM up to 4000 ft elev, occ farther e *(F. crinita)*   var. **crinita** (Rydb.) Hitchc.

  2b Lvs glaucous and ± bluish-green above, rather thick and not prominently veiny, upper surface not bulged between main lateral veins, gen glab, terminal tooth of lflets gen < adj lateral teeth; fls mostly exceeded by lvs; open woods and meadows to stream banks, plains and lower mts; Alas to Cal, mostly e Cas, e to Atl coast, Colo, and Ga; 2 vars.

3 F. virginiana Duchesne

  a1 Pubescence scanty, appressed on petioles and scapes; petals mostly 5–10 mm, narrowly obovate; RM, BC to NM, rare farther w, e to Mack and Alta; bluelf s. *(F. g., F. pauciflora)*   var. **glauca** Wats.

  a2 Pubescence gen abundant, spreading on petioles and scapes; petals mostly 8–12 mm, broadly obovate-orbicular; mostly e Cas, Alas to Cal, but occ w Cas, e to w Mont, Wyo, Colo, and Utah; broadpetal s. *(F. latiuscula, F. p., F. suksdorfii)*   var. **platypetala** (Rydb.) Hall

Geum L.   Avens

Fls perig, complete, mostly bracteate-cymose (solitary); calyx with entire to cleft bracteoles alt with 5 erect to spreading lobes, hypan saucerlike to cup-shaped; petals 5, yellow, pink, or purplish, < to > sepals; stamens ∝, borne with petals at edge of disc on hypan; pistils ∝, style straight to bent or geniculate and jointed, often elongating in fr when gen with deciduous terminal segm; frs achenes; per, often rhizomatous herbs with ± pinnate or pinnatifid basal lvs and few, alt or opp, mostly 3-lobed st lvs. (Old L name for one of the spp.). *(Acomastylis, Erythrocoma, Sieversia).*

1a Sepals green, reflexed in fl; hypan saucer-shaped, lined with glandular disc; styles strongly geniculate and jointed, persistent portion the longer, hooked at tip; petals yellow, spreading, 4–9 mm; sts 3–10 dm, hirsute to ± hispid

2a Lower segm of style eglandular, glab or slightly hirsute near base; terminal segm of basal lvs ± larger than main lateral lobes, but similarly cuneate-based; damp woods, stream banks, and marshes, e Cas; BC to Cal, e to RM, Alta to NM, e to Que and Newf; Eurasia; yellow a.

1 **G. aleppicum** Jacq.

2b Lower segm of style ± glandular-pubescent; terminal segm of basal lvs many times larger than main lateral lobes, gen rounded to subcordate at base; meadows, moist woods, and stream banks, from near sea level to subalp; Alas to Baja Cal, e in Can to GL area and NS; Asia; Ore a., largelvd a.; 2 vars.   2 **G. macrophyllum** Willd.

  a1 Terminal segm of basal lvs shallowly rounded-lobed and minutely 1–2-serrate-dentate; lflets of st lvs more nearly once–twice-toothed than -cleft; peduncles and peds scarcely glandular; faces of achenes hairy; Alas to s Cal, mostly w Cas, less common to se BC and w Mont; ne US, Can, and Asia   var. **macrophyllum**

  a2 Terminal segm of basal lvs lobed 1/3–1/2 length and again coarsely once–twice-toothed or -cleft; lflets of st lvs mostly shallowly cleft to deeply toothed; peduncles and peds ± strongly glandular; achenes glab to sparsely hairy on faces; Alas to Cal, mostly e Cas, e to Mich, common in RMS (*G. oregonense, G. p.*)   var. **perincisum** (Rydb.) Raup

1b Sepals often reddish or purplish, ascending to erect; hypan turbinate to shallowly bowl-shaped, lower half gen not disc-lined; styles sometimes neither geniculate nor jointed and hooked on the persistent portion; petals yellow to pinkish, often ascending to erect, 6–12 mm; sts 0.5–6 (8) dm, rarely hirsute and never hispid

3a Petals erect to convergent, fl thus ± bowl- or vase-shaped; calyx bracteoles linear to elliptic, simple to 2–3-cleft, < to > the (8–12 mm) sepals; st lvs 2 (4), opp, ± sheathing at base; style plumose, straight to tortuous, jointed near tip, terminal segm 3–6 mm, glab, deciduous or persistent; moister spots from sagebr plains and des foothills to subalp ridges; BC to Cal, mostly e Cas, e to Newf, NY, Ill, Neb, and RMS; old man's whiskers, prairie smoke a.; 3 vars.   3 **G. triflorum** Pursh

  a1 Terminal segm of style gen persistent, lower segm up to 5 cm; larger lfts of basal lvs cleft < half their length; common var. from e Alta to Newf and s; occ in RMS e of Cont Div   var. **triflorum**

  a2 Terminal segm of style gen deciduous, lower segm rarely > 3.5 cm; larger lfts of basal lvs mostly cleft, much > half their length, into ultimate linear or narrowly oblong segms

    b1 Petals gen > bracteoles and sepals; mont ecotype of OM and Saddle Mt, Clatsop Co, Ore   var. **campanulatum** (Greene) Hitchc.

    b2 Petals gen no > bracteoles and sepals; islands of PS to e Cas (*G. c., E. flavula, E. grisea*)   var. **ciliatum** (Pursh) Fassett

3b Petals gen spreading to ascending, fl thus ± rotate, but calyx bowl-shaped to broadly funnell, bracteoles simple, lanceolate to oblong, < the sepals; st lvs 1–several, alt

4a Style jointed, strongly geniculate at joint, upper segm 3–4 mm, hirsute, deciduous, lower segm persistent, 6–8 mm, hirsute below, hooked at tip, ca twice as long as achene; hypan bowl-shaped; basal lvs lyrate-pinnatifid, segms 7–15; pl (3) 4–8 dm; wet meadows, bogs, and edge of water; BC to Que and Newf, s to Okanogan Co, Wn, through RM to NM, and e to Mo, Ind, and NJ; Eurasia; water a., purple a. (*G. aurantiacum*)   4 **G. rivale** L.

4b Style neither jointed nor geniculate, glab, 3–4 mm, ca as long as achene; hypan shallowly and broadly funnell; basal lvs pinnatifid, segms 9–31; pl mostly < 3 dm; arctic-alp, Alas to Ore, Ariz, and NM, Asia; Ross' a.; 3 vars.   5 **G. rossii** (R. Br.) Ser.

  a1 Calyx gen deeply purple-tinged, lobes not heavily veined; hypan (pressed) gen < twice as broad as long; lflet segms gen lobed at least half length, ave 2–4 (5) mm broad; fls often several per st; cirques and alp talus slides to wind swept ridges; RM, c and w Mont to NM, disjunctly w to Wallowa Mts, Ore, and to Nev and Ariz (*A. gracilipes, Potentilla g., S. g., P. nivalis* Torr., *A. sericea, G. s., S. s.*)   var. **turbinatum** (Rydb.) Hitchc.

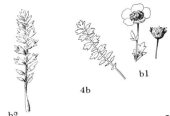

4b

b2

a2 Calyx gen green or only slightly purplish, lobes gen heavily veined; hypan (pressed) gen > twice as broad as long; segms of lflets mostly 3–6 mm broad; fls mostly solitary, occ 2 or 3

    b1 Petals gen (8) 10–12 mm; lflets greenish, subglab to ± pubescent or puberulent; arctic-alp, Alas and Yuk to VI, se BC, and Mission Mts, Lake Co, Mont   var. **rossii**

    b2 Petals gen 6–10 mm; lflets glandular-pubescent and villous-silky, grayish; Wen Mts, Wn   var. **depressum** (Greene) Hitchc.

Holodiscus Maxim.   Ocean-spray

Fls diffusely pan, perig, ⚥, rather small; calyx ebracteolate, deeply 5-lobed, hypan shallowly saucer-shaped; petals 5, white; stamens gen 20, inserted with petals just above disc-lining of hypan; pistils mostly 5, free of hypan, ovary 2-ovuled; fr gen 1-seeded achene, ± flattened, short-stip; style persistent; shrubs or small trees with alt, exstip, toothed to shallowly lobed, deciduous lvs. (Gr *holo*, whole, and *diskos*, disc, referring to unlobed disc lining hypan). (*Schizonotus, Sericotheca*). Orn of 2ndary value.

a2

a1

1b

1a

1a Lf bls gen > 3 cm, not decurrent along the (mostly 10–15 mm) petioles, shallowly lobed or toothed and gen with several 2ndary teeth, ± tomentose or crisply pilose beneath; ± erect shrub 0.5–3 m; pans diffuse, (6) 10–17 cm; fls ca 5 mm broad; styles ca 1.5 mm; coastal bluffs to moist woods and lower mts; BC to s Cal, e to w Mont, n Ida, and ne Ore, in our area more common w Cas; creambush o. (*Spiraea d., S. ariaefolia*)

                               1 H. discolor (Pursh) Maxim.

1b Lf bls 1–2 cm, ± decurrent along the (mostly 2–4 mm) petioles, shallowly to coarsely toothed, sometimes 2ndarily mucronulate, sessile-glandular to strongly pubescent (but never at all tomentose) beneath; low shrubs 0.1–1.5 (2) m; pans 3–10 cm; fls 4–5 mm broad; styles 1–1.5 mm; des valleys and hillsides to mont slopes; sc Ore to Cal, e to c Ida and nw Wyo, Colo, NM, Utah, Ariz, and Nev; gland o.; 2 vars.   **2 H. dumosus** (Hook.) Heller

    a1 Lvs strongly villous-pubescent and gen glandular beneath, ± ovate or elliptic, often toothed to below middle; c Ida s, occ w to n Cal

                                      var. **dumosus**

    a2 Lvs sparsely pubescent to glab, but also glandular, beneath, ± obovate, gen toothed only above middle; extreme se Marion Co to c and se Ore, s to ne Cal and nw Nev (*H. g., S. concolor*)

                         var. **glabrescens** (Greenm.) Hitchc.

Horkelia C. & S.   Horkelia

Fls cymose, complete, perig; calyx bracteolate, 5-lobed, hypan saucer- or cup-shaped; petals 5, white to pinkish; stamens 10, borne with stamens at edge of hypan; pistils mostly 10–25 (ours), style subterminal, mostly enlarged and glandular below, deciduous; achenes ± ovoid, gen smooth; per, mostly strongly pubescent, often glandular herbs; lvs alt, stip, pinnately compound, with mostly linear segms. (For Johann Horkel, 1769–1846, German physiologist).

1a

b1

al

1b

1a Basal lvs with filiform-lobed stips, segms of bl 5–11 (13), narrowly oblanceolate to linear-cuneate, 2–3-toothed at tip; petals cream, 5–6 mm; open sandy or rocky flats to open woods; WV, Ore, s to sw Ore; shaggy h. (*H. hirsuta, Potentilla c., Sibbaldia c.*)   **1 H. congesta** Dougl.

1b Basal lvs with mostly entire, adnate stips, segms of bl mostly (9) 11–19, cuneate to obovate-flabellate, several-lobed or -cleft; petals white or pink, 2.5–6 mm; damp meadows to rocky hillsides; c Wn to Cal, e Cas in our area, e to Ida, Wyo, and Nev; 4 vars.   **2 H. fusca** Lindl.

    a1 Lflets gen divided at least half length into linear segms; herbage green, puberulent to pubescent, not grayish-hairy; n Kittitas Co, Wn, s along Cas foothills to Hood River and Wasco cos, Ore; tawny h. (*H. tenuisecta, Potentilla douglasii*)   var. **fusca**

    a2 Lflets gen toothed or lobed much < half length, or more deeply lobed but silky or grayish-hairy, or both

        b1 Lflets gen broadly ovate to obovate; petals mostly > 4 mm; nw Ida to Blue and Steens mts, Ore, and nw Cal (*H. caeruleomontana, H. capitata, H. pseudocapitata*)   var. **capitata** (Lindl.) Peck

        b2 Lflets gen cuneate-oblanceolate or cuneate-obovate; petals gen < 4 mm

1b   c1

c2

c1 Lower surface of lvs puberulent and gen glandular, not grayish; Deschutes Co, Ore, s on e side Cas to the Sierra Nevada, Cal (*H. p., Potentilla andersonii*)   var. **parviflora** (Nutt.) Peck

c2 Lower surface of lvs strongly pubescent and gen grayish, slightly if at all glandular; n Deschutes Co, Ore, to Cal (*H. p.*)   var. **pseudocapitata** (Rydb.) Peck

Ivesia T. & G.   Ivesia

Fls cymose, ♀, perig; calyx linear-bracteolate between 5 lobes, hypan bowl-shaped to broadly conic, disc-lined; petals 5, yellow (ours); stamens 5 in ours; pistils mostly 2–6 (1–15), free of calyx, style subterminal, straight, deciduous; achenes mostly smooth; per, often glandular herbs with mostly basal, pinnately compound lvs and gen dissected lflets. (For Dr. Eli Ives, 1779–1861, Am botanist-physician).

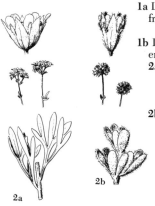

2a

2b

1a Lflets mostly < 20; infl open, dichotomously br; peds filiform, recurved in fr; se Ore to Nev and s Ida, not known quite within our area   1 **I. baileyi** Wats.

1b Lflets mostly > 20; infl congested, often subcapitate; peds gen stout and erect

    2a Petals = or > sepals; hypan shallowly bowl-shaped, 2–2.5 mm, < the sepals, broader than long (or deep); dry, open or wooded slopes to alp ridges; e side Cas from s Chelan to Yakima Co, Wn, also from near Coeur d'Alene Mts, Ida; Tweedy's i. (*Horkelia t.*)   2 **I. tweedyi** Rydb.

    2b Petals < the sepals; hypan turbinate-conic, (2.5) 3–4 mm, subequal to sepals, longer (or deeper) than broad; floodplains and riverbanks to alp ridges and talus; Table Mt, Chelan Co, to Mt Adams, Wn, Blue and Wallowa mts, Wn and Ore, s to Cal, e to w and sw Mont, Wyo, Utah, and Colo; Gordon's i. (*Horkelia g., I. alpicola*)   3 **I. gordonii** (Hook.) T. & G

Kelseya (Wats.) Rydb.   Kelseya

Fls complete, perig; calyx ebracteate, 5-lobed; petals 5, pink; stamens (7 to) 10; pistils gen 3, free of calyx; follicles gen 1 (2–4) -seeded; caespitose, cushionlike per with simple, alt, persistent, exstip lvs. (For Rev. F. D. Kelsey, 1849–1905, who first collected it).

K. uniflora (Wats.) Rydb. Solid-cushioned, intricately br shrubs adjusting to rock crevices, up to 8 dm broad and 3–8 cm thick; lfy brs greatly crowded, 1–3 (8) cm, covered with marcescent, often hardened and crusted lvs below, functional lvs imbricate, elliptic-oblanceolate to spatulate-obovate, 2.5–4 mm, finely sericeous and grayish-green; calyx camp in fl, becoming subglobose in fr, pinkish- or purplish-tinged, lobes oblong, ca 1.5 mm, ca = hypan; petals pinkish to purplish, fading to brown, oblong-elliptic, 2–3 mm; stamens gen slightly > petals, reddish-purple; styles ca = stamens; follicles ellipsoid, ca 3 mm, completely dehiscent on 2 sutures; limestone rock crevices at medium to high elev; Lewis & Clark and Meagher cos, Mont, Custer and Butte cos, Ida, and Big Horn Mts, Wyo (*Eriogynia u., Luetkea u., Spiraea u.*).

Luetkea Bong.   Partridgefoot; Luetkea

Fls complete, perig, white; calyx ebracteate, 5-lobed; petals 5; stamens ca 20, borne with petals at top of hypan; pistils 5 (4–6), free of calyx; follicles several-seeded; trailing evergreen semishrubs with ∝ ternately dissected lvs. (For Count F. P. Luetke, 1797–1882, Russian explorer). Very choice ground cover pl.

L. pectinata (Pursh) Kuntze. Matforming, rhizomatous and stoloniferous, upright lfy sts (5) 10–15 cm; lvs mostly in thick basal tufts, often marcescent, petioles narrowly wing-margined, 5–10 mm, blade = petiole, bi (tri) -ternately dissected into linear segms; calyx glab, lobes ca 2 mm, nearly twice as long as the obconic hypan; petals spatulate to obovate, 3–3.5 mm; follicles ca 5 mm, stipelike at base, freely dehiscent on ventral suture and tardily on the dorsal; moist or shaded areas, mostly where snow lies late in spring; Alas to n Cal, Cas and w, e to RM from s BC and sw Alta to e Ida and w Mont (*Eriogynia p., Spiraea p.*).

Oemleria Reichb.   Indian Plum; Osoberry

Fls ♂, ♀, perig, bracteate-racemose; calyx ebracteolate, lobes 5; petals 5, greenish-white, decid-uous; stamens 15, borne in 3 series near top of hypan; pistils gen 5, distinct; ovaries free of calyx, 2-ovulate; frs small 1-seeded drupes; deciduous shrubs with entire lvs and deciduous stips. (Gr *osme*, smell or fragrant, plus *Aronia*, another generic name). Pls of very slight orn value. (*Osmaronia*).

O. cerasiformis (H.&A.) Landon. Shrub to small tree 1.5–3 (5) m, bark pur-plish-brown, bitter; lvs with petioles 5–15 mm, bls narrowly oblong-lanceolate to -obovate, 5–12 × 2–4 cm, glab above, paler and often pu-bescent beneath; racemes axillary, gen pendent, bracts deciduous; calyx tur-binate-camp, 6–7 mm, lobes ca = hypan; petals elliptic-obovate, short-clawed, ascending, 5–6 mm, but somewhat smaller in ♀ fls; stigmas capi-tate, styles included in hypan; drupes 5 (1–4), 8–10 mm, bluish-black, bitter; stream banks and moist to rather dry woods; BC s, from coast to w slope Cas, through w Wn and Ore to Cal.

Peraphyllum Nutt.   Squaw Apple

Fls ⚥, epig; calyx turbinate, 5-lobed; petals 5, pale pink to rose; stamens 15 (20), inserted on top of hypan; pistil 1–2 (3) -carpellary, ovary inferior, 2 (3) -celled but falsely 4 (6) -celled by intrusion of 2 (3) parietal septa; styles 2 (3); fr several-seeded pome; unarmed, deciduous shrubs with sim-ple, minutely stip lvs. (Gr *pera*, leather pouch, and *phyllon*, lf). Attractive, but not successfully grown w Cas.

P. ramosissimum Nutt. Intricately and rigidly br shrub 0.5–2 m, bark dark grayish; lvs fascicled on short spurs, narrowly oblanceolate, abruptly acute, (1) 1.5–4 cm, short-petiolate, minutely serrulate or entire, ± ap-pressed-pubescent; stips tiny, reddish, quickly deciduous; fls 1–3 at ends of lateral spurs among fascicles of lvs; calyx lobes persistent, 3.5–5 mm, triangu-lar, ca = portion adnate to ovary and free hypan combined; petals spreading, obovate-oblong to obovate, 6–7 mm; pome yellowish to red, bitter, 8–10 mm; sagebr des to ponderosa pine woodl; nc Ore to ne Cal, e to s Ida, Utah, and Colo.

Petrophytum (Nutt.) Rydb.   Rockmat

Fls complete, perig, in crowded racemes at ends of bracteate-lfy peduncles; calyx ebracteolate, hypan ca 1 mm, turbinate to hemispheric, disc-lined, lobes 5; petals 5, white; stamens 20–40; pis-tils 5 (3–7), distinct, free of calyx; ovary 2–4-ovuled; style 2–4 times as long as ovary; follicles 1–2-seeded, dehiscent on 2 sutures; prostrate, matted shrubs with alt (tufted), simple, entire, ± spatulate, gen persistent and ± marcescent lvs. (Gr *petros*, rock, and *phyton*, pl, because pls grow in rock crevices). Attractive, but not easily grown rock garden pls.

1a Lvs 1-nerved on lower surface, sericeous, 5–12 (14) mm; peduncles 2–8 cm; stamens 20; styles ca 3 mm; foothills to alp ridges, mostly in rock crev-ices; ne Ore to Cal, e to Ida, Mont, SD, NM, and Tex; RM r. (*Eriogynia c.*, *Luetkea c.*, *Spiraea c.*)   1 **P. caespitosum** (Nutt.) Rydb.
1b Lvs 3-nerved on lower surface, glab to sericeous, 10–25 mm; peduncles 1–15 cm; stamens 20–40; styles 1–2 mm
  2a Stamens 20–25, ca twice as long as petals; styles ca 2 mm; hypan turbi-nate, grayish-sericeous; sepals lanceolate-triangular, erect; peduncles 5–15 cm; racemes often compound; peds 0.5–2 mm; basaltic cliffs along CR in Chelan Co, Wn; Chelan r. (*Luetkea c.*, *Spiraea c.*)
                                                              2 **P. cinerascens** (Piper) Rydb.
  2b Stamens 35–40; styles 1–1.5 mm; hypan hemispheric, sparsely hairy to glab; sepals oblong, reflexed; peduncles 1–5 (6) cm; racemes simple; peds 2–3 (5) mm; rocky cliffs, OM, Wn; OM r. (*Eriogynia h.*, *Luetkea h.*, *Spiraea h.*)   3 **P. hendersonii** (Canby) Rydb.

Physocarpus Maxim.   Ninebark

Fls complete, perig, in terminal corymbs; calyx ebracteate, hypan hemispheric to turbinate, disc-lined, lobes 5, mostly slightly > the hypan; stamens 20–40, borne with petals at edge of disc;

pistils 2–5 (ours), free of ovary; follicles several-seeded, ± inflated but gen ± compressed, dehiscent on 2 sutures; stellate-pubescent, deciduous shrubs with alt, mostly palmately 3–5-lobed lvs and deciduous stips. (Gr *physa*, bellows or bladder, and *carpos*, fr, the carpels inflated). (*Neillia*, *Opulaster*).

1a Pistils gen 2 (3–5); ovaries strongly stellate, connate at least half length; pl 0.5–2 m; canyons and hillsides, grassl to ponderosa pine and Douglas fir for; entirely e of Cas, sc BC through c and e Wn and Ore, e to sw Alta, Mont, Wyo, and Utah; mallow n. (*P. pauciflora*)

1 **P. malvaceus** (Greene) Kuntze

1b Pistils 3–5; ovaries glab to sparsely stellate along ventral suture, connate only at base; pl 2–4 (6) m; moist to wet places, lowl to lower mts; w Cas, Alas to Cal, also in n Ida; Pac n.   2 **P. capitatus** (Pursh) Kuntze

Potentilla L.   Cinquefoil; Five-finger

Fls single to cymose, perig, mostly complete (or ♀♂); calyx flaring into a shallow hypan, 5-lobed, bracteolate between lobes; petals mostly yellow (white, red, or purplish), deciduous; stamens mostly 10–∝, rarely fewer, borne at edge of hypan; pistils ∝, style apically to semibasally inserted, filiform to thickened, mostly deciduous; achenes smooth to reticulate; ann to per herbs or (1 sp.) shrubs with pinnately to digitately compound, alt, stip lvs. (L *potens*, powerful, referring to supposed medicinal value of some spp.). (*Argentina, Comarum, Dasiphora, Drymocallis, Hypargyrium, Pentaphyllum, Tridophyllum*). Several spp., but esp *P. fruticosa*, of hort merit.

1a Fls deep red to purple; rhizomatous pl of wet places, often with trailing sts; lvs pinnate, lflets (3) 5–7; lowl to subalp; Alas s along coast to Cal, e to Ida, Mont, Wyo, and to Lab and Ohio; purple c., marsh c.

1 **P. palustris** (L.) Scop.

1b Fls white, cream, or yellow; pl various
2a Pl shrubby, 1–16 dm; achenes hirsute, style midlaterally attached; lvs pinnately (3) 5 (7)-foliolate; petals yellow; foothills to subalp slopes; Alas s through Cas and OM to Cal, e through RMS and to Lab, NS, NJ, and Pa; Eurasia; shrubby c., yellow rose (*P. floribunda*)   2 **P. fruticosa** L.
2b Pl herbaceous; achenes glab
3a Pl ann or bien, rarely short-lived per without rootstocks; lvs not white-tomentose beneath; stamens 10–20   **Group I**, lead 9a
3b Pl per, mostly with rootstocks; lvs often white-tomentose beneath; stamens 10–40
4a Fls solitary on naked peduncles; pl strongly stoloniferous and "strawberrylike"; lvs multipinnate, lflets (7) 15–29; petals yellow
5a Achenes scarcely wrinkled on back; pl, other than lfts, sparsely appressed-hairy to glab; dunes to marsh edges; Alas to s Cal, very seldom far from coast; Pac silverweed   3 **P. pacifica** Howell
5b Achenes ± corky and wrinkled on back; pl lightly to strongly silky, with mostly spreading hairs; mud flats to meadowl; Alas to s Cal, mostly e Cas, e to Atl; Eurasia; common silverweed (*P. argentina*)   4 **P. anserina** L.
4b Fls gen several on lfy fl sts; pl nonstoloniferous; lvs often ternate to palmate, seldom with as many as 15 lfts
6a Style slenderly fusiform, mostly roughened slightly near midlength, attached on lower half of ovary; lvs pinnate, lflets 5–11; stamens ca 25 (40); petals mostly cream
7a Cymes narrow, strict, brs nearly erect; sepals 6–8 mm at anthesis; petals white or pale yellow, = or 1–2 mm > the sepals; pls gen > 4 dm; Alas to Ore, NM, Okla, Pa, and Que; tall, valley, or glandular c. (*P. convallaria, P. corymbosa*); ours the var. *convallaria* (Rydb.) Wolf, from Alas to s Ore, e Cas, e to RMS, Nev, Utah, and Ariz   5 **P. arguta** Pursh
7b Cymes open to glomerate, lateral brs not appressed; sepals often < 6 mm; petals pale to deep yellow, if pale then sepals mostly > petals or pl < 4 dm or infl open; BC to Baja Cal, e to RMS; sticky c., gland c.; several vars.   6 **P. glandulosa** Lindl.
a1 Calyx shallowly bowl-shaped, sepals ascending at anthesis; petals 2–4 mm > sepals, semi-erect, never spreading; washes to talus slopes, John Day Valley, e Ore
var. **campanulata** Hitchc.

a2 Calyx mostly subrotate, if sepals ascending then petals <
sepals or spreading
   b1 Petals < 0.5 mm > sepals, sometimes < they, often oblanceolate to narrowly obovate; hairs of st and lvs gen all
   glandular
      c1 Petals spreading to reflexed, as much as 1 mm < se-
      pals; bracts of infl reduced, nonlfy; Wasco Co, Ore, s (e
      Cas) to s Cal *(P. r.)*   var. **reflexa** Greene
      c2 Petals mostly ascending to erect, ca = sepals; infl
      mostly lfy-bracteate; coastal, s BC to Baja Cal, e in Wn
      and ne Ore to w Ida *(P. albida, P. amplifolia, P. rhom-
      boidea, P. valida, P. wrangelliana, Dr. oregana)*
                                              var. **glandulosa**
   b2 Petals gen at least 0.5 mm > sepals, oval to broadly
   obovate; hairs of st and lvs in part nonglandular
      d1 Petals lemon to deep yellow; pl 3–6 dm; nearly all
      hairs glandular; infl lfy-bracteate, congested, with few,
      ± erect brs; Mont to Colo, w occ to e side Cas in Wn
      and Ore *(Dr. foliosa, Dr. glabrata, Dr. pseudorupest-
      ris* var. *i., P. g.* ssp. *glabrata)*
                                       var. **intermedia** (Rydb.) Hitchc.
      d2 Petals canary yellow to nearly white; pl often < 3 dm,
      or with ∝ nonglandular hairs; infl seldom lfy-bracteate,
      brs often openly fl
         e1 Pl glandular almost throughout, lvs and sts scarcely
         pilose or hirsute; range largely that of var. *intermedia*
         (P. p., P. viscosa) var. **pseudorupestris** (Rydb.) Breit.
         e2 Pl ± pilose or hirsute, hairs eglandular in part; more
         s and w in range than var. *pseudorupestris (P. ashlan-
         dica, P. g.* var. *pumila, Dr. monticola)*
                                              var. **nevadensis** Wats.
6b Style tapered from base, or filiform, or attached near top of
ovary; lvs palmate or pinnate but lfets often > 11; stamens often
only 20
   8a Basal lvs mostly odd-pinnate, few if any digitate, sometimes a
   few ternate, but always some with at least 5 lflets
                                              Group II, lead 13a
   8b Basal lvs largely digitate but sometimes ternate
                                              Group III, lead 23a

Group I

9a Lflets (5) 7–15 (21), crowded, pinnately dissected, < 1 cm; sts prostrate to
ascending, seldom > 3 dm; petals white or cream, obovate, (4) 5–6 mm;
stamens 20; vernal pools and shores of lakes; sc Wn to ne Cal; Newberry's
c. *(Ivesia gracilis)*   **7 P. newberryi** Gray
9b Lflets often < 7 and not crowded, mostly > 1 cm; petals and stamens various
   10a Mature achenes with thickened and wedge-shaped appendage on
   adaxial side; stamens mostly 20 (sometimes fewer); lower lvs
   5–11-foliolate; petals yellow, 3–4 mm, obovate; moist flats and shores or
   banks of lakes and streams; chiefly e N Am, also in Asia, in our area at
   least in c Wn and in Klickitat Co, Wn; bushy c.   **8 P. paradoxa** Nutt.
   10b Mature achenes not prominently thickened on adaxial side; stamens
   10–20; lower lvs mostly 3 (5)-foliolate
      11a Sts stiffly hirsute proximally with 1-celled, eglandular, pustu-
      lar-based hairs; achenes mostly undulate-corrugate lengthwise; sta-
      mens (15) 20; petals mostly at least 3/4 as long as sepals; occ
      throughout our area, mostly in moist waste places; native in e N Am
      and widespread, but prob weedy in our area; Norwegian c. *(P. hirsuta,
      P. leurocarpa, P. monspeliensis)*   **9 P. norvegica** L.
      11b Sts soft-pubescent proximally, hairs often glandular or multicellular
      or semitomentose; achenes smooth or slightly striate; stamens gen
      10–15; petals gen < 3/4 as long as sepals
         12a Sts basally soft-pubescent, eglandular, often ± lanate with
         1-celled hairs; lower st lvs often 5-foliolate; calyx eglandular; moist
         areas as around ponds; BC s, e Cas mostly, to s Cal, CRG and WV, e

12b

to Sask, Mo, and NM; brook c., river c. *(P. leucocarpa, P. millegrana)*
                                                                    10 **P. rivalis** Nutt.
12b Sts basally pubescent with multicellular, moniliform, often glan-
    dular hairs; lower st lvs 3-foliolate; calyx mealy-glandular; waste
    places and in sandy soil of water courses and ponds; BC to Baja Cal,
    mostly e Cas, e to Sask, Daks, and Colo; biennial c. *(P. kelseyi, P.
    lateriflora)*   11 **P. biennis** Greene

Group II

13a Pl short-lived per, eglandular, grayish-hirsute, low and spreading but not
    rhizomatous; petals 4–6 mm, whitish; infl very lfy; style slender, thickened
    and ± glandular basally (see lead 9a)   7 **P. newberryi** Gray
13b Pl often glandular or upright; petals often yellow or > 6 mm; styles and
    infl various

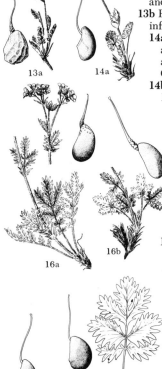

12b · 13a · 14a · 16a · 16b · 17a · 19a · a1 · c1 · 19b

  14a Pl finely glandular-puberulent, 5–10 (15) cm; style slender, attached ca
    at distal 3/4 of achene; lflets (3) 5 (7), crowded, orbicular-flabellate, cleft
    and crenate, mostly < 10 mm; petals yellow, 5–8 mm; high mont, ne
    Ore to c Ida, Wyo, and n Nev; short-lvd c.   12 **P. brevifolia** Nutt.
  14b Pl gen eglandular, mostly > 10 cm; style, lflets, and petals various
    15a Style subapical, thickened and glandular-roughened or papillose
      basally, tapered upward, slightly if at all > mature achene; lflets oblong
      to oblanceolate-obovate, lobed at least halfway to midvein; stips
      mostly deeply cleft; petals yellow, 4–6 mm
      16a Pl gen 2–5 dm; lflets 5–11, always > 5 on some lvs; pubescence of
        st mostly straight and spreading; grassl and sagebr des to mont
        ridges; Alas s, e Cas, to Nev, e to NH, Kans, and NM; prairie c. *(P.
        bipinnatifida, P. lasiodonta, P. virgulata)* 13 **P. pensylvanica** L.
      16b Pl gen < 2 dm; lvs ternate to 5-foliolate; lflets never > 5; pubesc-
        ence of st mostly tangled (lanate); gravelly drainages, subalp or alp;
        BC s in RMS to Colo and Utah, w to sc Ore, e to Sask; snow c.,
        five-lvd c. *(P. nivea* var. *pentaphylla)* 14 **P. quinquefolia** Rydb.
    15b Style slender, scarcely at all tapered, not glandular basally, much >
      the mature achene; lflets various; stips mostly entire or only slightly
      lobed
      17a Lflets obovate-cuneate, (1) 2–5 cm, greenish, dissected ca halfway
        to midvein, crowded; pl sparsely to fairly strongly hirsute-strigose,
        but greenish, gen > 3 dm; meadows to ridges, subalp to alp; BC s
        through Cas and OM to n Cal, e to sw Alta; Drummond's c. *(P.
        anomalofolia, P. cascadensis)*   15 **P. drummondii** Lehm.
      17b Lflets gen smaller and narrower, often deeply dissected or grayish,
        or both; pl often much < 3 dm
        18a Lflets mostly oblong, (1) 2–5 (12) cm, grayish-tomentose be-
          neath and grayish-strigose above, gen toothed < halfway to mid-
          vein; pl rarely < 2 and often > 3 dm
          19a Anthers 0.5–0.7 mm; lvs gen pinnate; lflets (5) 7–13, oblong to
            oblong-oblanceolate, lanceolate-toothed ca halfway to midvein;
            grassl and sagebr des to pine for; RM, Alta to NM and Ariz, e
            to Sask, Daks, and Neb; woolly c. *(P. argyrea, P. diffusa, P. ef-
            fusa, P. filicaulis, P. leneophylla)*   16 **P. hippiana** Lehm.
          19b Anthers > 0.8 mm; lvs subdigitate; lflets (5) 7–9, cu-
            neate-oblanceolate to oblong-elliptic or broadly oblanceolate,
            crenate to dissected; grassl and moist areas in des to subalp
            meadows; Alas to Baja Cal, e to Sask, Daks, Neb, and NM;
            several vars.   17 **P. gracilis** Dougl.
            a1 Calyx finely glandular-pubescent; lvs finely glandu-
              lar-pubescent and hirsute on both sides (but greenish); se Wn
              to e Ore, e to Ida and Mont and s *(P. b.)*
                                            var. **brunnescens** (Rydb.) Hitchc.
            a2 Calyx and lvs variously pubescent but seldom glandular
              b1 Lflets dissected 2/3 way to midvein into mostly linear
                segms grayish-hairy beneath
                c1 Lflets white-tomentose beneath but greenish above; BC
                  s along e base Cas to ne Cal, e to Mont *(P. ctenophora,
                  P. longiloba)*   var. **flabelliformis** (Lehm.) Nutt.

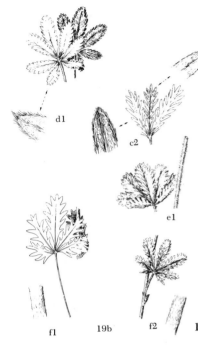

c2 Lflets silky to tomentose beneath, sericeous and grayish above; nc Ore to Cal, e to w Mont, Wyo, Colo, and NM; Elmer's c. *(P. candida, P. e., P. pecten, P. pectinisecta)*   var. **elmeri** (Rydb.) Jeps.

b2 Lflets gen dissected < 2/3 way to midvein, segms lanceolate, often greenish (though pubescent) beneath

d1 Lflets gen grayish beneath, serrate 1/5–1/2 way to midvein; bls often semipinnate; RMS, BC to Alta, s to n Mex, occ to e Wn and ne Ore; soft c. *(P. filipes, P. p.)*   var. **pulcherrima** (Lehm.) Fern.

d2 Lflets gen greenish beneath or coarsely and sharply toothed

e1 Pl gen hoary with spreading hairs; BC s, e Cas, through c Wn to Cal, e to Ida and Nev, occ to Mont and Colo *(P. dascia, P. hallii, P. longipedunculata)*   var. **permollis** (Rydb.) Hitchc.

e2 Pl strigose to sericeous, hairs scarcely spreading

f1 Lvs white-woolly beneath, lighter above; mostly coastal, the only var. w Cas, BC to Ore; slender c. *(P. macropetala)*   var. **gracilis**

f2 Lvs rarely woolly but gen ± greenish beneath and above; range of sp. as whole except not w Cas; esp common Alas to Alta and s to Ore, Ida, and Mont *(P. blaschkeana, P. chrysantha, P. dichroa, P. glomerata, P. grosseserrata, P. indiges, P. jucunda, P. rigida, P. viridescens)*
                    var. **glabrata** (Lehm.) Hitchc.

18b Lflets gen < 2 cm or otherwise not as above; pl often < 2 dm

20a Lflets gen 5–7, crowded, often in part digitate, obovate to oblong, 1–3 (5) cm, gen greenish on both surfaces; pl 1–4.5 dm, mont to alp; Yuk s chiefly in RMS to NM; diverse-lvd c., vari-lf c.; 3 vars.   18 **P. diversifolia** Lehm.

a1 Lvs primarily digitate; lflets deeply toothed to ± dissected; common var. throughout range *(P. dissecta, P. glaucophylla, P. vreelandii)*   var. **diversifolia**

a2 Lvs subpinnate to pinnate; lflets dissected at least halfway to midvein, segms linear to narrowly oblong

b1 Lvs coarsely strigose, grayish; segms ± linear, scarcely 1.5 mm broad; occ Ida and Mont to (common in) Utah and Nev *(P. m.)*   var. **multisecta** Wats.

b2 Lvs ± sericeous to glabrate, greenish; segms linear to oblong, up to 4 mm broad; esp common in Mont and se BC and se Alta *(P. decurrens)*
                    var. **perdissecta** (Rydb.) Hitchc.

20b Lfts often > 7, or tomentose; pl often < 1.5 dm

21a Lflets 5–7 (9), white-tomentose and often strigose-hirsute beneath, less so and paler above; pl low, spreading, rarely > 1.5 dm, from simple crown; prairies and foothills to alp ridges; Alta s in RMS to NM, w to Ida, Nev, and Ariz, e to Sask and Daks; early c.; 4 vars.   19 **P. concinna** Richards.

a1 Lvs more nearly digitate than pinnate; lflets gen 5

b1 Lflets not toothed > 1/2 way to midnerve; Alta to Man, s through e Mont and Daks to NM, occ in Ida and Nev   var. **concinna**

b2 Lflets dissected > 1/2 way to midnerve; range of var. *concinna* but farther w in Ida and Nev *(P. d., P. nivea* var. *dissecta)*   var. **divisa** Rydb.

a2 Lvs pinnate; lflets often 7 (9), sometimes either crowded and subdigitate, or lower ones scattered and distant from others

c1 Lflets grayish and slightly to strongly tomentose on upper surface; Alta to c Mont, mostly e RM *(P. m.)*
                    var. **macounii** (Rydb.) Hitchc.

c2 Lflets greenish on upper surface, but often strongly hirsute or strigose (not tomentose); mostly Colo, but occ n to Alta *(P. r.)*   var. **rubripes** (Rydb.) Hitchc.

21b Lflets (7) 9–21; pl often erect or with br crown and root-stocks, often > 1.5 dm

    22a Lflet segms gen > 1.5 mm broad; lflets (7) 9–11 (13), sparsely to strongly lanate beneath, 1–2 (2.5) cm; meadows and streambanks to alp slopes; occ in Cas from Lane Co, Ore, s to Sierran Cal, where common, e to se Ore and Nev; Brewer's c. (*P. plattensis* var. *leucophylla*)
                                                        20 **P. breweri** Wats.

    22b Lflet segms gen < 1.5 mm broad; lflets (7) 9–21, rarely > 1 cm, strigose-sericeous, but rarely at all lanate beneath; meadows to ridges and barren slopes, mont to alp; BC to Sask, s in RM to Ida, Mont, Wyo, Utah, Colo, and NM, also in e Ore and ne Cal; sheep c. (*P. klamathensis, P. monidensis, P. nelsoniana, P. pinnatisecta, P. versicolor, P. wyomingensis*)
                                                       21 **P. ovina** Macoun

Group III

23a Lflets of basal lvs gen 3, very rarely 5

    24a Lvs thin, glab or sparsely pubescent, greenish on both surfaces; styles not warty at base; calyx bracteoles elliptic to oval, often 2–3-toothed at tip; pl strongly rhizomatous; moist meadows to subalp ridges and talus slopes; BC s through Cas and OM to Sierran Cal, e to RM from Alta to Mont, common in Ida and Mont and in ne Ore; fan-lf c.
                                          22 **P. flabellifolia** Hook.

    24b Lvs mostly thick and leathery, grayish- or whitish-pubescent at least beneath; styles gen papillate or warty at base; calyx bracteoles gen entire

        25a Basal lvs in part 5-foliolate (see lead 16b)   14 **P. quinquefolia** Rydb.

        25b Basal lvs all 3-foliolate

            26a Calyx bracteoles broadly elliptic to oval, 1.5–2 (3) times as long as broad; sts 3–20 cm; petals 5–8 mm; alp talus slopes and rock crevices; Alas to Cas and OM, Wn, and in RM to s BC and sw Alta; Asia; ours the var. *parviflora* Hitchc.; villous c.   23 **P. villosa** Pall.

            26b Calyx bracteoles linear to narrowly elliptic or oblong, mostly at least 3 times as long as broad; petals 4–6 mm

                27a Petioles and lower portion of st without tomentum, rather strongly puberulent and with mixture of longer, coarse, spreading to antrorse hairs; arctic-alp; Alas to Can RM, reportedly to n Mont; Siberia; Hooker's c.   24 **P. hookeriana** Lehm.

                27b Petioles and lower portion of st lanate or pilose with slightly crisped hairs, but without fine puberulence

                    28a Petioles and basal part of st finely tomentose; arctic-alpine; Alas to Que, s in RM to Colo and e Utah, also in Nev; Eurasia; snow c.   25 **P. nivea** L.

                    28b Petioles and lower part of st pilose; arctic-alp, Alas s, mainly in RM, to Mont, Colo, and (prob) n Cas, Wn; reported from Wallowa Mts, Ore; one-fl c.   26 **P. uniflora** Ledeb.

23b Lflets of basal lvs gen 5 or more

    29a Basal lvs in part 3-foliolate, grayish-lanate beneath; lflets toothed 1/2–3/4 way to midvein; sts rarely > 2 dm; styles warty-papillose at base (see lead 16b)   14 **P. quinquefolia** Rydb.

    29b Basal lvs at least 5-foliolate, often neither grayish nor tomentose beneath; lflets often toothed < halfway to midnerve; sts often > 2 dm; styles not always warty-papillose

        30a Pl erect, hirsute-hispid and finely pubescent (not at all tomentose), per but short-lived; lvs mainly cauline; cymes flat-topped; mature achenes strongly reticulate and with slight dorsal keel; stamens 25 (30?); anthers at least 1 mm; petals yellow, obovate and emarginate, 5–12 mm; waste places mostly; e Wn and Mont, native to Eurasia; erect c.   27 **P. recta** L.

        30b Pl various, often tomentose; achenes often smooth; stamens gen 20; anthers often < 1 mm

            31a Fls gen < 1 cm broad, subfiliform-pedicellate in lfy-bracteate cymes; pl with many strongly lfy sts; lflets 0.5–2.5 cm, white-lanate beneath, greenish above, toothed > 1/2 way to midvein; style thickened and glandular-warty at base; sandy river banks to ponderosa pine for; very occ e Cas, Wn and Ida; native to Europe, more common in e Can and US; silvery c.   28 **P. argentea** L.

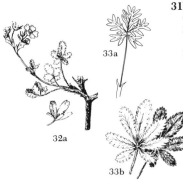

31b Fls gen > 1 cm broad, not lfy-bracteate; lvs mostly basal; lflets various; styles often filiform and not glandular-warty

    32a Pl low and spreading, mostly < 1 dm, but sts up to 2 dm; lvs white-tomentose beneath (see lead 21a)   19 **P. concinna** Richards.

    32b Pl gen erect or > 2 dm; lvs gen not white-tomentose beneath

        33a Pl alp or subalp, (1) 1.5–4.5 dm; lflets mostly 1–3.5 (5) cm, greenish or nearly equally grayish-sericeous on both surfaces; anthers mostly 0.4–0.5 mm (see lead 20a)

                18 **P. diversifolia** Lehm.

        33b Pl from lowl to midmont, up to 8 dm; lflets (2) 3–8 (12) cm, often tomentose beneath or much paler on lower than on upper side; anthers gen 0.8–1.3 mm (see lead 19b)

                17 **P. gracilis** Dougl.

Prunus L.   Plum; Cherry

Fls solitary to racemose or corymbose, complete, perig; calyx turbinate to camp; hypan disc-lined, ca = the 5 lobes, deciduous near base after anthesis; petals white to pink or red; stamens 20–30; pistil 1, simple, ovary superior, 2-ovuled; fr drupaceous, 1 (2)-seeded; trees or shrubs with alt, deciduous (ours), crenate to serrate lvs, often gland-toothed or glandular along petiole or lower part of bl. (Ancient L name for plum). (*Cerasus, Padus*).

1a Fls ∝, loosely racemose

    2a Lvs evergreen, thick and leathery; laurel cherry, an orn sometimes bird-disseminated w Cas, rarely reaching maturity   1 **P. laurocerasus** L.

    2b Lvs deciduous, firm but not leathery, elliptic to ovate-oblong or oblong-obovate, finely serrate, 3–10 cm, acute or acuminate; drupe red to black, sweet but astringent; widespread in s Can and much of US; common chokecherry; 2 vars.   2 **P. virginiana** L.

        a1 Lvs gen glab beneath or pubescent in axils or veins; pl small to medium-sized shrub 4 (6) m; drupe deep bluish-purple to black; e base Cas, BC to ne Ore, farther w in s Ore and Cal, e in Can to Alta and Daks, s in RMS to NM *(P. pinetorum)*   var. **melanocarpa** (Nels.) Sarg.

        a2 Lvs gen pubescent beneath; pl mostly 2–4 (6) m; drupe black; w Cas, BC to nw Ore *(P. d.)*   var. **demissa** (Nutt.) Torr.

1b Fls 1–several, umbellate or corymbose

    3a Fls corymbose on common axis of current season; rachis often lfy-bracteate; fls 10–15 mm broad; drupe 6–10 mm broad, not glaucous

        4a Corymbs lfy-bracteate at base; lvs oval to broadly elliptic-ovate, abruptly acute; rachis gen glab; drupe almost black, 6–8 mm thick; gen trees up to 10 m, used as budding stock and very occ escaped in e Wn and w Ida; Mahaleb c.   3 **P. mahaleb** L.

        4b Corymbs not lfy-bracteate at base; lvs various, often pubescent; fr often red, or > 8 mm thick; trees or shrubs

            5a Lvs gradually acuminate, finely serrate; drupe red, 4–7 mm; gen e of Cont Div; e BC to Colo, e to Lab, Newf, Va, and Ind; bird c., pin c.   4 **P. pensylvanica** L. f.

            5b Lvs rounded to acute, crenulate to serrate; drupe red to black, 8–12 mm, mostly w of Cont Div, BC s to s Cal, from coast inl to RMS; bitterc.; 2 vars.   5 **P. emarginata** (Dougl.) Walp.

                a1 Pl more treelike than shrublike, up to 15 m, bole up to 2.5 dm thick; calyx and lower surface of lvs thickly pubescent; mainly in lowl w Cas; BC to s Ore *(P. m., P. prunifolia, C. erecta)*   var. **mollis** (Dougl.) Brew.

                a2 Pl more shrublike than treelike, mostly several-std and 1–4 (8) m, glab to heavily pubescent; chiefly in and e Cas; BC to Ore, and s and e, also in OM *(P. corymbulosa, P. e., P. trichopetala)*

                    var. **emarginata**

    3b Fls single or umbellate at or near tip of shoot of previous year, often much > 15 mm broad; drupe gen > 10 mm broad or strongly glaucous

        6a Calyx strongly hairy externally; lvs rounded to obtuse, often ± cordate; thicket-forming shrubs 1–3 (8) m, often spinose; stream banks to woodl; WV to Cal, more common s; wild plum, Klamath p.

                6 **P. subcordata** Benth.

6b Calyx glab externally or if pubescent lvs acute to acuminate

  7a Fr pit subglobose (not flattened), smooth; fls 2.5–3.5 cm broad; peds mostly at least 2 cm; pl gen arborescent; lvs coarsely serrate, 5–15 cm; drupe not glaucous; cult cherries, occ escaped or persistent

    8a Lvs permanently hairy beneath, at least along main veins, mostly 8–15 cm; petals obovate; sweet c.   **7 P. avium** L.

    8b Lvs glab or glabrate beneath, mostly 5–8 cm; petals suborbicular; sour c.   **8 P. cerasus** L.

  7b Fr pit either strongly flattened or evidently rugose; fls often < 2.5 cm broad; peds gen < 2 cm; lvs often < 5 cm, or not coarsely serrate; fr gen glaucous

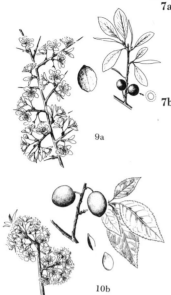

9a

    9a Fr pit turgid, rugose-pitted; fls mostly single from each node of spur brs; drupe 10–15 mm broad, bluish-black; fls 10–15 mm broad; rigid, thorny shrub 1–3 (4) m; cult in se Wn and escaped to moist draws and hillsides of se Wn and adj Ida, perhaps in ne Ore; blackthorn, sloe   **9 P. spinosa** L.

    9b Fr pit strongly flattened, smooth; fls 2-several per node of spur brs; drupe gen > 15 mm broad, yellow to bluish-black; fls 15–30 mm broad

      10a Fls 2–3 per node on spur shoots; lvs convolute in bud; fr gen yellow, green, or blue, glaucous; occ escape from cult esp in w Ida; cult p.   **10 P. domestica** L.

10b

      10b Fls gen 2–5 per node on spur shoots; lvs conduplicate in bud; fr gen red, slightly if at all glaucous; moist to dry places from plains into foothills; Alta to Utah and Ariz, w in Mont to Ravalli and Sanders cos? e to ne Can and US; wild p.   **11 P. americana** Marsh.

Purshia DC.

Fls complete, perig, solitary and terminal on lateral shoots; calyx 5-lobed; petals 5, deciduous; stamens ca 25, inserted with petals at top of hypan; pistil 1 (2), simple, 1-ovulate; fr achene, 1-seeded; rigidly br shrub with alt, deciduous, stip lvs. (For F. T. Pursh, 1774–1820, author of one of earliest floras of N Am). *(Tigarea, Kunzia)*. Excellent browse pl, and fair orn shrub e Cas.

P. tridentata (Pursh) DC. Antelope-brush, bitter-brush. Freely br shrub 1–2 (4) m; lvs cuneate, deeply 3-toothed at tip, 5–20 (25) mm, greenish above, grayish-tomentose beneath, margins gen ± revolute; calyx (5) 6–8 mm, stipitate-glandular and tomentose, lobes ovate-oblong, ca = the funnell-turbinate, gland-lined hypan; petals oblong-obovate to spatulate, yellow, 5–9 mm; stamens well exserted; achene cartilaginous, short-stipitate, ellipsoid-fusiform, ca 15 mm, puberulent, the thick style persistent; seed black, 6–8 mm; sagebr des to ponderosa pine for; BC to Cal, e Cas, e to Mont, Wyo, Colo, and NM, also in CRG.

Pyrus L.　Pear; Apple

Fls umbellate or corymbose on spur shoots, complete, epig, rather showy; calyx ebracteolate, with expanded free hypan above ovary, 5-lobed; petals white or pink; stamens 15– ∞, inserted with petals at top of hypan; pistil 1, 2–5-carpellary; ovary inferior; styles 2–5, free or basally connate; fr a pome, carpels imbedded in floral tube, seeds (1) 2 per carpel; shrubs or trees with alt, simple, toothed to lobed lvs. (L name for pear).

  1a Fr pear-shaped, containing grit cells; lvs ovate to elliptic, abruptly acuminate, crenate; fls white; cult pear, occ escaped   **P. communis** L.

  1b Fr applelike, not containing grit cells; lvs ± oblong-ovate, serrate to lobed; fls white or pinkish

    2a Carpels (and styles) gen 5; fr much > 2 cm thick; lvs not lobed; cult apple, occ escaped   **P. malus** L.

2b

    2b Carpels (and styles) gen 3 (4); fr < 2 cm in diam; lvs sometimes lobed; coastal bogs to mt slopes; s Alas s, w Cas, to Cal, up to 2500 ft elev in Cas; w crabapple *(Malus f., P. diversifolia, P. rivularis)*   **1 P. fusca** Raf.

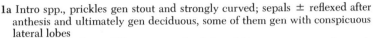

Rosa L.   Rose

Fls large, gen showy, complete, strongly perig, borne singly or in small (rarely large) cymes; sepals and petals gen 5, or petals ∝ in many cult double forms, light pink to deep rose (our native spp.), less often white or yellow; stamens ∝; pistils gen ∝ (rarely < 10); styles slender and ± exserted from mouth of the globose or ellipsoid to pyriform hypan; ovules solitary; achenes bony, enclosed within the hypan which becomes fleshy and gen reddish or purplish and berry-like in fr (hip); ± prickly shrubs or woody vines, often with a pair of prickles (infrastip prickles) at or just below each node; lvs alt, pinnate; lfts 3–11 (rarely more), gen ± toothed; stips well developed, gen green and lfy, adnate to petiole. (Classical L name).

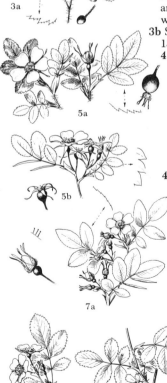

1a Intro spp., prickles gen stout and strongly curved; sepals ± reflexed after anthesis and ultimately gen deciduous, some of them gen with conspicuous lateral lobes

  2a Lower surface of lfts stipitate-glandular; foliage sweet-scented; sepals stipitate-glandular; petals 1.5–2 cm, bright pink; styles ± densely short-hairy; fr 1–1.5 cm, bright red; European sp., naturalized along roadsides and in pastures and more natural habitats; sweetbrier, eglantine *(R. rubiginosa); R. micrantha* Sm., differing in having glab styles, is reported from w Wn, but not known to be fully naturalized
                                                        1 R. eglanteria L.

  2b Lower surface of lfts glab or nearly so; foliage not sweet-scented; sepals not stipitate-glandular; petals 2–2.5 cm, white or pink; fr 1.5–2 cm, bright red; Eurasian, naturalized w Cas and in n Ida and e US, mostly along roadsides; dog rose   **2 R. canina** L.

1b Native spp., prickles stout or weak, but seldom much curved; sepals gen ascending or erect after anthesis and (except in no 3) persistent, seldom with lateral lobes

  3a Sepals, top of hypan, and styles deciduous together as fr matures; achenes gen 12 or fewer; sepals 5–12 mm; hypan glab, bright red and ca 1 cm when mature; sts 3–12 dm, bristly to nearly unarmed, the prickles slender; moist to dry woods, from near sea level to midmont; s BC s, in and w Cas, to Cal, also in nw Mont, w Ida, and ne Ore; baldhip r., little wild r. *(R. apiculata, R. dasypoda, R. helleri)*   **3 R. gymnocarpa** Nutt.

  3b Sepals and styles persistent; sepals often much > 12 mm; achenes gen 15–30 or more

    4a Sts ± bristly with slender prickles; infrastip prickles, if any, like the others

      5a Fls solitary (occ 2), on lateral brs of the season; low shrub, 2–12 dm, of for; lfts gen 5–7 (9); boreal conif for, s in cordillera to Ida (rarely) and Mont, and NM; prickly r. *(R. butleri)*
                                        4 R. acicularis Lindl.

      5b Fls gen several and cymose at end of main shoot of the season (also often on lateral shoots); semishrub to occ true shrub, 1–5 (15) dm; lfts (7) 9–11; prairies e RM, barely in our area; Arkansas rose *(R. blanda* var. *a., R. virginiana* var. *a.)*   **5 R. arkansana** Porter

    4b Sts mostly with well defined infrastip prickles, but sometimes nearly unarmed

      6a Fls relatively small, gen clustered; sepals gen 1–2 cm × 2–3.5 mm at base; petals 1.2–2.5 cm; hypan 3–5 mm thick at anthesis and ca 1 (1.5) cm thick in fr; pls mostly of lowl and hills

        7a Pls W Cas; sepals gen glandular-bristly on back; lfts rather finely toothed, inner margin of teeth mostly 0.5–1.0 mm; s BC to n Cal, gen where rather moist; clustered wild r., peafruit r. *(R. anacantha)*   **6 R. pisocarpa** Gray

        7b Pls E Cas; sepals rarely bristly or at all glandular on back; lvs (esp in var. *ultramontana)* more coarsely toothed, inner margin of teeth often > 1.0 mm; e Wn to s Cal, e to Wis, Mo, and Tex; 2 vars. in our area   **7 R. woodsii** Lindl.

          a1 Plains ecotype, up to ca 1 m tall, stiff and with crowded lvs; lfts gen small, 1–2 cm, ave less coarsely toothed, teeth sometimes gland-tipped; barely reaching our area in Mont; Wood's r. *(R. fimbriatula, R. macounii, R. sandbergii, R. w.* f. *hispida)*
                                           var. **woodsii**

         a2 Cordilleran ecotype in moist sites in otherwise dry habitats, lowl and foothills, gen 1–2 (3) m, often lax; lfts gen larger, up

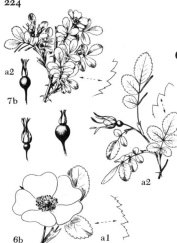

to 5 × 2.5 cm, teeth not gland-tipped; pearhip r. *(R. grosseser-rata, R. lapwaiensis, R. pyrifera, R. u., R. californica var. u.)*
var. **ultramontana** (Wats.) Jeps.

6b Fls relatively large, gen solitary; sepals gen 1.5–4 cm × 3–6 mm at base; petals gen 2.5–4 cm; hypan gen 5–9 mm thick at anthesis and (1) 1.2–2 cm thick in fr; pls of wooded or moist areas, often mont; widespread cordilleran sp.; 2 vars. in our area
8 **R. nutkana** Presl

a1 Lfts doubly serrate, the teeth glandular; infrastip prickles becoming much enlarged and much flattened toward base; lfts glandular beneath; lf rachis stipitate-glandular; mostly W Cas; Nootka r. *(R. durandii, R. muriculata, R. n. var. m., R. n. var. setosa)*
var. **nutkana**

a2 Lfts singly (seldom doubly) serrate, teeth gen not gland-tipped; prickles rarely enlarged and flattened; lfts and rachis glandular or not, otherwise glab or puberulent; chiefly E Cas; bristly Nootka r. *(R. anatonensis, R. caeruleomontana, R. columbiana, R. jonesii, R. macdougalii, R. megalantha, R. n. var. pallida, R. spaldingii, R. s. var. parkeri, R. s. var. chelanensis, R. s. var. h., R. rainierensis)*
var. **hispida** Fern.

Rubus L.   Bramble; Blackberry; Raspberry

Fls rather large, complete or sometimes ♀♂ or ♀, ♂, perig, solitary to clustered; calyx ebracteate, lobes 5 (6–7), persistent, hypan camp to saucer-like, disc-lined; petals white to red, as many as sepals; stamens (15) 40 to > 100, inserted with petals at edge of hypan; pistils ∝, on ± hemispheric, often fleshy recep; ovary 2-ovulate; fr an aggregation of weakly coherent drupelets, often remaining attached to the fleshy recep; per shrubs or vines, often strongly armed with prickles or bristles, with alt, simple to ternate or pinnate, deciduous or evergreen lvs, mostly with evident stips. (Roman name). *(Ametron, Batidaea, Bossekia, Cardiobatus, Comarobatia, Comaropsis, Dalibardia, Manteia, Melanobatus, Parmena, Psychrobatia, Rubacer).*

1a Pl unarmed
  2a Sts erect, woody, rarely < 0.5 m
    3a Petals red; lvs trifoliolate; fr salmon or yellowish, drupelets gen not coherent to recep; pl 1–3 (5) m; moist woods to stream banks or swamps, lowl to midmont; Alas to nw Cal from coast to Cas; frs edible, pl rather orn, but difficult to eradicate; salmonberry
1 **R. spectabilis** Pursh
    3b Petals white; lvs palmately lobed
      4a Fls solitary; lf bl mostly < 5 cm; styles hairy full length; twigs and peds puberulent but not stipitate-glandular; thickets on canyon sides along SR, Wallowa Co, Ore, and Idaho Co, Ida; Bartonberry
2 **R. bartonianus** Peck
      4b Fls 2–9, loosely cymose; lf bl 5–25 cm; styles glab distally; twigs stipitate-glandular; moist to dry, wooded to open areas from near sea level to subalp; Alas to s Cal, from coast to GL, Wyo, Colo, NM, and n Mex; thimbleberry, frs ± edible   3 **R. parviflorus** Nutt.
  2b Sts ± trailing, scarcely woody, prob always < 0.5 m
    5a Pl ♀, ♂; lvs broadly cordate-reniform to semiorbicular, 4–10 cm broad, shallowly (3) 5-lobed; fls solitary on erect lfy brs; circumboreal, s to VI, Me, and NY but not known to reach our area; cloudberry
4 **R. chamaemorus** L.
    5b Pl ♂; lvs compound (except in *R. pedatus*)
      6a Petals ± reddish-tinged, (8) 10–16 mm; pl not stoloniferous, but strongly rhizomatous; ann fl sts erect, 2–15 cm; mt meadows and bogs or woods to alp tundra; Alas to VI, e to Newf, Lab, and Minn, and s in RM to Mont, Wyo, and Colo; nagoonberry
5 **R. acaulis** Michx.
      6b Petals white, mostly < 8 mm; pl gen stoloniferous and trailing, but not rhizomatous (except *R. pubescens*)
        7a Lflets 5, or only 3 but basal pair deeply divided; pl mat-forming, erect sts scarcely 2 cm; filaments filiform; pistils 3–6, ovary glab; moist areas, open banks to dense for, from near sea level to near timberl; Alas to s Ore, from near coast to w Mont and n Ida; excellent trailing orn; fivelvd bramble, strawberry b.
6 **R. pedatus** J. E. Smith

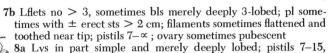

7b Lflets no > 3, sometimes bls merely deeply 3-lobed; pl sometimes with ± erect sts > 2 cm; filaments sometimes flattened and
toothed near tip; pistils 7–∝ ; ovary sometimes pubescent
    8a Lvs in part simple and merely deeply lobed; pistils 7–15,
ovaries strongly pubescent; filaments slender, not flattened; pl
non-rhizomatous; moist to dry woods, mont to subalp; BC to n
Cal, in both Cas and OM; excellent trailing orn; dwarf bramble
                                  7 R. lasiococcus Gray
    8b Lvs 3-foliolate; pistils 20–30, ovaries glab or only weakly pu-
bescent; filaments broad and flattened, with square shoulder or
2 teeth near tip; pl ± rhizomatous; clearings and burns to deep
for, gen where moist; BC to nc Wn, e to Newf and Lab, and to
Ia, Wisc, Ind, and in RM to n Colo, not known from Ida; dwarf
red blackberry (*R. transmontanus, R. triflorus*)
                                8 R. pubescens Raf.
1b Pl armed with bristlelike to stout and often curved prickles
    9a Lvs in part simple and cordate, evergreen, glossy, 3–6 cm; stips
ovate-lanceolate, slenderly acuminate; sts trailing, armed with small,
hooked prickles; petals gen pink to purple (white); stamens ca 15; pistils
4–9, drupelets red, 3–5 mm; open to shaded and gen moist slopes in the
mts; BC s in OM and Cas to sw Ore, e to Ida; snow bramble, snow dewberry (*R. pacificus*); excellent trailing orn shrublet   **9 R. nivalis** Dougl.
    9b Lvs mostly compound, if evergreen the lflets gen dissected; stips often
slender or adnate to petiole; petals mostly white; sts sometimes erect or
with large prickles; stamens and pistils mostly ∝
        10a Recep fleshy, forming part of the blackberry-like fr; sts trailing or
clambering, strongly armed with gen flattened or hooked prickles;
petals white or pale pink
            11a Pl partially or wholly ♀, ♂, the ♀ fls with rudimentary stamens,
the ♂ fls with nonfunctional pistils; sts slender, trailing, armed with
slender and scarcely flattened prickles, lvs trifoliolate (or some
simple), deciduous; abundant on prairies, burns, and clearings, but
also in open to rather dense woodl, from near the coast to midmont;
BC to n Cal, e to Ida; Douglasberry, Pacific blackberry, dewberry—
our only native blackberry (*R. helleri, R. macropetalus, R. vitifolius*); ours the var. *macropetalus* (Dougl.) Brown
                                10 R. ursinus Cham. & Schlecht.
           11b Pl ⚥; lvs gen 5-foliolate; sts thick, often clambering to erect, armed
with large, often flattened prickles; lvs various
               12a Lvs evergreen, greenish on both surfaces although hairy be-
neath; lflets laciniate to dissected; of European origin, widely
escaped w Cas, BC to Cal, occ e to Ida; evergreen blackberry
                                11 R. laciniatus Willd.
           12b Lvs deciduous or only partly evergreen, gen white beneath;
lflets merely toothed
               13a Lvs grayish- or white-tomentose beneath
                    14a Infl and esp peds stipitate-glandular; prickles straight; of
European origin, very sparingly estab w Cas, as in Grays
Harbor Co, Wn, and at Salem, Ore; European blackberry
                                12 R. vestitus Weihe & Nees
                  14b Infl and peds eglandular; prickles hooked; intro from Old
World and a serious pest in many areas, esp w Cas, from BC
to Cal, also along SR in Wn, Ore, and Ida; Himalayan blackberry (*R. fruticosus, R. procerus*)
                              13 R. discolor Weihe & Nees
                 13b Lvs soft-pubescent but not at all tomentose beneath,
3–5-foliolate; prickles mostly straight; European in origin,
sparingly intro and estab in w Wn; large-lvd blackberry
                        14 R. macrophyllus Weihe & Nees
        10b Recep dry or (*R. spectabilis*) slightly succulent, gen not forming part
of the raspberry-like fr, or if part of fr, petals pink; sts mostly erect or
arching; prickles mostly straight and not flattened
            15a Petals pink to red, gen > 1.5 cm; fr salmon-colored or red, drupe-
lets imperfectly coherent; sts erect, not vinelike, often armed only
near base; lvs not prickly (see lead 3a)   **1 R. spectabilis** Pursh

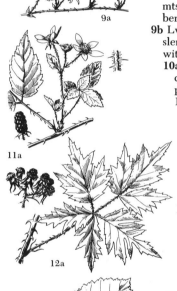

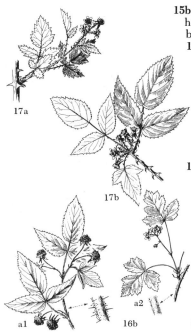

15b Petals white, gen < 1.5 cm; fr often black, drupelets mostly coherent; sts often trailing or arching, strongly armed; lvs often prickly beneath

  16a Main prickles flattened and often hooked; fr black to dark reddish-blue; fls gen ∝ in flat-topped clusters

    17a Lvs greenish and glab or glabrate beneath; pistils 25–40, ripening into barely coalescent, only moderately succulent, glab (puberulent) drupelets; moist hillsides along SR, Whitman Co, Wn; nw raspberry   **15 R. nigerrimus** (Greene) Rydb.

    17b Lvs grayish-tomentose beneath; pistils > 40, ripening into coalescent, succulent, finely tomentose drupelets; fields, canyons and lower mts; BC to s Cal, from coast inl to Mont, Utah, and Nev; black raspberry, blackcap   **16 R. leucodermis** Dougl.

  16b Main prickles not hooked, gen not flattened; fr yellowish or red; fls gen few in open, gen ± elongate clusters; wet or dry woods to open and often rocky mt slopes; over much of temp N Am and Eurasia; red raspberry; 2 vars.   **17 R. idaeus** L.

    a1 Lvs glab or subglab and greenish beneath; occ in ne Wn, n Ida, and nw Mont and n to Alas *(Ba. cataphracta, B. filipendula, B. unicolor, R. p., R. viburnifolia)*
                var. **peramoenus** (Greene) Fern.

    a2 Lvs permanently grayish-lanate beneath; widespread in much of N Am, Asia *(Ba. sandbergii, B. subcordata, R. acalyphaceus, R. melanolasius, R. strigosus, R. sachalinensis)*
                var. **gracilipes** Jones

Sanguisorba L. Burnet

Fls sessile in capitate to elongate, dense spikes, each subtended by a ± papery bract and 2 lateral bractlets, ♀ or ♀♂, apet, perig, white or greenish to red or purple; calyx 4-lobed, hypan urceolate, nearly closed by disc at top, sepals broad, petaloid; stamens mostly 4 (sometimes 2 or ca 12), inserted at top of hypan; pistils 1–2 (3), 1-ovulate; stigma papillate-discoid; achene enclosed in ± hardened, often winged, smooth to rugose hypan; ann to per herbs with alt, pinnate lvs; lfts toothed to pectinately dissected; stips adnate or free. (L *sanguis*, blood, and *sorbere*, to drink or absorb—some spp. thought to have blood-quenching property). *(Poteridium, Poterium)*.

1a Lflets pectinately pinnatifid; pl ann or bien, often semi-weedy; sts very lfy; stamens gen 2; fls 2–3 mm, ♂; grassy flats, sagebr des to lower mts; BC s (e Cas in Wn) to WV, Ore, and to s Cal, e to Ida and Mont; ann b. *(S. myriophylla)*   **1 S. occidentalis** Nutt.

1b Lflets merely toothed; pl per, often rhizomatous; sts mostly few-lvd; stamens 4 or more

  2a Stamens ca 12, filaments filiform; lflets 1–2 cm; fls mostly ♀♂, the ♂ below the ♀ in spike; hypan strongly verrucose-papillate in fr; intro from Europe, occ in w N Am, in our area only w Cas; garden b., small b. *(S. sanguisorba)*   **2 S. minor** Scop.

  2b Stamens 4, filaments stouter, often flattened or clavate; lflets mostly > 2 cm; fls ♂; hypan smooth or wrinkled in fr

    3a Calyx greenish or pinkish to rose-tinged; spikes 3–8 cm, tapered from base upward in fl; filaments strongly flattened and clavate, gen at least 3 times as long as sepals; wet places, mont; Alas to Ore, from Cas e to Ida and ne Ore; Sitka b. *(S. latifolia)*   **3 S. sitchensis** C. A. Meyer

    3b Calyx reddish to purple; spikes 1–3 (7) cm, not strongly tapered upward from base in fl; filaments sometimes terete, < 3 times as long as sepals

      4a Filaments flattened and clavate, ca twice as long as sepals; coastal bogs and marshes; Alas to OP, Wn; Menzies' b. **4 S. menziesii** Rydb.

      4b Filaments terete, not clavate, ca = sepals; ± circumboreal, in bogs and marshes; s from Alas chiefly along coast, to nw Cal, up to ca 5000 ft on Mr Hood, Ore; garden b. *(S. microcephala)*
                5 S. officinalis L.

Sibbaldia L. Sibbaldia

Fls complete, perig, bracteate-cymose; calyx bracteolate, 5-lobed; petals 5, pale yellow; stamens 5 (4–10); pistils 5–15 (20), free of calyx; achenes 1-seeded; per herbs with stip, 3-foliolate lvs. (For

Sir Robert Sibbald, 1641–1722, Professor of Medicine at Edinburgh).

S. procumbens L. Creeping s. Rhizomatous and mat-forming, strigillose per; lvs mostly on prostrate sts, slender-petiolate, lflets cuneate-obovate, 1–2 (3) cm, semitruncate and apically 3–5-toothed; fl sts 4–8 (15) cm, gen lfless below fls; cymes (1) 2–15-fld; sepals 2.5–3.5 (4) mm, > bracteoles and the shallow, gland-lined hypan; petals oval to spatulate, ca 1/2 as long as sepals; stamens < petals; style ca 1 mm; achenes pyriform-ovoid, short-stipitate, 2–2.5 mm; circumpolar, s in N Am, on open, dry to moist alp slopes, to Cal, Colo, Que, and NH (*Potentilla p., P. sibbaldii*).

Sorbus L.   Mountain-ash

Fls complete, corymbose, epig; calyx ebracteolate, 5-lobed, free hypan short; petals 5, white (cream); stamens 15–20; pistil 1, 2–5-carpellary; styles distinct; ovary inferior; fr fleshy, pomaceous, 2–5-celled, each cell 1–2-seeded; trees or shrubs with alt, deciduous, pinnate (ours) to simple, stip lvs. (L name for one of spp.). All attractive orn shrubs or trees.

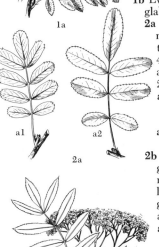

1a Lvs in part with > 13 lfts; winter buds grayish strigose-villose; infl grayish-lanate; styles 2–3; intro tree, 5–20 m, sometimes escaping or persisting; Rowan tree, European m. (*Pyrus a.*)   1 **S. aucuparia** L.
1b Lvs rarely or never (?) with > 13 lfts; winter buds rufous-pubescent to glab, sometimes glutinous; styles 3–5; shrubs, mostly 1–5 (6) m
  2a Winter buds and sometimes young growth and infl rufous-hairy; calyx mostly glab externally; lflets 7–11, ± truncate or rounded (subacute) at tip, rarely serrate > 3/4 length, gen at least 1/3 as broad as long; styles 4–5; frs red, but glaucous and with slight bluish cast; Alas s through Cas and OM to n Cal, e to Yuk, e BC, n Ida, and nw Mont, mont at 2000–10,000 ft elev; Sitka m.; 2 vars.   2 **S. sitchensis** Roemer
    a1 Lflets toothed mostly above midlength, sometimes entire; s BC s in Cas and OM to Cal (*S. occidentalis, S. pumilus*)
                                                                        var. **grayi** (Wenzig) Hitchc.
    a2 Lflets toothed 1/2–3/4 length; Alas and Yuk to sw Alta, nw Mont, and w BC, also in Wn Cas (*S. tilingii*)   var. **sitchensis**
  2b Winter buds ± glutinous and only sparsely whitish-hairy; young growth and infl grayish-strigillose-pilose; calyx gen ± pubescent externally; lflets (7) 9–13, acute or shortly acuminate, finely serrate almost full length, mostly < 1/3 as broad as long; styles 3–4; fr orange to scarlet, glossy; Alas s, in OM and Cas, to n Cal, e to w Alta, Daks, Wyo, Colo, and NM; Cascade m.; 2 vars.   3 **S. scopulina** Greene
    a1 Lflets in part 13 per lf; stips mostly shed before end of anthesis; widespread, Alas to Alta and Daks, s to NM, Utah, Ida, and (mainly e Cas) through Wn and Ore to n Cal (*S. angustifolia*)   var. **scopulina**
    a2 Lflets rarely if ever > 11; stips gen persistent until after anthesis; BC to Sierran Cal, mostly w Cas, but occ to ne Wn and Ore
                                                                        var. **cascadensis** (Jones) Hitchc.

Spiraea L.   Spirea; Spiraea

Fls complete, perig, in dense corymbs or pans, small but rather showy; calyx ebracteolate, turbinate to hemispheric, persistent, 5-lobed; hypan ca = sepals; petals 5, white, pink, rose, or purplish, deciduous; stamens 25–50 (ours), filaments slender, inserted with petals at edge of hypan; pistils 5 (3–7), distinct, stigma capitate-discoid; follicles leathery, ventrally dehiscent, 2–several-seeded; seeds fusiform; deciduous shrubs with simple, toothed to shallowly lobed lvs. (Gr *speira*, coil or wreath, the significance obscure). Our spp. 2nd rate as orns.

1a Fls in ± flat-topped corymbs; pl almost or quite glab (sometimes puberulent with tiny, straight hairs)
  2a Petals pink to red; infl and lower surface of lvs often puberulent; mts at 2000–11,000 ft, mostly where moist, but onto open rocky slopes; BC s in Cas and OM to Sierran Cal, e to se BC, Mont, Ida, and e Ore; subalp s.; 2 vars.   1 **S. densiflora** Nutt.
    a1 Pl nearly or quite glab except within hypan, or lvs and infl bractlets ciliate; common n phase, BC to Ore, e to e BC and Mont, transitional to var. *splendens* in Ida (*S. arbuscula*)   var. **densiflora**

a2 Pl finely puberulent in infl and often also on lower surface of lvs; common in s Ore and Cal, n to ne Ore, c Ida, and n Mont
var. **splendens** (Baumann) Hitchc.
2b Petals white or only slightly pinkish- or lavender-tinged; pl glab or merely ciliolate on lvs and bracts; wet places to open hillsides, from near sea level to perhaps 10,000 ft elev; BC s e Cas to nc Ore, e to Sask, SD, and Wyo; Asia; shiny-lf s. *(S. lucida);* ours the var. *lucida* (Dougl.) Hitchc.   2 **S. betulifolia** Pall.
1b Fls in rounded to much elongate corymbs; pl pubescent (at least in infl) with mostly crisp hairs
3a Petals pale pink or whitish; infl rounded to obconic, 1–2 times as long as broad; valleys to dry canyons; e side Cas, c BC to n Ore, e to Ida; pyramid s. *(S. tomentulosa)*   3 **S. pyramidata** Greene
3b Petals dark pink to rose; infl narrowly conic to much elongate, mostly several times as long as broad; swamps to damp meadows, from sea level to subalp; s Alas s, along coast and inl, to n Cal, e to se BC, n and c Ida, and ne Ore; 3 vars.   4 **S. douglasii** Hook.
a1 Lvs grayish-tomentose on lower surface; infl and calyces finely tomentose; s BC to nw Cal, mostly w Cas; Douglas' s.   var. **douglasii**
a2 Lvs glab to pubescent but not grayish-tomentose beneath; infl and calyces tomentulose to glab
b1 Infl and calyces tomentulose; lvs glab to pubescent beneath; s Alas to nw Ore, through Cas, e to n Ida and ne Ore; Menzies' s. *(S. m., S. subvillosa)*   var **menziesii** (Hook.) Presl
b2 Infl subglab to glab; lvs glab beneath; c Ida *(S. idahoensis, S. r.)*
var. **roseata** (Rydb.) Hitchc.

Waldsteinia Willd.

Fls complete, perig, cymose; calyx ebracteolate, 5-lobed; petals 5, cream or yellowish; stamens ∝, persistent in fr; pistils 2–6, ovaries 1-ovulate; fr an achene; per herbs with stip, mostly basal lvs and naked or bracteate scapes. (For Count F. A. Waldstein-Wartenburg, 1759–1823, Austrian botanist).

W. idahoensis Piper. Idaho barren strawberry. Hirsute, rhizomatous, per herb 1.5–4 dm; stips membranous, adnate to and forming broad base on the slender petioles, bls cordate-suborbicular, shallowly 3–5-lobed and coarsely bicrenate-dentate, mostly 3–7 cm broad; peduncles 1–3-bracteolate; fls 3–8 in loose, bracteate, raceme-like cymes; hypan 2–3 mm, obconic, lobes ca 4 mm, spreading; petals 4–5 mm, suborbicular; stamens ca 70; pistils 4–7, styles ca = filaments, soon deciduous; meadows and moist woods along streams; local in wc Ida.

LEGUMINOSAE   Pea Family

Fls gen ♂, hypog to slightly perig, solitary or gen racemose, capitate, or spicate; sepals gen 5, ± connate; corolla (ours) gen papilionaceous—strongly irreg, 1 petal (the uppermost) gen much the largest (banner), 2 lateral and ± horizontal (wings), gen smaller than banner, gen stuck to the lowermost, partially connate pair (keel) which enclose the stamens and pistil (in *Petalostemon* petals apparently 5 and much alike, all distinct); stamens (ours) gen 10 and monadelphous or diadelphous (9 in one group, 1 distinct), but occ all distinct, sometimes only 5; pistil 1, 1-carpellary; ovary superior, 1-celled, gen dehiscent on 2 sutures, but sometimes breaking crosswise into gen flattened, oval segms, very rarely indehiscent, sometimes spiny or burlike; seeds 1–several; ann to per herbs to trees, gen with alt, stip, pinnately (palmately) compound (occ simple), often tendril-bearing lvs. *Alhagi camelorum* Fisch., camelthorn, native to Asia Minor, is estab along Crab Creek, c Wn. It is a rhizomatous shrub ca 1 m, with needle-like axillary nodal spines, small entire lvs, small reddish fls, and pods strongly constricted between the seeds.

1a Pls intro trees or shrubs; brs either prickly, thorny, or greenish and strongly grooved, lvs then gen reduced and inconspicuous
2a Stips modified into thorns; large shrubs or trees with brs neither spine-tipped nor greenish and prominently ridged; fls white, in dropping racemes   **Robinia**
2b Stips not modified into thorns; shrubs, brs either spine-tipped or greenish and prominently grooved; fls gen yellow

3a Brs spine-tipped   **Ulex**

3b Brs not spine-tipped   **Cytisus**

1b Pls mostly native herbs or semishrubs; sts tending to die back, rarely at all prickly and seldom both greenish and strongly grooved; lvs mostly conspicuous

    4a Fertile stamens 5, alt with 4 petaloid staminodia; fls in dense spikes, small, not truly papilionaceous, all but 1 of the 5 apparent petals really staminodia   **Petalostemon**

    4b Fertile stamens gen 10, if fewer not alt with colored staminodia; fls papilionaceous

        5a Stamens distinct; lvs trifoliolate; fls yellow, racemose; pods several times as long as calyx   **Thermopsis**

        5b Stamens ± connate; lvs various, if trifoliolate then either fls not at once yellow and racemose or pods not much > calyx

            6a Herbage freely dotted with small glands; lflets either 3 (5) or frs covered with hooked spines

                7a Lvs pinnately many-foliolate; frs cocklebur-like, covered with hooked spines   **Glycyrrhiza**

                7b Lvs 3 (5) -foliolate; frs not spiny   **Psoralea**

            6b Herbage not glandular-dotted, or frs not burlike, or lflets other than 3

                8a Lvs (upper ones) simple, bl oblong, serrate; fls axillary, purplish or reddish; pl strongly pilose-villous   **Ononis**

                8b Lvs all compound or not serrate or fls not red or pl not pilose-villous

                    9a Lvs trifoliolate, lflets gen shallowly toothed; mature fr either spirally coiled or ± completely enclosed within the calyx

                        10a Pod falcate to coiled, strongly veined, sometimes prickly; fls yellow or purple, either few and in axillary, pedunculate, small heads, or more numerous and racemose   **Medicago**

                        10b Pod neither falcate or coiled nor spiny; fls often other than yellow or purple, mostly ∝ in heads or racemes

                            11a Fls in long narrow racemes, white or yellow; pl gen erect, 0.5–3 m   **Melilotus**

                            11b Fls in heads or short spikes; pl often decumbent or much < 0.5 m   **Trifolium**

        9b Lvs not trifoliolate, or lflets entire or fr projecting from calyx and not coiled

            12a Fr with short spiny teeth, 1–2-seeded, 6–8 mm long and nearly as broad; fls pinkish-lavender and prominently lined; wings scarcely half as long as keel; calyx lobes linear-lanceolate, much > tube; stips membranous, brown

                                                   Onobrychis

            12b Fr either not spiny or several-seeded and much longer than broad, or pl otherwise not as above

        13a Lvs palmately compound; lflets (4) 5–17

            14a Fls in elongate racemes; stamens monadelphous

                                               Lupinus

            14b Fls in heads or in tight spikes or racemes; stamens diadelphous   **Trifolium**

        13b Lvs not palmately compound

            15a Pods much constricted between seeds, not dehiscent but breaking crosswise into 1-seeded segms; keel sometimes = or considerably > wings

                16a Fls umbellate; petals strongly clawed; pod linear, 4-angled, not strongly flattened   **Coronilla**

                16b Fls racemose; petals not strongly clawed; pod strongly flattened   **Hedysarum**

            15b Pods not much constricted between seeds, dehiscent lengthwise; keel gen < wings

                17a Lvs even-pinnate, rachis prolonged as a slender bristle or a simple or br tendril; pl often scandent

                    18a Style filiform, ± equally hairy on all sides, but only on the terminal 1 mm   **Vicia**

                    18b Style flattened, hairy for much of length but only on the ventral side   **Lathyrus**

17b Lvs odd-pinnate or at least with terminal lflet, rachis not prolonged as a bristle or tendril, but terminal lflet sometimes confluent with rachis; pl not scandent   **Group I**

Group I

19a Fls axillary, solitary or in small pedunculate heads or umbels; free portion of filaments (or only of every other one) often dilated, sometimes broader than the anthers   **Lotus**
19b Fls gen spicate or racemose; filaments not dilated
    20a Lvs trifoliolate; pods pubescent; fls yellow   **Medicago**
    20b Lvs gen at least 5-foliolate; pods and fls various
        21a Calyx with 2 small deciduous bracteoles at base; style sparsely pubescent almost full length; fls in loose axillary racemes; peds slender, 3–5 mm; pl mostly > 5 dm   **Swainsona**
        21b Calyx gen ebracteolate, or style glab, or fls ∝ in long terminal or axillary racemes; peds often < 3 mm; pl often < 5 dm
            22a Keel petal abruptly narrowed to a beaklike point; pl gen without lfy st   **Oxytropis**
            22b Keel petal not abruptly beaked; pl gen with lfy st   **Astragalus**

Astragalus L.   Locoweed; Milk-vetch; Poison-vetch; Rattle-pod

Fls gen in pedunculate racemes (rarely single), papilionaceous, mostly showy, white or yellow to red or purple; calyx camp to tubular, with 5 tooth-like lobes; banner mostly well reflexed from wings, wings mostly exceeding the blunt to slightly beaked keel; stamens 10, diadelphous; pod sessile to long-stipitate, globose to linear, straight to coiled, membranous to leathery or fleshy (and woody when dried), often strongly inflated-bladdery, dehiscent to indehiscent, laterally to dorsiventrally flattened (compressed to obcompressed), 1-celled to completely 2-celled by intrusion of 1 or sometimes both sutures; ann to per herbs with odd-pinnate lvs, free or connate stips, and varied pubescence, the hairs mostly simple but sometimes attached subterminally or near the middle and thus 2-br (dolabriform or malpighiaceous). (Gr name, possibly derived from *astragalos*, ankle bone, in reference to shape of lvs or pod). (*Atelophragma, Cnemidophacos, Ctenophyllum, Cystium, Cystopora, Geoprumnon, Hamosa, Hesperonix, Homalobus, Kentrophyta, Microphacos, Onix, Orophaca, Phaca, Phacomene, Phacopsis, Pisophaca, Tium, Tragacantha, Xylophacos*). Common names are those listed by Barneby: Atlas of N Am Astragalus.

KEY TO ASTRAGALUS USABLE ON PLANTS IN FLOWER

(Key usable on plants in fruit on p. 246)

1a Lflets 5–11, linear-elliptic, all continuous with the rachis, sharp-pointed to spinose; raceme 1–3-fld, short-pedunculate to subsessile; fls small, banner 4–8 mm; pod < 1 cm, 1-celled; sandy des and badl to alp ridges and talus; s Alta to Colo, ne Ore, and ec Cal, to Daks and Neb; thistle m., kentrophyta; 3 vars.   **1 A. kentrophyta** Gray
    a1 Pubescence 2-br; banner whitish; Gr Pl, app our area in Mont (*K. montana*)   var. **kentrophyta**
    a2 Pubescence basifixed; banner sometimes blue or purplish
        b1 Banner white; ovules 2; pl of sagebr des to foothills; near Walla Walla, Wn   var. **douglasii** Barneby
        b2 Banner bluish to purplish; ovules 5–8; mont to alp, ne Ore to c Nev, e to RMS, Mont to NM (*A. aculeatus, A. tegetarius*)
                                    var. **implexus** (Canby) Barneby
1b Lflets various, but if continuous with rachis then racemes several-fld and long-pedunculate and fls larger
    2a Pl tufted, stless; lvs 3-foliolate; fls sessile among lvs, long and narrow, 15–30 mm; corolla whitish to yellow, sometimes purple-tipped; stips hyaline, acute, 6–15 mm; pubescence dolabriform (hairs sessile and 2-armed); plains along e base RM, Mont to Colo and e; plains orophaca (*A. triphyllus, Phaca caespitosa*)   **2 A. gilviflorus** Sheld.
    2b Pl mostly with lfy sts, if stless then either lvs pinnate or fls racemose and mostly < 15 mm

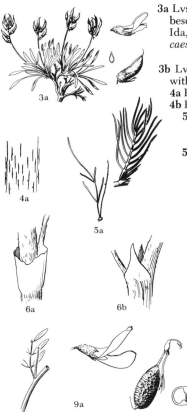

3a

4a

5a

6a   6b

9a

3a Lvs nearly all simple, oblanceolate; pl stless; fls purple, 7–9 mm; pubescence dolabriform, silvery; Gr Pl region, ec Alta and e Mont to se Ida, ne Utah, and n Colo, e to Sask, Daks, and w Neb; draba m. (*A. caespitosus* Gray, *Ho. brachycarpus, H. canescens*)

3 **A. spatulatus** Sheld.

3b Lvs mostly pinnate, if some reduced to simple rachis or bl then pl with lfy sts and pubescence basifixed

4a Pubescence of lvs and sts dolabriform   **Group I, lead 12a**

4b Pubescence of lvs and sts basifixed

5a Terminal lflet on at least some of upper lvs confluent (not jointed) with rachis, or lvs in part reduced to simple grasslike bl

Group II, lead 19a

5b Lflets all jointed to rachis; lvs odd-pinnate

6a Stips at lowest nodes, at least, fully surrounding st and connate opp the petiole—if lowest nodes bladeless then stips forming a low collar or sheath

7a Banner > 15 mm as measured along curvature over midvein   **Group III, lead 29a**

7b Banner not > 15 mm

8a Banner 10–15 mm   **Group IV, lead 37a**

8b Banner < 10 mm   **Group V, lead 51a**

6b Stips attached to petiole, sometimes ± clasping or decurrent on st, but not united opp petiole

9a Pl weedy, rhizomatous; sts fistulose, 3–7 dm; keel longer and broader than wings: fls off-white, 13–17 mm; pod pendulous, 2-celled, with stipe ca 2× length of calyx; intro in Clark Co, Ida; native to China; China m.   4 **A. chinensis** L. f.

9b Pl seldom either weedy or rhizomatous; sts mostly not fistulose; fls and pods various

10a Banner > 15 mm   **Group VI, lead 67a**

10b Banner not > 15 mm

11a Banner 10–15 mm   **Group VII, lead 98a**

11b Banner < 10 mm   **Group VIII, lead 117a**

Group I

14a

14b

c1

13b

12a Stips of lower nodes, at least, connate opp petiole

13a Banner 13–20 mm; keel obtuse, 9–15 mm; pod 2-celled

14a Fls declined, overlapping, white or off-white; keel gen purple-tipped; calyx 6–9 mm, tube ca twice length of teeth; pl extensively rhizomatous, sts arising singly or few together; BC to Atl, s to Cal, Colo, and Tex; Can m.; 2 vars.   **5 A. canadensis** L.

a1 Calyx with both black and white hairs, teeth subequal, 1.5–3 mm; pl greenish; ponderosa pine and lower mont for, occ into sagebr des; Okanogan Co, Wn, to Ida and Mont (*A. m., A. tristis*)

var. **mortonii** (Nutt.) Wats.

a2 Calyx with only white hairs, teeth 1–2 mm, upper 3 much the broadest but shortest; pls ± silvery-pubescent; water-courses or alkaline flats to lower open for; BC to Cal, e Cas, e to Daks and Colo (*A. b., A. spicatus*)   var. **brevidens** (Gandg.) Barneby

14b Fls erect or ascending; corolla white or purplish; calyx 5–9 mm, teeth 1/3–1/2 as long as tube, white- or black-strigillose; pl tufted, sts clustered; prairies to rocky foothills; c Wn to Alta, s to NM; standing m. (*A. striatus*)   **6 A. adsurgens** Hook.

13b Banner 6–13 mm; keel 6–9.5 mm, triangular and acute at tip; fls white to somewhat bluish or pinkish; calyx 3–4 (6) mm, teeth triangular, mostly < 1 (2) mm; grassl and foothills to alp meadows and dry ridges; s BC to n Wn, e to Alta, Daks, and RMS to Colo, Utah, and Nev; weedy m.; several vars.   **7 A. miser** Dougl.

a1 Pubescence at least in part dolabriform; lflets pubescent on both sides, ± silvery

b1 Herbage strigillose with straight, appressed or narrowly ascending hairs; pod strigillose

c1 Lflets narrowly linear to linear-elliptic; petals (except keel tip) whitish or straw-colored; ovules 7–11; sw Mont to c Ida, s to YNP

var. **praeteritus** Barneby

c2 Lflets broader, elliptic or oblanceolate; petals mostly pink-purple
or purple (white); ovules 12–18; s Mont s, e of divide, to sc Wyo
(A. decumbens, A. divergens, Ho. camporum)
var. **decumbens** (Nutt.) Cronq.

b2 Herbage villosulous with loose twisted hairs; pod minutely crisp-
villosulous; Bitterroot Mts, Ida and Mont
var. **crispatus** (Jones) Cronq.

a2 Pubescence all basifixed

d1 Lflets equally pubescent on both faces, silvery or cinereous; calyx
4.5–6 mm; keel 8–10.5 mm; pod gen pubescent; ne Wn and adj BC
to n Mont   var. **miser**

d2 Lflets mostly glab above or fls smaller or pod not pubescent

e1 Sts < 15 cm; keel mostly 8-10 mm; lflets broad, elliptic to oval;
pod glab; common in RM, Mont, nc Ida, and nw Wyo
var. **hylophilus** (Rydb.) Barneby

e2 Sts mostly at least 15 cm; keel 6–8 mm; lflets narrow; pod sometimes puberulent; s BC to Alta, s e Cas to Kittitas Co, Wn *(A.
palliseri, A. s.)*   var. **serotinus** (Gray) Barneby

12b Stips not connate opp petiole

15a Pl silvery-strigose, acaulescent; wing petals deeply cleft or toothed at
tip; fls white, keel often purple-tipped; calyx 6–9 mm, teeth 1/2–3/4
length of tube; pod 2-celled; grassl and sagebr des to lower mts; c Ida to
Wyo, s to Cal, Ariz, and NM; Torrey's m., matted m. *(A. brevicaulis)*
8 **A. calycosus** Torr.

15b Pl caulescent or (if acaulescent) wings entire or pod 1-celled

16a Banner 12–22 mm; keel 9–18 mm

17a Fls white (except purple-tip of keel), drying ochroleucous, 12–16
mm, nodding; racemes 10–30-fld; pod 2-celled; calyx 4–5 mm,
teeth triangular, < 1/4 length of tube; sagebr benchl to mont
ridges, often on limestone; c Ida to sc Mont, s and e to Teton Co,
Wyo; railhead m. *(A. reventoides)*   9 **A. terminalis** Wats.

17b Fls purple, ascending, 15–25 mm; racemes closely 3–9-fld; pod
1-celled; calyx 9–14 mm, teeth linear, 1/3–1/2 length of tube; Gr Pl
region, e foothills RM, Alta to NM and Tex; Missouri m.
10 **A. missouriensis** Nutt.

16b Banner 7–10 mm; keel 4–8 mm

18a Pl low, tufted, silvery-cinereous, almost stless; peduncles subscapose, 1–2 (8) cm; fls yellowish to ± purple-tinged; calyx teeth
linear, subequal to tube; pods 1-celled, ± lunate; Gr Pl region, e
base RM, Alta to NM and Tex; lotus m. *(Phaca cretacea, A. elatio-
carpus)*   11 **A. lotiflorus** Hook.

18b Pl 4–8 dm, greenish; fls ochroleucous; calyx teeth 1/3–1/2 length
of tube; pods 2-celled; intro in se Wn from the Caucasus; Russian sickle
m.   12 **A. falcatus** Lam.

Group II

19a Stips free; sts slender, prostrate to erect, 1–3 dm; lvs 4–15 cm; lflets 5–17,
linear, 3–10 × < 1.5 mm; fls 8–9 mm, whitish, purplish-lined or -tinged;
calyx 3.5–4 mm, teeth 1/3 length of tube; pod 1-celled; sagebr slopes, Baker
Co, Ore, to Nev and adj Ida; mourning m.; 2 vars.   13 **A. atratus** Wats.

a1 Lflets narrow and remote, terminal one continuous with rachis; pod
papery; Baker and Malheur cos, Ore, to sw Ida and ne Nev
var. **owyheensis** (Nels. & Macbr.) Jones

a2 Lflets more ample, less scattered, terminal one jointed with rachis; pod
leathery; stony flats where moist in spring; n edge SRP, Camas Co, Ida
var. **inseptus** Barneby

19b Stips connate at lower nodes, at least; pl various

20a Banner 15–24 mm; keel 11–16 mm; pl coarse, malodorous (seleni-
um-scented); pods woody

21a Petals white; lvs all pinnate; lflets of upper lvs not reduced in size;
pod 1-celled

22a Sts erect; pod erect, 9–18 × 2.7–3.5 mm thick, subterete, carinate
on upper suture; calyx 6.5–10.3 mm; fls 15–23 mm; ovules 14–23;
lflets (1) 2–5 pairs, narrowly oblong to oblanceolate, flat; adobe
plains, mainly; Carbon Co, Mont, to Carbon Co, Wyo; Gray's m.
14 **A. grayi** Parry

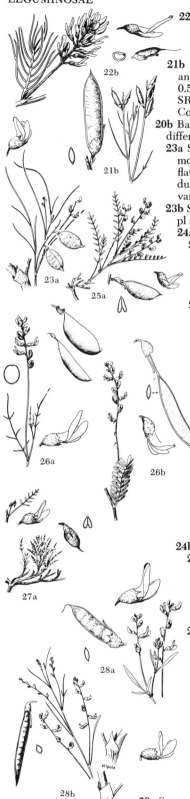

22b Sts diffuse or ascending; pod deflexed, oblong-ovoid, 10–20 mm; calyx 6–9 mm; fls 18–23 mm; ovules 22–32; lflets (2) 4–10 pairs, filiform to linear-oblanceolate, involute; Gr Pl region, e Cont Div, Mont to Colo, n to Alta; tine-lvd m. **15 A. pectinatus** Dougl.

21b Petals pink-purple; upper lvs reduced to a bare rachis; lflets (if any) only 3–30 mm; fls 15–20 mm; calyx 4.5–8 mm, teeth triangular, 0.5–2 mm; pod oblong-ellipsoid, 13–25 mm, ± 2-celled; low elev along SR, Fremont to Owyhee Co, Ida, to nw Utah and Nev and Malheur Co, Ore; Toano m. **16 A. toanus** Jones

20b Banner 4–14 mm (to 16.5 mm in *A. cusickii*—with bladdery pod and different range); keel 4–11 mm; pl scentless; pods papery

23a Sts single or few together, from buried brs with stip sheaths; lflets mostly lacking or 1–several, linear, 1–3.5 cm; pod membranous, inflated, purplish-mottled, oblong-ellipsoid, 2.5–3.5 cm, 1-celled; sand dunes to prairies; c Ida to e Mont and e and s; painted m.; ours the var. *apus* Barneby, from c Ida, only **17 A. ceramicus** Sheld.

23b Sts gen several, arising together from root crown or aerial caudex (if pl at all rhizomatous then pod not bladdery-inflated and 1-celled)

24a Ovary and immature pod stipitate

25a Lateral lflets mostly 9–12 pairs; fls cream-colored; banner often bluish-lined or -tinged, 6–8 mm; keel ca 5 mm; pod 3-cornered, 2-celled; calyx brown- or blackish-hairy in part, 3–4 mm, teeth ca = tube; sandy bluffs and talus along SR and Boise R, sw Ida, and adj Ore; Mulford's m. **18 A. mulfordiae** Jones

25b Lateral lflets mostly 4–9 pairs; banner 10–15.5 mm; keel 7–9.5 mm; pod 1-celled

26a Pod bladdery-inflated, slightly beaked, ± ellipsoid-obovoid, 2.5–4.5 cm × 6–22 mm (pressed); calyx 4–6.5 mm, teeth scarcely 1 mm; sagebr plains to rocky slopes; e Ore to wc Ida; Cusick's m.; 2 vars. **19 A. cusickii** Gray

a1 Lflets 7–11 (13), mostly strigillose on upper surface; pls 3–4 dm tall; calyx 5–6.5 mm; petals white; pod 12–22 mm broad (pressed), obovoid; SRC, from mouth of Grande Ronde R to Weiser, Ida var. **cusickii**

a2 Lflets 9–15 (19), mostly glab on upper surface; pls 3.5–7 dm; calyx 4–5 mm; petals white to pale purple, but keel purple-tipped; pod 6–12 mm broad (pressed), half obovoid or half ellipsoid; lower Salmon and Little Salmon R canyons, Ida and Adams cos, Ida var. **flexilipes** Barneby

26b Pod linear-oblong, not inflated, 1.5–2.5 cm × (3) 4–6 mm; calyx 4–6 mm, teeth scarcely 1 mm; petals cream; sagebr plains and foothills, sc BC through Ida and e Wn and Ore to ne Cal and much of Nev, s and Baja Cal; basalt m., threadstalk m. (*A. stenophyllus*, misapplied) **20 A. filipes** Torr.

24b Ovary and immature pod quite sessile

27a Pl densely lfy, tufted, ± suffruticulose at base and with rigid, persistent petioles; sts erect to decumbent, 2–20 cm; pod 5–8 mm, 2-celled; corolla 4–6 mm, cream (or bluish-lined); calyx 2.5–3 mm, teeth linear, subequal to tube; sandy or pumice soil, w Crook, Deschutes, and Klamath cos, Ore; Peck's m. **21 A. peckii** Piper

27b Pl dying back to root crown each year; sts often much > 20 cm; lvs not stiff-petioled; pod at least 1 cm, 1-celled; more e

28a Pod narrowly oblong, 3–4 mm broad; lflets mostly flat, thin, grasslike, 2–5 mm broad; st prostrate, flexuous; calyx teeth ca 1/2 as long as tube; moist, often alkaline soil; c Ida to n Utah, e to Wyo; meadow m. (*A. reclinatus*) **22 A. diversifolius** Gray

28b Pod linear to linear-oblanceolate, 2–2.5 mm broad; lflets involute, very narrow; st firm, rushlike, ascending to erect; calyx teeth scarcely 1/3 as long as tube; dry soil, grassl to sagebr des; SR drainage, se Ida, s to Utah, e to Wyo, also near Helena, Mont; lesser rushy m. (*A. junceus*) **23 A. convallarius** Greene

Group III

29a Sts and herbage villous-hirsute, hairs spreading, 1–2 mm; pod pendulous, stipitate, glab, 2-celled, cordate in section, 2–4 cm × 4–5 mm; corolla

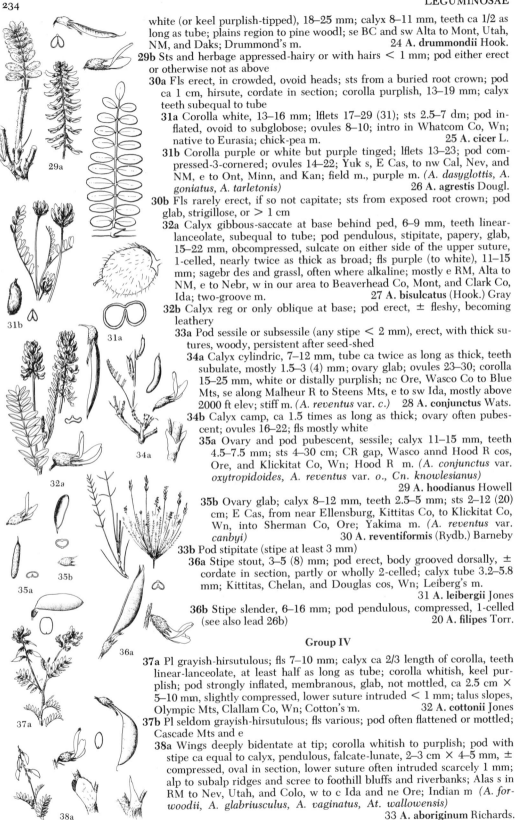

white (or keel purplish-tipped), 18–25 mm; calyx 8–11 mm, teeth ca 1/2 as long as tube; plains region to pine woodl; se BC and sw Alta to Mont, Utah, NM, and Daks; Drummond's m.   24 **A. drummondii** Hook.

29b Sts and herbage appressed-hairy or with hairs < 1 mm; pod either erect or otherwise not as above

    30a Fls erect, in crowded, ovoid heads; sts from a buried root crown; pod ca 1 cm, hirsute, cordate in section; corolla purplish, 13–19 mm; calyx teeth subequal to tube

        31a Corolla white, 13–16 mm; lflets 17–29 (31); sts 2.5–7 dm; pod inflated, ovoid to subglobose; ovules 8–10; intro in Whatcom Co, Wn; native to Eurasia; chick-pea m.   25 **A. cicer** L.

        31b Corolla purple or white but purple tinged; lflets 13–23; pod compressed-3-cornered; ovules 14–22; Yuk s, E Cas, to nw Cal, Nev, and NM, e to Ont, Minn, and Kan; field m., purple m. (A. dasyglottis, A. goniatus, A. tarletonis)   26 **A. agrestis** Dougl.

    30b Fls rarely erect, if so not capitate; sts from exposed root crown; pod glab, strigillose, or > 1 cm

        32a Calyx gibbous-saccate at base behind ped, 6–9 mm, teeth linear-lanceolate, subequal to tube; pod pendulous, stipitate, papery, glab, 15–22 mm, obcompressed, sulcate on either side of the upper suture, 1-celled, nearly twice as thick as broad; fls purple (to white), 11–15 mm; sagebr des and grassl, often where alkaline; mostly e RM, Alta to NM, e to Nebr, w in our area to Beaverhead Co, Mont, and Clark Co, Ida; two-groove m.   27 **A. bisulcatus** (Hook.) Gray

        32b Calyx reg or only oblique at base; pod erect, ± fleshy, becoming leathery

        33a Pod sessile or subsessile (any stipe < 2 mm), erect, with thick sutures, woody, persistent after seed-shed

            34a Calyx cylindric, 7–12 mm, tube ca twice as long as thick, teeth subulate, mostly 1.5–3 (4) mm; ovary glab; ovules 23–30; corolla 15–25 mm, white or distally purplish; nc Ore, Wasco Co to Blue Mts, se along Malheur R to Steens Mts, e to sw Ida, mostly above 2000 ft elev; stiff m. (A. reventus var. c.)   28 **A. conjunctus** Wats.

            34b Calyx camp, ca 1.5 times as long as thick; ovary often pubescent; ovules 16–22; fls mostly white

                35a Ovary and pod pubescent, sessile; calyx 11–15 mm, teeth 4.5–7.5 mm; sts 4–30 cm; CR gap, Wasco annd Hood R cos, Ore, and Klickitat Co, Wn; Hood R m. (A. conjunctus var. oxytropidoides, A. reventus var. o., Cn. knowlesianus)
                                      29 **A. hoodianus** Howell

                35b Ovary glab; calyx 8–12 mm, teeth 2.5–5 mm; sts 2–12 (20) cm; E Cas, from near Ellensburg, Kittitas Co, to Klickitat Co, Wn, into Sherman Co, Ore; Yakima m. (A. reventus var. canbyi)   30 **A. reventiformis** (Rydb.) Barneby

        33b Pod stipitate (stipe at least 3 mm)

            36a Stipe stout, 3–5 (8) mm; pod erect, body grooved dorsally, ± cordate in section, partly or wholly 2-celled; calyx tube 3.2–5.8 mm; Kittitas, Chelan, and Douglas cos, Wn; Leiberg's m.
                                      31 **A. leibergii** Jones

            36b Stipe slender, 6–16 mm; pod pendulous, compressed, 1-celled (see also lead 26b)   20 **A. filipes** Torr.

Group IV

    37a Pl grayish-hirsutulous; fls 7–10 mm; calyx ca 2/3 length of corolla, teeth linear-lanceolate, at least half as long as tube; corolla whitish, keel purplish; pod strongly inflated, membranous, glab, not mottled, ca 2.5 cm × 5–10 mm, slightly compressed, lower suture intruded < 1 mm; talus slopes, Olympic Mts, Clallam Co, Wn; Cotton's m.   32 **A. cottonii** Jones

    37b Pl seldom grayish-hirsutulous; fls various; pod often flattened or mottled; Cascade Mts and e

        38a Wings deeply bidentate at tip; corolla whitish to purplish; pod with stipe ca equal to calyx, pendulous, falcate-lunate, 2–3 cm × 4–5 mm, ± compressed, oval in section, lower suture often intruded scarcely 1 mm; alp to subalp ridges and scree to foothill bluffs and riverbanks; Alas s in RM to Nev, Utah, and Colo, w to c Ida and ne Ore; Indian m (A. forwoodii, A. glabriusculus, A. vaginatus, At. wallowensis)
                                    33 **A. aboriginum** Richards.

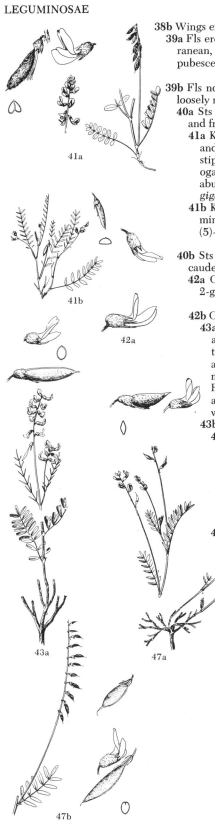

41a

41b

42a

43a

47a

47b

38b Wings entire or only shallowly emarginate; corolla and pod various
    39a Fls erect, subsessile, closely crowded in ovoid heads; caudex subterranean, with few to ∝ partially buried brs; calyx tube 4–7 mm; pods pubescent (see leads 31a–31b for distinction)   **26 A. agrestis** Dougl.
                25 A. cicer L.
    39b Fls not at once erect and crowded in dense heads, but rather often loosely racemose
        40a Sts arising singly or few together from slender, widely creeping and freely rooting, subterranean brs; pls of cool or moist soils in mts
            41a Keel broad, purple-tipped, = the purple-margined banner, > and wider than the narrow white wings; pod gen black-pilose, stipe 1.4–3.5 mm; racemes (5) 7–23-fld; circumboreal, s to Okanogan Co, Wn, Wallowa Mts, Ore, and in RM to ne Nev and NM, abundant in Mont and Ida; alpine m., purple m. *(A. andinus, A. giganteus)*   **34 A. alpinus** L.
            41b Keel narrow, lilac tipped, < the white wings and banner; pod minutely strigillose, stipe not > 1.5 mm; racemes mostly 2–3 (5)-fld; w Mont, ec Ida, and nc Colo; Park m. *(Phaca pauciflora)*   **35 A. leptaleus** Gray
        40b Sts arising together from root crown, if latter subterranean then caudex brs not rooting and pl not mont
            42a Calyx tube gibbous-saccate at base; pod pendulous, stipitate, 2-grooved on top (see also lead 32a)
                27 A. bisculatus (Hook.) Gray
            42b Calyx tube ± oblique but not gibbous at base; pod various
                43a Root crown subterranean; lflets 15–21, rounded to emarginate, 5–22 mm; calyx 4–5 mm, teeth slender, 1/4–1/3 as long as tube; pod with stipe scarcely 1 mm, body 12–21 × 2–3.5 mm, acute at each end, nearly terete, neither suture intruded; mostly e RM, esp in alkaline soil; extreme se BC and Alta s, e RM, to e Utah and Ariz, Colo and NM, and e to Man, Daks, and w Neb, w in Mont to Mo and Yellowstone rivers; pliant m., wiry m.   **36 A. flexuosus** (Dougl.) Don
            43b Root crown exposed or pls from far w of Mo R
                44a Immature pod erect and fleshy, often stipitate
                    45a Pods plainly stipitate; calyx tube 3–5.5 mm (see lead 36a)   **31 A. leibergii** Jones
                    45b Pods sessile or subsessile; calyx tube gen > 5.5 mm (see leads 34a–35b for distinction among)
                        28 A. conjunctus Wats.
                        29 A. hoodianus Howell
                  30 A. reventiformis (Rydb.) Barneby
              44b Immature pod pendulous, thin in texture
                46a Ovary and pod pubescent with (at least some) black hairs
                    47a Lflets 11–19, mostly elliptic and acute; pod compressed, sutures both prominent; ovules 2–6; alp or subalp or lower esp along rivers, often on limestone; RM, BC and Alta to Mont, w to Shoshone Co, Ida; Bourgeau's m.   **37 A. bourgovii** Gray
                    47b Lflets mostly 9–13 (19), oval or oblong, obtuse; pod obcompressed and obtusely 3-cornered in section; ovules 7–9 or rarely 3–6; Robbins' m.; 2 vars.
                        38 A. robbinsii (Oakes) Gray
                      a1 Pod (ours) 15–25 mm, 7–9-ovuled; stream banks to alp slopes; Alas and s BC to Alta, s to ne Wn, w Mont, c Ida, w Wyo, and Colo; Vt, Me, and NS *(A. blakei, A. macounii)*   var. **minor** (Hook.) Barneby
                    a2 Pod 8–11 mm, 3–5-ovuled; Wallowa Mts, Ore *(At. a., A. alpinus var. a.)*
                    var. **alpiniformis** (Rydb.) Barneby
                46b Ovary and pod glab or white-strigillose
                  48a Ovary and pod stipitate; sts all lfless at base
                    49a Pod linear-oblong, pendulous, compressed, 1-celled (see lead 26b)   **20 A. filipes** Torr.
                    49b Pod obovoid or half-ellipsoid, greatly inflated, bladdery

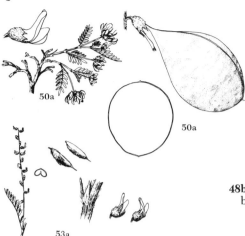

50a Sts mostly < 3 dm; pod symmetrically obovoid, balloon shaped, broadest just below the beakless apex, purple-mottled, 2–6 cm, 1-celled, lower suture not intruded; fls ochroleucous, lavender- or purple-tinged, ca 10 mm; open mont slopes to alp summits, often on serpentine; e slope Cas, Wn to ne Cal, e to sw Ida and nw Nev; balloon m. (*A. hookerianus, A. sonneanus*); ours the var. *sonneanus* (Greene) Jeps.   **39 A. whitneyi** Gray

50b Sts 3–7 dm; pod ovoid to half-ellipsoid, contracted at tip into a deltoid beak, not purple-mottled, 2.5–4.5 dm, 1-celled, lower suture not intruded; fls ochroleucous, at most lightly purple-tinged, 9–16.5 mm (see lead 26a)
                    19 A. cusickii Gray

48b Ovary and pod sessile or subsessile; sts in part lfy at base, forming lfy tuft (see lead 13b)   **7 A. miser** Dougl.

Group V

51a Keel petals 2.5–6 mm

   52a Herbage gray-villosulous; sts erect or ascending, often tomentulous at base; pod nearly or quite 2-celled, 5–9 mm, sessile or subsessile, broadly cordate-3-cornered in section

     53a Stips all free; pod 5–8 × 2–3.5 mm, sometimes not fully bilocular, with slightly convex lateral faces and tiny stipe scarcely 0.5 mm; sts uniformly pubescent, raceme-bearing almost half their length; sagebr plains and foothills; Douglas Co to Kittitas and Franklin cos, Wn; Lyall's m.   **40 A. lyallii** Gray

     53b Stips (at least basal ones) connate; pod 6–8.5 × 2–3 mm, completely bilocular, with flat or slightly concave lateral faces; sts more strongly tomentose at base than above, raceme-bearing only from the uppermost 2–5 nodes; dunes and hillsides, mostly in sagebr des; along lower Yakima and Columbia rivers, Wn, and along SR in se Ore and in Ida; buckwheat m.   **41 A. caricinus** (Jones) Barneby

   52b Herbage variably pubescent, if villosulous then sts prostrate and matted and pod nearly or quite 1-celled

     54a Ovules 2 (3); pls prostrate, very slender-std, grayish-strigillose; lvs 2–4 cm; lower stips connate, lflets (5) 7–11, 3–7 mm, sharp-pointed; pod 4–5 (6) mm, compressed, 1-celled; fls 3–7, 4–6 mm; corolla whitish, banner purplish-lined; ponderosa pine for; n Harney Co, possibly n to Grant Co, Ore; bastard kentrophyta, Deschutes m.
                42 A. tegetarioides Jones

     54b Ovules 4–11; pls various; mostly from e of Ore

       55a Sts arising mostly from buried rootcrown, subterranean for 1–7 cm; pod small, fleshy but becoming leathery and transversely rugulose, strongly obcompressed, 6–9 mm, 3.5–4 mm thick, scarcely half as broad, 1-celled, ± reniform in section; lflets mostly retuse, 5–13 × 1–3 mm; fls 7–8 mm, pale lavender to purple; grassl to juniper-covered hills; e foothills RM, Mont to NM and Tex, e to Daks, Neb, Kans, and Okla, also in Sask; slender m. (*A. microlobus*)
              43 A. gracilis Nutt.

       55b Sts from exposed rootcrown; pod mostly papery, not rugulose; mostly more w

         56a Immature pod gen in part brown- or blackish-hairy, pendulous, 8–12 mm, compressed but somewhat 3-cornered in section, lower suture intruded 0.5–1 mm; fls purplish, 6–9 mm; mont for and meadows; Alas, s in RM to Mont, extreme ne Ida, Wyo, ne Utah, and Colo, e across Can to Baffin I and Me; elegant m. (*Phaca elegans, P. parviflora*)   **44 A. eucosmus** Robins.

         56b Immature pod glab or white-strigillose

           57a Pod inflated, obovoid or subglobose, sessile, 8–12 mm, slightly compressed, 1-celled, lower suture not intruded; fls 5–12, 6–9 mm, pink to deep magenta-purple; stips mostly drying blackish; prairies and foothills to ponderosa pine for; ne Wn and adj BC, across Ida panhandle to near Helena, Mont; least bladdery m.   **45 A. microcystis** Gray

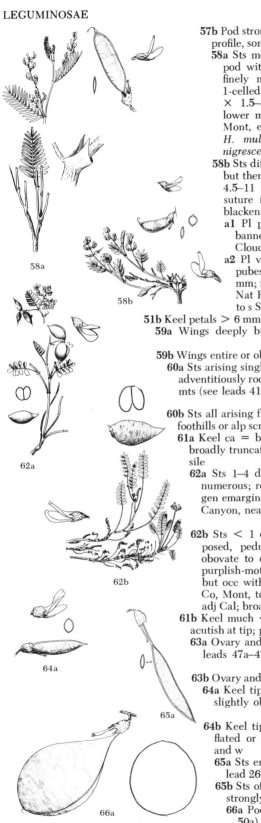

58a

58b

62a

62b

64a

65a

66a

57b Pod strongly compressed (not inflated), oblong or lentiform in profile, sometimes stipitate; stips often not drying blackish

58a Sts mostly erect; petals white (keel often purple-tipped); pod with stipe at least as long as calyx, body glab, often finely mottled, 7–15 × 3–5 mm, strongly compressed, 1-celled, neither suture intruded; lflets 11–31, (5) 10–20 × 1.5–5 mm; stips blackening; prairies and foothills to lower mts; Yuk s through e BC, mostly in RM and e, to Mont, e Ida, e Nev, Utah, and NM; pulse m. *(H. dispar, H. multiflorus, H. stipitatus, H. strigulosus, H. t., A. nigrescens)*   46 **A. tenellus** Pursh

58b Sts diffuse or prostrate, often matted, sometimes ascending but then petals often purplish; pod sessile, not mottled, (3.5) 4.5–11 × 2.5–3 mm, compressed and 1-celled, with neither suture intruded; lflets 7–13, 5–12 × 1–3 mm; stips not blackening; bent-fld m.; 2 vars.   47 **A. vexilliflexus** Sheld.

    a1 Pl prostrate, matted; herbage loosely silky-villosulous; banner ca 5–6 mm; pod 3.5–5 mm; alp stony crests, White Cloud Range, Custer Co, Ida   var. **nubilus** Barneby

    a2 Pl various, often not matted; herbage often appressed-pubescent; banner often 7–9 mm; pod mostly 5–11 mm; foothills to subalp slopes, RM, mostly on e side, Banff Nat Park, Can, s to Wyo, w in Mont to Beaverhead Co, e to s Sask and Daks *(A. amphidoxus)*   var. **vexilliflexus**

51b Keel petals > 6 mm

59a Wings deeply bidentate at tip (see lead 38a)
                        33 **A. aboriginum** Richards.

59b Wings entire or obscurely truncate-emarginate

60a Sts arising singly or few together from slender, widely creeping and adventitiously rooting, underground brs; pl of moist or cool soils in the mts (see leads 41a–41b for distinction between)   34 **A. alpinus** L.
                        35 **A. leptaleus** Gray

60b Sts all arising from root crown, if latter underground then pl of dry foothills or alp scree and ridges

61a Keel ca = banner, lower edge abruptly incurved and appearing broadly truncate at tip; pod sessile, bladdery-inflated, 2-celled, sessile

62a Sts 1–4 dm, diffuse or trailing, freely br, axillary peduncles numerous; root crown subterranean; lflets 9–13, oblong-obovate, gen emarginate, 5–10 mm; pod not mottled, 2–3.5 cm; Salmon R Canyon, near Challis, Custer and Lemhi cos, Ida; Challis m.
                  48 **A. amblytropis** Barneby

62b Sts < 1 dm, often almost lacking, tufted, the rootcrown exposed, peduncles few, subscapose; lflets (5) 13–21, oblong-obovate to oblanceolate, 3–8 (10) mm, rounded to acute; pod purplish-mottled, 2–3 cm; mont to alp, mostly on talus or scree, but occ with sagebr, in our area gen on limestone; Beaverhead Co, Mont, to Pahsimeroi R, nc Ida, more widespread in Nev, to adj Cal; broad-keeled m.   49 **A. platytropis** Gray

61b Keel much < banner or if subequal thereto then triangular and acutish at tip; pod mostly 1-celled, either not inflated or ± stipitate

63a Ovary and pod pubescent with (at least some) black hairs (see leads 47a–47b)   37 **A. bourgovii** Gray
              38 **A. robbinsii** (Oakes) Gray

63b Ovary and pod glab or white-strigillose

64a Keel tip bluntly rounded; pod linear-oblanceolate, terete or slightly obcompressed (see lead 43a)
           36 **A. flexuosus** (Dougl.) Don

64b Keel tip triangular, acute or acutish; pod either greatly inflated or compressed (not obcompressed); mostly mont, RM and w

65a Sts erect, 3–9 dm; pod stipitate, strongly compressed (see lead 26b)   20 **A. filipes** Torr.

65b Sts often prostrate, 1–5 dm; pods not at once stipitate and strongly compressed

66a Pod greatly inflated, balloon shaped, stipitate (see lead 50a)   39 **A. whitneyi** Gray

66b

67a

69a

71a

73a

75a

76a

a1

76b

66b Pod linear-oblanceolate, compressed, nearly or quite
sessile (see lead 13b)   **7 A. miser** Dougl.

Group VI

67a Wing petals 1–4 mm > banner, dilated upward and emarginate at or
below apex; fls 18–26 mm, pinkish, keel often purplish-tipped; calyx tubular, 9–15 mm, teeth linear-lanceolate, 1/2–2/3 as long as tube; pl hirsute;
pod glab, erect, sessile, cartilaginous, 3–4 cm × 5–8 mm, strongly compressed, narrowly cordate in section, 2-celled; sagebr des to lower foothills,
esp on dunes; Klickitat and Grant cos, Wn, to Umatilla and Gilliam cos,
Ore; crouching m., Columbia m.   **50 A. succumbens** Dougl.
67b Wing petals subequal to or < banner; pl and pod various
  68a Ovary and pod glab
    69a St and lvs hirsute, longest hairs 1–2 mm (see lead 29a)
                                    24 A. drummondii Hook.
    69b St and lvs mostly strigillose or glab, if loosely pubescent then hairs
      < 1 mm
      70a Ovary and pod stipitate or on a stipelike gynophore
                                Subgroup VIa, lead 79a
      70b Ovary and pod sessile   **Subgroup VIb**, lead 86a
  68b Ovary and pod pubescent
    71a Calyx gibbous-saccate at base behind and above ped (see leads
      82a–82b)   **59 A. collinus** Dougl.
                         60 A. curvicarpus (Sheld.) Macbr.
    71b Calyx not gibbous (though sometimes oblique) at base
      72a Racemes not > 10-fld; lflets uniformly and densely silvery-
      strigose or softly villous-tomentose on both sides
        73a Pl acaulescent; lflets silvery with straight, subappressed hairs;
        fls purple, 2–2.5 cm; calyx 10–16 mm, tube 3–4× as long as teeth;
        pod woolly, slightly obcompressed, 1-celled, 1.5–3 cm; foothills
        and lower mts, often in ponderosa pine for; c Crook Co, Ore, s to
        Cal and NM, e to s Ida and Utah; Newberry's m.
                              51 A. newberryi Gray
        73b Pl caulescent, or lvs villous-tomentose with curly hairs, or pod
        strigillose-villosulous
          74a Herbage villous with very fine, curly or sinuous, entangled
          hairs; pod hirsute, or hirsute and tomentose
            75a Pod thinly hirsute, not completely concealed by hairs,
            2–3.5 cm, distinctly beaked, lower suture not intruded; calyx
            11–15 mm, teeth subulate, (2) 2.5–3.5 (4) mm; fls white to
            purple, 20–25 mm; along SR and tributaries, n Malheur Co,
            Ore, and Payette Co, Ida, e along SR to s Elmore Co, Ida;
            cobblestone m.   **52 A. nudisiliquus** Nels.
           75b Pod densely tomentose and hirsute, concealed by hairs,
            1–3 cm, not beaked, lower suture sometimes strongly in-
            truded; calyx 6–18 mm, teeth lanceolate-subulate, 2–9 mm;
            fls white to purple, 12–30 mm
              76a Pl prostrate to ascending; sts 1–5 dm; lflets (13) 15–29,
              mostly acute, 7–18 mm; racemes 6–19-fld; fls rose-purple,
              20–30 mm; calyx 2/3–3/4 as long as corolla, teeth linear,
              1/2 as long to = tube; sagebr des to dry hillsides; c Wn to
              extreme ne Ore, e through Ida to c Mont; bent m., hairy
              m.   **53 A. inflexus** Dougl.
              76b Pl prostrate; sts 0.5–1 dm; lflets (3) 7–10, rounded to
              acute, 5–20 mm; racemes 3–10 (11) -fld; fls white or
              yellow to rose-purple, 10–30 mm; calyx 1/2–3/4 as long as
              corolla, teeth 1/3–1/2 as long as tube; pods 10–25 mm,
              straight to arcuate, lower suture not intruded to strongly
              intruded; prairies and sagebr des to lower mts e Cas; s BC
              to n Cal, e to Alta, Daks, Mont, and NM; woolly-pod m.,
              Pursh's m.; several vars.   **54 A. purshii** Dougl.
              a1 Fls 20–30 mm, ochroleucous to pale lavender, keel
                purple-tipped; lflets mostly 9–13, acute; pod 2–2.5 cm,
                nearly straight; sagebr valleys and foothills; ± gen e

b1

c1

76b

d1

74b

78b

79a

Cas, BC to ne Cal, e to Alta, Daks, and Colo *(Phaca mollissima, X. incurvus)*   var. **purshii**
a2 Fls 10–25 mm, reddish-purple; pod gen arcuate
  b1 Pod < 2 cm; fls 10–16 mm; Deschutes R, Deschutes
    and Crook cos, e to Harney Co, Ore, s to n Cal *(A. l.,*
    A. p. var. *lectulus, A. viarius)*
         var. **lagopinus** (Rydb.) Barneby
  b2 Pod and fls gen at least 2 cm
    c1 Pod very strongly sulcate, nearly 2-celled toward
    base; lflets gen oblanceolate-obovate, acute to
    rounded; gen in sagebr; e Wn and Ore to Nev and
    sw Ida *(A. allanaris, A. booneanus, A. g., A. lanocarpus)*   var. **glareosus** (Dougl.) Barneby
    c2 Pod not so strongly sulcate, sutures not contiguous
    within; lflets various
      d1 Lflets (5) 7–9, obovate to suborbicular; calyx
      slightly inflated, gen at least 5 mm broad when
      pressed; c Ida to wc Mont
            var. **concinnus** Barneby
      d2 Lflets (5) 9–13 (17), elliptic to obovate; calyx
      not inflated, < 4 mm broad when pressed;
      lower CRV, Wn, w, e Cas, to Nev and Cal *(A.*
      candelarius, A. ventosus)   var. **tinctus** Jones
74b Herbage pubescent with straight or merely incumbent hairs;
pod strigillose, woody, 15–25 mm, ca 8 mm thick, ± obcompressed; fls 20–25 mm, light to deep purple; calyx 9–12 mm,
teeth ca 1/3 length of tube; open sagebr hills to (often alkaline)
meadows and streams; SR, Ida, s to Nev and e Cal, e to Mont,
Wyo, and Colo; silver-lvd m.; 2 vars.   **55 A. argophyllus** Nutt.
  a1 Fls deep purple; keel 17–19 mm; pubescence appressed;
  SR, Ida, to Mont, s to Wyo, Utah, and Cal *(A. uintensis, A. a.*
  var. *cnicensis)*   var. **argophyllus**
  a2 Fls pale pink; keel 13–17 mm; pubescence spreading;
  sagebr des to juniper woodl; SR, Ida, and adj Mont and Wyo,
  s to Colo and Ariz, barely in our area *(A. a.* var. *pephragmenoides)*   var. **martinii** Jones
72b Racemes in part > 10-fld, if fewer-fld then herbage green and
lflets glab or medially glabrescent above
  77a Herbage and pod either hirsute or both softly hirsute and tomentose, some of hairs at least 1.3 mm
    78a Fls ascending; pod 1-celled, hirsute and tomentose, 1.5–3
    cm, hardened, arcuate, strongly obcompressed, lower suture
    intruded but not completely dividing cavity; fls rose-purple;
    calyx (9.5) 11–16.5 mm, teeth linear-lanceolate, 3.1–7 mm (see
    lead 76a)   **53 A. inflexus** Dougl.
    78b Fls declined; pod 2-celled, hirsute only, 2–3 cm, membranous-cartilaginous, somewhat arcuate, terete-cordate in section, lower suture deeply intruded and almost to completely
    partitioning cavity; fls deep magenta; calyx 8–12 mm, teeth
    linear-lanceolate, 1.5–4 (5) mm; sagebr des to for slopes; n
    Malheur and extreme s Baker cos, Ore, and adj Washington
    Co, Ida, s to ne and ec Cal and to Nev; shaggy m. *(A. obfalcatus)*   **56 A. malacus** Gray
  77b Herbage and pod strigillose or shortly villosulous, hairs all < 1
  mm   **Subgroup VIc, lead 91a**

Subgroup VIa

79a Calyx 9–14 mm, glab except for few hairs on teeth, teeth > 3.5 mm, gen
ca = tube; pod ± lunate, 1.5–3 cm, ± inflated, sometimes purplish-mottled, coriaceous, 1-celled, slightly obcompressed, almost 1 cm
thick, upper suture very prominent, lower one slightly intruded (> 2 mm);
lflets 15–25, 10–20 × 4–10 mm, rounded to retuse; Beckwith's m.; 2 vars.
                    57 A. beckwithii T. & G.
  a1 Pod rounded or broadly cuneate at base, deeply sulcate along lower suture; along Salmon R, Lemhi Co, Ida   var. **sulcatus** Barneby
  a2 Pod clavate or stipelike at base, only shallowly if at all sulcate along

lower suture; sagebrush desert to dry hillsides; near Kamloops, BC, and along or near SR, se Wn and ne Ore and adj Ida to se Ida
var. **weiserensis** Jones

79b Calyx various, gen pubescent but if glabrescent then the teeth much < 3.5 mm

  80a Calyx gibbous-saccate behind pedicel, sometimes almost symmetrical but with ped affixed at lower side; pod 1-celled

    81a Fls ascending, ochroleucous; calyx 8–10 mm, teeth 1–2 mm; pod erect or ascending, glab, ca 15 × 4–5 mm, nearly terete, both sutures prominent but neither sulcate; sagebr plains and foothills; Yakima Co, Wn, to ne Ore, near Columbia and lower Deschutes rivers; Tweedy's m.   **58 A. tweedyi** Canby

    81b Fls nodding or reflexed; pod pendulous, mostly pubescent

      82a Herbage green, nearly glab; reticulations of ripe pod sunken below gen surface, body of pod linear to lance-oblong or ± lunate, 7–25 × 2.5–4.2 mm, cuneate at each end, compressed and bicarinate by the prominent sutures, becoming leathery; hill m., hillside m.; 2 vars.   **59 A. collinus** Dougl.

        a1 Pod slenderly linear-oblong, 7–25 × 2.5–3.5 mm, straight or slightly curved, crisply villosulous, 12–18-ovuled; basaltic grassl and sagebr des; ne Ore, n to c Wn, e to wc Ida, also in s BC *(A. cyrtoides)*   var. **collinus**

        a2 Pod plumper, ave 8–15 × 3.3–4.2 mm, obliquely ovate-oblong, lunate, villosulous or glab, 8–12-ovuled; Morrow Co, Ore
var. **laurentii** (Rydb.) Barneby

      82b Herbage gray-pubescent; reticulations of ripe pod raised above gen surface, body 25–35 × 3–5 mm, coiled at least 1/2 turn, strongly compressed, upper suture very prominent, becoming leathery; sickle m., curvepod m.; 3 vars.
60 A. curvicarpus (Sheld.) Macbr.

        a1 Ovary and pod pubescent; banner 16–20 mm, with well-developed bl; keel 12–15 mm; lflets pubescent on upper surface; sagebr plains and foothills; sw Blue Mts, Ore, to SRP, s Ida, and s to Nev and Cal   var. **curvicarpus**

        a2 Ovary and pod glab or pl otherwise not as above

          b1 Calyx 6–8.5 mm; banner 13.5–15.5 mm; keel 9.5–11.5 mm; lflets pubescent on upper surface; dry hillsides and plains; upper forks of Deschutes R, Crook, Deschutes, and Jefferson cos, Ore   var. **brachycodon** Barneby

          b2 Calyx 9–13.5 mm; banner 13.9–19.5 mm; keel 11.5–14.5 mm; lflets nearly glab; rocky hillsides and sagebr flats, John Day and lower Deschutes rivers, Grant to Wasco cos, s to n Jefferson Co, Ore   var. **subglaber** (Rydb.) Barneby

  80b Calyx ± oblique at base but not gibbous or saccate; pod various

    83a Sts diffuse or decumbent; peduncles 2–6 cm; pod abruptly contracted at both ends, 2.5–4 cm, obcompressed, 6–8 mm wide and 10–15 mm thick, both sutures prominent and ± sulcate, lower one intruded and nearly dividing cavity; fls white, ca 2 cm; local in SRC, Washington and Adams cos, Ida, and adj Ore; Snake Canyon m.
61 A. vallaris Jones

    83b Sts erect or ascending; peduncles mostly > 8 cm; pod mostly < 8 (but rarely to 10) mm thick

      84a Stipe 6–15 mm, widely incurved-ascending; pod erect but distant from axis of raceme

        85a Pod 6–10 mm thick, truncate at base, corrugate-wrinkled, 15–20 mm, sulcate along both intruded sutures, lower suture forming a 3/4 complete partition; fls white to ochroleucous, ca 2 cm; calyx 8–10 mm, teeth lanceolate, ca 2 mm; sagebr plains and lower slopes; Beaverhead Co, Mont, to Lemhi Co, Ida; Bitterroot m.   **62 A. scaphoides** Jones

        85b Pod 3.5–7 mm thick, cuneately tapered to the stipe, mottled, not corrugate, 18–25 mm, sulcate only along lower suture, broadly cordate in section, partitioned at least 3/4 width; sagebr des to pinyon-juniper woodl; se Ore and sw Ida to Nev, nw Ariz, and sw Utah; hermit m. *(A. boiseanus, A. cusickii)*
63 A. eremiticus Sheld.

84b Stipe 2.5–6 mm, straight, erect in same plane as body of pod; pod subappressed to axis of raceme, body 14–28 × 2.5–4.5 mm, obtusely 3-angled in section, with convex lateral and shallowly grooved dorsal (lower) surfaces, keeled above by the suture, lower suture intruded to partially or completely divide the cavity; calyx 5–6.5 mm, teeth 1–2 mm; gravelly to sandy sagebr des and foothills, up to 8000 ft elev, Idaho Co (canyons of Salmon and SR) and nc Ida (upper Salmon R) n to Big Hole and Clark's Fork rivers, w Mont, e to Helena; Kelsey's m. (*A. kelseyi*; *A. stenophyllus*, misapplied)

64 **A. atropubescens** Coult. & Fish.

Subgroup VIb

86a Lfts 5–13, oblong-elliptic to narrowly oblong, 3–12 mm; sts diffuse, lfless at base, fl from 1–3 nodes below midlength; racemes 2–10-fld; corolla white, 18–19 mm; calyx 11–12 mm, black-strigillose, teeth ca 1/2 length of tube; pod curved 1/4–1/2 turn, 2.5–4 cm, ± obcompressed, 8.5–10 mm broad, beaked at tip; known only from Priest's Rapids, Yakima Co, Wn; Columbia m. (*A. casei*, misapplied)   65 **A. columbianus** Barneby

86b Lfts often > 13, or sts erect, or pod 2-celled

87a Peduncles 1–2 dm, erect; pod 1-celled; sts erect or incurved-ascending

88a Sts 2–4.5 dm, exceeding longest (gen lowest) peduncle and raceme together; pod erect, glab, fleshy and nearly solid, becoming coriaceous, 10–15 × 4–5 mm, cordate in section, lower suture sulcate; ovules 14–18; calyx 5–7 mm, teeth lanceolate, 1/4–1/3 length of tube; dry hillsides, often with sagebr; n edge of SRP, Camas and Custer to Elmore and Canyon cos, also in Cassia Co, Ida; Boise m.

66 **A. adanus** Nels.

88b Sts < 2 dm; either st < than longest peduncle and raceme or ovary and pod pubescent; ovules 20–36

89a Ovary and pod glab; pod ovoid-oblong or broadly oblong-ellipsoid, 7–10 mm thick, shallowly or not at all sulcate on lower suture, valves rarely at all inflexed, any septum < 1 mm high; Blue Mts, headwaters of Umatilla and Grande Ronde rivers, Ore, ne into extreme se Wash; Blue Mt m., longlf m.   67 **A. reventus** Gray

89b Ovary and pod rarely glab; pod narrowly oblong-ellipsoid, (4) 4.5–6.3 mm thick, openly sulcate on lower suture, almost bilocular, septum 1–1.6 mm high; lower Salmon R and SR, Wallowa Co, Ore, Idaho, Lewis, and Nez Perce cos, Ida, and s Asotin Co, Wn; Sheldon's m. (*A. reventus* var. *s.*, *A. conjunctus* var. *s.*)

68 **A. sheldonii** (Rydb.) Barneby

87b Peduncles 1–7 cm, mostly prostrate or ± incurved-ascending; sts often prostrate, to ascending; pod 2-celled

90a Pod subglobose, merely cuspidate at tip, not beaked, valves thickly fleshy, at least 2 mm thick; calyx closely subtended by gen 2 membranous bractlets, teeth 1/4–1/3 as long as tube; fls 22–30 mm, mostly on e side RM, se BC and Alta to NM and Tex, e to Man, Ill, and Ark; ground or buffalo plum, pomme de terre; ours the var. *paysonii* (Kelso) Barneby   69 **A. crassicarpus** Nutt.

90b Pod subglobose to lance-ellipsoid, always with a conspicuous, compressed, triangular or lance-acuminate, often incurved beak, leathery or thinly fleshy, valves much < 1 mm thick; calyx not subtended by bractlets, teeth ca 1/2 as long as tube; fls 8–18 mm; des flats to subalp slopes; throughout much of w US, in our area e Cas; freckled m., specklepod m.; several vars.   70 **A. lentiginosus** Dougl.

a1 Fls 13–18 mm; pod mostly 1.5–3 cm; nc Ore to ne Cal, e to Wyo and Utah (*A. merrillii, Cy. cornutum*)

var. **platyphyllidius** (Rydb.) Peck

a2 Fls mostly < 13 mm, if longer then pod only 1–2 cm

b1 Pod glab, shining, 1.5–2.5 cm; alkaline flats, c Ore to ne Cal, e to sw Mont and Utah (*A. heliophilus*)

var. **salinus** (Howell) Barneby

b2 Pod hairy, 1–2.3 (3) cm; mostly in sagebr, c Wn to Ore and Cal, e to Ida (*A. l.* var. *carinatus*)   var. **lentiginosus**

Subgroup VIc

91a Fls ascending at full anthesis (later sometimes declined)

    92a Pod long-stipitate, compressed, pendulous

        93a Pubescence of calyx and lflets mostly straight and appressed; lflets linear to linear-oblong; body of pod falcately or lunately incurved, 2–3.5 cm × 6–9 mm, strongly compressed, elliptic in section, 1-celled, both sutures prominent but neither intruded; ovules 30–36; calyx 5–7 mm, teeth broad, ca 1/4 length of tube; dunes and sandy des on both sides CR from The Dalles upstream to the Great Bend, n to Kettle Falls, Wn, and along the Okanogan R into BC; The Dalles m., stalked-pod m.   **71 A. sclerocarpus** Gray

        93b Pubescence of calyx and lflets incumbent or curly; lflets narrowly oblong to obovate-cuneate; body of pod < 6 mm broad; ovules 20–30

            94a Pod coiled into a ring, 3–5.5 mm broad, strongly compressed; lflets oblong and rounded to obovate and retuse or semi-obcordate, 2–10 (12) × 2–5 mm; fls white to lavender or bluish tinged; calyx 6–9 mm, teeth triangular, scarcely 1/4 as long as tube; sagebr des, esp near CR; Yakima, Kittitas, Klickitat, and Benton cos, Wn; medick m., curvepod m.   **72 A. speirocarpus** Gray

            94b Pod merely falcate, 24–27 × 4–5 mm, compressed; lflets obovate-cuneate to oblong-oblanceolate, truncate-retuse, 6–16 mm; fls white; calyx 9–11.5 mm, teeth narrowly triangular, 1/4–1/3 length of tube; rocky hillsides, apparently gen in sagebr; known only from Colockum Creek, Chelan Co, Wn; Whited m. (*A. whitedii*)   **73 A. sinuatus** Piper

    92b Pod sessile or subsessile, any stipe not > 1.5 mm, and gen concealed by calyx

        95a Pod fully bilocular, quite sessile; lflets glab beneath or occ with few hairs along midrib (see lead 90b)   **70 A. lentiginosus** Dougl.

        95b Pod 1-celled, obscurely stipitate, 2.5–3.5 cm, fleshy and drying cartilaginous, erect, obcompressed; lflets pubescent beneath; fls violet-purple to off-white, ascending; calyx 7–10 mm, teeth linear-lanceolate, 1/2–3/4 as long as tube; mainly in sagebr; c Ida to Mont, s to ne Nev, Utah, and nw Colo; browse m. (*A. arietinus*)   **74 A. cibarius** Sheld.

91b Fls nodding or declined at full anthesis

    96a Pod erect; pls of ne Ore and e

        97a Lflets 19–27; calyx teeth 1–2 mm; stipe 3–6 mm (see lead 84b)   **64 A. atropubescens** Coult. & Fish.

        97b Lets 23–39; calyx teeth 2–6 mm; stipe not > 1.5 mm (see lead 89b)   **68 A. sheldonii** (Rydb.) Barneby

    96b Pod pendulous, body 15–30 × 2.5–5 mm, compressed-cordate-triquetrous in section, lower suture almost completely intruded; fls ochroleucous; calyx 5–8.8 mm, teeth linear-lanceolate, 1/2 as long to nearly = tube; sagebr plains, lower Deschutes and John Day rivers, Wasco and Sherman cos, into Morrow and Umatilla cos, Ore; Howell's m.   **75 A. howellii** Gray

Group VII

98a Pls of Ore Coast Ranges; fls white; sts nearly glab; pod glab, triquetrous-compressed, 2-celled, falcate, linear or linear-lanceolate, 14–24 × 2.5–3.5 mm; calyx 5–7 mm, teeth 2–3 mm; dry woods; Yamhill Co, Ore, interruptedly to nw Cal; sylvan m., woodl m.   **76 A. umbraticus** Sheld.

98b Pls of interior, E Cas

    99a Cauline stips very large, foliaceous, becoming papery, several-nerved, deflexed; lflets 9–15, ample and thin-textured, 1.5–6 cm; pod pendulous, stipitate, bladdery, membranous, ellipsoid, 2–3 cm, nearly terete; pls of stream banks and mt woods to meadows; Alas s in RM through e BC and Mont to Colo, e to Ont and SD; Que; Am m., arctic m.

        77 A. americanus (Hook.) Jones

    99b Cauline stips not foliaceous; lflets mostly < 1.5 cm; pod sessile if inflated; pls mostly xerophytic

        100a Lflets 7–13, broadly ovate-oblong to oblong-elliptic, emarginate to retuse, thin, greenish, 3–18 mm; calyx 4–6 mm, teeth ca = tube; banner ca = wings, keel only 0.5–1 mm shorter; pod sessile, ca 15–17

100a

mm, 7–8 mm thick, slightly obcompressed, moderately inflated, shallowly sulcate on top, valves thin, becoming papery, inflexed as partial
septum ca 1.5 mm high; known only from Pass Creek Gorge, n of Leslie, Custer Co, Ida; Lost River m.   **78 A. amnis-amissi** Barneby

100b Lflets more ∝ or pl otherwise not as above
    101a Banner gen (because of its erectness) < wings although some-
        times only slightly so   **Subgroup VIIa**
    101b Banner (as measured along curved axis) > wings
                                                Subgroup VIIb, lead 106a

Subgroup VIIa

102a Calyx gibbous-saccate behind pedicel; fls nodding; pod pendulous, stipitate, 1-celled (see leads 82a–82b for distinction between)
                                                59 A. collinus Dougl.
                                    60 A. curvicarpus (Sheld.) Macbr.

102b Calyx sometimes oblique at base but never gibbous
    103a Lflets 21–33; petals ochroleucous; pods erect; se Wn and adj Ida
        104a Ovary and pod sessile, glab; pod oblong- or ovoid-ellipsoid, 15–25
            mm, 6–10 mm thick, slightly obcompressed, lower suture scarcely sul-
            cate; wings 1–4 mm > banner; banner abruptly recurved, 11.5–14
            mm, erose or undulate to lacerate on margins; calyx 7–10.5 mm, teeth
            2–5 mm; SR between mouth of Clearwater R and Whitman and Col-
            umbia cos, Wn, and Nez Perce Co, Ida; Piper's m.
                                                    79 A. riparius Barneby
        104b Ovary and pod stipitate, pubescent; pod narrowly oblong-
            ellipsoid, body 15–23 mm, 2.5–6.5 mm thick, obcompressed,
            keeled by upper suture, openly sulcate on lower side, and nearly com-
            pletely partitioned, broadly obcordate in section; calyx 5–6.5 mm,
            teeth ca 1.5 mm; sagebr flats to open pine for; Whitman Co, Wn, and
            adj wc Ida, also in ec Wn; Palouse m., hangingpod m. *(A. palousensis)*
                                                    80 A. arrectus Gray
    103b Lflets 9–19; petals whitish and ± purple-tinged; pods declined or
        deflexed (rarely ascending); Blue Mts, Ore, and se
        105a Racemes 2–8-fld; pod 4.5–6.5 mm thick, not quite so broad, ±
            mottled, somewhat arcuate, cordate-reniform in section, lower suture
            strongly sulcate and partially dividing cavity; sagebr-covered hills and
            valleys; Blue Mts, Baker, Harney, and Malheur cos, Ore, se to Owyhee
            Co, Ida; Trout Creek m., Salmon m.   **81 A. salmonis** Jones
        105b Racemes 6–15-fld; pod 3–4.2 mm thick, ca 3 mm broad, straight,
            purplish-mottled, slightly sulcate on lower suture, but suture not in-
            truded (see lead 19a)   **13 A. atratus** Wats.

104a

104b

105a

Subgroup VIIb

106a Pl densely villous-tomentose with fine entangled hairs; ovary and pod
silky-hirsute; fls pink or purple; sts very short, pls tufted or matted; mostly
around the s end of our area (see lead 76b)   **54 A. purshii** Dougl.

106b Pl mostly strigillose to nearly glab, if (as sometimes) villous-tomentose
then of more n range and fls ochroleucous or whitish and merely tinged
with lavender, and sts well developed
    107a Calyx villous; fl subsessile in heads or oblong spikes; ovules 4–10;
        pod 4–7 mm, compressed and 2-celled by complete intrusion of lower
        suture, mostly 1–2 seeded; calyx 5–8 mm, teeth subulate, ca half as long
        as tube
        108a Petals white tinged with lilac, glab on outer surfaces; lflets thin,
            gray-villous; sts 0.5–4 dm; grassl and sagebr des; Douglas and Lincoln
            cos to Yakima and Benton cos, Wn, e to Clearwater Valley, w Ida, s to
            Umatilla Co, and in c Baker Co, Ore; Spalding's m.
                                                    82 A. spaldingii Gray
        108b Petals pale yellow, all coarsely pubescent on outer surfaces; lflets
            apparently thick in texture because of dense gray tomentum; sts
            1–5.5 dm; Wasco Co, Ore; Tygh Valley m.   **83 A. tyghensis** Peck
    107b Calyx strigillose; fls loosely racemose; ovules and seeds more ∝
        109a Keel nearly as long as banner, narrowly triangular and acute at
            tip; pod erect, sessile, 2-celled, 15–25 × 3–4 mm; calyx 3–4 mm,

108a

108b

teeth ca 1/3 length of tube; sagebr slopes chiefly; s edge Blue Mts,
Deschutes, Grant, and Crook cos, Ore, s to ne Cal and n Nev, e to
Owyhee and s Elmore cos, Ida; arcane m.   84 **A. obscurus** Wats.
109b Keel (often greatly) < banner, rounded and obtuse at tip; pod and
calyx various
   110a Ovary and pod quite sessile
     111a Pod 1-celled, bladdery-inflated, 3–4 cm, ellipsoid, not mottled;
      calyx 6.7–8 mm, teeth narrowly subulate, ca = tube; shale and
      gravel banks; Custer and Lemhi cos, Ida; Lemhi m. *(A. wootonii,*
      misapplied; *A. w.* var. *a.)*   85 **A. aquilonius** (Barneby) Barneby
     111b Pod 2-celled; calyx teeth ca 1/2 length of tube (see lead 90b)
                                70 A. lentiginosus Dougl.

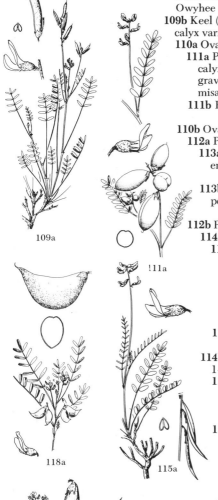

   110b Ovary and pod ± stip
     112a Pod erect or incurved to spreading
      113a Stipe 6–15 mm, widely incurved-ascending; body of pod
      erect but distant from axis of raceme (see also lead 85b)
                               63 A. eremiticus Sheld.
      113b Stipe 2.5–6 mm, straight, erect in the same plane as body of
      pod; pod subcontiguous to axis of raceme (see lead 84b)
                      64 A. atropubescens Coult. & Fish.
     112b Pod pendulous
      114a Fls nodding; pod ± triquetrous, 2-celled
       115a Lflets glab above; body of pod 3–4.5 cm, ca 1/10 as broad,
       compressed, triangular-cordate in section, nearly completely
       2-celled; calyx 5–7 mm, teeth oblong-triangular, ca 1/2
       length of tube; grassy hills and rocky meadows, on basalt; n
       foothills of Blue and Wallowa mts, Umatilla and Wallowa
       cos, Ore, Asotin Co, Wn, and Idaho and Nez Perce cos, Ida;
       Waha m., Arthur's m.   86 **A. arthuri** Jones
       115b Lflets pubescent above; body of pod ca 2–3 cm, at
       least 1/7 as broad (see lead 96b)   75 **A. howellii** Gray
      114b Fls ascending; pod laterally compressed, bicarinate,
      1-celled
       116a Lflets oblong to cuneate-obcordate, loosely pubescent
       with incumbent hairs; body of pod 3.5–5.5 mm broad,
       coiled into a ring or contorted through 1–2.5 turns; ovules
       20–30 (see also lead 94a)   72 **A. speirocarpus** Gray
       116b Lflets linear to linear-oblong, strigillose; body of pod 6–9
       mm broad, lunate or falcate; ovules 30–36 (see lead 93a)
                      71 A. sclerocarpus Gray

Group VIII

117a Pl ann or bien with slender taproot; lflets 3–13
   118a Lflets linear-oblong, terminal one > uppermost pair; pod lunately
   half-ellipsoid, inflated, 1-celled, with no septum; calyx 3–4 mm, teeth ca
   1/3 length of tube; sandy des, mostly on dunes; se Ore to Cal and Nev, e
   in SR drainage, Ida (where barely reaching our area) to Wyo and Utah,
   also in Walla Walla Co, Wn; Geyer's m.   87 **A. geyeri** Gray
   118b Lflets oval or obovate, terminal one no > rest; pod gen ±
   triquetrous, sulcate beneath and with a septum 0.5–1 mm high; calyx ca
   3.5–4.5 mm, teeth ca 2/3 as long as tube; Grant Co, Ore, down John Day
   R to The Dalles, Wasco Co, and across Columbia R in Klickitat Co, Wn;
   transparent m., John Day m. *(A. craigi, A. diurnus, A. drepanolobus)*
                            88 **A. diaphanus** Dougl.
117b Pl mostly per, if fl first year or characteristically bien, lflets always > 13
   119a Peduncles paired in at least some of upper axils; corolla 5–6 mm;
   calyx 3–4 mm, teeth 1.1–1.7 mm; pod 4–7 × 1.5–2.5 mm, compressed-
   triquetrous and 2-celled; e Cas, head Deschutes R, c Ore to ec Cal;
   Lemmon's m.   89 **A. lemmonii** Gray
   119b Peduncles solitary, or pl not otherwise as above
     120a Lflets 5–7 (in var. *lagopinus* as many as 11), densely pan-
     nose-tomentose; pod thickly hirsute-tomentose, 1-celled (see lead 76b)
                       54 **A. purshii** Dougl.
     120b Lflets > 7; ovary nearly or quite 2-celled if pl tomentose
      121a Herbage densely gray-villous or -tomentose; racemes 10–40-fld;
      pod 4–8 mm, 2-celled

128a

124a

129a

129b

130a

132a

122a Calyx 5–8 mm; banner at least 7 mm; racemes densely spicate in fl, interrupted only in fr; pods spreading or ascending (see leads 108a–108b for distinction between)   82 **A. spaldingii** Gray
83 **A. tyghensis** Peck

122b Calyx gen < 5 (but to 5.5) mm; banner gen < 6.5 (7.5) mm; racemes loosely fld; pods declined (see leads 53a–53b)
40 **A. lyallii** Gray
41 **A. caricinus** (Jones) Barneby

121b Herbage strigillose or thinly villosulous, if pubescence loose then racemes only 3–12-fld or pod either > 8 mm or 1-celled (or both)

123a Lflets (11) 17–27; pod pendulous, stipitate, triquetrous, 2-celled

124a Racemes 4–12-fld; fls veined with lilac, 6–7.5 mm; lflets 17–27, elliptic to oval or oblong, 1–7 mm; locally abundant in foothills of Sawtooth Mts, Blaine Co, Ida; Picabo m.
90 **A. oniciformis** Barneby

124b Racemes 10–25-fld; fls whitish-yellow, 6–12 mm; lflets 11–23, linear to linear-lanceolate or narrowly oblong, 5–10 mm (see lead 96b)   75 **A. howellii** Gray

123b Lflets 7–19 (23), if > 17 then ovary and pod sessile or 1-celled
Subgroup VIIIa

Subgroup VIIIa

125a Wings slightly > banner, keel much < either (see leads 105a–105b for distinction between)   13 **A. atratus** Wats.
81 **A. salmonis** Jones

125b Wings < banner or nearly as long but then keel also nearly = banner

126a Petals lilac or purple, often drying bluish; pls of RM (see 56a)
44 **A. eucosmus** Robins.

126b Petals white or ochroleucous, often drying yellowish, only the keel tip lilac (if banner faintly lilac- or pink-veined then pl often more w)

127a Keel 3.5–5.5 mm; ovules 4–10

128a Pod sessile, inflated, 1-celled, 10–15 mm, obliquely ellipsoid, sulcate on both sutures; pl villosulous; calyx hairy, ca 3–4 mm, teeth linear-lanceolate, ca 1/2 length of tube; dry open for, gen on volcanic soils, Klickitat Co, Wn, and ne Cal and adj Nev; Ames m.; ours the var. *suksdorfii* (Howell) Barneby   91 **A. pulsiferae** Gray

128b Pod shortly to prominently stipitate, ± lunate and nearly or quite 2-celled; pl more nearly strigillose; e Cas

129a Pl green and nearly glab; sts 20–45 cm; lflets 5–20 mm; pod declined, lunately linear-ellipsoid, 10–17 × 2.5–3.5 mm, tip cuspidate; fls white; open ridges in timber belt; rare and local, known only from w Wyo and from Clearwater Mts, Idaho Co, Ida; Payson's m.   92 **A. paysonii** (Rydb.) Barneby

129b Pl grayish-strigillose; sts 5–15 (20) cm; lflets 5–10 (14) mm; pod pendulous, lunately lance-elliptic, 12–25 × 2.5–4 mm, triquetrously compressed; pauper m.; 2 vars.   93 **A. misellus** Wats.

a1 Sts 1–2.5 dm long; lflets (9) 13–21, glab or nearly so above; e Cas, Deschutes and Jefferson to Grant Co, Ore (*A. drepanolobus* var. *aberrans, A. howellii* var. *a., A. h.* var. *m.*)
var. **misellus**

a2 Sts 2.5–7 cm; lflets (7) 11–13, pubescent above; near Ellensburg, Kittitas Co, Wn   var. **pauper** Barneby

127b Keel > 6 mm; ovules 14–40

130a Lflets 7–13, broadly ovate-oblong to oblong-elliptic, retuse or emarginate, thin and greenish, 3–18 mm; calyx mostly 4–6 mm, teeth ca = tube; pods sessile, strigillose (see lead 100a)
78 **A. amnis-amissi** Barneby

130b Lflets, calyx, and pod not at once as above

131a Peduncles 1–6.5 cm, < lvs; either tip of keel obtusely rounded or fls declined; pod inflated, neither linear-oblong nor strictly erect

132a Calyx tube broadly camp, 3.5–4 × 2.5–3.3 mm, teeth ca 3–4 mm; pod 1-celled (see lead 111a)
85 **A. aquilonius** (Barneby) Barneby

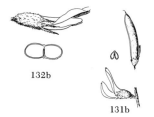

132b Calyx tube narrowly camp, ca 3–4 mm, but only 1.5–2.5 mm wide, teeth < 2.5 mm; pod fully 2-celled (see lead 90b)
                                           70 **A. lentiginosus** Dougl.
131b Peduncles 3–15 cm, mostly > lvs; tip of keel narrowly triangular, acutish; pod strictly erect, linear-oblong, 2-celled (see lead 109a)   84 **A. obscurus** Wats.

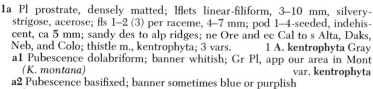

KEY TO ASTRAGALUS USABLE ON PLANTS IN FRUIT

1a Pl prostrate, densely matted; lflets linear-filiform, 3–10 mm, silvery-strigose, acerose; fls 1–2 (3) per raceme, 4–7 mm; pod 1–4-seeded, indehiscent, ca 5 mm; sandy des to alp ridges; ne Ore and ec Cal to s Alta, Daks, Neb, and Colo; thistle m., kentrophyta; 3 vars.   1 **A. kentrophyta** Gray

   a1 Pubescence dolabriform; banner whitish; Gr Pl, app our area in Mont (*K. montana*)   var. **kentrophyta**
   a2 Pubescence basifixed; banner sometimes blue or purplish
      b1 Banner white; ovules 2; sagebr des to foothills near Walla Walla, Wn
                                                    var. **douglasii** Barneby
      b2 Banner bluish or purplish; ovules 5–8; mont to alp; ne Ore to c Nev, e to RMS, Mont to NM (*A. aculeatus, A. tegetarius*)
                                              var. **implexus** (Canby) Barneby

1b Pl not at once prostrate, matted, and with linear-filiform, acerose lflets, 1–2 fls per raceme, and 1–4-seeded, indehiscent pods

   2a Infl, even in fr, a headlike spike nearly as broad as long; calyx often black- or brownish-hairy; corolla ca 15 mm; pod erect, ca 1 cm, ovoid, grayish- or blackish-pilose, cordate in section due to complete intrusion of lower suture

      3a Sts 5–30 cm; corolla mostly purple, if whitish then lavender tinged; lflets 13–21 (23); pod scarcely inflated, ovoid or oblong-ellipsoid, slightly compressed and 3-cornered; ovules 14–22; widespread e Cas, Yuk s through e Wn and Ore to ne Cal, e to Man, Minn, Kans, Colo, and NM; field m., purple m. (*A. dasyglottis, A. goniatus, A. tarletonis, A. virgultulus*)   2 **A. agrestis** Dougl.
      3b Sts 25–70 cm; corolla ochroleucous; lflets 17–31; pod inflated, ovoid or subglobose; ovules 8–10; lower stips connate-sheathing; European, rare in our area, intro in Whatcom Co, Wn; chick-pea m. 3 **A. cicer** L.

   2b Infl seldom capitate or as broad as long, when so pl not otherwise as above

      4a Pods woolly or villous with long white or grayish hairs that almost or quite conceal the surface, gen woody or coriaceous, lower suture intruded to ± completely divide the cavity into 2 cells
                                                       Group A, lead 15a
      4b Pods neither woolly nor with the surface concealed, glab or with dense but short, or less abundant but longer pubescence, varied in texture, lower suture often not at all intruded

         5a Pod fleshy, subglobose, 2–3 cm, 2-celled, glab; calyx with 2 (3) membranous scales at base; fls 22–30 mm, whitish but purplish-tinged, keel often purplish; prairies and foothills; mostly e RM, Alta and Sask to Tex, e to Mo and Tenn, in our range in Mont, where w as far as Flathead, Granite, and Beaverhead cos; ground or buffalo plum, pomme de terre (*A. carnosus, A. caryocarpus, A. prunifer, A. succulentus, A. s. var. paysoni*); ours the var. **paysonii** (Kelso) Barneby (var. *crassicarpus* in e Mont)   4 **A. crassicarpus** Nutt.
         5b Pod gen not fleshy, but when fleshy then otherwise not as above; calyx gen without scales at base
            6a Pl ann or bien
               7a Pod greatly inflated, 1-celled (lower suture not at all intruded), sessile, ca 2 cm; fls 6–8 mm, off-white to pale lavender; winter ann; sandy des, esp on dunes; se Ore to Cal and Nev, e along SR drainage to Wyo and in our area in Ida; Geyer's m.
                                                         5 **A. geyeri** Gray
               7b Pod not greatly inflated (but lower suture intruded), sessile, 1.5–2 cm; fls 4–8 mm, white but pinkish-lined and -tinged; prob bien; gravel bars, alluvial slopes, and thin soil overlying basaltic rocks along CR from Klickitat Co, Wn, and Wasco Co,

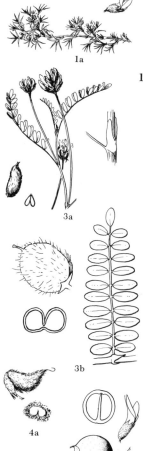

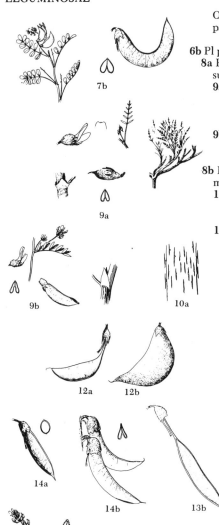

Ore, to mouth of John Day R and up river to Grant Co; transparent m., John Day m. (*A. craigi, A. diurnus, A. drepanolobus*)
6 **A. diaphanus** Dougl.

6b Pl per

8a Pod 4–7 mm, almost completely 2-celled by intrusion of lower suture; fls 4–6 mm; lflets not > 1 cm

9a Terminal lflet confluent with rachis, not noticeably different from it; pl grayish-strigillose; stips connate; sandy soil, often on pumice; w Crook, Deschutes, and Klamath cos, Ore; Peck's m.   7 **A. peckii** Piper

9b Terminal lflet unlike (gen much broader than) rachis, jointed with it; pl greenish; stips free; e Cas, from head Deschutes R, c Ore, to ec Cal; Lemmon's m. 8 **A. lemmonii** Gray

8b Pod either > 7 mm, or not 2-celled, or lflets > 1 cm; fls > 8 mm

10a Pl sub-acaulescent (*A. missouriensis* with lf sts 1–5 cm tall), silvery; pubescence dense, gen tightly appressed, dolabriform; lflets often lacking or not > 3   **Group B**, lead 23a

10b Pl gen caulescent or sub-acaulescent but either with basifixed pubescence or not silvery; lflets mostly several–∝

11a Pod inflated, thin and papery, not at all woody, from ovoid or globose to obovoid, rarely either strongly compressed or obcompressed

12a Pod narrowed to a stalklike base (stipe) at least = the intact calyx, and visible within or beyond it
Group C, lead 27a

12b Pod nonstipitate or with stipe < the intact calyx, and not visible within it   **Group D**, lead 32a

11b Pod not inflated or inflated but tough and often woody, gen elliptic to linear or oblong in outline, often strongly compressed or obcompressed

13a Pod not visibly stipitate; stipe (if any) < calyx and concealed by it

14a Pod 1-celled, with lower suture very slightly if at all intruded, either round in section or (more commonly) compressed   **Group E**, lead 42a

14b Pod either partially to completely 2-celled (lower suture at least somewhat intruded) or ± cordate, reniform, or obcompressed in section
Group F, lead 62a

13b Pod stipitate, body narrowed to stipe at least = calyx tube and visible at maturity   **Group G**, lead 89a

Group A

15a Pod 4–10 mm, villous but not woolly, cordate in section, lower suture completely intruded; fls 6–14 mm, white with purplish pencilling or spotting

16a Racemes dense; fls closely crowded, 10–14 mm, spreading to erect; pod ovoid, scarcely > calyx

17a Lflets thin, grayish-villous; sts 5–40 cm; petals white tinged with lilac, glab on outer surface; grassl and sagebr des; Douglas and Lincoln cos to Yakima and Benton cos, Wn, e to Clearwater Valley, w Ida, s to Umatilla Co, and in c Baker Co, Ore; Spalding's m.
9 **A. spaldingii** Gray

17b Lflets apparently thick in texture due to dense gray tomentum; sts 10–55 cm; petals pale yellow, all coarsely pubescent on outer surface; Wasco Co, Ore; Tygh Valley m.   10 **A. tyghensis** Peck

16b Racemes lax; fls not crowded, 6–9 mm, reflexed; pod twice length of calyx

18a Stips all free; pod 5–8 × 2–3.5 mm, sometimes not fully bilocular, with slightly convex lateral faces and tiny stipe scarcely 0.5 mm; sts uniformly pubescent, fl almost half their length; sagebr plains and foothills; Douglas Co to Kittitas and Franklin cos, Wn; Lyall's m.
11 **A. lyallii** Gray

18b Stips (at least basal ones) connate; pod 6–8.5 × 2–3 mm, with flat or slightly concave lateral faces, sessile, completely 2-celled; sts more

strongly tomentose at base than upward, fl only from the uppermost 2–5 nodes; dunes and hillsides, mostly in sagebr; lower Yakima and Columbia rivers, Wn, and along SR in se Ore and in Ida; buckwheat m.   12 **A. caricinus** (Jones) Barneby

15b Pod either 15–30 mm or woolly, partially to completely 2-celled; fls 15–30 mm, gen deep reddish or purplish

  19a Herbage hirsute; pod compressed, slightly falcate, 2-celled, hirsute, 2–3 cm, somewhat arcuate; fls declined; calyx 8–12 mm, teeth linear-lanceolate, 1.5–4 (5) mm; sagebr des to for slopes; n Malheur and extreme s Baker cos, Ore, and adj Washington Co, Ida, s to ne and ec Cal and to Nev; shaggy m. *(A. obfalcatus)*   13 **A. malacus** Gray

  19b Herbage nearer to being woolly or silky; pod obcompressed, incompletely 2-celled

    20a Sts gen > 15 cm; fls purple, 17–23 mm; calyx (9.5) 11–16.5 mm, teeth 3.1–9 mm, at least 1/2 as long as tube, linear-lanceolate; stips free; sagebr des to dry hillsides; c Wn to extreme ne Ore, e through Ida to c Mont; bent m., hairy m.   14 **A. inflexus** Dougl.

    20b Sts rarely > 10 cm; calyx teeth linear-lanceolate to triangular, gen < 4 mm and < half length of tube; fls white to purple

      21a Pubescence of lvs nearly straight, mostly appressed; sts almost lacking; pod 15–30 mm, densely hirsute to tomentose, strongly obcompressed but 1-celled, lower suture not intruded; calyx 10–16 mm, teeth 1/4–1/3 length of tube; foothills to ponderosa pine fir in our area; c Crook Co, Ore, to Cal and NM, e to s Ida and to Utah; Newberry's m.   15 **A. newberryi** Gray

      21b Pubescence of lvs more nearly tomentose or villous, not appressed; sts often evident

        22a Lflets mostly acutish or fewer than 11; pods tomentose, 10–25 mm, concealed by pubescence; pl gen subacaulescent; calyx 6–18 mm, teeth lanceolate to subulate, 2–3.5 mm; fls white to purple, 10–30 mm; prairies and sagebr des to lower mts e Cas; s BC to n Cal, e to Alta, Daks, Mont, and NM; woollypod m., Pursh's m.; several vars.   16 **A. purshii** Dougl.

        a1 Fls 20–30 mm, ochroleucous to pale lavender, keel purple-tipped; lflets mostly 9–13, acute; pod 2–2.5 cm, nearly straight; sagebr valleys and foothills; ± gen e Cas, BC to ne Cal, e to Alta, Daks, and Colo *(Phaca mollissima, X. incurvus)*
                                                                             var. **purshii**

        a2 Fls 10–25 mm, reddish-purple; pod gen arcuate

          b1 Pod < 20 mm; fls 10–16 mm; upper Deschutes R, Deschutes and Crook cos e to Harney Co, Ore, s to n Cal *(A. l. var. lectulus, A. viarius)*   var. **lagopinus** (Rydb.) Barneby

          b2 Pod and fls gen at least 20 mm

            c1 Pod very strongly sulcate, almost 2-celled near base; lflets mostly oblanceolate-obovate, acute to rounded; mostly in sagebr; e Wn and Ore to Nev and sw Ida *(A. allanaris, A. booneanus, A. g., A. lanocarpus)*
                                                          var. **glareosus** (Dougl.) Barneby

            c2 Pod not so strongly sulcate, sutures not contiguous within; lflets various

              d1 Lflets (5) 7–9, obovate to suborbicular; calyx slightly inflated, gen at least 5 mm broad when pressed; c Ida to wc Mont   var. **concinnus** Barneby

              d2 Lflets (5) 9–13 (17), elliptic to obovate; calyx not inflated, < 4 mm broad when pressed; lower CRV, Wn, s, e Cas, to Nev and Cal *(A. candelarius, A. ventosus)*
                                                                   var. **tinctus** Jones

        22b Lflets obtuse or rounded, 11 or more on at least some lvs; pods 20–35 mm, distinctly beaked, villous, not completely concealed by pubescence, lower suture not at all intruded; pl gen with evident sts; calyx 11–15 mm, teeth subulate, (2) 2.5–4 mm; fls white to purple, 20–25 mm; along SR and tributaries; n Malheur Co, Ore, and Payette Co, Ida, e to s Elmore Co, Ida; cobblestone m.
                                                             17 **A. nudisiliquus** Nels.

Group B

23a Lflets (at least on many lvs) 5 or more

  24a Pod nearly 2-celled by intrusion of lower suture, obcordate in section, oblong, arcuate, not at all inflated; lflets commonly 3–5 (7); pl silvery-strigose; wing petal deeply cleft or toothed at tip; grassl and sagebr des to lower mts; c Ida to Wyo, s to Cal, Ariz, and NM; Torrey m., matted m. (*A. brevicaulis*)   **18 A. calycosus** Torr.

  24b Pod 1-celled, gen obcompressed, ± conspicuously inflated, sometimes not arcuate; lflets mostly 7–15; wing petal not cleft at tip

    25a Racemes long-pedunculate, gen = lvs; fls purple, ascending, 15–20 mm; pod sparsely hairy; calyx 9–14 mm, teeth linear, 1/3–1/4 length of tube; Gr Pl region, e foothills RM, Alta to NM and Tex; Missouri m.   **19 A. missouriensis** Nutt.

    25b Racemes mostly very short-pedunculate; fls yellowish to ± purple-tinged, ca 10 mm, apparently nearly sessile in lf axils; pod fairly heavily pubescent with long crisp hairs, ± lunate; calyx teeth linear, subequal to tube; Gr Pl region, e base RM, Alta to NM and Tex; lotus m. (*Phaca cretacea, A. elatiocarpus*)   **20 A. lotiflorus** Hook.

23b Lflets no > 3

  26a Lvs all or nearly all simple; fls several in long-pedunculate racemes, borne above lvs, purple, 7–9 mm; Alta and GNP s, in e and c Mont, to se Ida, ne Utah, and n Colo, e to Sask, Daks, and w Neb; draba m. (*A. caespitosus* Gray, *Ho. brachycarpus, H. canescens*)   **21 A. spatulatus** Sheld.

  26b Lvs trifoliolate; fls in very short-pedunculate, mostly 1–2-fld racemes, apparently sessile in lf axils, whitish to yellow, 15–30 mm; along e base RM, Mont to Colo and e; plains orophaca (*A. triphyllus, Phaca caespitosa*)   **22 A. gilviflorus** Sheld.

Group C

27a Lflets 1.5–6 cm × 7–15 mm; stips lfletlike, often reflexed; fls ochroleucous; calyx lobes scarcely 1 mm; pl glab or subglab; pod pendulous, ellipsoid, 2–3 cm, nearly terete; stream banks and mt woods to moist meadows; Alas s in RM through e BC and Mont to Colo, e to Ont, SD, and Que; Am m., arctic m.   **23 A. americanus** (Hook.) Jones

27b Lflets mostly < 2 cm, either < 7 mm broad or pl otherwise not as above

  28a Lflets often lacking on many of lvs, linear, gen ca 1 mm broad; lf rachis 6–20 cm; calyx lobes 0.5–2 mm

    29a Fls gen > 10 mm, not purple-tipped, calyx lobes scarcely 1 mm; pod ± ellipsoid-obovoid, 2.5–4.5 cm, not mottled; lateral lflets 10–16; sagebr plains to rocky slopes; e Ore to wc Ida; Cusick's m.; 2 vars.   **24 A. cusickii** Gray

      a1 Lflets 7–11 (13), gen strigillose on upper surface; pl 3–4 dm; calyx 5–6.5 mm; petals white; pod 12–22 mm broad (pressed), obovoid; SRC, from mouth Grande Ronde R to Weiser, Ida   var. **cusickii**

      a2 Lflets 9–15 (19), mostly glab on upper surface; pl 3.5–7 dm; calyx 4–5 mm; petals white to pale purple, but keel purple-tipped; pod 6–12 mm broad (pressed), half obovoid or half ellipsoid; lower Salmon and Little Salmon R canyons, Idaho and Adams cos, Ida   var. **flexilipes** Barneby

    29b Fls gen < 10 mm, gen purple-tipped; calyx lobes linear, 1–2 mm long; pod mottled, oblong-ellipsoid, 2.5–3.5 cm; lflets lacking or 1–several and 1–3.5 cm; sand dunes to prairies; c Ida to e Mont and e and s; painted m.; ours the var. *apus* Barneby, from c Ida, only   **25 A. ceramicus** Sheld.

  28b Lflets present on all lvs, seldom < 2 mm broad, if narrower then rachis much < 6 cm, or calyx lobes > 2 mm

    30a Fls ochroleucous, 15–20 mm; stips not connate; herbage green; body of pod linear to lance-oblong or ± lunate, 7–25 × 2.5–4.2 mm, with reticulations sunken below the gen surface when mature; hill m., hillside m.; 2 vars.   **26 A. collinus** Dougl.

      a1 Pod slenderly linear-oblong, 7–25 × 2.5–3.5 mm, straight or slightly curved, crisply villosulous, 12–18- ovuled; basaltic grassl and sagebr des; ne Ore n to c Wn, e to wc Ida, also in s BC (*A. cyrtoides*)   var. **collinus**

      a2 Pod plumper, averaging 8–15 × 3.3–4.2 mm, obliquely ov-

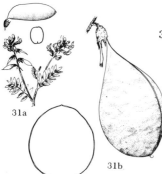

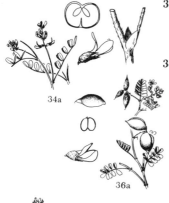

ate-oblong, lunate, villosulous or glab, 8–12-ovuled; Morrow Co, Ore var. **laurentii** (Rydb.) Barneby

30b Fls gen purple or at least purplish-tinged, 6–11 mm; basal stips gen ± connate

31a Body of pod < 3 cm, glab, not mottled, ± compressed, elliptic-oblong in outline, lower suture intruded (not > 1 mm); pl grayish-hirsutulous; alp talus slopes, OM only; Cotton's m.

27 **A. cottonii** Jones

31b Body of pod gen at least 3 cm, gen strigillose, often mottled, not compressed, elliptic-obovate in outline, lower suture not intruded; pl grayish-strigillose to greenish; open mont slopes to alp summits, often on serpentine; e slope Cas, Wn to ne Cal, e to sw Ida and nw Nev; balloon m.; ours the var. *sonneanus* (Greene) Jeps.

28 **A. whitneyi** Gray

Group D

32a Pod ca 1 cm, broadest at or above middle, hairy, not particularly oblique; stips gen drying blackish, connate; calyx teeth linear, subequal to tube; fls 6–9 mm long, pink to magenta-purple; open prairies and foothills to ponderosa pine for; ne Wn and adj BC across Ida panhandle to w Mont near Helena; least bladdery m. 29 **A. microcystis** Gray

32b Pod gen either much > 1 cm or oblique (or both); stips often free; calyx teeth often much < tube

33a Pod ± completely 2-celled, cavity often > 1/2 partitioned

34a Pod only slightly inflated, 15–17 × 7–8 mm, partially 2-celled, septum only ca 1.5 mm high; stips not connate; lflets 7–13, broadly ovate-oblong to oblong-elliptic, emarginate to retuse; calyx 4–6 mm, teeth ca = tube; known only from Pass Creek Gorge, n of Leslie, Custer Co, Ida; Lost R m. 30 **A. amnis-amissi** Barneby

34b Pod more strongly inflated, gen much > 8 mm wide, more completely 2-celled, the septum much > 1.5 mm high; stips often connate

35a Lvs grayish-hairy; lflets oblong-obovate, 5–10 mm; pod densely strigillose; keel nearly or quite = banner; pl sometimes almost acaulescent; stips connate

36a Sts 1–4 dm, diffuse or trailing, freely br; axillary peduncles ∝; root crown subterranean; lflets 9–13, oblong-obovate, gen emarginate, 5–10 mm; pod not mottled, 2–3.5 cm; Salmon R canyon, Custer and Lemhi cos, Ida, on shale or basalt; Challis m.

31 **A. amblytropis** Barneby

36b Sts < 1 dm, often almost lacking, tufted; rootcrown exposed; peduncles few, subscapose; lflets (5) 13–21, oblong-obovate to oblanceolate, 3–8 (10) mm, rounded to acute; pod gen purplish-mottled, 2–3 cm; mont to alp, mostly on talus or scree, but occ with sagebr, often on limestone; Beaverhead Co, Mont, to Pahsimeroi R, nc Ida, more widespread in Nev and adj Cal; broad-keeled m. 32 **A. platytropis** Gray

35b Lvs greenish, but sometimes ± hairy; larger lflets > 10 mm; pod glab or very sparsely hairy; keel much < banner; pl strongly caulescent; stips not connate; des flats to subalp slopes, throughout much of w US, in our area e Cas; freckled m., specklepod m.; several vars.

33 **A. lentiginosus** Dougl.

a1 Fls 13–18 mm; pod mostly 1.5–3 cm; nc Ore to ne Cal, e to Wyo and Utah (*A. merrillii, Cy. cornutum*)

var. **platyphyllidius** (Rydb.) Peck

a2 Fls mostly < 13 mm, if that long then pod only 1–2 cm

b1 Pod glab, shining, 1.5–2.5 cm; alkaline flats; c Ore to ne Cal, e to sw Mont and Utah (*A. heliophilus*)

var. **salinus** (Howell) Barneby

b2 Pod hairy, 1–2.3 (3) cm; mostly in sagebr; c Wn to e Ore and Cal, e to Ida (*A. l.* var. *carinatus*) var. **lentiginosus**

33b Pod strictly 1-celled, sutures slightly or not at all intruded

37a Fls 14–18 mm; calyx tube 4–7 mm (see lead 30a)

26 **A. collinus** Dougl.

37b Fls < 14 mm; calyx tube gen < 4 mm

38a Pod 8–20 mm; lvs 1.5–6 cm; fls < 10 mm; lflets always present

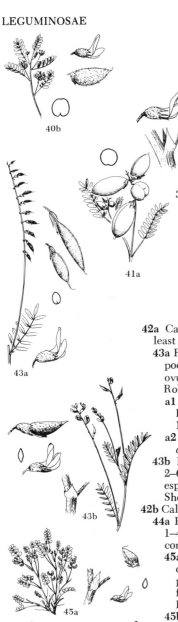

39a Pod gen > 15 mm, not strongly compressed; John Day Valley, Ore, n to CR (see lead 7b)   **6 A. diaphanus** Dougl.

39b Pod not > 15 mm, often compressed; not known from John Day Valley

    40a Stips connate; pod rather symmetrical, broadest at or above middle, lower suture not at all sulcate; corolla white to deep magneta-purple (see lead 32a)   **29 A. microcystis** Gray

    40b Stips not connate; pod obliquely ellipsoid, 10–15 mm, sulcate on both sutures; pl villosulous; calyx hairy, ca 3–4 mm, teeth linear-lanceolate, ca 1/2 length of tube; corolla ochroleucous or very pale purple; dry open for, gen on volcanic soils; Klickitat Co, Wn, and ne Cal and adj Nev; Ames' m.; ours the var *suksdorfii* (Howell) Barneby   **34 A. pulsiferae** Gray

38b Pod 25–40 mm; lvs in part often > 6 cm; fls ca 10 mm or sometimes shorter and then lvs in part without lflets

    41a Lflets present on all lvs; pod not mottled, 3–4 cm, ellipsoid; calyx 6.7–8 mm, teeth narrowly subulate, ca = tube; shale and gravel banks; Custer and Lemhi cos, Ida; Lemhi m. (*A. wootonii* var. *a.*)   **35 A. aquilonius** (Barneby) Barneby

    41b Lflets lacking on many lvs; pod mottled, oblong-ellipsoid, 2.5–3.5 cm; sts single or few together, from buried caudex brs with stip sheaths (see lead 29b)   **25 A. ceramicus** Sheld.

Group E

42a Calyx and pod very strongly pubescent with appressed black (or at least dark) hairs; pl not at all rhizomatous

    43a Racemes gen at least 20-fld; stips free or lowest ones slightly connate; pod often > 15 mm, obcompressed and obtusely 3-cornered in section; ovules 7–9 (rarely only 3–5); lflets 9–13 (19), oval or oblong, obtuse; Robbins' m.; 2 vars.   **36 A. robbinsii** (Oakes) Gray

    a1 Pod (ours) 15–25 mm, 7–9-ovuled; stream banks to alp slopes; Alas, s BC to Alta, s to ne Wn, w Mont, e Ida, w Wyo, and Colo; Vt, Me, and NS (*A. blakei, A. macounii*)   var. **minor** (Hook.) Barneby

    a2 Pod 8–11 mm, 3–5-ovuled; Wallowa Mts, Ore (*At. a., A. alpinus* var. *a.*)   var. **alpiniformis** (Rydb.) Barneby

    43b Racemes mostly 5–10-fld; stips strongly connate; pod ca 15 mm, 2–6-ovuled, compressed, both sutures prominent; alp or subalp or lower, esp along rivers, often on limestone; BC and Alta to Mont, w to Shoshone Co, Ida; Bourgeau's m.   **37 A. bourgovii** Gray

42b Calyx and pod both not black-hairy, or if so pl strongly rhizomatous

    44a Pod 4–10 (12) mm, compressed, 1-celled, elliptic-oblong, acute; lflets 1–4 mm broad, grayish-hairy; fls 4–8 (10) mm; stips (at least lower ones) connate

    45a Fls 4–6 mm, whitish, banner purple-pencilled; lflets sharp-pointed, obovate to obcordate, 3–7 mm; pod 4–6 mm; ovules 2 (3); seeds gen 1; pl prostrate, very slender-std, grayish-strigillose; dry ponderosa pine for; n Harney Co, possibly n to Grant Co, Ore; bastard kentrophyta, Deschutes m.   **38 A. tegetarioides** Jones

    45b Fls 5–8 (10) mm, gen ± purplish; lflets linear- to oblong-elliptic, 5–12 mm, not sharp-pointed; pod sometimes 7–11mm; ovules 4–7; seeds gen 2 or more; pl diffuse to ascending, sometimes not strigillose; bent-fld m.; 2 vars.   **39 A. vexilliflexus** Sheld.

    a1 Pl prostrate, matted; herbage loosely silky-villosulous; banner ca 5–6 mm; pod 3.5–5 mm; alp stony crests; White Cloud Range, Custer Co, Ida   var. **nubilus** Barneby

    a2 Pl various, often not matted; herbage often appressed-pubescent; banner often 7–9 mm; pod mostly 5–11 mm; foothills to subalp slopes; RM, mostly on e side, Banff Nat Park, Can, s to Wyo, w in Mont to Beaverhead Co, e to s Sask and Daks (*A. amphidoxus*)   var. **vexilliflexus**

    44b Pod > 10 mm, gen linear or oblong; lflets often either > 4 mm broad or not grayish-hairy; fls often > 10 mm

    46a Pl delicate, diffuse, sts arising at intervals from subterranean rootstock; lower stips connate; pod somewhat obcompressed, 8–14 × 2.5–4 mm; keel narrow, lilac-tipped, < white wings and banner; racemes mostly 2–3 (5) -fld; w Mont, ec Ida, and nc Colo; park m. (*Phaca pauciflora*)   **40 A. leptaleus** Gray

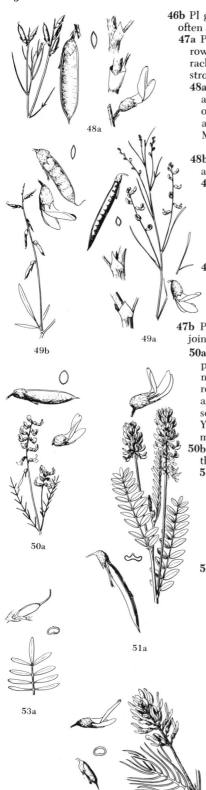

48a

49a

49b

50a

51a

53a

53b

46b Pl gen ascending to erect; sts mostly from woody, br caudex; stips often all free; pod gen compressed

47a Pl often rushlike or broomlike; lvs with comparatively few, narrow, much-elongate lflets, terminal lflet (if any) confluent with rachis; pod linear to narrowly oblong in outline, 2–5 mm broad, strongly compressed

48a Corolla 13–20 mm, pink-purple; pl erect; sts many, erect, stiff and broomlike; lvs few; lflets 0–9, very narrow, 5–30 mm; pod oblong-ellipsoid, 13–25 mm, both sutures prominent; low elev along SR, Fremont to Owyhee Co, Ida, to nw Utah and Nev and Malheur Co, Ore (s of our area entirely?); Toano m.

41 **A. toanus** Jones

48b Corolla not > 13 mm; pl often decumbent or otherwise not as above

49a Rachis of lvs 1–1.5 mm wide, attenuate toward tip; lateral lflets 2, 4, or more (sometimes 0), linear, 1–3 cm, often involute and scarcely as broad as rachis; pod linear, ca 2 mm broad; pl erect, chiefly on dry benchl; calyx teeth scarcely 1/3 length of tube; SR drainage, se Ida, s to Utah, e to Wyo, in our area near Helena, Mont; lesser rushy m. *(A. junceus)*

42 **A. convallarius** Greene

49b Rachis of lvs 1–2 mm broad, gen broadened into a terminal lflet, or (if not widened) lateral lflets 2–5 mm broad; pod 3–4 mm broad; pl reclining, in alkaline meadows; calyx teeth ca 1/2 length of tube; c Ida to n Utah e to Wyo; meadow m. *(A. reclinatus)*

43 **A. diversifolius** Gray

47b Pl not rushlike; lvs gen with many lflets, terminal lflet mostly jointed to rachis; pod various

50a Pod terete or slightly compressed, 12–21 mm, 2–3 mm thick, partially filled between seeds with criss-crossing fibrous strands, neither suture intruded; root crown subterranean; lflets 15–21, rounded to emarginate, 5–22 mm; calyx 4–5 mm, teeth 1/4–1/3 as long as tube; mostly e RM, esp in alkaline soil; extreme se BC and Alta s to Utah, Colo, and NM, w in Mont to Mo and Yellowstone rivers, e to Man, Daks, and w Neb; pliant m., wiry m.

44 **A. flexuosus** (Dougl.) Don

50b Pod gen either compressed or obcompressed, if terete then either > 3 mm thick or not filled with fibrous material

51a Pod strongly obcompressed, ca twice as thick as broad, very deeply furrowed on both sides of prominent upper suture, strictly 1-celled; wings and keel subequal; racemes gen > 30-fld; calyx gibbous-saccate behind ped; fls purple (to white), 11–15 mm; sagebr des and grassl, often where alkaline; mostly e RM, Alta to NM, e to Neb, w in our area to Beaverhead Co, Mont, and to Clark Co, Ida; two-grooved m.

45 **A. bisulcatus** (Hook.) Gray

51b Pod not bifurrowed on upper side, although sometimes obcompressed; pl various

52a Pl fleshy, malodorous; upper lflet confluent with rachis; fls white, ca 2 cm, ochroleucous; pod fleshy, nearly solid, drying to woody, subterete

53a Sts erect; pod erect, 9–18 mm, 2.7–3.5 mm thick; ovules 14–23; calyx 6.5–10.3 mm, white- (rarely blackish-) strigillose, teeth 1.2–2.5 mm; adobe plains e RM; Carbon Co, Mont, to Carbon Co, Wyo; Gray's m.

46 **A. grayi** Parry

53b Sts diffuse to ascending; pod deflexed, (10) 15–25 × 4.5–8 mm thick; ovules (23) 26–32; calyx 8–12 mm, black-strigillose, teeth 1.5–3 mm; e Cont Div, Alta to Colo, e to Kans; tine-lvd m.   47 **A. pectinatus** Dougl.

52b Pl not fleshy and malodorous; upper lflets gen not confluent with rachis; fls often < 2 cm, or purplish; pod not as above

54a Calyx tube 8–12 mm; fls gen purple, mostly 15–25 mm

55a Lvs appressed-villous or sericeous; sts gen < 10 cm; pod hirsute-strigose, 1.5–2.5 cm long, obcompressed, ca 8 mm thick; calyx 9–12 mm, teeth ca 1/3 length of tube;

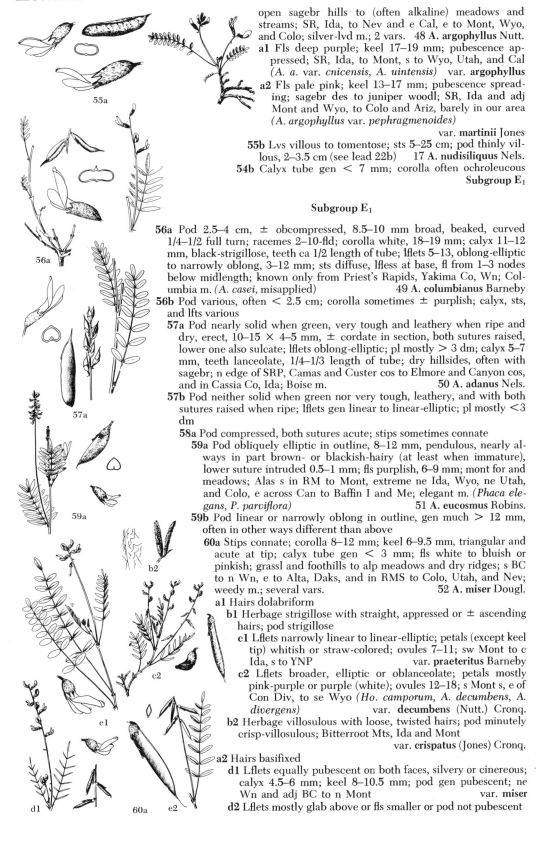

open sagebr hills to (often alkaline) meadows and streams; SR, Ida, to Nev and e Cal, e to Mont, Wyo, and Colo; silver-lvd m.; 2 vars. 48 **A. argophyllus** Nutt.

  a1 Fls deep purple; keel 17–19 mm; pubescence appressed; SR, Ida, to Mont, s to Wyo, Utah, and Cal (*A. a.* var. *cnicensis, A. uintensis*) var. **argophyllus**

  a2 Fls pale pink; keel 13–17 mm; pubescence spreading; sagebr des to juniper woodl; SR, Ida and adj Mont and Wyo, to Colo and Ariz, barely in our area (*A. argophyllus* var. *pephragmenoides*)

                                      var. **martinii** Jones

55b Lvs villous to tomentose; sts 5–25 cm; pod thinly villous, 2–3.5 cm (see lead 22b) 17 **A. nudisiliquus** Nels.

54b Calyx tube gen < 7 mm; corolla often ochroleucous

                                            Subgroup E₁

Subgroup E$_1$

56a Pod 2.5–4 cm, ± obcompressed, 8.5–10 mm broad, beaked, curved 1/4–1/2 full turn; racemes 2–10-fld; corolla white, 18–19 mm; calyx 11–12 mm, black-strigillose, teeth ca 1/2 length of tube; lflets 5–13, oblong-elliptic to narrowly oblong, 3–12 mm; sts diffuse, lfless at base, fl from 1–3 nodes below midlength; known only from Priest's Rapids, Yakima Co, Wn; Columbia m. (*A. casei*, misapplied) 49 **A. columbianus** Barneby

56b Pod various, often < 2.5 cm; corolla sometimes ± purplish; calyx, sts, and lfts various

  57a Pod nearly solid when green, very tough and leathery when ripe and dry, erect, 10–15 × 4–5 mm, ± cordate in section, both sutures raised, lower one also sulcate; lflets oblong-elliptic; pl mostly > 3 dm; calyx 5–7 mm, teeth lanceolate, 1/4–1/3 length of tube; dry hillsides, often with sagebr; n edge of SRP, Camas and Custer cos to Elmore and Canyon cos, and in Cassia Co, Ida; Boise m. 50 **A. adanus** Nels.

  57b Pod neither solid when green nor very tough, leathery, and with both sutures raised when ripe; lflets gen linear to linear-elliptic; pl mostly <3 dm

    58a Pod compressed, both sutures acute; stips sometimes connate

      59a Pod obliquely elliptic in outline, 8–12 mm, pendulous, nearly always in part brown- or blackish-hairy (at least when immature), lower suture intruded 0.5–1 mm; fls purplish, 6–9 mm; mont for and meadows; Alas s in RM to Mont, extreme ne Ida, Wyo, ne Utah, and Colo, e across Can to Baffin I and Me; elegant m. (*Phaca elegans, P. parviflora*) 51 **A. eucosmus** Robins.

      59b Pod linear or narrowly oblong in outline, gen much > 12 mm, often in other ways different than above

        60a Stips connate; corolla 8–12 mm; keel 6–9.5 mm, triangular and acute at tip; calyx tube gen < 3 mm; fls white to bluish or pinkish; grassl and foothills to alp meadows and dry ridges; s BC to n Wn, e to Alta, Daks, and in RMS to Colo, Utah, and Nev; weedy m.; several vars. 52 **A. miser** Dougl.

        a1 Hairs dolabriform

          b1 Herbage strigillose with straight, appressed or ± ascending hairs; pod strigillose

            c1 Lflets narrowly linear to linear-elliptic; petals (except keel tip) whitish or straw-colored; ovules 7–11; sw Mont to c Ida, s to YNP var. **praeteritus** Barneby

            c2 Lflets broader, elliptic or oblanceolate; petals mostly pink-purple or purple (white); ovules 12–18; s Mont s, e of Con Div, to se Wyo (*Ho. camporum, A. decumbens, A. divergens*) var. **decumbens** (Nutt.) Cronq.

          b2 Herbage villosulous with loose, twisted hairs; pod minutely crisp-villosulous; Bitterroot Mts, Ida and Mont

                          var. **crispatus** (Jones) Cronq.

        a2 Hairs basifixed

          d1 Lflets equally pubescent on both faces, silvery or cinereous; calyx 4.5–6 mm; keel 8–10.5 mm; pod gen pubescent; ne Wn and adj BC to n Mont var. **miser**

          d2 Lflets mostly glab above or fls smaller or pod not pubescent

61a

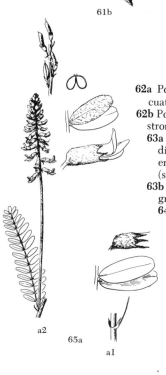

61b

a2
65a

a1

65b

el Sts < 15 cm; keel mostly 8–10 mm; lflets broad, elliptic
  to oval; pod glab; common in RM, Mont, nc Ida, and nw
  Wyo   var. **hylophilus** (Rydb.) Barneby
e2 Sts mostly at least 15 cm; keel 6–8 mm; lflets linear to
  narrowly elliptic; pod sometimes puberulent; s BC to Alta,
  s (e Cas) to Kittitas Co, Wn (*A. palliseri, A. s.*)
                                       var. **serotinus** (Gray) Barneby
  60b Stips free; corolla mostly 15–20 mm; calyx tube 4–7 mm (see
      lead 30a)   26 **A. collinus** Dougl.
58b Pod obcompressed, sulcate beneath; stips free
  61a Peds 3–6 mm; pod nearly straight, ca 3.5 (3–4.2) mm wide and
      thick, more membranous than woody, 1-celled, lower suture sulcate
      but not intruded; racemes 6–15-fld; fls 8–9 mm, whitish, purp-
      lish-lined; calyx 3.5–4 mm, teeth 1/3 length of tube; sagebr slopes,
      Baker Co, Ore, to Nev and adj Ida; mourning m.; 2 vars.
                                               53 **A. atratus** Wats.
    a1 Lflets narrow and remote, terminal one continuous with rachis;
        pod papery; Baker and Malheur cos, Ore, to sw Ida and ne Nev
                                   var. **owyheensis** (Nels. & Macbr.) Jones
    a2 Lflets more ample, less scattered, the terminal one jointed with
        rachis; pod leathery; stony flats where moist in spring; ne edge
        SRP, Camas Co, Ida   var. **inseptus** Barneby
  61b Peds 1–4 mm; pod ca 5.5 (4.5–6.5) mm in width and thickness,
      oblique or curved, woody (fleshy when immature), deeply sulcate on
      lower suture, nearly 2-celled; racemes 2–8-fld; sagebr-covered hills
      and valleys; Blue Mts, Baker, Harney, and Malheur cos, Ore, se to
      Owyhee Co, Ida; Trout Creek m., Salmon m.   54 **A. salmonis** Jones

Group F

62a Pod membranous, often somewhat transparent, strigillose, strongly ar-
    cuate; ann or bien (see lead 7b)   6 **A. diaphanus** Dougl.
62b Pod thick and opaque or sometimes thin but copiously dark-hairy; mostly
    strong per
  63a Pod inflated to turgid, ovoid to ovoid-lanceolate, gen terete or ±
      didymous in section, tapered to a distinct and prominent beak; beak lat-
      erally compressed to filiform and 1-celled; pl gen greenish; fls 8–18 mm
      (see lead 35b)   33 **A. lentiginosus** Dougl.
  63b Pod various, gen ± linear to oblong in outline, seldom beaked; pl often
      grayish-hairy
    64a Fls (15) 30–150 in spikelike congested racemes, 12–18 mm; ped ca 1
        mm; pubescence of calyx and sts appressed, 2-br; stips membranous,
        mostly connate; pod 8–20 mm, seldom > twice as long as calyx, lower
        suture deeply sulcate
      65a Pl strongly rhizomatous; pod 8–20 × 4–5 mm; fls declined, over-
          lapping, white or off-white, keel gen purple-tipped; BC to Atl, s to
          Cal, Colo, and Tex; Can m.; 2 vars.   55 **A. canadensis** L.
        a1 Calyx black- and white-hairy, teeth subequal, 1.5–3 mm; pl
            greenish; ponderosa pine and lower mont for, occ into sagebr des;
            Okanogan Co, Wn, to Ida and Mont (*A. m., A. tristis*)
                                          var. **mortonii** (Nutt.) Wats.
        a2 Calyx white-hairy, teeth 1–2 mm, upper ones much the broader
            and shorter; pl ± silvery-pubescent; water-courses or alkaline
            flats to lower open for; BC to Cal, e Cas, e to Daks and Colo (*A.
            b., A. spicatus*)   var. **brevidens** (Gandg.) Barneby
      65b Pl nonrhizomatous; pod 8–20 × 4–5 mm; fls erect or ascending,
          white or purplish; prairies to rocky foothills; c Wn to Alta, s to NM;
          standing m. (*A. striatus*)   56 **A. adsurgens** Hook.
    64b Fls mostly < 30, or not spicate-racemose, or < 12 mm; pubescence
        gen basifixed; stips mostly free; pod gen > 10 mm
      66a Lflets 3–10 mm, often filiform or linear, distant, the terminal one
          often confluent with rachis; sts either filiform and ± decumbent, or
          short (scarcely 5 cm); fls 7–10 mm; pod contracted to a short (0.5–2
          mm) stipe within the calyx, body 10–22 mm, cordate in section, 3–7
          mm thick (see leads 61a–61b)   53 **A. atratus** Wats.
                                               54 **A. salmonis** Jones

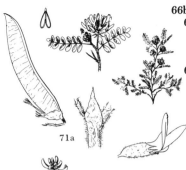

66b Lflets either > 10 mm, or pl (esp pods) not as above
    67a Pod strictly 1-celled, ± obcompressed, valves papery-membranous, not inflexed, strigillose with black or white hairs; stips connate; brs rhizome-like at base (see lead 46a)
        40 A. leptaleus Gray
    67b Pod either ± 2-celled due to intrusion of valves along the lower suture, or pl otherwise not as above
        68a Pod completely 2-celled to falsely 2-celled by intrusion of lower suture, partition always extending at least 3/4 width of cavity   **Subgroup F$_1$**
        68b Pod imperfectly 2-celled, the partition (or intruded lower suture) < 3/4 complete   **Subgroup F$_2$**, lead 81a

71a

Subgroup F$_1$

69a Fls 4.5–8 (9) mm, white or purplish-tinged; pods 5–9 mm, reflexed; lflets 5–20 mm, linear, grayish-hairy (see leads 18a–18b)
        12 A. caricinus (Jones) Barneby
        11 A. lyallii Gray
69b Fls either > 8 mm, or not whitish, or pods either > 10 mm or ascending to erect
    70a Sts villous to hirsute
        71a Pod glab, shining, not mottled; wing petal 1–4 mm > banner, dilated upward and emarginate at or below apex; fls 18–26 mm, pinkish, keel often purplish-tipped; calyx tubular, 9–15 mm, teeth linear-lanceolate, 1/2–2/3 as long as tube; sagebr des to lower foothills, esp on dunes; Klickitat and Grant cos, Wn, to Umatilla and Gilliam cos, Ore; crouching m., Columbia m.   **57 A. succumbens** Dougl.
72a
        71b Pod hirsute, dull, often mottled (see lead 19a)   **13 A. malacus** Gray
    70b Sts glab or strigose
        72a Calyx ca 4 (5) mm, teeth blunt, scarcely 1/4 length of tube; fls ca 15 mm, white but drying ochroleucous, keel purple-tipped; lfts gen oblong-oblanceolate to obovate; pod erect, glab, 15–20 × ca 4 mm, slightly obcompressed, broadly cordate in section, lower suture almost completely intruded; racemes 10–30-fld; sagebr benchl to mont ridges, often on limestone; c Ida to sc Mont, s and e to Teton Co, Wyo; railhead m. (*A. reventoides*)   **58 A. terminalis** Wats.
        72b Calyx teeth 1/3–4/5 length of tube, or pls otherwise not as above
            73a Pubescence dolabriform; stips not connate; calyx ca 5 mm; pod sessile, 2-celled; pl 4–8 dm, greenish; fls ochroleucous; intro from Caucasus and estab in se Wn; Russian sickle m.   **59 A. falcatus** Lam.

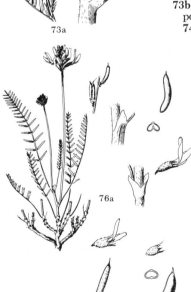

73a
            73b Pubescence basifixed; calyx either < 4.5 mm or much > 5 mm; pod sometimes stipitate; stips often connate
            74a Calyx 7.5–15 mm
                75a Lower stips connate (except sometimes in *A. hoodianus*), upper ones sometimes free; lflets 13–37
                    76a Calyx tube cylindric or subcylindric, 7–12 mm, tube ca 2× as long as thick, teeth subulate, mostly 1.5–3 (4) mm; ovary glab; ovules 23–30; corolla 15–25 mm, white or distally purplish; nc Ore, Wasco Co to Blue Mts, se along Malheur R to Steens Mts, e to sw Ida, mostly above 2000 ft elev; basalt m., stiff m. (*A. reventus* var. *c.*)   **60 A. conjunctus** Wats.
76a
                    76b Calyx camp, ca 1.5× as long as thick; ovary often pubescent; ovules 16–22; fls mostly white
                        77a Ovary and pod pubescent, sessile; calyx 11–15 mm, teeth 4.5–7.5 mm; pl 4–30 cm; CR gap, Wasco and Hood R cos, Ore, and Klickitat Co, Wn; Hood R m. (*A. conjunctus* var. *oxytropidoides*, *A. reventus* var. *o.*, *Cn. knowlesianus*)   **61 A. hoodianus** Howell
                        77b Ovary glab; calyx 8–12 mm, teeth 2.5–5 mm; e Cas, from near Ellensburg, Kittitas Co, to Klickitat Co, Wn, s to Sherman Co, Ore; Yakima m. (*A. reventus* var. *canbyi*)
                          62 A. reventiformis (Rydb.) Barneby
                75b Lower stips (as well as the upper) free although sometimes decurrent over halfway around st; lflets 23–41
                  78a Ovary and pod glab; pod ovoid-oblong or broadly oblong-ellipsoid, 7–10 mm thick, shallowly if at all sulcate on
77a   77b

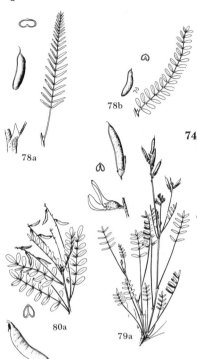

lower suture, valves rarely at all inflexed, any septum < 1 mm high; Blue Mts, headwaters of Umatilla and Grande Ronde rivers, Ore, ne into extreme se Wn; Blue Mt m., long-lf m.   63 A. reventus Gray

78b Ovary and pod rarely glab; pod narrowly oblong-ellipsoid, (4) 4.5–6.3 mm thick, openly sulcate on lower suture, almost 2-celled, septum 1–1.6 mm high; lower Salmon and SR, Wallowa Co, Ore, and Idaho, Lewis, and Nez Perce cos, Ida, and s Asotin Co, Wn; Sheldon's m.
                                    64 A. sheldonii (Rydb.) Barneby

74b Calyx rarely as much as 7 mm

  79a Pl silvery-strigillose; pod sessile, erect, 2-celled, 15–25 × 3–4 mm; calyx 3–4 mm, teeth ca 1/3 length of tube; keel nearly = banner, narrowly triangular and acute at tip; sagebr slopes chiefly; s edge Blue Mts, Deschutes, Grant, and Crook cos, Ore, s to ne Cal and n Nev, e to Owyhee and s Elmore cos, Ida; arcane m.   65 A. obscurus Wats.

  79b Pl pale green, nearly glab; pod with stipe 1–1.5 mm

    80a Banner ca 7 mm; pod declined, lunately linear-ellipsoid, body 10–17 mm; lflets 7–12; open ridges in timber belt, rare and local; known only from w Wyo and Clearwater Mts, Idaho Co, Ida; Payson's m.   66 A. paysonii (Rydb.) Barneby

    80b Banner 10–14 mm; pod declined, falcate, linear or linear-lanceolate, 14–24 × 2.5–3.5 mm; lflets 15–23; dry woods, Yamhill Co, Ore, interruptedly to nw Cal; sylvan m., woodl m.   67 A. umbraticus Sheld.

Subgroup F$_2$

81a Pod gray- or black-hairy, < 13 mm, pendulous; fls 6–9 mm (see lead 59a)
                                    51 A. eucosmus Robins.

81b Pod and fls not as above in all respects

  82a Fls < 10 mm; pod 6–9 mm, obcompressed, 3.5–4 mm thick, scarcely half as broad, grayish-hairy, gen 2-seeded, fleshy but becoming leathery and cross-rugulose; lflets mostly retuse, 5–13 × 1–3 mm; sts mostly from buried rootcrown; grassl to juniper-covered hills; e foothills RM, Mont to NM and Tex, e to Daks, Neb, Kans, and Okla, also in Sask; slender m. (A. microlobus)   68 A. gracilis Nutt.

  82b Fls or pods either much > 10 mm or pods not grayish-hairy

    83a Pod (2) 2.5–3.5 cm, distinctly arcuate, 7–10 mm thick, gen glab; lflets greenish, oblong-elliptic to obovate, glab on upper surface; fls violet-purple to off-white, ascending; calyx 7–10 mm, teeth linear-lanceolate, 1/2–3/4 length of tube; mainly in sagebr; c Ida to Mont, s to ne Nev, Utah, and nw Colo; browse m. (A. arietinus)
                                    69 A. cibarius Sheld.

    83b Pod either < 2 cm or straight, or lflets not oblong-elliptic to obovate and glab on upper surface

      84a Lflets 9–15, linear, 2–6 cm, at least the terminal one confluent with rachis; pods reflexed; calyx tube > 6 mm, gen black-hairy; e Cont Div in our area (see lead 53b)   47 A. pectinatus Dougl.

      84b Lflets either not linear and so much as 2 cm, or pods erect, or calyx tube > 6 mm

        85a Pl greenish; sts 3–7 dm; lflets 10–25 mm, oblong-elliptic, in part gen > 5 mm broad; pod fleshy and nearly solid when green, hardened and fibrous-filled when dry, 10–15 mm, lower suture intruded only slightly; calyx teeth 1.5–2 mm (see lead 57a)
                                    50 A. adanus Nels.

      85b Pl not at once greenish and with st, lflets, pods, and calyx as above

        86a Calyx teeth broadly triangular, ca 1 mm; pod partially septate; lflets oblong-oblanceolate to obovate, (2) 3–5 mm broad; fls white, drying ochroleucous (keel purple-tipped), 12–16 mm, nodding (see lead 72a)   58 A. terminalis Wats.

        86b Calyx teeth at least 2 mm long; pod not septate; lflets sometimes linear, mostly < 3 mm broad

          87a Wing petals 1–4 mm > banner; pod glab, oblong- or ovoid-ellipsoid, 15–25 × 6–10 mm, slightly obcompressed;

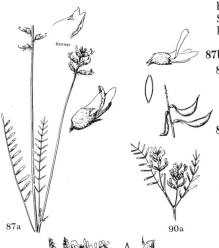

banner 11.5–14 mm, margins erose or undulate to lacerate; SR between mouth of Clearwater R and Whitman and Columbia cos, Wn, and Nez Perce Co, Ida; Piper m.

70 **A. riparius** Barneby

87b Wing petals < banner; pod often pubescent

88a Racemes 2–8-fld; fls 20–25 mm, purplish; calyx teeth ca 1/3 length of tube; pl silvery-sericeous to appressed-villous and grayish, low and caespitose, sts 2–10 cm; pod hirsute-strigose, slightly falcate, 15–25 × ca 8 mm (see lead 55a) 48 **A. argophyllus** Nutt.

88b Racemes 7–35-fld; fls (11) 13–23 mm, gen off-white but sometimes purplish; calyx teeth 1/3–4/5 length of tube; pl greenish to ± grayish, appressed-pubescent, gen ± erect, sts (2) 10–40 cm; pod glab to grayish- or blackish-strigose or -villous, straight to ± arcuate, 15–30 × 6–10 mm (see leads 75a–78b for distinctions among)

60 **A. conjunctus** (Rydb.) Barneby
61 **A. hoodianus** Howell
62 **A. reventiformis** (Rydb.) Barneby
63 **A. reventus** Gray
64 **A. sheldonii** (Rydb.) Barneby

Group G

89a Pod strongly curved, often to a complete turn or circle, but occ to < 1/2 turn—when always > 5 mm broad and with both sutures very prominent

90a Pod curved < 1/2 turn, ave at least 5 mm broad, 20–35 mm, strongly compressed, sutures very prominent; dunes and sandy des both sides CR from The Dalles upstream to near Kettle Falls, Wn, and along Okanogan R into BC; The Dalles m., stalked-pod m. 71 **A. sclerocarpus** Gray

90b Pod gen curved at least 1/2 turn, rarely as much as 5 mm broad, the sutures not prominent

91a Pod grayish-hairy, at least when young, 3–5.5 mm broad, rarely curved < 1 full turn; stipe 4–11 mm; calyx 5–9 mm, not gibbous-saccate at base behind ped; lflets mostly 2–10 mm; sagebr des, esp near CR, Yakima, Kittitas, Klickitat, and Benton cos, Wn; medick m., curvepod m. 72 **A. speirocarpus** Gray

91b Pod mostly not grayish-hairy, even when young, 2.7–4.5 mm broad, rarely curved as much as 1 full turn; calyx 6–13.5 mm, gibbous-saccate at base behind ped; lflets mostly 5–18 (21) mm; sickle m., curvepod m.; 3 vars. 73 **A. curvicarpus** (Sheld.) Macbride

a1 Ovary and pod pubescent; banner 16–21 mm, bl well developed; keel 12–15 mm; lflets pubescent on upper surface; sagebr plains and foothills; sw edge Blue Mts, Ore, to SRP, s Ida, s to Nev and Cal var. **curvicarpus**

a2 Ovary and pod glab or pl not otherwise as above

b1 Calyx 6–8.5 mm; banner 13.5–15.5 mm; keel 9.5–11.5 mm; lflets pubescent on upper surface; dry hillsides and plains; upper forks Deschutes R, Crook, Deschutes, and Jefferson cos, Ore var. **brachycodon** Barneby

b2 Calyx 9–13.5 mm; banner 13.9–19.5 mm; keel 11.5–14.5 mm; lflets subglab; rocky hillsides and sagebr flats; John Day and lower Deschutes R, Grant to Wasco cos, s to n Jefferson Co, Ore var. **subglaber** (Rydb.) Barneby

89b Pod nearly or quite straight, never curved as much as 1/2 a turn or circle, if strongly arcuate then (except in *A. beckwithii* and *A. vallaris*) not > 5 mm broad

92a Calyx at least 8 mm; pod either > 5 mm broad or thick, or stipe at least 15 mm or evidently jointed to pod

93a Fls 11.5–16.5 mm; pod erect, grooved above, ± cordate in section, partly or wholly 2-celled; stipe stout, 3–5 (8) mm; calyx tube 3.2–5.8 mm; Kittitas, Chelan, and Douglas cos, Wn; Leiberg's m.

74 **A. leibergii** Jones

93b Fls gen ca 20 mm; pl various

94a Pod 1-celled, body jointed to stipe at top of or barely within calyx tube, somewhat lunate, 1.5–3 cm, purplish-mottled, almost 1 cm

94a

95a

95b

98a

99b

101a

102a

thick; calyx 9–14 mm, teeth ca = tube; Beckwith's m.; 2 vars.
75 A. beckwithii T. & G.
a1 Pod rounded or broadly cuneate at base, deeply sulcate along
lower suture; along Salmon R, Lemhi Co, Ida
var. **sulcatus** Barneby
a2 Pod clavate or stipelike at base, only shallowly or not at all sulcate along lower suture; sagebr des to dry hillsides; near Kamloops, BC, and along or near SR, se Wn and ne Ore and adj Ida
to se Ida
var. **weiserenis** Jones
94b Pod nearly or quite 2-celled, the partition thick, at least 1/2 complete; stipe not jointed to body of pod; calyx 8–10 mm, teeth <
half as long as tube
95a Pod and st erect; body of pod 6–10 mm thick, truncate at base,
corrugate-wrinkled, 15–20 mm, both sutures grooved and intruded, lower one forming 3/4 partition; sagebr plains and lower
slopes; Beaverhead Co, Mont, to Lemhi Co, Ida; Bitterroot m.
76 A. scaphoides Jones
95b Pod pendulous; sts spreading; body of pod 6–8 mm wide and
10–15 mm thick, abruptly contracted at each end, both sutures
prominent and ± grooved, lower one intruded and almost completely partitioning cavity; local in SRC, Washington and Adams
cos, Ida, and adj Ore; Snake Canyon m.   **77 A. vallaris** Jones
92b Calyx < 8 mm, or pod much < 5 mm broad or with stipe much < 15
mm and not jointed to body
96a Pod from strongly compressed to round in section, strictly 1-celled,
lower suture rarely protruding into cavity, partition (if any) rudimentary and < 1/4 width of cavity   **Subgroup G₁**
96b Pod round, cordate, or triangular in section, or obcompressed, gen
with lower suture rather deeply sulcate, if pod round or compressed
then with a partition at least 1/3 width of cavity
Subgroup G₂, lead 103a

Subgroup G₁

97a Calyx tube 5–9 mm; pod oval to terete in cross section
98a Pod glab, erect, ca 15 × 4–5 mm, both sutures prominent but neither
grooved; fls spreading to erect, ochroleucous; sagebr plains and foothills,
Yakima Co, Wn, to ne Ore, near Columbia and lower Deschutes R;
Tweedy's m.   **78 A. tweedyi** Canby
98b Pod hairy, pendulous; fls sometimes reflexed
99a Calyx gibbous at base behind ped; fls reflexed (see lead 30a)
26 A. collinus Dougl.
99b Calyx not gibbous at base; fls ascending, white; pod 24–27 × 4–5
mm, compressed; lflets obovate-cuneate to oblong-oblanceolate, trun-
cate-retuse, 6–16 mm; rocky hillsides, apparently in sagebr; known
only from Colockum Cr, ca 2 mi w of Columbia R, Chelan Co, Wn, in
sagebrush; Whited m. *(A. whitedii)*   **79 A. sinuatus** Piper
97b Calyx tube < 5 mm, or pod strongly compressed
100a Pod black-hairy (see leads 43a–43b)   **36 A. robbinsii** (Oakes) Gray
37 A. bourgovii Gray
100b Pod not black-hairy
101a Rachis of lvs gen < 5 cm; lflets gen 2–5 mm broad; wings deeply
bidentate at tip; pod pendulous, falcate-lunate, 2–3 cm × 4–5 mm, ±
compressed; alp to subalp ridges and scree to foothill bluffs and riverbanks; Alas to Nev, Utah, and Colo through RM, w to c Ida and to ne
Ore; Indian m. *(A. forwoodii, A. glabriusculus, A. vaginatus, At. wallowensis)*   **80 A. aboriginum** Richards.
101b Rachis of lvs gen much > 5 cm; lflets often < 2 mm broad; wings
not deeply bidentate at tip; pod various, rather strongly compressed
102a Body of pod 7–15 × 3–5 mm, glab, often finely mottled; lflets
11–31, 1.5–5.5 mm broad; stips blackening upon drying; prairies
and foothills to lower mts; Yuk s through e BC and RM to Mont, e
Ida, e Nev, Utah, and NM, e to Man, Daks, Minn, and Neb; pulse
m. *(A. nigrescens, Ho. dispar, H. multiflorus, H. stipitatus, H. strigulosus, H. t.)*   **81 A. tenellus** Pursh
102b Body of pod 15–35 × (3) 4–6 mm, glab to strigillose, not mottled; lflets (5) 9–19 (23), linear, gen 7–20 (23) × 1–2 mm broad;

102b

103a

104a

108a

109a

110a

111a

stips not blackening; sagebr plains and foothills; sc BC through Ida and e Wn and Ore to ne Cal and much of Nev, s to Baja Cal; basalt m., threadstalk m. (*A. stenophyllus* of auth.)   82 A. filipes Torr.

Subgroup G₂

103a Pod woody, pendulous, ovoid or subglobose, body 1–2 cm, 2-celled; pl weedy, rhizomatous; keel > the narrow wings; sts fistulose, 3–7 dm; fls off-white, 13–17 mm; intro in Clark Co, Ida; China; China m.
83 A. chinensis L. f.

103b Pod not as above if pl rhizomatous; keel rarely > wings
  104a Pod and calyx black-hairy; pod pendulous, body 8–12 mm; pl rhizomatous; keel broad, purple-tipped, ca = purple-margined banner, longer and narrower than the white wings; circumboreal, s to Okanogan Co, Wn, Wallowa Mts, ne Ore, and RM to ne Nev and NM, abundant in Mont and Ida; alpine m., purple m. (*A. andinus, A. giganteus*)
84 A. alpinus L.

  104b Pod, calyx, and pl not at once as above
    105a Fls purple (occ lavender or nearly white); pod very strongly obcompressed and deeply 2-grooved on upper surface (see lead 51a)
45 A. bisulcatus (Hook.) Gray
    105b Fls either whitish or yellowish if pod obcompressed and sulcate on upper surface
      106a Lflets not > 10 mm; fls not > 9 mm
        107a Pod at least 4 mm broad and thick, mottled (see lead 61b)
54 A. salmonis Jones

        107b Pod either < 4 mm broad and thick or not mottled
          108a Lower stips connate; fls mostly 6–8 mm, whitish but drying cream-colored (± bluish-lined); calyx 3–5 mm; pod 10–16 × 3.2–4.5 mm, slightly curved, stipe 3–5 mm; sandy bluffs and talus along SR and Boise R, sw Ida and adj Ore; Mulford's m.
85 A. mulfordiae Jones
          108b Lower stips, as well as upper ones, not connate
            109a Stipe mostly 2–4 mm; pod 7–12 × 2.5–3.5 mm, only very slightly curved; fls 4–12, 5.5–7.5 mm, off-white, banner purplish-veined; calyx 3–4 mm; locally abundant in foothills of Sawtooth Mts, Blaine Co, Ida; Picabo m.
86 A. oniciformis Barneby
            109b Stipe 2.5–14 mm; pod 12–30 × 2.5–5 mm, often lunately curved; fls 5–25, 7–17 mm, whitish to yellowish; calyx 3.4–7.5 mm
              110a Pod 16–30 × 3–5 mm, stipe (4) 7–14 mm; fls 10–25, off-white, not purplish-veined, 10–17 mm; calyx 5–7.5 mm; sagebr and bunchgrass valleys and hillsides; John Day and Deschutes rivers, Wasco, Sherman, Morrow, and Umatilla cos, Ore; Howell's m.   87 A. howellii Gray
              110b Pod 12–25 × 2.5–4 mm, stipe 2.5–5 mm; fls 5–15, yellowish or greenish-white, 7–10 mm; calyx 3.4–4.8 mm; sagebr and grassl; c Wn and Ore; pauper m.; 2 vars.
88 A. misellus Wats.
                a1 Sts 10–25 cm; lflets mostly 13–21, glab or subglab on upper surface; Deschutes and John Day R valleys, Ore (*A. drepanolobus* var. *aberrans, A. howellii* var. *a.*)   var. misellus
                a2 Sts 2.5–7 cm; lflets mostly 11–13, pubescent on upper surface; near Ellensburg, Kittitas co, Wn)
var. pauper Barneby
      106b Lflets much > 10 mm, or fls at least 10 mm
        111a Pod body linear, 3.5–5 cm × 3–4 mm, strigillose, triangular-cordate in section; calyx 5–7 mm, teeth ca 1/2 length of tube; lflets glab on upper surface, 6–12 (14) mm; grassy hills and rocky meadows, on basalt; n foothills of Blue and Wallowa mts, Umatilla and Wallowa cos, Ore, n to Asotin Co, Wn, and to Nez Perce Co, Ida; Waho m., Arthur's m.   89 A. arthuri Jones
        111b Pod body not at once as above if lflets only 6–12 mm
          112a Pod glab, pendulous; pl grayish-villous with long soft hairs, (3) 4–8 dm; calyx 8–11 mm, teeth ca 1/2 length of tube; plains

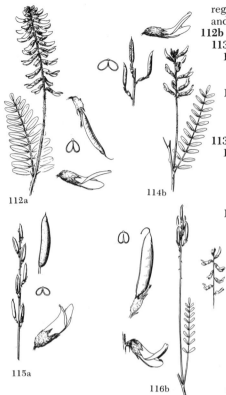

region to pine woodl; se BC and sw Alta to Mont, Utah, NM, and Daks; Drummond's m.   90 **A. drummondii** Hook.
112b Pod pubescent or erect or pl not otherwise as above
  113a Stipe 8–12 mm, ca twice length of calyx
    114a Pod compressed, almost completely 2-celled, gen pendent, softly short-strigose-lanate (see lead 110a)
                                   87 **A. howellii** Gray
    114b Pod obcompressed, imperfectly 2-celled, gen erect, glab to strigillose; sagebr des to pinyon-juniper woodl; se Ore and sw Ida to Nev, nw Ariz, and sw Utah; hermit m. (*A. boiseanus, A. cusickii*)   91 **A. eremiticus** Sheld.
  113b Stipe (2.5) 4–7 (8) mm, gen only slightly > calyx
    115a Fls ascending to erect; banner < wings; stips free; calyx 5–6.5 mm, teeth ca 1.5 mm; sagebr flats to open pine for; Whitman Co, Wn, and adj wc Ida, also in ec Wn; Palouse m., hangingpod m. (*A. palousensis*)
                                 92 **A. arrectus** Gray
    115b Fls nodding; banner > wings; stips sometimes connate
      116a Stips (at least lower ones) connate (see lead 93a)
                             74 **A. leibergii** Jones
      116b Stips all free; body of pod 14–28 × 2.5–4.5 mm, obtusely 3-angled in section, with convex lateral and shallowly grooved lower surfaces, keeled above by the suture; calyx 5–6.5 mm, teeth 1–2 mm; gravelly to sandy sagebr des and foothills, up to 8000 ft elev; Idaho Co (canyons of Salmon and SR) and nc Ida (upper Salmon R) n to Big Hole and Clark's Fork rivers, w Mont, e to Helena; Kelsey m. (*A. kelseyi, A. stenophyllus,* misapplied)   93 **A. atropubescens** Coult. & Fish.

Coronilla L.

Fls in long-pedunculate umbels; calyx 5-toothed, bilabiate; petals clawed; stamens 10, diadelphous; pod slender, terete to angled; per herbs (ours) with odd-pinnate lvs. (L *corona*, crown, the fls often spreading in a ring and suggesting a crown).

C. varia L. Crown vetch. Glab herb with spreading to diffuse sts up to 6 dm; lflets 9–15 (21), oblong or elliptic to obovate, mostly 1–2 cm, acute to truncate or retuse; peduncles gen exceeding the lvs; umbels 14–20-fld; calyx ca 2 mm; corolla white to pink, 10–12 mm; banner with an arched claw > the calyx; wings and keel strongly aur, the keel purple-tipped; loment linear, gen arcuate, up to 5 cm, 5–10-seeded; Old World orn, reported to be estab in parts of w Ore and in Ida.

Cytisus L.   Broom

Fls papilionaceous, axillary, umbellate, or racemose, white, yellow, or purple-tinged; calyx bilabiate, 5-toothed; stamens 10, monadelphous; pods flattened, several-seeded; seeds with wartlike strophiole; deciduous or evergreen shrubs with strongly angled greenish sts and alt, mostly 3-foliolate lvs, rarely lvs entire and merely scalelike; stips small, thickened. (Derivation obscure, perhaps from Cythrus, where the pl grew, or from Gr *kutisus,* the name of some leguminous pl).

1a Fls white, 1–2 (3) per lf axil, ca 1 cm; lowest lvs 3-foliolate, upper ones simple; pod appressed-pubescent; w Ore and Wn, native of s Europe; white Spanish b. (*Genista alba*)   1 **C. multiflorus** (Ait.) Sweet
1b Fls yellow or purplish
  2a Fls axillary, 1 (2–3) per axil; pod glab except on the margins; deciduous shrub up to 3 m, brs strongly angled; corolla ca 2 cm; common escape w Cas, BC to Cal, often planted along highways; Scot's b. (*Spartium s.*)
                               2 **C. scoparius** (L.) Link
  2b Fls racemose; pods hairy on valves and on margins; shrub to 3.5 m; fls 3–10 per raceme, light yellow, ca 1 cm; occ escape w Cas, BC to Cal; French b.   3 **C. monspessulanus** L.

Glycyrrhiza L.   Licorice; Licorice-root

Fls papilionaceous, ochroleucous (ours), spicate; calyx 5-toothed; stamens 10, diadelphous; pod indehiscent, 1-celled, few-seeded, burlike; glandular per herbs with odd-pinnate lvs and small deciduous stips. (Gr *glykys*, sweet, and *rhiza*, root; most spp. with sweetish licorice-flavored juice).

G. lepidota Pursh. Am l. Pl rhizomatous; sts 3–12 dm; lflets 7–15, 2–4 cm; fls 10–15 mm; pod 10–15 mm; waste places and low ground, BC to Ont and Minn, s to Cal, Ariz, NM, and Tex; 2 vars.

a1 Pl with stalked glands almost throughout; e Cas, BC to Cal, e to YNP (*G. g.*)   var. **glutinosa** (Nutt.) Wats.

a2 Pl with stalked glands only on calyx; s Cal ne to Ont and much of Mont, occ in Ida and Wn   var. **lepidota**

Hedysarum L.   Sweetvetch; Hedysarum

Fls ∝ in bracteate racemes, yellowish-white, pink, or purple, papilionaceous; calyx 5-toothed, 2-bracteolate at base; keel gen surpassing wings and banner; stamens 10, diadelphous; fr flat, breaking into 1-seeded, indehiscent segms; herbaceous per with odd-pinnate lvs, mostly gland-dotted; stips membranous, gen connate. (Gr name, *hedusaron*, meaning obscure).

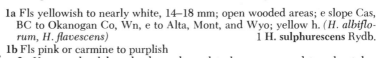

1a Fls yellowish to nearly white, 14–18 mm; open wooded areas; e slope Cas, BC to Okanogan Co, Wn, e to Alta, Mont, and Wyo; yellow h. (*H. albiflorum, H. flavescens*)   1 **H. sulphurescens** Rydb.
1b Fls pink or carmine to purplish
  2a Upper calyx lobes slender, subequal to lower ones and to calyx tube; basal lobe of wing petals broad, much < claw, not joined over ovary; pod plainly cross-corrugated, reticulations laterally elongated; Yuk to ne Ore, e to Newf and s in RM to NM and Ariz; northern h.; 3 vars.   2 **H. boreale** Nutt.
    a1 Racemes ± compact, 2–4 (6) cm; fls 5–15 (20), purple, 17–21 mm; pl gen glab (sparsely strigulose); upper surface of lvs glab to sparsely strigulose; segms of loments 3–6 (8); Yuk to Newf, s to BC, Alta, and Man, barely reaching our area (*H. carnosulum, H. pabulare, H. p.* var. *rivulare*)   var. **mackenzii** (Richards.) Hitchc.
    a2 Racemes often up to 15 cm; fls mostly 12–20, carmine, magenta, or purple, 11–17 mm; pl gen ± grayish-strigulose; segms of loments 2–5 (6)
      b1 Pl grayish-hairy; lvs pubescent on both surfaces; Ida and Mont to Wyo and Alta (*H. canescens)*   var. **cinerascens** (Rydb.) Rollins
      b2 Pl greenish; lvs sparsely hairy, often glab on upper surface; BC to Alta, s to Ida, Daks, and in RM to NM and Ariz, also in ne Ore
                                           var. **boreale**
  2b Upper calyx lobes broader but considerably < lower 3; basal lobe of wing petals slender, nearly = claw, the 2 petals weakly joined by these lobes over ovary; pod not strongly cross-corrugated, reticulations reg, not laterally elongated
    3a Pod 7–12 mm broad, margins winged for 1–2 mm; fls 16–22 (rarely < 18) mm; higher elev in mts; Ida and Mont s to Wyo and Colo, also in Cas and OM, Wn; western h. (*H. lancifolium, H. uintahense*)
                                       3 **H. occidentale** Greene
    3b Pod 3.5–6 mm broad, margins winged < 1 mm; fls 11–18 mm; circumboreal, BC to Alta, s to nc Mont; American h.; ours the var. *americanum* Michx.   4 **H. alpinum** L.

Lathyrus L.   Peavine; Sweet-pea; Vetchling

Fls papilionaceous, single to racemose in lf axils, white or yellow to pink, red, or purplish, relatively large and showy; stamens 10, diadelphous; pods 1-celled, dehiscent, several-seeded; ann or per herbs often with twining sts and tendril-bearing, pinnate lvs, lflets 2–8 (rarely lacking); tendrils simple to br, rarely reduced to nonprehensile bristles. (Ancient Gr name for this or related pl).

1a Lflets lacking; stips ovate-triangular and hastate, 1–4 cm; tendril prehensile, unbr; fls 1 (2) per axil, yellow, 10–14 mm; calyx lobes at least twice length of tube; European, intro in w Ore in our area; yellow v.

1 L. aphaca L.

1b Lflets 2 or more; stips various, seldom hastate; tendrils often br, rarely nonprehensile; fls variously colored; calyx lobes often < tube

    2a Lflets 2; ann or weedy per   **Group I**, lead 6a

    2b Lflets at least 4 on some lvs; native per (rarely the garden pea, *Pisum sativum* L., may be found persisting in waste areas, it will key here except that it is ann)

        3a Tendrils lacking, gen represented by terminal, simple, nonprehensile bristles   **Group II**, lead 14a

        3b Tendrils present, gen well developed, either br or prehensile

            4a Pls mostly along coast on sand dunes or in brackish sloughs; sts either winged or stips obliquely hastate or sagittate-ovate and subequal to or > lflets

                5a Sts winged; lflets mostly 6; stips mostly < half as long as lflets; fls 12–20 mm; Alas to ne Cal, mostly in brackish sloughs, also along Atl coast and in GL region; marsh p.   2 L. **palustris** L.

                5b Sts not winged; lflets 6–12; stips from only slightly < to > lflets; fls 17–30 mm; maritime p., beach p., sea p. (*L. maritimus*)

3 L. japonicus Willd.

            4b Pls rarely maritime, if from near coast then sts not winged and stips much smaller than lflets   **Group III**, lead 19a

Group I

6a Corollas 25–30 mm; pl ann

    7a Pods glab, 7–10 cm × 7–9 mm; fls rose-purple; occ escaped in Ore, native of Old World; Tangier p.   4 L. **tingitanus** L.

    7b Pods hairy, 3–6 cm × 4–7 mm; fls of various colors; common cult s.

5 L. odoratus L.

6b Corollas much < 25 mm; pl sometimes per

    8a Fls yellow; pl per, native of Europe, rarely escaped; meadow p., meadow v.   6 L. **pratensis** L.

    8b Fls not yellow; pl often ann

        9a Pl ann

            10a Peduncles 1-fld, rachis extending beyond ped of fl as a straight or curved bristle; pl (and pods) glab; Eurasian, well estab in parts of w Ore; grass p.   7 L. **sphaericus** Retz.

            10b Peduncles mostly 2-fld, if 1-fld then either rachis not prolonged as a bristle or pl hairy

                11a Pl hirsute; calyx lobes ca = tube; pods hairy, 5–8 mm broad; sts winged; Eurasian, estab in WV, Ore, and perhaps elsewhere; rough p., hairy v.   8 L. **hirsutus** L.

                11b Pl sparsely pubescent; calyx lobes ca twice length of tube; pods glab, 2–4 mm broad; intro from s US, in sw Ore, perhaps in our area; tiny p.   9 L. **pusillus** Elliott

        9b Pl per

            12a Sts wingless; fls 12–16 mm; lflets obovate to elliptic-oblanceolate, 2–4 cm, obtuse to rounded; Eurasian, occ escaped along roadsides; known from Okanogan Co, Wn, in our area; earth-nut p.

10 L. tuberosus L.

            12b Sts winged; fls 14–20 mm; lflets mostly acute and > 4 cm

                13a Pod 6–10 cm × 7–10 mm; stips 3–5 cm, ovate to broadly lanceolate; fls 15–20 mm, white to red; European, our most common weedy sweet pea, esp along railroads and highways; everlasting p., per p.   11 L. **latifolius** L.

                13b Pod 4–6 cm × 4–6 mm; stips 1–3 cm, linear-lanceolate; fls ca 15 mm, red; European, fairly commonly estab in Wn, Ore, and Ida; flat p., narrowlvd everlasting p.   12 L. **sylvestris** L.

Group II

14a Fls 1–2 per raceme, 8–13 mm; lower calyx teeth half again as long as tube; lflets elliptic to oval or ovate, apiculate; pl mostly villous throughout; open prairies and woods, mostly near coast; Pierce Co, Wn, to Santa Cruz Co, Cal; Torrey's p.   13 L. **torreyi** Gray

14b Fls either > 2 per raceme or pl otherwise not as above

  15a Pl densely villous, found only on coastal sand dunes; rachis of lvs flattened, 1.5–3 cm; fls 12–18 mm; VI to Monterey Co, Cal; beach p.
                                             14 L. **littoralis** (Nutt.) Endl.

  15b Pl either not villous or not coastal; rachis of lvs > 3 cm and not flattened

    16a Fls 8–13 (16) mm

      17a Lflets 2–4 (very occ 6); foothills, open parks or light woods; extreme e Wn and adj Ida (possibly Mont?); pinewoods p.
                                    15 L. **bijugatus** White

      17b Lflets very rarely < 6; ponderosa pine woodl to mont for; e side Cas, Wn to Cal, e to Ida and Utah; thick-lvd p.; 2 vars.
                                 16 L. **lanszwertii** Kell.

        a1 Tendrils lacking; fls 8–12 mm, almost white; Wn to Cal, e to extreme w Ida   var. **aridus** (Piper) Jeps.

        a2 Tendrils gen well developed; fls 13–16 mm, mostly pinkish to lavender; Wn to c Cal, e to c Ida and Utah (*L. coriaceus, L. oregonensis*)   var. **lanszwertii**

    16b Fls ave at least 18 mm

      18a Pls many-std from a thick taproot, not rhizomatous; rachis of lvs 2–5 cm, ± winged; lflets heavily veined; fls white; sagebr valleys and hillsides, Union Co, Ore, and Adams Co, Ida, to Modoc Co, Cal; rigid p., bushy p.   17 L. **rigidus** White

      18b Pls rhizomatous; rachis of lvs mostly > 5 cm, not winged; lflets not heavily veined; fls white to blue; mostly in woodl; several infraspecific taxa   18 L. **nevadensis** Wats.

      a1 Fls mostly > 17 mm, white to blue; tendrils bristlelike, or larger but rarely br; lflets 4–7 (8)

        b1 Corolla pinkish to fairly dark blue or reddish-purple (rarely white); Lane and Coos cos, Ore, to Humboldt and Fresno cos, Cal; Sierran p.   ssp. **nevadensis**

        b2 Corolla white, pinkish-veined; Blue and Wallowa mts, Ore, and adj Ida; Cusick's p.   ssp. **cusickii** (Wats.) Hitchc.

      a2 Fls mostly < 17 (13–20) mm, variously colored; tendrils gen present, often br; sts scandent; lflets mostly 8–14; Nuttall's p.; 3 vars.
                              ssp. **lanceolatus** (Howell) Hitchc.

        c1 Fls bluish to reddish-purple; mostly w Cas, BC to Cal, but e to Ida (*L. nuttallii, L. pedunculatus*)
                           var. **pilosellus** (Peck) Hitchc.

        c2 Fls white or pinkish; e Cas

          d1 Banner white, at most pinkish-lined; nc Ida
                        var. **parkeri** (St. John) Hitchc.

          d2 Banner pinkish-tinged; Wen Mts, Kittitas and Chelan cos, Wn   var. **puniceus** Hitchc.

Group III

19a Corolla white

  20a Bl of banner slightly < claw; lateral calyx lobes linear to lanceolate, not broadened above base; lflets (6) 8–12; fls 13–17 mm; WV and Umpqua R Valley, Ore; thin-lvd p.   19 L. **holochlorus** (Piper) Hitchc.

  20b Bl of banner = or > claw; lateral calyx lobes often broadened above base and ± ovate; lflets and fls various

    21a Lateral calyx lobes oblong-ovate to lanceolate, broadened considerably just above base

      22a Lflets mostly 6; e Cas, Okanogan Co, Wn, to n Ida, and Mont, across Can and s to much of c and ne US; cream-fld p.
                           20 L. **ochroleucus** Hook.

      22b Lflets 8–12; w Cas, Wn to Cal; several vars. of this sp. occur to the s of our area, all with bluish or purplish fls; ours, ssp. *ochropetalus* (Piper) Hitchc., the only white-colored phase, ranges from King Co, Wn, to sw Ore; Pac p.   21 L. **vestitus** Nutt.

    21b Lateral calyx lobes lanceolate, not broadened above base; e Cas, Chelan to Kittitas cos, Wn, e to se Wn and adj Ore and Ida (see also lead 18b)   18 L. **nevadensis** Wats.

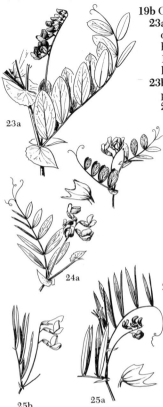

19b Corolla blue to red, sometimes pale and fading to brownish
  23a Lflets 10–16, scattered; stips sagittate-ovate, mostly at least 1/2 length of lflets; pl erect, rarely at all clambering, glab except on calyx; lower half or third of st gen lfless at anthesis; racemes 5–13-fld, secund; fls 16–20 mm; w Cas, Puget Trough to Lake Co, Cal, woods and prairies; lfy p.   22 **L. polyphyllus** Nutt.
  23b Lflets rarely > 10, if so then either stips smaller and narrower or pl pubescent or otherwise not as above
    24a Keel 2–4 mm < wing petals; calyx glab, or teeth ciliate, lowest tooth gen > tube; stips mostly broadly lanceolate to ovate and at least half as long as lflets (sometimes much smaller); grassl and sagebr slopes to ponderosa pine for; e Cas, Wn to s Cal, e to Ida, Utah, Colo, and Ariz; few-fld p.; several phases   23 **L. pauciflorus** Fern.
      a1 Fls mostly at least 18 mm; lflets mostly > 3 cm, linear to ovate
                                                       ssp. **pauciflorus**
        b1 Lflets linear to oblong-elliptic, mostly > twice as long as broad; Chelan Co, Wn, to c Ore, e to w Ida   var. **pauciflorus**
        b2 Lflets ovate to ovate-lanceolate, gen at least half as broad as long; Utah to sw Colo and nw Ariz, occ in Ida, Ore, and Wn
                                           var. **utahensis** (Jones) Peck
      a2 Fls mostly < 17 mm; lflets either linear or (mostly) obovate and not > 3 cm; sc Ore to s Cal, closely app by pls from Wasco Co, Ore
                                        ssp. **brownii** (Eastw.) Piper
    24b Keel gen subequal to wing petals or (if <) calyx often hairy; lowest calyx tooth often < tube; stips various
      25a Fls pale lavender-tinged to pinkish-violet or pinkish-orchid, 8–16 mm; lflets 4–12, linear to oblong-elliptic, mostly 3–8 times as long as broad, rarely if ever > 1 cm broad (see lead 17b)
                                        16 **L. lanszwertii** Kell.
      25b Fls more nearly rose, blue, or purple, 13–27 mm; lflets 4–10 but gen 6 or 8, often elliptic, gen > 1 cm broad and not > 3 times as long as broad (see lead 18b)   18 **L. nevadensis** Wats.

Lotus L.   Deervetch; Birdsfoot-trefoil; Lotus

Fls papilionaceous, sessile and axillary to umbellate, white or yellow and often tinged with reddish or purple; calyx tubular, teeth > to < tube; stamens 10, diadelphous; pods 1–∝ -seeded, gen dehiscent; glab to hairy ann or per herbs with pinnately 3–∝ -foliolate lvs and glandlike to lfletlike stips. (The Gr name). *(Acmispon, Anisolotus, Hosackia).*

1a Pl ann; stips mere blackish glands; fls 1–3, sessile or short-pedunculate in lf axils
  2a Fls subsessile in lf axils; rachis of lvs flattened; pods 8–15 × 3–4 mm, not constricted between seeds; calyx teeth nearly 2 × length of tube; mostly in sandy soil in open; e slope Cas to coast, s BC to n Cal; meadow l.   1 **L. denticulatus** (Drew) Greene
  2b Fls pedunculate; rachis of lvs not flattened; pods mostly much > 15 × < 3 mm, often constricted between seeds
    3a Lflets mostly 3, 3–10 (15) mm broad; fls 4–8 mm; calyx teeth mostly much > 2 mm, > tube; pod slightly or not at all constricted between seeds; sandy to rocky soil, open to wooded areas; both sides Cas, BC to Cal, e through Mont to C states, s to Mex; Spanish-clover (*L. americanus, L. unifoliolatus, H. elata, H. floribunda, H. mollis, H. pilosa*)
               2 **L. purshiana** (Benth.) Clements & Clements
    3b Lflets mostly 5, 1–4 mm broad; fls 4–5 mm; calyx teeth mostly < 1 mm and much < tube; pod mostly strongly constricted between seeds; sandy flats to open slopes; coast to mts w Cas, BC to Cal; small-fld d.   3 **L. micranthus** Benth.
1b Pl per; stips often well developed; fls in axillary, pedunculate umbels, mostly at least 3 per umbel
  4a Lvs apparently sessile; lflets 5, lowest pair basal, petiolulate, similar to the other 3, by their position resembling stips but true stips represented by tiny glands; fls 3–15 in compact umbels; European, escaped in moist areas and known from Ida and w Wn and Ore in our area; b.
                             4 **L. corniculatus** L.

4b Lvs petiolate; lflets 3–15, lowest pair not basal on petiole; stips gland-like or foliaceous but dissimilar to lflets and not petiolulate
  5a Stips glandlike, blacking when dried; lfts (3–4) 5; pods falcate, inde-hiscent, 1–3-seeded; common, both sides Cas in sandy or rocky soil; BC to Cal, e to Ida and Nev; Nev d.; ours the var. *douglasii* (Greene) Ottley (*L. d.*)   **5 L. nevadensis** (Wats.) Greene
  5b Stips membranous, not glandlike; lflets often 5–15; pods elongate, nearly or quite straight, dehiscent, mostly 5–20-seeded
    6a Bract of peduncle considerably below umbel, gen com-pound (simple); peds 1–4 mm; pods gen > 3 mm broad; nw Wn to s Cal, w Cas except in CRG, from lowl to mont; big d.; 2 vars.
          6 L. crassifolius (Benth.) Greene
      a1 Corolla whitish, tinged with red or purple; nw Wn s, w Cas, to s Cal (*H. rosea*)   var. **subglaber** (Ottley) Hitchc.
      a2 Corolla greenish-yellow, often spotted with deep purple; sw Wn to e Wasco Co, Ore, s to s Cal   var. **crassifolius**
    6b Bract of peduncle (if any) just below umbel, often simple; peds scarcely 1 mm; pods < 3 mm broad
      7a Bract of peduncle gen 3 (1, 5, 7)-foliolate; banner yellow; wings pink to rose; keel purplish-tipped; upper calyx teeth not connate > 1/2 length; moist soil, sea level to mont; sw Wn s, w Cas, to Monterey Co, Cal; seaside l.   **7 L. formosissimus** Greene
      7b Bract gen a mere tooth, but sometimes a single lflet, occ lacking; petals all cream to yellowish, with little if any pink; upper calyx teeth connate > 1/2 length; moist soil, nw Wn to c Cal, e along streams and lakes to Ida; meadow or bog d. (*L. bicolor*)
                      8 L. pinnatus Hook.

Lupinus L. Lupine

Fls papilionaceous, racemose, white, yellowish, pink, blue, or violet; calyx mostly bilabiate; stamens 10, dimorphic, monadelphous; pods flattened, hairy, (1) 2–12-seeded, dehiscent; ann or per herbs to semishrubs with palmately compound lvs and (3) 5–17 lflets. (L *lupus*, wolf, implication uncertain).

  1a Pl ann
    2a Ovules and seeds gen 2, sometimes only 1
      3a Keel ciliate on lower half; racemes verticillate, their peduncles ca = petioles; fls ca 15 mm; dry to moist soil, VI and coastal PS, s on e side Cas to Baja Cal; chick l.; 2 vars.   **1 L. microcarpus** Sims
        a1 Fls pale yellowish; VI and adj is of Wn   var. **scopulorum** Smith
        a2 Fls pink to reddish-purple; Yakima Valley, Wn, to Baja Cal
                      var. **microcarpus**
      3b Keel glab; racemes not verticillate; peduncles < petioles; fls ca 10 mm; sandy des; c Wn to Cal and Ariz, e to Alta, Daks, and Neb; rusty l., low l.; 2 vars.   **2 L. pusillus** Pursh
        a1 Fls mostly at least 10 mm; chiefly e RM, Alta to Colo, but along SR in Ida   var. **pusillus**
        a2 Fls mostly 7–10 mm; e Wn to Cal and Ariz, e to Ida, Nev, Wyo, Utah, and Ariz   var. **intermontanus** (Heller) Smith
    2b Ovules and seeds gen 4 or more, rarely only 3
      4a Fls 5–7 mm; banner subequal to wings but scarcely at all reflexed from them; keel blunt and not much upturned; peds gen 1–2 mm and rather stout; mostly w Cas, gravelly areas and "prairies;" BC to Cal, up CRG to ne Ore; field l., small-fld l.   **3 L. micranthus** Dougl.
      4b Fls gen > 7 mm (sometimes shorter); banner often much > wings and markedly reflexed from them; keel slender-tipped and gen rather sharply upturned; peds 2–5 (6) mm
        5a Fls (9) 10–13 mm; lflets broadly oblanceolate; banner gen as broad as long, not markedly < keel; s WV to Cal; fleshy l. (*L. carnosulus*)
                  4 L. affinis Agardh
        5b Fls 6–9 (10) mm; lflets narrowly oblanceolate; banner often longer than broad, frequently < keel; BC to s Cal, mainly w Cas and in CRG; two-color l. (*L. apricus, L. hirsutulus, L. strigulosus*)
                  5 L. bicolor Lindl.

  1b Pl per

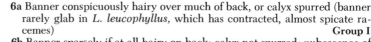

6a Banner conspicuously hairy over much of back, or calyx spurred (banner rarely glab in *L. leucophyllus*, which has contracted, almost spicate racemes)   **Group I**

6b Banner sparsely if at all hairy on back; calyx not spurred; pubescence of banner (if any) largely concealed by calyx, but sometimes extending above calyx as a line (in *L. sulphureus*)   **Group II**, lead 13a

Group I

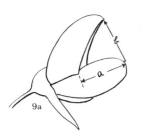

7a Calyx spurred at base above the ped; banner gen not pubescent on upper third; wing petals often hairy on upper half; fls at least 9 mm

  8a Wing petals pubescent near tip; upper calyx lip 1/5–1/4 as long as wings; sagebr and ponderosa pine country, e Cas, Wn to Cal, e to Ida, Mont, Utah, and Nev, w through CRG and into Rogue R drainage; spurred 1.; 3 vars.   6 **L. laxiflorus** Dougl.

    a1 Fls mostly white or cream to slightly or heavily tinged with pinkish, blue, or violet, rarely < 10 mm; mostly ne Ore to Sierran Cal, e to Ida, w Mont, and Utah, not uncommon in Wn (*L. c.*, *L. multitinctus*, *L. variegatus*)   var. **calcaratus** (Kell.) Smith

    a2 Fls mostly blue to violet or purple, 8–14 mm

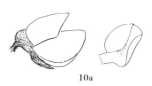

      b1 Lflets glab or subglab on upper surface, often obtuse or rounded; fls 8–11 mm; ne Wn to ne Ore, e to Mont (where common), occ elsewhere, esp along e base Cas, Wn (*L. laxispicatus*, *L. scheuberae*)   var. **pseudoparviflorus** (Rydb.) Smith & St. John

      b2 Lflets pubescent on upper surface, gen acute to obtuse; fls 9–14 mm; range of sp. as a whole (*L. arbustus*, *L. silvicola*) var. **laxiflorus**

  8b Wings glab; upper calyx lip 1/3–3/4 as long as wings; sagebr des to ponderosa pine for; e Ore to Cal, e to Mont and Colo; tailcup 1., Kellogg spurred 1. (*L. argentinus*, *L. hendersonii*, *L. lupinus*) 7 **L. caudatus** Kell.

7b Calyx not spurred on upper side at base (although gen ± asymmetrical and sometimes ± bulged); banner often pubescent on upper third; wing petals not pubescent on upper half; fls sometimes < 9 mm

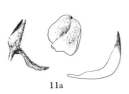

  9a Fls 10–18 mm, loosely racemose; banner pubescent to upper third, well reflexed, the index ave at least 15 (banner index, as here used, is an arbitrary unit of comparison; it is the product obtained by multiplying one factor, "a," the length, as measured in mm, of the exposed wing above the point where the banner is reflexed, by another factor, "b," also obtained by a measurement in mm, this the distance between the tips of the banner and wing petals); peds 4–11 mm

    10a Fls yellow (sometimes tinged with purple); banner very sparsely hairy; lflets gen > 6 × 1 cm; mostly in open ponderosa pine for; Blue Mts, se Wn and ne Ore; Sabin's 1.   8 **L. sabinii** Dougl.

    10b Fls gen blue, if yellow or yellowish then banner strongly hairy or lflets ave < 6 × 1 cm

      11a Keel slender, strongly curved, narrowed gradually to claw, tip gen exposed; racemes very loose, mostly long-pedunculate, lowest fl well above all lvs; fls 12–16 mm; banner gen yellow-centered; s WV, Ore, to foothill Cal; white-lvd 1.   9 **L. albifrons** Benth.

      11b Keel broader and not strongly curved, abruptly contracted to claw, tip rarely exposed; racemes not particularly loose, mostly short-pedunculate, lowest fls commonly not above uppermost lvs; fls mostly 10–12 mm; banner white-centered; sagebr des to lower mont for; e Cas, BC to Cal and Ariz, e to Alta and RMS to NM; silky 1.; 3 vars.   10 **L. sericeus** Pursh

        a1 Pl gen with only 1–3 (4) fl sts; fls white, banner only moderately pubescent (gen in median line); nonfor Asotin Co, Wn, from SR s, n into s Whitman Co where tending to be multistd
                                         var. **asotinensis** (Phillips) Hitchc.

        a2 Pl gen with > 3 sts; fls gen bluish or lavender; banner often strongly hairy on back

          b1 Banner only sparsely hairy; mostly along e base Cas, Okanogan Co, Wn, to Hood R Co, Ore var. **fikeranus** (Smith) Hitchc.

          b2 Banner strongly hairy over much of back; range of sp. as a whole (*L. flexuosus*, *L. leucopsis*, *L. obtusilobus*, *L. ornatus*)
                                         var. **sericeus**

12a

12b

14a

14b

16a

9b Fls either < 10 mm or tightly racemose; banner not much reflexed (index ave 10 or less), often not pubescent on upper half; peds 1–4 mm

12a Banner pubescent to upper third; racemes densely fld, mostly 15–20 cm; fls (7) 8–12 mm; foothills e Cas, c Wn to Mont, nw Wyo, Ida, and w Nev; velvet l.; 2 vars.   11 L. **leucophyllus** Dougl.

a1 Fls mostly 8–10 (12) mm; pressed racemes ave much > 2 cm broad; lflets gen > 7 mm broad; pubescence mostly long and loose; range of sp. as a whole (*L. canescens, L. cyaneus, L. erectus, L. macrostachys, L. plumosus, L. retrorsus*)   var. **leucophyllus**

a2 Fls 7–8 mm; pressed racemes ave < 2 cm broad; lflets gen < 7 mm broad; pubescence often short and appressed; sagebr and ponderosa pine regions; Lincoln Co, Wn, to c and se Ore and wc and sw Ida   var. **tenuispicus** (Nels.) Smith

12b Banner gen not pubescent to upper third; racemes interrupted below (at least), 5–15 cm; fls 5–7 (8) mm; sagebr slopes to ponderosa pine woodl; sc Ida, Adams to Blaine, Boise, and Owyhee cos, and adj Ore; little-fld l., silky l.   12 L. **holosericeus** Nutt.

Group II

13a Banner slightly or moderately reflexed from wings and keel, because of this, or because of the small fls, the banner index (for explanation see note under lead 9a) only 2–10; pl caulescent; sts rarely < 1 (gen > 2) dm, > the lower petioles

14a Basal lvs gen present at anthesis, petioles 3–5 times as long as lflets, much > petioles of cauline lvs; BC s, e Cas, to Cal, w through CRG to sw Wn and WV, Ore, e to se Wn, ne Ore, and prob adj Ida; sulfur l.; 3 vars.   13 L. **sulphureus** Dougl.

a1 Fls gen yellow but often white (not merely scattered albinos); Blue Mts, se Wn and ne Ore, app by pls of Okanogan Co, Wn, and elsewhere   var. **sulphureus**

a2 Fls mostly blue or purplish or scattered albinos

b1 Lflets pubescent on both surfaces; racemes mostly < 10 cm; fairly common in sagebr and ponderosa pine woodl; BC to c Ore, mostly along e base Cas and in e end CRG, e, but not reaching Ida (*L. bingenensis, L. mollis*)   var. **subsaccatus** (Suks.) Hitchc.

b2 Lflets glab on upper surface; racemes mostly 10–18 cm; WV to Douglas Co, Ore (*L. amabilis, L. oreganus*)   var. **kincaidii** (Smith) Hitchc.

14b Basal lvs gen lacking at anthesis, if present their petioles < 3 × as long as lflets; pine for to subalp ridges; c Ore to ne Cal, e to Alta, Daks, and in RMS to NM; silvery l.; 4 vars.   14 L. **argenteus** Pursh

a1 Fls (4) 5–7 (9) mm; lvs mostly glab on upper surface

b1 Lflets of basal lvs oblanceolate to almost obovate, rounded to obtuse; fls 4–6 mm; se Ida to Wyo and s to Utah and Colo (*L. alsophilus, L. floribundus*)   var. **parviflorus** (Nutt.) Hitchc.

b2 Lflets all narrowly lanceolate, acute or acuminate; fls (5) 6–9 mm; c Ida to s Mont, Utah, and NM (*L. leptostachyus*)   var. **stenophyllus** (Rydb.) Davis

a2 Fls (7) 9–12 mm; lvs various

c1 Pl subalp, grayish-hairy, 1.5–2.5 dm; racemes congested; c Ida to c Mont and Wyo (*L. evermannii, L. roseolus*)   var. **depressus** (Rydb.) Hitchc.

c2 Pl of various habitats, gen not grayish-hairy, often much > 2.5 dm; racemes not particularly congested; Ore to Alta, Mont, and SD (*L. adscendens, L. alpestris, L. corymbosus, L. decumbens, L. jonesii, L. lucidulus, L. laxus, L. maculatus, L. macounii, L. monticola, L. myrianthus, L. pulcherrimus, L. rubricaulis, L. spathulatus, L. tenellus*)   var. **argenteus**

13b Banner so well reflexed or fls so large the banner index ave > 10 and gen at least 15, if < 15 then pl subacaulescent, the sts rarely > 1 dm and gen < longer petioles

15a Fls gen yellow (to blue or purplish or rarely white), 13–18 mm; pl either shrubby, 1–2 m, and from along the coast (mostly on dunes) or from the Blue Mts, se Wn and ne Ore

16a Pl shrubby, 1–2 m, coastal; native to Cal but intro and estab along coast and along PS, Wn; tree l.   15 L. **arboreus** Sims

17a

20a

20b

19b

16b Pl herbaceous, local in ponderosa pine for (see lead 10a)
                                                        8 **L. sabinii** Dougl.
15b Fls rarely yellow, often < 15 mm; pl never shrubby, if coastal and on
dunes then gen prostrate
  17a Pls of immediate coast, prostrate and matted; sts and petioles mostly
    hirsute with rust-colored hairs 2–5 mm; BC to n Cal; seashore l.
                                                     16 **L. littoralis** Dougl.
  17b Pls not strictly coastal, often erect; sts and petioles seldom hirsute as
    above
    18a Sts gen br, lateral brs ending in racemes (lateral racemes often
      rudimentary as the terminal raceme starts to fl); lvs of main st gen
      ave 8 or more (rarely only 3–7); pls (except some on n side CRG)
      greenish, glab to moderately pubescent; lflets often glab on upper
      surface; peds gen at least 5 mm; rarely e Cas except in CRG
      19a Upper calyx lip similar to the lower, although shorter, gen at
        least 3 times as long as calyx tube; pls not villous; lowermost lvs
        deciduous by anthesis
        20a Keel ciliate, scarcely exposed; wings broad, almost com-
          pletely covering keel; margins of banner strongly reflexed;
          gravelly prairies, riverbanks, and open woods, always at low
          elev; BC to n Cal, w Cas; stream l., streambank l. (*L. amphi-
          bium*)   17 **L. rivularis** Dougl.
        20b Keel nonciliate, tip exposed; wings slender, not completely
          covering keel; margins of banner reflexed only slightly; lowl,
          PT s, w Cas, to Cal and w Nev; sickle-keeled l., pine l. (*L.
          andersonii, L. lignipes*)   18 **L. albicaulis** Dougl.
      19b Upper calyx lip much broader than the lower, gen not > 2.5
        times as long as calyx tube; pls sometimes villous; lowermost lvs
        sometimes remaining until anthesis; Cas to coastal mts, Alas to
        Cal, from lowl prairies to open subalp ridges; broadlf l.; 3 vars.
                                                  19 **L. latifolius** Agardh
        a1 Pl strigillose (rarely villous even in infl), gen > 2.5 dm; Cas to
          coast, BC to Sierran Cal (*L. ligulatus, L. viridifolius*)
                                                            var. **latifolius**
        a2 Pl gen strongly villous with whitish to reddish hairs, esp on
          calyx and infl, less often also on lvs and sts
          b1 Keel mostly nonciliate; pl ave well > 2.5 dm; CRG mostly,
            Wn and Ore (*L. suksdorfii*)
                                            var. **thompsonianus** (Smith) Hitchc.
          b2 Keel mostly ciliate; pl gen 1–2.5 dm; Alas s, on subalp
            ridges and mt meadows in Cas, to Mt. Rainier (*L. arcticus,
            L. volcanicus*)   var. **subalpinus** (Piper & Robins.) Smith
    18b Sts unbr or rarely br but then lateral brs not ending in racemes;
      lvs of main st gen < 8; pl often grayish; peds sometimes much < 5
      mm; more gen, often e Cas
      21a Pl low and spreading, gen < 3 dm; sts lfy, gen < peduncles
        and racemes; upper calyx lip bidentate for 1/2 length or more; both
        sides Cas, BC to Cal, e to Mont, Wyo, and Colo; prairie l.; 5
        vars.   20 **L. lepidus** Dougl.
        a1 Racemes sessile or very short-pedunculate, < lvs and largely
          concealed by them at anthesis
          b1 Wing petals slender, 7–8 mm, about 1/3 as broad; banner
            slender, width < 3/5 length; c Ore through c Ida and se Ore
            to w Mont and s to Colo and Utah (*L. caespitosus*)
                                                var. **utahensis** (Wats.) Hitchc.
          b2 Wing petals broader, gen > 8 mm and nearly 1/2 as broad;
            banner broader, width 3/5 of length; Blue Mts, Ore, and
            Okanogan Co, Wn (*L. abortivus, L. brachypodus, L. minuti-
            folius, L. volutans*)   var. **cusickii** (Wats.) Hitchc.
        a2 Racemes pedunculate, at least partially surpassing lvs
          c1 Pls prostrate and matted; lflets seldom > 15 mm; racemes
            mostly < 5 cm at anthesis; subalp; BC to s Cal, in Cas and
            OM, e to w Ida and Nev (*L. danaus, L. fruticulosus, L. lyal-
            lii, L. paulinus, L. perditorum*)   var. **lobbii** (Gray) Hitchc.
          c2 Pls not matted; lflets gen > 15 mm; racemes mostly at least
            5 cm; pl gen not high-mont

21a

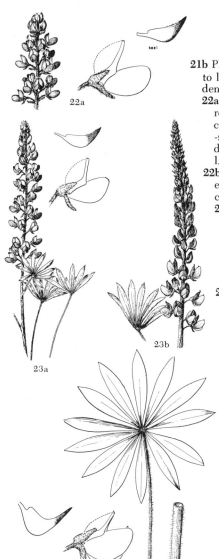

22a

23a

23b

23b

d1 Racemes gen partially concealed by longer lvs; fls mostly 9–11 mm; sc Wn to n Cal and Nev, e to sc Ida (*L. hellerae, L. piperi*)   var. **aridus** (Dougl.) Jeps.

d2 Racemes gen exserted well beyond longest lvs; fls 11–13 mm; s BC to nw Ore, w Cas (*L. minimus*)   var. **lepidus**

21b Pl gen much > 3 dm, mostly with lfy sts > peduncles and = to longest basal petioles; upper calyx lip subentire to deeply bidentate

22a Upper lip of calyx cleft at least 1/3 length; lflets oblanceolate, rounded to abruptly acute, 1.5–3 (4) cm; racemes mostly 5–10 cm at anthesis and 15 cm in fr; pls gen brownish-villous to -strigose on lvs and sts; mostly on basaltic rimrock in sagebr des or light pine woodl; c Wn and e Ore to n Cal, e Cas; rock l. (*L. subsericeus*)   21 **L. saxosus** Howell

22b Upper lip of calyx cleft < 1/3 length; lflets various but gen either acute or acuminate or over 4 cm; racemes mostly > 10 cm at anthesis; pl rarely brownish-villous

23a Lflets pubescent (gen strigose) on upper surface, mostly abruptly acuminate to apiculate, linear-oblanceolate to elliptic-oblanceolate, 2–4 (7) cm × 2–6 (14) mm; sts slightly or not at all fistulose; sagebr plains and valleys to mont or subalp for and open ridges; BC to Cal, e to Alta and through RMS to Colo; Wyeth's l. (*L. candicans, L. comatus, L. flavescens, L. humicola, L. rydbergii*)   22 **L. wyethii** Wats.

23b Lflets glab on upper surface, mostly rounded or acute, elliptic-oblanceolate, (3) 4–15 cm × 10–25 mm; sts often very fistulose; moist areas mostly, often along streams or meadows but into mts on shaded or open slopes; BC to Cal, e to Alta and RMS to Colo; biglf l., many-lvd l., large-lvd l.; 4 vars.
          23 **L. polyphyllus** Lindl.

a1 Sts strongly fistulose, gen > 6 dm; mostly in and w Cas

b2 Sts sparsely hirsute with long, white or reddish, stiff hairs; WV, Ore, to Thurston Co, Wn
          var. **pallidipes** (Heller) Smith

b2 Sts glab or strigillose; BC to Cal, from coast into Cas, occ in ne Ore and Ida (*L. cottoni, L. grandifolius*)
          var. **polyphyllus**

a2 Sts gen only slightly fistulose, ave < 6 dm

c1 Sts, petioles, and lower surface of lflets glab to strigillose, rarely with short, whitish, spreading pubescence; mainly e Cas, BC to Cal, e to Ida, Mont, and s; extremely variable in Wen Mts, Wn (*L. apodotropis*)
          var. **burkei** (Wats.) Hitchc.

c2 Sts, petioles, and lower surface of lflets strongly reddish-hirsute; mostly c and e Wn to adj Ida and ne Ore
          var. **prunophilus** (Jones) Phillips

Medicago L.   Medic(k)

Fls small, papilionaceous, in tight pedunculate racemes or heads, (white) yellow or bluish-purple; stamens 10, diadelphous; pods 1–several-seeded, often armed with 2 rows of prickles on the exposed, keeled margins, indehiscent; ann or per herbs. (Gr name, *medice*, apparently for alfalfa, introduced to Greece).

1a Pl per, deep-rooted; fls 6–10 mm; pod unarmed

2a Fls blue (rarely pink or white); pods coiled; lflets mostly 2–4 cm; European, escaped from cult in waste areas and along roadsides; alfalfa, lucerne   1 **M. sativa** L.

2b Fls yellow (violet); pods only slightly falcate; lflets mostly 1–2 cm; Eurasian, not often intro in N Am, but well estab near Midway, BC; yellow lucerne, sickle m.   2 **M. falcata** L.

1b Pl ann, shallow-rooted; fls < 6 mm; pod often armed

3a Pod 1-seeded, unarmed, reniform, curved to < 1 full spiral, only the style spiraled; fls 2–3 mm; European, estab in many waste places and along roadsides, esp on sandy or gravelly soil; black m., hop clover   3 **M. lupulina** L.

2a   2b   3a

4a

4b

3b Pod several-seeded, often prickly, spirally coiled as is the style; fls 4–5 mm
    4a Lflets mostly > 1.5 cm, with a central dark purplish spot; European, estab from Wn to Cal, mostly in waste places near coast; rarely fr not armed (ssp. *inermis* Ricker); spotted m.   **4 M. arabica** (L.) Hudson
    4b Lflets mostly < 1.5 cm, unspotted; sts glab or sparsely strigulose; stips lacerate; European, widely intro in w US, esp w Cas; bur-clover, hairy m.; the closely related *M. minima* (L.) Bart. (differing in having soft-hairy sts and subentire stips) has been collected once or twice as an escape, as has *M. turbinata* Willd., which has more strongly coiled pods with thickened, but not spiny margins *(M. polymorpha)*
                                       5 M. hispida Gaertn.

Melilotus Mill.   Sweet-clover; Melilot

Fls small, papilionaceous, ∝ in many spikelike racemes, white or yellow; stamens 10, diadelphous; pod ovoid, slightly > calyx, 1 (-4)-seeded, gen indehiscent; ann or bien, taprooted, strongly sweetish-odored herbs. (Gr *meli*, honey, and *lotos*, name for a cloverlike pl).

2a

2b

1a

1a Corolla white, ca 5 mm; pl 0.5–3 m; pod ca 4 mm; Eurasian; common weed in much of US and Can; white s.   **1 M. alba** Desr.
1b Corolla yellow, often < 5 mm; pl various; pod 2–3 mm
    2a Fls 4–6 mm; calyx teeth narrowly lanceolate-subulate; pls gen > 1 m; Eurasian; common weed in RMS, less common w, rarely w Cas; common yellow s.   **2 M. officinalis** (L.) Lam.
    2b Fls 2–3 mm; calyx teeth oblong-lanceolate, obtuse; pls gen < 1 m; European, sparingly intro in w Ore and w Wn, occ elsewhere; small-fld yellow s.   **3 M. indica** All.

Onobrychis Adans.

Fls papilionaceous, spicate-racemose; stamens 10, monadelphous but upper filament partially free; pods indehiscent, 1–2-seeded, mostly ± prickly; per herbs with odd-pinnate lvs. (Ancient Gr name).

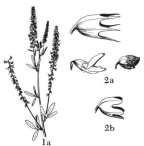

O. viciaefolia Scop. Saintfoin, sandfain, holy-clover. Pl 2–4 dm, strigose; stips reddish-brown, lanceolate; lflets 11–17 (27), narrowly elliptic to lanceolate, 1–2 cm, apiculate; fls 10–13 mm, pink to lavender, wing petal much smaller than keel and banner; frs ovate in outline, 6–8 mm, 1-seeded; European, intro in N Am, estab in BC, w Wn, in Powell Co, Mont, and prob elsewhere.

Ononis L.

Fls axillary, papilionaceous; stamens 10, monadelphous; pods few-seeded, dehiscent; ann or per herbs or shrubs with simple or 3-foliolate lvs and large stips adnate to petiole. (Ancient Gr name).

O. repens L. Restharrow. Prostrate per herb, long-villous and glandular on lvs and calyx; sts 2–5 dm; lvs 3-foliolate below, simple above; stips large, serrate; lflets oblong-obovate, (5) 10–20 mm, serrate distally; fls single in axils, 15–20 mm; calyx camp, lobes 2–3 times as long as tube; pod villous, 1–2 cm; European, occ escaped and estab as at Bingen, Klickitat Co, Wn, and near Linnton, Ore (where prob not persistent).

Oxytropis DC.   Crazyweed; Stemless-loco; Oxytrope

Fls spicate or racemose, papilionaceous, white to reddish or purple; calyx 5-toothed; keel prolonged into a point or tooth or a straight to curved beak; stamens 10, diadelphous; pod sessile or stipitate, often inflated, 1-celled to ± 2-celled, several-seeded; per herbs, stless or with short lfy sts and odd-pinnate lvs; lflets scattered to verticillate, stips free or connate around st. (Gr *oxys*, sharp, and *tropis*, keel, in reference to beaked keel). *(Aragallus, Spiesia)*.

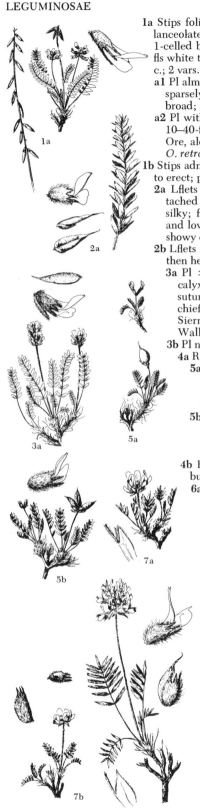

1a Stips foliaceous, adnate to base of petiole for only 1–3 mm, free portion
   lanceolate, 6–12 mm; pods pendulous, short-stipitate, 10–18 × 2–4 mm;
   1-celled but upper suture ± intruded; pls often with lfy sts 3–20 (30) cm;
   fls white to bluish-purple, 6–10 mm; N Am and Siberia, mont; pendent-pod
   c.; 2 vars.   1 **O. deflexa** (Pall.) DC.
   a1 Pl almost or quite acaulescent, greenish; racemes 7–10- fld, subcapitate,
      sparsely pilose; fls bluish-purple, ca 10 mm; sinuses between calyx lobes
      broad; Alta s in RMS to Colo, across Can   var. **foliolosa** (Hook.) Barneby
   a2 Pl with sts 3–20 cm, often grayish with pubescence; racemes elongate,
      10–40-fld; fls pale blue, 6–8 mm; calyx sinuses narrow and acute; BC to
      Ore, along e base Cas, e across Can and s in RMS to NM (*O. retroflexus,*
      O. retrorsa)   var. **sericea** T. & G.
1b Stips adnate to base of petiole for at least half their length; pods spreading
   to erect; pls acaulescent
   2a Lflets of all lvs fasciculate and apparently verticillate, 2 or more at-
      tached at a point on either side of rachis; herbage not glandular, densely
      silky; fls reddish-purple, 12–15 mm; gravel bars to meadowl, foothills
      and lower mts; RM, chiefly on e slope, Alas to NM, e to Ont and Minn;
      showy c. (*O. caudatus, O. richardsoni*)   2 **O. splendens** Dougl.
   2b Lflets subopp or scattered, or occ a few paired on one side of rachis and
      then herbage gen glandular
      3a Pl ± glandular-verrucose and often viscid, glands most obvious on
         calyx teeth; fls cream to purple, 10–15 mm; pods 10–15 mm, upper
         suture intruded half width of cavity; subalp to alp; Alas to Que, s
         chiefly in RM to Colo, also in Wallowa Mts, Ore, OM, Wn, and the
         Sierra Nevada; pls of OM with cream colored fls, elsewhere, and esp in
         Wallowa Mts, color purplish; sticky c. (*O. viscidula*) 3 **O. viscida** Nutt.
      3b Pl neither viscid nor glandular-verrucose
         4a Racemes 1–3-fld; fls purplish; pl alp; lvs gen < 5 cm
            5a Pod stipitate, bladdery-inflated, 15–25 × gen > 8 mm; lflets
               linear; fls 12–17 mm; ovary stipe 2–5 mm, but often concealed by
               calyx; alp ridges, RM, Alta to Colo, e to Atl coast; stalked-pod c.
               (*O. hallii*)   4 **O. podocarpa** Gray
            5b Pod sessile, oblong-ovoid, not greatly inflated, 15–23 × 4–8
               mm; lflets narrowly oblong; fls 7–10 mm; alp ridges, Wyo to NM,
               w to mts of c Ida (only area in our range), Nev, and Cal; Parry's c.
                                                         5 **O. parryi** Gray
         4b Racemes 4–∝ -fld, sometimes (depauperate specimens) fewer-fld
            but then fls ochroleucous, or pl not alp or lvs well > 5 cm
            6a Fls purple or reddish-purple; pod 1–2 cm, ca half partitioned by
               the intruded upper suture
               7a Calyx gen accrescent and not ruptured by the enlarging pod,
                  densely silky-villous and mostly concealed by hairs; scapes gen
                  < 10 cm; bracts membranous, shaggy pilose on back; lflets
                  mostly < 15 and seldom > 15 mm; sagebr plains to lower mts;
                  Ida and w Mont to Wyo; rabbit-foot c.; 3 vars.
                                                         6 **O. lagopus** Nutt.
                  a1 Lflets 5–9, ca = rachis of lf; calyx gen persistent until after
                     pod dehisces; c Mont, mostly e RM   var. **conjugens** Barneby
                  a2 Lflets mostly (9) 11 or more, rarely as long as rachis; calyx
                     gen deciduous with the enclosed pod before seed dispersal
                     b1 Corolla pinkish-purple; Ida and Mont (*O. blankenshipii*)
                                                         var. **lagopus**
                     b2 Corolla deep purple; Wyo and extreme s Mont
                                                         var. **atropurpurea** (Rydb.) Barneby
               7b Calyx not accrescent, gen ruptured by maturing pod, hisp-
                  id-hirsute, not concealed by hairs; scapes commonly at least 10
                  cm; bracts herbaceous, sparsely appressed-pilose; lflets often >
                  15 or > 15 mm; gravel benches, prairies, stream banks, and
                  lower mont slopes; c Ida and Mont to Colo; Bessey's c.; 3 vars.
                                                         7 **O. besseyi** (Rydb.) Blank.
                  a1 Pubescence of calyx appressed; Custer Co, Ida
                                                         var. **salmonensis** Barneby
                  a2 Pubescence of calyx spreading
                     b1 Scapes 10–20 cm; racemes loosely 8–20-fld; e slope RM, s
                        Sask to ne Wyo   var. **besseyi**

b2 Scapes 2–9 cm; racemes closely 3–10-fld; sw Mont and
   adj Ida   var. **argophylla** (Rydb.) Barneby
6b Fls white or yellowish, only the keel sometimes purple-tinged
  8a Pod fleshy, becoming hardened and bony, the dried wall
     nearly 1 mm thick; fls 15–27 mm; pls of RM, w in our area only
     to c Ida; prairies to subalp meadows and ridges; silky c; 2 vars.
                                             8 **O. sericea** Nutt.
   a1 Corolla cream to white, often pinkish tinged, keel often
      purple-tipped; w Mont to ne Nev, NM, and Tex *(A. albi-
      florus, A. condensatus, O. collina, O. pinetorum)*
                                                      var. **sericea**
   a2 Corolla lemon to sulfur yellow, keel mostly not purple-
      tipped; n BC to c Ida and n Wyo
                                        var. **spicata** (Hook.) Barneby
  8b Pod more membranous than fleshy, the dried wall scarcely 0.5
     mm thick; fls 10–20 mm; transcont in Can, s in RM to Colo, w
     to OM, Wn, submont to mont; slender c., field c.; 3 vars.
                                             9 **O. campestris** (L.) DC.
   a1 Corolla white, keel strongly purple-spotted; lflets gen
      fewer than 17 (up to 23); ne Wn and Flathead Lake region,
      Mont   var. **columbiana** (St. John) Barneby
   a2 Corolla ochroleucous, keel not spotted; lfts often > 17
      b1 Stips very hairy; scapes mostly > 15 cm; lflets gen > 17;
         w Wn to Alta and ND, s in RM to Colo *(A. cervinus, O.
         albertina, O. cascadensis, O. luteola, O. macounii, O.
         mazama, O. monticola, O. okanoganea, O. olympica, O.
         villosa)*   var. **gracilis** (Nels.) Barneby
      b2 Stips glab or glabrate; scapes rarely > 15 cm; lflets
         seldom > 17; range of var. *gracilis*, but not w Cas *(O. al-
         picola, O. paysoniana, O. rydbergii)*
                                             var. **cusickii** (Greenm.) Barneby

Petalostemon Michx.   Prairie-clover

Fls spicate, irreg but not papilionaceous, white, pink, or red; calyx camp, 5-toothed; petals apparently 5, largest one (prob the only true petal) broad-clawed and adnate to calyx, other 4 (really
staminodia?) narrow-clawed, adnate to staminal tube and alt with the 5 fertile stamens (monadelphous); pod 1–2-seeded, indehiscent, gen contained in calyx; glandular-punctate herbaceous per
(ours) with odd-pinnate lvs and setaceous stips. (Gr *petalon*, petal, and *stemon*, stamen, in reference perhaps to the petaloid stamens). *(Kuhniastera)*.

1a Fls white; calyx tube glab, ca 2.5 mm; dry plains and e slope RM, Sask and
   Mont to most of c and s US; white p.   1 **P. candidum** Michx.
1b Fls pink to purple (except occ albino-fld pls); calyx tube gen very hairy
  2a Lflets elliptic to obovate, at least some 3–8 mm broad; pl mostly glab
     (except calyces and sometimes stips); rocky or sandy places, often in
     sagebr; Yakima Co, Wn, to c Ore, e to w Ida; western p.
                                             2 **P. ornatum** Dougl.
  2b Lflets linear, 1–2 mm broad; pl subglab to grayish-hairy throughout; dry
     plains and foothills e RM; Sask and Man to Colo, Tex, and Ala; purple
     p. *(P. molle, P. violaceum)*   3 **P. purpureum** (Vent.) Rydb.

Psoralea L.   Bread-root; Scurf-pea; Psoralea

Fls papilionaceous, rather tightly racemose or spicate; calyx tubular to camp; corolla whitish to
purple; stamens (9) 10, monadelphous (diadelphous); pods 1-seeded, indehiscent (circumscissile
or irreg rupturing), ± globose, often beaked, mostly not > calyx; glandular-punctate, herbaceous per with 3–5-foliolate lvs. (Gr *psora*, itch or mange, referring to scablike glands of the pl).
(Hoita, Lotodes, Pediomelum, Psoralidium).

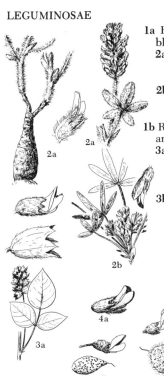

1a Roots enlarged and tuberlike; fls 10–16 mm, tightly spicate; calyx not black-hairy; RM and e

2a Pl long-hirsute; fls ca 15 mm; calyx tube ca 5 mm, lobes broadly lanceolate, > tube, all ca same size; prairies and foothills e RM; Alta to Okla, e to Wis, just on border of our area; Indian b.   1 **P. esculenta** Pursh

2b Pl appressed-pubescent; fls 10–13 mm; calyx tube 3–4 mm, lowest lobe much largest, 7–9 mm; Gr Pl region into RM foothills; Mont to NM and Okla, e to Neb; subterranean b.   2 **P. hypogaea** Nutt.

1b Roots not enlarged and tuberlike; either fls < 10 mm or calyx black-hairy and pl mostly from w Cas

3a Fls 9–12 mm, yellowish-green, keel purple-tipped; calyx black-hairy, strongly accrescent and as much as 12 mm in fr; lflets ovate-lanceolate, (2) 2.5–5 cm; mostly on disturbed or open land near woods w Cas; Wn to Cal, but reported from Ida; Cal-tea, Cal-tea s.   3 **P. physodes** Dougl.

3b Fls mostly 7–10 mm, mostly light to deep blue; calyx not black-hairy and not markedly accrescent; lflets linear to oblong, oblanceolate, or somewhat obovate

4a Lvs silvery-hairy, lower surface obscured by pubescence; fls 7–10 mm; e slope RM to Mo, Sask to NM; silver-lvd s. (*P. incana*)
          4 **P. argophylla** Pursh

4b Lvs greenish, upper surface glab; fls 4–7 mm

5a Fls blue, in loose interrupted racemes much > lvs; pod 7–8 mm; e side RM, Mont to Tex, e to Daks; slender-fld s. 5 **P. tenuiflora** Pursh

5b Fls white or keel blue, in short, congested racemes often < lvs; pod 4–5 mm; sandy soil, often with sagebr; e Cas, Wn to Cal and Nev, e to Neb; lance-lf s. (*P. elliptica, P. scabra*)
          6 **P. lanceolata** Pursh

Robinia L.   Robinia

Fls papilionaceous, showy; stamens 10, diadelphous; pods flattened, several-seeded, dehiscent; trees or shrubs with odd-pinnate lvs and gen with stips modified into large thorns. (For Jean and Vespasian Robin; the latter believed to have intro the genus into cult in Europe in the 16th century).

R. pseudo-acacia L. Black or yellow locust, false acacia. Tree to 25 m, well-armed; lflets 11–21, lanceolate to oblanceolate, 2–4 cm; racemes 10–14 cm, 30–70-fld, drooping; fls white, fragrant, 14–20 mm; calyx broadly camp, 5–6 mm; pod 6–10 cm; native in e US, often planted and rather widely escaped, esp near habitations or along rivers.

Swainsona Salisb.   Swainsona

Fls papilionaceous, loosely racemose; calyx 5-toothed, 2-bracteolate near base; stamens 10, diadelphous; pod stipitate, inflated, 1-celled, several-seeded; per herbs with odd-pinnate lvs. (For Isaac Swainson, 1746–1812, an English botanist).

S. salsula (Pall.) Taub. Rhizomatous per 4–9 dm; lflets 9–25, oblong-elliptic to elliptic-oblanceolate, 1–2 cm; stips connate basally; fls 10–15 mm, brick red or pinkish-brown, drying to violet or purplish; calyx camp, ca 5 mm, teeth subequal, the 2 subtending bracteoles scarcely 1 mm; pod 1.5–3 × 1–1.5 cm; stipe up to twice as long as calyx; Asiatic, escaped and well estab in places in c Wn and in e Ore in our area (*Astragalus iochrous, A. violaceus*).

Thermopsis R. Br.   Buck-bean; Golden-pea; Golden-banner; Thermopsis

Fls rather showy, racemose, papilionaceous, yellow; calyx camp, bilabiate; stamens 10, distinct; pod ∝-seeded, not inflated; per, rhizomatous herbs with trifoliolate lvs and large lfletlike stips. (Gr *thermos*, the lupine, and *opsis*, likeness or resemblance).

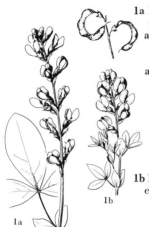

1a Pods erect or ascending to rarely somewhat spreading, rather straight; pl
   mostly 5–9 dm; Pac coast to RM; mt t.; 3 vars.   1 **T. montana** Nutt.
   a1 Lflets and stips narrow, 4–6 times as long as broad; pl densely pubes-
      cent; pods erect; racemes 10–60-fld; RM, Mont to Colo, occ w to e Ore
                                                              var. **montana**
   a2 Lflets and stips mostly < 4 times as long as broad; pubescence variable
      in amount; pods erect to spreading; racemes sometimes with < 10 fls;
      chiefly coastal, occ to RM
      b1 Pl, at least lflets, glab to subglab; racemes mostly 5–15-fld; pods
         sometimes spreading; Lane Co, Ore, to Cal, occ in Ida and Wn (*T.
         subglaber*)   var. **venosa** (Eastw.) Jeps.
      b2 Pl mostly densely pubescent; racemes mostly 10–60-fld; pods erect;
         coast to RM, Wn to Cal, common (*T. xylorhiza*)
                                                 var. **ovata** (Robins.) St. John
1b Pods spreading and gen recurved to ring-shaped; pl mostly 2–4 dm; mostly
   e RM, but w RM in Mont; round-lvd t.   2 **T. rhombifolia** Nutt.

Trifolium L.   Clover; Trefoil

Fls in pedunculate, ± capitate spikes or racemes, often invol, papilionaceous, white, yellow, or
pink to red or purple; calyx 5-toothed, teeth entire to 3-fid; corolla withering-persistent; stamens
10, diadelphous; pod globose to elongated, gen included in calyx, indehiscent, 1–several-seeded;
ann or per herbs, often rhizomatous, with palmately to semipinnately 3-foliolate (sometimes palmately 4–9-foliolate) lvs and membranous to foliaceous stips. (L name, referring to the trifoliolate
lvs).

1a Pl ann
   2a Heads invol   **Group I**, lead 11a
   2b Heads not invol   **Group II**, lead 18a
1b Pl per
   3a Fls subtended by an invol, bracts 1–4 (–12), distinct to connate, 0.5–2
      mm
      4a Heads 1–4-fld; pl 1–3 cm, densely matted, strongly taprooted; invol
         bracts gen 1–4; fls lilac-purple, aging brown, 15–22 mm; alp or sub-
         alp in RM; sw Mont to NM; dwarf c.   1 **T. nanum** Torr.
      4b Heads several-fld; pl mostly much > 3 cm
         5a Pl evidently pubescent, esp on calyx
            6a Calyx inflated and bladdery in fr; pl rhizomatous, not strongly
               pubescent except on calyx, 5–30 cm; fls 4–6 mm; European
               weed in waste places; Wn, Ore, and Ida in our area; strawberry c.
                                                         2 **T. fragiferum** L.
            6b Calyx not inflated or bladdery in fr; pl strongly pubescent, mat-
               ted, taprooted, 1–3 cm; fls 10–20 mm; alp and subalp slopes; RM,
               c Mont to Colo and e Utah; whip-root c. (*T. uintense*)
                                                    3 **T. dasyphyllum** T. & G.
         5b Pl glab or subglab, at least on calyx
            7a Pl matted, taprooted, 1–5 cm; heads on scapelike peduncles;
               invol bracts 6–12, distinct, entire-margined; fls 11–22 mm, dark
               reddish-purple, aging brown; subalp to alp meadows and stream
               banks; RM, Mont and adj Ida to NM and e Utah; Parry's c. (*T.
               montanense*)   4 **T. parryi** Gray
            7b Pl lfy-std, rhizomatous, mostly 10–80 cm; heads axillary on short
               peduncles of lfy sts; invol bracts 8–12, connate, margins toothed;
               fls 10–18 mm, reddish to purple, often white-tipped; meadows
               and stream banks to coastal dunes; BC to Cal and Mex, e to Ida,
               Colo, and NM; springbank c. (*T. fimbriatum, T. heterodon, T.
               spinulosum, T. willdenovii*)   5 **T. wormskjoldii** Lehm.
   3b Fls not subtended by a true invol, but sometimes stips of upper lvs ±
      invol-like
      8a Lflets commonly > 3; fls gen 30–100 per head, mostly > 18 mm;
         rachis of head not prolonged beyond upper fls
         9a Lflets linear to lanceolate, acute; calyx pubescent but not plumose;
            fls bright reddish-lavender to orchid, 18–22 mm; dry grassy hillsides
            just below ponderosa pine woodl; known only from Swakane Can-
            yon, Chelan Co, Wn; Thompson's c.   6 **T. thompsonii** Morton

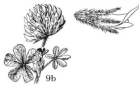

9b

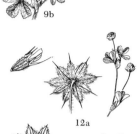

9b

12a

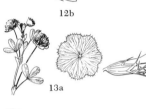

12b

9b Lflets oblanceolate to obcordate; calyx teeth plumose; fls pinkish to pinkish-rose, 22–28 mm; rocky places in sagebr des to ponderosa pine woodl; e Cas, sc Wn to Nev, e to w Ida; big-head c. *(T. megacephalum)* **7 T. macrocephalum** (Pursh) Poiret

8b Lflets mostly 3, if > 3 then either fls mostly < 30 per head and < 15 mm, or rachis of head extending well past upper fls

10a Calyx glab or only sparsely pubescent with scattered hairs **Group III, lead 28a**

10b Calyx strongly pubescent to villous or plumose **Group IV, lead 35a**

Group I

11a Invol long-villous, gen cup-shaped; calyx teeth simple; fls white to pinkish, 4–6 (7) mm

12a Calyx gen hairy, lower teeth at least = tube and often projecting past corolla; invol lobes gen subentire or entire; moist meadows to dryish hillsides; s BC to Baja Cal, e from the coast to Mont, Nev, and Ariz, often mont; woolly c., small-head c. **8 T. microcephalum** Pursh

12b Calyx glab or subglab, teeth all < tube and < corolla; invol lobes several-toothed; meadows and sandy to rocky soil w Cas; s BC to Cal; S Am; Valparaiso c., thimble c. **9 T. microdon** H. & A.

11b Invol glab or subglab, often flared and saucer-shaped; calyx teeth various

13a

15b

16a

17a

13a Lower calyx teeth mostly bi- or trifid; invol shallowly round-lobate, lobes finely erose-dentate, teeth < 2 mm; calyx 6–12 mm, ca = corolla; wet meadows to dryish sandy soil, e Cas; BC to Cal, e to Ida, prob intro along coast in few places; cup c. **10 T. cyathiferum** Lindl.

13b Lower calyx teeth entire, or invol with teeth > 2 mm

14a Corolla strongly inflated, narrowed toward tip; teeth of invol and calyx not stiffly spinulose; calyx tube 5–10-nerved

15a Fls 5–22 mm; stips 10–20 mm; sts 1–5 dm, mostly fistulose; s Ore and s, out of our range, but collected at least once in Seattle, Wn, where surely intro **11 T. fucatum** Lindl.

15b Fls 3–11 mm; stips 5–10 mm; sts 0.5–4 dm, not fistulose; c Ore to s Cal, occ and possibly always intro n to Wn and VI; poverty c. *(T. amplectens)* **12 T. depauperatum** Desv.

14b Corolla not strongly inflated; teeth of invol and calyx gen stiffly spinulose; calyx-tube 10–25-nerved

16a Sinuses of calyx equally cleft; calyx lobes narrowly lanceolate-subulate, entire; heads 1–2 cm broad; invol irreg lobed and gen cleft for at least half length; dry sandy soil to moist meadows mostly w Cas; BC to Cal, occ e to Mont and Utah; white-tip c. *(T. dianthum, T. rostratum)* **13 T. variegatum** Nutt.

16b Sinuses of calyx not equally cleft, upper one much the deepest; calyx-lobes often 3-toothed; heads often < 1 cm broad; invol sometimes lobed < half length

17a Heads 3–12 mm broad; fls 2–8 (15), 4–8 mm; meadows to dry rocky soil w Cas; s BC to Cal; few-fld c. *(T. pauciflorum,* misapplied; *T. variegatum* var. *p.)* **14 T. oliganthum** Steud.

17b Heads 10–30 mm broad; fls 6–60, (8) 10–17 mm; grassy hillsides and meadows, w Cas, BC to Cal; sand c., tomcat c. *(T. obtusiflorum)* **15 T. tridentatum** Lindl.

Group II

19a

17b

18a Calyx densely pilose-villous to strigose

19a Fls crimson, 12–15 mm; heads (2) 2.5–6 cm; stips sheathing at base, free portion obtuse to rounded; intro, occ escaped as in w Ida, well estab on wasteland in w Wn and Ore; crimson c. **16 T. incarnatum** L.

19b Fls white to pink or purple, 5–8 (12) mm; heads 0.5–2.5 cm; stips often nonsheathing, mostly acuminate to subulate, or at least acute

20a Calyx teeth < tube; calyx tube very prominently 10-striate and ovoid in fr; European; known from Linn Co, Ore, where prob not estab **17 T. striatum** L.

20b Calyx teeth mostly 2–4 times > tube; calyx tube not inflated nor prominently striate in fr

21a Corolla white or pink, ca 2/3 length of calyx; calyx teeth ca twice

length of tube; lflets linear-oblanceolate; pl erect, 1–4 dm; occ escape in old fields and waste places, mostly w Cas, intro from Europe; hare's-foot   18 **T. arvense** L.

21b Corolla purplish, often subequal to calyx or even longer; calyx teeth mostly > twice length of tube; lflets oblanceolate to obcordate; pl ascending to erect, 1–3 (5) dm; coastal to mont, BC to Baja Cal; rancheria c., Macrae's c.; 2 vars.   19 **T. macraei** H. & A.

 a1 Calyx rarely > corolla, teeth 2–3 times > tube; VI, w Wn, s Ore to Cal *(T. d.)*   var. **dichotomum** (H. & A.) Brew. & Wats.

 a2 Calyx mostly > corolla, teeth 3–6 times > tube; Wn and Ore to Baja Cal *(T. a.)* var. **albopurpureum** (T. & G.) Greene

18b Calyx glab or pubescent on teeth, the tube wholly glab or hairy only adj to teeth

 22a Calyx teeth strongly fimbriate-denticulate, upper 2 much larger than lower 3; fls white to purplish, 6–12 mm; w Cas, s Wn to Cal; foothill c., tree c. *(T. ciliatum* Nutt.)   20 **T. ciliolatum** Benth.

 22b Calyx teeth nearly or quite entire, upper 2 often smaller than lower 3

  23a Fls in part sterile; fertile (normal) fls (1) 2–7 per head, white or cream-colored; calyx 5–6 mm, ca half as long as corolla, teeth (lobes) filiform-setaceous, ca = tube; sterile fls gradually developing in center of head (above fertile ones), represented by rigid, terminally divaricate-spinulose processes (calyces?), becoming reflexed over maturing 1-seeded pods and transforming each head into a burlike structure which, by reflexion of peduncles, tends to become buried; European; occ intro for grazing purposes in US and now estab near CRG in Hood R Co, Ore, and adj Wn, n to PS region, Wn, and sw BC; subterranean c. or t.   21 **T. subterraneum** L.

  23b Fls all normal, gen ∝ (> 7), gen yellow or pink to purple; heads never burlike

   24a Fls white to pink or purplish; calyx 10-nerved; native spp.

    25a Calyx teeth lanceolate-acuminate, glab; fls 5–9 mm; calyx tube > 1.5 mm on upper side, upper teeth mostly much the largest; rachis of infl often prolonged above fls; grassy knolls w Cas; occ in Wn, common s to Baja Cal; slender c. 22 **T. gracilentum** T. & G.

    25b Calyx teeth slender, acicular, sparsely villous-pilose; fls 5–7 mm; calyx tube not > 1.5 mm on upper side, upper teeth scarcely larger than others; rachis of infl not prolonged beyond upper fls; mostly w Cas, Klickitat Co, Wn, s to s Cal; Pinole c. *(T. hallii);* ours the var. *decipiens* Greene 23 **T. bifidum** Gray

   24b Fls yellowish, fading to brown; calyx 5-nerved; intro weedy European spp.

    26a Pressed heads < 8 mm broad, gen with < 30 fls; corolla 3–3.5 mm; suckling c., least hop c.   24 **T. dubium** Sibth.

    26b Pressed heads mostly > 8 mm broad, gen with > 30 fls; corollla at least 4 mm

     27a Stalk of terminal lflet twice length of those of lateral lflets; heads mostly 8–11 mm broad; stips ovate, joined to petiole for ca half length, width of adnate portion nearly or quite = to length of free tip; hop c.   25 **T. procumbens** L.

     27b Stalk of terminal lflet ca = those of lateral lflets; heads mostly much > 1 cm broad; stips linear, often joined to petiole for 2/3 length, width of adnate portion much < length of free tips; yellow c.   26 **T. agrarium** L.

Group III

28a Heads 3–5 cm, gen not so thick; calyx tube with ca 20 prominent nerves, upper teeth curved downward; pl mostly > 5 dm, taprooted; fls reddish-purple, 14–20 mm; meadows and streambanks; Spokane Co, Wn, to Baker Co, Ore, e to adj Ida; Douglas' c. *(T. altissimum* Dougl.)
               27 **T. douglasii** House

28b Heads gen as thick as long; calyx ca (5) 10-nerved; pl mostly < 5 dm, sometimes rhizomatous or stoloniferous

 29a Fls 5–9 mm; heads axillary; European spp., intro and ± weedy

  30a Corolla white or slightly pinkish; pl stoloniferous; calyx glab; lflets mostly retuse to obcordate; stips 3–10 mm; waste areas to mt meadows; white c., Dutch c.   28 **T. repens** L.

30b Corolla gen pink or red (white); pl mostly with erect or ascending (but sometimes stolonous) sts; calyx with few hairs at base of teeth; lflets mostly rounded; stips 5–20 mm; gen estab in w US; alsike c.

29 **T. hybridum** L.

29b Fls at least 10 mm; heads often terminal and solitary; native spp.

31a Lflets (2) 2.5–5 cm broad; pl rhizomatous; sts 3–8 dm, fistulose; fls white or light pink, 11–15 mm; wet or shady places; w Cas, Lane Co, Ore, to n Cal; biglf c., Howell's c.   30 **T. howellii** Wats.

31b Lflets ave < 2 cm broad; pl various

32a Pl glab, caespitose or nearly so; lfy sts scarcely 5 cm; fls 13–17 mm; calyx 1/2–2/3 as long as corolla; alp or subalp meadows to ridges; RM, Carbon, Gallatin, and Park cos, Mont, to adj Wyo; Hayden's c.   31 **T. haydenii** Porter

32b Pl pubescent or with lfy sts > 5 cm

33a Rachis of infl prolonged and gen forked above fls; fls strongly reflexed, 11–18 mm; calyx scarcely 1/4 as long as corolla; corolla whitish, purple-tipped; moist to dry soil of for and open ridges; c Ore to Sierran Cal; elongated c., King's c. *(T. kingii* var. *p.)*

32 **T. productum** Greene

33b Rachis not prolonged; fls not reflexed, or calyx > 1/3 as long as corolla

34a Calyx gen < 1/2 as long as corolla, teeth mostly ca = tube; fls reddish or pale purple, 12–17 mm; mt meadows mostly; se Ore to Deer Lodge Co, Mont, s to Sierran Cal; Beckwith's c.

33 **T. beckwithii** Brew.

34b Calyx gen at least half as long as corolla, teeth mostly at least 2 × as long as tube; fls 11–18 mm, white to purple; lower mont valleys and meadows to subalp slopes; rather gen from nw Wn to s Cal, e to all the RMS; long-stalked c.; 5 rather poorly defined vars. in our area   34 **T. longipes** Nutt.

a1 Fls mostly or all reflexed or becoming reflexed in age

b1 Fls gen white, all strongly reflexed, the head gen as broad as long; ne Ore to Mont, s in RMS to NM *(T. l.* ssp. *r.)*

var. **reflexum** Nels.

b2 Fls gen light to rather dark purple, the lower ones reflexed more than the upper, the head often longer than broad; OM and c Cas, Wn, to n and c Ore *(T. oreganum* var. *multiovulatum* Hend. Madrono 3:231. 1936; *T. l.* var. *shastense*, misapplied; *T. caurinum; T. l.* ssp. *c.; T. covillei)*

var. **MULTIOVULATUM** (Hend.) C. L. Hitchc. hoc loc.

a2 Fls mostly ascending to erect, the lowermost sometimes reflexed in age

c1 Pl gen strongly rhizomatous; moist meadows, Cas, n Ore, to Cal and Nev *(T. l.* var. *nevadense, T. h., T. l.* f. *h., T. l.* ssp. *h.)*   var. **hansenii** (Greene) Jeps.

c2 Pl few-std from a taproot, only slightly if at all rhizomatous

d1 Tips of banner and wings slenderly acute; Wen Mts, Wn, to nc and ne Ore, s in e Ore to ne Cal

var. **longipes**

d2 Tips of banner and wings rounded to acute; mont, c Ida *(T. pedunculatum* Rydb. Bull. Torr. Club 30:254. 1903, *T. l.* ssp. *p.)*

var. **PEDUNCULATUM** (Rydb.) C. L. Hitchc. hoc loc.

Group IV

35a Heads 50–200-fld, sessile or with peduncles < subtending lvs; stips of upper lvs forming a false invol; fls deep red, 13–20 mm; European, often cult, now fairly widely intro; red c.   35 **T. pratense** L.

35b Heads sometimes few-fld, pedunculate, either peduncles > lvs or upper stips not invol

36a Peduncles mostly < lvs; lflets thick and leathery, gen not > 2.5 cm and with < 30 serrations; pl seldom > 1.5 dm; heads gen < 15-fld; fls white to pink, 8–14 mm; mostly dry soil of sagebr des to ponderosa pine

for; ne Ore to ne Cal, e to Mont, NM, and Ariz; hollylf c.; ours the var. *plummerae* (Wats.) Martin   36 **T. gymnocarpon** Nutt.

36b Peduncles gen > lvs; lflets either thin or > 2.5 cm or with > 30 serrulations; pl mostly > 1.5 dm; heads mostly with > 15 fls

37a Heads cylindric, gen nearly twice as long as thick; calyx densely villous-plumose, tube 20–35-nerved, teeth attenuate-aristate; lflets linear to linear-elliptic, mostly > 6 cm; fls whitish with pink to reddish tips, 14–20 mm; dry hillsides to meadowl, Blue Mts, Union Co, Ore, to Ida; plumed c.; 2 vars.   37 **T. plumosum** Dougl.

  a1 Lflets of basal lvs 2–5 (7) mm broad, acuminate; Blue Mts, se Wn, and Blue and Wallowa mts, ne Ore   var. **plumosum**

  a2 Lflets of basal lvs (8) 9–16 mm broad, acute; Washington Co, Ida   var. **amplifolium** Martin

37b Heads globose to hemispheric; calyx often merely ciliate or short-pubescent, tube 5–10-nerved; lflets seldom > 6 cm

38a Calyx plumose-villous; fls all strongly reflexed, curvature chiefly in calyx base, calyx tube thus ± gibbous; fls pinkish to red, 12–17 mm; Cas, sc Wn to n Cal and Nev, e to Ida and Utah; woolly-head c.; 3 vars.   38 **T. eriocephalum** Nutt.

  a1 Ovary with 2 (3) ovules; Klickitat Co, Wn, to n Cal (*T. e.* ssp. *cascadense, T. e.* var. *butleri, T. scorpioides*)   var. **eriocephalum**

  a2 Ovary with 4 ovules

    b1 Lflets linear, acuminate; Union Co, Ore, to se Ore and sw Ida (*T. harneyense, T. tropicum*)   var. **cusickii** (Piper) Martin

    b2 Lflets (at least basal ones) elliptic to oblong, rounded to retuse; Blue Mts, Wn and Ore, to adj Ida and Bitterroot Mts, Mont (*T. arcuatum, T. eriocephalum* f. or ssp. *a.*)
                                                                    var. **piperi** Martin

38b Calyx strigose to short-villous; fls erect to reflexed, curvature in peds, calyx tube not gibbous

39a Calyx lobes subplumose; lflets gen < 3 times as long as broad; peds often at least 2 mm; fls yellow to purple, 10–18 mm; meadows to rocky ridges; Wallowa Co, Ore, to Missoula Co, Mont; twin c. (*T. aitonii, T. orbiculatum*)
                                                          39 **T. latifolium** (Hook.) Greene

39b Calyx lobes short-pubescent; lflets mostly > 3 times as long as broad; peds mostly < 2 mm

    40a Wings and keel abruptly acuminate-apiculate; fls 12–19 mm, red to purple, lower ones reflexed; lflets gen <2 cm; dry slopes to subalp ridges; Wen Mts, Wn, and Wallowa and Steens mts, Ore; many-stalked c. (*T. longipes* ssp. *m.*)
                                                    40 **T. multipedunculatum** Kennedy

    40b Wings and gen keel not acuminate-apiculate; fls 11–18 mm, often ochroleucous or not reflexed; lflets mostly > 2 cm; widespread, BC s on both sides Cas to Cal, e to RM, Mont to Colo (see lead 34b)   34 **T. longipes** Nutt.

Ulex L.   Furze; Gorse

Fls papilionaceous, showy, yellow; calyx deeply 2-lobed; stamens 10, monadelphous; legume hairy, dehiscent; spiny, green-brd shrubs with rigid and prickly lvs. (Old L name for the pl).

U. europaeus L. Pubescent shrub 1–3 m; brs greenish, strongly angled, rigid, spine-tipped; fls borne along short lateral spinose brs; peds 3–6 mm, 2-bracteolate at tip; calyx 10–15 mm, hairy; corolla 15–20 mm; European, intro and freely escaped w Cas, in places a serious pest and fire hazard.

Vicia L.   Vetch

Fls solitary to racemose; calyx 5-toothed; corolla papilionaceous; stamens 10, mostly diadelphous; style filiform, densely bearded just below the stigmatic tip; pods flat, 2–several-seeded, dehiscent; ann or per herbs with trailing to climbing sts, pinnate lvs ending in simple to brd tendrils (rarely nontendrilar), and entire to sagittate stips. (Old L name for the plant). (*Ervum*).

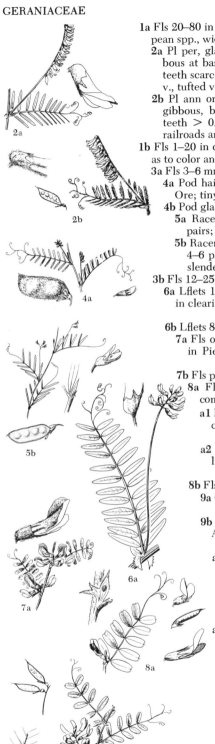

2a

2b

4a

5b

6a

7a

8a

9b

1a Fls 20–80 in dense, 1-sided, spikelike racemes, purplish, 10–18 mm; European spp., widely escaped and naturalized

    2a Pl per, glab to appressed-pubescent; fls 10–15 mm; calyx slightly gibbous at base, but bulge not extending behind point of insertion of ped, teeth scarcely 0.5 mm, lower ones gen no > tube; occ in RM and w; bird v., tufted v., cat peas, tinegrass   **1 V. cracca** L.

    2b Pl ann or bien, strongly hirsute-villous; fls 15–18 mm; calyx strongly gibbous, bulge extending behind point of insertion of pedicel, upper teeth > 0.5 mm, lower ones longer than tube; waste places and along railroads and highways; hairy, woolly, or winter v.   **2 V. villosa** Roth

1b Fls 1–20 in often loosely-fld racemes, sometimes only 1–2 in axils, various as to color and size; native or intro spp.

    3a Fls 3–6 mm; European spp., widely intro or escaped

        4a Pod hairy, 2 (3)-seeded; racemes 3–8-fld; fairly common, w Wn and Ore; tiny or hairy v.   **3 V. hirsuta** (L.) S. F. Gray

        4b Pod glab, 2–5-seeded; racemes often only 1–2 (to 5)-fld

            5a Racemes 2–5-fld; pods 10–16 × 5–8 mm, 2-seeded; lflets 8–10 pairs; widely intro, to be expected in our area   **4 V. disperma** DC.

            5b Racemes 1–2 (3)-fld; pods 10–15 × < 5 mm, (3) 4–5-seeded; lflets 4–6 pairs; occ in n Ida and w of Cas in Wn and Ore; smooth tare, slender v.   **5 V. tetrasperma** L.

    3b Fls 12–25 mm; native or intro spp.

        6a Lflets 19–29; fls white to orange; pl per, 1–2 m; sts fistulose; coastal, in clearings or along streams; Alas to Cal, inl to WV, Ore; giant v.   **6 V. gigantea** Hook.

        6b Lflets 8–20; fls various; pl ann or per, mostly < 1 m; sts not fistulose

            7a Fls ochroleucous; pl ann; European, occ escaped mostly w Cas, as in Pierce Co, Wn, and WV, Ore; Hungarian v.   **7 V. pannonica** Crantz

            7b Fls purplish; pl per

                8a Fls 1–3 per lf axil; European, long cult, now widely intro; common vetch, tare; 2 vars.   **8 V. sativa** L.

                    a1 Fls mostly at least 18 mm; wings often red; lflets mostly oblanceolate or broader; occ estab in Ida and w Wn and Ore   var. **sativa**

                    a2 Fls gen < 18 mm; wings seldom red; lflets variable, often linear; more common and more widely distributed   var. **angustifolia** (L.) Wahlb.

                8b Fls 4–10 (20) in pedunculate racemes

                    9a Calyx gibbous-based; fls 10–15 mm; pl ± weedy (see lead 2a)   **1 V. cracca** L.

                    9b Calyx not gibbous-based; fls 12–25; pl native, widespread; Alas to Cal and Mex, e to Ont, W Va, and Mo; Am v.; 3 vars.   **9 V. americana** Muhl.

                        a1 Lflets narrow, thick and coriaceous, densely pubescent with short curly hairs, lateral veins prominent, mostly unbr, departing from midrib at narrow angle; mostly e RM, but occ w in Mont and to Ida (*V. a.* var. *angustifolia, V. caespitosa, V. callianthema, V. dissatifolia, V. linearis, V. sparsifolia, V. sylvatica, V. trifida, V. vexillaris*)   var. **minor** Hook.

                    a2 Lflets various, more gen thin, glab, or with divaricate lateral veins

                        b1 Herbage, calyx, and sometimes pod copiously villous-pubescent; sts mostly zigzag; fls mostly not > 15 mm; Island Co, Wn, s, w Cas, to Cal (*V. californica*)   var. **villosa** (Kell.) Hermann

                        b2 Herbage and calyx glab to coarsely pubescent; pod mostly glab; sts gen not zigzag; fls ave > 15 mm; e and w Cas, BC to Cal, e to RM and plains region (*V. a.* var. *pallida, V. oregana*)   var. **truncata** (Nutt.) Brew.

GERANIACEAE   Geranium Family

Fls complete, hypog, reg (ours); sepals and petals 5 (ours), distinct; stamens gen in 1–3 series of 5 each, often basally connate, 1 or more whorls often without anthers; pistil 1, carpels 5 (ours), weakly united, ovary deeply 5-lobed and -celled; styles adnate to the much-elongate recep until fr mature; fr a caps, but often septicidally dehiscent into several segms, each segm 1–2-seeded and tipped with a slender, often spirally coiling, persistent style; ann or per herbs (ours) with alt or opp, stip, mostly palmately lobed or divided to pinnately compound lvs.

1a Lvs pinnately compound (ours); fertile stamens 5 **Erodium**
1b Lvs palmately lobed or divided; fertile stamens gen 10 **Geranium**

Erodium L'Her.   Alfilaria; Filaree; Clocks; Crane's-bill; Stork's-bill

Fls few, small, umbellate on axillary peduncles; sepals mostly setose, persistent; petals pinkish; filaments 10, distinct, alternately long and short, the shorter ones lacking anthers; styles spirally twisting at maturity; carpels sharp-pointed at base; ann herb (ours) with opp, pinnate lvs and interpetiolar stips. (Gr *erodios*, heron, in reference to the long-beaked fr.)

E. cicutarium (L.) L'Her. Lvs mostly basal, rosulate, pinnate-pinnatifid to pinnately divided with incised segms, ± reddish, nodes swollen; fls 10–15 mm broad; sepals mucronate to aristate; recep and styles 2.5–5 cm; Eurasian, intro and widely estab mostly e Cas, and esp in valleys and foothills; valuable range pl. Other spp. reported to be estab in our area are *E. moschatum* (L.) L'Her. ex Ait. and *E. aethiopicum* (Lam.) Brumh., both with less-finely divided lvs and nonaristate (rarely slightly mucronate) sepals.

Geranium L.   Crane's-bill; Geranium

Fls mostly showy, gen cymose (often only 2); stamens 10, all gen anther-bearing; styles gen recurving but spirally coiling at maturity of fr; ann or per herbs with alt or opp lvs and swollen nodes. (Gr *geranos*, crane, in reference to the long-beaked fr.)

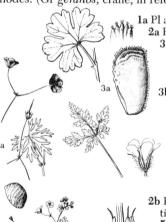

1a Pl ann or bien; petals < 12 mm; seeds smooth or reticulate-pitted
  2a Petals (5) 7–11 mm; carpels glab or hairy; sepals strongly bristle-tipped
    3a Lvs 5 (7)-lobed 1/2–3/4 length, the segms entire to shallowly 3-toothed at tip; petals 8–9 mm, pinkish; carpels puberulent, strongly (3) 5-ridged lengthwise esp over the top; European sp. recently recorded from Yamhill Co, Ore; shining c.   1 **G. lucidum** L.
    3b Lvs very deeply lobed, the main segms again ± dissected; carpels smooth to cross-rugose
      4a Calyx eglandular, sparsely appressed-pubescent; herbage dull green; petals 5–11 mm; carpels smooth; European, occ in c and e US, in our area in Whatcom Co, Wn, and along CR, Ore; long-stalked g.   2 **G. columbinum** L.
      4b Calyx slenderly-pilose; herbage light green; petals 7–10 mm; carpels cross-rugose, glab to densely puberulent; Eurasian, intro in w Wn and Ore; herb Robert, Robert g.   3 **G. robertianum** L.
  2b Petals mostly < 7 mm; carpels gen hairy, if glab then sepals not bristle-tipped
    5a Carpels glab, finely rugose; pl pilose-hirsute and ± glandular; petals emarginate, 3–5 mm; European; well estab esp in moist or waste places, mostly w Cas; dovefoot g.   4 **G. molle** L.
    5b Carpels gen hairy but not rugose; pls variously pubescent
      6a Fertile stamens 5; carpels pubescent with very fine, appressed hairs; seeds smooth; sepals not bristle-tipped; petals obcordate, 2.5–4 mm; European; estab throughout our lowl, in moist or waste places; small-fld c.   5 **G. pusillum** Burm.
      6b Fertile stamens 10; carpels pubescent with erect to spreading hairs; sepals bristle-tipped; seeds reticulate-pitted
        7a Beak of stylar column (including stigmas) 4–5 mm; stigmatic portion ca 1 mm; fr ped much > the calyx; petals retuse, 4–8 mm; native, BC to Newf, s to Wn, Mont, and Utah, and to ne US; Bicknell's g. (*G. nemorale, G. longipes*)   6 **G. bicknellii** Britt.
        7b Beak of stylar column, including stigmas, mostly < 3 mm; stigmas < 1 mm; fr ped mostly slightly or not at all > the calyx
          8a Pits of seeds prominent, nearly isodiametric; infl not much-congested; carpels rather evenly short-hirsute with

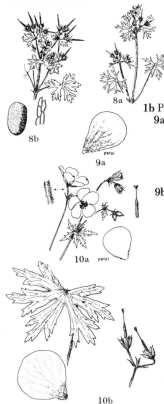

spreading hairs; petals retuse, 4–5 mm; European; scattered
weed along coast, BC to Cal; cut-lf g. *(G. laxum)*
7 **G. dissectum** L.

8b Pits of seeds not prominent, elongate; infl congested; carpels
long-hirsute with ascending hairs and shorter, spreading pubescence; petals rounded to retuse, 4–7 mm; much of e US and
Can, in our area as a widespread weed; Carolina g. *(G. ther-
male)*   8 **G. carolinianum** L.

1b Pl per; petals at least 12 mm; seeds reticulate

9a Petals white or pale pink, with pinkish or purple veins, 12–17 mm, pilose ca half the length on the inner face; infl pilose with glandular, purplish-tipped hairs; beak of stylar column no > the lobes of the stigma;
much of w N Am, e BC s, e Cas, to s Cal and NM, from foothills into the
mts, gen in shade; white g., sticky g. *(G. albiflorum* Hook., *G. hooker-
ianum, G. loloense)*   9 **G. richardsonii** Fisch. & Trautv.

9b Petals pink to deep magenta-purple, rarely white, 14–24 mm, gen pilose
on the inner face for not > 1/3 length; infl eglandular, or glandular with
yellowish hairs; beak of stylar column considerably > the stigmatic
lobes

10a Petals ciliate ca 1/5 their length but not pilose on inner face; stigmas
2–2.5 mm; w Cas, sw Wn to Cal, in meadows and woodl; western g.
(G. incisum)   10 **G. oreganum** Howell

10b Petals pilose on inner face; stigmas 4–5 mm; e Cas; sticky purple g.;
2 vars.   11 **G. viscosissimum** F. & M.

a1 Lower petioles and st hirsute and also glandular-puberulent, rather
evenly glandular-villous above; BC s to n Cal and n Nev, e to Sask
and n Wyo   var. **viscosissimum**

a2 Lower petioles and st glab, strigillose, appressed- to spreading-puberulent, or hirsute, but not glandular; range of var. *v.* but
somewhat more s, reaching Colo, Utah, and n Cal *(G. n., G. strigosum, G. canum, G. incisum)*   var. **nervosum** (Rydb.) Hitchc.

OXALIDACEAE   Oxalis or Wood-sorrel Family

Fls reg, ⚥, hypog, singly pedunculate to cymose or racemose; sepals 5, distinct; petals 5 (or lacking
in cleistogamous fls), distinct or basally connate; stamens gen 10 (ours), of 2 unequal lengths,
rarely only 5 fertile, basally connate; pistil 5-carpellary, ovary 5-celled; styles 5, stigmas mostly
capitate; fr caps, several–∞-seeded, loculicidal; herbs (ours) with watery, sour juice and alt, ternately (ours) to pinnately compound lvs and gen obcordate lfts.

Oxalis L.   Oxalis; Wood-sorrel; Lady's-sorrel

Petals showy, white or yellow to pinkish or violet; per herbs, often with corms or rhizomes and
long-petioled lvs. (Gr *oxys*, sharp [sour], in reference to the acrid juice of most species). *(Ceratox-
alis, Hesperoxalis, Xanthoxalis)*. Many spp. in cult, all apt to be weedy; *O. oregana* attractive and
(too) easily grown.

1a Pl without lfy st above ground; fls white or pinkish

2a Scapes 1-fld; petals mostly 13–20 mm; caps ovoid, scarcely 1 cm; moist
woods; along coast from OP to Cal, up CRG and onto w slope Cas in
Ore; Ore o., Ore w.   1 **O. oregana** Nutt.

2b Scapes 2–several-fld; petals ca 10 mm; caps linear, 2–3 cm; meadows
and moist woods; w slope Cas to coast, sw Wn to Cal, to 4000–6000 ft in
Cas; great o., trillium-lvd w.   2 **O. trilliifolia** Hook.

1b Pl with lfy st above ground; fls yellow

3a Petals 12–20 mm; peduncles 1 (2)-fld; moist coastal woods to rather dry
open slopes; w Cas, sw Wn to nw Cal; w yellow o. *(O. pumila)*
3 **O. suksdorfii** Trel.

3b Petals 4–9 mm; peduncles (1) 2–7-fld

4a Hairs of st and petioles gen at least in part septate, blunt at tip, gen
collapsing when dried or pressed; pl rhizomatous; stips lacking; widespread in N Am, in our area a very troublesome weed, esp in w Wn;
upright yellow w. *(O. ambigua, O. bushii, O. coloradensis, O. cymosa,
O. europaea, O. interior, O. rufa)*   4 **O. stricta** L.

4b Hairs of st and petioles nonseptate, pointed at tip, not collapsing when dried or pressed; pl more stoloniferous than rhizomatous; stips gen present, often greatly reduced
  5a Transverse corrugations of seeds white-ridged; pl caespitose, but outer brs sometimes rooting at base, erect part light green (or tan when dried), mostly antrorsely strigose; only occ as weed, chiefly e Cas; widespread in Can and c and e US to Mex; Dillen's w. (*O. stricta* of auth.)   5 **O. dillenii** Jacq.
  5b Transverse corrugations of seeds not white-ridged; pl with creeping and rooting main st, erect portion green to purple, glab or antrorsely to retrorsely strigose; Eurasian; common as weed in waste places and gardens; creeping yellow w. (*O. pusilla, O. repens, O. tropaeoloides*)   6 **O. corniculata** L.

LINACEAE   Flax Family

Fls complete, reg, hypog, polypet, cymose or racemose; sepals 5 (4), distinct or basally connate; petals 5 (4), often showy, gen quickly deciduous; fertile stamens 5 (ours) to 10 or more, gen connate basally, often alt with staminodia; pistil (2–4) 5-carpellary and mostly (2–4) 5–10-celled; styles as many as the carpels, mostly free; fr caps (ours), mostly with twice as many cells as styles; ann or per herbs (rarely suffrutescent) with simple, entire, alt (opp) lvs.

Linum L.   Flax

Fls white, yellow, pink, red, or blue; caps septicidal, each cell ± completely false-septate; seeds often mucilaginous when wet. (Original L name for flax). (*Cathartolinum, Hesperolinon*).

1a Petals yellow
  2a Pl per with woody base and ∝ sts; lvs ∝, crowded on sterile brs; high mts; se Ida to Wyo and Colo, prob not quite in our area   **L. kingii** Wats.
  2b Pl ann or erect short-lived per, gen simple-std; lvs scattered, basal ones often deciduous
    3a Lvs alt, entire; stigmas 4 or 5; petals 10–15 mm; pl of Gr Pl, Alta to Tex, possibly in our area in Mont; yellow f., large-fld y. f.
                                                                L. rigidum Pursh
    3b Lvs in part opp, upper bracts serrulate; stigmas 2; petals 3–6 mm; meadows and prairies e Cas; Wn and Ore to Sierran Cal; nw yellow f.
                                                                1 **L. digynum** Gray
1b Petals white, pinkish, or blue
  4a Petals white to pink, 4–7 mm; st gen pubescent; styles gen 3; dry open ground; Jefferson and Douglas cos, Ore, to Cal; small-fld white f.
                                                                2 **L. micranthum** Gray
  4b Petals blue; st gen glab
    5a Pl per; margins of sepals entire; lvs 1-nerved; petals 10–23 mm; prairies to alp ridges, gen on dry, well-drained soil; w N Am and Eurasia; 2 vars.   3 **L. perenne** L.
      a1 Fls of 2 kinds, 1 with styles > stamens, the other with stamens the longer; Eurasia; orn, occ garden escape; blue garden f.   var. **perenne**
      a2 Pls with fls all alike, native throughout much of w N Am; wild blue f. (*L. l., L. p.* var. *albiflorum*)   var. **lewisii** (Pursh) Eat. & Wright
    5b Pl ann (at most short-lived per); margins of inner sepals membranous and erose-serrulate or ciliate; lvs 3-nerved; petals sometimes < 10 mm
      6a Petals 5–8 mm; lvs linear-lanceolate, 1–2 cm × 1–2 mm; along roadsides, Lane Co, Ore, to coastal Cal; Europe; pale f., narrow-lvd f.   4 **L. angustifolium** Huds.
      6b Petals 10–14 mm; lvs linear, 1–2.5 cm × 1–1.5 mm; European, cult (for fibre and oil) and often escaping, but not tending to persist; cult f., common f., linseed   5 **L. usitatissimum** L.

ZYGOPHYLLACEAE   Caltrop Family

Fls reg, ♂, hypog, borne on axillary peduncles; sepals 5 (4), distinct; petals 5 (4, or rarely lacking), distinct; stamens 10–15, distinct, filaments often scale-appendaged; pistil 5 (4)-carpellary; styles 1 (to several); fr mostly a 4–5-celled caps or berry or a schizocarp of much-hardened carpels; ann or per herbs (ours) to trees with mostly opp, pinnatifid to pinnate lvs and persistent, often leathery or spinescent stips.

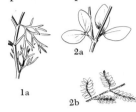

1a Lvs irreg pinnatifid-dissected, the ultimate segms linear; stamens 12–15; fr an ovoid caps   **Peganum**
1b Lvs pinnately compound or bifoliolate, the lfts not linear; stamens 10
  2a Lfts 2; filaments with a linear, 2-lobed scale adnate to the inner surface; fr an unarmed caps   **Zygophyllum**
  2b Lfts 8–14; filaments without scales; fr a spiny schizocarp   **Tribulus**

Peganum L.   Peganum

Fls white or pale yellow; pistil 2–3-carpellary; ann (ours) or per herbs. (Gr name for the pl, used by Theophrastus).

P. harmala L. Glab and glaucous ann with decumbent to erect, freely br st 2–5 dm; lfts irreg pinnatifid into ultimate segms 1–3 cm; stips setaceous, ca 1 mm, quickly deciduous; fls solitary on peduncles 1–2 cm; sepals linear, sometimes lobed, 1–2 cm; petals 14–18 mm, white; stamens 15; caps membranous, subglobose, 10–15 mm; European, very occ escaped in w US, as at Ephrata, Wn, and in Crook Co, Ore.

Tribulus L.   Puncture-vine; Land Caltrop; Ground Bur-nut

Fls solitary on axillary peduncles, 5-merous; petals yellow, borne at the edge of a 10-lobed disc; fr deeply lobed, hardened, separating ultimately into 5 2-spined segms; prostrate, freely br herbs with pinnately compound, opp, stip lvs. (L *tribulus*, thorn or thistle).

T. terrestris L. Pl ann, strigose-hirsute, matted, 3–10 dm broad; lfts 4–8 pairs, obliquely oblong-ovate, 5–15 mm; stips 1–2 mm; fls short-pedunculate; petals 3–5 mm; fr segms each with 2 large divergent spines 2–6 mm and with ∝ smaller spines forming a broad, dorsal, longitudinal row; intro in w US and rapidly spreading along roadsides and railways esp, mostly e Cas; much more common in sw and s US; particularly common on sandy soil in warm, dry regions, and a most serious pest, spoiling many areas for recreational use.

Zygophyllum L.

Fls yellow, mostly 5-merous; style and stigma 1; fr a 4–5-celled caps; ann or per herbs with opp, gen compound lvs and small, interpetiolar stips. (Gr *zygo*, yoke, and *phyllum*, leaf, referring to the bifoliolate lvs).

Z. fabago L. Syrian bean-caper. Glab, ± succulent ann with freely br, decumbent to ascending sts 2–5 dm; lfts obliquely oblong-obovate, 2–4 cm; stips oblong, 1–3 mm, sometimes fused; peduncles ca 1 cm; fls axillary, ca 1.5 cm broad; petals slightly > the sepals; stamens 10, their attached scales serrulate; caps 2–3 cm, oblong; native to the Old World, intro and perhaps estab in a few places as at Ephrata, Wn, and Minidoka, Ida.

SIMAROUBACEAE   Quassia Family

Fls mostly in large infls, ♂ to ♂ ♀ ♂ or ♀, ♂; calyx 3–8-merous, segms distinct to connate; petals mostly as many as sepals (lacking); disc gen present; stamens mostly = or 2× sepals; carpels mostly 5, weakly connate and sometimes baccate, but mostly ripening into separate drupes, samaras, or achenes; woody pls with alt, pinnately compound (simple) lvs and mostly very bitter bark.

Ailanthus Desf.   Tree-of-heaven

Pl subdioecious, fls small in ∝ pan cymules; sepals and petals 5 (6); stamens 10 (12); disc lobed; carpels nearly separate except for the common style, ripening into 5 (6) 1-seeded samaras; trees with large, pinnately compound, deciduous lvs. (Supposed oriental name meaning "tree of heaven").

A. altissima (Mill.) Swingle. Pl rapid-growing, smooth-barked; lvs 5–12 dm; lfts 10–25, lance-ovate, 5–15 cm, with 1–3 coarse teeth on each side near the base, each tooth with a gland on the lower surface; fls greenish, 6–8 mm; samaras 2.5–5 cm, often spirally twisted; Chinese, commonly grown orn, now often escaped, esp along highways (*A. glandulosa, Toxicodendron a.*).

EUPHORBIACEAE   Spurge Family

Fls ♂ ♀, sometimes borne separately, when often with a 5-merous calyx (rarely also with corolla) and often 10 stamens, but fls gen greatly reduced and borne in a cup-shaped, perianthlike invol; ♂ fls gen several, mostly included within the invol, stamen 1, filament jointed with the ped; ♀ fl single and terminal on a minute axis, gen exserted from the invol; pistil mostly 3 (1–4)-carpellary, gen separating (sometimes elastically) into 3 (1–4) 1-seeded segms, rarely the whole fr only 1-seeded; ♂ ♀ (♂, ♀) herbs, shrubs, or trees, often with milky juice and mostly opp (alt), simple to compound lvs.

1a Pl silvery-hairy and ± hirsute-hispid; lvs ovate, 1–3 cm; ♂ and ♀ fls separate, the ♀ naked   **Eremocarpus**
1b Pl not silvery-hairy, often glab; lvs various; ♂ and ♀ fls together in a calyx-like invol   **Euphorbia**

1a   1b

Eremocarpus Benth.   Doveweed; Turkey-mullein

Fls tiny, apet, ♂ and ♀ separate; ♂ ♀, prostrate or spreading, grayish-stellate ann with simple, petiolate lvs and nonmilky juice. (Gr *eremos*, lonely, and *karpos*, fr, referring to the solitary carpel).

E. setigerus (Hook.) Benth. Pl grayish-green, musky-scented, 2–10 cm tall, up to 5 dm broad; ♂ fls in terminal clusters, sepals 5–6, stamens 5–9; ♀ fls naked, solitary and axillary, style unbr, stigma 1; fr a 2-valved, 1-seeded caps; seeds 3–4 mm, smooth, grayish and mottled; dry, often rocky areas; e Cas in Wn, but e and w Cas in Ore, to Cal and Baja Cal, e to Nev (*Croton s., Piscaria s.*).

Euphorbia L.   Spurge; Euphorbia

Fls naked, in short clusters subtended by a camp to obconic, perianth-like invol; invol shallowly 5-lobed, bearing at the summit (4) 5 conspicuous glands sometimes with strongly colored borders; stamens (each a separate fl) several, included in the invol; ♀ fl single and terminal, 3-carpellary and mostly 3-lobed: fr caps, separating into 3 1-seeded segms; glab to ± hairy, ann or per herbs (ours) mostly with milky juice and opp lvs or bracts. (For Euphorbus, physician to King Juba II).

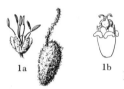

1a Pl mostly prostrate; lvs gen < 1.5 cm × < 5 mm, all opp; glands of invol petaloid (*Anisophyllum, Chamaesyce*)
2a Sts crisp-puberulent to pilose; lvs gen pilose on lower surface, larger ones sometimes > 15 mm; glands 4, broadly white- or pink-bordered
3a Pl prostrate; caps hairy; lvs 4–17 mm; seeds ca 1 mm, pinkish; native in e US, occ intro on Pac coast, known from Yakima Valley and from Walla Walla Co, Wn; milk s. (*E. maculata*, misapplied)

1 E. **supina** Raf.
3b Pl erect or ascending; caps glab; lvs 10–30 mm; seeds ca 1.3 mm, golden-brown; ND to Que, s to Fla and Mex; intro in Skamania Co, Wn; spotted s. (*E. nutans, E. preslii*)   2 E. **maculata** L.
2b St glab; lvs mostly glab, larger ones seldom > 15 mm; glands 4 or 5, with rather narrow whitish borders
4a Seeds coarsely transcorrugated; lvs thick-margined, linear-oblong, entire to crenulate-serrulate; invol ± turbinate; on dry and rather sandy soil; e Cas, BC to Cal, e to NB, Me, and s to Tex; corrugate-seeded s. (*E. greenei*)   3 E. **glyptosperma** Engelm.
4b Seeds smooth to wrinkled or pitted, not coarsely transcorrugated; lvs mostly obovate-oblong or ovate-oblong, margins serrulate but not

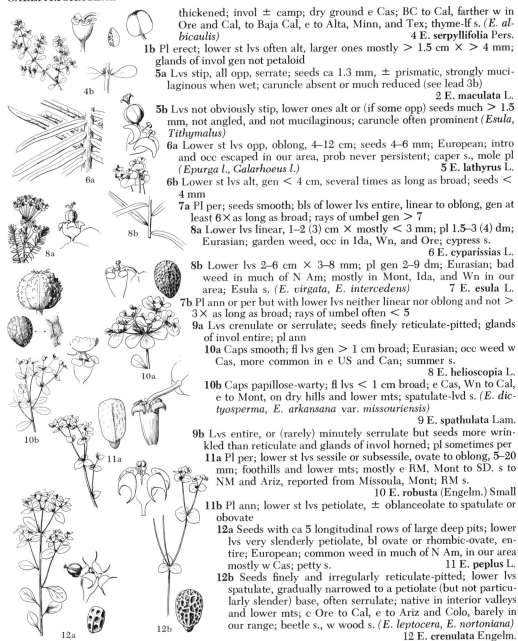

thickened; invol ± camp; dry ground e Cas; BC to Cal, farther w in Ore and Cal, to Baja Cal, e to Alta, Minn, and Tex; thyme-lf s. (*E. albicaulis*)   4 **E. serpyllifolia** Pers.

1b Pl erect; lower st lvs often alt, larger ones mostly > 1.5 cm × > 4 mm; glands of invol gen not petaloid

    5a Lvs stip, all opp, serrate; seeds ca 1.3 mm, ± prismatic, strongly mucilaginous when wet; caruncle absent or much reduced (see lead 3b)
                                                       2 **E. maculata** L.

    5b Lvs not obviously stip, lower ones alt or (if some opp) seeds much > 1.5 mm, not angled, and not mucilaginous; caruncle often prominent (*Esula, Tithymalus*)

        6a Lower st lvs opp, oblong, 4–12 cm; seeds 4–6 mm; European; intro and occ escaped in our area, prob never persistent; caper s., mole pl (*Epurga l., Galarhoeus l.*)   5 **E. lathyrus** L.

        6b Lower st lvs alt, gen < 4 cm, several times as long as broad; seeds < 4 mm

            7a Pl per; seeds smooth; bls of lower lvs entire, linear to oblong, gen at least 6× as long as broad; rays of umbel gen > 7

                8a Lower lvs linear, 1–2 (3) cm × mostly < 3 mm; pl 1.5–3 (4) dm; Eurasian; garden weed, occ in Ida, Wn, and Ore; cypress s.
                                            6 **E. cyparissias** L.

                8b Lower lvs 2–6 cm × 3–8 mm; pl gen 2–9 dm; Eurasian; bad weed in much of N Am; mostly in Mont, Ida, and Wn in our area; Esula s. (*E. virgata, E. intercedens*)   7 **E. esula** L.

            7b Pl ann or per but with lower lvs neither linear nor oblong and not > 3× as long as broad; rays of umbel often < 5

                9a Lvs crenulate or serrulate; seeds finely reticulate-pitted; glands of invol entire; pl ann

                    10a Caps smooth; fl lvs gen > 1 cm broad; Eurasian; occ weed w Cas, more common in e US and Can; summer s.
                                        8 **E. helioscopia** L.

                    10b Caps papillose-warty; fl lvs < 1 cm broad; e Cas, Wn to Cal, e to Mont, on dry hills and lower mts; spatulate-lvd s. (*E. dictyosperma, E. arkansana* var. *missouriensis*)
                                      9 **E. spathulata** Lam.

            9b Lvs entire, or (rarely) minutely serrulate but seeds more wrinkled than reticulate and glands of invol horned; pl sometimes per

                11a Pl per; lower st lvs sessile or subsessile, ovate to oblong, 5–20 mm; foothills and lower mts; mostly e RM, Mont to SD. s to NM and Ariz, reported from Missoula, Mont; RM s.
                        10 **E. robusta** (Engelm.) Small

                11b Pl ann; lower st lvs petiolate, ± oblanceolate to spatulate or obovate

                    12a Seeds with ca 5 longitudinal rows of large deep pits; lower lvs very slenderly petiolate, bl ovate or rhombic-ovate, entire; European; common weed in much of N Am, in our area mostly w Cas; petty s.   11 **E. peplus** L.

                    12b Seeds finely and irregularly reticulate-pitted; lower lvs spatulate, gradually narrowed to a petiolate (but not particularly slender) base, often serrulate; native in interior valleys and lower mts; c Ore to Cal, e to Ariz and Colo, barely in our range; beetle s., w wood s. (*E. leptocera, E. nortoniana*)
                      12 **E. crenulata** Engelm.

CALLITRICHACEAE   Water-starwort Family

Fls tiny, 1–3 per lf axil, naked or with 2 bracts, mostly ♂ ♀ ♂ ; stamens 1 (rarely more), pistil 2-carpellary and deeply 4-lobed, separating when mature into four 1-seeded frs; aquatic herbs, submerged or emergent and rooted in mud.

Callitriche L.   Water-starwort

Lvs exstip, either opp (rarely whorled), linear, and 1-nerved, or tufted at br ends and then commonly broadened and 3-nerved. (Gr *callos*, beautiful, and *trichos*, hair, in reference to the slender sts).

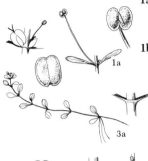

1a ♀ fls on distinct ped much > fr; lvs linear to linear-spatulate, 5–12 mm, joined at base by tiny winged ridges; frs ca 1 mm, wing-margined; vernal pools in w Cal; reported from near The Dalles and at Grants Pass, Ore; winged w.　　　　　　　　　　　　　　　　　　1 C. marginata Torr.
1b ♀ fls sessile or with ped scarcely 1/4 length of fr
　　2a Fr wing-margined all around, 1–1.5 mm, round to oblong-oval in outline
　　　　3a Fls with conspicuous white bracts nearly = the carpels; lf bases joined by small winged ridges; European, occ in Wn, Ore, and BC; more common in e US; pond w.　　　　　　　　　　2 C. stagnalis Scop.
　　　　3b Fls lacking conspicuous white bracts; lf bases connected by indistinct ridges; vegetatively not separable from C. hermaphroditica; near The Dalles, Ore; status uncertain, but doubtfully a valid sp.; Fassett's w.
　　　　　　　　　　　　　　　　　　　　　　3 C. fassettii Schotsman

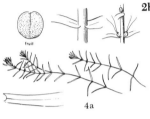

　　2b Fr winged only near the tip if at all
　　　　4a Lvs all linear, 1-nerved, 5–20 mm, bright lustrous green, not wing-joined at base; fl bracts lacking; widespread in N Hem, mostly emersed; autumnal w., northern w. (C. autumnalis, C. a. var. bicarpellaris, C. bifida)　　　　　　　　　4 C. hermaphroditica L.
　　　　4b Lvs varied, upper ones often broadened and several-nerved and joined at base by winglike ridge; fl bracts gen present
　　　　　　5a Carpel faces marked with tiny pits in rather reg longitudinal lines; fr gen slightly wing-margined at top, ca 1 mm, 1/5–1/3 longer than broad; submerged lvs linear, 5–20 mm, retuse, vein thickened but not protruding at tip; gen in US and s Can; Eurasia; spring w. (C. palustris var. v.)　　　　　　　　　　　5 C. verna L.
　　　　　　5b Carpel faces marked with pits irreg patterned, not in rows; fr scarcely at all winged, gen ca as broad as long
　　　　　　　　6a Fr gen widest above middle and ± obovate in outline; submerged lvs bidentate but with midvein barely perceptibly thickened at tip; emergent lvs often much > 5 mm broad; sts mostly 5–40 cm; different-lvd w.; 2 vars.　　6 C. heterophylla Pursh
　　　　　　　　　　a1 Fr gen < 0.9 mm; chiefly e US; occ Wn to Cal and S Am (C. austini, C. deflexa var.. a.)　　　　　var. heterophylla
　　　　　　　　　　a2 Fr gen > 0.9 mm; chiefly w US; Greenl (C. b., C. palustris var. b.)　　　　　var. bolanderi (Hegelm.) Fassett
　　　　　　　　6b Fr gen subrotund or oblong in outline; submerged lvs with midvein much-thickened and slightly protruding at tip; lvs all < 5 mm broad; sts scarcely 5 cm; gen submerged, when emergent lvs not strongly dimorphic
　　　　　　　　　　7a Fr wingless, oblong, ca 1–1.1 × 0.6–0.9 mm; carpels scarcely grooved along the outer edges; common in e N Am, and known from Alas, Wn, and Utah; two-edged w. 7 C. anceps Fern.
　　　　　　　　　　7b Fr narrowly wing-margined, subrotund, 1–1.5 mm broad; carpels grooved along the outer edges; mainly Cal, but reported also from Linn Co, Ore　　8 C. trochlearis Fassett

EMPETRACEAE　Crowberry Family

Fls axillary (ours), reg, hypog, apet, ☿ to ♂ ♀ or ♂, ♀; perianth segms 2–6, inner ones ± petaloid; stamens 2–4; pistil 2–9-carpellary and -celled; style 1; fr baccate, the "seeds" 2–9, with a hardened and stonelike surrounding shell; low evergreen shrubs with alt or whorled, entire, linear, heathlike lvs jointed between the bl and petiole.

Empetrum L.　Crowberry

Fls small, purplish; perianth apparently consisting of 9 segms; fr separating when ripe into 6–9 1-seeded pyrenes. (Gr en, upon, and petros, rock, the ancient Gr name for the pl).

E. nigrum L. Low spreading shrub up to 1.5 (3) dm, brs ± lanate; lvs subterete, some alt, some whorled in 4's, 4–8 mm, glandular-puberulent, revolute-margined, grooved beneath; fls ca 3 mm, closely subtended by ca 3 chaffey bracts similar to and almost as large as the sepals; inner (2) 3 perianth segms brownish-purple; stamens mostly 3; fr globose, ca 4–5 mm, purplish or black; stigma peltate, with 6–9 short lobes; ± circumpolar, s in Cas and along the coast in Wn and Ore to n Cal, mostly on exposed rocky bluffs, but also in peat bogs; hard to grow, but an interesting pl.

LIMNANTHACEAE   Meadow-foam Family

Fls complete, reg, hypog, pedunculate in the lf axils; perianth 3–5 (6)-merous; sepals semidistinct; petals distinct, white or yellowish (ours); stamens = or 2 × the petals; pistil 2–5 (6)-carpellary, ovary deeply divided into 2–5 globose segms maturing into 1-seeded frs, the styles free except basally; ann, mostly glab, juicy herbs with alt, pinnatifid to pinnate lvs, mostly of moist areas.

1a Perianth 4–5 (6)-merous; petals (3) 4–12 mm   **Limnanthes**
1b Perianth gen 3-merous; petals scarcely 2 mm   **Floerkea**

Floerkea Willd.   False-mermaid

Fls minute; stamens 3–6; carpels 2–3; fr papillose-warty; lvs pinnate. (For H. G. Floerke, 1764–1825, German botanist).

F. proserpinacoides Willd. Glab, weak, decumbent to erect ann 2–10 cm; lfts oval to elliptic or narrowly oblanceolate, 3–20 mm; sepals ca 3 mm; petals white, ca half as long as the sepals; stamens 3 (4), alt with the petals and each adnate to a scalelike gland, or twice as many and those opp petals without gland; carpels mostly 2 (3), separating at maturity; fr ca 2.5 mm; gen in wet places, esp under shrubs; BC to Cal, e through RMS to Atl coast *(F. occidentalis)*.

Limnanthes R. Br.   Meadow-foam

Fls rather showy; petals 4–5 (6), white or yellowish; stamens 8 or 10; carpels 4–5; lvs pinnately dissected. (Gr *limne*, marsh, and *anthos*, flower, referring to the habitat).

1a Fls 4-merous; petals scarcely 5 mm; rare and seldom collected; known
   only from Victoria, VI; Macoun's m.   1 **L. macounii** Trèl.
1b Fls 5-merous; petals 8–15 mm; reported from VI, but prob an escape, as pl
   not known natively from n of Douglas Co, Ore; Douglas' m.
                                                 2 **L. douglasii** R. Br.

ANACARDIACEAE   Sumac Family

Fls pan or thyrsoid, small, reg, mostly ♂ to ♂ ♀, ♂ ♂; calyx 5 (3–7)-parted; petals as many as sepals or lacking; stamens mostly = or 2× sepals (fewer or more numerous), borne on the outer edge of a perig disc; pistil gen 3-carpellary and 3-styled; ovary mostly 1-celled and 1-seeded; fr gen drupe; shrubs (ours) with alt, often pinnate (trifoliolate or sometimes entire), exstip lvs and acrid or milky juice.

Rhus L.   Sumac

Pl mostly ♂ ♀, ♂ ♂; fertile stamens 5; disc flat, lobed, fleshy; shrubs or woody vines with pinnately 3–∞-foliolate lvs. (Classical name for the pl). *(Philostemon, Schmaltzia, Sumac, Toxicodendron).* Lvs of most of the spp. color highly in the fall, esp *R. glabra*, hence pl often of hort value.

1a Fr strongly reddish-hairy; petals pilose on inner surface; lfts sometimes
   5–29
  2a Lfts 7–29; shrubs 1–3 m; fr reddish, 4–5 mm; e Cas, BC to Nev and
     Mex, e to NH and Ga; w or smooth s. *(R. occidentalis)*; widely rhizoma-
     tous and often "weedy," but a valued orn   1 **R. glabra** L.
  2b Lfts 3–5; shrubs to 2 m; fr reddish-orange, 6–8 mm; chiefly e RM, Alta
     to Mex, w to Ida and se Ore and Cal, e to Ia; lemonade s., smooth s.,
     squawbush, skunkbush *(Schmaltzia quinata, S. oxyacanthoides, R. o., S.
     pubescens, R. osterhoutii)*   2 **R. trilobata** Nutt.
1b Fr glab or subglab, white or yellowish; petals glab
  3a Lfts acute or acuminate; fr ca 4 mm; shrub up to 2 m, sometimes scan-
     dent; e Cas, c Wn and Ore, e to Atl coast and s to Mex; poison ivy,
     poison oak *(R. rydbergii, R. toxicodendron of auth., T. coriaceum, T.
     hesperium)*   3 **R. radicans** L.
  3b Lfts rounded, obtuse, or very abruptly acutish; fr ca 5 mm; shrub 1–2 m,
     sometimes vining and up to 15 m; mainly w Cas, Wn to Cal, e in CRG to
     Klickitat Co, Wn; poison oak, poison-ivy *(T. lobadioides)*
                                                 4 **R. diversiloba** T. & G.

CELASTRACEAE   Staff-tree Family

Fls single or cymose, small, greenish to red, reg, ♂ to ♀ ♂, ♀ ♀; sepals 4–5 (6), connate basally; petals as many as the sepals (rarely lacking); stamens = or 2× sepals and opp them, borne at the outer edge of a flattened or cuplike disc; pistil 2–5-carpellary; ovary 2–5-celled; style 1; fr a drupe, caps, samara, or follicle; seeds often with showy aril; shrubs (ours), vines, or trees with mostly opp simple lvs.

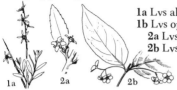

1a Lvs alt, entire; petals whitish, clawed; brs spinose at the tip **Glossopetalon**
1b Lvs opp, serrate; petals ± reddish, not clawed; brs not spinose
    2a Lvs persistent, 1–3 cm; fls 4-merous   **Pachistima**
    2b Lvs deciduous, gen much > 3 cm; fls gen 5-merous   **Euonymus**

Euonymus L.   Wahoo

Fls ♂; fr a 4–5-celled leathery caps; seeds arillate; shrubs or small trees with opp, deciduous (ours) lvs. (Gr *eu*, good, and *onoma*, name).

E. occidentalis Nutt. Western w., burning bush. Straggly shrub to 5 m; brs glab, striate; lvs thin, oblong-lanceolate, serrate, acuminate, 5–10 cm; stips minute or lacking; fls in mostly 3-fld axillary cymes; sepals rounded; petals ca 5 mm, greenish- and purple-mottled to reddish-purple; stamens 5, borne at the edge of a flattened disc partially surrounding the ovary, anthers transversely dehiscent, much > the filaments; seeds reddish-arillate; in woods w Cas, Lewis Co, Wn, to c Cal (*E. atropurpureus*); of little hort imp.

Glossopetalon Gray   Green-bush

Fls mostly ♂ or (ours) ♀ ♂ ♀, small, axillary; perianth 5 (4–6)-merous; stamens = or 2 × petals, inserted under the outer edge of a small, lobed disc; pistil 1–3-carpellary, ovary 1-celled, superior; fr a leathery follicle; greenish-barked shrubs with spinescent, angled and grooved sts and alt deciduous lvs. (Gr *glossa*, tongue, and *petalon*, petal, the petals narrow). *(Forsellesia)*.

G. nevadense Gray. Spiny g. Glab, erect to spreading, freely brd, glab shrub 1–3 m; brs spinose, strongly grooved; lvs entire, grayish-green, 3–15 mm, the slender petioles art with a fleshy, cushionlike, often purplish base; petals slender-clawed, deciduous, white, 4–9 mm; stamens gen opp and = in no to the sepals in the ♂ fls; fr ca 3 mm, grooved lengthwise, gen 1-seeded; rocky canyon walls; wc Ida to Cal, Ariz, and Utah; ours the var. *stipuliferum* (St. John) Hitchc.

Pachistima Raf.

Fls complete, reg, axillary, 4-merous except the pistil; ovary 2-carpellary and 2-celled; fr caps, 1–2-seeded; evergreen shrubs with leathery opp lvs. (Gr *pachus*, thick, and *stigma*, stigma).

P. myrsinites (Pursh) Raf. Mt-box, mt-lover, myrtle boxwood, Ore boxwood. Glab shrub 2–6 (10) dm; lvs glossy, oblong-lanceolate to oblanceolate, serrate, 1–3 cm; fls 3–4 mm broad; petals maroon; stamens borne at the outer edge of a flattened disc, the ovary sunken in the disc; caps 3–4 mm; seeds dark brown, ca 2/3 covered by a thin, lacerate, whitish aril; BC to Cal, e to RMS; mostly midmont, but to near the coast in w Wn; one of the nicest orn shrubs in nw; easily grown.

ACERACEAE   Maple Family

Fls corymbose to pan, ♂ to ♀, ♂, reg; sepals 4–5, distinct or ± connate; petals 4–5 (0), distinct; stamens 4–10 (mostly 8), inserted on and mostly exterior to a lobed disc; pistil 2-carpellary; styles 1 or 2; stigmas 2; ovary superior, 2-celled; fr with mostly 2 divergent wings (double samara), the halves ultimately separating, each 1-seeded; deciduous shrubs or trees with opp, palmately lobed to pinnately compound lvs.

Acer L.   Maple

Fls in axillary pans, racemes, or corymbs; petals small, sometimes lacking; lvs mostly palmately lobed, occ 3–5-foliolate. (L name for maple).

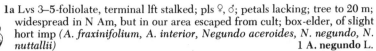

1a Lvs 3–5-foliolate, terminal lft stalked; pls ♀, ♂; petals lacking; tree to 20 m; widespread in N Am, but in our area escaped from cult; box-elder, of slight hort imp (*A. fraxinifolium, A. interior, Negundo aceroides, N. negundo, N. nuttallii*)   1 **A. negundo** L.

1b Lvs simple, palmately lobed or occ trifoliolate but lfts not stalked; petals gen present

  2a Fls 10–50, racemose or pan; fr bristly-hairy; large trees to 30 m; larger lvs > 15 cm broad; Alas to Cal, mostly w Cas, also in wc Ida; big-lf m., common m., Ore m.; little hort value   2 **A. macrophyllum** Pursh

  2b Fls mostly < 10, corymbose or umbellate; fr glab or sparsely pilose; pl often shrublike; larger lvs gen < 15 cm broad

    3a Lvs 7–9-lobed; stamens borne inside the disc; sepals red; shrub or small tree 1–8 m; Alas to n Cal, from e side Cas to coast; vine m., of great hort value, colors highly in fall   3 **A. circinatum** Pursh

    3b Lvs 3–5-lobed; stamens gen outside the disc; sepals green

      4a Sinuses of lvs narrowly acute, lobes ovate, sharply and finely bidentate; fls corymbose; petals gen present; shrub or small tree 1–10 m; less hort imp than vine m., but good; 2 vars.   4 **A. glabrum** Torr.

      a1 St reddish; lvs shallowly lobed, mostly > 6 cm broad; coastal and e to Ida and Mont; Douglas' m. (*A. d.*)

          var. **douglasii** (Hook.) Dippel

      a2 St mostly grayish; lvs more deeply lobed, mostly < 6 cm wide; mts of Mont and Ida to NM, e to Neb; RM m. (*A. subserratum*)

          var. **glabrum**

      4b Sinuses of lvs broad, lobes oblong, coarsely and sparingly sinuate-toothed; fls umbellate; petals lacking; pl 3–5 (6) m; e Ida, sc Mont, and w Wyo to Tex and Ariz; bigtooth maple (*Saccharodendron g.*)   5 **A. grandidentatum** Nutt.

BALSAMINACEAE   Balsam or Touch-me-not Family

Fls axillary, strongly irreg, complete or (some) cleistogamous and much-reduced, hypog; sepals 3 (5), petaloid, 1 gen much enlarged, saccate, and often spurred; petals 5, distinct or each lateral pair ± connate and apparently single and bilobed; stamens 5, often connate; pistil 5-carpellary, style single, very short, stigma 5-lobed; fr caps, elastically dehiscent (explosive), with axile placentation; our pls herbaceous and succulent, with simple, alt to whorled lvs.

Impatiens L.   Balsam; Touch-me-not; Jewelweed

Fls showy, ours yellow or orange to purplish or maroon; sepals apparently 4 and petals only 2, but sepals really 3 (1 saccate and gen spurred, other 2 much smaller) and petals 5, upper one gen notched at the tip, others connate in 2 pairs, 1 of each pair much the smaller; stamens mostly ± connate; pl of moist, gen shaded areas; lvs exstip. (L, meaning impatient, referring to the explosive frs).

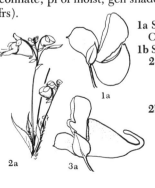

1a Saccate sepal not spurred; fls pale yellow to orange, unspotted, 1–2 cm; w Cas, Wn and Ore, and se BC to Mont; spurless b.   1 **I. ecalcarata** Blank.

1b Saccate sepal spurred; fls often spotted or > 2 cm or both

  2a Lvs in part opp or whorled, finely and closely serrate with 40 or more teeth; spur gen < 6 mm; fls purplish-pink or red, 2–3 cm; Asiatic orn, occ estab in BC and w Wn; policeman's helmet (*I. roylei*)

          2 **I. glandulifera** Royle

  2b Lvs alt; fls mostly yellowish or orange

    3a Fls scarcely 20 (mostly 10–15) mm exclusive of the recurved portion of the spur, orange, unspotted; extreme se Wn and n Ida to se BC and w Mont; orange b.   3 **I. aurella** Rydb.

    3b Fls mostly much > 20 (20–35) mm, yellow to orange, but ± spotted or mottled

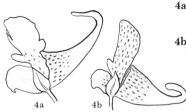

4a

4b

4a Fls 25–35 mm, yellow, sparingly flecked with brownish-purple on saccate petal esp; Alas and c BC to nw Wn; Eurasia; touch-me-not (*I. occidentalis*)   4 **I. noli-tangere** L.

4b Fls 20–25 (30) mm, orange, brown-spotted or -blotched, esp on saccate petal and the lower of the other 2 pairs; along CR in Ore and in extreme nw Wn and adj BC; common in e US and Can; orange b.   5 **I. capensis** Meerb.

RHAMNACEAE   Buckthorn Family

Fls reg, ♂ to ♂ ♀; calyx 5 (4)-lobed; petals 5 (4), rarely wanting, often clawed and with hooded bl; stamens as many as the sepals and alt with them (opp petals), borne at outer edge of a perig disc; pistil 2–4-carpellary, ovary superior or partially covered by the disc; fr caps, berry, or drupe, mostly with 1 (2) seeds per cell; ours deciduous to evergreen shrubs or small trees with alt or opp, simple and gen stip lvs.

1a   1b

1a Fr bluish-black, fleshy; fls greenish; petals short-clawed or lacking **Rhamnus**

1b Fr mostly green or brown caps, but often with hardened walls; fls white to blue; petals long-clawed **Ceanothus**

Ceanothus L.   Buckbrush; Buckthorn; Ceanothus; Wild-lilac

Fls in umbels or pans, showy en masse, 5-merous; fr separating into 3 1-seeded, dehiscent carpels; prostrate to erect shrubs, sometimes the brs spinose. (Gr pl name, the meaning uncertain). Most spp. imp orn shrubs, but our climate is a little too cold for most, and our native spp. are not the best.

2a

2b

3a

4b   4a

1a Lvs opp or whorled, persistent
  2a Pl prostrate; brs neither rigid nor spinose; lvs toothed, 1–2.5 cm; fls bluish to nearly white; e slope Cas, Yakima Co, Wn, to Cal and w Nev; squaw carpet, mahala mat, most attractive!   1 **C. prostratus** Benth.
  2b Pl erect; brs rigid, spinose; lvs rarely toothed, 0.5–2 cm; fls white; foothills, in dry areas, WV, Ore, to Baja Cal; common buckbrush, sedge-lf b., narrow-lf b.   2 **C. cuneatus** (Hook.) T. & G.
1b Lvs alt
  3a Lvs persistent, glutinous and shining on the upper surface, bl 5–10 cm; stips ca 1 mm, persistent; fls white; mt balm, sticky-laurel, greasewood, tobacco-brush; 2 vars.   3 **C. velutinus** Dougl.
    a1 Lvs glab beneath (at least on veins); w Cas, BC to n Cal
                var. **laevigatus** (Hook.) T. & G.
    a2 Lvs finely puberulent beneath; e Cas, BC to Cal and Nev, e to SD and Colo   var. **velutinus**
  3b Lvs deciduous, seldom either glutinous or shining on the upper surface; stips linear, 3–8 mm, deciduous; fls white to blue
    4a Caps crestless; lvs glandular-serrulate; fls white; both sides Cas, BC to Cal, e to Ida and w Mont; redstem c., Ore tea-tree (*C. oreganus*)   4 **C. sanguineus** Pursh
    4b Caps crested on back; lvs entire; fls white or blue; e Cas, Wn to Baja Cal, e to NM; deerbrush (*C. macrothyrsus, C. peduncularis*)   5 **C. integerrimus** H. & A.

Rhamnus L.   Cascara; Buckthorn

Fls in small axillary clusters, greenish-yellow, ♂ to ♂ ♀, ♂ ♂; calyx 4–5-lobed, camp, gland-lined at base; petals = sepals (lacking), scarcely hooded; ovary 2–4-celled; fr berrylike; shrubs or small trees with alt, deciduous, simple, prominently veined lvs. (Gr name for the pl).

1a   1b

1a Fls 8–40, in pedunculate umbels; hypan cup-shaped; petals present; main lateral veins of lvs gen > 8 per side; shrub or small tree up to 10 m; BC to Cal, mostly w Cas, but e to Ida and w Mont; bark widely collected for medicinal value; cascara, chittam bark   1 **R. purshiana** DC.

1b Fls 2–5 in sessile umbels; hypan discoid or saucer-shaped; petals gen absent; main lateral veins of lvs gen not > 8 per side; shrub 0.5–1.5 m; mont, e Cas, BC to Cal, e to Que and Me, and to Ida, Mont, and Wyo; alder b.   2 **R. alnifolia** L'Her.

VITACEAE   Grape Family

Fls small, reg, perig, cymose to pan, complete to ♂ ♀; calyx mostly 4–5-lobed; petals as many as sepals, white or greenish, free or ± connate and sometimes calyptrate; stamens as many as petals and opp them, inserted at the edge of an entire to lobed disc; pistil 2 (3–6)-carpellary; ovary 2 (3–6)-celled, superior; fr baccate; vining and tendril-bearing shrubs (ours) with alt, simple to compound, stip, deciduous or evergreen lvs and (often) swollen nodes.

Vitis L.   Grape -vine

Fls greenish; petals calyptrate; seeds 2–4; tendril-bearing vines with large, palmately-veined and -lobed lvs. (Ancient L name for the grape).

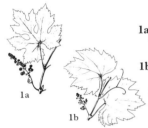

1a Lvs 5-lobed > half their length, tomentose beneath when young, floccose to glabrate in age; Eurasian, long-cult; known in our area from SRC, se Wn, and possibly from near Portland, Ore; European g.   1 **V. vinifera** L.
1b Lvs lobed < half their length, often glab or merely pubescent, strongly serrate-dentate; native in e US; sometimes grown for brilliance of the fall foliage: known from sc Mont and from near Portland, Ore; riverbank g.
2 **V. riparia** Michx.

MALVACEAE   Mallow Family

Fls mostly reg, hypog, ⚥ to ± ♂, ♀, mostly rather showy, white or yellow to pink or lavender; sepals 5, distinct to ± connate, sometimes bracteolate at base; petals 5, short-clawed, inserted on staminal tube slightly above recep; stamens ∝, monadelphous, forming an elongate tube, anthers separate or in groups of 2–6, reniform, 1-celled; pistil 2–∝-carpellary, ovary superior, 2–∝ -celled, segms gen ± loosely connate around the central axis; fr caps or a schizocarp (rarely baccate or samaroid); seeds 1–several per cell; herbs or shrubs (ours) with alt, stip, entire to palmately lobed lvs, gen stellate-pubescent, often with mucilaginous sap.

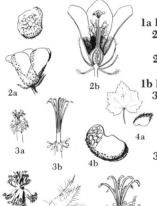

1a Petals yellow
  2a Carpels 1-seeded; petals ca 1 cm, not black-spotted within; calyx not becoming inflated and papery in fr   **Sida**
  2b Carpels 2–several-seeded; petals ca 2 cm, with a purplish-black basal spot within; calyx becoming papery and inflated in fr   **Hibiscus**
1b Petals white, pink, or lavender to reddish
  3a Stigmas terminal and capitate
    4a Petals mostly > 2 cm; lvs large and ± like those of the grape; carpels smooth on the sides, dehiscent full length   **Iliamna**
    4b Petals mostly < 2 cm; lvs smaller, often deeply lobed to incised; carpels strongly reticulate on the sides on the basal half   **Sphaeralcea**
  3b Stigmas extending full length of the style brs on the inner surface
    5a Stamens freed from top of the staminal tube in 2–4 series of connate groups of 2–6 each; calyx gen ebracteolate   **Sidalcea**
    5b Stamens freed singly or in pairs from the upper third of the staminal tube; calyx basally bracteolate   **Malva**

Hibiscus L.   Rosemallow

Stigmas 5, capitate; fr caps, 5-celled; herbs or shrubs with rather showy fls. (Ancient Gr name for mallow).

H. trionum L. Flower-of-an-hour, modesty. Hairy ann 3–6 dm; calyx subtended at base by an invol of several bracts, becoming distinctly 5-winged, papery, and inflated in fr; petals yellowish but with a large purplish-black spot near the base; filaments freed separately from most of the length of the staminal tube; native to Europe; often cult and sometimes escaped; apparently estab in WV, Ore.

Iliamna Greene   Globemallow

Fls racemose, pink to lavender, rather showy; calyx tribracteolate; petals ciliate-clawed; filaments freed separately from the upper 3/4 of the staminal tube; carpels fully dehiscent, hairy with long stiff hairs and small stellae, smooth on the sides; seeds 2–4 per carpel; herbaceous, stellate per with (3) 5–9-lobed lvs. (Gr, significance obscure).

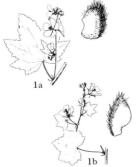

1a Sepals < 1 cm; peds stout, mostly < 1 cm; lvs 3–7-lobed, 5–15 cm; sepals obtuse (acute), 3–5 mm; carpels somewhat rounded at the tip; e Cas, BC to Ore, e to Mont and Colo; mostly in canyons or foothills; streambank g.; 2 vars.   1 **I. rivularis** (Dougl.) Greene
    a1 Lvs 3–5-lobed, truncate-based; e Ida to YNP (*Sphaeralcea r.* var. *d.*)
                                     var. **diversa** (Nels.) Hitchc.
    a2 Lvs 5–7-lobed, cordate-based; range of the sp. (*Malva r., Sphaeralcea r., Phymosia r., S. acerifolia*)   var. **rivularis**
1b Sepals ca 1.5 cm; peds slender, some gen > 1 cm; carpels obtuse or acute at the tip; sagebr-covered foothills to ponderosa pine woodl; lower levels e Cas, Kittitas Co to Chelan and Douglas cos, Wn; longsepal g. (*Sphaeralcea l., Phymosia l.*)   2 **I. longisepala** (Torr.) Wiggins

Malva L.   Mallow; Cheeseweed; Cheeses

Fls in axillary clusters or subterminal pan, inconspicuous to showy; calyx tribracteolate basally; petals whitish to lavender, retuse; stamens freed singly or in pairs; style brs 10–15, stigmatic most of length; carpels 10–15, 1-seeded, smooth to corrugated, glab to hairy, separating when ripe; ann to per herbs with long-petiolate, stip lvs, bl ovate to reniform, shallowly lobed to dissected; pubescence simple to stellate. (Ancient Gr name, meaning soft, the lvs supposedly being soothing to the skin).

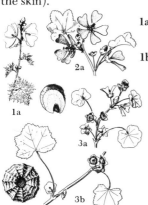

1a Upper st lvs dissected into linear segms; fls showy, petals 2–3 cm; European, escape from gardens, occ w Cas in Wn and Ore; musk m.
                                              1 **M. moschata** L.
1b Upper st lvs shallowly lobed; fls mostly not showy; petals mostly < 2 cm; intro European weeds
    2a Petals (1) 1.5–2 cm; calyx bracteoles ovate or oblong; common in much of N Am; in our area mostly w Cas; common m.   2 **M. sylvestris** L.
    2b Petals mostly < 1.5 cm; bracteoles linear
        3a Carpels rounded and smooth (except for some puberulence) on back; petals 2–4× as long as the calyx; common in much of US; dwarf m.
                                  3 **M. neglecta** Wallr.
        3b Carpels flattened and wrinkled on back; petals < to only slightly > calyx; rather common; cheeseweed, alkali m.   4 **M. parviflora** L.

Sida L.

Fls axillary, complete; calyx (ours) with 1–3 linear bracteoles; filaments freed singly from the staminal tube; carpels 1-seeded, ours indehiscent; densely stellate-pubescent per herbs with linear, deciduous stips. (Gr name, the meaning obscure).

S. hederacea (Dougl.) Torr. Alkali-mallow. Prostrate to erect, grayish-stellate, 1.5–4 dm; lf bls reniform, 1.5–3 cm; petals yellow, ca 1 cm, reddish-purple and stellate-pubescent on the edge exposed in the bud; carpels 5–12, thin but strongly reticulate on the contacting faces, stellate on the back; sandy (often alkaline) soil e Cas; Okanogan Co, Wn, to Cal, Ariz, Okla, and Mex (*Malva h., M. plicata, S. obliqua*).

Sidalcea Gray   Checker-mallow; Sidalcea

Fls white to deep pink or pinkish-lavender, racemose, ♀ to ♂, ♀, the ♂ fls largest; petals ciliate on the claws; stamens freed in 2–3 series, filaments of the outer 2 series mostly connate in groups of 2–6, the uppermost series gen with 10 partially connate pairs; carpels 5–10, tardily separating, 1-seeded; ann or per (ours) herbs, gen stellate-pubescent and often ± hirsute. (Gr *Sida*, plus *Alcea*, both names of other genera).

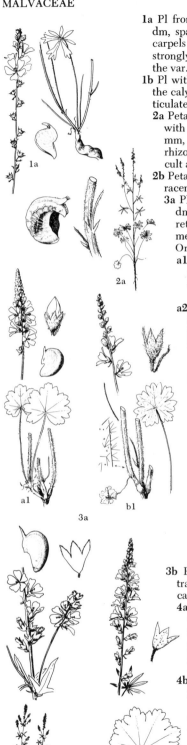

1a Pl from enlarged, ± fleshy, tap or fascicled root, not rhizomatous, 2–7 dm, sparsely hirsute, esp on the calyx; racemes elongate and loosely fld; carpels nearly smooth, ca 2.5 mm, with curved beak ca 0.6 mm; moist, strongly alkaline areas; ec Ore to Wyo, s to Cal, NM, and Mex; RM c.; ours the var. *crenulata* (Nels.) Hitchc. *(S. c.)*   1 **S. neomexicana** Gray

1b Pl with woody roots, often rhizomatous, gen strongly stellate, at least on the calyx; racemes often congested and spicate; carpels mostly strongly reticulate-alveolate

   2a Petals white to pale pink or pinkish-orchid; lower part of st gen hirsute with simple (forked) hairs; racemes elongate, loosely fld; carpels ca 3.5 mm, rather prominently reticulate-pitted on the sides; pl short-rhizomatous, 5–20 dm; dry fields, WV, Ore, from Portland s; often cult and occ escaped elsewhere; meadow s.   2 **S. campestris** Greene

   2b Petals gen deep pink to pinkish-lavender; lower part of st often stellate; racemes often closely fld and spikelike; pl sometimes nonrhizomatous

      3a Pl nonrhizomatous, often with decumbent but non-rooting brs, 2–15 dm; herbage often strongly glaucous; carpels from nearly smooth to reticulate-alveolate on the sides; mostly e Cas, from sagebr plains to meadows and ponderosa pine woodl; c Wn to Cal, e to Wyo and Utah; Ore c.; 5 vars.   3 **S. oregana** (Nutt.) Gray

         a1 Hairs of lower part of the st mostly simple; fls crowded and spicate, peds mostly 1–2 mm; calyx ca 5 mm at anthesis; carpels nearly or quite smooth; c Ore to Sierran Cal *(Callirhoe s., S. s., S. o. ssp. s.)*   var. **spicata** Jeps.

         a2 Hairs of lower part of st mostly stellate; fls not spicate nor crowded, at least after anthesis, peds ave 3 mm; calyx gen > 5 mm at anthesis; carpels reticulate-rugose

            b1 St pubescent near base with coarse, spreading, simple to stellate hairs; calyx mostly much > 6 mm; e Wn and adj Ida to ne Ore   var. **procera** Hitchc.

            b2 St variously pubescent or glab, if stellate the hairs mostly appressed; calyx various

               c1 Lvs glab or subglab, thick and fleshy; calyx sparsely stellate, lobes subglab on back but strongly ciliate with simple to stellate hairs 0.5–1 mm; sts glab at base or sparsely hairy with large, 4-rayed, appressed hairs; Wen Mts, Wn   var. **calva** Hitchc.

               c2 Lvs gen stellate; calyx evenly stellate, but lobes sometimes with cilia up to 0.5 mm; sts variously pubescent to glab

                  d1 Sts glab or sparsely stellate at base with 4–9-rayed, appressed hairs, often strongly glaucous; petals of ♂ fls gen > 15 mm; sc Ore to Wyo, Utah, and ne Cal *(S. nervata)*   var. **oregana**

                  d2 Sts ± strongly pubescent at base, nonglaucous or if glaucous then petals of ♂ fls mostly < 15 mm; c and s Ore to n Cal   var. **maxima** (Peck) Hitchc.

      3b Pl rhizomatous but rhizomes often short and thick; st sometimes trailing and rather freely rooting; herbage not particularly glaucous; carpels sometimes pitted; mostly w Cas

         4a St ± fistulose, 5–15 dm, glab or only sparsely pubescent at base with short, simple to forked hairs; racemes compound, closely fld and spikelike; calyx glab or subglab except for the cilia, purplish, much-enlarged in fr; carpels ca 4 mm, nearly or quite smooth, with beak gen at least 1 mm; coastal, gen on or near tidel; VI to Umpqua R, Ore; Henderson's c.   4 **S. hendersonii** Wats.

         4b St rarely fistulose (except occ in nos 6 and 7), mostly rather densely pubescent throughout; racemes often open; calyx gen strongly hairy; carpels sometimes < 3.5 mm, often prominently reticulate-alveolate, the beak mostly < 1 mm

            5a Calyx 4–6 mm, subglab to rather thickly pubescent with tiny stellae, gen purplish; petals 5–15 mm; carpels lightly reticulate on sides, beak < 0.5 mm; st glab to sparsely hirsute with short, appressed, simple hairs; gravelly, well-drained soil; WV, from Portland to Salem, Ore, w to e Tillamook Co; Nelson's c.   5 **S. nelsoniana** Piper

5b Calyx mostly > 6 mm, gen densely hairy; petals frequently > 15 mm; carpels often coarsely reticulate, beak commonly at least 0.5 mm; st gen very hairy

    6a St hirsute at base with rather stiff, simple or forked hairs 1–2.5 mm; racemes spikelike, congested, gen < 8 cm; calyx 9–15 mm, enlarged considerably in fr, finely stellate but also strongly hirsute with hairs 1–2 mm; carpels prominently reticulate-alveolate; coastal mts to bluffs along the ocean, but never on tideflats, n Lincoln to Tillamook and Clatsop cos, Ore, and Clark and Lewis cos, Wn; hairy-std c.   **6 S. hirtipes** Hitchc.

    6b St gen stellate; racemes elongate; calyx < 9 mm, or pl otherwise not as above

        7a Calyx lobes widened above base and ± ovate-lanceolate, prominently veined; rhizomes gen short and thick; st (4) 5–18 dm, often somewhat fistulose; racemes compounded, spikelike, closely fld; petals often > 5, truncate or erose, 10–18 mm; carpels ca 3 mm, smooth or only slightly reticulate on the sides; open fields, mostly in rather heavy soil, valleys of Coquille and Umpqua rivers, Ore; barely reaching our area; Cusick's c.   **7 S. cusickii** Piper

        7b Calyx lobes not widened above base, gen tapered rather evenly to the tip; st mostly < 10 dm, not fistulose, gen trailing and freely rooting at base, often truly rhizomatous; racemes commonly rather open; petals 5, rounded but retuse, 15–30 mm; carpels gen rather strongly reticulate-alveolate; fields and roadsides to grassy hillsides and lower mts, esp in moist meadows, WV to s Ore; rose c. (*S. malviflora* ssp. *v.*)   **8 S. virgata** Howell

Sphaeralcea St. Hil.   Globe-mallow

Fls racemose, complete, pink to red; calyx gen 3 (1–2)-bracteolate at base; filaments freed from staminal tube individually; carpels 8–12, 1–2-seeded, stellate-pubescent on the back, rugose-reticulate and indehiscent on the lower 1/3–4/5, the upper part dehiscent and smooth on the sides; stellate-pubescent per herbs, sometimes slightly frutescent, with thickish, simple to divided lvs. (Gr *sphaera*, sphere, and *alcea*, mallow, in allusion to the globose fr).

1a Lvs crenate to lobed < halfway to the midvein; calyx bracteolate; carpels gen 1-seeded, rugose-reticulate only on the lower, indehiscent third; petals 1–2 cm, apricot-pink to reddish; pl 2–8 dm; des plains to lower mts; sc BC to w Mont, Cal, and Utah; Munro's g., white-std g. (*Malva m., Nuttallia m., Malvastrum m., Malveopsis m.*)   **1 S. munroana** (Dougl.) Spach

1b Lvs divided nearly to the midvein; calyx often ebracteolate; seeds sometimes 2 per carpel

    2a Calyx gen ebracteolate; carpels 1-seeded, > 2/3 of the lateral faces rugose-reticulate; fls mostly in simple racemes; pl low and spreading, rhizomatous, 1–2 dm; petals 1–2 cm, rusty-red; plains region, extending into Granite and Powell cos, Mont; red g. (*Cristaria c., Malva c., Sida c., Malvastrum c.*)   **2 S. coccinea** (Pursh) Rydb.

    2b Calyx gen tribracteolate; carpels mostly 2-seeded, rugose-reticulate only on the lower half of the lateral faces; infl thyrsoid; pl erect, 3–7 (10) dm; petals 1.5–2 cm, reddish; c Ida to sc Wn, s to Utah and Nev; from open des to lower mts; gooseberry-lvd g. (*Sida g., Malvastrum g.*)   **3 S. grossulariifolia** (H. & A.) Rydb.

HYPERICACEAE   St. John's-wort Family

Fls cymose (rarely single), mostly showy, yellow (ours) or white, complete, reg, hypog; sepals 5 (4), mostly distinct; petals 5 (4), distinct, often glandular at base; stamens 15–100, filaments free or slightly connate into 3–8 separate groups; pistil 3–5-carpellary, styles distinct, ovary 1-celled with 3–5 parietal placentae or sometimes 3–5-celled; fr caps or berry; ann or per herbs with opp or whorled, simple, sessile (ours) or petiolate, finely translucent-dotted or black-dotted lvs.

Hypericum L.   St. John's-wort

Foliage and perianth mostly blackish- or purplish-dotted along the margins; sepals and petals 5, yellow; carpels and styles 3 in ours; fr caps, septicidal; ours per, glab, sessile-lvd herbs. (Gr name for the pl, the meaning obscure).

1a Petals gen < 6 mm, scarcely > the sepals, not black-dotted along the margins; stamens < 50, free or only slightly connate at base; caps 1-celled
  2a St erect, 1–5 dm; lvs 1–3.5 cm; wet ground, mostly in e Can and US, but collected in 1891 at Green Lake, Seattle, and recently in Pend Oreille Co, Wn   **1 H. majus** (Gray) Britt.
  2b St procumbent and matted, some upright and 0.5–1.5 dm; lvs 5–15 mm; moist ground, coastal to mont; BC to Baja Cal, e to Mont; bog s. (*H. bryophytum, H. tapetoides*)   **2 H. anagalloides** C. & S.
1b Petals either much > 6 mm (and gen much > the sepals) or blackish-glandular along the margins; stamens gen > 50, basally connate into 3 (4–5) groups
  3a Sepals linear-lanceolate, 3–5 times as long as broad, mostly acute; lvs lanceolate to obovate-oblanceolate or narrowly spatulate-oblanceolate; seeds brownish, ca 1.25 mm, strongly pitted in longitudinal rows, not striate; European, now widely intro and gen a serious pest on wastel and drier areas throughout much of US; esp common from Tacoma, Wn, to c Cal; Klamath weed, common s.   **3 H. perforatum** L.
  3b Sepals triangular to ovate-lanceolate, < 3× as long as broad, rounded to acute; lvs ovate-oblong to ovate; seeds yellowish, ca 0.8 mm, indistinctly reticulate-alveolate but not pitted, ± striate lengthwise; moist places, coastal to subalp; BC to Baja Cal, e to Mont, Wyo, and Mex; western s.; 2 vars.   **4 H. formosum** H.B.K.
    a1 Pl sturdy, mostly > 2 dm, often brd; lowl to midmont (*H. s.*)
                                   var. **scouleri** (Hook.) Hitchc.
    a2 Pl rather slender, gen simple-std, mostly < 2 dm; subalp (*H. n.*)
                                   var. **nortoniae** (Jones) Hitchc.

ELATINACEAE   Waterwort Family

Fls axillary, solitary to cymose, inconspicuous, complete, hypog, many often cleistogamous; sepals and petals 2–5, distinct (sepals sometimes slightly connate); stamens = or 2× petals, distinct; pistil 2–5-carpellary, styles short, distinct, ovary 2–5-celled; fr caps; ann (ours) aquatic or terrestrial herbs with opp (whorled), stip, simple lvs.

1a Fls 5-merous; sepals keeled on back; sts ± fibrous or woody at base, glandular-puberulent   **Bergia**
1b Fls 2–4-merous; sepals not keeled; sts herbaceous throughout, glab   **Elatine**

Bergia L.   Bergia

Sepals cuspidate, strongly ribbed; seeds lightly striate lengthwise; prostrate to erect herbs with glandular-pectinate stips. (For Peter Jonas Bergius, Swedish botanist, 1730–1790).

B. texana (Hook.) Seubert. Texas b. St 0.3–3 dm, ± woody at base; lvs glandular-denticulate, 2–4 cm; fls 1–several in the lf axils, ped; petals < sepals; stamens mostly 10, many fls cleistogamous; wet ground, esp around vernal pools, lower CR in our area; Cal to Ill, s to Tex (*Merimea t., Elatine t.*).

Elatine L.   Waterwort; Mud-purslane

Fls minute, mostly solitary; caps 2–4-celled; seeds reg reticulate in longitudinal rows; tiny prostrate herbs with entire stips, growing on mud flats or at edge of ponds, freely rooting at nodes. (Gr, meaning fir-like).

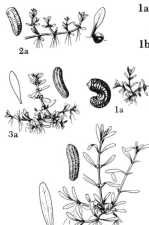

1a Caps 4-celled, ped; stamens 8; seeds curved at least to a 1/4 circle, pits 20–30 in each of ca 10 rows; Wn to Cal, e to Mont; Cal w. *(E. williamsii)*
                                                        1 E. californica Gray
1b Caps 2–3-celled, sessile; stamens 3–6; seeds only slightly curved, the pits often < 20 per row
  2a Pits of seeds mostly 10–15 per row; lvs rarely notched at tip; fls 2–3-merous; c Ore and s Cal, Colo, and c US; short-seeded w. *(Alsinastrum b., E. triandra* var. *b.)*   **2 E. brachysperma** Gray
  2b Pits of seeds mostly 18–27 per row; lvs often notched at tip; fls sometimes all 3-merous
    3a Lvs obovate, rounded at tip; sepals, petals, and stamens 3; mostly along Atl coast, possibly not in our area; scarcely to be told from the next; Am w. *(Peplis a., E. triandra* var. *a.)*
                                                        3 E. americana (Pursh) Arnott
    3b Lvs linear to narrowly oblanceolate, gen distinctly notched; sepals, petals, and stamens 2 or 3; gen in w US and s BC; Eurasia; three-stamen w.   **4 E. triandra** Schkuhr

TAMARICACEAE   Tamarisk Family

Fls small, reg, complete, in slender compound spikelike racemes; sepals and petals 4 or 5, distinct; stamens = or 2× petals, on a fleshy perig disc, filaments sometimes ± connate at base; pistil 3–5-carpellary, ovary superior, 1-celled with 3–5 parietal placentae; fr caps; shrubs or small trees with alt, scalelike, exstip lvs.

Tamarix L.   Tamarisk

Fls pinkish, only 2–4 mm, but showy en masse; shrubs or small trees of dry or saline areas. (Named for Tamaris River, Spain).

T. parviflora DC. Spreading shrub to 4 m; brs recurved or arching, slender; lvs overlapping, 4-ranked, 1–1.5 mm; fls 4-merous; European, estab in many moist spots in the des in w US; in our area in Ore along John Day R, Gilliam Co, and in Malheur Co. *T. gallica L.* and *T. pentandra* Pall. have been cited (but not seen) for our area; both are 5-merous, although otherwise very similar to *T. parviflora.*

VIOLACEAE   Violet Family

Fls complete, hypog to ± perig, 5-merous except the pistil, ours irreg, often cleistogamous; sepals nearly or quite distinct; petals distinct, the lowermost largest and spurred (ours), the other 4 in 2 dissimilar pairs; stamens connivent around pistil, filaments very short or lacking; pistil 3-carpellary, ovary 1-celled with 3 parietal placentae (ours), style 1, stigma ± globose; fr 3-valved caps; ours ann or per herbs with alt, simple or dissected, prominently stip lvs.

Viola L.   Violet

Fls mostly rather showy, blue, violet, yellow, or white, but some often cleistogamous and not showy; stamen connectives prolonged past the anther sacs, closely investing the ovary; caps explosively dehiscent. (L name for the pl). *V. douglasii* Steud. has been collected in Jefferson and Wasco cos, Ore, although more common s of our area. It will "key" with *V. hallii,* but differs therefrom because the 2 upper petals are yellow (rather than purple) on the inner surface.

1a Pl ann; stips large, laciniate into 5–9 linear segms, the terminal segm often nearly as large as main bl; fls whitish or light yellow with bluish tinge; European, widely cult and often escaping, prob never persistent; wild pansy, cult pansy   **1 V. arvensis** Murr.
1b Pl per; stips gen small and neither lf- nor lflet-like
  2a Lvs compound or deeply dissected into linear or oblong segms
    3a Lf segms leathery, ± prominently 3-nerved, glab, ± elliptic; petals bicolored, upper pair deep reddish-violet, lower 3 pale to fairly deep lilac, with yellowish or whitish base and purple blotching or pencilling; sagebr flats and rocky hillsides where moist early in spring; e Cas, Okanogan Co, Wn, to se Ore; sagebr v., 3-nerved v., Rainier v., des pansy *(V. beckwithii* var. *t., V. chrysantha* var. *glaberrima, V. t.* var. *semialba)*   **2 V. trinervata** Howell

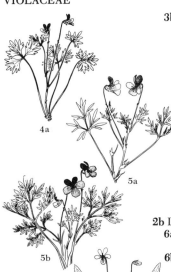

3b Lf segms not leathery, mostly 1-nerved, often pubescent, scarcely elliptic; petals various, lower 3 often mainly yellow

　4a Lf bl much broader than long, pedatifid; petals all yellowish on the inner surface, upper pair brownish-backed, lower 3 purple-pencilled, lateral pair sparsely clavate-bearded; chaparral or for, often under ponderosa pine; e Cas, from near Cle Elum, Kittitas Co, Wn, to Baja Cal; Shelton's v.　　3 **V. sheltonii** Torr.

　4b Lf bl mostly no broader than long, 2–3 times pinnatifid, upper petals sometimes purplish on inner surface

　　5a Pl glab or subglab; lower petals gen pale yellowish, purplish-penciled, upper pair purplish-blue on both surfaces; open woodl or light gravelly plains and foothills; WV, Ore, to nw Cal; Hall's.　　4 **V. hallii** Gray

　　5b Pl puberulent to strongly pubescent; lower petals mauve (white), yellowish only at base, purplish-pencilled, upper pair reddish-purple; sagebr hills and ponderosa pine woodl; ne Ore to Ida and Utah, s to Cal; Beckwith's v.　　5 **V. beckwithii** T. & G.

2b Lvs entire to crenate or serrate, neither compound nor dissected

　6a Petals mainly white, often with bluish or purplish shading, but never yellow　　**Group I**

　6b Petals mainly blue, violet, or yellow, rather than white

　　7a Petals bluish to purple, not yellow　　**Group II, lead 14a**

　　7b Petals partially or wholly yellow　　**Group III, lead 22a**

Group I

8a Lvs narrowly elliptic to elliptic-lanceolate, 0.5–1.5 (2) cm broad, 3–6 times as long, not at all cordate at base; pl 5–15 cm, stless or nearly so; fls (6) 8–11 mm; petals almost or quite beardless; bogs and moist meadows; native in c and e US, intro in Pierce and Pacific cos, Wn; lance-lvd v.　　6 **V. lanceolata** L.

8b Lvs reniform to oval, not > 2× longer than broad, mostly ± cordate-reniform

　9a Pl with ann, floriferous, short and tufted-lfy to elongate sts

　　10a Lateral petals violet-spotted at base, upper petals deep purplish on back; fl st 5–30 cm, puberulent; woodl of coast ranges, n Douglas Co, Ore, to Monterey Co, Cal, possibly not in our area; pinto v.　　7 **V. ocellata** T. & G.

　　10b Lateral petals not violet-spotted, upper petals often not purple on back

　　　11a Petals shaded with blue to purple on back, not at all yellowish; st 2–10 cm, sparsely pubescent; fls 1.5–2 cm; prairies and moist woods w Cas; s BC to nw Cal, e as far as Klamath Lake in s Ore; Howell's v.　　8 **V. howellii** Gray

　　　11b Petals gen shaded with some yellow, at least basally; st 10–40 cm, glab to puberulent; fls ca 1.5 cm; moist woodl, gen on loamy soil; Can v.; 2 vars.　　9 **V. canadensis** L.

　　　　a1 Stolons present, sometimes buried and not obvious; pl pubescent; lvs often wider than long, ciliate-margined; Alas to Ore and through RM to Colo, e occ to c US and s Appalachians; w Can v. (*V. geminiflora, V. rugulosa, V. rydbergii*)

　　　　　var. **rugulosa** (Greene) Hitchc.

　　　　a2 Stolons lacking; pl glab to puberulent; lvs mostly longer than broad, not ciliate; c and e US and Can, w occ to RM, Alta to NM and Ariz; e Can v.　　var. **canadensis**

　9b Pl without ann fl sts, fls pedunculate on main rhizome

　　12a Petals gen tinged with violet or blue on back, mostly 10–13 mm; stolons gen present; lvs 2.5–3.5 cm broad, glab; moist meadows and stream banks; BC to Cal, e to RM and to Lab and Me; Europe; marsh v. (*V. p.* var. *brevipes,* pure white-fld form)　　10 **V. palustris** L.

　　12b Petals pure white except for purplish pencilling, 5–15 mm; stolons often lacking; lvs often < 2.5 cm broad

　　　13a Stolons lacking; lvs mostly much > 3 (2–6) cm broad, reniform, often pilose beneath; petals 10–15 mm, all beardless; lowl for to subalp slopes; BC to ne US, s to extreme n Wn and in RMS to Colo; kidney-lvd v. (*V. brainerdii*)　　11 **V. renifolia** Gray

13b Stolons gen present; lvs mostly < 3 (1–3) cm broad, more cordate
than reniform, glab beneath; petals 5–10 mm, lateral pair gen
bearded; boggy or wet ground in mts of much of N Am; small white
v.; 2 vars.   12 V. **macloskeyi** Lloyd
  a1 Lf bl slightly crenate, mostly < 2.5 cm broad; BC to Alta, s in
  mts to s Cal   var. **macloskeyi**
  a2 Lf bl deeply crenate, gen > 2.5 cm broad; BC and Wn to Atl
  coast, Can to se US, and in RM to Colo
                                                 var. **pallens** (Banks) Hitchc.

Group II

14a Petals not uniformly colored, lateral pair mostly dark-spotted basally, all,
but esp upper pair, reddish-purple on back (see lead 10a)
                                                      7 V. **ocellata** T. & G.
14b Petals uniformly colored, or at least lateral pair not spotted basally
  15a Pl with very slender, elongate stolons, without erect lfy sts; lvs arising
  from the rhizome; growing on very moist to boggy ground (see lead 12a)
                                                    10 V. **palustris** L.
  15b Pl either without stolons or with erect lfy sts
    16a Aerial sts well developed, often > lvs, fl on upper 2/3; either lvs
    ± acuminate or pl subalp in OM
      17a Lvs reniform, 1.5–4 cm broad, purplish-green, rounded; pl 3–15
      cm, subalp in OM, Wn, mostly on talus or in rock crevices; Flett's v.
                                                    13 V. **flettii** Piper
      17b Lvs cordate, many gen > 4 cm broad, acute to acuminate, bright
      green; pl mainly > 15 cm; widespread, mostly in woodl (see lead
      11b)   9 V. **canadensis** L.
    16b Aerial sts lacking or < lvs, peduncles gen borne on rhizome; if, as
    rarely, st > lvs, then pl neither subalp in OM nor with acuminate lvs
      18a Petioles and peduncles sparsely to evidently hirsute; sepals ciliate;
      spur much < 1/2 length of bl of lowest petal; style glab; lower 3
      petals white-based and copiously white-bearded; moist open woods;
      ne N Am w to Ont; also in BC to near US border; northern v.
                                            14 V. **septentrionalis** Greene
      18b Petioles and peduncles glab, or sepals not ciliate, or pl otherwise
      not as above
        19a Style head not bearded; pl glab, caulescent or acaulescent
          20a Pl with evident erect lfy st; stips ovate to lanceolate, entire or
          glandular-denticulate; lf bl 2–5 cm broad; moist to boggy
          areas; Alas s along coast to s Ore, Asia; Aleutian v. *(V. simu-*
          lata)   15 V. **langsdorfii** (Regel) Fisch.
          20b Pl with scarcely evident erect st, almost acaulescent; stips
          linear-lanceolate, entire; lf bl up to 7 cm broad; meadows and
          stream banks; e Cas, BC to Cal, Ariz, NM, NY, and Newf; n
          bog v.; 2 vars.   16 V. **nephrophylla** Greene
            a1 Lvs thick, smooth, purplish-backed; rootstocks slender; all
            petals bearded; BC and Wn to RM, Alta to Colo *(V. c.)*
                                                var. **cognata** (Greene) Hitchc.
            a2 Lvs thinner, mostly not purple-backed; rootstocks thicker;
            upper 2 petals not bearded; BC to e Can, s along e side Cas
            in Wn to Colo and Ariz *(V. austinae, V. macabeiana, V. sub-*
            juncta)   var. **nephrophylla**
        19b Style head bearded; pl glab or hairy, gen caulescent
          21a Spur broad and pouched, much < half as long as bl of lowest
          petal; fls 15–20 mm; w Cas (see lead 11a)   8 V. **howellii** Gray
          21b Spur slender, mostly half as long as bl of lowest petal; fls
          5–15 mm; dry to moist meadows, woods, and open ground;
          common in much of N Am, in our area on both sides Cas, but
          more common e Cas; early blue v., hook v.; 4 vars.
                                                   17 V. **adunca** Sm.
            a1 Pl stless at anthesis of most normal fls, but later producing
            cleistogamous fls on well-developed but non-persistent sts;
            rare, or perhaps not detected; Okanogan Co, Wn, and Des-
            chutes Co, Ore; Cas v. *(V. c.)*
                                          var. **cascadensis** (Baker) Hitchc.

23a

a2 Pl caulescent, normal fls all on short aerial sts
  b1 Lvs mostly pubescent, considerably longer than broad; pl
    mostly > 5 cm; range of sp. as a whole (*V. drepanophora,*
    V. longipes, V. mamillata, V. montanensis, V. monticola,
    V. odontophora, V. oxyceras, V. oxysepala, V. retroscabra,
    V. verbascula, V. a. var. *glabra*)   var. **adunca**
  b2 Lvs glab; pl gen dwarf, scarcely 5 cm
    c1 Petals gen whitish at base, ca 5 mm; chiefly RM (*V. b.*)
                                          var. **bellidifolia** (Greene) Harr.
    c2 Petals not whitish at base, mostly much > 5 mm; local
      in s Ore, possibly not quite reaching our aea (*V. u.*)
                                          var. **uncinulata** (Greene) Hitchc.

Group III

22a Aerial sts naked below, lf- and fl-bearing only from near tip; lvs mostly
    large, thin, and cordate-based
  23a Petals mostly bluish, yellow only at base; alp talus and rock crevices
    (see lead 17a)   13 **V. flettii** Piper

24a

  23b Petals mostly yellowish or whitish-yellow, sometimes blue to brown on
    back
    24a Petals all clear yellow on both surfaces, 8–14 mm, the lower 3 pur-
      plish-pencilled within; lf bl reniform to ovate-cordate, abruptly acute;
      moist woods or stream edges; Alas s, on both sides Cas, to Sierran Cal,
      e to Mont; ne Asia; stream v., pioneer v.   18 **V. glabella** Nutt.
    24b Petals white with yellow base, all (but esp upper 3) ± purplish-
      tinged without (and sometimes also within); lf bl cordate, acute to
      strongly acuminate (see lead 17b)   9 **V. canadensis** L.

24b

22b Aerial sts lf- and fl-bearing most of length
  25a Lvs cordate to reniform, gen as broad as long
    26a Lvs finely dotted, flecked, or reticulately mottled with purple, rather
      firm and leathery, persistent, gen hairy; pl stoloniferous; w Cas, BC to
      Cal, mostly in moist woods, lowl to lower mont; redwoods v., ever-
      green v.   19 **V. sempervirens** Greene

26a

    26b Lvs not flecked with purple, thin, gen withering during winter, often
      glab; pl not stoloniferous; mont to alp; both sides Cas, BC to n Ore, e
      to Ida and Mont; round-lvd v., darkwoods v. (*V. sempervirens* var. *o.*,
      V. s. var. *orbiculoides*)   20 **V. orbiculata** Geyer
  25b Lvs neither truly cordate nor reniform, gen slightly to much longer
    than broad
    27a Lf bl coarsely veined, gen not > 4 cm, coarsely few-toothed or
      -lobed, not reg serrate or dentate, often glaucous and ± purplish, at
      least along veins; upper petals deep purple on back; caps puberulent;
      lowl, dryish areas to high mont, esp on open ridges or slopes; Chelan
      Co, Wn, to Cal and Ariz, e to Mont, Wyo, and Colo; goosefoot v., pur-
      plish v.; 2 vars.   21 **V. purpurea** Kell.

26b

      a1 Seeds with feathered and discoid caruncle often nearly 1 mm
        broad; Chelan Co, Wn, to e Sierra Nevada, e to Mont, Wyo, and
        Colo (*V. atriplicifolia, V. aurea* var. *v., V. nuttallii* var. *v., V. prae-*
        morsa var. *v., V. thorii*)   var. **venosa** (Wats.) Brain.
      a2 Seeds with rounded or amorphous caruncle < 0.5 mm broad; Cas,
        Deschutes Co, Ore, to Inyo Co, Cal (*V. pinetorum, V. p.* sspp. *di-*
        morpha and *geophyta*)   var. **purpurea**
    27b Lf bl not coarsely veined, entire to finely or obscurely serrate or den-
      tate, often much > 4 (2–10) cm, gen not glaucous and not purplish,
      even along veins; upper petals often not purple on back; caps glab or
      puberulent; BC to Cal, mostly e Cas, e to c US; 5 vars.
                                                        22 **V. nuttallii** Pursh
      a1 Lf bl narrowly lanceolate or elliptic-lanceolate, gen at least 3 times
        as long as broad, narrowed to petioles nearly or quite as long;
        chiefly e RM, Alta to Ariz, e to Mo, prob in our range in Mont; Nut-
        tall's v., yellow prairie v.   var. **nuttallii**
      a2 Lf bl various, gen < 3 times as long as broad, often truncate or
        subcordate at base; RM w
        b1 Upper petals not brownish-backed; fls 5–12 mm; lf bl mostly 2–5
          cm; caps glab; Cas, Mt Adams, Wn, to Cal; Baker v. (*V. b.*)
                                                        var. **bakeri** (Greene) Hitchc.

27a

b2 Upper petals gen brownish-backed; fls 8–15 mm; lf bl 3–10 cm; caps glab or hairy

    c1 Lvs glab to sparsely hairy, bl ovate to ovate-lanceolate, gen ± truncate or subcordate at base, mostly < 5 cm; caps glab; BC to Ore, e to RM; chiefly in sagebr or on sagebr-ponderosa pine woodl; valley yellow v. (*V. physalodes, V. subsagittifolia, V. v.*)   var. **vallicola** (Nels.) Hitchc.

    c2 Lvs densely hairy to glab, bl mostly at least 5 cm, seldom at all cordate; caps often hairy

        d1 Lvs mostly strongly hairy, bl thick and fleshy, ovate-lanceolate; mostly w Cas, valleys and prairies, BC to Cal; canary v., upland yellow v. (*V. p., V. p.* var. *oregana*)   var. **praemorsa** (Dougl.) Wats.

        d2 Lvs glab to moderately hairy, bl not noticeably fleshy; e Cas, mostly in ponderosa pine for and lower mts; Nuttall v. (*V. erectifolia, V. flavovirens, V. gomphopetala, V. linguaefolia, V. praemorsa* ssp. *arida, V. xylorhiza*)   var. **major** Hook.

LOASACEAE   Blazing-star Family

Fls solitary to cymose, reg, epig (ours), complete, with a short, often showy, gen flared, free hypan; calyx lobes 5 (4); petals 5 (4), distinct (ours); stamens ∝ (ours), distinct or basally connate into several groups, some often expanded into petaloid staminodia; ovary 1 (3)-celled; placentation mostly parietal (axile); fr caps; herbs or shrubs, gen scab or bristly with rough, barbed, sometimes stinging hairs; lvs alt to opp, exstip.

Mentzelia L.   Blazing-star; Mentzelia

Fls inconspicuous to showy, ours yellow to orange; ann or per herbs with alt, scab, barbellate-pubescent, brittle lvs readily attaching to any foreign object touched. (For C. Mentzel, 1622–1701, German botanist). (*Bartonia, Hesperaster, Acrolasia, Nuttallia, Touterea*). The 1st 2 spp. are beautiful-fld and desirable garden subjects for dry areas, esp e Cas.

1a Pl bien to per; fls showy; petals 1.5–8 cm; seeds flattened

  2a Petals apparently 10, inner 5 (really staminodia) slightly narrower than the outer 5; fl bracts adherent to the ovary; seeds thin-margined but not winged; plains and lower mts; e Mont to Mex, w in Mont to Ida; sand lily, evening star   **1 M. decapetala** (Pursh) Urb. & Gilg

  2b Petals 5; five of outer stamens sometimes flattened and ± petaloid, but much narrower than the true petals; fl bracts not adherent to the ovary; seeds distinctly wing-margined; des valleys and lower mts; BC and e Wn to Mont, s to Cal, Utah, and Wyo; blazing-star m.; 2 vars.

                        2 M. laevicaulis (Dougl.) T. & G.

    a1 Petals at least 4 cm; caps mostly > 2 cm; range of the sp. except largely replaced by next in areas mentioned (*M. acuminata*)   var. **laevicaulis**

    a2 Petals mostly 1.5–4 cm; caps mostly < 2 cm; s BC and valleys e Cas, Wn (*M. brandegei*)   var. **parviflora** (Dougl.) Hitchc.

1b Pl ann; petals < 1.5 cm; seeds not flattened

  3a Fl bracts mostly ovate-lanceolate to ovate; infl congested; caps linear; seeds in a single row the full length of the caps, prismatic, grooved on the vertical margins, very obscurely tuberculate-muricate and apparently smooth, even with 10× magnification; dry soil, plains to lower mts; e Cas, Wn to s Cal, e to Mont, Wyo, and Colo; bushy m., small-fld m. (*M. compacta, M. latifolia, M. pinetorum*)   **3 M. dispersa** Wats.

  3b Fl bracts narrowly to broadly lanceolate; infl not congested; caps mostly broadened upward; seeds obviously tuberculate-muricate under 10 × magnification, prismatic and often grooved on the vertical angles in the lower part of the caps, but arranged irreg in upper part of caps and irreg shaped and not grooved on the angles; dry, mostly sandy soil, des valleys to foothills; mostly e Cas, BC to s Cal, e to Mont and NM; white-std m., small-fld m. (*M. ctenophora, M. gracilis, M. parviflora, M. tenerrima, M. tweedyi*)   **4 M. albicaulis** Dougl.

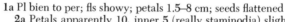

CACTACEAE   Cactus Family

Fls showy, solitary, reg, complete, epig; sepals and petals ∝, basally coalescent and adnate to one another and to the ovary, forming a distinctive free hypan bearing the ∝ stamens; pistil 3–10-carpellary, style 1, stigmas 3–10, ovary 1-celled, placentation parietal; fr baccate to dry and leathery; fleshy, herbaceous (ours) to ± woody, prickly or spiny pls with globose to cylindric or flattened, jointed, often ribbed or fluted sts, mostly with special cushions (areoles) bearing coarse woolly hair, strong spines, soft sharp bristles, and fls; lvs (ours) scalelike or lacking.

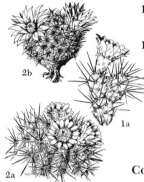

1a St jointed, joints (ours) ± flattened, without tubercles; areoles producing soft, sharp, barbed bristles as well as long spines; lvs scalelike, quickly deciduous **Opuntia**

1b St not jointed, ours ± globose, evidently tubercled; areoles gen with wool as well as strong spines, but without spinose, barbed bristles; lvs lacking

2a St with low, spiralled ribs running lengthwise; fl solitary near tip of tubercle at edge of spine-bearing areole; tubercles low, not cylindric **Pediocactus**

2b St without longitudinal ribs; fl solitary at base of groove on upper side of mature tubercles, not immediately adj to the areole; tubercles cylindric, rounded at the tip **Coryphantha**

Coryphantha (Engelm.) Lemaire   Coryphantha

Fls greenish-white to deep red or purple, borne near st tip; ovary not spinose; pl globose, spines produced from areoles at tip of spirally arranged, mammillate tubercles grooved on the upper side. (Gr *koryphe*, cluster, and *anthos*, flower).

1a Fls greenish-white, barely reddish-tinged; fr < 1 cm, subglobose, reddish; st gen 1, 3–5 cm, subglobose; valleys and foothills of des and grassl; mainly e RM, Man to Kans, but w to c Mont and Colo, and in c Ida, along Salmon R, Custer Co, Ida; nipple c. (*Mammillaria nuttallii, Neobesseya m., Neomamillaria m.*) 1 **C. missouriensis** (Sweet) Britt. & Rose

1b Fls reddish-purple; fr 1–2 cm, oblong, greenish; sts 1–several, subglobose, 3–6 cm; des valleys and foothills; Alta to se Ore, Colo, and Kans; cushion c., ball c. (*Cactus v., Escobaria v., Mammillaria v.*)
2 **C. vivipara** (Nutt.) Britt. & Brown

Opuntia Mill.   Cholla or Prickly-pear Cactus

Fls in areoles of previous year's st; sepals greenish, grading into the yellow to red petals; berry dry (ours) to juicy; st flattened, broad, jointed (ours). (Derivation uncertain).

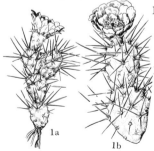

1a St joints not greatly flattened, often nearly as thick as broad, mostly < 5 cm, readily detached from pl; spines strongly barbed; areoles rather strongly white-woolly; fls yellow; pl low, matted, 0.5–2 dm; dry open ground, BC s, mostly e Cas, but also in n PT, to n Cal, e to Tex and Wisc; brittle c. 1 **O. fragilis** (Nutt.) Haw.

1b St joints much-flattened, larger ones 5–13 cm, not readily detached; spines only slightly barbed; areoles mostly rusty-woolly, or nonwoolly; fls yellow to reddish; pl 1–3 (4) dm, in rounded clumps or broader mats; plains and foothills to lower mts; BC to e Ore, e Cas, e to Alta, Ariz, Tex, and Mo; starvation c. (*O. columbiana, O. missouriensis* var. *microsperma, O. m. platycarpa*) 2 **O. polyacantha** Haw.

Pediocactus Britt. & Rose   Hedgehog-cactus

Small, lfless pls with 1–several globose to ovoid sts bearing low tubercles in 8–13 spiral longitudinal rows. (Gr *pedion*, field, and cactus).

P. simpsonii (Engelm.) Britt. & Brown. Mt h. St 7–12 cm thick; tubercles 12–25 mm; fls solitary in a specialized area near the tip of the tubercles at the edge of the spine-bearing areoles, yellow to purplish, 1.5–2 cm; free hypan short, camp; fr subglobose, 6–8 mm; seeds black, ca 3 mm; des valleys and low mts, e Wn to Nev, e to Wyo, Utah, and Colo (*Echinocactus s., E. simpsoni* var. *minor);* ours the var. *robustior* Coult.

ELAEAGNACEAE   Oleaster Family

Fls apet, ♂ to ♂, ♀, with tubular hypan which surrounds but does not adhere to the ovary; calyx 4-merous; stamens 4 or 8, borne near top of hypan, often alt with lobes of disc; pistil 1-carpellary; deciduous (ours) or evergreen trees or shrubs with alt or opp lvs,. lepidote to stellate hairs, and sometimes with spinose brs.

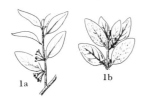

1a Lvs alt; pls apparently ♂-fld; stamens 4   **Elaeagnus**
1b Lvs opp; pls ♂, ♀; stamens 8   **Shepherdia**

Elaeagnus L.   Elaeagnus

Fls axillary; calyx lobed < half length, deciduous from top of fr; ovary wall becoming hardened and bony, fluted lengthwise, investing calyx, more mealy than juicy; alt-lvd shrubs, often of very arid regions. (Gr *elaia*, olive, and *agnos,* Gr name for the chaste-tree).

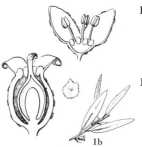

1a Unarmed shrub 1–4 m, strongly rhizomatous; lvs and brs silvery- to brownish-scurfy; lvs lanceolate to oblanceolate, 2–7 cm; fls apparently ♂, but functionally ♂ ♀, the ♂ fls with narrower tube and vestigial pistil; gravel benches, scabl, and water courses; BC to Que, s in RM to Ida, Mont, and Utah; silverberry, wolf-willow (*E. argentea*, misapplied)
                                                                1 **E. commutata** Bernh.
1b Spiny-br shrub or small tree to 8 m; lvs linear to linear-lanceolate; Eurasian, intro for use as wind break, and perhaps long-persistent or even estab in some places; russian olive   2 **E. angustifolia** L.

Shepherdia Nutt.   Buffalo-berry

Fls axillary; hypan of ♂ fls with 8-lobed disc in throat, becoming fleshy in fr; ovary wall not greatly hardened; opp-lvd shrubs (trees). (For John Sheppard, 1764–1836, English botanist). (*Hippophae, Lepargyraea*). Our spp. of little hort importance.

1a Lvs silvery on both surfaces, 2–5 × 0.5–1.5 cm; brs often spine-tipped; pl 2–6 m; fr reddish-yellow; mostly along water courses; BC to s Cal, e Cas, e to Minn; not known from Wn or Ida, and in Mont mostly e RM; thorny b.
                                                                1 **S. argentea** (Pursh) Nutt.
1b Lvs green above, brownish-lepidote beneath, 1.5–6 × 1–3 cm; brs unarmed; pl 1–4 m; fr yellow-red, bitter; open to wooded areas; Alas to Ore, e to Atl coast; Can b., russett b., soopolallie, soapberry (*Elaeagnus c.*)
                                                                2 **S. canadensis** (L.) Nutt.

LYTHRACEAE   Loosestrife Family

Fls complete or sometimes apet, ours reg, perig, axillary to racemose or cymose, with elongate hypan free of ovary; calyx lobes 4–7, often alt with toothlike appendages; petals distinct, mostly as many as sepals, borne at top of hypan, gen crumpled in the bud; stamens as many to 2× petals (occ more or fewer), borne on hypan; ovary (1) 2–6-celled, superior; style 1; stigma mostly 2-lobed; fr caps, mostly with axile placentation; ann to per glab herbs (ours) to shrubs or trees with entire, mostly opp or verticillate, exstip lvs.

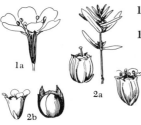

1a Hypan elongate-cylindric to conic, several times as long as thick, strongly nerved; petals 5–7, gen showy   **Lythrum**
1b Hypan camp to hemispheric, sometimes not strongly nerved; petals gen 4, inconspicuous (sometimes wanting)
2a Style mostly exserted from calyx, at least 1 mm; caps not dehiscing by reg sutures; lvs sessile, cordate-clasping at base, gen with > 1 fl per axil
                                                                Ammannia
2b Style not exserted from calyx, < 1 mm; caps dehiscent along reg sutures; lvs often petiolate, not at all cordate, gen with only 1 fl per axil
                                                                Rotala

Ammannia L.   Ammannia

Fls 4-merous (ours), in axillary, few-fld cymes; hypan strongly 8-ribbed and 4-angled, sinuses projecting as small teeth; petals small or wanting; caps 4-celled; small ann. (For Paul Ammann, 1634–1691, German botanist).

A. coccinea Rottb. Scarlet a. Simple to br ann 0.5–4 dm; lvs entire, oblong to oblong-lanceolate, 2–4 cm × 2–8 mm, cordate-clasping; fls 1–5 per axil, sessile or subsessile; hypan globose to camp, 2–3 mm but up to 5 mm in fr; petals purplish (occ wanting), the bl ca 1 mm; stamens 4; style slightly exserted, 1–1.5 mm; stigma discoid; caps membranous, subglosose, ca 4 mm; widespread in wet places, often where alkaline; known from much of N Am, but rare in w US; reported for our area from Klickitat Co, Wn (A. alcalina).

Lythrum L.   Loosestrife

Fls in axillary clusters or terminal spikes, often showy, complete; calyx tubular, 5–7-lobed and with linear processes between the lobes; petals 5–7, white to purple, inserted with stamens near top of hypan; stamens as many to 2× petals, varying in length, the fls often di- or trimorphic as to relative length of stamens and style; caps mostly 2-celled; ann or per herbs with opp (whorled, or very rarely alt), mostly sessile lvs. (Gr *luthron*, blood, name used by Dioscorides, prob referring to staining properties of pl, or to fl color).

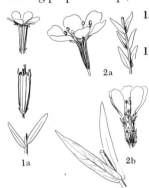

1a Pl ann, 1–4 dm; lower lvs mostly alt, petals white to rose, 2–3.5 mm; ± marshy places, mainly coastal in our area; Wn to Cal, e US; Europe; grass poly, hyssop l.   1 **L. hyssopifolia** L.
1b Pl per, mostly much > 4 dm; lower lvs gen opp; petals red or purple, 5–10 mm
  2a Main lvs 2–5 cm, not cordate; petals purple, ca 5 mm; pl 4–10 dm; perhaps not in our area, but known from BC; SC to Tex, e to Mass and Ga, on swampy to rather dry soil; winged l.   2 **L. alatum** Pursh
  2b Main lvs 3–10 cm, ± cordate-based; petals reddish-purple, 7–10 mm; pl 5–20 dm; marshes of PS area, Wn, also in c and e US, native to Europe; most colorful of our spp., suited to wet areas in the garden; purple l., long-purples, spiked willow-herb   3 **L. salicaria** L.

Rotala L.

Fls inconspicuous, ours 4-merous; calyx lobes minute; petals white, tiny; caps 4-celled, septicidally dehiscent; ann or per herbs with opp lvs. (L *rotula*, little wheel, said to refer to the whorled lvs of the type sp.).

R. ramosior (L.) Koehne. Toothcup. Glab, erect to procumbent, simple to br ann 5–15 cm; sts angled; lvs mostly short-petiolate, bl lanceolate to oblanceolate, 1.5–3 cm; fls 1 (2–3) per axil; hypan camp, 1.5–2.5 mm in fl but becoming globose and up to 3–4 mm in fr; petals white, ca 1 mm; stamens and style included in the hypan; caps globose, ca 3 mm; wet places, occ in Wn, s to Mex, e to Atl coast (Ammannia r., A. humilis).

ONAGRACEAE   Evening-primrose Family

Fls gen complete, rarely apet, mostly reg (irreg), epig, often showy, gen spicate or racemose, gen with a free hypan, this mostly short but sometimes > the ovary; calyx lobes mostly 4, occ 2, 3, or 5; petals white, yellow, or pink to lavender or purplish, gen as many as calyx lobes, stamens as many or twice as many, borne with petals at summit of hypan; ovary inferior, gen 4 (1, 2, or 5)-celled; style elongate; stigmas globose to (mostly) 4-lobed; fr a ∝-seeded caps or occ 1–2-seeded and nutlike; ann or per herbs with opp or alt, simple to pinnatifid lvs.

1a Fls 2-merous; fr with bristly, hooked hairs, 1–2-seeded, indehiscent
                                                  Circaea
1b Fls 4 (5)-merous; fr not bristly, gen several-seeded
  2a Petals and sepals 5 (ours); sepals persistent on caps; caps slender, elongate, eventually reflexed; stamens gen 10   **Jussiaea**

2b Petals gen 4, rarely lacking; sepals 4, often deciduous; caps often short and thickened; stamens gen 4 or 8

　3a Fr scarcely twice as long as thick, 4-angled and -sided, ∝-seeded; sepals persistent; free hypan almost nil　　　　　**Ludwigia**

　3b Fr mostly several times as long as broad, seldom 4-angled and -sided; free hypan often well developed and tubular (gen deciduous as fr matures)

　　4a Fr nutlike, hardened, indehiscent, 1–4-seeded, 4-angled and flat-sided　　　　　　　　　　　　　　　　　　**Gaura**

　　4b Fr caps, ∝-seeded, dehiscent, gen ± terete

　　　5a Seeds with conspicuous tuft of long hairs at the tip; pl often growing along streams or on moist soil; lvs often opp　**Epilobium**

　　　5b Seeds glab or very finely strigose-puberulent; pls mainly from dry hills or drying mud flats; lvs gen alt

　　　　6a Ovary 2-celled, slender; pl gen freely br; fls small, white to pinkish, petals 1–2 (5) mm　　　　　**Gayophytum**

　　　　6b Ovary 4-celled; habit and fls diverse, petals often > 5 mm

　　　　　7a Fls axillary, sessile or subsessile; petals 1–6 mm; calyx lobes erect; anthers basifixed; pl often villous-hairy

　　　　　　　　　　　　　　　　　　　　　　Boisduvalia

　　　　　7b Fls often ped, if sessile either petals > 6 mm, or calyx lobes reflexed, or anthers versatile

　　　　　　8a Petals yellow or white, sometimes aging reddish or purplish; anthers often versatile　　　　　**Oenothera**

　　　　　　8b Petals pink to purple; anthers erect, attached near base　　　　　　　　　　　　　　　　　　　　**Clarkia**

Boisduvalia Spach　Spike-primrose; Boisduvalia

Fls diurnal, pale pink to rose or purple, small; stamens 8; hypan short-funnelf; petals bilobed 1/3–1/2 length; caps elongate, 4-ribbed; seeds glab; ann, mostly soft-hairy herbs with alt, simple lvs. (For J. A. Boisduval, 1801–1879, French naturalist).

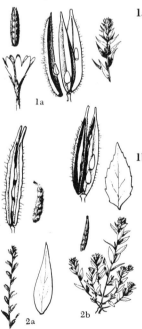

1a Petals (2.5) 3–8 (12) mm; caps slenderly fusiform, 6–10 mm, very short-beaked, straight, septa free of the valves, adherent to the placentae which form a persistent column after shed of valves and seeds; seeds mostly < 6 per locule; floral lvs crowded, ovate to ovate-lanceolate; pl gen densely strigose to pilose; BC s, both sides Cas, to Baja Cal, e to w Mont, Ida, and Nev; dense s.; 2 vars.　　　　　1 **B. densiflora** (Lindl.) Wats.

　a1 Pl with soft, spreading hairs, often ± glandular; petals dark, gen > 4 mm; range of sp. (*B. bipartita, B. imbricata, B. sparsifolia, B. d.* vars. *pallescens* and *montanus*)　　　　　　　　　var. **densiflora**

　a2 Pl mostly canescent with short, appressed hairs; petals mostly pinkish, 3–4 mm; with var. *densiflora* but less common (*B. salicina, B. sparsiflora*)　　　　　　　　　　　　　　　　　var. **salina** (Rydb.) Munz

1b Petals 1.5–4 mm; caps sometimes strongly beaked or curved, septa adherent to valves and shed with them, placentae gen fragmenting as seeds are shed; seeds mostly 6 or more per locule; floral lvs often linear and not crowded; pl sometimes glabrate

　2a Pl villous-pilose, gen canescent; fls not crowded; petals ca 2 mm; fl bracts linear to linear-lanceolate, entire; caps 6–11 mm, short-beaked, gen ± torulose, rather strongly curved; seeds 6–8 per locule; Yakima Co, Wn, s and on both sides Cas in Ore, to Cal, e to Ida and Nev; brook s., stiff s. (*B. diffusa, B. parviflora, B. torreyi*)　　　2 **B. stricta** (Gray) Greene

　2b Pl greenish, strigose to glabrate; fls gen crowded; petals 2–4 mm; fl bracts lanceolate to ovate-lanceolate, denticulate to dentate; caps ca 7 mm, nearly straight, pointed but not beaked; seeds often > 8 per locule; BC and e Wn to s Cal and Nev, e to Sask, Daks, and Utah; S Am; smooth s.　　　　　　　　　　　　　　　　3 **B. glabella** (Nutt.) Walp.

Circaea L.　Enchanter's Nightshade; Circaea

Fls small, white, in simple or compound racemes; sepals reflexed; hypan very short, deciduous after anthesis; petals notched; fr pear-shaped; seeds glab; small, juicy, opp-lvd per. (For Gr goddess *Circe*).

C. alpina L. Pl rhizomatous, clear green; sts arising from tuberous parts of the rhizomes, 1–5 dm, simple to freely br; lf bls cordate-ovate to ovate, 2–6 cm, subentire to sharply toothed, considerably > the petioles; racemes gen with 1–2 linear bracts at base, the fls minutely bracteolate; peds spreading or erect in fl, gen reflexed in fr, = or> fr; petals and sepals 1–1.5 mm; fr ca 2 mm, covered with short, hooked hairs; cool, damp woods; Alas to Cal, e to Newf and Ga; Eurasia *(C. pacifica, C. a. var. p.).* A nasty weed in the woodl garden.

Clarkia Pursh   Clarkia; Godetia

Fls reg to slightly irreg, open continuously or only during the day, buds nodding to erect; hypan well-developed, often with a ring of hairs within, deciduous as the fr matures; petals 4; stamens (4) 8; ovary elongate, slender to fusiform or clavate, terete or ± 4-sided, often beaked; seeds non-hairy but with a terminal crestlike border of elongate cells; ann with simple alt lvs. (For Capt. William Clark of the Lewis and Clark Expedition). Most spp. are attractive and easily grown anns, suitable for dry banks. *(Godetia, Phaeostoma).*

1a Petals clawed, gen ± 3-lobed; fls slightly irreg, not closing at night *(Clarkia)*
  2a Fertile stamens 4; petals very strongly 3-lobed, 5–25 mm, lavender to rose-purple; s BC s, e Cas, to se Ore, e to Ida and w Mont; pink fairies, deer horn, ragged robbin, elkhorns c.   1 **C. pulchella** Pursh
  2b Fertile stamens 8; petals narrowly rhomboidal, 5–10 mm, rose-purple, not lobed, but claw often toothed near base; s BC s, e Cas, in Wn, but e and w Cas in Ore, to s Cal, e to Ida, Utah, and Ariz; rhombic-petaled c., common c. *(C. gauroides)*   2 **C. rhomboidea** Dougl.
1b Petals neither clawed nor lobed; fls reg, closing at night *(Godetia)*
  3a Calyx lobes gen connate and turned to one side under fl, ovaries 4-grooved at anthesis; caps terete and 8-nerved, elongate, slender, often beaked, gen on ped 2–10 mm; stigmas 1–7 mm
    4a Buds and rachis of raceme recurved, the latter straightening and becoming erect as fls open; pl erect, 1.5–6 dm; petals pink to lavender, gen unspotted, 8–20 mm; stigma lobes 1–1.5 mm, cream; Klickitat and Walla Walla cos, Wn, s in Ore w Cas, to c Cal; slender g. *(G. amoena var. g., G. g.)*   3 **C. gracilis** (Piper) Nels. & Macbr.
    4b Buds and rachis of raceme erect; pl spreading to erect, 1–10 dm; petals pale pink to rose-purple, gen carmen-spotted in center, 10–40 mm; stigma lobes (1) 1.5–7 mm, yellow; farewell-to-spring; 3 vars.
          4 **C. amoena** (Lehm.) Nels. & Macbr.
      a1 Stigmas linear, gen much > 2 mm; petals mostly > 2 cm; caps not strongly curved; pl gen erect; occ from VI s *(G. a. var. l., G. grandiflora, C. superba)*   var. **lindleyi** (Dougl.) Hitchc.
      a2 Stigmas oval, gen not > 2 mm; petals mostly < 2 cm; caps often curved; pl often br from base and ± decumbent
        b1 Caps curved; petals rhombic-ovate, striped with 2 lunate median bands; pl gen basally br and ± decumbent; coastal bluffs of Lincoln Co, Ore, at one time near Port Angeles, Wn *(G. p., G. romanzovii)*   var. **pacifica** (Peck) Hitchc.
        b2 Caps straight; petals rounded to truncate, spotted or not; pl erect; occ from VI to s Ore, w Cas *(G. c., C. c.)*
            var. **caurina** (Abrams) Hitchc.
  3b Calyx lobes mostly distinct and closely reflexed; ovaries gen terete and not 4-grooved at anthesis; caps often ± 4-angled at maturity, gen sessile, never long-beaked; stigmas < 2 mm
    5a Fls mostly closely crowded; caps 15–25 × 3–5 mm, mostly much enlarged near center, largely concealed by fl bracts; stigmas scarcely 1 mm; petals 5–20 mm, crimson to purple, spotted or not; stigmas oval, scarcely 1 mm; dry ground, gen around vernal pools, rarely coll recently; Salem, Ore, to s Cal; purple g. *(G. p., G. p. var. parviflora, G. albescens, G. arnottii, G. decumbens, G. lepida)*
          5 **C. purpurea** (Curtis) Nels. & Macbr.
    5b Fls not crowded; caps 1.5–3 mm thick, not much enlarged near center, not concealed by bracts; stigmas often > 1 mm
      6a Hypan 6–12 mm, slender-based but flared above, considerably enlarged at top of ovary; petals 13–25 mm, rose-lavender to purple, often with large central or apical wine-colored blotch, sometimes

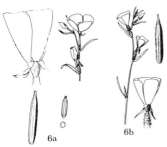

yellowish near base; stigma lobes ca 1.5 mm; very occ, Multnomah to Wasco cos, Ore, s, w Cas, to Cal, where common; twiggy g. (*G. v.*)   6 C. **viminea** (Dougl.) Nels. & Macbr.

6b Hypan 2–7 mm, ± uniform in taper, not enlarged at top of ovary; petals 5–15 mm, pale pinkish-lavender to deep rose-purple, with or without a carmine or purplish central spot; stigma lobes oval, ca 1 mm; very occ in open; OP to Tacoma, up CRG to Klickitat Co, Wn, and Hood R Co, Ore, and in WV, Ore, s (and more common) to Baja Cal; small-fld g. (*G. brevistyla, G. q., G. tenella*)

7 C. **quadrivulnera** (Dougl.) Nels. & Macbr.

Epilobium L.   Willow-herb; Willow-weed

Fls complete, white or yellow to rose-purple, mostly racemose; hypan short or lacking; stamens 8; stigma 4-lobed or oblong-clavate and not lobed; caps 4-celled, linear to fusiform; seeds strongly comose at the tip; ann or per herbs, a few somewhat shrubby, mostly spreading by stolons or rhizomes often having rosettes of new lvs at tips or with bulblike offsets (turions); lvs mostly opp, entire to dentate, often willowlike. (Gr *epi*, upon, and *lobos*, pod, referring to the inferior ovary). (*Crossostigma*). A few spp. showy but hard to grow, most others, including *E. angustifolium*, potentially bad weeds.

1a Stigma 4-cleft (shallowly in *E. obcordatum*); petals yellow, or at least 1 cm, or both

   2a Petals yellow

      3a Pl woody based; st 1–3 dm; lvs entire, 1–3 cm; caps fusiform, 1.5–2.5 cm; fls slightly irreg; petals mostly 5–9 (10) mm; gravelly stream banks to mont slopes; c Ida to Mont and Wyo; shrubby w. (*Cordylophorum*)

1 E. **suffruticosum** Nutt.

      3b Pl herbaceous throughout; st erect, 2–7 dm; lvs serrate-dentate, 3–8 cm; caps linear, 4–8 cm; fls reg; petals (10) 14–18 mm; moist soil; Alas to Cas, Ore, at middle and higher elev, w to OM and VI; yellow w. (*E. l.* var. *lilacinum*)   2 E. **luteum** Pursh

   2b Petals pink to purple (white)

      4a Free hypan 1–3 mm; ovary glandular-puberulent; petals 12–20 mm

         5a Pl ± prostrate, scarcely 1 dm; petals obcordate; lvs glab and glaucous, 1–2 cm; subalp or alp meadows, talus slopes, and ledges; c Cas, Ore to Cal and Nev, Blue Mts, Ore, and Sawtooth Mts, c Ida; rose w.   3 E. **obcordatum** Gray

         5b Pl erect, 8–12 dm; petals rounded to truncate or retuse; lvs pubescent, not glaucous, 3–8 cm; European, intro in swampy places w Cas, as at Bellingham and near Bingen, Wn; fiddle-grass

4 E. **hirsutum** L.

      4b Free hypan lacking, calyx cleft to top of ovary; ovary nonglandular; petals various (*Chamaenerion*)

         6a Lvs (5) 10–20 cm; pl mostly 1–2 (3) m, strongly rhizomatous: ˙tyle > stamens, hairy near base; fl bracts much-reduced, linear; racemes elongate, mostly at least 15-fld; lowl to high mont, often on old burns, Alas to Cal, e to Atl Coast; Eurasia; fireweed, blooming Sally

5 E. **angustifolium** L.

         6b Lvs (1) 2–6 cm; pl 0.5–4 dm, not rhizomatous; styles < stamens, glab; fl bracts similar to lvs or somewhat reduced; racemes mostly 3–12-fld; river bars, stream banks, and alp slopes, subalp–alp; Alas to Cal, in both Cas and OM, Blue-Wallowa mts, Ore, e to Ida and Mont and s in RM to Colo; Eurasia; red w.   6 E. **latifolium** L.

1b Stigma gen entire, but if short-lobed then petals never yellow (sometimes creamy white) and seldom as much as 1 cm

   7a Pl ann, taprooted, mostly on well-drained soil; epidermis on lower part of st gen exfoliating

      8a Lvs mostly opp below; seeds 0.5–0.9 mm, very faintly alveolate; st soft-pubescent; pl mostly < 4 (0.3–4.5) dm; petals 2–4 mm, white to rose; lowl to mont, mostly on gravelly, dry soil, widespread; BC to Cal, e to Mont; small-fld w. (*E. adscendens, Crossostigma lindleyi*)

7 E. **minutum** Lindl.

      8b Lvs all or all but the lowermost alt; seeds at least 1 mm, faintly spinose-papillate; pl 3–10 (20) dm; petals 3–12 mm, rose to pale pink;

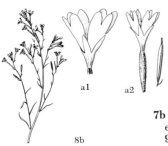

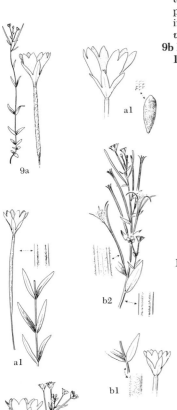

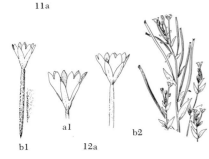

mostly on dry soil in open to wooded areas, often in ponderosa pine woodl, common in much of w US; autumn w., tall ann w.; 2 vars.

8 E. paniculatum Nutt.

a1 Petals 7–13 mm, mostly rose-colored; hypan 5–12 mm; coastal Wn to Ida and Cal (*E. j.*)   var. **jucundum** (Gray) Trel.

a2 Petals mostly < 7 mm, often light pink; hypan mostly < 5 mm; coastal to Daks, BC to s Cal and NM (*E. adenocladon, E. altissimum, E. apricum, E. fasciculatum, E. hammondi, E. laevicaule, E. subulatum, E. tracyi*)   var. **paniculatum**

7b Pl per, gen rhizomatous, mostly in moist areas; lower epidermis seldom exfoliating

9a Pls gen grayish-strigillose, at least above, producing turions; lvs linear to narrowly lanceolate, (1) 2–6 cm × 1–4 (8) mm; petals white to pink, notched, 3–5 mm; wet soil; Alas to Cas of c Wn, e to Atl coast, s in RM to Colo; Eurasia; swamp w., wickup (*E. leptophyllum, E. davuricum, E. densum, E. wyomingense*)   9 E. palustre L.

9b Pls not grayish-strigillose or if so otherwise not as above

10a Turions gen present; seeds often papillate; petals 3–10 mm, pale to deep pink or purplish; mostly midmont in moist places, Alas through Wn and Ore to Cal, e to n Atl coast, s in RM to Colo; Asia; common w.; 3 vars.   10 E. glandulosum Lehm.

a1 Petals 2–10 (but mostly > 6) mm; lvs sessile; st glab below but crisp-puberulent (mostly in lines) above; caps glandular-puberulent; range of sp. as whole (*E. drummondii, E. exaltatum, E. ovatifolium, E. palmeri, E. rubescens, E. sandbergii, E. saximontanum, E. stramineum*)   var. **glandulosum**

a2 Petals 3–5 mm; lvs sessile or petiolate; st pubescent, not always in lines; caps glandular to eglandular

b1 St not pubescent in lines; caps glandular-pilulose; lvs sessile or petiolate; Alas to s Cal, e to Colo (*E. brevistylum, E. delicatum, E. ursinum*)   var. **tenue** (Trel.) Hitchc.

b2 St pubescent in lines; caps strigillose, eglandular; lvs gen petiolate; Alas to Ore, e to Ida (*E. halleanum, E. leptocarpum, E. mirabile, E. paddoense*)   var. **macounii** (Trel.) Hitchc.

10b Turions lacking; seeds various

11a Pl glab or only with minute pubescence in the infl or on the ovary, mostly glaucous, often matted at base; lvs lanceolate to ovate, entire to denticulate, 1.5–5.5 cm; petals deep rosy-purple to light pink, 4–8 mm; mont in wet places; BC s through OM and Cas of Wn to s Cal, e to Ida, w Mont, and Utah; smooth w.; 2 vars.   11 E. glaberrimum Barbey

a1 Pl mostly > 3 dm, often br above; lf bls lanceolate to narrowly ovate, 2.5–4.5 cm, barely overlapping; ovaries hairy; Wn to Cal, e to Ida   var. **glaberrimum**

a2 Pl mostly 1–3 dm, simple; lf bls mostly 1–3 cm, ovate to lanceolate, closely crowded and overlapping; ovaries gen glab; common from w Wn to Cal, e to Mont and Utah (*E. atrichum, E. f., E. platyphyllum*)   var. **fastigiatum** (Nutt.) Trel.

11b Pl gen pubescent, not glaucous

12a St (2) 3–10 dm, freely br above midlength; rhizomes short or lacking; lvs (2.5) 3–7 cm, mostly evidently serrulate to serrate; seeds strongly crested-papillate in many parallel longitudinal lines; coma white; lowl to low mts, mostly where wet; gen in much of N Am; Watson's w.; 3 vars.   12 E. watsonii Barbey

a1 Petals 6–10 mm, deep purplish; pl gen reddish on lower lvs and st and on calyx; mostly near the coast, Wn to c Cal (*E. franciscanum*)   var. **watsonii**

a2 Petals commonly not > 6 mm, often pale; pl various

b1 Infl not glandular, often canescent with short, appressed to curved hairs; occ from RM to Wn and s to s Cal (*E. californicum, E. cinerascens, E. griseum, E. macdougalii, E. parishii*)   var. **parishii** (Trel.) Hitchc.

b2 Infl from sparsely to densely glandular; Alas to the Atl coast, s to Cal and Colo (*E. adenocaulon, E. americanum, E. occidentale, E. perplexans, E. praecox*)

var. **occidentale** (Trel.) Hitchc.

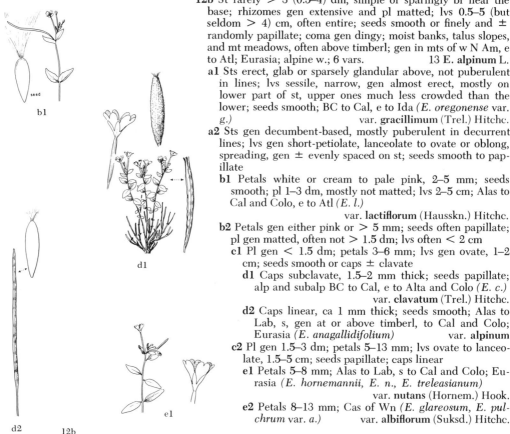

12b St rarely > 3 (0.5–4) dm, simple or sparingly br near the base; rhizomes gen extensive and pl matted; lvs 0.5–5 (but seldom > 4) cm, often entire; seeds smooth or finely and ± randomly papillate; coma gen dingy; moist banks, talus slopes, and mt meadows, often above timberl; gen in mts of w N Am, e to Atl; Eurasia; alpine w.; 6 vars.   13 **E. alpinum** L.

a1 Sts erect, glab or sparsely glandular above, not puberulent in lines; lvs sessile, narrow, gen almost erect, mostly on lower part of st, upper ones much less crowded than the lower; seeds smooth; BC to Cal, e to Ida *(E. oregonense* var. *g.)*   var. **gracillimum** (Trel.) Hitchc.

a2 Sts gen decumbent-based, mostly puberulent in decurrent lines; lvs gen short-petiolate, lanceolate to ovate or oblong, spreading, gen ± evenly spaced on st; seeds smooth to pap-illate

b1 Petals white or cream to pale pink, 2–5 mm; seeds smooth; pl 1–3 dm, mostly not matted; lvs 2–5 cm; Alas to Cal and Colo, e to Atl *(E. l.)*
var. **lactiflorum** (Hausskn.) Hitchc.

b2 Petals gen either pink or > 5 mm; seeds often papillate; pl gen matted, often not > 1.5 dm; lvs often < 2 cm

c1 Pl gen < 1.5 dm; petals 3–6 mm; lvs gen ovate, 1–2 cm; seeds smooth or caps ± clavate

d1 Caps subclavate, 1.5–2 mm thick; seeds papillate; alp and subalp BC to Cal, e to Alta and Colo *(E. c.)*
var. **clavatum** (Trel.) Hitchc.

d2 Caps linear, ca 1 mm thick; seeds smooth; Alas to Lab, s, gen at or above timberl, to Cal and Colo; Eurasia *(E. anagallidifolium)*   var. **alpinum**

c2 Pl gen 1.5–3 dm; petals 5–13 mm; lvs ovate to lanceo-late, 1.5–5 cm; seeds papillate; caps linear

e1 Petals 5–8 mm; Alas to Lab, s to Cal and Colo; Eu-rasia *(E. hornemannii, E. n., E. treleasianum)*
var. **nutans** (Hornem.) Hook.

e2 Petals 8–13 mm; Cas of Wn *(E. glareosum, E. pul-chrum* var. *a.)*   var. **albiflorum** (Suksd.) Hitchc.

Gaura L.   Gaura

Fls spicate or spicate-racemose, whitish to pink, slightly irreg; hypan narrowly cylindric-obconic; stamens 8; stigma 4-lobed; ovary 4-celled or sometimes 1-celled; ovules 1 or 2 per cell; fr hard-ened, indehiscent, ± nutlike, mostly fusiform, 1–4-seeded; ann or per herbs with alt lvs. (Gr *gau-ros*, proud, because of the erect or "proud" fls). Not suitable garden subjects, esp w Cas.

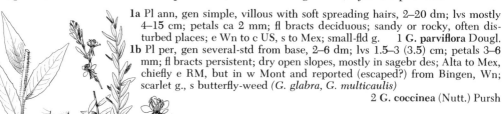

1a Pl ann, gen simple, villous with soft spreading hairs, 2–20 dm; lvs mostly 4–15 cm; petals ca 2 mm; fl bracts deciduous; sandy or rocky, often dis-turbed places; e Wn to c US, s to Mex; small-fld g.   1 **G. parviflora** Dougl.

1b Pl per, gen several-std from base, 2–6 dm; lvs 1.5–3 (3.5) cm; petals 3–6 mm; fl bracts persistent; dry open slopes, mostly in sagebr des; Alta to Mex, chiefly e RM, but in w Mont and reported (escaped?) from Bingen, Wn; scarlet g., s butterfly-weed *(G. glabra, G. multicaulis)*
2 **G. coccinea** (Nutt.) Pursh

Gayophytum Juss.   Groundsmoke; Gayophytum

Fls small, reg, racemose to spicate; free hypan almost lacking, calyx segms nearly distinct, gen reflexed; petals white to pinkish; stamens 8; stigmas capitate; caps linear to linear-clavate, ± torulose, 2-celled, 4-valved; seeds glab to puberulent, not comose, in 1 row per cell; freely br, slender-std ann with linear to lanceolate, alt (opp) lvs. (For G. Gay, author of flora of Chile, and Gr *phyton*, plant, "Gay's plant"). Of no hort imp.

1a Caps sessile or subsessile, slightly if at all constricted between seeds; pl basally br, very lfy above; lvs gen much > internodes

  2a Seeds rarely > 10, ca 1 mm, erect in the caps; ovary glab or nonglandular-pubescent, all valves free at maturity; petals scarcely 1 mm; open slopes, foothills to higher mont; mostly E Cas, Wn to s Cal, e to Mont and Colo; racemed g. *(G. caesium, G. helleri)*   1 **G. racemosum** T. & G.

  2b Seeds 15–20, 0.5–0.75 mm, obliquely aligned in the caps; ovary glab or ± glandular, 2 of valves attached to septum at maturity; petals ca 1 mm; foothills to medium elev in mts; E Cas, Wn to s Cal, e to Ida; Chile; dwarf g. *(G. nuttallii, G. pumilum)*   2 **G. humile** Juss.

1b Caps ped, gen constricted between seeds; pls gen br above, internodes often > lvs

  3a Petals ca 0.5 mm; caps 3–6 mm, < the gen sharply reflexed peds; pl glab; seeds glab, ca 0.5 mm; dry foothills and valleys to lower mts; e Wn to Cal, e to Mont and Colo; hairstem g.   3 **G. ramosissimum** Nutt.

  3b Petals 1–5 mm; caps 4–15 mm, often > the mostly erect or spreading peds; pl glab or pubescent; seeds glab or pubescent, 0.8–1.8 mm

    4a Petals 1–1.8 mm; caps mostly < 1 mm thick, slightly constricted between seeds; pl rather evenly br throughout; sandy or rocky, dry soil of des woodl to ponderosa pine for; sc Wn to se Cal, e to Mont and Colo; diploid (n=7) and scarcely distinguishable from *G. diffusum;* deceptive g.   4 **G. decipiens** Lewis & Szeyk.

    4b petals mostly 1.5–5 mm; caps mostly 1–1.5 mm thick, strongly constricted between seeds; pl br mainly above; open slopes, mostly on dry gravelly soil, valleys and foothills to timberl; gen in much of w US; tetraploid (n=14); spreading g. *(G. eriospermum, G. intermedium, G. lasiospermum, G. d.* var. *parviflorum)*   5 **G. diffusum** T. & G.

Jussiaea L.   Primrose-willow

Fls solitary in lf axils, 5 (4 or 6) -merous; free hypan almost nil; petals yellow (white), deciduous; stamens 10 (8 or 12); stigmas capitate (ours); per herbs (ours) with alt, simple, stip (ours), petiolate lvs. (For Bernard de Jussieu, 1699–1777, famous French taxonomist).

J. uruguayensis Camb. Glab or slightly hairy per; sts prostrate (or floating) to erect, reddish, often matted; lf bl entire, elliptic to spatulate or lanceolate, up to 12 cm; petioles 2–3 cm; sepals 6–13 mm; petals 12–20 mm, deep yellow; caps 10-nerved, up to 25 × 3–4 mm; seeds pendulous, ca 1.5 mm; swamps, lakes, and streams; S Am, intro in much of US; along CR near Portland.

Ludwigia L.   False Loosestrife

Fls sessile in upper axils, greenish, 4-merous; free hypan almost nil; petals lacking; stamens 4, filaments very short; stigma capitate; caps turbinate, ∝ -seeded; seeds glab; per herbs with opp lvs and prostrate sts (ours). (For C. G. Ludwig, 1709–1773, German botanist).

L. palustris (L.) Ell. Water-purslane. Glab, succulent per; lf bl 1–3.5 cm, ovate-elliptic to obovate, with petioles nearly as long; fls ca 2 mm; caps 2.5–4 mm, 4-sided, basally 2-bracteolate; marshes and bogs, widespread in N Am; Eurasia, Africa; with 2 vars.

  a1 Lf bl scarcely twice as long as broad; Cas to Atl coast

              var. **americana** (DC.) Fern. & Griscom

  a2 Lf bl > twice as long as broad; w Cas and Sierra Nevada

              var. **pacifica** Fern. & Griscom

Oenothera L.   Evening-primrose; Oenothera

Fls white or yellow, sometimes aging to pink or purplish, often nocturnal or fragrant, or both, racemose, spicate, or sessile among basal lvs; hypan often very prominent, sepals distinct to connate and turned to one side; petals clawless; stamens 8; stigma globose to deeply 4-lobed; caps woody to membranous, straight to ± coiled, dehiscent, 4-celled; seeds ∝, not hairy; ann to per, acaulescent to caulescent herbs with alt or basal, simple to pinnatifid lvs. (Gr name, used by Theophrastus, said to mean wine-scented). *(Calylophus; Camissonia; Chylismia; Jussiaea,* in part; *Lavauxia; Meriolix; Pachylophus; Sphaerostigma; Taraxia).* Most spp. are showy and several are widely cult, but mostly they are not suited to conditions w Cas.

1a Fls borne among basal rosettes of lvs; sts (if any) short and concealed by the lvs and fls; hypan several times as long as the ovary; per except *O. palmeri*   **Group I**, lead **5a**

1b Fls on lfy sts; hypan often < ovary; ann, bien, or per

    2a Petals white or pinkish, > 1 cm; fls gen opening in the early evening, fading by midmorning; stigma lobes linear, mostly 4–6 mm *(Anogra)*

        3a Hypan and sepals glandular-pubescent; pl rhizomatous per 4–10 dm; petals 1.5–2.5 cm; caps erect, linear, 2–3 cm; grassl and hillsides, mainly e Cont Div; RM, Can to Colo, e to Minn; white e., Nuttall's e. *(O. albicaulis* var. *n.)*   1 **O. nuttallii** Sweet

        3b Hypan not glandular-pubescent; pl, petals, and caps various

            4a Pl ± caespitose ann or bien, mostly 1–2 dm, strigillose to sublanate, brs prostrate to decumbent; lf bls oblanceolate-obovate, 2–7 cm, with slender petiole mostly = bl; petals (1) 2–4 cm; caps 2–7 cm × 3–5 mm; widespread in des; e Ore to s Cal, e to Utah and Ariz; var. *piperi* Munz, barely reaching the edge of our range; hairy e.   2 **O. deltoides** Torr. & Frem.

            4b Pl rhizomatous per 1–5 dm, glab to strigillose; lf bls linear to linear-lanceolate, mostly 2–5 cm, petiole scarcely 1/4 as long; petals 1.5–3 cm; caps 1.5–3.5 cm × 2–3 mm; sandy or gravelly soil, esp on dunes; e Cas, Wn to Ariz and e to Ida and NM; pale e., white-std e.; 2 vars.   3 **O. pallida** Lindl.

                a1 Pl grayish-strigillose, gen < 2 dm; Fremont Co, Ida   var. **idahoensis** Munz

                a2 Pl glab or subglab, gen > 2 dm; range of the sp. *(O. leptophylla)*   var. **pallida**

    2b Petals either yellow or < 1 cm; fls mostly open during the day; stigmas capitate, discoid, or shortly to linearly lobed   **Group II**, lead **10a**

Group I

5a Petals white or yellow, aging to pink or purple, 1–5 cm; stigma lobes linear, mostly > 3 mm; hypan deciduous from fr

    6a Petals yellow, aging purple, 1–2 cm; caps wing-margined, woody, ovoid, 1–2 cm; mostly in hard-packed soil of swales or margins of vernal pools, plains to lower foothills; Sask to Mex, w to Ida and Cal, disjunct (?) along Yakima R, Wn; long-tubed e.   4 **O. flava** (Nels.) Garrett

    6b Petals white, aging pink or purplish, 2.5–4.5 cm; caps woody, oblong-ovoid, 1–4 cm, angled but not winged; roadcuts, dry hills, and talus slopes to ponderosa pine for; widespread e Cas; rock-rose, des e., butte primrose; several vars.   5 **O. caespitosa** Nutt.

        a1 Caps short-ped, straight, 3-4 cm, tuberculate; pl mostly caulescent, villous-hirsute; CR, e Wn, to Cal, e to Ida and Colo *(O. m., P. m., P. cylindrocarpus)*   var. **marginata** (Nutt.) Munz

        a2 Caps sessile, gen curved, mostly < 3 cm, often smooth; pl often acaulescent

            b1 Pl almost or quite glab

                c1 Caps tuberculate on the angles, 1–2 cm; pl acaulescent; e Ore to Mont, Wyo, and Daks *(P. c., P. glabra)*   var. **caespitosa**

                c2 Caps not at all tuberculate, much > 2 cm; pl caulescent; Fremont Co, Ida *(P. p.)*   var. **psammophila** (Nels. & Macbr.) Munz

            b2 Pl pubescent, often villous

                d1 Caps tubercled, pubescent; pl canescent with appressed pubescence; e Wn to Nev *(P. canescens, O. marginata* var. *p.)*   var. **purpurea** (Wats.) Munz

                d2 Caps not tubercled, gen glab; lvs strigillose to villous on the veins; e Ore to Mont and Ida *(O. m., P. m., O. idahoensis)*   var. **montana** (Nutt.) Durand

5b Petals yellow, gen not aging to purple, < 2 cm; stigma globose or discoid; hypan filiform, persistent, flared only at the top *(Taraxia)*

    7a Pl ann; lvs 2–7 mm broad, linear, entire to denticulate; petals 3–5 mm; caps ovoid, 5–7 mm, hardened, 4-angled below, winged above; sandy des; se Ore to s Cal, e to Nev and Ariz; just within our range in n Malheur Co, Ore, also in wc Nev and more common in des s Cal; Palmer's e.   6 **O. palmeri** Wats.

    7b Pl per; lvs in part > 10 mm broad, gen lanceolate or oblong, dentate to pinnatifid; petals 6–16 mm; caps various, but not winged above; mea-

8a

9a

9b

11a

12a

13a

15a

15b

dows, swales, and river banks, grassl and sagebr des to midmont, where moist in spring but gen dry by midsummer

8a Lvs entire to dentate or lobed, glab or subglab (if pubescent, only so on margins and veins); caps glab, 12–18 mm, 4-angled, narrowed to a slender point; e Wn to Sierran Cal, e to Mont and Colo; long-lf e. (*O. heterantha, T. h.*) **7 O. subacaulis** (Pursh) Garrett

8b Lvs pinnatifid, strigose to hirsute; caps gen hairy

   9a Hypan 0.6–2.5 cm; style 3–7 mm, not > anthers; petals 6–8 mm; caps gen curved, 10–20 × 2–5 mm; drier meadowl and stream banks; e Ore to ne Cal, e to Sask, Mont, and Wyo; short-fld e. **8 O. breviflora** T. & G.

   9b Hypan (2) 2.5–6.5 mm; style 9–20 mm, > anthers; petals 10–20 mm; caps straight, 10–25 × 3–5 mm; Wn s, e Cas, to Sierran Cal, e to Ida and Mont; tansy-lf e. (*O. nuttallii* T. & G., *T. longiflora, T. tikurana*) **9 O. tanacetifolia** T. & G.

Group II

10a Hypan mostly at least 2 cm; petals yellow; stigmas distinct, linear, 4–6 mm; pl bien, mostly 3–12 dm (*Onagra*)

   11a Petals 10–20 mm, seldom fading to reddish; pl grayish-strigose and pustular-hirsute; meadows and stream banks, gen where moist, plains to lower mts, widespread; e Wn to Cal, e to Minn and Kans; common e. (*O. cheradophila, O. depressa, O. rydbergii, O. muricata* of auth.) **10 O. strigosa** Mkze. & Bush

   11b Petals mostly 25–40 (50) mm, gen fading to reddish; pl mostly crisp-puberulent or hirsute, or both

      12a Petals 25–40 mm, pale yellow, fading to orange-red; lvs plane to wrinkled; st lvs with bl 6–12 cm, scarcely 1/4 as broad; sts strongly pustulose-hirsute and finely strigose; widespread from the plains into the mts at medium elev, gen where moist; Hooker's e.; 2 vars. **11 O. hookeri** T. & G.

         a1 Sepals greenish; petals not fading to red; se Wn to c and sw Ida, esp near Boise (*O. o., O. h.* ssp. *o.*) **var. ornata** (Nels.) Munz

         a2 Sepals reddish; petals fading to red; se Ore to Cal, e to Nev, Utah, and Colo (*O. macbrideae, O. h.* ssp. *a.*) **var. angustifolia** Gates

      12b Petals 35–50 mm, deep yellow, gen fading to red; lvs strongly crinkled; st lvs with lf bl 5–10 cm, mostly at least 1/3 as wide; sts pustulose-hirsute and crisp-puberulent; garden hybrid, occ escaped, BC to Cal, w Cas; red-sepaled e. **× O. erythrosepala** Borb.

10b Hypan much < 2 cm; petals white or pink to yellow; stigma capitate to discoid or shallowly lobed

   13a Pl per, suffrutescent, many-std from the base; free hypan 6–15 mm; petals yellow, 8–25 mm; stigma discoid, very shallowly lobed; caps linear, 15–30 mm; prairies, fields, and sand dunes; mostly e RM, Alta and Man to Tex, in our area near Bozeman, Mont; shrubby e. (*O. leucocarpa*) **12 O. serrulata** Nutt.

   13b Pl ann; free hypan often < 6 mm; petals sometimes white or < 8 mm; stigma globose to capitate; caps various

      14a Caps slender-ped, not attenuate distally, but gen ± clavate; petals yellow (*Chylismia*)

         15a Petals 2–3 (4) mm; stamens = style and stigma; sts gen ± glandular-pubescent above, very slender; seeds ca 1.7 mm; sagebr des in our area, esp on sandy soil; e Ore to Cal, e to Ida, Wyo, and Colo; naked-std e. (*O. brevipes* var. *s., Ch. s.* var. *seorsa, O. s.* var. *seorsa*); ours the ssp. *brachycarpa* Raven **13 O. scapoidea** Nutt.

         15b Petals mostly 4–10 mm; stamens < style and stigma; sts glab or finely puberulent, fairly sturdy; seeds ca 1.2 mm; sandy des; e Ore to Cal, e to Ida and Utah; club-fr e. (*O. scapoidea* var. *c., O. c.* ssp. *citrina, Ch. cruciformis, O. c.*); ours is var. *cruciformis* (Kell.) Munz **14 O. claviformis** Torr. & Frem.

      14b Caps sessile or subsessile (never slender-ped), not clavate but gen attenuate distally, often contorted; petals often white (*Sphaerostigma*)

         16a Fls white (often aging to pink), opening in the evening; lvs in part often > 3 (sometimes up to 15) mm broad

            17a Petals 1–1.5 mm; style 1.2–3 mm, slightly > stamens; infl tip ± erect; pl fl from the base upward; caps linear, scarcely en-

a1

a2

18b

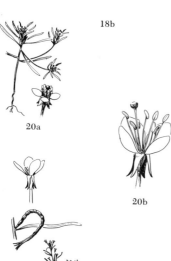

20a

20b

21a

larged at base, 12–25 mm, gen strongly contorted to coiled; lvs entire; e Cas, c Wn to n Cal, e across Ida to Wyo and s to Nev, Utah, and w Colo; small-fl e. (*O. alyssoides* var. *minutiflora, S. a.* var. *m., O. m.* var. *cusickii*)   15 **O. minor** (Nels.) Munz

17b Petals 1.5–9 mm; style 3.2–15 mm, < to much > stamens; tip of infl gen nodding; pl often not fl from the base; caps straight to contorted, 8–35 mm, often considerably thickened at base; lvs often irreg toothed

18a Petals 1.5–2.5 mm; stamens = style and stigma; seeds of lower part of caps strongly papillose, dark brown, those of upper part of caps nearly smooth, whitish; occ from c Wn to e Ore and adj s Ida; dwarf e. (*O. boothii* var. *p., S. b.* var. *p.*)
                                        16 **O. pygmaea** Dougl.

18b Petals 3–7.5 mm; stamens < style and stigma; seeds in gen all alike, none strongly papillose; e Cas, sc and se Wn to Cal and n Mex, e to Ida, Utah, and Ariz; 2 sspp.
                                        17 **O. boothii** Dougl.

a1 Pl strongly strigillose, rarely at all glandular, 0.5–3.5 dm; caps scarcely 1.5 mm thick at base, greatly contorted; lvs thin, gen entire; petals 3.5–5 mm; in our area in ec Ore and adj Ida, s to ne Cal, Nev, and Utah; alyssum-like e. (*O. a., S. a., S. implexum*)   ssp. **alyssoides** (H. & A.) Munz

a2 Pl villous and ± glandular, 1.5–6 dm; caps gen at least 1.5 mm thick at base, contorted; lvs thickish, gen coarsely toothed; petals 3.5–9 mm; sc and se Wn to e Ore and adj and se Ida, s to Nev and nw Ariz; Booth's e. (*O. gauraeflora* var. *b.*)   ssp. **boothii**

16b Fls yellow, (1) 1.5–4 (5) mm, gen opening in the morning; lvs rarely > 2.5 mm broad

19a Caps 4–15 mm, linear-fusiform to fusiform-lanceolate, mostly slightly thickened at base where 1–1.5 mm thick, straight or curved; br sts very slender, lfless except for the terminal lfy, crowded infl

20a Petals 1–2 (2.3) mm; style 1.7–3 mm; e Cas, sc BC to ne Cal, and Nev, e to Alta, w Mont, Ida, Wyo, and Utah; obscure e. (*O. a.* var. *anomala*)   18 **O. andina** Nutt.

20b Petals 2.5–5 mm; style 4.5–6 mm; sagebr des e Cas, Okanogan Co, Wn, s to CR in Klickitat Co, and to Multomah Co, Ore; Hilgard's e. (*S. andina* var. *h., O. a.* var. *h.*)
                                        19 **O. hilgardii** Greene

19b Caps linear, 20–40 × scarcely 1 mm, not thickened at base, arched or curved to much-contorted; br sts sturdy, gen ± lfy throughout, the infl not crowded at the tip

21a Caps sessile (seed-bearing almost to the subtending lf); pl rarely at all glandular, often strongly pubescent with short, appressed to spreading, stout hairs; petals mostly 2.5–5 mm; lvs often remotely dentate; s VI and is of PS, and from extreme sc Wn s to sw Ore and much of Cal, e to wc and sw Ida and extreme wc Nev; contorted-pod e. (*O. cheiranthifolia* var. *c., O. campestris* var. *cruciata, O. torulosa* f. *c.*)
                                        20 **O. contorta** (Dougl.) Kearney

21b Caps short-ped, seed-bearing to within 1–2 mm of the subtending lf; pl gen glandular-pilose at least in the infl; petals mostly 2–3.5 (1.9–3.6) mm; lvs entire or less strongly dentate; e Cas, Okanogan Co, Wn, to ne Cal, e across s Ida to Wyo, and through Nev and Utah to w Colo; hairy e. (*Camissonia pusilla, C. contorta* var. *p., S. c.* var. *p., O. contorta* var. *flexuosa* of Munz)   21 **O. pubens** (Wats.) Munz

HALORAGACEAE   Water-milfoil Family

Fls axillary (ours), very small, reg, epig, mostly ♀ ♂ or ♀, ♂; free hypan almost nil, calyx segms 2–4 (0); petals 0 or 2 or 4, small and soon deciduous; stamens 4 or 8; pistil mostly 4-carpellary, ovary 1–4-celled; ovules 1 per cell, pendulous; stigmas often plumose; fr drupaceous or nutlike, carpels gen separating as fr matures; ours aquatic per herbs with whorled, pinnately dissected lvs.

Myriophyllum L.   Water-milfoil

Fls 1 per axil, ♂ gen uppermost, all gen subtended by 2 tiny bracteoles; calyx lobes 4, small; petals 4, quickly deciduous (lacking); stamens 4 or 8; emersed lvs pectinately dissected to subentire or reduced, submersed lvs filiformly pinnatifid and sometimes stip. (Gr *myrios*, thousand, and *phyllon*, leaf).

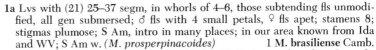

1a Lvs with (21) 25–37 segm, in whorls of 4–6, those subtending fls unmodified, all gen submersed; ♂ fls with 4 small petals, ♀ fls apet; stamens 8; stigmas plumose; S Am, intro in many places; in our area known from Ida and WV; S Am w. (*M. prosperpinacoides*)   1 **M. brasiliense** Camb.
1b Lvs mostly with 13–23 segm, in whorls of 3–6, those subtending fls modified, gen much-reduced in size, often emersed; fls various
   2a Bract lvs linear to narrowly oblong, 10–15 mm, 4–5 per whorl, serrate; petals tardily deciduous; stigmas very short, nearly smooth; stamens 4; ponds and slow streams; Wn and Ore, w Cas, to Mex and e to NY and se US; western w.   2 **M. hippuroides** Nutt.
   2b Bract lvs gen either ± ovate or much < 10 mm, 3–5 per whorl, sometimes dissected; petals soon deciduous; stigmas plumose; stamens 8
      3a Infl often forking; bracts oblong-ovate, 7–10 mm, 4–5 per whorl, pectinate to serrulate; bracteoles 1–2 mm, whitish, erose to pectinate; S Am and Australia; known from n of Bend, Ore, along Deschutes R; waterwort w.   3 **M. elatinoides** Gaud.
      3b Infl simple; bracts subentire to deeply pectinate, mostly much < 7 mm; bracteoles 1–1.5 mm, mostly greenish with whitish margins, entire to minutely erose; quiet, often brackish water; widespread in N Am, our common sp.; BC to Cal, e to Atl; Eurasia; spiked w.; 2 vars.
                                                                        4 **M. spicatum** L.
         a1 Bract lvs scarcely = frs, serrulate to entire; common (*M. e.*)
                                                          var. **exalbescens** (Fern.) Jeps.
         a2 Bract lvs mostly > frs, pectinate to serrate; rare in w US, more common e and in Eurasia (*M. verticillatum*)   var. **spicatum**

HIPPURIDACEAE   Mare's-tail Family

Fls gen ♀ (to ♀ ♂); perianth apparently lacking, but calyx adnate to ovary and very obscurely if at all lobed at tip; stamen 1; pistil 1-carpellary, style slender, stigmatic full length, in ♂ fls gen lying in groove between lobes of anther; fr nutlike, indehiscent, 1-celled and 1-seeded.

Hippuris L.   Mare's-tail

Glab, per, aquatic or amphibious herbs with creeping rhizomes; lvs entire, linear, sessile, in whorls of 4–12; fls single in axils of lvs of upper whorls, sessile. (Gr *hippos*, horse, and *oura*, tail).

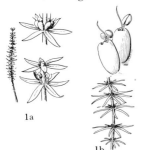

1a Lvs 5–8 per whorl, 5–10 × 0.5–1 mm; fls ♀ ♂, ♂ mostly in whorls below ♀; st 2–10 cm × scarcely 0.5 mm; mature ovary ca 1 mm; streams and mossy banks or wet meadows; OM and Cas, Wn, n to Alas, e to e BC; mt m.   1 **H. montana** Ledeb.
1b Lvs (4) 6–12 per whorl, (8) 10–50 × 1–2 mm; fls mostly ♀; st 5–30 cm × 1.5–2.5 mm; mature ovary ca 2 mm; streams and ponds to shallow lakes, gen at least partially emersed; ± circumboreal, s in N Am to Cal, NM, and c and e US; common m.   2 **H. vulgaris** L.

ARALIACEAE   Ginseng Family

Fls small, gen greenish, polypet, epig, ♂ (ours) to ♀ ♂, in globose-capitate umbels borne in corymbs, racemes, or pans; calyx lobes gen 5, very small; free hypan nil; petals 5 (3–10), often ± connate and calyptrate; stamens 5 (ours); pistil 5 (4–6) -carpellary, ovary covered by a fleshy disc, mostly 5 (1–15) -celled, ovules 1 per cell; styles as many as carpels, or only 1; fr gen baccate; per, often armed, sometimes climbing shrubs or trees with large, alt (ours), mostly palmately lobed to compound lvs.

1a Pls armed with slender spines   **Oplopanax**
1b Pls unarmed
  2a Lvs large, compound, basal; pl not scandent   **Aralia**
  2b Lvs simple, ∝ along the scandent sts   **Hedera**

2a   2b   1a

Aralia L.

Fls ⚥, pan to umbellate, 5-merous; styles distinct to base; fr baccate, 2–5-seeded; per herbs (ours) with large, alt, pinnately or ternately compound to decompound lvs. (Said to be derived from French-Canadian *aralie*, name under which the pl, collected in Quebec, was sent to Tournefort, who used name *Aralia* for first time).

A. nudicaulis L. Wild sarsaparilla. Rhizomatous; fl st sometimes barely reaching ground level; lvs gen single, 3–5 dm, biternate to ternate-pinnate, ultimate segms ± oblique, ovate-oblong, 5–12 cm, serrate to biserrate; infl 3–7 globose, corymbose to pan umbels, < lvs; petals greenish-white, ca 2.5 mm; fr dark purple, 6–8 mm; moist shade; e BC and ne Wn to Mont, s to Colo, e to e Can and Atl States; of slight value in the garden.

Hedera L.   Ivy

Fls ⚥, 5-merous; petals ± fleshy, greenish; style single; fr berry, 2–5-seeded; evergreen, mostly stellate, climbing shrubs with aerial roots and alt, leathery, long-petioled lvs. (Ancient classical name for ivy).

H. helix L. English i. Sts up to 30 m; lvs broadly ovate to triangular, entire to deeply 3–5-lobed, 4–10 cm; fls 5–7 mm, in 1–several racemose, globose umbels; fr deep bluish-black, 6–9 mm; cult European vine, freely disseminated by birds, often estab in w Wn and Ore; perhaps the best known evergreen vine in cult.

Oplopanax (T. & G.) Miq.

Fls mostly ⚥, greenish-white; ovary 2 (3)-celled, styles 2 (3), free; fr 2–3-seeded berry; thick-std, deciduous, fiercely armed shrubs with large, palmately lobed lvs. (Gr *hoplon*, tool or weapon, and *Panax*, another genus).

O. horridum (Smith) Miq. Devil's club. Pl 1–3 m, armed throughout with yellow spines 5–10 mm; sts punky, 1–4 cm thick; lf bl shallowly 7–9-lobed, 1–3.5 dm broad, cordate; fls 5–6 mm, subsessile in small capitate umbels in elongate pans or racemes as much as 2.5 dm; berry bright red, 5–8 mm; moist woods, esp near streams; Alas to Ore, e and w Cas, e to Ida and Mont, and in Mich and Ont (*Echinopanax h., Fatsia h., Panax h., Ricinophyllum h.*); interesting and attractive pl for the bog garden.

UMBELLIFERAE   Parsley Family

Fls epig, polypet, mostly reg and ⚥, occ some of them sterile or neutral and then often ± irreg; calyx teeth 5, or obsolete; petals 5, gen inflexed at tip, gen yellow or white, less often purple or other colors; stamens 5, inserted on an epig disk, alt with petals; ovary bicarpellate, 2-celled, each cell 1-ovuled; styles 2, often swollen at base to form a stylopodium; fr a dry schizocarp, consisting of 2 halves (mericarps) united by their faces (the commissure), each mericarp typically 5-nerved, gen with 1 or more vertical oil tubes in the intervals between the nerves and on the commissure; mericarps separating at maturity, revealing a slender central carpophore to which they are attached apically, the carpophore entire to deeply bifid, or sometimes obsolete by adnation of the separate halves to the commissural faces of the mericarps; ann or per herbs, rarely woody at base, often hollow-std, gen aromatic; lvs alt or rarely opp, or sometimes all basal, gen with sheathing petiole and large, compound to variously cleft or dissected bl, simple and entire in a few genera; fls typically in compound umbels, the umbel gen subtended by a few bracts (invol), and umbellets by bractlets (involucel), less often in simple or proliferating umbels, or in compound heads, the involucel sometimes dimidiate (asymmetrical, as though one half were missing). The fls are so uniform, and the vegetative features often so variable among closely allied spp. that the genera must often be separated largely by technical characters of the mature fr.

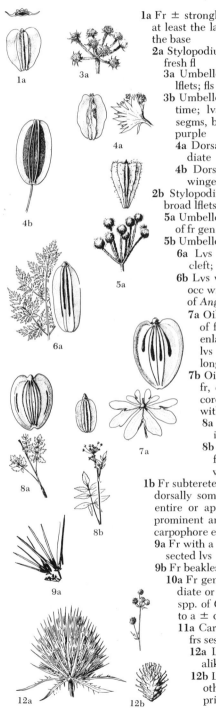

KEY EMPHASIZING CHARACTERS OF THE FRUIT
(Key Emphasizing Vegetative Characters on p. 317)

1a Fr ± strongly compressed dorsally (parallel to the commissure), unarmed, at least the lateral ribs gen winged, only occ wingless; carpophore bifid to the base

2a Stylopodium obsolete or nearly so, though sometimes ± evident in the fresh fl

3a Umbellets capitate, the fls sessile; maritime pls with definite broad lflets; fls white   **Glehnia**

3b Umbellets with ped fls, the peds sometimes very short; pls not maritime; lvs gen ± dissected, with small and often narrow ultimate segms, but sometimes with definite broad lflets; fls yellow to white or purple

4a Dorsal ribs of fr gen ± winged, rarely wingless; involucel dimidiate   **Cymopterus**

4b Dorsal ribs of fr gen filiform and wingless, occ very narrowly winged; involucel dimidiate or more often not   **Lomatium**

2b Stylopodium ± well developed; lvs in most spp. with well-defined broad lflets

5a Umbellets capitate, fls sessile; fr cuneate; dorsal as well as lateral ribs of fr gen winged; fls white   **Sphenosciadium**

5b Umbellets with ped fls; fr not cuneate; dorsal ribs winged or wingless

6a Lvs ± dissected, without well-defined lflets or with lflets ± cleft; dorsal ribs of fr narrowly winged; maritime pls   **Conioselinum**

6b Lvs with ± definite, entire or merely toothed lflets, or some lflets occ with a few irreg lobes; fr various; pls not maritime except 2 spp. of *Angelica*

7a Oil tubes reaching only part way from the apex toward the base of fr, readily visible through the pericarp; marginal fls of umbel enlarged and radiant, the outer corolla lobes commonly 2-cleft; lvs in our sp. trifoliolate, with very large lflets mostly 1–4 dm long and wide; fls white   **Heracleum**

7b Oil tubes extending all the way from the apex to the base of the fr, conspicuous or inconspicuous; marginal fls not enlarged, the corolla lobes not 2-cleft; lvs pinnately to ternately compound, with > 3 lflets, these gen < 1 dm wide

8a Intro bien weed with yellow fls and merely once-pinnate lvs; invol and involucel gen wanting   **Pastinaca**

8b Native pers, not weedy; lvs often 2–several times compound; fls white (pink) or in 1 sp. yellowish; invol and involucel often well developed   **Angelica**

1b Fr subterete or ± compressed laterally (at right angles to the commissure), dorsally somewhat compressed only in *Daucus*, which has armed fr and entire or apically cleft carpophore; ribs of fr varying from obsolete to prominent and corky-thickened, but seldom (except *Ligusticum*) winged; carpophore entire or variously bifid, or wanting

9a Fr with a conspicuous beak much > the body; intro ann weed with dissected lvs   **Scandix**

9b Fr beakless or with a short beak distinctly < body

10a Fr gen with conspicuous scales, or bristly hairs, or hooked or glochidiate or barbed prickles, or tubercles, glab and unarmed only in a few spp. of *Osmorhiza* and *Anthriscus* that have the fr narrowed at the tip to a ± distinct short beak

11a Carpophore wanting; calyx teeth well developed and prominent; frs sessile or nearly so, the ribs gen obsolete; our spp. per

12a Lvs simple, entire or merely toothed; infl capitate, the fls all alike and sessile; fr in our spp. covered with scales   **Eryngium**

12b Lvs compound or dissected; fls of 2 kinds, some ♂ and gen ped, others ♀ and sessile or nearly so; fr in our spp. with uncinate prickles   **Sanicula**

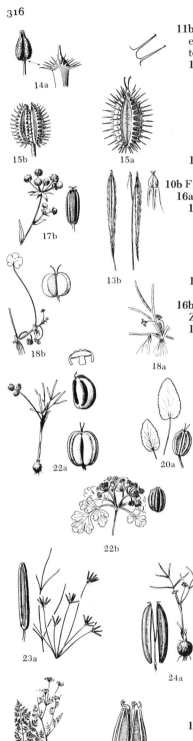

11b Carpophore well developed, entire or bifid at the apex; calyx teeth evident to more often minute or obsolete; frs ped; ribs of fr evident to obsolete

  13a Pls ann or bien, often weedy; lvs in our spp. dissected into small and narrow ultimate segms; native and intro spp.

    14a Fr shortly but distinctly beaked; ribs obsolete   **Anthriscus**

    14b Fr beakless; ribs evident

      15a Fr dorsally somewhat compressed, provided with glochidiate or barbed prickles   **Daucus**

      15b Fr laterally compressed, provided with uncinate prickles   **Caucalis**

  13b Pls per, native, not weedy; lvs with well defined broad lflets; fr merely bristly-hispid or glab, beakless or short-beaked **Osmorhiza**

10b Fr unarmed and commonly glab, not beaked

  16a Lvs all simple, entire or toothed to sometimes palmately lobed

    17a Umbels simple (sometimes proliferous, or aggregated into a scarcely umbellate infl); carpophore wanting; pls fibrous-rooted, growing in water or wet places

      18a Lvs reduced to narrow, hollow, septate phyllodes   **Lilaeopsis**

      18b Lvs with a distinct normal petiole and broad blade   **Hydrocotyle**

    17b Umbels compound; carpophore present, bifid to the base; pls taprooted, growing in upland habitats   **Bupleurum**

  16b Lvs compound or dissected (basal lvs merely toothed in our sp. of *Zizia*)

    19a Stylopodium obsolete or nearly so; pls (except *Zizia*) low, scapose or subscapose, not > ca 3 dm

      20a Pls per from a cluster of fleshy roots, mostly growing in wet places; calyx teeth well developed; carpophore bifid ca halfway to the base; fls yellow; our sp. with the basal lvs simple, merely toothed   **Zizia**

      20b Pls per from a taproot (with or without a br caudex), or from a thickened, cormlike root, mostly growing in dry places; calyx teeth, carpophore, and fl-color various; basal lvs compound or dissected

        21a Carpophore wanting

          22a Lateral ribs of fr developed into inflexed corky wings, a corky riblike projection also present along the length of the commissural face of each mericarp; lf segms elongate and narrow; root sometimes short and tuberous-thickened, sometimes more elongate   **Orogenia**

          22b Lateral and dorsal ribs of fr merely raised and thickened, with or without an irreg vestige of a wing; commissural face of the mericarp without a projection; primary lf segms 3–5, relatively broad and ± confluent, merely toothed or lobed; root thickened and elongate   **Rhysopterus**

        21b Carpophore present, entire to deeply cleft

          23a Frs linear, mostly 8–10 × 1–1.5 mm; n Ida and adj Wn (*L. orogenioides*)   **Lomatium**

          23b Frs linear-oblong or broader, 2–7 mm

            24a Calyx teeth in our spp. minute or obsolete; ribs of fr in our spp. filiform and inconspicuous, although fr sometimes grooved-sulcate; our spp. occurring in Wn   **Tauschia**

            24b Calyx teeth well developed and ± conspicuous; ribs of fr thickened and conspicuous; our spp. in Ida and Mont   **Musineon**

    19b Stylopodium ± well developed; pls often taller and lfy-std

      25a Carpophore entire; st purple-spotted; oil tubes ∝, small, ± confluent; robust taprooted bien weed with dissected lvs   **Conium**

      25b Carpophore bifid to base, or wanting; st not purple-spotted; oil tubes and habit various

        26a Pls taprooted; carpophore bifid to base, the halves distinct, not adnate to the mericarps

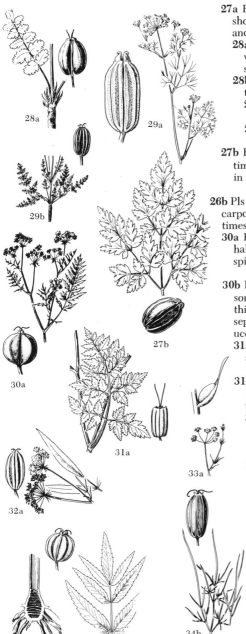

27a Ribs of fr not winged; intro; ± weedy biens or rather short-lived pers; lvs, or some of them, dissected into small and narrow ultimate segms

  28a Lowermost lvs (in our sp.) merely once compound and with well-defined broad lflets; fls white; our sp. a rather short-lived per   **Pimpinella**

  28b Lowermost lvs, when well developed, dissected like the others, without well-defined broad lflets

    29a Fls yellow; st glaucous; our sp. a robust, short-lived per, 1–2 m when well developed   **Foeniculum**

    29b Fls white (pink); st not glaucous; our sp. bien, up to ca 1 m   **Carum**

27b Ribs of the fr (except in 1 sp.) distinctly though sometimes rather narrowly winged; native pers, not weedy; lvs in most spp. less dissected and with evident lflets   **Ligusticum**

26b Pls with fibrous or tuberous-thickened roots, not taprooted; carpophore bifid to the base, or wanting, the halves sometimes adnate to the mericarps

  30a Ribs of fr inconspicuous; oil tubes ∝ and contiguous; halves of the carpophore adnate to the mericarps, inconspicuous; involucel well developed; roots not thickened   **Berula**

  30b Ribs of fr gen conspicuous and corky-thickened, except sometimes in spp. of *Perideridia* which have tuberous-thickened roots and have the halves of the carpophore separate from the mericarps; oil tubes, invol, and involucel various

    31a Calyx teeth well developed, forming a persistent conspicuous crown on the fr; carpophore wanting   **Oenanthe**

    31b Calyx teeth small and inconspicuous, or obsolete; carpophore bifid, sometimes deciduous or the halves adnate to the mericarps

      32a Invol of well-developed, subfoliaceous bracts; roots not tuberous-thickened and pls not bulbiliferous; halves of the carpophore (in our sp.) adnate to the mericarp   **Sium**

      32b Invol of inconspicuous, often scarious bracts, or wanting; gen either some of the roots tuberous-thickened or the upper lvs bearing axillary bulbils; halves of the carpophore distinct, not adnate to the mericarps, persistent or (in *Cicuta*) deciduous

        33a Pls with bulbils in the axils of at least the upper lvs   **Cicuta**

        33b Pls not bulbiliferous

          34a Base of st thickened, hollow, and with well-developed transverse partitions; lvs with well-defined broad lflets   **Cicuta**

          34b Base of st normal in structure, not modified as in *Cicuta*; lvs compound or dissected with mostly narrow and elongate ultimate segms   **Perideridia**

KEY EMPHASIZING VEGETATIVE CHARACTERS

1a Lvs all simple, entire or toothed (palmately lobed in *Hydrocotyle*)

  2a Infl densely capitate, without rays, fls and frs sessile; lvs often spiny-toothed   **Eryngium**

  2b Infl umbellate, fls and frs ± ped; lvs not spiny-toothed

    3a Pls aquatic or semi-aquatic, fibrous-rooted, with simple umbels

      4a Lvs with a broad, mostly rotund-reniform, toothed and gen lobed bl well differentiated from the petiole   **Hydrocotyle**

      4b Lvs reduced to long, narrow phyllodes, without a differentiated bl   **Lilaeopsis**

    3b Pls of uplands, with a br caudex; lvs narrow and entire, but well developed   **Bupleurum**

1b Lvs or most of them compound or very deeply cleft (basal lvs merely toothed in *Zizia*, but the cauline ones compound)

    5a Lvs or many of them with ± well-defined lflets, not dissected into small and narrow segms

        6a Basal lvs merely toothed   **Zizia**

        6b Basal lvs when well developed compound or deeply cleft

            7a Lflets 3, very large, mostly 1–4 dm long and wide   **Heracleum**

            7b Lflets gen > 3, seldom as much as 1 dm wide

                8a Pls per from fibrous or fleshy-thickened, fascicled roots, without a taproot or well-developed caudex, not maritime

                    9a Lvs palmately very deeply cleft or palmately once compound   **Sanicula**

                    9b Lvs pinnately to ternately 1–several times compound

                        10a Base of st thickened, hollow, and with well-developed transverse partitions; some of the roots gen tuberous-thickened; primary lateral veins of lflets tending to be directed to the sinuses between the teeth   **Cicuta**

                        10b Base of st without transverse partitions; roots not tuberous-thickened; veins not directed to the sinuses

                            11a Ribs of fr inconspicuous; calyx teeth minute or obsolete; widespread   **Berula**

                            11b Ribs of fr prominent, ± corky-thickened

                                12a Pls gen reclining or scrambling-ascending; primary lateral veins of lflets tending to be directed to the teeth; calyx teeth well developed; chiefly w Cas   **Oenanthe**

                                12b Pls erect; primary lateral veins of lflets bearing no obvious relation to the teeth or sinuses; calyx teeth minute or obsolete; widespread   **Sium**

                8b Pls ann, bien, or per from a taproot or stout caudex, or with fleshy-fibrous roots from a rhizome-caudex in 1 maritime sp. of *Conioselinum*   **Group I, lead 16a**

    5b Lvs ± dissected into mostly rather small and narrow ultimate segms, without well-defined lflets

        13a St purple-spotted; robust bien weed of moist places, 0.5–3 m, with white fls   **Conium**

        13b St not purple-spotted; habit and habitat various

            14a Pl a robust, short-lived per weed, 1–2 m when well developed, with glaucous sts, yellow fls, and finely dissected lvs, the ultimate segms distinctly < 1 mm wide   **Foeniculum**

            14b Pl distinctly otherwise, differing in 1 or gen more respects from the above

                15a Pls ann or bien, taprooted, often weedy; native and intro spp.; fls white, seldom pink or yellowish, never yellow

                              Group II, lead 27a

                15b Pls per, with or without a taproot; native spp., not weedy

                              Group III, lead 31a

Group I

16a Pls low, scapose or subscapose (or with a pseudoscape), up to 2 dm; inl, nonmaritime spp.

    17a Fr dorsally compressed and with evident lateral wings; lflets evidently toothed or cleft; fls ochroleucous to yellow; Wn and Ore, Cas w   **Lomatium**

    17b Fr slightly flattened laterally, very slightly or not at all winged; lflets and fls various

        18a Pls with a pseudoscape arising from the subterranean crown of a taproot; lflets toothed or lobed; fls white; e Ore   **Rhysopterus**

         18b Pls scapose or subscapose from a taproot and slightly br caudex; lflets mostly entire; fls yellow; Wn Cas   **Tauschia**

16b Pls (except maritime spp.) taller and gen ± lfy-std

    19a Ann or bien weeds

        20a Pl bien; fls yellow; fr unarmed   **Pastinaca**

        20b Pl ann; fls white or pink; fr prickly   **Caucalis**

    19b Pers, not weedy (except *Pimpinella*)

        21a Umbellets capitate, fls and frs sessile

            22a Pls low, subacaulescent, < 2 dm, maritime   **Glehnia**

22b Pls taller, mostly 5–18 dm, lfy-std, inl   **Sphenosciadium**

21b Umbellets not capitate, the fls and frs ped

  23a Fr dorsally flattened

    24a Stylopodium well developed; fls (except 1 sp.) white or occ pinkish (see also *Conioselinum*)   **Angelica**

    24b Stylopodium obsolete or nearly so; fls (except *L. martindalei,* which might occ be sought here) yellow   **Lomatium**

  23b Fr subterete or flattened laterally

    25a Fr linear or linear-oblong to clavate, not at all winged, 3–22 mm; lvs always with well-defined lflets   **Osmorhiza**

    25b Fr broader and often shorter, often some of the ribs winged; lflets not always well defined

      26a Fr 2–2.5 mm, the ribs filiform and rather obscure; Wn   **Pimpinella**

      26b Fr 3–7 mm, the ribs prominent and, except in 1 sp., winged; widespread   **Ligusticum**

Group II

27a Fr distinctly beaked

  28a Beak much > body of fr; pls mostly 1–3 dm   **Scandix**

  28b Beak < body of fr; pls mostly 4–15 dm   **Anthriscus**

27b Fr beakless

  29a Fr prickly

    30a Fr dorsally somewhat compressed, provided with glochidiate or barbed prickles or bristles   **Daucus**

    30b Fr laterally somewhat compressed, gen provided with uncinate prickles   **Caucalis**

  29b Fr unarmed, glab   **Carum**

Group III

31a Fr armed with hooked prickles   **Sanicula**

31b Fr unarmed

  32a Pls maritime   **Conioselinum**

  32b Pls not maritime

    33a Pls definitely taprooted, the taproot sometimes fleshy-thickened, but distinctly elongate; taproot often surmounted by a stout br caudex

      34a Fr essentially wingless, slightly compressed laterally

        35a Fr without a carpophore; Ore   **Rhysopterus**

        35b Fr with a definite carpophore; Mont and Ida   **Musineon**

      34b Fr with at least the lateral ribs produced into evident wing-margins when fully mature, these inflexed in *Orogenia*

        36a Carpophore wanting; W Cas in Ore   **Orogenia**

        36b Carpophore present, bifid to the base; widespread

          37a Body of fr subterete or slightly compressed laterally   **Ligusticum**

          37b Body of fr distinctly compressed dorsally

            38a Dorsal ribs of fr wingless or nearly so, only the lateral ribs distinctly winged; involucel dimidiate or not   **Lomatium**

            38b Dorsal ribs of fr gen winged, the wings sometimes narrower than those of the lateral ribs; involucel dimidiate   **Cymopterus**

    33b Pls with fibrous or fleshy-thickened roots; roots when fleshy-thickened either clustered or solitary, and when solitary always short, not > 2.5 cm

      39a Pls bearing bulbils in the axils of at least the upper lvs   **Cicuta**

      39b Pls not bulbiliferous

        40a Roots strictly fibrous, not at all fleshy; growing in water or very wet places   **Berula**

        40b Roots or some of them ± fleshy-thickened; wet or more often dry places

          41a Fr slightly to strongly flattened laterally, wingless or the mericarps with thickened, incurved lateral wings; fls white or pink

            42a Fr linear, mostly 8–10 mm; n Ida and adj Wn   **Lomatium**

            42b Fr linear-oblong or broader, 2–7 mm

              43a Pls low, up to ca 2 dm, with a solitary, globose, cormlike root

Figure labels: 22b, 24a, 25a, 24b, 26a, 26b, 28a, 28b, 29b, 30a, 30b, 31a, 32a, 35a, 35b, 36a, 37a, 38a, 38b, 39a, 40a, 42a

44a

44b

43b

41b

44a Fr linear-oblong, 5–7 mm, not winged; c Wn
                                           Tauschia
44b Fr oblong-elliptic, 3–4 mm, the lateral ribs developed
    into inflexed corky wings; widespread   **Orogenia**
  43b Pls taller, 2–12 dm, the thickened roots solitary or often
    clustered   **Perideridia**
41b Fr ± strongly flattened dorsally, the marginal ribs with thin
  spreading wings at maturity; fls yellow or white   **Lomatium**

Angelica L.   Angelica

Infl one or gen several compound umbels; invol and involucel of foliaceous or narrow and scarious bracts or bractlets, or wanting; fls white, seldom pink or yellowish; calyx teeth minute or obsolete; stylopodium broadly conic; carpophore bifid to base; fr elliptic-oblong to orbicular, strongly compressed dorsally, glab to scab, hispidulous, or tomentose, the lateral and often also the dorsal ribs evidently winged, or ribs all sometimes merely elevated and corky-thickened but scarcely winged; oil tubes few–∝ ; stout pers, gen single-std from stout taproot; lvs pinnately to ternately 1–3 times compound, with broad, toothed or cleft lflets, uppermost ones often merely bladeless petiolar sheaths. (L *angelus*, angel, referring to properties making some spp. of value in medicine or manufacture of cordials).

1a Pls maritime; invol mostly wanting but involucel present
  2a Lvs essentially glab; dorsal ribs or wings of fr similar to the lateral; oil
    tubes ∝ , adhering to the seed, which is free within the pericarp at maturity (fr unique among our spp.); beaches and coastal bluffs, apparently
    always maritime in our region, but inl elsewhere; Pac Basin, Siberia to n
    Cal, also from Lab to NY; sea-watch, seacoast a. (*Coelopleurum l., C.*
    maritimum, C. longipes)   **1 A. lucida** L.
  2b Lvs tomentose or woolly beneath (thus unique among our spp.); lateral
    wings of fr much better developed than the dorsal; oil tubes few, the
    seed adhering to the pericarp; bluffs and sand dunes along coast, s Wn to
    c Cal; Henderson's a.   **2 A. hendersonii** Coult. & Rose
1b Pls not maritime; invol and involucel various
  3a Fls distinctly yellowish; infl consisting of a single compound umbel
    (rarely 2); invol bracts conspicuous, foliaceous, often = rays of umbel, or
    rarely wanting; se BC and sw Alta to n Ida and nw Mont, extending e
    Cont Div in GNP, Mont; along steams to wet slopes in the mts; Dawson's
    a. (*Thaspium aureum* var. *involucratum*)   **3 A. dawsonii** Wats.
  3b Fls white or pinkish; infl gen 2–more compound umbels; invol gen
    wanting, occ a few small bracts or a single sheathing bract
    4a Rachis of lvs abaxially geniculate at point of insertion of the 1st pair of
      pinnae and commonly also at the points of insertion of successive pinnae, the primary pinnae gen deflexed; Alas to Cal, apparently wholly
      W Cas in US, but e to Selkirk Mts, BC, and to c Alta, in moist places;
      kneeling a.   **4 A. genuflexa** Nutt.
    4b Rachis of lvs not geniculate, the pinnae not deflexed; E Cas, except
      for the widespread *A. arguta*
      5a Lvs oblong to elliptic, pinnate to incompletely bipinnate; involucel
        wanting; sw Mont and adj Ida to Utah and NM; moist places, foothills to midmont; pinnate-lvd a.   **5 A. pinnata** Wats.
      5b Lvs more deltoid, ternate-pinnate; involucel present or absent
        6a Ovaries and fr scab-tuberculate or short-hairy; lflets 1.5–5.5 cm
          7a Involucel gen wanting; dorsal ribs of fr very narrowly winged,
            not much thickened; c Wn to n Ore, moist places; Canby's a.
                            6 A. canbyi Coult. & Rose
          7b Involucel of evident narrow bractlets; dorsal wings of fr thickened and fairly well developed, though not so large as lateral
            ones; w Mont and adj Ida to Utah and Colo; talus slopes and
            other dryish rocky places; mont, often at high altitudes; Rose's
            a.   **7 A. roseana** Hend.
      6b Ovaries and fr glab; lflets mostly 4–14 cm; involucel gen
        wanting; wet places, foothills and valleys to midmont; s BC, Wn,
        Ore, and n Cal, mostly e Cas in Wn, but common in w Ore in
        WV, e to sw Alta, Wyo, and Utah; Lyall's a., sharptooth a. (*A.*
        lyallii, A. piperi)   **8 A. arguta** Nutt.

2b

2a

3a

4a

5a

7b

7a

6b

Anthriscus Hoffm.   Chervil; Anthriscus

Infl of few-rayed compound umbels; invol gen wanting; involucel of several narrow, entire, gen reflexed bractlets; fls white; calyx teeth obsolete; styles short, stylopodium conic; carpophore entire or bifid at apex; fr ovoid to linear, short-beaked at tip, flattened laterally and often constricted at commissure, smooth or bristly, ribs obsolete; oil tubes obscure or obsolete; br, lfy-std anns or biens, seldom per, with petiolate, pinnately compound or dissected lvs. (Ancient Gr name of some umbellifer).

A. scandicina (Weber) Mansfeld. Bur c. Taprooted ann 4–9 dm, st gen glab; lvs basal and cauline, gradually reduced upwards, ± hirsute or hispidulous, bl pinnately dissected into small segms; peduncles slender, 2 cm or less, borne opp upper lvs; rays 3–6, 1–2.5 cm; involucel of a few small lanceolate bractlets; umbellets 3–7-fld, peds short, each with a ring of short flattened hairs at summit; fr ovoid, ca 4 mm, body covered with short thick uncinate prickles, beak short, unarmed; moist, often disturbed places; European, occ intro in e US and Pac states (Scandix anthriscus, A. vulgaris). A. sylvestris (L.) Hoffm., a bien, more robust weed (up to 15 dm), with short-beaked, ribless, smooth fr 5–8 mm, has been collected as an intro in Wn.

Berula Hoffm.   Berula

Infl of compound umbels; invol of evident, narrow, entire or toothed bracts; involucel of narrow bractlets; fls white; calyx teeth minute or obsolete; styles short, stylopodium conic; carpophore bifid to base, the halves inconspicuous, adnate to mericarps; fr elliptic to orbicular, somewhat compressed laterally, glab, ribs inconspicuous; oil tubes ∝ and contiguous; glab, lfy-std pers from fibrous roots; lvs pinnately compound, with entire to toothed or cleft lflets, or some lvs submerged and with filiform-dissected bl. (L name of some umbellifer).

B. erecta (Huds.) Cov. Cut-lvd water-parsnip, stalky b. Pl 2–8 dm, gen freely brd, often ± stoloniferous; submerged, filiform-dissected lvs sometimes present; aerial lvs ± dimorphic, the lower mostly with 7–21 ovate or lance-ovate to elliptic, crenate or crenately lobulate to occ serrate or laciniate lflets up to 5 ×3 cm, the upper with smaller and relatively narrower, mostly sharply toothed or often irreg incised or subpinnatifid lflets; umbels gen several or ∝ ; fr 1.5–2 mm; wet places, valleys and plains; s BC to Baja Cal, e to Ont, NY, and Fla, Europe (Sium e., S. pusillum, B. p.); ours the var. incisa (Torr.) Cronq.

Bupleurum L.   Bupleurum

Infl of compound umbels; invol of conspicuous lfy bracts, or wanting; involucel of broad, gen foliaceous, often connate bractlets; fls yellow or sometimes purplish; calyx teeth obsolete; styles short, the stylopodium depressed-conic; carpophore bifid to base; fr oblong to elliptic or orbicular, slightly flattened laterally and constricted at the commissure, evidently ribbed, otherwise smooth to ± roughened or tuberculate; oil tubes ± ∝ , or obscure or wanting; taprooted, glab and often glaucose, ± lfy-std anns or pers; lvs simple and entire. (Ancient Gr name, meaning ox rib, applied to some pl).

B. americanum Coult. & Rose. Am thorough-wax. Taprooted per with a br caudex, 0.5–5 dm, glab and glaucous; lvs narrow, elongate, entire, parallel-veined, basal ones up to 16 × 1 cm, cauline ones mostly sessile, shorter but often nearly or quite as wide as the basal, upper ones sometimes lanceolate or lance-ovate and clasping; fls yellow or sometimes purple; fr broadly oblong, 3–4 mm, glaucous, ribs prominently raised; rock outcrops and open grassl or meadows, foothills to high mont (above timberl); Alas and Yuk to Wyo, chiefly e Cont Div in our area, but entering Ida in Lemhi and Fremont cos (B. purpureum).

Carum L.

Infl of compound umbels on terminal and lateral peduncles; invol and involucel of a few narrow and inconspicuous bracts or bracteoles, or wanting; fls mostly white; calyx teeth obsolete; styles short, spreading, stylopodium low-conic; carpophore bifid to base; fr oblong to broadly elliptic-oblong, somewhat compressed laterally, glab, evidently ribbed; oil tubes solitary in the intervals, 2 on the commissure; glab, lfy-std pers or biens from a taproot, with pinnately dissected lvs. (Slightly modified name from ancient Gr and Roman names for some umbellifer).

C. carvi L. Caraway. Bien, 3–10 dm; basal and lower cauline lvs well developed, petiolate, the bl mostly 8–17 × 3–10 cm; middle and upper lvs few and ± reduced, but gen petiolate; fls white (reputedly rarely pink); rays of umbel gen 7–14, gen 1–3 cm at anthesis, up to 5 cm in fr; fr 3–4 mm, rather broadly oblong-elliptic; roadsides and meadows; European, ± estab in much of n US, and occ in our range.

Caucalis L.

Infl of compound umbels; invol a few entire or dissected often lfy bracts, or wanting; involucel a few entire to dissected, commonly somewhat scarious bractlets; fls white; calyx teeth evident; petals cuneate or obovate, with slender inflexed tip; styles short, stylopodium thick and conic; carpophore entire or bifid at apex; fr oblong or ovoid, somewhat compressed laterally, with stout, spreading, uncinate prickles along alternate ribs, merely bristly-hairy on the others; oil tubes solitary under alt ribs and 2 on the commissure; taprooted anns with lfy sts and petiolate, pinnatedly compound or dissected lvs. (Ancient Roman name). Several spp. of *Torilis*, a related genus, have been collected in or near the s boundary of our range in Ore, but do not appear to be estab. All have the tubercles or glochidiate or uncinate prickles of the fr gen distributed rather than in a few definite longitudinal rows. *T. nodosa* (L.) Gaertn. has sympodial sts with sessile or short-pedunculate simple capitate umbels opp the lvs. *T. japonica* (Houtt) DC. has compound umbels with 1 narrow bract to each ray. *T. arvensis* (Huds.) Link. and *T. scabra* (Thunb.) DC. have compound umbels without an invol, or with a single invol bract; *T. arvensis* has 2–10-rayed umbels with frs 3–5 mm; *T. scabra* has 2–3-rayed umbels with the fr 5–7 mm.

C. microcarpa H. & A. Cal hedge-parsley. Slender ann 1–4 dm, simple or br above, ± spreading-hirsute throughout; lvs chiefly cauline, bl mostly 2–6 × 2–5 cm, pinnately dissected into small, narrow ultimate segms; invol of several fairly well-developed, scarcely modified lvs; rays of umbels 1–9, 1–8 cm; involucel of several pinnatifid to entire bractlets; fr oblong, 3–7 mm; moist places; s BC s to Baja Cal, e and w Cas, e to Ida and n Mex. *C. latifolia* L., a stouter European sp. with less-dissected lvs (mostly only 1-pinnate, the segms of lflets 1–8 cm × 5–20 mm) and frs 10–12 mm, has been reported as a casual intro in Ore.

Cicuta L.   Water-hemlock

Umbels compound; invol wanting or a few inconspicuous narrow bracts; involucel of several narrow bractlets, rarely wanting; fls white or greenish; calyx teeth evident; styles short, spreading, stylopodium depressed or low-conic; carpophore bifid to base, deciduous; fr glab, elliptic or ovate to orbicular, compressed laterally, ribs gen prominent and corky; oil tubes solitary in intervals, 2 on the commissure; glab, violently poisonous pers; st tuberous-thickened at base and hollow with prominent transverse partitions; roots gen in part tuberous-thickened; lvs 1–3 times pinnate, lflets well-developed to mere linear segms. (Ancient L name of some poisonous umbellifer).

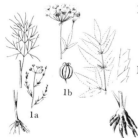

1a Pl bulbiliferous; bulbils borne in axils of at least the reduced upper lvs; segms of main lvs narrowly linear, mostly 0.5–4 cm × 0.5–1.5 mm; fr 1.5–2 mm; wet places or standing water; plains and lowl to mont; transcontinental in n US and s Can, but not common in our range; bulb-bearing w.
                                                                                1 C. **bulbifera** L.
1b Pl not bulbiliferous; lflets of principal lvs 3–10 cm × 5–35 mm; fr 2–4 mm; marshes, ditches, and wet low places; plains and lowl to mont; widespread cordilleran sp.; Douglas' w., western w.
                                                                2 C. **douglasii** (DC.) Coult. & Rose

Conioselinum Hoffm.   Hemlock-parsley

Umbels 1 or more, compound; invol a few narrow or lfy bracts, or wanting; involucel of well-developed, narrow, often ± scarious bractlets; fls white; calyx teeth obsolete; stylopodium conic; carpophore bifid to base or nearly so; fr elliptic or elliptic-oblong, dorsally compressed, glab, lateral ribs evidently thin-winged, dorsal ribs more narrowly so or low and corky; oil tubes 1–2 in the intervals, 2–4 on the commissure, often not reaching base of fr; per, sometimes with a caudex, gen single-std, st ± lfy; lvs pinnately or ternate-pinnately decompound, without well-defined lflets, or lflets ± deeply cleft. (Resembling *Conium* and *Selinum*).

C. **pacificum** (Wats.) Coult. & Rose. Pac h. St single, 2–10 (15?) dm, gen from short rhizome; pl glab except for some scabrosity on umbel rays, sometimes ± glaucous; lvs all cauline, pinnately 2–4 times compound; invol wanting or a few narrow bracts, seldom of lfy bracts; involucel of well-developed narrow bractlets gen = or > pedicels, these 4–8 mm; fr 5–8.5 mm, lateral wings well developed; bluffs and sandy beaches along seashore; Alas to Cal (*Selinum p.; C. fischeri*, misapplied; *S. hookeri*).

Conium L.   Poison-hemlock

Umbels compound; invol and involucel of several small, lanceolate to ovate bracts or bractlets; fls white; calyx teeth obsolete; styles reflexed, stylopodium depressed-conic; carpophore entire; fr glab, broadly ovoid, laterally ± flattened; oil tubes ∝ and small, ± confluent; glab biens. (Gr *koneion*, ancient name of our sp.).

C. **maculatum** L. Pls 0.5–3 m, coarse and freely brd, from a stout taproot, with purple-spotted, hollow st; larger lvs gen 1.5–3 dm, pinnately or ternate-pinnately dissected with rather small ultimate segms; umbels ∝, axillary and pedunculate, rays subequal, mostly 1–4 cm at maturity; fr 2–2.5 mm, with prominent, raised, often wavy and somewhat crenate, almost winged ribs; ditches and other disturbed sites; European weed now widely estab in N Am and esp common w Cas; highly poisonous.

Cymopterus Raf.   Cymopterus

Infl of compound, loose to capitate umbels on terminal peduncles or scapes; invol of scarious to herbaceous bracts, or wanting; involucel of large or small, ovate to filiform bractlets all on one side of the umbellet, often basally connate; fls white, yellow, or purple; calyx teeth evident to obsolete; styles slender, without a stylopodium; carpophore bifid to base; fr ovoid to oblong, somewhat flattened dorsally, the lateral and gen 1–more of the dorsal ribs conspicuously winged; oil tubes 1–∝ in the intervals, 2–∝ on commissure, and sometimes 1 under each wing; low, taprooted pers, caulescent or acaulescent, often with a pseudoscape; lvs ternately to pedately or more often pinnately ± dissected, with mostly small ultimate segms. (Gr *kyma*, wave, and *pteron*, wing). (*Pseudocymopterus, Pteryxia*).

1a Pls with 1–2 (seldom more) short to elongate pseudoscapes arising from the simple or occ few-brd subterranean crown of the taproot; lvs tending to form a flat rosette
  2a Fls white; fr umbel very dense and headlike; primarily Gr Pl and adj w intermont valleys, extending into the drier valleys of c Ida and through SRP to Harney Co, Ore; dry flats and hillsides; plains c. (*C. leibergii, C. glomeratus* var. *l.*)   1 C. **acaulis** (Pursh) Raf.
  2b Fls yellow; fr umbel fairly compact, but not at all headlike; mts of c Ida to nw Mont w Cont Div; open, gen gravelly or rocky slopes and flats from foothills to above timberl in mts; grayish c.   2 C. **glaucus** Nutt.
1b Pls caespitose, with ∝ lvs and several or ∝ peduncles or basally lfy sts arising from the surficial br caudex surmounting a taproot
  3a Fls white; calyx teeth minute and scarious, gen blunt; pls acaulescent
    4a Lvs subtripinnatisect, some of 2ndary segms gen again cleft, ultimate segms ∝ and crowded; rays of umbel, or some of them, often > 5 (to 17) mm at maturity; foothills to high mont, w Mont and Wyo to c Ida, and s to ne Nev and se Ore, in open, often rocky places; Hayden's c. (*Cynomarathrum macbridei*)   3 C. **bipinnatus** Wats.
    4b Lvs sub-bipinnatisect, some of distal primary segms gen entire, the others gen only once pinnatifid, ultimate segms fewer and less crowded than in *C. bipinnatus*; rays of umbel < 5 mm; local in rocky places, high mont; c Ida and ne Nev; snowline c.   4 C. **nivalis** Wats.

3b Fls yellow; calyx teeth evident, gen rather narrow and pointed, mostly green
  5a Lvs appearing very open and skeletonlike, segms remote, bl mostly 4–17 cm, pinnately or subternate-pinnately 2–3 times dissected; pl gen caulescent, 15–45 cm; open rocky places in dry or des regions; SRP, Ida, to se Ore, s to Cal and n Ariz, barely reaching s edge of our range; rock-loving c.   **5 C. petraeus** M. E. Jones
  5b Lvs with more dense, crowded segms
    6a Lvs ± ovate in outline, ternate-pinnately dissected; pls caulescent or sometimes acaulescent; widespread in cordillera e Cas; turpentine c.; 2 vars. in our area   **6 C. terebinthinus** (Hook.) T. & G.
      a1 Wings of fr ± strongly crisped, gen all ca alike and = or often broader than the body; infl at anthesis mostly 2–7 cm wide; ultimate segms of lvs gen relatively short and broad; Columbia Plateau, Wn and ne Ore, mostly at low elev, often in sand dunes
                                                              var. terebinthinus
      a2 Wings of fr scarcely or not at all crisped, dorsal ones often narrower than lateral ones, which are often no broader than the body; infl at anthesis mostly 1.5–3 cm wide; ultimate segms of lvs relatively long and slender; foothills to midmont, c Ida, e into Ravalli Co, Mont, and w into mts and foothills of ne and c Ore and extreme se Wn (*C. thapsoides, C. elrodi, C. foeniculaceus*)
                                                  var. foeniculaceus (T. & G.) Cronq.
    6b Lvs oblong in outline, pinnately dissected; pls acaulescent; rocky places in open, foothills to above timberl; Beartooth Mts, sw Mont, to n NM, w to c Ida, ne Nev, and s Utah; Henderson's c.
                                                  7 C. hendersonii (Coult. & Rose) Cronq.

Daucus L.   Carrot

Umbels compound; invol of ∝ entire or dissected bracts, or wanting; involucel of ∝ toothed or entire bractlets, or wanting; fls white, but central fl of umbel often purple, or rarely all fls pink or yellow; outer fls often with slightly enlarged and irreg corolla; calyx teeth evident to obsolete; styles short, stylopodium conic; carpophore entire or bifid at apex; fr oblong to ovoid, slightly compressed dorsally, evidently ribbed, beset with stout, spreading, glochidiate or barbed prickles along alt ribs, bristly or hairy on the others; oil tubes solitary under alt ribs, and 2 on the commissure; taprooted anns or biens with lfy sts. (Ancient Gr name of some umbellifer).

1a Relatively coarse bien 2–12 dm, ± hirsute throughout to subglab; invol bracts scarious margined below, segms firm, elongate, filiform-subulate; umbellets mostly (10) 20–∝ -fld; lf bls 5–15 × 2–7 cm, ultimate segms linear to lanceolate, 2–12 × 0.5–2 mm; roadsides and disturbed sites in moist places; European, now estab occ in much of our range; wild c., Queen Anne's lace   **1 D. carota** L.
1b Relatively slender ann 0.5–6 (9) dm, ± hirsute throughout; invol bracts not scarious-margined, segms linear or lanceolate, scarcely elongate; umbellets mostly 5–12-fld; lf bls 3–10 × 1.5–7 cm, ultimate segms linear, 1–5 × 0.5–1 mm; dry, open places at lower elev; s BC to Baja Cal, e to Mo and Fla, chiefly w Cas with us; Am c., rattlesnake weed (*D. microphyllus, D. c. var. m.*)   **2 D. pusillus** Michx.

Eryngium L.   Eryngo; Coyote-thistle

Infl of dense, bracteolate heads terminating brs, the bracteoles representing the involucels of the 1-flowered, sessile umbellets; invol of 1–more series of entire or variously-cleft or toothed bracts subtending the head; fls sessile, white to blue or purple, corolla lobes sometimes fimbriate; calyx lobes well developed and conspicuous, firm and sometimes spinescent; stylopodium and carpophore wanting; fr globose to obovoid, slightly or scarcely flattened laterally, with a broad commissure, variously covered with scales or tubercles, the ribs obsolete; oil tubes mostly 5; bien or per, caulescent or acaulescent, gen glab herbs; lvs gen firm, entire or toothed to deeply cleft and often ± spinose-toothed, petioles often hollow and septate-nodose. (Gr, but name of uncertain significance).

1a Bl of basal lvs almost = or > petiole, rounded or subcordate at base; heads blue or bluish; European weed, collected at Salem, Ore, but not known to be estab **E. planum** L.

1b Bl of basal lvs much < petiole, tapering to base, or basal petioles bladeless; native spp., not weedy

  2a Bls of larger lvs up to ca 1 cm wide; heads green or greenish; WV, Ore, extending up CR to e end of CRG; low ground, esp in places submerged in spring and dry by late summer; Ore e. or c. (*E. p.* var. *juncifolium*)

    1 **E. petiolatum** Hook.

  2b Bls of larger lvs mostly 1–3 cm wide; e Cas in our range

    3a Heads blue; corolla blue and bracteoles and calyx lobes bluish-tinged; pls mostly 3–10 dm; n Ida and adj Wn, irreg to c Cal; low gound along streams and lakes, often where submerged in spring; beefthistle e.

      2 **E. articulatum** Hook.

    3b Heads green or greenish; corolla white, the bracteoles and calyx lobes scarcely if at all bluish; pls mostly 1–3 dm; ne Cal and n Nev to Harney Co, Ore, prob not quite reaching our range

      E. alismifolium Greene

2a   3a

Foeniculum Adans.   Fennel

Umbels compound, the rays several to rather ∝ ; invol and involucel wanting; fls yellow; calyx teeth obsolete; styles very short, recurved, the stylopodium conic; carpophore 2-cleft to base; fr oblong, subterete or slightly flattened laterally, glab, with prominent ribs; oil tubes solitary in the intervals, 2 on the commissure; anns or pers, glab and glaucous, erect and lfy-std; lvs pinnately dissected with elongate, filiform ultimate segms; petioles broad and somewhat sheathing. (Diminutive of L *foenum*, hay, referring to the odor).

F. vulgare Mill. Sweet f. Stout per 1–2 m, with strong anise odor; st solitary, commonly br above; lvs (exclusive of petioles) sometimes as much as 3–4 dm long and broad, ultimate segms filiform, 4–20 × 1 mm; rays of umbel 10–40, unequal, mostly 2–8 cm at maturity; fr 3.5–4 mm; roadsides and waste places; Mediterranean sp. now widely intro, and found throughout much of US, estab at least w of Cas in our range (*Anethum f., F. officinale*).

Glehnia Schmidt   Glehnia

Umbels compound, on terminal peduncles; rays of primary umbel well developed; ultimate umbels capitate, the peds obsolete; invol a few narrow bracts, or wanting; involucel of several well-developed, lance-attenuate bractlets; fls white; calyx teeth minute; styles short, without a stylopodium; carpophore bifid to base; fr ovate-oblong to subglobose, somewhat compressed dorsally, ribs all broadly corky-winged; oil tubes several in the intervals, 2–6 on the commissure; stout, taprooted pers. (Presumably for Peter von Glehn, 1835–1876, curator of the Botanic Garden at St. Petersburg).

G. leiocarpa Mathias. American g. Pl nearly acaulescent, or with st and strongly sheathing petioles buried in the sand; lvs spreading, often prostrate, thick and firm, glab above, tomentose beneath, bl once or twice ternate or ternate-pinnate, lflets 2–7 cm, crenate or crenate-serrate with callous teeth, broadly elliptic to obovate or broader and deeply trilobed; peduncles < 1 dm; rays of umbel mostly 5–13, up to 4.5 cm at maturity; fr 6–13 mm, glab or with a few long hairs toward tip; dunes and sandy beaches along coast, Alas to nw Cal (*Cymopterus littoralis, G. l.* ssp. *leiocarpa*).

Heracleum L.   Cow-parsnip

Umbels compound, on terminal and axillary peduncles, rays unequal; invol wanting or a few deciduous bracts; involucel of ∝ slender bractlets, or rarely wanting; fls white or tinged with green or red; calyx teeth minute or obsolete; outer fls of at least the marginal umbellets gen irreg, the outer corolla lobes enlarged and often bifid; stylopodium conic, styles short, erect or recurved; carpophore bifid to base; fr orbicular to obovate or elliptic, strongly flattened dorsally, dorsal ribs narrow, lateral ones broadly winged; oil tubes 2–4 on the commissure, solitary in the intervals, extending only part way from the stylopodium toward the base of the fr, readily visible to the naked eye; coarse biens or pers. (Supposedly named for Hercules).

H. lanatum Michx. Single-std per, 1–3 m, thinly tomentose or woolly-villous to nearly glab; lvs once ternate, with broad, distinctly petiolulate, coarsely toothed and palmately lobed lflets 1–3 (4) dm long and wide; invol of 5–10 deciduous narrow bracts 0.5–2 cm; involucel of similar bractlets; rays up to 10 cm or longer, the terminal umbel 1–2 dm wide; fls white; fr obovate to obcordate, 7–12 × 5–9 mm; stream banks or moist ground, lowl to midmont; widespread in N Am; Siberia (*Pastinaca l., H. douglasii*).

Hydrocotyle L.   Marsh- or Water-pennywort

Infl gen a simple umbel, sometimes proliferous, peds ascending to reflexed; invol inconspicuous or wanting; fls white, greenish, or yellow; calyx teeth minute or obsolete; stylopodium conspicuously conic to depressed; fr orbicular or ellipsoid, ± flattened laterally, dorsal surface rounded or acute, with narrow, acute ribs, or the ribs obsolete; low pers with creeping or floating sts, and with petiolate, often peltate lvs. (Gr *hydor*, water, and *kotyle*, flat cup, referring to the peltate lvs).

H. ranunculoides L. f. Floating m. or w. Pl glab; sts floating or creeping; petioles weak, mostly 0.5–3.5 dm; lf bls rotund-reniform, gen 1–6 cm wide, not at all peltate, 5–6-lobed to about middle; peduncles axillary, much < petiole, gen 1–5 cm, recurved in fr; umbel simple, 5–10-fld, pedicels ascending or spreading, 1–3 mm; stylopodium depressed; fr suborbicular, 1–3 × 2–3 mm, dorsal surface rounded, ribs obsolete; marshes and wet ground; tropical Am, n to Del, Ark, Cal, and in PS region, Wn. *H. umbellata* L., differing in having distinctly peltate lvs, occurs in s Ore, but is not known to reach our range.

Ligusticum L.   Lovage; Licorice-root

Infl of compound umbels; invol and involucel wanting, or of a few inconspicuous narrow bracts or bractlets; fls white or sometimes pinkish; calyx teeth evident or obscure; styles short, spreading, stylopodium low-conic; carpophore bifid to base; fr oblong to ovate or suborbicular, subterete or slightly compressed laterally, glab, ribs evident, often winged; oil tubes 1–6 in the intervals, 2–10 on the commissure; taprooted pers with ternately or ternate-pinnately compound or dissected lvs. (Gr *Ligustikon*, classical name of an umbellifer).

1a Lvs dissected into ∝ ± linear ultimate segms gen 1–3 mm wide
  2a Pls relatively robust, gen 5–10 dm, with 1–more ± well-developed st lvs; basal lvs gen 10–25 cm wide; rays of main umbel gen (10) 12–20; fr 5–7 mm; Fremont Co, Ida, and Madison Co, Mont, s in mts of w Wyo and e Ida to Wasatch region, Utah; open or wooded, moist to dry slopes and ridges in mts; fern-lf l.   1 **L. filicinum** Wats.
  2b Pls relatively small and slender, 1–6 dm, scapose or more often subscapose, st lvs wanting or 1 but greatly reduced; basal lvs gen < 10 cm wide; rays of umbel gen 5–13; fr 3–5 mm; Wallowa and Blue mts, Ore, e across c Ida to sw Mont, also in Colo and ne Utah; marshes to wet or moist slopes, mont; slender-lvd l. (*L. filicinum* var. *t., L. oreganum*)   2 **L. tenuifolium** Wats.
1b Lvs less dissected, ultimate segms broader, ± toothed or cleft
  3a Ribs of fr narrowly winged; in and e Cas
    4a Lflets relatively large and broad, gen 3–8 × 2–5 cm; robust lfy-std pls gen 1–2 m; foothills and valleys, n Ida and adj Mont, moist woods to swampy areas; verticillate-umbel l. (*Angelica v.*)
                    3 **L. verticillatum** (Geyer) Coult. & Rose
    4b Lflets smaller, gen 1–5 × 0.5–2 cm; smaller, less lfy pls, up to ca 12 dm
      5a Rays of terminal umbel 15–40; pls 5–12 dm, with 1–more, ± well-developed st lvs; c and nc Wn to w Mont (w Cont Div), s to Valley Co, Ida, and Blue Mts, Ore; wet to moist, or occ dryish soil; Canby's l. (*L. leibergii*)   4 **L. canbyi** Coult. & Rose
      5b Rays of terminal umbel 7–14; pls 2–6 dm, scapose or with 1–2 strongly reduced st lvs; Cas-Sierran, e to c Ida and ne Nev; moist to dry, open or wooded mont slopes and drier meadows; Gray's l.

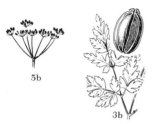

5b

(*Pimpinella apiodora* var. *nudicaulis, L. caeruleomontanum, L. cusickii, L. purpureum, L. tenuifolium* var. *dissimilis*)

5 L. grayi Coult. & Rose

3b Ribs of fr wingless; pls 4–15 dm, rays of umbel 12–30, gen 2.5–6 cm at maturity; fence rows and sparsely wooded slopes and prairies, lowl; w Cas, Wn to Cal; celery-lvd l., parsley-lvd l. (*Cynapium a.*)

6 L. apiifolium (Nutt.) Gray

3b

Lilaeopsis Greene   Lilaeopsis

Infl of simple, axillary, few-fld pedunculate umbels; invol of a few small bracts; fls white; calyx teeth minute; styles short, stylopodium depressed or obsolete; carpophore wanting; fr globose or ovoid, glab, ribs next to commissure corky-thickened, the dorsal ones narrow; oil tubes gen solitary in the intervals, 2 on the commissure; small rhizomatous pers with reduced lvs. (Named for its resemblance to the genus *Lilaea*).

L. occidentalis Coult. & Rose. Lvs tufted at intervals along rhizome, reduced to elongate, narrow, entire, hollow, transversely septate phyllodes 3–15 cm × 1–4 mm, with 5–11 partitions; peduncles slender, 0.5–3 (4.5) cm, ascending or often recurved above, gen much < the lvs, 5–12-fld; fr ovoid, terete or slightly flattened laterally, ca 2 mm, lateral ribs pale and prominent, dorsal ones inconspicuous; marshes, salt flats, and sandy or muddy beaches and shores along and near the coast, including PS; s VI to c Cal (*Crantziola o., L. lineata* var. *o.*).

Lomatium Raf.   Biscuit-root; Desert-parsley; Lomatium

Umbels compound; invol wanting or inconspicuous; involucels evident to inconspicuous, or wanting; fls gen white or yellow, less often purple; calyx teeth minute or obsolete; stylopodium scarcely developed, sometimes visible in fresh fls, but gen obscure at best in dried specimens; carpophore bifid to base; fr linear to orbicular or obovate, glab to hairy or granular, dorsally flattened and with ± well developed, thin or corky marginal wings, or wings rarely obsolete and fr narrow and scarcely flattened; dorsal ribs evident to obsolete, often raised, but only narrowly if at all winged; oil tubes solitary to ∞ in the intervals, 2–several on the commissure; taprooted pers, root often tuberous-thickened; lvs gen mostly oɩ all basal (some notable exceptions), pinnately to ternately or in part quinately compound to dissected. (Gr *loma*, a border, referring to the winged fr). (*Cogswellia, Leibergia, Leptotaenia, Peucedanum*).

3a

a1

4a

1a Ultimate segms of lvs relatively large, many or all at least 1 cm
  2a Ultimate segms of lvs forming ± definite lflets, these entire to deeply cleft, gen > 5 mm wide
    3a Lflets strongly toothed or cleft; fls white or yellowish-white to sometimes yellow; lvs chiefly basal or nearly so, glaucous, pinnately or ternate-pinnately 1–2 times compound, with < 60 ± lf-like ultimate segms; root often thickened below; rocky slopes to dry meadows, mont to subalp; Cas, s BC to s Ore, in OM, Wn, and coast range, Ore; Martindale's l., few-frd l.; 3 vars.   **1 L. martindalei** Coult. & Rose
      a1 Fls yellow; frs narrow, 2–3 times as long as wide; OM (*L. angustatum* var. *f.*)   var. **flavum** (G. N. Jones) Cronq.
      a2 Fls yellowish-white or very occ yellow; fr 1.5–3 times as long as wide
        b1 Frs gen 1.5–2 times as long as wide; Mt Hood, Ore, and Skamania Co, Wn, to s range of the sp.   var. **martindalei**
        b2 Frs gen 2–3 times as long as wide; Cas of s BC and Wn, s to Mackenzie Pass, Ore   var. **angustatum** Coult. & Rose
    3b Lflets mostly entire or shallowly toothed; fls yellow; widespread
      4a Herbage (at least sts and peduncles) ± hirtellous-puberulent; lvs ternately to ternate-pinnately (or the basal quinately) 2–3 times (seldom only once) cleft into long, gen narrow segms or lflets 1–10 (20) cm; open slopes and meadows, dry to fairly moist soil; lowl to mid-mont; s Alta and BC to Colo, Utah, and Cal; nine-lf l.; 2 sspp. and 2 vars, in our area   **2 L. triternatum** (Pursh) Coult. & Rose
        a1 Fr broadly elliptic, wings nearly or fully as wide as body; lvs rather reg dissected, more nearly ternate than pinnate, ult segms

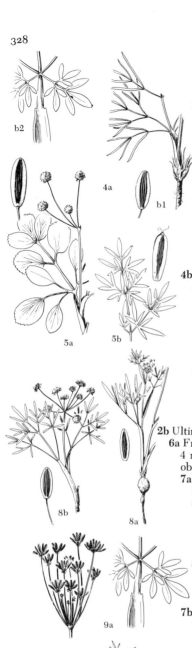

linear or nearly so, relatively few; pls more often scapose than distinctly caulescent; c and sw Colo to Mont (chiefly e Cont Div but w to Missoula Co), c Ida, e Ore, and drier parts of Wn, e Cas (*L. simplex, L. p.*)   ssp. **platycarpum** (Torr.) Cronq.

a2 Fr gen relatively narrow, wings seldom > half as wide as body; lvs less reg and more nearly pinnately dissected, although 1st division gen ternate or even quinate; ultimate segms of lvs gen more ∝, more markedly unequal, and more crowded, varying from linear to elliptic; pls distinctly caulescent or less often scapose; s Alta and BC to nw Mont, c Ida, s Ore, and n Cal
                                                                ssp. **triternatum**

b1 Ultimate segms of lvs as in ssp. *platycarpum*, gen linear or nearly so and ± strongly acute; range of ssp. (*L. t. f. lancifolium, Cogswellia brevifolia, L. t.* var. *b., L. t.* var. *macrocarpum, L. robustius*)   var. **triternatum**

b2 Ultimate segms of lvs mostly broader than linear, obtuse or rounded to barely acute; w Ida and adj Ore (*L. anomalum*)
                                            var. **anomalum** (M.E. Jones) Math.

4b Herbage essentially glab and often glaucous

5a Lflets gen 2–9 × (0.4) 1–6 cm; longer rays of umbel gen 6–20 cm in fr; lvs firm, ternately or ternate-pinnately 1–3 times compound, with 3–30 well-defined, veiny, often petiolulate ultimate lflets; dry, open to sparsely wooded places, lowl to mid-mont, often with sagebr or ponderosa pine; both sides Cas, s BC to c Cal, e to sw Alta, Ida, and w Utah; barestem l., pestle parsnip (*L. platyphyllum*)   3 **L. nudicaule** (Pursh) Coult. & Rose

5b Lflets gen 1–5 cm × 2–8 mm; longer rays 1.5–5 cm in fr, lvs not particularly firm, ternate-pinnately dissected; open or wooded slopes, foothills to fairly high mont; Wen Mts and Cas, Kittitas to Okanogan cos, Wn; Brandegee's l.
                                        4 **L. brandegei** (Coult. & Rose) Macbr.

2b Ultimate segms of lvs narrow and scarcely lfletlike, gen < 5 mm wide

6a Frs linear to narrowly oblong, gen (2.5) 3–8 times as long as wide, < 4 mm wide over-all, wings up to 1/3 (rarely 1/2) as wide as body, or obsolete

7a Invol gen wanting; peds elongate, gen 4–13 mm at maturity; fls yellow, or somewhat purplish in age

8a Pls distinctly caulescent and gen few-brd above; root globose-thickened or moniliform to sometimes more slender and elongate; widespread, open slopes and flats, lowl to midmont; s BC and Wn (E or barely W Cas) to ne Ore, e across Ida to Mont, Wyo, and Utah; swale d.   5 **L. ambiguum** (Nutt.) Coult. & Rose

8b Pls acaulescent or nearly so, with 0–2 st lvs; sts or scapes simple above the base; root elongate and not much thickened; open rocky slopes and dry meadows, mid-mont to subalp; c Ida; Ida l.   6 **L. idahoense** Math. & Const.

7b Invol present; peds short or elongate; fls yellow or white

9a Larger lf segms gen > 2 mm wide; st and peduncle finely hirtellous-puberulent, rarely subglab; peds 2–12 mm at maturity; fls yellow; widespread (see lead 4a for treatment of sspp. and vars.)
                                          2 **L. triternatum** (Pursh) Coult. & Rose

9b Larger lf segms mostly 1–2 mm wide; st and peduncle (or scape) glab to occ scaberulous or hirtellous; peds short, only 0.5–3 mm at maturity; fls yellow or white (*L. farinosum* and *L. hambleniae* might sometimes be sought here, except for the long peds)

10a Lf segms < 15; fls white; frs wingless; pl acaulescent; meadows and moist bottomlands; n Ida and adj Wn; Leiberg's l.   7 **L. orogenioides** (Coult. & Rose) Math.

10b Lf segms relatively ∝, many > 15; fls yellow, rarely white; frs very narrowly winged at maturity; pl caulescent or acaulescent; open slopes to meadows and swales, esp in heavy clay soils, foothills and lowl to midmont; Ore, e Cas, and se Wn, to ne Cal, e to Ida, and isolated (?) in s Wyo and adj Colo; slender-fr l., bicolor b. (*L. ambiguum* var. *l.*)
                                          8 **L. leptocarpum** (T. & G.) Coult. & Rose

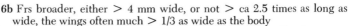

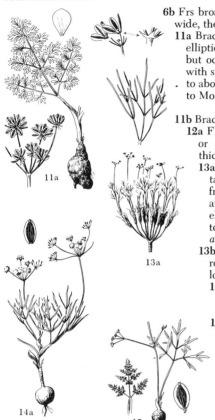

6b Frs broader, either > 4 mm wide, or not > ca 2.5 times as long as wide, the wings often much > 1/3 as wide as the body

11a Bractlets of involucel broadly oblanceolate to broadly obovate or elliptic; fls yellow; root gen short and strongly tuberous-thickened, but occ slender and elongate; dry, open, often rocky places, often with sagebr; esp common in foothills and lowl, but sometimes mont to above timberl; se Wn and ne Ore (w to Jefferson Co), e across Ida to Mont and n Wyo; cous, cous b. (*L. circumdatum, L. montanum*)
9 L. **cous** (Wats.) Coult. & Rose

11b Bractlets of involucel narrow, gen linear or lanceolate, or wanting

12a Fls white or yellowish-white to occ purple; low, often scapose or subscapose pls up to ca 3 (4) dm, often with tuberous-thickened roots

13a Pl when well developed gen with br caudex surmounting a taproot, the caudical brs often rather slender and elongate (and frequently broken off in herbarium specimens at the point of attachment to the root); peds 1–4 (5) mm; mont, at mid to high elevs, open or wooded, often rocky places; ne Ore across c Ida to sw Mont; fl June-July; Cusick's l. or d. (*C. altensis, C. brecciarum*)
10 L. **cusickii** (Wats.) Coult. & Rose

13b Pl with a globose-thickened or moniliform to more elongate root and a gen simple crown, without a br caudex; foothills and lowl (no 13 to midmont); fl Mar–May

14a Peds gen 6–22 mm in fr; fr 5–6.5 mm; scablands and rocky slopes, extreme e Wn, across n Ida to Missoula Co, Mont; Coeur d'Alene l.   11 L. **farinosum** (Hook.) Coult. & Rose

14b Peds short, gen < 4 mm; fr 5–12 mm

15a Pls relatively small, gen 1–1.5 dm at maturity; frs gen 5–7 mm; bractlets of involucel often < 2 mm; root subglobose; Kittitas, Lincoln, and s Spokane Co, Wn, and Kootenai Co, Ida, s to ne Cal; open slopes, scablands, and plains, often with sagebr; Gorman's l. or d. (*Peucedanum confusum, L. piperi, L. g. f. purpureum*)
12 L. **gormanii** (Howell) Coult. & Rose

15b Pls larger, gen 1.5–4 dm; frs gen 7–12 mm; bractlets of involucel 2–3 mm; root subglobose to more often elongate or moniliform; n to s BC from Kittitas, Lincoln, and Spokane cos, and Cleman Mt, Yakima Co, Wn, and from Kootenai Co, Ida; open slopes and flats; Geyer's l. or d. (*Orogenia fusiformis* var. *leibergii, O. l.*)
13 L. **geyeri** (Wats.) Coult. & Rose

12b Fls yellow or occ somewhat purplish, or wholly purple in some robust spp. mostly > 3 dm

16a Sts or scapes solitary or few from a simple or occ few-br root-crown

17a Pls essentially acaulescent (or with a pseudoscape), mostly 1–2 dm at maturity, glab, arising from a globose-thickened root; very similar to *L. farinosum* except for the yellow fls; scablands, often with sagebr; c Wn, Grant to Chelan and Yakima cos; Hamblen's l. or d.
14 L. **hambleniae** Math. & Const.

17b Pls either taller and often distinctly caulescent, or with a more elongate and not much thickened root, or often both; glab or hairy

18a Involucel gen absent; sts or scapes glab or merely scaberulous; midmont to subalp in c Ida (see lead 8b)
6 L. **idahoense** Math. & Const.

18b Involucel present; sts or scapes hirtellous-puberulent to sometimes merely scaberulous; widespread, gen in foothills and lowl, sometimes to midmont

19a Lvs irreg and ± distinctly pinnately dissected, ultimate segms seldom much > 1 cm; fr glab, elliptic, 5–8 mm; open slopes in and near the canyons of the Snake and lower Salmon rivers, e Ore and w Ida; Rollins' l. or d.   15 L. **rollinsii** Math. & Const.

19b Lvs ternately to ± ternate-pinnately or in part quinately dissected, ultimate segms gen much > 1 cm; fr

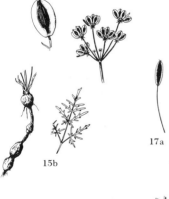

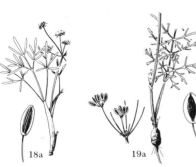

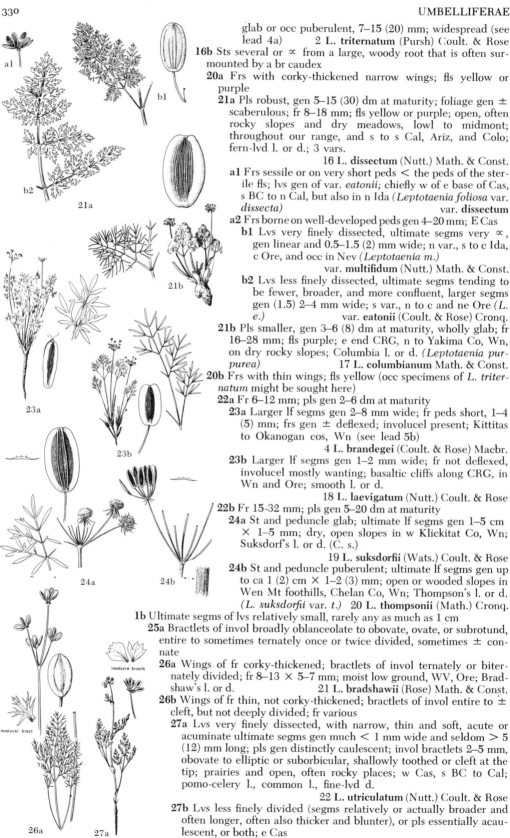

glab or occ puberulent, 7–15 (20) mm; widespread (see
lead 4a)   2 **L. triternatum** (Pursh) Coult. & Rose
16b Sts several or ∝ from a large, woody root that is often surmounted by a br caudex
  20a Frs with corky-thickened narrow wings; fls yellow or
  purple
    21a Pls robust, gen 5–15 (30) dm at maturity; foliage gen ±
    scaberulous; fr 8–18 mm; fls yellow or purple; open, often
    rocky slopes and dry meadows, lowl to midmont;
    throughout our range, and s to s Cal, Ariz, and Colo;
    fern-lvd l. or d.; 3 vars.
                    16 **L. dissectum** (Nutt.) Math. & Const.
      a1 Frs sessile or on very short peds < the peds of the ster-
      ile fls; lvs gen of var. *eatonii;* chiefly w of e base of Cas,
      s BC to n Cal, but also in n Ida *(Leptotaenia foliosa* var.
      dissecta)   var. **dissectum**
      a2 Frs borne on well-developed peds gen 4–20 mm; E Cas
        b1 Lvs very finely dissected, ultimate segms very ∝,
        gen linear and 0.5–1.5 (2) mm wide; n var., s to c Ida,
        c Ore, and occ in Nev *(Leptotaenia m.)*
                    var. **multifidum** (Nutt.) Math. & Const.
        b2 Lvs less finely dissected, ultimate segms tending to
        be fewer, broader, and more confluent, larger segms
        gen (1.5) 2–4 mm wide; s var., n to c and ne Ore *(L.
        e.)*   var. **eatonii** (Coult. & Rose) Cronq.
    21b Pls smaller, gen 3–6 (8) dm at maturity, wholly glab; fr
    16–28 mm; fls purple; e end CRG, n to Yakima Co, Wn,
    on dry rocky slopes; Columbia l. or d. *(Leptotaenia pur-
    purea)*   17 **L. columbianum** Math. & Const.
  20b Frs with thin wings; fls yellow (occ specimens of *L. triternatum* might be sought here)
    22a Fr 6–12 mm; pls gen 2–6 dm at maturity
      23a Larger lf segms gen 2–8 mm wide; fr peds short, 1–4
      (5) mm; frs gen ± deflexed; involucel present; Kittitas
      to Okanogan cos, Wn (see lead 5b)
                    4 **L. brandegei** (Coult. & Rose) Macbr.
      23b Larger lf segms gen 1–2 mm wide; fr not deflexed,
      involucel mostly wanting; basaltic cliffs along CRG, in
      Wn and Ore; smooth l. or d.
                    18 **L. laevigatum** (Nutt.) Coult. & Rose
    22b Fr 15–32 mm; pls gen 5–20 dm at maturity
      24a St and peduncle glab; ultimate lf segms gen 1–5 cm
      × 1–5 mm; dry, open slopes in w Klickitat Co, Wn;
      Suksdorf's l. or d. (C. s.)
                    19 **L. suksdorfii** (Wats.) Coult. & Rose
      24b St and peduncle puberulent; ultimate lf segms gen up
      to ca 1 (2) cm × 1–2 (3) mm; open or wooded slopes in
      Wen Mt foothills, Chelan Co, Wn; Thompson's l. or d.
      (L. suksdorfii var. *t.)* 20 **L. thompsonii** (Math.) Cronq.
1b Ultimate segms of lvs relatively small, rarely any as much as 1 cm
  25a Bractlets of invol broadly oblanceolate to obovate, ovate, or subrotund,
  entire to sometimes ternately once or twice divided, sometimes ± connate
    26a Wings of fr corky-thickened; bractlets of invol ternately or biter-
    nately divided; fr 8–13 × 5–7 mm; moist low ground, WV, Ore; Brad-
    shaw's l. or d.   21 **L. bradshawii** (Rose) Math. & Const.
    26b Wings of fr thin, not corky-thickened; bractlets of invol entire to ±
    cleft, but not deeply divided; fr various
      27a Lvs very finely dissected, with narrow, thin and soft, acute or
      acuminate ultimate segms gen much < 1 mm wide and seldom > 5
      (12) mm long; pls gen distinctly caulescent; invol bractlets 2–5 mm,
      obovate to elliptic or suborbicular, shallowly toothed or cleft at the
      tip; prairies and open, often rocky places; w Cas, s BC to Cal;
      pomo-celery l., common l., fine-lvd d.
                    22 **L. utriculatum** (Nutt.) Coult. & Rose
      27b Lvs less finely divided (segms relatively or actually broader and
      often longer, often also thicker and blunter), or pls essentially acau-
      lescent, or both; e Cas

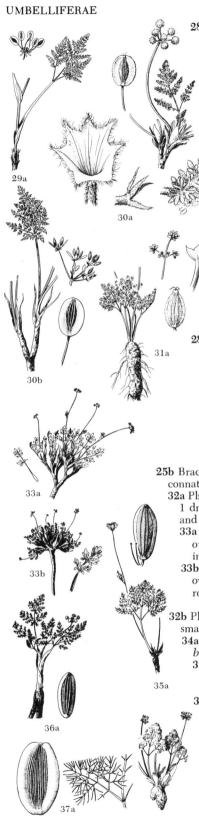

28a Fr peds ± elongate, gen 3–15 mm; taproot elongate and seldom much thickened

29a Lvs glab or merely granular-scaberulous; fr granular-roughened when young, not hairy; pls distinctly caulescent, 1.5–4 dm; invol scarcely dimidiate; fls yellow; open slopes and flats, foothills and valleys, often on heavy clays; Union, Grant, and Crook cos, Ore, s to Cal; broadsheath l.
23 L. vaginatum Coult. & Rose

29b Lvs evidently short-hairy; fr gen short-hairy at least when young; pls acaulescent or nearly so; invol dimidiate

30a Fls yellow; invol bractlets puberulent or villous-puberulent, often basally connate and tapered above; dry, open slopes, valleys and plains to midmont or higher; e and s from Mont, c Ida, and c Ore; fennel-lvd d.; 2 vars. in our area   24 L. foeniculaceum (Nutt.) Coult. & Rose

al Foliage and frs moderately to densely hirtellous; invol bractlets seldom much connate; w Cont Div (L. semise-pultum, L. m.)   var. macdougalii (Coult. & Rose) Cronq.

a2 Foliage and frs gen less densely hairy, but the hairs often longer; invol bractlets gen rather strongly connate; gen e Cont Div, but w into Powell Co, Mont (L. villosum)
var. foeniculaceum

30b Fls white; bractlets of the involucel glab or only very finely hirtellous; dry open slopes in foothills and plains; c Ore (Grant and Jefferson cos) to s Cal, Utah, Colo, Ariz, and n Mex; Nev d.   25 L. nevadense (Wats.) Coult. & Rose

28b Fr pedicels very short, gen 1–3 mm, or obsolete; taproot gen short and tuberous-thickened, sometimes more slender and elongate

31a Involucel dimidiate, and the bractlets gen connate below the middle or nearly to the tip; ovaries and young fr finely puberulent or glab; s Kittitas Co, Wn, to Jefferson Co, Ore, on open hillsides with sagebr; Watson's d. or l. (L. frenchii)
26 L. watsonii Coult. & Rose

31b Invol not dimidiate, and the bractlets distinct; fr glab or granular-roughened, not hairy (see lead 11a)
9 L. cous (Wats.) Coult. & Rose

25b Bractlets of invol narrow, gen linear or lanceolate, distinct or merely connate at the base, or wholly wanting

32a Pls dwarf alp or subalp pers from a taproot and much-brd caudex, < 1 dm, with very compact infl (rays of umbel all < 1 cm at maturity) and not much-dissected lvs; mts of ne Ore

33a Ultimate segms of lvs glab or slightly scaberulous, gen 3–8 mm; ovaries and fr glab; fr ca 3.5 × 2 mm; known only from 1 collection in Wallowa Mts, Ore; Greenman's l.   27 L. greenmanii Math.

33b Ultimate segms of lvs gen hirtellous-puberulent, gen 1–3 (6) mm; ovaries and fr hirtellous-puberulent; fr ca 5 × 2.5–3 mm; open rocky places high in the Wallowa and Blue mts, Ore; Ore l.
28 L. oreganum Coult. & Rose

32b Pls distinctly otherwise, gen larger and with looser mature infl, or if small then with a short, tuberous-thickened root

34a Wings of fr ± corky-thickened, gen narrow (or wider in L. columbianum); fls purple or yellow (or salmon-colored)

35a Foliage gen ± scaberulous, rarely glab; robust, mostly 5–15 (20) dm at maturity; throughout our range (see lead 21a)
16 L. dissectum (Nutt.) Math & Const.

35b Foliage glab; smaller, up to ca 6 (8) dm; more local in distribution

36a Fls salmon-yellow; fr 8–14 mm; open rocky slopes near Snake and Clearwater rivers, se Wn, wc Ida, and prob adj Ore; Salmon R l. or d.
28 L. salmoniflorum (Coult. & Rose) Math. & Const.

36b Fls purple (rarely yellow); more w or s spp.

37a Fr large, gen 16–28 mm; lf segms 3–20 mm; upper end CRG (see lead 21b)   17 L.columbianum Math. & Const.

37b Fr smaller, gen 9–16 mm; lf segms 1–8 mm; not of CRG

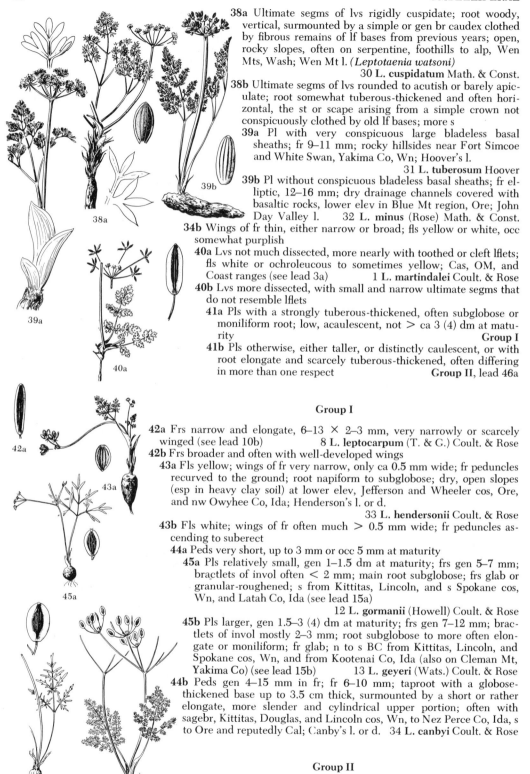

38a Ultimate segms of lvs rigidly cuspidate; root woody, vertical, surmounted by a simple or gen br caudex clothed by fibrous remains of lf bases from previous years; open, rocky slopes, often on serpentine, foothills to alp, Wen Mts, Wash; Wen Mt l. (*Leptotaenia watsoni*)
          30 L. **cuspidatum** Math. & Const.

38b Ultimate segms of lvs rounded to acutish or barely apiculate; root somewhat tuberous-thickened and often horizontal, the st or scape arising from a simple crown not conspicuously clothed by old lf bases; more s

 39a Pl with very conspicuous large bladeless basal sheaths; fr 9–11 mm; rocky hillsides near Fort Simcoe and White Swan, Yakima Co, Wn; Hoover's l.
        31 L. **tuberosum** Hoover

 39b Pl without conspicuous bladeless basal sheaths; fr elliptic, 12–16 mm; dry drainage channels covered with basaltic rocks, lower elev in Blue Mt region, Ore; John Day Valley l. 32 L. **minus** (Rose) Math. & Const.

34b Wings of fr thin, either narrow or broad; fls yellow or white, occ somewhat purplish

 40a Lvs not much dissected, more nearly with toothed or cleft lflets; fls white or ochroleucous to sometimes yellow; Cas, OM, and Coast ranges (see lead 3a) 1 L. **martindalei** Coult. & Rose

 40b Lvs more dissected, with small and narrow ultimate segms that do not resemble lflets

  41a Pls with a strongly tuberous-thickened, often subglobose or moniliform root; low, acaulescent, not > ca 3 (4) dm at maturity   **Group I**

  41b Pls otherwise, either taller, or distinctly caulescent, or with root elongate and scarcely tuberous-thickened, often differing in more than one respect   **Group II**, lead 46a

Group I

42a Frs narrow and elongate, 6–13 × 2–3 mm, very narrowly or scarcely winged (see lead 10b) 8 L. **leptocarpum** (T. & G.) Coult. & Rose

42b Frs broader and often with well-developed wings

 43a Fls yellow; wings of fr very narrow, only ca 0.5 mm wide; fr peduncles recurved to the ground; root napiform to subglobose; dry, open slopes (esp in heavy clay soil) at lower elev, Jefferson and Wheeler cos, Ore, and nw Owyhee Co, Ida; Henderson's l. or d.
         33 L. **hendersonii** Coult. & Rose

 43b Fls white; wings of fr often much > 0.5 mm wide; fr peduncles ascending to suberect

  44a Peds very short, up to 3 mm or occ 5 mm at maturity

   45a Pls relatively small, gen 1–1.5 dm at maturity; frs gen 5–7 mm; bractlets of invol often < 2 mm; main root subglobose; frs glab or granular-roughened; s from Kittitas, Lincoln, and s Spokane cos, Wn, and Latah Co, Ida (see lead 15a)
        12 L. **gormanii** (Howell) Coult. & Rose

   45b Pls larger, gen 1.5–3 (4) dm at maturity; frs gen 7–12 mm; bractlets of invol mostly 2–3 mm; root subglobose to more often elongate or moniliform; fr glab; n to s BC from Kittitas, Lincoln, and Spokane cos, Wn, and from Kootenai Co, Ida (also on Cleman Mt, Yakima Co) (see lead 15b) 13 L. **geyeri** (Wats.) Coult. & Rose

  44b Peds gen 4–15 mm in fr; fr 6–10 mm; taproot with a globose-thickened base up to 3.5 cm thick, surmounted by a short or rather elongate, more slender and cylindrical upper portion; often with sagebr, Kittitas, Douglas, and Lincoln cos, Wn, to Nez Perce Co, Ida, s to Ore and reputedly Cal; Canby's l. or d. 34 L. **canbyi** Coult. & Rose

Group II

46a Ovaries and young (often also mature) fr granular-scaberulous or ± hairy

 47a Herbage granular-scaberulous to subglabrous; fr granular-scaberulous at least when young; fls yellow

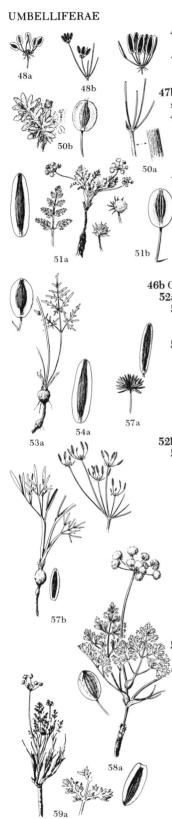

48a Peds 5–15 mm at maturity; fr 8–12 mm; c and ne Ore to Cal (see lead 29a)   23 **L. vaginatum** Coult. & Rose
48b Peds 2–5 (7) mm at maturity; fr 5–8 mm; open, rocky slopes and ridges, mid to high elevs; n Ida, nw Mont (chiefly w Cont Div), and extreme sw Alta; Sandberg's l. or d.   35 **L. sandbergii** Coult. & Rose
47b Herbage short-hairy (at least the scape or st and peduncle); fr short-hairy at least when young, not granular-scaberulous
49a Fls yellow
50a Fr large, gen 16–28 mm; pls gen 5–10 dm at maturity; c Wn (see lead 24b)   20 **L. thompsonii** (Math.) Cronq.
50b Fr small, mostly 5–10 mm; pls < 5 dm; not of Wn (see lead 30a)   24 **L. foeniculaceum** (Nutt.) Coult. & Rose
49b Fls white or somewhat purplish, very rarely yellow in no 36
51a Bractlets of involucel villous-puberulent, not markedly scarious-margined; fr (7) 10–20 mm, (1.8) 2–5 times as long as wide; open rocky hills and plains, scarcely mont; s BC to Cal, e to Man, ND, Wyo, and Utah, E Cas in our range; large-fr l. or d. *(L. flavum)*
                    36 **L. macrocarpum** (Nutt.) Coult. & Rose
51b Bractlets of involucel glab or very finely hirtellous, evidently scarious-margined; fr 6–10 mm, < twice as long as wide; c Ore and s (see lead 30b)   25 **L. nevadense** (Wats.) Coult. & Rose
46b Ovaries and fr glab
52a Fls ordinarily white or somewhat purplish, very rarely yellow in no 36
53a Herbage essentially glab; c Wn and n Ida to s BC (extreme specimens of *L. cusickii*, from ne Ore to sw Mont, might be sought here) (see lead 15b)   13 **L. geyeri** (Wats.) Coult. & Rose
53b Herbage evidently puberulent, villous-puberulent, or hirtellous
54a Bractlets of involucel villous-puberulent, not markedly scarious-margined; fr (7) 10–20 mm, (1.8) 2–5 times as long as wide; widespread E Cas (see lead 51a)
                    36 **L. macrocarpum** (Nutt.) Coult. & Rose
54b Bractlets of involucel glab or very finely hirtellous, evidently scarious-margined; fr 6–10 mm, > twice as long as wide; c Ore and s (see lead 30b)   25 **L. nevadense** (Wats.) Coult. & Rose
52b Fls yellow, rarely white in no 8
55a Taproot or tuberous-thickened root with a gen simple, often subterranean crown, sts solitary or few
56a Fr narrowly oblong, 1.5–3 mm wide, > 2.5 times as long as wide
57a Involucel well developed; peds 1–3 mm (see lead 10b)
                    8 **L. leptocarpum** (T. & G.) Coult. & Rose
57b Involucel none; peds 4–13 mm (see lead 8a)
                    5 **L. ambiguum** (Nutt.) Coult. & Rose
56b Fr elliptic, often broadly so, < 2.5 times as long as wide
58a Herbage glab; lvs blue-glaucous, mostly or all clustered at or near the base; c and e Ore, open, rocky or gravelly slopes and dry meadows to lower mont, not extending to SRC; Donnell's l. or d.
                    37 **L. donnellii** Coult. & Rose
58b Herbage scaberulous or crisp-puberulent to subglab; lvs cauline and basal, not glaucous; in and near canyons of Snake and lower Salmon rivers in n Ore and adj Ida (see lead 19a)
                    15 **L. rollinsii** Math. & Const.
55b Taproot gen surmounted by a br caudex, sts or scapes several or many (taproot sometimes with a simple crown in no 40, but still with clustered sts)
59a Lvs very finely dissected, gen with several hundred or > 1000 very narrow and often subterete ultimate segms that lie in numerous different planes so that the lf has "thickness"; lvs often evidently scaberulous; widespread E Cas, dry, open, often rocky places from foothills and lowl to midmont; c Wn to n Ida, s in e Ore and w Ida to ne Nev, irreg to se Ida, Wyo, and Colo; Gray's l. or d.
                    38 **L. grayi** Coult. & Rose
59b Lvs less finely dissected, with not > a few hundred flat, linear or oblong segms that tend to lie in nearly a single plane; lvs glab or very nearly so
60a Pls acaulescent; oil tubes solitary in the intervals; open, often rocky slopes in and near SRC, w Ida, e Ore, and extreme se Wn;

Snake Canyon l. or d. (*C. fragrans*)

39 **L. serpentinum** (M. E. Jones) Math.

60b Pls ± caulescent; oil tubes gen 2–3 in the intervals; W Cas, WV s to sw Ore, rocky crevices and bluffs in valleys and foothills; Hall's l. or d.   40 **L. hallii** (Wats.) Coult. & Rose

Musineon Raf.   Musineon

Infl a compound umbel; invol gen wanting; involucel of several distinct or basally connate bractlets; fls yellow or white; calyx teeth well developed; styles slender, spreading, without a stylopodium; carpophore entire to deeply cleft; fr ovoid to linear-oblong, laterally ± compressed, glab or scab, evidently ribbed; oil tubes 1–4 in intervals and sometimes 1 in each rib, 2–6 on commissure; low pers from thickened taproot; lvs chiefly at or near base (sometimes on a pseudoscape), pinnately or ternate-pinnately ± dissected. (From ancient Gr name for *Foeniculum* or some other umbellifer).

1a Lvs mostly subopposite, deeply pinnatifid, primary pinnae only shortly or scarcely stalked and gen appearing as ± deeply cleft or toothed segms; taproot with a single crown and pls often with a pseudoscape; open flats and valleys and foothills, primarily of high plains, Sask to Neb, w to Alta, Mont, c Ida, and e Nev; tolerant of alkali; leafy m. (*M. d.* var. *hookeri*, *M. h.*, *M. trachyspermum*, *M. angustifolium*)   1 **M. divaricatum** (Pursh) Nutt.
1b Lvs distinctly alt, ternate-pinnately dissected into ± linear ultimate segms, at least lowest pair of primary segms distinctly slender-stalked; taproot surmounted by a ± br caudex, and pls without a pseudoscape; rocky places, foothills and mts; Bridger Mts, Mont, to Bighorn Mt region, Wyo; seldom collected; sheathed m.   2 **M. vaginatum** Rydb.

Oenanthe L.   Water-parsley; Oenanthe

Infl of compound umbels; invol of narrow bracts, or wanting; involucel of ∝ small, narrow bractlets; fls white; calyx teeth evident; styles elongate, ± erect; stylopodium conic; carpophore wanting; fr oblong or elliptic, much broader than the stylopodium, terete or slightly compressed laterally, glab, ribs prominent, corky-thickened and wider than the intervals; oil tubes gen 1 in the intervals, 1–2 on the commissure; glab pers from fascicled fibrous or tuberous-thickened roots; lvs pinnately compound to dissected. (Ancient Gr name for some other pl).

O. sarmentosa Presl. Pac, or Am w. or o. Soft and weak, gen reclining or scrambling-ascending herbs, often rooting at the nodes, freely br, up to 1 m or more; lvs mostly bipinnate or subtripinnate, lflets 1.5–6 × 0.7–5 cm; primary lateral veins of lflets directed to marginal teeth; umbels pedunculate, leaf-opposed, rays 10–20, 1.5–3 cm at maturity; fr oblong, 2.5–3.5 × 1.5–2 mm; low wet places, w Cas, Alas to c Cal, up CR to w Klickitat Co, Wn, and in the Chilliwack Valley, BC.

Orogenia Wats.   Turkey-peas; Orogenia

Infl a small compound umbel with few, unequal rays; invol wanting; involucel minute or wanting; calyx teeth obsolete; petals white; anthers and top of ovary dark purplish; stylopodium and carpophore wanting; fr oblong to oval, glab, subterete or slightly flattened laterally, dorsal ribs evident or obscure, lateral ones developed into inflexed corky wings, a corky riblike projection also running the length of the commissural face of each mericarp; oil tubes several in the intervals and on the commissure; low, delicate vernal, scapose pers from a tuberous-thickened root; lvs 1–3 times ternate, with elongate, narrow, mostly entire ultimate segms, the petiole ± sheathing. (Gr *oros*, mt, and *genos*, race, referring to the habitat).

1a Root globose, up to ca 1.5 cm thick; fr 3–4 mm, oblong-elliptic, with evident dorsal ribs; open slopes and ridges, lower foothills to midmont; e Cas, s Wn to e Ore, e to Ravalli Co, Mont, s Ida, Utah, and w Colo; Mar-May; linear-lvd o., GB o.   1 **O. linearifolia** Wats.
1b Root elongate, several times as long as thick; fr as in no 1, but dorsal ribs tending to be obscure; lvs sometimes more ∝ than in no 1; open places, valleys to lower mont; Linn Co, Ore, to n Cal; May-July

2 **O. fusiformis** Wats.

Osmorhiza Raf.   Sweet-root; Sweet-cicely

Infl of loose compound umbels; invol wanting or of 1–few narrow, foliaceous bracts; involucel of several narrow foliaceous reflexed bractlets, or wanting; fls white, yellow, purple, or pink; calyx teeth obsolete; stylopodium conic to depressed; carpophore 2-cleft < half its length; fr narrow, linear or clavate, obtuse to constricted or short-beaked, rounded to caudate at base, somewhat compressed laterally, bristly-hispid to glab, ribs narrow, the oil tubes obscure or wanting; caulescent, thick-rooted pers with petiolate lvs, the bl ternately to pinnately 1–3 times compound, lflets lanceolate to orbicular, toothed to pinnatifid. (Gr *osme*, odor, and *rhiza*, root, from the pleasant odor of the original sp.). (*Glycosma, Washingtonia*).

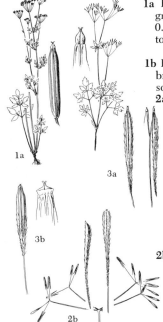

1a Fr glab, 12–20 mm, the base obtuse, exappendiculate; fls yellow or greenish-white; sts clustered; pl strongly licorice-odored; lflets gen 2–10 × 0.5–5 cm; thickets and open places, lowl to midmont; nw Wn to sw Alta, s to Cal and Colo; western s. (*G. o., G. ambiguum, O. a.*)
                                       1 **O. occidentalis** (Nutt.) Torr.
1b Fr attenuate at base into prominent, bristly appendages, the body often bristly as well; fls greenish-white or sometimes pink or purple; sts mostly solitary
2a Fr concavely narrowed to the summit, the terminal (0.5) 1–2 mm ± distinctly set off as a broadly beaklike apex (more marked in no 2 than in no 3); peds and rays of umbel gen ascending-spreading
3a Fr gen 12–22 mm; stylopdium ± conic, commonly at least as high as broad; fls mostly greenish-white; common in woodl, from near sea level to midmont; widespread cordilleran sp., also in ne N Am, and S Am; mt s. (*W. divaricata, O. nuda* var. *d., O. n., W. intermedia*)
                                      2 **O. chilensis** H. & A.
3b Fr gen 8–13 mm; stylopodium depressed, gen broader than high; fls pink or purple (sometimes greenish-white); relatively uncommon in our range, largely confined to moister mt areas; Alas to nw Mont, n Ida, and n Wn, and in Cas to s Ore and adj Cal; purple s. (*W. leibergii, O. l.*)   3 **O. purpurea** (Coult. & Rose) Suksd.
2b Fr convexly narrowed to the rounded or obtuse (to merely acutish) summit, the apex not at all beaklike; pedicels and rays of umbel gen widely divaricate; woodl, from near sea level to midmont; widespread cordilleran sp., wholly E Cas, also in ne N Am and S Am; blunt-fr s. (*W. obtusa, O. o.*)   4 **O. depauperata** Phil.

Pastinaca L.   Parsnip

Umbels compound; invol and involucel gen wanting; fls yellow or red; calyx teeth obsolete; styles short, spreading, the stylopodium depressed-conic; carpophore bifid to the base; fr glab, elliptic to obovate, strongly flattened dorsally, dorsal ribs filiform, the lateral ones narrowly winged; oil tubes solitary in the intervals and visible to the naked eye through the pericarp, 2–4 on the commissure; stout, lfy-std biens and pers, gen with well-developed taproot; lvs pinnately compound. (Ancient L name of the parsnip).

P. sativa L. Common p., wild p. Aromatic bien from a stout taproot, 3–10 dm; basal lvs up to 5 dm and nearly half as broad, lflets up to 13 × 10 cm, serrate and the lower ones sometimes also pinnately cleft or some lflets with a nearly separate large basal lobe; cauline lvs reduced upward; fls yellow; fr broadly elliptic, 5–6 × 4–5 mm; roadsides and other disturbed sites; European, now widely naturalized, but only occ in our area.

Perideridia Reichenb.   Yampah; False-caraway

Umbels compound, on ± elongate peduncles; invol of few–∞ small, ± scarious bracts, or wanting; involucel gen of scarious or colored bractlets, or obsolete; fls white or pink; calyx teeth well developed; styles short, with a ± conic stylopodium; carpophore bifid to base; fr linear-oblong to orbicular, laterally somewhat compressed, glab, ribs prominent to inconspicuous; oil tubes 1–5 in the intervals, 2–8 on the commissure; glab pers from ± tuberous-thickened, edible, often fascicled roots; lvs pinnately or ternate-pinnately compound or dissected, with mostly narrow ultimate segms. (Name unexplained).

1a

2b

2a

1a Principal lvs dissected, with ∝ ultimate segms, these ± distinctly dimorphic, some much longer than others; petioles markedly dilated; oil
   tubes 2–5 (gen 3) in the intervals; bractlets of involucel narrowly lanceolate
   to obovate; dry hillsides, ridges, and washes, foothills and high plains; e
   Wallowa Co, Ore, and w Ida Co, Ida, s through e Ore and Ida to ne Utah,
   Nev, and Sierran Cal; also in Jackson Hole, Wyo; Bolander's y. (*Podos-
   ciadium b., Eulophus b.*)   1 **P. bolanderi** (Gray) Nels. & Macbr.
1b Principal lvs merely 1–2 times pinnate or ternate, ultimate segms few and
   not markedly dimorphic; petioles not much dilated; oil tubes 1 in the intervals
   2a Fr orbicular or suborbicular, nearly or quite as wide as long; bractlets
      mostly setaceous, up to 0.5 mm wide, or obsolete; woodl and dry to wet
      meadows, lowl to midmont; s BC to s Cal, e to Sask, SD and Colo; Gaird-
      ner's y. (*Carum g., C. montanum, Atenia m.*); our pls have been de-
      scribed as ssp. *borealis* Chuang & Const.
                                       2 **P. gairdneri** (H. & A.) Math.
   2b Fr oblong-ovate, evidently longer than wide; bractlets broader and
      better developed, mostly 0.6–1 mm wide; moist or dry meadows to open
      slopes, valleys to rather high mont; Cas and w, sw Wn to Cal; eppaw,
      Ore y. (*Carum o., Ataenia o.*)   3 **P. oregana** (Wats.) Math.

Pimpinella L.   Pimpinella

Umbels compound; invol gen wanting; involucel of inconspicuous bractlets or wanting; fls gen
white or whitish, outer petals often larger than the inner; calyx teeth minute or obsolete; stylopodium ± conic; carpophore bifid to middle or to base; fr oblong to orbicular, narrowed at the apex,
rounded or cordate at base, somewhat flattened laterally, ribs equal, filiform to very narrowly
winged; pls mostly per, with br sts; lvs variously simple or compound to dissected. (L name prob
applied to a pl of this genus, said to be derived from *bipinnula*, referring to the lvs).

P. saxifraga L. Burnet-saxifrage. Pls 3–8 dm, shortly woolly-hirsute; lvs di- or
trimorphic, lowermost ones pinnately compound with gen 7–13 lflets 1–4 ×
up to 3 cm, the next above gen more dissected and without well-defined lflets,
the uppermost gen reduced to petiolar sheaths with vestigial or no bl; fr elliptic to orbicular, 2–2.5 mm, glab, the ribs filiform and rather obscure; oil
tubes gen 3 in the intervals, 2–4 on the commissure; Eurasian, occ estab weed
in US; known in our area from San Juan I, Wn, where represented by a hairy
phase called ssp. *nigra* (Mill.) Gaudin.

Rhysopterus Coult. & Rose   Rhysopterus

Umbels compact, compound, on terminal peduncles; invol wanting; involucel of well-developed,
subcoriaceous or firm-scarious, partly connate bractlets, all on one side of the umbellet; fls white;
calyx teeth small but evident, broadly rounded, persistent; styles short, without a stylopodium;
carpophore wanting; fr ovoid to orbicular; low, taprooted pers; lvs forming a rosette, not much
dissected. (Gr *rhysos*, wrinkled, and *pteron*, wing, referring to the fr).

R. plurijugus Coult. & Rose. Glab per with a slender pseudoscape arising to
the ground level from the subterranean crown of a thickened taproot; lvs petiolate, spreading, firm, bl gen 1.5–4 cm, not much dissected, the 3–5 primary
segms broad and ± confluent, toothed or lobed; fr somewhat flattened laterally, appearing to have strongly corrugated narrow wings when young, but at
maturity the ribs merely raised and thickened, with or without an irreg vestige of a wing; oil tubes solitary in the intervals and in the apex of each rib, 2
on the commissure; loose dry ground in se Ore, barely or scarcely entering
our range (*Cymopterus p.*).

Sanicula L.   Sanicle

Infl of several or many compact, headlike ultimate umbels arranged into a definite primary umbel
or dichasially or rather irreg arranged; fls white or greenish to yellow, purple, or blue, some ♂,
others ♀, each ultimate umbel with both types, or some wholly ♂; ♀ fls sessile or nearly so, ♂ fls gen
± ped; invol ± foliaceous, often appearing as opposite, sessile lvs in forms with dichasially ar-

ranged umbellets; involucel of several or ∝ prominent or inconspicuous bractlets; calyx teeth well developed, gen connate at least at base; stylopodium flattened and disclike, or wanting; fr oblong-ovoid to globose, ± compressed laterally, beset with prickles or tubercles, our spp. with uncinate prickles; ribs of fr gen obsolete; glab or subglab biens or more often pers with variously cleft or dissected lvs. (L *samare*, to heal).

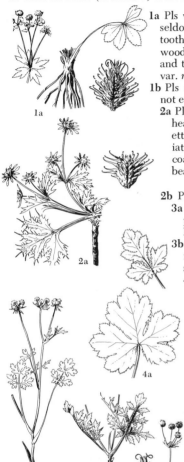

1a Pls with a cluster of fibrous roots from a short simple caudex or crown, seldom < 4 dm; lvs palmately cleft to palmately compound, lflets or segms toothed; fls greenish white; wholly e Cas, moist low ground to damp wooded slopes; Newf to Fla, w to e BC, n Ida, and prob extreme ne Wn, and to sw Mont and s RM; black snake-root, s. (*Caucalis m., S. canadensis* var. *m.*)   1 S. marilandica L.

1b Pls ± distinctly taprooted, often smaller than no 1, and except for no 5, not extending much e Cas; fls light yellowish to yellow or purple

   2a Pls prostrate or ascending, maritime; involucel conspicuous, gen > the heads; lvs somewhat succulent, often yellowish, basal ones rosette-forming, broadly petioled, bl 3-cleft, primary segms irreg laciniate-toothed or cleft and teeth softly bristle-tipped; coastal bluffs, VI and coastal Wn to s Cal, seldom collected in our area; footsteps of spring, bear's-foot s., snake-root (*S. howellii, S. crassicaulis* var. *h.*)
          2 S. arctopoides H. & A.

   2b Pls erect, not maritime, or only casually so; involucel inconspicuous

      3a Principal lvs 1–2 times pinnatifid, with a distinctly toothed rachis; fls purple; ± open slopes and drier meadows; VI and w Wn and Ore to Baja Cal, wholly w Cas in our area; purple s. 3 S. bipinnatifida Dougl.

      3b Principal lvs palmately or pinnipalmately lobed or divided to ternate-pinnate, without a toothed rachis; fls yellow or yellowish except sometimes in no 4

         4a Lvs palmately or pinnipalmately lobed or divided, without a narrow rachis, the primary divisions merely serrate or lobed; woods w Cas, s BC to Baja Cal, and through CRG to Klickitat Co, Wn; Chile; Pac s.; 2 vars.   4 S. crassicaulis Poepp.

            a1 Bl of basal lvs palmately lobed or cleft and ca as wide as long, or somewhat wider than long; fls yellow; general (*S. menziesii*)
                    var. **crassicaulis**

            a2 Bl of basal lvs more pinnipalmately cleft, often slightly longer than wide, central lobe ± elongate; fls often slightly purplish; PS, Wn, to WV, Ore, and up CRG to Klickitat Co, Wn (*S. t.*)
              var. **tripartita** (Suksd.) H. Wolff

         4b Lvs ± ternate-pinnate, primary divisions tending to be pinnatifid, the lowest pair of primary divisions separated from the terminal segm or segms by a narrow, entire rachis; widespread, ± open slopes and flats, lowl to midmont; s BC to s Cal, e to w Mont and nw Wyo; S Am; Sierra s. (*S. septentrionalis, S. nevadensis* var. *s., S. g.* var. *s., S. apiifolia*)   5 S. graveolens Poepp.

Scandix L.   Scandix

Infl of compound or simple umbels; invol wanting, or a lfy bract; involucel of several lobed or dissected bractlets; fls white; calyx teeth minute or obsolete; styles very short, stylopodium depressed; carpophore entire, or bifid at apex; fr ± linear, hispid or scab, body ± quadrangular to subcylindric, evidently ribbed, tipped by a long, subcylindric, nearly ribless beak; oil tubes solitary in the intervals; taprooted ann with lfy sts and pinnately dissected lvs. (Ancient Gr name of another umbelliferous pl).

·S. pecten-veneris L. Venus'-comb, lady's v.-c., shepherd's-needle. Pl 1–3 dm, gen br from base, ± hirsute to subglab; lvs scattered along st, petiole = or < bl, bl mostly 1.5–9 (15) × 1–6 cm, ultimate segms 1–3 × 0.5–1 mm; umbellets compact, < 1 cm wide at anthesis and not much wider at maturity, borne in 2's (or singly or in 3's); involucel bractlets several, gen bifid above; fr peds stout, erect, 2–5 mm; body of fr 6–15 × 1–2 mm, beak 2–7 cm; Eurasian, weedy sp. sparingly intro in US, reported from VI to Cal.

Sium L.   Water-parsnip

Umbels compound, congested; invol bracts subfoliaceous, entire or incised, often reflexed; involucel bractlets well-developed; fls white; calyx teeth minute or obsolete; styles short, reflexed, stylopodium depressed or rarely conic; carpophore bifid to the base, the halves adnate to the mericarps; fr glab, elliptic to orbicular, slightly compressed laterally and somewhat constricted at the commissure; ribs subequal, prominent, corky but scarcely winged; oil tubes 1–3 in the intervals, 2–6 on the commissure face; glab, lfy-std pers with mostly pinnately compound or decompound lvs. (Gr *Sion*, ancient name of some aquatic umbellifer, possibly a sp. of *Sium*).

S. suave Walt. Hemlock w. Crown erect, short, with fibrous roots sometimes originating also from 1 or 2 nodes above; st solitary, stout, ribbed-striate, gen 5–12 dm; lvs basal and cauline or all cauline, wing-petiolate, pinnately once compound with 7–13 lflets, these 2–9 cm × 1.5–10 (20) mm, merely serrulate to deeply pinnatisect and again cleft; primary lateral veins of lflets mostly br and inconspicuous, not bearing any obvious relation to marginal teeth; fr broadly elliptic to orbicular, 2–3 mm; swampy places to standing water; valleys and foothills; s BC to Cal, e to Newf and Va, throughout our area.

Sphenosciadium Gray   Swamp White-heads; Woolly-head Parsnip

Infl of loose compound umbels, umbellets capitate with sessile fls; invol wanting; involucel bractlets ∝, slender; calyx teeth obsolete; stylopodium small, broadly conic; carpophore bifid to base; fr tomentose, cuneate, strongly flattened dorsally; oil tubes small, 1 in intervals, 2 on commissure; per from simple or br caudex which may surmount a taproot; lvs petiolate, pinnately or ternate-pinnately 1–2 times compound. (Gr *sphen*, wedge, and *skias*, umbrella, presumably referring to frs and infl).

S. capitellatum Gray. Pl 5–18 dm; foliage scab to subglab; infl tomentose; lflets ± lanceolate, 2.5–8 × 0.5–2 cm; umbels 1–several, rays 1–5 cm; umbellets 6–12 mm wide; fls white or occ purplish; fr 5–8 × 3–5 mm, wings wider distally; wet meadows and moist low places, foothills to midmont; c Ida to Wallowa and Strawberry mts, e Ore, s irreg to Sierran Cal and w Nev.

Tauschia Schlecht.   Tauschia

Infl a compound umbel, gen with involucel but no invol; calyx teeth obsolete to prominent; petals yellow, white, or purplish; stylopodium nearly or quite wanting; carpophore 2-cleft to middle or below; fr linear-oblong to subglobose, slightly flattened laterally, glab, ribs ± evident but not winged; low, acaulescent or short-caulescent pers, our spp. glab; lvs pinnately to ternately cleft or dissected, rarely entire. (For Friedrich Tausch, 1793–1848, European botanist).

1a Lf segms linear, gen 1–2 mm wide; fr linear-oblong, 5–7 mm; fls white; sagebr scablands in Yakima Co, Wn; Hoover's t.
                                       1 **Tauschia hooveri** Math. & Const.
1b Lf segms lanceolate to ovate or elliptic, gen 4–10 mm wide; fr subglobose or broadly ellipsoid, 2–3 mm; fls yellow; meadows and moist slopes, 5000–6500 ft elev, Mt. Rainier, Wn; Strickland's t. *(Hesperogenia s.)*
                                       2 **T. stricklandii** (Coult. & Rose) Math. & Const.

1a   1b   **Zizia** Koch   Zizia

Infl of compound umbels; invol wanting; involucel a few inconspicuous bractlets; peds short and stout, central fl of each umbellet sessile or subsessile; fls bright yellow; calyx teeth well developed; styles slender, erect or spreading, without a stylopodium; carpophore bifid ca 1/2 length; fr oblong or broadly elliptic, glab, ± compressed laterally, ribs prominent but not much raised; oil tubes 1 per interval, 2 on commissure; pers from short caudex; lvs basal and cauline, simple or ternately compound. (For Johann Baptist Ziz, 1779–1829, German botanist).

Z. aptera (Gray) Fern. Sts gen clustered on a compact caudex, 2–8 dm; basal lvs rather long-petiolate, with broad, cordate, crenate or serrate bl 2.5–10 × 1.5–8 cm, occ some trifoliolate, with more elliptic, scarcely cordate lflets; cauline lvs few, mostly trifoliolate; infl compact and densely fld, gen 1.5–4 cm wide at anthesis, a little larger in fr, the rays up to 3.5 cm at maturity; fr 2–4 mm; moist or wet places, gen in low ground, tolerant of alkali; most of e and c US and adj Can, and in RM from s Alta to Colo, w to e Wn, nw Ore, and ne Nev; heart-lvd Alexanders; ours the var. *occidentalis* Fern.

CORNACEAE   Dogwood Family

Fls small, polypet, reg, ♂ to ♀♂ or ♀, ♂, epig, pan or cymose to capitate, often invol, 4- (ours) or 5-merous; sepals tiny, free hypan lacking; stamens as many as petals and alt with them, or twice as many; ovary 1–4-celled, ovules 1 per cell, pendulous; style 1; fr drupe or berry; trees or shrubs (1 sp. trailing and merely suffrutescent), with opp (ours), simple lvs.

Cornus L.   Dogwood

Fls ebracteate or subtended by 4–8 conspicuous whitish or pinkish bracts; petals white or greenish, often purple-tipped; stamens 4; pistil 2-carpellary, ovary 2-celled; drupe 2-seeded; suffrutescent to woody, ± malphghiaceous-strigillose shrubs or trees with simple lvs and prominent pinnate veins. (L, meaning horn or leather, referring to the hard wood, or perhaps the *cornus*, or orn knob of the roll on which manuscripts were kept). (*Chamaepericlymenum, Cornella, Cynoxylon*).

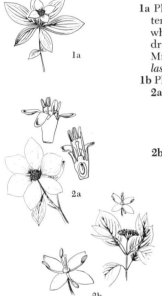

1a Pl low trailing subshrub 0.5–2 dm; lvs whorled at top of st, 2–8 cm; fls in terminal, pedunculate, condensed and semicapitate cyme, subtended by 4 white, pinkish, or purple-tinged ovate bracts 1–2.5 cm; petals 1–1.5 mm; drupes coral red, 6–8 mm; moist woods; Alas to Greenl., s to Cal, NM, Minn, and Pa; Asia; bunchberry (d.), puddingberry, dwarf cornel (*C. unalaschkensis*); an excellent garden subject   1 C. canadensis L.
1b Pl woody shrub or tree; lvs opp throughout
   2a Fls capitate, infl conspicuously bracteate, bracts 4–7, white or pinkish-tinged, 2–7 cm; pl shrublike to arborescent, 2–20 m; petals ca 2.5 mm; drupes red, ca 10 mm; open to rather dense for, esp along streams; BC s, in and w Cas, to s Cal, also in Idaho Co, Ida; Pac d., mt d., w flowering d.; one of finest orn in our flora   2 C. nuttallii Aud.
   2b Fls cymose, infl not conspicuously bracteate; pl many-std shrub 2–6 m; petals 2–4 mm; drupe white to ± bluish, 7–9 mm; creek d., red-osier d.; 2 vars.   3 C. stolonifera Michx.
      a1 Stone of fr smooth; petals 2–3 mm; styles 1–2 mm; pubescence strigose mainly; Alas to Newf, s in RM to Mex, e to Mo and Pa; the common phase in Mont, and Ida, occ in ne Wn   var. stolonifera
      a2 Stone grooved lengthwise; petals 3–4 mm; styles 2–3 mm; pubescence often spreading or curled; Alas to s Cal, e to Ida and Nev, occ in Mont   var. occidentalis (T. & G.) Hitchc.

GARRYACEAE   Silk-tassel Family

Fls small, apet, ♀, ♂, in pendent catkinlike racemes, in 3's (or singly) in axils of opp, ± cuplike bracts; ♂ fl with 4 elongate bractlike sepals alt with 4 stamens; calyx of ♀ fl adherent to ovary, lobes 2, 1, or lacking; carpels 2, styles 2, ovary 1-celled with 2 pendulous ovules; fr semi-baccate but becoming dry and ± caps, 1–2-seeded; evergreen shrubs with opp, leathery, simple, subentire lvs.

Garrya Dougl.   Silk-tassel

Shrubs or sometimes small trees with gen deep green lvs and drooping infl of long-persistent fls blossoming during winter or early spring. (For Nicholas Garry, a friend of Douglas). The spp. are choice orn pls.

1a Lvs ± strongly undulate-margined, thickly tomentose beneath, deep green and glab above, bl 5–8 cm, oblong-ovate to -obovate; pl 2–7 m; coastal bluffs and hills; s Lincoln Co, Ore, to Cal; wavy-lf s.   1 G. elliptica Dougl.
1b Lvs not undulate-margined, strigillose or glabrate beneath, glab above, rather yellow-green, bl elliptic-ovate to -oblong, 4–8 cm; pl 1–3 m; woodl and chaparral; CR, Wn and Ore, w side Cas, Lane Co, Ore, to Cal; Fremont s., bear-brush   2 G. fremontii Torr.

ERICACEAE   Heath Family

Fls often showy, white to red or purplish, polypet to gamopet (apet), reg or sometimes ± irreg, mostly 4–5-merous, hypog to epig, ♀; sepals distinct to connate; corolla rotate to funnelf or often urnshaped, occ lacking; stamens mostly = or 2 × corolla or calyx lobes (rarely more), almost or quite free of corolla; anthers often inverted, sometimes with awnlike projections from back or tip, opening by terminal or lateral pores or chinks, by longitudinal slits, or often by open terminal tubes, 2-celled; pistil 4–12-carpellary and -celled or occ 1–3-celled, placentation axile or rarely parietal; style 1, stigma capitate to lobed; fr mostly dry and caps, but sometimes with adherent fleshy calyx, sometimes a drupe, and occ a true berry; per, often glandular or hairy trees, shrubs, or herbs with alt, mostly leathery, gen evergreen lvs, but sometimes pl fleshy, without green color (white to brownish or red), and with lvs reduced to bracts.

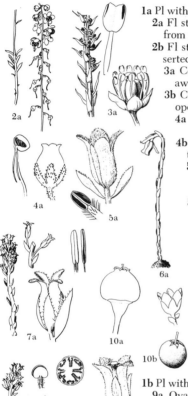

1a Pl without green lvs
  2a Fl st gen < 2 dm, with only 1–3 bracts below fls; styles long, exserted from fls; petals distinct, mostly spreading   **Pyrola**
  2b Fl st freely bracteate to infl, often much > 2 dm; either styles not exserted from fl or corolla lacking
    3a Corolla lacking; st pink- and white-striped lengthwise; anthers unawned, opening by basal (falsely terminal) pores   **Allotropa**
    3b Corolla present but calyx sometimes lacking; st not striped; anthers opening nearly full length by slits
      4a Corolla urn-shaped, gamopet; pl 3–10 dm; anthers awned on back   **Pterospora**
      4b Corolla not urn-shaped, gen polypet; pl 1–4 (rarely > 3) dm; anthers unawned
        5a Corolla gamopet, fimbriate, hairy within; placentation parietal; anthers oblong, ca 2 mm, dehiscent full length but not across the tip   **Hemitomes**
        5b Corolla polypet; placentation often axile; anthers various
          6a Fls single, waxy-white, drying black   **Monotropa**
          6b Fls 2–several, not waxy-white, darkening, but sometimes not to black, when dried
            7a Anthers 2–3 mm, linear, not much broader than filaments, dehiscent full length, but not across tip; filaments glab or minutely puberulent, flattened; corolla scarcely hairy within; caps subglobose, glab or subglab; placentation parietal   **Pleuricospora**
            7b Anthers ca 1 mm, oval, as broad as long, dehiscent by continuous slit over tip; filaments hairy; corolla gen strongly hairy within; caps oblong-ovoid, hairy; placentation sometimes axile
              8a Placentation axile; pl drying to brownish   **Hypopitys**
              8b Placentation parietal; pl drying blackish   **Pityopus**
1b Pl with (gen ∝) green lvs
  9a Ovary truly inferior or falsely inferior because of fleshiness of surrounding persistent calyx; fr berry or caps ± embedded in fleshy calyx
    10a Ovary truly inferior; mature fr never white; erect shrubs, mostly deciduous   **Vaccinium**
    10b Ovary superior, but apparently inferior when ripe because of surrounding (but scarcely adherent) calyx, if partially inferior the fr white; mostly prostrate to semierect evergreen shrubs or shrublets, but at least 1 sp. often erect   **Gaultheria**
  9b Ovary superior, free of, and not closely surrounded by calyx
    11a Fr fleshy; pl gen evergreen shrubs or trees with large leathery lvs, often with deep reddish to purplish bark; fls urnshaped
      12a Pl arborescent; fr warty, ∝ -seeded; lf bl 7–15 cm   **Arbutus**
      12b Pl shrubby; fr smooth to viscid, not warty, mostly 5-seeded; lf bl < 7 cm   **Arctostaphylos**
    11b Fr dry caps; herbs to large shrubs; corolla often not urnshaped; lvs often scalelike or ± needlelike
      13a Lvs scalelike, < 6 mm, mostly 4-ranked; anthers awned   **Cassiope**
      13b Lvs not scalelike, mostly > 6 mm if at all 4-ranked; anthers sometimes not awned
        14a Petals distinct or only very slightly connate, corolla mostly open and not at all tubular   **Group I**
        14b Petals connate, corolla often tubular   **Group II, lead 18a**

Group I

15a Pl suffrutescent, not truly woody, seldom > 3 dm; lvs large and coriaceous, evergreen, not revolute
 16a Lvs cauline, scattered or whorled (pl not scapose); style short, gen not
visible; filaments enlarged and hairy near midlength; fls corymbose
 Chimaphila
 16b Lvs mostly basal (pl scapose); style elongate, plainly visible, often
curved; filaments not enlarged and hairy; fls racemose or single **Pyrola**
15b Pl woody shrub, mostly > 3 dm; lvs sometimes deciduous or revolute
 17a Fls single and terminal on short axillary brs; petals 8–12 mm; lvs not
glandular **Cladothamnus**
 17b Fls in terminal corymbs; petals < 8 mm; lvs often glandular **Ledum**

Group II

18a Corolla saucerlike, with 10 shallow pouches partially enfolding anthers
before anthesis; lvs 1–4 cm, often strongly revolute; anthers dehiscent full
length **Kalmia**
18b Corolla neither saucerlike nor pouched to fit anthers, mostly bell- or
urn-shaped; lvs various, often < 1 cm or not revolute; anthers often dehiscent only at tip
 19a Lvs either deciduous or leathery and > 6 cm, not revolute; pl mostly
1–5 m
 20a Corolla urnshaped, 6–8 × 3–4 mm; lvs deciduous **Menziesia**
 20b Corolla shallowly camp to funnelf, 1–4 × 1–3 cm; lvs deciduous or
evergreen **Rhododendron**
 19b Lvs evergreen, < 3 cm, often revolute; pl < 1 m
 21a Anthers unawned; caps septicidal; lvs 4–16 × 1–2 (3) mm
 22a Stamens twice as many as corolla lobes; anthers opening by terminal pores; lvs alt, linear, 6–16 mm **Phyllodoce**
 22b Stamens as many as corolla lobes; anthers opening by full length
slits; lvs opp, ± elliptical, 4–8 mm **Loiseleuria**
 21b Anthers awned; caps loculicidal; lvs mostly > 15 × 3–5 mm
 Andromeda

Allotropa T. & G. Candystick; Sugarstick

Fls in elongate spikelike racemes, apet; sepals 5; stamens 10; pistil 5-carpellary; ovary superior; fr
a caps; fleshy, simple-std, saprophytic herb. (Gr *allos*, other, and *tropos*, turn, referring to the infl,
in contrast to *Monotropa*).

A. virgata T. & G. Pl 1–4 dm; st white- and pink-striped; raceme 5–20 cm; lvs
scalelike, linear-lanceolate, pinkish to yellow-brown; sepals distinct, white or
pinkish to brown, ca 5 mm; stamens purple, 1–2 × as long as the sepals, anthers unappendaged, opening by basal (falsely terminal) pores; ovary
5-celled; style very short, stigma shallowly 5-lobed; in deep humus of conif
for at lower elev; e slope Cas to near the coast, BC to Cal.

Andromeda L. Bog-rosemary; Moorwort

Fls few in terminal umbels, 5-merous; corolla urnshaped; stamens 10; fr 5-valved, subglobose;
small, glab, evergreen shrubs with alt, leathery, entire lvs. (For *Andromeda* of Gr mythology).

A. polifolia L. Pl spreading, 0.5–8 dm; lvs elliptic to broadly linear, 1–4 cm ×
1–6 mm, apiculate, whitish-glaucous beneath, ± revolute; peds 5–12 mm,
gen ± recurved; corolla 5–8 mm, pinkish; anthers slenderly 2-awned at tip,
opening by terminal pores; caps glab, 4–6 mm thick; acid bogs; circumboreal,
Alas to Lab, s to s BC and Alta, reported for both Wn and Ida; highly prized,
low-growing orn.

Arbutus L. Madroña; Madroño; Madrone

Fls in large, terminal, compound racemes, hypog, 5-merous; corolla white, gamopet; stamens 10;
fr a finely granular-tuberculate, ∞-seeded berry; lvs alt, leathery, shiny. (L name for another sp.
of the genus).

A. menziesii Pursh. Pac m. Tree 6–30 m; bark smooth, at first chartreuse then deep red, becoming dark brownish-red, exfoliating; lvs glab, 7–15 cm, ovate-oblong to elliptic, entire to serrate; corolla 6–7 mm; filaments pilose near base; anthers awned from back near tip, opening by terminal slitlike pores; berry ca 1 cm thick, subglobose, orange to red; BC to Baja Cal, in dry areas w Cas *(A. procera)*; very fine orn, but somewhat messy, shedding lvs and bark most of the year.

Arctostaphylos Adans.   Manzanita; Bearberry

Fls mostly pan (racemose), urnshaped, 5 (4)-merous, white or pink; stamens 10 (8), filaments broadened and hairy near base, anthers awned dorsally near tip, opening by terminal pores; ovary 5 (4–10)-celled, with 1 stony nutlet per cell; mostly evergreen shrubs with freely br sts; bark smooth, often reddish; lvs alt. (Gr *arktos*, bear, and *staphyle*, bunch of grapes, the frs eaten by bears). Most spp. are choice orn shrubs.

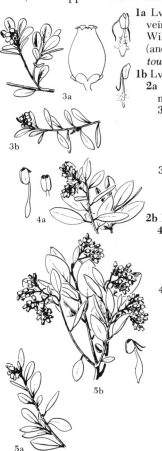

1a Lvs withering but persistent, thin, obovate, serrate, rugose and strongly veined, 2–4 (5) cm; fls 1–3 in lf axils; berry juicy, red (var. *ruber* Regd. & Wilson) or purplish-black (var. *alpina*); Alas to Greenl, s to Can Rockies (and possibly to our area in Mont) and mts of Me and NH; alpine b. *(Arctous a.)* **A. alpina** (L.) Spreng.

1b Lvs truly evergreen; fls mostly racemose or pan

  2a Pl prostrate, forming mats several m broad, rarely > 2 dm tall; lvs mostly obovate, 1.5–3.5 cm

    3a Lvs rounded to retuse, oblong to obovate or spatulate; berry bright red, 7–10 mm thick; pl trailing, seldom with erect br > 1.5 dm; Alas to Lab, s to coastal Cal, Ida, Mont, and in RM to NM, c and e US; Eurasia; kinnikinnick, bearberry, sandberry; one of the finest ground covers known, esp for dry banks **1 A. uva-ursi** (L.) Spreng.

    3b Lvs abruptly mucronate to apiculate, oblong to spatulate; berry rather brownish-red; pl not so completely prostrate, often 1–2 dm; mont, Wn Cas to Cal, e to Blue Mts, Ore; pinemat m., kinnikinnick; also an excellent ground cover **2 A. nevadensis** Gray

  2b Pl erect to spreading, rarely < 3 dm; lvs various, often ovate or elliptic

    4a Young sts grayish with long, stiff hairs mixed with finer tomentum or pubescence; lvs mostly grayish-puberulent, ovate or lanceolate to elliptic, bl 2–6 × 1–2.5 cm; pl erect or spreading shrub 1–3 (4) m; BC s, w Cas and mostly at low elev, to nw Cal; bristly m. *(A. tomentosa,* misapplied) **3 A. columbiana** Piper

    4b Young sts glab to uniformly pubescent; lvs glab to puberulent

      5a Lf bls gen oblanceolate to cuneate-obovate, ca 2.5 cm; berry red; pl spreading, 3-6 dm; occ hybrid between no 1 and no 3; with the parents at lower elev, esp on e side OP; propagated and widely used in nw as low-growing orn shrub; media m. **4 A. x media** Greene

      5b Lf bls gen ovate-lanceolate to oblong-elliptic, 2-5 cm; pl erect to spreading, 10-30 dm

        6a Young growth shortly glandular-puberulent; lf bls ovate-lanceolate to oblong-elliptic, 2-5 cm, blunt to rounded; Klickitat Co, Wn, s in Cas to Cal, e to Colo; green-lf m. *(A. obtusifolia)* **5 A. patula** Greene

        6b Young growth densely nonglandular-pubescent; lf bls oblong to narrowly elliptic, 1.5-3 cm, gen abruptly acute and ± mucronate; recently reported from near Lake Mary Ronan, Lake Co, Mont, this more s sp. is not otherwise known from our area; point-lf m. **6 A. pungens** HBK

Cassiope D. Don   Moss-heather; Moss-bush; Cassiope

Fls axillary or terminal, solitary or clustered, 5 (4)-merous; corolla white or pinkish, gamopet, camp; stamens 10 (8), included, anthers recurved-awned from back, opening by terminal pores; caps subglobose, loculicidal; low shrubs with 4-ranked, mostly scalelike lvs. (Name from Gr mythology). All attractive shrubs but not easily grown and seldom fl at all freely.

1a Lvs alt, spreading, short-petiolate, linear-oblanceolate, 3–5 mm; fls mostly single and terminal, erect, 5–6 mm; sepals erose-denticulate; filaments ± flared at base; alp meadows and bogs; Alas to Wn Cas; Alas c., h., or m. *(Harrimanella s.)* **1 C. stelleriana** (Pall.) DC.

1b Lvs opp, appressed, sessile, 2–5 mm; fls mostly several, lateral or subter-

minal, mostly pendent; sepals mostly smooth; filaments not flared at base

2a Lvs prominently grooved on back (beneath), puberulent and ciliolate; sepals entire; circumboreal and mont; Alas to n Cas, Wn, and to GNP, Mont; four-angled mt heather; ours the var. *saximontana* (Small) Hitchc.

2 C. **tetragona** (L.) D. Don

2b Lvs keeled but not grooved on back, glab or merely ciliate; sepals entire to erose-denticulate; high mts, Alas to Cal, e to RM and Nev; Mertens' mt heather; 2 vars.   3 C. **mertensiana** (Bong.) G. Don

a1 Sts and peds puberulent; lvs gen glab; Alas to s Ore, Can RM (*Andromeda m., A. cupressina*)   var. **mertensiana**

a2 Sts and peds glab; lvs gen minutely ciliate; ne Ore to Ida and Mont   var. **gracilis** (Piper) Hitchc.

Chimaphila Pursh   Pipsissewa; Prince's-pine

Fls pendent in small umbels or corymbs, 5-merous; sepals distinct; petals pink to rose, distinct, spreading, 5–7 mm; stamens 10, anthers awnless, opening by terminal pores; caps 5-celled, loculicidal; low, glab, rhizomatous subshrubs with slightly woody sts and leathery, ± whorled lvs. (Gr *cheima*, winter, and *philos*, loving, because of evergreen habit). Attractive but difficult to grow; should not be collected in the wild.

1a Fls 1–3; peduncles 2–5 cm; filaments hairy on expanded area; pl 0.5–1.5 dm; lvs ± elliptic, bl 2–6 cm; conif woods, BC to s Cal, e to Ida and Mont; little p. (*Pyrola m.*)   1 C. **menziesii** (R. Br.) Spreng.

1b Fls mostly 5–15; peduncles 5–10 cm; expanded area of filaments ciliate but not hairy; pl 1–3 dm; lvs more elliptic-oblanceolate, bl 3–7 cm; woods, mostly under conif; circumboreal, Alas to s Cal, e to RM of Colo, e US, Eurasia; prince's-pine, common p.; ours the var. *occidentalis* (Rydb.) Blake

2 C. **umbellata** (L.) Bart.

Cladothamnus Bong.

Fls 5-merous; corolla ± rotate, polypet; stamens 10 (occ fewer), anthers unawned, dehiscent most of length by lateral slits; style elongate, recurved; stigma discoid; caps 5 (6)-valved, septicidal; shrub with alt, deciduous lvs. (Gr *klados*, branch, and *thamnos*, bush).

C. **pyroliflorus** Bong. Copper-bush. Pl 0.5–3 m; lvs ± glaucous, elliptic-oblanceolate to oblanceolate, mucronate, entire, (1.5) 2–5 cm; sepals 7–10 mm; petals 10–15 mm, salmon or copper colored; style ca 1 cm; caps subglobose, 4–7 mm thick; moist for and stream banks; Alas to Clatsop Co, Ore, not common in our area but in both Cas and OM; a distinctive orn that is difficult to grow.

Gaultheria L.   Wintergreen; Salal; Gaultheria

Fls white or pinkish, solitary in the axils or in axillary or terminal racemes, 4–5-merous, gamopet; calyx mostly deeply lobed, becoming thickened and fleshy in fr; corolla camp to urnshaped; stamens 8 or 10, filaments basally expanded; anthers awned or unawned, opening by terminal pores; caps ∞-seeded, surrounded by persistent but scarcely adnate calyx, the whole berrylike; lvs persistent, leathery, mostly shining. (For Jean Francois Gaultier, 1708–1756, physician and botanist of Quebec). All excellent ground covers; *G. shallon* the best, *G. hispidula* the most difficult.

1a Lvs 4–10 mm, elliptic to obovate, entire, revolute; fls 4-merous, single in lf axils, subtended by 2 ovate bracts > calyx; corolla bell-shaped, scarcely 3 mm; anthers unawned but gen with 4 short terminal points; fr white, 3–5 mm thick, somewhat spicy and aromatic; in sphagnum bogs or deep conif woods; BC to Lab, s into n Ida and ne Wn; creeping snowberry, moxie-plum, maidenhair berry (*Vaccinium h., Oxycoccus h., Chiogenes h., C. serpyllifolia, G. s.*)   1 G. **hispidula** (L.) Muhl.

1b Lvs mostly much > 10 mm; fls mostly 5-merous, often racemose, subtending bracts, if any, much < calyx; anthers sometimes awned; fr red or bluish-black

2a Fls 5–15 in terminal and subterminal bracteate racemes, pinkish, 7–10 mm; anthers with 4 slender apical awns; creeping to erect shrub 1–20

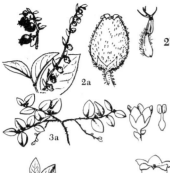

dm; lvs ovate to ovate-elliptic, 5–9 × 3–5 cm, sharply serrulate; fr purplish, 6–10 mm thick; moist to dry woods, lowl to lower mts; BC s, from
e base Cas to coast, to s Cal; salal, Ore w.   2 G. shallon Pursh

2b Fls solitary in lf axils, white or pinkish, 3–5 mm; anthers unawned; prostrate or sprawling shrublets scarcely 3 cm; lvs 1–4 × 0.5–3 cm; fr red,
5–8 mm thick

3a Calyx glab; lvs broadly ovate or oval to subelliptic, 1–2 (2.5) ×
0.5–1.5 cm, rounded or obtuse (acute), entire to crenulate or serrulate;
fls 3–4 mm; subalp to alp for and moist slopes; BC s, in OM and Cas,
to n Cal, e to Alta and s to Ida and Colo; alpine w., western w., matted
w. (*G. myrsinites*)   3 G. humifusa (Grah.) Rydb.

3b Calyx strongly brownish-reddish-pilose; lvs ovate, (1.5) 2–4 × (1)
1.5–3 cm, acute, more prominently serrulate; fls 3.5–5 mm; dryish
woods to moist subalp slopes, in drier habitats than *G. humifusa*; BC s,
in Cas and OM, to n Cal, e to Ida and perhaps to GNP; slender w.,
Oregon w.   4 G. ovatifolia Gray

Hemitomes Gray   Gnome-plant

Fls congested in short, terminal, bracteate spike, whitish to pink or pinkish-yellow; sepals distinct;
petals connate at least half their length; stamens gen 8; ovary superior, 1-celled with 7–9 parietal
placentae; style short, hairy; fr ± fleshy caps; fleshy saprophyte with simple st and thin, scalelike
lvs. (Gr *hemi*, half, and *tomias*, eunuch, one anther cell at first believed to be sterile). (*New-
berrya*).

H. congestum Gray. St 3–20 cm; lvs yellowish to brown, nongreen; sepals
2–4; petals 4, the calyx and corolla similar in texture, subequal, pinkish but
drying to brown, 12–20 mm; stamens included, filaments hairy, anthers ca 2
mm, fully dehiscent by slits, unawned; caps ovoid, hairy; conif for, in deep
humus; from sw side of Cas and the OP, Wn, to Monterey Co, Cal (*N. spicata,
N. longiloba, H. spicatum*).

Hypopitys Hill   Pinesap

Fls in bracteate terminal raceme, 4 (3–5) -merous; sepals oblanceolate, distinct; petals erose to lacerate-fimbriate, ± hairy on both surfaces; stamens twice as many as the petals; ovary superior,
4–5-celled with axile placentation below, but 1-celled above; fr caps, fleshy; white or yellowish to
pinkish saprophyte. (Gr *hypo*, beneath, and *pitys*, pine tree, in reference to the habitat).

H. monotropa Crantz. Fringed p. St unbr, 5–25 cm, pinkish to straw colored,
drying to black; lvs scalelike, nongreen, entire to fimbriate; raceme gen recurved at anthesis, erect in fr; peds 3–6 mm; sepals 5–9 mm; petals 9–18 mm,
hairy on one or both surfaces (glab), ± saccate at base; stamens < corolla;
style hairy, stigma slightly lobed; caps subglobose, 5–8 mm; in humus, mostly
under conif; BC to n Cal, e to Atl; Europe (*Monotropa h., H. h., M. fimbriata,
H. f., H. lutea* of auth., *H. latisquama, M. l., H. brevis*).

Kalmia L.   Laurel

Fls in lfy-bracteate terminal corymbs, 5-merous, light to deep pink; stamens 10 (8–12), anthers
unawned; caps 5-celled, septicidal; low evergreen shrubs with alt to whorled, leathery, entire lvs.
(For Peter Kalm, 1715–1779, a student of Linnaeus).

1a Pl low and spreading, rarely > 1.5 dm; lvs opp, dark green above, finely
granular-puberulent and gray beneath, ± oblong-elliptic, 1–2 (3) cm, 1/4–
> 1/2 as broad, slightly revolute or plane; corolla ave 10–12 mm broad;
calyx 5–7 mm broad; subalpine and alpine mt meadows and bogs; Alas s, in
Cas, to Cal, e to RM, s Can, and to Mont, Wyo, and Colo; small-lvd or
alpine l. or k. (*K. glauca* var. *m., K. polifolia* var. *m.*)
                                               1 K. microphylla (Hook.) Heller

1b Pl 1–4 dm; lvs as those of *K. m.*, but narrower and more often strongly
revolute, gen narrowly oblong-elliptic (partially due to the recurved margins),
1–3 (4) cm × 5–15 mm; corolla (12) 14–20 mm broad; calyx 8–10 mm
broad; s Alas s to nw Ore, W Cas in our area, gen in sphagnum bogs at low
elev; w swamp l. or k. (*K. polifolia* ssp. *o.; K. glauca* and *K. p.*, misapplied)
                                               2 K. occidentalis Small

Ledum L.   Labrador-tea

Fls in terminal racemes or corymbs, 5 (4–6) -merous; petals white, spreading to subrotate; stamens 10 (5–12), filaments slender, anthers unawned, opening by terminal pores; caps 5-valved, septicidal from base upward; evergreen shrubs with leathery, entire, often revolute lvs, often strongly glandular. (Gr *ledon*, mastic, a name used by the Greeks for another genus, *Cistus*, from which an aromatic resin was obtained).

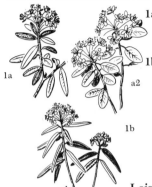

1a Lvs linear-elliptic, 2–6 cm, densely rusty-lanate beneath, strongly revolute; stamens 5–10, slightly > style; Alas to Greenl, s along coast, mostly in swamps and bogs, to nw Ore, possibly also to n Ida, e to Greenl and n Atl states; bog l. (*L. pacificum*)   1 **L. groenlandicum** Oeder

1b Lvs ovate to elliptic, 1.5–6 cm, mealy-puberulent to puberulent beneath, slightly to strongly revolute; stamens (5) 8–12, considerably > style; BC to Cal, e to RM; trapper's tea, smooth l., mt l.; 2 vars. 2 **L. glandulosum** Nutt.

a1 Lvs strongly revolute, 3–5 × scarcely l (1.5) cm; caps ovoid, 4–5.5 mm; Pacific Co, Wn, s along coast to Marin Co, Cal (*L. c., L. g.* ssp. *c., L. g.* ssp. *c.* var. *australe*)   var. **columbianum** (Piper) Hitchc.

a2 Lvs plane or slightly revolute, 1.5–3 (4) cm, gen at least half as broad; caps subglobose, 1–3 (4.5) mm; BC s, on e side Cas, to Wn, e in BC to RM, s to Mont and n Wyo, w to c Ida and ne Ore   var. **glandulosum**

Loiseleuria Desv.   Alpine Azalea; Trailing Azalea

Fls umbellately clustered from axils of upper st lvs, camp, 5-merous; stamens included; anthers subglobose; caps 2–3-celled and -valved, septicidal. (For J. L. A. Loiseleur-Deslongchamps, 1774–1849, a French botanist).

L. procumbens (L.) Desv. Diffusely br, decumbent shrub scarcely l dm; lvs opp, bright green, narrowly elliptic, 4–8 mm, revolute, evergreen; petioles 1–2 mm; peds 5–20 mm; corolla light to deep pink, 3–4 mm; stamens as many as corolla lobes, anthers unawned, opening full length by slits; caps ± ovoid, ca 4 mm; circumboreal, Alas to Greenl and ne US, s to BC and to at least one station (Trapper Peak, Skagit Co) in Wn (*Chamaecistus p.*); a most attractive orn, easily propagated by cuttings or layering.

Menziesia Smith   Menziesia

Fls in terminal clusters, appearing with lvs, 4 (5) -merous; calyx very short, shallowly lobed, glandular-ciliate; corolla greenish-white or pinkish to reddish-yellow (ours), finely puberulent inside; stamens 8 (ours), included; anthers unawned, opening by terminal pores; caps 4 (5) -celled, septicidal; alt lvd, deciduous shrubs. (For Archibald Menzies, 1754–1842, naturalist with Vancouver Expedition of 1790–95).

M. ferruginea Smith. Fool's huckleberry, mock azalea, rusty-lf, rusty m. Erect shrub 1–2 m; lvs ovate-elliptic to elliptic-obovate, 4–6 cm; ped 1–2 cm; pilose to glandular-bristly and sometimes also puberulent; corolla 6–8 mm; caps ovoid, 5–7 mm; moist woods and stream banks; Alas to Cal, e to RMS; rather desirable orn, esp because of fall coloration; 2 vars. in our area

a1 Lvs glandular-pubescent, gen apiculate or acute; calyx glandular-ciliate; ovary glandular-pilose but not puberulent; Alas s, from Cas to coast, to n Cal   var. **ferruginea**

a2 Lvs less glandular, finely puberulent on both surfaces, more rounded than acute; calyx mostly puberulent and glandular-ciliate; ovary puberulent as well as glandular-pilose; BC to Alta, s to Mont, Wyo, Ida, and e Wn and Ore, down CR to Mt Adams and Mt Hood (*M. g.*)
          var. **glabella** (Gray) Peck

Monotropa L.   Indian-pipe

Fls large, single and terminal; sepals prob absent, but upper 1–4 bracts of st proximal to fl and sepaloid, gen erose-lacerate, deciduous; petals 5 (4–6), distinct, glab outside, erect; stamens gen 10; ovary superior, 5-celled; fr caps, loculicidal; white to pinkish per saprophyte. (Gr *monos*, one, and *tropos*, direction, the fls pendulous and turned to one side).

M. uniflora L. Sts clustered, 5–25 cm, waxy white but turning black with age or upon drying; fls 1.5–2 cm, curved to one side or even drooping, narrowly camp, hairy within; petals ± saccate at base; stamens included, filaments ± hairy, anthers broader than long, unawned, dehiscent by curving slit; stigma discoid-lobed; caps subglobose, ca 6 mm; deep shady woods; Alas to n Cal, e to Atl coast (*M. morisoniana*).

Phyllodoce Salisb.   Mountain-heath: Mountain-heather

Fls single in lf axils, clustered near tip of st, 5-merous, camp to narrowly urnshaped; peds glandular-pubescent, slender; stamens 10 (7–9), included; caps septicidal from tip; dwarf evergreen alp shrubs with alt, linear lvs. (Gr name for a sea nymph). Attractive orn shrubs, but nearly impossible to grow and to get to fl.

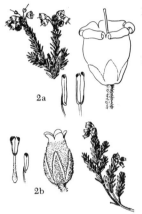

1a Corolla pale pink, narrowly camp, mostly sparsely glandular-pubescent outside; occ hybrid between next two, always with both parents present; hybrid m. (*P. hybrida*, × *P. intermedia*, *P. glanduliflora* var. *i.*)
           × P. intermedia (Hook.) Camp
1b Corolla greenish-white, dirty-yellowish, or deep pinkish-rose
   2a Corolla > twice as long as calyx, camp, pink to rose, glab outside, ca 7 mm, lobes ovate, recurved; sepals ovate to ovate-lanceolate, obtuse, ciliolate but not pubescent; filaments glab; mont to alp meadows and slopes; Alas s, in both OM and Cas, to Cal, e to Alta and s in RM to Colo, w to Ida; red m., pink m.   **1 P. empetriformis** (Sw.) D. Don
   2b Corolla narrowly urnshaped, scarcely twice as long as calyx, dirty yellowish to greenish-white, 5–7 mm, strongly glandular-pubescent outside, lobes spreading, ovate-lanceolate; filaments pubescent; mont to alp; Alas s to Ore, in OM and Cas in Wn, e to Can RM and s to Wyo and Ida; yellow m., cream m.   **2 P. glanduliflora** (Hook.) Cov.

Pityopus Small   Pine-foot; Pityopus

Fls in terminal, bracteate raceme, 4 (3–5) -merous; sepals distinct; stamens twice as many as petals; caps l-celled, with 7–10 parietal placentae; nongreen saprophytic per. (Gr *pitys*, pine tree, and *pus*, foot, in reference to the habitat).

P. californica (Eastw.) Copeland. Pl pinkish to yellow, drying black; st simple, 5–25 cm; lvs scalelike; fls 12–18 mm, petals distinct, gen hairy on both surfaces, pinkish to straw-colored, blotched with chocolate; filaments hairy; anthers ovoid, opening by a continuous slit full length and across the top; pistil 4–5-carpellary; caps fleshy, 5–8 mm; deep conif for; very rarely collected and known only from a few stations w Cas in Ore and from n Cal (*Monotropa c.*, *M. hypopitys* var. *c.*, *P. oregana*).

Pleuricospora Gray   Sierra-sap; Fringed-pinesap

Fls in terminal spikelike raceme, bracteate, mostly 4 (5–6) -merous; sepals and petals distinct; stamens 8 (10); caps fleshy, 1-celled with 4–6 parietal placentae; fleshy, whitish to yellow-brown saprophyte. (Gr *pleura*, side, and *sporos*, seed, in reference to the parietal placentation).

P. fimbriolata Gray. Pl gen glab, 3–12 (18) cm; fls 8–15 mm; bracts only slightly < fls, fimbriate-pectinate as the perianth segms; peds ultimately 1–2 cm; sepals oblong-ovate, ca half as long as the petals; petals oblong, slightly broadened above midlength; stamens included, filaments broadened and flattened, glab to puberulent; anthers 2–3 mm, linear, unawned, dehiscent lengthwise by slits; deep for, mostly just emerging from the duff; w Cas, OP, Wn, to nw and Sierran Cal (*P. densa*, *P. longipetala*).

Pterospora Nutt.   Pinedrops; Albany-beechdrops

Fls ∝ in elongate terminal raceme, 5-merous; corolla whitish to red, lobes short, spreading; stamens 10, anthers ovoid; caps 5-celled; tall yellow to reddish-brown saprophytes lacking chlorophyll. (Gr *pteron*, wing, and *sporos*, seed, the seeds with netlike wing on one end).

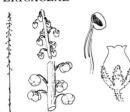

P. andromeda Nutt. Woodl p. Sts simple, 3–10 dm, reddish-brown, remaining as dried stalks through winter, glandular-hairy, narrow-bracteate; raceme gen at least as long as rest of st; fls 5–8 mm, pendulous; peds 5–15 mm, glandular-hairy; caps depressed-globose, 8–12 mm; deep humus in conif for, often under ponderosa pines; Alas to Cal and in RMS to Mex, e to Atl.

Pyrola L.   Pyrola; Shinleaf; Wintergreen

Fls reg to irreg, single and terminal to racemose, 5-merous; sepals persistent, ± connate at base; petals often ± unequal, mostly concave, deciduous; stamens 10, bent inward, unawned, dehiscent by terminal pores sometimes on short tubes; ovary 5-celled; style straight or strongly bent to one side, often with small collar near stigma; caps dry; glab per herbs with slender rhizomes and mostly rosettes of green lvs but sometimes largely saprophytic and almost or quite without green lvs. (L diminutive of *pirus* or *pyrus*, the pear, the lvs of some spp. being somewhat pear shaped). Many are interesting, but all hard to grow, and best left unmolested. (*Actinocyclus, Amelia, Erxlebania, Moneses, Orthilia, Ramischia, Thelaia*).

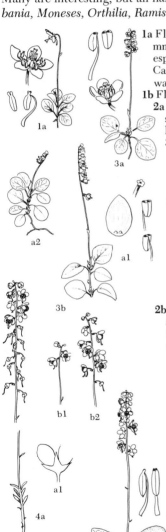

1a Fls 1.5–2.5 cm broad, white, solitary, terminal on scape 3–15 cm; style 2–4 mm, straight; lf bl ovate-elliptic to obovate, 1–2.5 cm; light to deep woods, esp where moist, or on rotting wood; lowl to mont; Alas s, e and w Cas, to Cal, e to e Can and Pa, and s in RMS to NM; woodnymph, single delight, wax-fl (*M. u., M. reticulata*)   **1 P. uniflora** L.

1b Fls 2–∝, racemose
  2a Style 1–4 mm, nearly or quite straight, without collar or ring below stigma; fls 5–7 mm broad; anthers without horns or tubes, opening by large terminal pores
    3a Style 1–2 mm; raceme 5–20-fld; fls not turned to one side, pink to rose, 5–7 mm broad; scape 1–2 dm; lvs basal and rosulate mostly, bl broadly elliptic to subrotund, 1.5–3.5 cm; woods, mostly under conif; Alas to s Cal, e to Atl, s in RM to Colo and to Ida and e Wn and Ore; Eurasia; lesser w., snowline p. (*P. conferta*)   **2 P. minor** L.
    3b Style 3–4 mm; raceme 6–20-fld; fls mostly turned to one side (secund), white, 5–6 mm broad; scape 5–15 cm; lvs ∝ and scarcely rosulate, bl ovate to ovate-elliptic, 1.5–6 cm; mostly in conif woods; Alas to s Cal, e to Newf and Atl, s in RMS to Mex; one-sided w., sidebells p.; 2 vars.   **3 P. secunda** L.
      a1 Lf bl mostly elliptic to ovate, acute; fls gen > 10; scape gen 8–15 cm; Alas to Cal and Mex., e to Atl; Eurasia   var. **secunda**
      a2 Lf bl subrotund, obtuse; fls gen < 10; scape 5–9 cm; Alas s, throughout RMS, e to Newf; Siberia   var. **obtusata** Turcz.
  2b Style mostly 3–7 mm, bent to one side, often with collar or ring below stigma; fls (8) 9–15 mm broad; anthers gen opening by small pores at the tip or on the side of short apical tubes
    4a Pl without detectable green lvs; several spp., esp *P. picta*, and less commonly *P. dentata*, *P. asarifolia*, and *P. chlorantha* may occur in this condition and not always be identifiable, although the following key may be helpful; leafless p.; traditionally they have been called
                               P. aphylla Smith
      a1 Petals pink to rose or purplish-red; sepals much longer than broad
                             P. asarifolia Michx.
      a2 Petals white to greenish-white, rarely pinkish; sepals not much longer than broad
        b1 Racemes mostly 2–8 (1–10) -fld; fls pale yellowish; sepals rounded, broader than long   **P. chlorantha** Sw.
        b2 Racemes mostly (6) 9–25-fld; fls white or greenish-white (yellowish); sepals acute, ca as long as broad
              P. picta Smith and **P. dentata** Smith
    4b Pl with green lvs on fl st or on sterile shoots attached to fl st
      5a Lvs deep green, mottled on upper surface with pale streaks above main veins, bls ovate to ovate-elliptic, coriaceous, 2–7 cm; sts reddish-brown, 10–25 cm; petals 6–8 mm, yellowish- or greenish-white to purplish; conif for; BC to s Cal, e to RMS, Mont to Colo; white vein p. (*P. conardiana, P. paradoxa, P. septentrionalis*)
                               4 P. picta Smith
      5b Lvs light to deep green, not mottled, bls various

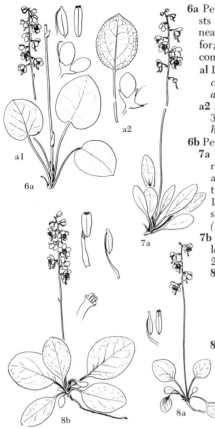

6a Petals pinkish to rose-purple, 5–7 mm; racemes (5) 10–25-fld; fl sts 1.5–4 dm; lf bl rotund to elliptic or obovate, (2) 3–8 cm and nearly as broad, petioles mostly = bls; moist ground in woodl to for; Alas to most of w US, across Can to ne N Am; alpine p., common pink w., liver-lf w.; 2 vars.   **5 P. asarifolia** Michx.

    a1 Lvs subentire, often cordate; sepals < 3.5 mm; range of sp. (*P. californica, P. elata, P. bracteata, P. a.* var. *b., P. incarnata, P. a.* var. *i., P. uliginosa, P. a.* var. *u.*)   var. **asarifolia**

    a2 Lvs serrulate, ± acute at least at one end; sepals gen at least 3.5 mm; range of sp. but more common w Cas (*P. bracteata* var. *hillii, P. rotundifolia* var. *p.*)   var. **purpurea** (Bunge) Fern.

6b Petals pale yellowish or greenish-white, often < 5 mm

    7a Lvs tapered to acute base, bl spatulate or oblanceolate to rhombic-elliptic, gen at least 3 (2–6) cm and not much > half as broad (seldom > 2.5 cm broad), pale green or bluish-green, thickish; petals cream to greenish-white, ca 6 mm; racemes 10–20-fld; conif woods, esp under ponderosa pine; BC s, both sides Cas, to Sierran Cal, e to Ida, Mont, and Wyo; toothlf p. (*P. picta* ssp. *d., P. d.* var. *integra*)   **6 P. dentata** Smith

    7b Lvs mostly ovate to orbicular and much > half as broad as long, many > 2.5 cm broad, often rounded at base; racemes (1) 2–20-fld

        8a Racemes (1) 2–8 (10) -fld; lf bls broadly elliptic or oblong-elliptic to rotund, 1–2.5 (3.5) cm, gen < petioles; sepals rounded, at least as broad as long; tubes at tip of anthers straight; moist conif for; Alas s, both sides Cas, to Cal, e in Can to Atl; Eurasia; greenish w., green w. (*P. virens, P. v.* var. *saximontana, P. c.* var. *s.*)   **7 P. chlorantha** Sw.

        8b Racemes (6) 9–20-fld; lf bls broadly elliptic to oblong or obovate, 3.5–7 cm, ca 3/4 as broad, gen > petioles; sepals mostly acuminate, longer than broad; tubes of anthers gen ± recurved; moist woods; BC and Can RM to ne N Am, s to w Mont, and prob to Ida; white w.   **8 P. elliptica** Nutt.

Rhododendron L.   Rhododendron; Azalea

Fls showy, ± irreg, 5-merous, gamopet; corolla funnelf to shallowly camp, deeply lobed; stamens (ours) 5 or 10; anthers unawned, opening by terminal pores; caps 5-celled; alt-lvd shrubs. (Gr *rhodon*, rose, and *dendron*, tree).

1a Stamens 5; lvs deciduous, bl elliptic to narrowly obovate, 3–9 cm; fls 5–20 in terminal corymbs, very fragrant, irreg, white to deep pink; corolla narrowly funnelf, 3–5 cm; pl 1–5 m; stream banks and moist areas in woods; sw Ore to s Cal, possibly not quite to our area, but widely planted and one of finest orn spp.; western a.   **1 R. occidentale** (T. & G.) Gray

1b Stamens 10; lvs various, sometimes persistent

    2a Lvs deciduous, bl elliptic-oblanceolate, 4–9 cm; fls in axillary clusters of 1–4, white or off-white; corolla 1.5–2 cm broad, shallowly camp, nearly reg; pl 1–2 m; stream banks to moist slopes, mont to subalp; BC to Ore, in most of our higher mts, e to w Mont; white rhododendron, Cas a., white-fld a.; attractive but too difficult to grow to be of value   **2 R. albiflorum** Hook.

    2b Lvs persistent, leathery, bl oblong-elliptic, 8–20 cm; fls ∝ in terminal corymbs, pale pink to deep rose-purple; corolla tubular-camp, 3–5-cm, somewhat irreg; pl 1–5 m; w slope Cas to coast, from sea level to lower mts; BC to n Cal; Pac r., Cal rose-bay, western r. (*R. californicum, R. m.* f. *album*); fine orn, easily grown   **3 R. macrophyllum** G. Don

Vaccinium L.   Bilberry; Blueberry; Huckleberry; Cranberry

Fls solitary to clustered terminally or in lf axils, 4–5 (6) -merous, polypet to gamopet, epig; calyx small, persistent to deciduous; corolla spreading and with reflexed segms to globose or urnshaped; stamens 2× as many as corolla lobes; anthers awned or awnless, dehiscent by pores at end of apical tubes; ovary 4–5 (6)-celled, ∝ -seeded; fr reddish to bluish berry; creeping to erect, deciduous to evergreen shrubs with alt lvs. (L name for the blueberry). *(Oxycoccus)*. Most spp. nice orns, esp *V. ovatum* and *V. parvifolium*.

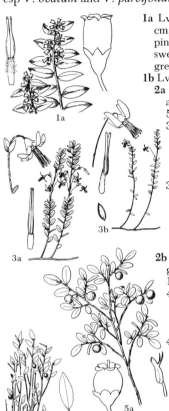

1a Lvs evergreen, glossy-green, serrulate, bl ovate to ovate-lanceolate, 2–5 cm; fls axillary in 3–10-fld clusters; corolla gamopet, narrowly camp, bright pink, ca 7.5 mm; anthers not awned; berry purplish-black, 4–7 mm thick, sweet and edible; pl 0.5–4 m; BC s, w side Cas to coast, to nw Cal; evergreen bl. or h., shot h. *(V. lanceolatum)* **1 V. ovatum** Pursh

1b Lvs gen either deciduous or entire or pls otherwise not as above

  2a Corolla polypet, petals 4, reflexed; stamens well exserted, anthers unawned; sts slender, prostrate; lvs persistent, ovate to elliptic-oblong, 5–15 mm *(Oxycoccus;* cranberries)

    3a Bracts lanceolate to oblong, often foliaceous, 3–10 mm, borne well above midlength of pubescent ped; fls lateral along st; petals 6–10 mm; filaments scarcely 1/3 length of anthers; berry 10–15 mm thick; native to ne N Am; cult and occ escaping and estab near coast in Wn; cult c. *(O. palustris* var. *macrocarpus)* **2 V. macrocarpon** Ait.

    3b Bracts linear or linear-lanceolate, not foliaceous, 1–2.5 mm, borne near or below midlength of glab or pubescent ped; fls l–several, terminal or lateral; petals 5–8 mm; filaments gen at least = anthers; berry 5–10 mm thick; native, circumboreal and transcont in Can, s in sphagnum bogs to Ore and Ida and to nc and ne US; wild c., swamp c. *(O. o., O. palustris, O. intermedius, V. o.* var. *i.)* **3 V. oxycoccos** L.

  2b Corolla gamopet, gen 5-merous, gen ± urnshaped or camp; stamens gen included, anthers awned; sts ± erect; lvs deciduous, often much > 15 mm

    4a Berry bright red; brs strongly angled, green or yellow-green; fls ca 4 (5) mm; lvs ovate-lanceolate to broadly elliptic, rarely broader in upper half than in lower, 8–30 mm, some often persistent through winter **Group I**

    4b Berry gen blue to blackish, sometimes reddish-blue; brs mostly not strongly angled; fls often > 5 mm; lvs various, if not > 2.5 mm then gen broader in upper half than in lower, all gen deciduous **Group II, lead 7a**

Group I

5a Pl erect, 1–4 m tall; brs not broomlike; lvs 10–25 mm, gen entire; pore-bearing tubes much < rest of anther; berry 6–9 mm thick; dryish to moist woods, esp on logs and stumps where bird-planted; lowl and lower mts, BC s, w Cas, to c Cal; red bi., red bl., a most attractive and versatile orn **4 V. parvifolium** Smith

5b Pl low and ± matted, 1–3 dm; pore-bearing tubes ca = to rest of anther

  6a Brs ∝, broom-like; lvs 8–15 mm; berry bright red, 3–5 mm thick; corolla ca 4 mm; subalp to alp woods on open slopes; BC s, mostly e Cas, to Cal, e to Alta and SD, and s to e Wn and Ore, Ida, and RMS to Colo; grouseberry, whortleberry *(V. myrtillus* var. *microphyllum, V. erythrococcum)* **5 V. scoparium** Leiberg

  6b Brs few, not broom-like; lvs 10–30 mm; berry dark red to bluish, 5–8 mm thick; corolla ca 5 mm; mont to subalp; BC s, e Cas, to Wen Mts, e to Alta, and to RMS, Mont to Colo; Eurasia; dwarf bi., low bi. *(V. oreophilum)* **6 V. myrtillus** L.

Group II

7a Pl gen 2–3 dm; brs ± angled and greenish; berry dark red to bluish, 5–8 mm thick; lvs ovate to elliptic-lanceolate, 1–3 cm, sharply serrulate; fls ca 5 mm, single in axils; mont to subalp (see lead 6b) **6 V. myrtillus** L.

7b Pls often > 4 dm; brs rarely both strongly angled and greenish; lvs often entire, > 3 cm, or broader above middle than below; berry mostly dark blue to blackish

  8a Calyx deeply lobed, lobes triangular, persistent in fr; fls 1–4 per axil; buds with 4–7 conspicuous scales at base of ped; brs not angled; lvs entire, 10-30 mm; pls 2–6 dm

    9a Lvs from not quite 1/2 to much > 1/2 as broad as long, firm in texture, strongly reticulate beneath; berry 6–8 mm; circumboreal, s in bogs along coast; Alas to n Cal, e in Can to Atl; Eurasia; bog bl., bog bi. *(V. u.* ssp. *occidentale* var. *o.)* **7 V. uliginosum** L.

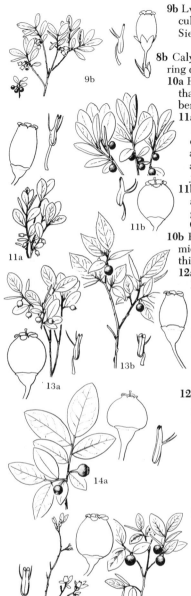

9b Lvs gen much < half as broad as long, glaucous and not strongly reticulate beneath; berry 4–5 mm thick; BC s, mostly e Cas, to nw and
Sierran Cal, e to Mont and n Utah; western h., w bog bl.
                                                      8 **V. occidentale** Gray

8b Calyx shallowly lobed, lobes rounded, deciduous; calyx tube forming
ring on top of fr; fls single in axils; brs often angled; bud scales 2

  10a Pls 2–3 (rarely to 6) dm; lvs ± serrate above midlength, rarely more
than indistinctly serrulate below this; twigs lightly angled to terete;
berry 5–8 mm thick

    11a Corolla narrowly urnshaped, 5–6 mm, ca half as thick; filaments
> anthers; lvs oblanceolate, strongly reticulate beneath, not glau-
cous, mostly 1–3 (but up to 5) cm; mont meadows and slopes, to
above timberl; Alas s, in Cas and OM, to Cal, e through Ida to RM
and in Can to Atl; dwarf bi., dwarf h. (*V. c.* vars. *arbuscula, angustifolium,* and *cuneifolium*)   9 **V. caespitosum** Michx.

    11b Corolla subglobular, 6–7 mm, much < half as thick; filaments <
anthers; lvs obovate to obovate-oblanceolate, faintly reticulate and
glaucous beneath, 1.5–5 (6) cm; mont to alp; s BC to n Ore, in both
Cas and OM; blue-lf h., Cas bl. or h.   10 **V. deliciosum** Piper

  10b Pls 4–20 (rarely < 5) dm; lvs sometimes entire or serrate to below
midlength; twigs sometimes rather strongly angled; berry 6–10 mm
thick

    12a Lvs sharply serrulate nearly full length

      13a Lvs oblong-obovate, rounded to abruptly acute, 2–4 (5) cm, ca
half as broad, strongly glaucous; berry bluish-purple, 6–8 mm
thick; lower and middle elev in mts; e Wn and Ore e through Ida
to Mont and Wyo; globe h.   11 **V. globulare** Rydb.

      13b Lvs ovate, ovate-oblong, or elliptic-obovate, gen long-pointed
or even ± acuminate, 2–5 cm, scarcely half as broad, not strongly
glaucous; berry reddish-purple or purple, 7–9 mm; mont slopes,
BC s, in Cas and OM, to Cal, e to Ida and Mont; thin-lvd bl.,
big h., tall bi. (*V. macrophyllum*)   12 **V. membranaceum** Dougl.

    12b Lvs entire or lightly serrulate, most or all serrulations on basal
half

      14a Peds mostly 10–15 mm in fr, slightly curved to straight, some-
what enlarged immediately below ovary; lvs sparsely glandular
along midnerve on lower surface, often puberulent, veins not
prominent, bl ovate-elliptic to elliptic-obovate, 2.5–6 cm; corolla
bronzy-pink, ca 7 mm, globose-urnshaped, at least as broad as
long; style gen exserted slightly; pl 5–12 dm; Alas s, from Cas to
coast, to nw Ore; lowl to mont; Alas bl. (*V. oblatum*)
                                                     13 **V. alaskaense** Howell

      14b Peds ca 5 (8) mm in fr, strongly recurved, not enlarged immedi-
ately below ovary; lvs glab, veins rather prominent, bl ovate-
elliptic, 2–4 (5) cm; corolla pinkish, tubular-urnshaped, ca 7
mm, longer than broad; style gen included; pl 4–10 dm; woods
and open slopes; Alas s, in Cas and OM, to Ore, e to Ida and
Mont, and in n Mich to Newf and NS; early bl., oval-lf h. (*V.
chamissonis*)   14 **V. ovalifolium** Smith

PRIMULACEAE   Primrose Family

Fls complete or rarely apet (*Glaux*), reg, hypog (ours), 5 (4–9)-merous; calyx shallowly to deeply
lobed; corolla gamopet, rotate and sometimes very deeply lobed to tubular or salverf and very
shallowly lobed; stamens as many as corolla lobes and opposite them; ovary 1-celled, placentation
free-central; style 1; stigma gen capitate; fr caps, valvate or circumscissile; ann or per herbs, often
scapose; lvs simple, mostly entire, alt or opp (whorled), exstip.

1a Fls mostly 6–7 (5–9)-merous; corolla rotate, lobed nearly to base; lvs reduced and often scalelike on lower part of st, larger ones crowded or
whorled near st tip   **Trientalis**

1b Fls mostly 5 (6)-merous; corolla various, if rotate and deeply lobed then st
lfless or equally lfy throughout

  2a Fls sessile or subsessile in lf axils, inconspicuous; corolla lacking or
membranous and much < the calyx

    3a Corolla lacking; sepals petaloid; per with mostly opp lvs; caps valvate
                                                                     Glaux

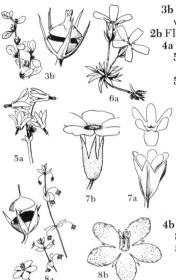

3b Corolla present, membranous, < the sepals; sepals not petaloid; ann
with mostly alt lvs; caps circumscissile   **Centunculus**
2b Fls stalked; corolla present, often showy
  4a Lvs in basal rosettes; fl st lfless
    5a Corolla lobes several times length of tube, sharply reflexed; stamens
      protruding full length; fls showy   **Dodecatheon**
    5b Corolla lobes < twice length of tube, not sharply reflexed; stamens
      not protruding; fls sometimes not at all showy
      6a Pl per with cushions of small, narrow, persistent lvs; fls showy,
        pink to violet; calyx ± keeled on and below lobes, glab or finely
        stellate, but not pilose   **Douglasia**
      6b Pl ann or per, if per and with cushions of basal lvs then fls white
        and calyx grayish-pilose, but not stellate, and not keeled along or
        below lobes
        7a Fls white, gen < 5 mm, sometimes up to 7 mm, but then pl
          grayish-pilose   **Androsace**
        7b Fls sometimes white but mostly pink to violet or purple, > 7
          mm; pl never grayish-pilose   **Primula**
  4b Lvs mainly on fl st
    8a Fls salmon pink; caps circumscissile; pl ann, decumbent   **Anagallis**
    8b Fls yellow; caps valvate; pl per, often erect   **Lysimachia**

Anagallis L.   Pimpernel

Fls axillary, ped, 5-merous; calyx deeply cleft; corolla rotate, deeply lobed; stamens exserted; low
prostrate ann with opp (ternate) lvs. (Gr name for pimpernel).

A. arvensis L. Scarlet p., poor man's weatherglass. Sts prostrate or ascending,
1–4 dm, glab; lvs ovate, 5–15 mm, sessile, entire; peds filiform, 1–4 cm; sepals
2–4 mm, acuminate; corolla 5–8 mm broad; caps globose; Eurasian, intro and
weedy; common in much of US, but only occ in our area from VI and w Wn,
reported from Ida.

Androsace L.   Fairy-candelabra; Rock-jasmine; Androsace

Fls in invol umbels, small, white or cream-colored, 5-merous; sepals connate ca half length; corolla
tubular-camp, contracted at throat, lobes spreading; stamens included; caps valvate; low ann or
per herbs with lvs in basal rosettes; fl sts several, lfless. (Gr name, *androsakes*, for some sea pl).

1a Pl per, grayish-pilose; scapes 1 from each rosette; corolla 5–7 mm broad,
white with yellow eye; lvs oblanceolate, 5–15 mm; peds < fls; arcticalp;
Alas to Big Snowy Mts, Mont; sweet-fld a. (*A. albertina, Drosace a., A.
chamaejasme* of auth.); attractive rock garden orn but not easily grown
                   1 A. lehmanniana Spreng.
1b Pl ann or bien, not pilose; scapes gen several per rosette; corolla mostly <
  5 mm broad, white; lvs often much > 15 mm; peds mostly much > fls
  2a Fls < 3 mm; calyx hemispheric, ca 2 mm; pl glab or sparsely glan-
    dular-puberulent above; lvs abruptly narrowed and petiolate; seeds yel-
    low, 0.2–0.3 mm; RM, Mont to Colo, w along CR to Klickitat Co, Wn,
    and WV, Ore; slender-std f. (*A. capillaris*)   **2 A. filiformis** Retz.
  2b Fls often > 3 mm; calyx ± turbinate; pl gen puberulent; lvs narrowed
    gradually and not distinctly petiolate; seeds dark brown, 0.7–1 mm
    3a Invol bracts oblong to oblong-obovate, 4–10 mm, ca 1/3 as broad;
      corolla scarcely exceeding calyx tube; calyx lobes ovate-lanceolate, ca
      = tube; RM, BC to NM, e to Miss Valley and w into Ariz and Cal;
      western f. (*A. platysepala, A. simplex*)   **3 A. occidentalis** Pursh
    3b Invol bracts lanceolate to linear, 3–6 mm, rarely much > 1 mm
      broad; corolla somewhat > calyx tube; calyx lobes < tube, ± del-
      toid; circumboreal, s in mts of w US to Cal, Ariz, and NM; northern a.
      or f. (*A. arguta, A. diffusa, A. gormani, A. puberulenta, A. subulifera,
      A. subumbellata*)   **4 A. septentrionalis** L.

Centunculus L.   Chaffweed

Fls subsessile in lf axils, inconspicuous, 4 (5)-merous; sepals greenish; corolla white or pinkish,
papery, nearly globose; delicate ann with mostly alt (or lowermost opp) lvs. (L diminutive of
cento, a patchwork).

C. **minimus** L. Sts decumbent to erect, 2–10 cm, glab; lvs obovate to spatulate or elliptic, 5–10 mm; fls single in axils, peds scarcely 1 mm; calyx 2–3 mm; corolla scarcely half length of calyx, drying and pushed upward by developing, subglobose caps; moist ground around vernal pools, coast to inl valleys, widespread in N Am; Europe.

Dodecatheon L.   Shooting Star

Fls few to many in terminal, invol umbels, showy, 4–5-merous; calyx divided nearly to base, lobes lanceolate; corolla with very short tube and long, strongly reflexed lobes, white to purple; stamens connivent around style, filaments short, free or connate; anthers slender, connective prominent, colored, smooth to transversely rugose; style slightly > stamens; caps 1-celled, valvate to tip or tip operculate; scapose per from rhizomes or short caudices, often with bulbils among roots, glab to strongly glandular-pubescent; lvs petiolate, entire to dentate. (Gr *dodeka*, twelve, and *theos*, god, the pls protected by the gods). *(Meadia)*. Several spp. excellent garden orns; *D. dentatum* and *D. jeffreyi* esp easily grown.

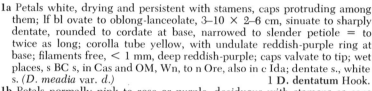

1a Petals white, drying and persistent with stamens, caps protruding among them; lf bl ovate to oblong-lanceolate, 3–10 × 2–6 cm, sinuate to sharply dentate, rounded to cordate at base, narrowed to slender petiole = to twice as long; corolla tube yellow, with undulate reddish-purple ring at base; filaments free, < 1 mm, deep reddish-purple; caps valvate to tip; wet places, s BC s, in Cas and OM, Wn, to n Ore, also in c Ida; dentate s., white s. *(D. meadia* var. *d.)*   1 **D. dentatum** Hook.
1b Petals normally pink to rose or purple, deciduous with stamens as caps matures; lf bls gradually narrowed to winged petioles, mostly entire or very shallowly toothed; caps sometimes with circumscissile tip
2a Stigma strongly capitate, gen at least twice as broad as thickness of style at midlength; filaments rarely > 1 mm; fls pink to reddish-purple, corolla tube ringed at tip with red or purple; connective cross-rugose, dark red to purple; wet places, mont to alp
3a Fls 4-merous; petals 10–18 mm; lvs glab, linear-oblanceolate, 3–10 cm × 3–15 mm; infl gen glab; caps valvate from tip, not operculate; mt streams and meadows; Wallowa Mts, Ore, to Cas of s Ore and Sierran Cal, e to Ariz and Utah; alpine s. *(D. meadia* var. *a., D. jeffreyi* var. *a.)*   2 **D. alpinum** (Gray) Greene
3b Fls 4- or 5-merous; petals 10–25 mm; lvs 5–40 × 1–6 cm, glab to glandular-pubescent; infl often glandular-pubescent; caps with operculate tip, remaining ovary wall valvate from tip; wet ground, esp meadows or stream borders; Alas s, in both Cas and OM, to s Sierran Cal, e to RM, and in both Ida and Mont; Jeffrey's s., tall mt s. *(D. crenatum, D. dispar, D. exilifolium, D. tetrandrum, D. viviparum)*   3 **D. jeffreyi** van Houtte
2b Stigma < twice as thick as style; filaments gen connate and forming tube > 1 mm
4a Lvs contracted abruptly to petiole, bl ovate to deltoid, 3–10 × 2–6 cm, length < 2.5 × width, petiole sometimes as long; bulblets present among roots at anthesis; infl glandular; stamen tube 2–4 mm, deep reddish-purple, gen cross-rugose; caps operculate at tip; prairies and open woods at lower elev; w side Cas, VI to s Ore; widespread in Cal; Henderson's s., broad-lvd s. *(D. atratum, D. cruciatum, D. latifolium, Meadia h.)*   4 **D. hendersonii** Gray
4b Lvs narrowed ± gradually to petiole, bl gen > 3× as long as broad; bulblets lacking; infl sometimes eglandular; stamen tube often < 1.5 mm; caps often not operculate
5a Filaments gen < 1 (up to 1.5) mm, free or connate, connectives cross-rugose; caps operculate; lvs 3–20 cm, bl lanceolate to oblanceolate, spatulate, or even obovate, entire; more moist areas, sagebr plains to mont meadows; e slope Cas, BC to Cal, e to Alta and Wyo; slimpod s., des s.; 2 vars.   5 **D. conjugens** Greene
a1 Pl glab or subglab; widespread; range of sp. *(D. acuminatum, D. albidum, D. campestrum, D. cylindrocarpum, D. glastifolium, D. pulchrum, D. hendersonii* var. *leptophyllum, D. c.* var. *l.)*
                                                                      var. **conjugens**
a2 Pl pubescent on lvs and often on scapes; more e part (se BC, and

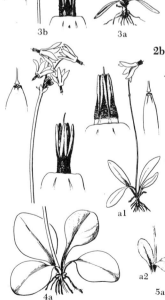

6a

7a

e Wn and Ore to sw Alta, Ida, and w Mont) of range of sp., with var. *conjugens* (*D. pubescens, D. v.*)   var. **viscidum** (Piper) Mason

5b Filaments gen > 1.5 mm, connectives often either smooth or wrinkled lengthwise; caps valvate to tip; lvs various

    6a Pl glab or very slightly puberulent; caps glab, walls comparatively thin and flexible, even toward tip; filament tube gen yellow or orange but sometimes purple, connectives smooth or ± wrinkled lengthwise; anthers yellow to reddish-purple; lvs entire to slightly denticulate; coastal prairies to saline swamps and mt meadows, sea level to above timberl; Alas to Mex, e to Pa; few-fld s., dark-throat s.; 3 vars.   **6 D. pulchellum** (Raf.) Merrill

        a1 Staminal tube reddish-purple to nearly black; pl gen > 6 cm; scapes several-fld; e Ore to ne Cal and Utah [*D. pauciflorum* (Durand) Greene var. *monanthum* Greene, Pittonia 2:73. 1890]

            var. **MONANTHUM** (Greene) C. L. Hitchcock, hoc loc.

        a2 Staminal tube yellow; pl various, sometimes dwarfed and scapes 1–2-fld

            b1 Pl dwarfed, mostly 2–5 cm; scapes mostly 1–2-fld; alp to subalp in RM and n Ore (*D. watsoni* Tidestrom, Proc. Biol. Soc. Wash. 36:183. 1923)

                var. **WATSONII** (Tidestrom) C. L. Hitchcock, hoc loc.

            b2 Pl rarely < 6 cm; scapes mostly several-fld; range of sp. as whole (*D. multiflorum, D. philoscia, D. radicatum, D. sinuatum, D. superbum, D. vulgare, D. pauciflorum* var. *alaskanum, D. p.* var. *shoshonensis, D. p.* var. *exquisitum, D. radicatum* ssp. *macrocarpum, D. meadia* var. *p., D.p., Exinia p., D. p. var. p.*)   var. **pulchellum**

    6b Pl strongly puberulent or glandular-puberulent; caps puberulent, walls comparatively hardened, esp toward tip, gen breaking rather than bending when pressed upon

        7a Filament tube 1.5–2 mm, purplish, connectives ± transversely rugose; lvs often strongly denticulate; grassy slopes to drier woodl, gen where moist in spring, often near seeps or springs along basaltic outcrops; from along Satus Cr, Yakima Co, to Klickitat and Skamania cos, Wn, and Wasco and Hood River cos, Ore; narcissus s.   **7 D. poeticum** Hend.

        7b Filament tube 1.5–2.5 (3) mm, yellow or occ reddish, connectives smooth or wrinkled lengthwise; lvs mostly entire (denticulate); grassl and for foothills; BC to e Wn, e Ore, e to Ida and Mont; Cusick's s., sticky s. (*D. puberulentum, D. pauciflorum* var. or ssp. *c.*)   **8 D. cusickii** Greene

Douglasia Lindl.   Douglasia

Fls solitary or gen 2–several in (mostly invol) umbels on naked peduncles, mostly 5-merous; sepals connate ca half length, carinate below lobes; corolla tubular-funnelf, tube ca = lobes, constricted at throat; caps 5-valvate; seeds 1–3; caespitose, matted to cushion-forming per with rosettes of small, entire to dentate lvs, gen finely stellate. (For David Douglas, 1789–1834, early pl explorer in our area). (*Gregoria*). Excellent rock garden pls, but more or less touchy.

    1a Fls 1 or 2, not invol, bright pink to rose-violet, 6–8 mm; corolla lobes ± retuse; lvs linear-lanceolate, 4–8 mm, minutely serrulate; foothills to open ridges and scree slopes; extreme se BC to Ida and n Wyo; RM d. (*Androsace uniflora, D. m.* var. *u., D. biflora, D. m.* var. *b.*)   **1 D. montana** Gray

1a

    1b Fls 2–10, in invol umbels; corolla lobes not retuse; lvs various

        2a Lvs grayish-stellate; invol bracts gen several times as long as broad; peds 3–40 mm; corolla bright red to magenta-purple; sagebr slopes to alp ridges and talus slopes; Wen Mts, Kittitas Co, n to Chelan and Douglas cos, Wn, also occ in ne Wn and in RM of BC and Alta; snow d. (*D. dentata, D. n.* var. *d., Androsace dieckeana*)   **2 D. nivalis** Lindl.

        2b Lvs glab or merely ciliolate; invol bracts gen not > 2 × as long as broad; peds 2–15 mm; corolla deep pinkish-rose; rocky alp ridges to moist coastal or river bluffs; w Cas; smooth d.; 2 vars.   **3 D. laevigata** Gray

            a1 Lvs glab or only very lightly ciliolate, mostly entire; umbels loose, peds 5–15 mm; CRG (*D. l.*)   var. **laevigata**

            a2 Lvs more strongly ciliolate, gen toothed; peds mostly 2–7 mm; w side Cas, Snohomish Co, and OM to Ore (Saddle Mt)   var. **ciliolata** Const.

2b

2a

Glaux L.   Saltwort; Sea-milkwort

Fls small, apet, single and sessile in lf axils, 5-merous; calyx camp, lobes ± petaloid; stamens free of calyx, alt with lobes; caps valvate; glab, fleshy, per herbs; lvs opp below, alt above. (Gr *glaucos*, bluish-green, in reference to color of pl). (*Glaucoides*).

G. maritima L. Pl rhizomatous, 3–30 cm; lvs oval to oblong or oblanceolate, sessile, 5–25 × 1.5–10 mm, jointed at base; calyx 4–5 mm, white or pinkish; caps ca 2.5 mm; coastal tidel to inl alkaline marshes and meadowl; widespread, in much of arctic and temp N Am; Eurasia (*G. m.* var. *obtusifolia*).

Lysimachia L.   Loosestrife

Fls ped, solitary in axils or in axillary or terminal racemes, yellow, often purple-dotted or -streaked, 5 (6–7)-merous; calyx lobed almost to base; corolla gen rotate, lobed at least half length; fertile stamens as many as corolla lobes, sometimes alt with toothlike staminodia, attached at base of corolla; caps valvate, subglobose, few-seeded; gen glab per with opp or whorled lvs. (Gr name, *lysimachos*, derived from *lysis*, a loosening, and *mache*, strife). (*Steironema, Nummularia*). Most spp. easily and commonly grown; all apt to become weedy.

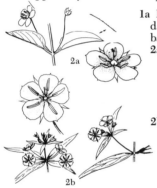

1a Fls solitary in axils; lvs lanceolate to ovate; pls erect, 3–12 dm, not dark-spotted; filaments distinct, staminodia present; corolla lobed to near base, lobes obovate, erose and gen abruptly mucronate (*Steironema*)
  2a Lvs at midstem ovate to ovate-lanceolate, 5–15 cm, gen > 3 (2.5–6) cm broad, mostly ciliate; petioles gen strongly ciliate; corolla lobes ca 1 cm; damp meadows, ponds, and streambanks; ± widespread, BC s, e Cas, to Ore, down CRG to Multnomah Co, e to NS, s to NM, in e Wn and ne Ore, and in Ida and Mont, and in much of c and e and se US; fringed l. (*S. laevigatum, S. c.* var. *occidentale, S. pumilum*)   1 L. **ciliata** L.
  2b Lvs at midstem linear to lanceolate, 5–10 × 1–3 cm, rarely at all ciliate; petioles sparsely ciliate; corolla lobes 6–9 mm; swamps and wet meadows to streambanks; common in upper Miss Valley and e US and Can; ours the var. *hybrida* Gray; western, in our area in Yakima Valley, Wn, more widespread in Ariz and NM as well as in Miss Valley; lance-lvd l.   2 L. **lanceolata** Walt.
1b Fls not solitary in lf axils or if so pl prostrate and lvs oval; pls gen reddish- or purplish-black spotted; filaments often connate at base, staminodia lacking
  3a Sts creeping; lvs short-petiolate, opp, bl suborbicular to oblong-ovate, 1–3 cm, nearly as broad; fls single in axils, peds 1–4 cm; corolla lobes 8–12 mm, oval-elliptic, dotted with red; filaments connate at base; European, freely escaped w Cas, Wn and Ore, much more common in c and e US; moneywort, creeping l., creeping jenny   3 L. **nummularia** L.
  3b Sts erect, 2–10 dm; lvs lanceolate to ovate-lanceolate, mostly 4–15 cm, rarely half as broad; corolla lobes mostly narrowly elliptic, sometimes not dotted; filaments sometimes distinct to base
    4a Pls hairy; lvs mostly in whorls of 3–4, broadly lanceolate; fls gen 3–5 in small, lfy, pedunculate groups in lf axils, peduncles becoming 1-fld above, but gen more ∝ than lvs; corolla not dotted or streaked with red or purple; filaments connate at base; Eurasian, grown as orn and occ escaping w Cas; possibly nowhere estab   4 L. **punctata** L.
    4b Pls glab; lvs opp, rather narrowly lanceolate; corolla ± streaked or dotted with red or blackish-purple; filaments sometimes distinct to base
      5a Racemes terminal, loose, ± lfy-bracteate, peds 1–2 cm; corolla rotate, deep yellow, lobes lanceolate-elliptic, streaked with dark lines; filaments connate at base, < corolla lobes; common in bogs and swamps in e Can and US, intro in cranberry bogs in w Wn and perhaps elsewhere; bog l.   5 L. **terrestris** (L.) B.S.P.
      5b Racemes axillary, closely fld, ebracteate, peds almost lacking; corolla lobes linear-lanceolate, ± clawed basally, ± erect, pale yellow, dotted or streaked near tip; filaments distinct, > corolla lobes; lakes, swamps, and ditches; Alas to NS, s to Cal, Col, and nc and e US; Eurasia; tufted l. (*Naumbergia t.*)   6 L. **thyrsiflora** L.

Primula L.   Primrose

Fls showy, umbellate, 5-merous; calyx persistent, deeply lobed; corolla salverf, lobes gen emarginate; stamens attached in upper third of corolla tube, included; style gen included; caps valvate; scapose, herbaceous (ours) per. (L *primus*, early or first, many spp. fl early in spring). Our spp. not satisfactory orn pl.

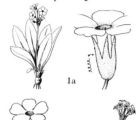

1a Lvs erect, 10–25 cm, strongly denticulate; corolla limb gen at least 1.5 cm broad, reddish-purple, eye yellow, tube 8–11 mm; pl viscid, rather ill-smelling; alp rock crevices, talus, and moist areas; RM, c Mont and Ida to NM, Utah, and Ariz; Parry's p. (*P. mucronata, P. p.* var. *brachyantha)*
1 **P. parryi** Gray

1b Lvs in flat basal rosettes, < 10 cm, entire to denticulate; corolla limb < 1.5 cm broad, bluish-violet to white

2a Scapes 2–9 cm, 1–3-fld; calyx 5–8 mm, lobes ca = tube; corolla ca 1 cm, bluish-violet to purplish, lobes ca 4 mm, shallowly retuse; foothills to subalp slopes, often on talus or near seepage; Blue and Wallowa mts, ne Ore, to c Ida; Cusick p., Wallowa p. (*P. brodheadae, P. angustifolia* var. *c.)* 2 **P. cusickiana** Gray

2b Scapes 6–40 cm, 3–12-fld; calyx 7–10 mm, lobed ca 1/3 length; corolla 8–11 mm, mostly lilac (white), lobes 2–3 mm, deeply retuse; moist meadows and streambanks; RM, s Mont to Colo, also in Utah, and in Lemhi Co, Ida (near Gilmore); mealy p.; Ida pls white-fld, others apparently always lilac (*P. americana, P. farinosa* var. *i.)* 3 **P. incana** Jones

Trientalis L.   Starflower

Fls axillary on slender, curved peds, 6–7 (5–9) -merous; calyx parted to near base, segms linearlanceolate; corolla white to pinkish-rose, rotate, divided nearly to base; stamens exserted, filaments shortly connate at base; caps globose, valvate; per rhizomatous herbs with short, thickened tubers and erect fl sts with crowded or whorled lvs near top and bracteate below. (L meaning having one-third of a foot, referring to height). Pls not unattractive but apt to become bad weeds.

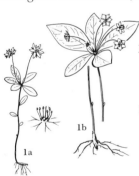

1a Fls gen white, 10–16 mm broad; peds > main lvs; reduced lvs scattered along st below crowded terminal cluster, main lvs 1.5–5 cm; tubers small, horizontal; bogs and swamps; Alas s in Cas and OM to n Ore and farther s along coast, e to Alta and n Ida; northern s. (*T. europaea* var. *a.)*
1 **T. arctica** Fisch.

1b Fls pinkish to rose, 8–15 mm broad; peds < lvs; lvs of terminal whorl 3–10 cm, all others reduced to tiny bracts; tubers 1–2 cm, ascending to erect; woods and prairies; s BC to Cal, e to Alta and Ida; western s., broad-lvd s. 2 **T. latifolia** Hook.

PLUMBAGINACEAE   Plumbago Family

Fls 5-merous, complete, reg, hypog, gen cymose or spicate to racemose or pan, sometimes invol; calyx lobed, prominently nerved, often papery and strongly plicate; corolla gamopet to almost polypet, scarious in ours; stamens on corolla tube, opp lobes; pistil 5-carpellary, styles 5, ovary 1-celled with single ovule; fr achenelike, often enclosed in calyx; scapose (ours) to lfy-std, herbaceous to shrubby per with alt or basal lvs.

Armeria Willd.   Thrift; Sea-pink

Fls subsessile in invol, capitate clusters, outer bracts reflexed and sheathing scape; calyx funnelf, dry and scarious; petals long-clawed, connate only at base, papery, gen pinkish; styles distinct, glandular-hairy at base; scapose per with dense tufts of narrowly linear, persistent basal lvs. (Derivation uncertain).

A. maritima (Mill.) Willd. Lvs (3) 5–10 cm × 1–3 mm, glab to ciliate or pub-erulent, persistent; scapes 1–5 dm, glab; outer invol bracts gen purplish; "heads" 1.5–3 cm broad; fls in clusters of 3, subtended by 2 transparent bracts; calyx 5–6 mm, 10-nerved, with 5 ribs ending in short teeth; corolla pinkish to lavender; beaches and coastal bluffs mainly, but inl occ as on Ta-coma prairies; circumboreal, s in N Am to s Cal and Newf; 2 vars. in our area
a1 Outer invol bracts triangular to lanceolate, often exceeding head; lvs glab;
  VI to Cal *(A. c.)*   var. **californica** (Boiss.) Lawrence
a2 Invol bracts ovate to obovate; lvs ciliate to pubescent; Wn to Alas; Europe
  (A. p., A. vulgaris ssp. *arctica*   var. **purpurea** (Mert. & Koch) Lawrence

OLEACEAE   Olive or Ash Family

Fls gen racemose to pan, hypog, reg, ♂ or occ ♂ ♂♀, or ♂♂, ♂♀; calyx 4-merous, rarely lacking; cor-olla gamopet (rarely polypet), gen 4-merous, or (ours) lacking; stamens 2 (3–4); pistil 2-carpellary; fr (ours) a samara; shrubs or trees with opp (rarely alt), simple to pinnately com-pound, gen exstip lvs.

Fraxinus L.   Ash

Fls gen preceding or developing with lvs, small, thyrsoid to glomerate; calyx much reduced, gen 4-lobed; corolla 2 (4)-merous or often lacking; filaments slender, anthers 2-celled; ovary 1- or 2-celled, gen 1-seeded; samara elongate; trees (ours) with odd-pinnate lvs. (L name for the ash tree).

F. latifolia Benth. Ore a. ♂, ♀ trees with trunks 10–20 × up to 1 m; bark rough, grayish-brown, at first gen puberulent to ± tomentose, but glabrate; lflets (3) 5–7 (9), gen 3–10 (15) cm, light green above, paler beneath; fls in crowded pans appearing with the lvs, ♂ with tiny bractlike calyx, stamens gen 2; ♀ fls with larger calyx, lobes 4, ± laciniate, ca 1/4 as long as ovary; samaras 3–5 cm × 3–9 mm; deep, fertile, gen moist soil along coast; BC s, w Cas, to Sierran and coastal Cal *(F. oregana, F. americana* var. *o., F. o.* var. *l.).*

BUDDLEJACEAE   Buddleja Family

Fls in large infls, complete, gamopet, hypog, 4–5-merous except pistil; stamens alt with corolla lobes; pistil 2-carpellary, ovary gen 2-celled and ∝ -seeded; fr caps or berry; shrubs (ours) with simple, opp lvs and interpetiolar stips.

Buddleja L.   Butterfly-bush

Fls pan to capitate; calyx 4-lobed; corolla ± funnell, 4-lobed; style single, stigma 2-lobed; caps 2-valved; shrubs with gen stellate pubescence and opp (ours) lvs connected at base by stip line. (For Adam Buddle, 1660–1715, English botanist).

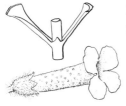

B. davidii Franch. Shrub up to 4–5 m; lvs ovate-lanceolate, up to 25 cm, grayish-tomentose beneath, green above; fls pan, corolla 7–9 mm, purplish with yellow orfice; orn that not infrequently escapes and persists along road-sides and esp along RR tracks in w Wn. Other spp. may also persist occ.

GENTIANACEAE   Gentian Family

Fls showy, gen ♂, reg, gamopet, hypog, 4–5 (7)-merous except pistil, solitary and terminal or var-iously cymose; calyx persistent, reg lobed to deeply cleft, or lobes sometimes lacking; corolla ro-tate to salverf or tubular, convolute in bud, often plicate in the sinuses, sometimes glandular esp at base of lobes; stamens inserted on corolla alt with lobes; ovary sessile to long-stipitate, 1-celled with 2 parietal placentae; style gen rather short and thick; caps gen valvate; seeds ∝ ; ann or per, mostly glab herbs with simple, mostly entire, opp to whorled, exstip lvs.

1a Petals spurred at base   **Halenia**
1b Petals not spurred
  2a Corolla salverform to tubular, lobes gen no > tube
    3a Anthers ± spirally twisted after dehiscence; fls gen pinkish or red
      (occ yellowish or white); calyx narrowly tubular-funnelf   **Centaurium**
    3b Anthers not twisted after dehiscence; fls gen bluish or bluish-tinged,
      or yellow but calyx other than narrowly funnelf
      4a Corolla yellowish, 4-lobed, gen < 8 mm, tube ca = calyx; pl ann
                                              Microcala
      4b Corolla mostly blue, rarely < 10 mm; pl ann or per   **Gentiana**
  2b Corolla rotate, lobes gen > tube
    5a Pl ann or bien; style lacking, the stigmas decurrent for half length of
      ovary   **Lomatogonium**
    5b Pl per; stigmas not decurrent on ovary, a style present
      6a Style thick, scarcely 1 mm; fls 5-merous   **Swertia**
      6b Style slender, at least 2 mm; fls 4-merous   **Frasera**

Centaurium Hill.   Centaury

Fls single or gen subsessile to long-pedunculate in few–∞-fld cymes, white to yellowish-pink or deep salmon, 5 (4)-merous except pistil; calyx slender, segms linear and often membranous-margined; corolla salverf, lobes narrow, mostly ca 1/2 length of tube; stamens inserted ca midlength of corolla tube, exserted, anthers oblong, spirally coiling after dehiscence; caps slender, 2-valved; ann (ours), strict, glab herbs with opp, sessile or clasping, entire lvs. (Old L name, *centaureum*, used by herbalists, referring to centaur, *Chiron*, who supposedly discovered medicinal properties of the pl). (*Erythraea, Centaurodes, Cicendia*).

1a Basal lvs several, 15–40 mm, often in rosette, rather prominently 2
  (5)-veined from base; fls gen ∞, crowded in compact clusters, subsessile,
  10–18 mm; anthers gen at least 1.5 mm; pl 1–5 dm; wastel, meadows, and
  prairies, gen where moist; estab from nw Wn to n Cal, gen w Cas, e to
  Ida; European c., common c. (*Gentiana centaurium, E. c.*)
                                                  1 C. **umbellatum** Gilib.
1b Basal lvs few, well spaced, not forming rosette, 5–40 mm, gen lightly
  1-nerved; fls gen few, often long-ped; anthers mostly < 1.5 mm
  2a Peds of central fls > the fls, gen > 2 cm; calyx lobes 2–3× > tube;
    moist places, esp around hot springs and alkaline lakes; e Cas, Wn, to
    Neb, s to Colo and Cal; western c.   2 C. **exaltatum** (Griseb.) Wight
  2b Peds all < 2 cm, gen much < the fls; calyx lobes not much > tube;
    moist soil; e Wn, Kittitas to Spokane cos, s to Nev and Cal, w through
    CRG to WV and s; Muhlenberg's c., Monterey c. (*C. curvistaminea, C.
    minima, C. m.* var. *albiflora*)   3 C. **muhlenbergii** (Griseb.) Wight

Frasera Walt.   Frasera

Fls 4-merous except pistil, rotate to shallowly camp, ∞ in a large loose to tight thyrse, withering-persistent; calyx cleft almost to base; corolla white or yellowish-green with purple spotting to bluish or purplish, deeply lobed, tube very short, often with short processes (squamellae) or with broader, scalelike, lacerate or fimbriate process (corona) fused to base of each corolla lobe and with 1 or 2 pits (foveae) near base of each segm; foveae ± completely surrounded by gen fimbriate membrane (hood); stamens alt with scalelike processes (crown scales) but scales sometimes reduced to squamellae or lacking; glab to puberulent per (bien) herbs with 1–several fl sts and opp or whorled, often basally connate-perfoliate, gen entire lvs. (For John Fraser, 1750–1811, English nurseryman). (*Leucocraspedum, Tessaranthium, Swertia* in part).

1a St lvs in whorls of 3–5 (6); fl st gen single, 5–20 dm; style < ovary; corolla
  lobes 10–25 mm, with 1 or 2 foveae
  2a Corolla greenish-yellow, spotted or blotched with purple, each lobe with
    2 elliptic foveae; fl st 10–20 dm; pl monocarpic; open or wooded foothills
    and valleys to subalp slopes; e Wn to Daks, s to Cal, NM, and Mex; giant
    f. (*F. angustifolia, F. s.* var. *a., F. macrophylla, S. radiata*)
                                                1 F. **speciosa** Dougl.
  2b Corolla ± clear blue or bluish-purple, each lobe with a single, nearly
    round fovea; fl st 5–15 dm; pl per; moist woods and meadowl; Blue Mts,

Ore and Wn, e to Idaho and Kootenai Cos, ne Ida; clustered f. *(F. thyrsi-*
flora)   2 **F. fastigiata** (Pursh) Heller

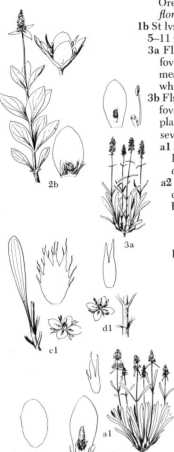

1b St lvs gen opp; fl sts gen $>$ 1, mostly $<$ 8 dm; style $>$ ovary; corolla lobes
5–11 mm, each with 1 fovea

  3a Fls white or cream; crown scales seldom $>$ 2 mm, entire to setiform;
foveae obovate, completely surrounded by a uniformly fringed hood; dry
meadowl to sagebr and wooded valleys and hillsides; c and wc Ida;
white f.   3 **F. montana** Mulford

  3b Fls blue (except for occ albino); crown scales 1–6 mm, lacerate to entire;
foveae oblong, hood smaller and scarcely fringed toward distal edge;
plains and lower mts; e Cas, s BC to Cal and Nev, e to Ida and w Mont;
several vars.   4 **F. albicaulis** Dougl.

    a1 Sts and lvs $\pm$ uniformly puberulent; crown scales mostly lacerate into
linear segms; c Wn to Ida and Mont; white-std f. *(F. a.* var. *alba, F.*
caerulea, F. nitida var. *a., S. watsonii)*   var **albicaulis**

    a2 Sts glab; lvs glab or puberulent only near base and along midrib;
crown scales various

      b1 Crown scales subentire, $\pm$ petaloid, 2.5–4.5 mm; lvs puberulent at
base and sometimes beneath midrib; Blue Mts, Grant and Union
cos, Ore, to Owyhee Co, Ida; Cusick's f. *(F. c., F. nitida* var. *c.)*
var. **cusickii** (Gray) Hitchc.

      b2 Crown scales lobed to lacerate, not petaloid, often $<$ 2.5 mm; pls
sometimes glab

        c1 Corolla pale blue, gen not darker-mottled; crown scales 2–6 mm,
gen deeply lacerate into linear segms; sheath of basal lvs gen not $>$
1.5 (2) cm, not bluish-tinged; Wallowa Mts, Baker Co, Ore, and
adj Adams Co, Ida; Ida f. *(S. i.)* var. **idahoensis** (St. John) Hitchc.

        c2 Corolla pale to dark blue, gen with darker mottling; crown scales
1–4 mm, from erose to lacerate; sheath of basal lvs gen $>$ 1.5 cm,
often bluish-tinged

          d1 Lower lf sheaths gen puberulent; crown scales (1) 3–4 mm;
near CR in Klickitat and Yakima cos, Wn, and adj Ore; Columbia f. *(F. nitida* var. *albida, S. c.)*
var. **columbiana** (St. John) Hitchc.

          d2 Lower lf sheaths often glab; crown scales gen $<$ 3 mm; se Ore
to c Cal, but also said to occur in our area; shiny f. *(F. n., S. a.*
var. *n.)*   var. **nitida** (Benth.) Hitchc.

Gentiana L.   Gentian

Fls solitary to ∞ in cymose clusters, 4–5 (6)-merous except the pistil, white or yellowish to blue or
purple, often $\pm$ strongly green-mottled; calyx from connate to near tip to lobate nearly to base,
sometimes lined with inner membrane projecting above base of lobes; corolla narrowly funnelf to
salverf, persistent, often either plicate in sinuses (plaits rounded to acute or lobed or toothed) or
with setaceous scales at base of lobes on inner surface; anthers versatile, erect or recurved; style
gen short and rather stout; ann, bien, or per, gen glab herbs from fleshy roots or slender rhizomes,
with opp, petiolate to sessile, sometimes clasping lvs. (For King Gentius of Illyris, who supposedly
knew of medicinal properties of the gentian). *(Amarella, Anthopogon, Chondrophylla, Dasyste-*
phana, Ericala, Gentianella, Gentianopsis, Pneumonanthe). Several spp. good rock garden subjects, but ours hard to grow.

1a Pl per; corolla gen 5-merous and much $>$ 2 (gen $>$ 2.5) cm, if shorter then
pl either $\pm$ prostrate, slenderly rhizomatous, and matted, or lvs connate
and long-sheathing at base and with bl $<$ 1 cm; stigmas never flabellate-erose

  2a Corollas gen much $>$ 2.5 cm; calyx with inner membranous lining extending above base of lobes; pl not slenderly rhizomatous

    3a Fls white or pale yellow, blotched and striped with purple; upper st
lvs mostly at least 3 $\times$ as long as broad

      4a Basal lvs linear-oblanceolate, 4–12 cm, $<$ 1/6 as broad; sts 5–20
cm; corolla 3.5–5 cm; alp bogs and meadows; Alas s through RM to
Colo, e Siberia; whitish g. *(G. romanzovii, G. a.* var. *r.)*
1 **G. algida** Pall.

      4b Basal lvs broadly spatulate to obovate, 2–6 cm, much $>$ 1/6 as
broad; sts 4–12 cm; corolla (2.5) 3–4 cm; moist mont meadows to

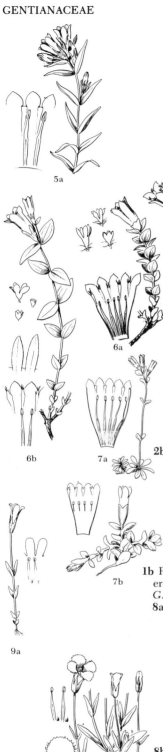

5a

6a

6b   7a

7b

9a

9b

subalp slopes; Cas, Deschutes Co, Ore, to Cal and w Nev; Newberry's g.   2 **G. newberryi** Gray
3b Fls blue or purplish or if, as sometimes, yellowish then upper st lvs scarcely 2× as long as broad
  5a Plaits or appendages of corolla sinuses truncate or rounded, entire; fls of lower axils sometimes with bractless ped; seeds fusiform, not winged; st 2–12 dm; larger st lvs oblong-lanceolate, 3–6 cm; pl of bogs and wet areas; BC to nw Cal, w Cas and mostly near coast; staff g., king's g. (*G. menziesii, G. orfordii*)   3 **G. sceptrum** Griseb.
  5b Plaits or appendages of corolla gen lacerate into narrow segms; all fls gen bracteate at top of ped; seeds sometimes winged; sts 0.5–8 dm; larger st lvs 1.5–5 cm; mostly mont
    6a Lvs ovate to obovate, mostly < 2× as long as broad; pl glab; fls gen solitary; seeds elongate, ± terete, not flattened or wing-margined; mont, gen alp or subalp meadows and stream banks to open slopes; BC to Cal, e to RM; explorer's g., mt bog g.; 3 vars.   4 **G. calycosa** Griseb.
    a1 Calyx mostly truncate at top, lobes very short or lacking; mainly in c Ida and adj Mont (*G. c.* ssp. *a., G. idahoensis*)
                                              var. **asepala** (Maguire) Hitchc.
    a2 Calyx lobed sometimes > half length
      b1 Calyx lobes large and similar to tube in texture, gen erect; RM, alta to Mont   var. **obtusiloba** (Rydb.) Hitchc.
      b2 Calyx lobes thicker and fleshier than tube, often spreading; pl gen somewhat taller and more often several-fld; mainly Cas and OM but also in RM (*G. c.* var. *stricta, G. c.* var. *monticola, G. c.* var. *xantha, G. cusickii, G. gormani, G. myrsinites, G. saxicola*)   var. **calycosa**
    6b Lvs gen at least 2× as long as broad, if at all ovate or obovate then st very finely pubescent in lines below lvs; lvs finely ciliolate; calyx lobes ± glandular-ciliolate; fls gen several; seeds flattened, broadly wing-margined; damp areas, valleys and foothills to upper mont, BC s, e side Cas, to Cal and Ariz, e to Alta and in RMS s to Mex; pleated g., prairie g. (*G. oregana*)
                                                   5 **G. affinis** Griseb.
2b Corollas < 2.5 cm; calyx not membranous-lined above base of lobes; pl gen slenderly rhizomatous, often ± matted
  7a Fls gen several (3–7), 10–20 mm, mostly 5-merous; st lvs 2–4 pairs, shortly connate at base; st 4–15 cm; alp meadows and tundra; Alas to BC and (reportedly) to n Cas, Wn, and in RM to Mont; Asia; glaucous g.   6 **G. glauca** Pall.
  7b Fls gen solitary and terminal, (8) 10–22 mm; st lvs ∝, long-sheathing; st 2–15 cm; alp bogs and high mont meadows; Alas s in RM to Colo, w to Utah, Nev, and c Ida; n Cal; Eurasia and S Am; moss g. (*G. fremonti*)   7 **G. prostrata** Haenke
1b Pl ann; corollas often < 2 cm, when longer gen 4-merous or with lacerate-fimbriate lobes or with broadly flabellate-erose stigmas; lvs (except in *G. propinqua*) not long-sheathing
  8a Corolla deep blue, gen > 2 (to 5) cm, 4-merous; stigmas broadly flabellate-erose
    9a Pl simple, st unbr; fls gen single, 2–4 cm; basal lvs gen not > 2; anthers 1–1.5 mm; style thick, 1–2 mm; mt bogs and meadows; Cas, Deschutes Co, Ore, to Sierran Cal, e through Ore to c Ida; one-fld g.
                                                      8 **G. simplex** Gray
    9b Pl rarely simple, st gen br; fls mostly several, 3.5–5.5 cm; basal lvs gen several; anthers 3–4 mm; style gen slender and 4–6 mm; circumboreal in meadows, bogs, and tundra; Alas to Newf, s to Cal, Ariz, Mex, SD, Ind, and NY, in our area only in Mont and c Ida; smaller fringed g. (*G. barbata, G. d.* var. *b., G. elegans, G. e.* var. *unicaulis, G. macounii, G. thermalis*); ours the var. *unicaulis* (Nels.) Hitchc.   9 **G. detonsa** Rottb.
  8b Corollas either < 2 cm, or not deep blue, sometimes 5-merous; stigmas not flabellate-erose
    10a Corolla sinuses plicate, often 2-toothed, base of lobes not fringed on inner side; calyx not lobed > half length
      11a Fls 5-merous, gen several per br; lvs neither long-sheathing nor overlapping; main sts gen one, but freely br from base, 5–20 cm;

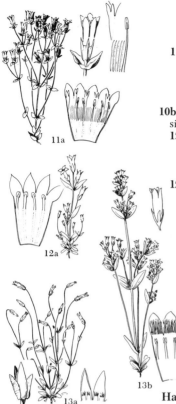

corolla 10–15 mm, tube white (drying yellowish), lobes blue on back; bogs, tundra, and lake margins; in our area at low to medium elev; Alas to VI; known from Snoqualmie Pass and Lake Ozette, Wn; swamp g. *(G. d.* var. *patens)*   10 **G. douglasiana** Bong.

11b Fls 4 (occ 5?) -merous, solitary on each st or br; lvs long-sheathing, lower ones overlapping; sts gen several from base and unbr above, 2–15 cm; corolla (8) 12–22 mm, from clear blue to purplish and often ± greenish mottled; high mont bogs and meadows (see lead 7b)   7 **G. prostrata** Haenke

10b Corolla sinuses not plicate, base of lobes sometimes fringed on inner side; calyx often lobed > half length

12a Corolla lobes not fringed within; fls mostly 4-merous, upper ones 15–22 mm, light purple; meadows and stream banks; Alas s in RM to BC, Alta, and Beaverhead Co, Mont, e to Que and Newf; four-parted g.   11 **G. propinqua** Richards.

12b Corolla lobes fringed within; fls 4–5-merous

13a Sepals distinct, slightly pouched at base, outer 2 broader and < inner; each corolla lobe with 2 scales within, scales fringed about half length; fls on long, naked peds, infl thus open; corolla 8–15 mm, white to bluish purple; mt ranges, where moist; circumboreal, s in N Am in RM to NM and Ariz, disjunct in Cal and Nev, known from YNP, reported from Ida, and to be expected in Mont; slender g. *(G. monantha, G. t.* var. *m.)*   12 **G. tenella** Rottb.

13b Sepals connate 1/5–1/3 length, ± unequal but not basally pouched; corolla lobes with inner scales fringed full length or nearly so; fls more often crowded in congested infl; corolla mostly 10–20 mm, from pale yellowish and lightly bluish-tinged to violet or lavender; moist areas; circumboreal, Alas s, in w N AM, to Mex, from coastal Wn to RM; northern g. *(G. acuta, G. anisosepala, G. heterosepala, G. plebeja, G. scopulorum)*
                                                        13 **G. amarella** L.

Halenia Borkh.   Spurred Gentian

Fls cymose, 4-merous (ours); calyx deeply lobed, segms foliaceous; corolla lobed to below middle, extended into short basal spurs; stamens freed at top of corolla tube; caps lanceolate; ann to per with opp or basal lvs. (For Jonas Halenius, 1727–1810, Swedish botanist).

H. deflexa (Sm.) Griseb. Ann; st erect, gen br above, mostly 1–5 dm; main lvs sessile, lanceolate to ovate, 2–5 cm; fls greenish-purple, ca 10 mm, spurs 1–5 mm (sometimes lacking); moist areas, BC to NS, known in our area only from Lincoln Co, Mont; ours the var. *d.*

Lomatogonium A. Br.   Marsh Felwort

Fls 4–5-merous except pistil; calyx deeply divided; corolla subrotate, lobed almost to base, segms with 2 small, scaly, basal appendages; style lacking, stigmatic surfaces decurrent as 2 lines on upper half or more of ovary; glab, opp lvd ann or bien. (Gr *loma*, fringe, and *gonos*, offspring, here referring to ovary).

L. rotatum (L.) Fries. St 10–25 cm; lvs oblanceolate or spatulate to lanceolate, 1–3 cm; fls mostly axillary on long, slender peds from upper lvs; sepals gen exceeding corolla; corolla blue (white), lobes elliptic-ovate, 6–15 mm; caps oblong-ovoid, ca = persistent corolla; wet, often saline soil, mont; Alas to Greenl, s in RM to NM, known from Ida in our area, but to be expected in Mont; Siberia *(Gentiana r., Narketis r., Pleurogyne r., Swertia r.).*

Microcala Hoffm. & Link   Timwort

Fls 4-merous except pistil, yellowish; corolla salverf; stamens freed above midlength of corolla tube; style slender, stigmas flabellate-lobed; seeds flattened, ovoid, coarsely reticulate; small, glab, opp-lvd ann. (Gr *mikros*, small, and *kalos*, beautiful).

M. quadrangularis (Lam.) Griseb. St slender, simple or ± br, 3–8 cm; lvs lanceolate to oblanceolate, 4–9 mm, mostly near-basal; fls 5–8 mm, terminal on main st or on elongate, axillary, gen naked peds; calyx broadly camp, strongly carinate on back of lobes, plicate-carinate between, teeth acute, scarcely 1/2 as long as tube; corolla tube ca = calyx, lobes ovate-rounded, ca 2/5 as long as tube, not plicate; caps ca = calyx; prairies and open moist flats, rare; WV and Umpqua Valley, Ore, s to much of ne Cal; S Am (*Exacum q., Gentiana q.*).

Swertia L.   Swertia

Fls cymose to thyrsoid, 5 (4)-merous except pistil; corolla deeply lobed, segms with basal pair of glands (foveae) surrounded by fringed hoods; crown scales lacking (ours); style short; glab ann or per herbs with opp, entire lvs. (For Emanuel Sweert, Dutch gardener and author of 16th century).

S. perennis L. Pl rhizomatous, fl sts gen single at ends of rhizomes, 5–50 cm; lvs mostly basal, (2) 5–12 (20) cm, obovate to oblong-elliptic; st lvs opp (sometimes few alt), much reduced except at first 2–4 nodes; corolla bluish-purple (occ albino), variously maculate with greenish or white, lobes narrowly oblong, entire to erose, 6–11 mm, tube with few small squamellae below foveae; foveae orbicular; style < 1 mm; caps compressed, 7–12 mm; Alas to Cal, e to RMS; Eurasia (*S. congesta, S. obtusa, S. ovalifolia, S. parallela*).

MENYANTHACEAE   Buck-bean Family

Fls racemose to cymose on long naked scapelike peduncles, 5 (4–6)-merous except pistil, reg, complete, gamopet; calyx adnate to lower 1/5–1/2 of ovary, deeply to shallowly lobed; corolla salverf to rotate, tube very short, lobes either with 3 fringed scales running lengthwise or strongly bearded with slender bristlelike scales; stamens 5, alt with corolla lobes; pistil 2–3-carpellary; ovary 1-celled with 2–3 parietal placentae; styles dimorphic, gen either short and thick or considerably elongate and slender; caps firm-walled, valvate or tardily dehiscent to indehiscent; glab, rhizomatous, per herbs with alt, sheathing-based, long-petioled, simple to trifoliolate lvs. Both our spp. are attractive and easily grown bog or pool pls.

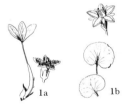

1a Lvs trifoliolate, lflets not crenate; corolla bearded with narrow hairlike scales   **Menyanthes**
1b Lvs simple, crenate; corolla crested with 3 fringed scales running full length of lobes   **Nephrophyllidium**

Menyanthes L.   Buckbean; Bogbean

Fls racemose; caps gen indehiscent; fleshy herbs with trifoliolate lvs. (Gr *meniaios*, monthly, and *anthos*, fl).

M. trifoliata L. Sts prostrate or fl brs ascending; petioles 1–3 dm, lflets short-stalked, elliptic-ovate to -obovate, 4–12 cm, entire to coarsely undulate-dentate; peduncles 5–30 cm; calyx 3–5 mm; corolla tube gen ca twice length of calyx, whitish, lobes gen purplish-tinged, 5–7 mm; anthers purple, ca 2 mm; bogs and lakes; Alas to Greenl, s, mostly w Cas, to s Ore and Sierran Cal, also in ne Ore; ranging through RM to Colo and in c and e US to Ind and Pa; Europe.

Nephrophyllidium Gilg   Deer-cabbage

Fls loosely cymose on long naked scapes; corolla rotate, lobes fringed; ovary at least half inferior; style < to > the stamens; caps ovoid-conic, valvate. (Gr, meaning like *Nephrophyllum*, another genus, which has kidney shaped lvs, from *nephros*, kidney, and *phyllon*, lf).

N. crista-galli (Menzies) Gilg. Petioles 2–3 dm, bls cordate-ovate, broadly reniform or reniform-obcordate, 3–12 cm broad, finely to coarsely crenate; peduncles 1–5 dm; cymes mostly open and loosely fld; floral tube turbinate; sepals lanceolate, spreading, 3–5 mm; corolla white, rotate, tube 2–4 mm, lobes ovate-lanceolate, 4–6 mm; anthers ca 3 mm; styles 1–5 mm; ovary > half inferior at anthesis, but caps elongating and superior portion becoming 10–18 mm; bogs, swamps, and wet meadows; OP, Wn, to Alas and Japan (*Fauria c., Menyanthes c.*).

APOCYNACEAE   Dogbane Family

Fls solitary or cymose (ours), complete, hypog, gamopet, reg or slightly irreg, 5-merous except pistil; calyx divided most of length; corolla camp to funnelf; stamens adnate to corolla alt with its lobes, anthers connivent around and sometimes adherent to style; pistil 2-carpellary, ovaries distinct and 1-celled (ours), gen surrounded by 5 nectaries; style 1, stigma enlarged; fr follicular (ours), ∝-seeded; seeds sometimes with tuft of soft hairs; per herbs (ours) with opp (alt or whorled), simple lvs and milky juice.

1a Fls greenish-white to pink, 2–10 mm; pl mostly erect   **Apocynum**
1b Fls blue (white), 3–5 cm; pl gen trailing   **Vinca**

Apocynum L.   Dogbane

Infl cymose, often much-compounded; calyx lobed 1/3–2/3 length; corolla tubular to camp, lobes oblong-lanceolate to ovate, erect to spreading or reflexed, tube with triangular subulate appendages alt with stamens; filaments freed at base of corolla tube, anthers narrowly lanceolate, sagittate at base, slightly convergent at tip and ± adherent to stigma; style very short, enlarged and broadly clavate-ovoid above; ovaries surrounded at base by 5 peglike nectaries; follicles elongate; seeds apically long-comose; rhizomatous per herbs with opp, subsessile to petiolate lvs and milky juice. (Gr *apo*, away from, and *kyon*, dog). (*Cynopaema*).

1a Fls 2–4.5 mm; corolla greenish-white to white, gen < twice as long as calyx, lobes erect or slightly spreading; lvs ascending
  2a Follicles gen > 12 cm; coma of seeds 2–3 cm; lvs yellow-green, oblong-ovate or -lanceolate, 5–11 cm, all except sometimes the basal distinctly petioled but not cordate-based; pl glab (ours), 3–10 dm; gen in wastel and seldom-cult areas, esp orchards, in most of US and much of Can; common d., hemp d.; 2 vars.   1 **A. cannabinum** L.
    a1 Calyx lobes ca half length of corolla; mostly in w US (*A. s.*)
                                 var. **suksdorfii** (Greene) Beg. & Bel.
    a2 Calyx lobes > half length of corolla; the more common in our area (*A. pubescens, A. c.* var. *p.*)   var. **glaberrimum** DC.
  2b Follicles < 12 cm; coma of seeds 1–2 cm; lvs of main st gen all sessile or subsessile and cordate-based; gen over much of US and Can; clasping-lvd d., Indian hemp; ours the var. *salignum* (Greene) Fern. (*A. hypericifolium*)   2 **A. sibiricum** Jacq.
1b Fls gen at least 5 (3.5–10) mm; corolla pinkish, often > twice length of calyx, lobes spreading to reflexed; lvs mostly drooping to spreading
  3a Calyx gen at least half as long as corolla, lobes narrowly lanceolate, acute to acuminate; lvs often ascending; gen in valleys and lower mts of w N Am to c US and Atl; western d. (*A. ciliolatum, A. convallarium, A. denticulatum, A. floribundum*)   3 **A. medium** Greene
  3b Calyx gen < half as long as corolla, lobes lanceolate to deltoid or ovate, often obtuse; lvs mostly spreading or drooping; gen on rather dry soil in valleys and foothills to semi-alp slopes; gen in much of Can and all but se US; spreading d., flytrap d., bitter-root; 2 vars.
                                 4 **A. androsaemifolium** L.
    a1 Corolla camp, 5–10 mm; follicles gen pendulous; cyme terminal, simple; range of sp. (*A. a.* var. *incanun, A. i., A. scopulorum, A. macranthum*)   var. **androsaemifolium**
    a2 Corolla more tubular, 4–7 mm; follicles gen erect; infl often larger and partially axillary as well as terminal; most of w US and Can (*A. a.* ssp. *detonsum, A. p., A. xylosteaceum*)   var. **pumilum** Gray

Vinca L.   Periwinkle

Fls axillary, long-ped; corolla salverf, crested in throat; stamens inserted below throat, filaments short; style elongate; stigma ovoid; seeds not comose; trailing per herbs. (L name, *vincapervinca*, from which comes periwinkle).

V. major L. Brs 1–6 dm, often trailing and freely rooting; lvs ovate to ovate-lanceolate, ciliolate, 3–9 cm; calyx lobes linear; corolla blue, 3–5 cm, tube ca 2 cm, lobes ± truncate; follicles 3–5 cm; European orn, often escaping w Cas.

ASCLEPIADACEAE   Milkweed Family

Fls complete, reg, gamopet, hypog, 5-merous except for the pistil; sepals free or basally connate; corolla gen lobed > half length, lobes reflexed; stamens adnate to base of corolla tube, monadelphous (rarely distinct), forming a "column" to which are attached 5 saccate "hoods" alt with corolla lobes, each gen with an elongate, inwardly curved appendage or "horn"; anthers basally attached, introrse, gen appendaged at tip with a scarious membrane or "corona," their hardened wing-margins forming elongate, raised, liplike "commissural grooves"; pollen of each anther sac coalescent into a single suspended mass or "pollinium," pollinia of pairs of anther sacs of adjacent anthers joined by an appendage ("translator arm"), the 2 arms attached above commissural groove to a single cleft gland opp lobe of stigma (glands fasten to legs of insects and are transported, with attached pollinia, to other fls); pistil 2-carpellary, ovaries and styles distinct, stigma 1, peltate-discoid, 5-lobed, adnate with stamens; fr follicular, ∞-seeded; seeds flattened, apically comose; per herbs (ours), often vinelike or shrubby, with opp or whorled (alt), simple lvs and milky juice.

Asclepias L.   Milkweed

Fls umbellate; sepals and petals in ours reflexed; 1 follicle often abortive; pl rhizomatous or with thickened, fleshy roots; lvs fleshy, gen with tiny caducous stips. (For *Asklepios*, legendary Gr physician and god of medicine).

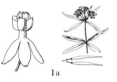

1a Lvs narrowly linear to linear-lanceolate, verticillate to scattered, 5–15 cm × 3–12 (16) mm; pl rhizomatous, 3–8 dm, glab to sparsely puberulent; sepals greenish to pinkish or purplish, ca 2 mm, puberulent; corolla lobes 2–4.5 mm, pale to dark pinkish-purple; hoods erect, 1–2 mm, subequal to stamens but much < the horns, attached well above base of staminal column; follicles erect, 6–12 cm, narrowly lanceolate-fusiform, smooth; dry to moist soil, esp along water courses; ne Wn to Utah, Ariz, and Baja Cal; w to Cas in sagebr and yellow pine region, reaching coast in Ore; Mexican m., narrow-lvd m. (*A. mexicana*, misapplied)   1 **A. fascicularis** Dcne.

1b Lvs gen lanceolate or broader, opp; hoods attached at base of staminal column

2a Hoods pink, not lobed, 9–12 mm, much > the stamens, ± divergent; follicles warty-spiny; pl rhizomatous, mostly grayish-lanate; petals pink to reddish-purple, ca 10 mm; loamy to sandy soil, esp where moist, as along waterways; e Cas, BC to Cal, e to Miss Valley; showy m., Greek m. (*A. douglasii*)   2 **A. speciosa** Torr.

2b Hoods pinkish, bilobed, 5–6 mm, barely exceeding anthers; follicles smooth; pl from woody, enlarged root, glab or subglab; petals pale greenish-yellow, often reddish-tinged on back, 8–12 mm; heavy clay to gravelly soil in hills and lower mts of Asotin Co, Wn, and Grant Co, Ore, to Payette Co, Ida, and Wyo, s to Cal, Colo, and Ariz; pallid m., Humboldt m. (*Acerates latifolia, Asclepias davisii, A. c.* ssp. *d.*)

3 **A. cryptoceras** Wats.

CONVOLVULACEAE   Morning-glory Family

Fls complete, gamopet, hypog, reg, gen 5-merous except the pistil, single and axillary to terminal and cymose; calyx often subtended by invol bracts, sepals distinct or basally connate; corolla often showy, camp or funnelf to salverf, entire-limbed to deeply lobed, mostly twisted in bud; stamens alt with corolla lobes; pistil 1, ovary 2 (1)-celled, gen 4-ovuled, often surrounded by small disc; styles 1–2; fr gen caps, 1–4 (6)-seeded; erect to procumbent or scandent herbs (ours) with alt, simple, exstip lvs and often milky juice.

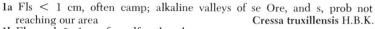

1a Fls < 1 cm, often camp; alkaline valleys of se Ore, and s, prob not reaching our area   **Cressa truxillensis** H.B.K.
1b Fls much > 1 cm, funnelf, rather showy
  2a Ann; stigmas capitate, often 3; occ escape from cult and temporarily estab but prob not persistent; common morning-glory
                                               Ipomaea purpurea (L.) Lam.
  2b Per; stigmas 2, elongate   **Convolvulus**

2b

Convolvulus L.   Bindweed; Morning-glory

Fls funnelf, large and showy, single or paired on axillary peduncles, 2-bracteate; bracts either slender, much smaller than calyx lobes, and gen distant from calyx, or broader and contiguous to calyx and gen = or > and largely concealing it; sepals subequal or outer ones considerably the broadest; corolla white to pinkish or ± purplish, twisted in bud; stamens included; style slender; ovary 2-celled, 4-ovuled; caps ovoid to globose, tardily 2–4-valvate, 4-seeded, or 1 or more seeds abortive; per (ours) herbs from slender rhizomes, gen with trailing to twining sts and simple, alt, often ± hastate lvs. (L *convolvere*, meaning to twine). (*Calystegia, Volvulus*).

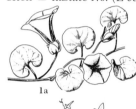

1a Lvs thick and fleshy, bl 1.5–4 cm broad, reniform, glab; calyx immediately subtended and almost = by broad, gen cordate bracts; corolla 3–5 (6) cm, pinkish-purple; confined to coastal beaches and sand dunes; BC to s Cal; Pac is; beach m., seashore m.   1 C. **soldanella** L.
1b Lvs mostly not fleshy or reniform, sometimes hairy; calyx sometimes not immediately subtended by bracts; corolla often white; pl rarely if ever on coastal beaches or dunes
  2a Fls double, bright pink; pl pubescent; sts twining; occ reported from Ida as an escape from cult, to be expected elsewhere; Cal rose
                                         C. japonicus Thunb.
  2b Fls single, gen white; pl often glab, sometimes not twining
    3a Bracts linear, often somewhat distant from calyx
      4a Pl pubescent; fls 1 per peduncle; bracts 3–12 mm, close to and overlapping base of calyx; corolla 3–4.5 cm, creamy-white or occ ± pinkish-streaked; dry valleys and hillsides, Maupin Valley, Wasco Co, Ore, and sw Ore to Cal; variable m., pale m.
                                  3 C. **polymorphus** Greene
      4b Pl often glab; fls often 2 per peduncle; bracts 2–4 mm, well below and not extending to base of calyx, sometimes lacking; corolla 1.5–2.5 cm, white to pinkish-purple; intro from Europe, one of most common and noxious weeds, esp in cult fields; field m., small b.
                                          4 C. **arvensis** L.
    3b Bracts broader, often cordate, gen adj to calyx and largely concealing it
      5a Pl glab, mostly on dry, rocky, open slopes or in ponderosa pine woodl; sts erect to trailing, but not twining; lf bl deltoid to hastate or cordate-hastate, 3–5 (7) cm; corolla white or pinkish-tinged, 3–5 cm; Mt Adams, Wn, s in Cas and WV to Cal; night-blooming m. (*C. atriplicifolia*)   5 C. **nyctagineus** Greene
      5b Pl glab to soft-pubescent, on ± moist soil, esp along river bottoms and coastal marshes, mostly w Cas; lf bl sagittate to hastate, 5–12 cm; corolla white to deep pink, 4–7 cm; intro from Europe, esp common in PS region; lady's-nightcap, bell-bind, Rutland beauty, hedge b. (*C. silvatica*)   6 C. **sepium** L.

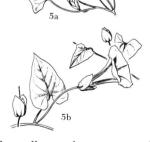

CUSCUTACEAE   Dodder Family

Fls small, complete, gamopet, hypog, gen pinkish (white), 5 (4–3)-merous except the 2-carpellary ovary; calyx deeply lobed, often fleshy at base; corolla narrowly tubular to camp, lobes erect to reflexed; stamens inserted just below sinuses of corolla lobes, portion fused with corolla tube often prominent and gen covered at base by mostly fimbriate-margined scales; ovary 2-celled, 4-ovuled; styles 2, distinct; stigmas capitate to linear-elongate; fr caps; twining, lfless, per herbs with very slender, pinkish-yellow to white, glab sts; lvs reduced to tiny scales; parasitic on many fl pls and sometimes causing considerable damage, esp to clover and alfalfa.

Cuscuta L.   Dodder; Love-tangle; Coral-vine

Fls sessile to short-ped in small cymules or aggregated into larger, often globular masses; caps membranous, circumscissile near base or indehiscent, 1–4-seeded, often rupturing corolla and pushing it upward as a calyptrate crown. (Derivation uncertain, word supposed to be Arabic). (*Epithymum*).

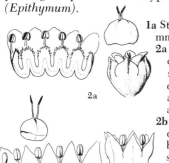

1a Stigmas attenuate, not capitate; caps irreg circumscissile near base; fls 2–3 mm
　2a Calyx ± fleshy, lobes as broad as long, tips turgid and slightly recurved; corolla gen not ruptured or pushed upward by developing caps, tube seldom any > calyx, lobes oblong-ovate; scales nearly = corolla tube, oblong, erose-fringed, rounded to slightly bifid; on leguminous crops, esp alfalfa, prob in all w US; Eurasia and Africa; clustered d., small-seeded alfalfa d. (*C. anthemi, C. gracilis*)　　　　1 C. **approximata** Bab.
　2b Calyx membranous, scarcely at all fleshy, lobes longer than broad, acute or acuminate, tips not turgid or recurved; corolla often pushed upward by caps, tube often exceeding calyx, lobes ovate-lanceolate to lanceolate; scales obovate-spatulate, ca 2/3 length of corolla tube, fringed at least above; native of Europe, widely intro on both sides Cas, on many hosts, but mainly on Leguminosae; common d., thyme d.
　　　　　　　　　　　　　　　　　　　2 C. **epithymum** Murr.
1b Stigmas capitate; caps not circumscissile; fls 1.5–6 mm
　3a Staminal scales lacking or greatly reduced and neither free-margined nor erose or pectinate
　　4a Corolla lobes reflexed, lanceolate, acute; anthers oblong, ca 0.7 mm; scales completely adnate, their outline barely noticeable; Wn to Baja Cal, on both sides Cas, on many hosts; Cal d.　3 C. **californica** Choisy
　　4b Corolla lobes spreading to erect; anthers < 0.5 mm
　　　5a Fls sessile; corolla tube = or > calyx; filament scales lacking or completely adnate, their outline barely visible; Wn to Cal, e to Colo, in most w states, on many hosts; western d.
　　　　　　　　　　　　　　　　　4 C. **occidentalis** Mills.
　　　5b Fls with peds 1–3 mm; corolla tube < calyx; filament scales sometimes almost lacking but gen very narrowly free-margined, entire or divergently 2-toothed at tip or with margins slightly erose; mont; s Wn to Cal, on *Aster, Spraguea,* and other hosts; Suksdorf's d., mt d. (*C. salina* var. *acuminata*)　　　　5 C. **suksdorfii** Yuncker
　3b Staminal scales present, their free margins gen erose to pectinate
　　6a Fls ca 2 mm, 4 (3–5)-merous, sessile or subsessile; sepals free almost to base, very unequal, outer ones strongly overlapping inner; corolla withering and calyptrate on caps, lobes obtuse, ca 1/3 length of tube; scales oblong, short-fringed to long-fringed near tips, ca 3/4 length of corolla tube; Wn and Ore to Atl coast, on many hosts; button-bush d.
　　　　　　　　　　　　　　　　　6 C. **cephalanthi** Engelm.
　　6b Fls either at least 2.5 mm or 5-merous and ped; sepals gen subequal and not strongly overlapping; corollas persistent or deciduous
　　　7a Fls gen 4-merous; caps globose or subglobose; corolla calyptrate on caps; scales slightly fringed or entire-margined (see lead 5b)
　　　　　　　　　　　　　　　　　5 C. **suksdorfii** Yuncker
　　　7b Fls gen 5-merous; caps often ovoid; corolla often persistent as caps matures; scales strongly erose to fimbriate-margined
　　　　8a Caps gen somewhat beaked or crested, longer than thick, gen globose-ovoid to ovoid-obconic; corolla not at all papillose
　　　　　9a Calyx lobes strongly erose-denticulate, oval to round; fls ca 2 mm; corolla tube barely > calyx; filament scales extending ca to base of free portion of filaments, obovate-oblong, subentire to irreg erose-dentate; des regions, on many shrubs; Lemhi Co, Ida, to Nev, Cal, and Utah; tooth-sepal d., des d.
　　　　　　　　　　　　　　　　　7 C. **denticulata** Engelm.

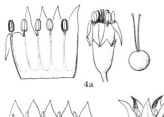

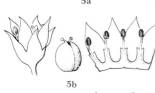

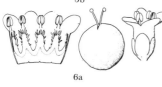

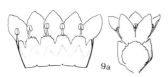

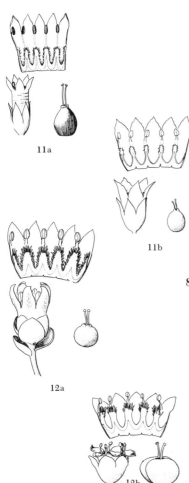

11a

11b

12a

12b

9b Calyx lobes gen acute, if rounded then not erose-denticulate; fls gen much > 2 mm; corolla tube often much > calyx

  10a Sepals almost or quite as broad as long, rounded; fls 2–3 mm; corolla lobes obtuse; seeds 3–4; se Mont to NM and e, not known from our range   **C. umbrosa** Beyrich

  10b Sepals considerably longer than broad, acute to acuminate; fls (2) 2.5–6 mm; corolla lobes acute; seeds gen only 1

    11a Fls 5–6 mm, often sessile; calyx ca half length of corolla tube, lobes strongly overlapping at base; corolla tube ca 2 × > lobes, cross-wrinkled; scales narrowly oblong, slightly > half length of corolla tube, free margins irreg fimbriate; w Ore to Mex, on many hosts, incl *Salix, Prunus, Ceanothus, Rhus,* and *Sambucus*; long-fld d., canyon d.   8 **C. subinclusa** Dur. & Hilg.

    11b Fls 2–4.5 mm, short-ped; corolla tube not cross-wrinkled, ca = corolla lobes, scarcely > calyx; calyx lobes scarcely overlapping at base; scales poorly developed, ca half length of corolla tube, narrowly oblong, free margins very narrow or lacking, entire to few-toothed, free tips gen with few teeth; mostly on Chenopodiaceae and Compositae, but also on other hosts; BC s along coast to Mex, inl to Utah and Ariz; salt-marsh d., alkali d.   9 **C. salina** Engelm.

8b Caps globose, gen not at all beaked, if ± pointed then corolla very slightly papillose

  12a Corolla 2–5 mm, very faintly papillose on margins and outer surfaces of the ± erect lobes; calyx lobed nearly to base, segms strongly overlapping; caps slightly crested-pointed; scales oblong to obovate, deeply fringed; wide-ranging in New World, on many hosts; gen in low, moist or irrigated areas; inelegant d.; ours the var. *neuropetala* (Engelm.) Hitchc., from s Ida to Cal, Ill, and Mex   10 **C. indecora** Choisy

  12b Corolla not papillose, lobes spreading or reflexed; calyx lobed ca 1/2 length; caps depressed at tip; scales oblong-obovate, strongly fringed above; in most of US and many other parts of the world, on many hosts, esp Leguminosae; field d., five-angled d.; 2 vars.   11 **C. pentagona** Engelm.

    a1 Fls gen < 2 mm; calyx lobes strongly overlapping (*C. arvensis* var. *p.*)   var. **pentagona**

    a2 Fls gen at least 2 mm; calyx lobes not strongly overlapping (*C. campestris*)   var. **calycina** Engelm.

POLEMONIACEAE   Phlox Family

Fls gamopet, ⚥, gen 5-merous as to calyx, corolla, and androecium; calyx lobes = or unequal, the tube often with alternating costae and hyaline intervals; corolla reg or seldom slightly irreg, contorted in bud; filaments alt with the corolla lobes, attached near the base or more often near the sinuses; ovary superior, with gen 3 (2–5) carpels; style simple, gen with separate stigmas; placentation axile; caps gen loculicidal, sometimes irreg or scarcely dehiscent; ovules and seeds few to ∞ ; embryo straight; endosperm present; ann to per herbs, less commonly shrubs or vines, with opp or alt, entire to pinnately compound or variously dissected lvs, the fls solitary or more often in open or compact (often headlike), variously modified cymes. All of our genera except *Phlox* and *Polemonium* have in the past often been referred to a broadly defined genus *Gilia.*

1a

3a

1a Lvs represented only by the persistent cotyledons and by a whorl of entire, often basally connate bracts just beneath the compact infl; small ann   **Gymnosteris**

1b Lvs ± well developed, either clustered at the base, or distributed along the st, or both; ann to per

  2a Calyx tube of essentially uniform texture throughout, accrescent, not ruptured by the developing caps; lvs not at once palmatifid and sessile

    3a Lvs pinnately compound, with definite lfts (these broader than linear except in 1 sp.); calyx tube herbaceous at anthesis; per (1 sp. ann), often with a mephitic odor   **Polemonium**

    3b Lvs entire to variously dissected, but without definite lfts; calyx tube

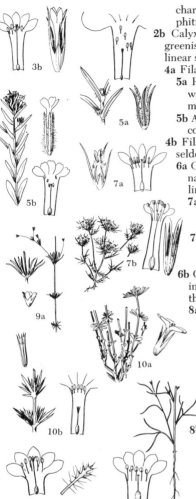

chartaceous at anthesis; ann (1 sp. per), only slightly or not at all mephitic **Collomia**

2b Calyx tube with green costae separated by hyaline intervals, or if
greenish essentially throughout, then the lvs sessile and palmatifid into
linear segms

   4a Filaments very unequally inserted; lvs entire, partly or wholly opp

      5a Per; lvs all (or nearly all) opp; seeds not becoming mucilaginous
when moistened; corolla showy, the lobes in most spp. well over 4
mm **Phlox**

      5b Ann; upper lvs alt; seeds becoming mucilaginous when moistened;
corolla relatively inconspicuous, the lobes only 1–4 mm **Microsteris**

   4b Filaments about equally, or occ somewhat unequally, inserted; lvs
seldom at once opp and entire

      6a Calyx lobes gen ± unequal; lvs gen pinnatifid to bipinnatifid with
narrow rachis and narrow, ± spinulose-tipped segms, varying to
linear and entire; ann with the fls in lfy-bracteate heads

         7a Heads distinctly tomentose with very fine, interwoven hairs; an-
thers well > 0.5 mm, nearly linear and deeply sagittate; lvs relatively soft **Eriastrum**

         7b Heads glab or glandular to rather coarsely villous, but not
tomentose; anthers small, up to ca 0.5 mm, not evidently sagittate; lvs gen rather firm and prickly **Navarretia**

      6b Calyx lobes = or nearly so; our spp. either per, or with the fls not
in heads, or with the heads essentially bractless, or with the lvs neither deeply pinnatifid nor entire

         8a Lvs sessile, palmatifid into linear segms

            9a Ann; lvs soft, mostly or all opp; seeds often becoming mucila-
ginous when moistened **Linanthus**

            9b Per; seeds remaining unchanged when moistened

               10a Hyaline intervals of the calyx very narrow and inconspic-
uous, not reaching the base; lvs rather soft, opp; pls woody
only at the base **Linanthastrum**

               10b Hyaline intervals of the calyx well developed and conspic-
uous; lvs firm and prickly, our sp. with only the lower ones
opp; pls shrubby **Leptodactylon**

         8b Lvs diverse, but not at once sessile and palmatifid; herbs or sub-
shrubs with alt (or all basal), entire to dissected lvs

            11a Lf segms or teeth very conspicuously bristle-tipped; ann
Langloisia

            11b Lf segms or teeth sometimes shortly and inconspicuously spi-
nulose, but not evidently bristle-tipped; ann, bien, or per **Gilia**

Collomia Nutt.   Collomia

Fls in terminal headlike clusters (these reduced to a single fl in 1 sp.) which may be overtopped by
their subtending brs so that the fls appear to be lateral or in the forks of the brs; calyx tube chartaceous, of nearly uniform texture throughout, not ruptured by the developing caps; calyx lobes
greenish and gen more herbaceous; corolla tubular-funnelf to nearly salverf, gen bluish or pinkish
to white (salmon or yellowish in 1 sp.); stamens equally or unequally inserted below the sinuses of
the corolla, of = or unequal length; seeds 1–3 per locule, those of the ann spp. becoming mucilaginous when moistened; ann or per herbs, gen taprooted, slightly or not at all mephitic, with alt (or
the lower opp), entire to variously dissected lvs, but without well defined lfts; fl spring-summer.
(Gr *kolla*, glue, in reference to the seeds when moist).

1a Pls per, with ± sprawling sts; talus slopes, often at high elev; alpine c.; 3
vars. 1 **C. debilis** (S. Wats.) Greene

   a1 Principal lvs entire or few-toothed to rather deeply 3–5-lobed, the lobes
entire; corolla gen 15–25 mm

      b1 Lvs relatively narrow and elongate, all entire or merely with a few
sharp, small teeth, ± strongly acute, or the lower ones more obtuse;
low elev, n Fork of Salmon R in Lemhi Co, Ida, n to Missoula, Mont
var. **camporum** Pays.

      b2 Lvs gen shorter and broader, varying from all entire to often some or
many of them ± deeply 3–5-lobed or cleft, the bl (when unlobed)
rounded to acutish; high elev, e of Cas, and in Cas of Wn var. **debilis**

a2 Principal lvs deeply and sometimes irregularly 3–7-cleft, with some or all of the principal segms again cleft; corolla gen 12–15 mm; high elev; OM, and Cas from Wn to n Cal *(C. l.)*   var. **larsenii** (Gray) Brand
1b Pls ann, often single-std, gen ± erect, seldom at high elev
  2a Lower lvs ± pinnatifid or subbipinnatifid, the middle and upper progressively reduced and less cleft, those subtending the fl clusters ± elliptic and entire or merely toothed; locules 2–3-seeded; corolla 8–17 mm, the lobes 3–4 mm; pls to 4 dm, in for openings and on loose streambanks at lower elev W Cas and in n Ida; varied-lf c.
                                                    2 C. **heterophylla** Hook.
  2b Lower lvs, like the others, entire, gen linear to lanceolate or lance-ovate; locules 1-seeded
    3a Corolla gen salmon or yellowish, gen 20–30 mm, the lobes 5–10 mm; pls 2–10 dm; dry, open places at middle and lower elev; large-fld c.
                                                   3 C. **grandiflora** Dougl.
    3b Corolla pink or lavender to blue or white, gen 4–15 mm, the lobes 1–3 mm
      4a Stamens unequally inserted; pls simple or brd, the compact, lfy-bracteate fl-clusters gen ∝ -fld and always appearing terminal to the main st or to the short or elongate brs on which they are borne; lvs often > 5 mm wide
        5a Calyx lobes merely acute, gen 3–4 mm in fr; pls gen 1–6 dm, widespread, chiefly E Cas; narrow-lf c.   4 C. **linearis** Nutt.
        5b Calyx lobes elongate, aristate-attenuate, the longer ones 5–11 mm in fr; pls to 1 dm, local in Gilliam Co, Ore; bristle-fld c.
                                      5 C. **macrocalyx** Leiberg
      4b Stamens equally inserted; pls tending to br just beneath the individual fls or fl clusters, so that many or all of the fls in well developed individuals appear to be borne in the forks or scattered along the brs; lvs to 5 mm wide; pls 1.5–2 dm, in dry, open places E Cas
        6a Fls, or most of them, borne in small clusters (2–5 together); corolla 8–14 mm; calyx lobes 2–3 mm at anthesis, sometimes to 4 mm in fr; yellow-staining c.   6 C. **tinctoria** Kell.
        6b Fls borne singly, or a few of them in pairs; corolla 4–6 mm; calyx lobes ca 1 mm at anthesis, to 2 mm in fr; diffuse c.
                                          7 C. **tenella** Gray

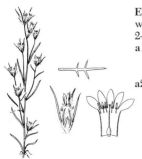

Eriastrum Woot. & Standl.   Eriastrum

Fls in lfy-bracteate, often rather small heads which are ± strongly tomentose with very fine, tangled hairs; calyx with broad, hyaline intervals between the more herbaceous costae, not much accrescent, gen ruptured by the developing caps; calyx lobes entire, scarcely to strongly unequal; corolla funnelf to salverf, the lobes often relatively longer than in *Navarretia;* filaments short or elongate, equally or subequally inserted at or shortly below the sinuses; anthers gen narrow and strongly sagittate, gen > 0.5 mm; ovary 3-carpellary, with 1–several ovules per locule; seeds gen becoming mucilaginous when moistened; ann with alt, remotely pinnatifid to linear and entire lvs which (including the segms) are spinulose-tipped, but less rigid than in *Navarretia.* (Gr *erion,* wool, and *aster,* star, referring to the pls and fls respectively).

E. sparsiflorum (Eastw.) Mason. Few-fld e. Slender, to 3 dm, freely brd when well developed; lvs 1–3 cm; corolla blue to white; anthers 0.8–1.3 mm; ovules 2–4 per locule; dry, often sandy places at lower elev E Cas; 2 vars.
a1 Lvs all entire, or occ some of them with a single pair of lateral lobes; corolla (boiled) gen 7–9 mm; c Ore to w Nev and s Cal, e on SRP to Twin Falls, Ida *(E. filifolium,* misapplied)   var. **sparsiflorum**
a2 Lvs, or many of them, with 1–3 pairs of lateral lobes; corolla (boiled) (8) 9–12 (13) mm; Grant Co, Wn, s through e Ore (where rare) to s Cal, e throughout SRP (and in Salmon R valley from near Challis to near Salmon) and to w Utah *(E. w.)*   var. **wilcoxii** (A. Nels.) Cronq.

Gilia R. & P.   Gilia

Fls borne in various sorts of basically cymose (determinate) infls; calyx slightly or scarcely accrescent, with prominent scarious or hyaline intervals between the more herbaceous costae or segms, gen ruptured by the developing caps; corolla camp to more often funnelf or salverf, variously colored; filaments gen = and equally (or occ unequally) inserted in the corolla tube or at the sinuses; seeds 1–∝ per locule (or some locules empty), in most spp. becoming mucilaginous when moistened; taprooted ann to per herbs with alt or all basal (or the lower opp), entire or toothed to more often pinnatifid or ternate (or sometimes palmatifid) lvs, without well defined lfts. (For the Spanish botanist Felipe Luis Gil). (*Ipomopsis*).

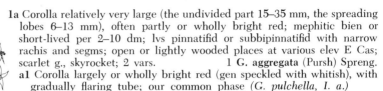

1a Corolla relatively very large (the undivided part 15–35 mm, the spreading lobes 6–13 mm), often partly or wholly bright red; mephitic bien or short-lived per 2–10 dm; lvs pinnatifid or subbipinnatifid with narrow rachis and segms; open or lightly wooded places at various elev E Cas; scarlet g., skyrocket; 2 vars.   1 **G. aggregata** (Pursh) Spreng.
  a1 Corolla largely or wholly bright red (gen speckled with whitish), with gradually flaring tube; our common phase (*G. pulchella, I. a.*)
                                           var. **aggregata**
  a2 Corolla white or yellowish, often speckled or tinged with red, the tube notably slender, scarcely flaring; s RM var., occ extending into our range in forms gen transitional to var. *aggregata* (*G. attenuata*)
                                     var. **attenuata** Gray
1b Corolla smaller (the undivided part not > 10 mm), never bright red
  2a Fls borne in 1 or more very dense, capitate or spicate-capitate clusters
    3a Pls distinctly per; corolla white; lvs linear and entire to more often trifid to pinnatifid or palmatifid with narrow rachis and segms, these occ again cleft; dry, open places from lowl to high elev
      4a Filaments < or barely = anthers, these 0.6–1.0 mm (dry); style distinctly < corolla tube, this 5–9 mm, well > calyx; corolla lobes 2.5–5 mm; sts simple and herbaceous throughout; Utah and c Ida to Gr Pls; spicate g.; ours is var. *orchidacea* (Brand) Cronq. (*G. cephaloidea*)   2 G. spicata Nutt.
      4b Filaments (1.5) 2–4 times as long as the anthers, these 0.4–0.6 mm (dry); style ca = or > corolla tube, this 3–4 mm, shortly or scarcely > calyx; corolla lobes ca 2 mm; sts basally br and somewhat woody; widespread E Cas; ballhead g., many-fld g.; we have 4 vars.
                                     3 **G. congesta** Hook.
        a1 Lvs green and glab or nearly so, gen many of them crowded on short, sterile shoots near the base; sts gen simple above the br base; middle to high elev in mts
          b1 Lvs all, or nearly all, entire; sw Mont to n and w Wyo
                                  var. **crebrifolia** (Nutt.) Gray
          b2 Lvs mostly trifid, often some of them pinnatifid or entire; c Ida to ne Ore (Wallowa Mts), and in ne Nev   var. **viridis** Cronq.
        a2 Lvs sparsely to copiously arachnoid-woolly, rarely glab; sterile shoots seldom much developed; sts often br above; foothills, valleys, and plains
          c1 Many of the lvs palmatifid or subpalmatifid; GB and Sierra Nevada var., disjunct to St. Anthony dunes and the Salmon R valley in c Ida   var. **palmifrons** (Brand) Cronq.
          c2 Lvs trifid to pinnatisect, or some of them entire, rarely any of them subpalmatifid; s margin of our range, and s and e
                                      var. **congesta**
    3b Pls ann; corolla bluish, 6–10 mm, the lobes and tube ca =; lvs with slender rachis and segms, the lower gen bipinnatifid; open places at lower elev W Cas and e to n Ida; bluefield g., globe g.; ours is var. *capitata*   4 **G. capitata** Sims
  2b Fls scattered in a ± open infl; ann, all E Cas
    5a Lvs mostly crowded at or near the base of the st, the principal ones ± strongly toothed to pinnatifid
      6a Lower part of the pl with some loose, cottony, ± deciduous pubescence; seeds well > 1 mm, becoming mucilaginous when moistened; lvs often ± strongly pinnatifid; corolla 3–11 mm; dry, open, often sandy places in the plains and foothills; GB and s RM sp, extending n to e Wn and the Salmon R valley in Ida; shy g., sinuate g.; we have 2 vars.   5 **G. sinuata** Dougl.
        a1 Corolla gen (6) 7–11 mm; seeds mostly 4–8 per locule; widespread (? *G. inconspicua*, an older name of doubtful application)
                                     var. **sinuata**

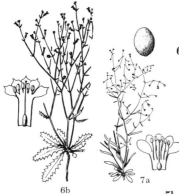

a2 Corolla gen 3–6 mm; seeds mostly 2–4 per locule; pls ave more slender and smaller-lvd, but often with relatively better developed cauline lvs; Salmon R valley of c Ida, s through Lost R region to SRP and s *(G. t.)*   var. **tweedyi** (Rydb.) Cronq.
6b Pl without cottony pubescence; seeds < 1 mm, remaining unchanged when moistened; lvs toothed or lobed, but scarcely pinnatifid; corolla 2–7 mm; dry, open places at low elev, esp in sandy soil; GB and s RM sp., n to e Wn and the Salmon R valley of Ida; 2 vars.
                                                          6 G. leptomeria Gray
  a1 Corolla gen 3–7 mm; seeds gen 8–12 per locule; pls up to 25 cm; peds ascending, gen 0.5–3 mm; widespread; GB g.
                                                                var. **leptomeria**
  a2 Corolla gen 2–3.5 mm; seeds gen 4–8 per locule; pls to ca 15 cm; peds more spreading, often > 3 mm; often in more alkaline soil than var. *l.*, barely entering our range *(G. m.)*
                                                      var. **micromeria** (Gray) Cronq.
5b Lvs well distributed along the st (sometimes also forming a basal rosette), mostly or all entire, gen linear or nearly so
  7a Corolla 1.5–3 mm; seeds gen solitary in each locule, 1–1.5 mm; filaments attached well below the sinuses; sts copiously stipitate-glandular, diffusely br, to 2 dm, sometimes with a rosette of basal lvs; open or lightly wooded slopes, from foothills to mid elev in mts; c Ida and sw Mont to ne Ore and s; delicate g.
                                                          7 G. tenerrima Gray
  7b Corolla (3.5) 4–8 (9) mm; lvs wholly cauline, not forming a basal rosette
    8a Filaments inserted just beneath the sinuses; seeds gen 2–5 per locule, 1–1.5 mm; sts copiously stipitate-glandular, openly br, to 3 (4) dm; meadows and open or lightly wooded slopes from foothills to mid elev in mts; c Wn to n and w Ida, s to Cal; smooth-lvd g.
                                                          8 G. capillaris Kell.
    8b Filaments inserted well below the sinuses; seeds 1 per locule, 2.5–3.5 mm; sts finely stipitate-glandular or glandular-scab, 1–6 dm, gen with a ± well defined central axis and ± elongate, ascending, fl but otherwise simple brs; dry, sandy places at low elev, often with sagebr; c and e Wn and adj BC to ne Ore, and e onto SRP; small-fld g.   **9 G. minutiflora** Benth.

Gymnosteris Greene   Gymnosteris

Fls sessile in a compact, terminal cluster which may be reduced to a single fl; calyx almost wholly scarious except for the gen more herbaceous teeth, a little inflated at anthesis, gen not ruptured by the growing caps; corolla white to yellow or pink, salverf, with slender tube, very short throat, and ± spreading lobes; filaments very short, inserted at the corolla throat; seeds several or rather ∞ in each locule, becoming mucilaginous when moistened; diminutive, essentially glab ann with persistent, connate cotyledons, short, naked st, and a whorl of entire, often basally connate lvs forming an invol just beneath the fl cluster. (Gr *gymnos*, naked, and *sterizo*, to support, referring to the lfless sts).

1a Corolla showy, the tube 6–10 mm, conspicuously exserted from the calyx, the lobes gen 3–6 mm; bracteal lvs linear or lanceolate; dry, open, often sandy places at low elev; Baker Co, Ore, to Blaine Co, Ida (n margin of SRP), and Ormsby Co, Nev; fl Apr–May; large-fld g.
                                                      1 G. nudicaulis (H. & A.) Greene
1b Corolla inconspicuous, the tube 2.5–5 mm, shortly or not at all exserted from the calyx, the lobes 0.7–1.5 mm; bracteal lvs narrowly lanceolate to ovate; open slopes, flats, and drier meadows, foothills and adj plains to mid or high elev in mts; c Ore to nw Wyo, s to Cal and Colo; fl May–July; small-fld g.   **2 G. parvula** Heller

Langloisia Greene   Langloisia

Fls sessile or nearly so in small, terminal, lfy-bracteate clusters; calyx with =, bristle-tipped segms, the hyaline intervals narrow and readily rupturing with the growth of the caps; corolla white or gen anthocyanic, tubular-funnelf, often ± bilabiate, filaments inserted at or shortly below the sinuses, = or unequal, < or slightly > corolla lobes; caps triangular, with gen 2–10 seeds per locule, these becoming mucilaginous when moistened; low, compact, much br ann with alt, deeply cleft or pinnatifid lvs; lf segms prominently bristle-tipped, the lower segms sometimes reduced to a bristle or cluster of bristles. (For Father Auguste Barthelemy Langlois, 1832–1900, amateur botanist of Louisiana).

L. setosissima (T. & G.) Greene. Bristly l. Pls forming low mats, to 1 (2) dm wide; lvs 1–2.5 cm, subcuneate, with 1 or 2 distal pairs of lateral lobes well developed, the proximal lobes often reduced to bristles; corolla gen light blue-lavender, reg or nearly so, 13–20 mm, the limb 7–12 mm wide; des, often in sandy soil, reaching the s margin of our range E Cas, and also in the Salmon R valley in c Ida.

Leptodactylon H. & A.   Leptodactylon

Fls sessile, solitary in the axils (terminal internodes sometimes foreshortened) or at the ends of short, lfy brs; calyx scarcely accrescent, the firm costae joined by prominent hyaline intervals, and each shortly exserted into a spinulose tip; corolla white to pink, lavender, or salmon, with long, slender tube and well developed, basally narrowed lobes; filaments ± equally inserted above the middle of the corolla tube, slightly, if at all, > the rather small anthers; seeds several or ∝ in each of the (2) 3 (4) locules, remaining unchanged when moistened; taprooted shrubs or subshrubs with small, alt or opp, sessile, rigid, palmatifid or sometimes pinnatifid lvs, these with narrow, spinulose-tipped segms. (Gr *leptos*, thin, and *daktylos*, finger, referring to the narrow lf segms).

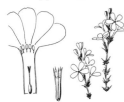

L. pungens (Torr.) Nutt. Prickly phlox. Sweetly aromatic shrub 1–6 dm; lvs ∝, alt or the lower opp, gen 5–12 mm, palmatifid into 3–7 segms and with axillary fascicles; corolla 15–25 mm, whitish (seldom yellowish) and gen washed or marked with lavender outside, nocturnal, the lobes 6–10 mm, loosely contorted-closed during the day; widespread cordilleran sp. of dry places from des to mid elev in drier mts E Cas.

Linanthastrum Ewan   Linanthastrum

Fls borne in lfy-bracteate, small, cymose, terminal infls which may be compact and headlike; calyx only slightly or scarcely accrescent, rather firm and subherbaceous throughout, the hyaline intervals narrow and inconspicuous, not reaching the base, or nearly wanting; corolla white or creamy, salverf; stamens ca= and equally inserted at the base of the short throat; locules with gen 2–4 ovules, but often maturing only 1 seed; valves of the caps persistent after dehiscence; seeds remaining unchanged when moistened; taprooted, per, woody-based herbs with sessile, opp, palmatifid lvs, the segms linear and elongate. (*Linanthus*, plus L *aster*, somewhat like or inferior to).

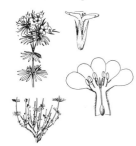

L. nuttallii (Gray) Ewan. Nuttall's l. Sweetly aromatic; sts ∝, slender, simple or sparingly br, to 3 dm; principal lvs 5–9-cleft into spinulose-tipped segms to 2 cm, firmer than in *Linanthus*, softer than in *Leptodactylon*, each gen with a smaller axillary fascicle; fls subsessile in dense infls; corolla tube woolly-puberulent, ca = the 6–9 mm calyx, the limb ca 1 cm wide; rocky slopes at middle and upper elev; Cas-Sierra region, e to c Ida and adj Mont, w Wyo, and s RM (*Gilia n., Leptodactylon n., Linanthus n.*).

Linanthus Benth.   Linanthus

Fls in open, gen dichasial infls, or in compact, headlike, terminal clusters; calyx only slightly or scarcely accrescent, gen with ± evident hyaline intervals between the more herbaceous costae, or the intervals occ ± reduced and the calyx almost wholly herbaceous; corolla camp to salverf; stamens gen = or nearly so and ca equally inserted in the throat or tube; valves of the caps persistent after dehiscence; seeds 1–several per locule, sometimes becoming mucilaginous when moistened; slender, taprooted ann to 3 dm with opp (or the uppermost alt), sessile, palmatifid lvs with gen linear and elongate segms, or the uppermost lvs entire. (Gr *linon*, flax, and *anthos*, fl).

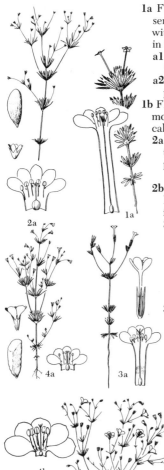

1a Fls clustered, subsessile; corolla 1–3 cm, salverf, with very narrow, exserted tube and abruptly spreading limb, the lobes 2–4 mm; calyx 5–10 mm,
with reduced intercostal membranes; open places at lower elev W Cas, and
in CRG; bicolored l.; 2 vars.   1 **L. bicolor** (Nutt.) Greene
  a1 Coastal ecotype with relatively small, often scarcely bicolored fls, the
tube 8–14 mm   var. **minimus** (Mason) Cronq.
  a2 Inl (and PS) ecotype, with larger, bicolored fls (throat yellow, lobes deep
pink or purplish to white), the tube 13–30 mm   var. **bicolor**
1b Fls evidently ped in an open infl; corolla to ca 1 cm, gen (except in no 3)
more funnelf or camp, with the tube scarcely or not at all exserted from the
calyx; calyx 2–5 mm; Cas and e
  2a Seeds solitary in each locule; corolla 1.5–2.5 mm, < 1.5 times as long as
the calyx, white (rarely pale blue?), glab, as also the filaments; open
places, foothills to mid elev in mts; Cas, e occ to c Ida; Harkness' l.
2 **L. harknessii** (Curran) Greene
  2b Seeds 2–8 per locule; corolla 2.5–10 mm, ca 1.5 times as long as the calyx, or longer, and gen either with an internal ring of hairs or with the
filaments hairy at the base
    3a Corolla lobes only ca half as long as the tube, which is gen exserted
from the calyx; filaments only 1–2 times as long as the anthers; ring of
hairs in the corolla tube borne well below the level of insertion of the
glab filaments; peds stipitate-glandular; corolla white to anthocyanic,
sometimes bicolored; dry, open places at lower elev; Cal, n rarely
along e side of Cas to Klickitat Co, Wn; fl Apr–June; Baker's l.
3 **L. bakeri** Mason
    3b Corolla lobes about = or more often distinctly > the tube, which is
scarcely or not at all exserted from the calyx; filaments several–∞
times as long as the anthers; peds not glandular; fl May–July
      4a Corolla 2.5–5 (6) mm, gen < 2 (often ca 1.5) times as long as the
calyx, white to light blue or lavender, the tube hairy within about at
the level of insertion of the glab or subglab filaments, or occ
subglab; widespread E Cas, gen in foothills, mts, and high plains;
northern l.   4 **L. septentrionalis** Mason
      4b Corolla (5) 6–10 mm, gen 2–3.5 times as long as the calyx, gen pale
blue with a yellow throat; filaments basally hairy, the corolla tube
gen glab or nearly so; E Cas from Wn and adj n Ida s to Cal, gen in
dry lowl, occ up to the ponderosa pine zone; thread-st l. (*L. liniflorus*
ssp. *p.*)   5 **L. pharnaceoides** (Benth.) Greene

Microsteris Greene   Microsteris

Fls terminal and paired (one subsessile, the other evidently ped), or partly solitary, but seemingly
scattered or loosely aggregated because of the br of the sts; calyx somewhat accrescent, with
prominent hyaline intervals, ruptured by the developing caps; corolla salverf, with slender tube
and short lobes; filaments short, very unequally inserted; valves of the caps disart completely on
dehiscence; seeds 1 per locule, becoming mucilaginous when moistened; ann with entire lvs, the
lower opp, the upper alt. (Gr *mikros*, small, and *sterizo*, to support, referring to the small size of
the pl?).

M. gracilis (Hook.) Greene. Pink m. Subsimple to much br, to 3 dm; lvs linear
or lance-linear to elliptic, or the lower obovate, to 5 cm × 8 mm; corolla with
white or yellowish tube and pink to lavender limb; dry to moderately moist,
open places, gen in foothills and lowl, the var. *gracilis* sometimes in meadows
or along streams; widespread in w N Am, and in temp S Am (*Collomia g.;
Gilia g.; Phlox g.*); 2 vars.
  a1 Primary st (up to the 1st fl) (5) 8–25 cm; pls br only above the
middle, or sometimes essentially throughout, but in any case higher
than broad; corolla (8) 9–15 mm, the lobes (1.5) 2–4 mm; gen W Cas,
also e, in less arid regions or habitats than var. *h.*, to Mont and Wyo (M.
micrantha)   var. **gracilis**
  a2 Primary st 1–5 (8) cm; pls at maturity much br, gen at least as broad as
high), corolla 5–8 (10) mm, the lobes 1–2 (2.5) mm; arid regions E Cas
(*M. humilis; M. micrantha*, misapplied)   var. **humilior** (Hook.) Cronq.

Navarretia R. & P.   Navarretia

Fls borne in dense, lfy-bracteate, glandular or glab to coarsely villous heads terminating the st and brs; calyx with broad hyaline intervals between the more herbaceous costae, not much accrescent, sometimes ruptured by the developing caps; calyx lobes gen ± unequal, gen spine-tipped (like the bracts and their segms), the larger ones often trifurcate; corolla salverf or funnelf with slender tube and gen short lobes; filaments elongate or very short, equally, or a little unequally, inserted, gen not far below the corolla sinuses; anthers small, elliptic, gen < 0.5 mm; stigmas 2 or 3, sometimes almost wholly connate; caps (1) 2–3-locular, the partitions sometimes imperfect, dehiscent or indehiscent; seeds 1–∝ per locule, becoming mucilaginous when moistened; prickly ann with alt (or the lower opp), gen pinnatifid or irreg bipinnatifid lvs (or the lvs all entire in depauperate pls), the segms narrow and spine-tipped. Fl measurements in the key are from boiled fls, in which the veins of the corolla are readily visible at 10× against a black background. (For Fr. Ferdinand Navarrete, Spanish physician).

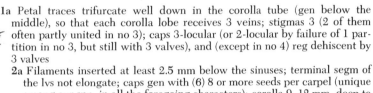

1a Petal traces trifurcate well down in the corolla tube (gen below the middle), so that each corolla lobe receives 3 veins; stigmas 3 (2 of them often partly united in no 3); caps 3-locular (or 2-locular by failure of 1 partition in no 3, but still with 3 valves), and (except in no 4) reg dehiscent by 3 valves

  2a Filaments inserted at least 2.5 mm below the sinuses; terminal segm of the lvs not elongate; caps gen with (6) 8 or more seeds per carpel (unique among our spp. in all the foregoing characters); corolla 9–12 mm, deep to pale blue; pls erect, to 4 dm, simple or moderately br, mephitic; open places W Cas; skunkweed *(N. pungens)*   1 **N. squarrosa** (Esch.) H. & A.

  2b Filaments inserted within 1.5 mm of the sinuses; terminal segm of lvs notably elongate; caps gen with 1–4 seeds per carpel; pls to 2.5 dm, often freely and widely br

    3a Corolla yellow (unique among our spp.), 5–8 mm; bracts distinctly pinnatifid; dry meadows, etc., often in shallow clay soil; foothills and high plains to mid elev in mts; c Wn to c Ida and w Wyo, s to Cal and Ariz; Brewer's n., yellow-fl n.   2 **N. breweri** (Gray) Greene

    3b Corolla pinkish or bluish to white

      4a Corolla 3.5–5 mm; filaments < anthers, up to ca 0.3 mm; brs tending to arise just beneath or even within the heads (unique among our spp. in the latter 2 characters); bracts ± palmatifid; widespread in dry, open places from foothills to mid elev in mts E Cas; fl June–Aug; mt n.   3 **N. divaricata** (Torr.) Greene

      4b Corolla 8–11 mm; filaments much > anthers, gen 1.5–2 mm; brs gen not closely associated with the heads; bracts distinctly pinnatifid; dry, open places at low elev; Klickitat Co, Wn, s to Cal; fl May–June; marigold n., northern n.   4 **N. tagetina** Greene

1b Petal traces simple up to the level of the sinuses, or beyond, so that each corolla lobe receives only 1 vein; stigmas 2, often almost wholly united, and caps ± 2-locular (rarely stigmas 3 and caps perhaps 3-locular); caps indehiscent, or tardily and irreg dehiscent by disintegration of the lower part of the lateral walls

  5a Corolla lobes only slightly or scarcely (to 1.5 times) longer than wide, with br midvein; ovules (2) 3–6 per carpel; lvs and bracts firm; corolla 4–11 mm; widespread in w N Am in moist to moderately dry sites up to mid elev in mts; fl June–Aug; needle-lf n.; 2 vars.

                                    5 **N. intertexta** (Benth.) Hook.

    a1 Corolla 7–11 mm, gen > calyx; lateral brs from midvein of corolla lobes gen again br; filaments 1.5–3.5 mm; infl conspicuously hairy; pls relatively robust, often 1–2.5 dm; W Cas, and in ne Ore, se Wn, and adj Ida   var. **intertexta**

    a2 Corolla 4–7 mm, = or more often < calyx; lateral brs from midvein of corolla lobes gen simple; filaments 0.5–2 mm; infl gen less strongly hairy; pls gen less robust, seldom > 1 dm; E Cas
                      var. **propinqua** (Suksd.) Brand

  5b Corolla lobes narrow, ca twice as long as wide, with unbr midvein; ovules 1 (2) per carpel; lvs and bracts relatively soft; corolla 4–6 mm; moist, open places at lower elev E Cas, from Klickitat Co, Wn, s; fl May–July; least n.   6 **N. minima** Nutt.

Phlox L.   Phlox; Wild Sweet-william

Fls borne in terminal cymes which are often reduced to a single fl; calyx scarcely accrescent, with prominent scarious or hyaline intervals between the more herbaceous costae, ruptured by the developing caps; corolla salverf, with slender tube and abruptly spreading lobes, white to pink, purple, or blue; filaments short and unequally placed in the corolla tube; anthers included, or often some of them partly exserted; seeds 1 (2–4) per locule, remaining unchanged when moistened; ann or gen (including all our spp.) per herbs, often ± shrubby at base, taprooted or less often fibrous-rooted, with opp (or the uppermost alt), entire, often narrow and needlelike, frequently basally connate lvs. (Gr, direct transliteration of word for flame, referring to the bright fls). Measurements of floral parts in the key are based on dry material, measurements of styles do not include the brs, and measurements of lvs do not include the connate part at the base.

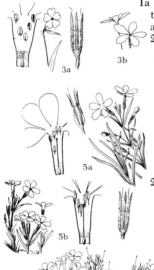

1a Pls ± erect, or loosely repent, with well developed internodes, and with the larger lvs often > 3.5 cm long or > 5 mm wide; fls gen slender-ped in a 3–∝ -fld terminal cyme (no 4 sometimes app the following group in habit)
   2a Style short, 0.5–2 mm, from barely > to more often distinctly < stigmas; pls 1.5–5 dm
      3a Corolla lobes nearly always notched at the tip; intercostal membranes of the calyx flat or nearly so; pls gen ± hairy or glandular, esp upwards, seldom wholly glab; lvs linear to broadly lanceolate, to 7 × 1 cm, some of them gen > 2 mm wide; widespread in sagebr and ponderosa pine areas E Cas in our range, s to Cal; showy p.

              1 **P. speciosa** Pursh
      3b Corolla lobes entire, not notched; intercostal membranes evidently carinate toward the base; pls wholly glab (except within the calyx); lvs linear, 2–7 cm × 0.5–2 mm; dry, open slopes and cliffs, SRC and adj parts of Salmon R canyon; SR p. 2 **P. colubrina** Wherry & Const.
   2b Style elongate, 6–15 mm, several times as long as the stigmas
      4a Lvs linear or nearly so, not > 5 mm wide; pls 0.5–4 dm
         5a Intercostal membranes of the calyx strongly and permanently carinate toward the base; herbage glab to glandular or hairy; dry, often rocky places from lowl to mid or occ high elev in mts; widespread E Cas in w US; long-lf p. (*P. linearifolia, P. viridis*)

              3 **P. longifolia** Nutt.
         5b Intercostal membranes of the calyx flat or nearly so; herbage glandular or hairy at least in the infl; open, gravelly or rocky places at mid elev in mts; local in se Wn, ne Ore, and adj Ida; sticky p. (*P. mollis*, a loosely woolly nonglandular form) 4 **P. viscida** E. Nels.
      4b Lvs lanceolate to ovate, elliptic, or nearly obovate, at least the larger ones (7) 10 mm wide or more
         6a Pls erect, gen 5–10 dm; sts solitary from the upturned end of a slender rhizome; larger lvs 5–9 cm, up to 3 cm wide; moist meadows and streambanks, Clearwater Co, Ida; Clearwater p.

            5 **P. idahonis** Wherry
         6b Pls repent or decumbent, to 3 dm, from an eventual taproot, often rhizomatous as well; larger lvs 1–3 cm × 7–18 mm; wooded slopes at mid elev in mts; Klamath region n, W Cas, to Linn Co, Ore; periwinkle p., woodland p. 6 **P. adsurgens** Torr.
1b Pls compact and tending to form mats, occ looser and up to 1 dm or even 1.5 dm; lvs short and crowded, rarely > 3 cm (to 3.5 cm in no 7), never > 5 mm wide; fls gen solitary (3), short-ped or sessile at the ends of the sts, scarcely forming distinct infls, though often largely covering the surface of the mat
   7a Intercostal membranes of the calyx gen ± carinate
      8a Keel of the intercostal membrane conspicuous, strongly bulged; style 5–10 mm; lvs 10–35 × 0.5–1.5 mm; corolla lobes 7–12 mm; dry, open, sometimes alkaline places, often with sagebr; SRP, barely entering our range; fl Apr–May; prickly-lvd p. 7 **P. aculeata** A. Nels.
      8b Keel of the intercostal membrane inconspicuous, low, linear; style 2–5 mm; lvs 8–15 (20) × 0.5–1.5 mm; corolla lobes 5–8 mm; foothills to mid or rather high elev in mts, gen avoiding the open plains; GB sp., n to c and ne Ore and c Ida; fl June–July; des p.

           8 **P. austromontana** Coville
   7b Intercostal membranes of the calyx essentially flat

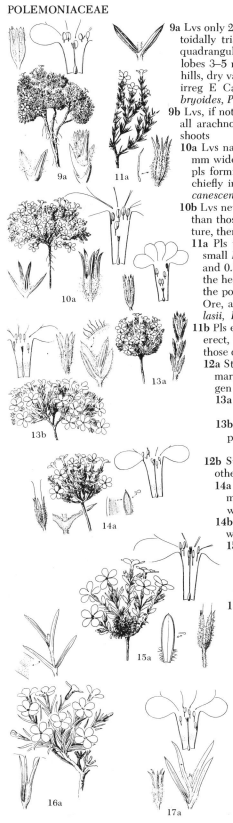

9a Lvs only 2–4 (5) mm, evidently arachnoid-woolly, narrowly to subdeltoidally triangular, closely crowded and appressed, tending to form quadrangular shoots; internodes woolly; corolla tube 5–10 mm, the lobes 3–5 mm; style 2–5 mm; dry, sometimes alkaline places in foothills, dry valleys, and plains, occ to mid elev in mts, often with sagebr; irreg E Cas, esp sw Mont and adj Ida; fl May–July; moss p. (*P. bryoides*, *P. lanata*)   **9 P. muscoides** Nutt.

9b Lvs, if not > 5 mm, either narrower and essentially linear, or not at all arachnoid-woolly, or both, never forming compact quadrangular shoots

 10a Lvs narrowly linear, gen (2.5) 4–10 (13) mm long and ca 0.5 (1) mm wide near the middle, firm and pungent, very often arachnoid; pls forming dense cushions; style 2–5 (6) mm; widespread E Cas, chiefly in foothills, valleys, and plains; fl Apr–June; Hood's p. (*P. canescens*)   **10 P. hoodii** Rich.

 10b Lvs never arachnoid, and gen longer, or wider, or distinctly softer than those of no 10, but if similar to those of no 10 in size and texture, then the pls looser and suberect

  11a Pls tending to be loosely erect, 5–15 cm, often resembling a small *Leptodactylon* in habit; lvs narrow, firm, gen 5–13 mm long and 0.5 (1) mm wide near the middle; calyx and peds (often also the herbage) gen glandular; style 3–7 (8) mm; chiefly in and near the ponderosa pine zone, from nw Mont to e and (occ) c Wn, ne Ore, and s BC; fl gen Apr–May; tufted p., clumped p. (*P. douglasii*, *P. rigida*)   **11 P. caespitosa** Nutt.

  11b Pls either more compact and mat-forming, or, if looser and suberect, then with the lvs distinctly longer, or wider, or softer than those of no 11; pubescence and styles diverse

   12a Styles 1–5 mm; lvs fairly firm, rarely as much as 1.5 cm, the margins somewhat thickened, but not strongly whitish; calyx gen glandular-hairy; moderate to high elev in mts; fl June–Aug

    13a Style 1–2 mm; Cas from c Wn to n Ore; Henderson's p.

     12 P. hendersonii (E. Nels.) Cronq.

    13b Style gen 2–5 mm; RM and GB sp., wholly E Cas; cushion p. (*P. caespitosa*, misapplied)

     13 P. pulvinata (Wherry) Cronq.

   12b Styles gen 5–12 mm, or occ only 4 mm in pls differing in other respects from the foregoing group

    14a Lvs 2–5 (7) mm, prominently white-margined; style 5–8 mm; moderate to high elev in mts of Mont and e Ida; white-margined p.   **14 P. albomarginata** M. E. Jones

    14b Lvs (at least the larger ones) (7) 10–25 mm, with or without white margins

     15a Lvs (at least the larger ones) gen 2–5 mm wide near the middle, (7) 10–25 mm, tending to be white-margined; style 6–12 mm; high plains, intermont valleys, and foothills, wholly e Cont Div; alyssum-lvd p.

      15 P. alyssifolia Greene

     15b Lvs (0.5) 1–2.5 mm wide near the middle, scarcely white-margined; style 4–8 (10) mm

      16a Lvs very finely and sparsely scaberulous, otherwise essentially glab, the better developed ones gen (10) 12–30 × 1–2 mm; pls never evidently glandular; mts from sw Mont and adj Ida s; many-fld p.

       16 P. multiflora A. Nels.

      16b Lvs not at all scaberulous, but often otherwise hairy or glandular, and gen ± ciliate-margined near the base

       17a Lvs either succulent, or evidently hairy (and often glandular) or both, the larger ones gen 10–25 mm; foothills, intermont valleys, and high plains, both sides Cont Div in Mont, e Ida, and Wyo; Kelsey's p.; we have 2 vars.   **17 P. kelseyi** Britt.

        a1 Pls ± succulent, occurring in alkaline meadows

         var. **kelseyi**

        a2 Pls more rigid, growing on open slopes (*P. m.*)

         var. **missoulensis** (Wherry) Cronq.

17b Lvs neither succulent nor hairy (except for the basally gen arachnoid-ciliate margins), often < 1 cm,
  gen not at all glandular; wholly w Cont Div, at mid
  to high elev in mts; spreading p.; ours is var. *longistylis* (Wherry) Peck   **18 P. diffusa** Benth.

Polemonium L.   Jacob's-ladder (taller spp.); Sky-pilot (dwarf alp spp.); Polemonium

Fls borne in diverse sorts of basically cymose infls; calyx essentially herbaceous, ± accrescent and
becoming chartaceous; corolla tubular-funnelf or subsalverf to nearly rotate, gen blue or white,
less often purple, yellow, or flesh-colored to salmon; stamens ca equally inserted; seeds 1–10 per
locule, sometimes becoming mucilaginous when moistened; per (1 sp. ann) herbs with alt, pinnately compound or very deeply pinnatifid lvs, gen with well defined lfts, these either entire or so
deeply 2–5-cleft as to appear verticillate; pls gen ± glandular and often strongly mephitic. (Said
to be named for Polemon, Gr philosopher, or from Gr *polemos*, strife).

1a Ann to 3 dm; fls terminal and solitary, but soon appearing lf-opposed because of the sympodial development of the st; corolla gen white, inconspicuous, 2–5 mm, < or merely = calyx; open plains and foothills, often with
  sagebr; widespread E Cas; fl Mar–May; littlebells p., ann p. *(Polemoniella
  m.)*   **1 P. micranthum** Benth.
1b Per; fls in terminal infls, more showy, the corolla much > calyx; fl
  May–Aug
  2a Corolla funnelf or tubular-funnelf to subsalverf, longer than wide, the
    lobes < tube; lfts (or their segms) 1.5–6 × 1–3.5 mm; dwarf (rarely to 4
    dm), alp and subalp, petrophilous, very strongly glandular and mephitic
    pls with ∝ sts from a taproot and much br (sometimes elongate) caudex;
    sky-pilot
    3a Lfts opp or offset, undivided; corolla gen 12–15 mm; Cas-OM, Wn,
      and adj BC; elegant p.   **2 P. elegans** Greene
    3b Lfts (or most of them) so deeply 2–5-cleft as to appear verticillate;
      corolla (13) 17–25 (30) mm; E Cas, and Coast Ranges of s BC; skunk
      p., sticky p.   **3 P. viscosum** Nutt.
  2b Corolla camp or broader, ca as wide as long, or wider, the lobes ca =
    tube, or > tube; pls of various habit and habitat, gen less strongly gland-
    ular and mephitic; Jacob's-ladder
    4a Fls large, the calyx 7.5–14 mm at anthesis, the corolla (dry) (15)
      18–28 mm, salmon or flesh-colored to yellow, white, or purple, only
      rarely blue; thickets, woodl, and for openings W Cas, up to mid elev in
      mts; great p., salmon p. *(P. amoenum)*   **4 P. carneum** Gray
    4b Fls smaller, the calyx up to 8 mm at anthesis, the corolla up to 17 mm;
      corolla gen blue (often with a yellow eye) or white
      5a Lfts narrowly linear; fls white or creamy; pls erect, 3–8 dm, with
        sts clustered on a taproot; moist bottoml in e Wn; Wn p.
                                                       5 P. pectinatum Greene
      5b Lfts broader; fls gen blue
        6a Sts solitary from the upturned end of a gen rather short and
          simple horizontal rhizome, erect, (1.5) 4–10 dm; very wet places
          at mid elev in mts; widespread cordilleran sp., chiefly E Cas;
          western p. *(P. caeruleum* spp. *o.)*   **6 P. occidentale** Greene
        6b Sts ± clustered from a br (sometimes elongate and rhizome-like)
          caudex which gen surmounts a taproot; pls lax, to 3 (5) dm, in
          moist to dry, often rocky places, from mid to high elev in mts;
          widespread cordilleran sp.; skunk-lvd p., showy p.; we have 2
          vars.   **7 P. pulcherrimum** Hook.
          a1 Relatively small and compact pls, gen of high elev, often above
            timberl, reg taprooted, seldom 2 (3) dm; lfts seldom as much as
            10 × 5 mm, gen all essentially distinct; calyx lobes gen < or
            ca = tube   var. **pulcherrimum**
          a2 Larger, laxer, and more robust pls of mid elev in mts, often 3
            dm and sometimes 5 dm, the taproot poorly or scarcely devel-
            oped; lfts fewer and larger, the larger ones seldom < 10 × 5
            mm, sometimes to 35 × 15 mm, the 3 terminal ones ±
            confluent; calyx lobes gen > tube, sometimes twice as long;
            Cas-Sierran region, e to c Ida and rarely w Mont *(P. colum-
            bianum)*   var. **calycinum** (Eastw.) Brand

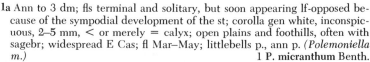

HYDROPHYLLACEAE   Waterleaf Family

Fls gamopet, ⚥, reg or nearly so, gen 5-merous as to calyx, corolla, and androecium; calyx cleft to the middle or more commonly to the base or nearly so; corolla imbricate (rarely contorted) in bud; filaments alt with corolla lobes, attached near the base or well up in the corolla tube, and very often flanked by a pair of small scales; ovary superior, gen 2-carpellary, with distinct or partly (rarely wholly) united styles and gen capitate stigmas; placentae 2, gen parietal, but often ± intruded, and sometimes meeting and joined, the ovary then 2-locular and the placentation axile; caps loculicidal (sometimes also septicidal), or irreg dehiscent; ovules few–∞ ; seeds 1–∞ ; embryo straight; endosperm present; herbs (rarely shrubs) with alt to sometimes partly or wholly opp, entire to cleft or compound lvs, the fls solitary or in variously modified (often helicoid) cymes.

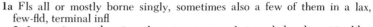

1a Fls all or mostly borne singly, sometimes also a few of them in a lax, few-fld, terminal infl
  2a Lvs entire or subentire; placentae ± strongly intruded and partitionlike
    3a Pls acaulescent, per, the fls on basal peduncles (habit unique in the family); mesophytes   **Hesperochiron**
    3b Pls caulescent, ours dichotomously br ann xerophytes with the fls sessile in the forks   **Nama**
  2b Lvs coarsely toothed to pinnatifid; placentae enlarged but not at all partitionlike, nearly filling the young ovary, and forming a lining for the mature caps
    4a Calyx provided with spreading or reflexed aurs at the sinuses; seeds with a partial or complete, persistent or deciduous outer covering (cucullus) over the seed coat (unique among our genera in both of these regards)   **Nemophila**
    4b Calyx exauriculate; cucullus wanting   **Ellisia**
1b Fls in definite infls, not solitary
  5a Style entire or nearly so; ± fibrous-rooted, mesophytic per with reniform-orbicular, chiefly basal lvs; placentae strongly intruded and partitionlike   **Romanzoffia**
  5b Style evidently (though sometimes shortly) lobed; habit otherwise
    6a Infl of subdichotomously br (sometimes capitate) cymes, without an evident central axis; placentae enlarged but not at all partitionlike, nearly filling the young ovary, and forming a lining for the mature caps; fibrous-rooted, per mesophytes with cleft lvs and exserted stamens   **Hydrophyllum**
    6b Infl of 1 or more sympodial, ± helicoid cymes which may be aggregated into a compound, often thyrsoid infl; placentae evidently intruded and partitionlike; taprooted ann or per, mesophytic or xerophytic, with entire to variously dissected lvs and included or exserted stamens   **Phacelia**

Ellisia L.   Ellisia

Fls solitary in or opp the axils, the st sometimes also terminating in a lax, few-fld cyme; calyx cleft to near the base, exappendiculate, accrescent and subrotate at maturity; corolla camp, white to partly or wholly lavender, ca = or slightly > calyx; filaments included, attached to the base of the obscurely appendiculate corolla tube; style < corolla, cleft up to half its length; ovules 2 on each of the 2 expanded, parietal placentae; caps 1-locular, dehiscent by 2 valves; seeds finely reticulate, without a cucullus; small, br ann with pinnatifid lvs, at least the lower ones opp. (For John Ellis, 1710–1776, English botanist).

E. nyctelea L. Nyctelea. Pls 0.5–4 dm, gen freely br, rather sparsely hairy, the petioles coarsely ciliate toward the base; bls to 6 × 5 cm, pinnatifid, with wing-margined rachis and gen 3–6 pairs of rather narrow, entire or few-toothed segms; peds gen < 1 cm at anthesis, sometimes to 5 cm in fr; calyx foliaceous-accrescent, the lobes becoming ca 1 cm; seeds gen 4, 2.5–3 mm; moist, shaded bottoms; e sp., barely entering our range (*Nyctelea americana*).

Hesperochiron Wats.   Hesperochiron

Fls borne singly on basal peduncles; calyx cleft to near the base; corolla camp to rotate, white,

often tinged or marked with lavender or purple, the tube gen evidently hairy within; filaments <
corolla, attached to the base of the tube, dilated at base, often unequal; style shortly 2-cleft; placentae shortly intruded; caps loculicidal, 1-celled, ∝ -seeded; small, scapose, per herbs from an
often caudexlike taproot, with petiolate, entire or subentire, oblanceolate to elliptic or ovate lvs,
the bl up to 7.5 × 2.5 cm. (Gr *hesperos*, evening, and *Chiron*, the centaur; significance uncertain). *(Capnorea).*

1a Corolla saucer-shaped or rotate, 1.5–3 cm wide, evidently wider than high
   even as pressed, reg, the lobes glab, (1.5) 2–4 times as long as the tube; fls
   gen 1–5 (8); lvs glab on the lower (or both) surfaces; pls ave less robust
   than no 2, the taproot more caudexlike, often premorse, and producing
   slender rhizomes from near the summit; meadows, swales, and moist, open
   slopes (seldom if ever in alkaline soil) from the valleys and plains to more
   gen in the foothills or well up in the mts; E Cas; dwarf h.
                                           1 H. pumilus (Griseb.) Porter
1b Corolla camp or funnelf, 1–2 (2.5) cm, as long as, or a little longer than,
   wide, often slightly irreg, the lobes often sparsely long-hairy within, gen
   0.8–1.5 times as long as the tube; fls ± ∝ , seldom as few as 5; lvs often
   short-hairy on both sides; taproot well developed, sometimes surmounted
   by a br caudex, without rhizomes; gen in ± alkaline meadows and flats in
   the plains, foothills, and intermont valleys; E Cas; Cal h.
                                      2 H. californicus (Benth.) Wats.

Hydrophyllum L.   Waterleaf

Fls borne in compact (often capitate), gen subdichotomously br cymes that lack a well developed
main axis; calyx divided to below the middle or near the base, gen exappendiculate; corolla camp
or a little broader, white to purple, gen 5–10 mm in our spp.; filaments exserted, attached to the
corolla tube at or near its base, each flanked by a pair of ciliate, linear corolla-appendages; style
exserted, rather shortly but distinctly 2-cleft; ovules 4; seeds 1–3; caps strictly 1-locular, dehiscent
by 2 valves; herbs, gen per and with fleshy-fibrous roots attached to a very short or ± well developed rhizome; lvs alt, best developed toward the base of the pl, variously cleft or pinnatifid. (Gr
hydor, water, and *phyllon*, leaf).

1a Lfts evidently toothed along the margins (teeth relatively few and large in
   no 3, which thus app or nearly matches no 4); rhizome gen ± evident,
   though often short; anthers 1.0–2.4 mm; pls (1.5) 2–8 dm, the peduncles
   gen 5–20 cm, seldom much, if at all, < lvs
2a Lf bls scarcely longer than wide, the 5 (7–9) segms or lfts gen approximate and pinnipalmately disposed, all but the lowermost pair gen confluent; peds gen 5–12 mm; moist woods at lower elev W Cas; Pac w.,
   slender-st w. *(H. viridulum)*   **1 H. tenuipes** Heller
2b Lf bls evidently longer than wide, the 7–15 lfts or segms pinnately disposed, the lower ones ± remote, only the upper ones confluent; peds gen
   2–7 mm
3a Lfts acuminate and sharply toothed, the teeth gen 4–8 to a side;
   thickets and moist, open places, from lowl to mid elev in mts; Fendler's w.; we have 2 vars.   **2 H. fendleri** (Gray) Heller
   a1 Pubescence relatively harsh and sparse, the lvs scab above and not
      white-hairy beneath, the sepals bristly-ciliate on the margins and
      glab to inconspicuously strigillose or strigillose-puberulent on the
      back, the st gen retrorsely hispid or hispid-hirsute; infl a little < to
      more often = or a little > lvs; lfts sometimes up to 15; Palouse-
      Blue Mt-c Ida region and s RM   var. **fendleri**
   a2 Pubescence relatively soft and copious, the lvs merely strigose
      above and softly white-hairy beneath, the sepals softly long-ciliate
      on the margins and villous-puberulent on the back, the st retrorsely
      strigose or hirsute-puberulent to hispid-hirsute; infl gen a little <
      lvs; lfts rarely > 11; OM-Cas-Sierra region and Palouse-Blue Mt-c
      Ida region *(H. a., H. congestum)*   var. **albifrons** (Heller) Macbr.
3b Lfts obtuse to abruptly acute, the teeth obtuse or acute, gen 2–4 to a
   side; s sp., barely reaching our range; Cal w., western w.; 2 vars.
                                               3 H. occidentale (Wats.) Gray
   a1 Pls relatively stiffly hairy; W Cas   var. **occidentale**

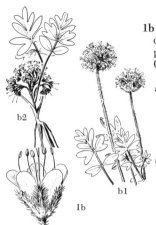

a2 Pls more softly hairy; GB, and disjunct in Elmore Co, Ida (*H. w.*)
var. **watsonii** Gray
1b Lfts entire or more gen some of them with 1 or 2 deep incisions (the segms directed forward) or distal teeth; rhizome scarcely evident; pls 1–4 dm, the peduncles, except in var. *thompsonii*, short, seldom > 5 cm; anthers 0.6–1.3 mm; widespread E Cas; ballhead w., wool breeches; 3 vars.
4 **H. capitatum** Dougl.
a1 St aerial as well as subterranean, the cymes somewhat elevated above the ground level; cymes gen capitate even in fr
b1 Infls = or > lvs, the peduncles elongate and erect, gen 5–20 cm; CRG and vic (*H. t.*)
var. **thompsonii** (Peck) Const.
b2 Infls well < lvs, the peduncles short, seldom any of them > 5 cm, often reflexed in fr; widespread in our range
var. **capitatum**
a2 St almost wholly subterranean, the cymes scarcely elevated above the ground level; cymes tending to be lax at least in fr, the peds often eventually recurved; GB var., reaching our range in c Ore var. **alpinum** Wats.

Nama L.   Nama

Fls solitary or in small cymes, ours solitary and terminal, immediately subtended by a lf, soon appearing to be in the forks of the brs, because of the development of a pair of axillary buds; corolla tubular to funnelf or narrowly camp, purple or bright pink to lavender or white; filaments included, unequally inserted, the adnate basal portion gen (but not in our spp.) with ± dilated, free margins or marginal appendages; style cleft only at the tip (ours) or more gen to the base; placentae ± intruded (and the ovary thus appearing 2-celled), but at least in our spp. apparently not joined; ovules and seeds ± ∝; caps loculicidal; herbs or subshrubs with alt, entire lvs, ours dwarf, prostrate, dichotomously br, hairy anns with oblanceolate lvs to 4 cm × 6 mm, occurring in dry sandy places in des and foothills, often with sagebr, sometimes extending into the ponderosa pine zone. (Gr *nama*, a spring, signficance doubtful). (*Conanthus*).

1a Corolla 8–14 mm, with bright pink to deep rose limb 5–10 mm wide, the tube gen yellow; style 2–6 mm; w GB sp., n to Yakima Co, Wn, and e on SRP to Gooding Co, Ida; purple n.
1 **N. aretioides** (H. & A.) Brand
1b Corolla 2.5–6 mm, the limb 1–3 mm wide, white to lavender; style 0.3–1.5 (2) mm; GB sp., n to Adams Co, Wn, and e on SRP to Gooding Co, Ida; matted n.; 2 vars.
2 **N. densum** Lemmon
a1 Pls densely spreading-hirsute throughout, with oblanceolate lvs seldom > 1.5 (2) cm; corolla gen 2.5–4 mm; style seldom > 1 mm; s var., barely reaching our range in c Ore (*C. d.*)
var. **densum**
a2 Pls less densely but more coarsely hairy, with longer, often relatively narrower (frequently linear-oblanceolate) lvs, the larger ones gen well > 1.5 (to 4) cm; corolla 3.5–6 mm; style to 1.5 (2) mm; range of the sp. (*N. p., C. p.*)
var. **parviflorum** (Greenm.) C. L. Hitchc.

Nemophila Nutt.   Nemophila

Fls solitary in the axils, or some of them in a loose, few-fld, racemelike, terminal cyme; calyx divided to near the base, and gen with a spreading or deflexed appendage (auricle) at each sinus; corolla camp to rotate, white to blue or purple; filaments < corolla, attached to the corolla tube near or well above the base, each gen flanked by a pair of small corolla-appendages; style shallowly to deeply cleft, < corolla; ovules 2–several on each of the 2 large parietal placentae; caps 1-locular, dehiscent by 2 valves; seeds with an obscure to evident, partial or complete covering (cucullus) external to the regular seed coat; delicate, taprooted ann with opp or less often alt, gen ± strongly pinnatifid or pinnatilobate lvs, the entire-margined cotyledons often enlarged and foliaceous, our spp. fl in spring and early summer. (Gr *nemos*, grove, and *philein*, to love, in reference to habitat).

1a Lvs all alt; corolla ca 2 mm, < calyx, lavender; seed gen solitary; style 0.5–1 mm, cleft only at the tip; sts beset with fine, short, retrorse prickles; foothills to mid elev in mts, widespread E Cas; GB n. 1 **N. breviflora** Gray
1b Lvs all opp or some of upper ones alt; corolla = or > calyx; seeds gen 2–22; sts variously glab or hairy, sometimes retrorse-bristly
2a Corolla 2–6 mm wide, gen camp; style 0.6–1.5 mm, cleft ca half way; widespread spp.

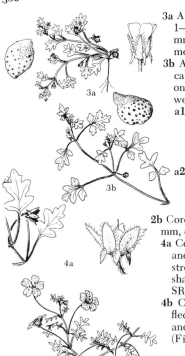

3a Aurs well developed, gen 1/2–3/4 (or fully) as long as the calyx lobes, 1–3 mm at maturity; lvs rather deeply pinnatifid, the rachis gen 1–4 mm wide; gen in moist, open places, esp in meadows and bottoml; meadow n., spreading n.   **2 N. pedunculata** Dougl.

3b Aurs relatively small and inconspicuous, seldom half as long as the calyx lobes, ca 1 mm or less at maturity; lvs often less deeply cleft; gen on wooded slopes or other ± shady places; small-fld n.; we have 2 well marked vars.   **3 N. parviflora** Dougl.

　　a1 Lvs relatively thin and deeply cleft, with prominently mucronate, often toothed lobes, the lower pair of sinuses in particular tending to app the midrib, and lower (or both) pairs of lobes tending to be narrowed at base; some of the upper lvs often alt; lower elev W Cas
                                                                   var. **parviflora**

　　a2 Lvs relatively firm, less deeply cleft, the lobes (or coarse teeth) only obscurely or scarcely mucronate and seldom evidently narrowed at base; lvs all opp; foothills and mid elev in mts E Cas (*N. a., N. inconspicua*)   var. **austiniae** (Eastw.) Brand

2b Corolla gen 7–25 mm wide, broadly camp to saucer-shaped; style 2–5 mm, cleft <half way, sometimes only at the tip

　4a Corolla gen 7–15 mm wide, gen blue-lavender toward the periphery and nearly white centrally, not flecked with dark purple; caps < the strongly accrescent fr calyx; seeds gen 2–4; rocky slopes, gen in partial shade, at low elev in SRC region, and up the Salmon R to Shoup, Ida; SRC n.   **4 N. kirtleyi** Hend.

　4b Corolla gen 15–25 mm wide, white or whitish and conspicuously flecked with blackish-purple; caps well > calyx; seeds gen 8–22; WV and Coast ranges of Ore, s to Cal; baby blue-eyes; ours is var. *atomaria* (Fish. & Mey.) Chandler   **5 N. menziesii** H. & A.

Phacelia Juss.   Phacelia

Fls borne in helicoid cymes, these often aggregated into a compound infl; calyx divided nearly to the base, with = or unequal, often accrescent lobes; corolla variously blue or purple to pink, white, or yellow, deciduous in most spp., tubular to rotate; filamens = or less often unequal, attached to the corolla tube near its base; style shortly (but evidently) to very deeply cleft; ovules 2–∞ on each of the 2 placentae, these slightly to strongly intruded; caps loculicidal; taprooted herbs with entire to pinnately dissected, chiefly or wholly alt lvs. (Gr *phakelos*, fascicle, referring to the congested infl). *(Miltitzia)*.

1a Fls gen 4-merous; corolla persistent, white or pale yellow, not > 2 mm; style < 0.5 mm; seeds gen 6–10; small, freely br ann with entire to shallowly lobed lf bls to 15 × 7 mm; alkaline flats and washes; w GB, n to Union Co, Ore; dwarf p. (*M. pusilla*)   **1 P. tetramera** J. T. Howell

1b Fls gen 5-merous; corolla 2.5 mm or more

　2a Corolla yellow (becoming whitish or even partly purplish in age), persistent and surrounding the fr; ovules and seeds 8–20; freely br low ann, often forming loose mats to 4 dm wide; lf bls to 2.5 × 1.5 cm; alkaline, gen clay flats and banks in des and foothills; irreg E Cas; yellow p.; we have 4 vars.   **2 P. lutea** (H. & A.) J. T. Howell

　　a1 Corolla 2.5–4 mm at anthesis, hispidulous outside, ca = calyx, slightly accrescent (to 5 mm) in fr, but then distinctly < calyx; style (0.5) 1–2.5 mm, gen hairy to near the middle; herbage shortly spreading-hairy, not glandular; longer filaments gen > sinuses; chiefly GB, but disjunct in Silver Bow Co, Mont (*P. s., M. s.*)
                                                                   var. **scopulina** (A. Nels.) Cronq.

　　a2 Corolla (4) 4.5–8 mm at anthesis, glab outside or somewhat hispidulous, gen > calyx, seldom definitely < calyx even in fr

　　　b1 Herbage glandular-villous; style 1.7–2.5 mm, hairy only at base; longer filaments often > sinuses; corolla ave a little < next 2 vars.; Ochoco-Blue Mt region of c Ore, notably in the alkaline, red clays derived from volcanic tuff   var. **purpurascens** J. T. Howell

　　　b2 Herbage gen not glandular, or glandular chiefly in the infl; style gen 2.5–3.8 mm

　　　　c1 Herbage wholly glab, or slightly glandular in the infl; longer filaments gen > sinuses; local in n Owyhee Co, Ida
                                                                   var. **calva** Cronq.

5a

6a

6b

c2 Herbage ± densely short-hairy, only occ also glandular; filaments < sinuses; se Ore and adj Nev, barely reaching our range *(M. l.)*   var. **lutea**

2b Corolla never wholly yellow (though sometimes ochroleucous, or with yellow tube and bluish limb), gen deciduous except in 2 per spp.

3a Lvs all entire, or in some spp. some of them with a large, entire terminal segm and 1 or several smaller, entire lobes or lfts below the middle

4a Pls bien or per; filaments strongly exserted; ovules 4; seeds 1–4; corolla 3–7 mm, whitish to bluish; the *P. magellanica* polyploid complex, with wholly confluent taxa too diverse inter se to be comfortably treated as a single sp.

5a Pls robust (5–20 dm), gen per and several-std, strongly spreading-bristly, with relatively large lf bls that gen bear 1 or more pairs of basal lfts or lobes; mesic, gen shady habitats at lower elev W Cas; woodl p., shade p.; our pls have been described as ssp. *oregonensis* Heckard   3 **P. nemoralis** Greene

5b Pls not presenting the foregoing combination of morphological characters, and gen either more e, or occurring in drier habitats, or both

6a Pls bien or short-lived per from a taproot, typically with a single erect st often well > 5 dm, or this surrounded by several lesser sts; some of the middle and lower lvs gen with 1–2 (4) pairs of lateral lobes or lfts at base of bl; herbage often ± griseous, but scarcely silvery, often markedly spreading-bristly; widespread cordilleran sp.; varileaf p., virgate p.; 2 vars.
                                                        4 **P. heterophylla** Pursh

a1 Small, weak ecotype, seldom 5 dm, with st(s) often curved at base, or ascending rather than erect, seldom virgate; mid elev in Cas-Sierra region *(P. nemoralis* var. *p., P. mutabilis)*
                                var. **pseudohispida** (Brand) Cronq.

a2 More robust series of ecotypes, to 12 dm; main st erect, often virgate; dry, open places in the foothills, valleys, and plains; widespread cordilleran var., commoner E Cas *(P. magellanica* var. *h., P. virgata)*   var. **heterophylla**

6b Pls per from a taproot which is gen surmounted by a br caudex, gen with several ± =, suberect to prostrate sts seldom > 5 dm; lvs all entire, or sometimes some of them with a pair of small lateral lobes near the base; herbage, esp in pls from lower altitudes, ± silvery with a short, dense pubescence, the longer bristles, if present, seldom very conspicuous; dry, open places at all elev, often in sand; widespread cordilleran sp.; whiteleaf p., silverleaf p.; we have 5 vars.   5 **P. hastata** Dougl.

a1 Pls of the foothills, valleys, and plains; sts ascending to suberect, gen > 15 cm; herbage gen silvery

b1 Lvs gen all entire, occ some of them with a pair of small lateral lobes; widespread E Cas, but not in the range of var. *hastata (P. l., P. magellanica* f. *l.)*
                                        var. **leucophylla** (Torr.) Cronq.

b2 Some of the lvs gen with a pair of small lateral lobes (this feature, sporadic and uncommon in var. *leucophylla,* is more stabilized here); calyx often very stiffly long-hispid; sands along CR e of CRG, n and s less often to Kittitas Co, Wn, and Deschutes and Grant cos, Ore   var. **hastata**

a2 Pls of mid to high elev in mts, and either with distinctly greener herbage than in the foregoing vars., or with the st either ± prostrate, or < 15 cm

c1 Fls gen whitish; sts often ± erect; lvs often with a pair of lateral lobes; herbage often more bristly than in var. *alpina*

d1 Dwarf, alp and subalp pls, gen not > 15 (20) cm, often prostrate; Cas-Sierra region from c Wn s *(P. leucophylla* var. *c., P. magellanica* f. *c.)*
                                        var. **compacta** (Brand) Cronq.

d2 Taller pls, gen 15–50 cm, gen of mid elev in mts, habitally much like var. *leucophylla,* but greener and more

bristly, and with the larger lvs often bearing a pair of lateral lobes; mts of Wn, ne Ore. n Ida, nw Mont, and s BC *(P. l.)*   var. **leptosepala** (Rydb.) Cronq.

    c2 Fls light lavender to dull purplish (unique in this sp. in this regard); sts gen prostrate or merely ascending at the tip; lvs gen all entire; mts of GB, n to Mont, c Ida, and ne Ore *(P. a., P. magellanica* f. *a., P. leucophylla* var. *a., P. heterophylla* var. *a.)*   var. **alpina** (Rydb.) Cronq.

4b Pls ann; filaments only shortly or not at all exserted; ovules (and gen seeds) > 4 except in *P. humilis*

   7a Corolla ± showy, the limb 4–18 mm wide; style (brs included) 3 mm or more; filaments ca = or shortly exserted from the corolla; herbage hairy, not glandular

     8a Ovules 4; seeds 1–4; corolla 4–7 mm wide; style cleft to the middle or below; lvs all entire, the bl gen 0.5–4 cm × 2–25 mm; pls 0.5–3 dm; open places at 1500–5500 feet E Cas in c Wn; se Ore to s Cal; low p.   **6 P. humilis** T. & G.

     8b Ovules gen 10–20 and seeds 6–15; corolla 8–18 mm wide; style entire to well beyond the middle; larger lvs often with 1–4 lateral segms below the middle; lf bls gen 1.5–11 cm × 1.5–12 mm (excluding segms); pls 1–5 dm; common in dry, open places at lower elev; cordilleran, gen E Cas; fl Apr–June; threadleaf p.   **7 P. linearis** (Pursh) Holz.

   7b Corolla inconspicuous, the limb 1.5–3 mm wide; style ca 1.5 mm or less; filaments included; lvs all entire; herbage ± glandular as well as hairy; ovules seldom < 10; uncommon spp.; fl July

     9a Principal lvs ± oblanceolate, 2–4 mm wide, the bl much > the short (to 4 mm) petiole; calyx segms conspicuously unequal in fr; mid elev in mts; c Ida, ne Ore (Wallowa Mts), and ne Nev; least p.   **8 P. minutissima** Hend.

     9b Principal lvs ± elliptic or oblong-ovate, 4–10 mm wide, the bl = or < the well developed (to 2 cm) petiole; calyx segms not very unequal; calcareous, dry slopes in foothills and at mid elev in mts; ec Ida, s irreg to Utah and Nev; hoary p.   **9 P. incana** Brand

3b Lvs coarsely toothed or pinnatilobate to more often pinnatifid or bipinnatifid, gen without a large, entire, terminal segm

  10a Filaments included; des or lowl anns, gen glandular, seldom (except no 12) as much as 3 dm, often without a well defined central axis

   11a Ovules and seeds ± ∝, the seeds cross-corrugated; style cleft up to 1/4 its length; calyx segms linear or linear-oblanceolate, not conspicuously accrescent; lvs subbipinnatifid, with narrow rachis and segms; mephitic pls of dry, sandy plains and hills, often with sagebr

     12a Corolla gen 6.5–12 mm (dry), the limb 5–10 mm wide; style (4) 5–8 mm; E Cas from c Ore to s Cal; two-color p.; ours is var. *leibergii* (Brand) Nels. & Macbr. *(P. l.)*   **10 P. bicolor** Torr.

     12b Corolla gen 4.5–6.5 mm, the limb 2–3.5 mm wide; style 2–3 mm; E Cas from c Wn to s Cal, and e through SRP to Wyo; sticky p. *(P. ivesiana* var. *g.)*   **11 P. glandulifera** Piper

   11b Ovules and seeds 4, or seeds fewer, these pitted-reticulate, not cross-corrugated; style cleft nearly half its length, or more; calyx segms broader and lvs less dissected than in the foregong group; corolla 3–4.5 mm, the limb 2–4 mm wide

     13a Pls rather thinly bristly-hispid throughout, as well as gen glandular; calyx segms unequally accrescent, the larger ones spatulate, not markedly firm and veiny; pls 1–6 dm; in thickets and under shelving rocks at lower elev; Wasco Co, Ore, to sw Ida and c Cal; fl June–July; Rattan's p.   **12 P. rattanii** Gray

     13b Pls merely glandular-hairy, not at all bristly-hispid; calyx segms lance-elliptic or ± oblong, conspicuously accrescent, firm and veiny in fr; pls 1–3 dm; heavy clay soil in des regions; n Malheur Co, Ore, s to sw Ida and ne Cal; fl May–June; hot-spring p.   **13 P. thermalis** Greene

  10b Filaments ca = or obviously > corolla; habit and habitat diverse

   14a Ovules and seeds 4 (or the seeds fewer); style cleft to well

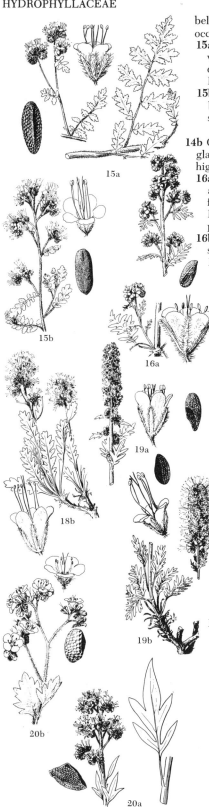

below the middle; strongly glandular-hairy and odoriferous pls, occurring in des and dry valleys and plains E Cas

    15a Per; sts prostrate or weakly ascending, 5–15 dm; corolla white to lavender or dull cream, the limb 6–12 mm wide; gen on basaltic talus or about ledges and cliffs; c Wn to Cal, sw Ida, and Ariz; br p.   **14 P. ramosissima** Dougl.

    15b Ann or bien; sts ascending or erect, 1–3.5 dm; corolla blue-violet, the limb 6–9 mm wide; loose banks and talus slopes; c Ida and sw Mont, s to Utah and Colo; glandular p.

                      15 P. glandulosa Nutt.

14b Ovules and seeds ± ∝ ; style not cleft beyond the middle; pls glandular or eglandular, occurring in more mesic habitats, or at higher elev, or both

    16a Ann or bien, single-std or with the central st surrounded by ascending lesser sts; corolla glab within, 6–9 × 8–12 mm; filaments only shortly or scarcely exserted; mid elev in mts; c Ida, n Wyo, and w Mont to L Superior and nw Can; Franklin's p.   **16 P. franklinii** (R. Br.) Gray

    16b Per, gen several-std (no 17 sometimes rather short-lived and single-std); corolla (except in no 21) gen ± hairy within

      17a Pls with basally disposed lvs, the basal and lower cauline ones well developed and gen persistent, the middle and upper ones gen ± reduced and less petiolate; helicoid cymes thyrsoidally or even capitately aggregated; infl not evidently glandular

        18a Lvs ± pinnatifid or bipinnatifid; pls scarcely glandular, sometimes silvery-hairy (almost always so when only 1–2.5 dm); corolla 4–6 mm wide and long, persistent and surrounding the caps

          19a Filaments barely exserted, < 1.5 times as long as the corolla; meadows, streambanks, and slopes in foothills and lower mts; c and n Ida; Ida p.

                  17 P. idahoensis Hend.

          19b Filaments long-exserted, (1.5) 2–3 times as long as the corolla; cordilleran; silky p.; 2 vars.

                  18 P. sericea (Grah.) Gray

            a1 Small, gen densely hairy (± sericeous) pls of high elev, 1–3 dm; lf segms gen relatively narrow and blunt; petioles seldom very strongly ciliate; widespread, but absent from Ore   var. **sericea**

            a2 Larger, less densely hairy (more strigose, often thinly so) pls of mid elev, (2) 3–6 (9) dm; lf segms ave broader, and sometimes more acute; petioles tending to be evidently ciliate; widespread, but absent from Wn (*P. c.*)   var. **ciliosa** Rydb.

        18b Lvs pinnatilobate or merely coarsely toothed; pls 0.5–2.5 dm, with green and gen glandular herbage; corolla 5–9 mm long and wide, deciduous; filaments well exserted; talus slopes and rock crevices at high elev; w Mont and adj Alta; Lyall's p.   **19 P. lyallii** (Gray) Rydb.

      17b Pls lfy-std, without tufts of basal lvs; helicoid cymes ± corymbosely aggregated; infl conspicuously glandular

        20a Corolla gen 5–9 mm wide, camp or open-camp, pale, gen greenish-white; filaments well exserted; pls erect, 5–20 dm, merely strigillose below the infl; seeds gen 12–16; meadows and open or semi-open slopes at mid elev in mts; Cas-Sierra region, and Blue Mt region of ne Ore and adj Ida; tall p.   **20 P. procera** Gray

        20b Corolla gen 10–20 mm wide, open-camp to subrotate, lavender to bluish or purplish; filaments ca = or slightly > corolla; pls sprawling to ± erect, seldom 1 m; herbage evidently spreading-hirsute at least in part; seeds gen 30–60; meadows and open, often loose slopes; coastal, c Ore and s, and disjunct in Wahkiakum Co, Wn; fl June–July; Bolander's p.   **21 P. bolanderi** Gray

Romanzoffia Cham.   Mistmaiden; Romanzoffia

Fls borne in loose or condensed, pedunculate, naked, racemelike, sympodial cymes; calyx divided nearly to the base; corolla camp, white, gen with a yellow eye, 6–11 mm long and wide; filaments ca =, attached to the corolla tube at or near its base; style simple, with a capitate, entire or obscurely lobed stigma; caps loculicidal, wholly or largely 2-celled; ovules ∝; low, per herbs with petiolate, reniform-orbicular, toothed or lobed, chiefly basal lvs, the cauline ones alt and mostly borne near the base; some of the lowest lvs often modified into conspicuous, brown-woolly, freely rooting tuberlike structures up to 1.5 × 1 cm. (Count Nikolai von Romanzoff, sponsor of Kotzebue's voyage to Cal).

1a Pls bearing well developed, brown-woolly, freely rooting "tubers" at base; petioles of foliage lvs somewhat enlarged and often bearing axillary bulbils, but not aggregated to form a bulbous base
   2a Pls condensed, 1 dm or less, the infl barely if at all > lvs. and the peds < 1 cm even in fr; herbage gen ± evidently glandular-villous; coastal bluffs, s VI and Wn to n Cal; Tracy's m.   1 **R. tracyi** Jeps.
   2b Pls laxer, gen 1–3 (4) dm, the infl well > lvs, and the peds (at least the lower ones) gen 1–4 cm in fr; herbage gen subglab, or finely glandular in the infl, seldom somewhat glandular-villous; wet cliffs and ledges; Coast Ranges and Cas of Ore and n Cal, not at high elev; California m. (*R. suksdorfii*, misapplied)   2 **R. californica** Greene
1b Pls with the petioles of the foliage lvs strongly dilated below and overlapping to form a bulbous base, sometimes also producing axillary, nonrooting bulbils in late season, but without the distinctive "tubers" of nos 1 and 2; pls otherwise much as in no 2, but ave a bit smaller, (0.5) 1–2 (3) dm; wet cliffs and ledges in the mts, sometimes above timberl, and descending to floor of CRG; n cordilleran sp, s to n Ore and rarely n Cal; Sitka m. (*R. suksdorfii*)   3 **R. sitchensis** Bong.

BORAGINACEAE   Borage Family

Fls gamopet, ⚥, gen 5-merous as to calyx, corolla, and androecium; sepals ± united or essentially distinct; corolla gen reg, gen bearing ± evident and often hairy appendages (the fornices) opp the lobes at the summit of the tube; stamens epipetalous, alt with the corolla lobes, included or sometimes exserted; ovary superior, basically 2-carpellary, each carpel 2-ovulate and with a 2ndary partition; fr of 4 nutlets which are typically attached individually to the short or elongate gynobase; style simple or sometimes cleft into 2 or even 4 segms (rarely wanting), typically attached directly to the gynobase and arising between the essentially distinct lobes of the ovary, less commonly attached to the summit of the entire or merely 4-lobed ovary which then separates only tardily into individual nutlets; seeds nearly or quite without endosperm; gen herbs (tropical members often woody), often rough-hairy, with exstip, alt or sometimes partly (rarely wholly) opp, gen entire lvs, with diverse sorts of basically cymose (determinate) infls, these most commonly sympodial, helicoid cymes which tend to elongate and straighten with maturity.

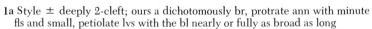

1a Style ± deeply 2-cleft; ours a dichotomously br, protrate ann with minute fls and small, petiolate lvs with the bl nearly or fully as broad as long
                                      Coldenia
1b Style entire or minutely lobed at the tip; pls not at once prostrate and with the lf bls ca as broad as long
   2a Ovary merely shallowly lobed, with the style wanting (ours) or borne on its summit, the stigma expanded and often as broad as the ovary; ours wholly glab, succulent pls with the white or faintly bluish fls crowded in naked, secund false spikes   **Heliotropium**
   2b Ovary deeply 4-parted, the style borne on the gynobase and arising between the essentially distinct lobes of the ovary; stigmas small; pls not both wholly glab and with the fls crowded in naked, secund, false spikes
      3a Base of the nutlet produced into a prominent, thickened rim which fits closely to the broad, low gynobase and surrounds the stipelike basal attachment, which latter fits into a distinct pit in the gynobase (body of the nutlet sometimes incurved so that the rim and attachment appear basilateral); ours intro, weedy spp. with blue or occ ochroleucous fls

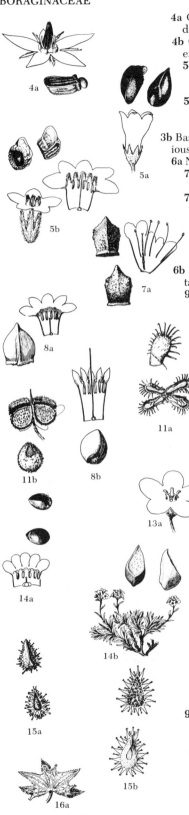

4a Corolla rotate; stamens exserted; anthers evidently appendiculate dorsally, connivent around the style   **Borago**
4b Corolla tubular-camp to funnelf or salverf, the stamens included, exappendiculate, not connivent
  5a Corolla tubular-camp, the well defined throat much > the short, erect or apically spreading lobes; style shortly exserted; basal rim of nutlets evidently toothed   **Symphytum**
  5b Corolla funnelf or salverf, the throat often poorly defined, not much if at all > the well developed, ± spreading lobes; style included; basal rim of nutlets ± entire   **Anchusa**
3b Base of the nutlet not produced into a thickened rim; attachment various, but rarely app that of the foregoing group
6a Nutlets broadly attached at the base to the broad, low gyno-base
  7a Corolla evidently irreg (the upper side distinctly the longer), gen blue; filaments (or some of them) strongly exserted   **Echium**
  7b Corolla reg, white or bluish-white to yellowish, yellow, or greenish; filaments inconspicuous, scarcely or not at all exserted
    8a Corolla lobes rounded and ± spreading; style in our spp. included   **Lithospermum**
    8b Corolla lobes acute or acuminate, erect; style conspicuously exserted   **Onosmodium**
6b Nutlets apically or medially to basilaterally or nearly basally attached, but the attachment, if nearly basal, always small
  9a Nutlets either evidently armed with glochidiate or uncinate bristles, or provided with a continuous, entire or cleft dorsomarginal ridge or wing that is complete around the base or across the back of the nutlet, or both armed and ridged
    10a Nutlets ± widely spreading in fr, the rather large scar apical or apicolateral, not extending below the middle of the fr, never elongate and narrow
      11a Nutlets armed with uncinate bristles; small, slender, white-fld ann   **Pectocarya**
      11b Nutlets armed with glochidiate prickles; coarse bien or per; fls gen blue or red   **Cynoglossum**
    10b Nutlets erect or incurved to somewhat spreading; scar distinctly otherwise
      12a Nutlets essentially unarmed, the dorsomarginal ridge entire or merely toothed or lacerate
        13a Scar medial; fornices conspicuous, exserted, outcurved; fr peds reflexed   **Dasynotus**
        13b Scar basilateral; fornices not outcurved; fr peds erect to spreading
          14a Dorsomarginal ridge complete around base of nutlet; pls sometimes caespitose, but not pulvinate   **Myosotis**
          14b Dorsomarginal ridge complete across back of nutlet well above the base, the part enclosed by the ridge (or wing) set at an angle to the basal part; pls pulvinate-caespitose   **Eritrichium**
      12b Nutlets armed with glochidiate prickles at least along the dorsomarginal ridge
        15a Nutlets narrowly attached to the elongate gynobase along the well developed median ventral keel, free near the base; peds erect or ascending in fr; "racemes" evidently bracteate; pls gen ann   **Lappula**
        15b Nutlets medially attached to the broad, low gynobase, the scar ± rounded, not elongate; peds recurved or deflexed in fr; "racemes" naked or nearly so; pls gen per (2 of our spp. ann or bien)   **Hackelia**
  9b Nutlets smooth or variously roughened, but without hooked or glochidiate prickles or bristles (except for some very small dorsal ones in spp. of *Plagiobothrys*), sometimes sharp-edged, esp distally, but not with the ridge complete across the back or around the base of the nutlet
    16a Fr calyx greatly enlarged and prominently veiny; nutlets obliquely compressed, with a small, ventromarginal scar above the middle; blue-fld, scrambling climbers with weak, retrorsely prickly-hispid sts   **Asperugo**

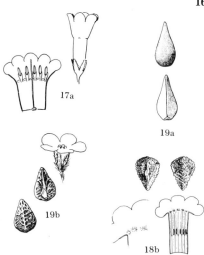

16b Fr calyx not greatly enlarged, seldom evidently veiny; nutlets otherwise, though sometimes obliquely compressed; fls various; habit otherwise
  17a Corolla blue or occ pink (white), tubular or tubular-funnelf to camp, never salverf; per, never pungently hairy **Mertensia**
  17b Corolla white to yellow or orange, often salverf; ann or per, often pungently hairy
    18a Corolla white, sometimes with a yellow eye, the throat ± closed by the fornices; cotyledons entire
      19a Nutlets with a closed or narrowly open ventral groove-scar running most of their length, this often expanded below into a depressed areola; pls ann to per, often pungently hairy **Cryptantha**
      19b Nutlets with a ventral keel extending to the middle or to near the base, the attachment elevated, caruncle-like, placed at the base of the keel, or rarely extending along the keel; ours ann, except *P. mollis,* and not pungently hairy **Plagiobothrys**
    18b Corolla yellow or orange, the fornices obsolete and the throat open except for no 3; cotyledons deeply 2-cleft, thus seemingly 4; ann, gen pungently hairy, with nutlets about as in *Plagiobothrys* **Amsinckia**

Amsinckia Lehm.   Fiddleneck; Tarweed; Fireweed; Amsinckia

Fls in sympodial, helicoid, gen naked false racemes or spikes which tend to elongate in age; calyx cleft essentially to the base, but some of the segms sometimes partly or wholly connate so that there may appear to be < 5 members; corolla yellow or orange, often marked with vermilion in the throat; corolla throat, except in no 3, open, the fornices obsolete; stamens included, the filaments short; nutlets with a well developed ventral keel extending from the tip to near or below the middle, often somewhat keeled dorsally as well, the scar small, placed at the end of the ventral keel, often elevated and caruncle-like; gynobase gen short-pyramidal; cotyledons deeply cleft, thus apparently 4; taprooted, bristly-hairy ann; lvs rather small and often narrow, entire or sometimes erose-dentate. (For William Amsinck, patron of the Hamburg botanical garden in 19th century).

1a Pls maritime; lvs tending to be erose-denticulate; 2 of the sepals gen partly united; corolla tube ca 10-nerved below the insertion of the stamens; seaside a. (*A. lycopsoides,* misapplied) 1 **A. spectabilis** Fisch. & Mey.
1b Pls not maritime; lvs entire; weedy spp.
  2a Sepals of many or all the fls unequal in width and reduced in no (gen 4, sometimes 2 or 3) by fusion, the broader one(s) often apically bidentate; corolla tube gen ca 20-nerved below the insertion of the stamens; chiefly E Cas; tessellate f. 2 **A. tessellata** Gray
  2b Sepals 5, essentially distinct, not very unequal in width; corolla tube ca 10-nerved below the insertion of the stamens
    3a Corolla throat obstructed by the well developed, hairy fornices; stamens inserted below the middle of the corolla tube (unique among our spp. in both of these respects); chiefly E Cas; tarweed f., bugloss f.
                                                3 **A. lycopsoides** Lehm.
    3b Corolla throat open, glab, the fornices scarcely developed; stamens inserted above the middle of the corolla tube
      4a St spreading-hispid and also evidently puberulent or strigose with shorter and softer, retrorse hairs; pubescence of the lvs tending to be ascending instead of widely spreading; corolla orange or orange-yellow, 5–8 mm, the tube not much, if at all, exserted, the limb gen 1.5–3 mm wide; chiefly E Cas; rigid f., harvest f.
                                                4 **A. retrorsa** Suksd.
      4b St spreading-hispid, nearly or quite without shorter and softer hairs below the infl; hairs of the lvs often widely spreading
        5a Corolla orange or orange-yellow, relatively large, 7–10 mm, the tube well exserted, the limb gen 4–6 mm wide; Cal, n to Klickitat Co, Wn, and perhaps n Ida, seldom collected in our range; ranchers f., fireweed f. 5 **A. intermedia** Fisch. & Mey.
        5b Corolla lighter yellow, smaller, 4–7 mm, the tube scarcely exserted, the limb only 1–3 mm wide; gen; Menzies' f., small-fld f. (*A. micrantha*) 6 **A. menziesii** (Lehm.) Nels. & Macbr.

Anchusa L.   Alkanet; Anchusa; Bugloss

Fls borne in terminal, sympodial, helicoid, bracteate false racemes which tend to elongate and straighten in age; peds persistently erect or ascending; calyx shallowly to deeply cleft; corolla funnelf or salverf, the throat often poorly defined, not much if at all > the well developed, ± spreading, apically rounded lobes; filaments inserted shortly below the level of the fornices, the anthers extending into the corolla throat; nutlets with a stipelike basal attachment which fits into a pit in the otherwise flattish gynobase, the attachment surrounded by a prominent, thickened rim (the basal margin of the nutlet) which fits close to the gynobase; ann or per, lfy herbs, often pungently hairy; ours blue-fld, taprooted per, spreading-hispid throughout, with oblanceolate, petiolate basal lvs and shorter-petiolate or sessile cauline lvs; intro from Mediterranean region. (Gr *anchousa*, ancient name for a related pl).

1a Corolla limb 6–11 mm wide; nutlets oblique, ca 2 mm high and 3–4 mm long, the tip directed inward; E Cas; common b.   **1 A. officinalis** L.
1b Corolla limb 12–20 mm wide; nutlets erect, 5–9 mm high and 3–5 mm thick; W Cas; Italian b. or an.   **2 A. azurea** Mill.

Asperugo L.   Madwort; Catchweed

Fls borne on short, stout, recurved peds in or near the axils of lvs or bracts, and in forks of brs; calyx 5-lobed to ca the middle, each lobe with a smaller tooth on each side near the base, the whole strongly accrescent, becoming compressed, firmly chartaceous, strongly reticulate-veiny, and shortly prickly-hispid with curved or hooked hairs; corolla small, blue, camp, with well developed fornices; anthers included; nutlets obliquely compressed, narrowly ovate, tessellate, attached to the elevated gynobase by a small scar just within the margin and above the middle; ann weeds with weak, climbing-scrambling, retrorsely prickly-hispid sts and opp or subopp to partly alt or partly whorled lvs that are subentire and often remote. (L *asper*, rough, referring to the pubescence).

A. procumbens L. St 3–12 dm; lvs thin, scab-hispid and irreg hispid-ciliate, the lower oblanceolate, petiolate, seldom to 10 × 2.5 cm, often deciduous, the others ± reduced, often broader in shape and subsessile; corolla 2–3 mm; fr calyx 1–2 cm wide; nutlets 2.5 mm, enveloped by the calyx; field weed, gen in fairly moist soil, intro from Eurasia; E Cas.

Borago L.   Borage

Fls borne in loose, terminal, modified sympodial cymes, these lfy-bracteate below, the elongate peds recurved in fr; sepals narrow, elongate, distinct or nearly so; corolla rotate, blue, with 5 elongate, acute lobes; fornices well developed; filaments inserted at the level of the fornices, each prolonged beyond the base of the anther into a prominent dorsal appendage; anthers elongate, conspicuous, connivent around the style; nutlets with a well developed, swollen, stipelike basal attachment which fits into a pit in the otherwise flattish gynobase, the attachment surrounded by a prominent, thickened rim (the basal margin of the nutlet) which fits close to the gynobase; ann or per, broad-lvd, coarsely hairy herbs. (L name, of unknown derivation).

B. officinalis L. Taprooted ann 2–6 dm; lower lvs petiolate, with broadly elliptic to ovate or merely oblanceolate bl 3–11 × 2–6 cm; cauline lvs progressively reduced but still ample, the upper often sessile and clasping; peds 1–4 cm; sepals densely bristly, 1–1.5 cm in fr; corolla 2 cm wide; anthers 5 mm, the linear appendages 3 mm; nutlets subcylindric, 4–5 mm, roughened; European sp., occ cult, estab as a casual weed W Cas.

Coldenia L.   Coldenia

Fls small, essentially sessile in few-fld, axillary clusters; calyx deeply (4) 5-cleft; corolla white (ours) or pink, (4) 5-lobed, often without fornices; stamens included; ovary entire or more often (incl our sp.) 4-lobed, separating into 4 (or fewer by abortion) nutlets at maturity; style terminal, not gynobasic (though the ovary in our sp. and some others is apically as well as laterally lobed), evidently 2-cleft, sometimes essentially to the base, each br with a capitate stigma; ann or per herbs or subshrubs. (For Dr. Cadwallader Colden, lieutenant-governor of NY, and correspondent of Linnaeus).

C. nuttallii Hook. Nuttall's c. Prostrate, dichotomously br, taprooted ann, forming rosettes to 3–4 dm wide, short-hairy throughout, and hispid-setose in the many small infls; lf bls elliptic to ovate or subrotund, strongly few-veined, 3–8 mm, ca = petioles; corolla minute, 1–2 mm wide; calyx 3–4 mm in fr; nutlets 1 mm, smooth; cotyledons strongly sagittate; sandy des; c Wn and s.

Cryptantha Lehm.   Cryptantha; White Forget-me-not

Fls gen borne in a series of sympodial, helicoid, naked or bracteate false spikes, these sometimes aggregated into a terminal thyrse, or rarely (no 8) the fls solitary in the axils; calyx cleft to the base or nearly so (except in no 8), ± accrescent; corolla white, the well developed fornices often yellow, the limb gen ± rotately spreading, often very small; filaments short, attached below the middle of the corolla tube; nutlets 4, or 1–3 by abortion, affixed to the somewhat elongate gynobase for much of their length, the scar narrow and gen appearing as an elongate, closed to narrowly open groove that is either forked at base or opened into a basal areola; taprooted, strigose to more often hirsute or partly hispid-setose herbs, ann or per, with gen narrow lvs. (Gr *kryptos*, hidden, and *anthos*, fl, referring to the cleistogamous fls of some spp.). *(Greeneocharis, Krynitzkia, Oreocarya, Piptocalyx)*. Lf length in the key includes the petiole.

1a Pls per or sometimes merely bien, relatively coarse, gen (except no 2) with a well developed tuft of basal lvs; spikes aggregated into a dense, terminal, irreg bracteate thyrse, often eventually elongating and distinct; corolla ± conspicuous, the limb gen 4–12 mm wide; calyx persistent *(Oreocarya)*
　2a Nutlets smooth; scar closed; pls per, 1.5–4 dm
　　3a Corolla tube > calyx at anthesis; corolla limb 5–9 mm wide; lvs all relatively slender and tapering gradually to an acutish tip, the lower ones 5–10 cm × 3–6 mm; dry plains of c Wn and adj Ore; gray c.
　　　　　　　　　　　　　　　　　　　　　1 **C. leucophaea** (Dougl.) Pays.
　　3b Corolla tube = or < calyx; corolla limb 8–11 mm wide; lower lvs more broadly rounded or obtuse, oblanceolate, gen broadly so, gen 4–10 cm × 6–12 mm; dry slopes and cliffs along Salmon R from Challis to Salmon, Ida; Salmon R c.
　　　　　　　　　　　　　　　　　　　2 **C. salmonensis** (Nels. & Macbr.) Pays.
　2b Nutlets ± roughened, at least on the back, the scar open or closed; pls bien or per, to 5 dm; corolla tube ca = or occ < calyx
　　4a Nutlets closely and unevenly rugose-reticulate on both sides; upper surface of basal lvs uniformly short-hairy, or with some poorly developed and inconspicuous longer setae, in any case not pustulate; corolla limb ca 8 mm wide; basal lvs 3–10 cm × 5–10 mm; dry, open hillsides in n Malheur Co, Ore, and adj Ida; Ida c.
　　　　　　　　　　　　　　　　　　　3 **C. propria** (Nels. & Macbr.) Pays.
　　4b Nutlets variously roughened, but scarcely as above; upper surfaces of lvs gen provided with well developed, gen pustulate-based setae in addition to the fine, short hairs
　　　5a Corolla relatively large and showy, the limb gen 8–12 mm wide; basal lvs relatively broad, gen broadly oblanceolate or spatulate, 2–8 cm × 4–15 mm, seldom conspicuously spreading-bristly; nutlets ovate or lance-ovate, ± evidently roughened on both sides; bien or per, 1–5 dm; dry lands of c Ore to s BC, e to Mont and Gr Pl; cockscomb c., northern c. *(C. bradburyana, C. sheldonii)*
　　　　　　　　　　　　　　　　　　　4 **C. celosioides** (Eastw.) Pays.
　　　5b Corolla smaller, the limb gen 4–8 mm wide (rarely to 10 mm in no 5); lvs various, often narrower and markedly spreading-bristly, esp in no 5; nutlets lanceolate or lance-ovate; per
　　　　6a Nutlets roughened ventrally as well as dorsally; scar closed or narrowly open; lvs to 6 cm × 8 mm; dry places at lower elev; SRP and adj Nev and Ore, n to Adams Co, Wn (rarely), and through ec Ida to w Mont and n; bristly c. *(C. macounii, C. spiculifera)*　　　　　　　　　　5 **C. interrupta** (Greene) Pays.
　　　　6b Nutlets smooth or nearly so ventrally, slightly or moderately roughened dorsally; pls gen of high elev in mts (no 7 sometimes descending to the foothills)

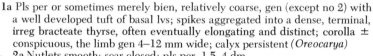

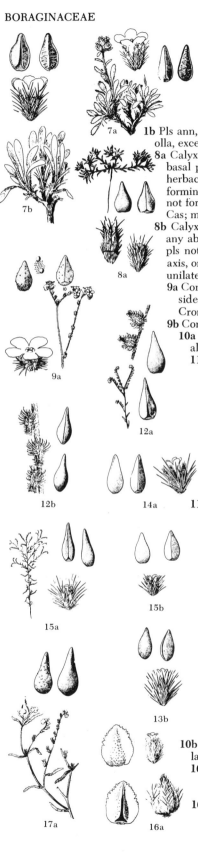

7a Pls dwarf, seldom 1.5 dm; lvs relatively short, the basal ones to 3.5 cm × 8 mm; scar gen closed; wholly E Cas in our range, and absent from Wn; Sierra c. *(C. hypsophila)*

6 **C. nubigena** (Greene) Pays.

7b Pls taller, gen 1–3 dm; lvs more elongate, the larger basal ones gen 4–7 cm × 5–10 mm; scar evidently open for most of its length; Wen Mts, from ponderosa pine belt upwards, often on serpentine; Thompson's c.   7 **C. thompsonii** Johnst.

1b Pls ann, relatively slender, without any conspicuous tuft of basal lvs; corolla, except in no 9, not > ca 2.5 mm wide

8a Calyx circumscissile a little below the middle, the persistent, cupulate basal portion scarious and obviously of different texture from the more herbaceous, deciduous portion; pls low, much br when well developed, forming cushions 1–6 cm broad; fls solitary in the crowded upper axils, not forming elongate spikes; dry, open, gen sandy places at lower elev E Cas; matted c.; ours is var. *c.*   8 **C. circumscissa** (H. & A.) Johnst.

8b Calyx divided essentially to the base, not circumscissile, and without any abrupt change of texture, in most spp. eventually deciduous intact; pls not cushion-forming, most spp. erect and with a ± evident central axis, or sometimes more bushy-br; fls borne in naked, eventually elongate, unilateral helicoid "spikes," these not closely aggregated *(Krynitzkia)*

9a Corolla relatively large, the limb 4–8 mm wide; dry, open slopes, both sides of Cas, e to w Ida; common c.; ours is var. *grandiflora* (Rydb.) Cronq. *(C. g., C. hendersonii)*   9 **C. intermedia** (Gray) Greene

9b Corolla small, the limb 0.5–2.5 mm wide; wholly E Cas

10a Nutlets all smooth, or finely and inconspicuously granular, not at all tuberculate or spiculate-papillate

11a Nutlets strictly solitary; hairs of the calyx tending to be uncinate or arcuate

12a Nutlets broadly truncate at base, the scar broadened below the middle into a definite, open areola; style reaching to the middle of the nutlet or gen beyond; dry, open places, Klickitat Co. Wn, s through e Ore to Cal; rare; beaked c. *(C. suksdorfii)*   10 **C. rostellata** Greene

12b Nutlets ± pointed or very narrowly truncate at base, the scar ± closed, not forming a definite areola; style scarcely reaching the middle of the nutlet; dry, open places, c and se Wn to w Ida and s Cal; common; weak-std c.

11 **C. flaccida** (Dougl.) Greene

11b Nutlets gen 4, in any case > 1; hairs of the calyx ± straight

13a Nutlets symmetrical, the scar median on the ventral face

14a Margins of nutlets prominent, sharply angled, esp above; scar opened at base to form a small areola; style nearly or fully = nutlets; dry, open places at lower elev, c Wn to w Mont and s, rare with us; Watson's c.

12 **C. watsonii** (Gray) Greene

14b Margins of nutlets rounded or obtuse, not prominent

15a Nutlets lanceolate, 0.5–0.7 mm wide; scar opening at base into an areola; style = or slightly > nutlets; sand dunes and very sandy soil; cordilleran sp., w to Franklin Co, Wn, and Umatilla Co, Ore, rare with us; Fendler's c.

13 **C. fendleri** (Gray) Greene

15b Nutlets ovate, 0.8–1.2 mm wide; scar closed; style somewhat < nutlets; common and widespread up to mid elev in mts, not in sand dunes; Torrey's c.

14 **C. torreyana** (Gray) Greene

13b Nutlets obliquely compressed, with a distinctly excentric scar near 1 margin; foothills to mid elev in mts E Cas; slender c.

15 **C. affinis** (Gray) Greene

10b Nutlets, or some of them, rough, with evident tubercles or spiculate papillae on the dorsal surface

16a Nutlets (except 1) with conspicuously winged margins; GB and s RM sp, n through e Ore to c Wn; winged c., wing-nut c.; ours is var. *p.*   16 **C. pterocarya** (Torr.) Greene

16b Nutlets not at all wing-margined

17a Nutlets distinctly heteromorphous, 1 nearly smooth and somewhat larger and more firmly attached than the other 3,

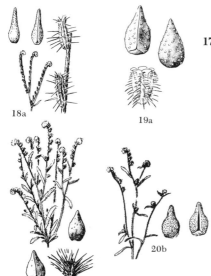

18a

19a

20b

20a

which are evidently tuberculate; dry, open plains, often in sandy soil; chiefly more e and s sp., reaching Helena, Mont, and Fremont Co, Ida; Kelsey's c.   17 C. **kelseyana** Greene

17b Nutlets all alike in size and texture when normally developed

18a Nutlets lanceolate, 0.5–0.7 mm wide, spiculate-papillate, esp distally; dry slopes and flats, often with sagebr, SRP and adj Ore and Nev, and in Salmon R valley of c Ida and apparently in Yakima Co, Wn; des c.   18 C. **scoparia** A. Nels.

18b Nutlets ovate, (0.7) 0.9–1.5 mm wide

19a Herbage closely strigose, essentially without spreading hairs; nutlets coarsely granular and with scattered low tubercles; with ponderosa pine; c Wn and possibly n Ida, s to Cal; pine woods c.   19 C **simulans** Greene

19b Herbage spreading-hirsute, at least in part

20a Nutlets with scattered low tubercles, gen also granular; dry, open places from lowl to mid elev in mts, widespread E Cas; obscure c.   20 C. **ambigua** (Gray) Greene

20b Nutlets densely spiculate-papillate; mid to rather high elev in mts, chiefly Sierran, disjunct in c Ida; prickly c.   21 C. **echinella** Greene

Cynoglossum L.   Hound's-tongue

Fls in sympodial, naked or subnaked false racemes, or in terminal, naked mixed pans, the peds ascending to more often spreading or even reflexed at maturity; calyx deeply cleft; corolla blue or violet to reddish, funnelf or salverf, with short, broad tube and ± spreading limb, the fornices well developed and often exserted; anthers included or borne at the corolla throat; nutlets ± widely spreading at maturity, attached to the broad, low gynobase by the apical end, the broad scar not extending below the middle of the ventral surface, the whole surface covered with short, stout, glochidiate prickles, or these sometimes restricted to the distal region of the free portion; entire-lvd, taprooted ann, bien, or more often per herbs, moderately to very robust. (Gr *kynos*, dog, and *glossa*, tongue, the lvs of some spp. very rough).

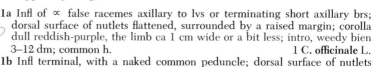

1a

1a

2a

3a

3b

1a Infl of ∝ false racemes axillary to lvs or terminating short axillary brs; dorsal surface of nutlets flattened, surrounded by a raised margin; corolla dull reddish-purple, the limb ca 1 cm wide or a bit less; intro, weedy bien 3–12 dm; common h.   1 C. **officinale** L.

1b Infl terminal, with a naked common peduncle; dorsal surface of nutlets broadly rounded, without a raised margin; native per, not weedy

2a Pls with all the lvs (or all but the reduced uppermost ones) petiolate; st glab or nearly so; corolla blue-violet, the limb gen 1–1.5 cm wide; woods at lower elev W Cas (and in CRG), s to Cal; Pac h.   2 C. **grande** Dougl.

2b Pls with only the lower lvs petiolate, the others sessile and often clasping; st evidently spreading-hairy; corolla limb < l cm wide

3a Sepals 1–2.5 mm in fl and fr; style 1–2 mm; nutlets 3.5–5 mm (unique among our spp. in all these respects); fls blue; woods; e Am sp., w to s BC; northern h.   3 C. **boreale** Fern.

3b Sepals 4–5 mm in fl, sometimes to l cm in fr; style 4–9 mm; nutlets 7–10 mm; fls dull red; with ponderosa pine; Cal n to Black Butte, Jefferson Co, Ore; western h.   4 C. **occidentale** Gray

Dasynotus Johnst.   Dasynotus

Fls borne in lax, few-fld false racemes that are basally lfy-bracteate but otherwise naked; peds elongate, ± deflexed in fr; calyx deeply 5-cleft; corolla white, subsalverf or broadly funnelf, with short tube and relatively large limb; fornices conspicuous, exserted, recurved, apically retuse; anthers borne near summit of corolla tube; nutlets somewhat spreading below, compressed, with an evident, erect, continuous, unarmed dorsomarginal ridge, short-hairy on the back, ventrally keeled toward the tip, the rather large scar medial or slightly supramedial; gynobase broadly pyramidal; erect per with ∝ sts and wholly cauline lvs. (Gr *dasys*, hispid, and *noton*, back, in reference to the nutlets).

D. daubenmirei Johnst. Pls 3–6 dm, from a stout but rather soft rhizome which may be shortened into a br caudex; lvs oblanceolate, sessile or nearly so, 9–17 × 1.5–3.5 cm; peds 1–3 cm (to 7 cm in fr); calyx 6–8 mm in fl, twice as long in fr; corolla tube 4–5 mm long and wide, the limb 2–2.5 cm wide; fornices 2–4 mm; nutlets 5–6 mm; forest openings at mid to rather high elev in mts of Idaho Co, Ida.

Echium L.   Viper's Bugloss

Fls borne in a series of sympodial, helicoid, bracteate cymes; calyx deeply cleft; corolla blue to purple or red (white), funnelf, irreg, the upper side evidently the longer, the lobes unequal; fornices wanting, the throat open; filaments slender, ± unequal, some or all of them strongly exserted; gynobase flat or nearly so; nutlets ± roughened, attached at the base, the large scar sometimes surrounded by a low rim; style exserted; herbs or subshrubs. (Gr *echion*, in turn from *echis*, viper).

E. vulgare L. Blueweed, common v.b., snake-fl., blue-thistle. Rough-hairy, taprooted bien 3–8 dm; basal lvs oblanceolate, 6–25 × 0.5–3 cm; cauline lvs progressively smaller; helicoid cymes ∝ and often short, aggregated into an elongate, often virgate infl; corolla bright blue (pink or white), 12–20 mm; 4 filaments long-exserted, the fifth scarcely so; roadside weed, intro from Europe.

Eritrichium Schrad.   False or Alpine Forget-me-not; Eritrichium

Fls borne in condensed, cymose clusters which may become elongate into sympodial, naked or ± lfy-bracteate false racemes or spikes terminating the often very short sts, the fr peds ± erect; calyx cleft essentially to the base; corolla blue, often with a yellow eye, salverf with short tube and abruptly spreading, evidently 5-lobed limb, the fornices well developed; filaments attached well down in the corolla tube, the anthers included; nutlets 1–4, basilaterally attached to the broad, low gynobase, provided with a prominent, ascending, entire to conspicuously toothed or lacerate dorsomarginal flange which is complete across the back of the nutlet well above the base, the flanged portion of the body tending to be set at an angle to the portion below the flange, more conspicuously so when the nutlets are only 1 or 2; dwarf, ± strongly pulvinate-caespitose per, often acaulescent, with small, ± hairy lvs densely crowded on the ∝ short shoots or toward the base of the more elongate fl sts. (Gr *erion*, wool, and *trichos*, hair, referring to the woolly pubescence of *E. nanum*).

1a Mature lvs densely and coarsely silvery-strigose, the surface gen hidden by the coarse, gen straight hairs, which do not form a pronounced apical tuft or fringe; corolla limb 5–9 mm wide; nutlets often hispidulous on the back; dry, open places, esp on limestone, from foothills to high elev in mts; e of Cont Div in Mont and n Wyo; fl May–July; Howard's a. f.

1a   1 **E. howardii** (Gray) Rydb.

1b Mature lvs loosely long-hairy, the surface readily visible between the hairs, which are often more ∝ distally and tend to form an apical tuft or fringe; corolla limb 4–8 mm wide; nutlets glab; rocky places at high elev; irreg circumboreal, s to s RM, in our range from Mont to ne Ore and in n Cas, Wn; ours is var. *elongatum* (Rydb.) Cronq. (*E. e.*); fl June–Aug; pale a. f.

                                                                 2 **E. nanum** (Vill.) Schrad.

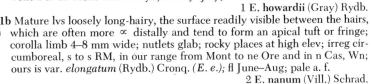

Hackelia Opiz   Stickseed; Wild Forget-me-not; Hackelia

Fls borne in sympodial, gen naked or only basally lfy-bracteate false racemes which tend to elongate in age, the peds deflexed in fr; calyx cleft nearly or quite to the base; corolla blue or white (sometimes ± ochroleucous), often with a yellow eye, salverf, with short tube and abruptly spreading, evidently 5-lobed limb, the fornices well developed; stamens included; nutlets > style, medially attached by a rather large scar to the broad, low gynobase, strongly glochidiate-prickly along the continuous dorsomarginal ridge and sometimes also across the back, the marginal prickles sometimes connate below into a cupulate border; taprooted per, or a few spp. bien or even ann. (For Joseph Hackel, 1783–1869, Czech botanist). (*Lappula*, in part).

1a Corolla limb only 1.5–3 mm wide; dorsal surface of nutlets 2–3 mm long; ann or bien, 1.5–10 dm; thickets and open woods; circumboreal, s to Wn, Ida, and s RM; nodding s.; ours is var. *americana* (Gray) Fern. & Johnst. (*L. a.*)   **1 H. deflexa** (Wahlenb.) Opiz

1b Corolla limb gen 4–20 mm wide; dorsal surface of nutlets gen 3–5 mm long; per, except no 3

  2a Corolla limb blue, sometimes withering pink

    3a Marginal prickles of nutlet united for 1/3–1/2 their length, forming a cupulate border; intramarginal prickles ± well developed; basal lvs seldom and cauline lvs never > 1 cm wide; pls 3–9 dm; corolla limb (5) 8–12 mm wide; sagebr and ponderosa pineland E Cas in c and n Wn; Okanogan s.   **2 H. ciliata** (Dougl.) Johnst.

    3b Marginal prickles, or many of them, free nearly or quite to the base, not forming a distinctly cupulate border; lvs often > 1 cm wide

      4a Pls robust, 3–10 dm, the st gen (2) 3–8 mm thick toward the base, the larger lvs often > 1.5 cm wide; fornices only very minutely papillate; widespread cordilleran spp. of thickets, meadows, streambanks, and for openings from foothills to fairly high elev in mts

        5a Pls relatively short-lived, bien or barely per, often single-std from a taproot and simple crown; intramarginal prickles wanting, or occ 1 or 2; corolla limb 4–7 mm wide; more nearly confined to moist places than no 4; chiefly RM, seldom in Wn and Ore; many-fld s.   **3 H. floribunda** (Lehm.) Johnst.

        5b Pls per, gen with several or ∝ sts from a taproot and br caudex; intramarginal prickles present, gen several; corolla limb (5) 7–11 mm wide; Cas-Sierran region, e to RM; blue s. (*H. jessicae, L. j.*) (Occ blue-fld forms of no 14 would key here, but for the papillate-puberulent fornices)   **4 H. micrantha** (Eastw.) J. L. Gentry

      4b Pls small and slender, 2–4 dm, the st gen 1–2 (3) mm thick toward the base, the larger lvs not > ca 1.5 cm wide; fornices evidently papillate-puberulent

        6a St appressed-hairy; middle and upper cauline lvs narrow, seldom as much as 1 cm wide, scarcely clasping, gen 4–10 times as long as wide; in shelter of juniper on drier mts, c Ore to n Cal; Cusick's s.   **5 H. cusickii** (Piper) Brand

        6b St spreading-hairy, at least below the middle; middle and upper cauline lvs markedly broad-based and clasping, 2–4 times as long as wide, many of them 1 cm wide or more; moist, rocky places at mid elev in mts of c Ida; Davis' s.   **6 H. davisii** Cronq.

  2b Corolla limb white to ochroleucous or greenish-tinted, sometimes marked or very lightly washed with blue

    7a Corolla relatively large and showy, the limb gen 13–20 mm wide, white; pls 2–4 dm; marginal prickles connate below to form an evident border; intramarginal prickles well developed, ca 15; with ponderosa pine in Chelan Co, Wn; showy s.   **7 H. venusta** (Piper) St. John

    7b Corolla smaller, the limb gen 4–12 mm wide; pls 2-10 dm

      8a Intramarginal prickles well developed, some of them gen > 1 mm and > half as long as the marginal ones, which are free nearly or quite to the base; basal lvs and lowermost cauline lvs ± reduced, the largest lvs borne shortly above base of st; for openings in and near Cas, at 4000–7000 ft; c Ore to c Cal; Cal s. (*H. elegans*)

                                 8 H. californica (Gray) Johnst.

      8b Intramarginal prickles less well developed, not > ca 1 mm, < half as long as the marginal ones; basal lvs well developed and often persistent, gen > cauline ones

        9a Corolla marked with blue; pubescence of the st largely or wholly appressed, sometimes very scanty; marginal prickles free nearly or quite to the base

          10a Sts sparsely antrorsely strigose above, glab below; fornices papillate or papillate-puberulent; with sagebr in n Malheur Co, Ore; Cronquist's s. (*H. patens* var. *semiglabra*)

                              9 H. cronquistii J. L. Gentry

          10b Sts antrorsely strigose above, retrorsely so below; fornices villous-puberulent; open places, often with sagebr, from foothills to near timberl; sw Mont and c Ida to ne Nev and s Utah, wholly e of no 9; spreading s. (*L. coerulescens*); ours is var. *patens*   **10 H. patens** (Nutt.) Johnst.

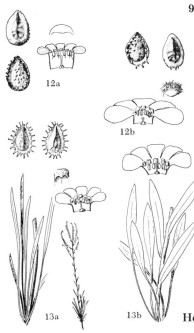

9b Corolla not marked, though occ more evenly and very lightly washed, with blue; pubescence of st appressed or often spreading; more n or w pls

11a Marginal prickles united for at least 1/3 their length, forming a distinct cupulate border to the nutlet

12a Fornices essentially glab; corolla ochroleucous or greenish-tinted, the limb gen 4–5 mm wide; cliffs and talus slopes; SRC vic, and Grand Coulee; rough s.

11 **H. hispida** (Gray) Johnst.

12b Fornices evidently papillate-puberulent; corolla white, the limb gen (5) 7–12 mm wide; gen on cliffs or talus, up to mid elev in mts; nw Mont and c Ida to c Wn, but not in the range of no 11; gray s.   12 **H. cinerea** (Piper) Johnst.

11b Marginal prickles many or all free nearly or quite to the base, not forming a distinct, cupulate border (or tending to form such a border in forms of no 13 transitional to no 12)

13a Larger lvs rarely as much as 1 cm wide; pubescence of st wholly appressed to partly spreading; dry slopes, esp with sagebr or ponderosa pine; c Wn; sagebr s.

13 **H. arida** (Piper) Johnst.

13b Larger lvs gen > 1 cm wide; pubescence of middle and lower part of st largely spreading; cliffs and talus slopes near e end of CRG, irreg n to s BC; diffuse s.

14 **H. diffusa** (Lehm.) Johnst.

Heliotropium L.   Heliotrope

Fls gen in terminal, naked or bracteate, helicoid false spikes or racemes; calyx shallowly to deeply cleft; corolla gen blue or white, salverf or funnelf, often with 5 small teeth alternating with the lobes; fornices wanting; anthers included, often connivent; ovary entire or merely shallowly lobed, the style terminal (or wanting and the stigma sessile); stigma with a broad, disklike base (often as broad as the ovary) that is gen surmounted by a gen short, entire or 2-cleft cone; fr separating at maturity into 4 nutlets, or the nutlets cohering in pairs; herbs or shrubs. (Gr *helios*, sun, and *tropos*, turn, the name only slightly modified from ancient Gr name).

H. curassavicum L. Salt h., seaside h. Glab, succulent, taprooted ann or short-lived per, prostrate or ascending, the sts 1–6 dm; lvs wholly cauline, oblanceolate to narrowly obovate, 2–6 cm × 6–18 mm; spikes 1–several on a short, naked common peduncle, to 6 (10) cm at maturity, naked; corolla white or faintly bluish, the limb gen 5–9 mm wide; stigma sessile, expanded and as broad as the ovary, the cone very short; saline places at low elev e Cas, s to tropical Am; ours is var. *obovatum* DC. (*H. spathulatum*).

Lappula Gilib.   Stickseed

Fls borne in sympodial, ± conspicuously bracteate, terminal false racemes which may elongate in age, the short peds erect or ascending in fr; calyx cleft nearly or quite to the base; corolla blue or occ white, gen small and relatively inconspicuous, ± funnelf, with definite fornices; stamens included; nutlets gen < style, narrowly attached to the elongate gynobase along the well developed median ventral keel (the lowermost part rounded and free), bearing 1 or more rows of glochidiate prickles along the continuous dorsomarginal ridge or cupulate border; entire-lvd, taprooted ann or winter ann or rarely bien, our spp. weedy, consisting of both native and intro forms, with rather ∝, linear or linear oblong to sometimes oblanceolate lvs. (Diminutive of L *lappa*, a bur).

1a Marginal prickles of the nutlets in 2 (3) rows, slender, not confluent at base; European s., bristly s. (*L. fremontii*)   1 **L. echinata** Gilib.

1b Marginal prickles in a single row; western s.; 2 vars.

2 **L. redowskii** (Hornem.) Greene

a1 Marginal prickles essentially distinct; the common form in our range

var. **redowskii**

a2 Marginal prickles on (2) 3 or all 4 of the nutlets connate below to form a ± prominent, cupulate, often swollen border; chiefly sw var., occ with us (*L. texana*, the more extreme form; *L. r.* var. *desertorum*, the less extreme form)   var. **cupulata** (Gray) M. E. Jones

Lithospermum L.   Stoneseed; Gromwell

Fls borne in modified lfy-bracteate cymes, or solitary in or near the upper axils, often heterostylic; fr peds gen erect or ascending; calyx deeply cleft; corolla yellow or yellowish, less often white or bluish-white, gen funnelf or salverf, the fornices present or absent; anthers included or partly exserted; gynobase flat or depressed; nutlets smooth to pitted or wrinkled, basally attached, the large scar often surrounded by a sharp rim, sometimes only 1 nutlet maturing; ann to per herbs, our spp. taprooted, strigose or rather softly hairy, with ± ∝, chiefly or wholly cauline, narrow, sessile or subsessile lvs seldom > 10 cm × 10 mm. (Gr *lithos*, stone, and *sperma*, seed, referring to the bony seeds).

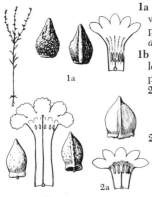

1a Ann, 1–7 dm, with 1–several sts, the central one gen the largest; corolla white or bluish-white, 5–8 mm, the limb 2–4 mm wide; nutlets wrinkled, pitted, and sometimes tuberculate; intro weed E Cas; corn g. *(Buglossoides a.)*
 1 **L. arvense** L.

1b Per, heavy-rooted, with clustered sts 0.5–6 dm; corolla yellowish or yellow, 8–40 mm, the limb 7–20 mm wide; nutlets shining, smooth or sparsely pitted; native, not weedy

2a Corolla pale yellowish, often greenish-tinted, the tube 4–6 mm, the limb 7–13 mm wide, the lobes entire or nearly so; nutlets smooth; common in open, fairly dry places up to mid elev in mts, chiefly E Cas; w g., Columbia puccoon
 2 **L. ruderale** Dougl.

2b Corolla bright yellow, the tube (12) 15–30 mm, the limb 10–20 mm wide, the lobes evidently erose; nutlets sparsely pitted; Gr Pl sp., entering our range in Mont and se BC; yellow g. *(L. angustifolium* Michx., a preoccupied name)
 3 **L. incisum** Lehm.

Mertensia Roth   Lungwort; Bluebells; Mertensia

Fls in modified, bractless, gen small cymes terminating the st and brs; calyx gen (except no 2) cleft at least to the middle, often to the base; corolla blue, or in occ individuals white or pink, shallowly 5-lobed, gen abruptly expanded at the throat and thus evidently divided into tube and limb (no 1 excepted), the fornices gen evident; filaments attached at or below the level of the fornices, often conspicuously expanded; nutlets attached laterally to the gynobase at or below the middle, gen rugose; glab to strigose or hirsute per herbs, the hairs not pungent. (For F. C. Mertens, 1764–1831, German botanist).

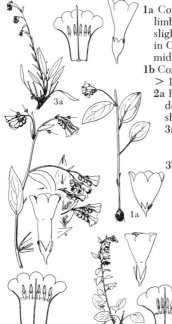

1a Corolla camp, flaring from near base, not sharply divided into tube and limb, 6–10 mm; filaments slender, attached ca 1 mm above corolla base, slightly > the 1 mm anthers; lvs with evident lateral veins; Klamath sp., n in Cas to c Ore, and in n Ida, in wet meadows and other moist places up to mid elev in mts; Ore l. or b.
 1 **M. bella** Piper

1b Corolla ± sharply divided into tube and limb (least so in no 2), often well > 10 mm

2a Pls relatively tall and robust (4–15 dm when well developed), with evident lateral veins in the cauline lvs (except often in no 2, marked by its short calyx lobes); pls gen growing along streams and in wet meadows

3a Calyx lobes distinctly < the well developed tube (unique among our spp. in this regard); corolla 15–20 mm; anthers 2.9–3.4 mm; c Ida; Ida b.
 2 **M. campanulata** A. Nels.

3b Calyx lobes distinctly > the gen very short tube; corolla 9–17 mm

4a Anthers 1.3–2.2 mm (dry); corolla limb gen 0.8–1.2 (1.5) times as long as tube; principal cauline lvs gen sessile or subsessile (except the lower) and tending to taper to the base; RM, GB, and Sierran sp., n to c Ore, c Ida, and wc Mont; ciliate b., broad-lf b.; ours is var. *ciliata*
 3 **M. ciliata** (Torr.) G. Don

4b Anthers 2.2–5.0 mm; corolla limb gen (1.0) 1.2–1.6 times as long as tube; principal cauline lvs gen rounded at base and ± petiolate

5a Anthers 2.2–3.4 mm; sts gen ∝ from a multicipital caudex or stout rhizome; boreal sp., s to wc Mont, c Ida, and c and sw Ore; in our range not extending w of Cas except in OM; tall b., pan b.; we have 2 vars.
 4 **M. paniculata** (Ait.) G. Don

a1 Pls more hairy, the lvs strigose above and hirsute or coarsely strigose beneath; widespread, but absent from Ore and all but the ne corner of Wn
 var. **paniculata**

a2 Pls less hairy, the lvs glab at least above, sometimes even glaucous; nw Mont and adj BC to Wn and Ore

var. **borealis** (Macbr.) Williams

5b Anthers 3.4–5.0 mm; sts gen arising singly from a stout, br rhizome; WV and s PS region, w, at low elev, to the coast; broadlvd b., western b.   **5 M. platyphylla** Heller

2b Pls smaller, seldom as much as 4 dm, gen without evident lateral veins in the cauline lvs, growing typically on open or lightly shaded slopes or ridges, less often in meadows, and fl as soon as snow and temperature permit

6a Filaments short, scarcely if at all > 1 mm, the base of the anthers not elevated beyond the fornices at the corolla throat; corolla 7–14 mm, the limb not much if at all < tube and tending to be widely flared; alp and subalp pls of Mont, c Ida, and s

7a Anthers included in corolla tube, their tips barely reaching the fornices; corolla tube glab within; alpine b.

6 M. alpina (Torr.) G. Don

7b Anthers exserted from corolla tube, their bases ca at level of the fornices; corolla tube with a ring of hairs inside below the middle, or sometimes with the hairs scattered over much of the inner surface; obscure b.   **7 M. perplexa** Rydb.

6b Filaments longer and more conspicuous, broad and flattened, 1.5–3 mm; base of anthers elevated well above the fornices; corolla diverse; pls gen of lowl to mid elev in mts, only no 8 extending to high elev

8a Corolla tube bearing a ring of hairs inside below the middle (or sometimes the hairs scattered over much of the inner surface), seldom much > limb; pls gen of mid to high elev, Mont to c Ida, c Ore, and s; green b. (*M. foliosa; M. cusickii*, a lower elev form of c and se Ore and adj states)   **8 M. viridis** A. Nels.

8b Corolla tube glab inside, obviously > limb; pls of valleys and foothills to mid elev in mts, widespread E Cas

9a Sts 1–2 (5) from a short, tuberous-thickened, easily detached, shallow root; basal lvs rarely developed at fl time; cauline lvs gen 1.5–4 times as long as wide; corolla tube gen (1.7) 2–3 times as long as limb; w Cont Div; small b., long-fld b.

9 M. longiflora Greene

9b Sts clustered on a stouter and more firmly attached, deeper-seated and scarcely tuberous root; basal lvs gen well developed; cauline lvs gen 2.5–7 times as long as wide; corolla tube gen 1.3–2 times as long as limb; widespread E Cas; lfy b. (*M. nevadensis; M. umbratilis* Greenm., a robust phase, to 6 dm, prob reflecting hybridization with no 4)

10 M. oblongifolia (Nutt.) G. Don

Myosotis L.   Forget-me-not; Scorpion-grass

Fls borne in terminal, naked, helicoid, eventually ± elongate, sympodial false racemes, or the lower ones scattered among the lvs, the fr peds erect or spreading; calyx 5-lobed, with a distinct tube; corolla gen blue, less often white, rarely yellow, salverf or broadly funnelf, with abruptly spreading, evidently 5-lobed limb, the fornices gen well developed; anthers gen included; nutlets 4, attached by a small, basilateral scar to the broad, low gynobase, smooth and shining, with an evident raised margin all the way around; herbs, glab to hirsute, but not hispid or setose, tending to be fibrous rooted. (Gr *mus*, mouse, *ous*, ear, from appearance of lvs of some spp.).

1a Calyx closely strigose, the hairs neither spreading nor uncinate; pls of moist soil and shallow water

2a Corolla limb 2–5 mm wide; style distinctly < nutlets; st often lax and decumbent at base, but scarcely creeping, and not at all stoloniferous; interruptedly circumboreal, in our area from e base of Cas to coast; small-fld f.   **1 M. laxa** Lehm.

2b Corolla limb 5–10 mm wide; style = or gen > nutlets; st often with creeping or stoloniferous base; European sp., occ intro in our range, up to mid elev in mts; common f. (*M. palustris*)   **2 M. scorpioides** L.

1b Calyx tube with some loose or spreading, uncinate hairs

3a Corolla rather showy, the limb 4–8 mm wide; mont to alp, circumboreal per, s to BC, c Ida, and n Wyo; wood f.; ours is var. *alpestris* (F. W. Schmidt) Koch (*M. a.*)   **3 M. sylvatica** Hoffm.

3b Corolla not showy, the limb 1–4 mm wide; ann or bien; from lowl to
   mid elev in mts, often weedy
   4a Calyx asymmetrical, sub-bilabiate, 3 lobes < the other 2; corolla
      white; fr peds evidently to barely < calyx; open, wet to rather dry
      places in foothills and lowl; widespread in e US; also in s BC and in
      Wn, Ore, and Ida; spring f., early f. (*M. virginica*, an illegitimate,
      prob misapplied name)   **4 M. verna** Nutt.
   4b Calyx symmetrical, with essentially similar lobes; corolla gen blue,
      sometimes white in no 5, or (in no 7) yellow when young; ± weedy
      spp., intro from Europe
      5a Fr peds = or gen > calyx; field f. or s.   **5 M. arvensis** (L.) Hill
      5b Fr peds distinctly < calyx
         6a Pls fl nearly to the base, the lower fls scattered among the lvs;
            style distinctly < nutlets; blue s.   **6. M. micrantha** Pall.
         6b Pls fl to not much if at all below the middle, the infl gen essen-
            tially naked; style somewhat < to more often = or distinctly >
            nutlets; yellow and blue f.   **7 M. discolor** Pers.

Onosmodium Michx.   False Gromwell

Fls borne in lfy-bracteate, helicoid terminal cymes which tend to elongate and straighten in age,
precociously sexual, the style exserted and the anthers dehiscent well before the corolla matures,
the style remaining exserted; calyx deeply 5-cleft, the narrow, often unequal segms eventually disart at base; corolla externally hairy, white to yellow, gen with ± greenish lobes, nearly tubular,
with rather narrow, erect, pointed lobes and thickened, basally inflexed sinuses; fornices wanting;
anthers barely or only partly included; nutlets turgid, without a prominent ventral keel, smooth or
merely pitted, broadly attached at base to the flattish or depressed gynobase, gen only 1 or 2 maturing; rather coarse per herbs, often rough-hairy, with largely or wholly cauline lvs that have
several prominent veins. (Named for its resemblance to *Onosma*, Old World genus of Boraginaceae).

O. molle Michx. Several-std per from a woody root, 3–7 dm, coarsely and
loosely hairy throughout; lower lvs reduced, the others rather ∝ and uniform,
lanceolate or narrowly ovate, 3–8 × 1–2 cm; corolla 12–16 mm, the sharply
acute lobes 2–4 mm; nutlets smooth and shining, broadly ovoid, without a
basal collar; open, moderately dry places; widespread e, and entering our
range in Mont; ours var. *occidentale* (Mack.) Johnst. (*O. o.*).

Pectocarya DC.   Combseed; Pectocarya

Fls borne in terminal, naked or irreg bracteate, sympodial false racemes, or the lower scattered
among the lvs, the short peds ascending to recurved; calyx deeply cleft; corolla minute, white;
anthers included; nutlets radially spreading from the broad, low gynobase at maturity, with relatively broad, apicolateral scar, uncinate-bristly at least in part, and with raised or winged margins;
low (to 2 dm), slender, taprooted, strigose ann with small, narrow lvs; our spp. wholly E Cas, in
dry, open places at lower elev, often with sagebr. (Gr *pektos*, combed, and *karyon*, nut, from the
row of bristles on the nutlet).

1a Nutlets isometrically spreading, ca 2.5 mm, the margins line-like or scarcely
   developed; sepals uncinate-bristly toward the tip; Cal n occ to Klickitat Co,
   Wn; little p.   **1 P. pusilla** (A. DC.) Gray
1b Nutlets spreading in pairs, at least 1 of each pair evidently wing-margined;
   hairs of the calyx not uncinate
   2a Nutlets = or more often > sepals, 1.5–2 mm, oblong, with upcurved or
      inflexed margins which are cleft (at least distally) into uncinate bristles;
      pls tending to be prostrate or weakly ascending; GB and Sonoran sp., n
      to s BC and c Ida; winged c. or p.; ours is var. *penicillata* (H. & A.) M. E.
      Jones (*P. p.*)   **2 P. linearis** (R. & P.) DC.
   2b Nutlets < sepals, 2–3 mm, orbicular-obovate, with ± scattered uncinate
      bristles, 1 nutlet of each pair with spreading wing margins, the other
      nearly marginless and partly hidden by the margined one; pls tending to
      be ascending or erect; GB and Sonoran sp., n rarely to Yakima Co, Wn;
      bristly c.   **3 P. setosa** Gray

Plagiobothrys F. & M.   Popcorn-flower; Plagiobothrys

Fls borne in a series of sympodial, helicoid, naked or irreg bracteate false racemes or spikes, these sometimes condensed into glomerules, but elongating with age in most spp.; calyx cleft to below middle or near base, sometimes moderately accrescent; corolla white, the well developed fornices sometimes yellow, the limb gen ± rotately spreading, not large; stamens included, the filaments short; nutlets 4, or 1–3 by abortion, tending to be keeled on the back, and with a well developed ventral keel extending from the tip to near the middle or near the base; scar gen elevated and ca- runclelike, gen small, lateral to basal, placed at one end of the ventral keel (rarely extending along it); gynobase gen short and broad; strigose or subglab to coarsely spreading-hairy herbs, ann or per, gen rather small (seldom > 2.5 dm, but to 4 dm in no 6 and 5 dm in no 3); lvs gen narrow, the lower often opp; our spp. all occurring in open places. (Gr *plagios*, placed sideways, and *both- ros*, pit, referring to the position of nutlet scar). *(Allocarya, Sonnea).*

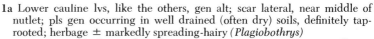

1a Lower cauline lvs, like the others, gen alt; scar lateral, near middle of nutlet; pls gen occurring in well drained (often dry) soils, definitely tap- rooted; herbage ± markedly spreading-hairy *(Plagiobothrys)*

  2a Nutlets 4, thick-cruciform (unique among our spp. in this regard); co- rolla limb gen 2–4 mm wide; basal lvs tufted and persistent, to 3 cm × 8 mm; sts erect or ascending, simple or few-brd; lower elev, chiefly E Cas, e to w Ida; slender p.   1 **P. tenellus** (Nutt.) Gray

  2b Nutlets 1–4, not all cruciform; habit various

    3a Corolla minute, the limb not > ca 1.5 mm wide; nutlets gen 1–2, hori- zontal; pls bushy–brd when well developed, the st evidently lfy and the basal lvs gen deciduous at maturity, the infls short and glomerate; GB sp., esp of sandy soil, reaching our range in c Ore; bristly p.

        2 **P. hispidus** Gray

    3b Corolla larger, the limb gen 4–9 mm wide; pls not at all bushy-brd, or somewhat so in no 4; basal lvs, except gen in no 4, tufted and per- sistent

      4a Scar short, scarcely longer than wide, placed at end of ventral keel; calyx eventually circumscissile; infls becoming elongate; nutlets 1–4; low elev at e end of CRG, and irreg s W Cas; rusty p.

        3 **P. nothofulvus** Gray

      4b Scar narrow and elongate, extending along crest of ventral keel; calyx not circumscissile; infls ± glomerate, not elongating; nutlets 3–4; GB sp., perhaps reaching our range in c Ore; Harkness' p.

        4 **P. harknessii** (Greene) Nels. & Macbr.

1b Lower cauline lvs, or some of them, opp; scar basal, basilateral, or lateral near the base; pls gen occurring in poorly drained soils, tending to be fibrous-rooted; herbage (except in no 5) gen appressed-hairy or subglab *(Allocarya)*

  5a Corolla relatively large, the limb gen 5–10 mm wide; scar lateral, near the base but often not reaching it

    6a Pls per (unique among our spp. in this regard), evidently spreading- hairy at least in part; sts lax, sometimes trailing; spikes gen not paired; GB sp., tolerant of alkali, perhaps reaching our range in c Ore; soft p.   5 **P. mollis** (Gray) Johnst.

    6b Pls ann, merely strigose; sts erect; spikes tending to be paired; nonal- kaline meadows, low ground, and moist fields W Cas and in CRG; fra- grant p. *(P. scouleri* and *A. s.,* misapplied)* 6 **P. figuratus** (Piper) Johnst.

  5b Corolla smaller, the limb 1–4 mm wide; scar basal or basilateral

    7a Scar basilateral, gen more lateral than basal; calyx lobes neither elon- gate nor much thickened, ± symmetrically disposed; pls prostrate to ascending or erect; widespread cordilleran sp., seldom in obviously alkaline places; Scouler's p.; 2 vars.   7 **P. scouleri** (H. & A.) Johnst.

      a1 Corolla limb gen 2–4 mm wide; pls more often ascending or erect; W Cas *(A. s., A. granulata, A. media,* and also all as *Plagiobothrys; A. fragilis)*   var. **scouleri**

      a2 Corolla limb gen 1–2 mm wide; pls more often prostrate to as- cending; E Cas *(A. cognata, A. cusickii, A. nelsonii, A. scopulorum,* and also all as *Plagiobothrys)*   var. **penicillatus** (Greene) Cronq.

    7b Scar very nearly basal; calyx lobes becoming elongate and slightly thickened, tending all to be directed towards the same side of the fr; pls gen prostrate; GB sp., n E Cas to n Ore, tolerant of alkali; slender- br p. *(A. orthocarpa, P. o.)*   8 **P. leptocladus** (Greene) Johnst.

Symphytum L.   Comfrey

Fls borne in several or ∝ small, sympodial, somewhat helicoid, naked, modified cymes; calyx shallowly to deeply cleft; corolla white or ochroleucous to pink or blue, tubular-camp, the well defined throat much > the short, erect or apically spreading lobes; fornices narrow and elongate, erect; filaments inserted at the level of the fornices; anthers included; style elongate, shortly exserted; nutlets incurved, ventrally keeled, smooth or finely wrinkled or tuberculate, with a well developed, stipelike basal attachment surrounded by a toothed rim (the basal margin of the nutlet) which fits close to the gynobase; broad-lvd per herbs, our spp. taprooted, 3–12 dm, with basal lf bls gen 15–30 × 7–12 cm and gradually reduced cauline lvs; escaped from cult and ± estab along roadsides etc, chiefly W Cas. (Gr *symphyton*, growing together, an ancient name for S. *officinale*).

1a St evidently winged by the conspicuously decurrent lf-bases; corolla ochroleucous or dull blue, ca 1.5 cm; common c.   1 **S. officinale** L.
1b St not winged, the lf bases not decurrent, or only shortly and inconspicuously so; corolla ave a bit larger, blue (pink before anthesis); rough c.
                                                                                2 **S. asperum** Lepech.

VERBENACEAE   Verbena or Vervain Family

Fls gamopet, gen ⚥; calyx gamosep, persistent, gen 4–5-lobed; corolla reg or more often irreg, the limb gen 4–5-lobed and often bilabiate; stamens (2) 4 (5), epipetalous, gen didynamous; ovary superior, scarcely or not at all lobed apically, 2 (4–5)-carpellary, or 1-carpellary by abortion, each carpel gen partitioned into 2 uniovulate segms; placentation axile; style slender, gen 2-cleft at the tip, the short brs often unequal; endosperm gen scanty or wanting; fr indehiscent and often drupaceous, or more gen a dry schizocarp separating into 4 nutlets; herbs, shrubs, trees, or woody vines, with opp (alt or whorled), exstip, simple or compound lvs, gen square sts, and diverse sorts of determinate or indeterminate infls.

Verbena L.   Verbena; Vervain

Fls in terminal, often densely ∝-fld spikes, each fl solitary in the axil of a narrow bract; calyx 5-angled and unequally 5-toothed, only slightly enlarged in fr; corolla salverf or funnelf, with flat, unequally 5-lobed, weakly bilabiate limb, anthocyanic or seldom white; only 1 lobe of the style stigmatic; ovary entire or very shallowly 4-lobed; fr gen enclosed in the calyx, dry, readily separating into 4 linear-oblong nutlets; endosperm wanting; herbs (ours) or shrubs with gen toothed to dissected lvs, our spp. strigose to hirsute.

1a Bracts of infl elongate, conspicuously > calyx; pls gen prostrate or decumbent, 1–6 dm; lvs, or many of them, ± deeply cleft, 2–5 × 1–2.5 cm; corolla inconspicuous, gen ± anthocyanic, the tube ca 4 mm, the limb 2–3 mm wide; taprooted ann or per; widespread native Am weed; bracted v. (*V. bracteosa*)   1 **V. bracteata** Lag. & Rodr.
1b Bracts of the infl inconspicuous, barely = calyx, or < calyx; pls erect, (2) 5–15 dm; lf bls gen merely toothed or very shallowly lobed, 4–15 × 1–5 cm
  2a Lvs evidently petiolate, the petiole (0.5) 1–2 cm, the bl gen lanceolate or lance-ovate; calyx 2–3 mm; corolla tube 3–4 mm, the limb 2.5–5 mm wide, blue or violet; spikes pan at top of st and gen ± ∝, 3–10 (15) cm; fibrous-rooted per of ditch banks and moist low ground throughout most of US; blue v., wild hyssop, simpler's joy   2 **V. hastata** L.
  2b Lvs sessile or subsessile, broadly elliptic or ovate; calyx 4–5 mm; corolla tube 6–7 mm, the limb 7–11 mm wide, deep blue or purple; spikes gen few or solitary, 6–30 cm; taprooted, short-lived pls of roadsides and other dry, open places, common on Gr Pl, and extending w through n Ida to ne Wn; hoary v., mullein-lvd v.   3 **V. stricta** Vent.

LABIATAE   Mint Family

Fls gamopet, ♂ (some pls sometimes ♀); calyx gamosep, reg or irreg, often bilabiate, 5 (10)-lobed, or the lobes rarely suppressed; corolla ± irreg, gen bilabiate, 5-lobed, or 4-lobed by fusion of the 2 lobes of the upper lip; stamens epipetalous, 4, didynamous, or 2 by abortion of either the upper or the lower pair; ovary superior, basically 2-carpellary, but ± deeply cleft into 4 uniovulate segms which ripen into hard nutlets, the style in most genera basal and surrounded by the basally attached, otherwise essentially distinct nutlets; style slender, gen 2 (4)-cleft at the tip, the short brs often unequal; ovules erect; seeds nearly or quite without endosperm; aromatic or sometimes inodorous herbs or shrubs, rarely trees or vines, with opp (whorled), exstip, entire or toothed to dissected lvs, gen square sts, and diverse sorts of basically cymose (determinate) infls, the fls very often verticillate, or in small cymes in the axils of lvs or bracts.

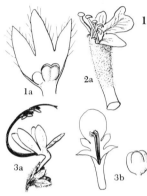

1a Ovary merely 4-lobed, not cleft to the base, the nutlets laterally attached; corolla (except in *Ajuga*) seemingly 1-lipped, or 2-lipped with the upper lip cleft to the base
  2a Upper lip of the corolla short but evident, not cleft beyond ca the middle; ours a mat-forming per   **Ajuga**
  2b Upper lip of the corolla not well defined, its lobes distinct to the base and sometimes well separated
    3a Lobes of the upper lip of the corolla adjacent, often slightly displaced toward the lateral lobes of the lower lip, but the sinus between them scarcely more prominent than the lateral sinuses, and the central lobe of the lower lip < twice as long as any of the other 4 corolla lobes; ours taprooted ann   **Trichostema**

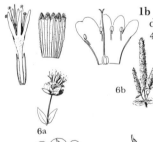

    3b Lobes of the upper lip of the corolla well separated, displaced onto the lateral margins of the lower lip; central lobe of the lower lip much larger than any of the other 4 corolla lobes; ours a rhizomatous per   **Teucrium**
1b Ovary cleft to the base, the nutlets basally attached; corolla either evidently 2-lipped or nearly reg
  4a Corolla only obscurely or scarcely bilabiate, either subequally 5-lobed with the lobes ± alike, or subequally 4-lobed with 1 of the lobes gen broader than the others and sometimes emarginate
    5a Stamens 4; pls aromatic; infls axillary or terminal
      6a Infl a dense terminal head subtended by conspicuous lfy or (in ours) dry bracts; corolla 5-lobed   **Monardella**
      6b Infl of axillary verticils, or terminal, spikelike, and inconspicuously bracteate; corolla gen 4-lobed, 5-lobed only in occ individuals   **Mentha**

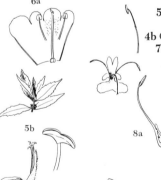

    5b Stamens 2; pls odorless; infls wholly axillary; corolla gen 4-lobed   **Lycopus**
  4b Corolla ± strongly bilabiate
    7a Fertile stamens 2
      8a Connective much elongate, jointed to and ± resembling the relatively short filament, the upper arm bearing a single pollen sac on the end, the shorter lower arm often wholly suppressed; herbs or shrubs (elongate connective unique among our genera)   **Salvia**
      8b Connective merely expanded at the end of the filament, bearing 2 pollen sacs adjacent to each other, end to end; herbs
        9a Upper lip of corolla narrow, galeate, in ours elongate and arcuate; infl (ours) a single, terminal, basally lfy-bracteate head; ours with the lvs gen > 1 cm wide   **Monarda**
        9b Upper lip of corolla short, straight, flat; infl a series of axillary verticils; ours with the lvs well < 1 cm wide   **Hedeoma**
    7b Fertile stamens 4 (or obsolete in ♀ pls)
      10a Calyx bilabiate, with entire lips, bearing a prominent transverse external appendage on the upper side; lower stamens each with only 1 functional pollen sac   **Scutellaria**
      10b Calyx reg or irreg, sometimes bilabiate, but in any case evidently toothed, and lacking any transverse appendage; each stamen with 2 pollen sacs, these sometimes confluent in dehiscence
        11a Infl appearing terminal, the verticils tending to be crowded and gen subtended by mere bracts that are evidently differentiated from foliage lvs

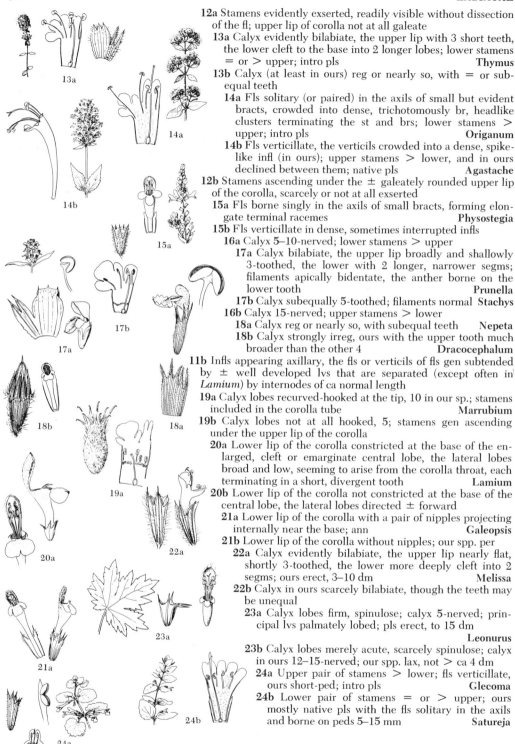

12a Stamens evidently exserted, readily visible without dissection of the fl; upper lip of corolla not at all galeate
  13a Calyx evidently bilabiate, the upper lip with 3 short teeth, the lower cleft to the base into 2 longer lobes; lower stamens = or > upper; intro pls   **Thymus**
  13b Calyx (at least in ours) reg or nearly so, with = or sub-equal teeth
    14a Fls solitary (or paired) in the axils of small but evident bracts, crowded into dense, trichotomously br, headlike clusters terminating the st and brs; lower stamens > upper; intro pls   **Origanum**
    14b Fls verticillate, the verticils crowded into a dense, spike-like infl (in ours); upper stamens > lower, and in ours declined between them; native pls   **Agastache**
12b Stamens ascending under the ± galeately rounded upper lip of the corolla, scarcely or not at all exserted
  15a Fls borne singly in the axils of small bracts, forming elongate terminal racemes   **Physostegia**
  15b Fls verticillate in dense, sometimes interrupted infls
    16a Calyx 5–10-nerved; lower stamens > upper
      17a Calyx bilabiate, the upper lip broadly and shallowly 3-toothed, the lower with 2 longer, narrower segms; filaments apically bidentate, the anther borne on the lower tooth   **Prunella**
      17b Calyx subequally 5-toothed; filaments normal **Stachys**
    16b Calyx 15-nerved; upper stamens > lower
      18a Calyx reg or nearly so, with subequal teeth   **Nepeta**
      18b Calyx strongly irreg, ours with the upper tooth much broader than the other 4   **Dracocephalum**
11b Infls appearing axillary, the fls or verticils of fls gen subtended by ± well developed lvs that are separated (except often in *Lamium*) by internodes of ca normal length
  19a Calyx lobes recurved-hooked at the tip, 10 in our sp.; stamens included in the corolla tube   **Marrubium**
  19b Calyx lobes not at all hooked, 5; stamens gen ascending under the upper lip of the corolla
    20a Lower lip of the corolla constricted at the base of the enlarged, cleft or emarginate central lobe, the lateral lobes broad and low, seeming to arise from the corolla throat, each terminating in a short, divergent tooth   **Lamium**
    20b Lower lip of the corolla not constricted at the base of the central lobe, the lateral lobes directed ± forward
      21a Lower lip of the corolla with a pair of nipples projecting internally near the base; ann   **Galeopsis**
      21b Lower lip of the corolla without nipples; our spp. per
        22a Calyx evidently bilabiate, the upper lip nearly flat, shortly 3-toothed, the lower more deeply cleft into 2 segms; ours erect, 3–10 dm   **Melissa**
        22b Calyx in ours scarcely bilabiate, though the teeth may be unequal
          23a Calyx lobes firm, spinulose; calyx 5-nerved; principal lvs palmately lobed; pls erect, to 15 dm   **Leonurus**
          23b Calyx lobes merely acute, scarcely spinulose; calyx in ours 12–15-nerved; our spp. lax, not > ca 4 dm
            24a Upper pair of stamens > lower; fls verticillate, ours short-ped; intro pls   **Glecoma**
            24b Lower pair of stamens = or > upper; ours mostly native pls with the fls solitary in the axils and borne on peds 5–15 mm   **Satureja**

Agastache Clayton   Giant-hyssop; Horse-mint; Agastache

Fls verticillate, the verticils crowded into a dense, spikelike infl (ours), or the infl sometimes more open; calyx with 15 or more prominent veins, equally 5-toothed, or the 3 upper teeth ± connate, often whitish or tinged with pink or blue; corolla pink to violet or white, bilabiate, with short lobes, 8–14 mm in ours; stamens 4, gen exserted, the upper pair > lower, and in ours declined between them; pollen sacs parallel or nearly so; per herbs with sts clustered on a br caudex which may surmount a taproot, the lvs petiolate or subsessile, entire or more often toothed, gen ovate to deltoid or subcordate. (Gr *agan*, much, and *stachys*, ear of grain, referring to the infl).

    1a Pls dwarf, only 1–2 dm, with lf bls gen 1–2 cm; dry, rocky places at mid elev in mts of c Ida, se Ore, and n Nev; Cusick's g. or h.; ours the var. *parva* Cronq.   1 **A. cusickii** (Greenm.) Heller
    1b Pls robust, 4–15 dm, ∝ of the lf bls > 3 cm (to 10 cm)
        2a Lvs very finely and closely tomentose-puberulent beneath; open, often rocky places on e slope of Wn Cas; western g. or h.
                                    2 **A. occidentalis** (Piper) Heller
        2b Lvs glab beneath, or often very minutely and obscurely hirtellous-scab, varying to occ loosely villous-puberulent; widespread cordilleran sp. of open places from foothills to rather high elev, wholly E Cas, and in Wn w only to Grand Coulee; nettle-lf h. or g.; ours the var. *u*.
                                    3 **A. urticifolia** (Benth.) Kuntze

Ajuga L.   Ajuga

Fls sessile in verticils of 2–6 axillary to bracteal lvs, forming a terminal lfy spike; calyx 10-nerved or irreg ∝-nerved, nearly reg 5-toothed; corolla ± distinctly bilabiate, the upper lip very short, entire to bifid about half way to the base, the lower lip with a large, 2-lobed segm and a pair of much smaller lateral segms, the whole corolla tending to persist in fr and become papery; stamens 4, the lower pair the longer, ascending under the upper lip and gen exserted; pollen sacs ± divergent or divaricate, confluent in dehiscence; ovary merely 4-lobed (not beyond the middle), the nutlets reticulate-rugose, laterally attached, gen connate to beyond the middle; style subequally 2-lobed; ann or more often per herbs, gen decumbent or stoloniferous, with coarsely toothed to rarely entire lvs. (Gr *a*, without, and *zugos*, yoke, the lower lip of the corolla not matched or yoked).

    A. reptans L. Bugle. Low per suggesting *Prunella vulgaris* in gen appearance, spreading by lfy stolons and forming mats; fl sts 1–3 dm; cauline lvs ovate to oblong-spatulate, 2–5 cm; calyx 4–6 mm, the triangular lobes < tube, 2 slightly < the other 3; corolla blue, 10–15 mm, the lower lip dilated, its lateral lobes reaching to the middle of the obcordate median lobe; European sp., occ escaped from cult W Cas.

Dracocephalum L.   Dragonhead

Fls verticillate in the axils of entire or more often spinose-dentate, small or lfy bracts; calyx 15-nerved, with oblique orifice, 5-toothed, the upper tooth much broader than the others, or the 3 upper teeth connate into a lip; corolla blue or purple to sometimes white, not much > calyx in ours, bilabiate, the upper lip emarginate and subgaleately rounded, the lower 3-lobed, with the large central lobe sometimes 2-cleft; stamens 4, ascending under the upper lip of the corolla, the upper pair the longer; pollen sacs divaricate at a wide angle, confluent in dehiscence; per or bien (ann) herbs with toothed or cleft lvs. (Gr *drakon*, dragon, and *kephale*, head, referring to fancied resemblance of fl). (Name previously often applied to *Physostegia*, but now conserved in the present usage). (*Moldavica*).

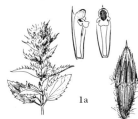

    1a Fls in dense, terminal, headlike clusters; bracts of the infl aristately few-toothed; calyx ca 1 cm, the tube ca = lobes; widespread native sp. of open, rather moist places from foothills to mid elev in mts, wholly E Cas; Am d. (*M. p.*)   1 **D. parviflorum** Nutt.
    1b Fls in loose, elongate, interrupted racemes; bracts entire or nearly so; calyx 6–8 mm, the tube obviously > lobes; Eurasian sp., occ with us; thyme-lvd d. (*M. t.*)   2 **D. thymiflorum** L.

Galeopsis L.   Hemp Nettle

Fls sessile, ∝ in remote or crowded axillary verticils, the lvs subtending the upper verticils ± reduced; calyx 5–15-nerved, with 5 equal or unequal, firm and apically subspinescent teeth; corolla bilabiate, the upper lip galeately rounded and entire; lower lip spreading, 3-lobed, with the central lobe sometimes cleft, bearing a pair of prominent projections or "nipples" on the upper side near the base; stamens 4, ascending under the upper lip, the lower pair the longer; pollen sacs contiguous in a nearly straight, transverse line at the summit of the apically expanded filament, the expanded portion representing the connective; each pollen sac transversely 2-valved, the upper valve ciliate, the lower larger and glab; ann with gen toothed lvs. (Classical Gr name).

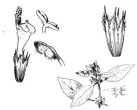

G. tetrahit L. Common h. n. Pls 1.5–7 dm, the st ± retrorse-hispid and often glandular-hairy; lvs petiolate, the bl 3–10 × 1–5 cm, coarsely blunt-serrate; verticils dense, gen several; calyx ca 1 cm; corolla gen 15–23 mm, purple or pink (white) with darker markings, the middle lobe of the lower lip nearly square, not emarginate; Eurasian sp., occ intro in meadows and other moist places; ours is var. *t.*

Glecoma L.   Ground Ivy

Fls in few-fld verticils axillary to normal lvs; calyx 15-nerved, oblique at the mouth, unequally 5-toothed; corolla gen blue or purple, bilabiate, the upper lip 2-lobed, ± concave, the lower lip spreading, with large central and smaller lateral lobes; stamens 4, the upper pair the longer, ascending under the upper lip of the corolla and scarcely exserted (sometimes abortive); pollen sacs divergent at about right angles, opening by separate slits; diffuse or creeping per herbs with toothed lvs. (Gr *glechon*, classical name of a sp. of *Mentha*).

G. hederacea L. Gill-over-the-ground, field balm, creeping Charlie. Fibrous-rooted from slender stolons or superficial rhizomes, 1–4 dm, pilose at the nodes; lvs wholly cauline, petiolate, with rotund-cordate to cordate-reniform bl 1–3 cm; calyx 5–6 mm, the upper teeth the longer; corolla blue-violet, purple-maculate, 13–23 mm, or in ♀ pls ca 1 cm or less; Eurasian weedy sp. of moist woods and thickets, often also in lawns; occ in our range (*Nepeta h.*).

Hedeoma Pers.   Pennyroyal

Fls verticillate in the axils of ordinary lvs; calyx ca 13-nerved, gen distended near the base and gen ± bilabiate, the 3 upper teeth ± connate, the 2 lower ones longer, narrower, and distinct; corolla ca 1 cm in our spp., bilabiate, the upper lip short, straight, flat, emarginate or 2-lobed, the lower spreading and 3-lobed; stamens 2, ascending under the upper lip of the corolla and often slightly exserted beyond it; pollen sacs separately dehiscent, contiguous in a nearly straight transverse line (or at a very wide angle) at the summit of the apically expanded filament, the expanded portion representing the connective; small, wiry-std ann or per herbs with entire or obscurely toothed, gen subsessile lvs; ours gen < 3 dm, with lvs up to 1.5 or 2 cm, growing in dry, open places at lower elev chiefly on and e of Gr Pl, encroaching into our range in Mont. (Gr *hedyosmon*, sweet odor, used for another member of the family).

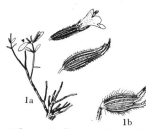

1a Pls per; calyx lobes subconnivent in fr, the whole calyx seeming to taper gradually from the base to (near) the tips of the lobes; Drummond's p.
                                       1 H. drummondii Benth.
1b Pls ann; upper calyx lobes ± strongly upcurved and divergent from the lower ones; rough p.   2 H. hispida Pursh

Lamium L.   Dead-nettle; Henbit

Fls verticillate in the axils of ordinary lvs or modified but well developed and lfy bracts; calyx inconspicuously nerved, the 5 pointed teeth = or the upper one the largest; corolla purple to white, hairy outside in ours, bilabiate, the upper lip entire or 2-lobed, galeately rounded, the lower spreading, constricted at the base of the enlarged, cleft or emarginate central lobe, the lateral lobes broad and low, terminating in a short tooth, seemingly borne on the corolla throat; stamens 4, the lower pair the longer, ascending under the upper lip; pollen sacs ± divergent or divaricate, conspicuously hairy; nutlets angularly 3-sided, truncate at the summit; ann or per herbs with toothed or pinnatifid, gen cordate lvs, ours all intro, ± weedy spp. (Ancient L name).

1a Upper lip of corolla 7–12 mm, the whole corolla gen 20–25 mm; pls per, fibrous-rooted from a creeping base; lvs all petioled; occ W Cas in our range; spotted d. or h.						**1 L. maculatum** L.

1b Upper lip of corolla 2–5 mm, the whole corolla < 20 mm; pls ann from a short taproot

  2a Lvs all petiolate; corolla with a ring of hairs inside near the base; red d. or h.					**2 L. purpureum** L.

  2b Lvs subtending the fl clusters mostly or all sessile and clasping; corolla glab inside; occ pls produce only small, cleistogamous fls; common d. or h.				**3 L. amplexicaule** L.

Leonurus L. Motherwort

Fls verticillate in the axils of ± reduced lvs; calyx 5 (10)-nerved, with 5 narrow, firm, spinulose, subequal lobes; corolla pink or white, bilabiate, the tube scarcely exserted from the calyx, the upper lip entire and galeately rounded, the lower spreading and 3-lobed, with obcordate central lobe and oblong lateral lobes; stamens 4, ascending under the upper lip and scarcely exserted, the lower pair = or > upper; pollen sacs gen parallel; nutlets triquetrous, truncate above; erect per herbs with the lower lvs palmately cleft, the upper gradually smaller, narrower, and less cleft, but still well > their axillary verticils of fls. (Gr *leon*, lion, and *oura*, tail).

L. cardiaca L. Sts clustered on a stout, br rhizome or caudex, 4–15 dm; lvs wholly cauline, the main ones palmately veined, palmately cleft and again coarsely toothed, 5–10 cm long and wide, ca = petiole, progressively reduced upward; calyx tube 3–4 mm, ca = lobes, the 2 lower lobes deflexed; corolla 1 cm, pale pink, upper lip white-villous. Asian sp., casual with us.

Lycopus L. Bugleweed; Water Horehound

Fls sessile, verticillate in the axils of scarcely reduced lvs; calyx evidently to obscurely 5-nerved, sometimes with additional lesser nerves, the 5 teeth ca =; corolla small, the tube short, internally hairy at the throat, the limb nearly reg 4-lobed; upper corolla lobe formed by fusion of the 2 lobes of the upper lip, tending to be broader than the other lobes, and often apically emarginate; fertile stamens 2, slightly exserted, the upper represented by small staminodes or obsolete; pollen sacs parallel; nutlets widened upward, bearing a corky ridge along the lateral angles and often across the tip, the outer surface smooth and nearly plane, the inner convex and gen glandular; rhizomatous, scarcely aromatic herbs with wholly cauline lvs, our spp. growing in marshes and low ground, often along streams and lakes. (Gr *lykos*, wolf, and *pous*, foot, translated from French name *patte du loup*).

1a Calyx lobes narrow, firm, slenderly subulate-pointed, distinctly > mature nutlets; lower elev

  2a Nutlets gen 1.6–2.1 × 1.4–1.8 mm, the outer apical margin truncate and often irreg toothed; lvs ± sessile, with abruptly contracted to occ more tapering base and rather coarsely but fairly evenly serrate margins; rhizome not tuber-bearing; pls tolerant of alkali, widespread E Cas; rough b. (*L. lucidus*, misapplied)		**1 L. asper** Greene

  2b Nutlets gen 1.0–1.4 × 0.8–1.2 mm, the outer apical margin smooth and broadly rounded; lvs tapering to a short petiole or petiolar base, irreg incised-toothed or subpinnatifid, rarely merely serrate; rhizome often (reg?) tuber-bearing; widespread sp. of nonalkaline sites; cut-lvd w. h.		**2 L. americanus** Muhl.

1b Calyx lobes broader, nearly ovate, soft, merely acutish, scarcely > mature nutlets; lvs gen tapering to a short petiole, sometimes more abruptly contracted and subsessile, the margins merely toothed; rhizomes tuber-bearing; gen in the mts, but descending to sea level W Cas; widespread sp.; northern b.			**3 L. uniflorus** Michx.

Marrubium L. Horehound

Fls in dense whorls axillary to normal lvs; calyx 5–10-nerved, the 5–10 gen pungent teeth gen recurved-hooked at the tip, at least in age; corolla white or purplish, bilabiate, the short tube included or barely exserted from the calyx, the upper lip flat or subgaleately rounded, entire or shortly bifid, the lower spreading, 3-lobed, the larger central lobe often emarginate; stamens 4, the lower pair the longer, included in the corolla tube; pollen sacs divaricate, soon confluent; per herbs, gen woolly, with toothed or incised, rugose lvs. (Classical L name, perhaps derived from Hebrew *marrob*, a bitter juice).

M. vulgare L. Sts gen several from a stout taproot, nearly prostrate to suberect, 3–10 dm, conspicuously white-woolly; lvs wholly cauline, ± canescent-woolly, petiolate, with broad, evidently crenate bl 2–5.5 cm; calyx tube 4–5 mm, > teeth, these 10, narrow, becoming widely spreading, the spinulose tips recurved from the first; corolla whitish; Eurasian weed, casual throughout our range.

Melissa L.   Balm

Fls borne in small verticils axillary to well developed lvs; calyx bilabiate, strongly 13-nerved, the upper lip nearly flat, shortly 3-toothed, the lower more deeply 2-cleft; corolla white to yellowish or light pink or blue, bilabiate, the upper lip nearly flat, emarginate, the lower spreading and 3-lobed; stamens 4, ascending under the upper lip, the lower pair the longer; pollen sacs ± strongly divaricate; herbs with toothed lvs. (Gr *melissa*, honey bee, the pl a source of honey).

M. officinalis L. Garden, lemon, or bee balm. Lemon-scented, fibrous-rooted per 3–10 dm; lvs petiolate, the lower soon deciduous, the middle cauline ones ovate to nearly deltoid, 4–9 × 2.5–5 cm, coarsely toothed; calyx 7–9 mm, the teeth firm, with subspinulose tip; corolla white or weakly anthocyanic, 10–15 mm, with short lips; Eurasian sp., cult and escaped to roadsides etc., occ in our range.

Mentha L.   Mint

Fls in verticils subtended by lvs or ± reduced bracts, the verticils crowded or remote, often forming spikelike infls; calyx 10-nerved, reg or subbilabiate, the 5 teeth = or unequal; corolla with short tube and nearly reg 4-lobed limb, the upper lobe formed by fusion of the 2 lobes of the upper lip, tending to be broader than the other lobes, and often apically emarginate; stamens 4, gen exserted; pollen sacs parallel; aromatic, rhizomatous per herbs with wholly cauline lvs. (L *menta*, Gr *minthe*, perhaps for the Gr nymph *Minthe*). Only nos 1, 2, 3, and 5 are natural spp., the others being hybrid cultigens.

1a Verticils of fls axillary, subtended by ± ordinary lvs and separated by internodes of normal length; pls of wet places
  2a Lvs small, gen 1–2 cm, with only 2–3 pairs of lateral veins; lvs of the infl tending to be deflexed and often scarcely > fl clusters; European sp., occ in disturbed sites W Cas; pennyroyal   **1 M. pulegium** L.
  2b Lvs larger, gen 2–8 cm, with several pairs of lateral veins; floral lvs spreading, conspicuously > fl clusters; common circumboreal sp.; corn m., field m.; ours is var. *glabrata* (Benth.) Fern. (*M. canadensis*)
                                                   2 M. arvensis L.
1b Verticils of fls crowded into terminal, inconspicuously bracteate, spikelike or headlike infls
  3a Lvs sessile or subsessile, the petiole, if any, not > ca 3 mm; calyx 1–2 mm; spikes relatively slender, gen 0.5–1 cm thick at anthesis
    4a Lvs gen 1–2 times as long as wide, acutish to more often obtuse or rounded, evidently hairy, esp beneath; chiefly W Cas, along roadsides and in waste places
      5a Lvs crenate-serrate, strongly rugose-reticulate, conspicuously villous-tomentose beneath, less so above; apple-scented m., round-lvd m.   **3 M. rotundifolia** (L.) Huds.
      5b Lvs sharply toothed, scarcely rugose and not strongly reticulate, often less densely hairy than no 3; woolly m.
                                    4 M. alopecuroides Muhl.
    4b Lvs gen 2–3.5 times as long as wide, ± acute, glab, or gen hirsute along the main veins beneath; occ on banks of streams and ditches throughout our range; spearmint, mackerel-mint   **5 M. spicata** L.
  3b Lvs evidently petiolate, many of the petioles 4 mm or more (to 15 mm);

6a

calyx 2–3 mm; spikes (or heads) relatively stout, gen 1–2 cm thick at anthesis; wet places

6a Pls with a characteristic peppermint odor; infl of several or ∝ verticils, forming a dense spike 2–7 × 1–1.5 cm; calyx teeth hispid-ciliate; occ throughout our range; peppermint   **6 M. piperita** L.

6b Pls with a characteristic lemon odor; infl of 1–3 principal verticils, forming an often ± elongate head ca 2 cm thick, sometimes with 1 or more additional remote verticils; calyx teeth scarcely ciliate; occ W Cas; bergamot m.   **7 M. citrata** Ehrh.

Monarda L.   Monarda

Fls in 1 or several dense, lfy-invol, terminal heads; calyx 13–15-nerved, the 5 teeth ca =; corolla strongly bilabiate, the upper lip narrow, entire, galeate, the lower spreading and 3-lobed; stamens 2, ascending under the upper corolla lip and gen exserted; pollen sacs end to end on the expanded summit of the filament (the connective), confluent in dehiscence; ann or more often (incl our spp.) per herbs with gen toothed lvs. (For Nicolas Monardes, 1493–1588, Spanish physician and botanist).

1a

1a Corolla lavender-purple, 2.5–3.5 cm; pls less robust, gen 3–7 dm, the lf bls 2.5–8 × 1–3 cm, the petioles seldom 1 cm; moist, open places up to mid elev in mts, chiefly RM and e; wild bergamot; ours is var. *menthifolia* (Grah.) Fern. (*M. m.*)   **1 M. fistulosa** L.

1b Corolla bright crimson, 3–4.5 cm; pls more robust, gen 7–15 dm, the lf bls gen 7–15 × 2.5–6 cm, the petioles 1–4 cm; e Am sp., occ escaped from cult W Cas; Oswego tea, Am bee-balm   **2 M. didyma** L.

Monardella Benth.   Monardella

Fls in dense, terminal heads subtended by conspicuous, lfy or dry bracts; calyx 10–15-nerved, with short, subequal teeth; corolla gen pink-purple, obscurely bilabiate, with 5 well developed, slender lobes; stamens 4, rather shortly exserted, subequal or the lower pair the longer; pollen sacs divergent or parallel; ann or per herbs with small, entire or serrate, sessile or petiolate lvs. (Diminutive of *Monarda*). (*Madronella*).

M. odoratissima Benth. Mt m. Sts clustered on a per taproot and br caudex, 1–5 dm; lvs shortly or scarcely petiolate, firm, entire, 1–3.5 cm × 3–12 mm; bracts conspicuous, 7–15 mm, rather dry, veiny, ± purplish; heads 1–4 cm thick; corolla pink-purple to whitish, 1–2 cm; open, often rocky places up to mid elev in mts, widespread E Cas; we have 2 vars.: var. *odoratissima*, with ± glab lvs, occurs in mts of c and ne Ore, ne Wn, and w and n Ida. Var. *discolor* (Greene) St. John (*M. d.*, *M. nervosa*), with the lvs ± densely short-hairy beneath, grows in the drylands of c Wn and c Ore, w onto the e slopes of Cas.

Nepeta L.

Fls in dense verticils crowded to form a terminal, inconspicuously bracteate infl, or sometimes borne in open, br, paniculiform infls; calyx 15-nerved, gen curved, oblique at mouth, 5-toothed, the teeth often unequal and sometimes forming 2 lips; corolla blue or white, bilabiate, the upper lip ± concave and often subgaleate, entire or bifid, the lower lip spreading, 3-lobed, the large central lobe sometimes notched; stamens 4, the upper pair the longer, ascending under the upper lip of the corolla and scarcely exserted; pollen sacs widely divaricate, confluent in dehiscence; ann or more often per herbs, ascending or erect, with toothed or cleft lvs. (Classical L name).

N. cataria L. Catnip, catmint. Taprooted, softly canescent per 3–10 dm; lvs wholly cauline, petiolate, the bl coarsely toothed, triangular-ovate, basally cordate, 2.5–7 × 1.5–5 cm; fls in short, dense, spikelike infls 2–8 × 1.5–2.5 cm; calyx 5–6 mm, tube > teeth; corolla whitish, gen purple-dotted, shortly exserted; Eurasian sp., now widespread in US, esp in disturbed sites.

Origanum L.   Wild Marjoram

Fls solitary (or paired) in the axils of small but evident bracts, crowded into dense, trichotomously br, terminal headlike clusters which collectively form a compact, corymbose-paniculiform infl; calyx ca 13-nerved, subequally 5-toothed, bearded internally at the throat; corolla bilabiate, the upper lip ± flat, emarginate or cleft, the lower lip spreading and 3-lobed; stamens 4, the upper pair ca = corolla, the lower pair exserted and divergent; pollen sacs strongly divergent, separately dehiscent; per herbs with rather small, toothed or entire lvs. (Gr *origonon*, name for several spp. of the family).

O. vulgare L. Rhizomatous per 3–7 dm, gen simple below the infl except for the axillary fascicles of reduced lvs; lvs wholly cauline, gen ovate or deltoid-ovate and 1.5–3 cm, ± entire, rather short-petiolate; bracts gen elliptic or rhombic to obovate, 3–5 mm, evidently purple-tipped; calyx 2–2.5 mm; corolla ca 5 mm (smaller in occ ♀ pls). Eurasian sp., now a roadside weed in ne US, and occ W Cas.

Physostegia Benth.   Physostegia

Fls borne singly in the axils of small bracts, forming elongate terminal racemes; calyx 10-nerved, equally or unequally 5-toothed; corolla pink or purple to white, elongate, bilabiate, the upper lip subgaleate and scarcely or barely notched, the lower lip ca = upper, spreading, 3-lobed; stamens 4, ascending under the upper corolla-lip, the lower pair the longer; pollen sacs nearly parallel; glab or finely puberulent, fibrous-rooted per herbs with toothed or sometimes entire lvs, recalling some of the larger spp. of *Penstemon* in aspect. (Gr *physa*, bladder, and *stege*, covering, the calyx becoming inflated and bladdery).

P. parviflora Nutt. Purple dragon-head. Sts solitary, 2–10 dm; lvs all cauline, gen 3–10 cm × 5–20 mm, sessile, linear-oblong to elliptic-oblong, the upper more lanceolate or lance-ovate; racemes 2–8 cm, closely fld, finely glandular-puberulent, shortly and unevenly toothed; corolla lavender-purple, 12–16 mm, the lips only 3–5 mm; moist, low places at lower elev, chiefly E Cas (*Dracocephalum nuttallii, P. n.*).

Prunella L.   Self-heal; All-heal

Fls in verticils, crowded into a dense, evidently bracteate terminal spike, the bracts sharply differentiated from the lvs; calyx irreg 10-nerved, bilabiate, the upper lip broad, shallowly 3-toothed, the lower deeply cleft, with 2 narrow segms; corolla blue or purple to white, bilabiate, the upper lip galeate and ± entire, the lower shorter and 3-lobed; stamens 4, ascending under the galea, the lower pair the longer, scarcely exserted (± reduced or obsolete in ♀ pls); filaments ± bidentate at the tip, the anther borne on the lower tooth; pollen sacs subparallel on the somewhat expanded connective, separately dehiscent; per herbs, erect to spreading, with entire to pinnatifid lvs. (Name of doubtful origin). (*Brunella*).

P. vulgaris L. Pls fibrous-rooted; sts 1–5 dm; lvs few, petiolate, the bl 2–9 × 0.7–4 cm; spikes short and dense, 2–5 × 1.5–2 cm; bracts depressed-ovate, abruptly short-acuminate, ca 1 cm, ± strongly ciliate; calyx 7–10 mm, the lips > tube, the teeth spinulose-tipped; corolla gen blue-violet, 1–2 cm (smaller in ♀ pls), with short lips; moist places, from sea level to mid elev in mts. The Eurasian var. *vulgaris* has the middle cauline lvs ca half as wide as long, with broadly rounded base; we have it as an occ weed in disturbed sites, where it is often dwarf and prostrate. The native Am var. *lanceolata* (Barton) Fern. is ascending or erect, with middle cauline lvs ca a third as wide as long and more tapering toward the base; it is common in both disturbed and natural habitats in our range.

Salvia L.   Sage

Fls in ± interrupted spikes (series of verticils) or in pans or racemes; calyx 10–15-nerved, bilabiate, the lips toothed or cleft to entire; corolla bilabiate, the lower lip 3-lobed, the upper lip tending to be subgaleate and with the 2 lobes sometimes suppressed; stamens 2, ascending under the upper lip, sometimes exserted; filaments rather short; connective elongate, art near the middle (or nearer the lower end) to the filament and bearing a pollen sac on each end, or the lower pollen sac or the whole lower arm of the connective ± suppressed; herbs or shrubs of diverse habit. (L name of the cult sage).

1a Pollen sacs 2, 1 at each end of the elongate connective, that on the lower arm < the one of the upper; suffrutescent per herb 2–6 dm, with petiolate, chiefly cauline lvs, the bl canescent, crenulate (seldom also few-lobed), 2–10 cm; Mediterranean sp., widely cult, and occ escaped in our range; garden sage   1 S. officinalis L.

1b Pollen sac 1, the lower arm of the connective short and sterile, or wholly suppressed

  2a Much br shrubs 2–5 dm; lvs ∝, closely silvery, entire, oblanceolate to obovate, (1) 1.5–3 (3.5) cm × 4–15 mm; bracts broad, ca 1 cm, often anthocyanic; corolla gen blue-violet, ca 1 cm; lower arm of connective suppressed; native sp. of dry, open places at lower elev, often with sagebr; GB, n E Cas to Wn; gray ball s.; ours is var. *carnosa* (Dougl.) Cronq. (*S. c.*)   2 S. dorrii (Kell.) Abrams

  2b Coarse herbs 2–15 dm with the largest lvs at or near the base; lower lf bls 6–25 cm, evidently toothed (sometimes also lobed); corolla gen 1.5–3 cm; lower arm of connective evident, but distinctly < upper; intro, weedy spp. of roadsides and other disturbed habitats

    3a Corolla pale yellow; herbage floccose-woolly, eventually partly glabrate, not evidently glandular; bracts broad, 1–2 cm; bien 2–7 dm; Mediterranean weed, intro in sw US and reaching our range in e Ore; African s.   3 S. aethiops L.

    3b Corolla blue to sometimes white, or marked with yellow; herbage more coarsely hairy to subglab, also glandular at least in the infl, not at all floccose-woolly; occ E Cas in our range

      4a Upper lip of calyx with short, approximate, inconspicuous teeth < 1 mm; bracts inconspicuous, gen < 1 cm; per 3–10 dm; meadow clary, meadow s.   4 S. pratensis L.

      4b Upper lip of calyx with the 2 lateral teeth prominent, aristate, 1.5–3 mm, and well separated from the smaller central tooth; bracts conspicuous, 1–3 cm; bien 5–15 dm; clary, clear-eye, see-bright   5 S. sclarea L.

Satureja L.   Savory

Fls verticillate (or solitary) in the axils of ordinary lvs, or the lvs subtending the upper (rarely all) whorls reduced; calyx 12–15-nerved in our spp., the 5 teeth subequal or arranged in 2 lips; corolla purple or yellow to white, 7–10 mm in our spp., bilabiate, the upper lip entire or emarginate, flat or nearly so; lower lip spreading, 3-lobed, the central lobe often > the others and sometimes emarginate; stamens 4, ascending under the upper lip, the lower pair often > upper; pollen sacs parallel or divergent, opening by separate slits; herbs or semishrubs with small, entire or toothed lvs. (L name of one of the spp., *S. hortensis*).

1a Per from a woody rhizome, the sts prostrate and often rooting, to 1 m, with short, ascending brs; lvs ovate to subrotund, 1–3.5 cm; fls solitary in the axils on peds 5–15 mm, the corolla white or purple-tinged; native in conif woods, chiefly W Cas, also e to n Ida; yerba buena   1 S. douglasii (Benth.) Briq.

1b Ann or nearly so, 1–2 dm, br from the base; lvs obovate-oblong to elliptic, 0.5–1.5 cm; fls in small axillary clusters on peds < 5 mm, the corolla bluish; European sp., occ in open places W Cas; basil-thyme (*Acinos arvensis*)   2 S. acinos (L.) Scheele

Scutellaria L.   Skullcap

Fls solitary (or less often verticillate) in the axils of normal or reduced lvs, sometimes forming ± definite racemes; calyx obscurely nerved, bilabiate, with entire lips, bearing a raised and gen prominent transverse appendage on or proximal to the upper lip; corolla variously colored, most often blue or violet, bilabiate, the upper lip gen ± galeate, the lateral lobes of the lower lip ± separated from the broad central lobe and marginally partly connate with the upper lip; stamens 4, the upper pair normal, the lower pair > upper, each with a single functional pollen sac, the other abortive; per (rarely ann) herbs or low shrubs. (L *scutella*, tray, referring to the appendage of the calyx).

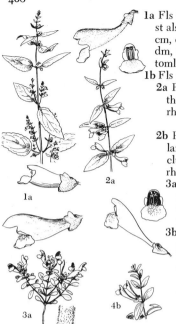

1a Fls in elongate racemes in the axils or terminating ordinary brs (the main st also terminating in a raceme); corolla 6–8 mm, blue; lvs gen 3–8 × 1.5–5 cm, ovate or lance-ovate, with broadly rounded or subcordate base; sts 2–8 dm, arising singly from slender rhizomes; transcontinental sp. of moist bottomlands; mad-dog s., blue s., madweed, hoodwort   **1 S. lateriflora** L.

1b Fls paired at the nodes (solitary in the axils); corolla (13) 15 mm or more

   2a Palate or corolla merely papillate; lvs mostly truncate-cordate at base, the larger ones gen (2) 2.5–5 cm; sts 2–8 dm, arising singly from slender rhizomes; circumboreal sp. of wet meadows etc; marsh s., willow-weed s.   **2 S. galericulata** L.

   2b Palate evidently long-hairy; lvs with ± tapering or rounded base, the larger ones gen 1–2.5 (3.5) cm; sts 0.5–3 dm, arising singly or in small clusters from slender to often moniliform-thickened or irreg tuberiferous rhizomes; w Am spp. of various, often drier habitats

      3a Corolla very pale in effect, the lower lip creamy, the upper gen dull pale-purplish; pubescence of the st downcurved; pls 0.5–1 dm; open, dry, often barren soil, often on basalt or andesite, in foothills and plains, c Ore to sw Ida, and s; dwarf s.; ours is var. *n*.   **3 S. nana** Gray

      3b Corolla deep blue-violet; pubescence of the st spreading or upcurved; pls (0.5) 1–3 dm; foothill and lowl to mid-mont pls, often in less severe habitats than no 3

         4a Corolla as pressed (13) 15–18 (22) mm, measured to the tip of the galea; c Ida to n Ore and s; snapdragon s.   **4 S. antirrhinoides** Benth.

         4b Corolla as pressed (22) 24–28 (32) mm; s BC to Cal, e to n and w Ida; narrow-lvd s.   **5 S. angustifolia** Pursh

Stachys L.   Hedge-nettle; Woundwort; Betony

Fls verticillate in the axils of the reduced upper lvs or bracts, gen forming terminal, often interrupted spikes; calyx 5–10-nerved, the 5 subequal teeth gen spinulose-tipped; corolla bilabiate, with entire or emarginate, subgaleate upper lip and spreading, 3-lobed lower lip, not much, if at all, exserted; pollen sacs strongly divergent, often confluent in dehiscence; ann or per herbs with toothed or entire lvs. (Transliteration of Gr word for ear of grain, referring to the infl). Ours all rhizomatous per with chiefly or wholly cauline lvs, growing in moist or wet places and fl June-Aug.

1a Lvs all evidently petiolate, the middle cauline ones with petioles gen 1.5–4.5 cm

   2a Corolla deep red-purple, relatively large, the tube 15–25 mm, the lower lip 8–14 mm; pls 7–15 dm; s BC to s Ore, from e slope Cas to coast; Cooley's h., great b. (*S. ciliata*, misapplied)   **1 S. cooleyae** Heller

   2b Corolla paler, gen pink or pink-purple, and smaller, the tube 8–13 mm; pls 3–8 dm; s BC to Cal, gen near the coast; great b., Mex b. (*S. emersonii, S. ciliata*)   **2 S. mexicana** Benth.

1b Middle and upper cauline lvs sessile, or some of them on short petioles not > 1 cm; corolla white-spotted on a purplish background

   3a Lvs becoming evidently petiolate toward base of st, some of the lower on petioles at least 1 (to 4) cm; calyx 5–6 mm, with relatively short, broad teeth; corolla 9–14 mm, the tube often evidently > calyx; Vancouver, Wn, s on both sides of Cas to Cal; rigid b. or h.; ours is var. *rigida*   **3 S. rigida** Nutt.

   3b Lvs all sessile or nearly so, the petiole, if any, gen well < 1 cm; calyx (6) 7–9 mm, the lobes a little < tube; corolla 11–16 mm, the tube only slightly (if at all) > calyx; widespread boreal sp., occurring nearly throughout our range; swamp h., marsh b.; ours is var. *pilosa* (Nutt.) Fern. (*S. p.*)   **4 S. palustris** L.

Teucrium L.   Germander; Wood Sage

Fls in terminal, bracteate spikes or racemes, or solitary in the axils of modified upper lvs; calyx 10-nerved, with 5 scarcely to evidently unequal teeth; corolla seemingly 1-lipped, the upper lip represented only by its 2 lobes, which are separated and displaced so as to arise from the margins of the well developed, declined, otherwise 3-lobed lower lip, of which the central lobe is much the largest; stamens 4, exserted, the lower pair the longer; pollen sacs strongly divergent, often confluent in dehiscence; ovary merely 4-lobed, the nutlets laterally attached and almost wholly united; herbs with entire to deeply lobed lvs. (Gr *Teukrion*, ancient name for some member of this or related genus).

T. canadense L. Rhizomatous per 2–10 dm, spreading-hairy, glandular in the infl; lvs cauline, short-petiolate, the bl 3–10 × 1–4 cm; infl a crowded, spiciform raceme 5–20 cm, with slender bracts ca 1 cm or less; calyx 5–7 mm, bilabiate; corolla purplish, 11–18 mm; widespread Am sp. of moist, low ground, occ in our range; ours is var. *occidentale* (Gray) McClintock & Epling (*T. o.*).

Thymus L.   Thyme

Fls verticillate in the upper axils; calyx 10–13-nerved, villous at the throat within, bilabiate, the broad upper lip 3-toothed, the lower lip more deeply cleft into 2 narrow lobes; corolla bilabiate, the upper lip nearly flat, the lower spreading and 3-lobed; stamens 4, gen exserted, subequal or the lower pair the longer; pollen sacs parallel or divergent, the connective expanded; small shrubs or subshrubs with small, entire lvs. (Gr *thymos*, ancient name of *T. vulgare*).

T. serpyllum L. Low per with br, widely divaricate or partly creeping, slender sts; lvs ± elliptic, seldom > 1 cm, short-petiolate or subsessile; verticils tending to be crowded to form a terminal, spicate-capitate or spiciform infl, the subtending lvs of the upper verticils ± reduced; calyx 3 mm; corolla purple, 4–5 mm; Eurasian sp., often cult as a ground cover, occ escaped in our range, esp W Cas.

Trichostema L.   Blue-curls; Trichostema

Fls in cymose axillary clusters; calyx 10-nerved, reg in ours; corolla irreg, the lowest lobe the largest, declined, the other 4 ca = and half as long as the lowermost one, the sinus between the 2 upper lobes scarcely deeper, and not much, if at all, more pronounced that the lateral sinuses; stamens 4, exserted, the filaments often arcuate; pollen sacs divaricate, often confluent in dehiscence; ovary rather deeply 4-lobed, the sculptured nutlets laterally attached, united ca 1/3 their length; ann or per herbs or near-shrubs with gen entire lvs. (Gr *trichos*, hair, and *stema*, stamen, the stamens long and exserted). Our spp. ann, 1–5 (10) dm, of moist, open places, often in disturbed soil.

1a Filaments 2.5–6 mm; corolla tube 2–3.5 mm, slightly arcuate, slightly or not at all exserted from the calyx; lowest corolla lobe 2–3.5 mm; lvs gen 1.5–5 cm × 5–20 mm and 1.7–3.5 times as long as wide, scarcely crowded, with normal pinnate venation; Wn and adj Ida to Cal and w Nev; mt b.
                                                                 1 **T. oblongum** Benth.
1b Filaments 10–20 mm; corolla tube 5–10 mm, evidently exserted from the calyx, abruptly bent upward at a right angle near the tip; lowest lobe of corolla 4–8 mm; lvs narrower, 2–7 cm × 4–20 mm and gen 3.5–7 times as long as wide, ± crowded, prominently (3) 5-nerved from below the middle; Wn to Baja Cal; vinegar weed   2 **T. lanceolatum** Benth.

SOLANACEAE   Potato or Nightshade Family

Fls gamopet, ♀, gen 5-merous as to calyx, corolla, and androecium; corolla gen reg, elongate to rotate; stamens epipetalous, alt with corolla lobes; ovary superior, gen 2-carpellary and 2-locular, with axile placentation, occ falsely 4-locular, rarely with several carpels and locules; style solitary, with capitate or slightly bilobed stigma; fr a caps or berry; seeds with well developed endosperm and subperipheral, often curved embryo; herbs, less often shrubs, vines, or even trees, with gen alt, exstip, simple to pinnately dissected lvs and diverse types of infls that appear to be mostly or wholly of eventually determinate origin.

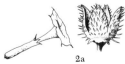

1a Fr caps, dehiscent; fls relatively large, the corolla funnelf to salverf, in ours 2–10 cm; heavy-scented, narcotic herbs, ours ann or bien
  2a Fls essentially terminal and solitary in origin, soon appearing to be in the forks of the brs; calyx circumscissile near the base; corolla (ours) 6–10 cm; fr gen spiny   **Datura**

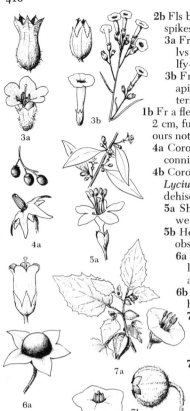

2b Fls borne in terminal, sometimes lfy, infls (pans, mixed pans, racemes, or spikes); calyx not circumscissile; corolla (ours) 2–6 cm; fr not spiny
  3a Fr calyx conspicuously accrescent, concealing the fr; caps operculate; lvs mostly sessile; fls in a terminal, secund raceme or spike that is lfy-bracteate at least below; corolla funnelf **Hyoscyamus**
  3b Fr calyx scarcely accrescent, not concealing the fr; caps dehiscent by apical valves; ours with mostly petiolate lvs borne in a terminal, sparsely or scarcely bracteate, mixed pan **Nicotiana**
1b Fr a fleshy or dry berry, indehiscent; fls smaller, ours (except in *Atropa*) < 2 cm, funnelf or camp to rotate, or with reflexed lobes; pls of diverse habit, ours not markedly heavy-scented
  4a Corolla lobes much > the short tube, sooner or later reflexed; anthers connivent, > filaments, opening by terminal pores **Solanum**
  4b Corolla lobes not much, if at all, > tube, not reflexed except casually in *Lycium,* or the corolla scarcely lobed; anthers free, ca = or < filaments, dehiscent by longitudinal slits
    5a Shrubs, often thorny; corolla with flaring, basally slender tube and well developed lobes > half as long as the tube **Lycium**
    5b Herbs, never thorny; corolla ± camp to subrotate, the lobes short or obscure, much < half as long as the tube
      6a Corolla gen 2.5–3 cm, distinctly longer than wide; robust, entire-lvd, per herbs to 15 dm, with ped, gen solitary fls at the upper axils **Atropa**
      6b Corolla < 2 cm, scarcely, if at all, longer than wide; ann or per herbs, ours to ca 6 dm
        7a Fr calyx bladdery-inflated, nearly closed, loosely investing and completely concealing the berry; corolla in ours light yellow or greenish-yellow, marked with purple or brown in the throat **Physalis**
        7b Fr calyx rather closely investing the berry, wide-mouthed, so that the top of the berry is exposed; corolla white or pale creamy, sometimes marked with purple or greenish **Chamaesaracha**

Atropa L.   Atropa

Fls ped and gen solitary at the upper axils; calyx herbaceous, deeply 5-cleft, ± accrescent and spreading in fr; corolla broadly tubular-camp, distinctly longer than wide, dirty purplish to greenish or yellowish, shallowly 5-lobed; filaments elongate, attached to base of corolla; anthers short, longitudinally dehiscent; fr a ∝-seeded black berry; seeds flattened, with strongly curved embryo; robust per herbs with entire, alt lvs, or the upper lvs in alt pairs, the members of each pair unequal and both on the same side of the st. (For *Atropos,* the Fate who severs the thread of life).

A. belladonna L. Dwale, deadly nightshade, belladonna. Per 6–15 dm, gen ± glandular-hairy above, otherwise ± glab; lvs short-petiolate, the bl ovate to elliptic, gen 8–16 × 4–8 cm; peds 1–2.5 cm; calyx segms ovate or broadly lanceolate, acuminate, 1–1.5 cm (or larger in fr); corolla 2.5–3 cm; berry 1–2 cm. European sp., cult for atropine, occ as a weed W Cas.

Chamaesaracha Gray   Chamaesaracha

Fls terminal and solitary (or in small clusters), but appearing axillary or eventually in the forks of the brs, because of the adjacent shoots developed from 1 or a pair of ± displaced axillary buds; calyx camp, 5-toothed or -lobed, ± accrescent but wide-mouthed, so that the rather closely invested berry is exposed above; corolla subrotate, woolly at the base within, slightly (if at all) 5-lobed; anthers = or < filaments, longitudinally dehiscent; fr a ∝-seeded berry; seeds flattened, with curved embryo; low per with entire to pinnatifid lvs. (Originally used as name of section of tropical genus *Saracha,* which was named for Isadore Saracha, a Spanish monk, with the addition of Gr prefix meaning on the ground).

C. nana Gray. Dwarf c. Rhizomatous, colonial per, seldom 1.5 dm; herbage and calyx strigose-puberulent; lvs petiolate, with narrowly to broadly ovate or rhombic, subentire bl 1.5–5 cm; corolla 1.5–2 cm wide, white or dingy, sometimes marked with purple; fr peds arching-recurved; berry ca 1 cm; dry, open places and open woods, esp in sandy soil; c Ore to Cal and s Nev.

Datura L.   Jimson-weed

Fls terminal and solitary in origin, but appearing (at least at maturity) to be borne in the forks of the brs, because of the shoots produced from a pair of ± displaced axillary buds; calyx cylindric or prismatic, gen circumscissile near the base, leaving a persistent flaring collar under the fr; corolla elongate-funnell, very large, the lobes well developed or represented by slender projections; stamens ± included, the anthers longitudinally dehiscent, much < filaments; fr a 2-carpellary, 4-celled, gen spiny caps, gen dehiscing by 4 valves; seeds ∝, flattened, with curved embryo; heavy-scented, narcotic, poisonous herbs, shrubs, or trees with large, entire to lobed lvs. (Name somewhat altered from Hindustani and Arabic cognate names for 1 or more of the spp.).

D. stramonium L. Stramonium, thorn apple, j. w. Coarse, often divaricately br ann to 1.5 m; lf bls coarsely few-toothed or sublobate, to 2 × 1.5 dm; calyx 3.5–5 cm; corolla 6–10 cm, the limb 3–5 cm wide; caps 3–5 cm; waste places; widespread weed, occ with us, commoner s; 2 vars.: Var. *stramonium* has green sts and white fls, with the lower spines of the fr gen < upper. Var. *tatula* (L.) Torr. (*D. t.*) has anthocyanic sts and fls, the lower spines of the fr ca = upper.

Hyoscyamus L.   Henbane

Fls. showy, in terminal, gen secund and ± lfy-bracteate racemes or spikes, the lower merely axillary; calyx camp or urceolate, 5-toothed, accrescent and enclosing the fr; corolla funnell, with an oblique, 5-lobed, slightly irreg limb; stamens gen exserted, anthers longitudinally dehiscent, much < filaments; caps ± 2-locular, circumscissile well above the middle; seeds ∝, flattened, tuberculate or roughened; embryo curved; heavy-scented, narcotic, poisonous herbs with ample, gen toothed to incised-pinnatifid lvs. (Gr *hyoskyamos*, from *hys*, sow, and *kyamos*, bean).

H. niger L. Black h., hog's bean. Coarse, viscid-villous ann to 1 m; lvs sessile, 5–20 × 2–14 cm, rather shallowly pinnatilobate; mature calyx 2.5 cm, urceolate, dry; corolla 2.5–4.5 cm, distally purple-reticulate on a pale background; fr 1–1.5 cm, with strongly thickened lid; European weed of roadsides and waste places, now casually estab over much of the US.

Lycium L.   Lycium

Fls axillary, 1–4 (∝) in an axil; calyx 3–6-lobed, camp to tubular, ruptured by the fr; corolla tubular to funnell, 4–7-lobed; anthers longitudinally dehiscent, much < the slender filaments, fr a berry; seeds 2–∝, ± compressed, with curved embryo; shrubs or small trees, gen thorny, with entire or minutely toothed, often fascicled lvs. (Gr *lykion*, the name of some thorny pl).

L. halimifolium Mill. Matrimony vine, box-thorn. Glab scrambling shrub 1–6 m, gen sparsely thorny; lvs short-petiolate, entire, to 7 × 3.5 cm, often much smaller; peds 1–3 per axil, 0.7–2 cm; corolla pale anthocyanic, 9–14 mm, the (4) 5 lobes < or ca = tube; fr ellipsoid or ovoid, 1–2 cm, red; Eurasian sp., occ escaped from cult in our range (*L. chinense*).

Nicotiana L.   Tobacco

Fls in terminal pans, mixed pans, or racemes; calyx toothed or cleft; corolla funnell or salverf, with gen spreading limb; stamens included or exserted, with long filaments and short, longitudinally dehiscing anthers; caps gen 2-locular and dehiscent by 2–4 apical valves; seeds ∝, scarcely flattened, with straight or curved embryo; narcotic herbs or shrubs with entire or merely toothed lvs, ours ann, 3–10 dm, with vespertinal, whitish fls and 4-valved caps ca 1 cm. (For Jean Nicot de Villemain, who was concerned with the intro of tobacco into Europe in the 16th century).

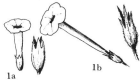

1a Longer calyx lobes 2–4 mm, distinctly < tube; corolla (2) 2.5–3.5 cm, the limb 8–14 mm wide when expanded; widespread native sp. of dry, open, often sandy places E Cas; coyote t.   1 N. attenuata Torr.
1b Longer calyx lobes 5–7 mm, = or > tube; corolla ave longer, to 6 cm, the limb 12–22 mm wide when expanded; S Am sp. of sandy stream banks, intro in Cal, rare with us; wild t. (*N. caesia*) 2 N. acuminata (Grah.) Hook.

1a   1b

Physalis L.   Ground-cherry

Fls actually terminal and gen solitary, but appearing axillary (or eventually in the forks of the brs) because of the adjacent shoots that develop from 1 or a pair of ± displaced axillary buds; calyx fairly small and 5-toothed or -lobed at anthesis, becoming bladdery-inflated (2–4 cm long in ours), 5–10-angled, and dry at maturity, loosely enclosing the fr, closed or barely open at the summit, the teeth remaining relatively small; corolla broadly camp to subrotate, shallowly or scarcely 5-lobed, gen (including ours) pale yellow or greenish-yellow, with purplish or brownish throat; anthers longitudinally dehiscent, ca = or < filaments; fr a ∞-seeded berry, ca 1 cm (or a bit more in ours), seeds flattened, with curved embryo; herbs with gen entire or merely toothed lvs, ours weeds 1–6 dm of cult fields and waste places, with lf bls 4–10 cm. (Gr word for bladder, referring to the inflated fr calyx).

1a Colonial per from running rhizomes (or roots); st glab or inconspicuously strigose; filaments flattened, app 1 mm wide; corolla 11–17 mm long and wide; lvs not cordate; chiefly or wholly E Cas; long-lvd g.; 2 vars.
              1 **P. longifolia** Nutt.
  a1 Lvs relatively thick and firm, gen ± entire, lanceolate or lance-elliptic to narrowly rhombic and ± tapering to the petiole; Gr Pl var., reaching our range in w Mont and casually intro elsewhere E Cas   var. **longifolia**
  a2 Lvs thinner and softer, gen ± sinuate-toothed, broader, more ovate, and more abruptly contracted to the often longer petiole; E Am var., casual with us (*P. s.*)   var. **subglabrata** (Mack. & Bush) Cronq.
1b Taprooted ann; st conspicuously spreading-villous; filaments slender, 0.2–0.3 mm thick; corolla 6–10 mm long and wide; lvs ± cordate or subcordate at base; tropical and e Am sp., intro W Cas; low hairy g.; ours is var. *grisea* Waterfall (*P. pruinosa*, perhaps misapplied)   2 **P. pubescens** L.

Solanum L.   Nightshade

Fls in diverse sorts of infls, extra-axillary in all ours, arising opp or between the lvs; calyx camp or rotate, toothed or cleft, sometimes accrescent; corolla with short tube and ± spreading, eventually reflexed, gen pointed lobes; fr a berry; seeds ∞, flattened, with annular embryo; herbs, shrubs, or vines, with simple to bipinnatifid lvs, ours all ± weedy and (except no 2) with a long fl season. (Classical L name for some pl).

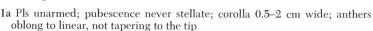

1a Pls unarmed; pubescence never stellate; corolla 0.5–2 cm wide; anthers oblong to linear, not tapering to the tip
  2a Rhizomatous per, woody below, tending to climb or scramble to 1–3 m; infls br, 10–25-fld; frs bright red; corolla blue or light violet, the lobes 5–9 mm, soon reflexed; lvs gen dimorphic, some simple, entire, and ovate-cordate, others with a prominent pair of basal lobes or lfts; Eurasian sp., widely intro in thickets, clearings, and open woods; bittersweet, climbing n., felonwort, blue bindweed   1 **S. dulcamara** L.
  2b Taprooted ann 1–6 dm, not climbing; infls simple, 2–8-fld; frs greenish or yellowish to black; corolla white or faintly bluish, 5–10 mm wide when expanded
    3a Lvs evidently pinnatilobate, gen 2–5 × 1–3 cm, the rachis seldom wider than the length of the lobes; fr greenish; calyx ± accrescent, to 6 mm; pls foetid; native weed E Cas; cut-lvd n.   2 **S. triflorum** Nutt.
    3b Lvs entire to merely toothed or wavy-margined, ovate to deltoid, 2–8 × 1.5–5 cm; pls relatively inodorous
      4a Calyx accrescent, becoming 4–6 (9) mm, cupping the lower half of the greenish or yellowish fr; sts spreading-hairy; S Am weed, now widespread in our range (*S. nigrum* var. *villosum*, misapplied), hairy n.   3 **S. sarrachoides** Sendt.
      4b Calyx scarcely accrescent, not cupping the fr, only 2–3 mm at maturity; fr black; sts glab or appressed-hairy; widespread weed; black nightshade; ours mostly the native Am diploid var. *virginicum* L. (*S. nodiflorum*), but the very similar European hexaploid var. *nigrum* has been found at Portland, Ore; black n., garden n.   4 **S. nigrum** L.
1b Pls ± spiny; pubescence of the lvs stellate; corolla gen (1.5) 2–3 (3.5) cm wide; anthers tapering to the tip; coarse herbs 3–10 dm
  5a Taprooted ann; calyx ± accrescent and enclosing the fr, the tube spiny even at anthesis; lvs deeply pinnately lobed

6a

6a Corolla yellow; infl stellate-hairy, not glandular-villous; 1 anther much > the others; Gr Pl sp., now widely intro elsewhere, and occ with us; buffalo bur 5 S. rostratum Dunal

6b Corolla blue or violet; infl glandular-villous as well as stellate-hairy; anthers =; S Am weed, occ with us; viscid n. 6 S. sisymbriifolium Lam.

5b Rhizomatous per; calyx not accrescent; weedy spp., occ with us

6a Lvs sinuate-lobed, gen at least 4 dm broad; pubescence not concealing surface beneath; horse-nettle 7 S. carolinense L.

6b Lvs entire to sinuate, gen < 3 cm broad; pubescence concealing surface beneath; silver-lf n., bull-nettle 8 S. elaeagnifolium Cav.

SCROPHULARIACEAE Figwort Family

Fls gamopet, ♂; calyx of 4 or 5 distinct or united, similar or dissimilar sepals, or fewer by reduction or fusion; corolla gen 4–5-lobed, tubular and bilabiate, less often rotate or reg or both, rarely wanting; stamens gen epipetalous, typically 4 and didynamous, sometimes only 2, or 5, with the 5th one sterile or rarely *(Verbascum)* fertile, basically alt with corolla lobes; ovary superior, 2-carpellary, 2-celled; placentation axile; style solitary, with distinct or united stigmas; fr caps; seeds with well developed endosperm, ∝ or less often few; herbs, or occ shrubs or vines, with opp or alt, exstip, simple or sometimes dissected lvs and diverse types of gen indeterminate infls.

1a Corolla galeate, i.e., the upper lip forming a hood or beak (galea) that tends to enclose the anthers, the teeth of the upper lip short or obsolete (galeate condition least marked in *Euphrasia);* infl of lfy-bracted spikes, spikelike racemes, or heads; stigmas wholly united; sepals ± connate at least below (sometimes in groups)

2a Lvs opp (lfy bracts subtending fls sometimes alt); ours ann

3a Calyx somewhat inflated at anthesis, accrescent and very conspicuously inflated in fr; anthers merely sagittate, the pollen sacs merely acutish to rounded at base; seeds flattened, winged **Rhinanthus**

3b Calyx not much, if at all, inflated; pollen sacs, or some of them, strongly acuminate-spurred to caudate at base; seeds turgid, wingless

4a Lvs entire (though the lfy bracts may have a few teeth); caps asymmetrical, curved, dehiscent only along the convex margin; seeds few, gen 1 or 2 per locule **Melampyrum**

4b Lvs (in our spp.) evidently toothed; caps symmetrical or nearly so in form and dehiscence; seeds ∝

5a Corolla strongly galeate, in ours yellow and 1.5–2 cm; intro, weedy pls **Parentucellia**

5b Corolla weakly galeate, in ours white to anthocyanic and not > 1 cm; native and intro pls **Euphrasia**

2b Lvs alt, sometimes nearly all basal; ann or per

6a Pollen sacs similar in size and position; calyx lobes gen 5, sometimes only 4 or 2; lvs often basal as well as cauline, the principal ones toothed to pinnately dissected, with gen > 5 teeth or primary segms to a side **Pedicularis**

6b Pollen sacs unequally set, 1 medifixed and appearing terminal on the filament, the other (sometimes reduced or obsolete) attached by its apex and pendulous or lying alongside the upper part of the filament; calyx lobes 2–4; lvs wholly cauline, entire or few-toothed or few-cleft, seldom with > 5 teeth or segms to a side

7a Galea ± strongly > lower lip; per, except for *C. exilis,* in which the galea is very conspicuously > lower lip **Castilleja**

7b Galea only slightly, or not at all, > lower lip; ann

8a Calyx not cleft to the base, the 4 lobes subequal or often partly connate in pairs to form 2 lateral segms; infls ∝–fld **Orthocarpus**

8b Calyx cleft to the base into a dorsal and a ventral segm; infls in our spp. with ca 5 or fewer fls **Cordylanthus**

1b Corolla not galeate, though often bilabiate, the upper lip, if differentiated, not forming a hood or beak; infls, sepals, and stigmas diverse

9a Filaments 5 (the 5th a mere projecting knob or scale on the upper lip in *Scrophularia);* stigmas wholly united

10a Anthers 5; lvs alt; corolla rotate, only slightly irreg **Verbascum**

10b Anthers 4; lvs opp (the cauline sometimes much reduced); corolla tubular, ± strongly bilabiate

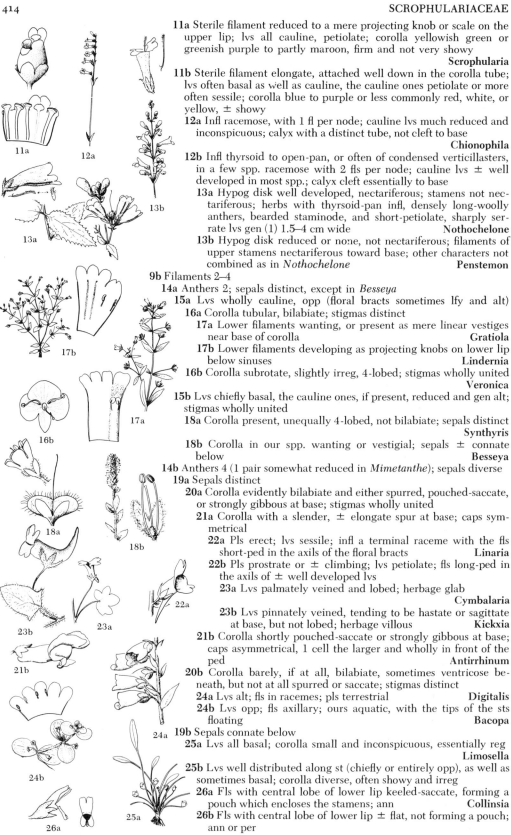

11a Sterile filament reduced to a mere projecting knob or scale on the upper lip; lvs all cauline, petiolate; corolla yellowish green or greenish purple to partly maroon, firm and not very showy **Scrophularia**

11b Sterile filament elongate, attached well down in the corolla tube; lvs often basal as well as cauline, the cauline ones petiolate or more often sessile; corolla blue to purple or less commonly red, white, or yellow, ± showy

  12a Infl racemose, with 1 fl per node; cauline lvs much reduced and inconspicuous; calyx with a distinct tube, not cleft to base **Chionophila**

  12b Infl thyrsoid to open-pan, or often of condensed verticillasters, in a few spp. racemose with 2 fls per node; cauline lvs ± well developed in most spp.; calyx cleft essentially to base

    13a Hypog disk well developed, nectariferous; stamens not nectariferous; herbs with thyrsoid-pan infl, densely long-woolly anthers, bearded staminode, and short-petiolate, sharply serrate lvs gen (1) 1.5–4 cm wide **Nothochelone**

    13b Hypog disk reduced or none, not nectariferous; filaments of upper stamens nectariferous toward base; other characters not combined as in *Nothochelone* **Penstemon**

9b Filaments 2–4

  14a Anthers 2; sepals distinct, except in *Besseya*

    15a Lvs wholly cauline, opp (floral bracts sometimes lfy and alt)

      16a Corolla tubular, bilabiate; stigmas distinct

        17a Lower filaments wanting, or present as mere linear vestiges near base of corolla **Gratiola**

        17b Lower filaments developing as projecting knobs on lower lip below sinuses **Lindernia**

      16b Corolla subrotate, slightly irreg, 4-lobed; stigmas wholly united **Veronica**

    15b Lvs chiefly basal, the cauline ones, if present, reduced and gen alt; stigmas wholly united

      18a Corolla present, unequally 4-lobed, not bilabiate; sepals distinct **Synthyris**

      18b Corolla in our spp. wanting or vestigial; sepals ± connate below **Besseya**

  14b Anthers 4 (1 pair somewhat reduced in *Mimetanthe*); sepals diverse

    19a Sepals distinct

      20a Corolla evidently bilabiate and either spurred, pouched-saccate, or strongly gibbous at base; stigmas wholly united

        21a Corolla with a slender, ± elongate spur at base; caps symmetrical

          22a Pls erect; lvs sessile; infl a terminal raceme with the fls short-ped in the axils of the floral bracts **Linaria**

          22b Pls prostrate or ± climbing; lvs petiolate; fls long-ped in the axils of ± well developed lvs

            23a Lvs palmately veined and lobed; herbage glab **Cymbalaria**

            23b Lvs pinnately veined, tending to be hastate or sagittate at base, but not lobed; herbage villous **Kickxia**

        21b Corolla shortly pouched-saccate or strongly gibbous at base; caps asymmetrical, 1 cell the larger and wholly in front of the ped **Antirrhinum**

      20b Corolla barely, if at all, bilabiate, sometimes ventricose beneath, but not at all spurred or saccate; stigmas distinct

        24a Lvs alt; fls in racemes; pls terrestrial **Digitalis**

        24b Lvs opp; fls axillary; ours aquatic, with the tips of the sts floating **Bacopa**

    19b Sepals connate below

      25a Lvs all basal; corolla small and inconspicuous, essentially reg **Limosella**

      25b Lvs well distributed along st (chiefly or entirely opp), as well as sometimes basal; corolla diverse, often showy and irreg

        26a Fls with central lobe of lower lip keeled-saccate, forming a pouch which encloses the stamens; ann **Collinsia**

        26b Fls with central lobe of lower lip ± flat, not forming a pouch; ann or per

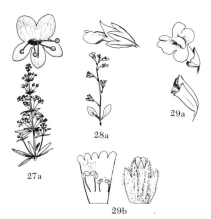

27a Corolla subrotate, slightly bilabiate, with short tube; stigmas wholly united; seeds gen 1–2 per cell **Tonella**

27b Corolla tubular, ± evidently bilabiate; stigmas gen distinct; seeds ± ∝

   28a Fls in terminal, minutely bracteate racemes; corolla blue-violet; calyx reg, the lobes = or > tube and with ± evident midrib **Mazus**

   28b Fls axillary to the upper lvs, these sometimes reduced, but the infl still gen ± lfy (fls falsely terminal in 1 sp. of *Mimulus*); corolla yellow to red, purple, or pink, but not blue-violet; calyx otherwise

      29a Calyx strongly 5-angled, the midribs to the = or unequal segms raised and prominent; deepest sinus of calyx extending much < half way to base **Mimulus**

      29b Calyx not angled, the midribs to the unequal segms obscure; ours with deepest sinus of calyx extending > half way to base **Mimetanthe**

Antirrhinum L.   Snapdragon

Fls in terminal racemes, or merely axillary; calyx of 5 essentially distinct sepals; corolla blue or purple to white or partly yellow, bilabiate, the lower lip with a prominent palate, the tube shortly pouched-saccate or strongly gibbous ventrally near the base; stamens 4, didynamous; stigma reduced to a small dot; caps asymmetrical, 1 cell larger or wholly in front of ped, the smaller cell opening by a single terminal pore, the larger one by 2 terminal pores, or the 2 pores confluent; style persistent and deflexed; seeds ∝; herbs, often glandular, with shortly or scarcely petiolate lvs, the lower opp, the upper gen alt; our spp. cult. (Gr *anti*, like, and *rhinos*, snout, in reference to the corolla).

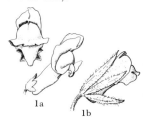

1a Corolla large, gen 2.5–4 cm, variously colored, gen several times as long as the ± ovate sepals, these 3–5 mm; lvs ± lanceolate, to 15 mm wide; pls fl the 1st year, but potentially per; common snapdragon, lion's-mouth; occ escaped from cult **1 A. majus** L.

1b Corolla smaller, gen 1–1.5 cm, gen pink-purple; sepals linear, elongate, 8–20 mm, often = or > corolla; lvs narrower, to 5 mm wide; ann; locally estab W Cas; weasel's snout, calf's snout, lesser s. **2 A. orontium** L.

Bacopa Aubl.   Water-hyssop

Fls axillary; sepals 5, distinct, dissimilar; corolla tubular-camp, 5-lobed, ± reg, the upper lobes external in bud; stamens 4; style shortly cleft, with distinct, terminal stigmas, rarely subentire; caps septicidal and loculicidal; seeds ∝; ± aquatic herbs with opp, gen sessile lvs. (L form of aboriginal name among natives of French Guiana).

B. rotundifolia (Michx.) Wettst. Fibrous-rooted per; sts 1–4 dm, distally floating and spreading-hairy; lvs palmately several-nerved, 1–3 cm, entire, glab; peds 0.5–2 cm; sepals 3–5 mm, the outer one broad, rotund-elliptic, the others much narrower; corolla white with yellow throat, 5–10 mm, the lobes = or a little < tube; mud-bottomed pools, chiefly c US, reaching our e border.

Besseya Rydb.   Besseya

Fls in terminal spikes or racemes; calyx of 2–4 segms, connate below; corolla wanting or vestigial in our spp.; stamens 2; stigma capitate; caps ± compressed, loculicidal; seeds ∝; fibrous-rooted per herbs with petiolate, toothed lvs and erect, lfy-bracteate sts, the bracts alt. (For Charles E. Bessey, 1845–1915, Am botanist).

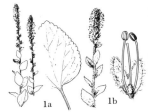

1a Calyx (3)4-lobed, the base cupulate and surrounding the ovary and stamens; lf bls gen 4–12 cm, tending to be deltoid or subcordate at base; open slopes and dry meadows in foothills and lowl E Cas; red b. (*Synthyris r.*) **1 B. rubra** (Dougl.) Rydb.

1b Calyx 2(3)-lobed, abaxial (external) to the ovary and stamens, and not surrounding them at base; lf bls gen 1.5–7 cm, seldom at all cordate; open slopes, from foothills to high elev; high plains and drier mts, w to e Ida; Wyo b. (*Synthyris w.*) **2 B. wyomingensis** (A. Nels.) Rydb.

Castilleja Mutis ex L. f.   Paintbrush; Indian-paintbrush; Castilleja
(Contributed by Noel H. Holmgren)

Fls several in bracteate, terminal spikes or spiciform racemes; calyx 4-cleft, the resulting segms subequal or in lateral pairs (primary lobes); corolla elongate and narrow, bilabiate, the upper lip (galea) beaklike, enclosing the anthers, the lower lip somewhat saccate at the base, the 3 lobes somewhat petaloid and nearly as long as the galea or reduced to teeth or rudimentary; stamens 4, didynamous, attached near or above the middle of the corolla tube, the anther-sacs 2, unequal in size and placement, the larger one medifixed, versatile, the smaller one apically attached, pendulous; stigma somewhat 2-lobed or capitate, projecting beyond the galea tip; caps loculicidal; seeds ∞ ; per or rarely ann herbs, hemiparasitic; lvs alt, entire or divided into 1–several pairs of lateral lobes. (Named for Domingo Castillejo, Spanish botanist).

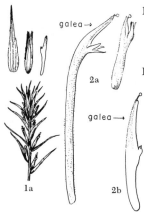

galea→

2a

galea→

1a   2b

1a Pls ann; lvs and bracts entire, linear-lanceolate; pl gen single-std; herbage villose, mainly with gland-tipped hairs; bracts red-tipped, calyx 15–25 mm, the segms 2–3 mm; corolla 15–25 (30) mm, the galea 5–10 (14) mm; saline marshes in the valleys; widespread in w US, e of the Cas from s Can s to n Ariz and nw NM; ann p.   1 C. **exilis** A. Nels.
1b Pls per; bracts 3–more-lobed, or if entire then broadly rounded; pl gen 2–several-std
  2a Galea short, 3–10 (13) mm (if > 10 mm the infl yellowish as occ in no 2, 20, 25, 26, and 27), rarely > half the length of the corolla tube; bracts gen yellowish, sometimes pinkish or dull red, rarely scarlet (as in no 19)   **Group I**
  2b Galea gen longer, (7) 10–20 mm (if < 10 mm the infl red or purplish as occ in no 32, 33, 39, and 42), > half the length of the corolla tube; bracts often showy, mostly red to purple, less often yellowish or whitish; lower lip much reduced, < 1/3 as long as the galea and with reduced incurved teeth   **Group II**, lead 29a

Group I

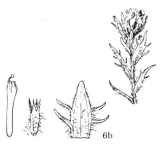

3a

3a Corolla tube long, (24) 30–45 mm, falcate, exserted from the calyx; calyx and bracts inconspicuous, wholly green or sometimes pink-tipped; herbage villose to somewhat lanate; lvs entire or the upper with a pair of narrow lateral lobes; calyx 25–40 mm; corolla 35–55 mm, the galea 9–12 mm; barely entering the region in nw Mont; dry habitats; Gr Pl from s Man & Ill, s to se Ariz, NM, n Mex, and w Tex; downy painted-cup
                                               2 C. **sessiliflora** Pursh
3b Corolla tube 8–19 (23) mm, gen enclosed by the calyx; calyx and bracts bearing most of the attractive coloration
  4a Calyx subequally cleft, the segms gen long, (3) 5–14 mm, linear to narrowly lanceolate, triangular in no 5, the entire calyx 10–18 (23) mm
    5a Herbage villose-hirsute to sometimes puberulent

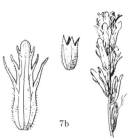

7b

      6a Bracts gen pale yellowish-green, sometimes with a purplish tint; sts ascending to erect, rarely decumbent, 1.5–3 (3.5) dm; galea (3) 4–5 (7) mm; mostly below 7000 ft elev
        7a Calyx 14–23 mm, the linear segms 6–12 mm; corolla (14) 16–24 (27) mm, the lower lip 3–5 (6) mm; sagebr and dry grassy meadows; barely entering the region in e Ore; s Ore, n Cal, and nw Nev; hairy p., white p. (*C. psittacina*)
                                    3 C. **pilosa** (Wats.) Rydb.
        7b Calyx shorter, 9–14 mm, the lanceolate segms 3–6 (9) mm; corolla (13) 15–17 (19) mm, the lower lip 2.5–4 mm; sagebr flats and slopes in dry gravelly soil; c Ida, nw Wyo, and adj Mont; white p.
                                      4 C. **longispica** A. Nels.
      6b Bracts purplish; sts decumbent, 0.2–0.7 (1) dm; galea 3.5–4 mm; calyx 10–13 mm, the lanceolate segms 3.5–6.5 mm; corolla 12–15 (17) mm; alp tundra; Wallowa Mts of ne Ore; purple alpine p.
                                     5 C. **rubida** Piper
    5b Herbage tomentose or arachnoid-lanate with woolly appressed hairs; bracts dull red, or red-orange to pale greenish-yellow
      8a Galea puberulent or only moderately pubescent on back with crisped hairs, 3–5 mm; bracts sparsely pubescent on back, dull red or red-orange, rarely greenish-yellow; sts decumbent, 0.8–2 (3) dm;

6b

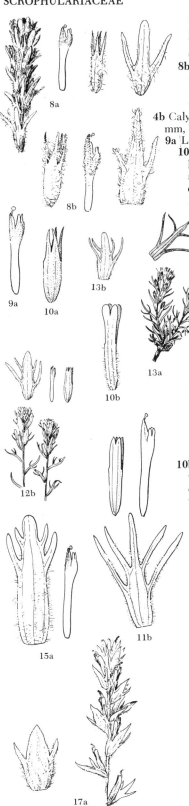

8a

8b

9a

10a

10b

13b

13a

12b

11b

15a

17a

calyx 12–19 mm; segms (5) 6–8 (10) mm; corolla (12) 15–17 (20) mm; dry sandy or gravelly slopes from mid to high elev; s Cas (from Three Sisters s), Klamath Mts, and n Sierra Nevada; cobwebby p., cotton p. (*C. payneae*)   6 **C. arachnoidea** Greenm.

8b Galea conspicuously shaggy-villose on back, 6–8 mm; bracts densely lanate, green or greenish-yellow; sts ascending, 0.5–1.5 dm; calyx 15–20 mm; segms 4–8 mm; corolla 18–25 mm; alp meadows; c and s Mont and adj Wyo; snow p.   7 **C. nivea** Pennell & Ownbey

4b Calyx clefts deeper medially than laterally, the ultimate segms 1–6 (8) mm, lanceolate to ovate, the entire calyx longer, (12) 14–32 mm

9a Lower corolla lip relatively long, (2) 3–6.5 mm, the lobes subpetaloid

10a Calyx segms acute (sometimes obtuse in no 11); herbage rather harshly pubescent or puberulent, not glandular; lower corolla lip ± strongly pouched and subequal to the galea (except in no 11); pls of dry habitats gen associated with sagebr

11a Herbage puberulent; lower corolla lip 3–7.5 mm, ± strongly pouched and 2/3 to as long as the galea

12a Calyx segms 4–7 mm, linear to lanceolate

13a Pls small, decumbent, 0.5–1.5 (1.7) dm; bracts strongly nerved; herbage puberulent with uniformly short recurved hairs; calyx segms 4–5 mm, lanceolate; galea 4–6.5 mm; c and sw Ida, se Ore, and ne Nev; dwarf pale i.
8 **C. inverta** (Nels. & Macbr.) Pennell & Ownbey

13b Pls taller, erect, (1) 1.2–2.5 dm; bracts less conspicuously nerved; herbage puberulent with various length hairs; calyx segms 5–7 mm, linear; galea 4–6 mm; ec Ore to wc Ida; pale Wallowa p.   9 **C. oresbia** Greenm.

12b Calyx segms shorter, 1–4 (6) mm, lanceolate to triangular; pls (0.5) 1–2.5 (3) dm; herbage puberulent with uniformly short recurved hairs; galea 4–6 (8) mm; 4500–9500 ft elev; w Mont, c Ida, and w Wyo; palish i.   10 **C. pallescens** (Gray) Greenm.

11b Herbage villous or hispid; lower corolla lip gen shorter, 2.5–4 (5) mm, less pouched and only 1/2–3/4 as long as the galea; pls 1–3 dm; calyx segms 1–5 (6) mm, broadly ellipsoid to triangular and broadly acute to sometimes obtuse or rounded; galea (4) 5–7 (8) mm; 2000–6000 ft elev; c Wn and s BC; Thompson's p.
11 **C. thompsonii** Pennell

10b Calyx segms obtuse to rounded, rarely acute; herbage villose to villulose with soft hairs, also obscurely to densely glandular; lower corolla lip little pouched and only 1/2 as long as the galea; pls of various habitats

14a Calyx segms 0.5–3 mm; galea (4) 5–7 (8) mm; herbage villose, sometimes viscid-villose; pls of moist habitats

15a Calyx (22) 23–28 mm; corolla mostly < the calyx, (19) 20–26 mm; pls erect to ascending, (1) 1.5–3 (4.5) dm; bracts > the fls, concealing them; valley to subalp meadows, 1000–9000 ft elev; lvs somewhat closely appressed, lanceolate to linear with 2 to 3 (4) pairs of narrow to subulate lateral lobes; bracts oblong-ovoid, obtuse; sw Alta, w Mont, w Wyo, and w through Ida to e Wn and ne Ore; Cusick's p. (*C. lutea*)
12 **C. cusickii** Greenm.

15b Calyx 12–21 (22) mm; corolla = or > the calyx, 12–22 mm; pls smaller, gen decumbent, 0.5–1.5 (2.5) dm; bracts gen < the fls; subalp to alp meadows, (6000) 7000–11,500 ft elev

16a Corolla 17–22 mm; stigma exserted beyond the tip of the galea; calyx lobes broadly obtuse or rounded

17a Bracts and calyces yellowish or at most faintly purplish, sometimes red; galea 6–8 mm; pls 0.9–1.8 (2.5) dm; lvs linear to lanceolate, entire or sometimes the upper with a pair of lateral lobes; bracts lanceolate to ovate, acute; 7000–9000 ft elev; Wallowa and Blue Mts of ne Ore; common Wallowa p., yellow p. (*C. ownbeyana*)
13 **C. chrysantha** Greenm.

17b Bracts and calyces purple or purplish, rarely yellowish; galea slightly shorter, 4.5–7 mm; pls gen smaller, 0.5–1 (1.7) dm; lvs linear to lanceolate with 1 (2) pair(s) of lat-

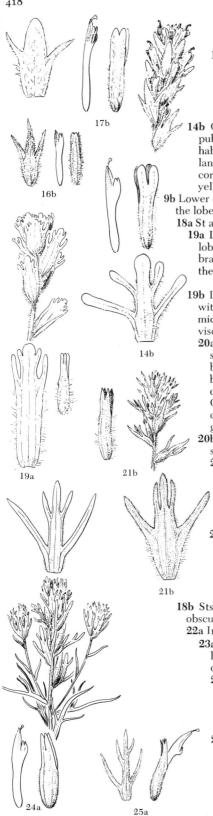

eral lobes; bracts lanceolate to ovate, rounded; (7000) 9000–11,500 ft elev; RM of Mont, Ida, Wyo, and ne Utah; showy i.   14 C. **pulchella** Rydb.

16b Corolla 12–16 mm, stigma not or tardily exserted from the galea; calyx lobes acute; lvs linear to lanceolate, entire or the upper with a pair of lateral lobes; bracts broader than the lvs, acute, purplish; at ca 6000 ft elev; endemic to Mt Rainier Nat Park, Wn; obscure i.
                                                                                    15 C. **cryptantha** Pennell & Jones

14b Calyx segms 3–7 mm; galea 7–8.5 mm; herbage glandular-puberulent in addition to the villose pubescence; pls of dry habitats; sts ascending to erect, 1–2.5 dm; lvs narrowly elliptic to lanceolate with 1 (2) pair(s) of lateral lobes; calyx (15) 20–27 mm; corolla 21–29 mm; 1000–2600 ft elev; John Day Valley of nc Ore; yellow-hairy i.   16 C. **xanthotricha** Pennell

9b Lower corolla lip much reduced, gen 2 mm or less, rarely up to 4 mm, the lobes reduced to teeth

18a St and lvs glandular-pubescent below the infl

19a Lvs and bracts oval, with 2–4 pairs of short, rounded, lateral lobes from near the apex; herbage viscid-villose; pls 1–5 dm; bracts golden yellow; calyx 15–18 mm; corolla 20–23 mm; lowl of the PT from VI, BC, s to WV, Ore; golden i.
                                                                                                17 C. **levisecta** Greenm.

19b Lvs and bracts linear to lanceolate, wavy-margined, entire or with 1 (2) pair(s) of slender, acute lateral lobes from ca the middle; herbage glandular-pubescent, often in addition to the viscid-villose pubescence; mts of ne Ore

20a Primary calyx lobes > 1/2 the calyx length, the ultimate segms 6–9 mm; galea stout and strongly decurved (i.e., dorsally bent forward from ca the middle); lvs short, 1–2 (2.5) cm; bracts gen greenish, the calyx yellowish-green to pink, orange, or reddish; galea 7–9 mm; dry gravelly slopes and summits; Ore, Tomalo Cr area (just e of the Three Sisters Peaks), Deschutes Co, and on Gearhart Mt of Lake and Klamath cos; green-tinged i.   18 C. **chlorotica** Piper

20b Primary calyx lobes < 1/2 the calyx length, the ultimate segms 2–6 mm; galea slender and only slightly arched forward

21a Corolla 18–22 (24) mm; calyx 13–17 mm; primary lobes 5–7.5 mm; pls 0.8–1.8 (3) dm; lvs short, 1.3–2.5 (3) cm; bracts and calyx yellowish-green, pink or red-orange to red; calyx segms 2–5 (6) mm; galea 7–9 mm; dry mt slopes, 6500–10,500 ft elev; Wallowa Mts of ne Ore, s through sw Ida to Nev; sticky i.   19 C. **viscidula** Gray

21b Corolla (20) 22–30 mm; calyx 17–22 mm; primary lobes (6) 7–11 mm; pls 1.2–2.5 (3) dm; lvs 2–3.5 (4.5) cm; bracts and calyx segms greenish-yellow, sometimes dull red; calyx segms 4–7 mm; galea 8–10 (12) mm; dry mt slopes, 7500–8500 ft elev; Blue Mts of ne Ore, sw to Strawberry Mt; glandular i.   20 C. **glandulifera** Pennell

18b Sts and lvs not glandular-pubescent below the infl (sometimes obscurely so in no 24)

22a Infl yellow, to yellow-orange, sometimes scarlet

23a Bracts narrowly lanceolate, not concealing the fls; infl puberulent to hirsute with whitish hairs; lvs linear, gen with 1–2 pairs of narrow to filiform lateral lobes, crisp-puberulent

24a Calyx subequally cleft medianly (in front and behind), 5–13 mm deep; calyx (12) 15–23 (27) mm, the segms (1) 2–4 mm; corolla (16) 19–24 mm, the galea 5–8 mm; dry sagebr slopes and plains; mts of c Ida and ne Ore; rural p., rustic p.   21 C. **rustica** Piper

24b Calyx more deeply cleft in front than behind, (6) 8–16 mm deep in front and 4–10 (14) mm deep in back

25a Pls 1.2–4 (5) dm; sagebr foothills, (5000) 6500–9000 ft elev; calyx 14–22 (28) mm, the segms 1–3 (5) mm; corolla 17–25 (30) mm, the galea (4) 6–9 mm; sw Mont, se Ida, w Wyo and adj Colo and Utah, recurring in sw Ida and ne Nev; yellow p.   22 C. **flava** Wats.

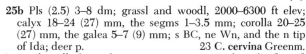

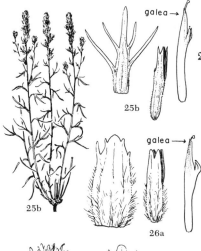

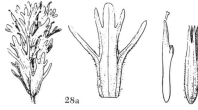

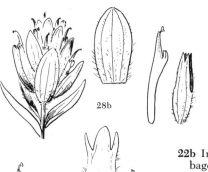

25b Pls (2.5) 3–8 dm; grassl and woodl, 2000–6300 ft elev; calyx 18–24 (27) mm, the segms 1–3.5 mm; corolla 20–25 (27) mm, the galea 5–7 (9) mm; s BC, ne Wn, and the n tip of Ida; deer p.   **23 C. cervina** Greenm.

23b Bracts broadly lanceolate to ovate, gen concealing the fls; infl villose, often with yellow hairs; lvs lanceolate, or if linear then entire, variously pubescent

  26a Pls small, 0.8–1.5 (2.5) dm; lvs lanceolate to narrow-lanceolate, gen all entire, but sometimes the uppermost with a pair of short lateral lobes, soft puberulent to viscid-villous; calyx (12) 15–23 mm, the primary lobes 5–10 mm, the segms 1.5–4 mm; corolla (15) 17–24 mm, the galea (5) 6–9 mm; subalp and alp meadows and talus, 7000–9000 ft elev; Can RM, s to GNP, recurrent in the RM of Colo; western p.
     24 C. occidentalis Torr.

  26b Pls relatively tall, 2.5–5 (8) dm

    27a Sts few to several, clustered on a stout woody caudex; corolla 20–30 mm

      28a Bracts with (1) 2 pairs of long, slender lateral lobes; herbage scab-puberulent; lvs entire or the upper with a pair of long lateral lobes; calyx (18) 20–25 (30) mm, the primary lobes 8–13 (18) mm, the segms 2–4 (5) mm, obtuse to acute; corolla 20–26 (28) mm, the galea (6) 7–9 (11) mm; low dry grassl and woodl, 1000–5500 (8000) ft elev; nw Mont, w across n Ida to extreme e Wn and ne Ore; yellowish p.   **25 C. lutescens** (Greenm.) Rydb.

      28b Bracts entire or with a pair of small lateral lobes; herbage glab to villose, sometimes slightly scab-puberulent; lvs entire, rarely the upper with a pair of short lateral lobes; calyx 15–22 (28) mm, the primary lobes 7–11 (13) mm, the segms 1–4 mm, obtuse to acute; corolla 20–27 (30) mm, the galea 7.5–10 (12) mm; moist meadows and slopes, 4000–12,000 ft elev; RM, s Alta, w Mont and adj Ida, s through Wyo to Colo, n NM, and Utah, also in the Black Hills of SD; sulfur p.   **26 C. sulphurea** Rydb.

    27b Sts gen solitary, rhizomatous; corolla shorter, (13) 15–21 mm; herbage glabrate to villose-puberulent; infl yellow, sometimes red to purple; calyx 13–20 mm, the primary lobes (5) 6–9 mm, the segms 1–3 (4) mm, triangular, obtuse to acute; galea 6–8 (10) mm; wet meadows and marshes, 5000–8000 ft elev; c Ida to nw Wyo and adj Mont; slender p.   **27 C. gracillima** Rydb.

22b Infl purple, crimson, rose, pink, or white, seldom scarlet; herbage glab to glabrescent below, villose in the infl; lvs broadly lanceolate to ovate, acute or obtuse, with 1(2) pair(s) of lateral lobes; subalp meadows; 3 vars. in our area **28 C. parviflora** Bong.

  a1 Corolla (16) 20–30 mm, the tube (9) 12–19 mm; calyx (16) 20–28 mm, the primary lobes (6) 8–14 mm, the segms (2)4–8 mm; pls 1.5–3 (3.5) dm; infl purple to crimson; Cas Range from Mt Rainier, Wn, s to the Three Sisters Peaks region of Ore; magenta p.   var. **oreopola** (Greenm.) Ownbey

  a2 Corolla shorter, 15–20 (23) mm, the tube 8.5–11 (14) mm; calyx 13–21 mm, the primary lobes 7–10 (11) mm, the segms 2–4 (6) mm

    b1 Infl purple to crimson; galea (6) 7–9 mm; pls 1–3 (4) dm; OM, Wn; OM p.   var. **olympica** (G. N. Jones) Ownbey

    b2 Infl pink to white, sometimes crimson; galea 5.5–7 mm; pls 1–2.5 (3) dm; Cas, from the vicinity of Cas Pass, Wn, n to BC; small-fld p.   var. **albida** (Pennell) Ownbey

Group II

29a Pls glandular-pubescent below the infl; lvs wavy-margined; 2 vars. in our area; Applegate's p., wavy-leaved p.   **29 C. applegatei** Fern.

  a1 Pls relatively tall, (2) 3–6 dm, if short (c Ida) the primary calyx lobes 4–11 mm; galea 9–12 (15) mm; lvs linear to lanceolate, sometimes all entire or the upper with a pair of slender lateral lobes; calyx (19) 20–24

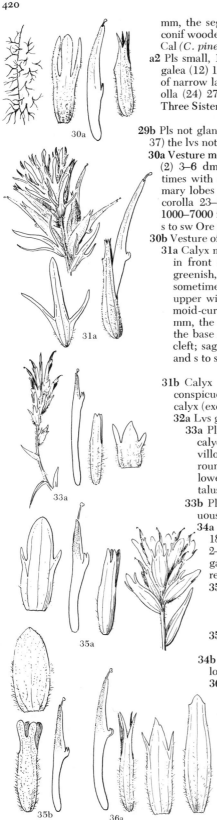

mm, the segms (3) 4–10 mm; corolla 23–30 (33) mm; sagebr and open conif wooded slopes; c Ida, across e Ore to the s Cas and s to n Nev and e Cal (*C. pinetorum*)   var. **fragilis** (Zeile) N. Holmgr.

a2 Pls small, 1–2.5 (3) dm, the primary calyx lobes gen longer, 9–15 mm; galea (12) 13–16 mm; lvs linear-lanceolate, at least the upper with a pair of narrow lateral lobes; calyx 18–25 mm, the segms (2) 3–5 (6) mm; corolla (24) 27–32 mm; open rocky or sandy pumice slopes; s Cas in the Three Sisters region, s to Crater Lake region and in the Paulina Mts
                                                                  var. **applegatei**

29b Pls not glandular pubescent below the infl or if glandular-pubescent (no 37) the lvs not wavy-margined

30a Vesture mostly of br (dendritic) hairs; pl sparsely pubescent to pruinose, (2) 3–6 dm; lvs mostly all entire, linear-lanceolate, the upper sometimes with a pair of lateral lobes; bracts red; calyx 15–25 mm, the primary lobes (5) 6–10 (15) mm, the slender, acute segms (2) 3–4 (7) mm; corolla 23–35 (38) mm, the galea (10) 14–19 mm; dry woodl or for, 1000–7000 ft elev; entering the NW around Mt Jefferson (Linn Co), Ore, s to sw Ore and n Cal; frosted p., pruinose p.   30 C. **pruinosa** Fern.

30b Vesture of simple hairs or none

31a Calyx much more deeply cleft in front than behind, 12–22 mm deep in front and 4–10 (18) mm deep in back; bracts reduced and gen greenish, the calyx conspicuously red colored; pls 3–7 (10) dm, glab to sometimes hispid below, hispid in the infl; lvs linear, entire or the upper with a pair of filiform lateral lobes; calyx 20–32 (39) mm, sigmoid-curved, the segms linear, 1–3 (4) mm; corolla (27) 30–42 (48) mm, the galea (11) 15–22 (24) mm, the tube bent forward from near the base so that the galea extends forward through the anterior calyx cleft; sagebr slopes and juniper woodl; s Mont, across s Ida to se Ore and s to s Cal, n Ariz, and nw NM; narrow-lvd p., Wyo p.
                                                       31 C. **linariaefolia** Benth.

31b Calyx ± subequally cleft medianly (in front and behind); bracts conspicuously colored and gen concealing the relatively inconspicuous calyx (except in no 32)

32a Lvs gen all entire, rarely the uppermost somewhat lobed

33a Pls small, 1–1.5 dm; infl coloration mostly from the bright red calyces, the bracts relatively inconspicuous; herbage viscid-villose to puberulent; calyx 15–20 mm, the segms 1–3 mm, rounded to acute; corolla 20–30 mm, the galea 8–12 mm, the lower lip relatively prominent, about 3 mm; alp meadows and talus; Wallowa Mts, Ore; fraternal p.   32 C. **fraterna** Greenm.

33b Pls > 2 dm; infl coloration various, mostly from the conspicuous bracts

34a Infl coloration purplish to crimson; pls gen < 3 dm; calyx 18–25 (28) mm, the primary lobes 8–12 (14) mm, the segms 2–4 (8) mm, acute to obtuse; corolla (20) 23–30 (34) mm, the galea 8–12 (15) mm (the following 2 spp. are very closely related perhaps varietally so)

35a Pl glabrate or only obscurely viscid-villose in the infl; RM from Alta and BC, s through Mont, Ida, Wyo, Colo, n NM, and Utah, also in ne Ore; rhexia-leaved p.
                                                       33 C. **rhexifolia** Rydb.

35b Pl strongly viscid-villose, at least in the infl; Wen Mts, n into s BC; Wen Mt p., Elmer's p.   34 C. **elmeri** Fern.

34b Infl coloration reddish (scarlet), sometimes red-orange to yellowish; pls gen > 3 dm

36a Bracts with a pair of long lateral lobes from about the middle, acute (except in some var. *dixonii*); galea (11) 15–20 mm, puberulent to villosulose on back; calyx segms (2) 4–8 mm, linear to lanceolate, acute; sts mostly erect; herbage glab to sparsely pubescent; calyx (18) 22–29 mm, the primary lobes 7–16 mm; corolla (25) 30–38 (42) mm; scarlet p., common p.; 2 vars. in our area   35 C. **miniata** Dougl.

a1 Lvs thin, attenuate, narrowly lanceolate to lanceolate; moist places at middle elev in the mts; widespread in w N Am, from s Alas, BC, and Alta s through all the mt States to NM, Ariz, and s Cal, missing the Coast Ranges of Ore and Cal   var. **miniata**

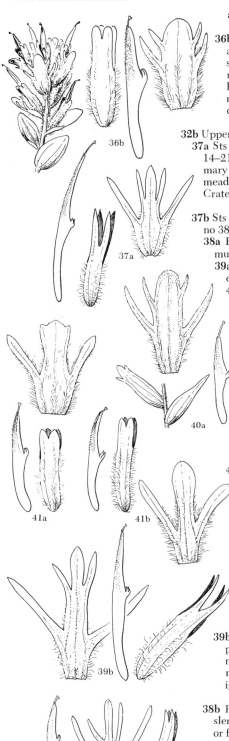

36b

37a

40a

41a   41b

39b

42a

a2 Lvs thick, lanceolate to ovate, blunt; s bluffs along the coast of Wn   var. **dixonii** (Fern.) Nels. & Macbr.

36b Bracts entire or with a small pair of lateral lobes near the apex, rounded to obtuse; galea gen shorter, 10–16 mm, shaggy-pubescent on back; calyx segms shorter, 1–3 (5) mm, rounded to obtuse; sts ± decumbent or ascending; herbage hispid-villose; calyx 18–25 (30) mm, the primary lobes 7–15 mm; corolla (23) 27–38 (42) mm; seaside bluffs; along the coast of Ore and n Cal; Pac p., Ore-coast p.

36 C. **litoralis** Pennell

32b Upper lvs gen with 1–3 pairs of linear, spreading lateral lobes

37a Sts solitary, rhizomatous; calyx segms (2) 5–12 mm; galea (12) 14–21 mm; pls (2.5) 3–5 (7.5) dm; calyx (20) 22–31 mm, the primary lobes (7) 12–18 mm; corolla (24) 30–40 mm; subalp meadows and for along streams; Cas from Mt Adams, Wn, to Crater Lake, Ore, n occ to Snohomish Co, Wn; Suksdorf's p.

37 C. **suksdorfii** Gray

37b Sts several to a cluster; calyx segms 1–5 (8) mm (longer in some no 38); galea (7) 10–15 (21) mm (gen 14–21 mm in no 43)

38a Pls (1.5) 2–3.5 (4.5) dm; herbage grayish-hispid with long multicellular hairs (except no 41); lvs linear-lanceolate or ovate

39a Corolla 20–30 (36) mm; pls of plains and foothills, gen fl early, Apr–June (Aug)

40a Lvs lanceolate to ovate-lanceolate; median clefts of the calyx subequal, 7–12 (14) mm deep; calyx 19–27 (29) mm; corolla 24–32 (36) mm, the galea (8) 10–14 (17) mm; grassy slopes and for openings from near the sea coast to 6000 (8000) ft elev; harsh p.; 2 vars.   38 C. **hispida** Benth.

a1 Lvs with 2–4 pairs of lateral lobes; pubescence ± villose; calyx segms gen short, 1–3 (8) mm, obtuse to rounded; Coast Ranges, extending inl along both sides of the international boundary to nw Mont and sw Alta   var. **hispida**

a2 Lvs with 1–2 pairs of lateral lobes; pubescence hispid-villose; calyx segms relatively long and narrow, (1) 3–5 (7) mm, acute; valleys in the RM of Ida and adj Mont, extending w to ne Ore and se Wn

var. **acuta** (Pennell) Ownbey

40b Lvs linear to linear-lanceolate; calyx slightly more deeply cleft in back (6–12 mm deep) than in front (4–7 mm deep); calyx segms 1.5–4 (5) mm, rounded to obtuse

41a Bracts and calyx segms purplish (rarely scarlet to yellowish); galea 6–12 mm; calyx 17–21 (23) mm; corolla 20–25 (27) mm; sagebr hills and plains; nw Wyo, adj Mont, and sw Ida, s to nw Utah; NW p.

39 C. **angustifolia** (Nutt.) G. Don

41b Bracts and calyx segms bright red, scarlet, orange, or yellow, never purplish; galea 10–18 mm; calyx (17) 20–26 mm; corolla 21–30 (36) mm; sagebr hills and plains; c and s Wyo, w Colo, nw NM, across s Ida, Utah, n Ariz, and Nev, to e Ore, s through e Cal; des p.

40 C. **chromosa** A. Nels.

39b Corolla 30–36 mm; pls of mts, fl later, July–Aug; herbage puberulent to obscurely villose; calyx 21–27 mm, the primary lobes 11–15 mm, the segms 3–5 mm; galea (10) 12–14 mm; extreme nw Wyo and adj Mont (possibly of hybrid origin involving nos 31 and 35); cocks-comb p.

41 C. **crista-galli** Rydb.

38b Pls small, 0.6–2 (2.5) dm; herbage green, villose, the hairs slender and entangled; lvs linear, with (1) 2–3 pairs of narrow or filiform lateral lobes; pls gen fl later, late June–Aug

42a Galea short, 7.5–13 mm, the tube longer, 14–20 (25) mm; pls 1–2 (2.5) dm; calyx 18–25 (30) mm, more deeply cleft in back (8–13 mm deep) than in front (5–9 mm deep), the segms 2.5–4 (6) mm; corolla (20) 22–30 (38) mm; rocky slopes and ridges, (6000) 7000–10,000 ft elev; RM of c Ida and adj Mont; RM p.   42 C. **covilleana** Hend.

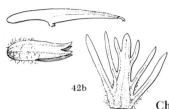

42b Galea longer, 14–21 mm, the tube short, (9) 11–14 (17) mm; pls 0.6–1.5 (2) dm; calyx 17–21 (26) mm, the median clefts subequal, 8–14 mm, the segms 1.5–6 (8) mm; corolla 27–35 (38) mm; crevices of rocks, 6000–7200 (4000 ft in BC) ft elev; Cas Range from c Ore to s BC; cliff p.

                                          43 C. rupicola Piper

Chionophila Benth.   Chionophila

Fls in a terminal, bracteate, secund raceme or spike; calyx with 5 teeth or lobes; corolla tubular, bilabiate, the upper lip external in bud; fertile stamens 4, the pollen sacs distally confluent; sterile filament slender, attached near corolla base, evidently < fertile ones; stigma capitate; seeds ∝, with an ariliform, reticulate coat; small, fibrous-rooted per with well developed, petiolate, basal lvs and ± reduced opp cauline lvs. (Gr *chion*, snow, and *philos*, beloved, referring to the alp habitat).

C. tweedyi (Canby & Rose) Hend. Slender, gen 0.5–2.5 dm, glab except the stipitate-glandular infl; basal lvs oblanceolate, 2.5–10 cm (petiole included) × 3–13 mm; cauline lvs much reduced, 1–3 pairs, seldom 2 cm; fls 4–10, short-ped, 9–14 mm, pale lavender; open places near timberl; c Ida and adj sw Mont.

Collinsia Nutt.   Blue-eyed Mary; Collinsia

Fls 1–several in axils of normal or ± reduced upper lvs (bracts); calyx with 5 subequal lobes; corolla strongly bilabiate, the upper lip 2-lobed, distally recurved, external in bud; lower lip 3-lobed, the central lobe < lateral ones, keeled-saccate, forming a pouch enclosing style and stamens; stamens 4, didynamous; pollen sacs confluent at tip; stigma ± capitate; caps dehiscent along 4 sutures; seeds 1–∝ per cell; slender ann 0.5–4 dm, with opp (or partly whorled), gen entire or toothed lvs. (For Zacheus Collins, 1764–1831, Am botanist).

1a Upper filaments pubescent; caps subglobose, 4–6 mm wide; seeds flattened, wing-margined, evidently cellular-reticulate; calyx 5–11 mm; corolla blue-lavender or white, 8–11 mm, the tube abruptly bent near the base; open slopes and swales E Cas, s Wn and n Ore to Cal; few-fld c. or b. m.; ours the var. *bruciae* (Jones) Newsom (*C. b.*)

                                       1 C. sparsiflora Fisch. & Mey.

1b Upper filaments, like the lower, glab except sometimes at point of insertion; caps ellipsoid, 2–4 mm wide; seeds, except gen in no 2, turgid, with thickened, inrolled margins, not evidently cellular-reticulate; corolla blue (pink), with ± white upper lip

  2a Corolla in the size range of no 3, but with the tube only slightly, if at all, bent; many of the hairs of the infl with expanded, glandular tip; open woods W Cas, s Wn to n Cal, and e rarely to John Day Valley, Ore; Rattan c.; ours, with flattened, thin-margined, cellular-reticulate seeds, have been called ssp. *glandulosa* (Howell) Pennell (*C. g.*)

                                       2 C. rattanii Gray

  2b Corolla tube evidently bent near the base so as to stand at an angle to the calyx and pedicel; hairs of infl eglandular or somewhat glandular, but seldom with expanded gland tip

    3a Corolla 4–7 (10) mm, the tube bent at an oblique angle to the calyx; widespread in vernally mesic habitats in w N Am; small-fld b. m.

                                     3 C. parviflora Lindl.

    3b Corolla 9–17 mm, the tube bent at about a right angle to the calyx; open places at middle and lower elev W Cas, and in CRG; large-fld b. m. (*C. diehlii*)   **4 C. grandiflora** Lindl.

Cordylanthus Nutt.   Birdbeak

Fls in small heads or compact spikes, or some single; calyx cleft to base into dorsal and ventral segms, or the lower segm obsolete; corolla tubular, bilabiate, the upper lip (galea) hooded and enclosing the anthers, with the lobes suppressed; lower lip ca = galea, a little inflated, external in bud, the 3 lobes short or obsolete; stamens 4 or 2, attached near middle of corolla tube; pollen sacs unequally set, 1 medifixed and appearing terminal on the filament, the other (sometimes reduced or obsolete) attached by its apex and pendulous or lying alongside upper part of filament; stigmas wholly united; caps loculicidal; seeds ∝ to rather few; ann with alt, entire to more often cleft or dissected, wholly cauline lvs; our spp. 1–6 dm, freely br when well developed. (L name from Gr *kordyle*, club, and *anthos*, fl). (*Adenostegia*).

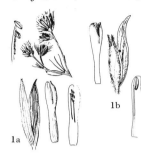

1a Stamens 4, each with 2 pollen sacs; upper calyx segm 5–7-nerved like the lower, and only minutely bidentate (to < 1 mm) at the tip; seeds ca 20; corolla dull yellowish, 12–17 mm; dry, open places E Cas, esp with sagebr; bushy b. (*A. ciliosa*)　　　　　　　　　　1 C. ramosus Nutt.

1b Stamens 2, each with a single pollen sac; upper calyx segm (the one next the galea) obviously only 2-nerved and obviously bifid (to 2–5 mm); seeds ca 8; corolla purplish, 10–15 mm; open places and dry woods E Cas, esp with sagebr; up to mid elev in mts; clustered or Yakima b. (*A. bicolor*)
　　　　　　　　　　　　　　　　　　2 C. capitatus Nutt.

Cymbalaria Hill

Fls long-ped, solitary in the axils; calyx reg, deeply 5-parted; corolla bilabiate, distinctly spurred at base, the upper lip external in bud, the throat closed by a prominent palate; stamens 4; caps globose, rupturing to form 2 terminal pores which later extend to the base; trailing herbs with alt, palmately lobed lvs. (From Gr word meaning cymbal).

C. muralis Gaertn. Ivy-lvd toadflax, Kenilworth ivy. Glab ann or bien 1–4 dm, trailing or twining, often climbing; lvs long-petioled, bl 1–3 cm, suborbicular in outline, with 3–7 shallow lobes; corolla 7–10 mm, blue with yellow palate, the spur obtuse, 2–3 mm; fr 3–4 mm thick; Eurasian sp., occ escaped from cult W Cas (*Linaria cymbalaria*).

Digitalis L.　Foxglove

Fls in terminal, bracteate racemes; calyx of 5 essentially distinct sepals; corolla tubular-camp, open, somewhat ventricose beneath, the lobes much reduced with the lowest one the largest, the upper external in bud; stamens 4, didynamous; stigmas 2, flattened; caps septicidal and loculicidal; seeds ∝; bien or per herbs with alt lvs. (L, pertaining to finger, or fingerlike, referring to the fls of *D. purpurea*).

D. purpurea L. Robust bien 5–18 dm, puberulent (or partly glab) and becoming viscid or glandular upward; lower lvs the largest, 1.5–5 dm (petiole included) × 3–12 cm, toothed; raceme elongate, secund, lfy-bracteate, the entire bracts unlike the foliage lvs; calyx foliaceous, 10–18 mm; corolla 4–6 cm, pink-purple, lower side paler and spotted; Eurasian sp., the source of digitalis, well estab in disturbed sites W Cas and occ farther e.

Euphrasia L.　Eyebright; Euphrasia

Fls in condensed or elongate, prominently lfy-bracteate terminal spikes; calyx 4-cleft, the lobes sometimes partly connate in lateral pairs; corolla bilabiate, weakly galeate, the upper lip only slightly hooded and often shortly 2-lobed; lower lip ca = or even > upper, evidently 3-lobed, external in bud; stamens 4, didynamous, converging under the galea; pollen sacs 2, =, 1 more strongly acuminate-spurred at base than the other; stigma entire; caps loculicidal, symmetrical in form and dehiscence; seeds gen ∝; slender ann or per, reputedly root-parasites; lvs opp, wholly cauline, entire to toothed or cleft, passing into the well developed, often alt bracts. (Gr, meaning delight, prob referring to reputed medicinal properties).

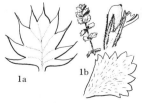

1a Bracteal lvs with subulate or bristle-tipped teeth; calyx and fr glab or nearly so; lvs gen glab; corolla 5–10 mm, the lower lip white with violet lines, the upper blue or purplish; European sp. of disturbed habitats, occ W Cas; hairy e. (*E. americana, E. canadensis*)　　　　1 E. officinalis L.

1b Bracteal lvs with acute or obtuse teeth; calyx and fr as well as lvs and bracts pubescent; corolla 3–6 mm, whitish with violet lines; circumboreal native sp. of wet places; northern e.; ours is var. *disjuncta* (Fern. & Wieg.) Cronq. (*E. d.*)　　　　　　　　　　　　　2 E. arctica Lange

Gratiola L.   Hedge-hyssop

Fls ped and solitary in the axils of ± reduced upper lvs; calyx of 5 distinct sepals, often closely subtended by a pair of sepal-like bracteoles; corolla tubular (ours with white or yellowish tube and small, purplish limb), bilabiate, the upper lip entire or shallowly lobed, external in bud, the lower 3-lobed; antheriferous stamens 2, the others wanting or represented by a pair of vestigial filaments near base of corolla tube on lower side; connective much enlarged and flattened, bearing the parallel pollen sacs on 1 face; stigmas 2, flattened; caps loculicidal and gen also septicidal (thus dehiscing by 4 valves); seeds ∝; ann or per herbs with opp, sessile, entire or toothed lvs, ours fibrous-rooted ann of shallow water, muddy shores, and other wet places in the valleys and plains, occurring ± throughout our range. (Diminutive of L *gratia*, favor, in allusion to its reputed healing properties).

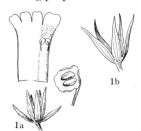

1a Ped bearing at the summit a pair of sepal-like bracteoles, the sepals thus apparently 7; pls ± stipitate-glandular upwards; corolla 7–10 mm; caps broadly ovoid, pointed, 5–7 mm; common Am h.   1 **G. neglecta** Torr.
1b Ped without bracteoles, the sepals evidently 5; pls glab or only obscurely glandular upwards; corolla 5–7 mm; caps more globose, not pointed, 4–5 mm; bractless h.   2 **G. ebracteata** Benth.

Kickxia Dumort   Cancerwort; Fluellin

Fls rather long-ped, solitary in the axils; calyx reg, deeply 5-parted; corolla bilabiate, distinctly spurred at base, the upper lip external in bud, the throat closed by the well developed palate; stamens 4; caps subglobose, each cell circumscissile on the side near its summit, nearly half of the valve being separated; ann with pinnately-veined, petiolate, alt lvs. (For Johann Kickx, Belgian professor).

K. elatine (L.) Dumort. Sharp-lvd f. Viscid-villous; sts prostrate, br, to 5 dm; lf bls 1–3 cm, broadly ovate to triangular-ovate, truncate at base and ± hastate, on a petiole 1–5 mm; peds 1–3 cm, glab, or villosulous at very base and very summit only; corolla 6–9 mm, yellow, upper lip purple within; spur 5 mm; Eurasian weed of arable land, intro W Cas in our range (*Linaria e.*).

Limosella L.   Mudwort

Fls on slender scapes that resemble elongate peds; calyx 5-toothed; corolla white or pinkish, inconspicuous, essentially reg, the 5 spreading lobes < tube, upper lobes external in bud; stamens 4, subequal, anther cells confluent; stigma capitate; caps septicidal, distally 1-celled; seeds ∝; glab per scapose herbs, the lvs entire and gen long-petiolate. (Latin *limus*, mire, and *sella*, seat, from the habitat).

L. aquatica L. Diminutive, fibrous-rooted from a small crown; lf bls rather narrowly elliptic, 5–18 × (1) 2–7 mm, much < the 1–8 cm petioles; scapes lax, 8–30 mm, 1-fld, or with a whorl of lvs at the top subtending several short-ped fls; calyx 2–3 mm; corolla tube nearly = calyx; widespread, circumboreal sp. of shallow water or wet mud at low elev.

Linaria Mill.   Toadflax

Fls in terminal racemes or spikes; calyx of 5 essentially distinct sepals; corolla yellow to blue or white, spurred at base, bilabiate, the upper lip external in bud, the lower lip often raised into a palate; stamens 4, didynamous; stigmas reduced to a point or capitate; caps cylindric or subglobose, rupturing irreg across the distal width of each cell; seeds gen ∝; herbs with sessile, alt (or the lower opp or ternate) lvs, ours glab, often glaucous. (L *linum*, flax, which several spp. resemble vegetatively).

2a

3a

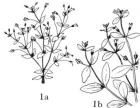

1b

1a Fls yellow with orange palate; frs 6–12 mm; intro, weedy Eurasian per from creeping roots, without a basal rosette

2a Lvs soft, linear, 2–5 cm × 2–4 mm, narrowed below to a short, petiole-like base; fls ∝ in a compact spike, 2–3 cm (incl spur); seeds flattened, winged; pls 2–8 dm, not esp robust, widely intro; butter and eggs
1 **L. vulgaris** Hill

2b Lvs firm, often broader, broadest toward the sessile, obtuse or rounded, often clasping base; racemes more elongate and loosely fld; seeds irreg wing-angled; pls robust, 4–15 dm

3a Fls 2.5–4 cm (incl spur); lvs relatively broad, 2–5 cm × (5) 10–25 mm, gen clasping; irreg throughout our range; Dalmatian t.
2 **L. dalmatica** (L.) Mill.

3b Fls 1.5–2 cm; lvs ave narrower, 1.5–4.5 cm × 2–10 (12) mm, not strongly clasping; occ W Cas; brown-lvd t.   3 **L. genistifolia** (L.) Mill.

1b Fls rather light blue, with paler, scarcely raised palate, to ca 2 cm (incl spur); frs 2.5–4 mm; ann or winter ann from a short taproot, with a basal rosette of prostrate, lfy sts and a slender, erect main st 1–5 dm; lvs of main st linear, 1–3.5 cm × 1–2.5 mm; moist, sandy places W Cas; blue t., wild t.; 2 vars.   4 **L. canadensis** (L.) Dumont

a1 Corolla gen 10–12 mm exclusive of the 5–9 mm spur; seeds densely tuberculate; s and w var., native W Cas (*L. t.*)
var. **texana** (Scheele) Pennell

a2 Corolla gen 8–10 mm exclusive of the 2–6 mm spur; seeds ± smooth; e Am var., occ intro W Cas   var. **canadensis**

Lindernia All.   Lindernia; False-pimpernel

Fls ped, solitary in the axils of the lvs; sepals distinct, 3-nerved in ours; corolla blue-violet, 6–10 mm in ours, bilabiate, the upper lip with short, acutish lobes, external in bud, the lower much longer and 3-lobed; filaments 4, didynamous, the upper short and antheriferous, the lower adnate to corolla tube most of their length, but visible as hairy ridges, the tips free and projecting upwards, lacking anthers in ours; stigmas distinct, flattened; caps septicidal, the septum persistent; seeds ∝ ; br ann, ours to 2 dm, with opp, entire or obscurely toothed, wholly cauline lvs (in ours sessile or subpetiolate) most or all of which subtend fls; ours widespread in the New World on moist banks, esp along the larger rivers, but uncommon with us. (For Franz Balthazar von Lindern, 1682–1755, German physician and botanist).

1a

1b

1a Peds (except sometimes the lowermost ones) conspicuously > their subtending lvs, gen 10–25 mm; lvs all, or all but the lowermost, broadly rounded at base   1 **L. anagallidea** (Michx.) Pennell

1b Peds < or only slightly > their subtending lvs, gen 5–15 mm; at least the lower lvs narrowed to base   2 **L. dubia** (L.) Pennell

Mazus Lour.   Mazus

Fls in terminal, nearly naked racemes; calyx 5-cleft to ca middle or a bit beyond, the narrow segms unequal, with ± evident midrib; corolla blue-violet, bilabiate, the short upper lip subacutely 2-toothed, external in bud, the lower lip much larger and 3-lobed; filaments 4, didynamous, all antheriferous; stigmas 2, flattened; fr loculicidal, dehiscing across the septum; seeds ∝ ; diffuse ann or bien with basal and opp lvs, or the uppermost lvs alt, as also the minute or obsolescent bracts of the raceme. (Gr *mazos*, papilla, referring to tubercles in the throat of the corolla).

M. japonicus (Thunb.) Kuntze. Japanese m. Br from base, to 15 cm, at least the sts with short, retrorse-spreading hairs; basal lvs spatulate or broader, to 4 × 1.5 cm, irreg toothed, cauline often smaller; racemes at maturity > the proper st, openly 4–10-fld; calyx 4–5 mm, to 9 mm in fr; corolla 7–10 mm, blue-violet, marked with yellow and white; lawns and wet bottom-lands; e Asian sp., intro W Cas.

Melampyrum L.   Cow-wheat

Fls in terminal, lfy-bracteate spikes or racemes (or solitary in the upper axils); calyx deeply 4-cleft; corolla bilabiate, galeate, the upper lip hooded and nearly lobeless; lower lip ca = upper, 3-lobed and with well developed basal palate, external in bud; stigma entire; caps loculicidal, curved, asymmetrical, dehiscent on the convex margin; seeds few, gen 1–2 per locule, hard; ann with opp, wholly cauline, entire lvs, the bracts often with divergent slender teeth near the base. (Gr *melas*, black, and *pyros*, wheat, referring to the color of the seeds).

M. lineare Desr. Narrow-lvd c. Slender, 1–3 dm, simple or few-br, glandular at least upwards on the st; lvs and bracts short-petiolate, linear or lanceolate, 2–5 cm × 1–8 mm; fls short-ped; corolla 5–10 mm, white or pinkish with yellow palate; seeds 3 mm; rich woods and wet meadows; widespread in e US and s Can, w to VI· and ne Wn; ours is var. *l.*

Mimetanthe Greene   Mimetanthe

Fls axillary; calyx ± 5-sulcate, but not angled, the midribs to the well developed, unequal segms obscure; corolla small, yellow or purplish, slightly bilabiate, the upper lobes external in bud; stamens 4, didynamous, the anthers of the lower pair tending to be reduced, or wanting; stigmas distinct; caps loculicidal and splitting across the septum distally; seeds ∝ ; ann with opp, narrow, sessile or subsessile lvs. (Gr *mimetes*, mimic, and *anthos*, flower, from resemblance to *Mimulus*).

M. pilosa (Benth.) Greene. Downy monkey-flower. Malodorous ann 1–4 dm, glandular and spreading-hairy throughout; lvs ± entire, 1–5 cm, to 1.5 cm wide; ped gen > calyx; calyx 5–6 mm at anthesis, the deepest sinus extending > half way to base; corolla 5–9 mm, yellow, often maroon-dotted; moist, low places in dry regions from e Wn to Baja Cal, e to sw Ida, Utah, and Ariz (*Mimulus p.*).

Mimulus L.   Monkey-flower

Fls axillary; calyx strongly 5-angled, the midveins to the gen rather short lobes prominent and often raised; corolla ± bilabiate, yellow to purple or red, the upper lobes external in bud; stamens 4, didynamous, all with well developed anthers; stigmas gen distinct, or sometimes marginally connate into a funnelf structure; caps loculicidal, sometimes splitting across the septum; seeds ∝ ; herbs (ours) with opp, entire or toothed lvs. (Diminutive of L *mimus*, a mimic).

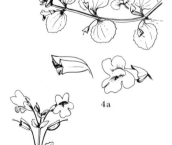

1a Pls per from rhizomes or stolons, gen of moist or wet places; fls long-ped, the peds > calyx; corollas dropping before withering; septum of caps remaining intact at maturity, or splitting only above the middle
  2a Corollas pink-purple, marked with yellow, 3–5.5 cm; calyx 1.5–2.5 (3) cm; erect, viscid-villous pls 3–10 dm; lvs sessile and several-nerved from base; middle and upper elev in all our higher mts; Lewis' m., great purple m.   1 **M. lewisii** Pursh
  2b Corollas yellow, sometimes marked with red or maroon, or in part washed with pale purple
    3a Upper calyx tooth conspicuously > others; lvs palmately or subpalmately veined, the 3–7 main veins arising at or very near base; corolla strongly bilabiate, with broad, strongly flaring throat
      4a Corolla 1–2 cm, with open throat; lateral and lower calyx teeth blunt and gen very short, the lower not folded upward; sts weak, gen decumbent to creeping or floating; Gr Pl and s sp., entering our range in Mont; glab m.; ours is var. *fremontii* (Benth.) Grant
                       2 **M. glabratus** H.B.K.
      4b Corolla (1) 2–4 cm, the throat nearly closed by the well developed palate; lateral and lower calyx teeth ± acute, the lower tending to fold upward in fr and partly close the orifice; sts weak and creeping to often erect; cordilleran spp., throughout our range
        5a Pls with definite, creeping, often sod-forming rhizomes, often stoloniferous as well; fls few (gen 1–5), large, the corolla gen 2–4 cm; low pls, gen not > 2 dm, of high altitudes; large mt m.; 2 vars. in our range   3 **M. tilingii** Regel
          a1 Lvs small, seldom > 1 cm; pls freely br, the brs ± stoloniform (*M. c.*)   var. **caespitosus** (Greene) Grant
          a2 Lvs larger, to 2.5 cm, the better developed ones gen 1 cm or more; pls less br, or the brs not stoloniform   var. **tilingii**
        5b Pls with stolons, but only rarely with definite creeping rhizomes; fls often > 5, and gen < 2 cm when few; pls very often > 2 dm;

2a

4a

5a

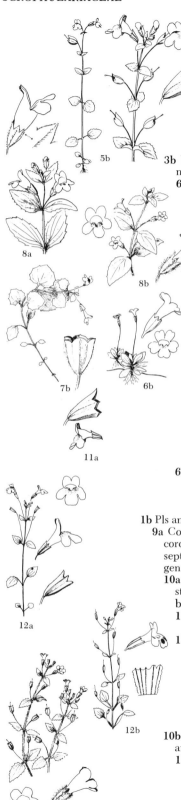

5b

8a

8b

7b

6b

11a

12a

12b

13a

sea level to mid elev in mts; yellow m.; we have 3 vars. (*M. langsdorfii, M. nasutus*) 4 **M. guttatus** DC.
a1 Pls very robust, stout-std, evidently hairy, on ocean cliffs from sw Wn to Cal var. **grandis** Greene
a2 Pls less robust, glab or hairy, not maritime
  b1 Pls ± depauperate, seldom > 3 dm, gen without stolons, with lvs 0.5–2 cm and fls only 1–2 cm, tending to occur in less distinctly hydric habitats than the next var. and often fl very early var. **depauperatus** (Gray) Grant
  b2 Pls larger, 2–8 dm, often stoloniferous; lvs well developed (sometimes to 8 cm); fls gen several or ∝, 2–4 cm; the common phase var. **guttatus**
3b Upper calyx tooth not evidently > others, or if so then the lvs pinnately veined
  6a Pls lfy-std; fls axillary; calyx tube hairy at least on nerves
    7a Calyx lobes 2–5 mm, acute or acuminate; rhizomes not producing cormlike resting buds, though sometimes moniliform in no 6; lvs pinnately veined
      8a Herbage ± hirsute, but scarcely viscid; calyx tube hirsute along the 5 ribs only; corolla 2.5–4 cm, strongly bilabiate, with broad, evidently expanded throat; W Cas; tooth-lvd m.
      5 **M. dentatus** Nutt.
      8b Herbage viscid-villous and tending to be slimy; calyx tube viscid-villous over the surface as well as along the ribs; corolla 1–3 cm, only slightly irreg, and with narrow, not much expanded, scarcely differentiated throat; musk-fl; musk-pl; widespread; 2 vars. 6 **M. moschatus** Dougl.
        a1 Pls relatively robust, to 7 dm, with large, often sessile lvs not much > half as wide as long; W Cas var. **sessilifolius** Gray
        a2 Pls smaller, seldom > 3 dm, with gen petiolate, shorter and relatively broader lvs, the bl seldom 4 cm and often > half as wide as long; widespread, but rare W Cas (var. *longiflorus*) var. **moschatus**
    7b Calyx lobes 1–2 mm, abruptly apiculate to rounded-mucronulate; rhizomes very slender, producing cormlike resting buds from which the sts arise; lvs ± palmately veined; herbage and calyx viscid-villous as in no 6; local at e end of CRG; liverwort m.
    7 **M. jungermannioides** Suksd.
  6b Pls mat-forming; lvs crowded at or near the ground; fls gen borne singly on axillary peds, these gen > st and appearing nearly terminal; calyx tube glab; lvs 3 (5)-nerved from base; widespread in mts E Cas; primrose m. 8 **M. primuloides** Benth.
1b Pls ann, without stolons or rhizomes; other features various
  9a Corollas yellow, often marked with red; peds elongate, gen > calyx; corollas (except sometimes in no 14) gen deciduous before withering; septum of caps remaining intact, or splitting only above the middle; spp. gen of moist or wet places of middle and lower elev (to high elev in no 14)
    10a Corolla strongly bilabiate, the lower lip evidently > upper and strongly deflexed from it; corolla at least 8 mm; principal lvs of fairly broad form and gen petiolate
      11a Upper calyx tooth obviously > others; ann forms of no 4 (esp var. *depauperatus*) keyed with the per spp.
      11b Upper calyx tooth ca = others, or smaller
        12a Corolla pubescent on the palate, sometimes finely red-dotted, but without a prominent blotch; calyx teeth all ± alike, gen acute; E Cas in Ore, Wn, and n Ida; Wn m. 9 **M. washingtonensis** Gand.
        12b Corolla glab, bearing a prominent maroon blotch at base of lower lip; upper 3 calyx teeth acute, lower 2 longer and rounded; shady places, esp in moss mats on cliffs, in and W Cas; chickweed m.
        10 **M. alsinoides** Dougl.
    10b Corolla only slightly bilabiate, the lower lip only slightly > upper and not much deflexed from it; corolla often < 8 mm; lvs diverse
      13a Lvs abruptly contracted to the petiole; herbage gen evidently viscid-pubescent, or sometimes only shortly and inconspicuously so; corolla 6–14 mm; gen E Cas; many RM pls, called var. *membranaceus* (A. Nels.) Grant, are small, few-fld, and thin-lvd, with relatively long petioles and inconspicuous pubescence; purple-st m.
      11 **M. floribundus** Lindl.

13b Lvs tapered to the petiolar or sessile base; herbage finely glandular-puberulent

14a Corolla 8–16 mm, gen 2–3 times as long as calyx; peds gen 1–1.5 cm at anthesis, tending to become arcuate-spreading or strongly divergent in fr; e base of Cas, from s Wn s; Pulsifer's m.
               12 **M. pulsiferae** Gray

14b Corolla 4–8 mm, slightly > calyx; peds shorter, < 1 cm at anthesis

 15a Lvs gen rather narrowly elliptic or rhombic-elliptic, gen short-petiolate; calyx teeth ± acute; fr peds tending to be loosely ascending; moist, open places in valleys and plains E Cas, not reaching Mont; short-fld m. 13 **M. breviflorus** Piper

 15b Lvs narrower, linear to narrowly oblong or oblanceolate, sessile or the lower short-petiolate; calyx teeth tending to be rounded-mucronulate; fr peds tending to be widely spreading, with suberect tip; open, moist to dry places at various elev; Mt Adams, Wn, and s; Suksdorf's m. 14 **M. suksdorfii** Gray

9b Corollas purple or red, gen marked in tube or throat with yellow or white; peds, except often in no 15, gen < calyx; corollas tending to persist for some time after withering (least so in no 15); spp. mesophytic to often xerophytic

16a Fls small and slender, 5–10 mm; septum splitting only above the middle; peds from a little < to evidently > calyx; slopes and meadows at mid elev in and E Cas; Brewer's m.
             15 **M. breweri** (Greene) Rydb.

16b Fls larger and more showy, 10–50 mm; septum splitting to base at maturity, the halves of the placenta adherent to the respective valves; peds < calyx (sometimes not much < in no 18)

 17a Caps symmetrical, dehiscent; corolla 1–3.5 cm; E Cas

  18a Lvs notably broad, evidently 3–5-nerved, the better developed ones gen 1–2.5 (3) cm wide, at least the upper ones broadly ovate or broader, with strongly acute to acuminate tip; corolla 2–3.5 cm; des and semides E Cas, s from Wasco Co, Ore; Cusick's m.
           16 **M. cusickii** (Greene) Piper

  18b Lvs narrower and more obscurely nerved, rarely any of them as much as 1.2 cm wide, seldom at all ovate, all obtuse or merely acutish; corolla 1–2.5 cm

   19a Fls subsessile, the peds only 1–3 mm; caps ovate, obtuse; pls of very dry places E Cas, becoming much br when well developed; dwarf purple m.   17 **M. nanus** H. & A.

   19b Fls evidently ped, the peds becoming 3–7 mm in fr; caps lance-linear; pls more mesophytic and gen simple or nearly so, from n Ida and adj Wn to s end of SRC; bank m.
           18 **M. clivicola** Greenm.

 17b Caps strongly oblique at base, ± woody, only tardily or not at all dehiscent; corolla 3–5 cm; wet clay soil, esp on sites of vernal pools, s WV and s; tricolored m.   19 **M. tricolor** Hartw.

Nothochelone Straw

Fls in verticillasters or in a thyrsoid-pan infl; calyx 5-cleft to base; corolla pink-purple, tubular, bilabiate, the lower lip > upper, the upper lobes external in bud; nectary a hypog disk, the stamens not nectariferous; fertile stamens 4, paired; anthers conspicuously long-woolly, deeply sagittate, the pollen sacs joined only near the tip, dehiscent throughout and eventually explanate; sterile filament inserted well down in corolla tube, well developed but < fertile ones, bearded throughout; stigma small, capitate; caps septicidal; seeds flattened, winged all around; per herbs (often a bit woody at base) with opp, toothed lvs. (Gr *notho*, false, plus *Chelone*, another genus).

N. nemorosa (Dougl.) Straw. Woodl beard-tongue. Sts (3) 4–8 dm, clustered on a br, woody caudex which tends to surmount a short taproot; lvs all cauline, the lower reduced, the others short-petiolate, conspicuously serrate, lanceolate to ovate, gen (2.5) 4–11 × 1.5–4 cm; infl glandular-hirsute; corolla 25–33 mm, glandular-hairy outside, glab inside except at base, strongly 2-ridged on lower side within; filaments long-hairy at base; caps 11–17 mm; seeds 2–3 mm; woods and moist, open, rocky slopes in and W Cas (*Penstemon n.*).

Orthocarpus Nutt.   Owl-clover

Fls in prominently bracteate, terminal spikes or spiciform racemes; calyx 4-cleft, the lobes often partly connate in lateral pairs; corolla elongate and narrow, bilabiate, the upper lip (galea) beak-like, its lobes united to tip and enclosing the anthers, the lower lip ± saccate-inflated, nearly or quite = galea, gen 3-toothed at tip, external in bud; stamens 4, didynamous, attached near summit of corolla tube; pollen sacs unequally placed, 1 apically attached and lying alongside upper part of filament, the other medifixed and versatile, or the lower 1 obsolete; stigma entire, penicillate; caps loculicidal; seeds ∝; slender ann (ours to 4 dm) with alt, entire to dissected, wholly cauline lvs. (Gr *orthos*, straight, and *karpos*, fruit, referring to the symmetrical caps).

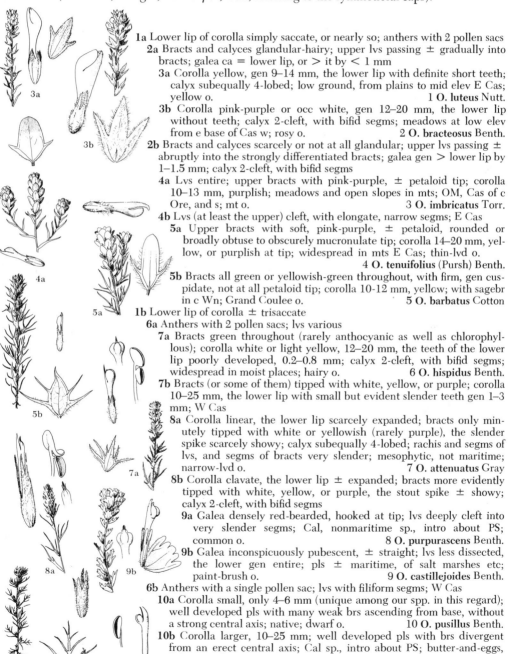

1a Lower lip of corolla simply saccate, or nearly so; anthers with 2 pollen sacs
  2a Bracts and calyces glandular-hairy; upper lvs passing ± gradually into bracts; galea ca = lower lip, or > it by < 1 mm
    3a Corolla yellow, gen 9–14 mm, the lower lip with definite short teeth; calyx subequally 4-lobed; low ground, from plains to mid elev E Cas; yellow o.   1 **O. luteus** Nutt.
    3b Corolla pink-purple or occ white, gen 12–20 mm, the lower lip without teeth; calyx 2-cleft, with bifid segms; meadows at low elev from e base of Cas w; rosy o.   2 **O. bracteosus** Benth.
  2b Bracts and calyces scarcely or not at all glandular; upper lvs passing ± abruptly into the strongly differentiated bracts; galea gen > lower lip by 1–1.5 mm; calyx 2-cleft, with bifid segms
    4a Lvs entire; upper bracts with pink-purple, ± petaloid tip; corolla 10–13 mm, purplish; meadows and open slopes in mts; OM, Cas of c Ore, and s; mt o.   3 **O. imbricatus** Torr.
    4b Lvs (at least the upper) cleft, with elongate, narrow segms; E Cas
      5a Upper bracts with soft, pink-purple, ± petaloid, rounded or broadly obtuse to obscurely mucronulate tip; corolla 14–20 mm, yellow, or purplish at tip; widespread in mts E Cas; thin-lvd o.
                                          4 **O. tenuifolius** (Pursh) Benth.
      5b Bracts all green or yellowish-green throughout, with firm, gen cuspidate, not at all petaloid tip; corolla 10-12 mm, yellow; with sagebr in c Wn; Grand Coulee o.   5 **O. barbatus** Cotton
1b Lower lip of corolla ± trisaccate
  6a Anthers with 2 pollen sacs; lvs various
    7a Bracts green throughout (rarely anthocyanic as well as chlorophyllous); corolla white or light yellow, 12–20 mm, the teeth of the lower lip poorly developed, 0.2–0.8 mm; calyx 2-cleft, with bifid segms; widespread in moist places; hairy o.   6 **O. hispidus** Benth.
    7b Bracts (or some of them) tipped with white, yellow, or purple; corolla 10–25 mm, the lower lip with small but evident slender teeth gen 1–3 mm; W Cas
      8a Corolla linear, the lower lip scarcely expanded; bracts only minutely tipped with white or yellowish (rarely purple), the slender spike scarcely showy; calyx subequally 4-lobed; rachis and segms of lvs, and segms of bracts very slender; mesophytic, not maritime; narrow-lvd o.   7 **O. attenuatus** Gray
      8b Corolla clavate, the lower lip ± expanded; bracts more evidently tipped with white, yellow, or purple, the stout spike ± showy; calyx 2-cleft, with bifid segms
        9a Galea densely red-bearded, hooked at tip; lvs deeply cleft into very slender segms; Cal, nonmaritime sp., intro about PS; common o.   8 **O. purpurascens** Benth.
        9b Galea inconspicuously pubescent, ± straight; lvs less dissected, the lower gen entire; pls ± maritime, of salt marshes etc; paint-brush o.   9 **O. castillejoides** Benth.
  6b Anthers with a single pollen sac; lvs with filiform segms; W Cas
    10a Corolla small, only 4–6 mm (unique among our spp. in this regard); well developed pls with many weak brs ascending from base, without a strong central axis; native; dwarf o.   10 **O. pusillus** Benth.
    10b Corolla larger, 10–25 mm; well developed pls with brs divergent from an erect central axis; Cal sp., intro about PS; butter-and-eggs, Johnny-turk o.   11 **O. erianthus** Benth.

Parentucellia Viviani   Parentucellia

Fls in a terminal, lfy-bracteate, spiciform raceme; calyx 4-lobed; corolla galeate, the upper lip sac-cate and lobeless, enclosing the anthers, the lower lip spreading, 3-lobed, ca = galea, its lobes external to the galea in bud; stamens 4, didynamous, the pollen sacs =, mucronate, woolly; stigma capitate, entire or nearly so; caps loculicidal, the cells slightly unequal; seeds ∝, minute, turgid, ± smooth; ann or bien, glandular-hairy herbs with sessile, toothed, wholly cauline lvs, these opp, or the upper alt or offset. (For Tomaso Parentucelli, founder of the Rome botanic garden).

P. **viscosa** (L.) Car. Yellow p. Erect, fibrous-rooted ann 1–7 dm, gen unbr, spreading-hairy, more stipitate-glandular upwards; lvs ovate or lanceolate, toothed, 1–4 cm, to 2 cm wide; fls subsessile, gen alt or offset; corolla yellow, 1.5–2 cm, > the eventually accrescent but not inflated calyx; Mediterranean weed of moist low ground, occ W Cas.

Pedicularis L.   Lousewort; Pedicularis

Fls in a lax and elongate to capitate, gen spiciform raceme; calyx reg (or less often irreg) cleft, with 5 (less often 4 or only 2) lobes; corolla yellow or white to purple or red, bilabiate, the upper lip (galea) hooded, enclosing the anthers, often extended into a beak, the lower lip gen shorter, gen 3-lobed, external to the upper in bud; stamens 4, didynamous; stigma capitate; caps glab, flat-tened, asymmetrical, loculicidal, gen arcuate and opening chiefly or wholly on the upper side; seeds several, often slightly winged; per (rarely apparently ann) herbs with alt, toothed to more often pinnatifid or pinnately dissected lvs. (L name, pertaining to lice, because of superstition that lice infestation followed its ingestion).

1a Lvs merely toothed, all cauline, gen 4–10 × 0.5–1.5 cm; calyx lobes 2; corolla 1–1.5 cm, the galea strongly arched and tapering into a slender, downcurved beak app or touching the prominent lower lip; widespread cordilleran sp., esp of conif woods; lfy l., sickletop l.; 2 vars.

1 P. **racemosa** Dougl.

  a1 Fls pink to purplish; in and W Cas   var. **racemosa**

  a2 Fls white; E Cas   var. **alba** (Pennell) Cronq.

1b Lvs, or many of them, pinnatilobate to bipinnatifid; calyx lobes 5 (except in no 2), often unequal or partly connate in groups

  2a St br, 1–4 dm; pls short-lived, perhaps ann; galea essentially beakless; calyx 2-cleft, with irreg lacerate or lobed segms; fls slender, 11–14 mm, purple; boreal Am sp. of muskeg etc, app our range in s BC and reput-edly s along coast to Ore; small-fld l.   2 P. **parviflora** Smith

  2b St simple; pls evidently per; other characters various

    3a Pls lfy-std, to 1 m, the basal lvs, if present, not markedly > cauline (though sometimes with longer petiole); bracts sharply differentiated from lvs; galea beakless or with a very short beak; calyx lobes 5, the upper one much the shortest, the others partly connate into 2 lateral segms; widespread cordilleran sp.; bracted l.; 8 vars. with us

3 P. **bracteosa** Benth.

      a1 Free tips of lateral sepals very slender and elongate, almost filiform, evidently glandular; corolla red or purple to partly or occ wholly yellow

        b1 Corolla dark blood-red to sometimes yellow; sepal tips tending to be evidently hairy as well as glandular, the glands dark, purple to blackish; OM (P. a.)

var. **atrosanguinea** (Pennell & Thomps.) Cronq.

        b2 Corolla purple to partly or sometimes wholly yellow; sepal tips gen without evident glandless hairs, the glands gen pale; BC and Alta to Mont, Ida, and Mt Adams, Wn   var. **bracteosa**

      a2 Free tips of sepals somewhat wider (linear to lanceolate or trian-gular) and often shorter, only very finely, or not at all, glandular; corolla yellow or yellowish, or in var. *latifolia* sometimes partly or wholly purple

        c1 Galea slightly beaked

          d1 Infl evidently arachnoid-villous; wc and nw Mont, and n Ida (P. c.)   var. **canbyi** (Gray) Cronq.

          d2 Infl glab or nearly so; sw Mont, nc Ida, and extreme se Wn (P. s.)   var. **siifolia** (Rydb.) Cronq.

c2 Galea beakless

e1 Free tips of lateral sepals evidently > portion which is connate above dorsal sinus

f1 Infl evidently hairy; free tips of lateral sepals gen at least twice as long as connate portion

g1 Galea not much raised above lower lip, sometimes even partly enfolded by it; roots tending to be tuberous-thickened; Wallowa and Blue Mt area of ne Ore and se Wn (*P. p.*)　　　var. **pachyrhiza** (Pennell) Cronq.

g2 Galea strongly raised above lower lip; roots gen not tuberous-thickened; sw Mont and e Ida s to s RM (*P. p.*) var. **paysoniana** (Pennell) Cronq.

f2 Infl glab or nearly so; free tips of lateral sepals < twice as long as connate portion; Klamath region, n in Cas to s Wn (*P. f.*)　　　var. **flavida** (Pennell) Cronq.

e2 Free tips of lateral sepals gen < portion which is connate above dorsal sinus; infl evidently hairy to sometimes subglab; Wn Cas, to s BC and n Ida (*P. l.*) var. **latifolia** (Pennell) Cronq.

3b Pls 0.5–7 dm, with the lvs basally disposed, the cauline lvs gen few and ± reduced (this habit least marked in no 9, which has the bracts, esp the lower ones, scarcely differentiated from lvs)

4a Galea essentially beakless, occ with an inconspicuous apiculation not > 1 mm

5a Corolla yellow or ochroleucous, sometimes faintly tinged with pink or purple

6a Pinnae of the lvs, or many of them, strongly incised

7a Fls large, the corolla (2) 2.5–3.5 cm; galea much > corolla tube; pls 0.5–1.5 dm; calyx lobes not very unequal; boreal sp., s at high alt to se BC; capitate l.　　4 P. capitata Adams

7b Fls smaller, the corolla ca 1.5 cm; galea ca = corolla tube; pls 1.5–4 dm; lateral lobes of calyx ± connate; Mt Rainier; Mt R l.　　5 P. rainierensis Pennell & Warren

6b Pinnae gen merely toothed; corolla 1.5–2.5 cm, the galea a little < tube; pls 0.5–2 dm; calyx lobes not very unequal; circumboreal sp., s in RM to nw Wyo; Oeder's l.　6 P. oederi Vahl

5b Corolla purple, gen 2–2.5 cm (or a bit shorter in no 7)

8a Pinnae, or many of them, deeply cleft; Mont and Wyo

9a Bracts not strongly differentiated from lvs, evidently pinnatifid or bipinnatifid at least distally; dwarf pls, > 1 dm; calyx lobes subequal; pretty dwarf l.　　7 P. pulchella Pennell

9b Bracts sharply differentiated from lvs, the distal portion elongate, narrow, and entire or merely toothed; taller pls, 1.5–4 dm; lateral calyx lobes connate in pairs; fern-lvd l.

8 P. cystopteridifolia Rydb.

8b Pinnae or lobes of the lvs gen merely toothed; lf rachis relatively broad, the lvs often merely pinnatilobate; pls 1–3 dm, ± lfy-std; boreal sp., s at high alt to Marble Mts, BC; Langsdorf's l.　　9 P. langsdorfii Fisch.

4b Galea evidently beaked, the beak 2 mm or more (or sometimes only 1 mm in no 11)

10a Beak straight, spreading, (1) 2–4 mm; corolla purple; calyx lobes not very unequal

11a Lvs subbipinnatifid, the pinnae deeply cleft and again toothed; infl capitate, sometimes with 1 or 2 smaller, separated lower clusters; corolla gen 1–1.5 cm; beak of galea gen 2–4 mm; Cas of Wn, and n; bird's-beak l.

10 P. ornithorhyncha Benth.

11b Lvs once-pinnatifid, the pinnae gen merely toothed; infl spicate-racemose, more evidently so in age; corolla gen 1.5–2 cm; beak of galea ca (1) 2 mm; s RM, n to sw Mont and c Ida; Parry's l.; ours is var. *purpurea* Parry　　11 P. parryi Gray

10b Beak strongly curved, often > 4 mm; corolla purple to white; calyx lobes variously subequal or unequal

12a Beak lunately downcurved, not much, if at all, exserted from the well developed lower lip; widespread sp. of wooded or open slopes and drier meadows at mid and upper alt in mts; white coiled-beak l.; 2 vars.　　12 P. contorta Benth.

a1 Pls glab throughout; fls ochroleucous or white, often finely
  marked with purple, seldom (as in Idaho Co, Ida) wholly
  pink; widespread   var. **contorta**
a2 Pls with the calyx and often also the bracts slightly villous at
  base; corolla wholly pink or purple; nw Wyo and adj sw
  Mont (*P. c.*)   var. **ctenophora** (Rydb.) Nels. & Macbr.
12b Beak strongly upcurved, the living fls (esp in no 13) reminis-
  cent of an elephant's head with trunk upraised; wet meadows
  and other wet places in mts (*Elephantella*)
  13a Pls essentially glab throughout; beak conspicuously ex-
    serted beyond the relatively small lower lip; widespread sp.
    (*P. surrecta*); pink elephants, elephant's head
                                       13 **P. groenlandica** Retz.
  13b Pls evidently villous in the infl; beak not extending much
    farther outward than the rather well developed lower lip;
    Cas of Ore, s to Cal; little elephant's head
                                       14 **P. attollens** Gray

Penstemon Mitch.   Beardtongue; Penstemon

Fls typically borne in verticillasters, these sometimes expanded to form an open-paniculiform infl,
or in a few spp. reduced to a pair of opp fls at each node of a then essentially racemose infl; calyx
5-cleft to base; corolla tubular, slightly to strongly bilabiate, gen blue or purple to lavender,
varying to pink, red, yellow, or white, the upper lobes external in bud; fertile stamens 4, paired,
the filaments of the upper ones nectariferous toward the base; anthers deeply sagittate, the pollen
sacs joined only near the tip; sterile filament well developed, inserted well down in the corolla
tube, > or < fertile ones, often bearded; hypog disk reduced or none; stigma small, capitate; caps
septicidal; seeds ∝, with a reticulate coat, not (or only slightly) winged; per herbs (less often
shrubs) with opp (rarely alt or partly ternate or quaternate), entire or toothed to rarely laciniate-pinnatifid lvs. (Gr *pente*, five, and *stemon*, thread, referring to the stamens).

Floral measurements below are from dried herbarium specimens. Measurements of the calyx are
taken at anthesis; the calyx tends to be ± accrescent in fr. Measurements of pollen sacs are taken
at full maturity, after dehiscence. When the pollen sacs open so widely after dehiscence as to form
essentially a plane, they are said to be explanate. The pollen sacs are joined (by the connective)
only at the apical end, and except in subg. Saccanthera they become ± divergent at maturity.
When the divergence is so strong that the 2 pollen sacs assume an end-to-end position in essentially a straight line, they are said to be opp. The ends of the pollen sacs which remain in contact
are here called the proximal ends, and the ends which become progressively more divergent are
called the distal ends.

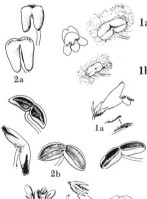

1a Anthers ± densely long-woolly with tangled hairs (conspicuously so to the
  naked eye); fls rather large, the corolla (2) 2.5–5 cm, glab outside, gen (except in no 5) long-hairy toward base of lower lip within, esp on the 2 internal ridges (subg. Dasanthera)   **Group I**
1b Anthers glab, or inconspicuously hispidulous with short, gen straight hairs;
  fls variously large or small, glab or hairy
  2a Pollen sacs opening across their confluent apices, the free tips remaining
    saccate and indehiscent, not becoming divaricate, the anther perma-
    nently horseshoe-shaped (subg. Saccanthera)   **Group II, lead 10a**
  2b Pollen sacs opening throughout their length, or remaining indehiscent at
    apex, almost always becoming divaricate (or even opp) after dehiscence
    (subg. Penstemon)   **Group III, lead 17a**

Group I

3a Lvs all cauline, the lower reduced, the sterile shoots none or few and similar to the fertile ones, the pls therefore without any tendency for the lvs to
  be clustered toward the base; pls wholly herbaceous, or woody at base
  4a Infl ± paniculiform, some or all of the axillary peduncles gen br and 2–
    several-fld; lvs relatively elongate and narrow, gen 3–13 × 0.3–1 (1.5)
    cm; pls 3–8 dm, ± erect; staminode glab; foothills to mid elev in mts of
    nw Mont, n Ida, and vic, seldom on talus; Lyall's b.   1 **P. lyallii** Gray

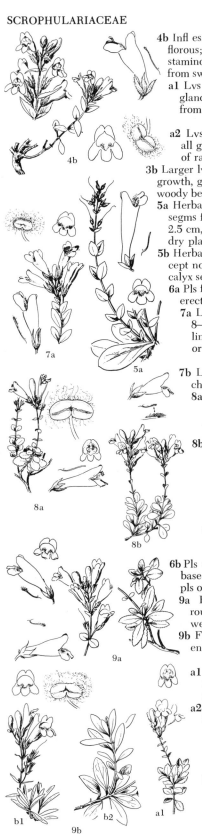

4b Infl essentially racemose, the axillary peduncles gen all simple and uniflorous; lvs shorter and broader, 1.5–5 × 0.7–2.5 cm; pls 1–3 dm, lax; staminode glab or bearded; gen on shifting talus at high elev; RM sp., from sw Mont and c Ida s; mt b.; 2 vars.   2 **P. montanus** Greene

 a1 Lvs entire or subentire, glab or rough-puberulent but gen not at all glandular, often glaucous; peds gen well < 1 cm at anthesis; c Ida, from w Custer to n Elmore and w Valley cos

           var. idahoensis (Keck) Cronq.

 a2 Lvs ± strongly toothed, gen rough-hairy, often glandular, not at all glaucous; peds often more elongate, sometimes to 2 cm; remainder of range of sp.   var **montanus**

3b Larger lvs tending to be clustered near base of pl or near base of season's growth, gen on short, sterile shoots as well as on fl sts; pls gen ± distinctly woody below, esp in larger forms; infl ± racemose, as in no 2

5a Herbage and infl glab and glaucous throughout, or the ± ovate calyx segms finely and inconspicuously glandular; lvs relatively large, to 8 × 2.5 cm, the larger ones gen > 1.5 cm wide; corollas lilac or rose-purple; dry places at e end of CRG; Barrett's b.   3 **P. barrettiae** Gray

5b Herbage gen ± hairy or glandular at least on sts or in infl, seldom (except no 5) at all glaucous; lvs rarely > 1.5 cm wide (to 1.8 cm in no 4); calyx segms often narrower

6a Pls forming dense mats on ground surface, with scattered, simple, gen erect fl sts not > ca 1.5 dm; lf bls 1–2.5 times as long as wide

 7a Lvs of erect fl sts relatively well developed, rarely < 1 cm; calyx 8–15 mm, the conspicuously glandular-hairy segms lance-linear to linear-oblong; lvs glab but not (or scarcely) glaucous, finely serrate or seldom entire; n RM sp.; elliptic-lvd b.

          4 **P. ellipticus** Coult. & Fisch.

 7b Lvs of erect fl sts small and often bractlike, seldom 1 cm; other characters not combined as in no 4; Cas and w

  8a Fls pink or nearly red to pink-lavender or rose-purple, glab inside, or with a few long hairs near base of lower lip; lvs evidently glaucous and often shortly spreading-hairy; cliff or rock p.

          5 **P. rupicola** (Piper) Howell

  8b Fls blue-lavender to purple-violet; sparsely or moderately hairy toward base of lower lip within; lvs glab but not glaucous; Davidson's p.; we have 2 vars.   6 **P. davidsonii** Greene

   a1 Lvs obscurely to evidently serrulate, tending to be broadest near (or even below) the middle, and sometimes acutish; pls ave smaller in all respects than the next var.; chiefly Wn and adj BC, s occ to c Ore *(P. m.)*   var. **menziesii** (Keck) Cronq.

   a2 Lvs entire, tending to be broadest above the middle, and more consistently rounded or obtuse at tip; chiefly Cal and Cas of Ore, occ in Wn   var. **davidsonii**

6b Pls ± ascending or erect, br above ground surface, the clusters of lvs at base of season's growth ± elevated above ground, not forming mats; pls often well over 1.5 dm; lf bls gen 2–10 times as long as wide

 9a Fls bright purple to deep blue-violet; lvs gen obtuse or round-tipped and with serrulate margins; pl 1–3 dm; W Cas; Cardwell's p.   7 **P. cardwellii** Howell

 9b Fls gen blue-lavender to light purplish; lvs gen acute or acutish, entire or toothed; pls 1.5–4 dm; E Cas; shrubby or bush p.; 3 vars.   8 **P. fruticosus** (Pursh) Greene

  a1 Lvs prominently serrate or dentate, relatively small, the bl gen 1–2.5 cm and 2–3.5 times as long as wide; corolla 3–4 cm; mts near SRC   var. **serratus** (Keck) Cronq.

  a2 Lvs otherwise, gen either longer, or subentire to entire, or both

   b1 Lvs ± toothed to entire, relatively very narrow, linear-oblanceolate or linear-elliptic, the larger ones with bl gen 2–5 cm × 3–5 (7) mm, gen 6–10 times as long as wide; corolla 3.5–5 cm; ne Wn, n Ida, and adj BC

         var. **scouleri** (Lindl.) Cronq.

   b2 Lvs entire or slightly serrulate, wider in shape and often also in measurement, to 6 × 1.5 cm, gen 2–7 times as long as wide; corolla (2.5) 3–4 cm; widespread and variable, but gen not in range of other 2 vars.   var. **fruticosus**

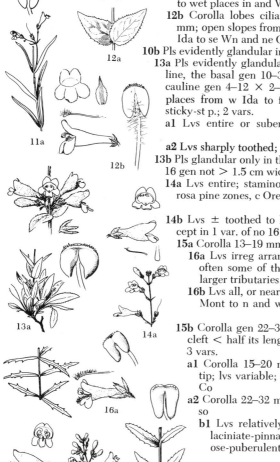

Group II

10a Pls not at all glandular; lvs all cauline

  11a Staminode glab; corolla 15–24 mm; lvs entire, < 10 (14) mm wide; herbage and infl finely hirtellous-puberulent; sagebr slopes and plains, esp on basalt; c Ore to sc Ida; Cusick's p.   **9 P. cusickii** Gray

  11b Staminode bearded; lvs evidently toothed or rarely subentire, often well > 10 mm wide; lvs glab

    12a Corolla lobes and fertile filaments glab; corolla 17-25 (28) mm; moist to wet places in and W Cas; coast p., Cas p.   **10 P. serrulatus** Menzies

    12b Corolla lobes ciliate; fertile filaments hairy above; corolla 25–38 mm; open slopes from foothills and valleys to mid elev in mts; n and w Ida to se Wn and ne Ore; lovely p., Blue Mt p.   **11 P. venustus** Dougl.

10b Pls evidently glandular in the infl

  13a Pls evidently glandular nearly or quite throughout; lvs basal and cauline, the basal gen 10–35 cm (petiole included) × 2.5–9 cm, the main cauline gen 4–12 × 2–5 cm; corolla 28–40 mm; staminode glab; open places from w Ida to foot of Cas in c Wn and n Ore; glandular p., sticky-st p.; 2 vars.   **12 P. glandulosus** Dougl.

    a1 Lvs entire or subentire; Chelan Co, Wn

            var. **chelanensis** (Keck) Cronq.

    a2 Lvs sharply toothed; rest of specific range   var. **glandulosus**

  13b Pls glandular only in the infl; lvs all cauline, < 10 cm, and except in no 16 gen not > 1.5 cm wide

    14a Lvs entire; staminode glab; corolla 15–25 mm; sagebr and ponderosa pine zones, c Ore and s; gay p.; ours is var. *roezlii* (Regel) Jeps.   **13 P. laetus** Gray

    14b Lvs ± toothed to laciniate-pinnatifid; staminode gen bearded, except in 1 var. of no 16

      15a Corolla 13–19 mm, the upper lip cleft > half its length

        16a Lvs irreg arranged, many of them ternate, or quaternate, and often some of them scattered or alt; low elev along SR and its larger tributaries; whorled p.   **14 P. triphyllus** Dougl.

        16b Lvs all, or nearly all, opp or subopp; lower parts of mts from wc Mont to n and wc Ida and se Wn; diphyllus p.

                15 P. diphyllus Rydb.

      15b Corolla gen 22–32 mm (15–20 mm in 1 local var.), the upper lip cleft < half its length; E Cas in BC, Wn, and Ore; Richardson's p.; 3 vars.   **16 P. richardsonii** Dougl.

        a1 Corolla 15–20 mm; staminode ± glab, evidently expanded at tip; lvs variable; s Gilliam and n Wheeler cos, Ore, n to s Wasco Co   var. **curtiflorus** (Keck) Cronq.

        a2 Corolla 22–32 mm; staminode gen bearded, often conspicuously so

          b1 Lvs relatively narrow and deeply laciniate-toothed to irreg laciniate-pinnatifid; upper surfaces of lvs glab to pruinose-puberulent; Wn and s BC, s barely across CR into Ore

                var. **richardsonii**

          b2 Lvs broader, ovate or lance-ovate, merely toothed, gen more reg so than in the foregoing var.; upper surfaces of lvs ± strongly puberulent; Wasco Co, Ore, to Union, Grant, and Crook cos   var. **dentatus** (Keck) Cronq.

Group III

17a Fls white or whitish, 8–20 mm; pls ± distinctly shrubby at base; lvs to 6 × 2.5 cm; widespread in dry places E Cas; hot-rock p.; we have 3 vars.   **17 P. deustus** Dougl.

  a1 Corolla essentially glab within, and only sparsely or scarcely glandular externally; pls tending to be compact and small-fld, with small and often relatively narrow lvs that are gen sharply toothed; Nev and ne Cal to c Deschutes Co, Ore *(P. h.)*   var. **heterander** (T. & G.) Cronq.

  a2 Corolla glandular-hairy inside and out

    b1 Lvs all opp, relatively broad and evidently toothed, those of fl sts gen sessile, 2–5 times as long as wide, rarely < 6 mm wide; staminode more often glab than bearded; widespread E Cas (and s WV) in our area, but not in range of other 2 vars.   var. **deustus**

21a

21b

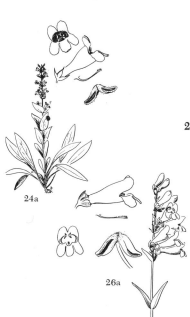

23a

24a

26a

b2 Lvs, or some of them, often ternate, or quaternate, or scattered, all narrow and irreg few-toothed to entire, often subpetiolate, > 5 times as long as wide and not > 6 mm wide; infl ave looser and fewer-fld than in var. *deustus;* staminode more often bearded than glab; Klickitat Co, Wn, to Deschutes, Grant, and Umatilla cos, Ore

var. **variabilis** (Suks.) Cronq.

17b Fls either not whitish, or the pls scarcely shrubby at base

  18a Pls not glandular (though sometimes glutinous in no 18); lvs entire or nearly so

    19a Seeds 1.8–4 mm and caps 7–13 mm; fls (except nos 18 and 19) large, the corolla 18–38 mm, the pollen sacs 1.1–3.0 mm; palate glab; caudex compactly br, not creeping, often surmounting a short taproot

      20a Pollen sacs essentially glab, 0.8–1.2 mm; corolla 13–21 mm; pls ± strongly glaucous

        21a Bearded portion of staminode gen 1–1.5 mm, the hairs ca 0.5 mm or less, or the staminode rarely glab; dunes and other dry, open places at low elev; E Cas in Wn, Ore, n Nev, and w Ida (to Gooding Co); sand-dune p., sharp-lvd p.   18 **P. acuminatus** Dougl.

        21b Bearded portion of staminode gen 2–4 mm, the hairs often well > 0.5 mm, sometimes > 1 mm; open places from plains to mid elev in mts; n Gr Pl sp., w to Lemhi Co, Ida; shining p.; 2 vars.

19 **P. nitidus** Dougl.

          a1 Lvs and bracts relatively narrow, the bracts gen lanceolate or lance-ovate; calyx gen 5–8 mm at anthesis; corolla gen with a few long hairs near base of lower lip within; Missoula, Powell, Granite, and Silverbow cos, Mont, and Lemhi Co, Ida

var. **polyphyllus** (Pennell) Cronq.

          a2 Lvs and bracts gen broader, the bracts gen clasping and ovate to subrotund; calyx gen 3–6 mm at anthesis; corolla gen glab; range of the sp., except that of previous var.   var. **nitidus**

      20b Pollen sacs evidently setose-dentate or dentate-ciliolate along the sutures, and sometimes hispidulous over the surface as well, 1.1–3.0 mm; corolla often well > 21 mm; pls glaucous or not; closely related spp., forming a geographical replacement series

        22a Pollen sacs 1.1–1.9 mm, straight or arcuate, becoming opp or upwardly divaricate after dehiscence, glab except along the sutures, or in no 20 often obscurely short-hairy toward the connective, often wholly dehiscent, esp in no 20; corolla 18–28 mm

          23a Middle and upper lvs relatively broad, esp toward base, gen lance-ovate or broader and ± clasping, (1) 1.5–4 cm wide; calyx 5–8 mm, segms with narrow to fairly broad, scarcely to evidently scarious-margined base and ± elongate, acuminate to subcaudate tip; c Ida, from Salmon R axis to n margin of SRP, chiefly toward w part of state, and reaching the Wallowa Mts of Ore; Payette p.   20 **P. payettensis** Nels. & Macbr.

          23b Middle and upper lvs narrow, gen lance-linear to linear-oblong, ca 1 (1.5) cm wide or less; calyx 3.5–5.5 mm, with broad, ± strongly scarious-margined, acutish to mucronate or abruptly short-acuminate segms; middle and lower SRP, hardly entering our range; very beautiful p.   21 **P. perpulcher** A. Nels.

        22b Pollen sacs 1.8–3.0 mm, tending to be sigmoidally twisted, downwardly divaricate (rarely becoming opp in no 23), their proximal (apical) portions remaining indehiscent; corolla 25–38 mm

          24a Lvs relatively broad, the middle and upper cauline gen 2.5–4 cm wide and 2–3 times as long; pollen sacs ± hispidulous toward the connective; calyx 6–9 mm, segms fairly narrow and evidently acuminate, the scarious margins not very broad; staminode bearded; Blue Mts of se Wn and adj Wallowa Co, Ore; Blue Mt p.   22 **P. pennellianus** Keck

          24b Lvs (except for the wider-lvd forms of no 23) relatively narrow, seldom any of them as much as 2.5 cm wide, the middle and upper cauline ones gen 3.5–10 times as long as wide

            25a Pollen sacs evidently hispidulous

              26a Calyx 4–7 mm, the segms very broad and with prominently erose-scarious margins, inconspicuously or scarcely pointed; staminode bearded; upper SRP, n to c Ida and sw Mont; dark-blue p.   23 **P. cyaneus** Pennell

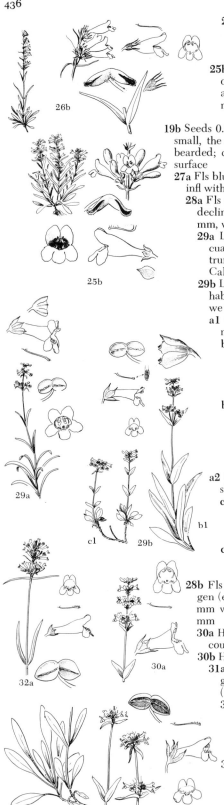

26b Calyx 7–11 mm, segms long-acuminate or subcaudate, less prominently scarious; staminode glab; n Lemhi Co, Ida, and adj Mont; Lemhi p.

24 **P. lemhiensis** (Keck) Keck & Cronq.

25b Pollen sacs glab, except along sutures; staminode glab or occ scantily bearded; calyx 4–8 mm, segms gen ovate and acutish to acuminate, with ± prominently erose-scarious margins; c Wn and e Ore to Cal and Utah; royal p., showy p.

25 **P. speciosus** Dougl.

19b Seeds 0.5–1.5 mm and caps 3–7 mm; fls (except in no 31) relatively small, the corolla 6–16 (18) mm, the pollen sacs 0.3–1.0 mm; palate bearded; caudex ± rhizomelike and tending to creep along ground surface

27a Fls blue or purple, rarely ochroleucous in forms of no 27; bracts of infl with entire, not evidently scarious margins

28a Fls relatively very small, the corolla 6–11 mm, with narrow, ± declined tube gen 2–3 mm wide at mouth; pollen sacs 0.3–0.7 mm, wholly dehiscent, eventually explanate

29a Lvs narrow, gen ± linear, rarely to 7 mm wide, often arcuate-recurved; calyx 1.5–2.5 mm, segms mucronate to subtruncate; dry foothills and lowl near e base of Cas, c Ore to Cal; ash p.   26 **P. cinicola** Keck

29b Lvs gen wider; calyx various; widespread, but gen in moister habitats or at higher elev, or both; tiny-bloom p., small-fld p.; we have 4 vars.   27 **P. procerus** Dougl.

a1 Calyx gen 3–6 mm, segms ± strongly caudate-tipped, or narrower and more gradually tapering

b1 Basal rosette gen only poorly or scarcely developed, the enlarged basal lvs gen few or none; pls gen (1) 2–4 (7) dm, gen with > 1 verticillaster of fls per st; corolla blue; widespread E Cas, up to mid (seldom high) elev in mts

var. **procerus**

b2 Basal rosette well developed; pls gen 0.5–1.5 (3) dm, with only 1 or less often 2–3 verticillasters of fls per st, the lvs ave shorter and relatively broader than in the previous var.; corolla blue or occ ochroleucous; mid to high elev in Cas of Wn and s BC, and in OM (*P. t.*)

var. **tolmiei** (Hook.) Cronq.

a2 Calyx gen 1.5–3 mm, segms obtuse or subtruncate to often shortly cuspidate; basal rosette gen well developed; fls blue

c1 Dwarf, alp and subalp pls, gen 0.5–1.5 dm, with relatively short, broad basal lvs and with infl gen reduced to 1 verticillaster; Wallowa and Strawberry mts of e Ore, and mts of Cal and Nev (*P. f.*)   var. **formosus** (A. Nels.) Cronq.

c2 Taller, gen mont to subalp pls, gen (1) 1.5–3 dm, the infl gen of > 1 verticillaster; Cas from Ore to Cal, and in Wallowa Mts (*P. b.*)   var. **brachyanthus** (Pennell) Cronq.

28b Fls larger, the corolla (10) 11–20 (22) mm, scarcely declined, gen (except no 29) with more expanded, ventricose tube gen 3–7 mm wide at mouth; pollen sacs more ovate or elliptic, 0.6–1.2 mm

30a Herbage glaucous throughout; Cas from s Wn to c Ore; glaucous p.   28 **P. euglaucus** English

30b Herbage not glaucous; gen not in Cas

31a Pollen sacs opening fully, confluent after dehiscence, the glab sutures becoming widely separated; corolla gen 11–15 (18) mm

32a Lvs all cauline, the lower strongly reduced, the st appearing relatively densely lfy; calyx 2–3.5 mm; corollas tending to ascend, the tube relatively very narrow, gen 2–3 mm wide at mouth; Blaine, Elmore, and Boise cos, Ida, n of SRP; loose p.   29 **P. laxus** A. Nels.

32b Lvs both basal and cauline, the basal often forming distinct rosettes, the st appearing less lfy than in no 29; calyx 3–7 (9) mm; corollas gen more spreading, the tube expanded, gen 3–5 mm wide at mouth; widespread, but not in range of no 29; Rydberg's p.; ours is var. *varians* (A.

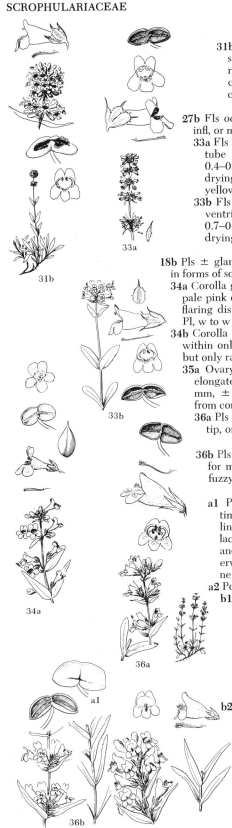

Nels.) Cronq. (*P. aggregatus, P. hesperius, P. oreocharis, P. vaseyanus*)   30 **P. rydbergii** A. Nels.

31b Pollen sacs incompletely dehiscent, their distal (free) ends shortly pouched, the partition between their proximal ends remaining intact, the gen setulose-dentate sutures not becoming very widely separated; corollas gen 15–20 (22) mm; c and n Ida, w to ne Ore; globe p.

31 **P. globosus** (Piper) Pennell & Keck

27b Fls ochroleucous or light yellow to slightly brownish; bracts of infl, or many of them, evidently scarious-margined and erose

33a Fls relatively small, the calyx 3–5 mm, the corolla 8–12 mm, its tube narrow, 2–3.5 mm wide at mouth, the pollen sacs gen 0.4–0.7 mm; lvs relatively thin and light green, not blackening in drying; widespread E Cas in our range, but not in range of no 33; yellow p.   32 **P. confertus** Dougl.

33b Fls larger, the calyx 5–9 mm, the corolla 12–16 mm, its tube ventricosely expanded, 3.5–6 mm wide at mouth, the pollen sacs 0.7–0.9 mm; lvs relatively thick and firm, tending to blacken in drying; Idaho Co, Ida, and adj Ravalli Co, Mont; pale yellow p.

33 **P. flavescens** Pennell

18b Pls ± glandular in the infl (or on the corollas), or occ essentially glab in forms of some spp. that gen have ± toothed lvs

34a Corolla glandular-puberulent near the mouth within, white, or with pale pink or bluish tube and white lobes, 1.5–2.5 cm overall, the tube flaring distally, the limb conspicuously expanded and spreading; Gr Pl, w to w Mont; white-fld p.   34 **P. albidus** Nutt.

34b Corolla gen either glab within or bearded on the palate, glandular within only in the more w, darker-fld no 43; corolla sometimes pale, but only rarely white

35a Ovary and caps gen glandular-puberulent near the top; calyx elongate, 7–13 mm, with herbaceous, entire segms; corolla 18–40 mm, ± inflated, 6–14 mm wide at mouth; staminode ± exserted from corolla mouth

36a Pls essentially glab below infl; staminode bearded only near the tip, or glab; s RM, n to sw Mont; Whipple's p.

35 **P. whippleanus** Gray

36b Pls ± hairy below infl, at least in part; staminode gen bearded for much of its length; widespread in foothills and lowl E Cas; fuzzytongue p., crested tongue p.; we have 4 vars.

36 **P. eriantherus** Pursh

a1 Pollen sacs relatively broad, becoming explanate, 0.9–1.4 times as long as wide when fully mature, with a relatively long line of contact at the proximal end; corollas lavender, pale lilac, or pale to medium purple-violet, tending to be ascending and the verticillasters not appearing well separated; habit otherwise variable; Gr Pl var., w through nw Mont to se BC and ne Wn   var. **eriantherus**

a2 Pollen sacs gen narrower

b1 Pollen sacs gen 1.4–2.0 times as long as wide when fully mature, scarcely to evidently explanate, gen with a fairly long line of contact at the proximal end; infl typically as in var. *eriantherus*, varying (esp toward the w) to sometimes as in the 2 following vars.; pls typically small (1–2 dm) and narrow-lvd, but sometimes (esp toward the w) exactly matching var. *whitedii* in habit; sw Mont and adj Ida, w across c Ida to Imnaha R in ne Ore (*P. whitedii* ssp. *tristis*)   var. **redactus** Pennell & Keck

b2 Pollen sacs gen 1.8–2.5 times as long as wide when fully mature, not explanate, often constricted at the proximal end and thus with a short line of contact; corollas spreading and the verticillasters appearing to be well separated

c1 Staminode densely bearded for most of its length (as in the 2 previous vars.); pls gen 2–4 dm, the cauline lvs gen broadest near the base and often clasping, to 2 cm wide; corollas light blue to orchid; c Wn (*P. w.*)

var. **whitedii** (Piper) A. Nels.

c2 Staminode sparsely bearded or nearly glab; habit of var. *whitedii*, or with narrower lvs; corolla deep red-purple to

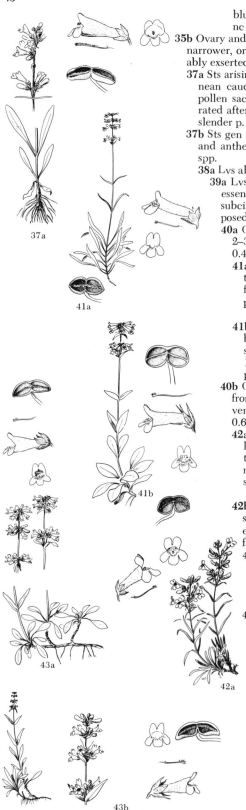

blue-purple; John Day and Deschutes drainages in c and nc Ore (*P. dayanus*)   var. **argillosus** M. E. Jones

35b Ovary and caps gen glab; calyx seldom > 7 (9) mm; corolla often narrower, or shorter, or both; staminode included in most spp. (notably exserted in no 48)

37a Sts arising singly or few together from a short, simple, subterranean caudex; corolla pale lilac or pale violet, whitish within; pollen sacs 1.0–1.5 mm, the sutures not becoming widely separated after dehiscence; Gr Pl, w into intermont valleys in Mont; slender p.   37 **P. gracilis** Nutt.

37b Sts gen arising from a ± superficial, br, woody caudex; corolla and anthers diverse, but not combined as in no 37; cordilleran spp.

38a Lvs all entire or nearly so

39a Lvs (except for often being glandular-hairy in the infl) all essentially glab, not at all cinereous (rarely a few of them subcinereous in no 40, marked by its narrow, basally disposed lvs)

40a Corolla relatively small, 8–12 mm, the narrow tube gen 2–3.5 mm wide at the mouth; pollen sacs subrotund, gen 0.4–0.6 mm

41a Lvs narrow, gen linear or nearly so, to ca 5 mm wide, the basal ones not much, if at all, developed, not forming rosettes; corolla pale purplish-blue to white; pls 2.5–7 dm; e slope of Ore Cas; Peck's p.
                                                    38 **P. peckii** Pennell

41b Lvs wider, many or all > 5 mm wide (to 18 mm), the basal ones gen well developed and forming obvious rosettes; corolla blue-purple, rarely ochroleucous; pls 1–2.5 dm; mts of Chelan and Okanogan cos, Wn; Wn p.   39 **P. washingtonensis** Keck

40b Corollas (except in forms of *P. attenuatus* var. *palustris* from c Ore) either longer, or broader and with evidently ventricose tube, or commonly both; pollen sacs ± ovate, 0.6–1.2 mm

42a Lvs all narrow, linear or lance-linear to narrowly oblanceolate, seldom any of them as much as 5 mm wide, the basal ones ∝ and well developed, forming prominent rosettes; verticillasters relatively loose and few-fld; sw Mont and adj Ida and Wyo; stiff-lf p.
                                                     40 **P. aridus** Rydb.

42b Lvs, or many of them, wider both in shape and measurement; verticillasters dense and ∝ -fld (the entire-lvd extremes of nos 49 and 54 would be sought here, except for their looser and fewer-fld verticillasters)

43a Pls 1–2.5 dm, with well developed basal rosettes; basal lvs (petiole included) to 6 cm, with ± elliptic blade; cauline lvs to 3.5 cm; corolla 10–13 mm, with evident guidelines in throat; Wallowa Mts; Wallowa p.   41 **P. spatulatus** Pennell

43b Pls otherwise, gen taller, or with larger basal lvs, or with longer corollas, or often differing in all of these respects; guidelines obscure or wanting; widespread E Cas, but not in Wallowa Mts; sulfur p., taper-lvd p. 4 vars.   42 **P. attenuatus** Dougl.

a1 Corolla relatively small, gen (7) 10–14 mm, the tube sometimes not much expanded; habit, anthers, and corolla color of var. *attenuatus;* Blue Mts, Ore, from s Morrow Co to Baker & Grant cos, often in wet meadows, not at high elev (*P. p.*)
                                          var. **palustris** (Pennell) Cronq.

a2 Corolla larger, gen 14–20 mm (or sometimes < 14 mm in var. *pseudoprocerus*), the tube evidently expanded distally

b1 Pollen sacs incompletely dehiscent, their distal (free) ends shortly pouched, the partition between their proximal ends remaining intact, the

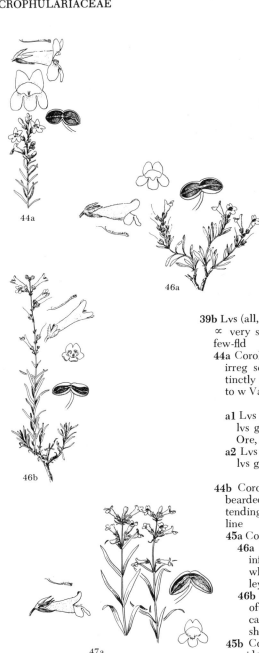

44a

46a

46b

47a

48a

sutures after dehiscence gen setulose-ciliate and not becoming very widely separated; corolla deep blue; habit gen of var. *attenuatus*, varying e to often nearly as in var. *pseudoprocerus;* mid to rather high elev in mts of c Ida from Salmon R axis to SRP, and barely entering Mont (*P. m.*)
    var. **militaris** (Greene) Cronq.

b2 Pollen sacs dehiscent throughout and eventually confluent, the glab sutures becoming widely separated

    c1 Pls gen robust, 3–9 dm; calyx lobes gen lanceolate and narrowly scarious; corolla 14–20 mm, various shades of blue, purple, or lavender to pale yellow or nearly white; drier meadows and moist, open or wooded slopes up to mid elev in mts; nw Mont (w of Cont Div) across n Ida (gen n of Salmon R axis) to e Wn and ne Ore, and along e side of Cas in Wn   var. **attenuatus**

    c2 Pls smaller, 1–4 dm; calyx lobes gen broader and more prominently scarious; corolla 12–18 mm, deep blue; drier meadows and open, often rocky places at mid to high elev in mts; sw Mont and adj Ida to Wyo (*P. p.*)
        var. **pseudoprocerus** (Rydb.) Cronq.

39b Lvs (all, or at least some of the cauline ones) cinereous with ∝ very short hairs; verticillasters gen relatively loose and few-fld

  44a Corolla glandular-hairy near the mouth within; lvs alt, irreg scattered, or opp, their bases gen remaining distinctly separated even when opp; E Cas in Ore and Wn, e to w Valley Co, Ida; Gairdner's p.; 2 vars.
        43 **P. gairdneri** Hook.

    a1 Lvs gen ± opp; pls ave relatively robust, 1.5–4 dm, the lvs gen 2–7 cm × 2–3 (5) mm; Union and Baker cos, Ore, to adj w Ida   var. **oreganus** Gray

    a2 Lvs gen alt or scattered; pls ave smaller, 1–3 dm, the lvs gen 1–4 cm × 1–3 mm; c Wn to c and se Ore
        var. **gairdneri**

  44b Corolla not glandular within, either glab or merely bearded on the palate; lvs all or mostly opp, their bases tending to meet around the st, or to be joined by a raised line

    45a Corolla glab within; lvs narrow, not > 6 mm wide

      46a Cauline lvs not very ∝, gen 1–5 pairs below the infl; pls < 2 dm, from a compactly br woody caudex which sometimes surmounts a short taproot; dry valleys of c Ida; dwarf p.   44 **P. pumilus** Nutt.

      46b Cauline lvs ∝, gen 6–12 pairs below the infl; pls often taller, the sts arising from a more loosely br caudex; dry plains and foothills, c Ore to sw Ida; short-lobed p.   45 **P. seorsus** (A. Nels.) Keck

    45b Corolla bearded on the palate; lvs often > 6 mm wide

      47a Lvs all cauline, the lower ones strongly reduced; sw Mont and adj Ida and s; matroot p.
        46 **P. radicosus** A. Nels.

      47b Lvs basal and cauline, the basal ones (or those on short, sterile shoots) ± well developed and persistent, tending to form distinct rosettes

        48a Calyx 2.5–6 mm, with lanceolate to ovate segms; corolla tube not strongly expanded distally, gen 2.5–6 mm wide at mouth; upper lip gen 2.5–5 mm, cleft > half way to base; sterile stamen shortly or not at all exserted, its beard hairs ca 0.5 mm; widespread in dry places at various elev between Cas-Sierra crest and Cont Div; lowly p.
        47 **P. humilis** Nutt.

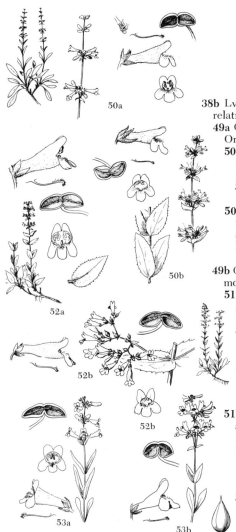

48b Calyx 6–10 mm, with lanceolate or narrower segms; corolla tube strongly inflated distally, gen (5) 7–12 mm wide at mouth; upper lip (5) 7–12 mm, seldom cleft > half way to base; sterile stamen conspicuously (though shortly) exserted, its beard hairs gen 0.7–1.5 mm or more; GB sp. of dry plains and foothills, n to Baker Co, Ore; golden-tongue p.   **48 P. miser** Gray

38b Lvs, or many of them, ± evidently toothed; verticillasters relatively loose and often few-fld

49a Corollas relatively small, gen 11–16 mm; c Wn and adj n Ore and s BC

50a Herbage essentially glab below the infl, or the st inconspicuously hirtellous-puberulent; lvs shallowly, irreg, and rather inconspicuously toothed, or occ all entire; pls gen 3–8 dm; sc Wn and adj Ore, at low elev; fine-toothed p.   **49 P. subserratus** Pennell

50b Herbage below the infl glandular-hirsute to merely hirtellous-puberulent, less often essentially glab; lvs gen sharply and conspicuously toothed; pls gen 1–4 dm; dry lowl in c Wn and adj BC; Chelan p.   **50 P. pruinosus** Dougl.

49b Corollas larger, gen (13) 15–23 mm; pls of either more e or more w distribution

51a Pls relatively robust, (3) 4–10 dm, with sharply toothed, gen relatively broad lvs, the larger cauline ones gen (0.8) 1.5–4 cm wide

52a St spreading-hirsute, often shortly so; lvs hairy like the st, at least along midrib beneath; woods at lower elev W Cas; broad-lvd p.   **51 P. ovatus** Dougl.

52b Pls gen glab below infl, or the st occ and the lvs rarely spreading-hirsute; open or wooded places from foothills to mid elev in mts of c and n Ida and adj ne Ore and nw Mont; Wilcox's p.   **52 P. wilcoxii** Rydb.

51b Pls smaller, gen 1–4 dm, with narrow, gen rather irreg and obscurely few-toothed lvs, the cauline ones up to ca 1.5 cm wide

53a Cauline lvs, or some of them, finely hirtellous-puberulent, all < 1 cm wide; overlooking SRC in Wallowa Co, Ore, and adj Ida; lovely p.   **53 P. elegantulus** Pennell

53b Cauline lvs, like the basal ones, glab, often > 1 cm wide; w Mont to c Ida, sw Alta, and se BC; Alberta p.   **54 P. albertinus** Greene

Rhinanthus L.   Rattle; Rattle-box

Fls in terminal, lfy-bracteate spikes; calyx ± inflated at anthesis, accrescent and very conspicuously inflated in fr, the 4 teeth relatively short; corolla yellow, bilabiate, galeate, the upper lip hooded and enclosing the anthers, its teeth short or obsolete; lower lip 3-lobed, < galea, external in bud; stamens 4, didynamous; pollen sacs =; stigma entire; caps flattened, orbicular, loculicidal; seeds ∝, flattened, orbicular, evidently winged; ann with opp, wholly cauline, sessile, toothed lvs. (Gr *rhinos*, snout, and *anthos*, flower, referring to the irreg fl).

R. crista-galli L. Yellow r., penny r. St 1.5–8 dm, thinly villous-puberulent on 2 of 4 sides; lvs firm, ± scab, lance-triangular to oblong, toothed, 2–6 cm × 4–15 mm, passing upwards into shorter, broader, more strongly toothed bracts; fr calyx 12–17 mm, nearly or fully as wide, reticulate-veiny; corolla 9–14 mm; circumboreal in moist places, s to nw Ore.

Scrophularia L.   Figwort

Fls ∝ in elongate, terminal, br, open, nearly naked infls; calyx deeply 5-cleft, the segms almost free; corolla greenish-yellow or greenish-purple to dark maroon, firm, in ours gen 9–14 mm, bilabiate, the upper lip 2-lobed, flat, projecting forward, external in bud, the lower shorter and with the central lobe deflexed; fertile stamens 4, the filaments expanded upward, the pollen sacs divergent; 5th stamen represented by a knob or scale < 2 mm on upper lip; stigma capitate; caps septicidal; seeds ∝, turgid; odorous per herbs with quadrangular sts (5–15 dm in ours) and opp, petiolate, toothed lvs, the bl in ours ± triangular-ovate, 5–15 × 2–7 cm, our spp. gen in moist low ground. (Named for its reputed value in treating scrofula).

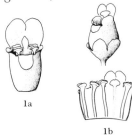

1a Sterile stamen purple or brown, gen 0.5–1 mm wide and a little longer than wide; corolla dark maroon, at least as to upper lip; coastal cos; Cal f.; ours is var. **OREGANA** (Pennell) Cronq. hoc loc. *(S. oregana* Pennell, Bull. Torrey Club 55: 316. 1928)   1 **S. californica** Cham. & Schlecht.

1b Sterile stamen yellow-green, gen 1–1.8 mm wide and a little wider than long; corolla yellowish-green, with a light maroon overcast esp above; widespread, noncoastal sp.; lance-lf f.   2 **S. lanceolata** Pursh

Synthyris Benth.   Kittentails; Synthyris

Fls in terminal racemes; calyx of 4 essentially distinct sepals; corolla blue (occ pink or lavender to white), camp to subrotate, unequally 4-lobed, the upper lobe the largest and internal to the others in bud; stamens 2; stigma capitate; caps ± compressed, sometimes notched at summit, loculicidal; seeds 2–∝ per locule; fibrous-rooted per herbs with well-developed, petiolate, basal lvs and erect or lax, naked or sparsely bracteate scapes or basal peduncles, the bracts (except in no 3) alt when present. (Gr *syn*, united, and *thyris*, door, referring to the caps valves).

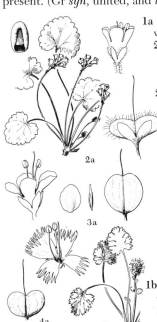

1a Lvs merely toothed or shallowly lobulate; frs, except in no 2, ca twice as wide as high, or even wider; corollas diverse

2a Corolla camp, the lobes a little < tube; peduncles weak and lax, ± reclining in fr; seeds with thick, strongly incurved margins; W Cas, and in CRG, gen at low elev in conif woods; snow-queen, round-lvd s.
                1 **S. reniformis** (Dougl.) Benth.

2b Corolla subrotate, the lobes much > tube, ± spreading; peduncles or scapes ± erect; seeds flat, thin-margined

3a Corolla lobes entire or slightly erose; caps wider than high, scarcely notched; n and wc Ida, se Wn, and ne Ore, from foothills to mid elev in mts, also in CRG [*S. missurica* var. *major* (Hook.) Davis, the larger form from lower elev, and *S. stellata* Pennell, the CRG phase, with more sharply toothed lvs and with better-developed bracts below infl, should perhaps be recognized as vars.]; mt k.
                2 **S. missurica** (Raf.) Pennell

3b Corolla lobes laciniately incised; caps nearly or fully twice as wide as high, strongly notched; W Cas, and in n Ida

4a Corolla 4–6 mm; lvs withering at end of 1st season; W Cas; fringed s.   3 **S. schizantha** Piper

4b Corolla 2.5–3.5 mm; lvs tending to persist until fl time of 2nd season; mts of nc Idaho Co, Ida; evergreen s.
                4 **S. platycarpa** Gail & Pennell

1b Lvs deeply cleft to pinnately dissected; frs ca as wide as high; corolla lobes ca equaling, or a little < tube

5a Lvs pinnipalmately or close-pinnately cleft to below middle, segms broadly confluent; Mission Range, Mont; Canby's s.   5 **S. canbyi** Pennell

5b Lvs pinnately dissected, with narrow rachis; more s or w; cut-lf s.; 3 vars.   6 **S. pinnatifida** Wats.

a1 Bracts gen obovate and rounded-obtuse; pls scantily villous or glab; caps glab; petioles often conspicuously woolly on margins at base; Utah and Wyo, perhaps reaching sw Mont   var. **pinnatifida**

a2 Bracts gen rhombic or ovate and acute to caudate-acuminate; pls often more hairy; petioles seldom conspicuously more hairy at base than above

b1 Herbage subglab to ± villous or tomentulose, gen glabrate in age; caps glab or hairy; sw Mont to c Ida *(S. cymopteroides, S. dissecta, S. hendersonii)*   var. **canescens** (Pennell) Cronq.

b2 Herbage permanently white-tomentulose, the tomentum gen a little finer than in the pubescent extreme of the previous var.; caps tomentulose; OM *(S. l.)*   var. **lanuginosa** (Piper) Cronq.

Tonella Nutt.   Tonella

Fls 1–several in axils of normal or ± reduced upper lvs (bracts); calyx with 5 subequal lobes > tube; corolla blue and white, subrotate, with short tube and spreading, ± bilabiate limb, upper lobes external in bud; stamens 4, =, exserted, filaments pubescent; pollen sacs confluent at tip; stigmas wholly united; caps dehiscing along 4 sutures; seeds 1–2 per cell, turgid; delicate ann < 4 dm with opp, gen tripartite or trifoliolate lvs. (Derivation unknown).

1a Corolla showy, 6–12 mm wide; peds and upper part of st evidently stipitate-glandular; se Wn, ne Ore, and adj Ida; large-fld t. 1 **T. floribunda** Gray
1b Corolla inconspicuous, 2–4 mm wide; peds and st glab or nearly so; W Cas and in CRG; small-fld t.   2 **T. tenella** (Benth.) Heller

Verbascum L.   Mullein

Fls in terminal racemes, spikes, or spikelike thyrses; calyx of 5 essentially distinct sepals; corolla yellow or occ white, rotate, slightly irreg 5-lobed, upper lobes external in bud and slightly < lower; stamens 5, all fertile, some or all filaments densely hairy; stigma capitate; caps ellipsoid to subglobose, septicidal; seeds ∝ ; coarse bien (ours) or per herbs with well developed basal and alt cauline lvs; ours intro weeds 4–20 dm. (The L name for some of the spp.).

1a Pls ± densely stipitate-glandular upward, essentially glab below, the lvs green; infl open, peds > bracts, 8–15 mm at anthesis; filaments covered with purple-knobbed hairs; basal lvs gen 5–15 × 1–3 cm; moth m.
                                                                      1 **V. blattaria** L.
1b Pls copiously tomentose throughout with br, glandless hairs; infl very dense, fls ± sessile; filaments with slender, yellow hairs, or some glab; basal lvs gen 10–40 × 3–12 cm, or larger; common m., flannel m.
                                                                        2 **V. thapsus** L.

Veronica L.   Speedwell

Fls in terminal or axillary racemes or spikes, or solitary in upper axils; calyx of (5) 4 essentially distinct sepals; corolla blue or violet to pink or white, subrotate, irreg 4-lobed, upper lobe largest and internal to others in bud, lower lobe smallest; stamens 2; stigma capitate; caps ± compressed, often notched or lobed at tip, loculicidal; seeds few–∝ ; ann or per herbs; lvs all cauline (at least in our spp.), entire or toothed, opp, or those of infl alt. (Derivation uncertain, perhaps for St. Veronica).

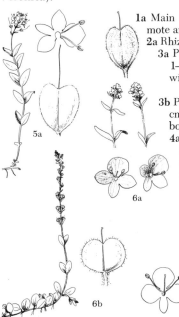

1a Main st terminating in an infl, its fls either densely crowded or more remote and axillary, the upper bract-lvs gen alt
  2a Rhizomatous per
    3a Pls 3–12 dm, erect; lvs sharply serrate, the larger gen 4–13 cm; fls in 1–several spikelike racemes 5–20 cm; European weed, occ W Cas with us; long-lf s. (*V. maritima; V. spicata*, misapplied)
                                                                      1 **V. longifolia** L.
    3b Pls 1–3 dm, erect or lax, or the sts more elongate and trailing; lvs to 4 cm, entire or with a few small teeth; infl gen either shorter or looser or both
      4a Bracts abruptly < foliage lvs, the fls thus in a terminal raceme or spike; lvs gen 1–4 cm, elliptic or lanceolate to broadly ovate or obovate; native and intro, sometimes weedy, but not of lawns
        5a Style elongate, (5) 6–10 mm; filaments 4–8 mm; lvs glab, entire; caps and habit nearly of no 3; native, not weedy; high elev in mts; Cusick's s.   2 **V. cusickii** Gray
        5b Style shorter, 1–3.5 mm; filaments 1–4 mm; lvs glab or hairy, entire or shallowly toothed
          6a Caps higher than wide; st (and often also the otherwise glab lvs) sparsely to densely villous-hirsute with loosely spreading hairs; sts simple, erect or merely decumbent at base; filaments 1–1.5 mm; mid to high elev in mts, native, widespread; Am alpine s. (*V. alpina* var. *w.*)
                                                                3 **V. wormskjoldii** Roem. & Schult.
          6b Caps wider than high; st finely and closely puberulent; lvs glab or nearly so; sts tending to creep at base, or to produce

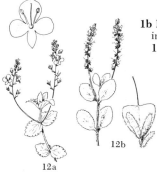

prostrate lower brs; filaments (1) 2–4 mm; lowl to rather high elev; thyme-lvd s.; 2 vars. **4 V. serpyllifolia** L.

  a1 Fls gen bright blue; infl more hairy than vegetative part of st, the peds with some spreading, viscid or glandular hairs; filaments 2–4 mm; widespread, native, lowl to high elev *(V. h.)* **var. humifusa** (Dickson) Vahl

  a2 Fls white or pale bluish, with darker blue lines; infl not evidently more hairy than vegetative part of st, the peds finely and closely puberulent; filaments 1–2.5 mm; European, ± weedy, occ intro W Cas **var. serpyllifolia**

4b Bracts only gradually reduced, the fls thus axillary, peds elongate, well > 1 cm; lawn weed with elongate, prostrate, rooting sts and small, rotund-cordate lvs gen < 1 cm, intro W Cas; thread-stalk s. **5 V. filiformis** Sm.

2b Ann

  7a Peds very short, only 1–2 mm even in fr; corolla only 2–3 mm wide; seeds 5 or more per locule

    8a Principal lvs linear-oblong to oblong or lanceolate, 3–10 times as long as wide; corolla white or whitish; style nearly obsolete, only 0.1–0.3 mm; seeds ∝; fibrous-rooted pls of moist or wet places; widespread; purslane s.; 2 vars. **6 V. peregrina** L.

      a1 St (and gen also sepals and caps) ± pubescent with short, gland-tipped hairs; native *(V. x.)*

        var. xalapensis (H.B.K.) St. John & Warren

      a2 Pls wholly glab; occ W Cas, presumably intro from e US

        var. peregrina

    8b Principal lvs ovate or broadly elliptic, 1–2 times as long as wide; corolla blue-violet; style 0.4–1.0 mm; seeds (5) 8–11 per locule; taprooted weed gen of fields and gardens, intro from Eurasia; wall s., common s. **7 V. arvensis** L.

  7b Peds elongate, gen 5–35 mm in fr; pls either with larger corolla or fewer seeds; ± taprooted, intro, weedy spp.

    9a Lvs, or ∝ of them, palmately 3–5-lobulate; caps slightly or scarcely notched; seeds 1–2 per locule; fr peds 0.5–1.5 cm; corolla ca 3 mm wide, pale bluish; ivy s. **8 V. hederaefolia** L.

    9b Lvs gen ± toothed, but not palmately lobulate; caps evidently notched; fls blue

      10a Peds gen 0.5–1.5 cm in fr; corolla 2–4 mm wide; seeds 2–4 per locule; caps barely wider than high, narrowly notched at least to middle; style evidently < the not very divergent lobes of caps; bilobed s. *(V. campylopoda)* **9 V. biloba** L.

      10b Peds gen 1.5–4 cm in fr; corolla 5–11 mm wide; seeds 5–10 per locule; caps conspicuously wider than high, broadly notched to not beyond middle; style = or > the strongly divergent lobes of caps; Persian s. **10 V. persica** Poir.

1b Main st never terminating in an infl, the lvs opp throughout and the fls all in axillary racemes

  11a Pls evidently pubescent; lvs relatively broad, 1–3 times as long as wide; seeds 6–12 to a locule, ca 1 mm; per mesophytes from creeping rhizomes, or lower part of st creeping

    12a Lvs coarsely toothed, with gen 5–11 teeth on a side, ± broad-based, sessile or nearly so; mature peds 5–9 mm, > subtending bracts; caps (seldom produced) < sepals; corolla 9–12 mm wide; weedy, often in lawns; European sp., occ intro, gen W Cas; Germander s.

      11 V. chamaedrys L.

    12b Lvs finely toothed, the larger ones gen with 12–20 teeth on a side, ± elliptic or elliptic-obovate, narrowed to a short petiole or subpetiolar base; mature peds 1–2 mm, < subtending bracts; caps > sepals; corolla 4–8 mm; weedy, but gen not in lawns; European sp., intro W Cas; common s., Paul's betony **12 V. officinalis** L.

  11b Pls essentially glab, or merely finely glandular in the infl (occ evidently hairy in no 16, which has linear or lanceolate lvs 3–20 times as long as wide); seeds various; pls of water or wet places

    13a Lvs all short-petiolate; caps and seeds ca as in spp. 14 and 15; Am brooklime **13 V. americana** Schwein.

    13b Lvs (at least the middle and upper ones of fl shoots) sessile

      14a Caps turgid, slightly or scarcely notched, slightly if at all wider

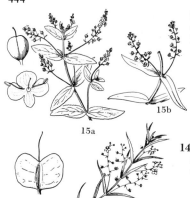

than high; seeds ∝ , 0.5 mm or less; lvs 1.5–5 times as long as wide; short-lived, perhaps sometimes bien

15a Lvs 1.5–3 times as long as wide; mature peds gen strongly ascending, or upcurved; caps ca as high as wide, or a little higher; fls blue or violet; water s., water pimpernel

14 V. anagallis-aquatica L.

15b Lvs (2.5) 3–5 times as long as wide; mature peds divaricately spreading; caps gen a little wider than high; fls white to pink or pale bluish; chain s. (*V. aquatica* Bernh., not S. F. Gray; *V. comosa, V. connata,* and *V. salina,* all misapplied)

15 V. catenata Pennell

14b Caps flattened, conspicuously notched, evidently wider than high; seeds 5–9 to a locule, 1.2–1.8 mm; lvs (3) 4–20 times as long as wide; rhizomatous per; marsh s., grass-lvd s., skullcap s.

16 V. scutellata L.

OROBANCHACEAE   Broomrape Family

Fls gamopet, ♀ or rarely some unisexual; calyx reg or irreg, 4–5-lobed, or the segms sometimes ± connate; corolla tubular, bilabiate or rarely reg, 5-lobed (or 2 lobes connate); stamens 4, didynamous, epipetalous; filaments often reverse-bent beneath anthers; ovary superior, 2–3-carpellate, gen 1-locular; placentae parietal, (1) 2 per carpel, intruded and often bifurcate, rarely confluent and the caps thus bilocular; style solitary, with a capitate or disciform, often 2–4-lobed stigma; fr a 2–3-valved, longitudinally dehiscent caps; seeds gen ∝ and small, with well developed endosperm and very small, few-celled embryo; herbaceous, often fleshy, root-parasites lacking chlorophyll, gen yellowish to brownish or purplish, with bract-like, alt lvs and spicate to corymbose or ± pan infl, or the fls solitary.

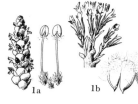

1a Pollen sacs blunt, barely or not at all mucronate; bracts of the densely spicate infl very broad, blunt, and conspicuous   **Boschniakia**

1b Pollen sacs pointed and ± mucronate at base; bracts narrower, less conspicuous, and gen more pointed; infl various   **Orobanche**

Boschniakia C. A. Mey.   Ground-cone

Fls in a dense, spicate infl, subtended by conspicuous, broad, blunt bracts; calyx irreg toothed or lobed, or truncate and lobeless; corolla curved, bilabiate, the upper lip often ± boat-shaped, shallowly if at all cleft, the lower lip often ± reduced; filaments ± hairy below, often densely so; pollen sacs blunt, seldom at all mucronate; ovary 2–3 (4)-carpellary, 1-locular, with 2–4 bifurcate, parietal placentae; stigma entire or 3–4-lobed; coarse and fleshy pls. (Boschniaki, Russian botanist).

B. hookeri Walpers. Vancouver g., small g. Pls yellow to dark red ór purple, 8–12 cm, to 3 cm thick in infl, often thicker at the cormlike base; sts conspicuously scaly-bracteate; calyx gen with 2–3 short lobes; corolla 1–1.5 cm, firm, constricted near or below middle, the lower lip < upper; filaments with a dense tuft of long hairs at base; caps 1–1.5 cm; parasitic on *Gaultheria shallon,* ± coastal (incl PS).

Orobanche L.   Broomrape; Cancer-root

Fls in spicate to pan or corymbose infls, or solitary; calyx irreg to nearly reg 4–5-cleft, or divided to base above and below; corolla bilabiate, both lips developed; filaments glab or sometimes hairy, esp below, but not with a basal tuft of long hairs; pollen sacs gen with well separated, pointed, ± mucronate bases; ovary 2-carpellary, 1-locular, with 4 parietal placentae; stigma entire or 2–4-lobed. (Gr *orobos*, vetch, and *anchein*, to choke). The interpretation of spp. no 3 and 4 is influenced by a manuscript by Lawrence R. Heckard, still unpublished as this treatment goes to press.

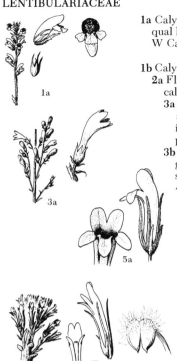

1a Calyx divided to base into 2 lateral segms, these gen cleft into 2 gen unequal lobes; fls sessile, without bractlets, 10–15 mm; Mediterranean sp. intro
  W Cas, parasitic esp on clover; clover b. *(O. columbiana)*
                                                        1 **O. minor** J. E. Smith
1b Calyx ± deeply cleft into 5 subequal lobes; native spp.
  2a Fls sessile or on peds up to ca 3 cm, with a pair of bractlets just beneath
    calyx, in addition to the subtending bract *(Myzorrhiza)*
    3a Calyx gen 5–8 mm, lobes ca = tube, or a little < tube; st gen with ∝
      short, slender brs, so that the yellowish fls are in a loose, paniculiform
      infl; anthers ± glab; widespread in coniferous woods of our range;
      pine b.   2 **O. pinorum** Geyer
    3b Calyx gen 10–20 mm, lobes much > tube; st simple or few-br; corolla
      gen purplish or pink; gen on open slopes or in meadows, often with
      sagebr
      4a Fls, esp the lower, gen ± ped, peds up to 3 cm; infl gen short and
        stout, often corymbose, but occ more elongate esp in forms of no 4;
        anthers woolly; corolla 16–35 or 40 mm
        5a Corolla relatively large, gen 25–35 or 40 mm, with long (10–14
          mm), widely spreading lips; coastal and sub-coastal, or in
          meadows inl, not occurring with sagebr; Cal b.; 2 vars. with us
                                          3 **O. californica** Cham. & Schlecht.
          a1 Pls strongly anthocyanic, the upper corolla tube and the lips
            dark violet, the calyx lobes, peds, and bracts violet-tinged; gen
            coastal and PS, occ inl to WV; parasitic esp on *Grindelia*
                                                        var. **californica**
          a2 Pls weakly or scarcely anthocyanic, the upper corolla tube and
            the lips white or yellowish to pinkish or purple-tinged, often
            with more deeply colored veins; calyx lobes, peds, and bracts
            pallid to purplish-tinged; gen in the mts of c and ne Ore (and s
            E Cas to Cal), also w through CRG and reg occ to PT *(O.
            grayana* Beck, Bibl. Bot. 4: (19) 79. 1890)
                                                        var. **grayana** (Beck) Cronq.
        5b Corolla gen smaller, 16–25 (30) mm, with shorter (4–9 mm),
          erect to ± spreading lips; drylands E Cas, parasitic esp on sage-
          br *(O. californica,* misapplied); the pls of Wn and s BC, often
          with more elongate, subspicate infl, have recently been annotated
          by Heckard as a new ssp., using the epithet *mutabilis;* flat-topped
          b.   4 **O. corymbosa** (Rydb.) Ferris
      4b Fls sessile or nearly so; infl spicate, tending to be elongate; anthers
        glab; corolla gen 15–22 mm; Suksdorf's b.; ours is var. *arenosa*
        (Suksd.) Cronq.   5 **O. ludoviciana** Nutt.
  2b Fls conspicuously long-ped, without bractlets; widespread *(Gymnocau-
    lis, Thalesia)*
    6a Peds gen 4–10, up to ca = the ± elongate st; calyx lobes = or <
      tube; foothills and valleys; clustered b.   6 **O. fasciculata** Nutt.
    6b Peds 1–3, much > the gen very short st; calyx lobes evidently >
      tube; lowl to mid elev in mts; naked b.; we have 2 vars.
                                                        7 **O. uniflora** L.
      a1 Anthers glab; corolla ochroleucous to purple, 1.5–2.5 cm, throat not
        much expanded (gen 3–5 mm wide as pressed), the relatively small
        limb not much spreading   var. **minuta** (Suksd.) Beck
      a2 Anthers gen ± woolly; corolla purple, 2–3.5 cm, with relatively
        broad throat (gen 5–9 mm wide as pressed) and expanded,
        spreading limb   var. **purpurea** (Heller) Achey

LENTIBULARIACEAE   Bladderwort Family

Fls gamopet, ⚥, irreg, calyx laterally 2-parted, often to base or nearly so, the segms entire, or sometimes the upper lip 3-lobed and the lower lip 2-lobed; corolla ± strongly bilabiate, the lips evidently lobed (upper 2, lower 3) or entire, the proper tube sometimes very short, the lower lip gen
prolonged into a basal spur; stamens 2, attached to corolla tube near its base, each with a single
pollen sac, but this, except in *Pinguicula*, transversely gen ± constricted; filaments often contorted; ovary superior, 1-locular, with free-central placenta and gen ∝ ovules; style short or obso-

lete; stigma 2-lobed, with 1 lobe smaller or obsolete; fr a caps, dehiscing by 2 or 4 valves, or bursting irreg; seeds gen ∝ and small, without endosperm, the embryo not much differentiated; ann or per herbs, aquatic or of wet soil, with alt (or all basal, or in part or wholly whorled), well developed to minute lvs, these when submersed gen dissected and producing ∝ small bladders that entrap crustaceans and other small aquatic animals; fls borne singly or more often in few-fld (rarely br) racemes, terminating erect scapes or peduncles.

1a Calyx 5-lobed, lobes partly connate into an upper lip of 3 and a lower lip of 2; fls solitary on bractless scapes; palate wanting; terrestrial herbs of wet places, with well developed, entire, rosulate basal lvs   **Pinguicula**
1b Calyx deeply 2-lobed; fls solitary or more gen racemose, each subtended by a bract; corolla tube in most spp. closed by a well developed palate at base of lower lip; aquatic pls with submersed, dissected lvs and emergent peduncles   **Utricularia**

1a   1b

Pinguicula L.   Butterwort

Fls solitary at the ends of erect, bractless scapes; calyx 5-lobed, lobes partly connate into an upper lip of 3 and a lower lip of 2; corolla bilabiate, upper lip 2-lobed, lower 3-lobed and prolonged into a basal spur, without a palate; caps 2-valved; fibrous-rooted, scapose, terrestrial herbs of wet places, with well developed, entire, rosulate basal lvs. (Diminutive of L *pinguis*, fat, from the greasy appearance of lvs of *P. vulgaris).*

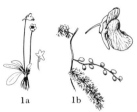

P. vulgaris L. Common b. Per, 5–15 cm; lvs succulent, broadly oblanceolate or subelliptic, short-petiolate, 2–5 cm × 7–18 mm, slimy on upper surface, digesting small insects that are caught; calyx ca 3 mm; corolla lavender-purple (nearly white), 1.5–2.5 cm, including the slender, 5–9 mm spur, with flaring throat and broad, rounded lobes; bogs and wet soil in mts; circumboreal, s to our range.

Utricularia L.   Bladderwort

Fls solitary or more gen several in a raceme ending an erect, emergent peduncle, each subtended by a bract; calyx deeply 2-lobed, lobes ± entire; corolla bilabiate, upper lip entire or shallowly 2-lobed, lower entire or 3-lobed, in most spp. elevated at base into a prominent palate, the tube prolonged at base into a spur or sac; rootless or apparently fibrous-rooted herbs, gen submersed and with much dissected lvs, the lvs alt, or partly or wholly whorled; submersed lvs with few to many small, buoyant, valve-lidded bladders which trap small crustaceans etc., or the traps borne on separate brs; our spp. all circumboreal, in standing or slowly moving water. (L *utriculus*, a little bag, referring to the bladders or traps).

1a Fls large, lower lip gen 10–20 mm; lvs gen 2-parted at base, then unequally and quasipinnately several times dichotomous, segms ± terete, ultimate ones filiform, strongly acuminate; bladders ∝, on ordinary lvs; pls free-floating, often in deep water; peds arcuate-recurved in fr; common b., greater b.   **1 U. vulgaris** L.
1b Fls smaller, lower lip gen 4–12 mm; lvs gen 3-parted at base, then 1–3 times dichotomous, often unequally, but not quasipinnately so, segms flat; pls of shallow water, st often creeping along bottom
2a Lf segms filiform; bladders all borne on the dissected lvs; scapes 1–3-fld; spur broadly conic, much < lower lip; common e and s of our area, but recently found by K. Chambers in Benton Co, Ore; humped bladderwort   **2 U. gibba** L.
2b Lf segms narrow, but flat, not filiform; bladders sometimes borne on separate specialized brs; scapes often > 3-fld; spur sometimes nearly = lower lip
3a Bladders gen borne on ordinary lvs; corolla with small or obsolete palate and reduced spur much < lower lip, this 4–8 mm; ultimate lf segms sharply acuminate; peds arcuate-recurved in fr; lesser b.   **3 U. minor** L.
3b Bladders borne on specialized brs distinct from the dissected lvs; corolla with palate well developed and with spur nearly = lower lip, this 8–12 mm; ultimate lf segms ± obtuse; peds erect in fr; mt b., flat-lvd b.   **4 U. intermedia** Hayne

PLANTAGINACEAE   Plantain Family

Fls reg (or calyx occ irreg), ♂ or rarely unisexual, gamopet, gen 4-merous as to calyx, corolla, and androecium, or stamens rarely only 1–3; corolla scarious, persistent; stamens epipetalous, alt with corolla lobes; ovary superior, gen 2-celled, with 1–∞ ovules in a cell, the cells sometimes each with a median, partition-like outgrowth from the septum; placentation gen axile; style solitary, with a simple, elongate stigma; fr gen a circumscissile caps, the top deciduous; endosperm present; herbs with entire or toothed, simple lvs, these gen basal in a close spiral, or in a few spp. cauline and opp; infl a bracteate spike.

Plantago L.   Plantain

Characters essentially of the family, as described above. (The classical L name, from *planta*, the sole of the foot). (*P. coronopus* L., a tooth-lvd European sp., has recently been collected on coastal strand in Pacific Co., Wn. The subpinnatifid lvs are distinctive.)

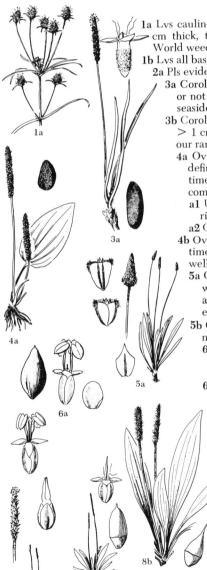

1a Lvs cauline, opp, linear, 2–8 cm × 1–3 mm; spikes 0.5–1.5 cm, nearly 1 cm thick, the lower bracts abruptly and firmly foliaceous-caudate; Old World weed, occ W Cas; sand p. (*P. indica*)   1 **P. psyllium** L.
1b Lvs all basal, the pls scapose
   2a Pls evidently per (some spp., such as no 3, may bloom the 1st year)
      3a Corolla tube pubescent (unique among ours); lvs ± linear, scarcely or not at all petiolate, seldom 1 cm wide; seeds 2–4; maritime; sea or seaside p.; ours is ssp. *juncoides* (Lam.) Hult.   2 **P. maritima** L.
      3b Corolla tube glab; lvs broader than linear, gen at least some of them > 1 cm wide, ± evidently petiolate; pls (except no 6) not maritime in our range
         4a Ovules and seeds 6–30 (unique among ours); lvs broad, the well defined bl broadly elliptic to cordate-ovate and gen 1.3–2.3 (3) times as long as wide; pls not woolly at base; seeds reticulate, 1 mm; common p., nippleseed p.; 2 vars.   3 **P. major** L.
            a1 Uncommon native, scarcely weedy, with ± succulent lvs, occurring in saline habitats (*P. nitrophila*)   var. **pachyphylla** Pilger
            a2 Common intro weed of lawns etc., not succulent   var. **major**
         4b Ovules 2–4; lvs narrower, the elliptic or narrower bl gen (2) 2.5–10 times as long as wide, or even longer; pls often woolly at base; seeds well over 1 mm, not reticulate
            5a Outer sepals (the 2 next the bract) connate, collectively entire or with a mere apical notch (unique among our spp.); bracts gen acuminate or caudate-acuminate; ovules 2; intro weed, common esp W Cas; English p., buckhorn p., ribwort   4 **P. lanceolata** L.
            5b Outer sepals, like the others, free; bracts obtuse to acute; native, not weedy
               6a Caps indehiscent (unique in the genus), 6–7 mm; seeds 2, 4–5 mm; cold, wet places, subcoastal but not maritime; Alas p.
                                                                        5 **P. macrocarpa** Cham. & Schlecht.
               6b Caps circumscissile at or below middle, 2–4.5 mm, (2) 3–4-seeded, seeds 1.7–3 mm
                  7a Corolla lobes 2–4 mm, erect after anthesis; spikes 5–25 cm, ca 1 cm thick; tropical Am sp., n on tidal flats and bluffs along coast to sw Wn; Mex p. (*P. subnuda*)
                                                                        6 **P. hirtella** H.B.K.
                  7b Corolla lobes 1.5 mm or less, spreading or reflexed; spikes 2–20 cm, < 1 cm thick; inl spp.
                     8a Pl densely and gen conspicuously brown-woolly at base; spike elongate, gen 5–20 cm at maturity; alkaline places esp on Gr Pl, w to intermont valleys of w Mont and c Ida; alkali p., saline p.   7 **P. eriopoda** Torr.
                     8b Pl sparsely and inconspicuously or scarcely brown-woolly at base; spike short, ca 2–7 cm at maturity; nonalkaline meadows at mid elev in RM, including w Mont and c Ida; Tweedy's p.   8 **P. tweedyi** Gray
   2b Pls ann (occ short-lived per in no 11, marked by the conspicuously long-exserted bracts)
      9a Caps 4 (6)-seeded; stamens gen 2; lvs puberulent or glab, seldom as

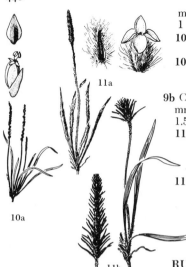

much as 3 mm wide; infl wholly glab; corolla very small, lobes <
1 mm; bracts ovate, 2 mm

10a Pls with a well developed taproot; widespread cordilleran sp. of
moist, ± saline sites; slender p.   9 **P. elongata** Pursh

10b Pls with a very short and quickly deliquescent taproot, appearing
fibrous-rooted; e Am sp. of nonsaline soils, occ intro W Cas; dwarf
p.   10 **P. pusilla** Nutt.

9b Caps 2-seeded; stamens 4; lvs ± woolly-villous to glab, often > 3
mm wide; infl sparsely to densely long-hairy; corolla larger, lobes gen
1.5–2 mm; bracts ± linear, gen longer; pls of nonsaline sites

11a Bracts gen inconspicuous and not much elongate, occ (esp lower
ones) moderately > the ± densely woolly spike, not blackening in
drying; lvs ± conspicuously woolly-villous; common native sp. E
Cas; Indian-wheat *(P. purshii)*   11 **P. patagonica** Jacq.

11b Bracts much elongate, conspicuously exserted from the gen not
very woolly spike, tending to blacken in drying; lvs gen not very
hairy, often glab; s US sp., occ weed w Cas, also in c and e Mont
where perhaps native; large bracted p.   12 **P. aristata** Michx.

RUBIACEAE   Madder Family

Fls reg, ♂ or rarely unisexual, gamopet, epig; calyx entire or toothed, or obsolete; corolla
(3)4–5-lobed; stamens epipetalous, alt with corolla lobes; pistil (1)2–10-locular, with 1–∞ ovules
in each cell, placentation axile, basal, or apical; style simple or divided, or styles separate; fr dry or
fleshy, dehiscent or indehiscent; endosperm copious to wanting; trees or shrubs, less often (but
including all our spp.) herbs, with opp (or whorled), simple and gen entire lvs; stips gen interpetiolar, sometimes reduced to a mere interpetiolar line, or sometimes enlarged and lflike, the lvs
thus appearing whorled; fls in various sorts of infls of cymose origin.

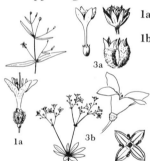

1a Lvs opp, with small stips; ovules basally attached; corolla salverf-funnelf
Kelloggia

1b Lvs whorled (or the uppermost sometimes merely opp); ovules laterally
attached (to the partition); corolla diverse

2a Corolla funnel-shaped, the tube not much, if at all, < lobes

3a Calyx teeth evident; fls in small heads with a basal invol of lflike
bracts (unique among our genera)   **Sherardia**

3b Calyx teeth obsolete; fls in an open infl   **Asperula**

2b Corolla rotate or nearly so, the tube much < lobes   **Galium**

Asperula L.   Asperula

Similar to *Galium,* and perhaps better included in that genus, but differing in its funnelf corolla,
the tube not much, if at all, < lobes; ours rhizomatous per. (Diminutive of Latin *asper,* rough,
presumably referring to the lvs).

1a Fr covered with hooked bristles; lvs gen 1.5–5 cm × 4–12 mm; pl vanilla-
scented; European sp., occ intro in woods W Cas; sweet woodruff, wald-
meister   1 **A. odorata** L.

1b Fr glab; lvs smaller, to 2 cm × 2 mm; pl inodorous; e Mediterranean sp.
casually intro at Bingen, Wn; spreading a.   2 **A. humifusa** (Bieb.) Bess.

Galium L.   Bedstraw; Cleavers

Fls ♂ or rarely unisexual, borne in small or large, basically cymose infls, or on axillary peduncles;
calyx lobes obsolete; corolla ± rotate, the 3–4 lobes much > tube, valvate; ovary 2-celled, with a
solitary ovule attached near middle of septum in each cell; styles 2, short; stigmas capitate; fr dry
(in all ours) or in some spp. fleshy, the carpels approximate or divergent, indehiscent; endosperm
well developed; herbs with square st and whorled, entire lvs (upper lvs merely opp in no 1),
exstip except insofar as some of the lvs of each whorl may be enlarged stips. (Gr *gala,* milk,
from the use of *G. verum* for curdling).

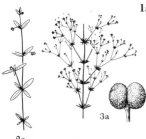

1a Pls ann from a short taproot

  2a Lvs 2–4 in a whorl, not cuspidate; st glab; fls solitary in axils on elongating peds; fr 2.5–3.5 mm, covered with hooked hairs; native, mont sp.,
gen E Cas; thinlf b., low mt. b.   1 **G. bifolium** Wats.

  2b Lvs gen 5–8 in a whorl, cuspidate; st retrorsely scab or rarely glab; fls on
3–5-fld axillary peduncles, or in more-open infls; frs diverse

    3a Frs very small, ca 1 mm, granular-roughened or with hooked hairs;
pls tending to be freely br and loosely ∝ -fld; European weed, occ
intro W Cas; wall b.; ours is var. *leiocarpum* Tausch *(G. anglicum, G.
divaricatum, G. parvifolium),* with merely granular-roughened frs
        2 **G. parisiense** L.

    3b Frs larger, (1.5) 2–4 (5) mm, pls gen not much br, and with relatively
few fls

      4a Frs with hooked hairs, or very rarely glab; peds divaricate or as-
cending, straight; peduncles = or gen > the subtending whorl of
lvs; lvs gen uncinate-hispid above; common cosmopolitan, rather
weedy sp.; cleavers, goose-grass; 2 vars.   3 **G. aparine** L.

        a1 Frs relatively small, up to ca 3 mm; our common phase *(G. vail-
lantii)*   var. **echinospermum** (Wallr.) Farw.

        a2 Frs larger, gen 3–4 (5) mm; occ with us   var. **aparine**

      4b Frs scab-muricate; peds strongly recurved in fr; peduncles = or <
the subtending whorl of lvs; lvs gen glab on upper surface (margins excepted); Eurasian weed, rare W Cas; rough-fr corn bedstraw
(G. tricorne, an illegitimate name)   4 **G. tricornutum** Dandy

1b Pls per from creeping rhizomes

  5a Frs provided with hairs that are hooked at tip (uncinate), or in no 7 the
fr sometimes merely muricate-scab

    6a Lvs 4 in a whorl, evidently trinerved; st glab, erect; W Cas

      7a St with gen 2–4 (5) whorls of lvs; lf-base evidently narrow, ± cu-
neate, the margins tending to be concavely rounded; infl few-fld,
the fls gen 2–3 (6) on each of the 1–3 terminal peduncles; boreal sp.,
s rarely to Cas of n Wn; boreal b., n wild licorice
        5 **G. kamtschaticum** Steller

      7b St with gen 5–8 (9) whorls of lvs; lf base broader, the margins
tending to be convexly rounded; fls more ∝ , each primary peduncle
tending to be cymosely br and several-fld; Cas and w; Ore b.
        6 **G. oreganum** Britt.

    6b Lvs gen 5–6 in a whorl, 1-nerved; st gen retrorsely scab on angles,
gen either prostrate, ascending, or scrambling on other vegetation;
widespread

      8a Bristles of fr very short, only 0.15–0.3 mm, or the fr sometimes
merely muricate-scab, with the short bristles very stout and
scarcely hooked; fls gen in loose, irreg br, ± lfy-bracteate infls at
ends of main st and brs; rough b.   7 **G. asperrimum** Gray

      8b Bristles of fr longer, gen 0.5–1.0 mm; fls gen borne in 3's at ends of
axillary peduncles, or the peduncles sometimes cymosely br and
several-fld; sweetscented b., fragrant b.   8 **G. triflorum** Michx.

  5b Frs without hooked hairs, not all scab-muricate

    9a Fls unisexual, the pls ♀, ♂; frs and young ovaries conspicuously pubes-
cent with long, spreading, straight or merely flexuous hairs; dry places
E Cas; shrubby b., many-fld b. *(G. bloomeri, G. watsonii)*
        9 **G. multiflorum** Kell.

    9b Fls ⚥; frs and young ovaries glab or in no 10 rather inconspicuously
short-hairy

      10a Fls ± ∝ in a terminal, compound and much br, rather showy infl;
sts prostrate or ascending to more often erect, glab or pubescent,
but not retrorsely scab on angles; corolla 4-lobed

        11a Lvs in whorls of 4, 3-nerved; frs short-hairy or
occ glab; common and widespread native circumboreal sp.;
northern b. *(G. septentrionale)*   10 **G. boreale** L.

        11b Lvs gen in whorls of 6–8 (12), 1-nerved, cuspidate; frs glab;
Eurasian weeds occ intro W Cas

          12a Fls white; st glab in infl; lvs broadly linear to linear-
oblong or oblanceolate; wild madder, white or great hedge b.
            11 **G. mollugo** L.

          12b Fls bright yellow (unique among our spp.); st gen puberulent

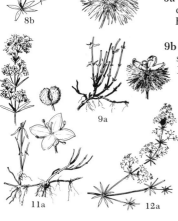

or hirtellous at least in infl; lvs narrowly linear; lady's b., yellow b.   **12 G. verum** L.

10b Fls solitary or few in small, rather inconspicuous infls; sts weak, tending to recline or scramble on other vegetation, gen retrorse-scab on angles; corolla 3-lobed or less often 4-lobed

13a Fls solitary or 2–3 at ends of axillary or terminal peduncles which may themselves be borne in 3's; corolla small, gen 1–1.5 mm wide; common, widespread circumboreal sp.; small b., small c.; ours is var. *pacificum* Wieg. *(G. brandegei*, a depauperate form; *G. subbiflorum)*   **13 G. trifidum** L.

13b Fls several in small, irreg br, basically cymose infls; corolla larger, gen 2–3 mm wide; gen W Cas and esp coastal (incl PS), occ inl to nw Mont; Pac b.   **14 G. cymosum** Wieg.

Kelloggia Torr.   Kelloggia

Fls ♂, in open, terminal, cymose infls, gen 4–5-merous; calyx teeth short; corolla funnel-salverf, valvate; ovary 2-celled, with a solitary, erect, basally attached ovule in each cell; style slender, bifid at tip; fr small, dry, indehiscent, covered with hooked bristles as in *Galium;* endosperm well developed; per herbs with opp, entire, sessile lvs and small interpetiolar stips. (Albert Kellogg, 1813–1887, Cal botanist).

K. galioides Torr. Glab, rhizomatous and sometimes also taprooted; sts ± clustered, 1–6 dm; lvs lanceolate or lance-linear, gen 1.5–5 cm × 2–15 mm; fls long-ped; corolla pink or white, 4–8 mm, the shortly hispid, ascending-spreading, narrow lobes nearly = the slender tube; stamens and style shortly exserted; fr 3–4 mm, the 2 halves readily separable; widespread at mid to high elev E Cas, also in OM.

Sherardia L.   Blue Field-madder; Herb Sherard; Spurwort

Fls ♂, borne in small heads with a basal invol of lflike bracts; calyx teeth 4–6, lanceolate, well developed; corolla funnelf, the slender tube evidently > the 4 valvate lobes; ovary 2-celled, with a solitary ovule attached near base of septum in each cell; style bifid at tip; stigmas capitate; fr dry, crowned by the persistent sepals, dicoccous, the carpels indehiscent; endosperm well developed; ann with habit of *Galium.* (William Sherard, 1659–1728, a patron of botany).

S. arvensis L. Sts slender, 0.5–3 dm, spreading-hairy or retrorsely scab; lvs in whorls of ca 6, 0.5–2 cm, stiffly hirsute above, the antrorsely scab, ± cartilaginous margins confluent distally into a firm point; heads on axillary and quasi-terminal, naked peduncles; invol of (7) 8 (10) shortly connate bracts 4–9 mm; corolla 3 mm, pinkish; fr scab-strigose, 2 mm exclusive of the prominent, pointed sepals; Mediterranean weed, intro W Cas.

CAPRIFOLIACEAE   Honeysuckle Family

Fls reg or irreg, ♂ (or the marginal ones sometimes neutral), gamopet, epig; calyx ± evidently 3–5-lobed; corolla gen 5-lobed, sometimes bilabiate, the tube sometimes spurred or gibbous; stamens epipetalous, gen 5 and alt with corolla lobes (only 4 in *Linnaea)*; pistil 2–5-locular, with 1–several pendulous ovules in each locule, sometimes only 1 locule fertile; stigma capitate or 2–5-parted, the style elongate or obsolete; fr indehiscent, gen fleshy (dry in *Linnaea);* endosperm copious; shrubs or woody vines, less often herbaceous or arborescent pls, with opp (rarely alt or whorled), gen exstip lvs, the stips when present gen small and adnate to petiole; fls in various sorts of infls of gen cymose origin.

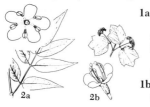

1a Style very short or none, stigmas sessile or nearly so; infl br and gen with ± ∝ fls, umbelliform to corymbiform or paniculiform; corolla rotate to shortly open-camp; fr drupaceous

2a Lvs pinnately or bipinnately compound; fr with 3–5 small, seedlike stones   **Sambucus**

2b Lvs simple, sometimes lobed; fr with 1 large stone   **Viburnum**

1b Style well developed, ± elongate; infl various, of paired fls ending a peduncle, or of short racemes or spikes, or of sessile verticils on an elongate or

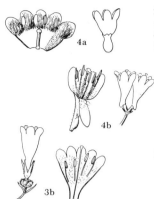

4a

4b

3b

foreshortened axis, but not umbelliform, corymbiform, or paniculiform; corolla short-camp to elongate and tubular, not at all rotate; lvs simple; frs diverse

3a Stamens as many as the corolla lobes, gen 5; fr fleshy; pls not at once trailing and with fls paired on terminal peduncles

    4a Corolla reg or merely ventricose (tube unequally expanded, ± bulged on one side); ovary with 2 fertile uniovulate locules and 2 sterile locules with several abortive ovules each; fr white, drupaceous, with 2 seedlike stones   **Symphoricarpos**

    4b Corolla evidently irreg, either 2-lipped, or spurred at base, or both; ovary gen 2–3-locular, with several ovules in each locule; fr a several-seeded berry, red to blue or black   **Lonicera**

3b Stamens 4; corolla lobes 5; fr dry; trailing pls with the fls gen paired at ends of terminal peduncles   **Linnaea**

Linnaea L.   Twinflower

Fls reg or nearly so, 5-merous as to calyx and corolla, ped nodding, paired at end of a long, naked peduncle; corolla funnelf to camp, pink or pinkish; stamens 4, inserted toward base of corolla, 2 > other 2; ovary 3-locular, 2 locules with several abortive ovules each, the other with a solitary pendulous normal ovule; style elongate; stigma capitate or obscurely lobed; fr small, dry, indehiscent, unequally 3-locular, 1-seeded; creeping, evergreen, herblike shrubs with small, exstip, few-toothed to entire lvs. (Named for Carl Linnaeus, 1707–1778, originally by Gronovius).

L. borealis L. Western t. Sts slender but woody, elongate, producing ∝ short, suberect, lfy sts gen < 10 cm, many of these terminating in a slender peduncle 3.5–8 cm; lvs short-petiolate, firm, rather broadly elliptic to subrotund, gen 7–25 × 5–15 mm; peduncle with a pair of minute bracts at summit, gen forking into a pair of peds 1–2.5 cm; corolla 9–16 mm, hairy within; circumboreal, s throughout our range in wooded regions; ours is var. *longiflora* Torr. (*L. l., L. americana*).

Lonicera L.   Honeysuckle

Fls reg or more often (incl all our spp.) irreg, 5-merous as to calyx, corolla, and androecium (or calyx lobes sometimes obsolete), borne on 2-fld axillary peduncles or in verticils (opp, 3-fld, sessile cymules) on terminal or terminal and axillary rachises; corolla reg 5-lobed, or often ± evidently bilabiate, with 4-lobed upper lip, the tube often gibbous or spurred near base; ovary gen 2–3-locular, each cell with 3–8 pendulous ovules; placentation axile; style elongate, with capitate stigma; fr a small, several-seeded berry; shrubs or woody vines with exstip or rarely stip, gen entire lvs. (Adam Lonitzer, 1528–1586, German herbalist).

2a

3a

4a

1a Fls paired on axillary peduncles; lvs all distinct; ± erect shrubs

  2a Bracts at top of peduncle enlarged, broad and foliaceous, often anthocyanic, forming an invol, the outer pair 8–15 mm or more; ovaries and frs wholly distinct; widespread cordilleran sp.; bearberry h., black twin-berry; 2 vars.   1 **L. involucrata** (Rich.) Banks

    a1 Shrub (0.5) 1–2(3) m; anthers ca = corolla or often slightly exserted; corolla seldom > 1.5 cm; widespread   var. **involucrata**

    a2 Shrub 1.5–4 m; anthers shortly included, sometimes only = sinuses of corolla; corolla to 2 cm; Cal Coast ranges, n along coast rarely to Lane Co, Ore (*L. l.*)   var. **ledebourii** (Esch.) Jeps.

  2b Bracts at top of peduncle narrow and gen small, relatively inconspicuous, < 5 mm except in forms of no 2

    3a Bractlets (inner bracts) wholly connate into a narrow-mouthed cup which tightly encloses the ovaries and grows with them into a fr, the ovaries thus appearing wholly united, but actually distinct within the cup; corolla yellow, 10–13 mm, obscurely or scarcely bilabiate, lobes ca = tube; sweet-berry h., bluefly h. (*L. cauriana*)   2 **L. caerulea** L.

    3b Bractlets small and inconspicuous, or obsolete, not enclosing the ovaries; ovaries and fr united at least at base, but obviously paired

      4a Fls ochroleucous or light yellow, 10–20 mm, obscurely or scarcely bilabiate, the slightly unequal lobes much < tube; ovaries and frs divergent, united only at base; shrub 1–2 m at mid to high elev in mts; Utah h. (*L. ebractulata*)   3 **L. utahensis** Wats.

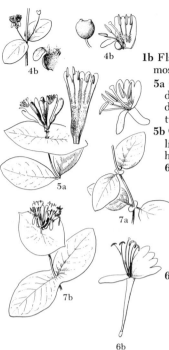

4b Fls dark reddish-purple, 8–11 mm, strongly and deeply bilabiate, lips > tube, upper lip shallowly 4-lobed; ovaries and frs united to middle or gen beyond; shrub 6–15 dm in Cas from Mt Adams s; purple-fl h., double h.   **4 L. conjugialis** Kell.

1b Fls in terminal or terminal and axillary infls of several or ∝ fls; uppermost lvs gen connate-perfoliate; vines or climbing shrubs

  5a Corolla slightly bilabiate, 2.5–4 cm, orange-yellow to orange-red, tube densely hairy within, ca 3–4 times as long as lips; filaments attached well down in tube; lvs glab except for ciliate margins; widespread native twining vine; trumpet, nw, or orange h.   **5 L. ciliosa** (Pursh) DC.

  5b Corolla strongly bilabiate, seldom orange, tube < to ca twice as long as lips; filaments attached nearly at orifice of corolla; lvs glab or often ± hairy

    6a Corolla 1.2–3 cm, tube densely hairy within; native

      7a Some lvs with well developed, connate stips (unique among our spp.); corolla 12–18 (20) mm, pink, or yellow tinged with pink, tube < to ca = lips; W Cas; Cal h., hairy h.

                                    6 L. hispidula (Lindl.) Dougl.

      7b None of the lvs stip; corolla to 30 mm, yellowish to purplish or reddish, tube from slightly > to ca twice as long as lips; e Am sp., w to se BC; smooth-lvd h., glaucous h., Douglas h.; ours is var. *glaucescens* (Rydb.) Butters (*L. g.*)   **7 L. dioica** L.

    6b Corolla 3–5 cm, pale yellowish, often tinged with purple, tube glab within, = or > lips; Mediterranean sp., estab in thickets along coast; Etruscan h.   **8 L. etrusca** Santi

Sambucus L.   Elderberry; Elder

Fls reg, gen 5-merous as to calyx, corolla, and androecium, borne in compound, umbelliform or paniculiform, basically cymose infls; calyx inconspicuous; corolla ± rotate; ovary 3–5-celled, with 1 pendulous ovule in each cell; style very short, almost obsolete, 3–5-lobed; fr berrylike, juicy, with 3–5 small, seedlike stones (each enclosing a seed); shrubs, or sometimes coarse herbs or small trees, with pithy sts and large, stip or exstip, pinnately or even bipinnately compound lvs and serrate lfts. (Classical L name for pl now called *Sambucus nigra* L.).

1a Infl ± flat-topped in life, its axis scarcely or not at all produced beyond the gen 4–5 subumbellately clustered principal brs; frs strongly glaucous, appearing pale powdery blue; widespread sp.; blue e.; ours is var. *c.*   **1 S. cerulea** Raf.

1b Infl pyramidal or strongly convex, paniclelike, its main axis extending well beyond the lowest pair of brs; frs not glaucous, gen red to black or purplish-black, seldom paler or even white; Am pls are ssp. *pubens* (Michx.) House, with 4 vars.   **2 S. racemosa** L.

  a1 Fr black or purplish-black; shrub gen 1–2 m; lvs glab or (esp when young) hairy beneath; the common phase E Cas in our range; black e. (*S. m.*)   var. **melanocarpa** (Gray) McMinn

  a2 Fr bright red, or occ yellow, chestnut, or even white

    b1 Large, sometimes arborescent shrubs, gen 2–6 m; lvs gen ± pubescent beneath, at least along midrib and main veins; the common phase W Cas; red e., coast red e. (*S. callicarpa*)

                            var. **arborescens** (T. & G.) Gray

    b2 Smaller shrubs, gen 0.5–3 m; E Cas

      c1 Lvs essentially glab; shrub 0.5–2 m; the common s RM phase, n rarely to Fremont Co, Ida; mt red e. (*S. m.*)

                var. **microbotrys** (Rydb.) Kearney & Peebles

      c2 Lvs gen evidently pubescent beneath, rarely glab; shrubs gen 1–3 m; the common phase of Can and e US, entering our range in se BC; red-berried e., eastern red e. (*S. pubens* var. *leucocarpa* T. & G., Fl. N. Amer. 2: 13. 1841, a name based on the rare, white-fr form, now unfortunately providing the epithet required by the Rules of Nomenclature when typical *S. pubens* is treated as a var. of *S. racemosa*)   var. **LEUCOCARPA** (T. & G.) Cronq. hoc loc.

Symphoricarpos Duhamel   Snowberry

Fls reg or corolla ± ventricose, 5-merous or occ 4-merous as to calyx, corolla, and androecium, borne in terminal or axillary short racemes or spikes, these sometimes reduced to single fls; corolla short-camp to elongate-camp or subsalverf, pink to white; ovary 4-locular, 2 cells containing several abortive ovules each, the other 2 each with a solitary, pendulous, normal ovule; style well developed, with capitate or slightly 2-lobed stigma; fr white (in our spp.), berry like, with 2 seed-like stones each enclosing a seed; erect or trailing shrubs with rather small (to 5 cm or in no 2 to 8 cm), exstip, simple, entire to occ coarsely toothed or even lobed lvs. (Gr *syn,* together, *phorein,* to bear, and *karpos,* fr, referring to the closely clustered frs).

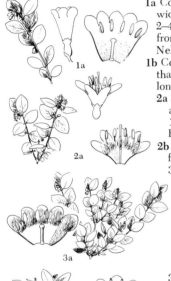

1a Corolla relatively long and narrow, elongate-camp, evidently longer than wide, not evidently ventricose, lobes gen 1/4–1/2 as long as tube; style gen 2–4 mm; erect shrubs 5–15 dm; widespread cordilleran sp. of open places from foothills to fairly high elev E Cas; ours is var. *utahensis* (Rydb.) A. Nels.; mt s. (*S. u., S. tetonensis, S. vaccinioides*)   1 **S. oreophilus** Gray

1b Corolla relatively short and broad, short-camp, not much, if at all, longer than wide (pressed), often bulged on one side (ventricose), lobes half as long to > tube

  2a Style elongate, (3) 4–7 (8) mm, ± exserted, gen long-hairy near middle; anthers 1.5–2 mm, evidently < filaments; corolla lobes gen = or a little > tube; erect shrubs 3–10 dm on open prairies and in moist low ground E Cas, esp in Mont; wolfberry, western s.   2 **S. occidentalis** Hook.

  2b Style short, 2–3 mm, glab; anthers ca 1.0–1.5 mm, not much if at all < filaments; corolla lobes often evidently < tube

    3a Erect shrubs (5) 10–20 (30) dm; corolla gen 5–7 mm; fr 6–15 mm; nutlets 4–5 mm; widespread sp. of thickets, woods, and open slopes, from lowl to mid elev in mts; common s.; 2 vars.   3 **S. albus** (L.) Blake

      a1 Atl slope phase, relatively small (to 1 m), with frs to 1 cm, the twigs and lower surfaces of lvs gen evidently short-hairy, seldom glab   var. **albus**

      a2 Pac slope phase, larger, to 2 (3) m, the larger frs 1–1.5 cm, the twigs and lower surfaces of lvs glab or inconspicuously short-hairy   var. **laevigatus** Fern.

    3b Trailing shrubs, the brs rising < 5 dm; corolla gen 3–5 mm; fr 4–6 mm; nutlets 2.5–3 mm; woods and open slopes up to mid elev in mts, W Cas, and e to n Ida; creeping s.; ours is var. *hesperius* (G. N. Jones) Cronq. (*S. h.*)   4 **S. mollis** Nutt.

Viburnum L.   Viburnum

Fls ♂, reg (or the marginal ones sometimes neutral and irreg), 5-merous as to calyx, corolla, and androecium, borne in umbelliform (all our spp.) or sometimes paniculiform, basically cymose infls; corolla open-camp to subrotate, gen white; ovary 3-locular, 2 locules reduced, sterile, and ± vestigial, the 3rd with a single pendulous ovule; stigmas minute, subsessile, 3, or apparently only 1; fr a unilocular, 1-seeded drupe with soft pulp, red in our spp.; shrubs or small trees, exstip or with small petiolar stips, with opp, simple, entire or toothed to lobed lvs. (The classical L name of *V. lantana* L.).

1a Infl with the marginal fls neutral and enlarged, their corollas gen 1.5–2.5 cm wide, the whole infl gen 5–15 cm wide at anthesis; lvs gen trilobed, the lobes coarsely few-toothed or subentire; circumboreal sp. of moist woods, s to CRG and n Ida; snowball, wild guelder-rose, cranberry-tree, high-bush cranberry; Am pls are var. *americanum* Ait. (*V. trilobum*)   1 **V. opulus** L.

1b Infl with fls all ♂ and alike, their corollas < 1 cm wide, the whole infl gen 1–5 cm wide at anthesis

  2a Infls small (gen 1–2.5 cm wide at anthesis) and with relatively few (< 50) fls; stamens inconspicuous, the filaments 1 mm or less; lvs, or many of them, tending to be trilobed as well as sharply toothed, the unlobed lvs gen acuminate; boreal Am sp., widespread in moist woods and swamps s to n Ore, n Ida, and Colo; moosewood v., high-bush cranberry, squash-berry (*V. pauciflorum*)   2 **V. edule** (Michx.) Raf.

  2b Infls larger (gen 2.5–5 cm wide at anthesis) and with more ∝ fls; stamens exserted, the filaments 3–5 mm; lvs coarsely and often rather bluntly toothed, not at all trilobed, acutish to rounded at tip; thickets and open woods W Cas; oval-lvd v., Ore v.   3 **V. ellipticum** Hook.

ADOXACEAE   Moschatel Family

Fls ± reg, ⚥, gamopet, semi-epig, the tips of the carpels free from the perianth; lateral fls gen with 3 sepals and 5 corolla lobes; terminal fls gen with 2 sepals and 4 corolla lobes; corolla rotate; stamens twice as many as corolla lobes, paired at the sinuses, each with only a single pollen sac; ovary gen 4–5-celled, the cells gen as many as the corolla lobes, with a single pendulous ovule in each cell; style short, distinct; fr a small, dry drupe with 4–5 nutlets; endosperm copious; exstip herbs with several basal lvs and a single pair of opp cauline lvs, the scales of the rhizome alt; fls in compact, headlike, gen 5-fld cymes.

Adoxa L.   Moschatel; Musk-root; Hollow-root

Characters of the family. (Gr *adoxos*, obscure, ignoble).

A. moschatellina L. Delicate herb with a musky odor, 5–20 cm, from a short rhizome; basal lvs long-petiolate, ternate, the discrete primary segms again once or twice ternate or parted; cauline lvs smaller and less dissected, gen borne a little above the middle; corolla inconspicuous, yellowish-green, 5–8 mm wide; circumboreal sp. of moist woods at upper elev, s in RM to Colo.

VALERIANACEAE   Valerian Family

Fls reg or irreg, ⚥ or unisexual, gamopet, epig; calyx segms either inrolled at anthesis and later expanded and pappuslike, or much reduced or obsolete; corolla gen 5-lobed, often ± bilabiate, the tube often spurred or gibbous; stamens epipetalous, 1–4, gen 3, alt with (but fewer than) corolla lobes; pistil basically 3-carpellary, 1 carpel fertile, the others sterile and sometimes obsolete; style with a simple, bilobed, or more often trilobed stigma; ovule solitary, pendulous; fr dry, indehiscent; endosperm wanting; opposite-lvd, exstip herbs with fls in various sorts of basically determinate infls.

1a Calyx segms ca 9–20, inrolled at anthesis, expanded on the mature fr; our spp. per, gen with some or all lvs ± pinnatifid to pinnately compound, fl (Apr) May–Aug   **Valeriana**
1b Calyx segms obsolete; ann with entire or merely toothed lvs, fl Mar–May (June)
　2a St dichotomously br above, the fls in cymose glomerules ending the brs; ovary 3-celled (only 1 cell fertile); stigma 3-lobed; intro, weedy spp. of moist, open, often disturbed sites   **Valerianella**
　2b St simple or with opp, axillary brs, the infl of terminal, subcapitate or interrupted-spicate cymose glomerules; ovary 1-celled, the sterile cells obsolete; stigma 2-lobed or occ 3-lobed; native spp. of vernally moist, open slopes and meadows at lower elev   **Plectritis**

Plectritis DC.   Plectritis

Infl of subcapitate or interrupted-spicate clusters of basically determinate nature, the st simple or with few, paired, axillary brs; corolla subequally 5-lobed or evidently bilabiate, the tube spurred or rarely merely gibbous at base; stamens 3; ovary unilocular, the 2 sterile carpels apparently obsolete; fr dry, indehiscent, gen with a pair of lateral wings, or these sometimes reduced or obsolete; ann with opp, entire or obscurely toothed, sessile or short-petiolate lvs, growing in vernally moist, open places at lower elev, fl spring. (Gr *plektos*, plaited, presumably referring to the complex infl). (*Aligera, Valerianella* in part).

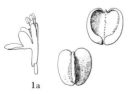

1a Convex side of fr keeled, not grooved; cotyledons transverse to ventral face of fr; wings of fr, when developed, tending to be connivent toward base and divergent above; hairs of fr, when present, ± pointed; in and W Cas; rosy p. (*P. anomala, P. aphanoptera, P. gibbosa, P. major, P. samolifolia*)
                                                                                   1 P. congesta (Lindl.) DC.
1b Convex side of fr broader, scarcely keeled, bearing a narrow groove down the center; cotyledons parallel to ventral face of fr; wings of fr, when present, tending to be about equally divergent above and below; hairs of fr,

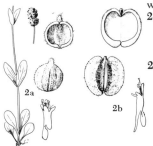

when present, clavate or long-cylindrical and blunt

2a Corolla gen white or pinkish, ca equally 5-lobed, with thick spur; hairs on fr, when present, without any definite arrangement at least on convex side; infl appearing terete in life; widespread in our range and s; long-horn p., white p. *(P. minor, A. grayi)*   **2 P. macrocera** T. & G.

2b Corolla gen pink or deep pink, evidently bilabiate, with slender spur; hairs of fr, when present, gen unevenly distributed, more ∝ in a band on each side of the groove on the convex face; infl tending to appear quadrangular in life; e end of CRG, and s W Cas to Cal; long-spurred p. *(V.c., A. macroptera, P. m.)*   **3 P. ciliosa** (Greene) Jeps.

Valeriana L.   Valerian

Fls in corymbiform to paniculiform or thyrsoid infls of basically determinate nature, ♂ or unisexual; calyx initially involute and inconspicuous, later enlarged and spreading, gen (incl all our spp.) with several or ∝ long, setaceous, plumose, pappuslike segms; corolla tube sometimes gibbous at base, the 5 lobes equal or subequal; stamens 3; ovary basically 3-carpellary, the 2 abaxial carpels vestigial; stigma 3-lobed; fr a nerved achene; ann or (all ours) per herbs with opp, entire to bipinnatifid lvs and a characteristic odor. (Name of L origin, possibly from *valere*, to be strong).

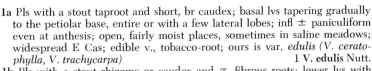

1a Pls with a stout taproot and short, br caudex; basal lvs tapering gradually to the petiolar base, entire or with a few lateral lobes; infl ± paniculiform even at anthesis; open, fairly moist places, sometimes in saline meadows; widespread E Cas; edible v., tobacco-root; ours is var. *edulis* (*V. ceratophylla, V. trachycarpa*)   **1 V. edulis** Nutt.

1b Pls with a stout rhizome or caudex and ∝ fibrous roots; lower lvs with sharply differentiated bl and petiole; infl ± corymbiform at anthesis, though often more expanded in fr

2a Lvs (except the reduced uppermost ones) all pinnately divided or compound, with 11–21 lfts or segms, the terminal one not much if at all > the lateral ones; pls gen 6–15 dm; European sp., occ escaped from cult in our range; garden heliotrope, garden v., all-heal   **2 V. officinalis** L.

2b Lower lvs undivided, or with 1–4 pairs of lateral segms or lfts, the terminal one evidently the largest; native spp.

3a Corolla small, gen 2–4 mm, the lobes not much if at all < tube; pls gyno-dioecious, some with chiefly ♂, others with chiefly ♀ fls

4a Relatively small pls, 1–4 (6) dm, not very lfy, the lateral lobes of the cauline lvs gen well < 1 cm wide; achenes lanceolate, glab; wet meadows etc. in the mts; circumboreal, s to nw Wn, c Ida, and nw Mont; northern v.; Am pls are called var. *sylvatica* (Rich.) Wats. (*V. s., V. septentrionalis*)   **3 V. dioica** L.

4b More robust pls, 3–9 dm, tending to be amply lfy, the lateral lobes of some of the cauline lvs often > 1 cm wide; achenes a little broader, gen lance-ovate, short-hairy or occ glab; RM from n Ida and adj Mont s, and w to e Ore; western v.   **4 V. occidentalis** Heller

3b Corolla larger, gen 4–18 mm, the lobes not > ca half as long as the tube; fls gen all ♂

5a Stamens gen > corolla lobes; corolla 4–9 mm; widespread

6a Robust pls, gen 3–12 dm, with ample cauline lvs, the basal lvs when present varying from smaller to a little larger than the cauline ones, the lf segms gen but not always coarsely crenate or wavy; frs gen ovate to oblong-ovate; moist places at mid and upper elev in mts in and E Cas; mt heliotrope, Sitka v.
                     5 V. sitchensis Bong.

6b Smaller, less lfy pls, gen 1–7 dm, the cauline lvs = or smaller than the well developed and persistent basal ones, the lf segms gen entire or nearly so, sometimes ± toothed; frs narrower, gen lance-oblong or lance-linear

7a Basal lvs, or some of them, gen pinnatifid; moist or wet places up to 4000 feet, gen W Cas; Scouler's v. (*V. sitchensis* var. *s.*)
                     6 V. scouleri Rydb.

7b Basal lvs gen lobeless; open, often rocky slopes at mid to high elev in mts, often near snowbanks; E Cas (*V. capitata* ssp. *a.*); ours is var. *pubicarpa* (Rydb.) Cronq.; downy-fr v. (*V.p.*)
                     7 V. acutiloba Rydb.

5b Stamens = or gen < corolla lobes; corolla 11–18 mm; open slopes in Wen Mts; Wenatchee v.   **8 V. columbiana** Piper

Valerianella Mill.   Valerianella

Infl dichotomously br; calyx lobes 3–6 or (in our spp.) obsolete; corolla tube often gibbous or minutely spurred, the 5 lobes ca =; stamens 3; ovary 3-carpellary; stigma 3-lobed; fr dry, 3-locular, the 2 abaxial locules sterile and gen with an evident groove between them; ann (ours) or bien herbs with opp, entire or toothed, gen sessile lvs, or the lower lvs often ± petiolate; our spp. occ weeds in moist, open places, intro from Europe. (Diminutive of *Valeriana*).

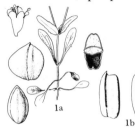

1a Fertile cell of fr with an enlarged corky mass attached to the back; groove between the sterile cells narrow, shallow, and relatively inconspicuous; lamb's lettuce, European corn-salad *(V. olitoria)* 1 **V. locusta** (L.) Betcke
1b Fertile cell of fr without a corky mass; groove between the sterile cells relatively wide, deep and conspicuous; keeled corn-salad
2 **V. carinata** Loisel.

DIPSACACEAE   Teasel Family

Fls ± irreg, ♀, gamopet, epig; calyx small, cupulate or ± deeply cut into 4 or 5 segms or into more ∝ teeth or hairs; corolla 4–5-lobed, often ± 2-lipped; stamens 4 or 2, epipetalous, alt with corolla lobes, exserted, not connate; style with a simple or 2-lobed stigma; ovary 1-celled, with a solitary pendulous ovule; fr dry, indehiscent, enclosed (except at tip) by a gamophyllous, apically cupulate-toothed or subentire involucel which may be adnate to the ovary below; seeds with thin or copious endosperm; herbs with opp or whorled, exstip lvs, the fls gen borne in dense, invol heads.

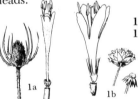

1a Recep chaffy; calyx cupulate, 4-angled or 4-lobed; st prickly   **Dipsacus**
1b Recep merely hairy; calyx of at least 8 elongate teeth or bristles, united below; st glab or hairy, not prickly   **Knautia**

Dipsacus L.   Teasel

Heads dense, ovoid or cylindric; invol bracts linear, spine-tipped, gen elongate; recep bracts ovate or lanceolate, acuminate into an awn > fls; involucel 4-angled, truncate or 4-toothed at summit; calyx short, often persistent at summit of achene; corollas all ca alike, with a long tube and 4 ± unequal lobes; stamens 4; stigma oblique, entire; coarse bien or per herbs with prickly sts and large, opp, sessile, often connate lvs. (The Gr and L name, presumably from Gr *dipsa*, thirst, referring to the accumulation of water in the cuplike base of the connate lvs).

D. sylvestris Huds. Gypsy-combs. Stout, taprooted bien 0.5–2 m; lvs ± prickly, esp on midrib beneath, otherwise ± glab, the basal oblanceolate, crenate, gen dying early in second season, the cauline lanceolate, to 3 dm, gen connate at base; heads erect, 3–10 cm on long, naked peduncles; invol bracts ± prickly, upcurved, unequal; recep bracts ending in a stout, straight awn; corolla pubescent, 10–15 mm, lobes 1 mm; European weed, widely intro in moist, low places.

Knautia L.

Heads dense, hemispheric; invol bracts lanceolate, herbaceous; recep densely hairy, without bracts; involucel compressed but ± 4-angled, the shortly cupulate apex gen obscurely 2-toothed; corollas ± unequally 4–5-lobed, those of the marginal fls often the largest; stamens 4; stigma emarginate; herbs with opp lvs. (Christian Knaut, 1654–1716, German physician and botanist).

K. arvensis (L.) Coult. Field scabious. Taprooted per 3–10 dm, ± hirsute; lowest lvs gen merely coarsely toothed, the others ± deeply pinnatifid with narrow lateral and broader terminal segms, to 2.5 dm, reduced upward; heads on long, naked peduncles, 1.5–4 cm wide; invol bracts 8–12 mm; calyx gen 3–4 mm, with 8–12 bristlelike teeth; corollas lilac-purple, 4-lobed; European weed, occ found in our range.

CUCURBITACEAE   Cucumber or Gourd Family

Pls monoecious or dioecious, the fls reg, gamopet or polypet, the ♀ ones epig; limb of calyx and corolla gen ± combined; stamens 1–5, gen 3, with two 2-celled and one 1-celled anther, free or variously monadelphous; style solitary, with thickened, entire or lobed stigma; placentas thickened, parietal, or often confluent and partitioning the 1–4-locular ovary; fr fleshy or sometimes dry, gen relatively large; seeds large, gen ± flattened, without endosperm; tendril-bearing vines with gen white or yellow to greenish fls and simple, alt, often lobed lvs.

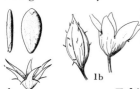

1a Ann; seeds flattened, roughened; germination epigaeous; fls gen 6-merous
                                                                        Echinocystis
1b Per from a much enlarged, woody root; seeds turgid, smooth; germination hypogaeous; fls 5–8-merous   **Marah**

Echinocystis T. & G.   Balsam Apple; Wild Cucumber

Pls monoecious, the ♂ fls in axillary, pedunculate, narrow pans, the ♀ ones from the same axils, solitary or sometimes paired; calyx lobes small and bristle-like, inserted at the sinuses of the corolloid perianth; stamens 3, the filaments and anthers ± connate; ovary 2-celled; ovules 2 in each cell, attached to the wall or to the partitions; style short, with a broad, lobed stigma; fr weakly spiny, ± bladdery-inflated, at length dry, bursting irreg at the apex, fibrous-netted within; seeds flattened; ann climbing vines with br tendrils and alt, lobed lvs. (Gr *echinos*, hedgehog, and *kystis*, bladder, referring to the frs).

E. **lobata** (Michx.) T. & G. High-climbing ann; lvs petiolate, 5–15 cm, scaberulous, cordate at base, palmately 5-lobed, the lobes gen triangular-acute and remotely toothed; perianth greenish-white, with short tube and spreading lobes gen 3–6 × 1 mm; fr ellipsoid or subglobose, 3.5–5 cm; seeds ± ovate, 1.5 cm; moist bottoms and thickets, widespread in e US, and w to Mont (*Micrampelis l.*).

Marah Kell.   Bigroot, Manroot

Pls monoecious, the ♂ fls in axillary, pedunculate racemes or narrow pans, the ♀ ones from the same axils, gen solitary; calyx lobes inconspicuous, inserted at the sinuses of the corolloid perianth; stamens 3, the filaments and anthers ± connate; ovary 2–4-celled; ovules 1–several in each cell, attached to the wall or to the partitions; style short, with a broad, lobed stigma; fr spiny or sometimes smooth, ± bladdery-inflated, at length dry, bursting irreg at apex, fibrous-netted within; seeds turgid, the cotyledons thickened; per from a much enlarged, woody root, the ann sts climbing or trailing, provided with br tendrils; lvs alt, gen lobed. (Hebrew *marah*, bitter, from the intensely bitter root).

M. **oreganus** (T. & G.) Howell. Ore b. High-climbing per; lvs petiolate, bl seldom > 2 dm, cordate at base, irreg palmately lobed, rough-hairy at least above; ♂ fls in racemes, perianth camp, 5–8-merous, whitish, tube 3–6 mm, lobes 3–8 [×] 2–5 mm; fr 3–8 cm, 2–4-locular, weakly or scarcely spiny; seeds 1–2 to a locule, 2 cm; bottomlands and open slopes at low elev W Cas, and e, rarely, to SRC (*Echinocystis o., Micrampelis o.*).

CAMPANULACEAE   Harebell Family

Fls reg or irreg, ⚥, gamopet, wholly epig, or with the apex of the ovary sometimes free, gen 5-merous as to calyx, corolla, and androecium; stamens gen (incl all our spp.) free from the corolla or nearly so, alt with the lobes; style 1; stigma lobes gen as many as the carpels; ovary 1–5-celled, with axile or parietal placentation and few to ∞ ovules, 2–5-carpellary; fr (in all ours) a caps, opening by valves or pores (or irreg), or sometimes a berry; ann, bien, or per herbs, or sometimes shrubs or trees, with gen alt (or all basal), simple, exstip lvs and often milky juice, the fls borne in diverse types of determinate or indeterminate infls.

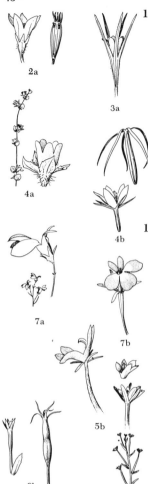

1a Filaments and anthers distinct; corolla reg (or wanting in some fls)

    2a Per (1 sp. bien); caps opening laterally (outside the sepals); cleistogamous fls wanting, the fls all with well developed corolla; filaments expanded and ciliate at base; our spp. all with evidently pedicellate or pedunculate fls   **Campanula**

    2b Ann; caps, fls, and filaments diverse; fls sessile or subsessile

        3a Caps opening only at apex, within the calyx; fls all with well developed corolla, technically solitary and terminal, but appearing irreg scattered because of the sympodial br of the st; filaments only slightly (if at all) expanded at base, not ciliate   **Githopsis**

        3b Caps opening laterally, sometimes near the summit, but outside the calyx; lower fls cleistogamous, with reduced or no corolla; filaments dilated and ciliate at base; infl ± spiciform or interrupted-spiciform

            4a Corolla shallowly lobed; infl a sympodial false spike, the bracts opp the fls; caps opening near base   **Heterocodon**

            4b Corolla deeply lobed (to below the middle); infl a monopodial false spike, the fls 1–several in the axil of each bract; caps opening near summit, or near middle   **Triodanis**

1b Filaments and anthers united into a tube, 2 of the anthers < others, the orifice of the tube thus oblique or appearing lateral; corolla distinctly irreg (or wanting in some fls)

    5a Fls ped (or sometimes subsessile in *Howellia);* hypan and fr fusiform to ellipsoid or globose; caps subapically or irreg dehiscent; duration and habit various

        6a Corolla well developed, in ours gen 7–20 mm; ovary and fr 2-locular, with axile placentation, the fr dehiscent by subapical valves; seeds ∝, minute; habit and habitat various, but not at once aquatic and with linear-filiform lvs

            7a Corolla tube cleft at least halfway to base on dorsal side (between the 2 lobes of upper lip), the sinus thus formed much deeper than the next adjacent lateral sinuses; our spp. per   **Lobelia**

            7b Corolla tube not deeply cleft dorsally, the depth of the dorsal sinus about = that of the lateral ones; ann   **Porterella**

        6b Corolla minute, scarcely 3 mm, or wanting; ovary and fr 1-locular, with parietal placentation, the fr irreg dehiscent by rupture of the thin lateral walls; seeds few and relatively large; aquatic ann, immersed or with floating brs, the lvs linear or linear-filiform   **Howellia**

    5b Fls sessile in axils of lvs or foliaceous bracts, but appearing long-stalked because of the elongate, narrow hypan; caps elastically dehiscent by long slits on the sides; ann, gen of moist places, but not truly aquatic   **Downingia**

Campanula L.   Harebell; Bellflower; Bluebells-of-Scotland

Fls typically ped in a racemiform infl, in which the terminal fl blooms first and the subsequent succession is centripetal, or the terminal fl abortive and the infl thus seemingly indeterminate, or not infrequently the fls solitary; corolla reg, ± camp, the 5 lobes short or ± elongate; stamens free from the corolla and from each other, the filaments short, expanded at the gen ciliate base; ovary gen 3-locular (5-locular in no 1), with axile placentation; stigma lobes as many as locules; caps short or elongate, opening by lateral pores that vary in position from near base to near apex; seeds ∝, compressed; per or seldom bien (as in no 1) or ann herbs with alt lvs and often milky juice. (Diminutive of L *campana*, bell, from the fl shape).

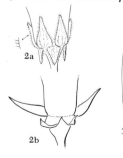

1a Garden and weedy spp. from Europe, occ intro W Cas; pls gen tall and lfy-std, seldom < 5 dm, with several or ∝ fls 2–5 cm in a ± elongate terminal false raceme

    2a Bien; calyx with a conspicuous, broadly cordate, reflexed appendage at each sinus; corolla large, gen 3.5–5 cm; stigmas 5 and caps 5-celled; Canterbury bell   1 **C. medium** L.

    2b Per; calyx without appendages; corolla smaller, gen 2–3.5 cm; stigmas 3 and caps 3-celled

        3a Lvs elongate and narrow, the principal ones ± linear to oblanceolate, < 1.5 cm wide and > 7 times as long as wide; fls relatively few, gen < 10; herbage glab; caps ± erect, opening above the middle   2 **C. persicifolia** L.

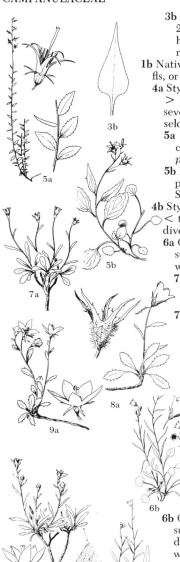

3b

5a

5b

7a

8a

9a

6b

10a   10b

3b Lvs broader, the principal ones with lance-ovate to cordate-ovate bl 2–5 cm wide and 1.5–3 times as long as wide; fls more ∝, often > 10; herbage glab or hairy; caps nodding, opening near base; creeping b., rover b.   3 **C. rapunculoides** L.

1b Native spp., all except some forms of no 11 either smaller, or with smaller fls, or both

4a Style well exserted from corolla; corolla lobes ± spreading or recurved, > tube; caps subglobose, 3–6 mm; lfy-std pls with toothed lvs and gen several or ∝ fls, occurring at low elev (up to 4000 feet) gen W Cas, seldom on e slope

5a Lvs all sessile or subsessile, gen 12–35 below infl; pls gen 3–8 dm; corolla lobes narrowly lanceolate; s WV to Cal; Cal h. or b. *(Asyneuma p.)*   4 **C. prenanthoides** Dur.

5b Lvs all petiolate, except the reduced upper ones, gen 4–10 below infl; pls gen 1–4 dm; corolla lobes broader, gen ovate-oblong; Alas to Cal; Scouler's h. or b.   5 **C. scouleri** Hook.

4b Style = or < corolla; corolla lobes slightly or scarcely spreading, > or < tube; caps diverse, often narrower and much longer; habit and habitat diverse, but not combined as above

6a Caps erect, opening by pores at or above middle; lvs never cordate or subrotund; pls gen dwarf, to 1.5 dm (or in no 10 rarely to 3.5 dm), with few or more often solitary fls, in alp and subalp habitats

7a Anthers 1–2.5 mm; caps 12–20 mm; corolla 6–12 mm, lobes = or > tube; lvs ± entire; circumboreal, s to Mont and Colo; arctic h. or b.   6 **C. uniflora** L.

7b Anthers 3–5.5 mm; caps 3–12 mm; corolla and lvs diverse

8a Hypan woolly-villous with long, loosely spreading hairs; corolla gen 18–30 mm, lobes much < tube; lvs slightly to strongly toothed; n Pac basin sp., s to se BC and Wn Cas; Alas h.   7 **C. lasiocarpa** Cham.

8b Hypan glab to merely scab or minutely hirtellous; corolla gen 6–16 mm, lobes > or < tube

9a Lvs sharply toothed; caps subglobose, 3–5 mm; OM; Olympic or Piper's h. or b.   8 **C. piperi** Howell

9b Lvs entire or nearly so; caps narrower; not of OM

10a Pls minutely spreading-hirtellous throughout; caps 5–7 mm; Cas, and rarely c Ida and w Mont; rough h. or b.   9 **C. scabrella** Engelm.

10b Pls glab or nearly so except the ciliate-margined bases of lower lvs; caps 7–11 mm; c and n Ida and adj Mont, and Wen Mts; Parry's h. or b.; ours is var. *idahoensis* McVaugh   10 **C. parryi** Gray

6b Caps nodding, opening by pores near base; basal lvs often cordate or subrotund, the others gen linear; pls when well developed taller (to 8 dm) and with several or ∝ fls, subalp specimens resembling no 10, but with smooth to merely scab lf-margins; widespread, circumboreal sp., gen at mid and lower elev; Scotch b., lady's-thimble *(C. petiolata)*   11 **C. rotundifolia** L.

Downingia Torr.   Downingia

Fls sessile, solitary in axils of middle and upper lvs (or forming a lfy-bracted spike); corolla bilabiate, inverted so that the 3-lobed, morphological upper lip is on the lower or abaxial side; filaments and anthers connate, 2 anthers < others; ovary 1-locular or 2-locular, with correspondingly parietal or axile placentation; stigma 2-lobed; fr elongate, slender, subcylindric, opening elastically by 3–5 longitudinal slits that extend nearly the length of the caps; seeds ∝; ± succulent ann with alt lvs, gen in wet meadows, vernal pools, or marshes, or at edges of ponds. (A. J. Downing, 1815–1852, Am horticulturist).

1a Corolla inconspicuous, scarcely if at all > calyx, gen 4–7 mm; caps 2-locular; tolerant of alkali; E Cas; GB d. *(D. brachyantha, Bolelia laeta)*   1 **D. laeta** Greene

1b Corolla showy, gen well > calyx, seldom < 7 mm; caps 1-locular

2a Anther tube scarcely exserted from corolla, not at all or only slightly incurved; corolla tube 3.5–6 mm; filament tube 2–4.5 mm; W Cas; Cas d., Willamette d.; ours is var. *major* McVaugh *(D. willamettensis)*   2 **D. yina** Appleg.

1a

2a

2b Anther tube evidently exserted from corolla, ± strongly incurved, gen standing almost at right angles to filament tube, the latter 4.5–10.5 mm; corolla tube gen short and widely flaring; n Ida to c Wn, s to Cal and Nev, extending W Cas in Ore; common or showy d. *(D. brachypetala* or *D. elegans* var. *b.,* a form app no 2)   3 **D. elegans** (Dougl.) Torr.

Githopsis Nutt.   Githopsis

Fls technically solitary and terminal, but appearing irreg scattered because of the ± sympodial br of the sts; corolla reg, tubular-camp, the 5 lobes = or < tube; stamens free from corolla and from each other, the filaments short, smooth, and only slightly (if at all) dilated at base; ovary 3-locular, with axile placentation; stigma 3-lobed; caps elongate-obconic, opening within the calyx by terminal pores; seeds ∝, angular; low, gen much br ann with alt, narrow, toothed lvs. ("Named in allusion to the resemblance of the flowers with those of *Githago segetum").*

G. specularioides Nutt. Common blue-cup. Pls to 3 dm; lvs oblong or narrower, to ca 15 × 3 mm; calyx divided to the hypan, its lobes elongate, ± linear, gen 5–15 (20) mm; corollas blue with whitish throat, dimorphic, sometimes < 1 cm and < calyx lobes, sometimes to 2 cm; caps prominently ribbed, 6–15 mm; dry, open places at lower elev, both sides of Cas, Wn to Cal.

Heterocodon Nutt.   Heterocodon

Fls subsessile in a lax, sympodial false spike, borne opp the bracts; lower fls cleistogamous, with reduced or abortive corolla; upper fls with normal, reg corolla, the 5 lobes < tube; stamens free from corolla and from each other, the filaments short, ciliate at the expanded base; ovary inferior, 3-locular, with axile placentation; stigma 3-lobed; caps short and broad, opening tardily by inconspicuous, irreg pores near base; slender ann with alt, sessile, toothed, short and broad lvs. (Gr *heteros,* different, and *kodon,* bell, referring to the 2 kinds of fls).

H. rariflorum Nutt. Lax, very slender, gen 0.5–3 dm, glab, or hispid on lf margins and st angles; lvs ± clasping, distant, rotund or rotund-ovate, small, seldom 1 cm; calyx lobes foliaceous, veiny, ovate or broader, 2–4 mm; corollas of upper fls blue, 3–6 mm, the others abortive; hypan gen spreading-hispid; moist open places at lower elev in our range.

Howellia Gray   Howellia

Fls axillary, ped or subsessile, both petaliferous and apet, the corolla when present bilabiate, with the tube deeply cleft dorsally; filaments and anthers connate, 2 anthers < others; ovary unilocular, with parietal placentation; stigma 2-lobed; fr irreg dehiscent by rupture of the very thin lateral walls; ovules few; seeds large, to 4 mm; ann aquatics, immersed or with floating brs. (Thomas and Joseph Howell, brothers, 1842–1912 and 1830–1912, respectively, pioneer resident botanists in our region).

H. aquatilis Gray. Pls rooted, glab, 1–6 dm, naked below, br above, the brs spreading or floating; lvs ∝, alt, or some opp or ternate, flaccid, linear or linear-filiform, 1–4.5 cm, to 1.5 mm wide; fls gen 3–10, axillary, on stout pedicels 1–4 (8) mm; corolla 2–3 mm, or wanting from earlier fls; fr 5–13 × 1–2 mm; seeds 2–4 mm; ponds and lakes in w Ore, w Wn, and n Ida.

Lobelia L.   Lobelia

Fls ped, borne in terminal racemes, or solitary in upper axils; corolla irreg, inverted so that the 3-lobed morphological upper lip is on the lower or abaxial side; corolla tube dorsally split (between the 2 lobes of the actual upper lip) to below middle, gen to near base, often fenestrate as well; our spp. with corolla blue or blue and white (or with a yellow eye) or wholly white, 7–20 mm; filaments and anthers connate, 2 anthers < others; ovary bilocular, with axile placentation; stigma 2-lobed; fr fusiform to ellipsoid or globose, dehiscent near top; seeds in our spp. roughened, < 1 mm; ann or (ours) per herbs, or shrubs, with alt (or all basal) toothed or subentire lvs. (Matthias de L'Obel, 1538–1616, Belgian botanist).

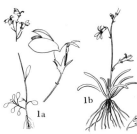

1a Lvs flat, linear to spatulate, cauline as well as basal, only the basal ones sometimes immersed; st solid; peds gen bibracteolate near middle; seeds acute at both ends; bogs, shores, etc; inl sp., w to Mont and ne Wn; brook l., Kalm's l.   1 L. **kalmii** L.

1b Lvs terete, hollow, all in a basal rosette; st hollow and essentially naked; peds without bractlets; seeds with square base at one end; aquatics with only the infl gen emergent; interruptedly circumboreal, W Cas with us; water l., water gladiole   2 L. **dortmanna** L.

Porterella Torr.   Porterella

Fls axillary, ped; corolla bilabiate, inverted so that the deeply 3-lobed morphological upper lip is on the lower or abaxial side; filaments and anthers connate, 2 anthers < others; ovary bilocular, with axile placentation; stigma 2-lobed; fr narrow, ± obconic, dehiscent near the apex; seeds ∝, smooth; fibrous-rooted, alt-lvd ann with the aspect of *Downingia*. (T. C. Porter, 1822–1901, Am botanist).

P. **carnulosa** (H. & A.) Torr. Glab, slightly succulent, 0.5–3 dm; lvs sessile, gen 0.5–2 (3) cm, the lower ± linear and often soon deciduous, the others broader, to 4 mm wide; fls solitary in middle and upper axils; corolla 8–15 mm, blue with white or yellowish eye; peds becoming 0.5–2 (3) cm in fr; fr 6–13 × 2–3 mm; wet meadows and pond edges; along our s border E Cas, and s *(Laurentia c.)*.

Triodanis Raf.   Triodanis; Venus'-looking-glass

Fls sessile or subsessile in axils of middle and upper lvs, 1–several in each axil, forming a dense or interrupted false spike on a monopodial axis, blooming from the bottom upward, except that the terminal fl develops before those immediately beneath it; lower fls cleistogamous, with reduced or abortive corolla; calyx lobes 5, or only 3 or 4 in cleistogamous fls; upper fls with normal, anthocyanic, reg corolla, the 5 lobes > tube; stamens free from corolla and from each other; filaments short, expanded and ciliate at base; ovary gen 3-locular, with axile placentation, varying to sometimes 1-locular and with parietal placentation; stigma gen 3-lobed, sometimes 2-lobed; caps linear to ellipsoid or clavate, opening by pores at or gen above middle; seeds ∝, ± lenticular; ann (ours 1–6 dm) with alt, petiolate or more often sessile, short or elongate, toothed lvs. (Name said by Rafinesque to refer to the 3 unequal teeth of the calyx). *(Specularia, in part.)*.

1a Middle and upper lvs (floral bracts) lanceolate or narrower, gen 5–10 times as long as wide; caps of cleistogamous fls gen 8–15 mm, 1-locular, of open fls 15–25 mm and 1-locular or 2-locular; corolla 7–10 mm, tube scarcely 2 mm; Gr Pl sp., reaching our range in Mont; western v.
                                                       1 T. **leptocarpa** (Nutt.) Nieuwl.

1b Lvs (and floral bracts) broader, gen rotund-ovate or broader, seldom > twice as long as wide; caps of cleistogamous fls gen 4–7 mm, of open fls to 1 cm, all 2- or 3-locular; corolla 8–13 mm, tube 2–4 mm; widespread, ± weedy Am sp.   2 T. **perfoliata** (L.) Nieuwl.

COMPOSITAE   Aster Family

Individual fls epig, ⚥ or unisexual, gamopet, reg or irreg, gen 5-merous, without definite calyx; stamens as many as the corolla lobes and alt with them, epipetalous, with elongate anthers united into a tube, or the anthers rarely distinct; ovary inferior, of 2 carpels, 1-locular, normally with a single erect anatropous ovule; style gen 2-cleft; fr an achene, unappendaged, or more gen crowned with a pappus consisting of hairs or scales; fls sessile in a close head on a common recep, sometimes individually subtended by a small bract (chaff), and nearly always collectively subtended by an invol of few to ∝ bracts; ann, bien, or per herbs, or less commonly woody pls, diverse in habit, foliage, and infl.

One of the largest families of fl pls, with > 15,000 spp., cosmopolitan.

The invol bracts are gen herbaceous or subherbaceous, varying to scarious, hyaline, or cartilaginous; they may be few and in a single row, or ∝ and imbricate, or modified into spines. The recep may be chaffy, with a bract subtending each fl, or may be covered with long bristles, or may be naked. There are (1)5–∝ fls in each head.

The fls are of several gen types. In one type they are ♀ (or functionally ♂) and the corolla is tubular or trumpet-shaped, with gen 5 short terminal lobes. This type of fl is a *disk fl;* a head composed wholly of disk fls is *discoid.* In another type the fl is ♀ or neutral (without a style), and the corolla is tubular only at the very base, above which it is flat, commonly bent to one side, and often 2–3-toothed at the tip. The flattened part of such a fl is a *ray* or *ligule,* and the fl bearing it is a *ray fl.* Ray fls occur only at the margin of the head, the center being occupied by disk fls (except in a few dioecious groups). A head with both ray and disk fls is *radiate.* In some spp. the lig of the marginal, ♀ fls fails to develop, so that the corolla is tubular, but gen more slender and not so evenly toothed as a disk corolla; such a head is *disciform.* A third type of fl superficially resembles the ray fl of a radiate head, but differs in being ♂ and in having gen 5 terminal teeth. Heads in the tribe Cichorieae are composed solely of fls of this type and are called *ligulate* heads. In our range this type of fl is found only in the Cichorieae. Some few members of the tribe Cynareae have the marginal disk fls enlarged, irreg, and transitional toward the lig type.

The pappus is highly diverse in structure. Phyletically a modified calyx, it may be composed of simple or plumose hairs in one or more series, or scales, or stout awns, or a mere projecting ring or crown, or combinations of these; or it may be lacking entirely.

The anthers are gen coherent by their lateral margins; their bases vary from truncate to slenderly caudate. The anthers dehisce introrsely, and the pollen is pushed out through the anther tube by the growth of the style. The style brs gen diverge above the anther tube, have various distinctive forms and textures, and tend to be stigmatic only on definite parts of their surface (the stigmatic lines). The characteristic style brs of the various tribes are to be sought only in the fertile disk fls; those of the rays are gen very similar in all groups, and those of the sterile disk fls are often reduced and undivided. The sterile disk fls, when present, are said to be functionally ♂; although the ovary is obsolete or nonfunctional, the style is still present and retains the function of acting as a plunger to push out the pollen.

TECHNICAL KEY TO THE GENERA, VIA THE TRIBES
(Alternate, artificial keys, leading directly to the genera, start on p 470)

1a Fls all lig and ♂; juice gen milky   **Cichorieae,** p 465

1b Fls, or some of them, tubular and elig; lig fls when present ♀ or neutral; juice gen watery

    2a Style with a ring of hairs, or sometimes merely with a thickened ring, below the brs, papillate thence to the tip (the brs sometimes connate); anthers tailed at the base (scarcely so in *Crupina*); heads discoid (sometimes falsely subradiate); recep densely bristly or sometimes naked; lvs alt; pls often prickly or spiny   **Cynareae,** p 466

    2b Style without any ring of hairs or distinct thickened ring below the brs; anthers (except in Inuleae) not tailed; pls seldom prickly; recep chaffy or naked, rarely *(Gaillardia)* bristly

    3a Style brs subterete or clavate, scarcely flattened, papillate, not hairy, stigmatic only near the base, the stigmatic portion gen not sharply differentiated in appearance; fls all tubular and ♂, never yellow; recep naked   **Eupatorieae,** p 466

    3b Style brs ± flattened, often minutely hairy, esp near the tip, the stigmatic portion often conspicuously defined; fls gen not all alike, some of them tubular and ♂ (sometimes sterile), others ♀ or neutral and often also lig, or, if occ all tubular and ♂, then nearly always yellow

      4a Anthers tailed at the base; ours all ± white-woolly herbs with the corollas (except in the yellow-rayed *Inula*) all tubular

                                              Inuleae, p 469

      4b Anthers truncate to evidently sagittate at the base, but scarcely tailed; heads gen radiate, sometimes disciform or discoid; pls variously hairy or subglab

      5a Style brs with well marked stigmatic lines and gen with a short or elongate appendage, the appendage glab within; lvs alt; recep naked; invol bracts gen but not always imbricate in several series and partly or wholly herbaceous   **Astereae,** p 463

      5b Style brs gen otherwise, though sometimes essentially as in the Astereae; characters of lvs, recep, and invol various, but rarely combined precisely as in the Astereae

        6a Pappus of capillary bristles (none in *Adenocaulon*); style brs

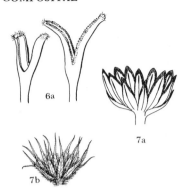

gen truncate, exappendiculate, and with a ring of hairs at the end, varying to sometimes as in the Astereae or Inuleae; lvs alt or opp; recep naked; invol chiefly of =, essentially uniseriate bracts, often with a few much smaller ones at the base

Senecioneae, p 469

  6b Pappus chaffy, or of awns, or none, never capillary; styles and recep various

    7a Invol bracts with scarious or hyaline margins, scarcely herbaceous, gen well imbricate; lvs alt; style brs gen exappendiculate, as in the Senecioneae   **Anthemideae, p 463**

    7b Invol bracts without scarious or hyaline margins, gen green and somewhat herbaceous, seldom much imbricate; lvs opp or alt; style brs commonly with elongate, pointed, hairy tip, with poorly differentiated stigmatic lines, varying to nearly as in the Astereae or Senecioneae   **Heliantheae, p 466**

Key to the genera of the Anthemideae

1a Recep chaffy, at least toward the middle

  2a Heads relatively large, terminating the brs; rays elongate; achenes terete, 4–5-angled, or compressed-quadrangular, not callous-margined

        Anthemis

  2b Heads small; infl corymbiform; rays short and broad; achenes strongly flattened, callous-margined   **Achillea**

1b Recep naked

  3a Infl corymbiform to capitate, or the heads solitary; heads variously large to small, radiate or rayless, with or without pappus

    4a Pappus of short but definite and discrete scales; style brs shortly appendiculate   **Hymenopappus**

    4b Pappus a minute crown, or none; style brs truncate, exappendiculate

      5a Achenes sessile; heads radiate or rayless; disk corollas 4-toothed or more often 5-toothed

        6a Recep flat or somewhat convex

          7a Heads rayless or nearly so; ♀ fls reg present, gen with short, tubular or somewhat ampliate corolla, the rays, if any, yellow and short   **Tanacetum**

          7b Heads radiate, the rays in most spp. white, or the ♀ fls rarely lacking and the heads thus discoid   **Chrysanthemum**

        6b Recep high, hemispheric or conic at least at maturity   **Matricaria**

      5b Achenes, esp the marginal ones, conspicuously stipitate, the stipe persistent on the recep; heads rayless; disk-corollas 4-toothed

        Cotula

  3b Infl spiciform, racemiform, or paniculiform; heads small, rayless, gen epappose   **Artemisia**

Key to the genera of the Astereae

1a Pappus of firm awns, or flattened, bristlelike scales, or of minute bristles with or without some firm longer awns, or none

  2a Pappus none; recep conic; pls scapose; rays white to pink or purple

        Bellis

  2b Pappus present; other characters various, but not combined as above

    3a Rays yellow, or wanting

      4a Heads relatively large, the disk gen ca 1 cm wide or more, the rays 10–45; pappus of 2– several firm, deciduous awns; lvs gen broader than linear; herbs   **Grindelia**

      4b Heads small, the disk ca 5 mm wide or less, the rays 10 or fewer; pappus of several scales or awns, sometimes united at the base; lvs linear; our sp. a shrub   **Gutierrezia**

    3b Rays white to pink, lavender, or blue, not yellow

      5a Recep flat; pappus (at least of the disk fls) of a single series of ∝ rigid, narrow, bristlelike scales; taprooted pls up to 3 dm, with rather few or solitary heads   **Townsendia**

      5b Recep conic or hemispheric; pappus of several minute bristles and (at least on the disk fls) 2 or sometimes 4 longer awns; fibrous-rooted pls, gen 3–20 dm, with ± ∝ heads   **Boltonia**

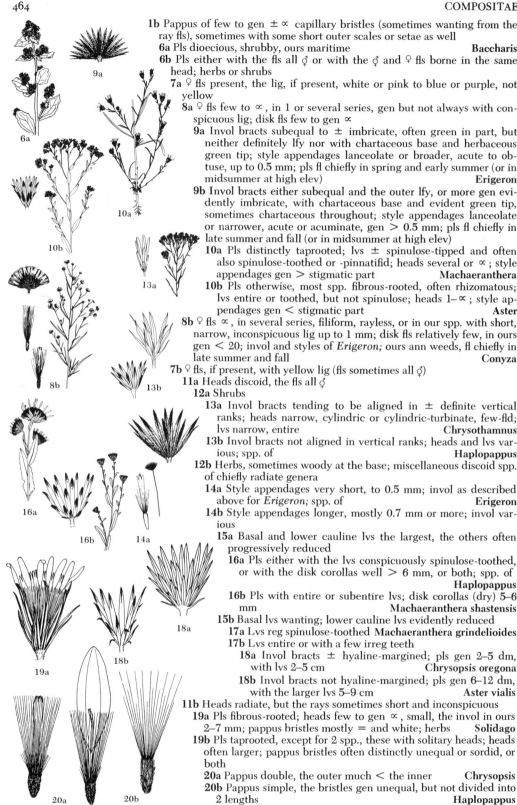

464 COMPOSITAE

1b Pappus of few to gen ± ∝ capillary bristles (sometimes wanting from the ray fls), sometimes with some short outer scales or setae as well

 6a Pls dioecious, shrubby, ours maritime   **Baccharis**

 6b Pls either with the fls all ♂ or with the ♂ and ♀ fls borne in the same head; herbs or shrubs

  7a ♀ fls present, the lig, if present, white or pink to blue or purple, not yellow

   8a ♀ fls few to ∝, in 1 or several series, gen but not always with conspicuous lig; disk fls few to gen ∝

    9a Invol bracts subequal to ± imbricate, often green in part, but neither definitely lfy nor with chartaceous base and herbaceous green tip; style appendages lanceolate or broader, acute to obtuse, up to 0.5 mm; pls fl chiefly in spring and early summer (or in midsummer at high elev)   **Erigeron**

    9b Invol bracts either subequal and the outer lfy, or more gen evidently imbricate, with chartaceous base and evident green tip, sometimes chartaceous throughout; style appendages lanceolate or narrower, acute or acuminate, gen > 0.5 mm; pls fl chiefly in late summer and fall (or in midsummer at high elev)

     10a Pls distinctly taprooted; lvs ± spinulose-tipped and often also spinulose-toothed or -pinnatifid; heads several or ∝; style appendages gen > stigmatic part   **Machaeranthera**

     10b Pls otherwise, most spp. fibrous-rooted, often rhizomatous; lvs entire or toothed, but not spinulose; heads 1–∝; style appendages gen < stigmatic part   **Aster**

   8b ♀ fls ∝, in several series, filiform, rayless, or in our spp. with short, narrow, inconspicuous lig up to 1 mm; disk fls relatively few, in ours gen < 20; invol and styles of *Erigeron;* ours ann weeds, fl chiefly in late summer and fall   **Conyza**

  7b ♀ fls, if present, with yellow lig (fls sometimes all ♂)

   11a Heads discoid, the fls all ♂

    12a Shrubs

     13a Invol bracts tending to be aligned in ± definite vertical ranks; heads narrow, cylindric or cylindric-turbinate, few-fld; lvs narrow, entire   **Chrysothamnus**

     13b Invol bracts not aligned in vertical ranks; heads and lvs various; spp. of   **Haplopappus**

    12b Herbs, sometimes woody at the base; miscellaneous discoid spp. of chiefly radiate genera

     14a Style appendages very short, to 0.5 mm; invol as described above for *Erigeron;* spp. of   **Erigeron**

     14b Style appendages longer, mostly 0.7 mm or more; invol various

      15a Basal and lower cauline lvs the largest, the others often progressively reduced

       16a Pls either with the lvs conspicuously spinulose-toothed, or with the disk corollas well > 6 mm, or both; spp. of   **Haplopappus**

       16b Pls with entire or subentire lvs; disk corollas (dry) 5–6 mm   **Machaeranthera shastensis**

      15b Basal lvs wanting; lower cauline lvs evidently reduced

       17a Lvs reg spinulose-toothed **Machaeranthera grindelioides**

       17b Lvs entire or with a few irreg teeth

        18a Invol bracts ± hyaline-margined; pls gen 2–5 dm, with lvs 2–5 cm   **Chrysopsis oregona**

        18b Invol bracts not hyaline-margined; pls gen 6–12 dm, with the larger lvs 5–9 cm   **Aster vialis**

   11b Heads radiate, but the rays sometimes short and inconspicuous

    19a Pls fibrous-rooted; heads few to gen ∝, small, the invol in ours 2–7 mm; pappus bristles mostly = and white; herbs **Solidago**

    19b Pls taprooted, except for 2 spp., these with solitary heads; heads often larger; pappus bristles often distinctly unequal or sordid, or both

     20a Pappus double, the outer much < the inner   **Chrysopsis**

     20b Pappus simple, the bristles gen unequal, but not divided into 2 lengths   **Haplopappus**

Key to the genera of the Cichorieae

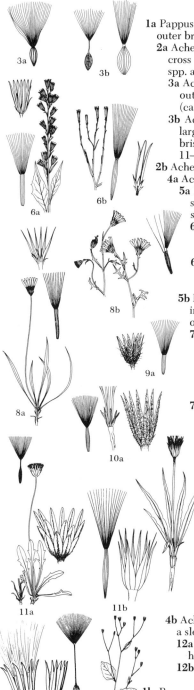

1a Pappus of simple (sometimes barbellate) capillary bristles only (some of the outer bristles sometimes evidently stronger than the others)

  2a Achenes ± strongly flattened; some of the pappus bristles < 5-celled in cross section at the base; corolla tube gen > half as long as the lig; our spp. all erect, lfy-std, and gen with several or ∝ heads

    3a Achenes beakless, without any enlarged pappiferous disk; some of the outer pappus bristles evidently stouter than the others; heads gen ∝ -fld (ca 85–250 fls in ours)   **Sonchus**

    3b Achenes beaked or less often beakless, in either case somewhat enlarged at the summit where the pappus is attached; none of the pappus bristles much stouter than the others; heads relatively few-fld (ca 11–56 fls in ours)   **Lactuca**

  2b Achenes terete or prismatic, scarcely flattened; other characters various

    4a Achenes smooth or nearly so, not evidently spinulose or muricate

      5a Fls pink or purplish to white; heads several or ∝ ; cauline lvs present, though sometimes reduced to scales; achenes beakless, though sometimes narrowed upwards

        6a Cauline lvs well developed, gen well > 1 cm wide; heads in a corymbiform or paniculiform to thyrsoid or subracemiform infl, in most spp. nodding   **Prenanthes**

        6b Cauline lvs narrow, < 1 cm wide, often reduced to mere scales, the pls ± rushlike; heads erect, gen terminating the brs   **Lygodesmia**

      5b Fls bright yellow to orange or red (sometimes drying pink or purple in *Agoseris*, which is scapose and monocephalous), or white in 1 sp. of *Hieracium*

        7a Pappus bristles (or most of them) ± connate at the base and tending to fall connected; achenes beakless

          8a Pappus bristles brownish, evidently barbellate; scapose, monocephalous per of wet places in the mts   **Apargidium**

          8b Pappus bristles whitish, scarcely or not at all barbellate; lfy-std des ann, with several or occ solitary heads   **Malacothrix**

        7b Pappus bristles all distinct, falling separately; achenes beaked or not

          9a Per from a very short rhizome, with ∝ fibrous roots; taproot wanting; achenes truncate or sometimes narrowed upwards, not beaked; pappus gen sordid or brownish   **Hieracium**

          9b Ann, bien, or more often per, from a taproot or several strong roots, without rhizomes (brs of the caudex atop the taproot may be elongate and rhizomelike in *Crepis nana*); achenes beaked or beakless; pappus gen white or nearly so

            10a Pls with at least a few cauline lvs; heads several to rarely solitary; invol ± calyculate   **Crepis**

            10b Pls scapose or nearly so, and with strictly solitary heads; invol bracts imbricate or subequal

              11a Achenes evidently beaked at maturity (except sometimes in 1 var. of *A. glauca* with conspicuously hairy herbage and invol)   **Agoseris**

              11b Achenes beakless; herbage and invol glab or nearly so   **Microseris**

    4b Achenes spinulose or muricate near the summit of the body, tipped by a slender beak; fls yellow

      12a Heads 7–15-fld; pls brd, with narrow cauline lvs and several or ∝ heads   **Chondrilla**

      12b Heads ∝ -fld; pls strictly scapose, with solitary terminal heads   **Taraxacum**

1b Pappus of plumose bristles, or bristles and scales, or scales, or minute awns, or none, the scales sometimes very slender and bristle-like in spp. of *Microseris*

  13a Pappus of minute scales, or a few minute awns, or none

    14a Pappus none, or a few minute awns; fls yellow; ann   **Lapsana**

    14b Pappus of minute scales; fls blue (white); per   **Cichorium**

  13b Pappus well developed

    15a Fls pink (white); principal invol bracts few, in our spp. gen 3–8; pappus plumose at least above   **Stephanomeria**

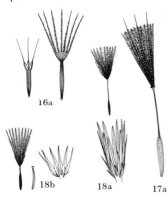

15b Fls yellow to orange or sometimes purple, not pink; principal bracts often more numerous

  16a Pappus of scales, these sometimes narrow, and often with a terminal (in 1 sp. plumose) awn, or in 1 sp. of intermingled capillary bristles and narrow, gradually attenuate scales; achenes sometimes narrowed above, but not beaked   **Microseris**

  16b Pappus (except sometimes of the marginal fls) of plumose bristles, sometimes with some shorter outer nonplumose bristles or scales; achenes beaked or beakless

    17a Plume brs of pappus interwebbed; invol uniseriate; lfy-std pls with ± grasslike lvs   **Tragopogon**

    17b Plume brs of pappus not interwebbed; invol imbricate or calyculate; our spp. scapose, with naked or merely bracteate st and scarcely grasslike lvs

      18a Recep chaffy-bracted   **Hypochaeris**

      18b Recep naked   **Leontodon**

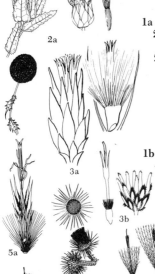

Key to the genera of the Cynareae

1a Achenes obliquely attached to the recep; marginal fls often enlarged

  2a Lvs prickly-margined; pappus of several short, horny teeth, 10 long awns or stout bristles, and 10 shorter ones; fls yellow   **Cnicus**

  2b Lvs not prickly; pappus various, but not as above; fls in most spp. not yellow

    3a Inner bracts tapering to a firm but scarcely spiny point; inner pappus ca 7–10 mm; recep chaffy with slender bracts   **Crupina**

    3b Inner bracts otherwise, gen either spine-tipped or some of them with broad, scarious or hyaline, erose to lacerate or pectinate tip or terminal appendage; pappus in our spp. (except *C. repens*) gen not > 5 mm, often wanting; recep bristly   **Centaurea**

1b Achenes basifixed; marginal fls not enlarged; fls not yellow

  4a Lvs prickly-margined, the pls thistlelike

    5a Heads 1-fld, united into 1 or more 2ndary heads   **Echinops**

    5b Heads ∝ -fld, discrete

      6a Recep fleshy, conspicuously honeycombed, not bristly or only sparsely and very shortly so   **Onopordum**

      6b Recep densely bristly, neither honeycombed nor obviously fleshy

        7a Lvs ± white-mottled along the veins; filaments united below; pappus not plumose   **Silybum**

        7b Lvs not white-mottled; filaments separate; pappus various

          8a Pappus bristles plumose (or those of the marginal fls sometimes merely barbellate); common, native and intro   **Cirsium**

          8b Pappus bristles merely barbellate; sparingly intro   **Carduus**

  4b Lvs not prickly-margined, the pls not thistlelike

    9a Invol bracts hooked at the tip; pappus bristles separately deciduous, not plumose; intro bien weeds   **Arctium**

    9b Invol bracts not hooked; pappus bristles plumose, deciduous in a ring (shorter, nonplumose outer pappus also present); native mont per   **Saussurea**

Key to the genera of the Eupatorieae

1a Achenes gen 5-angled, not ribbed   **Eupatorium**

1b Achenes gen 10 (8–20)-ribbed or -striate

  2a Invol bracts scarcely striate; infl spicate or racemiform; pappus in ours plumose   **Liatris**

  2b Invol bracts strongly striate-nerved; infl in ours corymbiform or paniculiform

    3a Pappus bristles merely barbellate or smooth   **Brickellia**

    3b Pappus bristles plumose   **Kuhnia**

Key to the genera of the Heliantheae

1a Recep chaffy, at least near the margin

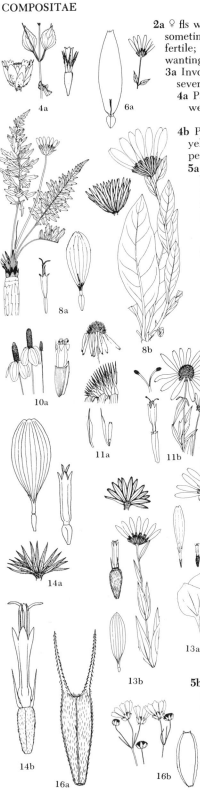

2a ♀ fls with an evident corolla, gen lig (very small in spp. of *Madia)*, or sometimes all the fls ♂; disk fls, except in some of the Madiinae, ♂ and fertile; anthers united; filaments distinct; pappus well developed to wanting

  3a Invol bracts not clasping the achenes, arranged in 1, or more often several, series; recep chaffy throughout; disk fls ± ∝

    4a Pappus of several or ∝ scales; rays white (in ours) or pink; ann weed with opp, petiolate, ovate or lance-ovate lvs (Galinsoginae)   **Galinsoga**

    4b Pappus a cup, or crown, or a few teeth or awns, or none; rays yellow or orange (sometimes discolored at the base) or pink in a few per spp., or none

      5a Achenes of the disk fls either flattened at right angles to the invol bracts, or quadrangular; recep bracts gen concave and clasping the achenes; invol bracts various, but scarcely biseriate and dimorphic

        6a Rays persistent and becoming papery, ♀ and fertile, subtended by recep bracts; recep conic, sometimes narrowly so; cauline lvs opp, well developed (Ziniinae)   **Heliopsis**

        6b Rays gen deciduous from the achenes at maturity (or sometimes persistent in spp. of *Balsamorhiza)*, ♀ or neutral, sometimes subtended by recep bracts, but gen not; recep and lvs various (Verbesininae)

          7a Rays ♀ and fertile; recep broadly convex, its bracts scarcely spinescent

            8a Pls subscapose, the cauline lvs much reduced or none   **Balsamorhiza**

            8b Pls with well developed, alt cauline lvs   **Wyethia**

          7b Rays neutral, with abortive ovary (or occ ♀ and fertile in *Echinacea*, which has conic recep with spinescent bracts)

            9a Recep enlarged, conic to columnar; cauline lvs all alt

              10a Rays subtended by recep bracts; achenes flattened, with 2 sharp and 2 very blunt angles; lvs pinnatifid   **Ratibida**

              10b Rays not subtended by recep bracts; achenes nearly equably quadrangular; lvs entire to pinnatifid

                11a Recep bracts spinescent, surpassing the disk fls; rays rose-purple (in ours); disk corollas slightly bulbous-thickened at base   **Echinacea**

                11b Recep bracts not spinescent, though sometimes shortly awn-pointed; rays yellow or partly brown-purple; disk corollas narrowed to a ± evident short tube at the base   **Rudbeckia**

          9b Recep flat to merely convex, or if definitely conic, then pls with opp lvs; at least lower lvs opp except in the scapiform genus *Enceliopsis*

            12a Disk achenes strongly compressed, thin-edged; scapose or lfy-std pls

              13a Pls scapose or subscapose   **Enceliopsis**

              13b Pls lfy-std   **Helianthella**

            12b Disk achenes equably quadrangular or moderately compressed, not thin-edged, lfy-std pls

              14a Pappus in our sp. none   **Viguiera**

              14b Pappus of 2 caducous awn-scales, rarely with a few shorter scales as well   **Helianthus**

      5b Achenes of the disk fls flattened parallel to the invol bracts (except in *Bidens beckii)*; recep bracts flat or only slightly concave; invol bracts biseriate and dimorphic (Coreopsidinae)

        15a Bracts of the invol all separate or nearly so

          16a Pappus of 2–6 awns or teeth, barbed or hispid, gen retrorsely so (very rarely smooth); achenes not wing-margined   **Bidens**

          16b Pappus of 2 short teeth or awns, barbed upwardly or not at all, or of a mere border, or none; achenes in most spp. wing-margined   **Coreopsis**

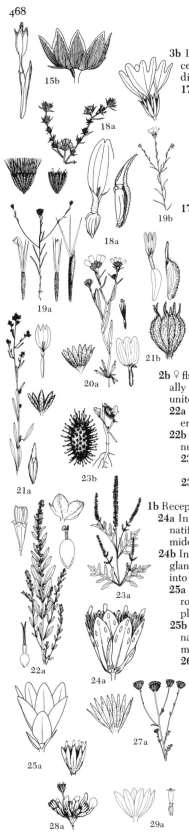

15b Inner bracts of the invol connate at least a third of their length   **Thelesperma**

3b Invol bracts clasping the achenes, gen uniseriate and =; recep, except in *Blepharipappus* and *Hemizonia*, chaffy only near the margin; disk fls 1–∝ (Madiinae)

 17a Invol bracts only partly clasping the achenes, nearly half the surface of the achene exposed

  18a Invol bracts (at least gen) all subtending rays; our sp. with spiny, gen laciniate or pinnatifid lvs, 15–35 rays, and convex, chaffy recep   **Hemizonia**

  18b Some of the invol bracts subtending disk fls; rays few, gen 3–8; lvs linear and entire, not spiny

   19a Rays inconspicuous, yellowish; recep naked except for the bracts immediately centripetal to the rays   **Rigiopappus**

   19b Rays showy, white; recep chaffy throughout   **Blepharipappus**

 17b Invol bracts wholly or largely enclosing the achenes, only the inner apex, or the inner margin, or none of the achene exposed

  20a Lvs, at least the lowermost, characteristically incised or pinnately lobed to bipinnatifid (occ all entire in depauperate pls); achenes compressed at right angles to a radius of the head; rays in ours white   **Layia**

  20b Lvs entire or merely toothed; achenes various; rays yellow or partly purplish

   21a Invol ± silky-villous, as well as often stipitate-glandular; rays purplish beneath; ray achenes compressed at right angles to a radius of the head; disk fls 6, sterile   **Lagophylla**

   21b Invol ± glandular, but not silky-villous; rays yellow throughout, sometimes minute or none; achenes and disk fls various   **Madia**

2b ♀ fls nearly or quite without corolla, ♂ fls with undivided style; functionally ♂ fls none; heads small, discoid, often unisexual; anthers scarcely united; filaments monadelphous; pappus none (Ambrosiinae)

 22a ♂ and ♀ fls in the same heads; invol of a few rounded, not at all tuberculate or spiny bracts   **Iva**

 22b ♂ and ♀ fls in separate heads, the ♂ gen uppermost; invol of ♀ heads nutlike or burlike

  23a Invol bracts of the ♂ heads united; invol of the ♀ heads with 1 or several series of tubercles or straight spines   **Ambrosia**

  23b Invol bracts of the ♂ heads separate; invol of the ♀ heads a bur with hooked prickles   **Xanthium**

1b Recep naked (or bristly in *Gaillardia*)

 24a Invol and gen also the lvs dotted with conspicuous oil glands; lvs pinnatifid, opp in ours; pappus scales ca 10–20, in ours each parted to the middle or beyond into 3–12 bristles (Tagetinae)   **Dyssodia**

 24b Invol and lvs not dotted with oil glands, although the lvs may be finely glandular-punctate; lvs various; pappus scales often fewer, not divided into bristles

  25a Invol bracts conspicuously imbricate in several series, broad, rounded; lvs opp; ours epappose, inconspicuously radiate, maritime pls (Jaumeinae)   **Jaumea**

  25b Invol bracts otherwise, seldom much imbricate, gen relatively narrow; lvs opp or alt (or all basal); heads radiate or discoid, ours not maritime except in 1 sp. of *Lasthenia*

   26a Invol bracts gen few, commonly 5–13 (15), or more in *Chaenactis*, gen subequal and uniseriate, tending to subtend the rays individually when the heads are radiate; achenes gen elongate, linear or clavate, at least 4 times as long as wide; heads radiate or discoid (Baeriinae)

    27a Heads discoid (marginal corollas occ enlarged and slightly irreg, but scarcely lig), the fls all tubular and ♂; recep flat; lvs alt or all basal   **Chaenactis**

    27b Heads gen radiate (discoid in rare forms of *Eriophyllum lanatum*), the rays sometimes very short and inconspicuous

     28a Achenes flattened, callous-margined, otherwise nerveless, the margins densely long-ciliate; diminutive woolly ann   **Eatonella**

     28b Achenes otherwise; habit various, but our spp. not as above

      29a Invol bracts carinate, distinct, permanently erect; lvs gen alt; pls floccose-tomentose, our spp. per   **Eriophyllum**

29b Invol bracts plane, distinct or connate below, sometimes reflexed in age; lvs gen opp; pls closely canescent to glab

    30a Recep flat or nearly so; our spp. per   **Bahia**

    30b Recep ± strongly conic; our spp. ann   **Lasthenia**

26b Invol bracts ∝, gen 20 or more, evidently in > 1 series, though seldom much imbricate, gen not individually subtending the rays; achenes, except in *Hulsea*, gen turbinate or obpyramidal, only 2–3 times as long as wide; heads gen (incl all our spp.) radiate; lvs alt or all basal (Heleninae)

    31a Achenes elongate, linear, at least 4 times as long as wide; recep flat   **Hulsea**

    31b Achenes short, only 2–3 times as long as wide; recep convex to conic or subglobose

        32a Recep bristly; style brs with a subulate appendage **Gaillardia**

        32b Recep naked; style brs truncate or nearly so

            33a Invol bracts appressed, permanently erect; ours either scapose or with pinnatifid or ternate lvs   **Hymenoxys**

            33b Invol bracts loose, sooner or later reflexed; st lfy, the lvs entire or merely toothed   **Helenium**

Key to the genera of the Inuleae

1a Heads with conspicuous yellow rays   **Inula**

1b Heads rayless

    2a Recep chaffy, at least near the margin (the chaff simulating an invol in *Filago*)

        3a Recep chaffy throughout; proper invol none   **Stylocline**

        3b Recep naked in the center, the inner fls bractless

            4a Pappus none

                5a Lvs opp; achenes with terminal or merely offset style; proper invol none   **Psilocarphus**

                5b Lvs alt; achenes with distinctly lateral style; invol present   **Micropus**

            4b Pappus present on inner fls, capillary; lvs alt   **Filago**

    2b Recep naked

        6a Pls dioecious or nearly so (♀ pl gen with a few central functionally ♂ fls in heads of *Anaphalis*), fibrous-rooted

            7a Pappus bristles of the ♀ fls united at the base, tending to fall off in a ring; basal lvs gen conspicuous, tufted, and persistent, the cauline ones gen reduced upwards and often few (or the pls acaulescent); strictly dioecious   **Antennaria**

            7b Pappus bristles distinct; basal lvs soon deciduous, scarcely if at all larger than the ∝ cauline ones; incompletely dioecious **Anaphalis**

        6b Pls with the fls all fertile, the outer ones in each head ∝ and ♀, the inner few and ♂; gen taprooted   **Gnaphalium**

Key to the genera of the Senecioneae

1a Pls with sterile disk fls and nearly or quite undivided style (subdioecious in *Petasites*); lvs large and chiefly basal or near-basal, the st otherwise naked or merely bracteate (lvs developing after the fls in *Tussilago*)

    2a Pappus of ∝ capillary bristles; anthers entire or minutely sagittate; ♀ fls with a ± evident ray

        3a Heads solitary, all alike, yellow; intro weed, fl before the lvs **Tussilago**

        3b Heads several, white or whitish, in some pls chiefly of ♀ fls, in others chiefly of sterile ♂ fls; native spp. of wet places, fl with the lvs **Petasites**

    2b Pappus none (unique among our genera of the tribe in this respect); heads all alike, whitish, the marginal fls ♀, with short, tubular corolla; anthers strongly sagittate   **Adenocaulon**

1b Pls with fertile disk fls and normal style; habit various

    4a Invol bracts only 2–6 and fls 2–9; heads strictly discoid

        5a Fls and invol bracts only 2 or 3, the fls white or anthocyanic; pappus bristles plumose, deciduous in a ring; diminutive ann **Dimeresia**

        5b Fls 4–9, yellow; invol bracts 4–6; pappus bristles separate, not plumose; shrubs with narrow lvs (< 5 mm wide) **Tetradymia**

    4b Invol bracts gen > 6 and fls > 9, or if not, then the heads radiate

        6a Style brs terminating in a definite, hispidulous to papillate-hairy appendage

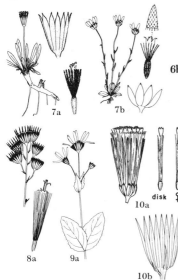

7a Style appendage slender; pappus strongly plumose; recep flat or
  nearly so; per, ours monocephalous and scapose or subscapose
                                                              Raillardella
7b Style appendage deltoid, flat; pappus not at all plumose; recep evidently conic; diminutive ann   **Crocidium**
6b Style brs otherwise
  8a Style brs with a thickened, very short and blunt, merely papillate
    appendage, not at all penicillate; heads strictly discoid   **Luina**
  8b Style brs exappendiculate or nearly so, with a tuft of short hairs all
    around at the end (the typical senecioneous style); heads radiate or
    less often discoid or disciform
    9a Cauline lvs opp (or occ some of the reduced uppermost ones
      offset)   **Arnica**
    9b Cauline lvs alt, or sometimes wanting and the lvs then all basal
      10a Heads disciform, the 2 or more marginal rows of fls ♀, with
        tubular-filiform, elig corolla   **Erechtites**
      10b Heads radiate or less often discoid, the ♀ fls, if present, lig
        (though sometimes shortly so), gen in a single row and as many
        as or > the principal invol bracts   **Senecio**

ARTIFICIAL KEYS TO THE GENERA OF THE COMPOSITAE

1a Fls all lig and ⚥; juice milky   The **Cichorieae;** see previous key, p 465
1b Fls not all lig; ray (lig) fls when present marginal (except in unisexual
  heads), either ♀ or neutral; juice gen watery
  2a Heads radiate
    3a Rays yellow or orange (sometimes marked with purple or reddish
      brown at the base)
      4a Pappus chaffy, or of firm awns, or none; recep chaffy, bristly, or
        naked   **Group I,** p 470
      4b Pappus partly or wholly of capillary (sometimes plumose) bristles;
        recep naked   **Group II,** p 472
    3b Rays white to pink, purple, red, or blue, not yellow or orange
                                                              Group III, p. 473
  2b Heads discoid or disciform, without rays (some pls with very small and
    inconspicuous rays are keyed here as well as in the radiate group)
    5a Pappus partly or wholly of ∝ capillary (sometimes plumose) bris-
      tles (wanting from the outer fls of *Filago;* bristles rather coarse and
      strongly plumose in *Saussurea*)   **Group IV,** p 474
    5b Pappus of scales, or awns, or very short chaffy bristles, or a mere
      crown, or none, never plumose (a few deciduous capillary bristles
      sometimes present on the central fls of the otherwise epappose genus
      Stylocline)   **Group V,** p 476

Group I

Rays yellow or orange; pappus chaffy, or of firm awns, or none

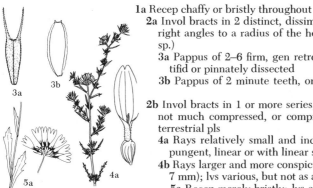

1a Recep chaffy or bristly throughout
  2a Invol bracts in 2 distinct, dissimilar series; achenes strongly flattened at
    right angles to a radius of the head (except in *Bidens beckii,* an aquatic
    sp.)
    3a Pappus of 2–6 firm, gen retrorsely barbed awns; lvs entire to pinna-
      tifid or pinnately dissected   **Bidens**
    3b Pappus of 2 minute teeth, or obsolete; lvs pinnately dissected
                                                              Coreopsis
  2b Invol bracts in 1 or more series, all ± similar in nature; achenes either
    not much compressed, or compressed parallel to a radius of the head;
    terrestrial pls
    4a Rays relatively small and inconspicuous, only ca 4–7 mm; lvs rigid,
      pungent, linear or with linear segms   **Hemizonia**
    4b Rays larger and more conspicuous, seldom much < 1 cm (rarely only
      7 mm); lvs various, but not as above
      5a Recep merely bristly; lvs alt, or occ all basal   **Gaillardia**

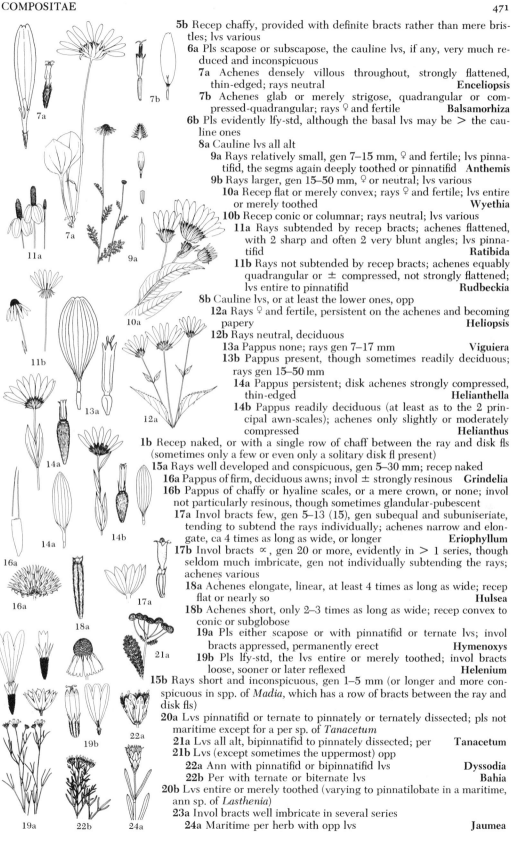

5b Recep chaffy, provided with definite bracts rather than mere bristles; lvs various

  6a Pls scapose or subscapose, the cauline lvs, if any, very much reduced and inconspicuous

    7a Achenes densely villous throughout, strongly flattened, thin-edged; rays neutral   **Enceliopsis**

    7b Achenes glab or merely strigose, quadrangular or compressed-quadrangular; rays ♀ and fertile   **Balsamorhiza**

  6b Pls evidently lfy-std, although the basal lvs may be > the cauline ones

    8a Cauline lvs all alt

      9a Rays relatively small, gen 7–15 mm, ♀ and fertile; lvs pinnatifid, the segms again deeply toothed or pinnatifid   **Anthemis**

      9b Rays larger, gen 15–50 mm, ♀ or neutral; lvs various

        10a Recep flat or merely convex; rays ♀ and fertile; lvs entire or merely toothed   **Wyethia**

        10b Recep conic or columnar; rays neutral; lvs various

          11a Rays subtended by recep bracts; achenes flattened, with 2 sharp and often 2 very blunt angles; lvs pinnatifid   **Ratibida**

          11b Rays not subtended by recep bracts; achenes equably quadrangular or ± compressed, not strongly flattened; lvs entire to pinnatifid   **Rudbeckia**

    8b Cauline lvs, or at least the lower ones, opp

      12a Rays ♀ and fertile, persistent on the achenes and becoming papery   **Heliopsis**

      12b Rays neutral, deciduous

        13a Pappus none; rays gen 7–17 mm   **Viguiera**

        13b Pappus present, though sometimes readily deciduous; rays gen 15–50 mm

          14a Pappus persistent; disk achenes strongly compressed, thin-edged   **Helianthella**

          14b Pappus readily deciduous (at least as to the 2 principal awn-scales); achenes only slightly or moderately compressed   **Helianthus**

1b Recep naked, or with a single row of chaff between the ray and disk fls (sometimes only a few or even only a solitary disk fl present)

  15a Rays well developed and conspicuous, gen 5–30 mm; recep naked

    16a Pappus of firm, deciduous awns; invol ± strongly resinous   **Grindelia**

    16b Pappus of chaffy or hyaline scales, or a mere crown, or none; invol not particularly resinous, though sometimes glandular-pubescent

      17a Invol bracts few, gen 5–13 (15), gen subequal and subuniseriate, tending to subtend the rays individually; achenes narrow and elongate, ca 4 times as long as wide, or longer   **Eriophyllum**

      17b Invol bracts ∝, gen 20 or more, evidently in > 1 series, though seldom much imbricate, gen not individually subtending the rays; achenes various

        18a Achenes elongate, linear, at least 4 times as long as wide; recep flat or nearly so   **Hulsea**

        18b Achenes short, only 2–3 times as long as wide; recep convex to conic or subglobose

          19a Pls either scapose or with pinnatifid or ternate lvs; invol bracts appressed, permanently erect   **Hymenoxys**

          19b Pls lfy-std, the lvs entire or merely toothed; invol bracts loose, sooner or later reflexed   **Helenium**

  15b Rays short and inconspicuous, gen 1–5 mm (or longer and more conspicuous in spp. of *Madia,* which has a row of bracts between the ray and disk fls)

    20a Lvs pinnatifid or ternate to pinnately or ternately dissected; pls not maritime except for a per sp. of *Tanacetum*

      21a Lvs all alt, bipinnatifid to pinnately dissected; per   **Tanacetum**

      21b Lvs (except sometimes the uppermost) opp

        22a Ann with pinnatifid or bipinnatifid lvs   **Dyssodia**

        22b Per with ternate or biternate lvs   **Bahia**

    20b Lvs entire or merely toothed (varying to pinnatilobate in a maritime, ann sp. of *Lasthenia*)

      23a Invol bracts well imbricate in several series

        24a Maritime per herb with opp lvs   **Jaumea**

24b Inl shrub or subshrub with alt lvs   **Gutierrezia**
23b Invol bracts essentially uniseriate
    25a Invol bracts united into a toothed or lobed cup; opp-lvd ann
        with well developed pappus   **Lasthenia**
    25b Invol bracts distinct nearly or quite to the base; lvs, duration,
        and pappus various
        26a Maritime ann with opp lvs and conic recep   **Lasthenia**
        26b Nonmaritime, ann or sometimes per herbs, with alt or opp lvs
          and small, flat or convex recep
          27a Achenes flattened, callous-margined, otherwise nerveless,
            the margins densely long-ciliate; recep naked; depressed ann
                  Eatonella
          27b Achenes otherwise; habit various, but scarcely as in *Eaton-ella*
            28a Invol bracts only partly clasping the achenes, nearly half
              the surface of the achene exposed; invol harshly puberu-
              lent, neither glandular nor silky-villous   **Rigiopappus**
            28b Invol bracts wholly or largely enclosing the achenes,
              only the inner apex, or the inner margin, or none of the
              achene exposed
              29a Invol ± silky-villous, as well as often stipitate-
                glandular; rays purplish beneath; ray achenes com-
                pressed at right angles to a radius of the head
                      Lagophylla
              29b Invol ± glandular, not silky-villous; rays yellow
                throughout; achenes various   **Madia**

Group II

Rays yellow or orange; pappus partly or wholly of capillary bristles

1a Shrubs   **Haplopappus**
1b Herbs
    2a Lvs (except sometimes the reduced uppermost ones) opp   **Arnica**
    2b Lvs alt (sometimes reduced to mere bracts), or all basal
        3a Recep strongly conic; delicate vernal ann with the invol bracts uni-
          seriate and individually subtending the rays   **Crocidium**
        3b Recep flat or nearly so; pls per, except a few spp. of *Senecio*
          4a Heads large, the invol ca 2–2.5 cm, the disk 3–5 cm wide; coarse,
            taprooted weed to 2 m, with the large lvs densely velvety beneath
                      Inula
          4b Heads obviously smaller, except for a few spp. of very different
            habit and vesture
             5a Invol bracts uniseriate, =, narrow, gen with a few very much
              shorter outer ones at the base
               6a Disk fls sterile, with undivided style; fls before the lvs, the st
                merely bracteate   **Tussilago**
               6b Disk fls fertile; style brs truncate, exappendiculate, minutely
                penicillate; fls with the lvs; st often lfy   **Senecio**
            5b Invol bracts in 2 or more series, = or imbricate, narrow or
             broad, style brs with short or elongate, externally hairy appen-
             dage
              7a Pls fibrous-rooted; heads few to gen ± ∝, small, the invol gen
               2–7 mm; pappus bristles gen = and white   **Solidago**
             7b Pls taprooted except for a few spp. that have solitary heads;
              heads often larger; pappus various
               8a Style appendages very short, up to 0.5 mm; invol bracts
                subequal or ± imbricate, often green in part, but neither
                definitely lfy nor with chartaceous base and herbaceous green
                tip; 4 spp. of   **Erigeron**
              8b Style appendages longer, gen 0.7 mm or more; invol various
                9a Pappus double, the outer much < the inner   **Chrysopsis**
                9b Pappus simple, the bristles gen unequal, but not divided
                  into 2 lengths   **Haplopappus**

Group III

Rays white to pink, purple, red, or blue

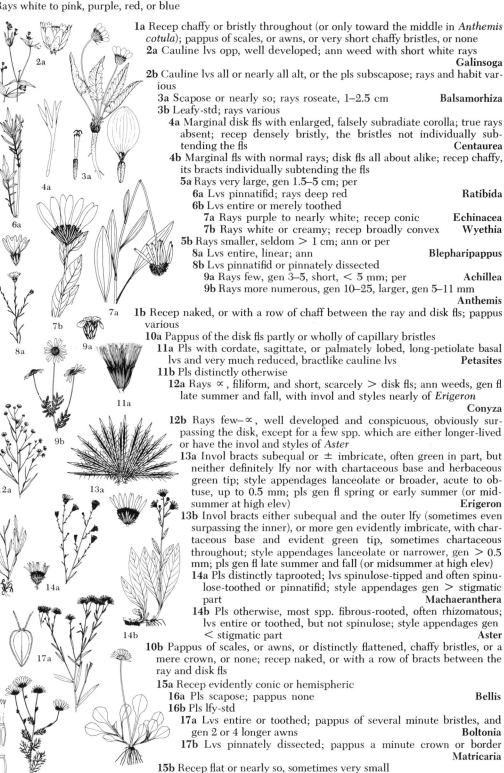

1a Recep chaffy or bristly throughout (or only toward the middle in *Anthemis cotula*); pappus of scales, or awns, or very short chaffy bristles, or none

  2a Cauline lvs opp, well developed; ann weed with short white rays   **Galinsoga**

  2b Cauline lvs all or nearly all alt, or the pls subscapose; rays and habit various

    3a Scapose or nearly so; rays roseate, 1–2.5 cm   **Balsamorhiza**

    3b Leafy-std; rays various

      4a Marginal disk fls with enlarged, falsely subradiate corolla; true rays absent; recep densely bristly, the bristles not individually subtending the fls   **Centaurea**

      4b Marginal fls with normal rays; disk fls all about alike; recep chaffy, its bracts individually subtending the fls

        5a Rays very large, gen 1.5–5 cm; per

          6a Lvs pinnatifid; rays deep red   **Ratibida**

          6b Lvs entire or merely toothed

            7a Rays purple to nearly white; recep conic   **Echinacea**

            7b Rays white or creamy; recep broadly convex   **Wyethia**

        5b Rays smaller, seldom > 1 cm; ann or per

          8a Lvs entire, linear; ann   **Blepharipappus**

          8b Lvs pinnatifid or pinnately dissected

            9a Rays few, gen 3–5, short, < 5 mm; per   **Achillea**

            9b Rays more numerous, gen 10–25, larger, gen 5–11 mm   **Anthemis**

1b Recep naked, or with a row of chaff between the ray and disk fls; pappus various

  10a Pappus of the disk fls partly or wholly of capillary bristles

    11a Pls with cordate, sagittate, or palmately lobed, long-petiolate basal lvs and very much reduced, bractlike cauline lvs   **Petasites**

    11b Pls distinctly otherwise

      12a Rays ∝, filiform, and short, scarcely > disk fls; ann weeds, gen fl late summer and fall, with invol and styles nearly of *Erigeron*   **Conyza**

      12b Rays few–∝, well developed and conspicuous, obviously surpassing the disk, except for a few spp. which are either longer-lived or have the invol and styles of *Aster*

        13a Invol bracts subequal or ± imbricate, often green in part, but neither definitely lfy nor with chartaceous base and herbaceous green tip; style appendages lanceolate or broader, acute to obtuse, up to 0.5 mm; pls gen fl spring or early summer (or midsummer at high elev)   **Erigeron**

        13b Invol bracts either subequal and the outer lfy (sometimes even surpassing the inner), or more gen evidently imbricate, with chartaceous base and evident green tip, sometimes chartaceous throughout; style appendages lanceolate or narrower, gen > 0.5 mm; pls gen fl late summer and fall (or midsummer at high elev)

          14a Pls distinctly taprooted; lvs spinulose-tipped and often spinulose-toothed or pinnatifid; style appendages gen > stigmatic part   **Machaeranthera**

          14b Pls otherwise, most spp. fibrous-rooted, often rhizomatous; lvs entire or toothed, but not spinulose; style appendages gen < stigmatic part   **Aster**

  10b Pappus of scales, or awns, or distinctly flattened, chaffy bristles, or a mere crown, or none; recep naked, or with a row of bracts between the ray and disk fls

    15a Recep evidently conic or hemispheric

      16a Pls scapose; pappus none   **Bellis**

      16b Pls lfy-std

        17a Lvs entire or toothed; pappus of several minute bristles, and gen 2 or 4 longer awns   **Boltonia**

        17b Lvs pinnately dissected; pappus a minute crown or border   **Matricaria**

    15b Recep flat or nearly so, sometimes very small

      18a Pappus of the disk fls of ca 10 or more flattened, bristlelike scales

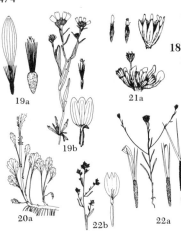

19a Rays essentially linear; lvs entire; recep naked   **Townsendia**
19b Rays broadly cuneate; lower lvs gen toothed or pinnatifid; recep with a row of bracts between the ray and disk fls   **Layia**
18b Pappus of ca 5 or fewer scales, or a mere crown, or none
  20a Per; lvs conspicuously toothed to pinnatifid   **Chrysanthemum**
  20b Ann; lvs entire or nearly so
    21a Achenes flattened, callous-margined, otherwise nerveless, the margins densely long-ciliate; recep naked; depressed ann
      Eatonella
    21b Achenes otherwise; recep with a row of bracts between the ray and disk fls; erect ann
      22a Invol bracts only partly clasping the achenes, nearly half the surface of the achene exposed; invol harshly puberulent, not glandular   **Rigiopappus**
      22b Invol bracts wholly enclosing the achenes; invol silky-villous and sometimes also stipitate-glandular
        Lagophylla

Group IV

Heads discoid or disciform; pappus capillary

1a Lvs either ± spiny and thistlelike, or heads with densely bristly (or elongate-chaffy) recep, or both
  2a Pls thistlelike, with ± spiny-margined lvs and gen also with spiny invol
    3a Lvs ± white-mottled along the veins; pappus not plumose   **Silybum**
    3b Lvs not white-mottled; pappus various
      4a Pappus bristles merely barbellate
        5a Recep densely bristly, neither honeycombed nor obviously fleshy   **Carduus**
        5b Recep fleshy, conspicuously honeycombed, not bristly or only sparsely and very shortly so   **Onopordum**
      4b Pappus bristles plumose (or those of the outermost row of fls sometimes merely barbellate); recep densely bristly   **Cirsium**
  2b Pls scarcely or not at all thistlelike, the lvs not spiny-margined; invol not spiny except in spp. of *Centaurea*
    6a Pappus bristles plumose, deciduous in a ring; achenes basifixed
      Saussurea
    6b Pappus bristles barbellate or occ plumose, but not deciduous in a ring; achenes obliquely attached
      7a Invol bracts tapering to a firm but scarcely spiny point; inner pappus ca 7–10 mm; recep chaffy with slender bracts   **Crupina**
      7b Invol bracts otherwise, gen either spine-tipped or some of them with broad, scarious or hyaline, erose to lacerate or pectinate tip or terminal appendage; pappus in our spp. (except *C. repens*) gen not > 5 mm, often wanting; recep bristly   **Centaurea**
1b Lvs not at all thistlelike, spiny only in spp. of the shrubby genus *Tetradymia*; recep naked (or with a few submarginal chaffy bracts in *Filago*)
  8a Shrubs
    9a Pls dioecious, maritime   **Baccharis**
    9b Pls with all ⚥ fls, not maritime
      10a Invol bracts 4–6, =   **Tetradymia**
      10b Invol bracts more ∝, imbricate, or at least in several series
        11a Invol bracts aligned in ± definite vertical ranks; lvs linear to oblong   **Chrysothamnus**
        11b Invol bracts not aligned in vertical ranks; lvs various
          12a Twigs white-tomentose; fls yellow; invol bracts not striate   **Haplopappus macronema**
          12b Twigs glab to puberulent or glandular, not tomentose; fls white or creamy to pink-purple; invol bracts striate (a rare form of *Haplopappus resinosus* might be sought here, except for the nonstriate bracts)   **Brickellia**
  8b Herbs (rarely subshrubby in *Luina*)
    13a Fls all ⚥ and fertile, the heads discoid
      14a Invol bracts and fls only 2 or 3 in each head; diminutive ann   **Dimeresia**
      14b Invol bracts and fls more numerous; habit and duration various
        15a Lvs opp; fls yellow or orange   **Arnica**

15b Lvs alt, or if rarely opp or whorled, then the fls not yellow or orange

  16a Principal invol bracts essentially = and uniseriate, although some very much reduced outer bracts may be present

    17a Pappus strongly plumose; pls scapose or subscapose   **Raillardella**

    17b Pappus merely barbellate; pls subscapose or more often lfy-std

      18a Pls either with palmately cleft lvs or with few (gen 5–29) fls in each head   **Luina**

      18b Pls with more numerously fld heads and not at all palmately cleft lvs   **Senecio**

  16b Principal invol bracts ± imbricate in 2–several series

    19a Style brs flattened, with well developed, introrsely marginal stigmatic lines and a short or elongate, externally papillate-hairy appendage; fls gen yellow, occ reddish; pappus merely barbellate

      20a Style appendages very short, up to 0.5 mm; 2 spp. of   **Erigeron**

      20b Style appendages longer, gen 0.7 mm or more

        21a Basal and lower cauline lvs the largest, the others often progressively reduced

          22a Pls either with the leaves conspicuously spinulose-toothed, or with the disk corollas well > 6 mm, or both; 2 spp. of   **Haplopappus**

          22b Pls with entire or subentire lvs; disk corollas (dry) 5–6 mm   **Machaeranthera shastensis**

        21b Basal lvs wanting; lower cauline lvs evidently reduced

          23a Lvs reg spinulose-toothed   **Machaeranthera grindelioides**

          23b Lvs entire or with a few irreg teeth

            24a Invol bracts ± hyaline-margined; pls gen 2–5 dm, with lvs 2–5 cm   **Chrysopsis oregona**

            24b Invol bracts not hyaline-margined; pls gen 6–12 dm, with the larger lvs 5–9 cm   **Aster vialis**

    19b Style brs otherwise; fls blue, purple, or pink to whitish, not yellow

      25a Style with a thickened, minutely hairy ring just beneath the short, papillate brs; anthers tailed; pappus bristles plumose, united at the base and falling connected   **Saussurea**

      25b Style without any thickened ring; style brs subterete and gen ± clavate, minutely papillate, without evident stigmatic lines; anthers truncate to sagittate, not tailed; pappus various, sometimes plumose, but the bristles separate

        26a Achenes gen 5-angled, not ribbed; pappus not at all plumose   **Eupatorium**

        26b Achenes gen 10 (8–20)-ribbed or -striate; pappus various

          27a Infl spiciform or racemiform; invol bracts scarcely striate; pappus plumose   **Liatris**

          27b Infl corymbiform or paniculiform; invol bracts strongly striate-nerved; pappus various

            28a Pappus bristles plumose   **Kuhnia**

            28b Pappus bristles merely barbellate or smooth   **Brickellia**

13b At least the outer fls ♀ (some dioecious pls, in which some heads are wholly ♀, and others functionally ♂, are keyed here)

  29a Basal lvs large, long-petiolate, either sagittate and toothed, or palmately cleft; cauline lvs very much reduced and bractlike   **Petasites**

  29b Basal and cauline lvs otherwise

    30a Herbage ± white-woolly; lvs simple and entire or nearly so; invol bracts gen with dry, scarious, thin, white to yellowish, brownish, or blackish-green tip (shortly or scarcely so in *Filago*)

31a Pls taprooted, ann or per; heads all with the outer fls ♀ and the central ones ♂ (or functionally ♂)

    32a Marginal recep bracts simulating an invol, but with a few inconspicuous ♀ fls borne between them and the shorter true invol bracts; recep otherwise naked; invol and recep bracts with strongly green midrib nearly or quite to the tip; ann **Filago**

    32b Recep naked; invol bracts, as in *Antennaria* and *Anaphalis,* with ± conspicuous scarious or hyaline tip; ann or per **Gnaphalium**

31b Pls fibrous-rooted, per, often with rhizomes or stolons, but without a taproot; dioecious or nearly so, the heads on at least some of the pls wholly ♂, or wholly ♀

    33a Basal lvs gen forming a conspicuous, persistent tuft (except gen in *A. geyeri*); st seldom very lfy, with or without stolons or rhizomes; strictly dioecious (♂ pls rare in some spp.) **Antennaria**

    33b Basal lvs soon deciduous, not markedly larger than the ∝ and well developed cauline ones; pls with rhizomes, but no stolons, ♀ pls gen with a few central, functionally ♂ fls in each head **Anaphalis**

30b Herbage not at all white-woolly (except in spp. of *Erechtites,* with lobed or pinnatifid lvs), though often otherwise pubescent; invol bracts not markedly scarious at the tip (or with a narrow scarious margin and tip in *Haplopappus carthamoides*)

  34a Invol a single series of = bracts, sometimes with some minute bracteoles at the base **Erechtites**

  34b Invol bracts in > 1 series, imbricate or sometimes subequal

    35a Per

      36a Heads large, the invol 1.5–3 cm, the disk corollas 10–14 mm **Haplopappus carthamoides**

      36b Heads much smaller, the invol only 4–9 mm, the disk corollas ca 3–5 mm; spp. of **Erigeron**

    35b Ann

      37a Invol 5–11 mm; herbage essentially glab except for the remotely hispidulous-ciliate margins of the lvs; the section Brachyactis of **Aster**

      37b Invol 3–4 mm, or if larger and up to 6 mm, then the herbage evidently pubescent **Conyza**

Group V

Heads discoid or disciform; pappus chaffy, or of awns, or none

1a Thistles with spiny-margined lvs

  2a Heads 1-fld, aggregated into 1 or more 2ndary heads **Echinops**

  2b Heads ∝ -fld, not aggregated into 2ndary heads **Cnicus**

1b Pls not thistlelike, the lvs not spiny-margined (but provided with tripartite axillary spines in *Xanthium spinosum*)

  3a Invols, or at least some of them, either armed with short, hooked prickles, or nutlike, or burlike and provided with tubercles or spines

    4a Heads all alike, with ± ∝ fls; invol bracts hooked at the tip; recep bristly; corollas ± evident **Arctium**

    4b Heads of 2 kinds; invol of the ♀ heads nutlike or burlike, provided with spines, tubercles, or hooked prickles; invol of ♂ heads unarmed; recep chaffy; corollas small and inconspicuous, or wanting

      5a ♀ invol with hooked prickles **Xanthium**

      5b ♀ invol with tubercles or straight spines **Ambrosia**

  3b Invols neither nutlike nor burlike nor provided with hooked prickles, all ca alike, spiny only in spp. of *Centaurea*

    6a Recep chaffy or bristly throughout, or a few of the central fls bractless

      7a Recep elongate, the large black heads thus shaped like the end of a finger or thumb **Rudbeckia**

      7b Recep flat or merely convex; heads not black (invol blackish in spp. of *Centaurea*)

        8a Pls ± white-woolly, slender, ann weeds with inconspicuous fls

          9a Recep chaffy throughout; proper invol none; lvs all or nearly all alt **Stylocline**

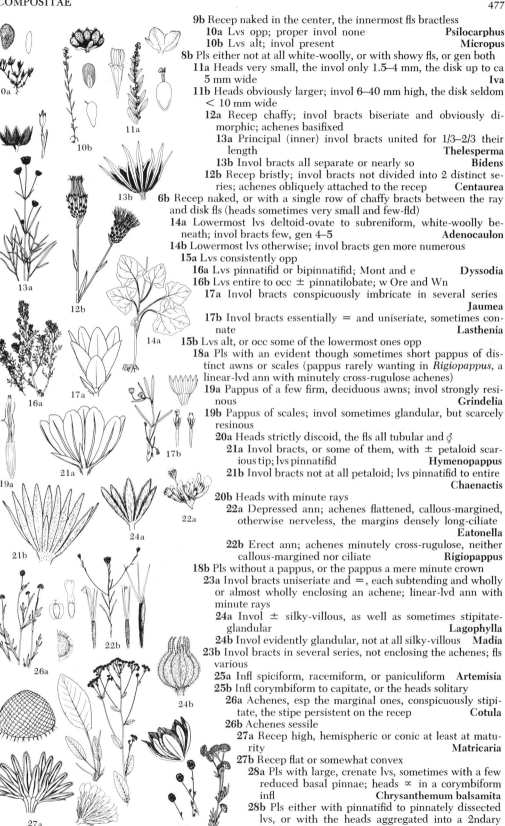

9b Recep naked in the center, the innermost fls bractless

    10a Lvs opp; proper invol none   **Psilocarphus**

    10b Lvs alt; invol present   **Micropus**

8b Pls either not at all white-woolly, or with showy fls, or gen both

    11a Heads very small, the invol only 1.5–4 mm, the disk up to ca 5 mm wide   **Iva**

    11b Heads obviously larger; invol 6–40 mm high, the disk seldom < 10 mm wide

        12a Recep chaffy; invol bracts biseriate and obviously dimorphic; achenes basifixed

            13a Principal (inner) invol bracts united for 1/3–2/3 their length   **Thelesperma**

            13b Invol bracts all separate or nearly so   **Bidens**

        12b Recep bristly; invol bracts not divided into 2 distinct series; achenes obliquely attached to the recep   **Centaurea**

6b Recep naked, or with a single row of chaffy bracts between the ray and disk fls (heads sometimes very small and few-fld)

    14a Lowermost lvs deltoid-ovate to subreniform, white-woolly beneath; invol bracts few, gen 4–5   **Adenocaulon**

    14b Lowermost lvs otherwise; invol bracts gen more numerous

        15a Lvs consistently opp

            16a Lvs pinnatifid or bipinnatifid; Mont and e   **Dyssodia**

            16b Lvs entire to occ ± pinnatilobate; w Ore and Wn

                17a Invol bracts conspicuously imbricate in several series   **Jaumea**

                17b Invol bracts essentially = and uniseriate, sometimes connate   **Lasthenia**

        15b Lvs alt, or occ some of the lowermost ones opp

            18a Pls with an evident though sometimes short pappus of distinct awns or scales (pappus rarely wanting in *Rigiopappus*, a linear-lvd ann with minutely cross-rugulose achenes)

                19a Pappus of a few firm, deciduous awns; invol strongly resinous   **Grindelia**

                19b Pappus of scales; invol sometimes glandular, but scarcely resinous

                    20a Heads strictly discoid, the fls all tubular and ♂

                        21a Invol bracts, or some of them, with ± petaloid scarious tip; lvs pinnatifid   **Hymenopappus**

                        21b Invol bracts not at all petaloid; lvs pinnatifid to entire   **Chaenactis**

                  20b Heads with minute rays

                        22a Depressed ann; achenes flattened, callous-margined, otherwise nerveless, the margins densely long-ciliate   **Eatonella**

                        22b Erect ann; achenes minutely cross-rugulose, neither callous-margined nor ciliate   **Rigiopappus**

            18b Pls without a pappus, or the pappus a mere minute crown

                23a Invol bracts uniseriate and =, each subtending and wholly or almost wholly enclosing an achene; linear-lvd ann with minute rays

                    24a Invol ± silky-villous, as well as sometimes stipitate-glandular   **Lagophylla**

                    24b Invol evidently glandular, not at all silky-villous   **Madia**

                23b Invol bracts in several series, not enclosing the achenes; fls various

                    25a Infl spiciform, racemiform, or paniculiform   **Artemisia**

                    25b Infl corymbiform to capitate, or the heads solitary

                      26a Achenes, esp the marginal ones, conspicuously stipitate, the stipe persistent on the recep   **Cotula**

                    26b Achenes sessile

                      27a Recep high, hemispheric or conic at least at maturity   **Matricaria**

                      27b Recep flat or somewhat convex

                          28a Pls with large, crenate lvs, sometimes with a few reduced basal pinnae; heads ∝ in a corymbiform infl   **Chrysanthemum balsamita**

                        28b Pls either with pinnatifid to pinnately dissected lvs, or with the heads aggregated into a 2ndary head or close, headlike cluster   **Tanacetum**

Achillea L.   Yarrow

Heads radiate or rarely discoid, the rays gen 3–12, ♀ and fertile (rarely neutral), white, sometimes pink or partly yellow, short and broad; invol bracts imbricate in several series, dry, with scarious or hyaline margins and often greenish midrib; recep conic or convex, chaffy throughout, the disk fls gen 10–75, ♂ and fertile; anthers entire at the base; style brs flattened, truncate, penicillate; achenes compressed parallel to the invol bracts, callous-margined, glab; pappus none; per herbs with alt, subentire to pinnately dissected lvs, and several to ∝ relatively small heads in a ± corymbiform infl. (Named for *Achilles*).

1a Heads very small, the invol only 2–3 mm; rays scarcely 1 mm, wider than long; otherwise much like the next; Eurasian weed, intro in nw Mont
1 **A. nobilis** L.
1b Heads larger, the invol gen 4–6 mm; rays 2–4 mm, as long as or longer than wide; aromatic per; lvs slender, pinnately dissected; heads in a short and broad, pan-corymbiform infl; rays gen 3–5, white (pink); disk fls 10–30; circumboreal, common, variable, not fully understood; common y., milfoil (incl *A. eradiata* Piper, a rayless phase with relatively few, broad, and distant ultimate lf-segms, known only from the original, at Mt Jefferson); several infraspecific taxa   2 **A. millefolium** L.
   a1 Tetraploid, widespread, nonmaritime; invol 4–5 mm; rays 2–3 mm (*A. l.*)   ssp. **lanulosa** (Nutt.) Piper
      b1 Lowl to midmont series of ecotypes, 3–10 dm; margins of invol bracts pale to brownish; doubtless further divisible when better understood
var. **lanulosa**
      b2 Alp and subalp ecotype, 1–3 dm; margins of invol bracts dark brown to nearly black (*A. a., A. subalpina*)   var. **alpicola** (Rydb.) Garrett
   a2 Hexaploid, maritime and up the CR to Bonneville; invol to 6 mm and rays to 4 mm; lvs larger, with slender ultimate segms longer and less crowded than in ssp. *lanulosa* (var. perhaps to be associated with other coastal hexaploids, such as the n *A. borealis* Bong., as a ssp.)
var. **californica** (Pollard) Jeps.

Adenocaulon Hook.   Trail-plant; Pathfinder; Adenocaulon

Invol small, of < 10 nearly = green bracts; heads disciform; fls whitish, tubular, the outer ca 3–7, ♀, the inner as many or a few more, ♂, with undivided style; recep naked; anthers strongly sagittate; pappus none; ann or per herbs with large alt lvs and ample subnaked infl, the brs and achenes ± stipitate-glandular. (Gr *aden*, gland, and *kaulos*, stem).

A. bicolor Hook. Fibrous-rooted slender per to 1 m; lvs mostly near the base, long-petiolate, thin, deltoid-ovate to subreniform, 3–15 cm wide, ± glab above, closely white-woolly beneath; invol bracts ca 2 mm, reflexed in fr and eventually deciduous; achenes clavate, 5–8 mm, coarsely stipitate-glandular above; moist, shady woods, W Cas, and e to nw Mont; Lake Superior.

Agoseris Raf.   False-dandelion; Mountain-dandelion; Agoseris

Heads lig, the fls all lig and ♂, yellow or orange, often turning pink or purple (respectively) in drying; invol camp, its bracts in 2–several series, subequal to strongly imbricate; recep gen naked; achenes terete or angular, prominently ca 10-nerved, ± strongly beaked at maturity (or beakless in 1 var. of *A. glauca*); pappus of ∝ capillary bristles; ann or per lactiferous herbs from a taproot, scapose or nearly so; lvs entire to pinnatifid; heads strictly solitary. (Gr *aix*, goat, and *seris*, chicory). *A. alpestris* is here referred to *Microseris*.

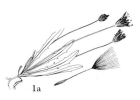

1a Pls ann; beak gen 2–3 times as long as body of achene; scapes to 4 dm; lvs to 15 × 1.5 cm; invol 5–13 mm in fl, 10–20 mm in fr; fls yellow; body of achene 2–5 mm; dry, open places in foothills and lowl, Pac and GB states; ann a.; ours is var. *heterophylla*   1 **A. heterophylla** (Nutt.) Greene
1b Pls per
   2a Achenes gen with short, stout, ± striate beak to ca half as long as the

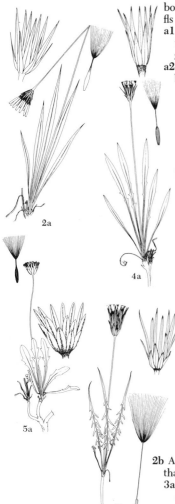

2a

4a

5a

6b

6a

5b

body (sometimes more in var. *laciniata,* or wanting in var. *dasycephala*); fls yellow; pale a., short-beaked a.; 5 vars.   **2 A. glauca** (Pursh) Raf.

a1 Pls glab, or merely sparsely ciliate on the lower part of the lvs and on the petiole; lvs strongly acute to acuminate, gen entire; widespread, gen at lower elev (*A. parviflora,* misapplied)   var. **glauca**

a2 Pls ± pubescent, at least on the invol or just below it

b1 Lvs gen oblanceolate or broader and ± obtuse, entire or sometimes weakly laciniate below; pls gen small, seldom > 25 cm, tending to be rather densely pubescent, conspicuously so on the invol, less so on the lvs; achenes sometimes beakless (Cas-Wen phase); n pls, in our range extending s to OM, Mt Rainier, and the mts of c Ida and w Mont, common and characteristically developed in the Cas and Wen Mts, e often less pubescent (*A. taraxacifolia, A. villosa*)

var. **dasycephala** (T. & G.) Jeps.

b2 Lvs more lanceolate, gen strongly acute or acuminate, gen (except often in var. *monticola*) laciniate; pls less densely hairy, often with the lf surfaces nearly or quite glab and the pubescence largely confined to the top of the peduncle and margins of the invol bracts

c1 Pls gen tall, seldom < 25 cm; outer invol bracts tending to be pinkish in part; pappus gen 14–16 mm; RM, w to e Wn and ne Ore, gen in foothills and at middle elev in mts, occurring at lower elev, in the area of range overlap, than var. *dasycephala*

var. **agrestis** (Osterh.) Q. Jones

c2 Pls short, gen < 25 cm; invol bracts gen not pinkish

d1 Invol bracts mostly in ca 2 length classes, the outer ones tending to be long-acuminate or slenderly acute; petioles and lower part of lvs arachnoid; pappus (12) 14–18 mm; pubescence wholly eglandular; GB and s RM, occ reaching the s fringes of our range, gen at lower elev and fl May–July (*A. parviflora; A. taraxacifolia,* misapplied)

var. **laciniata** (D. C. Eat.) Smiley

d2 Invol bracts mostly in ca 3 length classes, the outer conspicuously shorter, merely acute or acutish; lvs glab or evenly short-hairy, not arachnoid; pappus gen 10–11 (15) mm; pubescence consisting partly of glandular, translucent hairs; Cas from Mt Adams s, and at upper elev in mts of e Ore, fl July–Sept; seldom collected in our area (*A. m.*)

var. **monticola** (Greene) Q. Jones

2b Achenes with a long, slender, scarcely or not at all striate beak, more than half as long as the body of the achene

3a Beak < twice as long as body of achene

4a Fls burnt orange; body of achene 5–9 mm; lvs variable, to 35 × 3 cm; cordilleran, mont to alp; orange a.; 2 vars.

3 A. aurantiaca (Hook.) Greene

a1 Invol bracts narrow, gen pointed, slightly or not at all imbricate, sometimes finely purple-dotted, not mottled; widespread, the common phase with us (*A. gracilens*)   var. **aurantiaca**

a2 Invol bracts broader, blunter, conspicuously imbricate, strongly mottled or blotched with purple; s RM phase, n occ to s Mont (*A. p.*)   var. **purpurea** (Gray) Cronq.

4b Fls yellow; Cas and w

5a Heads small, the fr invol ca 1.5 cm, the body of achene gen 4–6 mm; lvs short, seldom > 10 cm; maritime, s Wn to Cal; seaside a.; ours are var. *maritima* (Sheld.) Q. Jones

4 A. apargioides (Less.) Greene

5b Heads larger, the fr invol gen 2–3 cm, the body of achene gen 8–10 mm; lvs 10–30 cm; nonmaritime, meadows and open woods up to middle elev in mts, Wn to Cal; tall a.

5 A. elata (Nutt.) Greene

3b Beak 2–4 times as long as the body of the achene; fls yellow

6a Achenes tapering to the beak; lvs antrorsely, divaricately, or irreg lobed, gen 10–25 × 1–3 cm; fr invol 1.5–4 cm; open places up to middle elev in mts, BC to Cal, w to n and c Ida, Nev, and (intro) Utah; large-fld a.   **6 A. grandiflora** (Nutt.) Greene

6b Achenes abruptly beaked from the truncate summit, lvs gen retrorsely lobed, 10–40 cm; fr invol 2.5–5 cm; dry, open woods, Cal n to sw Ida and Chelan Co, Wn; spear-lf a.

7 A. retrorsa (Benth.) Greene

Ambrosia L.   Ragweed; Bursage; Burweed

Heads unisexual, small, rayless; sterile (♂) heads in a spiciform or racemiform, bractless infl; invol subherbaceous, 5–12-lobed; recep flat, its bracts slender, filiform-setose; filaments monadelphous; anthers scarcely united; style undivided; fertile (♀) heads borne below the sterile ones, in the axils of lvs or bracts; invol closed, nutlike or burlike, beset with 1–several cycles of straight-tipped spines or tubercles; ♀ fls 1–several, without corolla; pappus wanting; coarse ann or per herbs (shrubs) with opp or alt, often lobed or dissected lvs and ∝ small, inconspicuous heads. (Ancient classical name of several plants). *(Franseria)*.

1a Invol with 2–several series of sharp, short spines, the bur 5–11 mm; lvs mostly alt
    2a Succulent, maritime, littoral per, 2–10 dm; BC to Cal; silver b., heath b.; with 2 coextensive vars.   **1 A. chamissonis** (Less.) Greene
      a1 Lvs merely toothed, densely silvery   var. **chamissonis**
      a2 Lvs bipinnatifid to tripinnatifid, gen less densely hairy
          var. **bipinnatisecta** (Less.) J. T. Howell
    2b Inl ann weed, not succulent, gen 1–8 dm; pls often chiefly ♂ or chiefly ♀, widespread in w US; bur r., ann b.   **2 A. acanthicarpa** Hook.
1b Invol with a single series of short spines or tubercles above the middle; at least the lower lvs opp; widespread weeds, occ with us
    3a Lvs palmately 3–5-lobed, or undivided, often 2 dm or more; pls coarse, often 2 m or more; fr invol 5–10 mm; giant r.; our pls var. *trifida*
          3 A. trifida L.
    3b Lvs mostly once or twice pinnatifid, to 1 dm; pls to 1 m; fr invol 3–5 mm
      4a Ann; lvs mostly petiolate, once or (more often) twice pinnatifid; fr invol with short spines; ann r., Roman wormwood *(A. elatior)*
          4 A. artemisiifolia L.
      4b Per from creeping roots; lvs thicker, short-petiolate or subsessile, gen only once pinnatifid; fr invol tuberculate; western r., common r. *(A. coronopifolia)*   **5 A. psilostachya** DC.

Anaphalis DC.   Pearly-everlasting

Dioecious or polygamodioecious, the ♀ heads sometimes with a few central ♂ fls; heads several or gen ∝, disciform or discoid, ∝-fld; invol bracts imbricate in several series, almost wholly dry and scarious; recep naked, flat or convex; anthers caudate; ♂ fls tubular, gen with undivided style; ♀ fls tubular-filiform, with bifid style; pappus, in both kinds of fls, of distinct capillary bristles, neither thickened at the tip nor conspicuously barbellate; white-woolly per with alt entire lvs. (Ancient Gr name of a possibly similar plant).

A. margaritacea (L.) B. & H. Common p. Rhizomatous, 2–9 dm, very lfy, loosely white-woolly; lvs ∝, lanceolate or narrower, to 12 × 2 cm, sessile, gen less hairy above, the basal ones soon deciduous; heads ca 1 cm wide or less, ∝, crowded in a short, broad infl; invol 5–7 mm, bracts pearly white, sometimes with a small basal dark spot; widespread, variable, boreal sp. *(A. subalpina)*.

Antennaria Gaertn.   Everlasting; Pussy-toes

Dioecious; heads solitary to ± ∝ in a gen congested infl, disciform or discoid, ∝-fld; invol bracts imbricate in several series, scarious at least distally, often colored; recep flat or convex, naked; anthers caudate; ♂ fls with gen undivided style and scanty pappus, the bristles ± clavate, or seldom only barbellate; ♀ fls with filiform-tubular corolla, bifid style, and ∝ capillary naked bristles that are slightly united at base; achenes terete or slightly compressed; white-woolly per with simple, gen entire, alt and basal lvs, the basal ones gen tufted and conspicuous, the cauline gen ± reduced upwards. (Name L, from the fancied resemblance of the ♂ pappus to insect antennae).

1a Heads solitary; low pls, gen < 1 dm; ♀ invol 7–15 mm; ♂ pappus merely barbellate; fl spring

  2a Pls with conspicuous filiform stolons to 1.5 dm; lvs linear, 1–3 cm, to 2 mm wide; dry mts and foothills, locally from c Wn and c Ore to n Wyo and sw Ida; stolonous e.   **1 A. flagellaris** Gray

  2b Pls without stolons, matted; lvs to 3 cm × 5 mm; dry, open places, esp in foothills and lowl, widespread in w US; low p. *(A. latisquama* Piper, a taller form of ne Ore and se Wn, to 10 cm) **2 A. dimorpha** (Nutt.) T. & G.

1b Heads several or ∝; pls often well over 1 dm

  3a Upper surface of basal lvs distinctly less pubescent than the lower, sooner or later glabrate

    4a Heads gen slender-pedunculate in an open, racemiform infl; upper part of st densely glandular, not tomentose; basal lf bls 1.5–8 × 1–5 cm; pls sexual, ♂ pls common, ♀ pls with stigmas = or > pappus; cool mt woods, sw Can and nw US; raceme p.   **3 A. racemosa** Hook.

    4b Heads in a crowded or even subcapitate cyme; st tomentose to the top, sometimes also glandular; basal lvs to 1.5 cm (or in var. *howellii* 2 cm) wide; pls gen apomictic, ♂ pls rare, ♀ pls with pappus > stigmas; open woods up to middle elev, widespread in N Am; field p.; 2 vars. in our range   **4 A. neglecta** Greene

      a1 Lvs thinly tomentose above when young
          var. **attenuata** (Fern.) Cronq.

      a2 Lvs glab above from the first *(A. h.)* var. **howellii** (Greene) Cronq.

  3b Upper surface of basal lvs nearly or quite as densely hairy as the lower, gen not becoming glabrate

    5a Pls with solitary sts 1–4 dm and no conspicuous tuft of basal lvs, perennating by conspicuous arching stolons; foothill meadows, n edge of SR plains to sw Wyo and ne Nev; arching p.   **5 A. arcuata** Cronq.

  5b Pls with clustered sts, the stolons, if present, prostrate

    6a Pls mat-forming, with ∝ lfy stolons

      7a Terminal scarious portion of invol bracts, at least the middle and outer ones, discolored and brownish to dirty blackish green; mont and alp

        8a Terminal scarious portion of invol bracts dirty blackish green throughout, the bracts gen sharp-pointed; sexual or apomictic; circumboreal, s in the mts, in our range alp and uncommon, passing into no. 7; alpine p.; ours is var. *media* (Greene) Jeps. *(A. m.)*   **6 A. alpina** (L.) Gaertn.

        8b Terminal scarious portion of the inner invol bracts whitish at the tip, or the whole scarious portion merely discolored and brownish; invol bracts gen blunt; chiefly apomictic; cordilleran and boreal, mont but seldom alp; umber p.
          7 A. umbrinella Rydb.

      7b Terminal scarious portion of invol bracts white or pink, sometimes with a basal dark spot; lowl to mont, but scarcely alp

        9a Invol bracts with a conspicuous black spot at the base of the scarious portion; basal lvs narrowly oblanceolate; gen sexual; invol 4–7 mm; meadows and moist, open woods, sw Mont and nc Ida to ec Ore and c Cal; meadow p., flat-topped p.
          8 A. corymbosa E. Nels.

        9b Invol bracts scarcely or not at all darkened at the base of the scarious portion; lvs gen somewhat broader; sexual or often apomictic

          10a Heads large, the invol 7–11 mm, the dry ♀ corollas gen 5–8 mm; plains sp., rarely to e Wn; Nuttall's p. *(A. aprica)*
            9 A. parvifolia Nutt.

          10b Heads smaller, the invol 4–7 mm, the dry ♀ corollas 2.5–4.5 mm; common cordilleran sp.; rosy p. *(A. arida, A. rosea)*   **10 A. microphylla** Rydb.

    6b Pls without stolons, mostly not mat-forming (except in *A. argentea),* though often with several or ∝ sts from a brd subterranean rhizome or caudex

      11a Invol scarious to the base, glab or nearly so, outermost bracts occ a little woolly at the base

        12a Invol dark brownish in aspect, although the tips of the bracts may be pale; infl very compact, gen 8–15 mm wide; lvs linear, gen 1–2 mm wide; c Wn to Nev and e to Blaine Co., Ida; narrow-lf p.   **11 A. stenophylla** Gray

12b Invol pale or nearly white (pink) in aspect; infl compact or open, often much > 1.5 cm wide; lvs gen broader, the larger ones gen 2–15 mm wide

    13a Pls with clustered sts from a brd caudex or system of short stout rhizomes; lvs relatively narrow, the basal gen linear-oblanceolate or occ oblanceolate, the middle cauline gen linear; open, not too dry places from foothills to middle elev in mts of nw US and s BC; woodrush p.

                           12 A. luzuloides T. & G.

    13b Pls vigorously rhizomatous, tending to form loose mats; lvs broader, the basal oblanceolate or broader, the middle cauline gen linear-oblanceolate; E Cas, c Ore to c Cal; silvery p.

                           13 A. argentea Benth.

11b Invol with a densely pubescent, not at all scarious lower portion, pubescence extending even to the inner bracts

    14a Largest lvs 1–3 cm; ♀ invols narrow, gen cylindric-turbinate, commonly somewhat pinkish, their dry corollas gen 4.5–8 mm; pls gen 5–15 cm; open ponderosa pine woods, e Wn to n Cal and e Nev; pinewoods p., Geyer p.   **14 A. geyeri** Gray

    14b Largest lvs 3–15 cm; ♀ invols broader, rarely pinkish, their corollas gen 2.5–4 (5) mm

        15a Pls gen 1–2 dm; invol blackish in aspect, although the inner bracts may be white at the tip; alp and subalp, mts from BC and Alta to Ore, Ida, and Wyo; woolly p.

                       15 A. lanata (Hook.) Greene

        15b Pls gen 2–5 dm

            16a Invol blackish in aspect, although some of the bracts may be white at the tip; cordilleran, alp and subalp, rare with us; showy p.   **16 A. pulcherrima** (Hook.) Greene

            16b Invol white or whitish in aspect, although the bracts may have a small dark spot at the base; open places at middle elev E Cas; tall p.   **17 A. anaphaloides** Rydb.

Anthemis L.   Chamomile; Dogfennel; Mayweed

Heads radiate or rarely discoid, the rays elongate, white or yellow, ♀ or neutral; invol bracts subequal or more gen imbricate in several series, dry, the margins ± scarious or hyaline; recep convex to conic or hemispheric, chaffy at least toward the middle; disk fls ∝, ♂, the tube at the base of the corolla gen cylindric, occ flattened; anthers entire or nearly so at the base; style brs flattened, truncate, minutely penicillate; achenes terete or 4–5-angled, or occ ± compressed, but not callous-margined; pappus a short crown, or more gen none; ann or per, gen aromatic herbs with alt, incised-dentate to pinnately dissected lvs and medium-sized to rather large, camp or subhemispheric heads terminating the brs. (Ancient Gr and L name of some of the spp.).

1a Rays 20–30, yellow, 7–15 mm; short-lived per, 3–7 dm; lvs pinnatifid, 2–5 cm, with winged rachis and deeply toothed or pinnatifid segms; recep chaffy throughout; pappus a very short crown; Eurasian sp., occ escaped from cult; yellow c.   **1 A. tinctoria** L.

1b Rays 10–20, white, 5–11 mm; ann, 1–6 dm; lvs 2–3 times pinnatifid, with very narrow segms; Eurasian weeds

    2a Rays ♀ and fertile; recep chaffy throughout; field c. or d., corn c.

                           2 A. arvensis L.

    2b Rays sterile and gen neutral; recep chaffy only toward the middle; mayweed c., stinking m.   **3 A. cotula** L.

Apargidium T. & G.   Apargidium

Heads lig, the fls all lig and ♂, yellow; invol calyculate; recep naked; achenes columnar, truncate, ca 10–12-ribbed; pappus of brownish, barbellate, capillary bristles, connate at the slightly thickened base; monocephalous, scapose per with milky juice. (Resembling *Apargia*—a synonym of *Leontodon*).

A. boreale (Bong.) T. & G. Glab, 1–5 dm, strictly scapose or with 1–2 small and inconspicuous bracts on the st, which is often curved at base; lvs elongate, 5–25 cm × 2–12 mm, entire or remotely runcinate-toothed, gradually acute; invol 10–13 mm, ± camp; achenes 5–6 mm; sphagnum bogs and wet meadows, Cas, Coast Range, and OM.

Arctium L.   Burdock

Fls all tubular and ♂, the corolla pink or purplish, with long, slender lobes; invol subglobose, its bracts multiserieate, narrow, appressed at base, with spreading, subulate, inwardly hooked tip; recep flat, densely bristly; anthers caudate; style brs linear, with a ring of hairs at base, or style pubescent below the brs; achenes oblong, slightly compressed, sub-3-angled, multinerved, truncate, glab; pappus of ∝ short, subpaleaceous, separately deciduous bristles; coarse per weeds of Eurasian origin, with large, alt, entire or toothed, gen cordate lvs and several or ∝ heads. (Gr *arktos*, bear, from the rough invol).

1a Infl of racemiform or thyrsoid brs, with short-pedunculate or subsessile
   heads; heads gen 1.5–2.5 cm thick; invol gen a little < the fls; pls seldom
   > 1.5 m; lf bls to 3 × 2.5 dm, lower petioles gen hollow; common b.
                                                 1 **A. minus** (Hill) Bernh.
1b Infl ± corymbiform, the heads gen long-pedunculate and 2.5–4 cm thick;
   invol gen = or > fls; pls to 3 m; lf bls to 5 × 3 dm, the petioles gen solid;
   occ; great b.
                                                 2 **A. lappa** L.

Arnica L.   Arnica

Heads radiate or discoid, the rays when present ♀, yellow or orange, relatively few and broad; invol bracts herbaceous, ± evidently biseriate, but subequal and connivent; recep convex, naked; disk fls ♂ and fertile, yellow; style brs flattened, truncate or nearly so, penicillate; achenes cylindric or nearly so, 5–10-nerved; pappus of ∝ white to tawny, barbellate to subplumose capillary bristles; fibrous-rooted per from a rhizome or caudex; lvs simple, opp; heads rather large (invol seldom < 1 cm), turbinate to hemispheric, 1–several or occ rather many. (Derivation uncertain).

1a Cauline lvs gen 5–12 pairs; rays gen 8–15, 1–2 cm; heads gen several
  2a Invol bracts obtuse or merely acutish, bearing a tuft of long hairs at or
    just within the tip; rhizome elongate, nearly naked; sts solitary, 2–10 dm;
    heads gen several; invol gen 8–12 mm; meadows and wet places, cordil-
    leran; lfy a., meadow a.; 4 vars. in our range   1 **A. chamissonis** Less.
    a1 Pappus tawny, subplumose; lvs gen toothed; hairs at base of invol
      with very prominent cross-walls; northern, s to c Wn and w c Mont,
      ours mostly transitional to spp. *foliosa*
                     ssp. **chamissonis** var. **interior** Maguire
    a2 Pappus stramineous or whitish, merely barbellate; lvs denticulate to
      more often entire; hairs at base of invol with less prominent
      cross-walls; w US and adj Can   ssp. **foliosa** (Nutt.) Maguire
      b1 Herbage conspicuously silvery-tomentose; ecotype of very wet
        places or in water, common from Cas of s Wn to Cal, occ e
                        var. **incana** (Gray) Hult.
      b2 Herbage less densely hairy, scarcely silvery; moist to occ dry or
        very wet places
        c1 Pls very robust, to 1 m, the lvs to 6 or 8 cm wide; local about
          Lower St. Mary Lake, GNP   var. **maguirei** (A. Nels.) Maguire
        c2 Pls less robust and with narrower lvs seldom > 4 cm wide; wide-
          spread in our range *(A. f.)*   var. **foliosa** (Nutt.) Maguire
  2b Invol bracts ± sharply acute, the tip not markedly more hairy than the
    body; rhizome freely rooting, short to ± elongate
  3a Lvs entire or nearly so, gen 5–12 × 1–2 cm, gradually acute or acu-
    minate; sts 3–6 dm, densely tufted, the rhizome gen shortened into a
    br caudex; invol 7–10 mm; pappus stramineous to tawny, barbellate or
    shortly subplumose; well drained sites about seeps or springs, and
    along cliffs and riverbanks; mts from s Alta to Wn, s to Cal and Colo;
    seep-spring a. *(A. myriadena)*   2 **A. longifolia** D. C. Eat.
  3b Lvs ± toothed, gen 5–12 × 1.2–6 cm; sts 3–8 dm, not much tufted,
    the rhizome more elongate; invol 9–15 mm; pappus tawny, subplu-
    mose; streambanks and moist woods, Alas to n Cal and w Mont;
    clasping a., streambank a.; 2 vars. in our range 3 **A. amplexicaulis** Nutt.

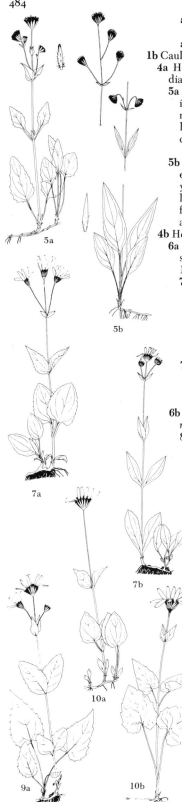

a1 Lvs very wide, to 6 cm; local in the CRG

var. **piperi** St. John & Warren

a2 Lvs narrower, seldom > 4 cm wide; widespread var. **amplexicaulis**

1b Cauline lvs gen 2–4 pairs, not including those (if any) of the basal cluster

4a Heads gen discoid (marginal corollas occ ampliate, rarely shortly radiate); sts 2–6 dm; invol 10–14 mm

5a Lower lvs gen ovate to subcordate, 4–8 × 1.5–5 cm, irreg toothed; invol bracts merely acutish or abruptly acute; heads not obviously nodding; rhizomes elongate, nearly naked; basal lvs well developed, long-petiolate; open woods, Cas region from Klickitat and Skamania cos, Wn, s to Cal; rayless a.; ours is var. *eradiata* (Gray) Cronq.

4 **A. discoidea** Benth.

5b Lower lvs gen lanceolate or lance-ovate to elliptic, 5–20 × 1.5–6 cm, evenly toothed or entire; invol bracts gradually and slenderly acute; young heads nodding; rhizomes nearly naked to freely rooting; basal lvs present or not; open woods, drier meadows, and moist slopes, foothills to middle elev in mts; BC and Alta to Colo and Cal; nodding a.; ours is var. *parryi*   5 **A. parryi** Gray

4b Heads gen radiate

6a Pappus subplumose, ± tawny; no evident tufts of basal lvs on the fl sts; rhizomes freely rooting, sometimes short; rays 1.5–2.5 cm; invol 10–16 mm

7a Heads narrow, ± turbinate; cauline lvs ovate or deltoid to broadly elliptic or lance-elliptic, sessile or shortly wing-petiolate, the middle ones gen largest, 4–8 × 2–6 cm; rocky places at middle and high elev; mts from Alas to Mont, Utah, and Cal; uncommon, gen apomictic, perhaps reflecting hybridization between *A. mollis* or *A. amplexicaulis*, on one hand, and *A. cordifolia* or *A. latifolia* on the other; sticky a.   6 **A. diversifolia** Greene

7b Heads broader, subhemispheric, with more ∝ fls; cauline lvs more variable; moist places at middle to high elev; mts from Alta and BC to Colo, Utah, and Cal, common and variable; hairy a.

7 **A. mollis** Hook.

6b Pappus merely barbellate (approaching a subplumose condition in *A. nevadensis*), gen white or nearly so

8a Lf blades relatively broad, the larger ones gen 1–2.5 (3) times as long as wide, pinnately or pinnipalmately veined; basal lvs sometimes persistent, but not densely tufted

9a Achenes gen glab below, or glab throughout; basal lvs (those on separate short shoots) seldom cordate when present, the cauline ones even less frequently so; lvs gen ± toothed; invol with few or no long hairs; common, variable; mts from Alas to Colo and Cal, with 2 vars.; mt a.   8 **A. latifolia** Bong.

a1 Small, tufted pls (1–3 dm) of rocky places at middle to gen high elev, the rhizome gen shortened into a slender, loosely br, scaly caudex; heads gen 3–9, narrow, small, the invol 7–13 mm; lvs seldom > 2.5 cm wide *(A. g.)*

var. **gracilis** (Rydb.) Cronq.

a2 Larger pls (1–6 dm) of moist woods, meadows, and moist open places in the mts (seldom at high elev); sts solitary or few together, the rhizomes elongate, gen nearly naked, but sometimes apically br and more scaly; heads gen 1–3, ave larger and broader, the invol 10–18 mm; lvs larger, gen 1.5–8 cm wide

var. **latifolia**

9b Achenes gen short-hairy (or glandular) nearly or quite to the base; lvs ± toothed to entire, often cordate; heads 1–3, occ more, with 10–15 rays 1.5–3 cm

10a Invol 10–15 mm, densely short-stipitate-glandular, with few or no long hairs; lvs entire or nearly so, the lower ones broad, rounded to subcordate at base, gen 3–7 × 2–4 cm; lateral setae of pappus bristles more prominent than in nos 8 and 10; high elev, OM and Cas, s to Cal, rare with us; Sierra a.

9 **A. nevadensis** Gray

10b Invol 13–20 mm, sparsely to gen copiously provided with long white hairs, esp below, gen also glandular; lvs toothed to occ entire, the basal and gen also the lower cauline ones gen ± strongly cordate (or scarcely so in alp forms); widespread cor-

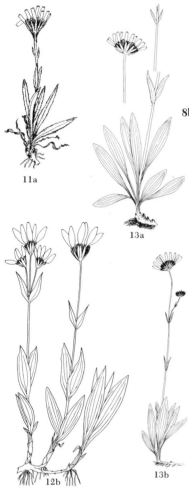

11a

13a

13b

12b

dilleran sp., common in woods from foothills to high elev; heart-lf a.; with 2 vars.   **10 A. cordifolia** Hook.

a1 Dwarf alp and subalp ecotype, gen < 2 dm, with relatively narrow, only slightly or scarcely cordate lvs 2–5 cm and with more glandular achenes (*A. humilis, A. p.*)
var. **pumila** (Rydb.) Maguire

a2 Foothill to midmont series of ecotypes, up to 6 dm, with the larger lvs gen 4–12 × 3–9 cm and ± strongly cordate
var. **cordifolia**

8b Lf blades narrower, gen 3–10 times as long as wide, with 3–5 main veins from the base; basal lvs often (not always) densely tufted; rhizomes short, often densely rooting; heads 1–3; achenes gen densely short-hairy throughout

11a Invol and herbage conspicuously woolly-villous; st solitary from a short rhizome or mere caudex, 0.5–2 dm; circumboreal arctic-alp sp., s in high mts to Alta, BC, and sw Mont; alpine a.; ours is var. *tomentosa* (Macoun) Cronq. **11 A. alpina** (L.) Olin

11b Invol and herbage ± pubescent, but not markedly woolly-villous

12a Heads relatively large, hemispheric or nearly so, with gen 10–23 rays 1.5–2.5 cm; pls 2–6 dm; lower cauline lvs gen petiolate; open places from the foothills to moderate elev in mts, BC to Sask, s to Cal and Colo

13a Old lf-bases with dense tufts of long brown wool in the axils; disk corolla with some spreading, glandless hairs, as well as stipitate-glandular; orange a., hillside a.
12 A. fulgens Pursh

13b Old lf-bases without axillary tufts, or the hairs few and white; disk corolla stipitate-glandular, gen not otherwise hairy; pls ave more slender than no 12, with less scaly, more slender, often longer rhizomes, and narrower and fewer lvs; twin a.   **13 A. sororia** Greene

12b Heads smaller, turbinate-camp, with gen 7–10 rays 1–2 cm; pls 1–3 dm; lower cauline lvs tending to be sessile; dry meadows and open slopes, gen at high elev; mts from BC and Alta to e Ore, n Cal, n Utah, and Colo; Rydberg's a., subalp a. (**A. louiseana** Farr, from Lake Louise, Alta, with nodding heads, more gradually and narrowly pointed invol bracts, and achenes glabrate below, may turn up in se BC)
14 A. rydbergii Greene

Artemisia L.   Mugwort; Sagebrush; Sagewort; Wormwood; Artemisia

Heads small, discoid or disciform, sometimes only with ♂ fls, sometimes the outer fls ♀, and the central ones then sometimes sterile; invol bracts dry, imbricate, at least the inner scarious or with scarious margins; recep flat to hemispheric, naked or densely beset with long hairs; style brs flattened, truncate, penicillate; achenes gen glab; pappus a very short crown, or gen none; ann, bien, or per herbs, or shrubs, gen aromatic, with alt, entire to dissected lvs and spiciform, racemiform, or paniculiform infl. (Ancient name of some of the spp., commemorating *Artemisia*, the wife of *Mausolus*).

4a

1a Marginal fls ♀; herbs or shrubs
  2a Disk fls fertile, with normal ovary
    3a Recep not hairy; lvs entire to dissected   **Group I**, lead 4a
    3b Recep beset with ∝ long hairs between the fls; lvs dissected
                                                          Group II, lead 16a
  2b Disk fls sterile, with abortive ovary   **Group III**, lead 18a
1b Fls all ♂; shrubs, all E Cas   **Group IV**, lead 21a

Group I

4a Shrubs 1–2 dm, with short, hyaline, irreg laciniate pappus; lvs canescent, 1–3 cm, once or twice trifid or irreg palmatifid, with narrow segms; infl racemiform, with 4–10 heads; invol 2–2.5 mm; dry, open, sometimes alkaline places, Blaine Co (Muldoon) to Owyhee Co, Ida, fl May–June; Owyhee s.
1 A. papposa Blake & Cronq.

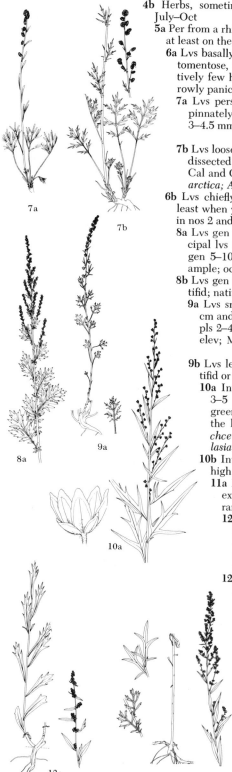

4b Herbs, sometimes a little woody at base; pappus wanting; fl (June) July–Oct

5a Per from a rhizome or caudex, or sometimes from a taproot; pls ± hairy, at least on the lower surface of the lvs when young

  6a Lvs basally disposed, sericeous to loosely villous or glabrate, scarcely tomentose, the lower gen 2–3 times cleft or divided; infl with relatively few heads, spiciform or racemiform, or in no 3 sometimes narrowly paniculiform; alp and subalp

    7a Lvs persistently sericeous, the bl 0.5–3 cm, ternately or ternate-pinnately to palmately cleft or dissected; pls 0.5–3 dm; invol 3–4.5 mm; OM and Cas of Wn; three-forked a.

      2 A. trifurcata Steph.

    7b Lvs loosely villous or eventually glabrate, the bl 2–10 cm, pinnately dissected; pls 2–6 dm; invol 4–7 mm; circumboreal, s in the mts to Cal and Colo; ours is var. *saxatilis* (Bess.) Jeps.; boreal w., mt s. (*A. arctica; A. s.*)

      3 A. norvegica Fries

  6b Lvs chiefly or all cauline, tomentose beneath (often also above) at least when young, scarcely villous or sericeous, often less divided than in nos 2 and 3

    8a Lvs gen with 1–2 pairs of stip-like lobes at base; divisions of principal lvs again toothed, cleft, or lobed; rhizomatous, 5–15 dm; lvs gen 5–10 × 3–7 cm, green above, white-tomentose beneath; infl ample; occ weed, intro from Europe; m. **4 A. vulgaris** L.

    8b Lvs gen without stip-like lobes at base, variously entire to bipinnatifid; native

      9a Lvs small and finely divided, the best developed ones gen 2–5 cm and bipinnatifid, often with the ultimate segms again toothed; pls 2–4 (7) dm; widespread cordilleran sp. of rocky places at high elev; Michaux m. (*A. discolor; A. vulgaris* sspp. *d.* and *m.*)

        5 A. michauxiana Bess.

      9b Lvs less divided and often larger, varying from entire to pinnatifid or occ subbipinnatifid

        10a Invol relatively broad and many-fld, wider than high, gen 3–5 × 4–8 mm; pls 3–15 dm; lvs tomentose beneath, gen green and shining above, entire to more often deeply lobed, or the lobes again cut; Aleutian m., mt w.; ours is var. *unalaschcensis* Bess. (*A. elatior, A. vulgaris* ssp. *t.*); passes to *A. douglasiana* and *A. ludoviciana* var. *latiloba* **6 A. tilesii** Ledeb.

        10b Invol narrower, higher than wide, or only about as wide as high

          11a Principal lvs relatively narrow, gen ca 1 cm or less wide exclusive of the lobes when present (rarely to 1.5 cm); pls rarely > 1 m

            12a Pls suffrutescent at base (sts not dying back to the ground each year), tending to be taprooted, without well developed rhizomes; below high-water mark, chiefly CR and its major tributaries; CR m., riverbank w. (*A. vulgaris* ssp. *l., A. prescottiana*) **7 A. lindleyana** Bess.

          12b Pls herbaceous to the base, not taprooted

            13a Pls freely spreading by creeping rhizomes, with solitary or loosely clustered sts; lvs entire to pinnatifid, seldom app the shape of no 9; widespread E Cas; western m., Louisiana s., prairie sage; 3 vars.

              8 A. ludoviciana Nutt.

            a1 Principal lvs entire or merely lobed; disk fls gen 6–21; plains and foothills, more common to the s of our range (*A. gnaphalodes*) var. **ludoviciana**

            a2 Principal lvs ± deeply parted or divided

              b1 Pls relatively heavily pubescent; heads relatively large; invol gen 3.5–4.5 mm, with 17–45 disk fls; lowl to moderate elev in mts (*A. candicans*) var. **latiloba** Nutt.

              b2 Pls relatively thinly pubescent and greenish, or even largely glabrate; heads smaller; invol gen 2.5–3.5 mm, with 15–30 (45) disk fls; lvs averaging smaller and more dissected than in var *latiloba*;

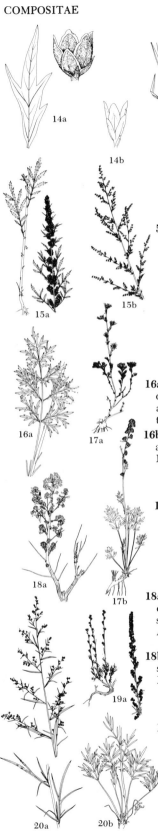

gen at moderate to high elev in mts (*A. i.*); passes to no 5   var. **incompta** (Nutt.) Cronq.

13b Pls with sts clustered on a woody caudex, without creeping rhizomes; lvs gen entire, narrow, and gradually tapering to the tip, rarely pinnately lobed; Gr Pl, barely reaching our range; long-lvd s. (*A. vulgaris* ssp. *l.*)   9 **A. longifolia** Nutt.

11b Principal lvs wider, gen 1–5 cm wide, exclusive of the lobes when present; pls often much > 1 m

14a Invol camp, tomentose; disk fls gen 10–25; W Cas in Ore and Cal, e through CRG and irreg across e Wn to n Ida; passes into *A. ludoviciana* var. *latiloba*; Douglas' s. (*A. vulgaris* ssp. *heterophylla*)   10 **A. douglasiana** Bess.

14b Invol cylindric or nearly so, glab and shining; disk fls gen 2–8; bluffs and beaches, W Cas, esp near the coast and about PS; Suksdorf's s., coastal m. (*A. vulgaris* ssp. *litoralis*)   11 **A. suksdorfii** Piper

5b Ann or bien from a taproot; lvs green, essentially glab

15a Infl dense, spikelike or of spiciform brs; heads crowded, obscurely pedunculate; invol 2–3 mm; pls nearly inodorous; lvs gen 5–15 cm, deeply pinnatisect into several narrow, gen toothed lobes, or the lower bipinnatifid; widespread native weed, esp in sandy soil; bien w.
    12 **A. biennis** Willd.

15b Infl loose, paniculiform; heads on evident slender peduncles; invol 1–2 mm; pls sweet-scented; lvs gen 2–10 cm, gen twice or thrice pinnatifid, with linear or lanceolate segms; Eurasian weed, occ with us; ann w.   13 **A. annua** L.

Group II

16a Pls robust, gen 4–12 dm; lvs relatively large, the better developed (lower) ones with bl 3–8 cm or more and with ultimate segms oblong or lanceolate and 1.5–4 mm wide; infl ample; invol 2–3 mm; Eurasian weed; w., absinthium   14 **A. absinthium** L.

16b Pls smaller, gen 1–4 dm; lvs smaller, the cauline with bl always < 3 cm and often with narrow ultimate segms

17a Pls mat-forming and subshrubby, with ∝ cauline lvs and gen ∝ heads; lf bls to 12 mm; dry, open plains and foothills; Gr Pl, to Ariz, Ida, e Wn (rarely), BC, and Alas; Siberia; pasture s., fringed s., prairie sagewort
    15 **A. frigida** Willd.

17b Pls strictly herbaceous and not at all mat-forming; cauline lvs few; heads few, gen 5–25; at least some of basal lvs with bl 1.5 cm or more; rocky places at high elev; mts of sw Mont to Colo, Utah, and NM; RM s.
    16 **A. scopulorum** Gray

Group III

18a Spinescent subshrub 0.5–5 dm, pungently aromatic; achenes and corollas cobwebby; larger lvs pedately 2–5-divided and again cleft into narrow segms, to 2 cm (petiole included), copiously villous; dry plains and hills, fl Apr–June; sw Mont and c Ida to se Ore, Cal, and NM; bud s., spring s.
    17 **A. spinescens** Eat.

18b Unarmed shrubs or herbs; achenes and corollas glab, or the corolla teeth short-hairy

19a Dwarf subshrub 0.5–1.5 dm; principal lvs 1–2 cm (petiole included), closely sericeous, once or twice ternate; heads few in a subracemiform or distantly spiciform infl; high plains and dry hills of Wyo, to Carbon Co, Mont; birdfoot s.   18 **A. pedatifida** Nutt.

19b Herbs 1–15 dm; principal lvs 2–10 cm; heads several or gen ∝ in a crowded-spiciform to ample and paniculiform infl

20a Lvs mostly entire, or the lower sometimes with 3–5 long, narrow segms; open, often dry places, plains to mid elev in mts; E Cas, widespread in w N Am; Eurasia; tarragon, dragon sagewort; ours is var. *dracunculus* (*A. dracunculoides*, *A. glauca*)   19 **A. dracunculus** L.

20b Lvs pinnatifid, bipinnatifid, or dissected, except the uppermost; open places, often in sandy soil; circumboreal, s to Ore, Ariz, Mich, and Vt; Pac s., northern w.; our pls 2 sspp.   20 **A. campestris** L.

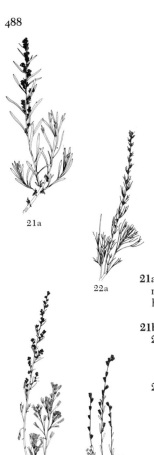

21a

22a

23a

24a

24b

a1 Per, several-std, with well developed rosettes of basal lvs each year, these gen persistent at least until anthesis; circumpolar ssp.; our pls in 3 vars.   ssp. **borealis** (Pall.) Hall & Clem.
  b1 Low pls, 1–4 dm, with narrow, spiciform-thyrsoid infl and relatively large heads; invol gen 3–4 mm
    c1 Herbage, esp lvs, densely and conspicuously sericeous; banks of CR, from Biggs to Hood R, and across the river in Wn (*A. ripicola*)   var. **wormskioldii** (Bess.) Cronq.
    c2 Herbage moderately to scarcely sericeous; high mts from n Wn to n Mont and Colo, and n (*A. p.*)   var. **purshii** (Bess.) Cronq.
  b2 Taller pls, gen 3–10 dm; infl larger and more open; heads smaller; invol 2–3 mm; widespread cordilleran var. at lower elev than var. *purshii* (*A. pacifica, A. campestris* ssp. *p., A. s.*)
                            var. **scouleriana** (Bess.) Cronq.
a2 Bien, single-std with basal rosettes only the 1st year; chiefly e and c US, occ with us, esp about PS (*A. c.*)
                          ssp. **caudata** (Michx.) Hall & Clem.

Group IV

21a Lvs all or nearly all entire, linear, 2.5–4 cm × 1–4 mm; pls otherwise much like the mont phase of no 25, but the upper lvs less reduced; mts and higher valleys, BC to Sask, s to n Cal, Utah, and NM; silver s., hoary s.
                                        21 **A. cana** Pursh
21b Lvs, or many of them, 3-toothed at the tip or 3–5-cleft
  22a Heads sessile in the axils, gen all surpassed by their subtending lvs; lvs deeply 3–5-cleft, deciduous; rounded shrub to 4 dm; dry, rocky places in the plains and foothills, w Mont to e Wn and c Ore; stiff s.
                              22 **A. rigida** (Nutt.) Gray
  22b Heads gen in a more brd and paniculiform infl, or spiciform with the upper heads not surpassed by their subtending lvs; lvs persistent, merely tridentate to deeply cleft
    23a Lvs, or many of them, deeply cleft into linear or linear-oblanceolate segms which may themselves be 3-cleft; fls gen 5–8; infl rather loosely paniculiform; pls gen 2–6 dm, sprouting freely after fire; dry plains and hills, often in moister or more favorable sites, or at slightly higher elev, than typical *A. tridentata*; BC to Cal, e to Mont and Colo; threetip s., cut-lf s. (*A. tridentata* ssp. *trifida*)   23 **A. tripartita** Rydb.
    23b Lvs gen merely toothed at the tip (occ more deeply cleft forms differ in other respects from no 23)
      24a Pls dwarf, gen 1–4 dm, with spikelike to narrowly paniculiform infl seldom 1.5 cm wide; lvs to 1.5 cm, sometimes more deeply cleft than in no 25; dry plains and hills, gen in less favorable sites than no 25; low s., dwarf s.; 2 vars.   24 **A. arbuscula** Nutt.
        a1 Invols canescent, 5–8-fld; typically w, from Wn to Cal, less commonly e to Wyo and Colo (*A. tridentata* ssp. *a.*) var. **arbuscula**
        a2 Invols glab or subglab, greenish-yellow, 3–5-fld; typically more e, chiefly from Mont and Ida to NM and Ariz (*A. n.; A. tridentata* ssp. *n.*)   var. **nova** (A. Nels.) Cronq.
      24b Pls taller, gen 4–20 dm, with more paniculiform infl gen 1.5–7 cm wide; invols canescent, 3–5-fld (or 5–8-fld in the mont phase); dry plains and hills, and to timberl in a mont ecotype; intolerant of alkali; BC to Baja Cal, e to ND and NM; big s. (*A. vaseyana*, the mont ecotype)   25 **A. tridentata** Nutt.

Aster L.   Aster; Michaelmas-daisy

Heads gen radiate, the few to ∝ ♀ fls gen with a lig, this variously blue, purple, or pink to white; in a few spp. the ♀ fls rayless and the heads thus disciform, in a few others the ♀ fls wanting and the heads thus discoid; invol bracts herbaceous throughout (the outer sometimes foliaceous and surpassing the inner), or more often chartaceous below, or sometimes chartaceous throughout, often strongly imbricate; recep flat or a little convex, naked; disk fls ± ∝, yellow or often reddish-purple; anthers entire or minutely sagittate at base; style brs flattened, with introrsely marginal stigmatic lines and lanceolate or narrower, acute or acuminate, externally minutely hairy appendage gen > 0.5 mm; achenes gen several-nerved (2-nerved in *A. alpinus*); pappus of capillary bristles, occ with a short outer series of minute bristles or scales; ann or gen per herbs with alt, entire or toothed, simple lvs and solitary to gen ± ∝, hemispheric to turbinate heads, fl mostly mid- or late summer and fall, except as noted. (Gr *aster*, a star, from the appearance of the heads). (*Brachyactis, Eucephalus, Sericocarpus*). Some spp. treated under *Aster* in the Flora are here referred to *Machaeranthera*.

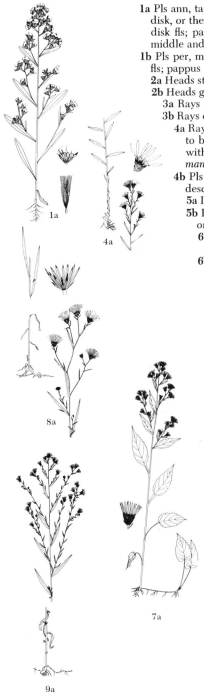

1a Pls ann, taprooted, with short, slender, inconspicuous rays only slightly > disk, or the ♀ fls with tubular-filiform, rayless corolla; ♀ fls more ∝ than the disk fls; pappus soft and copious, > disk fls; moist, gen saline ground at middle and lower elev *(Brachyactis)*   **Group I, lead 11a**

1b Pls per, most spp. with well developed rays; ♀ fls gen fewer than the disk fls; pappus = or < disk

   2a Heads strictly solitary   **Group II, lead 12a**

   2b Heads gen several or ∝

      3a Rays few (1–7), white; W Cas *(Sericocarpus)*   **Group III, lead 16a**

      3b Rays either more ∝, or colored, or both (or the ♀ fls none)

         4a Rays few, gen 5, 8, or 13 (21), or the ♀ fls none; invol bracts tending to be keeled, seldom strongly herbaceous; lower st lvs reduced; pls without well developed rhizomes (or shortly rhizomatous in *A. gormanii*) *(Eucephalus)*   **Group IV, lead 17a**

         4b Pls differing in 1 or more respects from the section Eucephalus as described above

            5a Invol and peduncles glandular   **Group V, lead 23a**

            5b Invol and peduncles, as well as the rest of the pl, without glands, or apparently so

               6a Disk corollas with the tube = or gen > the slender limb°; invol often ± suffused with anthocyanin   **Group VI, lead 27a**

               6b Disk corollas with the tube < limb; invol bracts sometimes purple-tipped, but normally not distinctly purple-margined or suffused with anthocyanin except in some forms of *A. foliaceus* and *A. junciformis*

                  7a Basal or lower cauline lvs with petiolate, cordate or subcordate bl 4–12 × 2–4 cm; pls rhizomatous; sts 2–12 dm; infl open, with very unequal, sparsely or scarcely bracteate peduncles; invol 5–8 mm, its bracts with long, narrow, acute or acuminate green tip; rays 12–25, blue, 8–15 mm; achenes glab; woods and clearings, Mack to BC, Wyo, and SD, e to Que and NY, rare with us; Lindley a. *(A. lindleyanus)*
                                          1 A. ciliolatus Lindl.

                  7b None of the lvs cordate or subcordate

                     8a Very slender pls to 8 dm, with elongate rhizomes seldom > 2 mm thick; lvs ± linear, gen 2–5 (9) mm wide; invol 5–7 mm, its slender, gen acute bracts ± imbricate, often with purple tips and margins; rays 20–50, white or pale bluish, 7–15 mm; cold bogs, Alas to Ida, e to Que and NJ, rare with us; rush a.   **2 A. junciformis** Rydb.

                  8b Pls otherwise

                     9a Pubescence of the st and brlets occurring in lines decurrent from the lf bases, gen neither uniform under the heads nor confined to the infl; infl gen large and lfy; sts 5–15 dm; lvs mostly sessile and linear to broadly lanceolate, 5–15 cm × 5–25 mm; invol 5–8 mm, with slender, green-tipped, acute bracts; rays 20–50, 6–14 mm, blue, less often pink or white; streambanks and ditches, plains to mid-elev in mts, Alta to Ida, e Nev, s Cal, and e to Wis and Mo, rare with us; relatively broad-bracted, white- or pink-rayed forms, often at higher elev, called *A. laetevirens* or *A. h.* var. *l.*, may reflect hybridization with *A. eatonii*; w willow a.; marsh a.
                                          3 A. hesperius Gray

                   9b Pubescence of the st uniform, or if in lines, then either uniform under the heads or very scanty and confined to the infl

                     10a Invol bracts, at least the outer, with loose or squar-

_____

° The disk corollas consist of a slender basal tube and a ± expanded terminal limb. The lobes are part of the limb.

rose, minutely spinulose-mucronate tip, or with loose, subulate, marginally inrolled tip; lvs narrow, seldom 1 cm wide; rays white   **Group VII, lead 28a**
10b Invol bracts neither spinulose-mucronate at the tip (rarely obscurely so in *A. chilensis*) nor with subulate marginally inrolled tip, but variously appressed or spreading; lvs narrow to broad; rays white to more often colored   **Group VIII, lead 30a**

Group I

11a Rays ca 2 mm, > styles; invol bracts obtuse or acutish; lvs seldom > 6 × 1 cm; e Wn to Cal, e to Utah, s Ida, and Wyo; short-rayed a., alkali a.   **4 A. frondosus** (Nutt.) T. & G.
11b Rays essentially wanting, the corolla of the ♀ fls tubular, < style; invol bracts definitely acute; lvs to 12 × 1 cm; BC and Alta to n Wn, Wyo, and Utah, e to Minn and Mo; Siberia; rayless alkali a. *(Brachyactis angusta)*   **5 A. brachyactis** Blake

Group II

12a Pappus distinctly double, with an outer series of very short bristles in addition to the principal bristles; lowest lvs reduced, the others ∝, narrow (to 4 mm wide), 1-nerved, uniform; pls 0.4–3 dm; at least the inner invol bracts scarious-margined and -tipped *(Ionactis)*
 13a Lvs very firm, gen 5–15 mm; sts 4–12 cm from a caudex that is sometimes evidently fibrous-rooted; invol 7–11 mm; rays 8, less often 13 or more; achenes densely silky; dry, open places, often with sagebr; w Mont to c Ida, se Ore, Nev, and Cal; crag a., lava a.   **6 A. scopulorum** Gray
 13b Lvs laxer, gen 15–40 mm; sts 15–30 cm; pls more definitely fibrous-rooted, and with the heads ave larger than in no 6; achenes less densely hairy; open places in the mts, at higher elev than no 6, c Ida and adj Mont to ne Wn and se BC; RM a.   **7 A. stenomeres** Gray
12b Pappus simple, or if inconspicuously double then the larger lvs gen well > 4 mm wide
 14a Achenes 2-nerved; sts 1–3 dm from a caudex or short rhizome; lvs basally disposed, the largest to 7 cm, oblanceolate or spatulate, apically rounded, ± trinerved; disk 2 cm wide or more; invol ca 1 cm, with herbaceous, loose, subequal bracts; rays white (violet), 1–1.5 cm or more; arctic-alp, circumboreal, s to Can border and rarely to Colo, fl June–July; boreal a.; Am pls are var. *vierhapperi* (Onno) Cronq.   **8 A. alpinus** L.
 14b Achenes several-nerved; fl July–Sept
  15a Sts subscapose from an erect, simple or slightly brd caudex that often surmounts a taproot; basal lvs 2–25 cm × 1–15 mm, the cauline few and much reduced; rays violet or lavender, 7–15 mm; alp and subalp; alpine a.; 3 vars.   **9 A. alpigenus** (T. & G.) Gray
   a1 Achenes gen hairy to the base; pls relatively robust, to 4 dm, the lvs to 25 × 1.5 cm; Cas of c Ore, s to Cal *(A. andersonii)*
        var. **andersonii** (Gray) Peck
   a2 Achenes glab below
     b1 Lvs gen oblanceolate, tending to be obtuse or rounded, the largest ones seldom < 5 mm wide (to 15 cm × 15 mm); pls fairly robust, to 2 (3) dm; OM, Cas from Wn to c Ore, and Wallowa Mts
        var. **alpigenus**
     b2 Lvs gen linear or linear-elliptic to linear-oblanceolate, tending to be acute, to 10 cm × 3 (5) mm; pls small, seldom 1.5 dm, with relatively narrow heads; w Mont and w Wyo, across Ida to the mts of e Ore, and in ne Nev *(A. h.)*   var. **haydenii** (Porter) Cronq.
  15b Sts obviously lfy, from a rhizome or caudex, with or without well developed basal lvs; various elev
        single-headed forms of spp. treated in **Groups III-VIII**

Group III

16a Rays 1–3 mm, < pappus, gen 2 (1–3) per head; pls relatively small and simple, 1–3 dm, gen with a single compact cluster of heads; sts from slender rhizomes; prairies, W Cas, s VI to sw Ore, chiefly in Wn; white-top a. *(Sericocarpus rigidus)*   **10 A. curtus** Cronq.

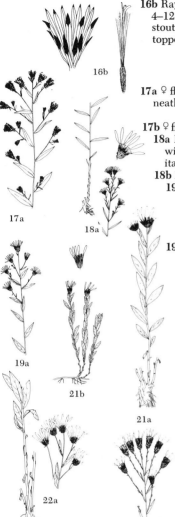

16b Rays 4–7 mm, > pappus, gen 5 (4–7) per head; pls larger and more brd, 4–12 dm, gen with several or ∝ clusters of heads; sts from a caudex or stout rhizome; woods, W Cas, sw Wn through w Ore to Cal; Ore white-topped a.   11 **A. oregonensis** (Nutt.) Cronq.

Group IV

17a ♀ fls none; sts 6–12 dm; lvs 5–9 × 1–3 cm, ± glab above, glandular be-neath; open woods, Lane and Douglas cos, Ore; wayside a.
                                                                12 **A. vialis** (Bradshaw) Blake
17b ♀ fls present, with well developed rays
  18a Herbage glab and glaucous; lvs 5–12 times as long as wide, to 15 mm wide; rays purple, 12–20 mm; sts 4–15 dm; Cas, Mt Adams to CR; Klick-itat a.   13 **A. glaucescens** (Gray) Blake
  18b Herbage ± pubescent to glandular or subglab, not glaucous
    19a Lvs large, gen 5–10 × 1.5–3.5 cm, 2.5–6 times as long as wide, glab, or somewhat hairy or glandular beneath; rays white (or pinkish in age or in drying), 15–25 mm; pls robust, gen 6–15 dm; open woods, foot-hills to high elev; E Cas, Wn and s BC, e to Alta, and s to Colo and ne Nev; Engelmann's a.   14 **A. engelmannii** (Eat.) Gray
    19b Lvs smaller, to 1.5 (2) cm wide; pls smaller, to 6 (8) dm
      20a Rays white, 8–15 mm; herbage glandular or glandular-hairy, nei-ther cottony nor markedly scabrous; heads few or solitary
        21a Invol bracts narrow, gen lance-linear; open slopes, OM and s VI; OM a.   15 **A. paucicapitatus** Rob.
        21b Invol bracts broader, gen lance-ovate; Cas of c Ore; Gorman's a.   16 **A. gormanii** (Piper) Blake
      20b Rays lavender-purple to deep violet; herbage not glandular except sometimes in the infl
        22a Lvs ± strongly cottony beneath, not scab, 3–7 × 0.5–2 cm; rays lavender-purple, 12–20 mm; meadows and open woods, up to timberl; Cas; Cas a.; 2 vars.   17 **A. ledophyllus** Gray
          a1 Rays gen ca 13 or even 21; pedundles glandular or glandless, gen with some cottony pubescence; Mt Jefferson and n
                                                                 var. **ledophyllus**
          a2 Rays gen ca 8, or even fewer; peduncles evidently glandular, not otherwise hairy; Mackenzie Pass and s *(Eucephalus c.)*
                                                        var. **covillei** (Greene) Cronq.
        22b Lvs ± scab-puberulent, not at all cottony, 2–5 × 0.3–1 cm; rays deep violet, 7–12 mm, gen ca 8 or ca 5; dry, open places, foothills to mid elev in mts; sw Mont to ne Ore, s to Nev and Utah; elegant a. *(Eucephalus elegans)*
                                                           18 **A. perelegans** Nels. & Macbr.

Group V

23a Lvs gen sharply toothed, thick and firm, the lowest ones ± reduced, the others elliptic (or the lower obovate), 6–18 × 2–8 cm; invol 9–12 mm, its bracts strongly imbricate, with broad, chartaceous base and loose or spreading, acute or acuminate green tip; rays 12–35, blue or violet, 10–15 mm; open woods, foothills to mid elev in mts; n Wyo, c Ida, and ne Ore, n to Sask, BC, and Yuk; showy a.   19 **A. conspicuus** Lindl.
23b Lvs entire (or rather distantly serrulate in *A. modestus*, which is thin-lvd and has subequal invol bracts), very often narrower than in *A. conspicuus*
  24a St equably lfy, the lvs all or nearly all sessile, the middle and upper ones ca as large as or larger than the often deciduous lower ones; infl typ-ically short and broad
    25a Lvs narrow, gen 2–8 cm × 2–8 (10) mm, only slightly or not at all clasping; pls 1–5 (7) dm; invol 5–8 mm; rays 15–20, light violet or purplish, 6–12 mm; open, often rather dry places, plains and valleys to mid elev in mts; s BC and e Wn to w Mont, s to Cal and Utah; w meadow a.; 2 vars.   20 **A. campestris** Nutt.
      a1 Herbage only slightly and inconspicuously pubescent to subglab (but glandular above); widespread   var. **campestris**
      a2 Herbage ± densely and conspicuously short-spreading-hairy (as well as glandular above); partly or wholly replacing var. *c.* along e foot of the Cas from Klickitat Co, Wn, s to Cal var. **bloomeri** Gray

25b Lvs wider, 5–13 cm × 10–40 mm, evidently clasping; pls 4–20 dm;
invol 6–11 mm; rays 20–∝, 8–15 mm

26a Sts 4–20 dm, clustered on a stout caudex or very short, thick rhizome with ∝ fibrous roots; rays 45–100, gen bright pink-purple;
achenes densely hairy; e Am sp., occ escaped in our range; New
England a.   21 **A. novae-angliae** L.

26b Sts 4–10 dm, scattered on elongate rhizomes; rays 20–45, purple
or violet; achenes sparsely hairy; widespread cordilleran sp. of
streambanks and moist woods; few-fld a., great n a.
                                                      22 **A. modestus** Lindl.

24b St rather inequably lfy, the lower or basal lvs either evidently petioled,
or larger than the ± reduced (though still often well developed) middle
and upper ones, gen 8–25 cm (petiole included), the others sessile and ±
clasping, 3–11 × 1–4 cm; infl ± elongate; invol 8–14 mm, outer bracts
green, 1–3 mm wide, inner narrower and purplish; rays 10–27, vio-
let-purple, 10–15 mm; drier meadows and open woods at mid elev in
mts; w Mont to se Wn, s to Cal and Colo; thick-std a., entire-lvd a.
                                                      23 **A. integrifolius** Nutt.

Group VI

27a Woodl pls of the Cas (and Ochoco) region and w, gen 2–7 dm and poly-
cephalous; lvs gen sharply and conspicuously toothed, to 10 × 4 cm; rays
white to purple; invol bracts with or without anthocyanin; rough-lvd a. *(A.
eliasii)*   24 **A. radulinus** Gray

27b Subalp and circumboreal pls of Mont, Ida, ne Ore, n Wn, and n, to 2 (4)
dm, lax and with 1–few (20) heads; lvs inconspicuously toothed to entire,
2.5–8 × 0.4–2.5 cm; rays purple; invol bracts anthocyanic; arctic a.; ours is
var. *meritus* (A. Nels.) Raup *(A. m.)*   25 **A. sibiricus** L.

Group VII

28a Green tips of invol bracts subulate and marginally inrolled, but soft, not
at all spinulose; sts clustered on a short caudex; e Am sp., rarely escaped in
w Wn; hairy a.; ours is var. *pilosus*   26 **A. pilosus** Willd.

28b Green tips of invol bracts flat, spinulose-mucronate; native spp.

29a Sts 3–20 dm, clustered on a caudex or very short rhizome; heads ∝,
tending to be secund on recurved-ascending brs, small; invol 4–6 (7)
mm, well imbricate; open, wet or dry places in the valleys and plains,
tolerant of alkali; e Wn and se BC to Mont, s to Utah, Colo, and Neb;
heath-lvd a., tufted white prairie a.   27 **A. pansus** (Blake) Cronq.

29b Sts 2–6 dm, scattered on long, creeping rhizomes; heads ave larger and
fewer, often solitary at the ends of the brs, these rarely at all recurved-
secund; invol 5–7 (8) mm, imbricate or not; dry, open places in the valleys and plains, widespread, but in our range not extending much w of
Cont Div; little gray a., creeping white prairie a. *(A. commutatus, A.
crassulus)*   28 **A. falcatus** Lindl.

Group VIII

30a Invol bracts strongly graduated, at least the outer bracts obtuse, markedly
shorter than the inner, and not foliaceous; widespread cordilleran sp. with
3 sspp.   29 **A. chilensis** Nees

a1 Infl conspicuously divaricate-bracteate, the bracts gen obtuse or acutish
and 2–4 times as long as wide; rays gen white; lvs linear; heads ave
smaller than in the other sspp.; W Cas, Ore and occ Wn; Hall's a. *(A. h.)*
                                                    ssp. **hallii** (Gray) Cronq.

a2 Infl not conspicuously divaricate-bracteate, the bracts, if present, gen
erect, > 4 times as long as wide, and markedly acute; rays gen blue or
pinkish, sometimes white; lvs linear or broader

b1 Innermost invol bracts gen acuminate or distinctly acute; middle st
lvs gen < 1 cm wide and > 7 times as long; RM, GB, and occ Blue
Mt reg; long-lvd a. *(A. a.)*   ssp. **adscendens** (Lindl.) Cronq.

b2 Innermost invol bracts gen obtuse or merely acutish; middle st lvs
often > 1 cm wide and < 7 times as long as wide; W Cas, Cal, n to
the s end of WV; common Cal a.   ssp. **chilensis**

30b Invol bracts not strongly graduated, or, if so, then the bracts markedly acute; invol bracts acute, or, if obtuse, then enlarged and foliaceous

31a Achenes glab or nearly so; pls glab, except sometimes for some short lines of puberulence in the infl, the herbage tending to be glaucous; lvs thick, the middle and upper sessile and clasping, even the bracts of the infl broad-based and clasping; sts gen clustered on a short rhizome or caudex; widespread E Cas, to the Atl; smooth a.; ours is var. *geyeri* Gray

30 **A. laevis** L.

31b Achenes ± pubescent, except occ in forms of *A. foliaceus;* herbage various, but not glaucous

32a Outer invol bracts with evident scarious margins near the base, the chartaceous part tending to be darkened, yellowish or brownish rather than white; lvs gen toothed; pappus gen reddish when dry; seashores, streambanks, and moist woods, chiefly W Cas, esp near the coast, but occ inl to Alta, Mont, n Ida, and ne Ore; Douglas' a. (*A. douglasii*)

31 **A. subspicatus** Nees

32b Outer invol bracts with inconspicuous or no scarious margins, the chartaceous part, if present, gen whitish or green-tinged; lvs gen entire

33a Infl a long, narrow, lfy pan with gen ∞ heads; st pubescence gen uniform and very short; lvs gen > 7 times as long as wide; rays gen pink or white; streambanks and other moist places, valleys to mid elev in mts; s BC to se Sask, s to Cal and NM; Eaton's a.

32 **A. eatonii** (Gray) Howell

33b Infl few-headed, or, if with ∞ heads, then shorter, more open, and cymose-pan, often with much-reduced lvs; rays gen blue or violet

34a Herbage and invol bracts ± densely and uniformly soft-hairy (or the lower parts glabrate); lvs not markedly aur-clasping; invol gen strongly graduated; sts 4–15 dm; streambanks and open places in the Palouse country, se Wn and adj Ida; Jessica's a., Pullman a.

33 **A. jessicae** Piper

34b Herbage gen glab or nearly so, except in forms of *A. foliaceus* with strongly clasping lvs and scarcely graduated invol; widespread and common cordilleran spp.

35a Lvs and invol bracts relatively small and narrow, the middle cauline lvs gen < 1 cm wide and > 7 times as long as wide, the invol bracts small and linear or linear-oblong, never enlarged and lfy; w mt a.; 2 vars.

34 **A. occidentalis** (Nutt.) T. & G.

a1 Relatively small, simple, few-headed pls, seldom > 5 dm, with gen 1–10 heads in a sparsely lfy infl; gen in mt meadows, widespread, but replaced by the next var. in e Wn and adj Ida var. **occidentalis**

a2 Larger, more-brd pls, with more ∞ heads in a copiously lfy-bracteate infl; not so strictly confined to meadows; E Cas in Wn and Ore, e to n Ida, at lower elev than var. *occidentalis* var. **intermedius** Gray

35b Lvs and invol bracts relatively large, the middle cauline lvs gen > 1 cm wide (narrower in reduced alp pls) and < 7 times as long as wide, some of the invol bracts often ± enlarged and lfy (sometimes all narrow and linear); lfy a., lfy-bract a.; 6 vars.

35 **A. foliaceus** Lindl.

a1 Heads solitary or sometimes several, with very lfy invol, the outer bracts nearly all foliaceous; sts erect, gen < 6 dm; Alas coast, s at increasing elev to OM, Cas of n Wn, and GNP var. **foliaceus**

a2 Heads gen several (except often in var. *apricus*); invol scarcely to strongly lfy; gen more s

b1 Pls to 2.5 dm, decumbent or ascending; invol bracts gen purple-tipped and -margined; alp and subalp sites, s BC to n Cal, e to Colo and Mont (*A. a.*) var. **apricus** Gray

b2 Pls gen > 2.5 dm, erect; invol bracts gen not purple-margined; gen not alp

c1 Middle and upper stem lvs very strongly aur-clasping; herbage and invol tending to be conspicuously soft-hairy at least in part, varying to subglab; achenes tending to be glab in age

35b

d1 Invol bracts very narrow and acute or acuminate, the foliaceus ones, if present, linear or narrowly lanceolate and long-pointed; mts of n Ida, less commonly to adj parts of Wn, Ore, Mont, and s BC *(A. l.)*

var. **lyallii** (Gray) Cronq.

d2 Invol bracts broader and often blunter, the foliaceous ones, if present, gen lanceolate or ovate, acute or obtusish; Wallowa Mts of ne Ore, less commonly to w Mont and extreme se Wn *(A. c.)*

var. **cusickii** (Gray) Cronq.

c2 Middle and upper stem lvs scarcely to moderately aur-clasping; herbage essentially glab except for the inconspicuously pubescent upper part of the st and the often ciliate lf margins; achenes persistently hairy; nearly throughout our range, but rare or absent in the principal area of the 2 previous vars.

e1 Invol bracts narrow, acute or acuminate, the foliaceous ones, if present, linear or narrowly lanceolate and very acute; lvs gen rather thin, the lower tending to be enlarged and persistent; gen along streams and other wet places in the mts; Cas of n Wn, to c Cal, e to Alta, Wyo, and NM *(A. frondeus)*

var. **parryi** (Eat.) Gray

e2 Invol bracts wider, obtuse or acutish, the foliaceous ones, if present, broadly lanceolate to ovate, rounded to obtuse or acutish; lvs gen rather thick, the lower seldom much enlarged and often deciduous; gen in drier sites than var. *parryi*, and more common se of our range *(A. c., A. burkei)* var. **canbyi** Gray

Baccharis L.   Baccharis

Dioecious; fertile heads with ± ∝ ♀ fls with tubular-filiform corolla; sterile heads with ± ∝ functionally ♂ fls, the ovary abortive, the style brs sometimes connate; fls white to yellowish or greenish; invol bracts subequal to strongly imbricate, chartaceous or subherbaceous; recep flat or merely convex, naked; pappus of ∝ capillary bristles, those of the sterile heads fewer and shorter; achenes gen ± compressed and ribbed; shrubs or subshrubs with alt lvs. (Name from Bacchus, originally applied to some different shrubs).

B. pilularis DC. Chaparral broom. Much-brd shrub 3–15 dm; lvs thick, glutinous, oblanceolate or cuneate to subrotund, coarsely few-toothed or angled, to 5 × 2.5 cm, sessile or nearly so; heads ∝, in small clusters; invol imbricate, 3–5 mm; bluffs and thickets along the coast, Tillamook Co, Ore, to s Cal, fl Aug–Sept; our pls are var. *consanguinea* (DC.) Kuntze *(B. c.)*.

Bahia Lag.   Bahia

Heads radiate or rarely discoid, the rays yellow or sometimes white, ♀ and gen fertile; invol bracts few, loosely erect, nearly =, thin and herbaceous; recep flat or nearly so, naked; disk fls ♂ and fertile; anthers minutely sagittate; style brs flattened, with introrsely marginal stigmatic lines and short, blunt, externally merely papillate-hairy appendage, or the appendage nearly obsolete; achenes elongate, slender, narrowed below; pappus of several short, scarious or hyaline scales with firmer, ± callous-thickened base which may be extended into an evident midrib, or rarely obsolete; per herbs with opp (alt) entire to dissected lvs. (For Juan Francisco Bahi, prof. of botany at Barcelona).

B. oppositifolia (Nutt.) DC. Pls 1–2.5 dm, several-std, finely canescent; lvs 1–4 cm, opp (except upper ones), deeply tripartite into slender, sometimes again tripartite segms; heads several; invol 4–7 mm; disk 4–12 mm wide; rays ca 5 or 8, gen 2–5 cm; achenes and tube of disk corolla glandular; pappus scales with ± evident midrib; plains sp., entering our range in Mont *(Picradeniopsis o.)*.

Balsamorhiza Nutt.   Balsamroot

Heads solitary or few, large, radiate, the rays ♀ and fertile, yellow (roseate); invol of several series of subequal or imbricate, ± herbaceous bracts, or the outer bracts enlarged and foliaceous; recep broadly convex, chaffy, its bracts clasping the achenes; disk fls ∝, ♂ and fertile, gen yellow; anthers sagittate; style brs slender, smooth below, the introrsely submarginal stigmatic lines fading into the elongate, hispidulous terminal part, or hispidulous throughout and the stigmatic lines obscure; pappus gen none; disk achenes compressed-quadrangular; scapiform per with a rosette of large basal lvs and often with 1 or more much reduced, gen bractlike cauline lvs; fl spring. (Gr *balsamon*, balsam, and *rhiza*, root).

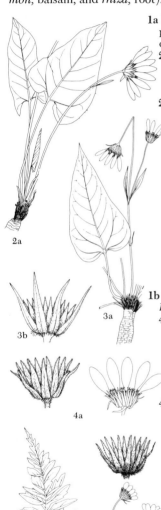

1a Lvs deltoid or sagittate, entire or merely crenate; rays permanently yellow; pls with a large, deep-seated woody root surmounted by a multicipital caudex (Sect. Artorhiza)

  2a Pls relatively densely and softly hairy, the invol gen woolly-tomentose, the lvs silvery (esp beneath) with a fine, close, feltlike tomentum of short hairs when young, greener and often glabrate in age; lowl to mid elev in mts, widespread E Cas; arrowlf b.   1 **B. sagittata** (Pursh) Nutt.

  2b Pls relatively sparsely hairy, the invol only slightly or scarcely woolly, the lvs green, their hairs fewer, coarser, and less conspicuous

    3a Rays only tardily deciduous, tending to persist on the achene and become ± papery; heads several; E Cas, s BC to c Ore; Carey's b.; in 2 vars.   2 **B. careyana** Gray

      a1 Achenes hairy; central head scarcely larger than the others; c Wn, s to n Umatilla and n Morrow cos., Ore   var. **careyana**

      a2 Achenes glab; pls app *B. deltoidea* in having a larger central head, some evidently enlarged and foliaceous outer bracts, and often some blunt teeth on the lvs; replacing var. *careyana* along the e side of the Cas, n to the Int Boundary and s to Bend, and in the n part of the John Day Valley   var. **intermedia** Cronq.

    3b Rays soon deciduous, not becoming papery; lvs thinner and less veiny than in *B. careyana*, often crenate; heads 1–several, the lateral ones, when present, obviously smaller than the c one; achenes glab; PT, s VI to s Cal; Puget b., deltoid b.   3 **B. deltoidea** Nutt.

1b Lvs ± cleft or pinnatifid (or some or all of them merely sharply serrate in *B. serrata*, or even crenate in *B. rosea*, which has roseate rays)

  4a Well developed pls with a large taproot from which several erect brs arise underground, thus somewhat cespitose and with the root transitional between sects. Artorhiza and Balsamorhiza; pls robust, the lvs gen 3–6 dm, with broad, gen entire or few-toothed segms 5–12 cm; lvs sparsely or evidently long-hairy but not at all tomentose; GB, reaching our range along the edge of upper SR plains, and in adj Mont; large-lvd b.   4 **B. macrophylla** Nutt.

  4b Pls with a smaller, somewhat carrotlike taproot surmounted by a simple or occ few-brd crown, often also producing deep-seated, slender, shortly creeping roots from which new pls arise; lvs either smaller, or more dissected, or tomentose (Sect. Balsamorhiza)

    5a Lvs and st silky-tomentose, with long, soft, tangled hairs; primary segms of lvs entire or with a few coarse teeth or segms, broader than in *B. hookeri* and *B. hirsuta*, sometimes 4 cm wide; heads large, the rays 3–6 cm; meadows and slopes at mid and lower elev in mts; ne Ore and se Wn to se Mont and wc Wyo; hoary b.   5 **B. incana** Nutt.

    5b Lvs and st scab to hirsute, hispid, or strigose, often glandular as well (or sericeous-strigose in some small-headed pls with the rays gen < 3 cm)

      6a Rays becoming roseate in age, 1–2.5 cm; achenes strigose; lvs crenate to pinnatifid; local E Cas in Wn; rosy b.

           6 **B. rosea** Nels. & Macbr.

      6b Rays permanently yellow, often > 2.5 cm; achenes glab

        7a Lvs varying from sharply serrate to deeply pinnatifid, even on the same pl, but gen at least some of them merely serrate; lvs scab-hispid, strongly reticulate; dry, rocky knolls and outcrops, E Cas in Ore and Wn; serrate b., toothed b.

           7 **B. serrata** Nels. & Macbr.

        7b Lvs deeply pinnatifid, less scab and less reticulate

          8a Middle and outer invol bracts with an ovate base narrowed to

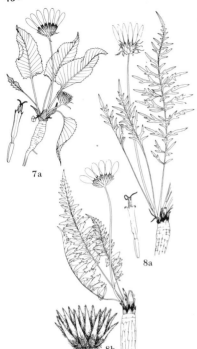

a subcaudate, gen reflexed appendage; reduced cauline lvs relatively well developed and conspicuous, pinnatifid or bipinnatifid; meadows and slopes in foothills and at mid elev in mts, local in e Ore, esp Union Co; hairy b.   **8 B. hirsuta** Nutt.

8b Invol bracts variously shaped, but seldom with subcaudate appendages; reduced cauline lvs smaller and inconspicuous, gen mere linear bracts, sometimes toothed or pinnatifid, or wanting; irreg distributed E Cas, to Cal and Colo, but absent from Ore; Hooker's b.; ours chiefly in 4 copiously intergrading vars.   **9 B. hookeri** Nutt.

   a1 Invol strigose or sericeous to hirsute or hispid, scarcely woolly

      b1 Lvs and invol sericeous or sericeous-strigose, the lvs often only once pinnate; pls dwarf; Klickitat Co, Wn

                                 var. **hookeri**

      b2 Lvs glandular and hispidulous, gen > once pinnate; invol sparsely hispid or hirsute-hispid; pls larger; chiefly SR plains, to Nev and Utah, but also near Yakima, Wn

                    var. **hispidula** (Sharp) Cronq.

   a2 Invol ± woolly

      c1 Pls relatively large and robust, with broad, gen uncut lf segms; Adams, Gem, and Wn cos, Ida

                var. **idahoensis** (Sharp) Cronq.

      c2 Pls smaller, and with more-dissected lvs, but not so small as var. *hookeri*; e Wn, but not Klickitat Co

              var. **lagocephala** (Sharp) Cronq.

Bellis L.   Bellis; Daisy

Heads radiate, the rays ♀, white to pink or purple; invol bracts herbaceous, =; recep conic, naked; disk fls ♂, yellow; style brs flattened, with introrsely marginal stigmatic lines and short, deltoid or ovate, externally short-hairy appendages scarcely longer than broad; achenes compressed, gen 2-nerved; pappus wanting; ann or per herbs, scapose or nearly so, with solitary heads. (L *bellus*, pretty).

B. perennis L. English d., lawn d. Spreading-hairy per; lf bls elliptic or obovate to orbicular, toothed, to 4 × 2 cm, = or > petiole; scape 5–20 cm; disk 5–10 mm wide; rays ∝, ca 1 cm or less; weed in lawns and waste places, adventive and ± estab across n US, in our range chiefly W Cas; fl Mar–Sept.

Bidens L.   Beggar-ticks; Beggars-tick; Bur-marigold; Sticktight

Heads radiate or discoid, the rays if present neutral (♀), yellow, seldom white or pink; invol bracts biseriate and dimorphic, the outer ± herbaceous, often very large, the inner membranous, often striate; recep flat or a little convex, chaffy throughout, its bracts narrow, flat or nearly so; disk fls ♂; anthers entire or minutely sagittate; style brs flattened, with externally hairy, gen short appendages, without well marked stigmatic lines; achenes flattened parallel to the invol bracts, not winged, often merely compressed-quadrangular, occ almost reg tetragonal, rarely *(B. beckii)* subterete; pappus of 1–6 (gen 2–4) awns or teeth, gen retrorsely barbed, sometimes antrorsely barbed or even barbless, rarely obsolete; ann to per herbs (shrubs), with opp, simple to dissected lvs; fl summer, into fall. (L, meaning two teeth, referring to the pappus).

1a Aquatic per with the submersed lvs filiform-dissected, the emersed ones gen merely serrate; achenes subterete, 10–14 mm, < the 3–6 pappus awns; rays 1–2 cm; ponds and slow streams; e Am sp., intro in Ore, Wn, and s BC; water marigold   **1 B. beckii** Torr.

1b Terrestrial or semi-aquatic ann, without filiform-dissected lvs; achenes flat or compressed-quadrangular, > the 2–4 pappus awns

   2a Lvs, except sometimes the lowermost, sessile; outer invol bracts gen spreading or reflexed; rays 4–15 mm, or none

      3a Principal lvs deeply tripartite; achenes truncate; rays < 1 cm; VI; VI b.   **2 B. amplissima** Greene

      3b Lvs all merely toothed to subentire; summit of mature achenes convex and cartilaginous (unique among our spp. in this regard); rays to 1.5 cm, or none; widespread; nodding b., s.   **3 B. cernua** L.

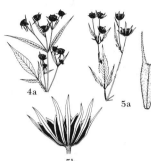

2b Lvs with distinct (sometimes winged) petiole gen 1–6 cm; outer invol bracts ascending or erect; rays none, or inconspicuous and up to 4 mm; wet to dry places

4a Lvs simple, though often deeply tripartite; widespread Eurasian and e Am weed, occ intro in our range, as at Bingen and Portland; three-lobed b. **4 B. tripartita** L.

4b Lvs pinnately compound, with 3–5 lfls, at least the terminal 1 slender-petiolulate; common and widespread weeds

5a Outer invol bracts gen 5–8; disk orange; lfy b., s. **5 B. frondosa** L.

5b Outer invol bracts gen 10–16; disk rather pale yellow; tall b., western s. **6 B. vulgata** Greene

Blepharipappus Hook.   Blepharipappus

Heads radiate, the 2–7 broad rays white, ♀ and fertile; invol bracts subequal, few, with narrow hyaline margins infolded around the subtended achene; recep naked, chaffy throughout, its bracts thin and narrow; disk fls rather few, ♂, white; style short-hairy below the brs, 2-cleft at the summit only, the short rounded brs stigmatic within; achenes silky-villous, somewhat flattened at right angles to a radius of the head; pappus of several or rather ∝ narrow, sometimes awnlike, fimbriate scales, with thickened axis, or sometimes reduced or wanting; ann with alt, entire, linear lvs. (Gr *blepharis*, eyelash, and *pappus*, seed down, referring to the fringed scales of the pappus).

B. scaber Hook. Slender ann 1–3 dm, scab-puberulent and often spreading-hairy, becoming glandular above; lvs ∝, 6–25 × up to 1.5 mm; heads 1–∝, terminating the brs; invol 4–7 mm; rays 4–10 mm, broad, 3-cleft; style purple, anthers darker; bunchgrass prairies and grassy foothills, chiefly E Cas; se Wn and adj Ida to Cal and nw Nev; Eugene; ours is var. *scaber*.

Boltonia L'Her.   Boltonia

Heads radiate, the rays white to pink or blue, ♀; invol bracts subequal to strongly imbricate, scarious-margined, with green or greenish midrib or tip; recep hemispheric or conic, naked; disk fls yellow, ♂; anthers entire at base; style brs flattened, with introrsely marginal stigmatic lines and short, lanceolate, externally short-hairy appendages; achenes obovate, strongly flattened, ± wing-margined; pappus of several minute bristles and 2 (4) longer awns, these gen reduced or wanting in the ray achenes, sometimes also in the disk; glab, often stoloniferous, short-lived per with relatively narrow, gen entire, alt lvs and several to very ∝, small to medium-sized hemispheric heads. (For James Bolton, English botanist of the 18th century).

B. asteroides (L.) L'Her. Fibrous-rooted, gen 5–15 dm; lvs to 15 × 2 cm; invol bracts ± imbricate; disk 6–10 mm wide; rays 5–15 mm; moist, low places; e Am sp., intro (?) here and there with us; fl summer into fall; ours chiefly var. *recognita* (Fern. & Grisc.) Cronq., with linear, acute invol bracts. Some specimens app var. *latisquama* (Gray) Cronq., with more-spatulate, rounded-obtuse bracts.

Brickellia Ell.   Brickellia; Brickellbush; Thoroughwort

Heads discoid, the fls all tubular and ♂, 3–∝ in each head; invol bracts striate, imbricate; recep naked; corollas white or creamy to pink-purple, never yellow; anthers minutely rounded-sagittate at base; style brs with short stigmatic lines and an elongate, papillate appendage; achenes 10-ribbed; pappus of 10–80 barbellate or nearly smooth to rarely subplumose bristles; per (ann) herbs or shrubs, most spp. fibrous-rooted; lvs alt or opp, simple. (John Brickell, 1749–1809, physician and botanist of Savannah, Ga). *(Coleosanthus)*.

1a Lvs lance-linear to oblong or elliptic-oblong, entire or nearly so, sessile; herb or subshrub 1–6 dm, with ∝ sts from near base; invol 10–20 mm; dry, often rocky places, foothills and lowl, cordilleran, E Cas, fl summer; narrow-lvd b. or t.; 2 vars. **1 B. oblongifolia** Nutt.

a1 Achenes glandular or glandular-hispidulous; widespread with us var. **oblongifolia**

a2 Achenes hispidulous, scarcely glandular; s border of our range, and s *(B. l.)* var. **linifolia** (D. C. Eat.) Rob.

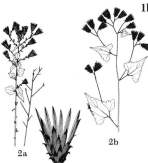

1b Lvs narrowly to broadly deltoid to ovate, rotund, or subcordate, toothed, evidently (though sometimes shortly) petiolate; invol 7–12 mm
  2a Lf bls gen 0.5–2 cm, with petiole 1–3 mm; outer invol bracts, like the inner, without appendages; shrub or subshrub 3–6 dm; fls 17–24 in a head; dry, rocky foothills; extreme se Wn to Cal, Wyo, and NM, fl Aug-Oct; small-lvd b.; ours is var. *microphylla*
  2 B. microphylla (Nutt.) Gray
  2b Lf bls gen 2–11 cm, with petiole 3–70 mm; outer invol bracts tipped with a well developed slender appendage; herb 2.5–7 dm; fls 20–40 (70); many habitats; cordilleran and plains sp., chiefly E Cas, fl July–Aug; large-fld b. or t.
  3 B. grandiflora (Hook.) Nutt.

Carduus L.   Thistle

Similar to *Cirsium,* differing chiefly in the merely minutely barbellulate, rather than plumose pappus bristles; st gen winged by decurrent lf bases; our spp. intro Eurasian ann and bien weeds with purple fls; fl summer, into fall. (Ancient L name for thistle).

1a Heads camp to hemispheric; disk 1.5–8 cm wide as pressed; invol bracts without the tiny hairs of no 3
  2a Invol bracts conspicuously broad, gen 2 mm wide or more, sometimes to 8 mm; heads gen solitary and nodding at br tips, often very large, the disk (1.5) 4–8 cm wide as pressed; pls to 2.5 m; widely but sparingly intro; musk t.
  1 C. nutans L.
  2b Invol bracts narrow, rarely any of them as much as 2 mm wide; heads small (disk 1.5–2.5 cm wide), erect, often clustered; pls to 1.5 m; adventive in n Ida, and to be expected elsewhere; acanthus t.
  2 C. acanthoides L.
1b Heads cylindric or cylindric-turbinate, clustered at the br tips, relatively narrow, the disk to 1.5 cm wide; invol bracts copiously beset with tiny, firm, forward-pointing hairs, esp on midrib; s WV and s; Italian t., compact-headed t. (*C. tenuiflorus* Curtis, similar but without these hairs, may also be expected along our s border)
  3 C. pycnocephalus L.

Centaurea L.   Knapweed; Star-thistle; Thistle; Centaurea

Heads discoid, the fls sometimes all tubular and ♂, or more gen the marginal ones sterile, with enlarged, irreg, falsely subradiate corolla; invol bracts imbricate in several series, either spine-tipped or more often some of them with enlarged, scarious or hyaline, erose to lacerate or pectinate appendage; recep nearly flat, densely bristly; corollas purple or blue to yellow or white, with slender tube and long, narrow lobes; anthers shortly to strongly caudate; style with a thickened, often hairy ring and an abrupt change of texture at the base of the appressed, gen ± connate brs; stigmatic lines marginal, extending nearly or quite to the tip; achenes obliquely or laterally attached to the recep, seldom evidently nerved; pappus of several series of graduated bristles or narrow scales, often much reduced, or wanting; ann, bien, or per herbs with alt or all basal, entire to pinnatifid lvs and solitary to ∞, small to large heads. (Gr *kentaurion*, plant of the Centaurs, applied by herbalists to several very different genera). Our spp. all intro from Eurasia and ± weedy; fl summer, into fall.

Carthamus tinctorius L., safflower, which is readily distinguished from *Centaurea* by its foliaceous outer invol bracts and bright orange fls, may be expected to escape from cult.

1a Invol bracts, or some of them, evidently spine-tipped; ann or bien; marginal fls not enlarged
  2a St merely angled, not winged; pappus none; lvs pinnatifid, with narrow segms, or the upper entire
    3a Central spine of each invol bract slender, gen 1.5–4 mm; fls creamy (purplish); invol 8–10 mm; local, sometimes abundant; tumble k., bushy k.
    1 C. diffusa Lam.
    3b Central spine stout, the larger ones gen 10–30 mm; fls purple; invol 10–18 mm; uncommon s., purple s.
    2 C. calcitrapa L.
  2b St evidently winged by the decurrent lf bases; pappus present, at least in the central fls; lower lvs toothed to lyrate-pinnatifid, the middle and upper smaller and becoming linear and entire; fls yellow; invol to 15 mm
    4a Larger spines of the invol bracts gen 11–22 mm; herbage persistently

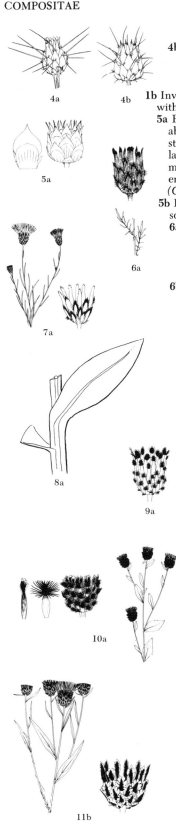

4a   4b

5a

6a

7a

8a

9a

10a

11b

tomentose; marginal fls epappose, the others with pappus gen 3–5 mm; chiefly E Cas; yellow s., St Barnaby's t.   3 C. **solstitialis** L.

4b Larger spines of the invol bracts gen 5–9 mm; herbage lightly arachnoid when young, not tomentose; marginal fls, like the others, with evident pappus 1.5–3 mm; chiefly W Cas; tocalote, maltese s., Napa t.
                                                         4 C. **melitensis** L.

1b Invol bracts not at all spiny, gen ± lacerate or fringed; ann, bien, or per, with or without enlarged marginal fls

5a Pappus well developed, gen 6–11 mm, the larger bristles subplumose above; bushy-brd per with deep-seated creeping roots; invol greenish-stramineous, 9–15 mm, the middle and outer bracts broad, glab, with large, broadly rounded, subentire hyaline tip, the inner narrower and more tapering, with plumose-hairy tip; fls purple, the marginal ones not enlarged; noxious weed; chiefly E Cas; Turkestan t., Russian s. or k. *(C. picris)*   5 C. **repens** L.

5b Pappus shorter, gen 1–3 mm, or none; ann, bien, or per, not (except to some extent in *C. montana*) creeping below ground

6a Lvs all (except some of the reduced ones of the infl) pinnatifid, with narrow segms; bien or short-lived per, tending to be taprooted; invol 10–13 mm, the middle and outer bracts with short, dark, pectinate tip; fls pink-purple (white), the marginal ones enlarged; chiefly E Cas; spotted k.   6 C. **maculosa** Lam.

6b Lvs entire or toothed, or some of the larger ones few-lobed

7a Ann or winter ann; lvs linear or nearly so, gen < 1 cm wide; invol 11–16 mm, its bracts with a narrow, often dark, pectinate or lacerate fringe near the tip; fls anthocyanic or white, the marginal ones enlarged; cosmopolitan, sometimes cult; cornflower, bachelor's button   7 C. **cyanus** L.

7b Per; lvs broader, some of the lower ones, at least, much > 1 cm wide (even without including any lobes that may be present)

8a St conspicuously winged by the decurrent lf bases; heads solitary (few), large, the invol 15–25 mm, the marginal fls much enlarged, so that the fl head is 4–8 cm wide; escaped from cult W Cas; mont s.   8 C. **montana** L.

8b St not winged; heads gen several or ± ∝, smaller, the invol 12–18 mm, the marginal fls often enlarged, but the head not > ca 4 cm wide; widespread weeds

9a Scarious tips of the invol bracts small, gen 1–3 mm, conspicuously blackish at least in part, the middle and outer ones deeply and rather reg pectinate; heads relatively narrow, pressed invol at least as high as broad; marginal fls enlarged; pappus gen none; occ weed; short-fringed k.   9 C. **dubia** Suter

9b Scarious tips of the invol bracts larger, gen 3–6 mm, sometimes paler and less reg cleft; heads wider, the pressed invol wider than high

10a Scarious tips of the invol bracts conspicuously blackish at least in part, the middle and outer ones deeply and fairly reg pectinate, with the laciniae gen 2–3 times as long as the width of the undivided central part of the appendage; seldom any of the bracts markedly bifid; marginal fls not enlarged; pappus gen evident, to 1 mm; black k., lesser k., hardheads, horse-knops   10 C. **nigra** L.

10b Scarious tips of the invol bracts gen paler, tan to dark brown or rarely blackish, less deeply and often less reg cleft, the inner (esp in *C. jacea*) often enlarged and bifid; marginal fls enlarged

11a Scarious tips of the middle invol bracts fairly reg pectinate-fringed, but not so deeply so as in *C. nigra*, the laciniae about as long as the width of the undivided body of the bract; pappus minute but gen present; a common (with us) intermediate form between *C. nigra* and *C. jacea*, presumably of eventual hybrid origin, but apparently self-perpetuating and here for convenience given a binomial; other intermediate or introgressant forms also exist; meadow k.   11 C. **pratensis** Thuill.

11b Scarious tips of the middle bracts irreg lacerate rather than pectinate; pappus gen none; brown k.   12 C. **jacea** L.

Chaenactis DC.   False-yarrow; Chaenactis

Heads discoid, the fls all tubular and ☿, though the outer are sometimes enlarged and shortly sublig; invol bracts narrow, herbaceous or subherbaceous, = or occ ± imbricate; recep naked; fls white, pink, or yellow; anthers shortly sagittate; style brs slightly compressed, externally hairy nearly or quite to the base, with introrsely submarginal, sometimes obscure stigmatic lines extending nearly to the tip; achenes clavate, subterete or somewhat compressed; pappus of 4–20 hyaline scales, or these rarely obsolete; ann, bien or per taprooted herbs with alt, entire to more often pinnately or irreg dissected lvs. (Gr *chaino*, to gape, and *actis*, ray, referring to the enlarged orifice and irreg, subradiate limb of the marginal disk corollas of ∝ spp.).

1a Pls ann or winter ann; invol < 1 cm; fl spring
  2a Fls yellow; pappus of ca 10 minute, vestigial scales; lvs pinnatifid; pls 1–3 dm, glandular-puberulent and arachnoid-villous; barren clays derived from volcanic tuff, nc Ore, esp John Day Valley; John Day c.
                                                                           1 C. **nevii** Gray
  2b Fls white or pinkish; pappus well developed, of 10–14 scales < corolla; lvs entire or nearly so, subglab, linear or oblanceolate, 1–4 cm × 2–5 mm; pls to ca 1 dm; dry, open places, low elev, e Ore and sw Ida; Cusick's c.   2 C. **cusickii** Gray
1b Pls bien or per; invol often > 1 cm; fls pink or whitish; pappus well developed; herbage sparsely to densely tomentose or arachnoid, often also glandular; fl summer *(C. douglasii* also in spring)
  3a Lvs pinnatilobate rather than pinnatifid, with gen 2–4 pairs of lobes, plane, 1–5 cm; mat-forming per with ∝ subnaked peduncles to 1 dm; talus slopes, high mts of c Ida; Evermann's c.   3 C. **evermannii** Greene
  3b Lvs pinnatifid to pinnately dissected, the gen more ∝ segms so shaped or oriented (except in *C. thompsonii*) that the lf does not appear plane; pls not mat-forming
    4a Alp per with 1 or gen several rosettes, and naked axillary peduncles to 1 dm, otherwise stemless; talus slopes, RM, w to ne Ore; alpine c. *(C. douglasii* var. *a.*)   4 C. **alpina** (Gray) Jones
    4b Lfy-std bien or per, with or without rosettes, seldom alp
      5a Sts ∝, lax; lvs gen only once pinnatifid; Wen Mts
        6a Per from a taproot and brd caudex; lvs flat; Thompson's c.
                                                                   5 C. **thompsonii** Cronq.
        6b Bien or short-lived per from a taproot; lf segms somewhat curled as in *C. douglasii;* branching c.   6 C. **ramosa** Stockwell
      5b Sts few or solitary, gen erect; lvs gen bipinnatifid or pinnately dissected; widespread and variable, with 4 confluent vars. in our range; hoary c. or f.   7 C. **douglasii** (Hook.) H. & A.
        a1 Dwarf but still lfy-std, often per ecotype, seldom > 1.5 dm, gen several-std from the base, often with brd caudex; mid to fairly high elev; RM *(C. humilis, C. cineria)*   var. **montana** Jones
        a2 Taller, gen 1.5–6 dm; lowl to mid elev in mts
          b1 Bien, gen single-std, 1.5–4 (5) dm; invol 8–12 mm
            c1 Peduncles and invols merely stipitate-glandular; herbage greener than in the next var.; fls pink; replacing the next var. in much of ne Ore, extreme se Wn, and adj Ida
                                                                   var. **glandulosa** Cronq.
            c2 Peduncles and invols glandular-puberulent and often ± tomentose; fls white to pink; the common and widespread phase of the sp. at mid and low elev *(C. a.)*
                                                       var. **achilleaefolia** (H. & A.) A. Nels.
          b2 ± per, often with brd caudex, more robust (to 6 dm) and more densely hairy, with invols to 16 mm; replacing the preceding var. at low elev along e foot of Cas, Yakima Co, Wn, to Jefferson Co, Ore   var. **douglasii**

Chondrilla L.   Skeleton-weed

Fls all lig and ☿, yellow, 7–15 in a head; invol cylindric, calyculate; achenes multinerved, glab, muricate above, and with a beak which is expanded distally into a pappiferous disk; pappus of ∝ capillary bristles, gen white; lactiferous, br, often rushlike herbs with well developed, gen pinnatifid basal lvs and reduced, gen scattered and entire cauline lvs. (Name used by Dioscorides for some gum-bearing pl).

C. **juncea** L. Gum-succory, Rushlike. taprooted per; st 3–15 dm, spreading-hispid below; basal lvs runcinate-pinnatifid, 5–13 × 1.5–3.5 cm, often deciduous; cauline lvs linear, 2–10 cm × 1–8 mm; invol 9–12 mm, white-tomentose; body of achene 3 mm, with a circle of scales at base of long, slender beak; Eurasian weed, occ with us; fl July–Sept.

Chrysanthemum L.   Chrysanthemum

Heads radiate or rarely discoid, the rays, when present, uniseriate, ♀ and fertile, white in our spp.; invol bracts ± imbricate in 2–4 series, dry, becoming scarious or hyaline at least at the margins and tip, the midrib sometimes greenish; recep flat or convex, naked; disk fls tubular and ⚥, the corolla (4)5-lobed; anthers ± entire at base; style brs flattened, truncate, minutely penicillate; achenes subterete or angular, 5–10-ribbed, or those of the rays with 2–3 wing-angles; pappus a short crown, or none; per (ours) or less often ann herbs with alt, entire or toothed to pinnatifid lvs and solitary to ∝ small to large heads; our spp. all intro, fl summer-fall. (Gr *chrysanthemon*, golden fl).

1a Heads solitary or few, naked-pedunculate, relatively large, the disk 1–3 cm wide, the rays 1–3 cm or more; basal or lower cauline lvs largest, gen oblanceolate or spatulate, crenate and (in no 1) often ± lobed or cleft
    2a Heads middle-sized, the rays 1–2 cm, the disk 1–2 cm wide; common in disturbed sites with adequate moisture; marguerite, oxeye-daisy, moon-daisy *(Leucanthemum vulgare)*   **1 C. leucanthemum** L.
    2b Heads larger, the rays 2–3 cm, the disk 2–3 cm wide; pls more robust and with more closely and shallowly toothed lvs; occ W Cas; Shasta daisy   **2 C. maximum** Ramond.
1b Heads several or ∝ in a terminal infl, smaller, the disk gen 4–9 mm wide, the rays when present < 1 cm; occ escaped
    3a Lvs pinnatifid with incised to again pinnate segms, chiefly cauline, the bl to 8 × 6 cm; rays 10–20, or ∝; featherfew, feverfew *(Matricaria p.)*   **3 C. parthenium** (L.) Bernh.
    3b Lvs merely crenate, or with a few reduced basal pinnae, the basal ones the largest, with bl 8–25 × 2.5–8 cm, on petioles same length; rays wanting, or occ few and < 1 cm; costmary, mint geranium *(Tanacetum b.)*   **4 C. balsamita** L.

Chrysopsis (Nutt.) Ell.   Goldaster; Golden-aster

Heads radiate or occ discoid, the rays yellow, ♀ and fertile; invol bracts imbricate, wholly scarious to subherbaceous; recep flat or nearly so, naked; disk fls yellow, ⚥ and fertile; anthers ± entire at base; style brs flattened, with introrsely marginal stigmatic lines and externally hairy, gen elongate appendage; achenes ± flattened; pappus double, the inner of capillary, gen unequal and often sordid bristles, the outer of short coarse bristles or scales (sometimes obscure); per or ann, gen ± hairy herbs (ours 1–5 dm, several-std from an often ± woody base, the lowest lvs reduced and deciduous, the others rather ∝ and uniformly distributed, seldom > 5 cm), with alt, gen entire lvs and hemispheric to camp, medium-sized to rather large heads, ours in a corymbiform infl; fl summer. (Gr *chrysos*, gold, and *opsis*, aspect, from the color of the heads). Genus sometimes submerged in *Heterotheca*, which when more strictly defined has ray achenes lacking pappus.

1a Heads discoid; outer pappus obscure, invol bracts evidently subhyaline-margined; sand and gravel bars along rivers, e base of Cas to coast; Ore g.   **1 C. oregona** (Nutt.) Gray
1b Heads with 10–25 rays 6–10 mm; outer pappus more evident; invol bracts (at least the outer ones) scarcely or less evidently margined; widespread, but chiefly E Cas; hairy g.; 3 vars. with us   **2 C. villosa** (Pursh) Nutt.
    a1 Pubescence of the lvs and invol of spreading hairs, or the hairs partly replaced by glands; dry, open places, widespread E Cas in our range and se *(C. h.)*   var. **hispida** (Hook.) Gray
    a2 Pubescence of the lvs appressed or subappressed, seldom at all glandular; pubescence of the invol appressed to spreading
        b1 Lvs relatively broad, gen broadly oblong to elliptic or obovate and scarcely petioled, gen some of them rather closely subtending the heads; Gr Pl to our range in Mont   var. **foliosa** (Nutt.) D. C. Eat.
        b2 Lvs narrower, more oblanceolate or spatulate, and mostly sub-petiolate; heads gen more evidently pedunculate; Gr Pl, and extending w throughout our range along streambanks   var. **villosa**

Chrysothamnus Nutt.　Rabbit-brush

Heads strictly discoid, narrow, with gen 5 (–20) fls; invol bracts imbricate, tending to be keeled, and arranged in 5 ± distinct vertical ranks, chartaceous or coriaceous except for the sometimes herbaceous tip; recep naked; fls ⚲, yellow; style brs flattened, with introrsely marginal stigmatic lines and ± elongate, externally short-hairy appendage; pappus of ∝ capillary bristles; brd shrubs with alt, sessile, narrow (gen linear), entire lvs and ± ∝ heads; fl mid- or late summer to fall. (Named for its "affinity to *Chrysocoma*, and brilliant golden yellow flowers").

1a Twigs covered with a feltlike tomentum, this sometimes so close as to escape casual observation

2a Outer involucral bracts gen prolonged into a slender herbaceous tip or appendage; shrubs 1–6 dm; lvs green and often ± viscid; infl tending to be elongate and subracemiform; chiefly s RM and GB sp.; Parry's r.; with 2 outlying, still undescribed vars. in our range　1 **C. parryi** (Gray) Greene

　a1 Fls ca 5–6 in each head; pls gen 3–6 dm; rocky slopes along Salmon R from near Clayton to below Challis (*C. parryi* var. *attenuatus*, misapplied)　　　　　　　　　　　　　　　　　　　　　var. **a.**

　a2 Fls 9–10 in each head; pls dwarf, < 2 dm; dry rocks at 9500 ft, Red Conglomerate Peaks, Clark Co, Ida (*C. parryi* var. *parryi* and var. *howardii*, misapplied)　　　　　　　　　　　　　　　var. **b.**

2b Outer invol bracts reg shortened, not herbaceous-tipped; dry, open places from lowl to occ mid elev in mts, chiefly E Cas; common r., gray r.; 6 vars. with us　　　　　　　　　2 **C. nauseosus** (Pall.) Britt.

　a1 Achenes glab; habit and pubescence nearly of var. *artus*, the lvs ave a bit wider and hairier; cliffs and basalt reefs in Clark Co, Ida　　　　　　　　　　　　　　　　　　var. **petrophilus** Cronq.

　a2 Achenes obviously pubescent

　　b1 Small pls, gen 1–6 dm, woody only at base

　　　c1 Disk corollas gen 6.5–8 mm (fresh or boiled); outermost invol bracts gen puberulent; pls 2–6 dm; chiefly Gr Pl, reaching the edge of our range in Mont　　　　　　var. **nauseosus**

　　　c2 Disk corollas gen 8–10 mm; invol gen glab throughout; pls 1–3 dm, on dry rocky ridges and cliffs at 4000–6500 ft in Blue and Wallowa mts of ne Ore and extreme se Wn　var. **nanus** Cronq.

　　b2 Larger, more woody pls, gen 4–20 dm

　　　d1 Invol bracts, at least the outer, ± tomentose-puberulent (sometimes only on the margins); tomentum of the lvs and twigs ave denser and more persistent than in the next 2 vars., inclined to be grayish or sometimes white; common and widespread (*C. speciosus; C. n.* ssp. *s.*)　　　var. **albicaulis** (Nutt.) Rydb.

　　　d2 Invol bracts wholly glab; tomentum of the lvs thin and ± deciduous, of the twigs close and gen light yellowish green

　　　　e1 Lvs very narrow, gen 0.5–1 mm wide; infl tending to become elongate; gen in alkaline meadows; Mont, across c Ida to e Ore and s (*C. consimilis, C. n.* ssp. *c.*)　var. **artus** (A. Nels.) Cronq.

　　　　e2 Lvs wider, gen 1–3 mm wide; infl gen not elongate; drier, better drained, scarcely or not at all alkaline soil; chiefly Gr Pl, but w occ to c Ida (*C. graveolens; C. n.* ssp. *g.*)　　　　　　　　　　　　　　var. **petrophilus** Cronq.

1b Twigs glab or minutely spreading-puberulent, not at all tomentose; widespread in des and dry foothills E Cas

3a Style short, included within the corolla, or merely the tips of the brs exserted; style appendages ca = stigmatic part; dwarf pls, only 1–2 (3) dm, with plane or only slightly twisted lvs 1–2.5 cm × 1–2 mm; Nev and ne Cal, n to the edge of our range in c Ore; Truckee green r. (*C. viscidiflorus* ssp. *h.*)　　　　　　　　　3 **C. humilis** Greene

3b Style longer, the stigmatic lines as well as the appendages exserted at maturity; style appendages < stigmatic part; larger pls, 2–12 dm, often with ± twisted lvs; green r.; 3 vars.　4 **C. viscidiflorus** (Hook.) Nutt.

　a1 Lvs and twigs ± densely puberulent, esp in the infl, the lvs (1) 2–6 mm wide and often 3–5-nerved (*C. l.*)　var. **lanceolatus** (Nutt.) Greene

　a2 Lvs and twigs glab, or the lvs often with scabrociliate margins

　　b1 Low shrubs, gen 2–4 dm, with linear-filiform lvs ca 1 mm wide or less; fls in each head gen 3 or 4, occ 5; GB var., perhaps not reaching our range (*C. s.*)　　var. **stenophyllus** (Gray) Hall

　　b2 Taller shrubs, (2) 4–12 dm, with lvs gen 1–4 mm wide, or occ wider; fls gen 5, occ only 4　　　　　　var. **viscidiflorus**

2a

2b

3b

Cichorium L.   Chicory

Fls all lig and ♂, blue (white), invol bracts biseriate, the outer shorter; recep naked; achenes glab, striate-nerved, sub-5-angled, or the outer slightly compressed; pappus of 2–3 series of scales, sometimes minute; brd, lfy-std herbs with milky juice, alt lvs, and several to ∝ ± showy heads. (Name latinized from the Arabic).

C. intybus L. Wild succory, blue-sailors. Glab to hirsute per 3–17 dm from a long taproot; lower lvs oblanceolate, petiolate, toothed or more often pinnatifid, 8–25 × 1–7 cm, the others progressively reduced, becoming sessile and often entire; heads to 4 cm wide in fl, 1–3 together in axils of much-reduced upper lvs, often on long racemiform brs; invol 9–15 mm; achenes 2–3 mm; pappus minute; cosmopolitan weed from Eurasia, common esp W Cas; July–Oct.

Cirsium Mill.   Thistle

Heads discoid, the fls all tubular and ♂, or pls sometimes dioecious by abortion; invol bracts in several series, imbricate or seldom subequal, gen some or all of them spine-tipped, ∝ spp. also with a thickened, glutinous dorsal ridge; recep flat to subconic, densely bristly; corollas purple or reddish to yellowish or white, with slender tube, expanded throat, and long narrow lobes; filaments gen papillose-hairy; anthers caudate at base; style with a thickened, minutely hairy ring and an abrupt change of texture below the brs, these connate almost throughout; stigmatic lines marginal, extending nearly or quite to the scarcely differentiated free tips; achenes glab, basifixed or nearly so, quadrangular or flattened, 4–∝ -nerved; pappus of plumose bristles (or that of the outermost fls merely barbellate), deciduous in a ring; ann, bien, or per spiny herbs with alt, toothed to more often pinnatifid lvs and 1–∝ smallish to large heads; fl summer. (Gr *kirsos*, a swollen vein, for which thistles, called *kirsion*, were a reputed remedy). Measurements of corollas given below are based on dry material.

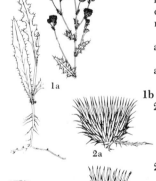

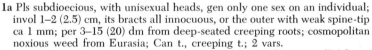

1a Pls subdioecious, with unisexual heads, gen only one sex on an individual; invol 1–2 (2.5) cm, its bracts all innocuous, or the outer with weak spine-tip ca 1 mm; per 3–15 (20) dm from deep-seated creeping roots; cosmopolitan noxious weed from Eurasia; Can t., creeping t.; 2 vars.

1 C. arvense (L.) Scop.

    a1 Lvs merely toothed or shallowly lobed, weakly spiny; not common with us   var. **arvense**

    a2 Lvs deeply pinnatifid, strongly spiny; our common phase

var. **horridum** Wimm. & Grab.

1b Pls with ♂ fls; invol and habit not combined as in no 1

  2a Lvs scab-hispid above; st spreading-hirsute to arachnoid, conspicuously winged by the spiny decurrent lf bases; invol bracts all spine-tipped, without a glutinous dorsal ridge; widespread bien weed from Eurasia; common t., bull t., spear t. (*C. lanceolatum*, misapplied)

2 C. vulgare (Savi) Tenore

  2b Lvs arachnoid-villous to floccose, tomentose, or glab above, not at all scab or hispid; lf bases not decurrent, or only shortly so

    3a Some or all invol bracts with a well developed, thickened, glutinous dorsal ridge; middle and outer invol bracts gen appressed except for the often abruptly spreading prickle; ± strongly spiny, not at all succulent pls with an open infl (heads solitary in small pls)

      4a Pubescence mostly relatively thin and loose, floccose-tomentose to even partly (esp on the st) arachnoid-villous, much of it sooner or later deciduous; bien or short-lived per, not creeping below ground; fls gen whitish; foothills and dry meadows, sw Mont and nw Wyo, across c Ida to ne and c Ore; gray-green t.

3 C. canovirens (Rydb.) Petr.

      4b Pubescence of the lower surfaces of the lvs and often also of the st closer and more persistent, not at all arachnoid-villous; taprooted per with a ± well developed tendency to spread by short-lived creeping roots

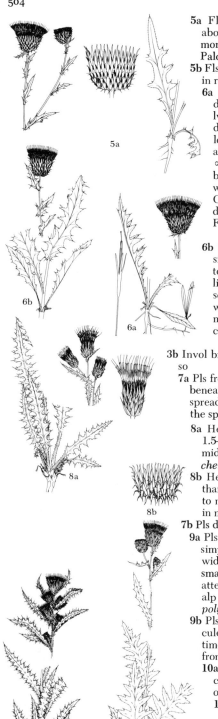

5a

6b

6a

8a

8b

9a

12a

5a Fls white or ochroleucous; lvs early green and subglabrate above; achenes gen 5–5.5 mm, stramineous or pale brown, with a more yellow apical band; grassl, se Wn, ne Ore, and adj Ida; Palouse t. *(C. palousense)*   4 **C. brevifolium** Nutt.

5b Fls lavender-purple to bright red-purple, sometimes very pale, or in rare individuals white

   6a Relatively slender pls 3–8 dm of swales and other poorly drained sites, vigorously spreading by creeping roots; rosette lvs subentire and merely spiny-margined to less often lobed, distinctly greenish above; cauline lvs gen deeply pinnatifid, the lobes typically < 7 mm wide; invol 2–2.5 (3) cm, not invaginated at base, its bracts appearing darker, narrower, and more ∝ than in no 6, the spine tip often < 3 mm; fls relatively bright and deep purple; achenes 3–5 mm, yellowish brown, with an apical yellow or ochroleucous band ca 0.5 mm wide; Gr Pl, and adj intermont valleys in Mont; biologically sharply distinct from no 6, but individual characters often overlapping; Flodman's t. *(C. canescens,* misapplied)

   5 **C. flodmanii** (Rydb.) Arthur

   6b Stouter pls 3–12 dm of dry hillsides and other well drained sites, weakly spreading by creeping roots; rosette lvs densely tomentose on both sides, strongly lobed to occ subentire; cauline lvs with lobes gen > 7 mm wide; invol 2.5–4 (5) cm, somewhat invaginated at base; middle and outer invol bracts with spine tip 3–5 mm; fls paler lavender-purple; achenes 5–7 mm, brown, without an apical pale band, or the band inconspicuous and < 0.5 mm; wavy-lvd t.

   6 **C. undulatum** (Nutt.) Spreng.

3b Invol bracts without any thickened, glutinous dorsal ridge, or nearly so

   7a Pls freely brd, strongly spiny, not at all succulent; lvs ± tomentose beneath; middle and outer invol bracts tending to have a loose or spreading upper portion which includes part of the body as well as the spine tip; open slopes in e Ore, c Ida, and s

      8a Heads relatively small and few-fld (gen 50–100 fls), the invol 1.5–2 (2.5) cm and scarcely broader; fls gen pale pink-purple; middle elev, mts of w Wyo to ne Ore; Jackson's Hole t. *(C. pulcherrimum,* misapplied)   7 **C. subniveum** Rydb.

      8b Heads larger and with more ∝ fls; invol 2–4 cm, gen broader than high (as pressed); fls white or pale pinkish; chiefly GB, but n to ne Ore and extreme se Wn on dry, open slopes at middle elev in mts; Utah t. *(C. wallowense)*   8 **C. utahense** Petr.

   7b Pls differing in 1 or more respects from the above

      9a Pls very strongly spiny, not at all succulent, 1.5–8 dm, strict and simple, crisp-arachnoid to subglab; lvs slender, gen < 4 cm wide; heads sessile in a gen elongate, lfy, spiciform infl, or in small pls in a subcapitate cluster; inner invol bracts slender, attenuate, often innocuous, the others tapering to an erect spine; alp and subalp; sw Mont, nw Wyo, and adj Ida; Tweedy's t. *(C. polyphyllum)*   9 **C. tweedyi** (Rydb.) Petr.

      9b Pls otherwise, from slightly to moderately spiny, sometimes succulent, short or often taller, variously simple to openly brd, sometimes with broader lvs, or fringe-tipped invol bracts, or both; gen from lowl to middle elev in mts

         10a Invol bracts relatively broad, the middle and outer gen lanceolate or ovate, ± strongly imbricate, slightly if at all hairy, often some of them with dilated, fringed tip

            11a All or nearly all of the invol bracts with dilated, fringed tip; sts not succulent, or slightly so in no 11, in any case tapering above and slender under the heads

               12a Lvs relatively broad, the larger ones gen 2–3 (4) times as long as wide (measured across the widespread lobes) and often > 5 cm wide; pls often with an openly brd infl and long, subnaked peduncles; W Cas, and e to Crook Co, Ore; mt t. *(C. centaureae);* ours the var. *oregonense* (Petr.) J. T. Howell   10 **C. callilepis** (Greene) Jeps.

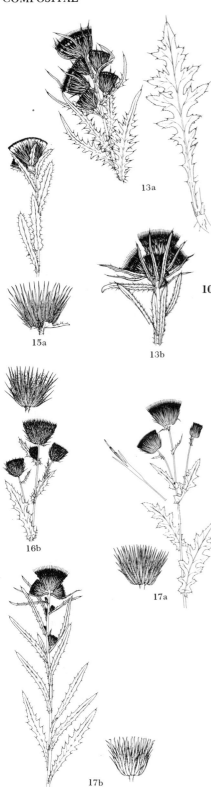

13a

15a

13b

16b

17a

17b

12b Lvs narrower, gen well > 4 times as long as wide and not > ca 5 cm wide; infl gen more congested (but not as dense as in no 13), the lateral heads often on short, ascending brs; Little Belt Mts and adj valleys, Mont; long-styled t.   11 **C. longistylum** Moore & Frankton

11b Only the inner (or none) of the bracts with dilated, fringed tip; heads pedunculate or sessile in a gen congested infl; sts ± succulent

   13a Pls gen 10–20 dm, the st tapering and becoming slender under the heads; infl congested, but the heads often short-pedunculate; corollas pink-purple, with the deepest sinuses gen 8–10 mm; wet alkaline meadows at n edge of SRP in Gem, Payette, and Blaine cos, Ida; showy t.

                     12 **C. magnificum** (A. Nels.) Petr.

   13b Pls gen < 10 (12) dm (often nearly acaulescent), the st gen thick and succulent throughout, seldom much tapering upwards; heads gen sessile or nearly so in a dense, terminal (occ elongate) cluster; corollas white to rather pale pink-purple, the deepest sinuses gen 5–8 mm; meadows and other moist (sometimes alkaline) places, foothills to well up in mts; widespread E Cas in w US; elk t. (*C. drummondii,* misapplied; *C. foliosum,* misapplied; *C. acaulescens; C. kelseyi*)   13 **C. scariosum** Nutt.

10b Invol bracts relatively narrow and gradually tapering, gen narrowly lance-linear, not much imbricate, gen ± strongly arachnoid-villous (least so in no 17), esp on the margins, rarely any of them (except sometimes a few in no 17) with dilated, fringed tip

   14a Fls purple (rare albino forms may be expected)

      15a Style < the corolla, or exserted only ca 1 mm; corolla tube 12–18 mm, 2.5–4 times as long as the short (4–6 mm) throat; corolla lobes gen 2–4 mm, subequal (proportions of parts of corolla and style unique among our spp.); anthers 3.5–4.5 mm; pls 6–30 dm; widespread in our range, in moist places up to middle elev in mts, but gen at lower elev than no 16, where the ranges overlap; short-styled t. (*C. edule,* misapplied)   14 **C. brevistylum** Cronq.

      15b Style exserted (2) 3–8 mm beyond the corolla lobes; corolla tube 7–11 mm, up to twice as long as the 4.5–13 mm throat; corolla lobes (3) 5–10 mm, ± unequal; anthers > 6 mm; pls 4–20 dm, chiefly W Cas

         16a Some of the outer invol bracts (as distinguished from the lvs closely subtending the head) with a few distinct, short, marginal spines; heads small, the corolla scarcely 20 mm, the throat ca 4.5–6 mm; WV and coast range of Ore; Hall's t.   15 **C. hallii** (Gray) Jones

         16b Invol bracts gen without marginal spines, though these may be present on the reduced lvs subtending the heads; heads ave larger, the corolla often > 20 mm, the throat (5) 6–13 mm; Cas to coast in Wn, to c BC and to Saddle Mt, Clatsop Co, Ore; Indian t., edible t. (*C. macounii*)   16 **C. edule** Nutt.

   14b Fls creamy white (rarely light pinkish?)

      17a Pls very weakly spiny, brd above and gen with open infl when well developed, 3–15 dm; lvs relatively large and thin in well developed pls, the larger cauline ones gen (3.5) 5–20 cm wide; invol only moderately hairy; lowl from e base of Cas to coast, Wn and s; weak t.

                  17 **C. remotifolium** (Hook.) DC.

      17b Pls moderately spiny, gen simple-std and with congested infl, 4–8 (15) dm; lvs narrower, the cauline to 4 cm wide; invol moderately to very strongly hairy; moist bottoms, open slopes, and fields; E Cas, BC and Alta to n Wn, n Ida, and nw Mont; Hooker's t., white t.

                  18 **C. hookerianum** Nutt.

Cnicus L.   Blessed-thistle

Heads discoid, the fls all tubular and ♂, yellow; invol of several series of broad, firm, spine-tipped bracts, the spines of the inner ones pinnatifid; recep flat, densely bristly; anthers shortly tailed at base; style with a ring of hairs and an abrupt change of texture at the base of the very short, scarcely divergent brs; achenes obliquely attached to the recep, terete, strongly ∝ -ribbed, glab, with firm, 10-toothed crown; pappus biseriate, the outer of 10 firm, smooth awns about = achene, alternating with as many much shorter, minutely hairy and sparsely pectinate inner ones; prickly ann thistle with alt lvs. (Ancient L name of the safflower, *Carthamus tinctorius*, from the Gr *knekos*).

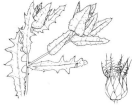

C. benedictus L. St brd, 1.5–8 dm, spreading-villous; lvs toothed or pinnatifid, to 15 × 5 cm, the lower petiolate, the middle and upper sessile, scarcely or not at all decurrent; heads terminating the brs, closely subtended by ovate or lance-ovate foliage lvs; invol 3–4 cm; achenes 8 mm; Mediterranean weed sparingly estab here.

Conyza Less.   Conyza

Heads disciform or minutely radiate, the ∝ ♀ fls with slender, tubular-filiform corolla that is in our spp. produced into a short, narrow, inconspicuous, whitish ray slightly if at all > pappus; invol bracts ± imbricate, scarcely herbaceous; recep flat or nearly so, naked; disk fls few, in ours not < ca 21, much like those of *Erigeron;* achenes 1–2-nerved, or nerveless; pappus of capillary bristles, sometimes with a short outer series; ann or per herbs with alt lvs and several – ∝ gen rather small heads, ours ann weeds to 1 m with ∝ narrow lvs to 10 × 1 cm. (Name used by Dioscorides and Pliny for some kind of fleabane, supposed to come from *konops*, a flea).

1a Invol glab, 3–4 mm; ♀ fls gen 25–40, with lig 0.5–1.0 mm, very slightly > pappus; heads ± ∝ in a gen elongate infl with a single main axis; widespread Am weed; horseweed, Can fleabane *(Erigeron c.);* ours is var. *glabrata* (Gray) Cronq., with sparsely hairy or subglab st
1 **C. canadensis** (L.) Cronq.
1b Invol copiously short-hairy, (3.5) 4–6 mm; ♀ fls gen 50–100 or more, with lig to 0.5 mm, = or < pappus; infl as in no 1, or often with 1–several elongate brs overtopping the main axis; tropical Am weed, occ with us near Portland; s Am c. *(Erigeron b., E. floribundus, C. f.)*
2 **C. bonariensis** (L.) Cronq.

Coreopsis L.   Tickseed; Coreopsis

Heads radiate, the rays conspicuous, broad, gen neutral, gen yellow, sometimes reddish-brown at base; invol bracts biseriate and dimorphic, all joined at base, the outer narrower, gen shorter, and more herbaceous than the inner; recep flat or slightly convex, its bracts thin and flat; disk fls tubular and ♂; anthers entire or sagittate at base; style brs flattened, with short or elongate, subtruncate to caudate, externally hairy appendage, gen without well marked stigmatic lines; achenes flattened parallel to the invol bracts, winged in most spp., not beaked; pappus of 2 smooth or upwardly barbed awns or short teeth, or a minute crown, or obsolete; ann to per herbs or subshrubs, with opp, entire to pinnatifid or ternate lvs. (Gr *koris*, bug, and *opsis*, resemblance, from the appearance of the achene).

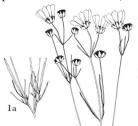

1a Achenes 2–3 mm, narrowly winged; glab ann or bien 3–12 dm; foliage rather scanty; lvs pinnatifid or bipinnatifid, with linear or linear-elliptic segms; heads ± ∝ on long, slender peduncles; rays 1–2 cm, orange-yellow with reddish-brown base; pappus teeth minute or obsolete; moist banks along major rivers, fl summer; Columbia c.   1 **C. atkinsoniana** Dougl.
1b Achenes wingless; otherwise much like no 1; Gr Pl sp., widely cult and occ escaped; calliopsis   **C. tinctoria** Nutt.

Cotula L.   Cotula

Heads disciform, the outer 1 or several series of fls ♀, with short tubular corolla or (ours) no corolla; invol bracts slightly unequal, ca 2-seriate, membranous or subherbaceous, gen with narrow scarious margins; recep flat to conic, naked; disk fls gen fertile, with 4-toothed yellow corolla; anthers entire or nearly so at base; style brs flattened, truncate, minutely penicillate; achenes, esp of the outer fls, ± stipitate (the stipe persisting on the recep), compressed parallel to the invol bracts, gen minutely striate, otherwise 2–4-nerved or nerveless (in our spp. the outer ones broadly winged, the others wingless or nearly so); pappus a short crown or (ours) none; ann or per herbs with alt, entire to gen pinnatifid or pinnately dissected lvs and small or medium-sized, ± pedunculate heads. (Gr *kotule*, a small cup).

1a Per, glab, ± succulent, 5–30 cm, often trailing; lvs 1–6 cm, with sheathing base, narrow, entire or with a few coarse teeth or spreading narrow segms; ♀ fls in one series; disk 5–11 mm wide; tidal flats, or occ other moist places, but seldom far inl; S Afr sp., intro along the coast from BC to Cal, and up the CR to Cas; brass buttons   1 C. **coronopifolia** L.

1b Ann, sparsely hairy, not succulent, 3–20 cm; lvs 2–4 cm, pinnatifid or bipinnatifid, not sheathing; ♀ fls in 2–3 series; disk 4–6 mm wide; Australian weed of disturbed sites, intro in trop Am and n occ to WV (Portland); Australian c.   C. **australis** (Sieb.) Hook.f.

Crepis L.   Hawksbeard

Fls all lig and ♂, yellow (sometimes reddish on the outer face); invol cylindric or camp, the principal bracts in 1 or 2 series, the reduced outer ones few to ∝; recep naked; achenes terete or subterete, fusiform or nearly columnar, often beaked, 10–20-ribbed; pappus of ∝ white or whitish capillary bristles, all > 4-celled in cross section at base; herbs with milky juice and well developed, entire to bipinnatifid basal lvs (cauline lvs alt, gen ± reduced, the uppermost bractlike) and few to ∝ small or large, few –∝ -fld heads in an open, corymbiform or paniculiform infl, our spp. all with a taproot or several strong roots, without rhizomes. (Name used by Pliny for some pl, from Gr *krepis*, a boot or sandal). Infraspecific taxa in *Crepis* are here treated consistently as sspp., to avoid the necessity for several new combinations in varietal status.

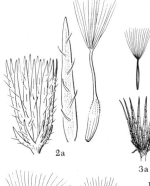

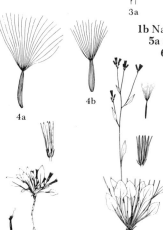

1a Intro ann or bien weeds of lawns and waste places, 1–11 dm; heads ± ∝ in well developed pls, with 20–∝ fls; fl May–Nov

2a Achenes with a distinct, slender beak 1–2.5 mm, 3.2–5 mm overall; st, lvs, and esp invol hispid-setose with yellowish bristles, not glandular; W Cas, esp upper WV; rough h.   1 C. **setosa** Hall f.

2b Achenes ± strongly narrowed upwards, but not beaked, 1.5–4.5 mm; pubescence gen finer and shorter, sometimes glandular

3a Inner invol bracts finely pubescent within; mature achenes dark purplish-brown, 2.5–4.5 mm; occ with us; ann h., rooftop h.
                                                      2 C. **tectorum** L.

3b Inner invol bracts glab within; mature achenes gen tawny or pale brown

4a Achenes 2.5–4 mm; invol 8–10 mm; recep finely ciliate; occ W Cas, French h.   3 C. **nicaeensis** Balb.

4b Achenes 1.5–2.5 mm; invol 5–8 mm; recep glab; common W Cas; smooth h.   4 C. **capillaris** (L.) Wallr.

1b Native per spp., not weedy

5a St and lvs glab or ± hispid, but not at all tomentose

6a Fls gen 6–12 in each head; pls glab and glaucous; fl July–Aug

7a Achenes beakless or with a very short beak; ribs of the achenes broad, rounded, smooth or slightly rugulose; lvs entire; dwarf alp and subalp pls of w N Am and n Asia; dwarf h.; 2 sspp.
                                                      5 C. **nana** Rich.

a1 Pls depressed, to 1 dm, the heads borne among the lvs, the taproot surmounted by a short, stout, erect caudex; the widespread typical phase   ssp. **nana**

a2 Pls taller, more br, to 2 dm, the heads borne well above the basal lvs, the caudex or its brs gen slender and elongate; less rigorous (but still subalp) habitats in s part of range of the sp.
                                                      ssp. **ramosa** Babc.

7b Achenes with a delicate beak ca 1/4 as long as the body; ribs of the achene narrow, scab-hirtellous at least above; lvs entire to coarsely dentate; larger pls 1–3 dm; taproot surmounted by a short, simple or slightly br caudex; riverbanks and sandbars, or occ on dry plains and bluffs, at lower elev than no 5; elegant h.   6 C. **elegans** Hook.

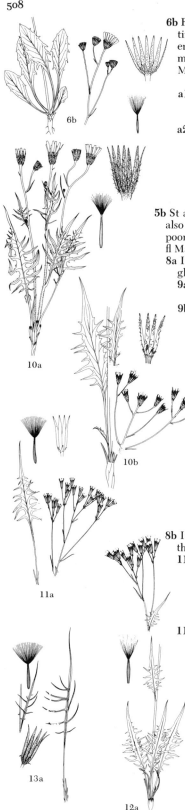

6b Fls gen 20–50 in each head; pls gen 2–7 dm, subscapose, glab (sometimes also glaucous) to ± setose or glandular-setose, esp on invol; lvs entire to runcinately toothed or pinnatifid; moist, often alkaline meadows in mts and foothills, widespread in w US, E Cas; fl May–July; dandelion h., meadow h.; we have 3 sspp.

  7 C. runcinata (James) T. & G.

  a1 Invol not at all glandular; gen in alkaline places; GB and Colo Plateau ssp., extending n occ to e Ida and w Mont *(C. g.)*

  ssp. **glauca** (Nutt.) Babc. & Stebb.

  a2 Invol with at least a few gland-tipped hairs

    b1 Basal lvs gen 0.5–3.5 cm wide, 4–8 times as long; heads (1) 3–7 (12); gen in nonalkaline meadows; Gr Pl ssp., extending into w Mont and occ to Ida ssp. **runcinata**

    b2 Basal lvs gen 3–8 cm wide, 2–4 times as long; heads (9) 12–25 (30); gen in alkaline meadows; the characteristic ssp. in most of our range ssp. **hispidulosa** (Howell) Babc. & Stebb.

5b St and lvs ± tomentose or puberulent, at least when young, sometimes also setose or glandular-hispidulous; polyploid-apomictic complex of poorly defined spp., in dry, open places up to middle elev in mts E Cas; fl May–July

  8a Invol, or lower part of st, or both, conspicuously setose, but not at all glandular

    9a Lvs deeply pinnatifid, with narrow, entire segms (see lead 13a)

  12 C. atrabarba Heller

    9b Lvs otherwise, either less deeply cleft, or with broader, again toothed or cleft segms

      10a Pls 1–3 dm, with 1–9 heads bearing 10–60 fls each; inner invol bracts gen ca 13 (10–15); achenes greenish-black to deep reddish-brown; sexual and apomictic forms; low h., Siskiyou h.; 2 sspp.

  8 C. modocensis Greene

        a1 Setae of st and petioles stiff, yellowish, those of invol blackish, all straight or slightly curved; achenes merely narrowed to the summit; Mont to ne Ore and s, ours all polyploid *(C. scopulorum)* ssp. **modocensis**

        a2 Setae all or nearly all whitish, elongate, and conspicuously curled or crisped, those on the invol gen very dense; achenes gen with a short, stout beak; Wn and s BC, diploid and polyploid *(C. r.)* ssp. **rostrata** (Coville) Babc. & Stebb.

      10b Pls gen 3–8 dm, with 6–70 heads bearing 8–25 fls each; inner invol bracts gen ca 8 (6–10); achenes olive or yellowish, merely narrowed to the summit; a series of polyploid apomicts derived by hybridization among nos 8, 10, and 12; Ore and Wn to n Ida; bearded h. **9 C. barbigera** Leib.

  8b Invol and st sparingly or not at all setose, or, if evidently setose, then the setae gland-tipped

    11a Invol gen glab or nearly so, 9–12 (15) mm, with 5–6 (8) inner bracts; heads gen 20–100 or more in well developed pls, with only 5–6 (10) fls, the corolla gen 10–15 (18) mm; lvs pinnately lobed, gen with broad central rachis and narrow, entire or sometimes toothed or cleft segms; widespread in w US, ours chiefly sexual; long-lvd h., tapertip h.; ours is ssp. *acuminata* **10 C. acuminata** Nutt.

    11b Invol gen tomentulose, often also setose; heads gen larger, or with more ∝ fls or invol bracts; lvs various

      12a Fls gen 7–12 (16) in each head; inner invol bracts 7–8 (12); heads gen 10–60; a series of polyploid apomicts derived by hybridization of no 10 with no 13 and sometimes nos 8 or 12; gray h. *(C. acuminata var. i.)* **11 C. intermedia** Gray

      12b Fls gen 10–40 in each head; inner invol bracts 8–14; heads gen 2–30 (40)

        13a Lf segms linear or narrowly lanceolate, gen entire, or the lvs rarely all narrow and entire; achenes gen greenish; slender h.; 2 sspp. **12 C. atrabarba** Heller

          a1 Smaller pls, 1.5–3.5 dm, with 3–18 heads, and with some black glandless setae on the invol bracts; polyploid and gen apomictic, tending to occur at slightly higher elev than the next ssp. *(C. exilis)* ssp. **atrabarba**

          a2 Larger pls, 3–7 dm, with 10–30 (40) heads, the invol nearly

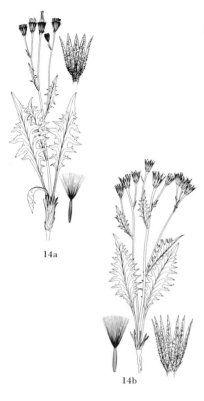

14a

14b

or quite devoid of setae; diploid and polyploid, often sexual
ssp. **originalis** Babc. & Stebb.
13b Lf segms broader, gen lanceolate or deltoid, some of them
gen toothed; achenes gen yellowish or brownish
14a Herbage gray-tomentulose; peduncles not markedly expanded upwards; western h.; widespread sp., with 4 poorly
marked sspp. 13 **C. occidentalis** Nutt.
a1 Invol with at least some gland-tipped hairs
b1 Largest heads with 10–13 inner bracts and 18–30 fls
ssp. **occidentalis**
b2 Largest heads with 8–9 inner bracts and 12–14 fls
ssp. **costata** (Gray) Babc. & Stebb.
a2 Invol completely devoid of glandular pubescence
c1 Pls low, gen 0.5–2 dm, br below and with
long-pedunculate heads; longest outer bracts gen >
half as long as inner; lvs deeply pinnatifid, with remote,
entire or coarsely few-toothed lobes
ssp. **conjuncta** Babc. & Stebb.
c2 Pls taller, gen 1–4 dm, with a well defined primary
axis; longest outer bracts gen < half as long as inner;
lvs gen pinnatifid, but with more closely set and strongly
toothed or pinnatifid lobes
ssp. **pumila** (Rydb.) Babc. & Stebb.
14b Herbage green, not markedly tomentulose; peduncles expanded near the tip; ours polyploid and often apomictic;
Baker's h.; we have 2 sspp. 14 **C. bakeri** Greene
a1 Outer invol bracts deltoid, the longest < half as long as
inner; invol relatively narrow; pappus > achene; lvs shallowly pinnatifid; Nez Perce Co, Ida
ssp. **idahoensis** Babc. & Stebb.
a2 Outer invol bracts lanceolate, the longest half or > half
as long as inner; invol relatively broad; pappus = achene;
lvs deeply pinnatifid; e slope of Cas, c Wn to Cal
ssp. **bakeri**

Crocidium Hook.   Spring-gold; Crocidium

Heads radiate, the rays ♀ and often fertile, yellow; invol a single series of rather broad, herbaceous,
= bracts; recep strongly conic, naked; disk fls ♂ and fertile, yellow; anthers entire or nearly so at
base; style brs flattened, with marginal stigmatic lines and a well developed, flat, deltoid, externally minutely papillate-hairy appendage; achenes covered with thick, papillalike hairs, becoming
mucilaginous when wet; pappus of ± ∝, very fragile, deciduous white bristles, sometimes
wanting from the ray fls; delicate ann with small, alt and basal, entire or few-toothed lvs and
rather small, long-pedunculate heads. (Diminutive derived from Gr *kroke*, loose thread or wool,
referring to the persistent axillary tomentum).

C. multicaule Hook. Gen several-std, to 1.5 (3) dm, with loose tufts of axillary
wool, otherwise ± glab; lvs slightly fleshy, the basal oblanceolate or broader,
to 2.5 × 1 cm, the cauline few and reduced; rays (5–) 8 (–13), 4–10 mm, individually subtended by the invol bracts, these 3–7 mm; disk to 1 cm wide;
sand plains, cliff ledges, and other dry, open places at low elev, VI to Cal,
both sides of Cas; Mar–May (*C. pugetense*).

Crupina Cass.   Crupina

Heads discoid, the marginal fls few and neutral, the others ♂ and fertile; invol bracts imbricate in
several series, wholly chaffy, tapering to a firm but scarcely spiny point; recep flat, beset with ∝
long, slender, thin, chaffy bracts; corollas anthocyanic, with slender tube and long, narrow lobes;
anthers sagittate and only shortly caudate-acuminate; style with a thickened, hairy ring and an
abrupt change of texture at base of the appressed, connate brs; stigmatic lines marginal, extending
nearly to tip; achenes ± obliquely attached to recep, thickened, hairy, obscurely nerved; pappus
members ∝, in several series, outer of short, blunt, narrow scales, inner of firm, slender, elongate
bristles; ann herbs with alt, gen pinnatisect lvs and few–several heads in an open-corymbiform,
subnaked infl.

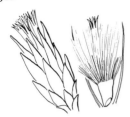

C. vulgaris Cass. Slender, 3–12 dm; lowest lvs ± entire, oblanceolate to rotund, 1–3 cm, often deciduous, the others to 7 cm, pinnatifid or bipinnatifid, with remote segms, margins glandular-scab or stipitate-glandular, the upper reduced; heads naked-pedunculate; invol ca 1.5–2 cm; achenes 4–5 mm; inner pappus 7–10 mm; field and roadside weed of e Medit Region, recently intro into w Ida (Grangeville).

Dimeresia Gray   Dimeresia

Heads discoid, 2–3-fld, the fls tubular and ♂; invol of 2–3 herbaceous bracts, these united at base, each broadly rounded on the back and loosely embracing a fl; anthers sagittate, but scarcely caudate; style brs flattened, slightly broader upwards, papillate-puberulent over the outer surface, the stigmatic lines ventromarginal, extending completely around the rounded apex; pappus of ca 20 coarse, sparsely long-plumose bristles, united at base and deciduous in a ring; achenes glab, striate; compact little ann with entire lvs. (Gr *dimeres*, of two parts).

D. howellii Gray. Nearly acaulescent, cushionlike, somewhat arachnoid at base, more glandular upwards; lvs spatulate to elliptic or ovate, to 3 × 1 cm (petiolar base included), aggregated around the 1–several compact clusters of heads; invol 4–6 mm; fls whitish or purplish; foothills, high des, and dry mts, ne Baker Co, Ore, to ne Cal and nw Nev.

Dyssodia Cav.   Dyssodia

Heads radiate or rarely discoid, the rays ♀ and fertile, yellow or orange; invol bracts conspicuously gland-dotted, arranged in 1 or 2 series, the inner joined at base or nearly throughout; recep flat or nearly so, naked or with poorly developed bristles; disk fls ∝, tubular and ♂; anthers entire or nearly so at base; style brs flattened, with introrsely marginal stigmatic lines, truncate or with elongate hairy appendages; achenes narrow, ± striate; pappus of ca 10–20 scales, each divided to the middle or beyond into several or ∝ bristles; herbs with gland-dotted, gen opp, entire to pinnately dissected lvs. (Name of doubtful origin).

D. papposa (Vent.) Hitchc. Fetid marigold, false dogfennel. Ill-smelling ann, gen much br, 0.5–4 dm; lvs 2–5 cm, pinnatifid or bipinnatifid, with narrow segms; heads ∝, disk 4–10 mm wide; invol 6–8 mm; outer bracts linear, 1/2–3/4 as long as inner, these broader and joined at base; rays few, erect, hardly > 1 mm; pappus scales deeply divided into 5–10 bristles; chiefly Gr Pl, reaching our range in w Mont; fl July–Sept.

Eatonella Gray   Eatonella

Heads discoid or inconspicuously radiate, the rays when present ♀ and fertile; invol of 5–13 ± uniseriate bracts, reflexed in age; recep flattish, naked; disk fls ♂ and fertile, yellow; anthers sagittate at base; style brs flattened, with introrsely marginal stigmatic lines and very short and blunt, externally minutely papillate-hairy appendages; achenes flattened parallel to invol bracts, nerveless except for the callous-thickened and densely long-villous-ciliate margins; pappus of 2–4 scales; dwarf, tomentose ann with simple, gen alt (or all basal) lvs and small heads. (For Daniel Cady Eaton, Am botanist, 1834–1885, who first described our sp.).

E. nivea (Eat.) Gray. White e. Depressed, white-woolly, brd at base; lvs ∝, linear-oblanceolate to spatulate, to 1.5 cm; heads sessile or on filiform axillary peduncles to 4 cm; invol 5 mm; rays scarcely > disk, yellow or purplish; pappus of 2 erose, shortly awn-tipped scales; GB des, disjunct along Salmon R from Clayton to Salmon City, Ida, and in c Wn near CR; fl May–July.

Echinacea Moench   Coneflower

Heads radiate, the rays neutral or occ ♀ and fertile, large, purple to nearly white or sometimes yellow, ± drooping; invol bracts in 2–4 subequal or slightly imbricate series, with firm base and spreading or reflexed green tip; recep conic, its bracts clasping the achenes, firm, with stout, spinescent tip conspicuously > disk corollas; disk corollas slightly bulbous-thickened at base; anthers slightly sagittate; style brs flattened, without well marked stigmatic lines, the slender, acuminate appendages hispidulous esp externally; achenes quadrangular, glab or sparsely pubescent on the angles; pappus a short, toothed crown; per herbs with simple, alt, entire or toothed, gen trinerved lvs and solitary or few, gen long-pedunculate heads. (Gr *echinos*, hedgehog or sea urchin, from the spinescent receptacular bracts).

E. pallida Nutt. Pale purple c. Taprooted, 1–6 dm, ± spreading-hirsute; lvs linear to lanceolate or lance-elliptic, blade to 15 × 2.5 cm, the basal long-petiolate; heads solitary; disk 1.5–3 cm wide; rays 2–4 cm, purple to nearly white; dry, open prairies and plains, c and e US, entering our range in Mont; ours is the Gr Pl phase, var. *angustifolia* (DC.) Cronq.; fl June–Aug (*E. a.*).

Echinops L.   Globe-thistle

Heads 1-fld, ∝, closely aggregated into globose 2ndary heads, discoid, the fls all tubular and ♂; 2ndary heads often with a common invol of a few narrow, reflexed bracts; invol proper gen subtended by a tuft of capillary bristles, its bracts imbricate in several series, firm, often awn-tipped; corolla with slender tube and long, narrow lobes, blue or purple to white; anthers tailed at base; filaments glab; style with a ring of hairs and an abrupt change of texture beneath the short brs; achenes elongate, quadrangular or subterete, gen hairy; pappus of ∝ short setae or narrow scales, free or ± united and forming a crown; spiny per herbs, gen ± white-tomentose, with alt, toothed or pinnatifid lvs. (Gr *echinos*, hedgehog or sea urchin, and *opsis*, appearance, from the spiny herbage and heads).

E. ruthenicus M. Bieb. Coarse, brd thistle to 1.5 m; st arachnoid-tomentose, nearly or quite eglandular; lvs to 40 × 20 cm, subbipinnatifid, densely white-arachnoid-tomentose beneath, green and with scattered viscid hairs above; 2ndary heads naked-pedunculate, 3–4 cm; invol and fls light blue-lavender; pappus a short crown; European weed, occ intro in se Wn (*E. sphaerocephalus*, misapplied, the true *E. s.* glandular as well as tomentose).

Enceliopsis (Gray) A. Nels.   Enceliopsis

Heads solitary, gen radiate, the rays large, yellow, neutral; invol bracts subherbaceous, ± imbricate; recep flat or convex, chaffy, its bracts soft and scarious, clasping the achenes; disk fls ∝, fertile, yellow; anthers minutely sagittate; style brs flattened, with short, externally hairy appendages, without well marked stigmatic lines; achenes gen densely villous, strongly flattened at right angles to the invol bracts, thin-edged; pappus of 2 short, subulate awns and some minute, often confluent scales, or none; ± scapose per herbs with simple lvs. (Name from the resemblance to *Encelia*).

E. nudicaulis (Gray) A. Nels. Naked-std sunray. Caespitose from a taproot and br caudex, 1–5 dm, densely tomentose-canescent; lvs basal, bl broadly rounded, 1–6 cm, = or < petiole; disk 2–4 cm; rays gen ca 21, 2.5–4 cm; pappus inconspicuous; GB des, and disjunct along Salmon and Lemhi rivers from Clayton to Salmon City, Ida.

Erechtites Raf.   Burnweed; Fireweed

Heads disciform, dull yellow or whitish; invol a single series of narrow, =, ± herbaceous bracts, sometimes with a few bracteoles at base; recep flat, naked; outer fls ♀, filiform-tubular, elig, in 2–several series; inner fls ♂ but sometimes sterile, the corolla narrowly tubular, 4–5-toothed; anthers entire or slightly sagittate; style brs flattened, minutely penicillate about the subtruncate or very shortly appendiculate tip; achenes 5-angled or 10–20-nerved; pappus of ∝ capillary bristles; herbs with alt, entire to pinnately dissected lvs and cylindric to ovoid heads, ours all ann weeds to 2 or 2.5 m, with ∝ lvs well distributed along the st, fl late summer and fall. (Name given by Dioscorides to a pl perhaps related to this).

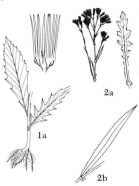

1a Invol 10–17 mm, swollen at base in life; pl glab or ± spreading-hairy; lvs to 20 × 8 cm, sharply serrate and sometimes irreg lobed, lower oblanceolate to obovate, often ± petiolate, others sessile and often aur; widespread Am weed, occ intro W Cas; eastern b. or f.; ours is var. *h.*

1 **E. hieracifolia** (L.) Raf.

1b Invol 5–7 mm, nearly cylindric; Australian spp., occ intro along our Pac coast

  2a Lvs lobed or pinnatifid, often also toothed, to 15 × 4 cm, lower gen ± petiolate, upper scarcely so and often aur; herbage thinly and ± deciduously villous-tomentulose; cut-lvd coast f. or b.　2 **E. arguta** DC.

  2b Lvs finely and sharply dentate, not at all lobed or pinnatifid, evidently aur; ave less hairy than no 2, and often with larger infl; toothed coast f. or b. (*E. prenanthoides,* misapplied)　3 **E. minima** (Poir.) DC.

Erigeron L.　　Daisy; Fleabane; Erigeron

Heads gen radiate, the few to gen ± ∝ ♀ fls gen bearing evident and often narrow rays, these gen anthocyanic to white, or in a few spp. yellow; in a few spp. the ♀ fls rayless and the heads thus disciform, in a few others the ♀ fls wanting and the heads thus discoid; invol bracts narrow, varying from herbaceous and = to scarcely herbaceous and evidently imbricate, the loss of herbaceousness either uniform throughout their length, or more prominent toward the tip; recep flat or nearly so, naked; disk fls ± ∝, yellow; some spp. with rayless ♀ fls between the disk and ray fls; anthers entire or nearly so at base; style brs flattened, with introrsely marginal stigmatic lines and short (to 0.5 mm), externally minutely hairy, lanceolate and acute to more often broadly triangular and obtuse appendages, or the appendage obsolete *(E. annuus);* achenes 2–∝-nerved; pappus of capillary and often fragile bristles, with or without a short outer series of minute bristles or scales; herbs with alt (or all basal) lvs and 1–∝ hemispheric to turbinate heads. (Gr *eri*, early, and *geron*, old man, prob referring to the early fl and fr of most spp.).

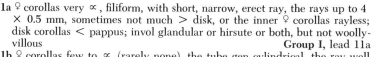

1a ♀ corollas very ∝, filiform, with short, narrow, erect ray, the rays up to 4 × 0.5 mm, sometimes not much > disk, or the inner ♀ corollas rayless; disk corollas < pappus; invol glandular or hirsute or both, but not woolly-villous　**Group I**, lead 11a

1b ♀ corollas few to ∝ (rarely none), the tube gen cylindrical, the ray well developed and spreading, or occ reduced or absent, but not short, narrow, and erect except in a few high-northern spp. with woolly-villous invol; disk corollas in most spp. > pappus

  2a Pappus of the ray and disk fls conspicuously unlike, that of the disk fls composed of long bristles and short outer setae, that of the ray fls lacking the long bristles; weedy ann (bien) 3–15 dm; rays 50–125, white; disk corollas 1.5–2.8 mm; fl all summer　**Group II**, lead 12a

  2b Pappus of the ray and disk fls alike, of long bristles, sometimes also with short outer setae or scales; ann or bien to more often per, most spp. not weedy; rays various; disk corollas often > 3 mm

    3a Internodes very ∝ and short; lvs linear or narrowly oblong and essentially uniform from base to near top of pl, without any enlarged basal cluster; invol bracts clearly and reg imbricate; fl summer

Group III, lead 13a

    3b Internodes neither excessively ∝ nor gen very short; lvs variously shaped, sometimes linear, but the basal ones gen obviously larger than the cauline; invol bracts = to ± imbricate; most spp. fl spring or early summer, unless at high elev where there is only one blooming season

      4a Pls fibrous-rooted from a rhizome or simple to br caudex, well developed individuals tall and ± erect (often > 3 dm), somewhat Asterlike, the cauline lvs gen ample, lanceolate or broader; achenes 2–7-nerved (*E. glaucus* and *E. peregrinus* are keyed only here, although small forms, with reduced cauline lvs, would be sought in the opposing group except for the 4 or more nerves or angles on the achenes)　**Group IV**, lead 14a

      4b Pls otherwise, either ± distinctly taprooted, or smaller (< 3 dm), or with narrower cauline lvs, often differing in 2 or all 3 of these respects; achenes 2-nerved

        5a Lvs, or some of them, ± lobed, divided, parted, or coarsely toothed; st < 3 dm　**Group V**, lead 23a

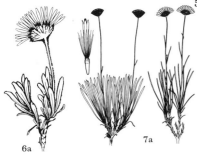

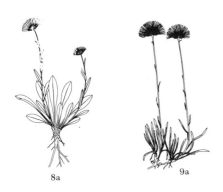

6a

7a

5b Lvs entire, or rarely slightly toothed, occ some of them trilobed at the tip in *E. lanatus*
  6a Invol woolly-villous with multicellular hairs (sparsely so in *E. flettii)*; monocephalous alp and subalp pls, < 3 dm
        Group VI, lead 28a
  6b Invol variously pubescent or glandular to glab, but not woolly-villous; habit and habitat various
    7a Rays yellow, or < the disk and inconspicuous, or lacking; st gen < 3 dm   **Group VII, lead 33a**
    7b Rays evident, blue or purple to pink or white
      8a Cauline lvs relatively well developed, lanceolate to ovate or ovate-oblong; invol bracts subequal; pls 5–25 cm from a stout, br caudex which sometimes surmounts a taproot; rays 20–50, 6–12 × 1–2 mm; disk corollas 3–4.5 mm; pappus bristles 12–20; Cas and w **Group VIII, lead 39a**
      8b Cauline lvs gen not very well developed, gen linear or oblanceolate, sometimes narrowly lance-oblong, those of some spp., while narrow, not much smaller than the lowermost ones (lvs sometimes broader in *E. caespitosus*, an E Cas sp. with evidently imbricate invol); some spp. > 3 dm
        9a Pls with ∝ fibrous roots from a simple to rhizomatously br caudex, without a taproot or stout central underground axis   **Group IX, lead 40a**
        9b Pls with an evident taproot or stout central underground axis, without ∝ fibrous roots
          10a Pubescence of the st appressed or ascending, or wanting   **Group X, lead 42a**
          10b Pubescence of the st widely spreading
                Group XI, lead 54a

8a

9a

Group I

11a

11b

11a Rayless ♀ fls present between rays and disk fls; infl corymbiform, the peduncles arcuate or obliquely ascending, or the head solitary; invol bracts acuminate to long-attenuate; circumpolar sp., s in mts to Cal and Colo; bitter f., northern d.; 3 vars. in N Am **1 E. acris** L.
  a1 Peduncles and invols nearly or quite glandless; gen in swampy places, mainly n of our range *(E. e.)*   var. **elatus** (Hook.) Cronq.
  a2 Peduncles and invols ± glandular; more often in rocky places
    b1 Pls relatively tall, gen at least 3 dm, with several or ∝ heads; circumpolar, s to n Ore and n Utah *(E. a., E. droebachensis, E. elongatus)*
          var. **asteroides** (Andrz.) DC.
    b2 Pls low, rarely 3 dm, with few or solitary heads; cordilleran *(E. d.)*
          var. **debilis** Gray
11b Rayless ♀ fls wanting; infl racemiform, the peduncles erect or nearly so, or the head solitary; invol bracts acute or acuminate, gen purple-tipped; meadows and other moist places, gen in mts; cordilleran sp., widespread E Cas; spear-lf f., short-rayed d. **2 E. lonchophyllus** Hook.

Group II

12a

12b

12a Foliage ample, the cauline lvs broadly lanceolate or gen broader, gen toothed; pls gen 6–15 dm; st long-spreading-hairy below the large and gen lfy infl; invol finely glandular and sparsely long-hairy; widespread Am weed of moist ground and waste places; ann f. **3 E. annuus** (L.) Pers.
12b Foliage sparse, the cauline lvs linear to lanceolate and gen entire; pls gen 3–7 dm; st and invol often with shorter pubescence, that of the st often more appressed; infl scarcely lfy, of few–∝ heads; widespread Am weed, often in drier places than no 3; daisy f., branching d.; we have 2 vars. *(E. ramosus)* **4 E. strigosus** Muhl.
  a1 St with the hairs short and appressed, except near the base; hairs of the invol < 1 mm, not flattened; common *(E. ramosus)* var. strigosus Muhl.
  a2 St with the hairs longer and more spreading, much as in no 3; hairs of the invol obviously flattened, gen > 1 mm; pls ave a bit more robust and lfy than var. *strigosus*; phase intermediate toward no 3, only occ with us and in our area perhaps reflecting recent hybridization
          var. **septentrionalis** (Fern. & Wieg.) Fern.

Group III

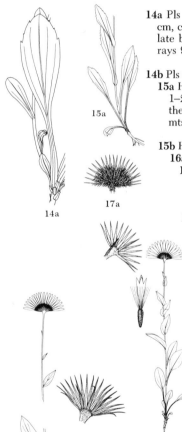

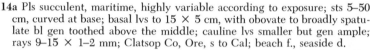

13a Heads with 15–65 rays, these 5–15 mm, gen blue; lvs 1–8 cm × 0.5–4 mm; dry, often rocky places, Cas and w, Cal n to c Ore (Mt Jefferson); lfy e.; we have 2 vars.   **5 E. foliosus** Nutt.
 a1 Pls 2.5–5 dm; herbage subglab or sparsely strigose; heads several *(E. h.)* var. **hartwegii** (Greene) Jeps.
 a2 Pls 1–2.5 dm; herbage often strigose-hirsute or even spreading-hirsute; lvs more crowded; heads few or solitary *(E. c.)*
                                                       var. **confinis** (Howell) Jeps.
13b Heads strictly discoid, without ♀ fls; lvs to 5 cm × 2–6 mm; dry, often rocky places in foothills, valleys, and plains; se and sc Wn through e Ore to Cal and w Nev; Cal rayless d.; ours is var. *inornatus*   **6 E. inornatus** Gray

Group IV

14a Pls succulent, maritime, highly variable according to exposure; sts 5–50 cm, curved at base; basal lvs to 15 × 5 cm, with obovate to broadly spatulate bl gen toothed above the middle; cauline lvs smaller but gen ample; rays 9–15 × 1–2 mm; Clatsop Co, Ore, s to Cal; beach f., seaside d.
                                                            7 E. glaucus Ker
14b Pls neither succulent nor maritime
 15a Hairs of invol with black crosswalls near the base; rays white, 9–24 × 1–2 mm; achenes 2-nerved; st spreading-hirsute at least above; at least the lower lvs gen toothed; meadows and streambanks, gen well up in mts, n Ida and n Ore to Cal and NM; Coulter's d. or f.
                                                    8 E. coulteri Porter
 15b Hairs of invol, when present, without black crosswalls
  16a Rays gen 2–4 mm wide; pappus gen simple
   17a Achenes 2–(occ) 4-nerved; lvs hirsute; invol hirsute on the lower 1/4–3/4, glandular thence to the tip; disk corollas gen 3–4 mm; rays 10–15 mm, white to pink-purple; mts W Cas; Alice f., Eastwood's d.
                                                9 E. aliceae Howell
   17b Achenes 4–7-nerved; lvs glab to occ hairy; invol glandular to less often hirsute, not hirsute below and glandular above; disk corollas gen 4–6 mm; rays (8) 10–25 mm
    18a Basal lf bls abruptly contracted to the petiole; cauline lvs thin, conspicuously clasping; rays white; disk corollas more flaring than in no 11; moist, often rocky places, s side of CRG; Howell's d.   **10 E. howellii** Gray
    18b Basal lf bls gen tapering to the petiole; cauline lvs various, but rarely at once thin and strongly clasping; rays colored or sometimes white; widespread cordilleran sp., at higher elev (or more boreal) than no 10; subalp d., wandering d. or f.; we have 2 sspp., each with 3 vars.   **11 E. peregrinus** (Pursh) Greene
     a1 Invol bracts villous on the back, or sometimes merely ciliate on the margins and glutinous on the back, not at all glandular; rays gen rather pale or even white; herbage often soft-pubescent, the st gen sparsely villous, the peduncular hairs rather loose; lvs often toothed; coastal and Cas mts, c Wn n
                                                    ssp. **peregrinus**
      b1 Invol bracts hairy on the back, often also ± long-ciliate on the margins; rays more often colored; foliage ample to rather scanty
       c1 Upper cauline lvs reduced and distant; n Wn to Unalaska
                                           var. **dawsonii** Greene
       c2 Upper cauline lvs either ample or closely set; chiefly from Alas panhandle n, but occ in BC   var. **peregrinus**
      b2 Invol bracts merely glutinous on the back, ciliate on the margins; rays gen white; foliage scanty; bogs on s fringe of OM *(E. t.)*   var. **thompsonii** (Blake) Cronq.
     a2 Invol bracts densely glandular on the back, rarely with a few long hairs as well; rays gen rich rose-purple or darker; herbage gen glab except for the closely villosulous peduncles; lvs gen entire; Cas and e *(E. salsuginosus,* misapplied)
                                           ssp. **callianthemus** (Greene) Cronq.
      d1 Reduced alp pls, < 2 dm, with relatively ample, apically

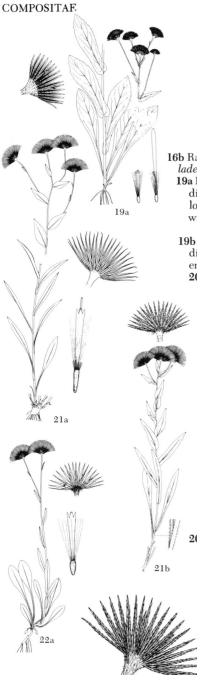

19a

21a

21b

22a

22b

24a

rounded or obtuse basal lvs, and very much smaller cauline lvs, often subscapose; widespread

var. **scaposus** (T. & G.) Cronq.

d2 Larger, gen subalp pls, to 7 dm (smaller in alp phases of var. *angustifolius,* but then with narrow, acute basal lvs)

e1 Lvs narrow, the basal oblanceolate or narrower, the cauline linear or lanceolate; Cas and w, s BC to Cal, uncommon in our range   var. **angustifolius** (Gray) Cronq.

e2 Lvs more ample, the basal oblanceolate or broader, the cauline gen ovate and not greatly reduced; common and widespread   var. **eucallianthemus** Cronq.

16b Rays gen ca 1 mm wide or less; pappus gen double except in *E. philadelphicus*

19a Rays very ∝ , gen 150-400, and very narrow, 5–10 × 0.2—0.6 mm; disk corollas gen 2.5–3.2 mm; bien or short-lived per, long-spreading-hairy to occ subglab; lvs often toothed or lobed; widespread Am weed of moist places; Phila. d. or f.

12 **E. philadelphicus** L.

19b Rays fewer, gen 65–150 (to 175 in *E. glabellus*), 8–18 × ca 1 mm; disk corollas gen 3.5–5.5 mm; true per, except *E. glabellus;* lvs gen entire, except often in *E. glabellus;* not weedy

20a Pls rather equably lfy, the upper lvs gradually reduced, the middle lvs gen as large as or > the often deciduous lowermost ones

21a Lvs glab or nearly so except for the ciliate margins; st glab below the infl or merely with a few scattered hairs; invol with few or no long hairs; widespread cordilleran sp. of open woods and clearings in foothills and at middle elev in mts; showy f.; 2 vars.   13 **E. speciosus** (Lindl.) DC.

a1 Lvs relatively narrow, the uppermost lanceolate; pls relatively a little more hairy, the st often sparsely hairy above, the uppermost lvs tending to be strongly ciliate all around and often with a few hairs on the surface, the invol gen with a few long hairs; Wn, Ore, n Ida, and w Mont var. **speciosus**

a2 Lvs relatively wider, the uppermost ovate; pls less hairy, the st gen glab except directly under the heads, the lvs smooth on the surface and only rarely strongly ciliate all around, the invol without hairs; chiefly more s and e, but extending to c Ida and sw Mont *(E. m.)*   var. **macranthus** (Nutt.) Cronq.

21b Lvs, st, and invol ± long-hairy; widespread cordilleran sp., gen in slightly drier habitats than no 13, but intergrading with it in Wn; three-veined f.; ours is var. *conspicuus* (Rydb.) Cronq. *(E. c.)*   14 **E. subtrinervis** Rydb.

20b Pls rather inequably lfy, the uppermost lvs strongly reduced, the middle ones gen smaller than the gen persistent lowermost ones

22a St and invol glandular or viscid, sometimes also hairy; st gen curved at base; per; s RM sp. of meadows and open ground in mts, often at high elev, n irreg to sw Mont; beautiful d. *(E. viscidus)*   15 **E. formosissimus** Greene

22b St and invol ± hairy, scarcely glandular or viscid; st gen erect; bien or per; GP and RM sp. of meadows and moist open ground, reaching our range in Mont and n and ec Ida; smooth d.; 2 vars.   16 **E. glabellus** Nutt.

a1 Pubescence of the st appressed; chiefly s var., the common phase with us   var. **glabellus**

a2 Pubescence of the st spreading; chiefly n var., s occ to Helena, Mont   var. **pubescens** Hook.

Group V

23a St lfy, not scapose or subscapose; herbage evidently hairy; Ore and Wn, fl May–Oct

24a Pappus bristles characteristically curled and twisted for at least the upper half; sts lax, simple; basal lvs tufted, spatulate to obovate, coarsely toothed or incised, to 9 × 2.5 cm; cauline lvs broadly lanceolate to elliptic or ovate, entire, to 4 × 1 cm; CRG; gorge d.   17 **E. oreganus** Gray

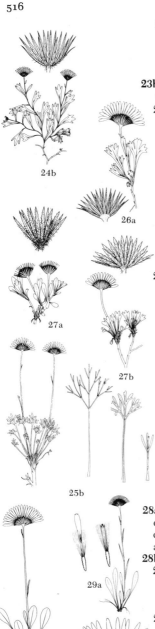

24b Pappus bristles straight or nearly so; sts sprawling or pendent, br, lfy esp distally, but without a basal cluster; lvs cuneate to obovate, to 4 × 1.5 cm, ± deeply and often irreg trilobed, the lobes broad, often again slightly lobed; cliff crevices in basaltic canyons at low elev, Yakima Co, Wn; basalt d.   18 E. basalticus Hoover
23b St scapose or subscapose (or ± lfy in *E. flabellifolius,* of Wyo and Mont, which has no long hairs)
  25a Caudex divided into several or ∝ long, slender, rhizome-like brs; lvs ternate to flabellate, with broad lobes, or some of them entire; pls not of Wn
    26a Invol and herbage glandular-puberulent, not at all villous or hirsute; lvs flabellate to deeply 3-lobed, the lobes sometimes again shallowly lobed; loose talus slopes at high elev, mts of nw Wyo and adj Mont; fan-lvd d.   19 E. flabellifolius Rydb.
    26b Invol and herbage ± hirsute or villous, often also glandular; lvs deeply 3-lobed to (some of them) entire
      27a Rays narrow, erect, and relatively short, to 5 × 0.8 mm; invol villous; mts of se BC and sw Alta; pale d.   20 E. pallens Cronq.
      27b Rays well developed, spreading, 4–7 × 1–2 mm; invol hirsute; shifting talus at high elev, s RM to Cal and n into our range in Wallowa Mts, Ore; loose d.   21 E. vagus Payson
  25b Caudex stout, surmounting a taproot, occ br, but the brs gen relatively short and stout, not slender and diffuse (a few pls from Cas of Wn app the preceding group in development of the caudex); lvs trifid to ternately dissected, with relatively narrow lobes; rays to 12 × 2 mm, or often reduced or wanting (the heads then disciform); widespread sp.; cut-lvd d., dwarf mt f.; we have 3 vars.   22 E. compositus Pursh
    a1 Best developed lvs gen 3–4 times ternate, often irreg so, with very long and linear segms; pls ave larger in all parts than the other 2 vars.; sandy riverbanks at low elev in Wn, Ore, and adj Ida   var. compositus
    a2 Best developed lvs gen 1–3 times ternate, the segms gen not so long and narrow; rocky places at moderate to high elev in mts, gen not on loose talus (cf. nos 19–21); mainly apomictic complex, not well sorted out into vars.
      b1 Lvs gen 2–3 times ternate; common and widespread, gen at middle to fairly high elev in mts   var. glabratus Macoun
      b2 Lvs gen only once ternate; pls gen smaller than the other vars., sometimes pulvinate; chiefly s RM, but occ at high elev in our range (*E. trifidus*)   var. discoideus Gray

Group VI

28a Rays yellow, 6–9 × 1.4–2.5 mm; basal lvs petiolate, with elliptic to obovate or subrotund, gen broadly rounded or obtuse bl to 13 mm wide; cauline lvs few and reduced; Cas of Wn, n to s Alta and s BC; golden f., alpine yellow d. (*Haplopappus brandegei*)   23 E. aureus Greene
28b Rays white to pink, blue, or purple, of various sizes; lvs various
  29a Rays very narrow and short, 3–6 × 0.5–1.0 mm; hairs of invol with dark, blackish-purple crosswalls; invol gen heavily tinted with blackish-purple; boreal sp., s in mts to GNP; arctic-alpine d. (*E. unalaschkensis*)   24 E. humilis Graham
  29b Rays relatively well developed, 7–15 × 1.0–2.5 mm; hairs of invol with clear or occ bright reddish-purple cross-walls; invol green to ± anthocyanic, but less deeply so than in no 24
    30a Pappus bristles gen 25–35; pls essentially scapose, to 5 cm, loosely long-woolly-villous, esp upwards; caudex slenderly br; nw Mont to adj Alta and (?) BC; also Colo; woolly d.   25 E. lanatus Hook.
    30b Pappus bristles gen 10–20; pls with some fairly well developed cauline lvs, gen with at least 1 lf above the middle of the st
      31a Rays gen 25–50, white; invol sparsely to moderately woolly-villous; OM; OM d.   26 E. flettii G. N. Jones
      31b Rays gen 50–125, blue or pink, rarely white; invol moderately to densely woolly-villous; E Cas
        32a Disk corollas gen 4.0–5.0 mm; lvs strongly hairy; pappus bristles 15–20; BC and Alta to Alas; large-fld d.
                                                                        27 E. grandiflorus Hook.
        32b Disk corollas gen 3.0–3.6 mm; lvs glab or very moderately hairy; pappus bristles 10–15; Mont to ne Ore, s to Ariz and NM; alpine d., one-std f.   28 E. simplex Greene

Group VII

33a Rays yellow

34a Pubescence·of the st and lvs fine and appressed; bases of sts and of basal lvs conspicuously indurate and ± enlarged, stramineous to sometimes purplish; lvs ± linear, 1.5–9 cm × 0.5–3 mm; dry, often rocky plains and foothills, to moderate elev in mts; widespread E Cas; line-lf f., desert yellow d. (*E. peucephyllus*)   29 **E. linearis** (Hook.) Piper

34b Pubescence of the st, or lvs, or both, obviously spreading

35a Pubescence of the st mostly appressed, esp above; lvs ± linear, flexuous, to 4 cm × 1.5 mm, not all basal; dry, open places, often with sagebr; CR plains of sc and se Wn; Piper's d. (*E. curvifolius*)

30 **E. piperianus** Cronq.

35b Pubescence of the st spreading; lvs narrowly oblanceolate, straight or slightly arcuate, to 9 cm, 1–3 mm wide, all or nearly all in a basal cluster; dry, open places, often with sagebr; se Wn to SRP, s to n Cal and n Nev; dwarf yellow f., golden d.; 2 sspp. 31 **E. chrysopsidis** Gray

a1 Rays well developed and conspicuous; the common phase in our range, with 2 vars.   ssp. **chrysopsidis**

b1 Pls of high altitudes in Wallowa Mts, relatively small and compact, with small heads (invol 4–5 mm) and often partly appressed pubescence   var. **brevifolius** Piper

b2 Pls of middle and lower elev, extreme se Wn, and E Cas in Ore, gen more robust and with larger heads (invol 5–7.5 mm) and more uniformly spreading pubescence   var. **chrysopsidis**

a2 Rays short and inconspicuous, scarcely or not at all > disk, or the ♀ fls often tubular and rayless; more s ssp., barely reaching the s edge of our range (*E. a.*)   ssp. **austiniae** (Greene) Cronq.

33b Rays absent, or inconspicuous and < disk

36a ♀ fls present; herbage spreading-hairy; pappus bristles 7–25 (30)

37a Achenes only sparsely or moderately hairy; outer pappus, if present, not obscured by the hairs of the achene

38a Outer pappus setose and obscure, or wanting; sts subnaked (see lead 35b)   31 **E. chrysopsidis** ssp. **austiniae**

38b Outer pappus of evident (sometimes narrow) scales; sts ± lfy; GB sp. of hot, dry places in foothills and des, barely reaching the s fringe of our range; basin rayless d.; ours is var. *aphanactis*

32 **E. aphanactis** (Gray) Greene

37b Achenes densely long-hairy, the hairs completely covering the surface of the achene and sometimes obscuring the outer pappus; rare discoid form (see lead 60a)   55 **E. poliospermus** Gray

36b ♀ fls absent; herbage finely strigose; lvs linear, 2–7 cm × 0.7–2 mm, all or nearly all in a basal cluster; pappus bristles 25–40; dry, often rocky places in foothills and mts, c Wn to c Ida, s to Cal and Nev, E Cas in our range; scabland f.; ours is var. *bloomeri*   33 **E. bloomeri** Gray

Group VIII

39a Lvs, invol, and gen st with obvious long hairs, as well as stipitate-glandular; rays gen blue or purple; rocky places at middle to high elev, Cas and Wen Mts of c and n Wn; Leiberg's f.   34 **E. leibergii** Piper

39b Lvs, invol, and st glab to sparsely glandular, sometimes also with a few long hairs; rays gen white; rocky places at middle to high elev, Cas and Calapooia Mts, Ore, 43rd to 45th parallel; Cas d.   35 **E. cascadensis** Heller

Group IX

40a Sts erect from a simple or slightly br caudex, 1–6 dm; heads 1–15, with ca 125–175 rays (see lead 22b)   16 **E. glabellus** Nutt.

40b Sts gen ± curved at base, 0.5–2.5 dm; heads solitary, with ca 30–100 rays

41a Invol glandular and spreading-hirsute; basal lvs oblanceolate, acute or acutish to sometimes obtuse or rounded, up to 12 cm × 11 mm, but gen much smaller; meadows and slopes at alp and subalp stations; wc Mont and e Ida to Colo, Ariz, and e Nev; Bear R f.   36 **E. ursinus** D. C. Eat.

41b Invol strigose or strigose-hirsute with appressed or slightly loose hairs, scarcely or not at all glandular; lvs ave narrower and a little hairier than in no 36; meadows in foothills and at mid elev in mts; sw Mont, ec Ida, and w Wyo; slender f.   37 **E. gracilis** Rydb.

Group X

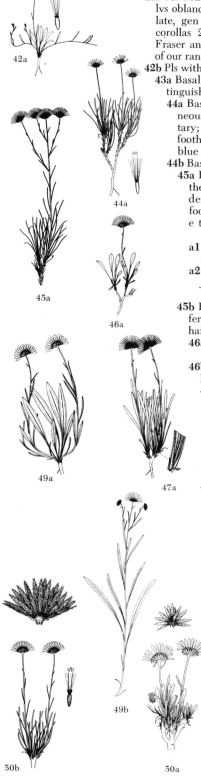

42a Pls with long, trailing, sparsely lfy stolons; bien or short-lived per; basal lvs oblanceolate, to 5 cm × 8 mm, cauline ± linear; heads long-pedunculate, gen solitary on erect, sparsely lfy or subnaked sts 5–40 cm; disk corollas 2.5–3.5 mm; pappus bristles 10–15; ± open places, valley of Fraser and Thompson rivers, BC, disjunct from principal area to s and e of our range; trailing f.; ours is var. *f.*   38 **E. flagellaris** Gray
42b Pls without stolons; true per
  43a Basal lvs narrow, linear or rather narrowly oblanceolate, the bl, if distinguishable, tapering gradually to the petiole
    44a Base of st conspicuously enlarged, shining, and ± indurate, stramineous or purplish; lvs chiefly basal, 1.5–6 cm × 0.5–1 mm; heads solitary; rays 15–20, blue or pink; open, often rocky places in plains and foothills, esp on basaltic or volcanic rock; ne Ore to n Cal; volcanic d., blue dwarf f.   39 **E. elegantulus** Greene
    44b Base of st not conspicuously enlarged, shining, and indurate
      45a Basal lvs with base neither enlarged nor of different texture from the bl; lvs linear or linear-filiform, 1–8 cm × 0.3–3 mm; st more densely hairy toward the base than above; dry places in plains and foothills, often with sagebr; s BC to nw Mont, s E Cas to n Cal, and e through SRP nearly to Wyo; thread-lf f.; 2 vars.
                                      40 **E. filifolius** Nutt.
        a1 Heads gen several, with 15–50 (75) rays; common and widespread   var. **filifolius**
        a2 Heads gen solitary, with 50–125 rays; pl stouter than var. *filifolius*; Kittitas Co, Wn, to Wasco and Gilliam cos, Ore
                                        var. **robustior** Peck
      45b Basal lvs with base ± enlarged, membranous or indurate, of different texture from the bl; lvs linear or broader; st not more densely hairy at base than above
        46a Invol glandular, not hairy; lvs glab (see lead 52b)
                                   46 **E. leiomerus** Gray
        46b Invol ± hairy, sometimes also viscid or glandular; lvs evidently hairy, or glandular, or both
          47a Petioles or margins of at least the basal lvs with some coarse spreading hairs unlike the other hairs of the lvs; lvs not triple-nerved, the basal to 10 cm × 4 mm; heads solitary or few, naked-pedunculate; rays gen white; woods, meadows, and open hillsides from foothills to mid elev in mts; ours is var. *davisii* Cronq., of Lewis and w Ida cos, Ida, and Wallowa Co, Ore, the typical var. disjunct to the e of our range; Engelmann's d.
                                 41 **E. engelmannii** A. Nels.
         47b Petioles of basal lvs without coarse spreading hairs
           48a Basal lvs gen triple-nerved, (4) 6–25 cm × 2–10 mm; outer pappus setulose and obscure, or wanting
             49a Cauline lvs gen abruptly reduced and smaller than the basal ones; pls 0.5–3 dm; E Cas; Eaton's d.; we have 2 vars.   42 **E. eatonii** Gray
               a1 Invol bracts distinctly imbricate and evidently glandular, gen only slightly or moderately hirsute; sw Mont, Wyo, s Ida, and s   var. **eatonii**
               a2 Invol bracts subequal, conspicuously white-villous-hirsute esp toward base, only slightly if at all glandular; pubescence of herbage gen longer, finer, looser, and more copious than in var. *eatonii*; c and w Ida to Wn and s Ore   var. **villosus** Cronq.
             49b Cauline lvs only gradually reduced; pls 1.5–7 dm; open places in WV; Pac f., WV d.; ours is var. *d.*
                             43 **E. decumbens** Nutt.
           48b Basal lvs not at all triple-nerved, 1–12 cm × 0.5–5 mm; outer pappus gen more conspicuous, of scales or thick setae; RM and e
             50a Pls nearly or quite scapose; disk corollas 2.3–3.0 mm; pappus bristles 6–12; lvs to 2 cm (see lead 59a)
                             53 **E. radicatus** Hook.
           50b Pls with some evident (sometimes small) lvs on the st; disk corollas 2.8–4.3 mm; pappus bristles 10–20; lvs to 12

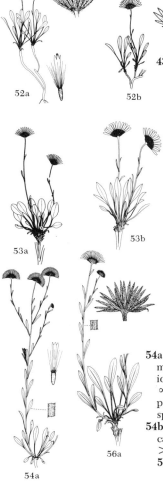

52a
52b

53a
53b

54a

56a

56b

58a

59a

59b

cm; dry plains and barren places in mts, sometimes at high elev; e of Cont Div, Mont to Alta and Gr Pl; buff f.; 2 vars.

44 **E. ochroleucus** Nutt.

   a1 Pls relatively tall and robust, 8–40 cm, with well developed (though narrow) cauline lvs   var. **ochroleucus**

   a2 Pls relatively low and slender, 1–6 (10) cm, with few and small cauline lvs *(E. s.)*

var. **scribneri** (Canby) Cronq.

43b Basal lvs relatively wide, broadly oblanceolate or gen broader, the bl well defined, gen ± abruptly contracted to the petiole

   51a Lvs glab, the basal spatulate or obovate to merely oblanceolate, rounded or retuse at tip, to 7 cm × 15 mm; high mts

      52a St essentially scapose; rays white; pappus of 25–35 very unequal bristles; shifting talus, c Ida and adj Mont; Evermann's d.

45 **E. evermannii** Rydb.

      52b St with several well developed (though small) lvs; rays deep blue to occ nearly white; pappus bristles 15–25; more stable rocky outcrops; ec Ida, s to RM; smooth d.   46 **E. leiomerus** Gray

   51b Lvs ± hairy, with well marked, acute to round-tipped bl

      53a Invol ± strigose, not glandular; herbage finely and densely canescent-strigose; basal lf bls obovate to elliptic or suborbicular, to 25 × 13 mm; dry places at middle elev in mts; sw Mont and nw Wyo to c Ida; Tweedy's d.   47 **E. tweedyi** Canby

      53b Invol ± glandular, gen also hirsute with short, spreading hairs, not strigose; herbage moderately or sparsely strigose; basal lf bls oblanceolate to obovate or rhombic, gen acute or acutish, 7–25 × 2–7 mm; rocky places and dry soil in mts, sometimes at high elev; w Wyo to e Ore, c Cal, and c Utah; slender d.   48 **E. tener** Gray

Group XI

54a Pls freely br, 1–7 dm, gen bien or short-lived per; disk corollas ca 2–3 mm; herbage covered with spreading hairs gen well < 1 mm; basal lvs petiolate, with oblanceolate or spatulate bl to 2.5 cm, gen deciduous, cauline ∝, smaller, often linear; heads gen ∝; rays 75–150, 5–10 × 0.5–1.2 mm; pappus bristles 5–12; dry and waste places in valleys and foothills; widespread E Cas; diffuse or spreading f.; ours is var. d. 49 **E. divergens** T. & G.

54b Pls simple or sparingly br (except *E. pumilus*); true per, often with br caudex; disk corollas > 3 mm except in *E. radicatus;* pappus bristles gen > 12 except in *E. radicatus;* rays often wider and often fewer

   55a Basal lvs in well developed pls evidently triple-nerved

      56a Invol bracts evidently thickened on the back; basal lvs gen rounded to obtuse at tip; st rarely purplish at base; cauline lvs ovate-oblong to linear; dry, open, often rocky places; widespread E Cas, but absent from Ore; gray d., tufted f.   50 **E. caespitosus** Nutt.

      56b Invol bracts only slightly or obscurely thickened on the back; basal lvs acute; st gen purplish at base; cauline lvs linear or lance-linear; open, dry places, often with sagebr; widespread E Cas in our range; long-lf f., foothill d.   51 **E. corymbosus** Nutt.

   55b Basal lvs not triple-nerved, or only very faintly so

      57a Heads turbinate, solitary or sometimes 2; pubescence of lvs short and fine; mt pls

         58a Lf surfaces uniformly pubescent with short (< 0.5 mm) spreading hairs; basal lvs broadly oblanceolate or broader, to 8 cm × 13 mm; pls lax, 2–20 cm; rays 10–25, gen deep blue or violet; disk corollas 4.2–6.3 mm; pappus bristles 20–30; middle to high elev; c Ida and ne Nev; rough f.   52 **E. asperugineus** (Eat.) Gray

         58b Lf surfaces sparsely and finely hairy (but the hairs often > 0.5 mm), or subglab; basal lvs linear or narrowly oblanceolate, to 5 cm × 3 mm; pls compact, ± tufted, to 8 cm; high mts

            59a Disk corollas gen 2.3–3.0 mm; pappus bristles gen 6–12; rays 20–50, white; rare, s Alta and (?) BC; taprooted f.

53 **E. radicatus** Hook.

            59b Disk corollas gen 3.7–5.0 mm; pappus bristles gen 15-20; rays 15-35, violet or white; sw Mont and adj Wyo; Rydberg's d.

54 **E. rydbergii** Cronq.

60a

61a

b1   61b   b2

57b Heads hemispheric; lvs conspicuously pubescent with coarse spreading hairs (> 0.5 mm); gen in foothills and plains

  60a Achenes densely long-hairy, the hairs completely covering the surface of the achene and sometimes obscuring the outer pappus; rays 15–45, anthocyanic, 5–14 × 1.3–3.6 mm; pappus bristles 20–30; dry, open places, often with sagebr; E Cas, s BC to c Ore, e to w Ida; cushion f., hairy-seeded d.; 2 vars.   **55 E. poliospermus** Gray

    a1 Herbage densely hairy and only obscurely or moderately glandular (invol sometimes more glandular than hairy); sts gen simple and monocephalous; widespread   var. **poliospermus**

    a2 Herbage and invol densely glandular and only sparsely or moderately hairy; sts often br and several-headed; Kittitas Co and adj parts of Grant and Chelan cos. Wn   var. **cereus** Cronq.

  60b Achenes only sparsely or moderately hairy, the surface gen exposed between the hairs, the outer pappus not at all obscured by the hairs

    61a Rays 30–60 (100), 1.5–2.3 mm wide, white (or pinkish in age); disk corollas coarser and less indurate than in *E. pumilus;* sts 3–12 cm; heads solitary; SR vic from extreme se Wn to Owyhee Co, Ida; SR d.   **56 E. disparipilus** Cronq.

    61b Rays 50–100, 0.7–1.5 mm wide, blue, pink, or white; disk corollas 3.5–5 mm, slender, the limb pale and indurate below; sts 5–50 cm, with 1–∝ heads; widespread E Cas; shaggy f.; we have 2 sspp.   **57 E. pumilus** Nutt.

      a1 Gr Pls ssp., encroaching into our range in intermont valleys of Mont; rays white; outer pappus setose; principal pappus bristles 15–27; disk corollas glab   ssp. **pumilus**

      a2 N RM ssp., the common phase in our range; rays anthocyanic to occ white; outer pappus setose-squamellate; principal pappus bristles 13–20; indurate part of disk corollas often slightly puberulent (*E. concinnus*, misapplied); 2 vars.

                    spp. **intermedius** Cronq.

      b1 Pls relatively robust, the sts gen > 1.5 mm thick at base and bearing 5–∝ heads; chiefly Wn, n Ida, and adj Mont and Ore   var. **euintermedius** Cronq.

      b2 Pls smaller and more slender, the larger sts seldom > 1.5 mm thick, rarely with as many as 5 heads; chiefly c and s Ida and adj parts of Ore, Mont, and Wyo   var. **gracilior** Cronq.

Eriophyllum Lag.   Eriophyllum

Heads radiate or rarely discoid, the rays few, ♀ and fertile, yellow; invol of 1 or apparently 2 series of firm, permanently erect, carinate bracts which partly embrace the ray achenes; recep flattish to low-conic, naked; disk fls few to ∝, ☿ and gen fertile; anthers minutely sagittate at base; style brs flattened, with introrsely marginal stigmatic lines and short, externally papillate-hairy appendages; achenes slender, 4-angled; pappus of nerveless chaffy scales, gen erose or fimbriate, seldom obsolete; herbs or subshrubs with woolly herbage and gen alt, entire to toothed or cleft lvs. (Gr *erion*, wool, and *phyllon*, foliage).

a1

b2

E. lanatum (Pursh) Forbes. Common e., woolly sunflower. Tomentose per 1–6 dm, gen with several sts; lvs 1–8 cm, entire to pinnatifid or ternate; heads long-pedunculate; pappus of 6–12 hyaline scales, or a mere toothed crown, or obsolete; dry, open places, lowl to mid elev in mts, BC to Cal, e to Mont and Utah; we have 3 vars.

  a1 Heads gen large, invol 9–12 mm; rays 1–2 cm, gen ca 13 (8); lvs pinnatifid to entire; pls gen robust, 2.5–6 dm, occ smaller; W Cas, and along their e front, also in se Wn, ne Ore, nc and wc Ida, and extreme w Mont   var. **lanatum**

  a2 Heads gen smaller, invol 6–10 mm; rays 5–12 mm, gen ca 8 (5)

    b1 Lvs, or some of them, gen ± ternate-pinnatifid, often opp; pls gen tall as in var. *lanatum;* Cal, n to Lane Co, Ore *(E. a.)*

                    var. **achillaeoides** (DC.) Jeps.

    b2 Lvs gen entire or apically trilobed, all or nearly all alt; pls gen 1–2.5 dm; widespread E Cas *(E. i.)*   var. **integrifolium** (Hook.) Smiley

Eupatorium L.   Boneset; Thoroughwort; Eupatorium

Heads discoid, the fls all tubular and ♂, few to ∝ (9–22 in ours) in each head; invol bracts imbricate or subequal; recep flat to conic, naked; corollas pink or purple to blue or white; anthers obtuse and entire at base, or minutely sagittate; style brs with short stigmatic lines and an elongate, papillate, obtuse, often clavate appendage; achenes 5-angled, ours glandular; pappus of ± ∝ capillary bristles; gen per herbs (ours fibrous-rooted) or (extralimital spp. only) shrubs, with entire or toothed (ours) to occ dissected, often glandular-punctate, gen opp, sometimes whorled or alt lvs (ours short-petiolate) and small to large heads in a gen corymbiform infl; our spp. fl summer. (Named for Mithridates Eupator, 132–63 B.C., king of Pontus).

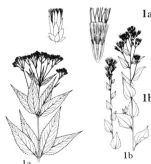

1a Lvs whorled (gen in 3's or 4's), bl lance-elliptic to lanceolate or lance-ovate, 6–20 × 2–7 cm, densely covered beneath with short, spreading, curly hairs; invol 6.5–9 mm, its bracts conspicuously imbricate; pls 6–15 (20) dm; heads in a large, corymbiform, rather flat-topped infl; swamps and other moist places, widespread in e US, w. on high plains and in intermont valleys to s BC, Utah, and NM; Joe-Pye weed; ours is var. *bruneri* (Gray) Breitung *(E. b.)*   1 **E. maculatum** L.
1b Lvs alt, bl narrowly to broadly deltoid or deltoid-ovate, 1.5–6.5 × 1–5 cm, gen atomiferous-glandular beneath; invol 3–5 mm, its bracts subequal; pls 1.5–7 dm; heads in smaller, more rounded (or more elongate) clusters; rocky uplands, cordilleran, E Cas; western e.   2 **E. occidentale** Hook.

1a   1b

Filago L.   Filago

Heads ∝, glomerate, disciform, white or whitish; invol scanty, its bracts resembling those of the cylindric or obconic to merely convex recep; outer fls ♀, fertile, with tubular-filiform corolla, in several series, the outermost of these epappose and subtended and partly enclosed by concave or boat-shaped bracts, the others bractless and with a pappus of capillary bristles, or some of them nearer the outside with more open subtending bracts; central fls 2–5, appearing ♂, but often sterile, bractless, and with capillary pappus; anthers short-tailed; achenes small, nerveless; white-woolly ann with alt, entire lvs. (L *filum*, a thread).

F. arvensis L. Field f. Pls simple or br, 0.3–5 dm, weedy; lvs erect, ± linear, to 4 cm × 5 mm; heads gen 3–5 mm; proper invol very scanty; recep bracts woolly, simulating an invol, somewhat concave-clasping, obscurely hyaline-margined and -tipped; recep nearly flat; European weed, becoming common on overgrazed ranges E Cas in our region, fl summer.

Gaillardia Foug.   Gaillardia

Heads radiate or occ discoid, the rays yellow, or partly or wholly purple, broad, 3-cleft, gen neutral, sometimes ♀ and fertile; invol bracts 2–3-seriate, herbaceous except for the chartaceous base, ± spreading, becoming reflexed in fr; recep convex to subglobose, provided with ∝ soft or more often chaffy or spinelike setae which do not individually subtend the disk fls; disk fls ♂ and fertile; anthers aur at base; style brs flattened, with well developed introrsely marginal stigmatic lines and ± elongate, externally hairy to subglab appendage, this gen with a more evident tuft of hairs across the base; achenes broadly obpyramidal, partly or wholly covered by a basal tuft of long, ascending hairs; pappus of 6–10 awned scales; herbs with alt (or all basal), entire to pinnatifid lvs and rather large, gen long-pedunculate heads. (Named for Gaillard de Marentonneau, French botanist).

G. aristata Pursh. Blanket-fl. Per from a slender taproot, 2–7 dm, with 1 or gen several simple or subsimple sts; lvs to 15 cm (petiole included) × 2.5 cm, entire to coarsely toothed or even subpinnatifid, hairy like the st, occ all basal; heads 1–few, long-pedunculate; disk 1.5–3 cm wide, gen ± purple; rays 6–16, often purple-based, 1–3.5 cm; disk corollas densely woolly-villous distally; open places, cordilleran, gen E Cas.

Galinsoga Cav.   Quickweed; Garden Pest

Heads radiate, the rays few, short, broad, only slightly > disk, white or pink, ♀ and fertile; invol bracts few, relatively broad, ± membranous, but greenish at least in part, several-nerved, each subtending a ray and gen joined at base with 2 adj recep bracts, a few shorter and narrower but otherwise similar outer bracts often present; recep conic, chaffy, its bracts membranous, rather narrow, nearly flat; disk fls ♂; anthers minutely or scarcely sagittate at base; style brs flattened, with short, externally minutely hairy appendage, without well marked stigmatic lines; achenes 4-angled, scarcely compressed, or, esp the outer ones, ± flattened parallel to the invol bracts; pappus of several or ∝ scales, often fimbriate or awn-tipped, that of the rays often reduced or wanting; ann herbs with opp, simple lvs and small heads. (Named for Mariano Martinez Galinsoga, Spanish physician and botanist).

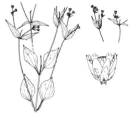

G. ciliata (Raf.) Blake. Freely br, 2–7 dm, ± spreading-hairy esp upwards, the hairs of the peduncle and often also of the invol gland-tipped; lvs petiolate, bl ovate or lance-ovate, 2–7 × 1–4 cm, toothed; heads in lfy cymes; disk 3–6 mm wide; rays white, with pappus scales ca = tube; disk pappus tapering to awn tips, ± fimbriate-ciliate; widespread trop Am weed, W Cas in our range.

Gnaphalium L.   Cudweed; Everlasting

Heads disciform, the ∝ outer fls slender and ♀, the few inner ones coarser and ♂; invol ovoid or camp, bracts slightly to evidently imbricate, scarious at the tip or nearly throughout; recep naked; fls yellowish or whitish; style brs of ♂ fls flattened, truncate, without sharply differentiated stigmatic portion; anthers caudate; pappus of capillary bristles, sometimes thickened distally, sometimes united at base; achenes small, nerveless; ann to per, gen ± white-woolly herbs with alt, entire lvs. (Gr *gnaphalion,* anciently applied to these or similar pls).

1a Pappus bristles ± united at base, tending to fall in a ring; ann or bien, to 4 dm; lowest lvs largest, spatulate or oblanceolate, to 10 × 2 cm; invol 3–5 mm, woolly at base

    2a Infl narrow, dense, spiciform-thyrsoid, occ br; pappus bristles firmly united at base; invol bracts imbricate, gen acute or acuminate, light brown, often tinged with anthocyanin; widespread Am weed, often in sandy soil, not wholly restricted to disturbed habitats; in our range gen W Cas; purple c.; ours is var. *p.*   **1 G. purpureum** L.

    2b Infl of 1 or gen several cymosely arranged terminal glomerules; pappus bristles only weakly united at base, the ring readily fragmented; invol bracts gen more obtuse or rounded, brownish-stramineous; European weed, locally intro, chiefly W Cas; weedy c.   **2 G. luteo-album** L.

1b Pappus bristles distinct, falling separately; lvs chiefly cauline

    3a Heads small, invol gen 2–4 mm, densely woolly below, not much imbricate; glomerules of heads lfy-bracted; ann, gen much br, to 1.5 (3) dm, ± weedy

        4a Lvs oblanceolate to oblong, often broadly so; tomentum loose; invol bracts gen brown with whitish tip; native, semi-weedy, in moist, open places, often in dried beds of vernal pools; lowl c.   **3 G. palustre** Nutt.

        4b Lvs linear or narrowly oblanceolate; tomentum close; invol less involved in tomentum, gen darker, bracts often greenish or brownish to the tip; European sp., W Cas with us; marsh c.   **4 G. uliginosum** L.

    3b Heads larger, invol gen 4–7 mm, woolly only at base if at all; glomerules not conspicuously lfy-bracted; ann or per, 2–9 dm, simple or ± br

        5a Herbage ± glandular-hairy, at least as to upper lf surface, sometimes also ± tomentose; ann or bien; lvs narrowly decurrent

            6a Invol bracts from merely acutish to more often obtuse or broadly rounded, tending to be pearly white; st gen evidently tomentose and only obscurely or scarcely glandular; Cal sp., n along coast to Lincoln Co, Ore; Cal c., green c.   **5 G. californicum** DC.

            6b Invol bracts ± strongly acute, gen yellowish or ± dingy; st gen evidently glandular-hairy, seldom tomentose below infl; widespread n and e of no 5; sticky c. (*G. decurrens, G. ivesii, G. macounii*)

                                                     6 G. viscosum H.B.K.

        5b Herbage ± tomentose, not at all glandular

            7a Per from a taproot; lvs narrowly decurrent, gen 3–10 cm × 2–10 mm, broadly linear or the lower oblanceolate; heads gen ∝ in small glomerules, forming a broad, open infl; widespread cordilleran sp. of open, gen dry places, esp burned-over for land; slender c.,

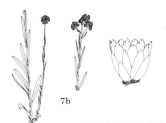

white c.; ours is var. *thermale* (E. Nels.) Cronq. *(G. t.)*
7 **G. microcephalum** Nutt.
7b Ann or bien; lvs merely adnate-aur, often broader than in no 7, sometimes oblong or lanceolate; pls more copiously and loosely tomentose than no 7; invols gen with a more yellowish cast, tending to be grouped in 1 or a few dense glomerules; widespread cordilleran sp., sometimes weedy, gen in moister soil than no 7; cotton-batting pl   8 **G. chilense** Spreng.

Grindelia Willd.   Gumweed; Gumplant; Resinweed; Grindelia

Heads radiate or occ discoid, the rays gen 10–45, ♀ and fertile, yellow; invol ± resinous or gummy, its bracts firm, herbaceous-tipped, imbricate or subequal; recep flat or convex, naked; disk fls yellow, the inner and often also the outer sterile; anthers entire or nearly so at base; style brs flattened, with marginal stigmatic lines and an externally hairy, lance-linear or occ very short appendage; achenes compressed to subquadrangular, scarcely nerved; pappus of 2–several firm deciduous awns, often ± serrulate; ann, bien, or per herbs, ours all gen < 1 m, with alt, ± resinous-punctate lvs and several or ∝ medium-sized to rather large heads, the disk in ours 1–3 cm wide; our spp. gen fl midsummer to fall. (David Hieronymus Grindel, 1776-1836, Russian botanist).

1a Heads discoid; bien or short-lived per, glab; lvs entire or with some small teeth, narrow, seldom > 10 × 1.5 cm; invol nearly of no 5; gen in gravelly or sandy places along streams; c Wn and extreme n Ida, and down the CR to Portland; CR g. *(G. nana* var. *discoidea)*
1 **G. columbiana** (Piper) Rydb.
1b Heads radiate
  2a Tips of invol bracts loose or spreading, but not reg reflexed; invol only slightly or moderately glutinous; herbage ± villous to sometimes glab; per; rays gen 10–35; W Cas; WV g., PS g.; 2 vars.   2 **G. integrifolia** DC.
    a1 Rays short, gen 8–12 mm; various nonmaritime habitats in PT, from s VI to s WV   var. **integrifolia**
    a2 Rays longer, gen 12–20 mm; heads ave a bit larger than in var. *integrifolia*; salt marshes and rocky shores, coastal, Alas to n Cal, common about PS *(G. stricta)*   var. **macrophylla** (Greene) Cronq.
  2b Tips of invol bracts (at least the middle and outer) reg reflexed; invol often ± strongly resinous; herbage glab except in no 3; chiefly E Cas
    3a St glandular and ± villous, at least in infl; prob bien; heads of no 5; n Ida and adj Mont, rare; Howell's g.   3 **G. howellii** Steyerm.
    3b St essentially glab; widespread spp.
      4a Rays gen 12–25, rarely more on some heads of robust pls; lvs entire or sharply toothed, not at all callous-serrulate; invol seldom strongly resinous; achenes gen with 1 or more short knobs on apical margin; per; common in e Ore and Wn and adj Ida, less common in Mont; low g.; we have 2 vars.   4 **G. nana** Nutt.
      a1 Heads relatively small, the invol to ca 1 cm high, the disk to 1.5 cm wide; invol bracts relatively short-tipped; pls relatively small and narrow-lvd; e Wn and adj Ida, s to Union Co, Ore   var. **nana**
      a2 Heads larger, the invol 1–1.7 cm high, the disk to 2.5 cm wide; green tips of invol bracts longer; more generally distributed, but not in range of var. *nana*   var. **integrifolia** Nutt.
    4b Rays gen 25–40, occ fewer on some heads of small pls; lvs reg callous-serrulate to sharply toothed or entire; invol strongly resinous; achenes gen without apical knobs; bien or short-lived per; common in Mont and Ida, less common w; resin-weed, curly-cup g.; we have 3 vars.   5 **G. squarrosa** (Pursh) Dunal
      a1 Lvs entire or remotely serrulate, or (especially the lower) coarsely and irreg toothed or incised; gen short-lived per; native *(G. perennis)*   var. **quasiperennis** Lunell
      a2 Lvs closely and evenly serrulate or crenate-serrulate; gen bien; intro in our range
        b1 Upper and middle lvs 2–4 times as long as wide, gen ovate or oblong   var. **squarrosa**
        b2 Upper and middle lvs 5–8 times as long as wide, gen linear-oblong to oblanceolate *(G. s.)* var. **serrulata** (Rydb.) Steyerm.

Gutierrezia Lag.   Matchbrush; Matchweed; Snakeweed

Heads radiate, the rays few and short, ♀, yellow; invol bracts strongly imbricate, herbaceous-tipped; recep small, naked but often deeply pitted; disk fls few, fertile or sterile; anthers entire or nearly so at base; style brs flattened, with introrsely marginal stigmatic lines and an externally short-hairy, ± elongate appendage; pappus of several scales or awns, sometimes united at base, or wanting; achenes terete, gen several-nerved; taprooted herbs or low shrubs with alt, linear, often punctate lvs and small, ∝ heads in a gen flat-topped infl. (Pedro Gutierrez, Spanish botanist).

G. **sarothrae** (Pursh) Britt. & Rusby. Broom s. Shrub or subshrub, 2–6 dm, with ∝ erect, slender, brittle brs; lvs punctate, 2–4 cm × 1–2 mm; heads ∝, tending to be in small glomerules; invol glandular-glutinous, subcylindric to narrowly obconic, 3–4.5 mm; rays gen 3–8, 2–3 mm; disk fls gen 3–8, fertile; pappus of narrow scales; dry, open places, gen at lower elev; Gr Pl and RM, w to e Ore and extreme se Wn; fl July–Sept.

Haplopappus Cass.   Bristleweed; Goldenweed

Heads radiate, with few to moderately ∝, ♀ or neutral, yellow (rarely pale or even white) rays, or the rays sometimes wanting and the heads thus discoid; invol bracts, as in *Aster*, gen either ± lfy, or green-tipped and imbricate, varying to wholly chartaceous, or sometimes narrow and subherbaceous throughout as in *Erigeron*; recep flat or a little convex, naked; disk fls few to more often ∝, yellow, or occ white or purplish; anthers entire or nearly so at base; style brs flattened, with introrsely marginal stigmatic lines and well developed, obtuse to more often acuminate, externally minutely hairy appendage often > stigmatic portion; achenes angled or striate to smooth; pappus of ± ∝, distinctly unequal, often sordid capillary bristles; taprooted herbs or shrubs, rarely with creeping rhizomes as well, with gen alt (or all basal) lvs and solitary to moderately ∝, hemispheric to cylindric heads (the disk fls gen larger, and in the herbaceous spp. gen also more ∝, than in *Solidago*). (Gr *haplous*, simple, and *pappos*, seed-down). (*Ericameria, Hesperodoria, Macronema, Pyrrocoma, Sideranthus, Stenotus, Tonestus*). *Haplopappus nuttallii* is here treated as *Machaeranthera grindelioides*.

1a Pls br low shrubs, 1–6 (9) dm, not at all cespitose; achenes elongate; fl (June) July–Sept (*Macronema*)

2a Rays uniformly wanting; twigs densely white-tomentose; lvs strongly glandular, 1–3 cm × 3–6 mm, margins often crisped; disk fls gen 10–25, corollas 8–11 mm; rocky slopes at high elev, c Ida to s RM, GB, and Cal; discoid g.; ours is var. *m.* (*M. discoideum*)   1 H. **macronema** Gray

2b Rays 1–8, or wanting from some heads; twigs glandular or glab to occ gray-tomentose

3a Heads relatively large and broad, gen with 20–40 fls; lf margins tending to be crisped; herbage glandular; rays showy, to nearly 2 cm; disk corollas 7–11 mm; rocky places at high elev, sw Mont to ne Ore, s to Nev and Cal; shrubby g., big-head g.
                                             2 H. **suffruticosus** (Nutt.) Gray

3b Heads narrow and in gen smaller, with gen 10–20 fls; lf margins plane or revolute, not crisped

4a Small-headed cliff-crevice pls gen of plains and foothills; lvs gen 0.5–3 cm × 0.5–2 mm; rays 2–6 mm; disk corollas 4.5–7.5 mm; invol 4–8 mm; herbage glab, resinous

5a Fls pale yellow to nearly white; rays 3–6 mm; disk fls ca 13 (11–16); corolla lobes 1–2 mm; E Cas, Wn and n Ore and adj Ida; gnarled g., Columbia g.   3 H. **resinosus** (Nutt.) Gray

5b Fls bright yellow; rays 2–3 mm; disk fls 4–10; corolla lobes < 1 mm; GB and SRP, n to Beaverhead Co, Mont; rubber-weed, dwarf g.   4 H. **nanus** (Nutt.) Eat.

4b Larger-lvd, larger-headed pls, gen not of cliff-crevices; lvs gen 1.5–6 cm × 0.5–7 mm; rays 6–15 mm; disk corollas 7–11 mm; invol 7–12 mm; herbage glandular or glandular-glutinous to subglab or ± tomentose-puberulent

6a Pls of foothills and moderate elev in mts, with gen linear or narrowly oblanceolate lvs, and with the infl tending to be elongate in well developed individuals; Cas and mts of c Ore and ne Wn; rabbitbrush g.   5 H. **bloomeri** Gray

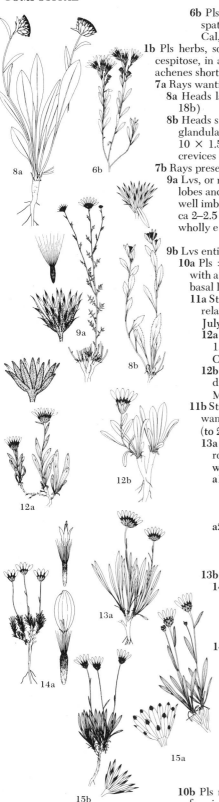

6b Pls gen of high elev in mts, with broader, more oblanceolate or spatulate lvs, and with short, compact infl; Cas from n Wn to n Cal, e to c Ida; Greene's g. *(H. mollis)*   **6 H. greenei** Gray

1b Pls herbs, sometimes with a shortly aerial woody caudex, often densely cespitose, in any case the fl sts herbaceous and of only one year's duration; achenes short or elongate

7a Rays wanting (see also *Chrysopsis oregona*)

8a Heads large; invol 1.5–3 cm; disk fls ∝, corolla 10–14 mm (see lead 18b)   **17 H. carthamoides** (Hook.) Gray

8b Heads smaller; invol ca 1 cm; disk fls 25–40, corolla 6–8 mm; herbage glandular-pubescent; lvs sharply toothed, the lower oblanceolate, to 10 × 1.5 cm, the upper becoming sessile and more often oblong; rock crevices in mts of c Ida; Ida g.   **7 H. aberrans** (A. Nels.) Hall

7b Rays present, gen conspicuous

9a Lvs, or most of them, pinnatifid or bipinnatifid, 1.5–6 cm, with narrow lobes and narrow rachis < 3 mm wide; invol 5–8 mm, its narrow bracts well imbricate, gen bristle-tipped; rays 15–50, 8–10 mm; achenes short, ca 2–2.5 mm; dry, open places in plains and foothills, chiefly Gr Pl, and wholly e of Cont Div in our range; fl July–Sept; spiny g.; ours the var. *s.*
          8 H. spinulosus (Pursh) DC.

9b Lvs entire or merely toothed; achenes elongate

10a Pls ± densely cespitose and gen mat-forming or cushion-forming, with a much brd caudex (or even creeping rhizomes in no 9), ∝ tufted basal lvs, and ± ∝ monocephalous sts rarely > 2 dm

11a Sts relatively lfy, the cauline lvs not greatly reduced; invol bracts relatively loose and herbaceous; rays 10–35; alp and subalp pls, fl July–Sept *(Tonestus)*

12a Herbage strongly glandular-puberulent; sts 3–15 cm; basal lvs 1.5–7 cm × 4–12 mm; BC and Alta to Wn, ne Ore, ne Nev, and Colo; Lyall's g.   **9 H. lyallii** Gray

12b Herbage subglab, or more viscid-villous upwards, scarcely glandular; sts 1–6 cm; basal lvs 1–5 cm × 1.5–5 mm; s RM, n to sw Mont; dwarf g.   **10 H. pygmaeus** (T. & G.) Gray

11b Sts only sparsely lfy or subnaked, the cauline lvs much reduced or wanting; invol bracts firmer, closer, and less herbaceous; rays 6–15 (to 20 in no 11); fl May–July (Aug) *(Stenotus)*

13a Lvs soft, floccose-tomentose to glandular or subglab; open, gen rocky or gravelly places; c Wn and e Ore to Ida and w Mont; woolly g.; 2 vars.   **11 H. lanuginosus** Gray

a1 Lvs green and gen ± glandular, only sparsely or scarcely tomentose; invol bracts relatively narrow and herbaceous, gen 1–2 mm wide; mts of Mont and c Ida, at moderate to high elev *(Stenotus a.)*   var. **andersonii** (Rydb.) Cronq.

a2 Lvs evidently gray-tomentose, though not densely so, seldom at all glandular; invol bracts broader and more scarious, gen 2–3 mm wide; Ore and Wn and adj Ida, plains to rather high elev   var. **lanuginosus**

13b Lvs rigid, glab to scab-puberulent or glandular

14a Lvs linear, to ca 2 cm × 2 mm, glandular-scabrid; lfy caudical brs more elongate than in the next 2 spp., forming a loose mat; dry, rocky soil, often with sagebr; plains and foothills, E Cas, c Wn to Nev and n Cal, e, rarely, to c Ida; narrow-lf g.
          12 H. stenophyllus Gray

14b Lvs oblanceolate, gen 2–10 cm × 2–7 mm, seldom glandular

15a Invol bracts broadly rounded or obtuse, conspicuously imbricate; herbage glab to sometimes resinous-glandular; dry plains and hills, esp on shifting, rocky soil or badlands; Gr Pl to s RM and GB, and e of Cont Div in our range; thrift g.   **13 H. armerioides** (Nutt.) Gray

15b Invol bracts acute or acuminate, moderately or scarcely imbricate; herbage glab to densely scab-puberulent; dry, open places, foothills and higher valleys to rather high elev in mts, widespread E Cas; stless g., cushion s. *(Stenotus caespitosus; S. falcatus; H. a.* var. *glabratus* Eat., the glabrate phase)   **14 H. acaulis** (Nutt.) Gray

10b Pls not densely cespitose and not at all mat-forming or cushion-forming, the caudex simple or moderately br, the fl sts several or

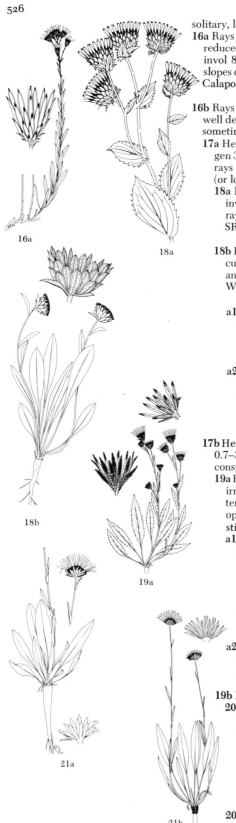

16a

18a

18b

19a

21a

21b

solitary, lfy or subnaked, with 1–∝ heads, often much > 2 dm

16a Rays few, gen ca 5 or 8; lvs ∝ and nearly all alike except for the reduced and often deciduous lowermost ones, 2–5 cm × 3–11 mm; invol 8–11 mm, rather narrow, resinous, strongly imbricate; dry slopes overlooking CR toward e end of gorge, and locally s in Cas to Calapooia Mts; fl Aug–Oct; Hall's g., hesperodoria
　　　　　　　　　　　　　　　　　　　　　　　　15 H. hallii Gray

16b Rays more ∝, gen 10–50 (or fewer only in no 17); basal lvs gen well developed and persistent; cauline lvs gen smaller, though still sometimes good-sized; fl (June) July–Aug (Sept) *(Pyrrocoma)*

17a Heads large, invol gen 1.5–3 cm, invol bracts relatively lfy and gen 3–8 mm wide (or even much wider), disk corollas 10–15 mm; rays relatively short and inconspicuous, seldom much > pappus (or longer in no 16), sometimes few or none

18a Pls very robust, basal lvs 5–20 cm wide, disk 3–4 cm wide, invol 2.5 cm, bracts ovate-oblong, imbricate, pale-margined; rays ca 34, 6–12 mm; dry hillsides in and near s end of SRC; SR g. *(H. carthamoides* var. *maximus)*
　　　　　　　　　　　　　　16 H. radiatus (Nutt.) Cronq.

18b Pls less robust, basal lvs 0.5–4 cm wide; rays 0–30, inconspicuous (rarely to 7 mm); meadows and open hillsides, valleys and plains to mid elev in mts; E Cas, Wn to nw Mont, s to Wyo, Nev, and Cal; large-fld g., Columbia g.; 2 vars.
　　　　　　　　　　　　　　17 H. carthamoides (Hook.) Gray

a1 Invol bracts tending to be broadly oblong, imbricate, and obviously scarious-margined, rarely with some of the outer enlarged, lfy, and spine-toothed; heads relatively broad, gen camp-hemispheric; e end of CRG, n and e to n Wn and nw Mont *(P. rigida)*　　　　　　　　var. **carthamoides**

a2 Invol bracts gen narrower and tapering from near the base to the more acute point, looser and seldom at all imbricate, obscurely or not at all scarious-margined; heads ave narrower, gen turbinate-camp; Blue Mts of e Ore and extreme se Wn, e to c Ida, and s to Nev and Cal *(P. c.)*
　　　　　　　　　　　　　　　　　var. **cusickii** Gray

17b Heads smaller, invol gen 0.5–1.5 cm, invol bracts narrower, gen 0.7–3 mm wide, disk corollas 5–10 mm; rays well developed and conspicuous

19a Herbage ± glandular, sometimes also woolly; lvs sharply and irreg toothed; heads several or sometimes solitary, the infl tending to be elongate in well developed pls; meadows and open slopes, foothills to mid elev in mts; c Wn to w Ida and s; sticky g., hairy g.; 3 weak vars.　　　　**18 H. hirtus** Gray

a1 Invol gen 9–15 mm, its bracts little if at all imbricate, with loose, often spreading tip, the green tip longer than in var. *lanulosus*

b1 Relatively robust pls of mesic habitats in Wn, ne Ore, and adj Ida, the lvs to 4 cm wide, the disk to 2.5 cm wide *(P. s.)*　　　　　　var. **sonchifolius** (Greene) Peck

b2 Smaller pls of drier places from ne Ore to Nev and Cal, the lvs to 2 or 2.5 cm wide, the disk to 2 cm wide *(P. h.)*
　　　　　　　　　　　　　　　　　var. **hirtus**

a2 Invol 8–11 mm, its bracts evidently imbricate, appressed, merely green-tipped; pls ave more strongly villous-tomentose and less glandular than in the other vars.; s Grant Co, Ore, s to Cal *(P. l.)*　var. **lanulosus** (Greene) Peck

19b Herbage variously hairy to glab, but not glandular

20a Heads gen solitary, occ 2 or 3 in robust pls; wet or dry, often alkaline meadows, occ also on open slopes

21a Invol gen 10–14 mm; disk corollas gen 7–10 mm; rays gen 1–2 cm; w Mont and adj Clark and Fremont cos, Ida; entire-lvd g.　　　　　**19 H. integrifolius** Gray

21b Invol gen 5–10 mm; disk corollas 5–7 mm; rays to ca 1 cm; c Ida and w Mont to Sask, Colo, se Ore, and Cal; one-fld g.; ours is var. *uniflorus*
　　　　　　　　　　20 H. uniflorus (Hook.) T. & G.

20b Heads gen several or ∝, occ solitary in depauperate pls; rays gen 5–12 mm

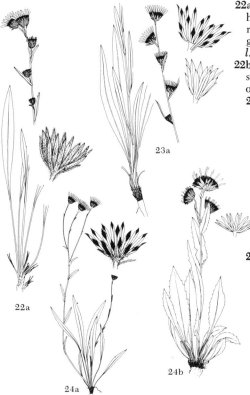

22a Lvs conspicuously pubescent (sometimes glab in age); heads relatively large, the invol 10–16 mm, the disk corollas 7–10 mm; grassy hillsides and prairies of Palouse region, se Wn and adj Ida; Palouse g. *(H. integrifolius* ssp. *l., P. scaberula)* 21 **H. liatriformis** (Greene) St. John

22b Lvs ± glab, except sometimes in no 23, which has smaller heads; wet or dry, often alkaline meadows, or occ on open slopes; not in se Wn and adj Ida

  23a Infl spicate-racemiform, the heads all short-pedunculate or subsessile; invol 8–14 mm; disk corollas 5–7 mm; n Ore and adj Ida to s Cal and Nev; racemed g.; we have 2 vars.

    22 **H. racemosus** (Nutt.) Torr.

    a1 Heads relatively broad, hemispheric or camp-hemispheric, the disk to 2.5 cm wide, the rays to 35; WV, apparently in scarcely alkaline soil

      var. **racemosus**

    a2 Heads narrower, gen turbinate-camp, the disk seldom much > 1 cm wide; the rays seldom > 15; alkaline meadows in e Ore and adj Ida *(P. glomerata, P. duriuscula)* var. **glomerellus** Gray

  23b Infl more corymbiform, the heads gen ± long-pedunculate (or subsessile in the more e *H. insecticruris)*

    24a Heads relatively small, the invol 5–10 mm, the disk corollas 5–7 mm; n Gr Pl, w to c Ida, e Ore, and Cal; lance-lf g. *(H. tenuicaulis)*

      23 **H. lanceolatus** (Hook.) T. & G.

    24b Heads larger, the invol 10–13 mm, the disk corollas 7–10 mm; Big Camas Prairie of Camas Co, Ida, extending into Blaine Co; bug-leg g. *(H. integrifolius* ssp. *i.)* 24 **H. insecticruris** Henderson

Helenium L.   Sneezeweed

Heads gen radiate, the rays ♀ or neutral, yellow or partly or wholly purple, cuneate, 3-lobed, few; invol bracts in 2–3 series, subequal or the inner shorter, narrow, herbaceous or subherbaceous, soon deflexed; recep convex or conic, naked; disk fls very ∝, ⚥; anthers minutely sagittate; style brs flattened, the tip dilated, subtruncate, minutely penicillate, the stigmatic lines well developed and introrsely submarginal; achenes truncately obpyramidal, 4–5-angled, with as many intermediate ribs; pappus of several series of hyaline, often awn-tipped scales; ann or per herbs with alt, glandular-punctate, gen decurrent, simple lvs. (Ancient Gr name of some plant said to be named for Helen of Troy).

H. autumnale L. Poisonous per to 12 dm; lvs ∝, lanceolate, narrowed to a sessile or subpetiolar base, 4–15 cm × 5–40 mm; heads several or ∝, hemispheric or subglobose, disk yellow, 1–2 cm wide; rays gen 10–20, yellow, soon deflexed; pappus gen < 2 mm; widespread Am sp. of moist low ground; 2 vars. in our range

a1 Rays gen ca 1 cm, sometimes to 1.5 cm; pl gen 1.5–6 dm, occ taller; E Cas *(H. m.)* var. **montanum** (Nutt.) Fern.

a2 Rays gen 1.5–2.5 cm; pl gen 4–12 dm; CRG and W Cas *(H. g.)*

    var. **grandiflorum** (Nutt.) T. & G.

Helianthella T. & G.   Little-sunflower; Helianthella

Heads radiate, the rays large, yellow, neutral; invol bracts ± herbaceous, in several series, subequal or imbricate, or the outer enlarged; recep flat or convex, chaffy throughout, its bracts clasping the achenes; disk fls ∝, fertile, yellow in our spp.; anthers sagittate; style brs flattened, without well developed stigmatic lines, and with short, blunt, externally hairy appendage; achenes strongly compressed at right angles to the invol bracts, thin-edged; pappus of several confluent, fimbriate, short scales, and gen 2 slender awns, persistent; taprooted, lfy-std per herbs with few or solitary, rather large heads; lvs simple and entire, at least the lower opp. (Name a diminutive of *Helianthus).*

1a Invol bracts ovate or lanceolate; rays pale yellow; recep bracts soft and
    scarious; lvs sparingly hairy, gen with 2 prominent pairs of lateral veins, the
    lower long-petiolate, larger than the often progressively reduced middle
    and upper; drier meadows and moist slopes at moderate to rather high elev
    in mts, e Ida and sw Mont s to Mex; nodding h.
                                            1 **H. quinquenervis** (Hook.) Gray
1b Invol bracts lance-linear or sometimes linear-oblong; rays bright yellow;
    recep bracts firm; lvs scab, gen ± triple-nerved, the lowest ± reduced;
    hillsides and open woods E Cas; RM h.; 2 vars.
                                            2 **H. uniflora** (Nutt.) T. & G.
  a1 Invol bracts conspicuously hirsute-ciliate, otherwise not very hairy; st
     gen spreading-hairy; lvs often broad-based; pl larger than var. *u.*, disk
     2–2.5 (3) cm wide, rays 3–4 cm; nw var., c and e Wn (and adj BC) to c
     Ore and c Ida (and sw Mont?) (*H. d.*)   var. **douglasii** (T. & G.) Weber
  a2 Invol bracts scarcely or only inconspicuously ciliate; sts appressed-
     puberulent; lvs all narrow-based; pls smaller, disk 1.5–2 (2.5) cm wide,
     rays 2–3 cm; se var., c Ida and sw Mont to Nev and NM   var. **uniflora**

Helianthus L.   Sunflower

Heads radiate, the rays large, yellow, neutral; invol bracts subequal or sometimes evidently imbricate, gen green and ± herbaceous; recep flat to convex or low-conic, chaffy throughout, its bracts
clasping the achenes; disk fls ♂ and fertile; anthers entire or minutely sagittate at base; style brs
flattened, externally (and sometimes distally internally) hispidulous, the short or elongate appendage hispidulous on both sides, the marginal or introrsely submarginal stigmatic lines poorly developed; achenes thick, moderately compressed at right angles to the invol bracts, with 2 evident
and gen 2 obscure angles, gen glab or nearly so; pappus of 2 readily deciduous awns with enlarged, thin, paleaceous base, rarely with some additional short scales; coarse ann or per herbs
with simple lvs, the lowest ones opp, the others sometimes alt. (Gr *helios*, sun, and *anthos*, fl).

1a Pls ann, fl June–Sept
  2a Central recep bracts conspicuously white-bearded at tip; invol bracts
     lanceolate, tapering, not long-hairy, scarcely ciliate; st seldom > 1 m;
     lvs narrower than in no 2, more often entire, rarely cordate; Gr Pl sp.,
     now widespread as a weed, in our range E Cas and less common than no
     2; prairie s.; ours is var. *p.*   1 **H. petiolaris** Nutt.
  2b Central recep bracts inconspicuously hairy, not at all bearded; invol
     bracts gen ovate or ovate-oblong and abruptly contracted above the
     middle, gen ciliate and with some long hairs on the back; st 0.4–2 m;
     lower lvs often cordate; open, dry to moderately moist soil, esp in waste
     places, in the valleys and foothills, native in w US and now widespread
     as a weed, also cult, chiefly in forms with a single very large head; com-
     mon s. (*H. aridus*, *H. lenticularis*)   2 **H. annuus** L.
1b Pls clearly per
  3a Invol bracts strongly imbricate, broad (ovate or lance-ovate), firm, ap-
     pressed, sharply acute to obtuse; lobes of disk corollas gen red or purple;
     Gr Pl and e Am sp., entering our range in the more e intermont valleys of
     Mont; showy s. (*H. laetiflorus*, misapplied, the name properly belonging
     to hybrids of *H. rigidus* and *H. tuberosus*); ours is var.
     SUBRHOMBOIDEUS (Rydb.) Cronq. hoc loc. (*H. subrhomboideus*
     Rydb. Mem. N. Y. Bot. Gard. 1: 419. 1900)   3 **H. rigidus** (Cass.) Desf.
  3b Invol bracts narrow, seldom much imbricate, some or all with loose,
     acuminate or attenuate to subcaudate tip; lobes of disk corollas yellow
     4a Lf bls broadly lanceolate to broadly ovate, gen 10–25 × 4–12 cm, on
        conspicuous petioles 2–8 cm; rhizomes gen tuber-bearing; st 0.7–5 m,
        evidently spreading-hairy; invol bracts gen rather dark, esp near base;
        sp. of c and e US, occ intro in our range; Jerusalem artichoke
                                            4 **H. tuberosus** L.
     4b Lf bls lanceolate or narrower, seldom > 3.5 cm wide, gen
        short-petiolate or sessile (or with longer petiole in no 7); rhizomes
        (when present) not tuber-bearing; st variously hairy or glab; invol
        bracts greener
        5a Sts 2–12 dm, nearly prostrate to erect, clustered on the crown of a
           thickened, soft, often ± turnip-shaped taproot; herbage rough-hairy
           or scab, or the st subglab; lvs entire, mostly or all opp; dry, open

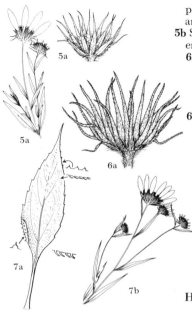

plains and foothills, e Ore and adj sw Ida and ne Cal n, between Cas and CR, to Ellensburg, Wn; Cusick's s. **5 H. cusickii** Gray

5b Sts 5–30 dm, erect, borne on short or elongate rhizomes with thickened, often fleshy roots

6a St pubescent with fine, appressed, white hairs; lvs entire or nearly so, not trinerved at base, gen at lease some of them conduplicate, often falcate; invol bracts gen long-attenuate or subcaudate; sp. of c US, occ intro in our range; Maxmilian's s.

6 H. maximilianii Schrad.

6b St glab (and often glaucous) or with a few coarse, spreading hairs; lvs flat, tending to be ± trinerved at base; invol bracts merely acuminate

7a Lvs glab or only slightly scab above, more softly hairy beneath, gen most of them alt, on a distinct petiole gen (1) 2–5 cm, evidently toothed to occ subentire; st gen glaucous; sp. of c and e US, sporadically intro in our range; saw-tooth s.

H. grosseserratus Martens

7b Lvs ± scab or scab-hispid on both sides, from mostly opp to mostly alt, on a short petiole seldom > 2 (3) cm, entire or nearly so; st glaucous or not; widespread cordilleran sp., common in bottoml, meadows, and other moist places up to moderate elev in mts, E Cas; Nuttall's s.; ours is var. *n.*

7 H. nuttallii T. & G.

Heliopsis Pers.   Heliopsis

Heads radiate, the rays yellow, ♀, fertile, persistent on the achene and becoming papery; invol bracts in 1 or 2 subequal series, herbaceous at least toward the tip; recep conic, sometimes narrowly so, esp in fr, chaffy throughout, its bracts clasping the achenes, subtending the rays as well as the disk fls; disk fls ☿ and fertile, the corolla tubular, scarcely narrowed at base; anthers entire or nearly so at base; style bulbous-thickened at base; style brs flattened, with short, externally hairy appendage, without well marked stigmatic lines; achenes nearly equably quadrangular (or those of the rays 3-angled) or occ subterete; pappus none, or of a short, irreg crown or a few teeth; per herbs with opp, petiolate lvs. (Named for its resemblance to the sunflower).

H. helianthoides (L.) Sweet. Ox-eye, false sunflower. Fibrous-rooted per 4–15 dm; lf bls ovate or lance-ovate, broad-based, serrate, 4–13 × 2–7 cm, petiolate, scab on both sides; heads 1–several, disk 1–2.5 cm wide; rays 8–15, rather pale yellow, 1.5–3 cm; dry woods, prairies, and waste places, widespread, chiefly more e sp., extending w to se BC; ours is var. *scabra* (Dunal) Fern. (*H. s.*).

Hemizonia DC.   Tarweed; Hemizonia

Heads radiate, the rays 3–45, yellow or white, ♀ and fertile; invol bracts herbaceous or subherbaceous, rounded on the back, each clasping the outer half of a ray achene, the inner margins of these achenes free; recep flat to subconic, sometimes chaffy throughout, with the individual bracts clasping the disk achenes, sometimes naked except for a single, cuplike row of bracts between disk and ray fls; disk fls fertile or more often sterile; anthers minutely sagittate; style brs of fertile disk fls flattened, externally hairy, with short, rather broad, introrsely submarginal stigmatic lines, the appendage sometimes hairy within; style brs of sterile disk fls gen more reduced, sometimes consisting merely of the short, hairy appendage; achenes not much compressed, often short-beaked; pappus of several short scales, or none; herbs with linear and entire to toothed or pinnatifid lvs, at least the lowermost gen opp. (Gr *hemi*, half, and *zone*, girdle, referring to the ray achenes).

H. pungens (H. & A.) T. & G. Common spikeweed. Coarse ann 1–10 dm; lvs ∝, alt, rigid, pungent, linear or with linear segms, the lower gen 2–6 cm and laciniate or pinnatifid, the upper smaller and often entire, often with reduced axillary fascicles; heads ± ∝, invol 3–5 mm; bracts shortly spine-tipped; rays 15–35, rather pale yellow, 4–7 mm; recep convex, chaffy; disk achenes mostly or all sterile; Cal sp. of roadsides, fields, and waste land, intro in Ore and Wn. Var. *p.*, known from a few scattered stations in our range, has scab lvs. Var. **SEPTENTRIONALIS** (Keck) Cronq. hoc loc. (*H. pungens* ssp. *septentrionalis* Keck, Aliso 4: 110. 1958), now a common weed about Walla Walla and in adj Umatilla Co, Ore, has the lvs smooth except for scattered long hairs.

Hieracium L. Hawkweed

Fls all lig and ⚥, gen yellow, sometimes red-orange or white; invol cylindric to hemispheric, the bracts evidently to obscurely imbricate; recep naked; achenes terete or prismatic, gen narrowed toward the base, truncate or occ narrowed toward the summit, ± strongly ribbed and sulcate; pappus of ∝ whitish to more often sordid or brownish capillary bristles; herbs with milky juice, fibrous-rooted from a short or elongate rhizome; lvs alt or all basal, entire or toothed; heads 1–∝ in a corymbiform or paniculiform, sometimes narrow and elongate infl; most spp. with at least a few stellate hairs on the herbage or invol. (Gr *hierax*, a hawk).

1a Basal and lowermost cauline lvs small and soon deciduous, the others, except for the reduced upper ones, gen rather ∝ and nearly alike in size and shape, sessile; invol with few or no long hairs; lvs not markedly setose-ciliate; fls 40–110 to a head; widespread spp. of thickets, woodl, and moist, open places, fl July–Sept

  2a St without long spreading hairs; lvs to 10 × 2 cm, gen 4–12 times as long as wide, not at all clasping, often narrowed to the base, provided with ∝ well developed, short, stout, subconic hairs, at least near the margins; circumboreal, s to n Ida and nw Ore; narrow-lvd h. (*H. scabriusculum*, the Am phase)   **1 H. umbellatum** L.

  2b St gen pubescent below with long, spreading hairs; lvs to 12 × 4 cm, gen 2–5 times as long as wide, tending to be broadly rounded and ± clasping at base, without subconic hairs, or these few and poorly developed; N Am sp., more common e of our range, but reaching Mont and less commonly BC, Wn, and reputedly Ore; Can h. (*H. columbianum*)
    2 H. canadense Michx.

1b Basal or lower cauline lvs markedly larger than the progressively reduced (and often few) middle and upper ones (this habit least pronounced in no 6), or the pl scapose; pubescence various

  3a Pls (except sometimes no 4) stoloniferous and with ± elongate rhizomes; st naked or with only 1–2 strongly reduced lvs; intro, weedy spp.

    4a Heads solitary or rarely 2–3 and then long-pedunculate (unique among our spp. in this regard); pls to 2.5 (4) dm; lvs tawny-tomentose with stellate hairs beneath, and with some long setae as well, green and glab above except for the very long setae; W Cas; mouse-ear h.
      3 H. pilosella L.

    4b Heads 5–30 in a more compact, corymbiform or umbelliform infl; pls to 9 dm; lvs long-setose on both sides or nearly glab above, not tomentose

      5a Fls yellow; locally intro in ne Wn; meadow h. **4 H. pratense** Tausch.

      5b Fls red-orange (unique among our spp.); W Cas, also in Flathead Co, Mont; king devil, orange h., golden mouse-ear h.
        5 H. aurantiacum L.

  3b Pls without stolons and with a short, caudexlike or praemorse rhizome; st gen (except in no 11) with several lvs

    6a Fls yellow; stellate hairs always present on invol or herbage or both (though sometimes obscured by the longer bristles)

      7a Lvs gen ± long-hairy, at least marginally (except sometimes in no 8), the lower ones often much > 10 cm; gen rather stout pls, 3–13 dm, occurring (except no 6) in dry places

        8a St appearing amply lfy, the lowermost lvs reduced and soon deciduous, those next above the largest (to ca 12 cm), the middle and often also the upper ones well developed and often only gradually reduced; lvs thin, often toothed, gen conspicuously setose-ciliate but only rather sparsely if at all hairy over the surface; invol gen evidently long-setose; CRG; long-beaked h. (*H. piperi*)
          6 H. longiberbe Howell

        8b St appearing less lfy, the basal or lower cauline lvs the largest (often much > 12 cm), the others progressively and strongly reduced; lvs firm (except no 7), variously hairy or subglab

          9a Lvs thin, some or all of them coarsely few-toothed, only rather sparsely setose (or more densely so on the midrib beneath), the basal ones rather abruptly tapered to the petiole; fls 40–80; European weed, intro at Portland and to be expected to spread W Cas; common h. (*H. lachenalii*, *H. strumosum*)
          7 H. vulgatum Fries

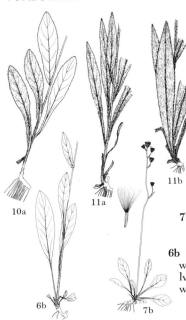

10a 11a 11b 6b 7b

9b Lvs firm, entire, variously hairy or glab, the basal ones more gradually tapered to the petiolar base; fls 15–50; widespread native spp.

10a Herbage sparsely or moderately setose below, subglab (or merely inconspicuously stellate) and often glaucous above; invol often glandular but seldom evidently long-setose; Cascade-Sierran sp., less common elsewhere in our range; woolly-weed **8 H. scouleri** Hook.

10b Herbage evidently long-setose above as well as below, not at all glaucous; gen E Cas

11a Invol evidently glandular and rather sparsely setose, varying to moderately or not at all setose; herbage moderately to rather sparsely or rather densely setose; houndstongue h. (*H. griseum*) **9 H. cynoglossoides** Arv.-Touv.

11b Invol and herbage very densely and conspicuously long-setose, scarcely or obscurely glandular; western h.
10 H. albertinum Farr

7b Lvs glab, or sometimes inconspicuously short-hairy, to 10 (13) cm; slender, subscapose pls of moist places in mts, gen 0.3–3.5 dm; cordilleran; slender h., alpine h. **11 H. gracile** Hook.

6b Fls white (unique among our spp. in this regard); stellate hairs wanting; pls tending to be long-hairy below, more glab above; lowest lvs largest; cordilleran sp. of fairly moist slopes and open woods; white-fld h. **12 H. albiflorum** Hook.

Hulsea T. & G.   Hulsea

Heads ∝ -fld, radiate, the 10–60 relatively narrow rays ♀ and fertile, yellow; invol bracts herbaceous, ± ∝, subequal in 2–3 series; recep flat, minutely horny-toothed, otherwise naked; disk fls yellow, ♂ and fertile; anthers minutely sagittate at base; style brs flattened, with well developed introrsely marginal stigmatic lines which sometimes extend nearly to the tip, and with short, poorly differentiated, merely externally minutely papillate-hairy appendage; achenes linear, hairy; pappus of several hyaline, nerveless scales, connate at base into a cartilaginous ring; aromatic, glandular herbs with alt, entire to pinnatifid lvs, our spp. tufted pers with short, monocephalous sts and ± succulent, glandular-villous to ± woolly lvs. (For Dr. G. W. Hulse, of the US Army).

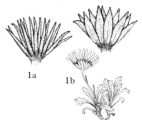

1a 1b

1a Invol bracts lance-linear, to 2 (2.5) mm wide; ± lfy-std pls, 1–4 dm, the lower lvs to 15 cm; rays gen 25–55; high mts E Cas (and Sierra Nevada); alpine h. (*H. carnosa*) **1 H. algida** Gray

1b Invol bracts lance-oblong, the wider ones gen 2–4 mm wide; gen scapose or subscapose pls to ca 1 dm, the lvs seldom 5 (8) cm, rarely taller and more lfy; rays gen ca 21, or occ fewer; cinder cones, moraines, and pumice plains, Cas; dwarf h. (*H. larsenii*, the more woolly phase) **2 H. nana** Gray

Hymenopappus L'Her.   Hymenopappus

Heads discoid, the fls all tubular and ♂; invol bracts few, appressed, in 2–3 gen subequal series, at least the inner with broad, obtuse or rounded, scarious or hyaline, ± petaloid, yellowish, whitish, or pinkish tip, more herbaceous below; recep small, flat, naked; corolla yellow or white; anthers minutely sagittate; style brs flattened, with introrsely marginal stigmatic lines which often extend into the short, externally papillate-hairy appendage; achenes rather narrow and elongate, 4–5-angled and less obviously 15–20-nerved; pappus of several or ∝ membranous or hyaline scales, or obsolete; herbs with alt or all basal, gen pinnatifid lvs. (Gr *hymen*, membrane, and *pappos*, seed-down).

H. filifolius Hook. Columbia cut-lf. Taprooted, several-std per, thinly tomentose (densely so at base); basal lvs tufted, once or twice pinnatifid, with linear segms; cauline lvs similar or smaller or none; infl corymbiform; invol 7–10 mm; pappus to 1 mm; widespread E Cas in dry places at lower elev; 3 vars.

a1 Tall (4–9 dm), lfy, freely brd, often with notably long lf segms; c Wn to c Ore **var. filifolius**

a2 Smaller (1–5 dm), less lfy, and less brd

b1 St with several lvs; Gr Pl and intermont valleys of Mont (*H. p.*) **var. polycephalus** (Osterh.) Turner

b2 St subscapose, with 0–2 lvs; pl stouter than var. *p.*; Salmon and Lemhi valleys, Ida **var. idahoensis** Turner

Hymenoxys Cass.   Hymenoxys

Heads radiate, the rays ♀ and fertile, broad, yellow, gen 5–35; invol bracts partly or wholly herba-ceous, in 2–3 similar or sharply differentiated series; recep naked, hemispheric to merely convex; disk fls ± ∝, yellow, ♂ and fertile; anthers entire or merely sagittate at base; style brs flattened, truncate, minutely penicillate, with introrsely marginal stigmatic lines; achenes turbinate, hairy, gen 5-angled; pappus of a few (gen 5) hyaline, often aristate or awned scales; per (all ours) or ann aromatic herbs with alt or all basal, entire to pinnatifid or ternate, gen punctate lvs. (Gr *hymen,* membrane, and *oxys,* sharp, referring to the pappus). (*Actinea* and *Actinella,* misapplied).

1a Lvs entire, essentially all basal, strongly sericeous; scapes monocephalous, to 3 dm; disk 8–20 mm wide; invol bracts similar, distinct; rays 5–20 mm; widespread inl sp. of high plains and dry slopes, reaching our range in Mont; Ariz h., stless h.; ours is var. *a.* (*Tetraneuris a.*)
1 H. acaulis (Pursh) Parker
1b Lvs cauline and basal, many or all of them ternately cleft or pinnatifid
  2a Heads small, rays ca 8 or 13, 7–15 mm, disk 7–17 mm wide, invol 4–8 mm, strongly biseriate, the firm, keeled outer bracts connate below; pls 1–2 dm, with gen 1–3 heads, the larger lvs ternate; Gr Pl, s RM, and GB sp., reaching our range in the intermont valleys of Mont; Richardson's h.; ours is var. *r.*
2 H. richardsonii (Hook.) Cockerell
  2b Heads large, rays 15–35, 15–30 mm, disk 15–35 mm wide, invol 9–15 mm, its bracts loose, slender, herbaceous, similar; pls 1–3 dm, with 1–few heads; lvs once or twice pinnatifid, or merely ternate; talus slopes and other rocky places or gravelly meadows at high elev, c Ida and sw Mont to Colo and Utah; old-man-of-the-mt (*Rydbergia g.*)
3 H. grandiflora (T. & G.) Parker

Hypochaeris L.   Cats-ear

Similar to *Leontodon,* from which it is distinguished primarily by its chaffy-bracted recep; our spp. European weeds with some or all of the achenes long-beaked, intro chiefly W Cas, fl May–Oct. (Name used by Theophrastus for this or some other cichorioid genus).

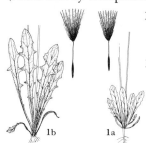

1a Ann or winter ann, taprooted, essentially glab; heads opening only in full sun, not very showy, the ligs about = invol and only ca twice as long as wide; invol 8–10 mm at anthesis, to 17 mm in fr; disturbed and waste places, esp in sandy soil, wholly W Cas; smooth c.   1 H. glabra L.
1b Per from a caudex, fibrous-rooted or more often several of the roots en-larged; lvs hispid; heads opening in bright or dull weather, showy, the ligs > invol and ca 4 times as long as wide; invol 10–15 mm at anthesis, to 25 mm in fr; lawns, pastures, and other disturbed sites, chiefly W Cas, but also in n Ida; spotted c., hairy c., gosmore   2 H. radicata L.

Inula L.   Inula

Heads gen radiate, the rays gen yellow, ♀ and fertile; invol hemispheric or camp, its bracts imbri-cate, the inner gen narrow and scarious, the outer broader and often herbaceous; recep flat or convex, naked; disk fls tubular and ♂, yellow; anthers caudate; style brs flattened, slightly broader upwards, the stigmatic lines ventromarginal, extending completely around the rounded apex; pappus of capillary bristles, often unequal; achenes prominently to obscurely 4–5-ribbed or -angled; gen per herbs, ± glandular or hairy, with alt or basal lvs. (The ancient L name).

I. helenium L. Elecampane. Coarse per to 2 m; st finely spreading-hairy; lvs ± toothed, densely velvety beneath, sparsely spreading-hairy or subglab above, the lower long-petiolate, with elliptic bl to 5 × 2 dm, the upper be-coming ovate, sessile; heads few, large, disk 3–5 cm wide; invol 2–2.5 cm, outer bracts herbaceous; rays slender, > 1 cm; European sp., occ cult, and occ as a roadside weed W Cas.

Iva L.   Poverty-weed; Marsh-elder

Heads disciform, the ♀ fls few, with tubular or obsolete corolla; invol of 1–3 = or imbricate series of bracts, sometimes with a shorter inner series subtending the achenes; recep small, chaffy, with linear or spatulate bracts; ♂ fls with undivided style; filaments monadelphous, anthers obtuse at base, scarcely united; achenes compressed parallel to invol bracts; pappus none; herbs (our spp.) with opp (or the upper alt) lvs and small heads (invol 1.5–4 mm in our spp.) of greenish-white fls. (Name unexplained).

1a Ann, 1.5–20 dm; lvs long-petiolate, with ovate, coarsely toothed bl gen 5–20 × 2.5–20 cm; infl large, paniculiform, the heads not individually subtended by lvs; invol with a shorter inner series of bracts subtending the achenes; bottoml and rather moist waste places, cordilleran, E Cas, fl Aug–Oct; tall m. (*Cyclachaena x.*)   1 I. xanthifolia Nutt.

1b Per from deep-seated creeping roots, 1.5–6 dm; lvs sessile or nearly so, oblong or ovate to broadly linear, entire, 1–5 cm; heads axillary, nodding; invol without shorter inner bracts; dry, often alkaline places at lower elev, sometimes a field weed; cordilleran, gen E Cas, fl May–Sept; p-w., deep-root   2 I. axillaris Pursh

Jaumea Pers.   Jaumea

Heads gen radiate, the rays ♀ and fertile, yellow; invol of rather few, broad, ± herbaceous bracts, well imbricate in several series; recep flat or conic, naked; disk fls ☿ and fertile; anthers sagittate at base; style brs flattened, with introrsely marginal stigmatic lines, the appendage short and blunt, minutely papillate-hairy inside and out, or obsolete; achenes elongate, 10-nerved; pappus of rather ∝ narrow scales or bristles, or none; per herbs or subshrubs with opp, narrow, entire lvs. (Named for I. H. Jaume St. Hilaire, French botanist).

J. carnosa (Less.) Gray. Fleshy j. Succulent, lax, rhizomatous per to 3 dm; lvs 2–6 cm × 2–6 mm; heads gen solitary; invol 8–12 mm; rays narrow and inconspicuous, 3–5 mm; recep conic; style art above a bulbous swelling near the base; achenes with a cartilaginous apical ring; pappus none; tidal flats and marshes, s VI to s Cal.

Kuhnia L.   Kuhnia

Heads discoid, the fls all tubular and ☿; invol bracts striate, obviously imbricate, the outer several series gen much < or differently shaped (or both) than the only slightly unequal inner ones; recep flat or nearly so, naked; corollas creamy-white to dull yellow or red; anthers gen partly or wholly separate by fl time, obtuse at base; style brs with an elongate, clavate, papillate appendage; achenes 10-ribbed; pappus of 10, 15, or 20 uniseriate, =, plumose bristles; per herbs from a stout taproot, with opp, offset, or scattered lvs and 1–∝ subcylindric to camp, medium-sized heads. (Named for Adam Kuhn of Philadelphia, 1741–1817, who brought the original sp. to Linnaeus).

K. eupatorioides L. False-boneset. Pls 3–10 dm, densely puberulent; lvs ± sessile, gland-dotted beneath, gen 2–10 cm × 4–40 mm; heads gen in small, corymbiform clusters at br tips; invol 9–14 mm, outer bracts with slender, attenuate tip; fls 14–33, creamy; dry, open places, widespread in c and e US, reaching our range in Mont; ours is var. *corymbulosa* T. & G., of the prairies and plains.

Lactuca L.   Lettuce

Fls all lig and ☿, yellow, blue, or white, the corolla tube gen > half as long as the lig; invol cylindric, often broadening at base in fr, calyculate or more often with the bracts imbricate; recep naked; achenes compressed, winged or strongly nerved marginally, with 1–several lesser nerves on each face, beaked or beakless, but in any case expanded at the summit where the pappus is attached; pappus of ∝ capillary bristles, none markedly larger than the others, at least some of them no > 4-celled in cross section at base; lfy-std herbs with milky juice, alt, entire to pinnatifid lvs, and gen ∝ rather few-fld heads (5–56 fls in our spp.) in a gen paniculiform infl, our spp. fl July–Sept (no 7 June–Sept). (The ancient L name, from *lac*, milk, referring to the milky juice). The common cult lettuce, *L. sativa* L., which may occ escape or persist after cult, may be distinguished from any of the following spp. by its very broad, merely toothed lvs.

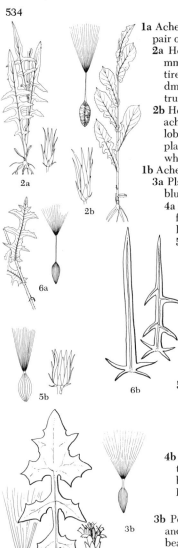

1a Achenes with only a median nerve on each face, or occ with an additional pair of very obscure ones.

  2a Heads relatively small, with gen 13–22 yellow fls; fr invol gen 10–15 mm; achenes gen 4.5–6.5 mm (beak included); pappus 5–7 mm; lvs entire to pinnately lobed, not prickly-toothed; weedy ann or bien, 3–25 dm; widespread in c and e US, and occ intro in our range; Can wild l., trumpet fireweed   **1 L. canadensis** L.

  2b Heads larger, with gen 20–56 yellow or blue fls; fr invol gen 15–22 mm; achenes gen 7–10 mm; mature pappus 7–10 mm; lvs pinnately lobed or lobeless, prickly-toothed; short-lived per, 3–15 dm, of open, rather moist places; Gr Pl, w to Mont and Lemhi Co, Ida, and casually intro elsewhere in our range; western l.   **2 L. ludoviciana** (Nutt.) Riddell

1b Achenes evidently to very prominently several-nerved on each face

  3a Pls ann or bien; heads relatively small; invol 9–15 mm in fr; fls yellow, blue, or whitish

    4a Fls gen 8–34 (55) to a head; lvs variously entire or toothed to pinnatifid, but not with an ivy-like terminal segm; middle and upper cauline lvs only gradually reduced

      5a Achenes with a filiform beak from nearly as long to twice as long as the body; pappus white; fls yellow, often drying blue

        6a Lvs prickly margined, pinnately lobed or (in var. *integrata* Gren. & Godr.) lobeless (the var. perhaps reflecting introgression from *L. sativa*); achenes evidently spinulose above, at least at maturity; fls (13) 18–24 (27) to a head; European weed 3–15 dm, of fields and waste places, now naturalized throughout most of the US (*L. scariola*, a variant spelling); prickly l.   **3 L. serriola** L.

        6b Lvs linear and entire or with scattered, sometimes slightly toothed lobes, but not prickly margined; achenes merely scab above; fls 8–16 to a head; European weed, 3–10 dm, of waste places, occ intro W Cas; willow l., least l.   **4 L. saligna** L.

      5b Achenes with a short stout beak < half as long as the body, or beakless; pappus brownish; fls gen 13–34 (55) to a head, bluish to white, less often yellow; lvs pinnatifid or merely toothed, but not prickly-margined; robust ann or bien weed of moist places, 6–20 dm, with notably narrow and elongate infl; tall blue l. (*L. spicata*, misapplied)   **5 L. biennis** (Moench) Fern.

    4b Fls 5 to a head, yellow; lower lvs pinnatifid, with broad, ± ivy-like terminal segm; middle and upper lvs few and reduced; slender ann or bien 3–9 dm; beak of achene short, slender, < half as long as body; European weed of moist places, occ W Cas; wall l.   **6 L. muralis** (L.) Fresen.

  3b Per from deep-seated creeping roots, 2–10 dm; heads relatively large and showy; invol 15–20 mm in fr; fls 18–50, blue; achenes with a stout beak up to half as long as the body; widespread cordilleran sp. of meadows, thickets, and other moist places; blue l. (*L. tatarica* spp. *p.*)

                        7 L. pulchella (Pursh) DC.

Lagophylla Nutt.   Hareleaf; Rabbitleaf

Heads radiate, the ray fls 5, ♀ and fertile, with broad, 3-cleft or -parted lig; invol bracts thin, herbaceous, deciduous with the completely enclosed ray achene; recep small, with a bract subtending each of the outer disk fls; disk fls 6, sterile, with slender, abortive achene; ray achenes compressed at right angles to a radius of the head, glab; pappus none; slender ann with narrow, entire or glandular-toothed lvs, the lower opp, the upper alt. (Gr *lagos*, hare, and *phyllon*, lf, referring to the canescent lvs).

L. ramosissima Nutt. Slender h., common h. Ann 1–10 dm, ± canescent, esp upwards; lvs entire or nearly so, the lowest oblanceolate or spatulate, gen deciduous, the others smaller, lanceolate to linear, sessile; invol 4.5–8 mm, gen short-stipitate-glandular as well as silky-villous; rays 2–5.5 mm, pale yellow, turning purple beneath; dry plains and foothills, Wn and adj Ida to Cal, chiefly E Cas.

Lapsana L.   Lapsana

Fls all lig and ⚥, yellow; invol cylindric-camp, minutely calyculate, the inner bracts subequal, uniseriate, often keeled; recep naked; achenes narrow, subterete or slightly compressed to ± angular, often curved, narrowed to both ends, ca 18–30-nerved, 5 or 6 nerves gen stronger; pappus none, or occ of some short, inconspicuous, recurved-divergent, firm awns; ann with milky juice, alt, entire to pinnatifid lvs, and several or ∞ small, rather few-fld heads. (Gr *lapsane*, applied by Dioscorides to a crucifer).

1a Erect, single-std pls 1.5–15 dm; principal invol bracts gen 8, persistently erect, evidently keeled below; lvs cauline, petiolate, with ovate to subrotund, obtuse or rounded, toothed or occ basally lyrate bl 2.5–10 × 2–7 cm; fls gen ca 13; achenes curved; Eurasian weed, occ encroaching into woodl, now widely intro in US, in our range gen W Cas; fl June–July; nipplewort
1 **L. communis** L.

1b Prostrate pls with several sts 1–3 dm; principal invol bracts gen 5, scarcely keeled, becoming stellate-spreading at full maturity; lvs chiefly basal, 4–10 cm (petiole included) × 1–2.5 cm, often lyrate; fls gen 5 or 8; achenes straight; e Asian weed of cult fields and disturbed sites, W Cas in our range; fl May; Japanese l. 2 **L. apogonoides** Maxim.

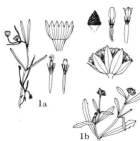

Lasthenia Cass.　Lasthenia

Heads radiate, the rays yellow, ♀ and fertile, in our spp. short and inconspicuous; invol of a few =, gen ± herbaceous, distinct to partly or wholly connate bracts, often with a tendency to embrace the subtended achenes; recep conic, varying to subulate or hemispheric, naked; disk fls ♂ and fertile, yellow; anthers minutely or scarcely sagittate at base; style brs flattened, with introrsely marginal stigmatic lines, with or without an externally papillate-hairy appendage; achenes elongate; pappus of awns, or scales, or both, or sometimes wanting; slender ann (our spp.) or per herbs gen < 4 dm, with opp, sessile or subsessile, entire to laciniate-pinnatifid, often narrow lvs. (Named for Lasthenia, a courtesan who was a student of Plato; of no obvious application). *(Baeria)*.

1a Invol bracts united into a toothed cup; pappus of 5–10 firm, erose-laciniate to shortly awn-tipped scales; lvs linear, entire, 2–8 cm; wet or muddy ground W Cas; s Wn to Cal; fl May–July; smooth l. 1 **L. glaberrima** DC.

1b Invol of separate bracts; pappus of 3–7 awns and as many alternating, laciniate, shorter scales; lvs narrowly oblong, entire to slenderly pinnatilobate, 1.5–5 cm; coastal, s BC to Cal; fl July–Sept; hairy l.; ours is var. **MARITIMA** (Gray) Cronq. hoc loc. *(Burrielia maritima* Gray, Proc. Am. Acad. 7: 358. 1868; *Baeria maritima)* 2 **L. minor** (DC.) Ornduff

Layia H. & A.　Tidytips; Layia

Heads radiate, the 1–13 yellow or white, 3-toothed or -lobed, broad rays ♀ and fertile; invol bracts herbaceous, flattened on the back below, with abruptly dilated thin margins infolded around the achene; recep broad and flat, bearing a series of thin chaffy bracts between the ray and disk fls; disk fls ♂ and fertile; anthers slightly sagittate; style brs flattened, externally hairy above, with broad, introrsely submarginal stigmatic lines extending nearly or quite to the slender tip; ray achenes flattened at right angles to a radius of the head, without pappus; disk achenes gen with a pappus of 10–35 bristles or slender scales; ann herbs; lvs chiefly alt, at least the lowest ones laciniate or pinnately lobed to bipinnatifid in well developed pls. (Named for Thomas Lay, naturalist on Beechey's voyage).

L. glandulosa (Hook.) H. & A. White l., white daisy t. Brd, spreading-hairy, 0.5–4 dm, ± stipitate-glandular esp above; lvs 1–6 cm, narrow, the lower gen toothed or pinnatifid; invol 6–9 mm; rays broad, white, 4–15 mm; disk achenes appressed-hairy, with ca 10 flattened white pappus bristles which are long-woolly-villous toward the base within; dry, open places at low elev, esp in sandy soil; E Cas.

Leontodon L.　Hawkbit

Fls all lig and ♂, yellow; invol ovoid or oblong, imbricate or calyculate; recep ± alveolate or fimbriately villous, but not chaffy-bracted; achenes narrow, subterete, several- or ∞-nerved, long-beaked or merely narrowed upwards; pappus of plumose bristles, sometimes with some shorter nonplumose outer bristles or scales, or that of the marginal achenes sometimes wholly of the latter type; ann or (ours) per lactiferous herbs with well developed basal lvs and a naked or scaly-bracted scape, monocephalous or with the heads terminating the brs. (Gr *leon*, lion, and *odous*, tooth, referring to the toothed lvs).

1a Scape scaly-bracted and gen several-headed; pappus wholly of plumose bristles; lvs glab or hirsute with simple hairs; Eurasian weed of roadsides, pastures, etc., now well estab in e US, and casual with us; fl July–Sept; fall dandelion or h.   1 **L. autumnalis** L.

1b Scape monocephalous and gen naked; pappus with some shorter outer merely barbellate bristles or scales, some of the marginal achenes wholly without plumose bristles; lvs hispid with shortly forked hairs; European weed of lawns and waste places, estab W Cas, esp along the coast; hairy h.; fl June–July; ours is called ssp. *taraxacoides* (Vill.) Schinz & Thell. (*L. leysseri*)   2 **L. nudicaulis** (L.) Merat

Liatris Schreb.   Liatris; Blazing-star

Heads discoid, the fls all tubular and ⚥, 3–100 or more to a head; invol bracts imbricate, often with petaloid scarious margins; recep naked; corollas pink-purple (white); anthers minutely sagittate at base; style brs with short, rather obscure, marginal stigmatic lines and an elongate, papillate, often clavate appendage; achenes ca 10-ribbed, pubescent; pappus of 1 or 2 series of strongly barbellate to plumose capillary bristles; per herbs, gen from an evident corm, or sometimes from a more elongate caudex or stout rhizome, with alt, entire, ± punctate lvs, the lower gen the largest, and smallish to middle-sized heads in a gen spiciform or racemiform infl. (Name of uncertain derivation).

L. punctata Hook. Pl 1–4 dm, glab except the often ciliate lf-margins; lvs ∝, ± linear; invol subcylindric, 10–18 mm, its bracts mucronate-acuminate; fls (3) 4–6 (9) to a head; corolla tube hairy toward the base within; pappus evidently plumose; Gr Pl sp., often of sandy soil, reaching our range in Mont; fl July–Sept.

Luina Benth.   Luina

Heads discoid, the fls all ⚥ and fertile, yellow or yellowish; invol a single series of firm, =, scarcely herbaceous to subherbaceous bracts; recep naked; anthers entire or minutely sagittate at base; style brs flattened, externally merely papillate or papillate-puberulent, with broad, introrsely marginal stigmatic lines and a thickened, very short and blunt, papillate appendage; pappus of ∝ capillary bristles; achenes prominently several-nerved; per herbs with simple, entire to deeply cleft, alt lvs; fl June–Oct. (Name an anagram of *Inula*). (*Cacaliopsis, Rainiera*).

1a Principal lvs with broad, palmately cleft blade to 2 × 2.5 dm, abruptly contracted to the long petiole, basally disposed, the cauline lvs few and progressively reduced; heads several, large; invol 10–17 mm; disk 12–30 mm wide; fls > 30; meadows and open woods, crest and e slope of Cas in Wn and n Ore, irreg s and w; silvercrown l. (*Cacaliopsis n.*); ours is var. *glabrata* (Piper) Cronq.   1 **L. nardosmia** (Gray) Cronq.

1b Principal lvs entire or slightly toothed, sessile or tapering to the petiole, < 1 dm wide; heads smaller; invol 5–10 mm; disk to ca 1 cm wide; fls < 30

2a Infl short, corymbiform or subumbelliform; lvs white-tomentose esp on the lower surface; st rather equably lfy; basal lvs wanting

3a Principal lvs petiolate, narrowly elliptic to lanceolate, 7–13 cm (petiole included), 5–11 times as long as wide; invol 8–10 mm, of 10–17 (ave 13) bracts; fls 15–29 (ave ca 21) to a head; open, rocky, serpentine slopes, Grant Co, Ore; colonial l.   2 **L. serpentina** Cronq.

3b Lvs all sessile, rather broadly elliptic or ovate, 2–6 cm, 1.5–3.5 times as long as wide; invol 5–7 (8) mm, of ca 8–10 bracts; fls 10–17 (ave 13) to a head; rocky places, Cas and w; silverback l. 3 **L. hypoleuca** Benth.

2b Infl elongate, thyrsoid-racemiform; lvs glab, the basal and lowermost cauline ones 1.5–3.5 dm (including the petiolar base) × 2–7 cm, persistent, the middle and upper ones progressively reduced; heads with ca 5 fls and 5–6 invol bracts; meadows and moist open slopes at high elev, Cas of c Wn to c Ore; tongue-lf l. (*Psacalium s., Rainiera s.*)   4 **L. stricta** (Greene) Rob.

Lygodesmia D. Don   Rush-pink; Skeletonweed; Skeletonplant

Fls all lig and ♀, pink or purple (white); invol cylindric, of 4–8 principal bracts, with a few ± reduced outer ones; recep naked; achenes linear, subterete, prominently several-nerved or nerveless, glab, gen narrowed above and sometimes also below; pappus of ∝ capillary bristles; ann or (ours) per, ± rushlike herbs with milky juice and alt, gen linear or subulate lvs (these sometimes reduced to mere scales, or the lowest ones in some extralimital spp. more ample). (Gr *lygos*, a pliant twig, and *desme*, bundle, from the habit of the pl).

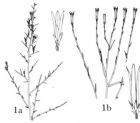

1a Brs spine-tipped, rigid, divaricate; sts several from a taproot and br caudex, bearing tufts of pale or brownish wool at base; heads short-pedunculate, borne laterally on the brs; dry, open, gen rocky places at lower elev; E Cas, rare and local in our range; spiny s. 1 **L. spinosa** Nutt.

1b Brs not spine-tipped; pls single-std from a deep-seated creeping root, without wool; heads terminating the brs; dry, open, often sandy places; Gr Pl sp., w occ to e Wn; rush-like s.   2 **L. juncea** (Pursh) D. Don

1a   1b

Machaeranthera Ness   Aster

Heads gen radiate, rays ♀, anthocyanic to white, or occ none and the heads discoid; invol bracts imbricate in several series, pale and chartaceous or coriaceous toward the base, distally ± herbaceous and often squarrose-reflexed, often glandular or canescent or both; recep flat or nearly so, naked but often strongly alveolate; disk fls ± ∝, yellow to red; anthers ± entire at base; style brs flattened, with introrsely submarginal stigmatic lines and a well developed, externally papillate-hairy appendage that in our spp. is ± narrowly acute and = or > the stigmatic lines; achenes gen several-nerved; pappus of ± ∝ markedly unequal barbellate bristles, often brownish; taprooted herbs or shrubs (ours all ann or bien to per herbs) with alt, spinulose-tipped, entire to more often spinulose-dentate or pinnatifid to pinnately dissected lvs and in our spp. several or ± ∝ heads terminating the brs; our spp. fl July–Oct. (Gr *machaira*, sword, and *anthera*, anther, from the anther appendages).

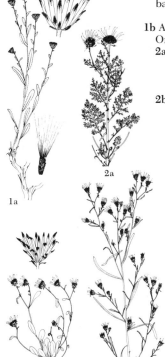

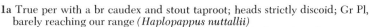

1a True per with a br caudex and stout taproot; heads strictly discoid; Gr Pl, barely reaching our range (*Haplopappus nuttallii*)
                            1 **M. grindelioides** (Nutt.) Keck & Cronq.
1b Ann, bien, or short-lived per; heads gen radiate (discoid in some pls from c Ore)
  2a Ann with pinnately incised to tripinnatifid lvs; achenes relatively short and broad, 2–3.5 mm, densely hairy; rays 12–36, 10–16 mm; Gr Pl sp., entering our range in Mont; tansy a. (*Aster t.*)
                            2 **M. tanacetifolia** (H.B.K.) Nees
  2b Bien or short-lived per with entire or merely toothed lvs; achenes broadly linear or clavate, 3–5.5 mm, rather sparsely hairy or glab; rays 8–35 (50), 5–12 mm
    3a Lvs green and glab; st glandular in the infl, otherwise subglab; resembling no 4 in habit and invol, gen not > 2 dm; mts of c Ida and ne Nev; vivid green a. (*Aster leiodes;* not *A. laetevirens*)
                            3 **M. laetevirens** Greene
    3b Lvs and st canescent, often glandular in the infl
      4a Invol bracts relatively broad, gen 1–2 mm wide; pls bien or often short-lived per
        5a Invol bracts gen 3–4-seriate, somewhat or not at all squarrose; rays gen ca 8 or ca 13, or none; dry, open places; mts of e Ore, w to Cas of c Ore, and s; Shasta a.; we have 3 vars. (*Aster s.*)
                            4 **M. shastensis** Gray
          a1 Rays wanting; habit gen of var. *glossophylla;* Cas from c Ore s
                            var. **eradiata** (Gray) Cronq. & Keck
          a2 Rays well developed
            b1 Lvs narrow, the lower gen linear-oblanceolate, not infrequently toothed; pl nearly or quite without a caudex, gen erect or stiffly ascending, to 4 dm; plains and foothills to moderate elev in mts, c and ne Ore, s to Cal
                            var. **glossophylla** (Piper) Cronq. & Keck
            b2 Lvs broader, the lower gen oblanceolate or broader, gen entire or subentire; pl tending to have the taproot surmounted by a slender brd caudex, gen 1–2 dm and decumbent or as-

b2   b1
        5a

cending; alp and subalp in Wallowa Mts and Cas of c Ore
var. **latifolia** (Cronq.) Cronq. & Keck
5b Invol bracts gen 4–8-seriate and ± strongly squarrose-tipped;
rays gen > 13; mts of c Ida to Utah and Wyo; mixed a.
5 M. commixta Greene
4b Invol bracts narrower, up to about 1 mm wide, gen 4–8-seriate
and evidently squarrose-tipped; pls bien or seldom per, gen br and
several-std, 1–5 (10) dm; lvs gen toothed; dry, open places, gen in
plains and foothills, cordilleran, widespread E Cas but rare or absent from most of Blue Mt region of Ore; hoary a. (*Aster c.; A. leu-
canthemifolius*, misapplied) **6 M. canescens** (Pursh) Gray

Madia Mol.   Tarweed; Madia

Heads radiate, the rays ♀ and fertile, yellow, relatively broad, often short, or rarely wanting
and the heads thus discoid; invol bracts subherbaceous, uniseriate, =, enfolding and gen completely enclosing the ray achenes, the invol gen appearing deeply sulcate; recep flat or convex,
with a single series of bracts between ray and disk fls, otherwise naked or merely hairy; disk fls ♂
but sometimes sterile; anthers minutely sagittate; style brs externally hairy at least above, those of
the fertile disk fls, at least, with ± evident, introrsely marginal stigmatic lines extending to near
the tip; achenes of the rays finely striate, gen incurved, in most spp. compressed parallel to a
radius of the head, those of the disk similar, or abortive; pappus gen none, or sometimes a short
crown or a few scales; ± glandular and heavy-scented (typically tar-scented) ann or occ bien or
per herbs with entire or slightly toothed lvs, at least the lower opp, the upper (except in no 2) gen
alt. (Name from *Madi*, the Chilean name of *M. sativa*).

1a Heads very small, invol 2–4.5 mm, rays minute; recep bracts united into a
cup about the solitary fertile disk fl (1–4 additional, gen fertile disk fls
rarely present); fl May–July
2a Achenes compressed parallel to a radius of the head; middle and upper
lvs gen alt; heads on filiform naked peduncles; dry, open woods and
grassl in the plains and foothills, occ up to mid elev in mts; widespread
in our range, s to Nev and Baja Cal; little t.
1 M. exigua (J. E. Smith) Gray
2b Achenes ± compressed at right angles to a radius of the head (unique in
the genus in this respect); lvs all or mostly opp; heads in the forks of the
st and in small, cymose clusters terminating the brs; dry, open ponderosa
pine woods, and in prairies W Cas; Wn and adj n Ida and s BC, s to Cal;
small-head t. (*Hemizonella m., H. durandii*) **2 M. minima** (Gray) Keck
1b Heads larger, invol 5–12 mm (or only 4 mm in some spp. with conspicuous
rays); disk fls gen several
3a Disk fls fertile; recep bracts gen distinct, each enclosing an achene; fl
June–Sept
4a Heads fusiform (unique in the genus in this respect), 2–5 mm wide as
pressed, glomerate; rays 1–5, ca 2 mm, or on some heads none; widespread cordilleran sp. of dry, open places from foothills and valleys to
mid elev in mts, often along roadsides, avoiding the drier plains;
cluster t., mt t. **3 M. glomerata** Hook.
4b Heads ovoid or broadly urn-shaped, the invol gen 6–12 mm wide as
pressed, not glomerate except in forms of no 5; rays 3–7 mm
5a Herbage mainly hairy, becoming stipitate-glandular chiefly in the
infl, the glands rarely extending below the middle of the st; heads
not glomerate; odor ± spicy; rays 5–13, typically 8; widespread
cordilleran (and Chilean) sp. of dry, open places, often along roadsides, from the valleys to mid elev in mts, avoiding the dry plains;
slender or common t., gum-weed (*M. dissitiflora*)
4 M. gracilis (J. E. Smith) Keck
5b Herbage strongly stipitate-glandular as well as hairy, the glands
extending nearly or quite to the base of the st; heads often glomerate; odor heavy and unpleasant; rays gen ca 13, occ only 8; dry,
open places, often along roadsides, W Cas from n Wn to Cal, and in
Chile; Chile t., coast t.; 2 vars. **5 M. sativa** Mol.
a1 Heads short-peduncled, scattered or subglomerate, the subtending bracts narrowly lanceolate, short **var. sativa**

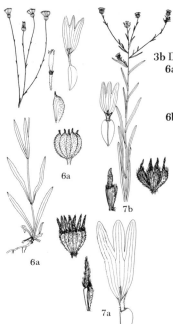

6a

7b

6a

7a

a2 Heads congested in terminal glomerules, the subtending (mature) bracts deltoid-lanceolate and gen > invol; well developed pls stouter and more divaricately br than var. *s. (M. capitata)*
                                                                                             var. **congesta** T. & G.
3b Disk fls sterile, the recep bracts ± united into a cup surrounding them
   6a Bien or short-lived per, gen with a short rhizome; disk fls with evident chaffy pappus (unique among our spp. in both these respects); recep glab; rays gen ca 8 (13); open woods W Cas, fl June–July; woodland t.
                                                                                   6 **M. madioides** (Nutt.) Greene
   6b Ann; disk fls without pappus; recep with ∝ erect straight hairs
      7a Rays gen ca (8) 13 (21), relatively large, gen 10–17 mm (large size unique in genus); achenes compressed; pls not lemon-scented; dry, open places W Cas, often becoming a roadside weed; showy t.; we have 2 vars.   7 **M. elegans** D. Don
         a1 Summer-fl mont ecotype 2–8 dm; basal lvs few or in a small rosette, the lower st lvs gen not so conspicuously crowded as in var. *d.*   var. **elegans**
         a2 Fall-fl lowland ecotype to 25 dm; basal lvs ∝ in a large rosette, the lower st lvs crowded and closely imbricate
                                                                                        var. **densifolia** (Greene) Jeps.
      7b Rays gen 5–8, smaller, gen 5–10 mm; achenes rounded-trigonous; pls lemon-scented (unique among our spp. in these last 2 respects); dry hillsides E Cas, se and sc Wn to n Cal, fl May–June; lemon-scented t.   8 **M. citriodora** Greene

Malacothrix DC.   Malacothrix

Fls all lig and ♂; invol bracts imbricate or calyculate, scarious-margined; recep setose or naked; achenes columnar, truncate, ribbed, with an entire or denticulate border or crown at the top; pappus of ∝ white capillary bristles, ± united at base and tending to fall off in a ring, often with 1–several persistent, stiffer bristles as well; herbs with milky juice and alt, toothed to pinnatifid lvs, the cauline ones sometimes much reduced. (Gr *malakos*, soft, and *thrix*, hair, doubtless referring to the pappus).

M. torreyi Gray. Torrey m. Ann 1–3 dm, arachnoid at first, soon subglab; basal lvs pinnatilobate and toothed, gen < 10 × 2 cm, the cauline few and ± reduced; heads few or solitary, yellow, showy; invol 8–13 mm, outer bracts short; achenes 3–4 mm, 15-ribbed, 5 ribs the stronger; 1–several persistent pappus bristles gen present; dry plains and hills, GB and s RM, and along Salmon R near Challis, Ida.

Matricaria L.   Matricaria

Heads radiate or discoid, the rays white, ♀ and gen fertile, or none; invol bracts dry, 2–3-seriate, not much imbricate, with ± hyaline or scarious margins; recep naked, hemispheric to conic or elongate; disk corollas yellow, 4–6-toothed; anthers entire or nearly so at base; style brs flattened, truncate, minutely penicillate; achenes gen several-nerved on the margins and ventrally, nerveless dorsally, glab or roughened; pappus a short crown or none; herbs with alt, pinnatifid or pinnately dissected lvs and small or middle-sized heads terminating the brs or in a corymbiform infl. (Name given by the herbalists, from L *mater* or *matrix*, to herbs of reputed medicinal value).

1a Heads with 10–25 white rays 4–13 mm; disk corollas 5-toothed; disk 6–15 mm wide
   2a Achenes with 2 marginal and 1 ventral, strongly callous-thickened, almost winglike ribs, minutely roughened on the back and between the ribs; recep hemispheric, rounded; pls ann to short-lived per, nearly inodorous; European weed, now estab in e US and occ in our range; typical *M. m.* occurs in saline places, esp along the seashore, and tends to be more br and longer-lived, with shorter and broader lf segms than the more weedy, shorter-lived, inl phase known as *M. inodora* or *M. m.* var. *agrestis;* it is uncertain whether the differentiation is ecotypic or reflects a direct response to the environment; scentless may-weed
                                                                                            1 **M. maritima** L.

2a

2b Achenes with 2 nearly marginal and 3 ventral, raised but not at all wing-like ribs, otherwise smooth; recep conic, pointed; pls ann, aromatic; European weed, sporadic with us; wild chamomile   2 **M. chamomilla** L.
1b Heads discoid; disk corollas 4-toothed; disk 5–10 mm wide; recep strongly conic, pointed; pineapple-scented ann; achenes with 2 marginal and 1 or sometimes more rather weak ventral nerves; native cordilleran weed; pineapple weed (*M. suaveolens, M. discoidea*)

3 **M. matricarioides** (Less.) Porter

Micropus L.   Micropus

Heads small, disciform, the outer fls slender and ♀, the inner coarser and functionally ♂, with abortive ovary; invol of a few scarious bracts, < bracts of the recep; ♀ fls few, each enclosed by a firm, saccate, woolly recept bract, this bearing a beaklike or hyaline appendage anteriorly below the summit, open anteriorly from the appendage to the basal annulus, only the style exserted; central ♂ fls few, gen bractless; anthers shortly sagittate-tailed; pappus none, or the central fls with a few caducous capillary bristles; achenes small, nerveless, with lateral style; slender, woolly ann with alt, entire lvs. (Gr *micros*, small, and *pous*, foot, referring to the small, soft-woolly heads).

M. californicus Fisch. & Mey. Pls 3–30 cm, gen simple; lvs erect, linear or linear-oblong, to 15 × 3 mm; heads in the upper axils; ♀ fls gen 4–6, their enclosing bracts ca 2 mm, excluding the short, erect, hyaline or ± beaklike tip; ♂ fls bractless, epappose; dry, open slopes, W Cas from Marion Co, Ore, to Cal; slender cottonweed; ours is var. *c*.

Microseris Don   Microseris

Fls all lig and ☿, yellow; invol bracts subequal, imbricate, or calyculate; recep naked; achenes columnar to fusiform, but scarcely beaked, with a whitish basal callosity, prominently ca 8–10-ribbed or -nerved; pappus of 5–∝ members, these gen with scalelike base and slender, bristlelike, naked or plumose tip, varying to very narrow and elongate bristlelike scales (in 1 sp. the scales so narrow as to be merely flattened bristles, in another sp. the narrow scales intermingled with capillary bristles); taprooted herbs with milky juice, alt (or all basal), entire to pinnatifid lvs and 1–∝ heads on long, naked peduncles; our spp. fl spring and early summer (no 7 in midsummer). (Gr *micros*, small, and *seris*, chicory). (*Calais, Nothocalais, Ptilocalais, Scorzonella, Uropappus*).

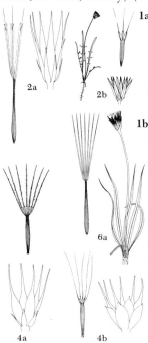

1a Ann; pappus members 5
  2a Pappus scales bifid at the tip, the awn inserted in the notch; heads relatively large, the invol gen 15–30 mm, the blackish achenes 9–13 mm; dry, open places in foothills, valleys, and plains, E Cas in Wn and Ida, s to RM and GB; Lindley's m. (*M. linearifolia, U. l.*)

1 **M. lindleyi** (DC.) Gray
  2b Pappus scales tapering to the awn, not notched; heads smaller, the invol gen 6–15 mm, the brownish achenes 3.5–6 mm; open, rather moist places W Cas; coast m.   2 **M. bigelovii** (Gray) Schultz-Bip.
1b Per; pappus members 8–∝
  3a Pls ± caulescent, often br and several-headed; pappus of 8–20 members, these chaffy at base, tipped with a definite, slender awn
    4a Pappus of 15–20 members, the awn evidently plumose; gen in open, rather moist places, lowl to fairly high elev in mts; widespread E Cas; nodding m. (*P. major*)   3 **M. nutans** (Geyer) Schultz-Bip.
    4b Pappus of ca 10 (or only 8) members, the awn merely barbellate; fairly moist meadows or sometimes drier slopes; Wn to Cal, e foothills of Cas to coast; cut-lvd m. (*M. leptosepala*)

4 **M. laciniata** (Hook.) Schultz-Bip.
  3b Pls scapose (occ with a small lf or bract on the scape in no 7), simple and monocephalous, resembling *Agoseris* in habit; pappus of 10–80 very narrow, gradually attenuate scales or bristles or both, these not clearly divided into a chaffy base and awn tip (*Nothocalais*)
    5a Pappus of 10–30 very slender, gradually attenuate scales
      6a Lvs narrow, gen 20–50 times as long as wide, the margins gen ± crisped or wavy; invol bracts lanceolate or linear-lanceolate, with or without dark dots; dry, open places in foothills and lowl E Cas, to Mont and Utah; false-agoseris   5 **M. troximoides** Gray

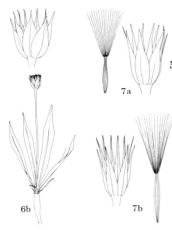

6b Lvs broader, gen 5–20 times as long as wide, plane, the margins scarcely or not at all crisped or wavy; invol bracts ovate or broadly lanceolate, finely and conspicuously dotted with blackish-purple; meadows and moist slopes in mts of c Ida, w Mont, and n Wyo; black-hairy m.   **6 M. nigrescens** Henderson

5b Pappus of 30–80 members, some or all of them so narrow as to be essentially capillary bristles

7a Pappus of 30–50 unequal, ± flattened capillary bristles; some or all lvs coarsely toothed or laciniate, or occ all lvs entire; achenes 5–8 mm; meadows and open slopes at middle and upper elev; Cas from Mt. Rainier s to Sierra Nevada; alp lake agoseris (*Agoseris a.*)

7 M. alpestris (Gray) Q. Jones

7b Pappus of 40–80 mixed capillary bristles and very slender, gradually attenuate scales; lvs entire; achenes 8–10 mm; Gr Pl, barely entering our range in Mont; toothed m. (*Agoseris c.*)

8 M. cuspidata (Pursh) Schultz-Bip.

Onopordum L.   Thistle

Similar to *Cirsium*, differing chiefly in the recep, which is flat, fleshy, and honeycombed, often with short bristle tips on the partitions, but not densely bristly; pappus bristles naked or plumose; spiny bien herbs, gen tomentose or woolly. (Name Latinized from *onopordon*, the ancient Gr name referring to a supposed effect of causing flatulence in donkeys).

O. acanthium L. Scotch t., cotton t. Coarse, strongly spiny, to ca 2 m, with broadly winged st; lvs toothed or slightly lobed, sessile and decurrent, or the lower petiolate, the bl to 6 × 3 dm, often much smaller; heads 2.5–5 cm wide; invol bracts all spine-tipped; achenes transversely rugulose; pappus merely barbellate; Eurasian weed, sporadic E Cas, esp SRC vic.

Petasites Mill.   Coltsfoot; Butterbur

Pls subdioecious, the fls in the ♀ heads all or nearly all ♀ and fertile, with or without rays, those in the ♂ heads chiefly or entirely ⚥ but sterile; invol a single series of ± herbaceous bracts, sometimes with a few bracteoles at base; recep flat, naked; ♀ fls with filiform corolla, with or without a lig; ⚥ fls tubular, with 5-cleft limb; anthers entire or slightly sagittate at base; style puberulent, undivided or nearly so; achenes linear, 5–10-ribbed; pappus of ∝ capillary bristles, elongating in fr, that of the sterile fls ± reduced; ± white-tomentose or woolly per herbs with large basal lvs, merely bracteate sts (the bracts alt), and several or ∝ medium-sized whitish (our spp.) to purple or rarely yellowish heads; our spp. fl May–July. (Gr *petasos*, a broad-brimmed hat, referring to the large basal lvs).

In addition to the following spp. we have a local population in Wen Mts, described as *P. warrenii* St. John, which may represent a vestigial colony of a hybrid between *P. sagittatus* and *P. frigidus* var. *palmatus*. The oldest binomial which prob applies to this hybrid combination is *P. vitifolius* Greene. Many such putative hybrids in Can are much like *P. frigidus* var. *nivalis*, which now forms a distinctive cordilleran high-altitude population. The pls called *P. warrenii* have only slightly lobed lvs and approach var. *frigidus*, a now more strictly n var. with coarsely few-toothed lvs, the teeth 5–15 on each side.

1a Lvs ± strongly lobed; meadows, swampy places, and moist woods; circumboreal, in our range only in and W Cas; sweet c., alpine or arctic b.; we have 2 vars.   **1 P. frigidus** (L.) Fries

a1 Lvs pinnipalmately lobed and veined, seldom > 2 dm wide, seldom evidently wider than long; cordilleran, s at high elev to OM and in Cas to n Ore   var. **nivalis** (Greene) Cronq.

a2 Lvs palmately lobed and veined, tending to be broader than long, often very large (to 4 dm wide); in and W Cas, at lower elev than var. *nivalis* (*P. p., P. speciosus*)   var. **palmatus** (Ait.) Cronq.

1b Lvs varying from merely a little wavy and callous-denticulate to more gen conspicuously dentate with 20–45 teeth on each side; wet places, often in shallow standing water; widespread in Can and Alas, in our range wholly E Cas, ne Wn across n Ida to w Mont; arrowlf or arrowhead c.

2 P. sagittatus (Banks) Gray

Prenanthes L.   Rattlesnake-root; White Lettuce

Fls all lig and ⚥, white or creamy to pink, purple, or pale yellow; invol of 4–15 principal bracts (ca 8 in our spp.) and several much reduced outer ones, or the outer occ better developed and almost passing into the inner; recep naked; achenes elongate, cylindric or slightly tapering to the summit, glab, gen ± ribbed-striate; pappus of ∝ capillary bristles, all > 4-celled in cross section at base; per herbs with milky juice, slightly to strongly tuberous-thickened roots, well developed alt lvs (our spp. with at least the lower lvs petiolate and deltoid to sagittate or hastate), and corymbiform to thyrsoid infl, the heads often nodding; our spp. occurring gen along streambanks and in other moist, often shaded places, fl July–Sept. (Gr *prenes*, drooping, and *anthe*, fl, referring to the nodding heads). *(Nabalus)*.

1a Infl narrow and elongate, thrysoid-spiciform; gen only the lower lvs deltoid or sagittate to hastate, the others few, narrower, and often much reduced; n Ida and w Mont to s Alta; arrow-lf r.   1 **P. sagittata** (Gray) A. Nels.
1b Infl broader and more open, ± corymbiform, not elongate; both the middle and the lower lvs deltoid to more often sagittate or hastate, only the upper narrower and reduced; W Cas; western r.

2 **P. alata** (Hook.) D. Dietr.

Psilocarphus Nutt.   Woolly-heads

Heads disciform, small, spheroid, terminal, but gen subtended by a pair of brs, thus appearing to be borne in the forks (or solitary in depauperate pls); proper invol none, although the heads are gen subtended by several foliage lvs; recep subglobose to truncately obpyriform; ♀ fls in several series, with short, filiform-tubular corolla, each loosely enclosed by a saccate, woolly recep bract, this bearing a hyaline appendage anteriorly below the summit, open anteriorly from the appendage to the basal annulus, only the style exserted; central fls few, bractless, functionally ♂, with abortive ovary; pappus none; achenes small, smooth, nerveless, turgid or compressed; small, woolly ann with opp (or occ a few alt), entire lvs, fl late spring and summer. (Gr *psilos*, bare, and *karphos*, chaff, "in allusion to the membranous bracteal scales," presumably referring to the hyaline appendages of the recep bracts).

1a Well developed recep bracts gen ca 3 (2.5–4) mm at maturity
  2a Achenes oblanceolate, with offset style; lvs tending to be broadest near the base; pls varying from dwarf, erect, and monocephalous, to more often prostrate and br, or occ erect and br, seldom 10 cm; dried beds of vernal pools; Cal and S Am, and widespread (but rare) E Cas in our range; dwarf w. *(P. globiferus)*; ours is var. *b*.   1 **P. brevissimus** Nutt.
  2b Achenes narrowly oblong or elliptic-oblong, with essentially apical style; lvs gen linear-oblong; pls tending to be erect and br, to 15 cm, or monocephalous and only 1 cm when depauperate; dried beds of vernal pools, and other open, moist or vernally moist places; W Cas, and mts of ne Ore, se Wn, and adj Ida; tall w.   2 **P. elatior** Gray
1b Well developed recep bracts gen ca 2 (1.3–2.7) mm at maturity; well developed pls gen depressed and much br
  3a Lvs linear or linear-oblanceolate, gen 6–12 times as long as wide; achenes narrowly oblong or elliptic-oblong; dried beds of vernal pools; E Cas in Wn, Ore, and adj Ida, s to Cal; Ore w.   3 **P. oregonus** Nutt.
  3b Lvs spatulate, oblanceolate, or oblong, gen 2–6 times as long as wide; achenes broadly oblanceolate or narrowly obovate; dried beds of vernal pools, and other open, moist or vernally moist places; chiefly Cal and sw Ore, isolated in n Ida and at e end of CRG; slender w.; ours is var. *t*.
                                  4 **P. tenellus** Nutt.

Raillardella Benth.   Raillardella

Heads discoid, or sometimes with ± well developed, fertile rays; invol bracts subequal, subherbaceous, ± uniseriate, in our spp. < disk; recep naked; anthers ± truncate at base; style brs flattened, externally hispidulous, with introrsely marginal stigmatic lines and a slender, short-hairy appendage; pappus of ± ∝ strongly plumose bristles, these ± flattened toward the base; achenes ± compressed, several-nerved; per herbs with simple, entire, alt or basal lvs and solitary or few yellow heads; our spp. scapose, monocephalous alp pls, chiefly Cal, but extending n in Cas to c

Ore, fl middle and late summer. (Named for its similarity to *Raillardia*).

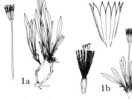

1a Lvs green and ± glandular, not at all tomentose, gen 3–15 cm × 3–7 mm; scapes 5–40 cm; heads discoid or occ with irreg developed short rays; green-lvd r.   **1 R. scaposa** Gray
1b Lvs grayish, silky-tomentose, 2–5 cm × 2–6 mm; scapes 1–12 cm; heads strictly discoid; slender r.   **2 R. argentea** Gray

Ratibida Raf.   Coneflower

Heads radiate, the rays neutral, large, yellow or sometimes partly or wholly purple; invol a single series of green, subherbaceous, linear or lance-linear bracts; recep columnar, its bracts subtending the rays as well as the disk fls, ± clasping the achenes; disk fls ♂ and fertile, their corollas short, cylindric, scarcely narrowed at base; anthers sagittate; style brs flattened, with short, externally hairy appendage, without well developed stigmatic lines; achenes compressed at right angles to the invol bracts, often also evidently quadrangular, glab except the sometimes ciliate margins; pappus coroniform, with 1 or 2 prolonged awnlike teeth, or of awn teeth only, or none; bien or per herbs with alt, pinnatifid lvs and naked-pedunculate heads. (Name unexplained).

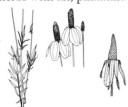

R. columnifera (Nutt.) Woot. & Standl. Prairie c. Taprooted, several-std per 3–12 dm, coarsely strigose or shortly hirsute; heads gen several; disk columnar, dark, gen 1.5–4 cm, 1/4–1/2 as wide; rays gen 3–7, 1.5–4.5 cm, very broad, spreading or reflexed; pappus of 1 or 2 unequal awn-teeth; Gr Pl sp., w to Mont and se BC, and occ intro elsewhere E Cas (*R. columnaris; R. c.* f. *pulcherrima*, the form with purple or partly purple rays).

Rigiopappus Gray

Heads radiate, the rays ♀ and fertile, short; invol ± turbinate, its bracts narrow, subequal, partly clasping the outer achenes; recep flat, naked except for a partial or complete row of bracts between the ray and disk fls; disk fls ♂ and fertile, yellow, reputedly sometimes turning purple; anthers not tailed; style brs flattened, with introrsely marginal stigmatic lines and an externally hairy appendage; achenes linear, ± compressed parallel to a radius of the head; pappus gen of 3–5 awnlike scales; slender ann with alt lvs. (Gr *rigios*, stiff, and *pappos*, seed-down).

R. leptocladus Gray. Bristle-head. Short-hairy to subglab ann to 3 dm, br when well developed, the br gen overtopping the main st; lvs narrowly linear, erect, 1–3 cm; heads terminating the brs, 8–80-fld; invol 4–7 mm; rays gen 3, 5, or 8, yellowish, 1.5–2 mm; achenes cross-rugulose; dry, open places at low elev E Cas, gen in grassl or sagebr; c Wn and w Ida to Utah and Cal.

Rudbeckia L.   Rudbeckia; Coneflower

Heads radiate or discoid, the rays neutral, large, yellow, orange, or partly purple; invol bracts 2–3-seriate, subequal or irreg unequal, ± herbaceous, gen spreading or reflexed; recep enlarged, conic or columnar, chaffy throughout, its bracts clasping the achenes; disk fls ♂, fertile, narrowed to a ± distinct tube at base; anthers obtuse or sagittate at base; style brs flattened, with short or elongate, externally hairy appendage, without well marked stigmatic lines; achenes equably quadrangular, or somewhat flattened at right angles to the invol bracts, glab; pappus a short, often toothed or irreg crown, or none; ann to more often per herbs with alt, entire to pinnatifid lvs. (Named for the botanists Olof Rudbeck, father and son, 1630–1702, 1660–1740, predecessors of Linnaeus at Uppsala).

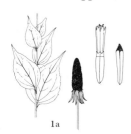

1a Heads discoid, the disk black, becoming thick-columnar, 2–6 cm in fr; coarse per 5–20 dm; lvs ample, broadly ovate or elliptic, to 25 × 15 cm; style appendages very short and blunt; pappus an evident short crown; streambanks and woods E Cas; black head; 2 vars.   **1 R. occidentalis** Nutt.
    a1 Lvs merely toothed to subentire; herbage subglab or inconspicuously hairy, seldom more evidently hairy; widespread, but not in Wen Mts
                                            var. occidentalis
    a2 Lvs, or some of them, irreg lobed or pinnatifid; herbage evidently short-hairy; Wen Mts (*R. a.*)   **var. alpicola** (Piper) Cronq.
1b Heads radiate; disk hemispheric or ovoid to sometimes (in no 2) short-columnar

2a Lvs large and broad (the larger ones gen 1–2.5 dm wide), laciniate-pinnatifid with broad segms; coarse per to 2 m, glab and often glaucous, or the lvs somewhat hairy; disk yellow or grayish; rays 6–16,
drooping, yellow, 3–6 cm; style appendages short and blunt; pappus a
short crown; streambanks and other moist places; widespread e Am sp.,
w to Mont, s Ida, and Ariz; tall c., green-headed c.; ours is var. *ampla* (A.
Nels.) Cronq.   2 **R. laciniata** L.
2b Lvs narrower (< 5 cm wide), ± entire, the larger (lower) ones oblanceolate to elliptic; bien or short-lived per 3–10 dm, ± rough-hairy
throughout; disk dark purple or brown; rays 8–20, spreading, orange,
2–4 cm; style appendages subulate; pappus none; disturbed places or
open meadows, widespread in E Am, casually intro with us; black-eyed
Susan; ours is var. *pulcherrima* Farw., the widespread, weedy phase of
the sp.   3 **R. hirta** L.

Saussurea DC.   Saussurea

Heads discoid, the fls all tubular and ☿; invol bracts imbricate in several series; recep flat or convex, densely paleaceous-bristly to naked, even in the same sp.; corolla blue or purple (white), with
slender tube and long narrow lobes; anthers tailed; style with a thickened, minutely hairy ring just
beneath the short, papillate brs; achenes glab, variously nerved, thick or flattened; pappus bristles
plumose, united at base and falling connected, gen with a shorter outer nonplumose series which
may or (our spp.) may not be attached to the ring of longer bristles; per herbs with alt, entire or
toothed to pinnatifid lvs. (Named for Theodore (the son) and Horace Benedict (the father) Saus-
sure, eminent Genevese naturalists).

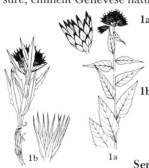

1a Pls robust, gen 5–12 dm; lvs relatively broad, the principal ones gen triangular-ovate to triangular-cordate, to 15 × 8 cm, sharply toothed; infl corymbiform; fls gen ca 13, sometimes 8 or 21; moist meadows and moist,
open, sometimes rocky slopes and middle and upper elev in mts; n cordilleran, s to OM, Cas of s Wn, ne Ore, c Ida, and nw Mont; Am sawwort
                                                                 1 **S. americana** Eat.
1b Pls dwarf, gen 0.5–2 dm; lvs narrower, gen narrowly elliptic and gradually
tapering at base, to 8 × 2 cm, irreg toothed; infl capitate; fls gen ca 21;
rocky slopes in mts; sw Alta and se BC, and reputedly VI; dwarf s.; perhaps
better as var. *densa* (Hook.) Hult. of the more n and Eurasian *S. nuda*
Ledeb.   2 **S. densa** (Hook.) Rydb.

Senecio L.   Groundsel; Ragwort; Butterweed

Heads radiate or less often discoid, the rays ♀ and fertile, yellow to orange or occ reddish or white,
or wanting; invol bracts herbaceous or subherbaceous, ± =, uniseriate or by lateral overlapping
subbiseriate, often with some smaller bracteoles at base; recep flat or convex, naked; disk fls ☿ and
fertile, yellow or orange to reddish; anthers entire to minutely sagittate; style brs flattened, truncate, penicillate, with introrsely marginal stigmatic lines extending to the tip; achenes subterete,
5–10-nerved; pappus of ∝, gen white, entire or barbellulate bristles; ann, bien, or per herbs (all
ours) or shrubs, vines, or even trees, with alt (or all basal), entire to variously toothed or dissected
lvs, and solitary to ∝ gen small to middle-sized, cylindric or camp to hemispheric heads; our spp.
gen fl summer. (L *senex*, old man, prob referring to the white pappus or hoary pubescence of
some spp.).

1a Pls ann, ± taprooted, ± lfy throughout, with very short and inconspicuous rays, or without rays; pappus = or > disk corollas; weeds of Eurasian origin
2a Rays none; bracteoles black-tipped; principal invol bracts ca 21; pl not
malodorous; widespread, but commoner W Cas; old-man-in-the-spring,
common g.   1 **S. vulgaris** L.
2b Rays gen present, minute (< 2 mm); bracteoles not black-tipped; principal invol bracts ca 13; pl powerfully malodorous; chiefly or wholly W
Cas; wood g.   2 **S. sylvaticus** L.
1b Pls otherwise, most spp. clearly per and fibrous-rooted, the short-lived or
taprooted spp. gen with conspicuous rays; pappus seldom > disk corollas
3a Cauline lvs well developed, only gradually reduced upwards; no well
developed tuft of basal lvs present, except often in no 8, keyed under
both leads

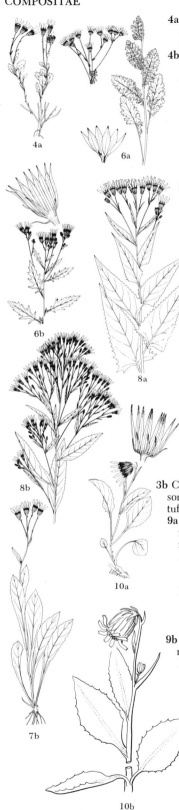

4a Dwarf, freely br, gen ca 1 (2) dm; lvs to 4 × 2 cm, ± toothed, short-petiolate or sessile; rocky places at high elev; cordilleran; dwarf mt b.; ours is var. *f.*   3 **S. fremontii** T. & G.
4b Taller, gen 2–15 dm or more, the st simple below the infl; lf bls, at least of the larger lvs, gen > 4 cm
   5a Lvs laciniate to bipinnatifid; bien or short-lived per with a poorly developed to evident taproot
      6a Lvs gen 2–3 times pinnatifid; poisonous Eurasian weed of pastures and other disturbed sites, estab W Cas; tansy r.
                                                      4 **S. jacobaea** L.
      6b Lvs merely laciniate-pinnatifid; native cordilleran sp. of thickets and woods, wholly E Cas, rarely found in our range and not known from Wn or Ore; dryland r.; ours is var. *e.*
                                                      5 **S. eremophilus** Rich.
   5b Lvs entire to coarsely toothed, but not at all laciniate or pinnatifid; true per, all fibrous-rooted
      7a Cauline lvs, except sometimes the reduced upper ones, petiolate or tapering to a narrow, petioliform base; heads few–∝; pls 3–15 or 20 dm
         8a Lvs, at least the lower, triangular, with deltoid to cordate base, often > 4 cm wide; rays 7–13 mm; pls 3–15 dm; widespread cordilleran sp. of streambanks and other moist places, often at low elev W Cas, but E Cas confined to upper and middle elev; arrowlf g.; 2 vars.   6 **S. triangularis** Hook.
            a1 Ecotype of sphagnum bogs at low elev near coast; pls small, slender, with relatively narrow lvs, the upper strongly reduced and becoming linear   var. **angustifolius** G. N. Jones
            a2 Series of ecotypes of various habitats; pls gen larger and broader-lvd   var. **triangularis**
         8b Lvs all tapering to the base, not at all triangular, gen 1–4 cm wide; rays 5–8 mm; pls 5–15 or 20 dm; meadows and other moist places from foothills to mid elev in mts; widespread in US cordillera E Cas; butterweed g., tall b.; ours is var. *s.*
                                                      7 **S. serra** Hook.
      7b Cauline lvs, or some of them, sessile, with fairly broad, ± clasping base; lvs up to 4 (5) cm wide; heads few or solitary, rarely > 12; pls 2–7 dm; rays 6–13 mm; open places from foothills to well up in mts; sw Mont irreg to ne Ore, and s; thick-lvd g.
                                                      8 **S. crassulus** Gray
3b Cauline lvs gen strongly and progressively reduced upwards (or the st sometimes scapose); basal or lower cauline lvs well developed, often tufted
   9a Heads 1–2 (4), nodding; alp talus pls 0.5–2 (4) dm with entire or merely dentate lvs and a well developed, often br caudex or short rhizome
      10a Lvs gen thinly arachnoid-villous at fl time, at least beneath; heads large, the invol 11–17 mm, the disk 1.5–2.5 cm wide; OM; OM b. (*S. websteri* Greenm. not Hook.)   9 **S. neowebsteri** Blake
      10b Lvs gen glab or nearly so at fl time; heads ave smaller, the invol 9–15 mm, the disk 1.2–2 cm wide; sw Mont, s to s RM; clasping g.; ours is var. *holmii* (Greene) Harrington (*S. h.*)
                                                      10 **S. amplectens** Gray
   9b Heads 1–∝, erect (several on flexuous peduncles, sometimes slightly nodding, in no 29); habit and habitat various
      11a Pls glab from the first, or sometimes lightly floccose-tomentose when young; if tomentose when young, then glab by fl time except occ for a little inconspicuous tomentum at the base and in the axils
         12a Lvs entire or denticulate to sharply dentate, not at all pinnatifid, lobed, wavy, or crenate; pls, except sometimes no 8, with ∝ fibrous roots from a very short, erect, short-lived crown, without any more-elongate caudex or rhizome (the glab extremes of some of the pubescent spp. might be sought here, except for the well developed rhizome or caudex)
            13a Heads few, seldom > 12, apparently always radiate; pls of dry to moderately moist habitats (see lead 7b)
                                                      8 **S. crassulus** Gray
            13b Heads more numerous, seldom < 12 in well developed pls,

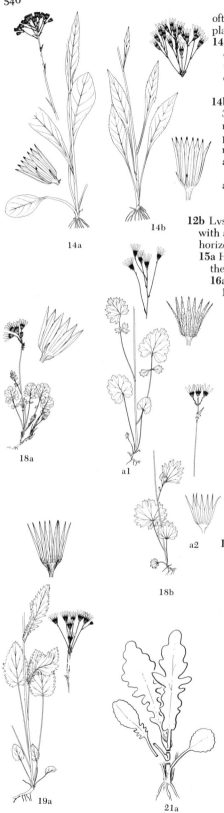

14a

14b

often very ∝, radiate or more often discoid; pls gen of wet places

14a Robust, gen ± glaucous, ± fleshy pls, 4–20 dm; lvs entire or occ irreg toothed; swampy places in valleys and foothills, tolerant of alkali and standing water; widespread in US cordillera E Cas; alkali-marsh butterweed
        11 S. hydrophilus Nutt.

14b Less robust, scarcely glaucous and scarcely fleshy pls, 3–10 dm; lvs saliently toothed or occ subentire; pls of wet meadows and wet hillsides, but not of swamps or alkaline places; widespread E Cas in our range, s to Cal; sweet-marsh b.; 2 vars.
        12 S. foetidus Howell

  a1 Infl congested; heads more often discoid; sts often ± clustered; more common w
        var. **foetidus**

  a2 Infl open; heads more often radiate; sts arising singly; more common e (S. h.)
        var. **hydrophiloides** (Rydb.) T. M. Barkley

12b Lvs, or some of them, pinnatifid, lyrate, crenate, or wavy; pls with a more evident, longer-lived and more-woody, ascending or horizontal caudex or rhizome

15a Heads characteristically radiate (rare discoid forms of most of these spp. occur, gen in company with the normal radiate pls)

16a Heads gen several

17a Basal lvs either lyrate-pinnatifid, or often tending to be shallowly palmately lobed, with the lobes gen again toothed and the bl gen ± cordate; Cas and W Cas

18a Pls gen 0.5–2 (4) dm, with subnaked st; basal lvs more often lyrate-pinnatifid, but sometimes as in no 14 instead; open, rocky places, esp on talus; OM, and about Mt Rainier, s to Clatsop Co, Ore; Flett's g.
        13 S. flettii Wieg.

18b Pls gen 2–6 dm, with several cauline lvs; basal lvs more often shallowly palmately lobed, but sometimes nearly as in no 13; moist, rocky woods, banks, bluffs, and beaches; CRG, and W Cas from sw Wn to Cal; Bolander's g.; 2 vars.
        14 S. bolanderi Gray

  a1 Invol (5) 6–8 mm, gen with some coarse, conspicuously multicellular hairs; lvs relatively thick and firm; along the coast, from CR estuary s, and, in forms often transitional to var. h., in Ore coast range well back from ocean
        var. **bolanderi**

  a2 Invol 4–5 (6) mm, glab; lvs relatively thin and lax; Cas from CRG to Lane Co, Ore (S. h.)
        var. **harfordii** (Greene) T. M. Barkley

17b Basal lvs gen neither palmately lobed nor lyrate-pinnatifid; Cas and e

19a Basal lvs, or some of them, cordate or subcordate, sharply toothed; cauline lvs laciniate-pinnatifid, at least toward the lf-base; lvs thin; streambanks, wet meadows, and moist woods, in and near the mts; widespread cordilleran sp.; streambank b.; ours is var. p.
        15 S. pseudaureus Rydb.

19b Basal lvs not cordate or subcordate, though sometimes subrotund (or subreniform and crenate to entire in forms of no 16); cauline lvs variously toothed or entire to pinnatilobate or laciniate-pinnatifid

20a Lvs relatively thick and firm, the basal ones gen elliptic to subrotund, broadly obovate, or subreniform, coarsely crenate to shallowly lobulate or entire

21a Cauline lvs well developed, the middle and lower clasping and as large as or even larger than the basal ones; basal lvs gen broadly ovate to orbicular or subreniform, 1–1.5 times as long as wide; alp and subalp meadows and rocky slopes; chiefly s RM sp., n to extreme sw Mont and adj Ida; Payson's g.; ours is var. *paysonii* T. M. Barkley
        16 S. dimorphophyllus Greene

14a

18a

a1

a2

18b

19a

21a

21b Cauline lvs progressively reduced, not clasping; basal lvs gen elliptic to obovate or subrotund, typically ca twice as long as wide; woods and moist to moderately dry open places at mid to high elev in mts; widespread cordilleran sp.; RM b., cleft-lf g. *(S. cymbalarioides* Nutt., not Buek; *S. suksdorfii)*
               17 **S. streptanthifolius** Greene

20b Lvs relatively thin and lax, the basal ones gen elliptic or oblanceolate, crenate or serrate to subentire; moister or wetter places than no 17, often at lower elev; widespread cordilleran and e Am sp.; balsam g., Can b. *(S. balsamitae, S. flavovirens, S. multnomensis);* cordilleran pls tend to have relatively large heads (invol 6–9 mm) and have been distinguished, perhaps on insufficient grounds, as var. *thomsoniensis* (Greenm.) Boiv.   18 **S. pauperculus** Michx.
16b Heads solitary or rarely 2; alp and subalp pls, to 3 dm

  22a Pls with 1 or more ± well developed cauline lvs, seldom < 1 dm tall; rhizome short; widespread spp.

    23a Rhizome very slender; heads yellow, not orange; pls ave more slender and often taller than no 20; wet meadows; few-lvd g., alp meadow b. *(S. subnudus)*
             19 **S. cymbalarioides** Buek

    23b Rhizome short and thick; heads tending to be ± orange or even orange-red; pls 0.5–2 dm; drier, more rocky and exposed habitats; dwarf arctic b.
             20 **S. resedifolius** Less.

  22b Pls essentially scapose, < 1 dm, the peduncle merely bearing 1 or 2 minute bracts; rhizome elongate, br; open, rocky sites; Wallowa Mts; Porter's b. 21 **S. porteri** Greene
15b Heads characteristically discoid (rare radiate forms of some of these spp. occur, gen in company with the normal discoid pls)

  24a Heads (1) 2–6 (12), orange or reddish; lvs relatively thick and firm; alp and subalp pls, 1.5–4 dm, of meadows and moist cliffs; Can, s to n Wn and n Ida, and disjunct in Cal and nw Wyo; rayless alp b. *(S. discoideus)*
             22 **S. pauciflorus** Pursh

  24b Heads more numerous, gen 6–100 or more, yellow or sometimes orange; pls 1–8 dm, not alp or subalp

    25a Lvs thick and firm, the basal ones entire to more often crenate or lobulate, the cauline sublyrate to pinnately divided, with deep, rounded sinuses and narrow, obtuse or rounded lobes; infl a corymbiform cyme; moist meadows and bottoml, esp in alkaline places; c Ida and sw Mont to Colo; weak b.   23 **S. debilis** Nutt.

    25b Lvs thin, the basal serrate or sometimes incised, the cauline sharply incised-pinnatifid, the lobes gen again irreg few-toothed; infl tending to be subumbelliform; streambanks and moist woods; Can and Alas to n Wn, w Mont, and nw Wyo, and disjunct in n Cal; rayless mt b. *(S. idahoensis)*   24 **S. indecorus** Greene
11b Pls gen ± pubescent at fl time

  26a Pubescence of the lvs of 2 types, the coarse, flattened, multicellular hairs overlain by a ± deciduous, arachnoid-villous tomentum; basal lvs scarcely petiolate, 3–7 × 0.8–2.2 cm; rays orange, 8–15 mm; arctic-alp sp. 1–2 dm, in our range only in the Beartooth Mts of Mont; twice-hairy b. *(S. bivestitus)*
             25 **S. fuscatus** Hayek

  26b Pubescence crisp-villous to tomentose, but of essentially uniform quality on each individual; basal lvs gen ± evidently petiolate

    27a Heads 1–4, relatively very large, the invol gen 12–16 mm, the disk gen 1.5–2.5 cm wide; rays gen 15–25 mm; lvs entire to callous-denticulate; open, rocky places at middle and upper elev in mts of Mont and n Ida to Alta and BC; large-headed b.
             26 **S. megacephalus** Nutt.

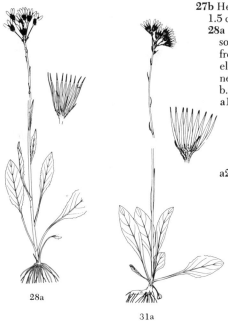

28a

31a

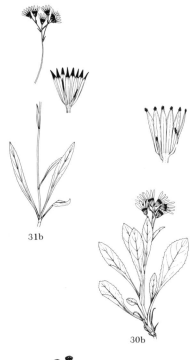

31b

30b

32a

27b Heads 1–∝, smaller, the invol gen 5–10 (12) mm, the disk < 1.5 cm wide, the rays up to ca 15 mm

28a Pubescence loosely crisp-villous or arachnoid-villous, sometimes very sparse by fl time; pls with ∝ fibrous roots from a very short, erect, short-lived crown, without a more elongate caudex or rhizome; open places from the valleys to near timberl; cordilleran and Gr Pl sp.; western g., one-std b.; 4 vars. in our range ..... 27 S. integerrimus Nutt.

a1 Invol bracts relatively narrow, only minutely and irreg black-tipped, or not at all black-tipped; peduncle of the terminal head as long as the others, or sometimes shortened and thickened; pl early glabrate, the pubescence sparse and inconspicuous by fl time; Gr Pls var., barely reaching our range in Mont ..... var. **integerrimus**

a2 Invol bracts relatively broader, more evidently and conspicuously black-tipped; peduncle of the terminal head consistently thickened and < the others; pls gen (not always) more persistently hairy, though still tending to be glabrate in age; cordilleran vars.

b1 Heads discoid; middle and upper elev in Wen-Cas area of c Wn, s irreg to s Ore, and in Wallowa Mts (S. v.) ..... var. **vaseyi** (Greenm.) Cronq.

b2 Heads gen radiate; more widespread

c1 Rays white or ochroleucous; basal lvs tending to have a deltoid or subcordate bl, although the bl is sometimes as narrow as in the next var.; gen in slightly more mesic habitats than the next var., often in woodl; w Mont, across n Ida to Wn Cas, and s in Cas to n Cal ..... var. **ochroleucus** (Gray) Cronq.

c2 Rays bright yellow; basal lvs gen oblanceolate or elliptic to occ subrotund, or even subcordate as in the preceding var.; more often in open places, widespread, encompassing the range of the 2 previous vars. ..... var. **exaltatus** (Nutt.) Cronq.

28b Pubescence finer, more tomentose or floccose (or scarcely so in no 30); pls with an evident, ± woody, ascending or horizontal caudex or short rhizome

29a Lvs sharply dentate (with divergent teeth) to merely callous-denticulate or even (in forms of no 28) entire; pls, except no 30, with the sts gen arising singly from a short, horizontal rhizome

30a Pls gen single-std from a short rhizome (occ several-std in no 29), thinly tomentulose when young, later often ± glabrate; pls typically of wet meadows

31a Invol bracts with minute black or brownish tip, gen ca 21, rarely only 13; pls 3–8 dm, of moderate elev in and near mts, Mont to ne Ore (Wallowa Mts), s to Colo and Nev, and disjunct in s WV; mt-marsh b. ..... 28 S. **sphaerocephalus** Greene

31b Invol bracts with very conspicuous black tip, gen ca 13, rarely up to 21; boreal sp., 1–5 dm, extending s at high elev to OM and mts of Mont and n Wyo; black-tipped b. (S. glaucescens) ..... 29 S. **lugens** Rich.

30b Pls with several sts 1–3 dm from a well developed br caudex, subtomentosely arachnoid-villous at first, only thinly so (or even glabrate) at fl time; heads gen 3–13 on flexuous peduncles, often somewhat nodding; talus slopes and other rocky places at high elev; Wen and Cas Mts from c Wn to s BC; Elmer's b. (sp. related to nos 9 and 10) ..... 30 S. **elmeri** Piper

29b Lvs entire to irreg subpinnately lobed, but not dentate or denticulate; pls gen with several sts arising from a br caudex which sometimes surmounts a taproot

32a Pls dwarf, gen 2–15 cm, with ± scapiform sts, only slightly or obscurely tomentose at fl time; heads 1–6, gen rather long-pedunculate, radiate or often discoid; talus slopes and other rocky places at high elev in mts;

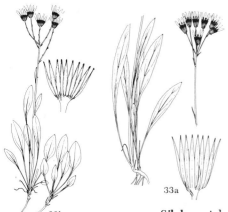

Mont and c Ida to Colo, Ariz, and Cal; rock b. *(S. petrocallis, S. saxosus)* 31 S. werneriaefolius Gray

32b Pls gen more robust and > 15 cm, or more evidently tomentose, or both; heads often > 6, gen radiate; various habitats, seldom at such high elev as no 30

   33a Tomentum relatively thin or even obscure; robust pls, gen 2–7 dm, with relatively long, slender lvs; open woods and dry, open places W Cas, from VI to s Ore; Puget b. *(S. fastigiatus* Nutt., not Schwein.) 32 S. macounii Greene

   33b Tomentum relatively dense, though sometimes partly deciduous; smaller pls, gen 1–3 (4) dm; lvs various; cordilleran sp., widespread E Cas in dry, open, often rocky places from the plains and foothills to timberl or above; woolly g. *(S. howellii, S. purshianus)* 33 S. canus Hook.

Silybum Adans.   Thistle; Milk-thistle

Heads discoid, the fls all tubular and ⚥; invol bracts imbricate in several series, broad, firm, most of them spiny-margined and strongly spine-tipped; recep flat, densely setose; corollas purple, with slender tube and long, narrow lobes; filaments glab, connate at least toward the base; anthers shortly tailed; style with an abrupt change of texture and often with a ring of hairs below the brs, which gen remain connate to the tip; achenes basifixed, flattened, glab; pappus of ∝ slender, unequal, subpaleaceous bristles, deciduous in a ring; spiny winter ann or bien with alt lvs and large, globose heads terminating the brs. (Name used by Dioscorides for this or some similar pl).

S. marianum (L.) Gaertn. Glab or slightly tomentose, gen 6–15 dm; lvs spiny-margined, ± marked with white along the main veins, the lower petiolate, to 6 × 3 dm, pinnately lobed, the upper smaller, sessile, aur-clasping; disk 3–6 cm wide; achenes 6–7 mm; Mediterranean weed, in our range mainly in WV.

Solidago L.   Goldenrod

Heads radiate, the rays ♀ and fertile, yellow, small, seldom > 5 mm; invol small, in ours 2–7 mm, its bracts ± imbricate in several series or rarely subequal, ± chartaceous at base, gen with ± herbaceous green tip; recep small, flat or a little convex, naked; disk fls ⚥ and fertile, yellow; anthers entire or nearly so at base; style brs flattened, with introrsely marginal stigmatic lines and an externally minutely hairy, gen lanceolate appendage; achenes subterete or angled, several-nerved; pappus of ∝ equal or sometimes unequal capillary bristles, gen white; fibrous-rooted per herbs from a rhizome or caudex, bearing simple, alt, entire or toothed lvs and few to often ∝, gen small, camp to subcylindric heads; fl July–Oct. (L *solidus* and *ago*, to make whole, referring to reputed healing properties). *(Euthamia, Oligoneuron)*.

  1a Pls with well developed creeping rhizomes; st varying from densely and rather equably lfy, without well developed basal lvs (in most spp.) to sometimes more sparsely and inequably lfy, with enlarged lower and basal lvs (in forms of no 4)

    2a Lvs punctate (sometimes rather obscurely so), linear or lance-linear, sessile, 3–7-nerved, to 13 cm × 12 mm; heads sessile in small glomerules, forming a corymbiform or diffuse infl; rays gen 15–30; pls (3) 5–20 dm, in moist, low ground in the valleys and plains *(Euthamia)*

      3a Infl gen ample, copiously lfy-bracteate, tending to be interrupted and elongate, or in small pls occ app that of no 2; widespread cordilleran sp.; western g.   1 S. occidentalis (Nutt.) T. & G.

      3b Infl gen compact and flat-topped, the glomerules fairly closely aggregated, or the lateral clusters even overtopping the central ones; invol bracts ave broader and blunter than in no 1; chiefly more e sp., rare in our range; bushy g., fragrant g.; ours is var. *major* (Michx.) Fern., with middle cauline lvs 7–11 times as long as wide

          2 S. graminifolia (L.) Salisb.

    2b Lvs not punctate, variously shaped and nerved; infl corymbiform to more often paniculiform, not glomerulate; rays gen ca 8 or ca 13, rarely to 17

5a

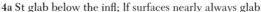

4a St glab below the infl; lf surfaces nearly always glab

  5a St glaucous, becoming definitely puberulent in the infl; pls 5–20 dm; rays gen ca 13 (10–17); lvs ∝ and equably distributed; infl paniculiform, with recurved-secund brs; widespread sp. of moist, gen open places in the valleys and plains; smooth g., late g.; ours is var. *serotina* (Kuntze) Cronq. *(S. s.)*   3 **S. gigantea** Ait.

  5b St glab or very nearly so throughout, not glaucous; pls gen 2–9 dm; rays gen ca 8, rarely 13; habit varying from that of the S. *spathulata* group to nearly that of S. *gigantea* and S. *canadensis;* widespread sp. of rather dry, open places from the valleys and plains to fairly high elev in mts; Mo g.; 4 poorly defined vars.

                                              4 **S. missouriensis** Nutt.

    a1 Pls tending to be tall and rather lfy-std, gen 4–9 dm, the basal and lower cauline lvs gen deciduous; infl evidently secund; Gr Pls var., w occ to Grand Coulee, Wn *(S. glaberrima)*

                                    var. **fasciculata** Holz.

    a2 Pls gen shorter, seldom > 5 dm, the lowermost lvs gen persistent, the middle and upper ones gen fewer and more reduced; infl moderately to not at all secund; the characteristeric cordilleran vars.

      b1 Pls of the prairies W Cas, with relatively large heads (invol 4–5 mm), narrow bracts, and somewhat secund infl *(S. t.)*

                              var. **tolmieana** (Gray) Cronq.

      b2 Pls strictly E Cas, and with the characters of invol and infl seldom combined precisely as in var. *tolmieana*

        c1 Heads relatively large, the invol gen 4–5 mm; infl seldom at all secund; typically in the more mesic habitats of moderate to rather high elev in the mts, less common than the next var. *(S. concinna)*   var. **extraria** Gray

        c2 Heads gen smaller, the invol 3–4 (5) mm; infl tending to be a little secund; gen in the foothills and valleys

                            var. **missouriensis**

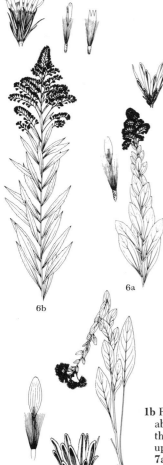

6a

6b

4b St puberulent, at least above the middle; lvs puberulent to subglab

  6a Rays ca 8, 3–4 mm; middle cauline lvs gen 2.5–4 times as long as wide; lvs canescent on both sides; pls 1–6 dm; Gr Pl sp., entering our range in Mont; velvety g.   5 **S. mollis** Bartl.

  6b Rays ca 13 (10–17), 1–3 mm; middle cauline lvs gen > 4 times as long as wide; lvs glab to canescent; pls 4–20 dm; transcontinental sp., gen of moist soil; Can g., meadow g.; 3 vars. in our range

                                      6 **S. canadensis** L.

    a1 Invol not much imbricate, the outer bracts > half as long as the inner; infl short and compact, not at all secund, sometimes scarcely surpassing the lvs; pls gen 4–8 dm; pubescence as in var. *salebrosa;* coastal region of BC and s Alas *(S. lepida)*

                          var. **subserrata** (DC.) Cronq.

    a2 Invol more evidently imbricate; infl gen larger

      b1 Lvs densely short-hairy on both sides (the hairs a little shorter above), st densely short-hairy to near the base; brs of the infl obviously secund; pls 4–8 (12) dm; Gr Pls var., extending into the intermont valleys of Mont and apparently to the Grand Coulee, Wn *(S. pruinosa)*   var. **gilvocanescens** Rydb.

      b2 Lvs less densely hairy or even subglab; st less densely hairy, often glab below; infl typically elongate and scarcely secund *(S. elongata)*, varying to like that of var. *gilvocanescens* or nearly like that of var. *subserrata;* pls 4–20 dm; common nearly throughout our area, sometimes extending well up into the mts

                        var. **salebrosa** (Piper) Jones

1b Pls with a gen short and stout rhizome, or a mere caudex; st rather inequably lfy, the basal or lower cauline lvs larger and differently shaped than those above, tending to be persistent, the cauline lvs conspicuously reduced upwards and not very ∝

  7a Lvs densely and finely canescent with short spreading hairs; spp. of dry, open places at lower elev

    8a Invol bracts longitudinally few-striate; achenes glab except gen for a few short loose hairs near the tip; basal lvs gen 2–8 cm wide; infl dense, corymbiform; heads relatively large and ∝-fld, the disk gen 5–10 mm wide, the invol 5–7 mm high, the rays 2.5–5 mm; wide-

8a

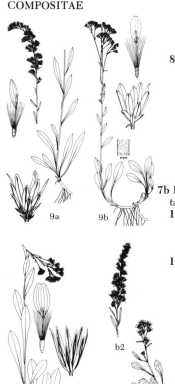

9a 9b

10a b1 b2 10b

spread e sp., entering our range e Cont Div in Mont *(Oligoneuron)*; stiff g., hard-lvd g.; ours is var. *humilis* Porter *(O. canescens)*
7 **S. rigida** L.

8b Invol bracts not striate; achenes short-hairy throughout; basal lvs gen 0.5–3 cm wide

9a Infl either with ± secund brs, or elongate and nodding at the tip; disk fls gen 5–9, about as many as or more often fewer than the rays; e sp., reaching the edge of our range in Mont; gray g., field g.; ours is var. *longipetiolata* (Mack. & Bush) Palmer & Steyerm. *(S. pulcherrima, S. n.* var. *decemflora)*   8 **S. nemoralis** Ait.

9b Infl corymbiform and gen relatively broad, neither nodding nor with evidently secund brs; disk fls gen 8–16, gen more ∝ than the rays; foothills, valleys, and plains from w Mont and c Ida s; low g.
9 **S. nana** Nutt.

7b Lvs glab except for the sometimes ciliate margins; spp. of various habitats

10a Lowermost lvs with ciliate-margined petiole; rays gen ca 13 (or more on the terminal head); invol bracts not much imbricate; pls 0.5–5 dm; boreal sp., s in the cordillera at high elev, missing from Ore; northern g.; ours is var. *scopulorum* Gray *(S. ciliosa)*   10 **S. multiradiata** Ait.

10b None of the lvs with ciliate-margined petiole; rays gen ca 8; invol bracts evidently imbricate; widespread sp. of various open, not too wet habitats; dune g.; 3 well marked vars. in our range
11 **S. spathulata** DC.

a1 Coastal sand dune pls, 1–6 dm, very strongly resinous and aromatic, with spiciform-thyrsoid, gen elongate infl and gen spatulate or obovate basal lvs   var. **spathulata**

a2 Inl pls, less strongly resinous and aromatic

b1 Alp and subalp pls, 0.5–1.5 dm, with short, compact infl and gen spatulate or obovate basal lvs *(S. decumbens)*
var. **nana** (Gray) Cronq.

b2 Valley and mont, but scarcely alp pls, 1.5–8 dm, with more-elongate, spiciform-thyrsoid to subracemiform infl, and gen with oblanceolate basal lvs *(S. glutinosa)*
var. **neomexicana** (Gray) Cronq.

Sonchus L.   Sow-thistle

Fls all lig and ⚥, yellow, few to more often ∝ (in our spp. gen 120–160, occ 85-250); invol ovoid or camp, its bracts gen imbricate in several series, often basally thickened and indurated in age; recep naked; achenes flattened, gen 6–20-ribbed, merely narrowed at the apex, beakless, often transversely rugulose, otherwise glab; pappus of ∝ white, capillary, often ± crisped bristles which tend to fall connected, gen with some stouter outer bristles which fall separately; some of the pappus bristles no > 4-celled in cross section at base; herbs with milky juice, alt or all basal, entire to pinnatifid or dissected, gen aur, often prickly-margined lvs, and solitary to gen several or ∝, medium-sized to rather large heads in an irreg corymbose-paniculiform to subumbelliform infl, our spp. cosmopolitan weeds of European origin, fl July–Oct. (The ancient Gr name).

2a 2b

3a 3b

1a Per with deep vertical roots, extensively spreading by horizontal, rhizome-like, often deep-seated roots; heads relatively large, gen 3–5 cm wide in fl, the fr invol gen 14–22 mm high; achenes with 5 or more prominent longitudinal ribs on each face, strongly rugulose

2a Invol and peduncles with coarse, spreading, gland-tipped hairs; per s., field milk-thistle   1 **S. arvensis** L.

2b Invol and peduncles glab, or obscurely tomentose; marsh s. *(S. a.* var. *glabrescens)*   2 **S. uliginosus** Bieb.

1b Ann from a short taproot; heads relatively small, gen 1.5–2.5 cm wide in fl, the fr invol gen 9–14 mm high

3a Achenes transversely tuberculate-rugulose at maturity, as well as several-nerved; lvs soft, scarcely prickly-margined, runcinate-pinnatifid to occ merely toothed, prominently aur, the aurs with well rounded margins but eventually sharply acute; common s.   3 **S. oleraceus** L.

3b Achenes merely several-nerved, not rugulose; lvs firmer, finely ± prickly-margined, pinnatifid, or often obovate and lobeless, provided with rounded, not acute aurs; prickly s.   4 **S. asper** (L.) Hill

Stephanomeria Nutt.   Rush-pink; Skeletonweed; Wirelettuce; Stephanomeria

Fls all lig and ⚥, pink or occ white; invol gen cylindric, the few principal bracts nearly =, the others shorter and often merely calyculate; recep naked; achenes ± 5-angled or 5-ribbed, sometimes with intermediate ribs; pappus of 10–∝ slender bristles, plumose at least above, often chaffy-flattened and connate toward the base; ann or per, often rushlike herbs with milky juice and alt, gen small and often scalelike lvs, growing in dry, gen open places, fl summer or late summer. (Gr *stephanos*, a wreath, and *mereia*, a division; of doubtful significance). (*Ptiloria*).

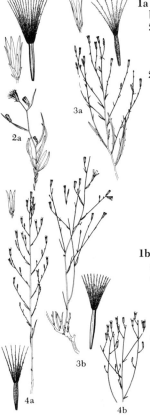

1a True per; pappus white or nearly so, the bristles plumose essentially to the base; achenes, except in no 2, smooth
    2a Heads relatively large, the invol with gen (5) 8 principal bracts, the fls gen (7) 10–21; pls 1–3 dm, gen single-std from deep-seated creeping roots; dry pine woodl, n Deschutes Co, Ore, s to Cal and Nev; large-fld w.   1 **S. lactucina** Gray
    2b Heads smaller, the invol gen with 5 principal bracts, the fls gen 5; pls several-std from a taproot which may be surmounted by a stout, br caudex, sometimes also with creeping roots
        3a Pls low, gen 1–3 dm; principal lvs runcinate-pinnatifid; achenes pitted and tuberculate; Gr Pl sp., reaching our range in Mont; runcinate-lvd s.   2 **S. runcinata** Nutt.
        3b Pls taller, gen 2–7 dm, with linear or filiform, entire or toothed lvs; achenes smooth or nearly so (except for being longitudinally ribbed); cordilleran, E Cas; narrow-lvd s., bush w.; 2 vars.
                3 **S. tenuifolia** (Torr.) Hall
            a1 Invol 7–11 mm; fls and principal bracts gen 5 each; pls moderately br; our common phase   var. **tenuifolia**
            a2 Invol 5–8 mm; fls and principal bracts 5 or often only 3 each; pl more slender, intricately br; GB phase, app or barely reaching our s border (*S. m.*)   var. **myrioclada** (Eat.) Cronq.
1b Ann or bien; pappus diverse, but not both white and plumose to the base; achenes rugose-tuberculate and pitted
    4a Pappus ± tawny or brownish, its bristles plumose nearly or quite to the base; infl paniculiform; pls 2–10 dm, br upwards, the brs relatively short, so that the pl appears narrower than in no 5; fls and principal bracts gen 5–8; E Cas, Wn, Ore, and adj Ida, s to Cal; stiff-br w.
                4 **S. paniculata** Nutt.
    4b Pappus white or nearly so, its bristles plumose only on the upper 1/2 or 2/3; heads terminating the brs, not arranged in a paniculiform infl; pls 1–6 dm, moderately to intricately bushy-br; principal bracts 3–5; fls (3) 5 (8); s sp., n to Crook Co, Ore; small w.   5 **S. exigua** Nutt.

Stylocline Nutt.   Stylocline

Heads disciform, white or whitish, small, ovoid or spheroid, ± glomerate; proper invol none, although the glomerules of heads are gen loosely lfy-bracted; ♀ fls with short, filiform-tubular corolla; central fls functionally ♂, with abortive achene; recep cylindric, chaffy throughout, the bracts subtending the ♀ fls boat-shaped or saccate, with scarious tip; anthers tailed; pappus none, or the few ♂ fls with some slender deciduous bristles; achenes small, ± compressed; small, white-woolly ann (ours seldom > 1 dm) with alt (or the lower rarely opp), entire, narrow lvs rarely 2 cm, growing in dry, open places, ours fl Apr–June. (Gr *stylos*, a column, and *kline*, a bed, referring to the form of the receptacle).

1a Bracts subtending the central, ♂ fls inconspicuous, = or < the others, not at all hooked; ♀ fls 8–13; GB sp., reaching our s border in Ore; Peck's s.
                1 **S. psilocarphoides** Peck
1b Bracts subtending the central, ♂ fls conspicuous, > the others, each with a rigid, incurved, hooked point; ♀ fls 5–9; E Cas in Ore, s to Cal and ne Nev; northern s., hooked s.   2 **S. filaginea** Gray

Tanacetum L.   Tansy

Heads gen disciform or shortly radiate, the outer fls ♀, with short, tubular corolla which in some spp. is expanded into a short yellow ray; invol bracts imbricate, dry, the margins and tips (at least of the inner) gen ± scarious; recep flat to low-conic, naked; disk fls ♂, with 5-toothed tubular yellow corolla; anthers entire or nearly so at base; style brs flattened, truncate, often minutely penicillate; achenes gen 5-ribbed or 5-angled, gen glandular; pappus a short crown, or none; aromatic ann or (all ours) per herbs, sometimes suffrutescent, with alt, pinnately dissected to ternate or even entire lvs and small or middle-sized, hemispheric to camp heads in a corymbiform to capitate infl. (Name of uncertain derivation). *(Sphaeromeria).*

2a

2b

a1   a2

3a

4b   4a

1a Robust, lfy-std pls, 2–20 dm, from creeping rhizomes; cauline lvs well developed, pinnately dissected, 0.5–2 dm, the larger ones gen > 1 dm

   2a Lvs glab or nearly so, punctate; pinnae with broadly winged rachis; disk 5–10 mm wide; Eurasian sp., cult and escaped in N Am esp along roadsides, fl Aug–Oct; common t.   **1 T. vulgare** L.

   2b Lvs ± pubescent, obscurely or scarcely punctate; pinnae with scarcely or very narrowly winged rachis; disk 8–15 mm wide; coastal sand dunes, BC to n Cal, fl June–Sept; n dune t., western t. *(T. camphoratum,* misapplied)   **2 T. douglasii** DC.

1b Smaller pls, without rhizomes, gen 0.5–3 dm, with gen basally disposed lvs, the cauline lvs ± reduced, seldom > 0.3 dm, and always much < 1 dm; fl May–July

   3a Basal lvs ± pinnatifid; receptacle white-hairy; disk 5–11 mm wide; meadows, esp alkaline meadows; Grant Co, Ore, and Elmore Co, Ida, s to c Nev and e Cal; cinquefoil t. *(Vesicarpa p.);* 2 well marked vars.

                                       3 T. potentilloides Gray

      a1 Pls gen 1–3 dm, with moderately lfy st; basal lvs gen pinnately dissected, the primary pinnae again deeply cut; heads gen several or ∝ ; Ore and Cal   var. **potentilloides**

      a2 Pls gen 0.5–1.5 dm, with more nearly naked st; basal lvs less dissected, gen merely pinnatifid; heads gen 1–4; Ida and Nev

                                      var. **nitrophilum** Cronq.

   3b Basal lvs entire to more often apically tridentate or ternate (or even palmatifid), the segms sometimes again ternate; receptacle glab; disk 3–7 mm wide; dry, open places in foothills and valleys

      4a Heads sessile in a very dense, globose head; basal lvs deeply trifid or palmatifid, the larger ones gen with the segms again trifid; Wyo and s Mont; cluster-headed t.   **4 T. capitatum** (Nutt.) T. & G.

      4b Heads pedunculate or subsessile in a close corymbiform cluster or rather loose head; basal lvs merely tridentate, sometimes strongly so, varying to entire; c Ida and adj Mont to Wyo; Nev; chicken sage *(S. argentea)*   **5 T. nuttallii** T. & G.

Taraxacum Hall.   Dandelion

Fls all lig and ♂, yellow, gen ∝ ; invol bracts biseriate, the outer gen < inner and often reflexed; inner invol bracts in some spp. corniculate (i.e., with a hooded appendage near the summit); recep naked; achenes columnar or thickly fusiform, terete or 4–5-angled, longitudinally sulcate or ribbed, gen muricate or tuberculate at least above, gen topped by a smooth, conic or pyramidal cusp which tapers into a long or short beak; pappus of ∝ capillary bristles, white to violet-brown; taprooted per, strictly scapose herbs, with erect, solitary heads and rosulate, entire to pinnatifid or subbipinnatifid lvs. (Name of doubtful origin, perhaps from Gr *tarassein,* to stir up, referring to reputed healing qualities). *(Leontodon,* misapplied).

2a   2b

1a Intro, weedy spp., fl throughout the season

   2a Achenes becoming red to reddish-brown or reddish-purple at maturity, the beak gen 1–3 times as long as the body; lvs tending to be deeply cut their whole length, without an enlarged terminal segm, the lobes narrow; outer invol bracts appressed to loose or sometimes reflexed; inner invol bracts gen corniculate; less common than no 2; red-seeded d. *(T. erythrospermum)*   **1 T. laevigatum** (Willd.) DC.

   2b Achenes olivaceous or stramineous to brown, the beak gen 2.5–4 times as long as the body; lvs gen less deeply cut, with an enlarged terminal lobe, but sometimes as in no 1; outer invol bracts reflexed; inner invol bracts not corniculate; common d. *(T. vulgare, L. taraxacum)*

                                         2 T. officinale Weber

1b Native, unagressive spp. of the high mts; lvs less dissected than in no 1; outer invol bracts appressed to lax or somewhat spreading, not reflexed; fl summer

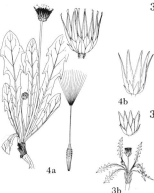

3a Achenes olivaceous or stramineous to brown or reddish, beak much >
body; pls of meadows and other moist places, often larger than no 5
4a Achenes olivaceous or stramineous to brown, only obscurely or
scarcely quadrangular; inner invol bracts often corniculate; circumboreal, s to Cal and NM; horned d. (*T. lapponicum, T. montanum*)
3 **T. ceratophorum** (Ledeb.) DC.
4b Achenes becoming red or reddish-brown to reddish-purple at maturity, tending to be sharply quadrangular; inner invol bracts seldom
corniculate; cordilleran; RM d. (*T. olympicum*) 4 **T. eriophorum** Rydb.
3b Achenes blackish, rarely slightly reddish at the summit, not quadrangular, beak about = body; inner invol bracts seldom corniculate; dwarf pls
of rocky places, seldom > 15 cm, with lvs seldom > 15 mm wide, and
with small heads (invol 7–18 mm); arctic Am and e Asia, s in the cordillera; dwarf alp d. (*T. scopulorum*)   5 **T. lyratum** (Ledeb.) DC.

Tetradymia DC.   Horse-brush

Heads discoid, yellow, 4–9-fld; invol of 4–6 erect = bracts; recep small, naked; corolla lobes >
throat; anthers strongly sagittate, almost caudate; style brs varying from as in *Luina* to nearly as in
Senecio; achenes terete, obscurely 5-nerved, glab to densely long-hairy; pappus of ∝ white or
whitish capillary bristles; ± canescent, br low shrubs (ours to 12 dm) with alt (and often fascicled) narrow entire lvs, occurring in dry, open places in the foothills and plains, fl May–Sept. (Gr
tetradymos, fourfold, referring to the teramerous heads of several of the spp.).

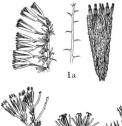

1a Primary lvs forming rigid, divaricate or recurved spines 5–13 mm, bearing
axillary fascicles of linear, subterete, green and glab lvs to ca 1 cm; fls 5–9;
heads pedunculate, scattered along the brs; invol conspicuously
white-tomentose; GB sp., n to n Malheur Co, Ore, c Ida, and sw Mont; cot-
tonhorn h., catclaw h.   1 **T. spinosa** H. & A.
1b Pls without spreading spines; fls 4; heads in small cymose clusters terminating the brs
2a Primary lvs appressed or closely ascending, 6–10 mm, tending to be
weakly spinescent; 2ndary lvs fascicled in the axils, blunt, linear, to ca 1
cm; twigs often rather persistently tomentose in lines, the herbage and
invol otherwise only thinly and deciduously tomentose or essentially
glab; GB sp., n to c Ida and Wasco Co, Ore; little-lf h. 2 **T. glabrata** Gray
2b None of the lvs spinescent; primary lvs linear or oblanceolate, 1–3 cm ×
1–4 mm, sometimes with shorter axillary fascicles; lvs, invols, and twigs
conspicuously and closely white-tomentose; cordilleran, widespread E
Cas; gray h., spineless h. (*T. inermis*)   3 **T. canescens** DC.

Thelesperma Less.   Thelesperma

Heads gen radiate, the rays neutral, broad, yellow or red-brown, or sometimes (incl our sp.)
wanting and the heads discoid; invol bracts biseriate and dimorphic, the outer ± herbaceous, gen
short and spreading, the inner membranous, with hyaline or scarious margins, united for at least
the lower third; recep flat, chaffy throughout, its bracts flat, or concave and somewhat clasping the
achenes; disk fls ⚥ and fertile; anthers entire or barely sagittate at base; style brs flattened, the
broad stigmatic lines nearly covering the inner surface below the externally short-hairy appendage; achenes ± flattened parallel to the invol bracts, linear or linear-oblong, glab, sometimes
tuberculate; pappus of 2 gen retrorsely barbed awns, or obsolete; herbs with opp (or the upper
alt), pinnately dissected to entire lvs and long-pedunculate heads. (Gr *thele*, nipple, and *sperma*,
seed, from the papillosity of some of the achenes).

T. subnudum Gray. Per, 1–2 dm from creeping roots, ± glab; principal lvs
crowded below, irreg pinnatifid or subbipinnatifid with few and long lobes
1.5–5 cm × 1–3 mm; heads 1–3, discoid (in ours), yellow; invol 6–9 mm; dry,
open places at lower elev; e Cont Div in Mont and adj Can, s to s RM and
Colo Plateau; ours is var. *marginatum* (Rydb.) Melchert (*T. marginatum*
Rydb. Mem. N. Y. Bot. Gard. 1:421. 1900).

Townsendia Hook.   Daisy; Townsendia

Heads radiate, the rays ♀ and fertile, white or anthocyanic; invol bracts narrow, ± imbricate, gen with a green central stripe and lacerate or fringed margins; recep flat, naked; disk fls ∝, ☿; anthers entire or nearly so at base; style brs flattened, with introrsely marginal stigmatic lines and short or elongate, externally hairy appendage; achenes flattened, 2-nerved (or those of the rays triangular), gen pubescent, but occ glab; pappus of the disk fls a single series of ∝ rigid, narrow, barbellate, bristlelike scales, that of the rays similar or sometimes reduced and chaffy; taprooted herbs with alt or all basal, entire lvs and gen rather few or solitary heads. (Named for David Townsend, 1787–1858, amateur botanist of West Chester, Pa.).

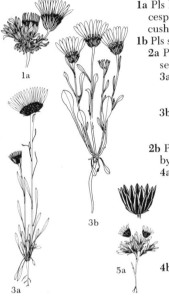

1a Pls loosely woolly-villous; pappus readily deciduous at maturity; compact, cespitose, short-lived, alp and subalp per with white or pinkish rays; cushion t. (*T. spathulata*, misapplied)   1 **T. condensata** Eat.
1b Pls strigose or hirsute-strigose to subglab; pappus persistent
 2a Pls ann, bien, or short-lived per with a mere crown on the taproot, gen several-std from the base (the sts lfy), seldom < 5 cm
  3a Rays lavender or purplish to blue, distinctly bluish when dried; heads large, the disk gen 1.5–3.5 cm wide, the invol 9–16 mm high; mont and subalp, n RM sp., w to Wallowa Mts; Parry's t.   2 **T. parryi** Eat.
  3b Rays pinkish; heads smaller, the disk gen 1–2 cm wide, the invol 7–10 mm high; plains and foothills E Cas, often with sagebr; showy t. (*T. watsoni*)   3 **T. florifer** (Hook.) Gray
 2b Pls cespitose, acaulescent per, seldom > 5 cm, the taproot surmounted by an evident br caudex
  4a Mont or alp spp. with blue or violet to occ white rays, glab or subglab achenes, and sparsely strigose or glabrate herbage
   5a Invol bracts broadly lanceolate to ovate or elliptic, obtuse or acute; sw Mont across c Ida to Wallowa Mts, Ore, s to Wyo and Utah; mt t.; ours is var. *m.*   4 **T. montana** Jones
   5b Invol bracts linear to lanceolate, strongly acute; ours morphologically app no 4 and often growing with it; c Ida and sw Mont, s to s RM and GB; common t.   5 **T. leptotes** (Gray) Osterh.
  4b Plains and valley spp. (entering our range e Cont Div in Mont) with white or pinkish rays, the herbage rather sparsely to densely strigose, the achenes persistently hairy, at least toward the base
   6a Invol bracts with a tuft of tangled cilia at the tip, linear-acuminate; heads relatively small, the invol gen 7–12 mm, its bracts ca 1 mm wide or less; rays gen 8–12 mm; disk corollas gen 5–8 mm; lvs linear or nearly so, gen 1–2 mm wide; Hooker's t. (*T. mensana*, misapplied; *T. sericea*, an illegitimate name)   6 **T. hookeri** Beaman
   6b Invol bracts without a tuft of tangled cilia, merely acute; heads relatively large, the invol gen 12–16 mm, its larger bracts gen 1.5–2.5 mm wide; rays gen 12–18 mm; disk corollas gen 8–12.5 mm; lvs oblanceolate, the larger ones gen 2–4 mm wide; Easter daisy
    7 **T. exscapa** (Rich.) Porter

Tragopogon L.   Goatsbeard; Salsify

Fls all lig and ☿, yellow or purple; invol cylindric or camp (conic in bud), the bracts uniseriate and =; recep naked; achenes linear, terete or angled, 5–10-nerved, narrowed at base, slender-beaked, or the outer occ beakless; pappus a single series of plumose bristles, united at the base, the plume-brs interwebbed, several of the bristles gen > the others and naked at the apex; taprooted herbs with milky juice, our spp. all bien or occ winter ann; lvs alt, linear, entire, clasping, gen ± grasslike; heads solitary at the br tips. (Gr *tragos*, goat, and *pogon*, beard, presumably from the conspicuous pappus).

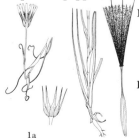

1a Peduncles not enlarged in fl, scarcely so in fr; achenes gen 15–25 mm (beak included); fls wholly yellow, the outer ligs = or > the invol bracts; invol bracts gen ca 8, 12–24 mm at anthesis, elongating to 18–38 mm in fr; lvs obscurely floccose at first, soon glabrate, with cirrhose-recurved tip; widespread diploid weed of European origin, gen in slightly moister habitats than no 3; meadow s., Jack-go-to-bed-at-noon   1 **T. pratensis** L.
1b Peduncles enlarged and fistulous above in fl and fr; achenes gen 25–40 mm; fls yellow or purple, the ligs slightly to very much < the invol bracts; lvs not cirrhose-recurved at tip except in no 2; invol bracts and pubescence various
 2a Heads yellow, or with a yellow eye; invol bracts gen ca 13, or only 8

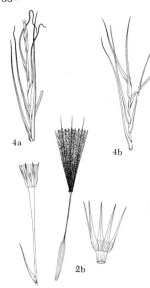

on last-formed heads or on depauperate pls

3a Heads wholly yellow; ligs much < invol bracts

   4a Lf apices cirrhose-recurved; lvs obscurely floccose when young, soon glab; recent amphiploid of nos 1 and 3, now becoming estab in e Wn and adj Ida, and in sw Mont and adj Wyo; full natural range not yet estab; hybrid g.   2 **T. miscellus** Ownbey

   4b Lf apices straight or nearly so, not cirrhose-recurved; lvs evidently floccose when young, later ± glabrate except gen in the axils; invol 2.5–4 cm in fl, elongating to 4–7 cm in fr; widespread diploid weed of European origin, in our range often also in relatively little-disturbed sites; yellow s. *(T. major)*   3 **T. dubius** Scop.

3b Heads purple or brownish-purple with a yellow eye; outer ligs only slightly < invol bracts; lvs obscurely floccose when young, soon glab; recent amphiploid of nos 3 and 5, now becoming estab in e Wn and adj Ida; full natural range not yet estab; remarkable g.

                          4 **T. mirus** Ownbey

2b Heads wholly purple; ligs slightly to much < invol bracts; lvs glab; invol bracts gen ca 8, gen 2.5–4 cm in fl, elongating to 4–7 cm in fr; weedy cult pl of European origin, occ with us along roadsides and in waste places, gen in rather moist soil; salsify, vegetable oyster

                          5 **T. porrifolius** L.

Tussilago L.   Coltsfoot

Heads radiate, the rays yellow, ♀ and fertile, ∝ in several series; invol a single series of =, ± herbaceous bracts, sometimes with a few basal bracteoles; recep flat, naked; central fls ♂ but sterile, with undivided, merely lobed style; anthers entire or minutely sagittate at base; achenes linear, 5–10-ribbed; pappus of ∝ capillary bristles, that of the sterile fls ± reduced; per with large basal lvs, scaly-bracted sts (the bracts alt), and medium-sized solitary heads. (L *tussis*, cough, for which the pl was a reputed remedy).

T. farfara L. Rhizomatous per 0.5–5 dm; bracts of the st ca 1 cm; lvs basal, developing well after the fl sts, long-petioled, the bl cordate to suborbicular, with deep, narrow sinus, toothed and shallowly lobed, gen 5–20 cm, glab above, persistently white-tomentose beneath; invol 8–15 mm; rays narrow, not much > invol and pappus; Eurasian weed, now widespread in e US and occ found W Cas in our region.

Viguiera H.B.K.   Viguiera

Heads radiate, the rays neutral, yellow; invol bracts narrow, ± herbaceous, 2–several-seriate; recep ± conic or convex, chaffy throughout, its bracts clasping the achenes; disk fls ∝, ♂ and fertile, yellow; anthers sagittate; style brs flattened, externally hairy above, without well marked stigmatic lines, the appendages shortly and minutely hairy, less so within; achenes quadrangular, somewhat flattened at right angles to the invol bracts; pappus persistent, of 2 awns and several short scales, or none. (Named for D. A. Viguier, librarian and botanist, of Montpellier, France). *(Heliomeris)*.

V. multiflora (Nutt.) Blake. Nw v. Taprooted, several-std per 3–13 dm, strigose to scab-puberulent; lvs lance-ovate to lance-linear, ± entire, gen 3–8 cm × 2–25 mm, short-petiolate, all but the uppermost opp; invol 5–10 mm, its bracts ± linear; disk 6–14 mm wide; rays 10–14, 7–17 mm; pappus none; open, gen dry hillsides; GB sp., n to sw Mont; ours is var. *m*.

Wyethia Nutt.   Mule's-ears; Wyethia

Heads large, solitary or several, radiate in all our spp., the rays ♀ and fertile, yellow or white; invol of several series of herbaceous or coriaceous bracts, subequal, or the outer enlarged; recep broadly convex, chaffy throughout, its bracts clasping the achenes; disk fls ∝, ♂ and fertile, light yellow; anthers sagittate; style brs slender, hispidulous to the base, without well marked stigmatic lines; pappus gen coroniform; disk achenes compressed-quadrangular; fragrant, taprooted, lfy-std per herbs; lvs simple, alt, the basal in our spp. enlarged, elliptic to elliptic-ovate or lance-elliptic, short-petiolate, the cauline smaller and often sessile; our spp. fl May–June. (Named for Nathaniel Wyeth, 1802–1856, western explorer, who first collected it).

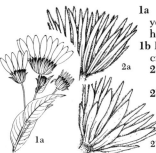

1a Herbage and invol entirely glab, resinous-varnished; rays sunflower yellow; cordilleran, E Cas, on open slopes and in dry meadows from foot-hills to mid elev in mts; northern w. or m.   1 **W. amplexicaulis** Nutt.

1b Herbage ± pubescent, not resinous-varnished; invol bracts conspicuously ciliate

2a Rays white or pale creamy; moist or wet meadows at mid elev in mts E Cas; white-rayed w., white-head w. or m.   2 **W. helianthoides** Nutt.

2b Rays sunflower yellow; meadows and moist, open hillsides W Cas and in CRG; narrow-lf w. or m.   3 **W. angustifolia** (DC.) Nutt.

Xanthium L.   Cocklebur

Heads small, unisexual; ♂ heads uppermost, ∝ -fld; invol of separate bracts in 1–3 series; recep cylindric, chaffy; filaments monadelphous; anthers free, obtuse at base; pistil vestigial, the style simple; invol. of ♀ heads of united bracts, completely enclosing the 2 fls (except for the exserted style brs), forming a conspicuous 2-chambered bur with hooked prickles; corolla none; achenes thick, solitary in each chamber of the bur, germinating in successive years; pappus none; coarse ann weeds to 2 m, fl Apr–Oct, with alt lvs and solitary or clustered axillary heads. (Ancient Gr name of some pl producing a yellow dye).

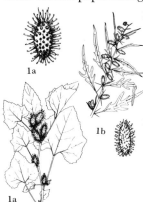

1a Lvs broadly ovate to suborbicular, gen cordate or deltoid at base, long-petiolate, without spines; burs 1.5–3.5 cm, with 2 beaks; widespread weed; common c.; we have 2 vars.   1 **X. strumarium** L.

a1 Bur gen 2–3.5 cm; lower part of prickles conspicuously spreading-hairy, as well as ± stipitate-glandular; our common phase (*X. italicum, X. pensylvanicum, X. saccharatum*)   var. **canadense** (Mill.) T. & G.

a2 Bur gen 1.5–2 (2.5) cm, merely atomiferous-glandular or finely glandular-puberulent to subglab; casual in our range (*X. chinense, X. orientale*)   var. **glabratum** (DC.) Cronq.

1b Lvs lanceolate, tapering to the short-petiolate base, provided with a conspicuous tripartite axillary spine 1–2 cm; burs ca 1 cm or a little more, beakless or with a single short beak; casual with us, more common in warm, dry regions; spiny c. (*Acanthoxanthium s.*)   2 **X. spinosum** L.

BUTOMACEAE   Flowering-rush Family

Fls ♂; sepals 3, gen greenish; petals 3, gen showy; stamens 6–∝ ; pistils (2) 3–21, whorled, distinct or slightly connate basally; fr follicular, ovary ovuliferous over much of inner surface; herbaceous aquatic or marsh-inhabiting per with linear to broad and flat lvs.

Butomus L.   Flowering-rush

Fls ∝ in terminal invol umbel; stamens 9; pistils 6; scapose, marsh-loving per with ∝ linear basal lvs. (Gr *boutomus*, name for some aquatic pl, derived from *bous*, ox, and *temno*, to cut).

B. umbellatus L. Pl fleshy-rhizomatous, up to 1 m; lvs erect, 2-ranked, narrow, ensiform, 3-cornered near base, ca = scape; perianth rose-colored, 2–2.5 cm broad, persistent, sepals slightly more greenish than the petals; anthers red; follicles inflated but hardened, long-beaked by the persistent style, body ca 1 cm, dehiscent ventrally; seeds ∝ ; in or at edge of shallow water; Eurasian, intro and widely estab in N Am, reported in our area from near Moscow, Latah Co, and Idaho Falls, Bonneville Co, Ida.

ALISMATACEAE   Water-plantain Family

Fls gen ♂ but sometimes ♂♀, umbellate, racemose, or pan, reg; sepals 3, greenish, gen persistent; petals 3, white to pinkish or purplish, deciduous (rarely lacking); stamens 6–∝ (3); pistils gen several–∝ , distinct or ovaries ± connate basally; style persistent and often beaklike; ovary 1–several-seeded; fr gen an achene, sometimes semidehiscent; per (ours), aquatic, mostly scapose herbs with erect to floating lvs gen sheathing at base.

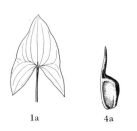

1a

4a

3b

4b

1a Lvs in part sagittate or hastate **Sagittaria**
1b Lvs neither sagittate nor hastate
  2a Pl submersed; lvs in part with threadlike petioles < 1 mm thick and up to 3–6 dm; bls floating, elliptic-lanceolate, 1–3 (4) × < 1 cm, but most lvs gladiate and tufted; nonfl, deep-water form found in Hicks Lake, Thurston Co, Wn; unidentifiable, but believed to be either the fairly common **Sagittaria latifolia** Willd.
    or the otherwise unknown (in our area)
    Echinodorus ranunculoides (L.) Engelm.
  2b Pl emersed or submersed but lvs neither gladiate nor with petioles < 1 mm thick; bls various
    3a Pistils in 1 whorl; stamens gen 6; fls ♂
      4a Petals incised-fimbriate; beak of achene erect, subequal to achene body **Machaerocarpus**
      4b Petals subentire; achene not beaked **Alisma**
    3b Pistils spirally arranged in a ± globose mass; stamens > 6; fls gen ♀♂ **Sagittaria**

Alisma L.   Waterplantain

Fls pan; pan brs whorled; stamens 6 (9), opp petals; pistils (5) 10–25, style subterminal, short (ca 1 mm in ours), ovary 1-celled and 1-ovuled; fr achene, strongly compressed; per, scapose, aquatic or palustrine, cormose herbs.

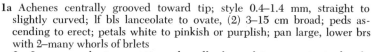

2b

1b

1a Achenes centrally grooved toward tip; style 0.4–1.4 mm, straight to slightly curved; lf bls lanceolate to ovate, (2) 3–15 cm broad; peds ascending to erect; petals white to pinkish or purplish; pan large, lower brs with 2–many whorls of brlets
  2a Lvs narrowly to sometimes broadly lanceolate, cuneate to barely rounded at base; petals pink to purple; style 0.4–0.9 mm, almost terminal on ovary; anthers < twice as long as broad; European, rare in N Am except in Cal; reported for our area from Eugene, Ore
    A. lanceolatum With.
  2b Lvs broadly lanceolate to ovate, rounded to cordate at base; petals either white or style 0.7–1.4 mm, style slightly lateral on ovary; anthers at least twice as long as broad; marshy areas, sometimes largely submerged; s BC to NS, s to s Cal, Ariz, Colo, Mo, NY, and Me; throughout much of Old World; Am w.; 2 vars. **1 A. plantago-aquatica** L.
    a1 Petals pink to purple; style 0.4–1.5 mm, slender, gen straight; common in Old World, but only occ in N Am, recorded in our area from Seattle, Wn (*A. michaletii*) var. **plantago-aquatica**
    a2 Petals white or only slightly pinkish; styles 0.4–0.8 mm, stout, gen slightly curved; range of sp. in N Am (*A. brevipes, A. trivialis*) var. **americanum** Schul. & Schul.
1b Achenes centrally ridged and with 2 grooves toward tip; style scarcely 0.5 mm, gen curved to at least 1/2 turn; lf bls narrowly lanceolate or spatulate to linear, mostly 0.5–2 (occ up to 3 or 4) cm broad; peds in part widely spreading to recurved; petals gen pinkish; infl small, lower brs with gen only 1 whorl of brlets; marshes and edges of lakes and ponds, occ almost completely submersed; sw BC to Que, s to n Cal, Ida, Colo, SD, Minn, and NY; Old World; narrowlf w.; 2 vars. **2 A. gramineum** Gmel.
  a1 Lvs stiffly erect, rarely submersed, bls mostly 0.5–2 (4) cm broad; range of sp. (*A. geyeri, A. validum*) var. **angustissimum** (DC.) Hendricks
  a2 Lvs gen submersed or floating, linear, almost bladeless, but pl sometimes emergent and with narrowly elliptic bls scarcely 0.5 cm broad; range of sp., perhaps only a deep-water ecological variant (*A. g.* var. *graminifolium*) var. **gramineum**

Machaerocarpus Small   Fringed or Star Waterplantain

Fls ♂, umbellate or pan, rather showy; petals white to pink, incised-fimbriate, spreading, deciduous; stamens 6; pistils (6) 9–15, in 1 whorl; fr achene, 1-seeded; per, scapose herb with long-petioled lvs. (Gr *machaira*, dagger, and *carpos*, fr, referring to the dagger-like beak of the achene).

M. californicus (Torr.) Small. Pl mostly 2–4 (6) dm; lvs erect to floating, clustered on the short rhizome, ca = infl, bl linear to oblong-lanceolate, rounded to subcuneate at base, 3–8 cm; infl a simple invol umbel or a 2–4-whorled pan; peds 2–8 cm, spreading to recurved; sepals 4–5 mm; petals 7–10 mm; stamens opp sepals, filaments flattened-deltoid, anthers linear, ca 2 mm, twisting after dehiscence; achenes flattened, divergent, strongly ribbed on each margin and depressed between, body and the erect beak both ca 4 mm; sloughs, marshy fields, and ditches; Wheeler Co, Ore, to c Cal, sw Ida, and w Nev (*Alisma c., Damasonium c.*).

Sagittaria L.   Arrowhead

Fls in whorls of gen 3 in a simple, bracteate raceme, lower ones gen ♀ (⚥), upper ones ♂, or sometimes all either ♂ or ♀; sepals persistent; petals white, longer than the sepals, gen witheringpersistent; stamens 7–∝; pistils ∝, receptacle ± globose; achenes laterally flattened and gen wing-margined, short-beaked; scapose, per, palustrine to aquatic herbs, gen rhizomatous and often tuber-bearing; lvs mostly sheathing, long-petioled, bls hastate or sagittate to lanceolate, at least some lvs (in submersed pls sometimes all) bladeless, elongate, and broadly to very narrowly subulate. (L *sagitta*, arrow, referring to shape of lf bls). Several spp., esp S. *latifolia*, are sometimes planted by "sportsmen," as they produce large, starchy, edible tubers (called "wapato" by the Indians) of which ducks are fond.

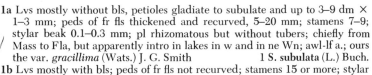

1a Lvs mostly without bls, petioles gladiate to subulate and up to 3–9 dm × 1–3 mm; peds of fr fls thickened and recurved, 5–20 mm; stamens 7–9; stylar beak 0.1–0.3 mm; pl rhizomatous but without tubers; chiefly from Mass to Fla, but apparently intro in lakes in w and in ne Wn; awl-lf a.; ours the var. *gracillima* (Wats.) J. G. Smith   1 **S. subulata** (L.) Buch.
1b Lvs mostly with bls; peds of fr fls not recurved; stamens 15 or more; stylar beak 0.2–1.5 mm; pls tuber-bearing
  2a Pls ♂♀, ♀ fls with much shorter peds than the ♂; bracts of infl (5) 10–30 mm, lanceolate, acute to acuminate; mature achenes 2–2.5 mm, beak (in fl the style) 0.2–0.4 mm, terminal on achene, pointing forward; sagittate lf bls mostly 6–15 × 2–10 cm; stamens 15–25; ponds and ditches; n BC s, apparently e Cas in Wn and Ore, to s Cal, e to NS, NY, Ill, Utah, NM, and n Tex; arumlf a., wapato (*S. paniculata, S. suksdorfii, S. arifolia* var. *c., S. a.* vars. *aquatilis, stricta,* and *tenuor*)   2 **S. cuneata** Sheld.
  2b Pls sometimes ♂, ♀, occ largely ♂ but fls with a few pistils, or with both ♂ and ♀ fls; ♀ fls with peds almost or fully as long as the ♂; bracts of infl mostly 5–10 (15) mm, blunt, hooded; mature achenes ca 3 (2.5–4) mm, beak (in fl the style) ca 1 mm, arising in line with the upper margin and pointing upward at ca right angle to body of achene; sagittate lf bls up to 25 cm and nearly as broad; stamens gen 25–40; swamps, ditches, and ponds and lakes; abundant in c and e US and common on Pac Coast from VI to c Cal, almost limited to w side Cas and CRG, but reported also from se Wn and n Ida; broadlf a., wapato (*S. esculenta, S. sagittifolia* var. *l., S. s.* vars. *macrophylla* and *vulgaris*)   3 **S. latifolia** Willd.

HYDROCHARITACEAE   Frog's-bit Family

Pls mostly ♂, ♀, but sometimes ♂♂♀ or (rarely) ⚥-fld; fls reg, 1–several within a spathe of mostly 2 (1–3) ± connate bracts; spathes sessile to pedunculate; sepals gen 3 (2), greenish; petals gen 3 (2), mosty white, often reduced or lacking; ♂ fls gen > 1 per spathe, stamens (1–2) 3–12; ♀ fls 1 per spathe, ovary inferior, 1-celled with 3–6 parietal (sometimes much-intruded) placentae, styles gen 3–6 and each sometimes 2–3-brd; fr dry to baccate, indehiscent, ∝-seeded; gen submerged aquatics of fresh or salt water, with anchored or floating roots and simple, tufted, alt, opp, or whorled lvs.

1a Lvs ∝, opp or whorled, borne on well-developed sts, not tape-like **Elodea**
1b Lvs few, alt, tufted on a very short st, long and tape-like   **Vallisneria**

Elodea Rich. in Michx.   Waterweed; Ditchmoss

Spathes sessile to pedunculate, 2-lobed, 1 (2–3)-fld; sepals and petals 3; ♂ fls with (3 to) 9 stamens; ♀ fls solitary in tubular spathes, with ped-like hypan gen elongating to bring rest of fl to the water surface, stigmas 3, gen bilobed, tending to float, the styles slender; fr cylindric; slender, submersed, ♂, ♀ (ours), per, with br and gen nodally rooting sts and 1-nerved, sessile, (alt) opp or whorled, gen ± denticulate lvs, gen in fresh (brackish) water of streams, ponds, sloughs, and ditches. (Gr *elodes*, marshy, referring to the habitat). (*Anacharis, Philotria*).

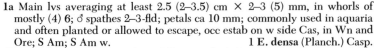

1a Main lvs averaging at least 2.5 (2–3.5) cm × 2–3 (5) mm, in whorls of mostly (4) 6; ♂ spathes 2–3-fld; petals ca 10 mm; commonly used in aquaria and often planted or allowed to escape, occ estab on w side Cas, in Wn and Ore; S Am; S Am w.   1 **E. densa** (Planch.) Casp.

1b Main lvs rarely as much as 2.5 cm, only 2–3 per node; ♂ spathes 1-fld; petals scarcely 5 mm

   2a Most of lvs (other than the basal) opp, larger ones gen (17) 20–26 × 1–2 (2.5) mm; ♂ fls with slender hypan up to 3 dm, sepals ca 4 mm, petals ca 5 mm; ♀ fls with hypan 2–3 dm, fls slightly smaller than the ♂; RM area, Alta to ND, s to Mont, Wyo, Colo, NM, and Utah; long-sheath w.

                                               2 **E. longivaginata** St. John

   2b Most of lvs (other than the basal) in whorls of 3, largest ones rarely > 15 (17) mm

      3a Lvs 1–4 (ave 2) mm wide, 6–17 mm long, tapered abruptly to a blunt point; ♂ fls with slender hypan not deciduous at anthesis, sepals ca 4 mm, petals slightly longer; ♀ fls with hypan up to 15 cm, sepals and petals ca 2.5 mm; often where alkaline; BC to Cal, e to Que and most of w and c US; Can or RM w. (*E. planchonii*)

                                        3 **E. canadensis** Rich. in Michx.

      3b Lvs 0.3–1.5 (ave not > 1.5) mm wide, 6–13 mm long, tapered to a slender point; ♂ fls sessile, deciduous from pl at anthesis and floating on water surface, sepals ca 2 mm, petals smaller (lacking); ♀ fls with hypan to 10 cm, sepals and petals ca 1.5 mm; fresh to slightly brackish water; common in e US, occ in Wn, Ore, n Ida, and elsewhere in w US; Nuttall's w.   4 **E. nuttallii** (Planch.) St. John

Vallisneria L.   Tapegrass; Wild Celery

♂ fls clustered in a subglobose, (2) 3-parted spathe, deciduous and free-floating at anthesis, sepals 3 (unequal), petals rudimentary, stamens gen 2 (1–3); ♀ fls tubular, on long slender scapes, solitary, borne at water surface, sepals and petals 3; stigmas 3, 2-lobed; fr cylindric; ♂, ♀, scapose per with linear, tufted, long, submersed lvs. (Named for Antonio Vallisnieri de Vallisnera, 1661–1730, an Italian botanist who was esp interested in aquatic pls).

V. americana Michx. Am w. Pl rhizomatous; lvs thin, flat, 1–5 (20) dm × (1) 3–10 mm, with ∝ longitudinal veinlets and cross-septa; ♂ fls several hundred per spathe, only ca 1 mm broad; ♀ fls 2–3 cm, solitary, the corolla white, ca 5 mm wide, the spathe with peduncle up to 1–2 m (tending to coil and retract after pollination); fr 5–10 cm; ponds and lakes and quiet streams; native from Que to Tex and Fla, intro in Dry Falls Coulee, Grant Co, and in several lakes w Cas, in Wn, and prob also in Ore. *V. spiralis* L., a European sp., has been reported as estab in places in w Wn, but has not been seen.

SCHEUCHZERIACEAE   Scheuchzeria Family

Fls few in terminal bracteate racemes, ⚥, reg, trimerous; perianth segms 6, in 2 similar series, persistent; stamens 6, filaments rather short, anthers linear; carpels gen 3 (occ 4–6), free or ± connate at base; stigma subsessile, papillate; ovaries gen with 2 (1–3) ovules; frs follicles, 1–2-seeded; per herbs with linear, basally sheathing lvs.

Scheuchzeria L.   Scheuchzeria

Pl rushlike; lvs broadly sheathing at base and with a prominent lig at juncture of sheath and bl, the bl semiterete. (For Johan Jakob Scheuchzer, 1672–1733, a Swiss botanist).

S. palustris L. Pl rhizomatous; fl sts 1–4 dm, covered with marcescent lvs at base; basal lvs 1–4 dm; st lvs gradually reduced upward; lig 1–10 mm; bl erect, 1–3 mm broad; racemes 3–12-fld, peds up to 25 mm in fr; perianth greenish-white, segms oblong, 1-nerved, ca 3 mm; follicles 5–8 (10) mm, compressed, divergent, light greenish-brown, connate only at base, the stylar beak 0.5–1 mm; seeds 4–5 mm; in bogs and along lake margins; s Alas to Lab and Newf, s in BC and Wn to n Cal and e to Ida, Wis, Ia, Ind, and NJ; Eurasia; ours the var. *americana* Fern.

JUNCAGINACEAE   Arrow-grass Family

Fls racemose to spicate, small; perianth parts sometimes lacking but gen 6, greenish to purplish, biseriate; stamens 1, 4, or 6, the anthers subsessile in the axils of the perianth segms; carpels 1, 3, 4, or 6, free to connate, styles lacking to greatly elongate, stigmas capitate to ± plumose; ovule 1 per carpel; frs from several and simple to 1 and compound; ann to per, scapose, often rhizomatous, ⚥-fld to ⚥♀♂ herbs with terete and onionlike to sheathing, rushlike lvs.

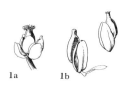

1a Perianth 6-parted; fls ebracteate, all ⚥ and in terminal racemes; carpels obviously 3 or 6; styles lacking   **Triglochin**
1b Perianth lacking; fls bracteate, of 3 types, ⚥, ♀, and ♂, only the ♀ borne in the lf axils, the ⚥ and ♂ in pedunculate spikes; carpels apparently 1 (or 3 and completely fused); styles of basal ♀ fls greatly elongated   **Lilaea**

1a   1b

Lilaea Humb. & Bonpl.   Flowering Quillwort

Fls naked, the ♀ in part single in the sheathing lf axils and thick-walled, with long filiform style and capitate stigma, but mostly in pedunculate spikes, where mixed with ⚥ fls or below them, bractless, the style short or lacking, the pistil 1-celled and 1-ovuled, but mostly 3-lobed at tip; ⚥ and ♂ fls in the pedunculate spikes only, the ♂ consisting of a single stamen, borne mostly above the ♀ fls, each attached to a caducous bract (bract sometimes interpreted as an expanded connective), the anther subsessile in the axil of the bract, the 2 pollen sacs widely separated; ⚥ fls 1-bracteate, consisting of 1 stamen and 1 pistil; frs strongly ribbed, woody, indehiscent, those of the basal lvs 3-angled and 3-horned at tip, those of the spikes flattened, 2-winged, and short-beaked; acaulescent, ann, aquatic pls with terete, tufted, onion- or quillwort-like lvs. (For Alire Raffeneau Delile, 1778–1850, a French botanist).

L. scilloides (Poir.) Hauman. Acaulescent ann; lvs tufted, terete, (3) 5–35 cm × 1–4 mm, acute to subulate, sheathing at base; basal pistils becoming triquetrous in fr, 4–6 mm, the style 5–25 cm; fl spikes with peduncles 3–20 cm, spike varying from 0.5–4 cm depending upon age; ♂ fls with caducous bract 2–3 mm; ♀ fls bractless or with small, recurved, caducous bract; fr winged, (2) 3–5 mm, lower ones with beaks 1–2 mm, upper ones almost beakless; gen in shallow water but often stranded in mud at edge of receding ponds; coastal tideflats to interior valleys; BC to Baja Cal, e to Alta, Mont, Ida, and Nev; S Am (*Heterostylus gramineus, L. subulata*).

Triglochin L.   Arrow-grass

Fls ebracteate, ⚥, 3-merous, racemose; anthers 6, almost sessile in the axils of the 6 greenish, 1.5–2 mm perianth segms; carpels 3 or 6, ± strongly connate, the 3 or 6 stigmas sessile; fr follicular, 1-seeded, carpels tardily separating from base upward and often remaining attached to or suspended from a terete to winged central axis; gen rhizomatous herbs of wet places, often where saline or alkaline. (Gr *treis*, three, and *glochis*, point; the fr of some spp. is 3-pointed). (*Hexaglochin, Juncago*).

1a Carpels and stigmas 3; fr linear-clavate, 6–8 mm, carpels separating upward from the 3-winged axis and remaining suspended from the top, subulate at the base; pl 1.5–6 dm; lig 0.5–1.5 mm, parted to the base; lf bls 1–2 mm broad; coastal bogs to inl meadows, mud flats, and gravelly stream margins, often where brackish or alkaline; Alas to Cal, e to Lab, Ill, NY, and NM; S Am and Eurasia; marsh a.   **1 T. palustre** L.
1b Carpels and stigmas gen 6; fr oblong or ovoid-prismatic, completely deciduous from the terete axis, not subulate at the base

1a

2a Lig entire or only slightly bilobed, (1) 1.5–5 mm; lf bls ± obcompressed, mostly 1.5–2.5 (4) mm broad; pls gen at least 3 (to 12) dm; marshy to moist and gravelly areas, mostly where saline or alkaline; Alas to Baja Cal and Mex, e to Lab, Ill, Ind, and Pa; S Am, Eurasia; seaside a.   **2 T. maritimum** L.

2b Lig bilobed full length, 0.5–1 mm; lf bls subterete, 0.5–1 (1.5) mm broad; pls often < 3 (0.5–4) dm; dry meadows and marshes, always where brackish or saline to alkaline; extreme sw BC s along coast to Baja Cal, e to Daks, Wyo, Colo, and Ariz; S Am; graceful a.; 2 vars.
    3 T. concinnum Davy

    a1 Pls mostly < 2 dm; lvs gen as long as scape; coastal, BC to Baja Cal
    var. **concinnum**

    a2 Pls mostly 2–3 (1.5–4) dm; lvs much shorter than scape; e Cas in Ore and mainly e Sierra Nevada in Cal, e to beyond RM, ND to Colo; known in our area from Teton Co, Mont, but to be expected elsewhere (*T. maritimum* var. *d.*, *T. d.*)   var. **debile** (Jones) Howell

NAJADACEAE   Water-nymph Family

Fls ♂♀; ♂ fls with 1 stamen; ♀ fls naked, pistil 1, 1-celled and 1-ovuled, but stigmas 2–4; fr achene; submersed ann with slender, br sts and narrow, opp lvs.

Najas L.   Water-nymph

Fls tiny, axillary, the ♂ enclosed in a membranous, hyaline, saccate bract surrounded by a firmer, entire to 4-lobed, perianth-like structure, filament almost lacking; stigmas sessile, considerably elongate; lvs often apparently 4 per node, toothed to entire, dilated and semi-sheathing at base, gen clustered toward st tips and concealing fls, but frs plainly visible as brs elongate. (Gr *naias*, a water-nymph).

1a Lvs 2–4 mm broad, prominently dentate, the teeth plainly visible to the unaided eye, at least 0.5 mm; pl ♂, ♀; wideranging Old World sp., intro in US where known from ND, Utah, Nev, and Cal, but apparently not reported for our area   **N. marina** L.

1b Lvs scarcely 1 mm broad, minutely serrulate, the teeth scarcely discernible to the unaided eye, much < 0.5 mm; pls ♂♀

  2a Seeds ca 3 mm, the coat smooth and shining; lvs 1–3 cm, tapered at least from midlength to a long slender point; sts 3–15 dm; anther 1-celled; fresh to slightly brackish water; BC to Cal, e to Ida and from Daks to ne Can and Mo; n Europe; wavy w.   **1 N. flexilis** (Willd.) Rost. & Schmidt

  2b Seeds 2.5–3 mm, the coat coarsely pitted, dull; lvs 1–2 cm, tapered mainly for distal 2–3 mm to a somewhat rounded or obtuse but ultimately finely pointed tip; sts rarely > 7 dm; anther 4-celled; widespread in fresh water in the New World; in our area in Wn, Ore, and Ida; Guadalupe w.   **2 N. guadalupensis** (Spreng.) Morong

POTAMOGETONACEAE   Pondweed Family

Fls ⚥, 4-merous, sessile in pedunculate spikes, gen in several approximate to distinct whorls; perianth segms 4, clawed, each with an expanded, upturned, gen oval bl; stamens fused with claws of the perianth segms, anthers 2-celled, sessile; pistils 4, sessile, 1-carpellary, stigma sessile or on a short, persistent, often curved style; ovule 1; fr achene, gen with a ± prominent dorsal keel, often beaked by the persistent style; per, rhizomatous herbs of fresh (rarely ± brackish) water, with terete or slightly flattened, simple or br sts and alt (opp), prominently stip lvs, the stips sometimes sheathing.

Potamogeton L.   Pondweed

Pls submersed or partially floating; lvs all alt or the uppermost sometimes opp, those submersed gen much thinner, shorter-petioled, and narrower than those (if any) floating; fr semifleshy, becoming hardened. (Gr name, *Potamogeiton*, for some aquatic pl; derived from *potamos*, river, and *geiton*, neighbor, in allusion to the habitat). (*Buccaferrea, Spirillus*).

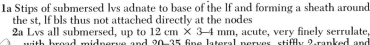

1a Stips of submersed lvs adnate to base of the lf and forming a sheath around the st, lf bls thus not attached directly at the nodes

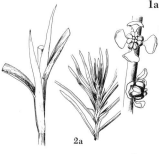

  2a Lvs all submersed, up to 12 cm × 3–4 mm, acute, very finely serrulate, with broad midnerve and 20–35 fine lateral nerves, stiffly 2-ranked and crowded on st; stips adnate for (5) 10–15 mm, free portion longer, white, many-nerved, ultimately shredding into white fibers; sts terete, stout, gen br, 1–4 (10) dm; achenes obovoid, 3.5–4 mm, scarcely beaked; quiet water, mostly in lakes, esp w Cas; BC to Cal, e to Que, Ida, Mont, Wyo, and to Ind and Pa; Robbins' p.   1 **P. robbinsii** Oakes

  2b Lvs sometimes in part floating, those submersed entire, < 3 mm broad, gen not crowded on st, mostly 1- or 3-nerved

    3a Adnate portion of stips 1–3 (5) mm, free portion ca twice as long; floating lvs sometimes present, these elliptic or oblong-elliptic, 1–3 cm, slender-petiolate; submersed lvs delicate, narrowly linear, 2–4 cm × scarcely 1 (1.5) mm; sts freely br, terete, very slender, up to 5 dm; achenes 1.1–1.5 (1.8) mm, semiorbicular, style tiny; shallow creeks and ponds in most of US, in our area known from s BC, Ida, and Mont; it is in the Steens Mts and possibly also in our area in Ore; diverse-lvd p. (*P. hybridum* Michx.)   2 **P. diversifolius** Raf.

    3b Adnate portion of stips gen > 5 mm, often > the free portion; floating lvs absent

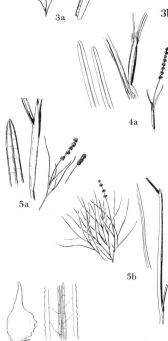

      4a Main stip-sheaths subtending 2 or more brs, strongly inflated, gen brownish, 2 or > times as broad as the st, the free stip tips forming a short lig; lvs rounded to retuse at tip, 2–8 (10) cm × 0.5–2 mm, 1 (3) -nerved; spike with (4) 5–9 evenly spaced whorls of fls; achenes obliquely obovate in outline, ca 3 mm; shallow to fairly deep, standing to swiftly moving water; Alas s, e Cas, to Ore, e to Newf, Ida, Mont, Wyo, Wis, and NY; Eurasia; sheathing p. (*P. monili-formis*)   3 **P. vaginatus** Turcz.

      4b Main sheaths subtending 1–2 brs, not strongly inflated, gen green-ish, the free stip tips projecting upward as a slender lig; lvs acute or acuminate; spike with mostly 2–5 whorls of lvs

        5a Stigma sessile, the fr not beaked; sheaths at least partially closed unless mechanically ruptured, remaining around the st; sts br below but largely unbr above, slender, subterete, mostly < 4 (5) dm; lvs up to 10 cm × scarcely 1.5 mm, 1-nerved; achenes ca 2 mm; shallow, standing or slow-moving water; Alas s, mainly e Cas, to Cal, Ariz, and Colo, e to Greenl, Mich, and Pa; wide-spread in Old World; slender-lvd p. (*P. borealis, P. interior*)
                                                                4 **P. filiformis** Pers.

        5b Stigma on a very short style (ca 0.5 mm) which persists as a tiny beak on fr; sheaths open to the base; sts gen dichotomously br to near tips, filiform, terete, up to 4 dm; lvs filiform, up to 12 cm × scarcely 1 mm, 1 (3)-nerved; achenes ca 3–3.5 mm; shallow, fresh to brackish water; one of commonest spp. of N Am, throughout our area, on both sides Cas; Eurasia; fennel-lvd p. (*P. columbianus*)   5 **P. pectinatus** L.

1b Stips of submersed lvs free of the rest of the lf, lf bls or petioles attached directly at the nodes of the st

  6a Lvs all submersed, narrowly oblong, crispate and finely serrulate, 3–8 cm × 3–10 (12) mm, sessile and ± clasping, 3–5 (7) -nerved, midrib bordered by lacunae and as much as 1 mm broad; achene body ca 3 mm, tapered uniformly into a conical beak 2–3 mm; ponds and streams, often forming mats; native to Old World; intro widely in N Am and known in our area from w Wn and w Ore, more common in Cal and in c and ne US and adj Can; curled p.   6 **P. crispus** L.

  6b Lvs sometimes in part floating, those submersed never serrulate and often < 3 or > 12 mm broad; achene beak < 2 mm

    7a Sts flattened, winged, (1) 2–3 mm broad, often half as broad as lvs; lvs all submersed, linear, up to 20 cm × (2) 3–5 mm, with (10) 15–30 or more very slender veins, not clasping; achenes 4 (4.5) mm; style ca 0.7 mm; mostly in lakes in our area; BC s, on both sides Cas, to n Cal, e to Ida and Mont and in most of c and ne US and adj Can; eel-grass p. (*P. compressus* and *P. zosterifolius* of Am auth.)   7 **P. zosteriformis** Fern.

    7b Sts subterete or only slightly flattened; lvs sometimes dimorphic and some floating, if linear gen < 2 or > 5 mm broad or with < 10 veins

      8a Lvs all ± cordate and clasping at base, submersed, rarely < 10 (to 30) mm broad

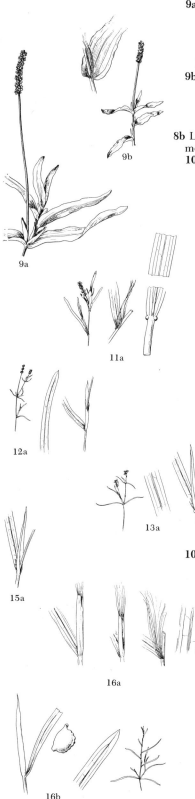

9a Sts often zigzag; stips gen persistent, mostly at least 3 (3–10) cm, whitish; lvs gen at least 10 cm, often > 2 (1–3) cm broad; peduncles 1–3 dm; achenes 4–5 mm, with a short but prominent beak; mostly in deep water of lakes in our area; Alas s on both sides Cas, to Cal, e to Newf and s to Utah, Colo, Neb, Ind, and NY; Europe; long- or white-stalked p.   8 **P. praelongus** Wulf.

9b Sts rarely at all zigzag; stips quickly disintegrating into fibers 1–2 cm; lvs gen < 10 × < 2 cm; peduncles mostly 0.2–2 dm; achenes 2.5–3.5 mm, the beak nearly 1 mm; widespread in N Am, both sides Cas; Richardson's p. *(P. perfoliatus* var. *lanceolatus)*

9 **P. richardsonii** (Bennett) Rydb.

8b Lvs not cordate and clasping at base, often dimorphic, those submersed often < 10 mm broad

10a Nodes with paired, globose, gen yellowish glands situated on either side of lf bases (glands often lacking on some nodes)

11a Stips prominent, 7–15 (20) mm, strongly fibrous, white, ultimately shredding into persistent fibers; lvs 1.5–3 (3.5) mm broad, with 1 broad midvein and (2) 4–6 indistinct lateral nerves; sts and peduncles flattened; stips freely br, 0.5–1 mm broad; achenes (1.5) 2–2.5 mm, with stylar beak ca 0.6 mm; in rather shallow, brackish to fresh water, chiefly in lakes; Alas to Skagit and Okanogan cos, and is of PS, Wn; in Can to Newf, also in c and e US; Europe; flat-stalked p.   10 **P. friesii** Rupr.

11b Stips not strongly fibrillose, mostly not white, often quickly deciduous; lvs mostly < 1.5 mm broad; sts and peduncles seldom flattened

12a Stips connate at least half length before rupture, surrounding internode, 6–15 mm, membranous and gen disintegrating from the tip into slender fibrils; lvs 6–7 cm × 0.5–2 mm, 3 (5) -nerved, the lateral nerves obscure; achenes 2–2.5 mm, style slightly recurved, scarcely 0.5 mm; shallow ponds and streams; BC s, on both sides Cas, to Cal, e through most of US and Can to Atl; Eurasia; small p. *(P. panormitanus)*

11 **P. pusillus** L.

12b Stips with free margins, often erect and forming a tube, but open to the base on one side, gen quickly deciduous

13a Lvs mostly < 1.5 (0.5–2) mm broad, up to 7 cm long, gen acute; achene 2–2.5 mm, stylar beak almost terminal, scarcely 0.5 mm; shallow, acidic to saline water; Alas s, on both sides Cas, to Cal, e to Newf and most of US to the Atl; Europe; Berchtold's p. *(P. pusillus* of Am auth.)

12 **P. berchtoldii** Fieb.

13b Lvs (1.5) 2–3 mm broad, gen obtuse or rounded; achene 3–3.5 mm, stylar beak ca 0.5 mm; c and ne US and adj Can, reported from VI, but not known from our area

P. obtusifolius Mert. & Koch

10b Nodes lacking glands

14a Lvs < 3 mm broad, all submersed, linear

15a Stips with free margins, often erect and forming a tube, but open down one side, often deciduous (see lead 13a)

12 **P. berchtoldii** Fieb.

15b Stips connate, forming a tube around st, often rupturing and shredding, persistent or deciduous

16a Stips firm, strongly fibrous, the fibers free at the stip tips as prominent cilia 2–3 mm, persistent, but rest of the stip rather quickly disintegrating; lvs narrowly linear, 3–6 cm × ca 1.5 (2) mm; shallow lakes and streams; sw Wn e through e Ore and Ida to YNP, seldom collected; fibrous-stip p.   13 **P. fibrillosus** Fern.

16b Stips delicate, not particularly fibrous, tending to disintegrate completely; lvs linear, up to 10 cm × 1–1.5 (2.5) mm; standing to slow-moving, gen shallow water; BC s, on both sides Cas, to Cal and Mex, e to Newf and most of US; close-lvd p.; 2 vars.   14 **P. foliosus** Raf.

a1 Lvs 1.5–2.5 mm broad, 3–5-nerved; sts 2–10 dm, sparingly br; range of the sp. *(P. pauciflorus* var. *californicus, P. f.* var. *c.)*   var. **foliosus**

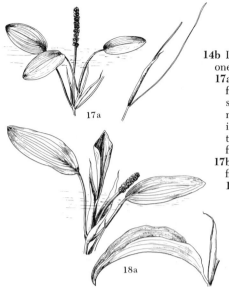

17a

18a

20a

21a

21b

floating leaf 19b

a2 Lvs mostly < 1.5 mm broad, (1) 3-nerved; sts 1–6 dm, gen freely br; range of sp., but perhaps the more common phase in w and nw N Am *(P. curtisii)*
var. **macellus** Fern.

14b Lvs in part > 4 mm broad, often dimorphic, the broader ones gen floating

17a Submersed lvs narrowly linear, 10–20 cm × 1–2 mm; floating lvs ovate-lanceolate to ovate-elliptic, gen rounded to semicordate at base, 5–10 × (1.5) 2.5–6.5 cm; achenes 3–5 mm (incl the stylar beak up to 1 mm); shallow, mostly standing, often brackish water; Alas s, on both sides Cas, to Cal, e to Newf and most of w, nc, and ne US; Europe; broad- or floating-lvd p.   **15 P. natans** L.

17b Submersed lvs either < 10 cm long or > 2 mm broad; floating lvs mostly acute at base

18a Submersed lvs 2–5 cm broad, folded and strongly falcate, 8–20 cm, acute at each end, mostly 25–45-nerved (the more deeply emersed lvs narrower, with fewer veins, and sometimes neither folded nor falcate, gen disintegrating by anthesis of pl); stips mostly > 5 (to 10) cm, free of petiole, gen persistent; achenes 4–5 mm incl the stout 0.5–1 mm style; floating lvs ovate to ovate-elliptic, acute to rounded at base, 5–10 × 2–4 cm; gen in fairly deep water; BC s, esp common w Cas, to Cal, e to w Mont and from Daks to Ark and e to Atl from Newf to Va; large-lvd p.   **16 P. amplifolius** Tuckerman

18b Submersed lvs mostly < 2 cm broad, rarely folded, never strongly falcate, mostly with < 25 veins; stips often < 5 cm

19a Stips < 3 cm (except sometimes in *P. alpinus*); submersed lvs sessile

20a Floating lvs (if any) not markedly different from the submersed, bl obovate or oblanceolate to elliptic-oblanceolate, 4–6 × 1–2.5 cm, 7–15-nerved; submersed lvs linear-lanceolate to linear-oblong, 7–20 × 0.5–2 cm, 7 (9) -nerved; pl gen reddish-tinged; achenes 3.5–4 mm incl the ca 0.5 mm subterminal curved stylar beak; streams, ponds, and lakes; Alas s, on both sides Cas, to Cal, e to Newf and s to Utah, Colo, Gr Lakes region, and Pa; Eurasia; northern or reddish p. *(P. montanense, P. tenuifolius* var. **subellipticus)**   **17 P. alpinus** Balbis

20b Floating lvs markedly dissimilar to the submersed, the latter often < 7 cm; pl greenish

21a Submersed lvs linear, flaccid, (5) 10–20 cm × (1) 3–10 mm, 5–7 (13) -nerved, midnerve cellular-reticulate and bordered on either side (on dorsal or lower surface) by air chambers, the median stripe 1–1.5 mm broad; floating lvs long-petioled, bls elliptic to oblong-elliptic, 2–8 × 0.5–2 cm, ca 15–25 (41) -nerved; shallow to deep water; Alas s, on both sides Cas, to Cal, e to Ida, Mont, and Colo and from Man to Newf and s to c, e, and ne US; ribbon-lf p. *(P. nuttallii, P. n.* var. **ramosus,** *P. e.* var. *r.)*   **18 P. epihydrus** Raf.

21b Submersed lvs mostly < 10 (2–11) cm, 3–10 (12) mm broad, the midvein much < 1 mm broad; floating lvs narrowly to broadly elliptic or oblong-elliptic, mostly 2–5 × 1–2 (2.5) cm, 13–19-nerved; standing or running water; Alas s, on both sides Cas, to Cal, Ariz, and Colo, e to most of c and ne US; Eurasia; grass-lvd p.
                      19 P. gramineus L.

19b Stips in part at least 3 cm; submersed lvs often petiolate

22a Submersed lvs (1.5) 2–4 (5) cm broad, often sessile or only short-petiolate, petioles rarely as much as 2

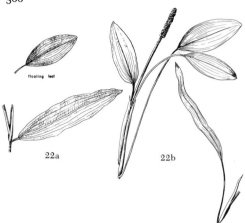

floating leaf

22a   22b

cm; petioles of floating lvs gen < bls, bls mostly broadly elliptic or oblong-elliptic, 4–12 × 2–6 cm, 13–29-nerved; stips 2.5–7 cm; achenes ca 4 mm, incl the subapical 0.5 mm beak; ponds and lakes, gen in rather deep water; BC s, on both sides Cas, to s Cal, e to most of RMS and to c and ne US and adj Can; Illinois p.   20 **P. illinoensis** Morong

22b Submersed lvs rarely > 2 (1–3) cm broad, gen with petioles 2–10 cm; petioles of floating lvs gen > bls; bls mostly elliptic to oblong-elliptic, 5–12 ×(1.5) 2–4 cm; stips (3) 4–8 cm; achenes 3–4 mm, the stylar beak < 0.5 mm; shallow to rather deep water; BC s, on both sides Cas, to Mex, e in most of US to Atl and to ne Can; widespread in other parts of the world; Loddon p., long-lvd p. (*P. fluitans* of auth., *P. occidentalis*, *P. rotundatus*)   21 **P. nodosus** Poir.

RUPPIACEAE   Ditch-grass Family

Fls sessile on a short spadix enclosed in sheathing lf bases, gen 2 per spike, naked, each gen consisting of 2 stamens and 4 pistils, but occ 1 stamen lacking or pistils as many as 8; 1 stamen typically above and 1 below the pistil(s) of each fl, the filament and broadened connective very short, the anther sacs ca 3/4 encircling the spadix; pistils sessile at anthesis, stigma sessile or subsessile, peltate-umbonate, ovary 1-celled and 1-ovuled; fr small, indehiscent, drupelike, borne on a slender stipe that progressively elongates as fr matures and often becoming twisted, the peduncle also becoming much elongated and often spirally coiled; submersed aquatic per herbs.

Ruppia L.   Ditch-grass

Pls of saline or brackish water; lvs mostly alt, linear, stip. (For Heinrich Bernard Ruppias, 1688–1719, a German botanist).

R. maritima L. Sts up to 8 dm, filiform, br; lvs semicapillary, up to 12 cm × scarcely 0.5 mm, bordered at base by adnate, ± sheathing stips 5–15 mm, stip tips sometimes free for 1–2 mm; spikes axillary to and sheathed by uppermost lvs of st, peduncle elongating as frs mature, ultimately 3–30 cm, straight or strongly coiled; fr ovoid or ± pyriform, symmetrical to strongly asymmetrical, 1.5–3 mm; Pac coast from Alas to Baja Cal, inl in ditches, ponds, and lakes throughout Can and US to Atl; Eurasia (*R. curvicarpa*, *R. occidentalis*, *R. pectinata*, *R. spiralis*, *R. m.* vars. *curvicarpa*, *occidentalis*, *rostellata*, and *spiralis*).

ZANNICHELLIACEAE   Horned Pondweed Family

Fls ♂♀, naked (ours); stamens 1 in ours, the filament slender, the anther (1) 2-celled; ♀ fls with (1) 2–9 distinct carpels, the ovary 1-ovuled; fr achene; pl ♀♂ or ♀, ♂, submersed, rhizomatous; lvs narrow, opp (ours).

Zannichellia L.   Horned Pondweed

Fls axillary, ♂ and ♀ gen but not always in same axil; ovaries subsessile, becoming obliquely oblong achenes. (For Gran Girolamo Zannichelli, 1662–1729, an Italian botanist).

Z. palustris L. Submersed per with slender rhizomes; sts almost filiform, freely br; lvs capillary, 1-nerved, 2–10 cm, with membranous, free stips; ♀ fls sessile to short-pedunculate, with short hyaline subtending bract < the pistils; filament slender, much > the pistils; achene gen short-stipitate, falcate, 3–4 mm, compressed, keeled, smooth to serrulate or dentate on the margins, stylar beak slender, ca 1 mm; almost throughout coastal and inl N Am, in fresh to brackish water (*Algoides p.*).

ZOSTERACEAE   Eel-grass Family

Fls naked, ♀♂, sessile on 1 side of a flattened and ± fleshy axis (spadix) surrounded by a tardily ruptured sheath (spathe); ♂ fls in same spike with the ♀ or on separate pls, each a single, sessile, 1-celled anther; ♀ fl a single 2-carpellary pistil with 1 style and 2 stigmas, the ovary 1-celled and 1-ovuled; fr utricular, 1-seeded, ultimately rupturing; submersed or partially floating pers with extensive rhizomes, compressed and simple or br sts, and alt, 2-ranked, linear lvs sheathing at base.

1a Pls ♀, ♂; spadix bordered by conspicuous flaplike projections   **Phyllospadix**
1b Pls ♀♂, the ♀ and ♂ in the same infl; spadix with inconspicuous projections, if any   **Zostera**

Phyllospadix Hook.   Surf-grass

Infl pedunculate, 1-sided, the ♀ and ♂ alike, but on separate pls; fls in 2 rows on the spadix, both the anthers and pistils sessile; ♀ spadix bordered on both sides by a series of oblong-ovate flaps (retinacula) covering the fls; style short, stigmas capillary; fr beaked, caudate to sagittate at base; linear-lvd, rhizomatous, submersed pers with lfy ann sts. (Gr *phyllon*, leaf, and *spadix*, the infl).

1a ♀ infls gen in pairs at 2–several nodes; lvs ca 1.5 mm broad; sts often br, up to 12 dm; at intertidal or (mostly) subtidal levels along coast from Ore to Baja Cal, just barely reaching our area? Torrey's s.   **1 P. torreyi** Wats.
1b ♀ infl only 1 per st; lvs 3–5 dm × 2–4 mm; sts mostly unbr, 0.5–4 dm × ca 2 mm; intertidal and subtidal levels along coast from Alas to s Cal; Scouler's s.   **2 P. scouleri** Hook.

Zostera L.   Eel-grass

Infl 1-sided, the ♀ and ♂ fls sessile and alt in 2 rows on each spadix; style short, stigmas 2, capillary; fr beaked, flask-like, rounded at base; submersed pers with extensive rhizomes and lfy ann sts with linear, basally sheathing lvs. (Gr *zoster*, a belt or band, referring to the narrow, elongate lvs).

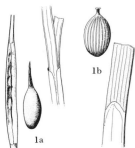

1a Lvs scarcely 1.5 mm broad, 3-nerved (lateral nerves marginal); lf sheath open to the base; spadix 3–6 cm, with tiny flaplike projections along the margin; sts mostly unbr, 1–3 dm; fr ca 2 mm; mud flats between low and semi-high tide levels; intro in our area in Pacific Co, Wn; Europe; dwarf e.
                                                                   1 Z. nana Roth
1b Lvs (2) 3–12 mm broad, up to 12 dm long, 5–∞ -nerved; lf sheath closed at base; spadix 3–8 cm, without flaplike projections; sts br, up to 2.5 m; fr ca 4 mm; mostly subtidal; Alas to s Cal, Atl coast; Europe; e., grass-wrack (*Z. oregana*, *Z. pacifica*, *Z. m.* var. *latifolia*)   **2 Z. marina** L.

JUNCACEAE   Rush Family

Fls small, capitate to cymose or pan; perianth greenish to brown or purplish-brown, the segms (tepals) in 2 rather similar series of 3 each; stamens 3 or 6 (rarely only 1 or 2); pistil 3-carpellary, stigmas linear; style 1; ovary superior, 1- or 3-celled; fr caps, 3-valved, with 3 or with ∞ seeds; seeds often appendaged at 1 or both ends; ann or per, often rhizomatous herbs with simple, mostly terete sts and alt, terete to grasslike lvs with open or closed sheathing bases.

1a Seeds 3; lvs with closed sheaths, the bls often with long straight marginal hairs   **Luzula**
1b Seeds ∞ ; lvs with mostly open sheaths, the bls rarely at all hairy   **Juncus**

Juncus L.   Rush

Fls few (rarely only 1) to ∞ in a terminal, essentially cymose, open and pan-like to greatly congested infl, sometimes in 1 − ∞ capitate clusters, lowest (invol) bract of infl from greatly elongate to reduced, flat to terete, when terete the infl often seemingly lateral on the st; stamens gen 6 or 3, rarely only 1 or 2, filaments from < to much > the anthers; caps 1-celled with 3 parietal placentae or placentae intruded and caps 3-celled; seeds ∞ , gen ± fusiform, often appendiculate at 1 or both ends, faintly areolate or reticulate; ann or per, caespitose to strongly rhizomatous herbs with terete to flattened, lfless to lfy fl sts; lvs sheathing at base, the sheath often extending upward on the sides at the juncture with the bl, the projections rounded to pointed (termed auricles, but rarely extending outward as do the aurs of the grass lf); bls from terete to laterally or dorsiventrally flattened, often septate within. (Classical Gr name for the rush.)

1a Pls ann, gen diminutive, much < 1 dm (except *J. bufonius)*; lvs narrow, often involute, rarely > 1 mm broad

 2a Pl rarely as much as 4 cm, scapose, the fls 1–2 (3) in a terminal head on a naked fl st; stamens 3

  3a Bracts lacking at base of the single terminal fl; known from Harney and Lake cos, Ore, and Gooding Co, Ida, and not improb within our range in Ore or Ida   **J. abjectus** Herm.

  3b Bracts (1 or 2) present at base of the 1–3 fls

   4a Bracts single, truncate, scarcely 0.5 mm (but much broader), nearly surrounding base of the solitary fl; tepals gen at least 3 mm; presently not known n of Lake Co, Ore, but possibly reaching our area   **J. uncialis** Greene

   4b Bracts mostly 2, if single then acute to acuminate and not surrounding base of the perianth, at least 1 of them gen almost 1 mm

    5a Bracts very unequal, upper one ca 1 mm, lower one much reduced or even lacking; seeds very indistinctly reticulate to smooth; caps thin-walled, ca 0.5 mm > the narrowly lanceolate tepals; fls 1 per peduncle; mud flats, vernal pools, and moist to wet meadows; Klickitat Co, Wn, and from Columbia Co, Ore, s through WV to Josephine and Lake cos, widespread and largely mont in Cal; dwarf r. *(J. uncialis*, misapplied)

               1 J. hemiendytus Herm.

    5b Bracts subequal, mostly 1–1.5 mm; seeds rather prominently reticulate or ridged; caps rarely > tepals; fls often > 1 per peduncle

     6a Seeds prominently ridged lengthwise and less strongly crosslined; caps ca = tepals; tepals 2.5–3.5 mm, abruptly acute or short-acuminate, not minutely roughened; damp or wet areas from open fields to mont meadows at medium elev; VI, along CR, Klickitat Co, Wn, and Columbia and Hood R cos, Ore, s through WV and sw Ore to s Cal; Kellogg's r. *(J. brachystylus, J. triformis* var. *b.)*   **2 J. kelloggii** Engelm.

     6b Seeds not prominently ridged lengthwise; caps scarcely 3/4 as long as tepals; tepals gradually narrowed to acicular, spreading and minutely roughened tips; known in Ore only from Harney Co, but possibly n into our area   **J. capillaris** Herm.

 2b Pls gen > 4 (3–30) cm; fls gen lateral as well as terminal on st; stamens 6; tepals 3–7 mm; seeds apiculate at each end; moist areas gen, from near sea level to midmont, throughout much of N Am; Eurasia; a bad garden weed w Cas; toad r. *(J. sphaerocarpus*, misapplied)

               3 J. bufonius L.

1b Pls per, gen at least 1 dm; lvs often much > 1 mm broad

 7a Infl apparently lateral, lowest invol bract terete, erect, apparently a continuation of the st

  8a Fls gen 1–4 (6–7) per st; invol bract rarely as much as 5 cm; caps ca = or slightly > the perianth; pl alp (or subalp)

   9a Uppermost of the basal sheaths of the st with a well-developed bl mostly 2–7 cm; caps sometimes acute

    10a Caps retuse; tepals acute, 4–5 mm; pl 2–3 (4) dm; RM, sc Mont to Colo; Hall's r.   **4 J. hallii** Engelm.

    10b Caps acute; tepals acuminate, 6–7 mm; pl 0.5–3 dm; BC s, in Cas and OM of Wn, to Cal, e to RM, sw Alta to Colo; Parry's r. *(J. drummondii* var. *p.)*   **5 J. parryi** Engelm.

   9b Uppermost of the basal sheaths bladeless or with a bl scarcely 1 cm; caps retuse; tepals (4) 5–7 mm; Alas to Cal, in both OM and Cas, e to RM from Alta to NM; Drummond's r.; 2 vars.

              6 J. drummondii E. Meyer

    a1 Mature caps gen at least 1 mm > perianth; perianth 5–6 (4–7) mm; Alas to Cal, e to Ida and e BC *(J. d.* var. *longifructus, J. pauperculus, J. s.)*   var. **subtriflorus** (Meyer) Hitchc.

a2 Mature caps ca = or only slightly > perianth; perianth mostly 6–7 mm; Rocky Mts   var. **drummondii**

8b Fls seldom fewer than 8 per st; invol bract gen > 5 cm; pl often other than alp or subalp

11a Anthers mostly < 1 (very rarely > 1.2) mm, from < to only slightly > the filaments

12a Stamens 3 (occ 6); tepals 2–3.5 mm; caps ca = tepals, slightly obovoid, distinctly 3-cornered above; invol bracts gen < half as long as st; pl strongly rhizomatous, 2–10 (1.5–13) dm; sheaths bladeless or with a short awnlike bl vestige; coastal tideflats to mont meadows and ridges; Alas to Baja Cal, e to Newf and most of c and e US, Europe; soft r., common r.; 3 vars.   7 **J. effusus** L.

a1 Tepals deep brownish, gen with green midvein, not rigid when dry, mostly 2.5–3 mm; sheaths mainly deep brown; infl small, rather loose, 1.5–4 cm, often subglobose; Alas to Cal, not e Cas, but also not along the immediate coast, up to 4000 ft elev (*J. e.* var. *bruneus*)   var. **gracilis** Hook.

a2 Tepals gen greenish, with little or no brownish border, stiff to rigid when dry, mostly < 2.5 or > 3 mm; sheaths often greenish, or brown below (sometimes throughout); infl sometimes > 4 cm

b1 Tepals rarely > 2.5 mm; pan compact, 1–4.5 cm; intro? on VI and prob elsewhere (Ida and Wn) in our area; native to e Can and ne US, and Europe (*J. bogotensis* var. *c.*, *J. e.* var. *caeruleomontanus?*)

var. **compactus** Lejeune & Court.

b2 Tepals 2.5–3.5 mm; pan loose, 2.5–15 cm; sheaths gen brown; coastal bogs and tidal mudflats to w slopes Cas

var. **pacificus** Fern. & Wieg.

12b Stamens 6; tepals often > 3.5 mm; caps sometimes either subglobose or only slightly if at all 3-cornered; invol bract sometimes > half as long as st

13a Perianth > 4.5 mm; infl small, rather tight; Alas to BC, not believed to reach our area (*J. balticus* var. *haenkei, J. a.* ssp. *sitchensis*)   **J. arcticus** Willd.

13b Perianth not > 4.5 mm; infl often open

14a Invol bract at least half as long as st; sts slender, 0.5–3 dm × 1.5 mm; infl small, gen tight, mostly < 2 cm or with < 20 fls; coastal bogs and lakeshores to mont marshes and meadows, from near sea level up to ca 5000 ft elev in our area; Alas to n Ore, e to Newf, and in RM to Ida, Mont, Wyo, and Utah; Eurasia; thread r.   8 **J. filiformis** L.

14b Invol bract < half as long as st; sts 2.5–9 dm, not particularly slender, gen > 1.5 thick; infl sometimes lax and > 2 cm and with > 20 fls

15a Seeds long-appendaged at each end, each appendage almost = body of seed; perianth (3.5) 4–4.5 mm; caps cylindric, obtuse, 0.5–1.5 mm > tepals; wet places, BC to Que, s to Ida and to Colo and Ia, known from our area only in Kootenai Co, Ida; Vasey's r.   9 **J. vaseyi** Engelm.

15b Seeds not long-appendaged, merely white-apiculate; perianth 2.5–3 mm; caps subglobose, subequal to the tepals; wet places, lowl valleys to low mont for w Cas, WV and Cas, Ore, to s Cal; reported for Wn; spreading r.

10 **J. patens** E. Meyer

11b Anthers rarely < 1.2 mm, at least twice as long as the filaments

16a Anthers ca 2.5 (2.2–2.8) mm, ca twice as long as the filaments; tepals (5) 6–7 mm; infl gen globose; sts 2–6 dm; coastal marshes and dunes; s BC to Cal; salt r. (*J. balticus* var. *l.*)

11 **J. lesueurii** Boland.

16b Anthers 1.2–2 (2.2) mm, mostly 3–5 times as long as the filaments; tepals 4–5 (3.5–6) mm; infl various; sts 1.5–8 (1–12) dm; wet places, often where saline or alkaline, over much of temp and subarctic N Am and Eurasia; Baltic r.; 3 vaguely delimitable vars.

12 **J. balticus** Willd.

a1 Tepals 5–6 mm, dark brown-margined, outer series acuminate and ca 0.5 mm > the inner; pan diffuse, mostly 5–15 cm; sts

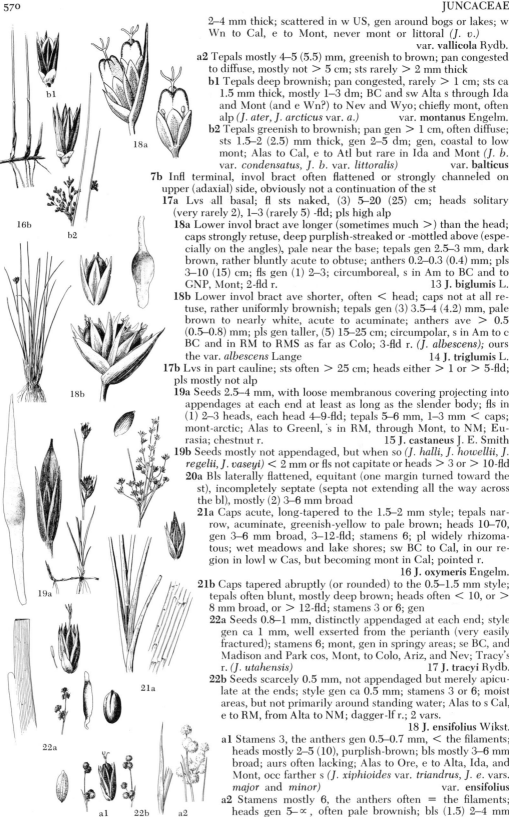

2–4 mm thick; scattered in w US, gen around bogs or lakes; w Wn to Cal, e to Mont, never mont or littoral *(J. v.)*
var. **vallicola** Rydb.

a2 Tepals mostly 4–5 (5.5) mm, greenish to brown; pan congested to diffuse, mostly not > 5 cm; sts rarely > 2 mm thick

b1 Tepals deep brownish; pan congested, rarely > 1 cm; sts ca 1.5 mm thick, mostly 1–3 dm; BC and sw Alta s through Ida and Mont (and e Wn?) to Nev and Wyo; chiefly mont, often alp *(J. ater, J. arcticus var. a.)* var. **montanus** Engelm.

b2 Tepals greenish to brownish; pan gen > 1 cm, often diffuse; sts 1.5–2 (2.5) mm thick, gen 2–5 dm; gen, coastal to low mont; Alas to Cal, e to Atl but rare in Ida and Mont *(J. b. var. condensatus, J. b. var. littoralis)* var. **balticus**

7b Infl terminal, invol bract often flattened or strongly channeled on upper (adaxial) side, obviously not a continuation of the st

17a Lvs all basal; fl sts naked, (3) 5–20 (25) cm; heads solitary (very rarely 2), 1–3 (rarely 5) -fld; pls high alp

18a Lower invol bract ave longer (sometimes much >) than the head; caps strongly retuse, deep purplish-streaked or -mottled above (especially on the angles), pale near the base; tepals gen 2.5–3 mm, dark brown, rather bluntly acute to obtuse; anthers 0.2–0.3 (0.4) mm; pls 3–10 (15) cm; fls gen (1) 2–3; circumboreal, s in Am to BC and to GNP, Mont; 2-fld r.   13 **J. biglumis** L.

18b Lower invol bract ave shorter, often < head; caps not at all retuse, rather uniformly brownish; tepals gen (3) 3.5–4 (4.2) mm, pale brown to nearly white, acute to acuminate; anthers ave > 0.5 (0.5–0.8) mm; pls gen taller, (5) 15–25 cm; circumpolar, s in Am to c BC and in RM to RMS as far as Colo; 3-fld r. *(J. albescens)*; ours the var. *albescens* Lange   14 **J. triglumis** L.

17b Lvs in part cauline; sts often > 25 cm; heads either > 1 or > 5-fld; pls mostly not alp

19a Seeds 2.5–4 mm, with loose membranous covering projecting into appendages at each end at least as long as the slender body; fls in (1) 2–3 heads, each head 4–9-fld; tepals 5–6 mm, 1–3 mm < caps; mont-arctic; Alas to Greenl, 's in RM, through Mont, to NM; Eurasia; chestnut r.   15 **J. castaneus** J. E. Smith

19b Seeds mostly not appendaged, but when so *(J. halli, J. howellii, J. regelii, J. vaseyi)* < 2 mm or fls not capitate or heads > 3 or > 10-fld

20a Bls laterally flattened, equitant (one margin turned toward the st), incompletely septate (septa not extending all the way across the bl), mostly (2) 3–6 mm broad

21a Caps acute, long-tapered to the 1.5–2 mm style; tepals narrow, acuminate, greenish-yellow to pale brown; heads 10–70, gen 3–6 mm broad, 3–12-fld; stamens 6; pl widely rhizomatous; wet meadows and lake shores; sw BC to Cal, in our region in lowl w Cas, but becoming mont in Cal; pointed r.
16 **J. oxymeris** Engelm.

21b Caps tapered abruptly (or rounded) to the 0.5–1.5 mm style; tepals often blunt, mostly deep brown; heads often < 10, or > 8 mm broad, or > 12-fld; stamens 3 or 6; gen

22a Seeds 0.8–1 mm, distinctly appendaged at each end; style gen ca 1 mm, well exserted from the perianth (very easily fractured); stamens 6; mont, gen in springy areas; se BC, and Madison and Park cos, Mont, to Colo, Ariz, and Nev; Tracy's r. *(J. utahensis)*   17 **J. tracyi** Rydb.

22b Seeds scarcely 0.5 mm, not appendaged but merely apiculate at the ends; style gen ca 0.5 mm; stamens 3 or 6; moist areas, but not primarily around standing water; Alas to s Cal, e to RM, from Alta to NM; dagger-lf r.; 2 vars.
18 **J. ensifolius** Wikst.

a1 Stamens 3, the anthers gen 0.5–0.7 mm, < the filaments; heads mostly 2–5 (10), purplish-brown; bls mostly 3–6 mm broad; aurs often lacking; Alas to Ore, e to Alta, Ida, and Mont, occ farther s *(J. xiphioides var. triandrus, J. e. vars. major and minor)* var. **ensifolius**

a2 Stamens mostly 6, the anthers often = the filaments; heads gen 5–∞, often pale brownish; bls (1.5) 2–4 mm

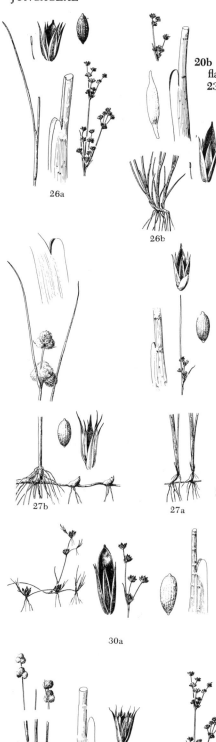

broad, commonly with aurs; BC to Cal, e to Ida and Mont and s to NM and Ariz (*J. brunnescens, J. saximontanus, J. s.* var. *robustior, J. xiphioides* vars. *macranthus* and *montanus*)   var. **montanus** (Engelm.) Hitchc.

20b Bls not laterally flattened and equitant, but dorsiventrally flattened or terete, mostly < 3 mm broad

  23a Lvs septate, terete

    24a Caps tapered almost from the base into a long, slender, tardily dehiscent or nondehiscent beak exceeding the perianth; tepals greenish-brown or tawny, narrowly lanceolate, acuminate-subulate; sts often arising from slender rhizomes at much-swollen, ± fusiform nodes

      25a Stamens mostly 3; heads 5–30 (2–50); caps gen dehiscent to the tip; rhizomes lacking or very short

        26a Pl (3) 4–8 dm; perianth 3–3.5 mm; seeds ca 0.4 mm, not appendaged; caps ca = perianth; wet places, sw BC s, mostly w Cas, to Cal, e to sw Ida and Ariz, and in much of e N Am; tapered r. (*J. bolanderi* var. *riparius*)   19 **J. acuminatus** Michx.

        26b Pl 2–3 (possibly 4) dm; perianth 3.5–4mm; seeds ca 1 mm, appendaged at each end, the appendages scarcely 1/5 as long as the body; caps 0.5–1 mm > perianth; mont bogs and margins of rivers and lakes; sc Mont, near YNP, and adj Ida, s in w Wyo to ne Utah; Tweedy's r. (*J. kuntzei, J. canadensis* var. *k.*)
                                                        20 **J. tweedyi** Rydb.

      25b Stamens 6; heads mostly 1–15; caps rarely dehiscent through the stylar beak; rhizomes often tuberous; e Cas in our area

        27a Tepals (3) 3.5 (4) mm; heads scarcely 10 mm broad when pressed; rhizome tubers, if any, much < 1 cm; ditches, pond margins, and wet places gen; BC to Cal e Cas, e to Newf and s to NM; tuberous r.
                                                        21 **J. nodosus** L.

        27b Tepals (4) 4.5–5.5 mm; heads 10–15 mm broad when pressed; tubers ca 1 × 0.5 cm; wet places, esp around lakes and along streams, from lowl into the lower mts; BC to Cal, e to Ida and Mont and to Ont, Me, NY, and Ala; Torry's r. (*J. megacephalus, J. nodosus* var. *m.*)
                                                        22 **J. torreyi** Cov.

    24b Caps rather abruptly narrowed above, dehiscent through the stylar beak, if any; tepals often deep brownish or brownish-purple, mostly acute to acuminate, rarely subulate; tuber-bearing rhizomes lacking

      28a Stamens 3, anthers much < filaments; tepals acuminate and often subulate, greenish to brown, but never deep brownish-purple; heads 3–∞

        29a Caps (0.5) 1–1.5 mm > tepals; pls 1–3 (4) dm; heads 3–10-fld

          30a Pl slenderly rhizomatous, proliferous from the upper nodes; seeds 0.6–0.7 mm; w Cas in marshes, ponds, and ditches, often where submerged early in the season; Alas s to nw Cal; spreading r. (*J. oreganus*)   23 **J. supiniformis** Engelm.

          30b Pl with very short rhizomes if any, not proliferous above; seeds ca 1 mm; RM (see lead 26b)
                                                        20 **J. tweedyi** Rydb.

        29b Caps subequal or ca = to the perianth; pls 3–8 dm; heads mostly (5) 10–30-fld

          31a Pl nonrhizomatous; heads mostly (2–3) 5–30, (5) 10–20-fld, 5–8 mm broad when pressed; infl gen diffuse, (3) 5–15 cm, very rarely heads globose and > 20-fld; tepals light brown or greenish (see lead 26a)
                                                        19 **J. acuminatus** Michx.

          31b Pl rhizomatous; heads (1) 3–9, mostly > 20-fld, gen > 8 mm broad (pressed), congested into 1–2 (3–5) approximate to widely spaced clusters; tepals light to

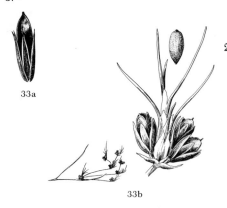

33a

33b

35a

35b

36a

37a

37b

al

dark brown, ca 3.5 mm; marshes and river bottoms, from tidel into the lower mts; s BC s, w Cas, to coastal c Cal; Bolander's r.   24 **J. bolanderi** Engelm.

28b Stamens gen 6, the anthers often = or > the filaments; tepals often blunt or deep brownish-purple; heads sometimes only 1–2

32a Pl gen growing partially to completely submerged during spring; sts and lvs often capillary, sts ± strongly proliferous above, rooting at the nodes; caps 0.5–1.5 mm > the pale brown or greenish-brown tepals, rounded above

33a Tepals 3–4 (4.5) mm, narrowly lanceolate, acute to acuminate-subulate; caps 1–1.5 mm > tepals (see lead 30a)   23 **J. supiniformis** Engelm.

33b Tepals ca 2.5 (3) mm, obtuse or the outer series acute; caps only slightly > tepals; known in our area from Snohomish Co, Wn; mainly Eurasian; bulbous r.   25 **J. supinus** Moench

32b Pl rarely either decumbent and nodally rooting or proliferous above; caps often no > the sometimes deep brownish-purple tepals

34a Tepals 2–3 mm; anthers 6, < the filaments; heads ∝, small, 3–12-fld

35a Caps uniformly tapered to the tip, > the tepals; tepals (2) 2.5–3 mm, ca equal or inner ones the longer, all slightly acuminate; brs of infl mostly spreading; wet places, tidel to streams or lake margins in mts; s BC to Cal, e to RMS and to nc and ne US and Newf; Eurasia; jointed r. (*J. a.* var. *obtusatus, J. amblyocarpus*)   26 **J. articulatus** L.

35b Caps distinctly rounded, = or slightly > tepals; tepals 2–2.5 mm, inner ones the shorter, blunt or rounded; brs of infl stiffly ascending; wet places, esp around ponds and lakes; Alas to Newf, s to Wn, Ida, Mont, Colo, Minn, and Pa; Eurasia; northern r. (*J. affinis, J. nodulosus*)
     27 **J. alpinus** Vill.

34b Tepals gen > 3 mm, if not then anthers either 3 or > filaments or heads single or > 12-fld

36a Heads gen 1 but very rarely 2, > 12-fld; tepals deep brown; anthers rarely > 2/3 as long as the filaments; caps slightly obovoid, the valves broadened above midlength and then abruptly rounded to almost truncate, the tip often slightly retuse; pls mont to alp, in wet places and along streams and lakes; Alas s, in Cas and OM, to Cal, e to w Alta and s in RM to NM; Mertens' r. (*J. m.* var. *filifolius*)   28 **J. mertensianus** Bong.

36b Heads (2) 3–∝, mostly < 12-fld; tepals sometimes light brown; anthers > 2/3 as long as the filaments (gen fully as long); caps gen cylindric or at least not thickened above midlength, the valves ± oblong, gradually rounded to acute at the tip; pls often in lowl

37a Tepals light brown, mostly no > caps; anthers gen only 3 in at least some fls, oblong, < the filaments (see lead 26a) 19 **J. acuminatus** Michx.

37b Tepals often deep brown, gen somewhat > the caps; anthers 6, gen linear and at least = the filaments; wet places, as near streams and lakes, from near sea level to lower elev in the mts; s BC to Cal, e to RM, Mont to NM; sierra r.; 4 vars.   29 **J. nevadensis** Wats.

al Heads 5–12 in a single dense cluster, each 8–12 mm broad; tepals fairly dark brown, 3.5–4.5 (5) mm; anthers slightly > the filaments; known only from along Ore coast in Lane and Lincoln cos (*J. i.*)
     var. **inventus** (Hend.) Hitchc.

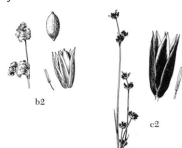

b2

c2

c1

37b

41a

39a

41b

44a

44b

a2 Heads few–∝, mostly < 8 mm broad, gen discrete but sometimes aggregated into 2 or more clusters (pls of e Wn and Ore rarely with heads in 1 cluster, but these with pale brown perianth); tepals pale to deep brown, sometimes < 3.5 mm; anthers sometimes much > the filaments; rarely along Ore coast

> **b1** Tepals rather dark brown; heads mostly solitary on ascending brs of the infl
>
>> **c1** Heads mostly 2–5; tepals ca 3 mm; anthers only very slightly > the filaments; e Wn and Ore to Mont, Wyo, Colo, and n NM (*J. b., J. truncatus*)
>>
>>> var. **badius** (Suksd.) Hitchc.
>>
>> **c2** Heads gen > 5; tepals 3–5.5 mm; anthers gen much > the filaments; range of the sp. as a whole (*J. phaeocephalus* var. *gracilis, J. mertensianus* ssp. g., *J. suksdorfii*)
>>
>>> var. **nevadensis**
>
> **b2** Tepals light brown; heads often in part aggregated in clusters on short, rather stiff brs of the infl; lower CR and e Wn and Ore to Mont (*J. c., J. mertensianus* ssp. *gracilis* var. *c.*)
>
>> var. **columbianus** (Cov.) St. John

23b Lvs not septate, either ± dorsiventrally flattened or semiterete but with a flat or grooved surface facing the st

> **38a** Seeds appendaged at each end, 1 mm or longer
>
>> **39a** Caps retuse; anthers scarcely 1 mm (see lead 10a)
>>
>>> 4 **J. hallii** Engelm.
>>
>> **39b** Caps rounded to acute; anthers gen at least 1 mm
>>
>>> **40a** Fls in heads; anthers 1–2.6 mm
>>>
>>>> **41a** Seeds ca 1 mm, the appendages < the body; aurs well developed, 1–3 mm; anthers 1.8–2.6 mm, much > the filaments; heads gen 3–9, mostly < 12-fld; moist areas in the mts; chiefly Cal, but specimens from ne Ore and wc Ida seem to be referable here; Howell's r.   30 **J. howellii** Herm.
>>>>
>>>> **41b** Seeds ca 1.5 mm, the appendages often as long as the body; aurs lacking or < 1 mm; anthers < 1.5 mm, often barely > the filaments; heads mostly 1–3 and > 12-fld; wet mont to subalp areas; Cas of s BC and Wn to Cal, e to Ida, Mont, Utah, and possibly Wyo; Regel's r.   31 **J. regelii** Buch.
>>>
>>> **40b** Fls in a congested to open pan-like cyme; anthers < 1 mm (see lead 15a)   9 **J. vaseyi** Engelm.
>
> **38b** Seeds < 1 mm, unappendaged (sometimes merely because of immaturity) or with only short appendages   **Group I**

Group I

42a Fls in heads; tepals gen minutely roughened on the outside

> **43a** Pls rarely as much as 2.5 dm, growing mainly along coast (inl to Cas in Wn); either caps 1–2 mm > the 3–4 mm tepals or the seeds pyriform and lustrous
>
>> **44a** Caps subglobose, < the (4) 4.5–6 mm tepals, leathery; seeds ca 0.7 mm, obliquely pyriform, loosely covered with a shining (lustrous), reticulate membrane; style barely 0.5 mm; coastal swamps and tideflats back into the sand dunes area; Alas to s Cal; Asia; sickle-lvd r. (*J. menziesii*); ours the var. *sitchensis* Buch., ranging from Alas to Ore   32 **J. falcatus** E. Meyer
>>
>> **44b** Caps cylindric-ovoid, up to 1 mm > perianth, not leathery; seeds ca 0.3 mm, not lustrous, faintly reticulate; style ca 1 mm; wet places, esp around lakes; VI to Cal, from coast to Cas, apparently also in Ida (Idaho Co); Coville's r.; 2 vars.   33 **J. covillei** Piper
>>
>>> **a1** Tepals dark brown, inner ones rounded; caps dark brown, gen ca 1 mm > tepals; lowl areas, rarely in mts (*J. falcatus* var. *paniculatus, J. latifolius* var. *p.*)   var. **covillei**

46a

46a

46b

46b

47a

47b

48a

49a

50a

52a

50a

52b

a2 Tepals pale brown, inner series obtuse to subacute; caps pale brown, only slightly > tepals; mont, both sides Cas, Wn to Cal, also in Ida (*J. o.*)   var. **obtusatus** Hitchc.

43b Pls mostly 2–6 dm, growing mainly in the mts and often e Cas; caps rarely > tepals; seeds never pyriform and lustrous

45a Sheaths projecting upward into prominent, truncate to rounded aurs at the point of juncture with the bls

46a Tepals broadly lanceolate, medium brown with a broad greenish midstripe and broad, membranous, smooth or minutely roughened, silvery margins, 5–6 mm; heads (2) 3–5, discrete or ± aggregated (sometimes into a single cluster); seeds 0.4–0.5 mm, slightly apiculate at each end, strongly striate lengthwise; foothills and lower mts, where moist; BC s, along e base Cas, to Sierran Cal, e to Ont, Neb, Wyo, Colo, and NM; long-styled r.   34 **J. longistylis** Torr.

46b Tepals lanceolate-acuminate, medium- to chestnut-brown with a broad greenish midstripe, minutely papillose toward the tip, not silvery-margined, 5–6.5 mm; heads (2) 3–9, discrete; seeds ca 1 mm, covered with a strongly reticulate membrane that forms a conspicuous appendage at each end (see lead 41a)   30 **J. howellii** Herm.

45b Sheaths forming small, linear, acute aurs (if any)

47a Heads 1–3 (4), ∝ -fld, mostly 1–2 cm broad when pressed; seeds 1.3–1.8 mm; anthers 1–1.4 mm (see lead 41b)   31 **J. regelii** Buch.

47b Heads (2) 3–12 (16), mostly 5–10-fld and > 1 cm broad when pressed; seeds ca 0.6 mm; anthers (1) 1.6–3 mm; wet places in mts; OM and s Cas, Wn, to s Cal, e to se Wn, ne Ore, and w Nev; straight-lvd r. (*J. longistylis* var. *latifolius, J. o.* var. *congestus*)
                                                                35 **J. orthophyllus** Cov.

42b Fls borne singly in pans or umbels, but not in heads; tepals not roughened on the outside

48a Fls 1–10, very short- ped; caps retuse, > the tepals; seeds ca 1 mm; infl apparently lateral, invol bract terete, lf-like and gen exceeding fls, but occ exceeded by infl which then appears to be terminal (see lead 10a)
                                                                4 **J. hallii** Engelm.

48b Fls often > 10, or long-ped; caps not retuse, often < tepals

49a Caps cylindric, 0.5–1 (1.5) mm > the tepals; seeds 0.8–1.2 mm (see lead 15a)   9 **J. vaseyi** Engelm.

49b Caps ovoid-cylindric to subglobose, rarely > the tepals; seeds < 0.5 mm

50a Tepals blunt, gen hooded at the tips, 3–4 mm; anthers ca 1.5 mm, much > the filaments; uppermost lf bl from above midlength of st; coastal salt marshes; VI, BC, and PS area, Atl coast of N Am; Eurasia; mud r. (*J. bulbosus* var. *g., J. fucensis*)   36 **J. gerardii** Loisel.

50b Tepals acute to acuminate, not hooded at the tips; anthers < 1 mm, gen < the filaments; uppermost lf bl from well below midlength of st

51a Caps 3-celled; aurs membranous, gen conspicuous; tepals 3.5–5 mm

52a Fls few, in a congested infl 1–2 cm, invol bract gen several times as long; tepals 3.5–4 mm, tawny or brownish with a broad greenish midstripe and membranous margins, the tip obtuse or acute to shortly acuminate, but not pungent; lvs < 1 mm broad; common on moist soil at lower elev; se BC and e Wn and Ore to Cal, e to sw Alta and s in RMS to Colo; Colo r.
                                                                37 **J. confusus** Cov.

52b Fls ∝ in a terminal, loose, cymose infl 2–7 cm, slightly exceeded by the invol bract; tepals ca 5 (4.6–5.3) mm, pale greenish, only narrowly membranous-margined, long-acuminate and ± pungent; lvs ca 1 (1.5) mm broad; moist areas at lower elev; e Wn and Ore and se to Ida, Colo, NM, and Ark; not in Mont? shortlvd r.   38 **J. brachyphyllus** Wieg.

51b Caps 1-celled or if, as occ, 3-celled then margins of sheath not extending upward into membranous aurs; tepals sometimes < 3.5 mm

53a Pl mostly > 5 dm; tepals 3–4 mm; caps almost 3-celled; margins of sheath rounded above, not projecting as aurs; reported for our area, but all material seen under this name is referable elsewhere, chiefly to *J. tenuis*   **J. interior** Wieg.

53b Pl gen < 5 (1.5–8) dm; tepals (3,5) 4–5 mm, greenish to tawny with a green midrib, membranous-margined, acute to

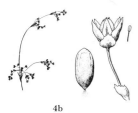

subulate; caps 1-celled; margins of sheath projecting upward into either long, thin, and membranous, or short and cartilaginous aurs; moist places; Alas to Mex, e over most of s Can and US; widespread in other continents; slender r.; 3 strongly intergradient vars.   39 J. tenuis Willd.

a1 Aurs cartilaginous, often yellowish to brown, ca 0.5 mm; fls few, gen congested into a tight infl 1–5 cm; sc BC s, e Cas, to Cal, e to Me, most of c US, and Colo (*J. d.*)

var. **dudleyi** (Wieg.) Herm.

a2 Aurs membranous, gen greenish or whitish, often > 0.5 mm; fls few to ∝, the infl often open

b1 Infl congested, mostly < 3 cm; tepals gen at least 4 mm, brownish with a broad green midstripe; Ore to Cal, e to Mont, Wyo, and Utah (*J. occidentalis, J. t.* var. *o., J. t.* var. *c.*)   var. **congestus** Engelm.

b2 Infl open, gen > 3 cm; tepals often < 4 mm, greenish; range of the sp.   var. **tenuis**

Luzula DC.   Woodrush

Infl spikelike to umbellate or pan-like; perianth greenish to brown or purplish-brown, often scarious; stamens 6; caps 1-celled, 3-seeded; seeds gen basally attached by cottony fibers, sometimes with a fleshy basal appendage (caruncle), gen smooth and shiny; grasslike per herbs; lvs with closed, sheathing base and long flat bl fringed with slender hairs. (From *Gramen Luzulae,* ancient L name for the plant, meaning grass of light, supposedly in reference to its tendency to shine when wet with dew). (*Juncoides, Juncodes*).

1a Fls solitary or occ in 2's at the ends of the brs of an open pan

2a Brs all divaricately spreading; peds stiff; tepals ca (2) 2.5 mm, light chocolate brown, ovate-lanceolate and acuminate; anthers 0.6–0.9 (1.2) mm, mostly ca = filaments; rhizomatous per 1.5–5 (8) dm; st lvs 2–5, bls 3–10 mm; wooded to open mont or subalp slopes, in dry to moist soil; rare in our area, known from Mt St Helens, Wn, and Multnomah Co, Ore, s to Cal, where more common; spreading w. (*L. parviflora* ssp. *d.*)

1 L. **divaricata** Wats.

2b Brs rarely stiffly divaricate but when occ so the tepals either dark purplish-brown or at least 3 mm; anthers various

3a Anthers 0.8–1.5 mm, much > the filaments; style ca 1 mm; stigmas ca 2 mm; tepals (2.5) 3–3.5 mm, dark brown or purplish-brown; seeds ca 1.5 (1.8) mm; st lvs 2–6 cm × (3) 4–10 mm; mont for to subalp or alp meadows and slopes; s BC s, in OM and Cas, to s Ore; e to sw Alta, Ida, Mont, and n Wyo; smooth w. (*J. arcuata* var. *major; Juncoides majus; L. glabrata,* misapplied; *L. piperi* of many auth.)

2 L. **hitchcockii** Hamet-Ahti

3b Anthers 0.3–0.7 mm, gen < the filaments; style rarely > 0.5 mm; tepals mostly < 2.5 mm, often pale; seeds mostly < 1.5 mm; pl often of lowl

4a St lvs mostly 2–3, gen 2–3 mm broad; bracts and bracteoles of infl finely lacerate-fimbriate; perianth deep purplish-brown; anthers gen ca = filaments; pl strongly caespitose but short-rhizomatous, 1.5–4 dm; arctic-alp; Alas s to OM and Mt Rainier, Wn, e in Can to Que and s to Mont; Piper's w. (*L. wahlenbergii* of auth., *L. spadicea* var. *w., L. parviflora* var. *intermedius, Juncoides piperi*)

3 L. **piperi** (Cov.) Jones

4b St lvs 2–4, mostly 3–10 mm broad; bracts and bracteoles mostly erose or only shallowly lacerate and not fimbriate; perianth greenish to purplish-brown; anthers gen < filaments; pl not rhizomatous, 1.5–5 dm; moist to fairly dry areas, from coastal rain for to alp slopes; Alas s to Cal, e to Newf, Minn, NY, Ariz, and NM; Eurasia; common throughout our area; smallfld w. (*Juncus p., J. melanocarpus, Juncoides piperi, L. piperi* Jones, *L. fastigiata, L. p.* var. *f., L. melanocarpa*)   4 L. **parviflora** (Ehrh.) Desv.

1b Fls in 1–more semicapitate spikes or in a spike-like pan

5a Infl rather stiffly erect, never drooping; invol bract conspicuous, often > infl; seeds 1.4–1.75 mm, with a whitish, basal, spongy, cellular caruncle up to half as long as the body; anthers from slightly < to several times

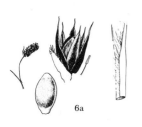

5a

6a

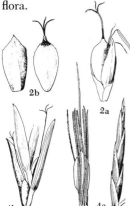

6b

> the filaments; pls often in the lowl on open gravelly prairies to deeply shaded, often moist areas; Alas to s Cal, e to most of Can and US; field w., Sweep's brush; 3 vars.   **5 L. campestris** (L.) DC.

a1 Perianth mostly 3–4.5 mm; anthers (1) 1.5–2.2 mm, at least twice as long as the filaments; VI and gravelly prairies of PT, s on w side Cas to Cal *(J. c., L. comosa* var. *c., L. multiflora* ssp. *c., L. comosa* var. *subsessilis, L. s.)*   var. **congesta** (Thuill.) E. Meyer

a2 Perianth mostly 2–3 (3.5) mm; anthers < 1.5 mm, gen only slightly if at all > filaments

> **b1** Caps = or slightly > the 2–2.5 (3) mm perianth; heads (spikes) gen with < 10 fls, up to not quite twice as long as thick; Alas to OM, Wn, and Mt Hood, Ore; ne N Am and Europe *(L. c.* var. *columbiana, L. f., L. multiflora* var. *f.)*   var. **frigida** Buch.

> **b2** Caps < the 2.5–3 (3.5) mm perianth; heads gen with > 10 fls, some gen > twice as long as thick; widespread, in all our area *(Juncus intermedius, L. i., Juncus m., L. m., L. i.* var. *m.)*
>
>   var. **multiflora** (Ehrh.) Celak.

5b Infl tending to droop, much > invol bract; seeds < 1.4 mm, the caruncle, if any, much < half as long as the body; anthers much < the filaments; pls subalp to alp

> **6a** Fls crowded into a single, continuous to interrupted, spikelike pan 1–3 cm; tepals 2–3 mm, dark brown, slenderly acuminate; pl 0.5–4 dm; open slopes, moraines, and stream banks; Alas and BC s to Cal, in OM and Cas, Wn, e to c Ida and to Mont, Wyo, and Colo; Eurasia; spiked w. *(Juncus s., L. cusickii, L. orestera)*   **6 L. spicata** (L.) DC.

> **6b** Fls in 2–several small, capitate spikes borne on almost filiform spreading but gen drooping brs; tepals light to dark brown, ca 2 mm, acute but scarcely acuminate; pl 0.5–2 (3) dm; rocky or gravelly soil, gen on moraines or above timberl; Alas, BC, w Mont, and Mt Rainier, Wn; Eurasia; curved w. *(J. a., L. a.* var. *unalaschkensis, L. u.)*
>
>   **7 L. arcuata** (Wahlenb.) Wahlenb.

CYPERACEAE   Sedge Family

Fls much reduced, ♂ or ♂ ♀ (pls then monoecious or seldom dioecious), sessile in spikes or spikelets, apparently or truly axillary to small bracts (scales); perianth of 1–∝ (often 6) short or elongate bristles, or lacking; stamens gen 3, or only 1–2, exserted at anthesis (pls wind pollinated); ovary superior, tricarpellate or less often bicarcellate, the style correspondingly 3-cleft or 2-cleft; ovary 1-celled, 1-ovuled; fr an achene; fibrous-rooted, often rhizomatous herbs, often grasslike, the culm triangular to terete, solid or seldom hollow; lvs mostly 3-ranked and with closed (rarely open) sheath and parallel-veined, typically elongate and grasslike blade, or some or all lvs with reduced or no blade.

Bulbostylis capillaris (L.) Clarke, a small, tufted ann with filiform lvs and few spikelets with spirally arranged scales, no perianth, and 3-angled achenes with a minute tubercle, is a widespread sp. s of our area, and has been reported for Ore, but is not clearly known to be an element of our flora.

1a Achene enclosed or enwrapped in a small bract (perigynium—abbreviation peri) as well as subtended by a scale

> **2a** Peri (gynium) open, with unsealed margins, merely wrapped around the achene   **Kobresia**

> **2b** Peri closed (except at the tip), though often showing a dorsal suture distally   **Carex**

1b Achene not enclosed or enwrapped in a peri

> **3a** Scales of spikelet arranged in 2 vertical ranks (distichous)

> > **4a** Perianth of 6–9 bristles; pls notably lfy-std, with ∝ well distributed, rather short lvs, the lower ones reduced, the upper ones bearing axillary infls   **Dulichium**

> > **4b** Perianth none; pls with fewer, more-basally disposed, often more-elongate lvs. the infl terminal and subtended by a cluster of sheathless, ± lfy bracts   **Cyperus**

> **3b** Scales of spikelet spirally arranged

> > **5a** Style thickened toward base, the thickened part persistent on achene

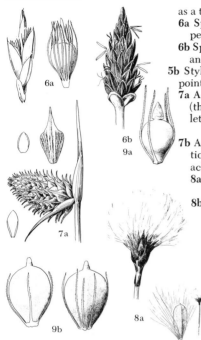

as a tubercle, gen sharply differentiated from body of the achene
6a Spikelets several–∝; fertile fls or achenes 1 (occ 2) per spikelet; perianth bristles in our sp. gen 10–12 **Rhynchospora**
6b Spikelet solitary; fertile fls or achenes several–∝ per spikelet; perianth bristles 0–6 **Eleocharis**
5b Style not thickened, although achene sometimes with a persistent, pointed stylar tip
7a Achene gen subtended by an inconspicuous, hyaline, adaxial scale (this rarely obsolete) in addition to the ordinary scale of the spikelet; small anns with 1 (2) stamens and lenticular achenes **Hemicarpha**
7b Achene subtended by 1–∝ bristles (these rarely obsolete) in addition to the ordinary scale of the spikelet; per; stamens (2) 3; achenes lenticular or often trigonous
8a Perianth bristles ∝ (> 10), conspicuous and much elongate **Eriophorum**
8b Perianth bristles 0–6, in most spp. not exceeding the scales
9a Spikelets single and strictly terminal, not subtended by modified bracts of any kind, although the lowest 1 (or sometimes more) scales gen empty; 3 spp. with styles only slightly tubercle-like at base **Eleocharis**
9b Spikelets often several–∝, when solitary lowest 1–several scales gen prolonged and ± unlike the fl-bearing scales; achene sometimes with a short, slender stylar tip but without a tubercle **Scirpus**

Carex L. Sedge

Pls ♂ ♀ or seldom ♂, ♀; fls naked, borne in spikes, each subtended by a gen small and scarious bract (scale); spikes 1 − ∝, each gen subtended by a large and lfy to much-reduced and inconspicuous bract which may or may not be sheathing at base, or this bract wanting; individual spikes sessile or pedunculate, racemosely arranged in a loose to tight terminal infl, sometimes some of them well removed from the others and axillary to lvs near base of st; spikes unisexual or bisexual, when bisexual the ♂ fls terminal (spikes androgynous) or basal (spikes gynaecandrous) or seldom ± intermingled with the ♀ ones; stamens 3 or occ 2; ♀ fls each enclosed by a sac-like bract called a peri (gynium), from the mouth of which the style or stigmas protrude, and subtended by an open scale (the pistillate scale); stigmas 2 or 3 (rarely 4), achene accordingly 2 or 3 (4)-sided; grasslike per herbs with 3-ranked lvs, closed sheaths, and triangular to round, solid sts. (The classical L name).

The peri is a highly modified bract on the adaxial side of the ♀ fl; it is wrapped around the fl, its margins being connate, forming an enclosing sac with a minute apical opening. The side of the peri next to the scale (the abaxial side) is called the dorsal side, and the side next to the spike axis (the adaxial side) is the ventral side.

The lowermost lvs are frequently very much reduced, with a short, nongreen bl that is often firm, pointed, and ± sheathing, the true closed sheath then often being short or absent. When the lowermost lvs are of this type, with the foliage lvs being borne farther up the st, the pl is said to be aphyllopodic; when the lowest lvs are normally developed, the pl is said to be phyllopodic. The distinction is not absolutely sharp, many spp. being slightly aphyllopodic, with 1–2 reduced lvs at the bottom but with foliage lvs still arising near the base of the culm.

Many characteristically tristigmatic spp. occ have a few distigmatic fls; such specimens should be keyed as being tristigmatic. With few exceptions (esp *C. saxatilis*), our distigmatic spp. only rarely produce any tristigmatic fls.

All students of *Carex* admit that the spp. must often be recognized by small technical differences, esp in the details of the structure of the peri. Often therefore precise measurement of the mature peri and achene may be necessary for accurate identification; immature specimens often are not keyable.

In addition to the spp. formally treated here, the European sp. *C. arenaria* L. has been collected several times on ballast at Portland, Ore. In the key it will fall between *C. siccata* and *C. sartwellii*, differing from both in having the bracts subtending the lower spikes strongly setose-prolonged and surpassing the spike.

1a Spike solitary, terminating the culm; peri often containing a vestigial rachilla as well as the achene; pls (except no 1) seldom < 3 dm
                                                                            Group I (lead 8a)
1b Spikes 2–∝ (in no 35, 1 spike terminal and 1 or more basal or nearly so);
peri never containing a rachilla; small to large pls
  2a Stigmas 2 or 3 (4) and the achene accordingly lenticular or trigonous (or
    quadrangular); if stigmas 2 then the spikes elongate and cylindric (the
    larger ones seldom under 1.5 cm), or at least some of them evidently
    pedunculate, or both; peri gen without a dorsal suture
    3a Style continuous with the achene and of the same bony texture, not
      withering, often becoming flexuous or contorted; stigmas 3 and achene
      trigonous, except in no 20; lvs often septate nodulose
                      Group II (lead 25a)
    3b Styles deciduous; other characters various
      4a Stigmas 3 (4) and the achene trigonous (quadrangular); a few len-
      ticular, distigmatic achenes seldom intermingled with the others
        5a Peri pubescent *(C. scirpoides*, a ± ♂, ♀ sp., might be sought here,
        but more properly belongs with Group I, where it is keyed)
                      Group III (lead 32a)
        5b Peri glab (glandular-papillate in *C. californica)*
                      Group IV (lead 41a)
      4b Stigmas 2 and achenes lenticular   **Group V** (lead 75a)
  2b Stigmas 2 and achenes lenticular; spikes sessile and relatively short
    (seldom > 1.5 cm), not elongate and cylindric; peri gen with a ± evi-
    dent dorsal suture, at least distally
    6a Spikes androgynous, or pl ± ♂, ♀ and most or all of the spikes uni-
    sexual   **Group VI** (lead 88a)
    6b Spikes gynaecandrous, or some of the lateral ones wholly ♀
      7a Peri planoconvex, often with raised margins, but not thin-edged
                      Group VII (lead 107a)
      7b Peri planoconvex or flattened, evidently thin-edged or wing-
      margined   **Group VIII** (lead 117a)

Group I

8a Achene relatively very large, gen 4–5 mm; peri few (1–3), 5–6 mm, beakless or nearly so; lvs elongate, nearly or quite = culms, flat, 1.5–3 mm
wide; sts ± clustered, 1.5–5 dm; woodl, open slopes, and dry meadows,
foothills to midmont; BC and Alta to n Cal, Utah, and Colo, mainly E Cas;
elk s., Geyer's s.   **1 C. geyeri** Boott
8b Achene smaller, not > 3.5 mm; peri various, often strongly beaked
  9a Rachilla well developed, at least half as long as the achene; lvs slender
  and often wiry, not > 1.5 mm wide
    10a Pl ± densely tufted, without rhizomes except in no 2, a short-
    rhizomatous sp. with 2 stigmas
      11a Peri planoconvex, with sharply angled thin margins (esp distally),
      not wholly filled by the achene, obscurely serrulate on margins dis-
      tally, or wholly glab; scales not much if at all wider than the peri,
      not tending to wrap around a large part of the spike; stigmas 2 or 3;
      arctic and alp spp.
        12a Pls aphyllopodic, tufted but not very densely so, the culms
        1–3.5 (4) dm, crowded on short rhizomes; peri relatively broad,
        gen broadly ovate and ± rounded toward base; stigmas 2; high
        mont, gen above timberl; circumboreal, s in w cordillera to s BC
        and Alta, and irreg to Cal, sw Mont, Utah, and n Mex; capitate s.
        (C. arctogena)   **2 C. capitata** L.
        12b Pls phyllopodic, very densely tufted, without rhizomes, culms
        0.2–1.5 dm, old sheaths persistent and very conspicuous; peri
        narrower, elliptic or lanceolate to oblong-obovate, tapering to
        the base; stigmas 2 or 3; open, often rocky places, gen above
        timberl; circumboreal, s in w cordillera to s Wn, Ida, Nev, and
        Colo; spikenard s. *(C. hepburnii*, the cordilleran phase)
                            3 C. nardina Fries
      11b Peri ± turgid, not sharp-margined, wholly filled by the mature
      achene, glab or short-hairy; scales evidently broader than peri,
      tending to wrap around a large part of the slender spike
        13a Peri glab or with only a very few inconspicuous short hairs near

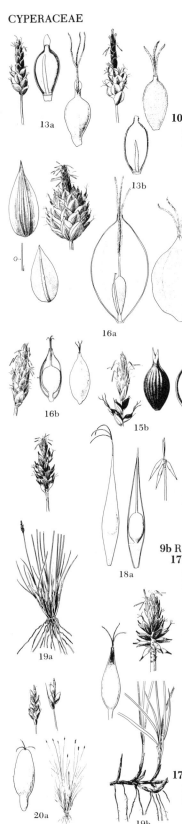

the beak; pl alp, gen above timberl; culms up to ca 1.5 dm; open dry slopes, sw Mont (n to Little Belt Mts) to Colo, w to Nev; Kobresia-like s.   **4 C. elynoides** Holm

13b Peri finely short-hairy, at least above the middle; dry-land sp., often with sagebr, plains to lower valleys, only occ high mont; culms (0.5) 1–3 dm; cordilleran and Gr Pl sp., wholly E Cas, thread-lvd s.   **5 C. filifolia** Nutt.

10b Pl not densely tufted, culms arising singly or few together from long, creeping rhizomes; stigmas 3

 14a ♀ scales gen long-persistent, sometimes deciduous, but peri then remaining appressed-ascending; rachilla included within peri

  15a Peri thin-walled, not at all coriaceous, gen rather ∝, (5) 10–40 or more; alp spp.

   16a Peri relatively large, (4) 4.5–6.5 mm, much larger than the achene; spike stout, gen 6–12 mm wide and 1.2–2.5 times as long as wide; culms 0.5–3 dm; dry to wet soil or talus, near or above timberl; Cas, s BC to Sierran Cal and adj Nev; Colo and Utah to Wyo and reputedly Mont; Brewer's s.; 2 vars.
                  6 C. breweri Boott

    a1 ♀ scales ± distinctly 3-nerved, lateral nerves often broad and pale; lvs terete, or subterete and shallowly channeled, rather than involute; terminal ♂ portion of spike short to elongate, gen well developed but < ♀ part; pl ave slightly more robust and gen with larger and broader peri; Cas-Sierran from Mt Adams, Wn, s   var. **breweri**

    a2 ♀ scales 1-nerved; lvs tending to be involute or deeply channeled; terminal ♂ portion of spike always short and relatively inconspicuous; Cas from Mt Adams n, disjunct to RM (*C. p.*, *C. engelmannii*)   var. **paddoensis** (Suksd.) Cronq.

   16b Peri smaller, 2.5–3.5 (4) mm, somewhat (but not conspicuously) larger than the achene; spike more slender, gen 4–6 mm wide and 2.5–3 times as long as wide; culms 0.5–2 dm; moist meadows, near or above timberl; Sierran Cal, and scattered stations in e Ore, c Ida, Wyo, and Utah; dark alp s.
                7 C. subnigricans Stacey

  15b Peri thick-walled and coriaceous, few, gen 1–6; dry, often grassy places, plains and foothills to fairly high mont; Eurasia and w N Am, s to Utah and NM, in our range wholly e Cont Div; blunt s.   **8 C. obtusata** Lilj.

 14b ♀ scales deciduous as the peri approach maturity; peri 3–5 mm, becoming spreading or reflexed, rachilla elongate, > style, exserted from peri; alpine bog sp., circumboreal, s occ to Mont and Colo
             9 C. microglochin Vahl

9b Rachilla obsolete; lvs various

 17a ♀ scales deciduous as the peri appoach maturity; peri becoming spreading or reflexed

  18a Peri 6–7.5 mm, shrunken and spongy for about 1–2 mm at the base, the achene borne above the base; pls of spagnum bogs; circumboreal, s near coast in w N Am to VI and Whatcom Co, Wn; few-fld s.   **10 C. pauciflora** Lightf.

  18b Peri 3–4.5 mm, not spongy-based; pls high-mont or alp, not in sphagnum bogs; widespread in our range

   19a Pl densely cespitose, without evident rhizomes; lvs slender, 0.5–1.5 mm wide, gen 2–4 (occ more) per culm; ♀ spike ave 4–7 mm thick; peri gen less widely spreading than in no 12; meadows, ledges, and rock crevices; Eurasian and cordilleran, s to Cas of Wn and Wallowa Mts of Ore, and to Utah and Colo; Pyrenaean s.   **11 C. pyrenaica** Wahl.

   19b Pl evidently rhizomatous, only loosely or not at all cespitose; lvs wider, mostly (1) 1.5–3 mm wide, gen 4–9 per culm; ♀ spike ave thicker (6–10 mm thick); moist to wet places; cordilleran, s to Cal and Colo; black alp s.   **12 C. nigricans** Retz.

 17b ♀ scales gen long-persistent, sometimes deciduous, but peri then remaining appressed-ascending

  20a Peri rounded or emarginate at tip, not at all beaked, with a narrow, often substipitate, spongy base 0.5–1 mm; lvs flat but slender, 0.7–1.2 mm broad; sphagnum bogs and other wet places, lowl to

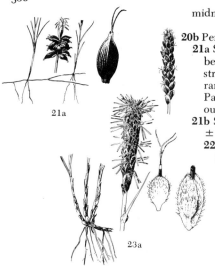

midmont; widespread N Am sp.; bristle-stalked s.

13 C. **leptalea** Wahl.

20b Peri ± distinctly (though often only shortly) beaked

  21a Stigmas 2; rhizomes well developed but very slender; peri soon becoming widely spreading, 3–3.5 mm, lightly to strongly multistriate; sts 0.5–2 dm; sphagnum bogs and marshy places, in our range rare and midmont to subalp; circumboreal, s in N Am to Pa, Colo, Nev, ne Ore, and s BC, but not in Wn; yellow bog s.; ours is var. *gynocrates* (Wormsk.) Ostenf. *(C. g.)* 14 C. **dioica** L.

  21b Stigmas gen 3, rarely 2; rhizomes either stouter or absent; peri ± ascending or appressed

    22a Spikes nearly or quite unisexual; peri ± ∝, at least > 15

      23a Peri evidently pubescent; open places high in mts, often above timberl; irreg circumboreal, s to NY, Colo, and Cal; Can single-spike s.; 3 vars. in our area

15 C. **scirpoidea** Michx.

        a1 Peri relatively long and narrow, gen 3–4.5 mm and 2.5–4 times as long as wide; ♀ scales also relatively narrow; pl phyllopodic; moister mts, Wn, ne Ore, n Ida, and nw (and Big Snowy Mts) Mont, n to Alas; chiefly w Cont Div *(C. s.)*

var. **stenochlaena** Holm

        a2 Peri relatively short and broad, gen 2–3.5 mm and 1.5–2.5 times as long as wide

          b1 Pls chiefly or wholly phyllopodic, fl culms arising from vegetative shoots of previous year, on which basal lvs have ± degenerated and cannot be distinguished readily; common form in s RM, w occ to Cal and n to Little Belt Mts, Mont, mts of c Ida, and se Ore *(C. p.)*

var. **pseudoscirpoidea** (Rydb.) Cronq.

          b2 Pls distinctly aphyllopodic, fl culms arising from vegetative shoots of the current year, on which basal scale lvs are conspicuous; common form of sp. over most of range except where other vars. are dominant, mostly not in our range, but reaching nw Mont and se BC, often at lower elev and in moister habitats than var. *pseudoscirpoidea (C. scirpiformis)*   var. **scirpoidea**

      23b Peri glab or nearly so, 1.9–3.0 mm; meadows and moist areas in gen, prairies and plains to lower midmont; occ pls (see lead 54a)   45 C. **parryana** Dewey

    22b Spikes evidently androgynous; peri few, gen 5–15 *(C. pyrenaica* might sometimes seem to key here except for the more numerous peri)

      24a Pl densely tufted, not rhizomatous; lvs narrow, 0.5–1 (1.3) mm wide; peri 4.5–6 mm, lanceolate or ± spindle-shaped; coastal bluffs and slopes, OM, Wn, to Alas; coiled s.

16 C. **circinata** C. A. Mey.

      24b Pl rhizomatous, only loosely if at all tufted; lvs gen 1–3 mm wide; peri 3–4 mm, elliptic; open slopes, high mont, often above timberl; circumboreal, s in Am to Alta and Que, and irreg to Colo and ne Utah; known from our area in w Mont; curly s. *(C. drummondii)*   17 C. **rupestris** All.

Group II

25a Sheaths distinctly short-hairy, at least on ventral surface; style straight; peri glab or hairy

  26a Peri short-hairy, 5–8 mm, teeth of the beak (0.5) 1–1.5 mm; ♀ scales narrowly acute or shortly awned; moist meadows and streambanks; foothills to lower mont; ne Ore to c Ida, Utah, and ne Cal; Sheldon's s.

18 C. **sheldonii** Mack.

  26b Peri glab, 7–10 mm, teeth of the beak 1.5–2.5 (3) mm; ♀ scales acute, with an awn tip 1–5 (9) mm; wet ground to rather deep water; lowl to midmont; circumboreal, s in Am to s Ore, Colo, and NY; awned s.

19 C. **atherodes** Spreng.

25b Sheaths and peri glab; style (except in no 21) gen becoming flexuous or contorted as the achene matures

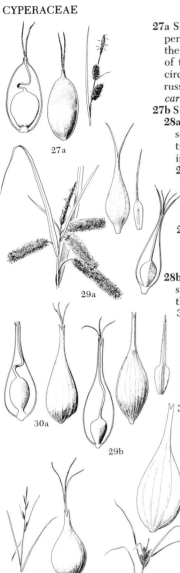

27a Stigmas gen 2 and achenes lenticular, seldom 3 and achene trigonous; peri with 2 marginal nerves only, or with a few faint additional nerves on the dorsal surface, scarcely inflated, the achene filling the lower 2/3–3/4 of the body; wet meadows and edge of water, midmont to ca timberl; circumboreal, s in Am to Wn (Clallam Co), Nev, Utah, Colo, and Lab; russet s.; cordilleran plants are var. *major* Olney *(C. physocarpa, S. vesicaria* ssp. *s.)*   20 C. **saxatilis** L.

27b Stigmas 3 and achenes trigonous; peri evidently several–∝ -nerved

  28a ♀ spikes (at least lower ones) ± nodding on slender peduncles; ♀ scales with a short, largely scarious or hyaline body only 1–2 mm, tipped by a firm awn 2–6 mm; peri lanceolate or lance-ovate, not much inflated, strongly (12) 15–20-nerved, 5–7.5 mm

    29a Teeth of peri beak elongate, 1.2–2.3 mm, arcuate or divergent; peri firm-textured, shortly stipitate; style gen straight or nearly so, seldom contorted; marshes and wet meadows; e US and adj Can; disjunct in Pac States, where rarely found from Wn to Cal and n Ida; bristly s.   21 C. **comosa** Boott

    29b Teeth short, 0.2–0.9 mm, straight, not divergent; peri thinner, not stipitate; style becoming flexuous or contorted; wet ground near streams, lowl to midmont; Wn to Cal, e to the Atl, seldom collected in our area; porcupine s.   22 C. **hystricina** Muhl.

  28b ♀ spikes erect or stiffly ascending, subsessile or (esp lowest) slender-pedunculate but still ± erect; ♀ scales and peri otherwise, the peri distinctly inflated; wet soil or shallow water

    30a Peri ascending, lanceolate to lance-ovate, tapering gradually to the often poorly defined beak, infl loose as in no 24; sts ± clustered on rather short rhizomes; lowl to midmont; circumboreal, s in Am to Cal, NM, Mo, and Del; inflated s.; 2 vars.   23 C. **vesicaria** L.

      a1 Peri 7.5–10 (11) mm, pl in gen more robust; humid coastal and subcoastal regions, Alas to Cal, occ e to nw Mont *(C. v.* var. *lanceolata, C. exsiccata)*   var. **major** Boott

      a2 Peri 5–8 mm; more gen, esp E Cas   var. **vesicaria**

    30b Peri ± strongly spreading at maturity (or lowest ones reflexed), body ellipsoid or ovoid to subglobose, ± abruptly contracted to the conspicuous beak

      31a Lfy bract subtending lowest spike only 1–2 times as long as the whole infl, infl ± elongate and spikes remote; peri 4–7 mm; pls sod-forming, sts arising singly or few together from long rhizomes; lowl to midmont, very common and widespread; circumboreal, s in Am to Cal, NM, Neb, and Del; beaked s. *(C. inflata, C. utriculata, C. u.* var. *minor)*   24 C. **rostrata** Stokes

      31b Lfy bract subtending lowest spike 2–several times as long as the whole infl, spikes crowded or only lowest one remote; peri 7–10 mm; pls densely tufted from very short rhizomes; foothills and lowl; E Cas from s BC to ne Ore, e to Colo and the Atl; retrorse s., knot-sheath s.   25 C. **retrorsa** Schw.

Group III

32a Achene filling body of peri; spikes all rather short, ♀ ones not > ca 1.5 cm, ♂ ones not > 2.5 cm

  33a Spikes all borne close together, toward the end of the culm, none basal

    34a Peri with a very short beak up to ca 0.5 mm; bracts subtending the spikes all very much reduced and scarcely lflike, not > ca 1 cm

      35a Stigmas 3; ♂ spike small, only 3–7 mm; ♀ scales conspicuously < the peri; pls gen 5–15 cm; lvs 1–3 mm wide, from < to nearly = sts; mont, in conif woods; Yuk to Newf, s to s BC, Wallowa Mts, Ore, YNP, Wyo, and to Colo, SD, and Mich; low northern s.   26 C. **concinna** R. Br.

      35b Stigmas gen 4; ♂ spike larger, 8–20 mm; ♀ scales from slightly < to slightly > the peri; pls larger, gen 15–35 cm; lvs 2–5 mm wide, sometimes much < the sts and only 5–10 cm, sometimes up to 20 cm or more; in and about conif woods; BC to n Cal, e to ne Ore, w Mont, and Alta; nw s.   27 C. **concinnoides** Mack.

    34b Peri with a prominent beak 0.5–1.5 mm; bract subtending lowest ♀ spike often longer (to 3 cm) and somewhat lf-like; stigmas 3; ♂ spike 10–25 mm; open rather dry woods and prairies; widespread in e US

and on n half of Gr Pl, uncommon in RM region, common in and w Cas, s BC to n Cal; long stolon s.; 2 vars.   28 **C. pensylvanica** Lam.
  a1 ♀ spikes short-oblong, up to ca 1.5 cm; bract subtending lowest spike gen 1–3 cm; characteristic var. of Cas and w *(C. inops, C. verecunda, C. vespertina)*   var. **vespertina** L. H. Bailey
  a2 ♀ spikes shorter, typically globose; bract subtending the lowest ♀ spike often shorter; n plains and e slope of RM   var. **digyna** Boeck.
33b Some of the ♀ spikes borne on peduncles which originate near the base of the culm, well removed from the other spikes
  36a ♂ spike gen 10–25 mm, > bract which subtends the lowest nonbasal ♀ spike; pl (1) 2–5 dm, tufted and also with creeping rhizomes; (see lead 34b) forms of   28 **C. pensylvanica** Lam.
  36b ♂ spike gen 5–12 (15) mm, often < bract which subtends lowest nonbasal ♂ spike; habit various
    37a Sts arising from an extensive system of compactly br rhizomes, pls gen forming low, turflike patches with the inconspicuous sts seldom > 1 (1.5) dm; dry soil near coast, s VI to Cal; short-std s.
                                                                                  29 **C. brevicaulis** Mack.
    37b Sts closely clustered, pls tuft-forming, gen without creeping rhizomes, 0.5–3 (4) dm; dry to fairly moist sites in conif for or adj meadows and prairies, from near sea level to near timberl; widespread in w cordilleran region, and irreg e to Mich; Ross s. *(C. brevipes, C. diversistylis, C. farwellii)*   30 **C. rossii** Boott
32b Achene loose in the peri, the upper part of peri empty; spikes gen longer, ♀ spikes often > 1.5 cm, ♂ often > 2.5 cm
  38a Peri only sparsely and inconspicuously hirtellous, except for the serrulate beak and distal margins, thin-walled, ± strongly compressed, not turgid; bract subtending lowest ♀ spike with a well developed (1–4.5 cm) sheath and a short (1–8 cm) bl which is much < infl; bogs and wet places in gen, foothills to near timberl; s BC to Cal, e to sw Mont, w Wyo, and Utah; woodrush s. *(C. ablata, C. fissuricola)* 31 **C. luzulina** Olney
  38b Peri densely velvety or velvety-sericeous, thick-walled and ± turgid, not strongly compressed; bract subtending lowest ♀ spike with short (up to 1.5 cm) or no sheath and with a ± well developed bl which is often > 8 cm and gen > infl
    39a Pl strongly phyllopodic, only 1.5–4 dm, lvs rather ∝ and crowded toward base of culm, gen 2–5 mm wide; dry, open or thinly wooded slopes and dry meadows, often on pumice or lava rock; Cas, from Mt. Adams, Wn, to n Cal; Hall's s. *(C. oregonensis)* 32 **C. halliana** Bailey
    39b Pl distinctly aphyllopodic, often much > 4 dm, lvs fewer and more evenly distributed along culm, widespread spp. of wet places or shallow water
      40a Lvs upwardly inrolled or folded along midrib, often appearing terete, 1–1.5 (2) mm wide as folded; circumboreal, s to Cas of s Wn, n Ida, Ia, and Pa; slender s. *(C. filiformis, misapplied)*; Am pls are var. *americana* Fern.   33 **C. lasiocarpa** Ehrh.
      40b Lvs flat or nearly so, the larger ones gen 2–5 mm wide; lowl to midmont; widespread N Am sp.; woolly s. *(C. filiformis var. latifolia, C. lasiocarpa var. or ssp. lanuginosa)*   34 **C. lanuginosa** Michx.

Group IV

41a Peri very large, gen 10–15 mm, conspicuously lacerate-margined; maritime, gen ♂, ♀ pls with ♀ spikes very closely bunched into a large, cylindric-oblong head; beaches and dunes, coastal from Lincoln Co, Ore, to Alas, China and Japan; large-headed s., bighead s. *(C. anthericoides)*
                                                                                  35 **C. macrocephala** Willd.
41b Peri smaller, rarely approaching 10 mm, not at all lacerate-margined; habit and habitat various
  42a Lower ♀ scales much enlarged and lf-like, gen 2–7 cm; achene ca 3 mm, without a stylar apiculus; moist woods or thickets, lowl to midmont; s BC, n Wn (Okanogan Co), and ne Ore (Wallowa Mts), e to Utah, Colo, NY and Que, infrequent in our range; Back's s. *(C. durifolia, C. saximontana)*   36 **C. backii** Boott
  42b Lower ♀ scales much smaller and not at all lf-like, gen < 1 cm; achene gen smaller and with a stylar apiculus (i.e., the base of the style persistent as a short beak)

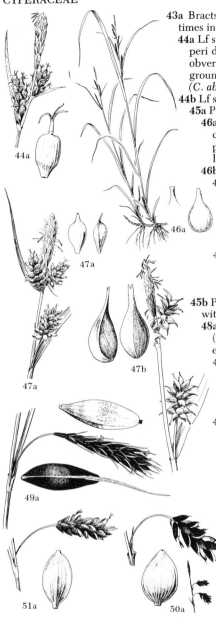

43a Bracts subtending spikes all sheathless or nearly so, except sometimes in *C. raynoldsii* (see lead 56b)

44a Lf surfaces pubescent; spikes all short and approximate, 5–15 mm; peri distinctly short-beaked (beak 0.2–0.4 mm), broadly ellipsoid or obversely trigonous-pyramidal, evidently 15–25-nerved; moist low ground; n Gr Pl sp., barely reaching our area in Mont; Torrey's s. (*C. abbreviata*) 37 **C. torreyi** Tuckerm.

44b Lf surfaces glab; spikes often longer and more remote

45a Peri with a conspicuous beak 0.8–2.3 mm

46a Sts 5–10 dm, arising singly or few together from long, creeping rhizomes; lvs very broad, gen 8–20 mm wide; wet places, lowl to midmont; s BC to Cal, e to Ida and se Ore; bigleaf s., ample-lvd s. 38 **C. amplifolia** Boott

46b Sts 1–8 dm, clustered, not rhizomatous; lvs 1–5 mm wide

47a Peri small, gen 2.2–3.3 mm, spreading, straight or nearly so; lvs narrow and tending to be channeled, gen 1–3 mm wide; wet, low ground, from sea level to midmont; circumboreal, s in Am to nw Cal, s Nev, n NM, and NJ; ours the var. *viridula* (Michx.) Kuek., green s. (*C. viridula*, *C. flava* var. *v.*, *C. f.* var. *recterostrata*) 39 **C. oederi** Retz.

47b Peri larger, gen 3.7–6.2 mm, most of them strongly recurved-falcate; lvs wider and flat, gen 2.5–5 mm; wet places, lowl and foothills to midmont; circumboreal, s in Am to s BC, c Ida, Mont, and NJ; yellow s. 40 **C. flava** L.

45b Peri beakless, or with beak no > 0.6 mm, except *C. aboriginum* with beak up to 1.0 mm

48a Roots covered with a yellowish brown felt; lateral spikes all (or at least the lower ones) nodding on slender peduncles; peri, except in *C. macrochaeta*, densely and conspicuously papillate

49a ♀ scales conspicuously awn-tipped, at least some of the awns in each spike 2 mm or more (up to 1 cm); peri not notably papillate; moist open places, often near the beach; coastal from Aleutian Is to s BC, reputedly also in CRG; large-awn s. 41 **C. macrochaeta** C. A. Mey.

49b ♀ scales awnless or with short awn-tip up to ca 1.5 mm; peri densely and conspicuously papillate

50a ♀ scales black or very nearly so, tending to wrap around the lower part of the peri; bract subtending the lowest ♀ spike 5–20 mm, awn-tipped, gen < the subtended peduncle and always < the spike; wet places, near coast; Alas to nw Wn; Siberia; several-fld s. (*C. stygia* and *C. rariflora* var. *s.*, misapplied) 42 **C. pluriflora** Hulten

50b ♀ scales merely brown to straw-colored, not wrapping around peri; lowest ♀ bracts gen well developed, lf-like, much > 2 cm, but sometimes shorter and setaceous

51a Pls strongly aphyllopodic; lvs tending to be channeled, 1–2 mm wide; ♀ spikes 1–2.5 cm, sometimes with a few ♂ fls at the tip, but never at the base; sphagnum bogs; circumboreal, s in mts to Ore, Cal, Nev, and Utah; mud s., shore s. 43 **C. limosa** L.

51b Pls phyllopodic and with old lvs ± persistent at base; lvs flat, 1–3 mm wide; ♀ spikes ave fewer-fld, 0.7–1.5 cm, often with a few ♂ fls at the base, never at the tip; sphagnum bogs; circumboreal, s in mts to (reputedly) Wn, n Ida, ne Utah, and Colo; Eurasia; poor s.

44 **C. paupercula** Michx.

48b Roots glab or inconspicuously hairy, not felty-tomentose; lateral spikes sessile or pedunculate, erect to nodding; peri (except in no 51) rather inconspicuously or not at all papillate, but cellular reticulum sometimes readily visible

52a Terminal spike ♂ or androgynous; bisexual spikes, if any, always androgynous **Subgroup IVa** (lead 53a)

52b Terminal spike gynaecandrous, or in no 45 often wholly ♀ or with ± intermingled ♂ and ♀ fls; bisexual spikes never androgynous **Subgroup IVb** (lead 58a)

43b Bracts not all sheathless, at least the one subtending lowest spike with a well-developed sheath gen at least 5 mm

Subgroup IVc (lead 65a)

Subgroup IVa

53a Peri small, only 1.9–3.3 mm; achenes 1.4–1.9 mm, only slightly smaller than peri cavity; spikes all erect or closely ascending

    54a ♀ scales light to dark brown; pl loosely tufted and with shortly creeping rhizomes; peri ± distinctly obovate, up to 3.0 mm; style soon deciduous, not notably exserted; meadows and moist low ground, plains and foothills, chiefly e Cont Div; extending onto Pac slope in c and e Ida and n Utah; Parry s. *(C. hallii, C. idahoa, C. elrodi; C. aboriginum*, misapplied*)*
                                        45 C. parryana Dewey

    54b ♀ scales purplish-black; pls densely tufted on short, stout rhizomes; peri elliptic to sometimes elliptic-obovate, up to 3.3 mm; style conspicuously exserted-persistent in early maturity but eventually deciduous; near the coast, but not strictly maritime, Clallam Co, Wn, n around continent to Lab and Que; long-styled s.   **46 C. stylosa** C. A. Mey.

53b Peri larger, gen 3.0–5.5 mm; achenes either > 2.0 mm or notably smaller than peri cavity, or both; lateral spikes erect or nodding

    55a Peri somewhat inflated; spikes all erect or closely ascending; pls phyllopodic or only slightly aphyllopodic

        56a Peri ca 5 mm, elliptic or ovate, becoming compressed near the beak; beak rather prominent (to 1 mm), evidently bidentate; gumbo soil, where wet in spring, Washington Co, Ida; Indian Valley s.
                                 47 C. aboriginum M. E. Jones

        56b Peri (3.0) 3.3–4.4 mm, elliptic or elliptic-obovate, inflated distally and abruptly contracted to a short (not > 0.5 mm) beak, this shortly or scarcely bidentate; moist or dry meadows and open or wooded slopes, foothills to fairly high mont; BC to Sierran Cal, e to Alta and Colo; Raynolds' s. *(C. lyallii)*   **48 C. raynoldsii** Dewey

    55b Peri strongly flattened, though often distended around achene; lateral spikes nodding to less often erect; pls distinctly aphyllopodic to sometimes phyllopodic

        57a Peri narrowly elliptic or elliptic-ovate, up to ca 1/2 as wide as long, gen (2.6) 2.9–5.0 mm, including the well defined, slender beak which is 0.2–0.5 mm, the 2 evident nerves prominent, marginal or rarely submarginal, the faces nerveless or obscurely few-nerved; meadows and moist or wet places, midmont to alp; Alas and Yuk s to Cal, e to Alta, n Ida, and w Mont; e Asia; showy s. *(C. nigella, C. tolmiei, C. montanensis, C. venustula)*   **49 C. spectabilis** Dewey

        57b Peri broadly elliptic to elliptic-ovate or subrotund, gen 3.3–4.3 mm, well over half as wide as long, abruptly contracted to the short (0.2–0.5 mm), shallowly to ± bidentate beak, the "marginal" nerves distinctly inframarginal on the abaxial side, the actual margins nerveless, the surfaces otherwise nerveless or lightly few-nerved; moist meadows to rocky slopes, high mont, often above timberl; sw Alta to w Wyo and Utah, w across c Ida to ne Ore and e Nev; Payson s. *(C. tolmiei*, misapplied; *C. podocarpa*, misapplied*)*   **50 C. paysonis** Clokey

Subgroup IVb

58a Pls vigorously rhizomatous, the sts of the season arising singly or few together and not surrounded at base by dried sheaths of previous years; ♀ scales surpassing the peri and shortly awn-tipped, awn mostly 0.5–3 mm; peri densely and conspicuously papillate; peat bogs and wet places in gen, from near sea level to midmont, widespread, but seldom collected ; circumboreal, s to c Cal, Utah, Colo, and NC; Eurasia; Buxbaum's s. *(C. holmiana)*
                                        51 C. buxbaumii Wahl.

58b Pls with short or no creeping rhizomes, the sts of the season clustered and surrounded by dried sheaths of previous years; ♀ scales otherwise; peri less conspicuously papillate, or not papillate

    59a Peri small, gen 1.9–3.0 mm; spikes ± approximate and erect or closely ascending

        60a Spikes relatively short, terminal one 6–14 mm; ♀ scales very dark, blackish purple or brownish black; midmont to alp, in moist to wet places; circumboreal, s to BC, Ida, NM, and Que, absent in Pac states; Scandinavian s. *(C. media)*   **52 C. norvegica** Retz.

        60b Spikes longer, the terminal one (10) 15–30 mm; ♀ scales brown or straw-colored, not at all blackish; foothills and plains (see lead 54a)   **45 C. parryana** Dewey

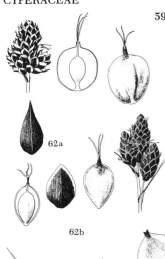

62a

62b

64a

64b

65a

66a

66b

67a

59b Peri larger, gen 3.0–5.5 mm (as little as 2.7 in spp. with spikes in a head)

  61a Spikes ± closely packed in a dense head; pls strictly alp and subalp

    62a Spikes all sessile or very nearly so and very closely crowded, lateral ones rather widely ascending or spreading; achenes tending to be stipitate; pl (1) 2–6 dm; streambanks to moist slopes, high mont, often above timberl; c Ida, w Mont, and Wyo to Nev, Utah, and NM; new s. *(C. nelsonii, C. pelocarpa)*   53 **C. nova** Bailey

    62b Spikes somewhat less crowded, more closely ascending, lowest one often short-pedunculate; achenes nearly or quite sessile; pl 1–3 dm; open, often dry or rocky slopes, gen above timberl; BC to Cas and OM, Wn, e to Alta, Mont, Utah, Colo, and Ariz, rarely in Sierran Cal; black-and-white scaled s.   54 **C. albonigra** Mack.

  61b Spikes ± approximate, but not closely aggregated into a dense head; pls 1.5–8 dm, mont to alp

    63a Peri ± inflated and with a prominent (to 1 mm), evidently bidentate beak; local in Wash Co, Ida (see lead 56a)

                           47 **C. aboriginum** M. E. Jones

    63b Peri ± strongly flattened and with a short (not > 0.5 mm), scarcely cleft beak; widespread spp.

      64a Spikes gen 2–5, erect or nodding; principal nerves of peri marginal; lowest bract ca = or < infl; wet meadows to rather dry slopes, midmont to alp, sometimes above timberl; circumboreal, s to Cal, n Ariz, Colo, and Vt; blackened s.; 2 vars. in our area

                               55 **C. atrata** L.

        a1 Peri elliptic-obovate to broadly elliptic or subrotund, strongly flattened, much wider than the achene, only obscurely or scarcely papillate-roughened, broadly rounded distally and abruptly beaked, 3.0–4.2 mm, including the 0.3–0.6 mm beak, 1/2–7/8 as wide as long, ca = to somewhat > the rather narrow scales, typically with pale green margins and darker, anthocyanic center; Sierran Cal to Mt Adams, Wn, e to nw Mont, n Ida, and Utah *(C. epapillosa, C. heteroneura)*

                           var. **erecta** Boott

        a2 Peri more narrowly lance-ovate to elliptic, more tapering toward the short (0.2–0.4) mm beak, rougher-textured, 2/5 to barely > 1/2 as wide as long, body tending to be golden brown, narrower in proportion to the achene and thus more distended by it; ♀ scales broader; mainly Can RM to nw Mont and occ to n Cas, Wn, and c Ida *(C. atrosquama, C. apoda)*   var. **atrosquama** (Mack.) Cronq.

      64b Spikes gen (5) 6–10, nodding on slender peduncles; principal nerves of peri evidently inframarginal on the abaxial side; lowest bract lfy, > infl; wet places, low elev to fairly high mont, but not reaching timberl; Alas and Yuk s to Cal, c Ida, and w Mont; Mertens' s. *(C. columbiana)*   56 **C. mertensii** Prescott

Subgroup IVc

65a Peri strongly compressed, narrow, lanceolate or lance-ovate, gen ciliolate-serrulate on the margins distally; pl tufted, not rhizomatous

  66a Terminal spike ♂ or androgynous; lvs 3–9 mm wide; mont but scarcely alp (see lead 38a)   31 **C. luzulina** Olney

  66b Terminal spike gynaecandrous, or sometimes wholly ♀; lvs 1–3 (4) mm wide; meadows and open slopes above timberl; circumboreal, s in RM to Mont, Colo, and Utah; few-fld s. *(C. fuliginosa* var. or ssp. *m.)*

                              57 **C. misandra** R. Br.

65b Peri scarcely or not at all compressed, variously shaped, not ciliate-serrulate; habit various

  67a Lvs very broad, larger ones gen 6–14 mm wide; peri 20–25-nerved; achenes relatively large, gen 2.7–3.2 mm; pl tufted, not rhizomatous; swampy places to wet mont W Cas, lowl to lower mont; isolated in Idaho Co, Ida; Henderson's s.   58 **C. hendersonii** Bailey

  67b Lvs narrower, gen 1–5.5 mm wide; peri with not > ca 15 nerves; achenes smaller, gen 1.2–2.6 mm; habit various

    68a One or more of the bracts elongate and conspicuously > infl; peri crowded, widely spreading or reflexed, distinctly beaked, borne in compact, erect to stiffly ascending spikes; pls densely tufted, not rhizomatous

69a Peri small, gen 2.2–3.3 mm, spreading, straight or nearly so; lvs narrow and tending to be channeled, gen 1–3 mm wide (see lead 47a) 39 **C. oederi** Retz.

69b Peri larger, gen 3.7–6.2 mm, gen strongly recurved-falcate; lvs wider and flat, gen 2.5–5 mm wide (see lead 47b) 40 **C. flava** L.

68b None of the bracts much > infl; other characters various, but not combined as above

    70a Peri very long-beaked, beak 1.7–4.0 mm, from slightly < to distinctly > the body, whole peri 4–8 mm; sts clustered on short to somewhat elongate rhizome; moist to wet places at relatively low elev; widespread in ne N Am, w to e base RM, s Alta to Colo, entering our range in e intermont valleys of Mont and along Fraser R, BC; Sprengel's s. 59 **C. sprengelii** Dewey

    70b Peri short-beaked or beakless, any beak not > ca 1.2 mm and distinctly < body; habit various

        71a Culms densely tufted, strongly phyllopodic; perigynia small, only 2.4–3.3 mm; achenes 1.2–1.5 mm; wet places, midmont up to ca timberl; circumboreal, s to se BC, ne Ore, ne Nev, n NM, and NY; apparently not in Wn; hair s. 60 **C. capillaris** L.

        71b Culms arising singly or few together from creeping rhizomes, phyllopodic or aphyllopodic; peri and achenes larger (except often in *C. crawei*), the peri gen 3.3–5.0 mm and the achenes gen 1.6–2.6 mm

            72a Peri short-beaked (beak up to ca 0.4 mm) or essentially beakless, not at all papillate-glaucous; ♀ spikes (5) 15–50-fld; pls 1–3 (4) dm; wet places at lower elev, often with limestone or in marl-bogs; se BC (and reputedly Wn) to sw Mont and Utah, e to Que and Ala; Craw's s. 61 **C. crawei** Dewey

            72b Peri either with a definite beak 0.5–1.2 mm, or densely papillate-glaucous; pls either with spikes only 5–15-fld or with relatively tall culms gen 2–7 dm

                73a Peri beakless or with very short beak up to ca 0.2 mm, densely papillate-glaucous; ♀ spikes 5–15-fld; lvs 1–3.5 mm wide, often channeled; bogs and swamps at low elev; ± circumboreal, s to w Wn (Grays Harbor Co), nw Mont, Mich, and NJ, disjunct in nw Cal; pale s. (*C. limosa* var. *l.*) 62 **C. livida** (Wahl.) Willd.

                73b Peri with a definite beak 0.5–1.2 mm, glab to papillate-glandular, scarcely glaucous; lvs 2–5 mm wide, flat

                    74a Pl strongly aphyllopodic, foliage lvs distinctly cauline; peri, midstripe of ♀ scales, and lower surface of lvs densely and finely glandular-papillate or very finely stipitate-glandular, ♀ spikes (10) 15–35-fld; wet prairies and brushy slopes near sea level to midmont; n Wn to n Cal, W Cas, reported from n Ida; Cal s. (*C. polymorpha* var. *c.*) 63 **C. californica** Bailey

                    74b Pl only slightly or not at all aphyllopodic, principal foliage lvs basal or nearly so; peri, ♀ scales, and lvs neither stipitate-glandular nor glandular-papillate, or peri sometimes very finely and obscurely glandular-papillate distally; ♀ spikes 5–12-fld; muskeg and moist to wet spruce woods; circumboreal, s to s BC (Selkirk Mts), Mich, and NY; sheathed s. (*C. altocaulis, C. saltuensis*) 64 **C. vaginata** Tausch.

Group V

75a Bract subtending lowest spike with a well-developed sheath at base; peri hardly at all compressed, ellipsoid to obovoid or obovoid-globose, ± rounded and beakless distally, 1.7–3.0 mm, greenish when young, gen turning golden or yellow-brown and becoming somewhat fleshy; pls 0.3–4 dm; moist or wet places, lowl to near timberl; widespread in N Am; golden s., golden-fruited s. (*C. garberi, C. hassei*) 65 **C. aurea** Nutt.

75b Bract subtending lowest spike sheathless or nearly so; other characters various

    76a Peri very firm and thick-walled; achenes sometimes fiddle-shaped, with a median constriction; ♀ spikes, or at least the lower ones, gen either

nodding on slender peduncles or with a flexuous axis and nodding tip; Cas and w

77a Very coarse, stout pls (3) 6–15 dm, inl as well as coastal; ♀ spikes elongate, larger ones gen 5–12 cm, seldom conspicuously pedunculate, but gen with flexuous axis and nodding tip; lower sheaths gen breaking and becoming fibrillose; peri 2.4–3.1 mm, with marginal nerves only; achene often with a ± deep median constriction; wet ground or standing water, esp along rivers or coastal swamps; Cas to the coast, s BC (VI) to Cal; slough s.   **66 C. obnupta** Bailey

77b Less coarse pls 1.5–10 dm, strictly maritime; ♀ spikes shorter, 1.5–5 cm, lower ones, at least, gen nodding on slender, often elongate peduncles; sheaths not becoming fibrillose; peri 2.2–3.5 mm, obscurely to evidently nerved on both faces; achene in our area gen not constricted; ± circumboreal along seacoast, s to Que and n Cal; Lyngby's s.; ours the var. *robusta* (Bailey) Cronq. *(C. cryptocarpa, C. scouleri, C. qualicumensis)*   **67 C. lyngbyei** Hornem.

76b Peri thinner-walled and ± membranous, except in *C. nebrascensis,* an inl sp. with erect spikes; achenes never fiddle-shaped; spikes erect, except in *C. sitchensis*

78a Lower sheaths breaking and becoming conspicuously fibrillose; peri lightly 3–several-nerved on each face, or the nerves sometimes obscure or obsolete

79a Lowest bract short, gen 1–8 cm, distinctly < infl; lvs flat or channeled, 1–4 mm wide; rocky beds of fast-flowing streams, below high-water mark, W Cas and CRG, s Wn to c Cal; torrent s. *(C. hallii, C. pulchella, C. suborbiculata, C. tenacissima)*   **68 C. nudata** W. Boott

79b Lowest bract gen 8–30 cm, = or > infl; lvs flat, 2–5 (7) mm wide; wet meadows and waters edge, in and near Cas, s Wn to n and Sierran Cal, apparently disjunct in mts of n and c Ida; wide-fr s. *(C. oxycarpa, C. egregia)*   **69 C. eurycarpa** Holm

78b Lower sheaths remaining intact or breaking irreg, not becoming notably fibrillose; peri nerved or nerveless on faces

80a Peri evidently nerved on both faces as well as on the margins; lowest bract gen = or > terminal spike, seldom somewhat <

81a Small, slender, arctic-alp pls, only 1–3 dm; lvs 1–2.5 mm wide; spikes ± capitate-crowded, small, only (0.7) 1–2 cm, terminal one gynaecandrous; wet meadows and streambanks; e Asia, and irreg from Yuk to GNP, Mont; goose-grass s. *(C. eurystachya, C. plectocarpa)*   **70 C. eleusinoides** Turcz.

81b Larger, coarser pls (1) 2–10 dm, occurring from lowl to near timberl, but scarcely arctic-alp; lvs gen 2–10 mm wide; spikes remote to ± aggregated but not capitate-crowded, gen 1.5–7 cm, the terminal one ♂ or occ gynaecandrous

82a Pls ± densely tufted, without long rhizomes; peri membranous, 1.9–3.0 (4) mm, the beak entire, 0.1–0.35 mm; lvs relatively narrow, seldom > 4 mm wide; wet places, from sea level to high mont, but below timberl; widespread N Am sp.; 2 vars. in our area   **71 C. lenticularis** Michx.

a1 Peri (2.4) 3–4 mm, scales often = peri; seashore pls, Alas to n Cal *(C. hindsii)*   var. **limnophila** (Holm) Cronq.

a2 Peri 1.9–3.0 mm, scales < peri *(C. kelloggii)*   var. **lenticularis**

82b Pls scarcely tufted, vigorously rhizomatous; peri coriaceous, (2.5) 2.8–3.9 mm, the beak (0.2) 0.4–0.6 mm, ± distinctly bidentate; lvs broader, the larger ones gen (3) 4–10 mm wide; wet places, often where alkaline, lowl to midmont; Wn to Cal, wholly e Cas, e to SD, Kans, and Neb, and through RMS to NM; Neb s. *(C. jamesii, C. j. var. ultriformis, C. u., C. n. var. erucaeformis)*   **72 C. nebrascensis** Dewey

80b Peri nerveless or nearly so on both faces, or occ some with 1–2 irreg nerves; lowest bract variously > or < the terminal spike

83a Peri distinctly turgid-inflated, only slightly compressed, gen pale coppery and often purple- or reddish-brown-dotted, 2.1–3.0 mm, with very short stipe and very short (0.2–0.4 mm) beak; wet low ground, esp on floodplains and lake-shores; s BC to nw Ore, e to Ida and nw Mont; Columbia s. *(C. turgidula)* **73 C. aperta** Boott

83b Peri ± strongly compressed, only slightly if at all inflated

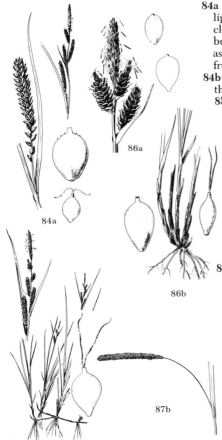

84a Peri very small, broadly obovate to broadly elliptic or elliptic-ovate, 1.4–2.0 mm, including the 0.2–0.4 mm beak,
closely investing the achene; rocky banks and beds of streams,
but also in other wet, low places; lower reaches of CR, e as far
as w Klickitat Co, s in Ore, w Cas, to Douglas Co; green-
fruited s. (*C. verticillata* Boott)   74 **C. interrupta** Boeck.
84b Peri larger, gen (1.8) 2.0–3.5 mm, more loosely enveloping
the achene, distal part of the cavity gen empty
  85a Lowest bract short, up to ca 7 cm, < terminal spike
    86a Pls 1–5 (6) dm, sod-forming, with freely br rhizomes
and scattered sts, not strongly aphyllopodic; larger lvs
(0.5) 1–3 (4) dm; ♀ spikes 1–2.5 (3) cm; widespread, wet
meadows and lakeshores to open slopes, midmont to alp,
often above timberl; s BC to Cal, e to w Mont, Wyo, and
Colo; Holm's RM s. (*C. stylosa* var. *virens, C. campylo-
carpa, C. spreta, C. accedens, C. chimaphila, C. gymno-
clada, C. miserabilis*)   75 **C. scopulorum** Holm
    86b Pls 4–9 dm, densely tufted and only shortly rhizomatous,
strongly aphyllopodic; larger lvs mostly 3–6 dm; ♀ spikes
1–3 (5) cm; wet to moist places, gen midmont, c and n
Ida and adj nw Mont, ne Wn, and ne Ore; saw-lvd s.
          76 **C. prionophylla** Holm
  85b Lowest bract longer, gen 7–50 cm, gen = or > terminal
spike
    87a Pls with well-developed, deep-seated, horizontal rhi-
zomes; spikes all erect, even when pedunculate; peduncles often all < 3 cm; lowest bract moderately elongate,
gen 7–25 cm; in shallow water or wet places, foothills to
near timberl; circumboreal, s to Cal, NM, and NJ, in and e
Cas; water s. (*C. substricta, C. suksdorfii, C. aperta* f. *con-
cinnula*)   77 **C. aquatilis** Wahl.
    87b Pls tufted, with very short if any rhizomes; lower spikes
tending to nod on flexuous, gen ± elongate peduncles,
lowest peduncle seldom < 3 cm; lowest bract elongate,
gen 20–50 cm; swamps and other wet places, Alas to Cal
(mainly in and w Cas), e to n Ida; Sitka s. (*C. howellii, C.
dives, C. aquatilis* var. *d., C. pachystoma, C. a.* var. *p.*)
          78 **C. sitchensis** Prescott

Group VI

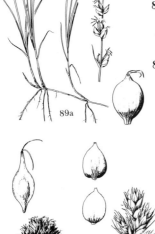

88a Pls ± rhizomatous, culms arising singly or only loosely clustered
  89a Infl interrupted-linear, individual spikes gen well separated from each
other, very small and few-fld, ca 5 mm or less, with gen 1–3 peri and ca
as many ♂ fls; soft, slender pls of swamps and streambanks; circumboreal,
s to Cal, NM, and NJ, largely E Cas, but in OM; soft lvd s.
          79 **C. disperma** Dewey
  89b Infl ± headlike, though sometimes elongate, the individual spikes, or
most of them, ± closely aggregated, often > 5 mm and with more fls
than *C. disperma;* various habitats, but seldom in swamps or along
streambanks
    90a Spikes very closely aggregated into a compact head, not individually
distinguishable to the unaided eye; peri 3.3–4.8 mm, including the
slender, stipitate base and prominent beak; dwarf, alp and subalp pls
up to ca 2.5 dm, moist or wet places; Mt. Adams, Wn, to Sierran Cal,
Nev, ne Ore, irreg to Utah, Colo, and Wyo; Europe; ours the var. *vernacula* (Bailey) Kuek.; foetid s. (*C. vernacula*)   80 **C. foetida** Allioni
    90b Spikes not so closely aggregated, lower ones, at least, readily distin-
guishable to the unaided eye; gen taller pls of more moderate elev
      91a Pls phyllopodic to somewhat aphyllopodic, lvs all clustered at or
near the base; peri essentially nerveless ventrally (except sometimes
at the base), often sharp-edged but not thin-margined
        92a Peri very small, 1.7–2.4 (2.6) mm, the beak 0.2–0.4 (0.6) mm,
the body very firm and thick-walled, esp proximally; wet meadows, foothills to midmont; E Cas, s Wn to Cal, e to s Alta, Mont,
NM; short-beaked s., analogue s.   81 **C. simulata** Mack.
        92b Peri larger, 2.6–4.6 mm, the beak 0.5–1.5 mm, the perigynial
wall sometimes fairly firm but not esp thick

93a Spikes gen all unisexual and plants dioecious; spikes gen
10–15 mm, the head gen 1.5–4.5 cm; peri 3.5–4.6 mm, with a
prominent beak gen 1.0–1.5 mm; dry to wet places, from valleys and plains to midmont, tolerant of alkali; s BC and Wn to
Cal, wholly e Cas, e to NM, Neb, Ia, and Man; Douglas' s. (*C.
nuttallii, C. irrasa*)   82 **C. douglasii** Boott
93b Spikes (or most of them) ordinarily androgynous, rarely all
unisexual and pls dioecious, the spikes, heads, and peri then
smaller than in *C. douglasii*
    94a Rhizome slender, brownish; pls slender, 0.5–2 dm, with
very narrow lvs only 0.3–1.5 mm wide; peri 2.6–3.3 (3.5)
mm; open, dry to moderately moist, often grassy places,
plains to mont, but not up to timberl; Ia and Neb w to e base
RM, scattered farther w to se BC, Crook Co, Ore, and to
Nev, Ariz, and Yuk; Eurasia; narrow-lvd s. (*C. eleocharis*)
                                       83 **C. stenophylla** Wahl.
    94b Rhizome coarse, black or brownish-black; pls gen either
taller, or with somewhat wider lvs, or both; peri often larger
        95a Maritime pls, 0.5–3 dm; infl gen ovoid and 1–2 cm
thick; sandy beaches and dunes along coast, Wn to Cal;
sand-dune s.   84 **C. pansa** Bailey
        95b Inl pls, not approaching the sea, (1) 3–7 dm; infl com-
monly more elongate and ca 1 (1.5) cm thick or less; open,
moist or wet, often alkaline places; plains and lowl to
mid-mont; common and widespread E Cas from s BC to
Cal, e to Man, Ia, and Tex; clustered field s. (*C. marcida,
C. camporum, C. douglasii* var. *brunnea, C. alterna, C. siccata* f. *obscurior*)   87 **C. praegracilis** W. Boott
91b Pls strongly aphyllopodic, the lvs not forming a basal cluster; peri
± evidently nerved ventrally, the body tending to be thin-margined
distally
    96a Spikes ∝ , 20 or more; peri small, gen 2.3–4.0 mm, the beak <
half as long as the body; pls gen rather tall and coarse, 3–8 dm,
the lvs fairly well distributed along at least the lower half of the
st, the sheaths elongate and relatively firm even on the ventral
side; gen in moist or wet places at lower elev; mainly n Gr Pl, w
occ to Flathead Lake, Mont; Sartwell's s. (*C. disticha* var. *s.*)
                                       86 **C. sartwellii** Dewey
    96b Spikes relatively few, gen 6–12; peri larger, gen 4.5–6.2 mm,
the beak > half as long as the body; pls more slender and with
most or all the lvs on lower 1/4 of the st; sheaths less elongate and
distinctly thin-hyaline ventrally; ± open, often grassy slopes,
base of mts up nearly to timberl; Wen Mts, Wn, and sw Alta to
Ariz and Mack, e to the Atl; silvertop s., dry s.
                                       87 **C. siccata** Dewey
88b Pls ± densely tufted, gen with short or no creeping rhizomes (stout rhizome often elongate in no 93, but the pls still tufted)
    97a Peri variously shaped (but not lance-triangular) gen broadest distinctly
above the base, the body ± abruptly tapered or contracted into the
beak, which may be of different texture from the body
    98a Spikes few, gen 10 or less; sheaths not red-dotted
        99a Body of peri strongly planoconvex, broadly rounded on the convex
side (marginal nerves tending to appear ventromarginal), very abruptly contracted to the beak and wholly filled by the mature achene, not evidently serrulate-margined (except often on the beak) at
10× magnification; beak of peri obliquely cleft, only very shallowly
or not at all bidentate; spikes loosely aggregated into an irreg, oblong-cylindric head; foothills to midmont, often with sagebr or
aspen; c Ore to c Ida and extreme sw Mont, Wyo, and SD, s to Cal,
Utah, and Mex; valley s. (*C. vagans, C. phaeolepis, C. brevi-
squama*)   88 **C. vallicola** Dewey
        99b Body of peri moderately planoconvex (marginal nerves only occ
appearing ventromarginal), less abruptly contracted to the beak, the
upper part (except sometimes in *C. tumulicola*) not wholly filled by
the achene, ± evidently serrulate margined above the middle; beak
of peri ± evidently bidentate
        100a Spikes loosely aggregated into an irreg or interrupted, ob-

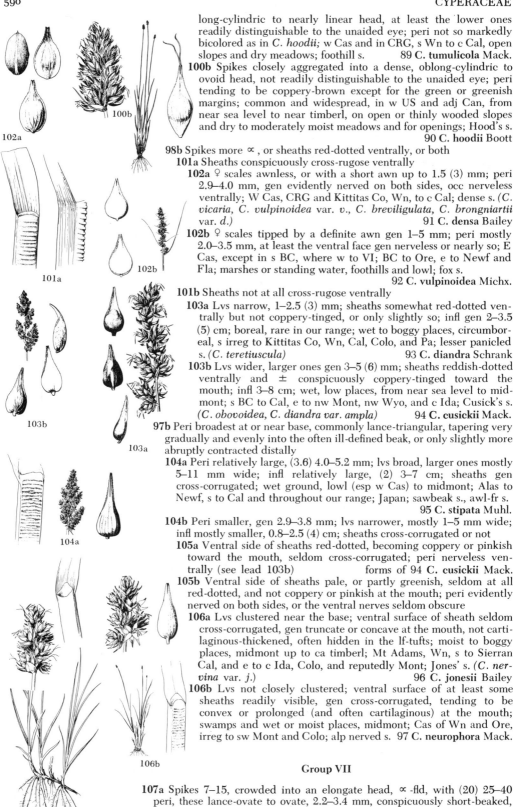

long-cylindric to nearly linear head, at least the lower ones readily distinguishable to the unaided eye; peri not so markedly bicolored as in *C. hoodii;* w Cas and in CRG, s Wn to c Cal, open slopes and dry meadows; foothill s.   89 **C. tumulicola** Mack.

100b Spikes closely aggregated into a dense, oblong-cylindric to ovoid head, not readily distinguishable to the unaided eye; peri tending to be coppery-brown except for the green or greenish margins; common and widespread, in w US and adj Can, from near sea level to near timberl, on open or thinly wooded slopes and dry to moderately moist meadows and for openings; Hood's s.
90 **C. hoodii** Boott

98b Spikes more ∝ , or sheaths red-dotted ventrally, or both

101a Sheaths conspicuously cross-rugose ventrally

102a ♀ scales awnless, or with a short awn up to 1.5 (3) mm; peri 2.9–4.0 mm, gen evidently nerved on both sides, occ nerveless ventrally; W Cas, CRG and Kittitas Co, Wn, to c Cal; dense s. *(C. vicaria, C. vulpinoidea* var. *v., C. breviligulata, C. brongniartii* var. *d.)*   91 **C. densa** Bailey

102b ♀ scales tipped by a definite awn gen 1–5 mm; peri mostly 2.0–3.5 mm, at least the ventral face gen nerveless or nearly so; E Cas, except in s BC, where w to VI; BC to Ore, e to Newf and Fla; marshes or standing water, foothills and lowl; fox s.
92 **C. vulpinoidea** Michx.

101b Sheaths not at all cross-rugose ventrally

103a Lvs narrow, 1–2.5 (3) mm; sheaths somewhat red-dotted ventrally but not coppery-tinged, or only slightly so; infl gen 2–3.5 (5) cm; boreal, rare in our range; wet to boggy places, circumboreal, s irreg to Kittitas Co, Wn, Cal, Colo, and Pa; lesser panicled s. *(C. teretiuscula)*   93 **C. diandra** Schrank

103b Lvs wider, larger ones gen 3–5 (6) mm; sheaths reddish-dotted ventrally and ± conspicuously coppery-tinged toward the mouth; infl 3–8 cm; wet, low places, from near sea level to midmont; s BC to Cal, e to nw Mont, nw Wyo, and c Ida; Cusick's s. *(C. obovoidea, C. diandra var. ampla)*   94 **C. cusickii** Mack.

97b Peri broadest at or near base, commonly lance-triangular, tapering very gradually and evenly into the often ill-defined beak, or only slightly more abruptly contracted distally

104a Peri relatively large, (3.6) 4.0–5.2 mm; lvs broad, larger ones mostly 5–11 mm wide; infl relatively large, (2) 3–7 cm; sheaths gen cross-corrugated; wet ground, lowl (esp w Cas) to midmont; Alas to Newf, s to Cal and throughout our range; Japan; sawbeak s., awl-fr s.
95 **C. stipata** Muhl.

104b Peri smaller, gen 2.9–3.8 mm; lvs narrower, mostly 1–5 mm wide; infl mostly smaller, 0.8–2.5 (4) cm; sheaths cross-corrugated or not

105a Ventral side of sheaths red-dotted, becoming coppery or pinkish toward the mouth, seldom cross-corrugated; peri nerveless ventrally (see lead 103b)   forms of 94 **C. cusickii** Mack.

105b Ventral side of sheaths pale, or partly greenish, seldom at all red-dotted, and not coppery or pinkish at the mouth; peri evidently nerved on both sides, or the ventral nerves seldom obscure

106a Lvs clustered near the base; ventral surface of sheath seldom cross-corrugated, gen truncate or concave at the mouth, not cartilaginous-thickened, often hidden in the lf-tufts; moist to boggy places, midmont up to ca timberl; Mt Adams, Wn, s to Sierran Cal, and e to c Ida, Colo, and reputedly Mont; Jones' s. *(C. nervina* var. *j.)*   96 **C. jonesii** Bailey

106b Lvs not closely clustered; ventral surface of at least some sheaths readily visible, gen cross-corrugated, tending to be convex or prolonged (and often cartilaginous) at the mouth; swamps and wet or moist places, midmont; Cas of Wn and Ore, irreg to sw Mont and Colo; alp nerved s. 97 **C. neurophora** Mack.

Group VII

107a Spikes 7–15, crowded into an elongate head, ∝ -fld, with (20) 25–40 peri, these lance-ovate to ovate, 2.2–3.4 mm, conspicuously short-beaked, and ± spreading; wet places, lowl to near timberl; Yuk to Que, s to Wn

and Ore (in and w Cas), nw Cal, n Ida, nw Mont, and Minn and NY; n clustered s. *(C. canescens* var. *oregana, C. heleonastes* var. *scabriuscula)*

98 **C. arcta** Boott

107b Spikes 2–8 (10), seldom at once > 6 and all crowded into a head; peri (1) 5–30

  108a Peri ascending or ascending-spreading at maturity *(C. integra,* lead 127b, and *C. leporinella,* lead 137a, might sometimes fall in this category)

    109a Peri 1.7–3.4 mm (to 3.8 mm in *C. laeviculmis),* short-beaked or nearly beakless, the beak often < 0.5 mm, less often up to 1.0 mm, only inconspicuously or not at all serrulate-margined; lvs 1–2.5 (3.5) mm wide

      110a Spikes 2–4, all approximate; ♀ scales ca = or > peri (or slightly < the beak); peri 2.4–3.4 mm, the prominent dorsal suture 0.7–1.5 mm; pls alp, 0.5–3 dm, rare with us; wet places, circumboreal, s to BC, Alta, and n Mont, disjunct in Utah and Colo; two-parted s. *(C. lachenalii)*

99 **C. bipartita** Allioni

      110b Spikes 4–8 (10), lower ones (or all) often ± remoté; ♀ scales tending to be < body of peri; peri gen 1.7–2.5 (to 3.8 mm in *C. laeviculmis)*

        111a Spikes relatively small and few-fld, with only 5–10 (15) peri; wall of distal part of peri-body very thin and easily ruptured; ventral surface of peri nerveless or obscurely nerved; dorsal suture of peri prominent, extending the length of the beak-apiculation and encroaching onto the distal part of the body

          112a Peri small, gen 1.7–2.5 mm, with a short beak-apiculation up to 0.5 mm; wet places in the mts; circumboreal, s to Cas of s Ore, and to Utah, Wyo, and Va, seldom collected in our range; brownish s. *(C. canescens* var. *b.)*

100 **C. brunnescens** (Pers.) Poir.

          112b Peri larger, gen 2.5–3.8 mm, with a well-developed beak gen 0.5–1.0 mm; wet places, wooded regions from lowl to midmont; Kamtchatka to n Cal, mainly W Cas, but also inl to se BC, Mont, and ne Ore; smooth-st s.

101 **C. laeviculmis** Meinsh.

        111b Spikes larger and more numerously-fld, with gen (10) 15–30 peri; wall of the distal part of the peri not esp thin and fragile; ventral surface of the peri evidently nerved

          113a Spikes 4–6, all closely aggregated, brownish in aspect; dorsal suture gen extending the length of the beak-apiculation, but scarcely encroaching onto the body of the peri; pls 1–3 dm; wet places at high elev; s Wn (Simcoe Mts) s to Sierran Cal, e to Colo and nw Wyo, rather seldom collected; teachers' s.

102 **C. praeceptorum** Mack.

          113b Spikes 4–8, silvery green to pale grayish or stramineous in aspect, the lower ones, at least, gen somewhat remote, so that the infl is interrupted and not head-like; dorsal suture of peri short and poorly developed, gen restricted to the distal part of the beak-apiculation; pls often > 3 dm; wet places, sometimes in sphagnum bogs, lowl to near timberl; circumboreal, s to Cal, Ariz, and Va; gray s.   103 **C. canescens** L.

    109b Peri (3.1) 3.5–4.8 mm, relatively long-beaked, the beak gen (0.8) 1.1–1.9 mm and serrulate-margined; lvs gen 2–5 mm wide; streambanks and moist woodl or for openings, from near sea level to near timberl; widespread in N Am and e Asia, and found throughout our range; Dewey's s. *(C. bolanderi, C. leptopoda)* 104 **C. deweyana** Schw.

  108b Peri widely spreading at maturity, or the lower ones reflexed

    114a Peri strictly entire-margined, not serrulate on either the body or the beak, 2.5–3.2 mm; wet places, high mont, often above timberl; OM and Cas, Wn, and adj BC, s to Sierran Cal, e to Mont and Colo; sheep s., small-headed s.   105 **C. illota** Bailey

    114b Peri ± evidently serrulate on the beak (and often also the distal part of the body) under 10× magnification

      115a Beak of peri short and stout, 1/4–1/2 as long as body, up to ca 1 mm, only very shortly and inconspicuously bidentate; peri mostly 2.2–3.2 mm; swamps and other wet places, lowl to midmont; widespread N Am sp., found throughout our range; inl s.

106 **C. interior** Bailey

107a

110a

112b

112a

113a

113b

109b

114a

115a

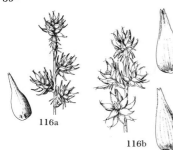

116a

116b

115b Beak of peri longer and relatively more slender, 1/2 as long to almost = the body, gen 1–2 mm; peri often > 3.2 mm

    116a Peri relatively small, gen 2.8–3.5 (4.0) mm; lvs narrow, gen 1–2 mm wide; swamps, bogs, and other wet places, from near sea level to midmont; ± circumboreal, s to Cal, Colo, and NC; muricate s. (*C. stellulata, C. sterilis, C. angustior, C. cephalantha, C. ormantha*)   107 C. muricata L.

    116b Peri larger, gen (3.1) 3.5–4.8 mm; lvs wider, gen 2–3 mm wide; pls appearing coarser than *C. muricata*; wet places, often in sphagnum bogs, W Cas, mainly near coast but not truly maritime, Alas to n Cal; coastal stellate s.

                              108 C. phyllomanica W. Boott

Group VIII

118a   119a

119b

117a One or more of the lower bracts of infl elongate, = or gen much > infl

    118a Peri very narrow and elongate-beaked, (4.5) 4.8–6.5 × 0.8–1.0 mm, 5–7 times as long as wide; lowest bract 8–20 cm and distinctly lf-like, the infl seemingly surrounded by lvs; moist or wet low ground; Ont and NY to Alta, Mont, and SD, occ w to BC (Kamloops) and Okanogan Co, Wn; seldom collected in our range; many-headed s.

                          109 C. sychnocephala Carey

    118b Peri shorter and relatively broader, 3.0–4.5 (4.7) × 0.9–1.8 mm, 2–4.5 times as long as wide; lowest bract often < 8 cm and scarcely lf-like, the head in any case obviously terminal or apparently lateral, not appearing to be surrounded by lvs

        119a Beak of peri ill-defined, evidently margined and serrulate all the way to the tip or to within < 0.5 mm of the tip; lowest bract (2.5) 5–15 cm, often erect and appearing like a continuation of the st; moist or wet places at low elev, BC to Cal, mainly W Cas; one-sided s.

                          110 C. unilateralis Mack.

        119b Beak of peri more evident and more narrowly margined, the distal 0.5–0.9 mm marginless, subterete, and gen entire; lowest bract (1) 2–8 (10) cm, setaceous, not appearing like a continuation of the st; widespread cordilleran sp. of moist or wet places, lowl to midmont; slenderbeaked s. (*C. macloviana* var. *pachystachya* f. *involucrata*)

                          111 C. athrostachya Olney

117b Bracts all short and inconspicuous, or the lowest 1 seldom somewhat setaceous-prolonged and up to ca = the infl, rarely a little longer

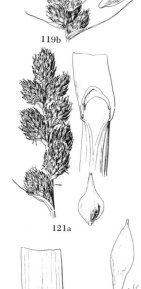

121a

121b

123a

    120a Lig elongate (but largely joined to bl), gen (2) 3–8 mm as measured from level of margin of lf collar; lf sheath often but not always conspicuously prolonged ventrally and > the collar by 3–13 mm; infl pale green to straw-colored or light brown, (2) 3–6 cm; Cas-Sierran region and w

        121a Ventral side of lf sheath firm and green all the way up to the level of the collar, except sometimes for a short, hyaline triangle within 5 mm of the collar (unique among our spp. of Group VIII in this regard); lf sheath not always prolonged beyond the collar; beak of peri flattened, serrulate, and narrowly margined to within ca 0.2 mm of the tip, or throughout; marshes to meadows, lowl to lower mont; W Cas and CRG; greensheathed s.   112 C. feta Bailey

        121b Ventral side of lf sheath fragile and white-hyaline at least in upper half; lf sheath conspicuously prolonged 3–13 mm beyond collar on ventral side; peri beak slender, the distal 0.5 mm smooth and marginless; wet to dry soil, from foothills to near timberl; Cas from Chelan Co, Wn, s to Cal; fragile-sheathed s. (*C. specifica*, misapplied)

                          113 C. fracta Mack.

    120b Lig short, seldom approaching 3 mm; lf sheath not conspicuously prolonged ventrally, not > level of collar by > ca 2 mm; infl and distribution various

        122a Peri ± evidently (seldom obscurely) multinerved dorsally (nerves 10 or more) and gen also ventrally, distinctly planoconvex, (4.1) 4.5–6.2 mm; spikes crowded in a narrow to globose-ovoid head

            123a Lvs narrow and firm, 0.5–2 (2.5) mm wide; pls 0.5–3 (4) dm, widespread in mts of our range, near or gen above timberl; dunhead s., mt hare s. (*C. petasata* var. *pleiostachya, C. eastwoodiana*)

                          114 C. phaeocephala Piper

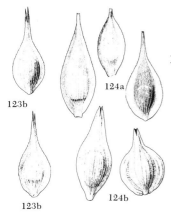

123b

123b

124a

124b

127a

127b

128b

128a

131a

132a

132b

123b Lvs wider and softer, gen 2–3.5 mm wide; pls gen 1.5–6 dm; meadows, streambanks, and open slopes, midmont or occ up to near timberl; Cas of n Wn, s to Cal, e occ to Ida and Mont; many-ribbed s.   115 **C. multicostata** Mack.

122b Peri otherwise, either fewer-nerved, or of a different size, or strongly flattened, except occ in some spp. with looser, more-spicate infl

124a Peri evidently slender-beaked, beak rather narrowly margined and serrulate in the proximal portion only (or not at all), the distal 0.5 or more being subterete, marginless, and gen entire or nearly so; infl variously pale to dark   Subgroup VIIIa (lead 125a)

124b Peri with a flattened, broadly margined and serrulate, often ill-defined beak, sometimes margined and serrulate throughout, or sometimes with a minute, subterete, marginless and entire tip < 0.5 mm; infl always relatively pale, greenish or straw-colored to light brown   Subgroup VIIIb (lead 139a)

Subgroup VIIIa

125a Peri small, mostly 2.5–3.2 (3.6) mm, planoconvex
  126a Peri sharp-edged but scarcely wing-margined, entire or nearly so
    127a Peri widely spreading, their short beaks standing out from the body of the spike and giving it a somewhat ragged outline; infl rather dark brown to greenish-black, very small and compact, mostly 8–15 mm; widespread (see lead 114a)   105 **C. illota** Bailey
    127b Peri appressed-ascending, their beaks not standing out from the body of the spike; infl gen paler, gen light brown to medium brown, often somewhat longer and looser than in *C. illota;* Cas-Sierran, from Mt Jefferson, Ore, to Cal; meadows and open slopes, midmont; smooth-beaked s.   116 **C. integra** Mack.
  126b Peri distinctly wing-margined and serrulate, at least on distal part of the body
    128a Infl small and compact, gen 8–15 mm, the spikes scarcely distinguishable by the unaided eye, medium brown to more often dark brown or greenish-black; ventral surface of peri gen nerveless or nearly so; pls smaller and slender, mostly 1.5–4 dm; moist to wet places, midmont; Mt Rainer, Wn, and Wallowa Mts, Ore, to Alta and Wyo, s to Nev and Colo, widespread, but seldom collected; pond s.
                                                      117 **C. limnophila** Hermann
    128b Infl larger and looser, gen 15–30 mm, the spikes readily distinguishable to the unaided eye, light brown or greenish to sometimes medium brown; ventral surface of peri gen nerved, seldom essentially nerveless; pls sometimes larger and coarser, up to 7 dm; moist to dry soil, midmont; Cas-Sierran axis n to c Ore and reputedly to Chelan Co, Wn, e to Utah and NM; rusty s. *(C. macloviana* var. *s., C. tenerae-formis)*   118 **C. subfusca** W. Boott

125b Peri larger, gen 3.2–7.9 mm, planoconvex or flattened
  129a ♀ scales distinctly < and narrower than the peri, largely exposing at least the distal margins as well as the beak of the peri (in *C. proposita* scale much narrower than, but often nearly = peri)
    130a Peri ± distinctly planoconvex, the cavity nearly filled by the plump achene
      131a Peri relatively large, gen 5.8–7.9 mm; wet places, plains to mont, sometimes to timberl; BC to Sask, s to Cal and Colo, wholly E Cas; Liddon's s. *(C. liddoni, C. rufovariegata, C. constanceana)*
                                                      119 **C. petasata** Dewey
      131b Peri smaller, gen 3.3–5.1 mm; n and nw spp., chiefly of humid regions
        132a Peri very slender, only 0.8–1.0 mm wide and 3.5–4.8 times as long as wide; infl pale greenish to straw-colored; moist or wet places, lower mont to midmont; BC to Newf, s to Snoqualmie Pass, Wn, and n Ida, in our range, and to Mich and NJ; Crawford's s.   120 **C. crawfordii** Fern.
        132b Peri broader, 1.4–2.1 mm wide and 1.5–3 times as long as wide; infl sometimes pale as in *C. crawfordii,* but more often darker brown; open, often rather dry slopes, to wet places, from near sea level to timberl; Alas to Ore and Cal, esp in and w Cas, e to sw Alta, sw Mont, Wyo, and ne Ore and adj Ida; thick-

headed s. *(C. macloviana* var. *p., C. mariposana, C. preslii, C. multimoda, C. olympica, C. platylepis)*
                                                                            121 C. **pachystachya** Cham.
130b Peri ± strongly flattened except where distended by the relatively small achene; inl spp., seldom reaching as far w as Cas
  133a Lvs 2–4 mm wide, flat or nearly so; infl a dense, typically globose-ovoid or broadly ellipsoid head
    134a Peri middle-sized, gen 3.2–5.0 mm, lightly to evidently several-nerved on both sides, gen paler than the ♀ scales and the spike thus conspicuously bicolored; pls gen (1.5) 2–6 (8) dm, occurring in moist or wet places from foothills to near timberl; common and widespread; E Cas in w US and adj Can; small-winged s. *(C. festiva* var. *viridis, C. festivella, C. stenoptila)*
                                                                            122 C. **microptera** Mack.
    134b Peri larger, gen (4.5) 5.0–6.2 mm, often nearly or quite as dark as the scales, ventral side tending to be nerveless or nearly so; pls gen 1–3 (4) dm, occurring near or above timberl; wet to moderately dry places; s BC to sw Alta, s to Cal, c Nev, and Colo; chiefly e Cas; Hayden's s. *(C. festiva* var. *h., C. macloviana* var. *h.)*   123 C. **haydeniana** Olney
  133b Lvs gen 0.5–2 mm wide, often folded or channeled; infl looser and narrower, forming a compact spike or loose head gen distinctly longer than wide; pls 1–3 dm, local in mts of c Ida, Wen Mts, Wn, and Sierran Cal; open rocky slopes and ridges, often on talus, near or above timberl; Smoky Mt s. 124 C. **proposita** Mack.
129b ♀ scales nearly or quite as long and wide as peri and ± completely concealing them in dorsal view, only the tip of the beak sometimes exposed
  135a Peri large, gen 5.8–7.9 mm, planoconvex, nerved on both sides, 2.6–4.0 times as long as wide; gen lower- and mid-mont, often in dry lands, E Cas (see lead 131a)   119 C. **petasata** Dewey
  135b Peri smaller, 3.2–5.7 (6.0) mm, variously shaped and nerved
    136a Peri strongly flattened, only slightly or not at all planoconvex, gen 1.8–2.6 times as long as wide, gen evidently nerved on both sides; infl stiff and compact, the 2 lowest internodes each mostly 2–5 (7) mm; pls mostly (3) 4–8 dm, occurring from the lowl to midmont, s BC to Cal, chiefly w Cas, but inl to n Ida and se BC, Eurasia; hare s. *(C. tracyi)*   125 C. **leporina** L.
    136b Peri ± evidently planoconvex, gen 2.6–4.0 times as long as wide (only 2.2 times in *C. phaeocephala,* an alp or subalp sp. only 1–3 (4) dm), nerved or more often nerveless or nearly so on the inner face; infl various, often looser
      137a Peri relatively small, gen 3.2–4.0 × 0.8–1.2 mm, and 3.3–4.0 times as wide, sharp-edged but scarcely wing-margined, the beak rather ill-defined; infl compact, the 1st internode gen 2–6 (9) mm, the 2nd 1–5 (6) mm; moist meadows, rather high mont, sometimes above timberl; s Wn to Sierran Cal, e across ne Ore and c Ida to w Wyo, and to Utah; Sierra-hare s.
                                                                            126 C. **leporinella** Mack.
      137b Peri larger, gen (3.6) 4.0–5.7 (6.0) × 1.3–2.0 mm and 2.2–4.0 times as long as wide, ± distinctly wing-margined at least toward distal part of body, the beak often but not always well defined
        138a Pls of lowl to midmont, often > 4 dm; infl relatively loose and flexuous, the 2 lowest internodes each 5–10 (15) mm; lvs (1) 2–4 mm wide; moist to wet meadows, streambanks, and moist woods, lowl to midmont, seldom app timberl; Alas to Greenl, s to nw Cal, ne Ore, c and se Ida, Utah, Colo, and ND; meadow s. *(C. pratensis* Drejer, *C. p.* var. *furva, C. praticola* var. *f.)*   127 C. **praticola** Rydb.
        138b Pls of high elevs, near or above timberl, only 1–3 (4) dm; infl more compact, the 2nd internode mostly 2–5 (6) mm, the 1st similar or sometimes up to 10 mm; lvs 0.5–2 (2.5) mm wide (see lead 123a)   114 C. **phaeocephala** Piper

Subgroup VIIIb

139a ♀ scales distinctly shorter and narrower than the peri, largely exposing at

134a

134b

133b

135a

136a

137a

138b

138a

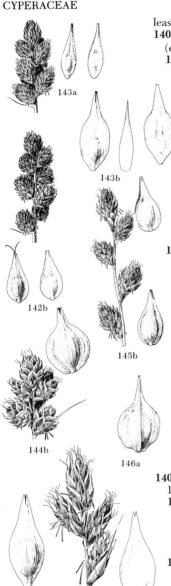

least the distal margins as well as the beak of the peri

140a Dorsal surface of the peri relatively few-nerved, the nerves evident (except sometimes in *C. crawfordii*) but < 10

141a Peri 2.5–5 times as long as wide; achenes very narrow, only 0.6–0.8 mm wide

142a Peri either at least 4 mm or > 3 times as long as wide; beak of peri becoming slender and subterete in the distal 0.2–0.5 mm

143a Peri 3.3–4.0 × 0.8–1.0 mm, 3.5–4.8 times as long as wide, planoconvex and not much wider than the achene; achene 1.0–1.3 mm (see lead 132a)   **120 C. crawfordii** Fern.

143b Peri 4.1–5.5 × 1.5–2.0 mm, 2.5–3.0 times as long as wide, strongly flattened and much wider than the achene; achene 1.3–1.8 mm; moist to wet ground at low elevs, chiefly where humid; widespread N Am sp., W Cas with us; pointed broom s.
128 C. scoparia Schk.

142b Peri 2.8–3.7 × 1.1–1.5 mm, up to 2.7 (3.0) times as long as wide, planoconvex; beak flattened to tip; wet places, valleys and lowl to midmont, mostly in moist regions; widespread N Am sp.; Bebb's s.
129 C. bebbii Olney

141b Peri 1.7–2.5 (3.0) times as long as wide; achenes, except in *C. bebbii*, at least 0.9 mm wide

144a Peri 1.1–1.9 mm wide, (1.5) 1.7–2.5 times as long as wide, plano-convex or somewhat flattened, tapering to an often ill-defined beak; achene 0.6–1.3 mm wide

145a Achene only 0.6–0.8 mm wide; spikes ± closely aggregated; peri 1.1–1.5 mm wide (see lead 142b)   **129 C. bebbii** Olney

145b Achene 0.9–1.3 mm wide; spikes more distant in a moniliform to interrupted infl, less often aggregated as in *C. bebbii;* peri 1.4–1.9 mm wide; wet, low ground in the plains and foothills; widespread in e US and adj Can, w to nw Mont; slender s. (*C. tincta*)   **130 C. tenera** Dewey

144b Peri 2.2–3.4 mm wide, (1.2) 1.3–1.7 (1.8) times as long as wide, strongly flattened, body suborbicular and ± abruptly contracted to the well-defined beak; achene 1.3–1.9 mm wide; streambanks, meadows, and swales, valleys and plains to midmont; widespread sp. of e US and adj Can, w to ne Wn and se BC, but chiefly e Cont Div; short-beaked s. (*C. straminea* var. *b., C. festucacea* var. *b.*)
131 C. brevior (Dewey) Mack.

140b Dorsal surface of peri obscurely to conspicuously multinerved, nerves 10 or more; peri 1.5–2.9 times as long as wide

146a Spikes borne in a compact head, 1st 2 internodes collectively only 4–7 mm; open slopes, often near persistent snowbanks, gen near or above timberl; n Wn, s irreg to Sierran Cal and adj Nev; disjunct in parts of c Ida, Mont, and Utah; Mt Shasta s. (*C. straminea* var. *congesta*)   **132 C. straminiformis** Bailey

146b Spikes borne in a more elongate, spiciform infl, 1st 2 internodes collectively gen 10–18 mm; grassl, open slopes, and mt parks, high plains to spruce-fir zone in mts; Gr Pl and s RM sp., to be expected at e margin of our range in Mont; dryland s. (*C. foenea* var. *x.*)
133 C. xerantica Bailey

139b ♀ scales nearly or quite as long and as wide as the peri and ± concealing them in dorsal view, the tip of the beak sometimes exposed

147a Peri 1.6–2.8 mm wide, 2.0–3.0 times as long as wide, often > 4 mm, always evidently wing-margined at least distally; pls 3–7 dm

148a Infl stiff; peri ± closely appressed, gen 4.5–7.0 mm, not notably darkened below (see lead 146b)   **133 C. xerantica** Bailey

148b Infl loose and gen flexuous; peri looser, not closely appressed, gen 3.6–4.9 mm, often notably darkened below the middle; moist or wet places, mostly where humid; valleys and plains to midmont; rather infrequent, widespread inl and e Am sp., w occ to our range in Mont; bronze s. (*C. foenea* var. *a.*)   **134 C. aenea** Fern.

147b Peri 0.9–1.2 mm wide, 3.3–4.0 times as long as wide, 3.2–4.0 mm, sharp-edged but scarcely wing-margined; pls 1–3 dm (see lead 137a)
126 C. leporinella Mack.

Cyperus L.   Flatsedge; Cyperus

Spikelets several–many, in 1 or gen several capitate to spicate clusters, the terminal cluster gen sessile or nearly so, the others borne on ± elongate rays from axils of sheathless, lfy invol bracts; scales of spikelet arranged in 2 vertical rows; fls ♂, borne singly in axils of the scales; perianth none; stamens 3 (1–2); style bifid or more often trifid, achene accordingly lenticular or trigonous; ann or gen per herbs with mostly triangular, solid sts, lvs with closed sheath and gen elongate, grasslike bl. (*Gr Kypeiros*, the ancient name).

1a Spikelets borne in very short, ± capitate clusters with a very short rachis; pl ann; rachilla wingless; stamens 1 or 2, rarely 3

2a Pistil bicarpellate; stamens 2 (3); scales gen 2–2.5 mm, straight, blunt; pl 0.5–2 (3) dm; wet places, lowl, tolerant of alkali; widespread in US and s Can, but more common e, and rarely collected in our area, s to S Am; shining f. or c.   **1 C. rivularis** Kunth

2b Pistil tricarpellate; stamen 1; scales sharp-pointed or awn-tipped, often outcurved distally, often < 2 mm

3a Scales 3-nerved, tending to be shortly recurved-acuminate, but not awn-tipped, 1.5–2 (2.5) mm, closely set, the tip of each scale surpassing that of the scale next below it by 0.4–0.7 mm; wet, low places in valleys and lowl, tolerant of alkali; ND to Ga, w to Pac, seldom collected in our range; short-pointed f. or c.
2 C. acuminatus Torr. & Hook.

3b Scales (5) 7–9-nerved, with a slender, squarrose-recurved, short but distinct awn-tip 0.3–1 mm, more loosely set, the tips gen > 0.7 mm apart; wet places, valleys and lowl, cosmopolitan except at high latitudes; awned f. or c. (*C. inflexus, C. a.* var. or f. *i.*) **3 C. aristatus** Rottb.

1b Spikelets borne in ± elongate and more open, cylindric spikes (except in no 7); pls per (except no 4); rachilla winged (or scarcely so in no 7); stamens 3; fls tricarpellate

4a Ann; scales small, gen 1.2–1.5 mm, closely set, the tip of each surpassing the one below it by 0.4–0.7 mm, the keel prominent, green, minutely excurrent; achene 0.7–1.0 mm; wet places, valleys and lowl, esp along banks of streams and major rivers; widespread in US and s Can, but seldom collected in our area; red-rooted f. or c. (*C. washingtonensis*)
4 C. erythrorhizos Muhl.

4b Per; scales larger, 2–6 mm, less closely set, the tips 1–3 mm apart; achenes 1.3–2.5 mm

5a Wings of the rachilla clasping each achene toward the base; pls wholly without rhizomes; spikelets deciduous at maturity, the rachilla disart just above the base, the individual scales not falling away from the rachilla; wet places at lower elev, as along banks of large streams; widespread; straw-colored f. or c.   **5 C. strigosus** L.

5b Wings of rachilla narrow, not clasping the achenes, or rachilla scarcely winged; pls rhizomatous; scales falling from the rachilla at maturity, the rachilla sometimes (*C. schweinitzii*) itself deciduous thereafter

6a Clusters of spikelets elongate and open, ± cylindric; rhizomes elongate, slender, with a terminal tuber; rachilla evidently hyaline-winged; moist, low ground along streams and ditches, or sometimes in drier ground, often becoming weedy; widespread, but seldom collected in our area; yellow nut-grass   **6 C. esculentus** L.

6b Clusters of spikelets relatively short and dense, subcapitate; rhizomes short, thickened at intervals, but scarcely tuber-bearing; rachilla narrowly or scarcely winged, the wing when present firm, not hyaline; sandy places, wet or dry, in valleys and lowl; Schweinitz f. or c.   **7 C. schweinitzii** Torr.

Dulichium Pers.   Dulichium

Spikelets in short, axillary spikes, in 2 distinct rows on the rachis; scales of spikelet arranged in 2 vertical rows; fls ♂, borne singly in axils of the scales; perianth 6–9 retrorsely barbellate bristles; stamens 3; style bifid; achene lenticular; rhizomatous per herbs with terete or obtusely triangular, hollow sts; lvs ∝, well distributed along sts, those of lower 1/3–1/2 of st reduced to bladeless sheaths, others with grasslike bl. (Name of uncertain origin).

D. arundinaceum (L.) Britt. Sts 3–10 dm, arising singly from often deep-seated rhizome; well-developed lvs with firm bl ave 4–15 cm × 2.5–8 mm; spikes with gen 7–10 spikelets each, internodes of rachis ca 2 mm; spikelets 1–2.5 cm, slender, ca 2 mm wide or less; scales 5–8 mm, several-nerved, acute or acuminate; achene narrow, gen 2.5–3 mm, exclusive of the short (to 1 mm) stout stipe, surpassed by the brownish bristles; marshes and wet meadows, lowl and lower mts, sw BC to Cal, inl to n Ida and nw Mont, and widespread farther e *(Cyperus a.).*

Eleocharis R. Br.   Spike-rush

Spikelet solitary and terminal, without subtending bracts; scales spirally arranged (ours) or sometimes distichous, largely scarious or hyaline, the lowest one(s) in most spp. empty; fls ♂, borne singly in axils of scales; perianth of 0–6 (9) bristles, these < to somewhat > the achenes, sometimes reduced or obsolete; stamens 3 (sometimes fewer); style bifid or trifid, thickened toward base, the thickened part persistent on the achene as a tubercle which is typically sharply differentiated from the body of the achene, but which in a few spp. is confluent with it and poorly differentiated from it; achene lenticular to planoconvex or ± trigonous; herbs with angular to terete or flattened sts, the lvs all basal or nearly so and reduced to mere sheaths or scarcely sheathing scales. (Gr *helos,* a marsh, and *charis,* grace).

1a Stigmas 3; achenes ± distinctly trigonous; fls gen 2–30 in each spikelet
  2a Tubercle confluent with the achene, not forming a distinct apical cap
    3a Achenes and scales small, achenes 0.9–1.3 mm (including tubercle), scales gen 1.5–2 (2.5) mm; pls very small, matforming, mostly 0.2–0.6 (1) dm; spikelets 2.5–4 (6) mm, with 2–9 (20) fls; lowest scale empty; wet, saline or alkaline sites, Newf to VI, s to n S Am; Europe; small s. (*E. pygmaea, E. p. var. anachaeta, Scirpus nanus* var. *a., E. leptos* Svenson, *S. coloradoensis, E. l.* var. *c., S. p.);* 2 vars.
                         1 **E. parvula** (R. & S.) Link
      a1 Perianth bristles commonly = or > achene; achene smooth; salt marshes along both coasts (VI to Cal; Newf to Cuba and Mex) and at scattered stations inl   var. **parvula**
      a2 Perianth bristles much-reduced or obsolete; achenes often cellular-roughened; mostly e and se US and Mex, but known from a single collection (Canyon Co, Ida) in our range
                    var. **anachaeta** (Torr.) Svenson
    3b Achenes and scales larger, achenes 0.9–2.8 mm, scales gen 2.5–5.5 mm; larger plants, seldom < 1 (0.5) dm
      4a Lowest scale, like the others, subtending a fl; sts slender and short, (0.5) 1–3 (4) dm, seldom as much as 1 mm wide, not flattened, not proliferous; spikelets 4–8 mm, with mostly gen 3–9 fls; wet (often boggy) places, lowl to above timberl, tolerant of salt and alkali; circumboreal, s to Cal, NM, Ill, and NJ; few-fld s. (*Scirpus p.*)
                    2 **E. pauciflora** (Lightf.) Link
      4b Lowest scale empty; sts coarser, gen (2) 4–10 dm or more, ± flattened at least distally and gen 1–2 mm wide, some of them gen proliferous; spikelets (5) 8–13 mm, with (5) 10–20 (25) fls; salt marshes along coast, and in alkaline or highly calcareous places inl, often around hot springs; VI to NS, s to S Am; beaked s. (*Scirpus r., E. suksdorfiana, E. pauciflora* var. *s.)*   3 **E. rostellata** Torr.
  2b Tubercle forming a distinct apical cap well differentiated from the body of the achene
    5a Achene whitish or pale gray to ochroleucous, longitudinally many-ribbed, with ∝ fine cross-ridges, forming ladder-like configurations; scales gen with green or straw-colored midvein, not notably pale-tipped, lowest one floriferous and not markedly broader or different from the others; sts filiform, 1–12 cm
      6a Per, rhizomatous (but still densely tufted); anthers 0.7–1.3 mm (dry); scales (1.3) 1.5–2.2 mm; spikelets 2.5–7 (9) mm; marshes and other wet places, lowl to rather high mont; circumboreal, s to Fla and Mex, and found throughout our range; needle s. (*Scirpus a., E. a.* var. *occidentalis)*   4 **E. acicularis** (L.) R. & S.
      6b Ann, rarely rhizomatous; anthers 0.25–0.4 mm; scales 1.0–1.3 mm; spikelets 1.5–3 mm; borders of lakes and marshes and in muddy to

springly places, lowl to midmont; Wn to n Ida and to s Cal and NM, wholly e Cas in our range; delicate s. *(E. acicularis* vars. *bella* and *minima)*   **5 E. bella** (Piper) Svenson

5b Achene golden yellow (reputedly sometimes black), merely 3-ribbed, the surface finely cellular-roughened; scales (except the broader, often suborbicular, and gen sterile lowest ones) largely or wholly dark (often blackish-purple) except for the conspicuously pale and hyaline tip; sts slender but not filiform, 5–40 cm

7a Rhizomes ± elongate, the sts not very densely clustered; achene (0.7) 1–1.5 mm, the tubercle gen depressed and with an apiculate center; wet places, mostly e N Am, but w occ to Alta, Mont, and BC; slender s. *(Scirpus t., S. capitatus, E. nitida, C. compressa* var. *borealis);* ours the var. *borealis* (Svenson) Gleason, which is trans-continental   **6 E. tenuis** (Willd.) Schultes

7b Rhizomes very short and freely rooting, with densely clustered sts; achene ca 1.5 mm, tubercle broad and depressed, sometimes appearing almost as an apical scar with a central mucro; wet places, foothills to midmont; Cas and Wallowa Mts, Ore, to Cal, e to se Ida and Utah; Bolander's s. *(E. montevidensis* var. *b.)*
                                                      7 E. bolanderi Gray

1b Stigmas 2 (sometimes 3 in *E. ovata,* an ann with ∝ fls in each spikelet); achenes lenticular

8a Achenes gen 1.5–2.5 mm (including tubercle); anthers gen 1.3–2.5 mm; scales gen 2–4.5 mm; per, rhizomatous; wet places from sea level to midmont, tolerant of alkali; widespread in temp and cold-temp regions of N Hem, and common in most of our range; common or creeping s. *(Scirpus p., S. uniglumis, E. u., S. p.* var. *u., E. calva)*
                                                      8 E. palustris (L.) R. & S.

8b Achenes gen 0.5–1.5 mm; anthers gen 0.25–1.0 mm; scales gen 1.0–2.5 mm; anns, except no 11

9a Achenes black to dark cherry red, 0.5–0.6 mm; scales gen 1.0–1.5 mm; diminutive ann gen < 1 dm; wet places, pantropical, n irreg to s US, and sporadically in our range, as at Lake Chelan, Wn, and Lake Osoyoos, BC; purple s. *(Scirpus a.; E. capitata,* misapplied)
                                                      9 E. atropurpurea (Retz.) Kunth

9b Achenes straw-colored to dark brown or purplish brown, 0.9–1.5 mm; scales gen 1.7–2.5 mm; ann or per, often > 1 dm

10a Fls ∝ (seldom < 40) in each spikelet; tubercle flattened, broad-based, gen > 0.5 mm wide; tufted ann 0.5–5 dm; marshes and other wet places, from sea level to midmont; widespread in N Hemisphere, and throughout our area; ovoid s. *(Scirpus ov., Bulbostylis ov., S. obtusus, E. ob., E. monticola* var. *pallida)*
                                                      10 E. ovata (Roth) R. & S.

10b Fls few (< 20) in each spikelet; tubercle narrower and more conic, ca 0.25 mm wide; slenderly rhizomatous per, tufted, 0.3–1 dm; wet places; mainly tropical, n to s US, intro in Cal, isolated stations n, as at YNP and adj sw Mont, where prob confined to margins of thermal streams and pools; yellow s. *(Scirpus f.);* ours the var. *thermalis* (Rydb.) Cronq.   **11 E. flavescens** (Poir) Urban

Eriophorum L.   Cotton-grass

Spikelets 1–several in a terminal infl subtended by 1–several elongate and lfy to small and scale-like invol bracts, these sometimes (esp in spp. with only a single spikelet) ± resembling the scales of the spikelet and then referred to as sterile scales; scales spirally arranged, scarious, not awned; perianth of ∝ (> 10) conspicuous, persistent bristles, these white to rufous, much elongate, at least in fr, together forming a cottony tuft gen 2–4 cm; stamens 3; style trifid; achene unequally 3-sided, often with a short, slender stylar apiculus; herbs with triangular or terete, gen solid sts, lvs with closed sheath and elongate, grasslike bl, upper sheaths often bl-less. (Gr *erion,* wool, and *phoros,* bearing).

1a Spikelets 2 or more

2a Foliaceous bracts 2 or more, at least the longest one gen > (or at least =) infl; lvs flattened to well beyond middle, triangular only near the tip; achenes blackish, broadly oblanceolate to obovate, gen 2–3 times as long as wide; peduncles compressed, smooth or minutely hairy

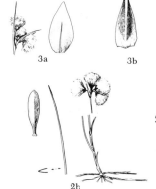

3a

3b

2b

5b

5a

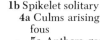

4b

3a Midrib of scale attenuated distally, not reaching the very thin end of the tawny to brownish or blackish-green scale; common sp. of cold swamps and bogs at mid to high elevs; circumboreal, s to Me, n NM, ne Utah, Ida, and c Ore (Mt. Jefferson); many-spiked c. (*E. ocreatum, E. angustifolium, Scirpus a., E. p.* var. *a.*)   1 **E. polystachion** L.

3b Midrib of scale expanded (or at least not attenuated) distally, reaching the end of the blackish-green scale; rare sp. of cold swamps and bogs, mid- to high-mont; Newf to Alas, s to NY, Colo, and n Ida; green keeled c. (*E. latifolium* var. *v., E. polystachion* var. *v.*)
2 **E. viridicarinatum** (Engelm.) Fern.

2b Foliaceous bract solitary, gen < infl; distal half (or more) of each foliage lf triangular or strongly channeled; achene straw-colored or light brown, rather narrowly elliptic-oblong or narrowly oblanceolate, gen 3–4.5 times as long as wide; peduncles subterete, short-hairy; scales blackish-green to greenish-brown; wet places, mid- to high-mont; circumboreal, s to Pa, Colo, and c Cal; slender c. (*Scirpus ardea*)
3 **E. gracile** Koch

1b Spikelet solitary

4a Culms arising singly from creeping rhizomes; bristles white to often rufous

5a Anthers gen 0.5–1 mm; scales narrow, tapering, blackish-green, not pale-margined; pls seldom > 3 dm; bristles white; rare with us and strictly alp; circumboreal, s to Newf and s Alta, and apparently isolated in high mts in Mont (Beartooth), and in Utah and Colo; Scheuchzer's c. (*E. leucocephalum, Scirpus l.*)
4 **E. scheuchzeri** Hoppe

5b Anthers gen 1–2.5 mm; scales broader and blunter, pale distally and along margins; pls gen 3–7 dm; bristles often reddish; common sp., from sea level to midmont, in swamps and other wet places; circumboreal, s to NB, Minn, nw Wyo, and along coast to Lane Co, Ore; Chamisso's c. (*Scirpus c., E. altaicum* var. *neogaeum, E. russeolum*)
5 **E. chamissonis** C. A. Mey.

4b Culms densely clustered, without rhizomes, gen 3–6 dm; bristles white or off-white; anthers up to ca 1 mm; cold, wet places, mont; circumboreal, s to Newf and s BC and reputedly to Mont; short-anthered c. (*Scirpus b.*)
6 **E. brachyantherum** Trautv. & Mey.

Hemicarpha Nees   Hemicarpha

Infl 1–several small spikes at the summit of the culm, subtended by 2–3 unequal invol bracts, the longest of which resembles a continuation of the culm, the infl thus appearing lateral near the summit rather than evidently terminal; spikes compound, of very ∝, spirally arranged, sessile spikelets, each spikelet with a single ♂ fl subtended by a well-developed outer scale and a smaller, very thin (often much-reduced or obsolete) inner scale (between the fl and the spike axis) that has sometimes been interpreted as a modified perianth, perianth otherwise lacking; stamen 1 (2); achene bicarpellate, dorsiventrally ± compressed; style deeply bifid, deciduous at maturity, leaving a small mucro tipping the achene; delicate, glab anns, with clustered sts up to 2 dm, the lvs few and slender, with closed sheaths. (Gr *hemi*, half, and *karphos*, chaff, referring to inner scale of spikelet).

1a

1b

1a Scales tapering to the prominent, elongate, firm awn-tip, tips of the middle ones gen 0.5–1.0 mm, those of the lowest ones somewhat longer and gen reaching or surpassing the middle of the relatively short, gen subglobose spike; wet low ground, Sierran Cal, n occ to Klickitat Co, Wn; western h.
1 **H. occidentalis** Gray

1b Scales abruptly contracted to a short mucro or minute awn-tip, the tips of the middle ones gen 0.1–0.5 mm, those of the lowest ones sometimes somewhat longer, but not reaching the middle of the more-ovoid spike; beaches, sand bars, and wet bottomland; tropical Am, n to s Wn, s Ida, s Wyo, and Me; small-fld h. (*H. intermedia, H. aristulata*) 2 **H. micrantha** (Vahl) Britt.

Kobresia Willd.   Kobresia

Fls ♂♀, naked, solitary in axils of small, scarious bracts (scales), arranged in small spikelets aggregated into 1 or more spikes, each spikelet subtended by a scale on the axis of the spike (this scale not directly subtending a fl) and having a single proximal ♀ fl and 1 or more distal ♂ fls, or some of the spikelets unisexual, with only the single ♀ fl (and often a prolonged rachilla) or 1 or more ♂ fls; scale subtending ♀ fl borne with its back to the axis of the spike, loosely wrapped about the trigonous achene, resembling and homologous with the peri of *Carex*, but open, with unsealed margins; stamens 3; stigmas 3; densely caespitose, grass-like per herbs with narrow lvs, closed sheaths, and obtusely triangular, solid sts. (For von Kobres, naturalist and patron of botany, contemporary of Willdenow).

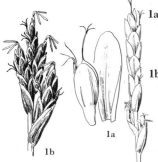

1a Spikes solitary, 1–3 cm; lf bls of previous year gen deciduous before fl time; open, dry or somewhat moist places, high mont, often above timberl; circumboreal, s to Newf, Que, Can RM, and irreg to Colo, Utah, Ore (Wallowa Mts), and Sierran Cal; Bellard's k. *(Carex m., K. bellardii)*
                                                                                1 **K. myosuroides** (Vill.) Fiori
1b Spikes several, gen 3–12, 0.5–1.5 cm; some or all of the dried lf bls of previous year still persistent at fl time; bogs and other wet places, mont, but gen not above timberl; circumboreal, s to Newf, Que, BC, and irreg to Colo, Ida (Teton Co), and Ore (Wallowa Mts), not recorded from Wn or Mont; simple k. *(Carex s.)*   2 **K. simpliciuscula** (Wahl.) Mack.

Rhynchospora Vahl   Beakrush

Spikelets rather small, several–many in open to very compact, axillary or terminal, cymose infl; scales spirally arranged, scarious, lower 2–3 comparatively small and empty, the succeeding 1–10 each subtending a ☿ fl, the terminal 1–2 empty or subtending ♂ or rudimentary fls; perianth of several (up to 20) bristles, or wanting; stamens 1–12, commonly 3; style shortly bilobed to deeply bifid; achene bicarpellate, crowned with a conspicuous tubercle consisting of the broad, persistent, indurated base (or greater part) of the style; ann or per, grasslike herbs with narrow lvs, closed sheaths, and ± triangular, solid sts. (Gr *rhynchos*, beak, and *spora*, seed, referring to the beaked achene).

R. alba (L.) Vahl. White b. Culms densely tufted, (0.8) 1.5–5 dm; lowest lvs ± reduced, others slender and elongate, up to 1 mm wide; infl 1–3 compact, head-like clusters gen 5–15 mm wide; spikelets gen 3.5–5 mm, 2 (3)-fld, maturing 1–2 achenes; bristles 10–12, > body of achene, achene 1.5–2 mm, exclusive of the elongate, narrow tubercle; sphagnum bogs and other wet places, lowl to midmont; interruptedly circumboreal, Newf to NC, inl to Gr Lakes region, and from Alas to c Cal, chiefly W Cas, but also inl in n Ida *(Schoenus a.).*

Scirpus L.   Bulrush; Clubrush

Spikelets 1–many in capitate to umbelliform or paniculiform infls subtended by 1–several large and lfy to small and scale-like bracts, the spikelet sometimes solitary and with the subtending bract(s) not much different from the scales except for being empty; scales spirally arranged, scarious, with or without an excurrent awn-tip; fls ☿, borne singly in axils of scales; perianth of 2–6 bristles, these < to much > the achene, sometimes conspicuously exserted from the spikelet or sometimes ± reduced or obsolete; stamens (1–) 3; style bifid or trifid, and achene lenticular or ± trigonous; achene commonly with a short, slender, stylar apiculus (or this wanting in some spp.), but without an apical tubercle, the style deciduous; herbs with triangular or terete, gen solid sts; lvs with closed sheath and elongate grasslike bl, or bl sometimes much reduced or obsolete. (Classical L name for some of the spp.).

1a Pls ann, dwarf, up to 2 (3.5) dm; bristles absent; achene scarcely apiculate; spikelet solitary, 2–5.5 mm; fresh or gen brackish or saline marshes and beaches or shores; near coast and about PS, s BC to n Mex, and irreg cosmopolitan; low c. *(S. pygmaeus, S. c.* var. *p.)*   1 **S. cernuus** Vahl
1b Pls per, often larger; most spp. with evident perianth bristles, or with a stylar apiculus on the achene, or both; spikelets 1–many
2a Invol of merely 2–3 empty, slightly modified lowest scales of the solitary spikelet, which appears to be strictly terminal
3a Culms triangular, scab; perianth bristles much elongate, conspicuously > scales; spikelet 5–7 mm; sphagnum bogs etc; circumboreal, s

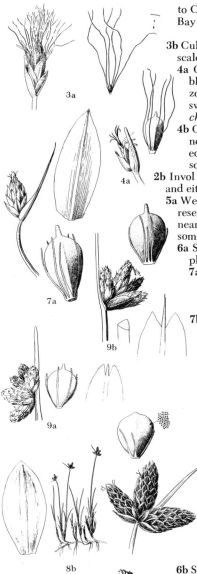

3a

4a

7a

9b

9a

8b

10a

11a

to Conn, Sask, and s BC, and reported by Hulten for Mont; Hudson's Bay b. or c. *(Eriophorum h., E. alpinum, Scirpus trichophorum)*
2 S. **hudsonianus** (Michx.) Fern.

3b Culms subterete, smooth; perianth bristles shorter, scarcely if at all > scales, or obsolete

    4a Outermost scale with the raised midrib prolonged into a broad, blunt awn at least 1 mm; pls 1–4 dm, densely tufted on a short rhizome; spikelet 4–6 mm; sphagnum bogs etc; circumboreal, s to NC, sw Mont, ne Utah, c Ida, w Wn, and Cas of c Ore; tufted c. *(Eleocharis c.)* 3 S. **cespitosus** L.

    4b Outermost scale with the midrib only very minutely (< 0.5 mm) or not at all exserted; pls rhizomatous, not densely tufted; interruptedly circumboreal, known from Colo, Cal, and Can RM and to be sought in our area S. **pumilus** Wats.

2b Invol very different from the scales, at least 1 of the bracts ± elongate and either lf-like or resembling a prolongation of the culm

    5a Well developed invol bract solitary, green, lfy or more often erect and resembling a prolongation of the culm, the infl thus apparently lateral near the tip of the culm rather than truly terminal; smaller invol bracts sometimes also present, but scalelike and not chlorophyllous

        6a Spikelets solitary or few (up to 10–15), sessile in a sessile cluster; pls, except no 5, seldom > 1 m

        7a Spikelet strictly solitary; pl gen aquatic and with flaccid, distally floating sts and lvs in quiet, shallow water gen 2–8 dm deep; Alas to s Ore, chiefly W Cas, but also inl to n Ida and nw Mont; Utah; e US and adj Can; water c. 4 S. **subterminalis** Torr.

        7b Spikelets gen 2 or more, occ solitary on some culms; pls sometimes growing in shallow water, but culms stiff and emergent for most of their length

            8a Achenes evidently apiculate, not cellular-reticulate even at 20 ×; culms ± distinctly triquetrous

            9a Bract solitary; culms gen 5–15 dm, very sharply triquetrous, with notably concave sides; achenes 1.8–2.5 × 1.4–1.7 mm; marshes and other wet, low places, tolerant of alkali; Wn (Grant Co) to NS, s to S Am; Olney's b. *(S. chilensis,* prob misapplied) 5 S. **olneyi** Gray

            9b Bracts 2–3, 2nd and 3rd resembling enlarged scales of the spikelet, but not subtending fls; culms gen 1.5–10 dm, not very sharply triquetrous, the sides plane to slightly concave or slightly convex; achenes 2.2–3.3 × 1.6–2.3 mm; marshes and wet low ground, tolerant of alkali; widespread in US and s Can, and throughout our range, as well as on other continents; Am b., three-square b. 6 S. **americanus** Pers.

            8b Achenes not at all apiculate, evidently cellular-reticulate at 10 ×; culms subtetete, 1–4 dm; moist or wet, alkaline sites in des or semi-des regions; c Wn (e Cas) to Cal, e to Sask, ND, Wyo, and Utah; Argentina; Nev c. or b. 7 S. **nevadensis** Wats.

        6b Spikelets ± ∝ in a br infl, often sessile in clusters at br ends; culms terete, gen 1–3 m

            10a Pistil in most or all fls tricarpellate; spikelets all or nearly all individually pedunculate; uncommon, seldom collected in our range; margins of freshwater lakes and streams, Wn and Ore e to Que and Mass, s in Gr Pl to Okla; pale great b. or c. 8 S. **heterochaetus** Chase

            10b Pistil in most or all fls bicarpellate; spikelets mostly or all sessile in small clusters; common spp.

                11a Spikelets appearing dull gray-brown, individual scales (at 10 ×) with prominent red-brown short lines on a pale, gray-white background, at least the lower and middle scales gen (3) 3.5–4 mm; marshes and muddy shores, sometimes in water up to 1 m deep; tolerant of alkali; widespread in temp N Am, commonest in w US, and throughout our area; hardstem b., viscid b. *(S. occidentalis, S. a.* var. *o., S. lacustris* var. *o., S. malheurensis, S. a.* var. *m.)* 9 S. **acutus** Muhl.

                11b Spikelets appearing more reddish-brown, the individual scales with a brownish or tawny ground-color, the red-brown lines thus gen not prominent; scales gen (2) 2.5–3 (3.5) mm;

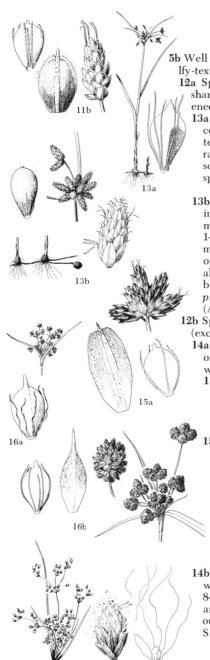

marshes and muddy shores at lower elevs, tolerant of alkali; widespread in temp N Am, s into tropical Am, and throughout our area, but less commonly than no 9; tule, softstem b., Am great b. *(S. lacustris* var. *v.)* 10 S. validus Vahl

5b Well developed invol bracts 2 or more, gen very unequal, but green, lfy-textured, and ± spreading, the infl thus evidently terminal

12a Spikelets 12–20 mm, not very numerous (up to ca 50); culms sharply triquetrous; rhizomes extensive, tuber-bearing or ± thickened at the nodes

13a Pistil tricarpellate; culms 1–1.5 m; secondary rootlets ∝ and conspicuous; spikes mostly 10–50, some of them in a compact terminal cluster, others borne singly or in small clusters at ends of rather lax brs 3–7 cm; fresh or brackish marshes and quiet water sometimes > 1 m deep, often forming dense colonies; widespread in temp N Am, but seldom collected in our range; river b. 11 S. fluviatilis (Torr.) Gray

13b Pistil bicarpellate; culms 0.2–1.5 m; secondary rootlets few, inconspicuous; spikes 3–20 or more, all sessile in a compact terminal cluster, or the principal cluster somewhat surpassed by 1–more short peduncles bearing each a subsidiary cluster; marshes, wet meadows, and margins of ponds, esp where alkaline or saline; widespread but rather irreg distributed in N Hem; esp along seacoasts, found essentially throughout our range; seacoast b. *(S. campestris, S. robustus* var. *c., S. paludosus, S. robustus* var. *p., S. campestris* var. *p.);* ours the rather ill-defined var. *paludosus* (A. Nels.) Kuek. 12 S. maritimus L.

12b Spikelets smaller, gen 3–8 mm, ± ∝ (> 50) in a br infl; culms (except in no 15) only obscurely triquetrous

14a Bristles only slightly if at all surpassing the mature achenes, < or barely = the scales; lvs (4) 6–20 mm broad; pls (except no 15) with well developed creeping rhizomes

15a Pistil bicarpellate; sheaths anthocyanic; pls robust, gen 6–15 dm; widespread, wet ground, lowl to midmont; throughout w cordilleran region, e to the Atl; small-fr b. *(S. sylvaticus* var. *m., S. lenticularis, S. s.* var. *digynus, S. macounii, S. m.* var. *longispicatus)* 13 S. microcarpus Presl

15b Pistil tricarpellate; sheaths not anthocyanic; habit various

16a Midrib of scales not exserted, or only very shortly and inconspicuously so; pls relatively small, gen 1.5–7 dm; Cas-Sierran sp., barely reaching our range at Mackenzie Pass, Ore; Congdon's b. *(S. c.* var. *minor)* 14 S. congdonii Britt.

16b Midrib of scales conspicuously exserted, forming a short, stout awn gen ca 0.5 mm; pls larger, (4) 6–15 dm or more; inl sp., lowl to midmont, in wet places, se Wn and ne Ore to RMS and to Minn and Mo; pale b. *(S. atrovirens* var. *p.)* 15 S. pallidus (Britt.) Fern.

14b Bristles elongate, > scales, conspicuously so in fr; pls tufted, without long rhizomes; pistil tricarpellate; lvs 2–6 mm broad; pls 8–15 dm; wet low ground, Newf to Fla, w to Minn, Mo, and Tex, and occ across s Can to se BC and reputedly to Wn; wool-grass; ours the mainly n var. *brachypodus* (Fern.) Gilly *(S. atrocinctus, S. a.* var. *b.)* 16 S. cyperinus (L.) Kunth

GRAMINEAE Grass Family

Fls mostly ⚲, but sometimes ♂, ♀, or sterile, sessile along a shortened, jointed axis (the rachilla) in 1–several-fld reduced spikes (spikelets); spikelets sometimes sessile in 1–several spikes, but gen borne on individual peds, each normally subtended by 2 opp arranged bracts (glumes), 1 (rarely both) of which may be reduced or lacking; individual fls (gen called florets and abbreviated here to flts) greatly variable, each gen borne within 2 bracts that may or may not be similar to the glumes in texture, size, and nervation (the tiny veins of the floral bracts are referred to as nerves; these often are slightly raised on the exposed surface, but gen are detectable only under at least 10 × magnification), the abaxial or outer member (lemma) gen partially enfolding the adaxial or inner one (palea), at maturity the basal portion of the lemma often (esp in 1-fld spikelets) elongate past the point of attachment to the rachilla and ± pointed and sometimes hardened, forming a so-called "callus"; stamens mostly 3, but sometimes 1, 2, or 6; pistil gen with 2 (rarely 3 or only 1) distinct styles and plumose stigmas; fr 1-seeded, indehiscent, the seed gen firmly attached to the ovary wall; perianth apparently lacking, but supposedly represented by 2 (rarely 3 or 0) greatly modified tiny structures (locidules; abbreviated here to lods) on the abaxial side of the fl (between the ovary and the lemma), which gen become distended at anthesis, forcing apart the lemma and palea; flts, if > 1 per spikelet, borne alternately on opp sides of the rachilla, the basal or the upper flts often reduced, sometimes each represented by an empty lemma only, the rachilla sometimes prolonged as a bristle behind the palea of the uppermost flt; ours ann or per herbs with gen terete (or flattened or grooved), hollow or (less commonly) solid, rarely persistent or woody, erect sts (culms) and ± swollen nodes, often some sts stolonous or rhizomatous; lvs alt, 2-ranked, the lower portion (sheath) surrounding the st, the sheaths with margins overlapping (open) or joined (closed), the upper margins (throat) often more hairy than the rest of the sheath; lf bls gen narrow and elongate, mostly flat, but often folded or rolled, the young lvs of the new shoots (innovations) either equitant (folded in the bud) or enclosing one another with the margins overlapping (rolled in the bud); lower margins of the bls sometimes free as small, rounded to acuminate projections on one or both sides (auricles; abbreviated to aurs), the abaxial (dorsal) area of the juncture of bl and sheath (the collar) gen marked by less-heavy venation, or by a difference in color or pubescence, the adaxial (ventral) point of juncture gen with a structure (ligule; abbreviation lig) varying from membranous to membranous-based and ciliate-fringed, or entirely hairy-fringed, very rarely the lig absent.

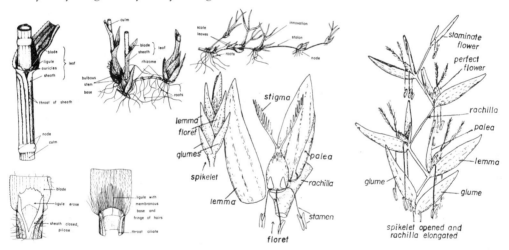

KEY TO MAIN GROUPS AND MISCELLANEOUS GENERA

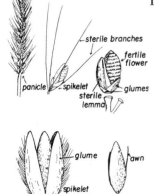

1a Spikelets shed as a unit, the glumes not persistent on the pl (rarely the spikelets without glumes to begin with), the point of disarticulation gen being just below the glumes, but not rarely along the axis of the infl (rachis) or brs of the rachis. Note—at maturity the grass infl gen breaks apart (disarticulates; abb. disart) readily, when it easily can be seen whether or not empty glumes remain. However, it may be necessary to dry other than mature grasses, since time must be allowed for the development of an abscission area, where disart occurs. When material is dry, the place of normal disart can be determined by rubbing or rolling part or all of an infl in the hands. In some cases it may be more satisfactory to test the point of disart on individual spikelets. To do this it is essential that the uppermost flt of the spikelet be grasped and it is perhaps best to use tweezers for the purpose. Included in Group I are the following kinds of grasses—1) a few genera with the glumes wanting or so small as to be overlooked even when persistent; 2) a few with the spikelets borne in spikes, and disart above the glumes, but the infl coming apart, gen along the rachis or at the base of the spikelets which are shed as units (in *Cenchrus* these are modified into spiny burs); 3) a few spp. with 1-fld spikelets disart

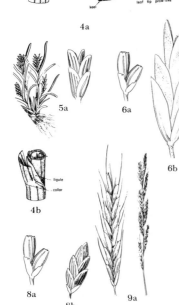

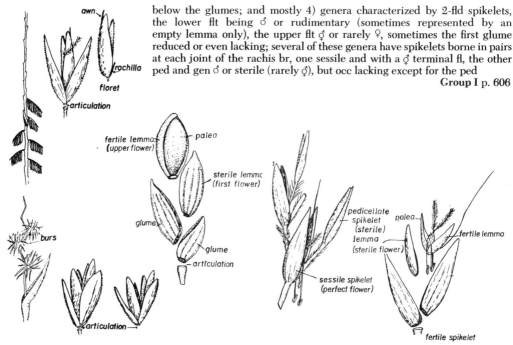

below the glumes; and mostly 4) genera characterized by 2-fld spikelets, the lower flt being ♂ or rudimentary (sometimes represented by an empty lemma only), the upper flt ♀ or rarely ♀, sometimes the first glume reduced or even lacking; several of these genera have spikelets borne in pairs at each joint of the rachis br, one sessile and with a ♀ terminal fl, the other ped and gen ♂ or sterile (rarely ♀), but occ lacking except for the ped

Group I p. 606

1b Spikelets not shed as a unit, but disart above the glumes, the empty glumes persistent on the pl

  2a Ligs mainly a fringe of hairs, sometimes with a membranous base but the terminal fringe at least as long as the membranous portion

Group II p. 608

  2b Ligs mainly or entirely membranous, often ciliate, erose, or lacerate, but any fringe < the membranous portion

    3a Sheaths closed at the base for at least 1/4 their length or lf tips hooded or prow-like, rather than flat (Care must be taken in examining for the 1st feature to distinguish between naturally open sheaths, in which the margins will gen tend to overlap, and naturally closed sheaths that have been ruptured, in which case the margins ordinarily do not overlap, but will fit together along the line of fracture. In *Poa*, the sheaths are open above but gen closed below, one margin sometimes overlapping the other, even though fused with it)

    4a Lvs folded in the bud, the tips prow-like; lemmas unawned

      5a Spikelets sessile or subsessile on one side of the rachis, in 1 or more spikes; pl ann   **Sclerochloa**

      5b Spikelets ped in open or contracted panicles; pl mostly per

        6a Lemmas prominently 3-nerved, the nerves not convergent, the tip erose-truncate; flts mostly 2 per spikelet   **Catabrosa**

        6b Lemmas obscurely to rather prominently 5–∝ -nerved, the nerves convergent, the tip pointed; flts gen > 2 per spikelet

                                                Poa

    4b Lvs rolled in the bud or the tips not prow-like; lemmas sometimes awned

      7a Lemmas unawned, prominently nerved, the nerves not converging at the tip

        8a Lemmas 3-nerved; flts mostly 2 per spikelet   **Catabrosa**

        8b Lemmas 5–9-nerved; flts mostly at least 3 per spikelet

                                            Glyceria

      7b Lemmas either awned or obscurely convergent-nerved (or both)

        9a Spikelets few in a loose raceme, 7–13-fld; lemmas awned, rather prominently parallel-nerved; keels of the palea winged

                                      Pleuropogon

        9b Spikelets pan, often < 7-fld; lemmas often convergent-nerved or unawned; keels of the palea not winged

          10a Callus (the basal, elongated, often pointed portion of the lemma) strongly bearded; lemmas prominently 7–13-nerved,

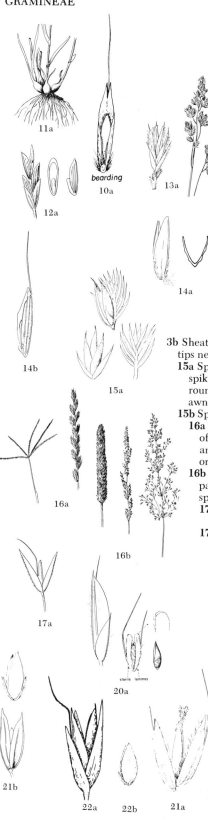

11a

bearding

10a

12a

13a

14a

14b

15a

16a

16b

17a

20a

sterile lemmas

21b

22a   22b   21a

awned from a deeply bifid apex; culms never bulbous at the base **Schizachne**

10b Callus not bearded or if bearded the lemmas obscurely 5–11-nerved or unawned, or both; culms sometimes bulbous at the base

    11a Culms bulbous-based   **Melica**

    11b Culms not bulbous-based

        12a Upper 2–4 flts sterile and represented by empty lemmas that enfold one another; aurs lacking; sheaths often closed their full length; pl per   **Melica**

        12b Upper 1 or 2 flts sometimes sterile but the lemmas not enfolding one another, a palea present in each; aurs sometimes present; sheaths never closed their full length; pls sometimes ann

            13a Spikelets subsessile in dense 1-sided clusters at the tips of comparatively few stiff pan brs; glumes keeled, the 1st lopsided, gen 2 (3)-nerved, stiffly ciliate on the keel, the 2nd narrower, gen 1-nerved, both with a soft awnlike tip; lemmas gen awn-tipped; sheaths compressed-keeled   **Dactylis**

            13b Spikelets not borne in dense 1-sided clusters; glumes and lemmas various

                14a Lemmas keeled on the back, never bifid at the tip, often with cobwebby callus, never awned   **Poa**

                14b Lemmas rounded on the back, often bifid at the tip, not cobwebby on the callus, often awned   **Bromus**

3b Sheaths closed for < 1/4 their length (gen open their full length); lf tips neither hooded nor prow-like

    15a Spikelets of 2 kinds, borne in pairs or clusters in a single, secund, spikelike, terminal pan, gen dimorphic—some fertile and often surrounding others which are sterile and consist only of narrow, empty, awned, scalelike lemmas   **Cynosurus**

    15b Spikelets all alike or at least not dimorphic and in part sterile

        16a Spikelets sessile or subsessile in spikes, either turned to one side of the rachis or in 2 rows on opp sides of it; spikes either single and terminal (if so the lvs often with aurs), or > 1 and racemose or digitately arranged   **Group III p. 609**

        16b Spikelets ped (sometimes very shortly so) in open to contracted pan, the pan (when contracted and pseudospicate) bearing the spikelets all the way around the main axis

            17a Spikelets 1-fld (occ in part 2-fld in *Calamagrostis* and *Cinna*)   **Group IV p. 610**

    17b Spikelets 2–∝ -fld

        18a Spikelets 2- or 3-fld, the uppermost flt ♀ and with both a lemma and a palea, the lower flts either ♂ or sterile, sometimes reduced to small, ± bristle-like remnants (1 of which might readily be mistaken for a prolonged rachilla)

            19a Spikelets 3-fld

                20a Glumes very unequal, the 1st 1-nerved and only ca 1/2 as long as the 2nd; lower 2 flts sterile, the lemmas awned from the back; lods lacking   **Anthoxanthum**

                20b Glumes subequal, both gen 3-nerved; lower 2 flts either ♂ or sterile, but with unawned lemmas; lods gen present

                    21a Lower 2 flts ♂, the lemmas sometimes awned; fertile lemma membranous; aurs lacking   **Hierochloe**

                    21b Lower 2 flts sterile, the lemmas reduced to tiny awnless vestiges; fertile lemma ± coriaceous; aurs gen present   **Phalaris**

            19b Spikelets 2-fld

                22a Lower flt ♂, the lemma awned from the back; culms sometimes bulbous at base   **Arrhenatherum**

                22b Lower flt sterile, the lemma reduced to a tiny awnless vestige; culms not bulbous at base   **Phalaris**

    18b Spikelets 2–∝ -fld, reduced flts (if any) either above, or both above and below the ♀ ones

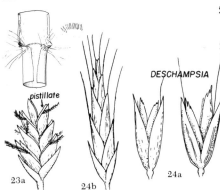

23a

24b

24a

DESCHAMPSIA

pistillate

23a Culms solid; pl strongly rhizomatous per; sheaths with
long stiff hairs near the throat, the collar gen pilose
Distichlis

23b Culms gen hollow, if (as rarely) solid then the sheaths
and throat without long stiff hairs; pls often not rhizomatous

24a Glumes (1 or both) > the 1st lemma; lemmas gen
awned, the awn arising from the back of the lemma or
sometimes from the sinus of a bifid apex, but never from
a pointed tip **Group V p. 611**

24b Glumes (both) < the 1st lemma; lemmas gen awnless
but sometimes awned from the pointed tip or from near
the tip **Group VI p. 612**

Group I

Art below the glumes, i.e., at spikelet base, or along the rachis or the brs of the rachis, the spikelets not breaking apart in the infl

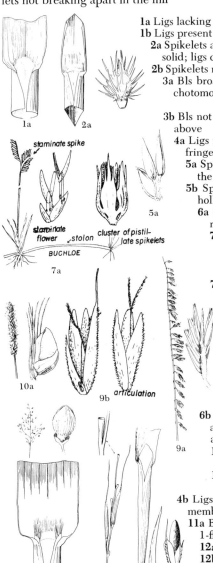

1a

2a

staminate spike

5a

staminate
flower
stolon

cluster of pistil-
late spikelets

BUCHLOE

7a

10a

9b

articulation

9a

10b

12a

13a

1a Ligs lacking **Echinochloa**

1b Ligs present

2a Spikelets aggregated into spiny burs that are shed as units; pl ann; culms
solid; ligs composed mainly of hairs **Cenchrus**

2b Spikelets not aggregated into spiny burs; pls various

3a Bls broad, cordate-clasping at base; pl ann; spikelets sessile in a dichotomously brd infl; known from our area only as a casual intro
Arthraxon hispidus (Thunb.) Makino

3b Bls not cordate-clasping at base; pl ann or per; spikelets not borne as
above

4a Ligs mainly a fringe of hairs, any basal membranous portion < the
fringe

5a Spikelets in 2's or 3's at each joint of the infl, 1 sessile and ☿-fld,
the other(s) ped and ♂; culms solid, rarely < 1 m **Sorghum**

5b Spikelets rarely in 2's or 3's, if so then all alike or the culms
hollow

6a Spikelets sessile in 1–several racemosely or paniculately arranged spikes, borne in 2 rows on 1 side of the rachis

7a Spikelets of 2 kinds, some ♂, 2-fld, and borne severally in
racemosely arranged spikes, others ♀, borne in clusters surrounded by sheathing lvs, and deciduous as a whole; pl per,
♂, ♀, solid-std, sod-forming **Buchloe**

7b Spikelets all alike

8a Pl ann; 1st glume adnate to the rachilla joint; known in
our area only from a collection made on ballast near Portland, Ore **Eriochloa villosa** (Thunb.) Kunth

8b Pl per, gen strongly rhizomatous; glumes well developed,
the 1st one not adnate to the rachilla joint

9a Culms solid; spikelets themselves art above the glumes,
but the entire spikes individually deciduous **Bouteloua**

9b Culms hollow; spikelets art below the glumes and shed
separately, but the spikes not individually deciduous
Spartina

6b Spikelets ped in open to contracted panicles, if subsessile the
apparent spike single and terminal, the spikelets arising all
around the rachis

10a Spikelets subsessile in an erect pan, subtended and gen
exceeded by ∝ sterile, bristle-like brs **Setaria**

10b Spikelets ped in open pan, not subtended by bristles
Panicum

4b Ligs mainly membranous, the terminal fringe, if any, < the basal
membranous portion

11a Both glumes lacking or greatly reduced and vestigial; spikelets
1-fld

12a Flts ♂♀; stamens 6; pl ann, aquatic **Zizania**

12b Flts ☿; stamens 2 or 3; pl either per or nonaquatic

13a Pl ann; lemma 1-nerved, ca 1 mm; stamens 2; rare on sand
bars along the CR **Coleanthus**

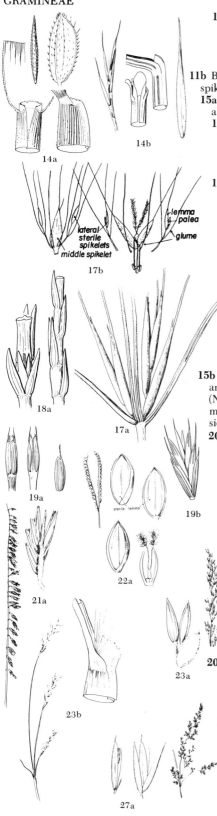

lateral sterile spikelets

middle spikelet

17b

lemma
palea
glume

18a

17a

19a

sterile lemma

19b

22a

21a

23a

23b

27a

13b Pl per; lemma 2- or 5-nerved, 5–10 mm; stamens 3

   14a Pl aquatic, rhizomatous; lemma ca 5 mm, 5-nerved, bristly-pubescent   **Leersia**

   14b Pl nonaquatic, tufted; lemma 7–10 mm, 2-nerved, nonbristly   **Nardus**

11b Both glumes not greatly reduced, at least 1 well developed; spikelets often > 1-fld

  15a Spikelets sessile or subsessile in a single terminal spike, borne alternately on opp sides of the rachis

    16a Spikelets ave 2–3 per node

      17a Spikelets all alike, gen 2 per node   **Sitanion**

      17b Spikelets dissimilar, gen 3 per node, the central one sessile and fertile, the 2 lateral ones ped and gen sterile   **Hordeum**

    16b Spikelets mostly 1 per node

      18a Flts mostly 1 (2) per spikelet; spikelets sunken in the rachis of the spike; glumes asymmetrical, borne in front of the flts; lemmas hyaline, shorter than the glumes, unawned; known in our area only through a collection made on ballast grounds at Albina, near Portland, Ore   **Parapholis incurva** (L.) Hubbard

      18b Flts 2–12 per spikelet; spikelets gen not sunken in the rachis, or glumes symmetrical, or the lemmas membranous and awned

        19a Lemmas toothed at the tip, the nerves not convergent; pl ann; spikelets terete, ± sunken in the rachis   **Aegilops**

        19b Lemmas gen not toothed at the tip, the nerves convergent; pl gen per; spikelets flattened, not sunken in the rachis   **Agropyron**

  15b Spikelets ped in an open to greatly congested pan, or sessile and spicate but only on one side of the rachis, the spikes > 1 (Note—a congested pan is apt to be mistaken for a single terminal spike, but will be seen to have spikelets arising on all sides of the rachis, rather than in 2 opposed rows)

    20a Culms solid; pl per; spikelets either all sessile or in 2's, 1 of each pair sessile

      21a Spikelets sessile, in small spikes, the spikes ∝ in a terminal raceme, ultimately separately deciduous   **Bouteloua**

      21b Spikelets sometimes in part ped, if all sessile the spikes gen not racemose and not separately deciduous

        22a Fertile (upper) lemma hardened, firmer than the lower (sterile) one and the glumes; spikelets all sessile in 2 rows on 1 side of the rachis in 2 (ours) terminal spikes   **Paspalum**

        22b Fertile (upper) lemma less firm than either the lower one or the glumes; spikelets in 2's in ∝ small racemes, 1 of each pair sessile, the other ped

          23a Bls in part much > 10 mm broad; racemes 1–5-jointed, borne in a large pan; lig fringed   **Sorghum**

          23b Bls rarely as much as 10 mm broad; racemes gen > 5-jointed, mostly digitate or single at each node of the rachis; lig almost entirely membranous   **Andropogon**

    20b Culms hollow; pls often ann; spikelets often all ped

      24a Callus long-bearded, the hairs at least twice as long as the glumes; rhizomatous per, collected near Portland, Ore, but prob not estab anywhere in our area   **Imperata cylindrica** (L.) Beauv.

      24b Callus not long-bearded, the callus hairs (if any) rarely > half as long as the glumes; pls various

        25a Spikelets ped in open to contracted pan, not confined to 1 side of the rachis

          26a Spikelets 1-fld (some of spikelets occ 2-fld in *Cinna*); sterile lemmas lacking

            27a Pan loose and open; lemma awnless to shortly awn-tipped; glumes unawned   **Cinna**

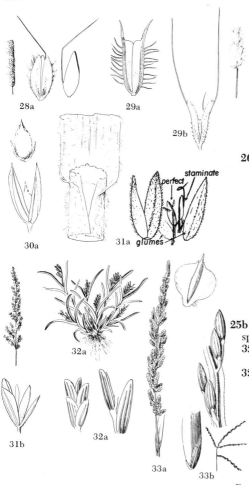

27b Pan contracted and ± spikelike; lemma often long-awned; glumes often awned

  28a Glumes unawned; lemma awned from the back   **Alopecurus**

  28b Glumes terminally awned; lemma unawned or subterminally awned

    29a Glumes ± stiffly long-ciliate on the keel, the awn stout, scarcely 3 mm; lemma unawned   **Phleum**

    29b Glumes not stiffly long-ciliate on the keel, the awn 3–10 mm; lemma awned subterminally   **Polypogon**

26b Spikelets 2–3-fld, 1–more of the flts sometimes sterile or ♂

  30a Spikelets 3-fld, terminal flt ♀, the lower ones represented by tiny bristle-like vestiges of the lemmas   **Phalaris**

  30b Spikelets 2-fld, lower flt ♀, the upper one sometimes ♂

    31a Glumes alike in width, > either lemma; lower lemma unawned; upper flt ♂, the lemma gen short-awned   **Holcus**

    31b Glumes very unlike in width, the 2nd much broader than the 1st, from slightly < to slightly > the 1st lemma; upper flt gen ♀; neither lemma awned   **Sphenopholis**

25b Spikelets sessile or subsessile in spikelike racemes or spikes, confined to 1 side of the rachis

  32a Spikelets 3-fld, the uppermost flt sterile; glumes 3–7-nerved   **Sclerochloa**

  32b Spikelets 1-fld or if 2-fld the lower flt gen sterile; glumes gen 1- or 3-nerved

    33a Glumes =, keeled-inflated, semicircular in outline, abruptly acute; spikelets 1-fld, the lemma less hardened than the glumes   **Beckmannia**

    33b Glumes unequal, the 1st greatly reduced; spikelets 2-fld, only the upper flt fertile, its lemma firmer than the glumes and the sterile lemma   **Digitaria**

Group II

Spikelets disart above the glumes; lig mainly a fringe of hairs, sometimes with a membranous base but the fringe at least = the membranous portion

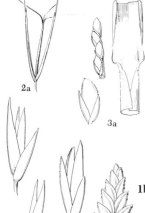

1a Spikelets 1-fld; sterile lemmas lacking

  2a Lemma with a prominent, terminal, trifid awn   **Aristida**

  2b Lemma unawned or with a small, simple awn

    3a Spikelets in several digitate spikes, all sessile in 2 rows confined to one side of the rachis   **Cynodon**

    3b Spikelets in open to contracted pan, not confined to one side of the rachis

      4a Callus with copious long hairs at least half as long as the lemma   **Calamovilfa**

      4b Callus not obviously bearded, the hairs (if any) < 1/4 as long as the lemma

        5a Pl ann, 5–15 cm; culms hollow; infl a dense spikelike pan 1–3 cm × 5 mm, gen well exserted from sheath at maturity; spikelets ca 2 mm   **Heleochloa**

        5b Pl ann or per, if ann gen > 15 cm, culms gen solid or pith-filled, and infl an open to contracted pan often largely contained in sheath; spikelets mostly > 2 mm   **Sporobolus**

1b Spikelets > 1-fld, gen with at least 2 ♀ flts, but if with only 1 ♀ flt then with 1 or 2 sterile lemmas

  6a Pl ann; lemmas 3-nerved, the nerves prominent

    7a Spikelets in an open pan, not closely subtended by the lvs; nerves not protruding at the tip of the lemma   **Eragrostis**

    7b Spikelets aggregated, 2 or 3 per node, on one side of a short rachis,

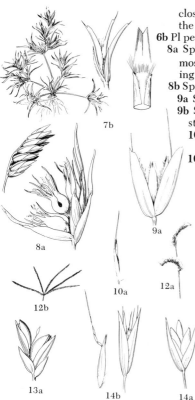

closely subtended by foliage lvs; at least 1 of the nerves protruding at
the tip of the lemma   **Munroa**
6b Pl per; lemmas often > 3-nerved, the nerves mostly obscure
  8a Spikelets of 2 kinds (♂ and ♀), the ♂ 2-fld, borne severally in race-
    mosely arranged 1-sided spikes, the ♀ in clusters surrounded by sheath-
    ing lvs; pls ♂, ♀   **Buchloe**
  8b Spikelets all alike; pls not ♂, ♀
    9a Spikelets 3-fld, the 2 lower flts ♂, the uppermost ♀   **Hierochloe**
    9b Spikelets 2–∝ -fld, the fls gen either all ♀ or the uppermost ♂ or
    sterile, but rarely the lowest lemma sterile and glume-like
      10a Hairs of the rachilla = or > the lemmas; stout reeds gen > 1.5
      m   **Phragmites**
      10b Hairs of the rachilla (if any) much < the lemmas; pls gen < 1.5
      m
        11a Spikelets in racemose or digitately arranged spikes, sessile in
        2 rows confined to one side of the rachis
          12a Spikes racemose; spikelets with 1♀ flt below 1 or 2 promi-
          nently 3-awned, sterile ones   **Bouteloua**
          12b Spikes digitate; spikelets mostly 1-fld, the sterile lemmas, if
          any, not 3-awned   **Cynodon**
        11b Spikelets ± pan
          13a Glumes < the 1st lemma (in the comparison of length of
          glumes and lemmas, only the length of the body is consi-
          dered, exclusive of any awn); lemmas awnless, not bifid
                                                                      Molinia
          13b Glumes > the 1st lemma; lemmas awned or mucronate
          from a slightly to strongly bifid apex
            14a Lemmas rather shallowly trifid, not truly awned
                                                                        Sieglingia
            14b Lemmas deeply bifid, with a large, twisted, geniculate
            awn from just below the sinus   **Danthonia**

Group III

Spikelets sessile or subsessile in 1 or more spikes, disart above the glumes; lig mainly or entirely
membranous; lvs often aur

1a Spikes single and terminal, the spikelets (except in *Nardus*) not confined to
one side of the rachis; aurs often present
  2a 1st glume lacking, the 2nd sometimes very small; spikelets either turned
  with their edges toward the rachis (rather than flatwise) or confined to
  one side of the rachis and partially sunken in it
    3a Spikelets several-fld, turned with one edge to the rachis, the 1 glume
    well developed; aurs gen present   **Lolium**
    3b Spikelets 1-fld, not obviously turned with one edge to the rachis, the 1
    glume minute; aurs lacking   **Nardus**
  2b 1st glume as well as the 2nd gen present; spikelets borne alternately on
  opp sides of the rachis and flatwise to it
    4a Spikelets ave 2 or 3 per node
      5a Spikelets gen (1) 2 per node, all alike and each gen with at least 2
      fertile fls
        6a Spikelets 2 at almost all nodes, 2–6-fld; glumes 1–5-nerved, nar-
        row, entire to cleft   **Elymus**
        6b Spikelets 1 at many nodes, 3–12-fld; glumes 3–9-nerved, gen ±
        lanceolate   **Agropyron**
      5b Spikelets gen 3 per node, the middle one sessile and with only 1
      fertile flt, the other (lateral) 2 ped and gen sterile   **Hordeum**
    4b Spikelets ave 1 per node
      7a Stamens 1; lemmas < the glumes, short-awned from a bifid apex;
      spikelets 1-fld   **Scribneria**
      7b Stamens 3 (2); lemmas gen = or > the glumes, awnless or awned
      mostly from the tip; spikelets mostly 2–∝ -fld
        8a Glumes stiff, subulate, 1-nerved; spikelets 2-fld   **Secale**
        8b Glumes gen at least 3-nerved; spikelets often > 2-fld
          9a Lemmas slightly toothed as well as gen awned, the nerves
          prominent, nonconvergent; pl ann
            10a Spikelets subterete, not flattened, fitting into the curvature

7b

8a

9a

12a

10a

12b

13a

14b

14a

3a

3b

6a

5b

6b

7a

8a

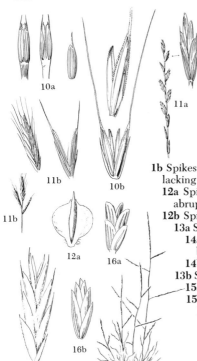

of the rachis; glumes and lemmas both rounded on the back
Aegilops
10b Spikelets flattened, not fitting into the rachis; glumes and
gen also the lemmas keeled   **Triticum**
9b Lemmas not toothed, awnless or awned, the nerves gen rather
obscure, convergent toward the tip; pls mostly per
11a Pl per; aurs gen present; spikelets mostly sessile and
obviously borne in 2 rows on opposite sides of the rachis in a
spike; glumes 1–9-nerved; common   **Agropyron**
11b Pl ann or per; aurs lacking; spikelets subsessile to
short-pedicellate, gen not borne in 2 clearly marked rows on
opposite sides of the rachis; glumes 5–7-nerved; very occ,
mainly (only?) in w Ore in our area   **Brachypodium**
1b Spikes gen > 1, the spikelets gen confined to one side of the rachis; aurs
lacking
12a Spikelets 1 (2)-fld; glumes strongly compressed, ca =, semicircular,
abruptly acute; spikes ∝, racemose-pan   **Beckmannia**
12b Spikelets often > 2-fld; glumes never semicircular; spikes often few
13a Spikelets 1-fld; lf bls white-margined; sheaths keeled
14a Pl ann, stoloniferous; lemmas awned; known in our area only from
ballast ground near Portland, Ore   **Chloris radiata** (L.) Swartz
14b Pl per, caespitose; lemmas unawned   **Schedonnardus**
13b Spikelets 2–∝ -fld; lf bls rarely white-margined; sheaths not keeled
15a Lemmas short-awned from a bifid apex   **Leptochloa**
15b Lemmas unawned
16a Glumes 3–7-nerved; spikelets 3-fld; lf bls often ± prow-shaped
at the tip (see lead 5a, Key to Main Groups, where this genus
normally will separate)   **Sclerochloa**
16b Glumes apparently only 1-nerved; spikelets > 3-fld; lf bls not
prow-shaped at the tip   **Eleusine**

Group IV

Spikelets 1-fld, disart above the glumes, ped in open to contracted pan; ligs mainly or entirely
membranous

1a Lemma with a terminal trifid awn   **Aristida**
1b Lemma unawned or with a simple awn
2a Glumes =, > the lemma, strongly compressed-keeled, truncately narrowed at the tip and with the midnerve ± extended into an abrupt
point; lemma awnless; spikelets in a cylindric, spikelike pan; culms
sometimes bulbous at base   **Phleum**
2b Glumes unequal, or < the lemma, or not strongly compressed-keeled,
mostly gradually tapered; lemma sometimes awned; spikelets often in
open pan; culms never bulbous at base
3a Lemma hardened, much firmer than the glumes, gen terete to obcompressed, awned; callus sometimes hard and sharp
4a Awn deciduous, bent but mostly not twisted; lemma plump, rarely
> 5 times as long as thick; callus rarely at all sharp   **Oryzopsis**
4b Awn persistent, gen twisted as well as bent; lemma mostly at least
6 times as long as thick; callus often hard and sharp   **Stipa**
3b Lemma membranous, gen no firmer than the glumes, ± compressed,
often unawned; callus not hard and sharp
5a Pl ann; glumes 1-nerved, 2–4 times as long as the lemma, swollen,
rounded, and ± saccate at base, the tips linear-lanceolate; lemma
pubescent, much < the terminal awn; pan spikelike   **Gastridium**
5b Pl ann or per; glumes often > 1-nerved, mostly < the lemma, but
occ as much as 2–3 times their length, gen tapered from near the
base; lemma often glab, unawned or short-awned; pan often open
6a Callus with bearding 1/4 as long to = the lemma; lemma awnless or awned from midlength or below; rachilla prolonged behind
the palea as a bristle mostly (0.5) 1–2 mm; palea well developed,
gen subequal to the lemma; glumes 3.5–12 mm
7a Lemma unawned, 8–10 mm; pl strongly rhizomatous; ligs
10–24 mm   **Ammophila**
7b Lemma awned, < 7 mm; pl often caespitose; ligs gen < 10
mm   **Calamagrostis**

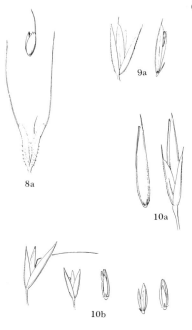

8a

9a

10a

10b

6b Callus not bearded or with bearding mostly < 1/4 as long as the lemma or rarely with the bearding longer, but then the lemma unawned or awned from above midlength; rachilla rarely prolonged as much as 0.5 mm; palea often < 1/2 as long as the lemma, sometimes lacking; glumes often < 3.5 mm (both *Cinna* and *Polypogon* have spikelets art below the glumes, and should separate under lead 26a in Group I, but this feature cannot be detected readily in fresh material)

    8a Glumes 2–3 × as long as the lemma, terminally awned, the awn slender and at least = the body of the glumes; lemma erose, awned from near the tip   **Polypogon**

    8b Glumes often < 2 × as long as the lemma, often short-awned or awnless; lemma not both erose and awned from near the tip

        9a Flt short-stipitate; rachilla prolonged, bristle-like, ca 0.6 mm; lemma gen with the vestige of a subapical awn; lf bls 7–15 mm broad; palea subequal to the lemma   **Cinna**

        9b Flt not stipitate; rachilla rarely prolonged and bristle-like; lemma often awnless or awned from the back; lf bls often < 7 mm broad; palea often much < the lemma

            10a Lemma mostly terminally awned or awn-tipped, if unawned then considerably > the glumes; rachilla not prolonged behind the palea; culms often solid; palea well developed, mostly subequal to the lemma   **Muhlenbergia**

            10b Lemma unawned or awned from the back, if unawned then often < the glumes; rachilla sometimes prolonged; culms never solid; palea often reduced or lacking **Agrostis**

Group V

Spikelets 2–∝-fld, art above the glumes, borne in open or contracted pans; ligs mainly or entirely membranous; 1 or both glumes > 1st lemma; lemmas occ awnless but gen awned from the back or from the sinus of a bifid apex, never from a pointed tip

1a Spikelets never > 2-fld or > 1 cm; 1 or both lemmas awned

    2a Pl ann, delicate; rachilla not prolonged beyond the 2nd flt   **Aira**

    2b Pl per; rachilla often prolonged beyond the 2nd flt

        3a Lemmas 1-nerved, awned from near the base, the awn segmented near midlength, the lower segm reddish, the upper segm greenish   **Corynephorus**

        3b Lemmas 4–5-nerved, awned from near midlength or above (in *Calamagrostis* sometimes below), the awn not segmented

            4a Lower fl ♀, the lemma awnless; upper fl ♂, the lemma awned from near the tip; culms pubescent at the nodes (spikelets ultimately disart below the glumes, and therefore *Holcus* separable under lead 31a, Group I, but this feature often overlooked with fresh material)   **Holcus**

            4b Lower fl similar to the upper; culms mostly glab

                5a Lemmas erose-truncate, awned from (and rounded on) the back; rachilla prolonged   **Deschampsia**

                5b Lemmas acute or bifid, but not erose at the tip, often awned from the sinus, gen keeled on the back; rachilla sometimes not prolonged

                    6a Lemmas bifid at the tip and gen awned from the sinus, but sometimes awnless; rachilla prolonged beyond the 2nd flt   **Trisetum**

                    6b Lemmas not bifid at the tip, always awned, the awn mostly from midlength or below; rachilla not prolonged beyond the 2nd flt (it is only rarely that *Calamagrostis* will key here, since the spikelets are almost always 1-fld)   **Calamagrostis**

1b Spikelets > 2-fld or > 1 cm or lemmas unawned

    7a Spikelets ave > 1 cm

        8a Glumes ave at least 20 mm, 7–9-nerved   **Avena**

        8b Glumes ave 10–15 mm, 3–5-nerved   **Helictotrichon**

    7b Spikelets ave < 1 cm

        9a Pl per, strongly rhizomatous; lemmas awnless, not bifid   **Scolochloa**

        9b Pl ann or a caespitose per; lemmas gen awned or bifid

            10a Lemmas erose-truncate at the tip, awned from the back   **Deschampsia**

2a

3a

2a

4a

2a

6b

5a

6a

8b

8a

10a

9a

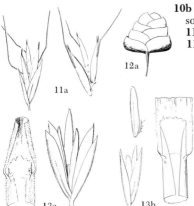

10b Lemmas acute or bifid at the tip, mostly awned from the sinus, but sometimes awnless
  11a Lemmas with an exserted, bent awn   **Trisetum**
  11b Lemmas unawned or with a short, straight awn
    12a Spikelets strongly compressed and ± cordate-ovate in outline; glumes spreading, broadly and deeply prow-like at the tip, and although gen slightly > the 1st lemma, neither covering nor largely concealing it; pl ann   **Briza**
    12b Spikelets neither strongly compressed nor cordate-ovate in outline; glumes not prowlike at the tip, gen largely concealing the 1st lemma; pl gen per
      13a Ligs mostly 0.5–2 (rarely 3) mm; pan greatly condensed; lf bls folded in the bud; rachilla joints glab   **Koeleria**
      13b Ligs (2.5) 3–4 mm; pan somewhat open; bls rolled in the bud; rachilla joints bearded   **Trisetum**

Group VI

Spikelets 2–∞-fld, art above the glumes, borne in open to contracted pan; ligs mainly or entirely membranous; both glumes < 1st lemma; lemmas either awnless or terminally or subterminally awned

1a 1st lemma entire, awned from the tip; 2nd lemma deeply bifid, awned from the back; glumes 6–9-nerved; callus bearded   **Ventenata**
1b 1st lemma essentially like the 2nd; glumes often only 1- or 3 (5)-nerved; callus mostly not bearded
  2a Spikelets cordate-ovate in outline, snail-shell-like, < 5 mm; lemmas awnless, cordate-based, saccate, papery, horizontally spreading; pan open; pl ann   **Briza**
  2b Spikelets neither cordate-ovate in outline nor shell-like, often > 5 mm; lemmas various; pan often compressed; pl often per
    3a Nerves of the lemma 5–7, parallel and not convergent above at the tip of the lemma, gen fairly prominent; lemmas unawned
      4a Pl ann; glumes 3- and 7-nerved respectively (except for atypical specimens, this genus will ordinarily separate elsewhere in the key)   **Sclerochloa**
      4b Pl per; glumes mostly 1- and 3-nerved, respectively   **Puccinellia**
    3b Nerves of the lemma often other than 5–7, tending to converge at the tip of the lemma, gen obscure; lemmas often awned
      5a Lf bls folded in their immature state in the innovations; lvs often turned up at the tip and hooded or ± like the prow of a canoe (*Poa* will ordinarily be separated earlier in the key, under lead 6b, Key to Main Groups)
        6a Sheaths compressed-keeled; spikelets borne in 1-sided clusters on long, stiff, ascending pan brs; pl per, the lf tips not prow-like; lemmas awn-tipped   **Dactylis**
        6b Sheaths gen terete or only slightly flattened, if keeled the lf tips prow-like, the pan open to diffuse, or the lemmas not awn-tipped
          7a Lemmas rounded on the back; rachilla joints persistent and stipelike on the individual flts; pl ann   **Scleropoa**
          7b Lemmas mostly ± keeled on the back; rachilla joints not forming a stipe on the individual flts; pl ann or per
            8a Lemmas unawned; lf tips gen prow-like; mostly the sheaths glab or pl rhizomatous   **Poa**
            8b Lemmas often awn-tipped; lf tips not prow-like; sheaths pubescent; pls never rhizomatous   **Koeleria**
    5b Lf bls rolled in their young state in the innovations; lf tips pointed, not prow-like
      9a Aurs well developed on at least some lvs
        10a Pl ann   **Elymus**
        10b Pl per
          11a Spikelets pan, all ped   **Festuca**
           11b Spikelets semispicate, at least some sessile   **Elymus**
      9b Aurs lacking
        12a Pl ♂, ♀, per, caespitose; lemmas awnless   **Hesperochloa**
        12b Pl mostly ♀-fld, often ann or rhizomatous; lemmas sometimes awned

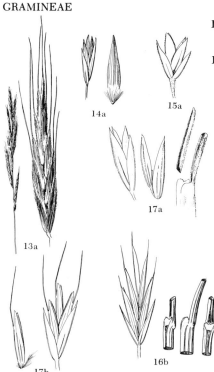

13a Spikelets 1–5 (10) per culm, subsessile in a terminal raceme, 7–13-fld; glumes 5–7-nerved, hardened
                                  Brachypodium

13b Spikelets gen > 5 per culm, borne in spikelike to open pans; glumes rarely > 3-nerved

  14a Callus strongly bearded; 1st glume 3-nerved, the 2nd 5-nerved; lemmas awnless; pl per, rhizomatous **Scolochloa**

  14b Callus rarely if ever bearded; glumes gen 1- and 3-nerved, respectively; lemmas often awned; pl ann or per, seldom rhizomatous

    15a Spikelets never > 2-fld, the rachilla not prolonged beyond the 2nd fl; culms often solid; lemmas not bifid (this genus is normally 1-fld and nearly all specimens will be separated under lead 10a, Group IV)
                                  Muhlenbergia

    15b Spikelets (at least in part) with > 2 fls, or the rachilla prolonged; culms never solid; lemmas sometimes bifid

      16a Lemmas ± bifid at the tip, the awn (if any) arising from the sinus between the teeth and therefore never truly terminal; lig rarely higher on the sides than at the back (opp the throat)

        17a Lig mostly 0.5–2 (rarely to 3) mm; rachilla joints glab   **Koeleria**

        17b Lig (2.5) 3–4 mm; rachilla joints bearded
                                    Trisetum

      16b Lemmas not bifid, the awn (if any) terminal; lig often higher on the sides than at the back   **Festuca**

Aegilops L.   Goatgrass

Infl a rather compact, terminal, gen cylindric spike tardily disart near the base and sometimes also at the upper nodes; spikelets 1 per node, flatwise to the rachis, 2–5-fld; glumes both present, gen awned; lods 2; stamens 3; ann with hollow culms, narrow, flat, aur or non-aur, gen hairy bls, short membranous ligs, and open sheaths. (The ancient Gr name).

A. cylindrica Host. Ann, (1.5) 3–5 (7) dm; sheaths glab to gen sparsely long-hirsute; bls flat, 2–3 mm broad, gen long-hirsute; lig ca 0.5 mm; spikes cylindrical, 4–10 cm, the spikelets sunken into the curvature of the rachis; glumes 10–14 mm, prominently 6–8-veined, ± lopsidedly stronger-nerved or keeled into a single awn 0.5–3 (8) cm, the abaxial tip with a short toothlike lobe; lemmas unawned or the upper ones awned and lobed as the glumes; European weed, occ in much of w US, but known in our area only from Mont and se Wn and adj Ida, chiefly as a weed in wheat *(Triticum c.)*. A. ovata L. and *A. triuncialis* L., with several-awned glumes, are well estab in parts of w US, but are not known to occur in our area.

Agropyron Gaertn.   Wheatgrass

Spikelets in a terminal spike, gen sessile and solitary but occ 2 or rarely even 3 per node (central spikelet then sometimes short-talked), borne flatwise to the continuous or rarely tardily disart rachis, (3) 4–12-fld; rachilla disart above the glumes; glumes = or subequal, 3–9 (rarely 1–2) -nerved, acute or acuminate (occ obtuse or rounded), awnless to short-awned, mostly subequal to the lemmas; lemmas rounded on the back, very indistinctly to strongly several-nerved, occ obtuse but gen acute, awnless to strongly awned; lods 2; stamens 3; anns or (as all our native spp.) caespitose to strongly rhizomatous pers with hollow (rarely pithy) culms, open sheaths, flat to involute bls rolled in the bud, mostly well-developed aurs, and short, membranous or membranous-ciliolate ligs. (Gr *agrios*, wild, and *pyros*, wheat). Our native per spp. are mostly valuable range grasses, esp A. *spicatum*. Several are important soil binders, but 1 of the rhizomatous spp., A. *repens*, is among the most troublesome weeds with which gardeners have to contend.

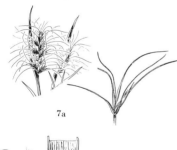

1a Pl ann, 1–3 dm; spikes ovate-oblong, 1–1.5 cm; aurs very slender, nearly 1 mm; lig erose-fimbriate, ca 1 mm; bls flat, 2–3 mm broad; spikelets 11–19, greatly crowded, divergent, 6–8 mm; a weedy sp. from Eurasia, sparingly intro in drier regions of w N Am, known from Wn, Ore, Ida, and Mont
1 **A. triticeum** Gaertn.

1b Pl per; spikes gen slender, at least 2 cm

    2a Spike flexuous; bls flat, 5–10 mm broad; lemma awns 2–3 cm; lig long-ciliate; known only as an intro on ballast near Portland, Ore
A. semicostatum (Steud.) Nees

    2b Spike stiff, not flexuous; bls various; lemmas awned or awnless; lig not long-ciliate

        3a Pls bunch grasses, rhizomes lacking or very poorly developed

            4a Spikelets crowded, at least some as much as 4✕ as long as the internodes of the rachis, strongly divergent; pl 3–10 dm; aurs slender, ca 1 mm; bls flat, gen pubescent above, 2–7 (10) mm broad; glumes and lemmas with awns 2–4 mm; widely intro from the Old World for forage purposes and often estab (*A. pectiniforme* R. & S. [*A. desertorum* Fisch.] and *A. sibiricum* Willd. are closely related spp. also intro for forage, but scarcely distinguishable from *A. cristatum*); crested w.
2 **A. cristatum** (L.) Gaertn.

            4b Spikelets distant to crowded, rarely > 3✕ as long as the internodes of the rachis (exclusive of the awns), not strongly divergent

                5a Glumes and lemmas blunt or rounded; anthers 5–6 mm; pl gen rhizomatous; bls stiff, involute, glab or subglab, 2–5 mm broad; aurs large; spikes slender, 1–2 dm, rachis not disart; European, intro for forage purposes and occ escaping; collected in Mont and in se Wn; intermediate w.
3 **A. intermedium** (Host) Beauv.

                5b Glumes acute to acuminate, or blunt or rounded but anthers < 5 mm; native grasses, gen abundant

                    6a Anthers ca 5 (4–6) mm; spikelets gen 1 per node, rather distant, from < to only slightly > the internodes of the rachis; pl (3) 4–10 dm, glab to puberulent, gen strongly caespitose, but occ ± rhizomatous; spikes loose, 8–15 (20) cm; lemmas 7–11 mm, glab to puberulent, awnless to strongly and divergently awned; widespread E Cas, Alas to Cal, e to Alta, Daks, and NM, plains to middle elev in mts; bluebunch w.; 2 vars.
4 **A. spicatum** (Pursh) Scribn. & Smith

                        a1 Pl glab to finely puberulent-strigillose overall; range of the sp. as a whole (*A. divergens, A. d.* var. *tenuispicum, A. s.* var. *inerme, A. vaseyi*)
var. **spicatum**

                        a2 Pl copiously short-pubescent overall; Wen Mts, Chelan and Kittitas cos, Wn (*A. s.* ssp. *puberulentum*); occ pls of *Elymus ambiguus* var. *salinus,* from Salmon R C, c Ida, may key here also, but differ in having 1-nerved, broader glumes
var. **pubescens** Elmer

                  6b Anthers 1–2.5 mm; spikelets crowded, mostly 2–3 ✕ > the internodes of the rachis

                    7a Rachis tending to disart at maturity; glumes very narrowly lanceolate, mostly 1–3 (5)-nerved, ending in a slender, gen divergent awn at least 1 cm; lemmas attenuate to a strongly divergent awn 2–3 (4) cm, glab to scab; mont, mostly above timberl, gen on open slopes or talus; s BC and n Cas, Wn, s through e Ore to Ariz and NM, e to Mont and Wyo; spreading w. (*A. bakeri, Elymus s., Sitanion marginatum*)
5 **A. scribneri** Vasey

                    7b Rachis not disart; glumes narrowly to broadly oblong-elliptic, (3) 4–7-nerved, acute to slender-awned; lemmas glab to soft hairy, awnless or with straight to geniculate awn up to 2.5–3 cm; from lowl to high in the mts, Alas to Mex, e to Atl, common in much of inl N Am; Eurasia; awned or bearded w.; several vars.
6 **A. caninum** (L.) Beauv.

                      a1 Glumes narrow and gen rather rigid, 3 (4–5)-nerved; anthers ave 2 mm; Eurasian, perhaps occ in N Am as an escape (reported from Portland, Ore); cutting w.
ssp. **caninum**

                      a2 Glumes mostly rather broad and ± membranous-margined, gen not rigid, mostly 4–7 (not rarely 3)-nerved;

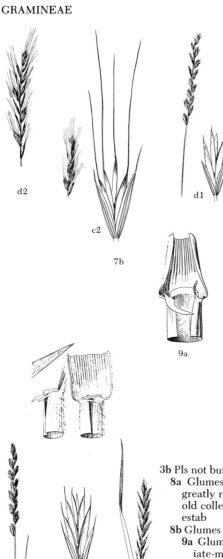

d2

c2

7b

10a

12a

d1

9a

anthers 1–1.6 mm; native to N Am; 4 vars. of
ssp. **majus** (Vasey) Hitchc.

b1 Spikelets 2.5–3× > the internodes, the spikes therefore broad and compact, gen 4–8 (10) cm; lemmas and awns often purplish; pl gen in mts, where often alp, mostly 2–5 (6) dm; aurs gen lacking

   c1 Lemmas awnless or with awn up to 5–6 mm; spikes often purplish, glab to soft-pubescent; pl mont to alp, Yuk s through e Wn and Ore to Cal, e to Que and Me, and s through RMS; broadglumed w. *(Triticum pauciflorum; A. p.,* misapplied; *Triticum trachycaulum; A. t.; A. trachycaulum* vars. *hirsutum* and *pilosiglume; A. violaceum* var. *latiglume; A. l.; A. trachycaulum* var. *l.)*
var. **latiglume** (Scribn. & Smith) Hitchc.

   c2 Lemmas with relatively long awns (mostly 8–20 mm); pls mont at lower elev; Yuk to Sierran Cal, e to RMS s to Colo, Utah, and Ariz; bearded w. *(Triticum subsecundum, A. s., A. caninum* var. *pubescens, A. violaceum* var. *a., A. a.)*
var. **andinum** (Scribn. & Smith) Hitchc.

b2 Spikelets gen (1) 2–2.5 × as long as the internodes, the spike relatively slender, 8–20 (2–25) cm; lemmas and awns rarely purple; lowl into the mts, but not primarily mont, gen much > 5 dm; aurs gen present

   d1 Lemmas unawned or with awns up to 5 mm; Alas to Cal, e to Newf, Mich, Minn, and NY and s to NM and n Mex; slender w. *(A. tenerum, A. repens* var. *t., A. trachycaulum* var. *t., A. t.* var. *longifolium, A. violaceum* var. *major, A. tenerum* ssp. *m., A. trachycaulum* var. *m.)*
var. **majus**

   d2 Lemmas with straight to divergent awns 10–25 (6–30) mm; Yuk to Wn and Ore, e to Me, Ia, Minn, NY, and Utah and Colo *(A. richardsoni* var. *ciliatum, A. trachycaulum* var. *c., A. u., A. trachycaulum* var. *u., A. c.* f. *violacescens, A. v.)*
var. **unilaterale** (Vasey) Hitchc.

3b Pls not bunch grasses, rhizomes present and gen well developed

8a Glumes 9-nerved; lowest 2–5 lvs with partially closed sheaths, greatly reduced bl (if any), and no lig; known in our area only from old collections from near Portland, Ore, where intro and prob never estab
A. junceum (L.) Beauv.

8b Glumes gen (1) 3–7-nerved; all lvs gen with bl and lig

   9a Glumes blunt; lemmas rounded to blunt; sheaths strongly ciliate-margined (see lead 5a) 3 **A. intermedium** (Host) Beauv.

   9b Glumes acute to acuminate; lemmas mostly acute; sheaths ciliate or nonciliate; pls mostly native or well estab intro

      10a Bls normally flat, 5–10 mm broad; lower sheaths gen hirsute-pilose; pl gen not strongly glaucous; lemma awn (if any) straight, up to 10 mm; intro from Eurasia, widespread, esp common in lawns, waste areas, and meadowl, not occurrent on dry des; quack, quitch, or couch grass, a bad weed *(Elymus r., A. leersianum* sensu Rydb.) 7 **A. repens** (L.) Beauv.

      10b Bls normally strongly involute or considerably < 5 mm broad; lower sheaths gen glab or strigillose (sometimes hirsute-pilose); pl often strongly glaucous; awn of lemma, if any, often divergent

         11a Culms pith-filled; pls intro in our area where known only from an old collection at Linnton, near Portland, Ore
A. pungens (Pers.) R. & S.

         11b Culms hollow, not pith-filled; pls native and widespread E Cas

            12a Spikes loose and open, some of the spikelets gen < internodes of the rachis; rachilla gen plainly visible; lemmas glab or puberulent, never pubescent; anthers ca 5 (4–6) mm; rhizomes short (see lead 6a)
4 **A. spicatum** (Pursh) Scribn. & Smith

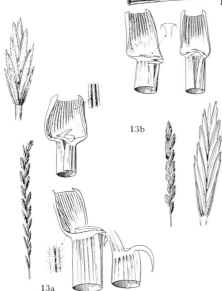

12b Spikes compact, the upper spikelets gen 2–3 × as long
  as the internodes of the rachis; rachilla gen almost concealed; lemmas often pubescent; anthers 3–5 mm; rhizomes gen extensive
13a Glumes mostly 5–7-nerved, oblong and lanceolate,
    broadest at or above midlength, < 1st lemma, rarely
    awn-tipped, gen hairy (glab); lemmas gen copiously
    hairy; lower sheaths finely strigillose to pilose; anthers
    ca 5 mm; sand dunes to heavy alkaline flats from sagebr
    des or grassl into lower levels of the mts; BC to ne Cal,
    e Cas, e to Alta, Mich, Ill, Colo, and Nev; thick-spiked
    w., downy w. (*A. albicans, A. a.* var. *griffithsii, A. g., A.
    elmeri, A. lanceolatum, A. pseudorepens, A. subvil-
    losum, A. riparium*) 8 **A. dasytachyum** (Hook.) Scribn.
13b Glumes mostly 3–5-nerved, lanceolate, tapered from
    near base, often = the 1st lemma, frequently
    awn-tipped, mostly glab but sometimes pubescent;
    lemmas glab to moderately pubescent; lower sheaths
    often glab; anthers 3–4 (4.5) mm; grassl and sagebr des,
    often on heavy moist soil, into the foothills and lower
    mts; BC s, e Cas, to Ore, Nev, and Ariz, e to Ont, NY,
    Tenn, and Tex; bluestem w. (*A. spicatum* var. *molle*,
    A. m., A. smithii* var. *m.*)   9 **A. smithii** Rydb.

Agrostis L.   Bentgrass

Spikelets 1–fld, rather small, art above the glumes, in open to contracted pan; glumes = to unequal, bluntly acute to acuminate or aristate, 1-nerved, gen scab on the keel; lemmas from slightly
to considerably < the glumes, thin, (3) 5-nerved, the nerves often projecting into minute teeth,
the midnerve sometimes extending into a subbasal to subterminal, fragile to rather stiff, straight to
bent and sometimes twisted awn; callus from minutely bearded (rarely glab) to bearded with hairs
half as long as the lemma; palea lacking, rudimentary, or well developed and subequal to the
lemma, nerveless or 2-nerved; rachilla sometimes prolonged as a short stub or bristle behind the
palea; stamens 3; lods 2; anns or tufted to rhizomatous or stoloniferous pers with hollow sts, open
sheaths, membranous and ± puberulent ligs, and flat to folded or involute bls rolled in the bud.
(Gr name for some grass). (*Anemagrostis, Apera, Podagrostis*). One of the more important genera
of range grasses; native spp. comprise an appreciable element of most mt meadows, and certain
intro spp., esp. *A. alba* and *A. tenuis*, are gen important elements in permanent pastures, meadows, and lawns.

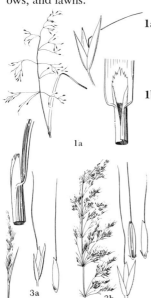

1a Lemma awned from just above the base (lowest fifth), the awn twisted,
  bent, ca 5 mm, well exserted from the glumes; pl lax, tufted, 3–4 (6) dm; lig
  2–3.5 (4.5) mm, very delicate, blunt to acute, erose, and gen strongly lacerate; pan open, 1–2 (3) dm, the brs spreading, flexuous; palea lacking; on
  moist rocks, s side CRG, Hood River and Multnomah cos, Ore; Howell's
  b.   1 **A. howellii** Scribn.
1b Lemma awnless or awned from near midlength or above, the awn often
  straight and < the glumes
  2a Pl ann; lemma subapically awned, the awn 4–7 mm; palea nearly = the
    lemma, 2-nerved
    3a Anthers ca 0.5 mm; pan brs appressed, short, spikelet-bearing to the
      base; European, now well intro and ± weedy, esp in dry areas on
      semi-waste land e Cas; BC to Ore, e to Ida and Mont, also near Port-
      land, Ore; interrupted apera (*A. spica-venti* ssp. or var. *i.*)
                                                           2 **A. interrupta** L.
    3b Anthers ca 1.2 mm; pan brs slender, spreading, not spikelet-bearing
      toward the base; European, well estab in various places in e US,
      known in our area only from collections near Portland (Albina,
      Linnton), Ore; silky apera   3 **A. spica-venti** L.
  2b Pl per or if ann the awn (if any) arising below the top 1/3 of the lemma;
    palea often much reduced, rarely 2-nerved
    4a Callus bearding 1–2 mm; anthers 1.6–1.8 mm; pl per, rhizomatous,
      4–10 (12) dm; ligs 2–7 mm, acute, gen deeply lacerate; bls flat, 2–5
      mm broad; palea lacking; dry woods to meadowl, nw Ore s, along and

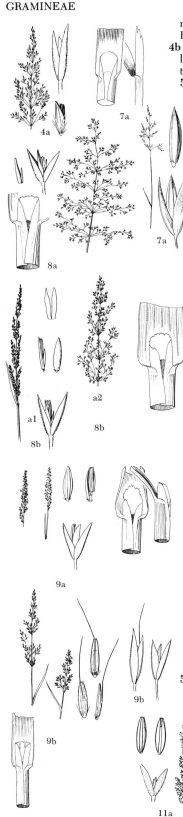

near the coast, to s Cal; in n Ore chiefly in lower mts on w side WV; Hall's b. *(A. occidentalis)* 4 **A. hallii** Vasey

4b Callus bearding lacking or gen scarcely as much as 0.5 mm, rarely longer but then anthers much < 1.6 mm; pl often ann or nonrhizomatous

5a Anthers at least 1 mm

6a Palea well developed, gen at least half as long as the lemma

7a Rachilla prolonged behind the palea as a short stub or bristle 0.6–1 mm; rhizomatous per 3–8 dm; lig 1–4 (5) mm; bls flat to weakly involute, 1.5–3 mm broad; bogs and wet places near the coast; Alas to VI and nw end of OP (Lake Ozette), Wn; Alas b. *(A. canina* var. *a.)* 5 **A. aequivalvis** (Trin.) Trin.

7b Rachilla not prolonged behind the palea

8a Pan open, the brs filiform, spreading, not spikelet-bearing to the base; ligs mostly 1–2 (3) mm; pl 3–5 (2–8) dm; Eurasian, a common lawn grass in much of temp N Am, and mainly w Cas in our area, but occ also e to n Ida and w Mont; colonial b. *(A. alba* var. *sylvatica, A. vulgaris* var. *aristata, A. t.* var. *a., A. stricta)* 6 **A. tenuis** Sibth.

8b Pan compressed or (if open) the brs often stiff and spikelet-bearing to the base; ligs of upper culm lvs mostly 3–6 mm; pl 5–13 dm; prob native to Europe and intro in N Am, where common as an escape from cult, esp from lawns; 3 vars. 7 **A. alba** L.

a1 Pan compressed, gen not > 15 mm broad when pressed; pl more stoloniferous than rhizomatous; common on w coast from s Alas to n Cal and to Ida, Utah, Colo, and NM, also in ne US and adj Can; creeping b. *(A. exarata* var. *stolonifera, A. reptans, A. palustris, A. stolonifera* var. *p.)* var. **palustris** (Huds.) Pers.

a2 Pan more open, gen > 15 mm broad when pressed, the brs tending to spread or ascend; pl often rhizomatous

b1 Bls often > 4 mm broad; pl mostly stoloniferous, gen not > 6 dm; Alas to Wn, w Cas, e to se Can and ne US, Europe; fiorin *(A. s., A. vulgaris* var. *s.)* var. **stolonifera** (L.) Smith

b2 Bls rarely > 4 mm broad; pl rhizomatous (also sometimes stoloniferous), often up to 12 dm; in most of nonarctic Can and temp, moist US, common throughout our area from lowl to lower mts; redtop *(A. stolonifera* var. *alba)* var. **alba**

6b Palea lacking

9a Pan strongly compressed, 3–10 cm × scarcely 15 mm when pressed; rhizomes long and wiry; bls 1.5–3 (4) mm broad, mostly involute; ligs firm, obtuse, 1–3 mm; coastal sand dunes or adj woods, Grays Harbor Co, Wn, to San Francisco; dune b. *(A. exarata* var. *littoralis)* 8 **A. pallens** Trin.

9b Pan moderately open, (3) 5–15 (20) cm × often > 15 mm when pressed; rhizomes short, sometimes barely evident; bls 1.5–3 (4) mm broad, mostly flat; ligs delicate, 1–4 mm, semi-truncate; dry to moist slopes and meadows from near sea level almost to treel in the mts; BC to Cal, on both sides Cas, e to Mont, Ida, and Nev; thin b., lfy b. *(A. foliosa, A. diegoensis* var. *f., A. pallens* var. *vaseyi, A. canina* var. *stolonifera)* 9 **A. diegoensis** Vasey

5b Anthers < 1 mm

10a Palea well developed, gen at least half as long as the lemma

11a Glumes blunt, scab over the back, ca 2 mm; pan compact, the brs spikelet-bearing to the base; culms decumbent based and freely rooting at the nodes, 3–8 dm; moist areas at lower elev, ± weedy; occ in Wn and Ore, more common in sw US; Eurasia; water b. *(A. verticillata, Nowodworskya v., Phalaris semiverticillata, Polypogon s., Nowodworskya s.)* 10 **A. semiverticillata** (Forsk.) Christ.

11b Glumes acute, scab on the keel only, sometimes > 2 mm; pan often open, the brs rarely spikelet-bearing to the base

12a Lemma with a bent awn ca 2.5 (3) mm; upper ligs 5–7 (12)

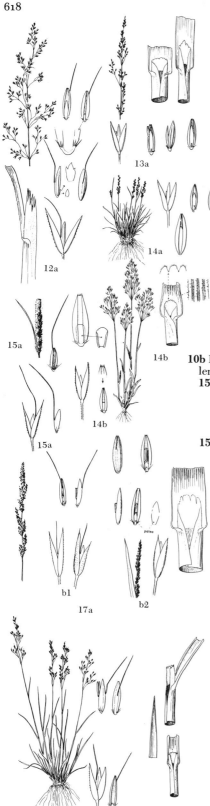

mm; pl tufted, 4–7 dm; palea mostly 0.2–0.5 (but up to 1) mm; moist to wet areas, gen in swamps or bogs; mostly coastal, Tillamook Co, Ore, to Cal, but also near Salem, Marion Co, Ore; Pac b.   11 **A. longiligula** Hitchc.

12b Lemma unawned or with a short straight awn; ligs 1–2.5 (3) mm

13a Pl gen > 15 cm, short-rhizomatous, per; pan (3) 5–10 cm, ± open at anthesis; palea almost = lemma; rachilla vestige 0.1–0.3 mm; lig mostly 1–3 mm; alp meadows and slopes to lake and stream margins, mostly alp or subalp, but down to ca 1000 ft elev near the coast; BC and s Alas to Cal, e to RMS as far s as Colo; Thurber b. (*A. atrata*)   12 **A. thurberiana** Hitchc.

13b Pl often < 15 cm, nonrhizomatous, sometimes ann; pan 1–6 cm, often compressed at anthesis; palea not > 3/4 as long as the lemma; rachilla vestige often lacking

14a Palea 2/3–3/4 as long as the lemma; pl per; pan greatly compressed, scarcely 5 mm broad when pressed; glumes scarcely 2 mm; lemma subequal to the glumes; subalp or alp meadows, streambanks, and moist slopes; Cas, BC, Wn, and Ore, e to Ida, Mont, Utah, and Colo; alpine b.   13 **A. humilis** Vasey

14b Palea scarcely half as long as the lemma; pl ann; pan not greatly compressed, > 5 mm broad when pressed; glumes 2–2.5 mm, the 1st the longer; lemma ca 1.5 mm; around warm springs, creeks, and geyser overflows, YNP, Wyo; Ross' b. (*A. exarata* var. *r.*)   14 **A. rossiae** Vasey

10b Palea lacking or poorly developed, always < half as long as the lemma

15a Pl ann, glab, 0.8–3 (4) dm; lemma with a bent awn 3–7 mm; lig 1–3 (5) mm; glumes (3) 3.5–4.2 mm, acuminate or with awn-tip up to 1 mm; moist soil in the open, VI s, w Cas, to Baja Cal, also in CRG; small-lvd b. (*A. inflata, A. exarata* var. *m.*)   15 **A. microphylla** Steud.

15b Pl per or (if ann) with unawned lemma or with awn straight or < 3 mm

16a Pan relatively strongly congested, several times as long as broad when pressed, the brs ascending, gen spikelet-bearing to near the base; pls mostly either alp or subalp or with glumes scab over the back

17a Glumes scab over the back, often aristate or awn-tipped; pl gen > 2 (2–12) dm; lvs rarely < 2 (2–10) mm broad; lig mostly 3–8 (2–13) mm; moist areas, from near sea level to midmont, n Alas to Cal, e to Alta, SD, Neb, Tex, and n Mex; spike b.; several phases   16 **A. exarata** Trin.

a1 Glumes long-acuminate to aristate, the awnlike tip sometimes as much as 1 mm (including the awn, if any), the overall length rarely < 3 mm; 2 vars.   ssp. **exarata**

b1 Lemma awned from above the middle, the awn (2) 3–4 (6) mm, bent; VI to s Cal, rarely in w Nev, collected also in Elmore Co, Ida (*Polypogon alopecuroides, A. ampla, A. inflata, A. ampla* f. *m.*)   var. **monolepsis** (Torr.) Hitch.

b2 Lemma unawned or very rarely with a short straight awn barely exceeding the lemma; the common phase from Alas to s Cal, mostly w Cas in Wn and Ore (*A. asperifolia, A. grandis, A. scouleri*)   var. **exarata**

a2 Glumes ± sharply acute but rarely aristate, mostly < 3 mm; Alas to s Cal, e to Alta, Daks, Neb, and NM (*A. albicans*)   ssp. **minor** (Hook.) Hitchc.

17b Glumes rarely scab over the back, neither aristate nor awn-tipped; pl gen not > 2 dm; lvs mostly < 2 mm broad

18a Lemma with a prominent, geniculate awn 2–3 mm; glumes mostly ca 3 mm; pl 1.5–4 dm; lf bls 1–3 mm broad; pan narrow but not greatly congested, 3–12 × gen at least 1 cm; moist arctic-alp areas; Alas to Newf, s

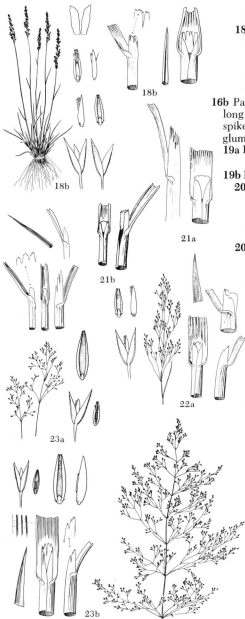

in RMS to Colo and Utah, reportedly to BC and Wn; northern b. *(A. bakeri)*   17 **A. borealis** Hartm.

18b Lemma unawned or very occ with a weak straight awn; glumes ca 2.5 mm; pl 0.8–1.5 (2) dm; lf bls 1–2 (2.5) mm broad; pan much-congested, 2–6 × scarcely 1 cm; subalp to alp meadows and open ridges; BC to s Cal, e to Mont, Wyo, and Colo; variant b. *(A. varians* Trin.; *A. rossae*, misapplied)   18 **A. variabilis** Rydb.

16b Pan relatively loose and open, often not > 2–3 times as long as broad when pressed, the brs often spreading or spikelet-bearing only above midlength; pls often not subalp, glumes rarely scab over the back

  19a Pl ann, local around hot springs in YNP (see lead 14b)

                    14 **A. rossiae** Vasey

  19b Pl per, gen in distribution

    20a Lemma gen with a bent awn (1.5) 2.5–3 mm

      21a Lig of culm lvs 3–7 (12) mm; pl 4–7 dm; coastal (see lead 12a)   11 **A. longiligula** Hitchc.

      21b Lig of culm lvs gen < 3 mm; pl not > 4 dm; alp (see lead 18a)   17 **A. borealis** Hartm.

    20b Lemma gen unawned or with a straight awn < 2.5 mm, if awned the pan brs rarely spikelet-bearing below midlength

      22a Pan gen purplish, (1) 1.5–3 dm, diffuse, the brs capillary, erect to spreading, rarely brd and never spiklet-bearing below midlength; glumes often acuminate to semi-aristate, mostly ca 2.3 (1.8–3) mm, sometimes scab over the back; lemmas sometimes awned; moist to dry areas from near sea level to subalp, Alas to Newf, s to all but se US; Asia; winter b., rough hair-grass, tickle-grass *(A. geminata, A. hiemalis* var. *g., A. hi(y)emalis* of most auth., *A. nootkaensis, Vilfa scabra, A. scabriuscula)*   19 **A. scabra** Willd.

      22b Pan often contracted, if diffuse the main brs forked and gen spikelet-bearing below midlength; glumes often much < 2.3 mm, not scab over the back; lemmas unawned

        23a Pl mostly 1–3 (4) dm; pan 3–10 cm, often contracted; glumes 1.6–2.4 (2.6) mm; damp to swampy areas, mostly at medium elev in the mts, but up to 10,000 ft s of our area; s BC to s Cal, e to Mont, Wyo, Colo, and NM; Ida b. *(A. tenuiculmis; A. tenuis*, misapplied)   20 **A. idahoensis** Nash

        23b Pl gen > 4 (3–8) dm; pan mostly 10–30 cm, open, the brs stiff, spreading to ascending; glumes 2–2.3 (3.5) mm; wet places in the mts at medium elev; VI s to Cal and w Nev, mostly in Cas and w, also in nc Ida, but prob not in Mont and Wyo as reported; Ore b. *(A. attenuata, A. schiedeana* var. *armata)*   21 **A. oregonensis** Vasey

Aira L.   Hairgrass

Spikelets pan, art above the glumes, 2-fld; glumes subequal, keeled, rather broad, 1 (3)-nerved, gen exceeding the upper flt; lemma rounded on the back, obscurely 3 (5)-nerved, slenderly bifid at the tip, minutely bearded on the callus, both awned from below midlength or the 1st awnless; awns twisted and once-bent, gen exserted; stamens 3; lods 2; slender delicate anns with hollow culms, open sheaths, membranous and gen puberulent ligs, and filiform, involute bls scarcely 1 mm broad; aurs lacking. (Ancient Gr name for some unknown weed). *(Airopsis, Aspris, Caryophyllea)*. The pls have no forage value.

1a Pan compact, ± spikelike, 1–3 cm; glumes ca 3 (2.7–3.5) mm; lemmas ca 3 (2.5–3.2) mm, both awned; near coast, on gravelly prairies, sea bluffs, and dunes; s VI to Cal, also in e US; weed intro from Europe; early h., little h. *(Agrostis p., Avena p., Trisetum p.)*   1 **A. praecox** L.

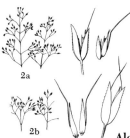

1b Pan open, often > 3 cm; glumes mostly < 3 mm; lemmas 1.5–2.3 mm, the
1st sometimes not awned
  2a Lemmas ca 1.5 (1.9) mm, the 1st unawned; lig very slightly scab or glab;
  sparingly intro from Europe in dry, mostly disturbed or overgrazed
  areas; w Cas, Wn to Cal, coast from Tex to Md; elegant h., diffuse h.
  (*Avena capillaris*)   2 **A. elegans** Willd.
  2b Lemmas 2–2.3 mm, both awned; lig strongly scab; moist to dry, gravelly
  soil; intro from Europe and common w Cas, sw BC to Cal, occ to
  common in se and e US, in our area sometimes a weed in rock gardens;
  silver h. (*Avena c.*)   3 **A. caryophyllea** L.

Alopecurus L.   Foxtail; Meadow-foxtail; Alopecurus

Spikelets 1-fld, art below the glumes, borne in cylindrical spikelike pan, strongly flattened; glumes
= or subequal, ± connate up to half their length, the tips acute to blunt or rounded; lemmas
from slightly < to barely > the glumes, strongly flattened, very indistinctly 3- or 5-nerved,
awned from the back, the margins ± connate, sometimes to above midlength; awn arising from
just above the base to about midlength of the lemma, straight to strongly twisted and bent, from
< the glumes to > twice their length; palea lacking; rachilla not prolonged; stamens 3; lods
wanting; ann or per with hollow culms and open sheaths; ligs membranous, truncate to acute,
erose-ciliolate to entire or lacerate, ± scab-puberulent; bls gen flat, rolled in the bud. (Gr *alopex*,
fox, and *oura*, tail). The spp. are highly palatable and nutritious, but rarely abundant enough to
constitute an appreciable element of the range forage.

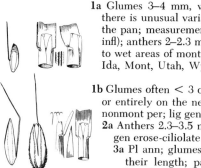

1a Glumes 3–4 mm, woolly-silky over their entire surface, (in *Alopecurus*
there is unusual variation in the length of the spikelets at various levels in
the pan; measurements here are for spikelets from near the middle of the
infl); anthers 2–2.3 mm; lig truncate, 1–3 mm; tufted per (1) 2–8 dm; moist
to wet areas of mont to alp slopes and meadows; Alas to Newf, s in RM to
Ida, Mont, Utah, Wyo, and Colo; alpine f. (*A. occidentalis*)
                                                                        1 **A. alpinus** Smith
1b Glumes often < 3 or > 4 mm, silky to villous, but the pubescence mostly
or entirely on the nerves and keel; anthers often < 2 mm; ann or mainly
nonmont per; lig gen acute or > 3 mm
  2a Anthers 2.3–3.5 mm; glumes rarely < 4 (4–7) mm, acute; lig truncate,
  gen erose-ciliolate
    3a Pl ann; glumes short-ciliate on the nerves and keel, connate 1/3–1/2
    their length; pan tapered and ± acute at each end; bls 2–4 mm
    broad; intro from Europe; known in our area from collections from
    Wn and Ore; black twitch   2 **A. myosuroides** Huds.
    3b Pl per; glumes villous on the nerves and keel, often connate for < 1/3
    their length; pan rounded at each end; bls 3–10 mm broad; intro from
    Europe, and not uncommon in much of N Am, in swampy places; occ
    with us in much of our area; meadow f.   3 **A. pratensis** L.
  2b Anthers gen < 2 mm, if that long then the glumes obtuse or < 4 mm
  and the lig ± acute and not erose-ciliolate
    4a Awn arising from near the middle of the lemma, rarely as much as 1.5
    mm > the glumes; glumes 2–2.3 (1.8–2.5) mm; anthers (0.5) 0.6–0.7
    (0.8–1.0) mm; pl per; wet places, often submerged; Alas to Newf, s to
    most of US and n Mex, both sides Cas; Eurasia; shortawn f., little m.
    (*A. geniculatus* var. *robustus*, *A. caespitosus*)   4 **A. aequalis** Sobol.
    4b Awn arising from the lower third of the lemma, often from just above
    the base, gen at least 1.5 mm > the glumes; glumes often much > 2.5
    mm or anthers < 0.5 or > 1.0 mm; pl ann or per
      5a Anthers gen 0.3–0.5 (0.6) mm; glumes 2–2.4 (2.6) mm; pl ann or
      winter ann; waste areas, old fields, and roadsides where moist; na-
      tive to much of c and se US, prob intro in our area where sporadic
      in s BC, Wn, Ore, Ida, and Mont; Carolina f. (*A. geniculatus* var.
      caespitosus, *A. macounii*)   5 **A. carolinianus** Walt.
      5b Anthers at least 0.6 mm; glumes often > 2.6 mm; pl ann or per
        6a Glumes 3.5–5 mm, ± acute; anthers 0.6–1.2 mm; pl ann, 1–3
        dm; awn gen (7) 8–10 mm; moist ground, often in vernal pools; e
        Wn to nc Cal; Pac m., saccate f. (*A. howellii*) 6 **A. saccatus** Vasey
        6b Glumes ca 2 (2.5–3.5) mm, obtuse; anthers 1.2–2.2 mm; pl per,
        (2) 3–5 dm; awn 4–5 (6) mm; wet places, often in standing water;
        Alas s to Cal and Ariz, e to Newf and Atl coast, and through RM
        to Colo; water f. (*A. pallescens*)   7 **A. geniculatus** L.

Ammophila Host   Beachgrass

Spikelets in a much-compressed pan, 1-fld, art above the glumes; glumes strongly keeled, narrow, subequal, the 1st 1-nerved, the 2nd gen 3-nerved; lemma keeled, 5-nerved, unawned, the callus sparsely short-bearded; rachilla prolonged as a short bristle beyond the palea; stamens 3; lods 2; strongly rhizomatous per with wiry, hollow culms, open sheaths, prominent membranous ligs and non-aur bls, rolled in the bud. (Gr *ammos*, sand, and *philos*, loving, in reference to the habitat).

A. arenaria (L.) Link. European b. Culms in tufts, up to 11 dm, connected by tough, elongate rhizomes; sheaths smooth; lig 10–25 mm, puberulent, acute, the margin entire but sometimes lacerate; bls involute, tough and fibrous, smooth, 2–4 mm broad; pan spikelike, (10) 15–30 cm × 15–20 mm when pressed; glumes pale, 10–14 mm, subequal, the 2nd the longer; lemma 1–3 mm < the 1st glume, gen with the midnerve barely excurrent just below the tip, the callus bearding 2–3 mm; rachilla vestige ca 1.5 mm; excellent sand-binder, native to Europe, intro along Pac coast.

Andropogon L.   Bluestem; Beardgrass

Spikelets semi-racemose, borne in 2's at the nodes, 1 sessile, the other ped; ped spikelet sterile or ♂, often considerably reduced in size and no of nonessential parts, the ped gen villous; sessile spikelet 2-fld, the upper flt ♀, the lower one an empty lemma; glumes rather coriaceous, narrow, awnless, the 1st gen ± flattened and often strongly 4-nerved and minutely bifid, the 2nd slightly compressed; sterile lemma like the glumes in texture, but shorter, gen bifid; fertile lemma papery, < the glumes, gen awned; palea much reduced, papery; stamens 3; lods 2, cuneate- flabellate, finely ∝ -nerved, often with a few long terminal hairs; strong, course per with solid and often grooved culms, open sheaths, membranous ligs, and flat or folded non-aur bls rolled or folded in the bud, the infl a pan, the brs bearing 1–∝ racemes, each with an ultimately disart rachis. (Gr *andr*, a combining form from *aner, andros*, man, and *pogon*, beard, in reference to the hairy ped spikelets that are gen ♂).

1a Sheaths strongly keeled; bls mostly < 5 mm broad; racemes single at the tip of each br of the pan, the joints of the rachis hollow and cuplike at the tip; culms gen 8–10 (5–15) dm; prairies and foothills; Alta to Que, s through most of US e of RM, w in Mont to Flathead Lake (record questionable) and at the edge of our area elsewhere; little b., broom b. *(Sorghum s.)*   1 **A. scoparius** Michx.

1b Sheaths not strongly keeled; bls mostly (3) 4–10 mm broad; racemes 2 or more on each br of the pan, the joints of the rachis only slightly if at all hollow at the tip; app our area in e Mont

2a Awn of fertile spikelet rarely < 10 mm; rhizomes very short or wanting; joints of the racemes gen with sparse pubescence mostly < 2 mm; Sask to Que, s along e base RM to Ariz and n Mex and most of the rest of the US; big b. *(A. furcatus)*   2 **A. gerardii** Vitman

2b Awn of fertile spikelet rarely > 5 mm; rhizomes well developed; joints of the racemes mostly with copious pubescence 2–4 mm; mostly in sandy areas, esp dunes; e Mont and ND s to Ariz, NM, and Tex; sand b. *(A. paucipilus)*   3 **A. hallii** Hack.

Anthoxanthum L.   Vernalgrass

Spikelets in small congested pan, art above the glumes, 3-fld, the lower 2 flts represented by empty lemmas; glumes unequal, acute, the 1st 1-nerved, exceeding the flts, the 2nd ca twice as long as the 1st; sterile lemmas pilose-hirsute, brownish, strongly compressed, rounded and bifid at the tip, awned from the back; fertile lemma firmer than the sterile ones, completely enfolding the palea, unawned, lightly several-nerved; stamens 2; lods lacking; small ann or per with hollow culms, open sheaths, gen short rounded aurs, membranous ligs, and flat bls rolled in the bud. (Gr *anthos*, flower, and *xanthos*, yellow, in allusion to the tawny infl).

1a Pl ann, 1–3 dm; bls 1–2 mm broad; lig 0.5–2 mm; aurs scarcely 0.5 mm (or lacking); pan 1.5–4 cm; rather arid, often waste areas; VI to Cal, both sides Cas, native to Europe; ann v. *(A. odoratum* var. *a.)*   1 **A. aristatum** Boiss.

1b Pl per, 3–6 dm, vanilla-scented, esp when dry; bls (2) 3–7 mm broad; lig (1) 2–3 mm; aurs ca 1 mm; pan 2–9 cm; intro from Europe for haying purposes, now widespread in lawns, pastures, and wasteland, gen where ± moist; Alas to Cal, mostly w Cas; sweet v.   2 **A. odoratum** L.

Aristida L.   Threeawn; Aristida

Spikelets 1-fld, pan, art above the glumes; glumes narrow, 1–3-nerved, acute to acuminate, sometimes awn-tipped; flt hardened, terete, the callus hardened, ± sharp-pointed, gen bearded; lemma convolute, lightly 3-nerved, entire to slightly bifid, with a terminal, trifid (ours), gen twisted awn; lods 2, oblong; ovary glab; ann or mostly per, tufted pls with hollow culms, open sheaths, short fringed-membranous ligs, and narrow, convolute to involute, non-aur bls. (L *arista*, awn). The pls are of little forage value and may cause mechanical injury because of the sharp callus.

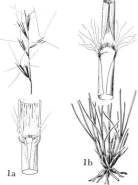

1a Pl ann, up to 6 dm; culms pithy; bls 1–1.5 mm broad; lig ca 0.5 mm, the fringe not quite so long as the basal membrane; dry soil, often where disturbed; WV, Ore, to Cal and Ariz; also in the Miss Valley and e to Atl coast; prairie t. *(A. o.* var. *nervata)*   1 A. oligantha Michx.

1b Pl per, strongly tufted, mostly 3–4 dm; culms hollow; bls 7–25 cm × 1–2 mm; lig scarcely 0.5 mm, the frontal bl margins with hairs 2–3 mm; des plains and grassl into the foothills; BC and e Wn to the Daks, s to Ore, Ariz, Tex, and n Mex *(A. purpurea* var. *l.)*; ours mostly referable to var. *robusta* Merrill—var. *longiseta* is a shorter grass with somewhat shorter, more-curved bls so similar to those of *A. fendleriana* Steud. (a pl of the Gr Pl region, with crisped bls scarcely 5 cm) as to suggest that the 2 taxa are not specifically distinct; red t.   2 A. longiseta Steud.

Arrhenatherum Beauv.   Oatgrass

Spikelets in large, rather narrow pan, art above the glumes, 2-fld, the 1st flt gen ♂, the 2nd ♀, the rachilla prolonged as a bristle behind the upper palea; glumes papery, keeled, unequal, the 1st 1-nerved, subequal to the 1st flt, the 2nd 3-nerved, gen = the 2nd flt; lemmas bearded on the callus, 7-nerved, entire to shallowly bifid, that of the 1st flt with a stout, twisted, and strongly bent awn arising near midlength, the 2nd with a short, straight, subterminal awn; stamens 3; lods 2; tall per with hollow culms, open sheaths, membranous ligs, and flat, non-aur bls rolled in the bud. (Gr *arrhen*, masculine, and *ather*, awn, referring to the awned ♂ flt).

A. elatius (L.) Presl. Tall o. Culms 8–15 dm, sometimes pubescent at the nodes, frequently enlarged and bulbose at the base, but erect and not rhizomatous; lig 1–2 (3) mm, truncate, erose-ciliolate, finely puberulent; bls 4–8 (10) mm broad; pan shining, narrow, 1–3 dm; glumes mostly 5–7 and 7–10 mm, respectively; awn of 1st lemma 10–20 mm, that of the 2nd mostly < 6 (15) mm or rarely lacking; native to Europe, widely intro as a meadow grass and now estab from sw BC to Cal, mostly w Cas, as well as in many other states; pls with bulbous bases, var. *bulbosum* (Willd.) Spenner, as well as those with nonbulbous bases are to be found in our area.

Avena L.   Oat

Spikelets 2–3-fld, gen 2–3 cm, art above the glumes, pendulous in large, open pan; glumes subequal, membranous, mostly > the flts, the 1st 7-nerved, the 2nd 9-nerved; lemmas coriaceous, mostly 7-nerved, glab or pilose on the lower half, gen awned at about midlength, the awn gen stout, twisted, and bent, but sometimes reduced and straight, occ lacking; rachilla prolonged; stamens 3; lods 2; moderately tall ann with hollow culms, open sheaths, rather short membranous ligs, and gen flat, non-aur bls rolled in the bud. (The ancient L name).

1a Tip of lemma bifid-setaceous, the teech ca 4 mm, bristle-tipped; culms to 1 m; anthers 2–3 mm; bls 3–6 mm broad; a weedy European sp. common in sw US, occ in w Ore, and collected also in Klickitat Co, Wn; slender o.
                                                                                    1 A. barbata Brot.

1b Tip of lemma with short, acute teeth only 1 (2) mm; anthers 3–5 mm

2a Flts gen 2 (3), not readily separable from one another, gen only the 1st flt awned, the awn straight or curved but not bent; lemmas greenish and thickened at the tip, glab; intro from Europe, found chiefly along railways and roadsides and on fallow soil, where not believed to be persistent for long *(A. fatua* var. *s.)*; common o.   2 A. sativa L.

2b Flts gen 3, readily separable from one another, the 1st 2 awned, the awns strongly bent; lemmas gen hirsute, thin and membranous at the tip; European, widely intro and common in our areas on waste ground and as a weed in grain fields; wild o. *(A. f.* var. *glabrata)*   3 A. fatua L.

Beckmannia Host   Sloughgrass; Beckmannia

Spikelets 1-fld (ours), art below the glumes, sessile, borne closely aggregated in 2 rows on one side of the rachis in ∝ short, appressed to spreading, racemose-pan spikes; glumes strongly compressed and inflated, ca equal, 3-nerved, semicircular, abruptly acute; lemma lightly 5-nerved, lanceolate, much narrower than the glumes, ± acuminate, = the glumes but only slightly larger than the 2-nerved palea; stamens 3; lods 2; rather large ann with hollow culms, open sheaths, large membranous ligs, and broad, flat, non-aur bls rolled in the bud. (For Johann Beckmann, 1739–1811, a German botanist).

B. syzigachne (Steud.) Fern. Am s. Pl up to 1 m, scab but otherwise glab, often stoloniferous; lig acuminate, 6–11 mm, entire or lacerate, gen pubescent; bls flat, 5–10 mm broad; infl narrow, crowded, up to 3 dm; spikes mostly 1–2 cm; rachilla not prolonged; glumes and lemma ca 3 mm; anthers 1–1.4 mm; lods almost 1 mm; marshes, wet meadows, and ponds and ditches, lowl to mid-mont, Alas to Cal, e to Man, Gr Lakes region, and NY, Mo, and NM; ne Asia (*B. erucaeformis* of Am auth.).

Bouteloua Lag.   Grama; Mesquite-grass

Spikelets with 1 ⚥ fl below 1 or 2 greatly reduced sterile flts, art above the glumes, sessile in 2 rows on one side of the rachis in (1) 2–∝ short, racemose spikes; glumes 1-nerved, the 1st much the smaller and narrower; fertile lemma gen at least as long as the glumes, 3-nerved, the nerves ending in awns; sterile lemmas small, with 3 strong awns gen > those of the fertile lemma; stamens 3; lods 2; ours pers with solid or pith-filled culms, open sheaths, short ligs composed chiefly of a fringe of straight hairs, and flat to involute bls, rolled in the bud. (For Claudio [1774–1842] and Esteban [1776–1813] Boutelou; Claudio was a professor of agriculture in Madrid). The genus includes some of the most common and palatable of range grasses.

1a Spikes (20) 30–80, each 10–15 (20) mm, consisting of 5–8 spikelets and deciduous as a unit; culms 4–8 dm; bls (2) 2.5–4 mm broad; pl widely rhizomatous, forming open patches; plains to lower mts, often on scabland or rocky slopes; Mont s to most of w US except Ida, Ore, and Wn, e to Ont and Me, in Mont w to Beaverhead Co; Mex, S Am; sideoat g.
                                        1 **B. curtipendula** (Michx.) Torr.
1b Spikes gen 2 (1–3), 1.5–4 (5) cm, each consisting of 30–40 spikelets and not deciduous as a unit, the individual spikelets art above the glumes; culms 2–4 (5) dm; bls 1–2 mm broad; pl short-rhizomatous and forming dense, thick mats; a major component of the shortgrass prairie e RM, Alta to Man, s to Tex, Mex, and Cal, w in Mont to Beaverhead, Beartooth, Deerlodge, and Helena Nat Forests; blue g.   2 **B. gracilis** (H.B.K.) Lag.

Brachypodium Beauv.   False-brome

Spikelets few, semi-spicate (the pedicels very short or lacking), several-fld, art above the glumes; glumes < their respective subtended flts, unequal, 5–7-nerved, hardened; lemmas with 7 converging nerves, rounded on the back, awned from the tip; paleas subequal to the lemma, 2-nerved, strongly ciliate; lods 2; stamens 3; ann or per with hollow culms, open sheaths, membranous ligs, and non-aur bls. (Gr *brachys*, short, and *podion*, foot, in reference to the short pedicels).

1a Pl ann, 0.5–3.5 dm; spikelets 1–5, each 1.5–3 cm, 7–13-fld; bls 1–3 (4) mm broad; intro from Europe, occ in N Am, esp near the coast; known in our area only from near Portland, Ore (*Bromus d.*)   1 **B. distachyon** (L.) Beauv.
1b Pl per, (4) 5–7 dm; spikelets mostly 5–10, each 1.5–4 cm, 7–17-fld; bls 4–10 mm broad; intro from Europe, occ cult (for orn purposes), and known in our area from near Corvallis, Ore (*Festuca s.*)
                                        2 **B. sylvaticum** (Huds.) Beauv.

Briza L.   Quaking-grass

Spikelets in open pan, several-fld, art above the glumes, compressed and ± cordate-ovate in outline; glumes gen > 1st flt, but not covering it, broad, rounded on the back, broadly and deeply prow-like at the tip, obscurely 3-nerved, papery-margined; lemmas awnless, obscurely nerved, ovate, obtuse at the tip, cordate at the base, gen saccate, papery; stamens 3; lods 2; glab ann with hollow culms, open sheaths, membranous ligs, and flat, non-aur bls. (Ancient Gr name for some grass or grain, gen supposed to have been rye). The spikelets are dainty and shell-like and dry without shattering, therefore the pls are prized as components of dry bouquets.

B. minor L. Little q. Glab ann up to 5 dm; sheaths much > in back than on the margins, the lig correspondingly higher in the center (opp the throat), but the free membrane only 1–4 mm; bls 2–8 mm broad, the margins freed from the sheath at different levels; pan pyramidal, 5–15 cm; spikelets broadly triangular, (3) 4–6-fld; glumes at right angles to the rachilla, ca 2 and 2.5 mm; lemmas cordate-based, ca 2 × 1.5 mm; European, prob intro for orn purposes, and well estab in dry areas of e N Am, sw Ore, and Cal, occ in sw Ida and nw Ore and collected near Nanaimo, VI.

Bromus L.   Brome; Bromus; Brome-grass; Cheat; Chess

Spikelets several-fld, borne in open to much-contracted pan (occ racemose), compressed to semi-terete, art above the glumes; glumes unequal, each < its subtended flt, the 1st 1- or 3-nerved, the 2nd 3- or 5 (7)-nerved, acute to blunt, rarely awn-tipped; lemmas keeled to rounded on the back, with 5–9 (11) rather faint, converging nerves, gen bifid but sometimes rounded, awnless or commonly awned from just below the tip (gen from between the teeth of the bifid apex); lods 2; stamens gen 3 (2), the anthers sometimes included and greatly reduced (flts cleistogamous), or exserted and much larger (flts chasmogamous); pl ann, bien, or per, rarely rhizomatous; culms hollow; sheaths closed (gen to near the top), mostly pubescent; aurs present or absent; ligs membranous, entire to erose or lacerate, relatively short (mostly 1–6 mm); bls flat to ± involute, rolled in the bud. (Gr *bromos*, ancient name for the oat, denoting food). *(Ceratochloa)*. The per spp. are palatable and readily grazed; a few, esp *B. inermis*, are widely used in hay-meadows. The ann spp. are mostly weedy and of little forage value except for a short period before maturity; when mature some spp. become potentially injurious because of the hardened, sharp, awned frs.

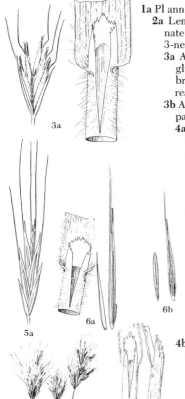

1a Pl ann
  2a Lemmas narrowly acute to acuminate, the teeth acuminate to acuminate-aristate, at least 2 mm; 1st glume 1-nerved, the 2nd mostly 3-nerved; awns gen at least 12 mm
    3a Awns bent, strongly twisted near the base; pan brs at least = the glumes, bearing ∝ spikelets; pl 3–10 dm, ± pilose; bls 2–10 mm broad; intro from S Am, well estab in sw US n to e Ore, possibly reaching our area; Chilean c. or b. *(Trisetum t.)*   1 **B. trinii** Desv.
    3b Awns rarely bent, and never strongly twisted, if at all twisted then the pan brs gen < the glumes or bearing only 1 or 2 spikelets
      4a Pan brs slender, some at least = the glumes, the pan itself rather open and loose
        5a Spikelets 15–20 mm; 2nd glume gen ca 10 mm; ligs ave barely 2 (1.2–3) mm; European, widely intro in dry areas in much of the world; common in waste or disturbed (esp overgrazed) areas mostly e Cas but throughout our range; perhaps the most common of all our grasses; cheat grass, downy c. 2 **B. tectorum** L.
        5b Spikelets at least 25 mm; 2nd glume gen at least 13 mm; ligs ave > 2 mm
          6a 1st and 2nd glumes 15–20 and 25–33 mm; lemmas 22–30 mm; pl 4–10 dm; bls 3–12 mm broad; European, widely intro, gen on waste ground and along railroads and highways, mostly w Cas, but e to Ida and RMS; ripgut *(B. gussoni, B. r.* var. *g.)*
                             3 **B. rigidus** Roth
          6b 1st and 2nd glumes 8–10 and 12–15 mm; lemmas 16–20 mm; pl 4–10 dm; bls 2–3 (4) mm broad; sparingly intro from Europe, mostly along roadsides and in waste places, s BC to Cal, e to s RMS; barren b.   4 **B. sterilis** L.
    4b Pan brs stout, rarely = the glumes, mostly erect, the pan itself strongly contracted; spikelets subsessile, stiffly erect; pl 1–4.5 dm; bls 1.5–3 mm broad; glumes ca 8 and 10–11 mm; lemmas mostly 13–15 mm; intro from s Europe, now common esp on overgrazed land e Cas, occ collected w Cas, as at Portland, Ore; foxtail c. or b. *B. madritensis* L., a similar Old World pl but with a looser pan

(lower peds up to 15 mm) is occ in sw US, and has been collected on
ballast near Portland, Ore 5 **B. rubens** L.

2b Lemmas rounded at the tip, gen shallowly bidentate, the teeth rarely as
much as 2 mm; 1st glume 3 (5)-nerved, the 2nd 5 (7, 9)-nerved; awns
often much < 12 mm

7a Spikelets gen awnless or merely awn-tipped (tip scarcely 1 mm),
9–15-fld, 15–30 mm; ligs pubescent; fr curved in cross-section and
enwrapped by the lemma; pl 2–6 dm; bls 2–5 mm broad; intro from
Europe, estab in waste areas, along roadsides, and on overgrazed land;
of considerable orn value and sometimes cult; rattle grass or c., rattlesnake g. 6 **B. brizaeformis** Fisch. & Mey.

7b Spikelets mostly with awns at least 2 mm, occ awnless but then otherwise not as above

8a Lemmas rounded, the edges rolled around the curved (lunate in ✕
section) fr; pan loose, spikelets ultimately much-flattened, the rachilla gen visible; awns often divaricate or zig-zag; rachilla not readily
disart, the flts tardily deciduous; palea only slightly if at all < the
lemma; bls 3–8 mm broad; intro from Europe, occ in our area along
roadsides, and in dry meadows and wasteland, or as a weed in
grainfields; ryebrome, chess, cheat *(B. velutinus, B. mollis* var. *s.)*
7 **B. secalinus** L.

8b Lemmas seldom if ever with the margin rolled around the fr; pan
various, often congested; spikelets sometimes not strongly flattened,
the rachilla mostly hidden, often readily disart; awns gen straight;
palea sometimes 1–3 mm < the lemma

9a Awn gen divergent or geniculate when dried naturally; pan
open, the lower brs often recurved or flexuous; sheaths
long-pilose; lemmas glab to scab, 1.5–3 (4) mm > the paleas;
intro from Eurasia, now often a weed of roadside or wasteland e
Cas, but also on ballast near Portland, Ore; Japanese b. or c.
8 **B. japonicus** Thunb.

9b Awns mostly neither divergent nor geniculate; pan open to congested, the brs sometimes all erect; sheaths pubescent to subglab;
lemmas often pubescent, sometimes < 1.5 mm > the paleas

10a Nerves of the lemma prominent, the area between the nerves
gen depressed and ± concave; infl mostly congested and erect,
in gen most of the spikelets < their peds; lemmas 6–8 (9) mm;
pl 2–7 (1–9) dm, gen soft-pubescent throughout; intro from
Europe and now a common weed from Alas to Baja Cal and
throughout our area; soft b. or c. *(B. racemosus, B. intermedius,
B. hordeaceus* of Am auth.) 9 **B. mollis** L.

10b Nerves of the lemmas only slightly or not at all elevated, the
area between the nerves plane; infl rather open, the brs mostly
ascending to spreading, most of the spikelets < their peds;
lemmas rarely < 8 (to 11) mm

11a Pan brs mostly slender and recurved or flexuous; lemmas
pilose; an Australian sp. now rather widely intro in sw US
and reported for Ore, where possibly estab, but seen from
our area only as an old collection from near Portland, Ore;
Australian b. 10 **B. arenarius** Labill.

11b Pan brs mostly ascending to spreading, rarely at all flexuous; lemmas mostly glab or scab, rarely pubescent; intro
from Europe and now widely distr, mostly e Cas; meadow or
hairy b., hairy c. 11 **B. commutatus** Schrad.

1b Pl per (occ fl the 1st year)

12a Spikelets strongly compressed; lemmas ± keeled (rather than rounded)
on the back

13a Pan large and open, the brs few (gen only 2–3 at the 1st node),
spreading to drooping; sheaths and bls soft-pubescent; lemmas
soft-pubescent; ligs (2) 3–5 mm, copiously hairy; bls lax, mostly 4–10
(12) mm broad; culms 6–15 (20) dm, sometimes rooting at the lower
nodes; moist woods, meadows, and lake or stream margins; s Alas and
VI to Lincoln Co, Ore, mostly along the coast; Pac b. *(B. magnificus)*
12 **B. pacificus** Shear

13b Pan often narrow and contracted, if large and open the pl otherwise
not as above, the ligs sometimes shorter, or sheaths and lemmas glab,
or bls narrower or broader

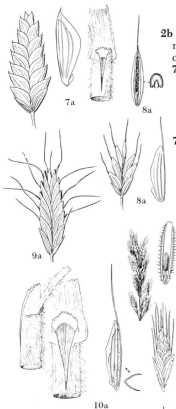

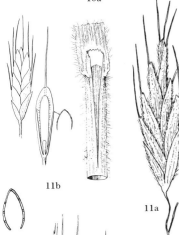

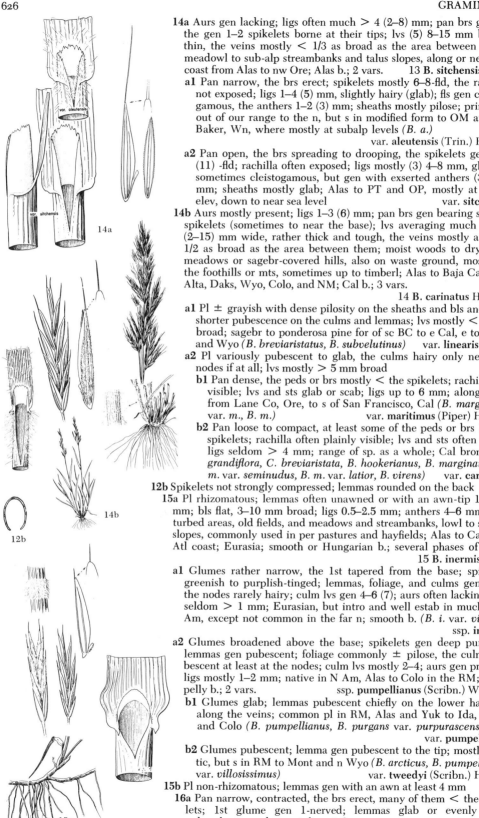

14a Aurs gen lacking; ligs often much > 4 (2–8) mm; pan brs gen > the gen 1–2 spikelets borne at their tips; lvs (5) 8–15 mm broad, thin, the veins mostly < 1/3 as broad as the area between them; meadowl to sub-alp streambanks and talus slopes, along or near the coast from Alas to nw Ore; Alas b.; 2 vars.   13 **B. sitchensis** Trin.

  a1 Pan narrow, the brs erect; spikelets mostly 6–8-fld, the rachilla not exposed; ligs 1–4 (5) mm, slightly hairy (glab); fls gen cleistogamous, the anthers 1–2 (3) mm; sheaths mostly pilose; primarily out of our range to the n, but s in modified form to OM and Mt Baker, Wn, where mostly at subalp levels (*B. a.*)

                                  var. **aleutensis** (Trin.) Hultén

  a2 Pan open, the brs spreading to drooping, the spikelets gen 4–7 (11) -fld; rachilla often exposed; ligs mostly (3) 4–8 mm, glab; fls sometimes cleistogamous, but gen with exserted anthers (3) 4–6 mm; sheaths mostly glab; Alas to PT and OP, mostly at lower elev, down to near sea level   var. **sitchensis**

14b Aurs mostly present; ligs 1–3 (6) mm; pan brs gen bearing several spikelets (sometimes to near the base); lvs averaging much < 10 (2–15) mm wide, rather thick and tough, the veins mostly at least 1/2 as broad as the area between them; moist woods to dry open meadows or sagebr-covered hills, also on waste ground, mostly in the foothills or mts, sometimes up to timberl; Alas to Baja Cal, e to Alta, Daks, Wyo, Colo, and NM; Cal b.; 3 vars.

                                   14 **B. carinatus** H. & A.

  a1 Pl ± grayish with dense pilosity on the sheaths and bls and with shorter pubescence on the culms and lemmas; lvs mostly < 5 mm broad; sagebr to ponderosa pine for sc BC to e Cal, e to Mont and Wyo (*B. breviaristatus*, *B. subvelutinus*)   var. **linearis** Shear

  a2 Pl variously pubescent to glab, the culms hairy only near the nodes if at all; lvs mostly > 5 mm broad

    b1 Pan dense, the peds or brs mostly < the spikelets; rachilla not visible; lvs and sts glab or scab; ligs up to 6 mm; along coast from Lane Co, Ore, to s of San Francisco, Cal (*B. marginatus* var. *m.*, *B. m.*)   var. **maritimus** (Piper) Hitchc.

    b2 Pan loose to compact, at least some of the peds or brs > the spikelets; rachilla often plainly visible; lvs and sts often hairy; ligs seldom > 4 mm; range of sp. as a whole; Cal brome (*C. grandiflora, C. breviaristata, B. hookerianus, B. marginatus, B. m.* var. *seminudus, B. m.* var. *latior, B. virens*)   var. **carinatus**

12b Spikelets not strongly compressed; lemmas rounded on the back

  15a Pl rhizomatous; lemmas often unawned or with an awn-tip 1–2 (4) mm; bls flat, 3–10 mm broad; ligs 0.5–2.5 mm; anthers 4–6 mm; disturbed areas, old fields, and meadows and streambanks, lowl to subalp slopes, commonly used in per pastures and hayfields; Alas to Cal, e to Atl coast; Eurasia; smooth or Hungarian b.; several phases of

                                 15 **B. inermis** Leys.

    a1 Glumes rather narrow, the 1st tapered from the base; spikelets greenish to purplish-tinged; lemmas, foliage, and culms gen glab, the nodes rarely hairy; culm lvs gen 4–6 (7); aurs often lacking; ligs seldom > 1 mm; Eurasian, but intro and well estab in much of N Am, except not common in the far n; smooth b. (*B. i.* var. *villosa*)

                                   ssp. **inermis**

    a2 Glumes broadened above the base; spikelets gen deep purplish; lemmas gen pubescent; foliage commonly ± pilose, the culms pubescent at least at the nodes; culm lvs mostly 2–4; aurs gen present; ligs mostly 1–2 mm; native in N Am, Alas to Colo in the RM; pumpelly b.; 2 vars.   ssp. **pumpellianus** (Scribn.) Wagnon

      b1 Glumes glab; lemmas pubescent chiefly on the lower half and along the veins; common pl in RM, Alas and Yuk to Ida, Mont, and Colo (*B. pumpellianus, B. purgans* var. *purpurascens*)

                                 var. **pumpellianus**

      b2 Glumes pubescent; lemma gen pubescent to the tip; mostly Arctic, but s in RM to Mont and n Wyo (*B. arcticus, B. pumpellianus* var. *villosissimus*)   var. **tweedyi** (Scribn.) Hitchc.

  15b Pl non-rhizomatous; lemmas gen with an awn at least 4 mm

    16a Pan narrow, contracted, the brs erect, many of them < the spikelets; 1st glume gen 1-nerved; lemmas glab or evenly scab-puberulent or pubescent; ligs < 2 mm

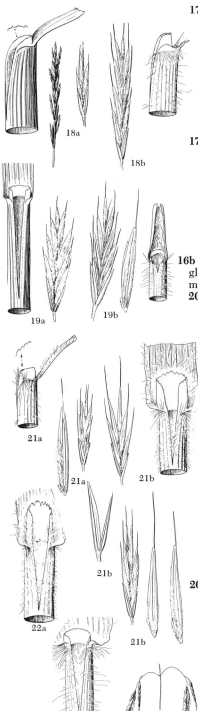

17a Bls flat, (4) 6–12 mm broad; native, gen in the mts, never w Cas
   18a Sheaths and bls glab; awn scarcely 4 mm; pan brs erect; anthers ca 3 mm; open meadows and slopes at medium to subalp levels; s Cas, Wn, and Cas, Blue, Wallowa, and Siskiyou mts, Ore, to s Sierran Cal and w Nev; Suksdorf's b.
                                        16 B. suksdorfii Vasey
   18b Sheaths long-pilose, at least at the throat; awn 5–7 mm; pan brs mostly stiffly spreading; anthers 3–4 mm; dry woods, esp under ponderosa pine, or in chaparral; Cas from near Mt Rainier, Wn, to s Cal, also in ne Ore and Ariz; Orcutt b. (*B. brachyphyllus*)
                                 17 B. orcuttianus Vasey
17b Bls stiff, gen involute, 2–5 mm broad; pls sometimes w Cas
   19a Lemmas rounded on the back, 10–13 mm, strongly pubescent; anthers 2–3 mm; bls 3–5 mm broad; dry to moist meadows and slopes from plains and foothills up to near timberl; RM, Alta to n Mex, e to Daks, w to Ida, Nev, Cal, and Ariz; nodding b. (*B. kalmii* var. *occidentalis*)
                                 18 B. anomalus Rupr.
   19b Lemmas slightly keeled, 8–11 mm, glab to scab-puberulent; anthers 5–6 mm; bls < 3 mm broad; intro from Europe and now estab in scattered areas, as in BC and Wn; meadow b. (*B. macounii*)
                                     19 B. erectus Huds.
16b Pan open, the brs gen ± spreading, mostly > the spikelets; 1st glume gen 3-nerved; lemmas often unevenly pubescent; ligs often much > 2 mm
   20a Lemmas pubescent unevenly, often only along the lateral nerves, the tip glab or the lower portion with much longer hairs than the upper; 1st glume often 1-nerved; pan open, the brs slender, spreading to drooping; bls 5–10 (15) mm broad
      21a Ligs ave ca 1 mm; awn mostly 2–4 (5) mm; anthers 1.5–2 mm; riverbanks, lake margins, and wet to semidry rocky slopes; c BC s through e Wn and Ore to Baja Cal, e to Lab, Mass, Tenn, Tex, and n Mex; fringed b. (*B. purgans* var. *longispicata, B. p.* var. *pallidus, B. richardsoni, B. inermis* var. *c.*)
                                      20 B. ciliatus L.
      21b Ligs (2) 3–5 mm; awn 3–8 mm; anthers 3–5 mm; shaded or open woods to moist or dry banks, from near sea level to dry, rocky slopes up to ca 6000 ft elev; BC s along the coast to Cal, e to sw Alta and Mont and Wyo; Columbia or narrow-fld b.; 2 vars.
                                21 B. vulgaris (Hook.) Shear
         a1 Sheaths gen glab; 1st glume gen 3-nerved; awns mostly 4–6 mm; ligs 2–3 (4) mm; occ from s Cas, Wn, to n Sierra Nevada, Cal, also in ne Ore (*B. ciliatus* var. *glaberrimus, B. laevipes, B. e.*)
                                  var. eximius Shear
         a2 Sheaths pilose or if glab the 1st glume 1-nerved or awns > 7 mm long or ligs at least 4 mm; range of the sp. as a whole (*B. ciliatus* var. *pauciflorus, B. eximius* spp. *umbraticus, B. e.* ssp. *robustus, B. purgans* var. *v., B. ramosus* sensu Chase)
                                        var. vulgaris
20b Lemmas pubescent to scabrid-puberulent evenly over the entire surface except sometimes the margins with longer hairs; 1st glume 3 (5)-nerved (1-nerved in *B. pacificus*); pan sometimes with short, stiffly spreading brs
   22a Ligs 2–5 mm; pan mostly > 15 cm, very loose and open; bls rarely < 6 mm (see lead 13a)
                                  12 B. pacificus Shear
   22b Ligs 1–1.5 (2) mm; pan rarely as much as 15 cm, often rather strict; bls often < 6 mm broad
      23a Bls very rarely as much as 6 mm broad; pan brs slender, gen drooping (see lead 19a)
                                18 B. anomalus Rupr.
      23b Bls mostly 5–10 (4–12) mm broad; pan brs short and stiffly spreading (see lead 18b)
                                17 B. orcuttianus Vasey

Buchloe Engelm.   Buffalo-grass

Spikelets of the ♂ pls 2-fld, sessile on one side of the rachis in short, racemosely arranged spikes, the glumes unequal and 1-nerved, the lemmas 3-nerved, > the glumes; spikelets of ♀ pls in a cluster surrounded by several ± modified and partially sheathing lvs, the aggregation deciduous as a whole, the spikelets gen in 2's in 2–3 separately ped groups, each pair surrounded by greatly modified and thickened structures that are lobed at the summit and more lf-like than glumaceous, the true glumes small and membranous, subequal, the single lemma 3-nerved and 3-lobed; low, creeping and sod-forming, ♂, ♀ per with solid culms and largely exposed internodes, open sheaths, short ligs composed chiefly of a fringe of hair, and short, curly, flat, non-aur bls rolled in the bud. (Gr *boubalos*, buffalo, and *chloe*, grass).

B. dactyloides (Nutt.) Engelm. Matted per, gen with pilose-hirsute sheaths and bls, the hairs of the throat 1–2 mm; culms 5–20 cm; bls short and curled, 1–2 mm broad; ♂ spikes 2–5, ca 1 cm, the spikelets 8–20, ca 5–6 mm; ♀ clusters ca 10 mm, the thickened, hardened subtending bracts of the spikelet pairs 5–7 mm; short-grass prairies from c Mont to Minn, s to e Ariz, NM, and n Mex; an important grazing element of the area between the RM and the Miss R; to be expected along our e border in Mont.

Calamagrostis Adans.   Reedgrass; Calamagrostis

Spikelets art above the glumes, borne in open to contracted or spikelike pan, gen 1-fld but a 2nd fl sometimes present (notably in *C. purpurascens*), the rachilla (ours) prolonged behind the palea and ± strongly bearded; glumes ca =, acute to acuminate, ± keeled, the 1st gen 1-nerved, the 2nd 3-nerved; lemma sometimes nearly = the glumes but gen <, thin, mostly 5-nerved, the 4 lateral nerves often extending into minute teeth, the midnerve prolonged into a straight to bent awn freed anywhere from just above the base to near the tip of the lemma; palea well developed, 2-nerved, from barely > 1/2 as long to fully as long as the lemma; stamens 3; lods 2; per of wet to dry places, gen rhizomatous, with open sheaths, prominent, membranous, externally ± scab-puberulent ligs, and flat to involute, non-aur bls rolled in the bud. (Gr *calamos*, reed, and *agrostis*, grass). *(Deyeuxia)*. The spp. are mostly palatable and nutritious, although not among the best of range grasses, seldom being particularly abundant.

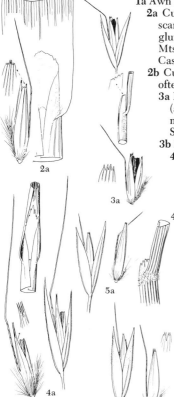

1a Awn bent, exserted (1) 1.5–10 mm beyond the glume tips
  2a Culm lvs flat, 5–13 mm broad; ligs mostly 6–15 mm; callus bearding scarcely 1 mm; pl (4) 6–15 dm, glab; pan 8–16 × ca 2 cm when pressed; glumes 5.5–7 (4.5–9) mm; subalp slopes and moist meadows, Chelan Mts, Chelan and Kittitas cos, Wn, and in Idaho Co, Ida; gen in timber; Cascade r.   1 C. **tweedyi** (Scribn.) Scribn.
  2b Culm lvs gen folded or involute, rarely as much as 5 mm broad; ligs often much < 6 mm; bearding of the callus often 1.5–4 mm
  3a Bls filiform, scarcely 1 mm broad; ligs barely 1.5 mm; glumes 3.3–4.3 (ave not > 4) mm; stream banks, lake margins, and moist subalp to alp meadows; Mt Hood, Clackamas Co, Ore, and Salmon-Trinity Mts and Sierra Nevada, Cal; shorthair r.   2 C. **breweri** Thurb.
  3b Bls gen > 1 mm broad; ligs > 1.5 mm; glumes ave > 4 mm
    4a Awn exserted 7–11 mm from the glumes (often the tip broken off); pan open, pyramidal, the brs spreading; ligs acute, 3–6 mm; bls 1–2.5 (3) mm broad, tough and wiry, nearly = culms; pl 3–6 dm; glumes 6–7 mm; rocky banks and crevices of cliffs in CRG, Wn and Ore; Howell's r.   3 C. **howellii** Vasey
    4b Awn exserted for not > 6 mm; pan compact to moderately open, oblong, the brs ascending to erect; ligs gen truncate or obtuse
      5a Collar crisp-pubescent; glumes 4–5 mm; awn strongly bent, gen exserted < 1.5 mm; pan tight to somewhat loose, 4–15 × 1–2.5 cm when pressed; open sagebr flats to timbered slopes of both dry and moist mont for; BC s in Cas of Wn and Ore to s Cal, e to Alta, and in RMS to Colo; pinegrass (*C. cusickii, C. suksdorfii, C. s.* var. *luxurians, C. l.*)   4 C. **rubescens** Buckl.
      5b Collar glab or only scab-puberulent, not crisp-pubescent; glumes 5–8 mm; awn gen not strongly bent, mostly exserted at least 1.5 mm
        6a Pan compact, mostly ca 10 (15–25) mm broad when pressed, scarcely interrupted; awn exserted 1–2 (3) mm; collar strongly scab-puberulent; glumes scab over the back; culms 3–6 (9) dm; valleys and foothills to subalp rocky slopes, often in parks in ponderosa pine for, typically on dry rocky soil in the open; Alas s, in both OM and Cas, to Cal, e to Que and s to Colo through

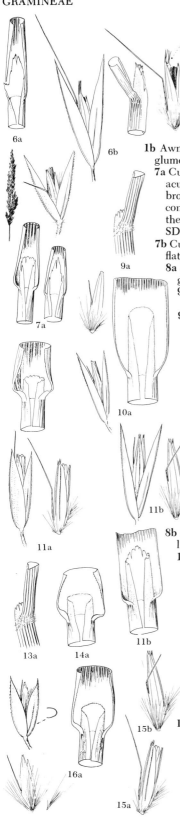

RMS and to SD and Minn; Greenl, Asia; purple r. or pinegrass (*C. vaseyi, Deschampsia congestiformis*)

5 C. **purpurascens** R. Br.

6b Pan somewhat loose and interrupted, mostly > 15 mm broad when pressed; awn exserted 3–5 mm; collar glab or at most lightly scab; glumes scab on the keel only; culms 1.5–4 (6) dm; on cliffs and shaded to open rocky mont to subalp slopes in our area, mostly at elevs of 3500–6000 ft; OM and Cas and Wen Mts, Kittitas to Skamania Co, Wn; Aleutian Is to ne Asia; one-and-a-half-fld r. (*Trisetum s.*)

6 C. **sesquiflora** (Trin.) Kawano

1b Awn straight or sometimes bent, but not exserted > 1 mm beyond the glume tips

7a Culms mostly 1.5–3.5 (6) dm, scab just below the pan; ligs 3–5 mm, gen acute and ± strongly lacerate; lvs strongly involute, erect, 1.5–3 mm broad; awn strongly bent, projecting sidewise from the spikelet; pan congested, 1–1.5 cm broad when pressed; grassl and sagebr benchl into the lower mts, esp on heavy soil; se BC and Alta to Man, s to Mont, Colo, SD, and Minn; plains r. 7 C. **montanensis** (Scribn.) Scribn.

7b Culms mostly either > 3.5 dm or not scab, or ligs not lacerate; lvs often flat; awn often straight; pan sometimes open

8a Awn bent, exserted sidewise slightly beyond the glumes; callus hairs gen < half as long as the lemma

9a Collar of at least some lvs crisp-pubescent (see lead 5a)

4 C. **rubescens** Buckl.

9b Collar never crisp-pubescent

10a Pl strongly rhizomatous, mostly (4) 8–15 dm; collar thickened and almost nerveless; bls (4) 6–12 mm broad; pl coastal, often in sand dunes, from along the beaches into the adj low mts; Alas to Cal; Pac r. (*C. albicans, C. aleutica, D. breviaristata*)

8 C. **nutkaensis** (Presl) Steud.

10b Pl scarcely rhizomatous, rarely as much as 8 dm; collar not thickened, the nerves evident; bls 2–6 mm broad; pl not coastal, mostly e Cas

11a Awn attached near the base of the lemma; bls mostly 2–4 mm broad; ligs 1.5–3.5 (4) mm; dry ridges and talus slopes to mt meadows, sw Ore to Cal, and in Teton Co, Ida, to nw Wyo, said to occur n to Wn, but not seen from n Ore or Wn; fire r. (*C. densa*) 9 C. **koelerioides** Vasey

11b Awn attached near or above midlength of the lemma; bls 3–7 mm broad; ligs 3–8 mm long; moist to dry soil of mont to fir, often on open slopes; doubtfully in Mont as reported, Wyo and Utah to NM and Ariz; cliff r. 10 C. **scopulorum** Jones

8b Awn nearly or quite straight, not projecting sidewise out of the spikelet; callus hairs often > half as long as the lemma

12a Callus hairs rarely > half as long as the lemma

13a Collars hairy, the hairs > those of the adj bl or sheath (see lead 5a) 4 C. **rubescens** Buckl.

13b Collars glab or only puberulent, the hairs, if any, not > those of the sheath or bl

14a Collar thickened, smooth (veins obscure or lacking); bls thick and tough, gen (4) 6–12 mm broad; pl coastal, mostly (4) 8–15 dm (see lead 10a) 8 C. **nutkaensis** (Presl) Steud.

14b Collar unthickened or at least with evident veins; bls 2–7 mm broad, not particularly thickened; pl of the dry interior, 4–8 dm

15a Awn from near the base of the lemma; bls 2–4 mm broad; lig 1.5–3.5 mm (see lead 11a) 9 C. **koelerioides** Vasey

15b Awn from near midlength of the lemma; bls 3–7 mm broad; lig 3–8 mm (see lead 11b) 10 C. **scopulorum** Jones

12b Callus hairs gen at least 3/4 as long as the lemma

16a Glumes thickened, rounded rather than keeled toward the base, ca 1/4 as broad as long; awn attached ca midlength of the lemma; ligs 1–3 mm; bls tough, flat to involute, 3–5 mm broad; rare in coastal swampland from Kodiak I to Cal, but known from a few stations on VI, 2 in Wn (Whatcom Lake and Lake Ozette), and from Mendocino Co and Pt Reyes Peninsula, Cal; thickglume r., Thurber's r. (*C. neglecta* var. *c.*) 11 C. **crassiglumis** Thurb.

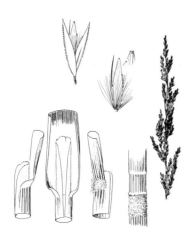

18a

18b

16b Glumes membranous, gen keeled to the base, mostly < 1/4 as broad as long; awn sometimes attached much below or above midlength of the lemma; ligs often much > 3 mm; bls mostly either tightly involute or lax and flat, often much > 5 mm broad

17a Pan relatively congested, rarely >2 cm broad when pressed, the individual brs erect or ascending, gen obscured by the spikelets; glumes mostly 3–4 mm; bls gen 1.5–4 mm broad, often involute; callus hairs often only 1/2–3/4 as long as the lemma; awn stout

18a Ligs of upper culm lvs mostly 2–4 mm, seldom lacerate; bls mostly 1.5–3 mm broad, flat to involute, scab chiefly on the margins; lemma very obscurely scab-puberulent; wet meadows, stream and pond margins, and moist mont slopes; Alas s, e Cas, to e Ore and n Nev, e to Greenl, nc and ne US, and s in RMS to Colo; Eurasia; slimstem r. (*C. laxiflora* Kearney, *C. lucida);* ours the var. *neglecta*

12 **C. neglecta** (Ehrh.) G.M. & S.

18b Ligs of upper culm lvs gen (4) 5–10 mm, often lacerate; bls often > 3 mm broad, stiff, gen strongly scab; lemma rather prominently scab-puberulent; wet to rather dry meadowl or streambanks, and other moist areas in the foothills and mts; Alas s to Cal and Ariz, e to Newf, Va, Ill, Neb, and NM; narrow-spiked r.; 2 vars.   13 **C. inexpansa** Gray

a1 Collars glab; pan rarely > 1.5 (2) cm broad when pressed; range of sp. except not in w Wn; marsh or northern r. (*Deyeuxia americana, C. neglecta* var. *i.*)

var. **inexpansa**

a2 Collars hairy; pan 2–2.5 cm broad when pressed; occ near sea level in Kitsap, Mason, and Grays Harbor cos, w Wn

var. **barbulata** Kearney

17b Pan almost always open, mostly > 2 cm broad when pressed, the brs tending to spread, rarely obscured by the spikelets; glumes often > 4 mm; bls sometimes > 4 mm broad, gen glab; callus hairs gen almost to fully as long as the lemma; awn delicate; wet places from near sea level to fairly high in the mts; Alas to Que, s to all but se US; bluejoint r.; several vars.

14 **C. canadensis** (Michx.) Beauv.

a1 Glumes (4.5) 5–6 mm, rather strongly scab; pan loose and open, rarely < 15 cm; bls flat, ave 4–10 mm broad; collars not hairy; Alas to Que, s to our area

b1 Awn geniculate, rather stout, gen slightly > the 4.5–5 mm glumes; callus hairs 1/2–3/4 as long as the lemma; sporadic, w side Mt Baker, Wn, and Siskiyou Mts, Ore (*C. l., C. langsdorffii* var. *l.*)   var. **lactea** (Beal) Hitchc.

b2 Awn almost or quite straight, weak, < the 5 mm glumes; callus hairs mostly = the lemma; Alas to Que and NH, s to OM and Cas, w Wn, and n Sierra Nevada, Cal (*C. c.* var. *langsdorfi* of auth., *C. oregonensis, C. scabra*)

var. **scabra** (Kunth) Hitchc.

a2 Glumes mostly < 4.5 mm, or collar hairy, or pan < 15 cm, or lvs narrow

c1 Collar hairy, the pubescence sometimes lacking on some lvs and often sparse on all; occ in the mts; Yuk to Mont, Wyo, and Colo, w to Ida, Wn, and Ore (*C. dubia, C. scribneri, C. langsdorfii* var. *s.*)   var. **robusta** Vasey

c2 Collar glab or scab but not pubescent

d1 Glumes mostly at least 3.8 mm, narrow, acuminate; awn from below the terminal 1/3 of the lemma

e1 Pan mostly diffuse and open, > 10 cm; glumes purple to green, rarely brownish-margined; widespread, but primarily RM, Alta to Colo, w to Pac from Alas to Cal (*C. langsdorffii* var. *a.*)

var. **acuminata** Vasey

e2 Pan rather narrowly pyramidal, gen tapered from the base, rarely > 10 cm; glumes purplish, not brownish-margined; n Can and Alas to BC, reoccurrent on Mt Adams, Wn (*C. anomala, C. scribneri* var. *i.*)   var. **imberbis** (Stebbins) Hitchc.

b1

e2   b2

17b

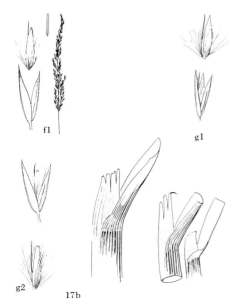

fl

gl

g2

17b

d2 Glumes ave < 3.8 mm, often abruptly acute, if longer the awn sometimes from the terminal 1/3 of the lemma; pan sometimes rather congested and narrow
 f1 Awn arising from near the tip of the lemma (at least on the terminal 1/4, gen slightly > the glumes; pan often straw-colored or light green, rather narrow; mostly n RM, Alta to se BC and YNP, but occ from Ida to se Wn and ne Ore *(C. blanda, C. p.)*
           var. **pallida** (Vasey & Scribn.) Stebbins
 f2 Awn arising on the lower 2/3 of the lemma, seldom > the glumes; pan gen purple, wide and lax
  g1 Glumes mostly < 3 mm, very abruptly acute, scarcely > the acute lemma; rare or lacking in w US, although apparently in a much-modified form as far w as Ida and Crook Co, Ore; absent in Wn; common from Gallatin Co, Mont, to Daks, Ill, and Ia *(C. c.* var. *campestris, C. m.)*
         var. **macouniana** (Vasey) Stebbins
  g2 Glumes mostly > 3 mm, gradually acute or acuminate, distinctly > the blunt lemma; common var. in most of c and e N Am, w to Pac coast and n to Alas, rare or lacking in Cal *(Arundo c.)*
              var. **canadensis**

Calamovilfa Hack.

Spikelets borne in rather large, narrow to open pan, 1-fld, art above the glumes, the rachilla not prolonged behind the palea; glumes slightly to strongly unequal, 1-nerved, keeled, rather sharply acute; lemma unawned, gen > the 1st glume but < the 2nd, with a strongly bearded callus; stamens 3; lods 2; wiry per with semi-solid culms, open sheaths, and ligs of short hairs, the collar and basal margin of the bl long-hairy, the bls non-aur, tough, involute, rolled in the bud. (Gr *kalamos*, reed, and *Vilfa*, another grass genus). Our pl is of little value as forage, although grazed to some extent esp by horses, but it is a sand-binder and helps to check wind erosion.

C. **longifolia** (Hook.) Scribn. Prairie sandgrass. Pl strongly rhizomatous, culms 6–15 dm; sheaths glab to soft-villous, the hairs at the throat 2–3 mm; ligs barely 1 mm; bls strongly involute, 3–8 mm broad, narrowed to long slender tip; pan narrow to somewhat spreading, up to 4 dm; glumes (3.5) 4.5–7 mm, acuminate, the 1st 1.5–2.5 mm < the 2nd; lemma glab, acute to blunt, the callus hairs abundant, ca half as long as the lemma; anthers 3.5–4.5 mm; prairies and foothills; se BC and Alta s to n Ida and in RM to Colo, e to Ill and Mo *(Calamagrostis l., Ammophila l.)*.

Catabrosa Beauv.   Brookgrass; Water Whorlwort

Spikelets in open pan, mostly 2 (1–5)-fld, art above the glumes; glumes membranous, nerveless, gen ± erose, < the 1st flt; lemmas with 3 prominent, nonconvergent nerves, unawned, the tip membranous, truncate and ± erose; lods membranous, flabellate and minutely lobed; stamens 3; rhizomatous per with hollow culms, at least partially closed sheaths, membranous ligs, and flat, non-aur bls folded in the bud, the tips ± prow-like. (Gr *catabrosis*, a devouring, referring to the erose glumes and lemmas). Our sp. is palatable, but of little forage value because of its aquatic habitat and comparative scarcity.

C. **aquatica** (L.) Beauv. Culms 1–6 dm, erect to decumbent; sheaths sometimes open half their length, sometimes closed to near the summit; ligs 2–8 mm, erose to subentire, minutely scab but not puberulent; bls flat, 2–13 mm broad; pan open, (5) 7–20 cm; glumes 0.7–1.3 and 1.5–2.2 mm, respectively; spikelets gen 2-fld, but not rarely only 1-fld; lemmas 2.5–3 mm; rachilla joint gen at least 1 mm, the upper flt gen deciduous considerably before the lower; wet meadows, stream banks, and edges of standing water; Alta to Newf, s through Ida and e Ore to Nev, Ariz, Colo, Ia, and Wis *(C. a.* var. *laurentiana)*.

Cenchrus L.   Bur-grass; Hedgehog-grass; Sandbur

Spikelets tightly enclosed within (and deciduous with) a spiny bur formed by the basal union of ∞ sterile brs, 1–several per bur, each 2-fld; 1st glume considerably reduced, narrow, 1 (3)-nerved; 2nd glume gen < 1st lemma, acute, 3-nerved; 1st (lower) flt sterile but gen with both a lemma and a reduced to well-developed palea; upper flt ♂, the lemma firmer in texture than the sterile lemma but not coriaceous, enfolding the palea; stamens 3; lods lacking in ours; ann or per, often matted, with solid culms (ours), open sheaths, ligs mainly of hairs, and flat to folded bls rolled in the bud, the burs subglobose, few to several in terminal spikelike racemes, individually deciduous. (Gr name, *kegchros*, for some millet).

C. longispinus (Hack.) Fern. Prostrate, often extensively matted ann; sheaths keeled, glab; collar constricted, sometimes pubescent; ligs 1–1.5 mm, the fringe 2–4 times as long as the basal membrane; bls flat or folded, 2–6 mm broad; burs 5–10 in a short, crowded raceme, the body 4–6 mm broad when pressed, finely pubescent, the spines flattened, ± connate at base; spikelets gen 2 per bur, 5–6 mm; e Wn and Ore to Mex and S Am, e to Ont and most of the US (except Mont?), gen in sandy soil; a serious pest for man and beast, esp for picnickers, campers, or bird hunters (*C. tribuloides* and *C. pauciflorus*, misapplied).

Cinna L.   Woodreed; Wood Reed-grass

Spikelets in large pan, 1-fld (very occ some 2-fld), flattened, art below the glumes; glumes (ours) 1-nerved, strongly keeled; flt stipitate, the rachilla prolonged as a bristle behind the palea; lemma much-flattened, 3-nerved, awnless or gen with a short, subterminal, straight awn; callus not bearded; stamens 3; lods 2; per, mostly of wet places, with hollow culms, open sheaths, membranous ligs, and broad, lax, non-aur bls rolled in the bud. (Gr *kinni*, an old name for some grass). Our sp. is palatable, but seldom abundant enough to furnish much forage.

C. latifolia (Trevir.) Griseb. Rhizomatous, mostly 7–20 dm; sheaths glab or finely scabridulous; ligs pubescent, 3–8 mm, erose and intact to ± strongly lacerate; bls flat, 7–15 mm broad; pan 15–30 cm, loose, the brs spreading to drooping; glumes slender, acuminate, (2) 3–4 mm, the 2nd slightly the longer, scab-puberulent on the keel and often over the back; lemma strongly compressed, 2–3.2 mm, puberulent over the back, awnless or awned, sometimes in the same pan, the awn subterminal, up to barely 1 mm; rachilla bristle ca 0.6 mm; moist woods, meadows, and wet areas along streams, from near sea level to subalp; Alas to Newf, s on both sides Cas, to Cal, Nev, NM, Gr Lakes, and NC (*Agrostis l., C. expansa, C. pendula* vars. *glomerula* and *mutica*).

Coleanthus Seidel   Moss-grass

Spikelets 1-fld, in a pan of rather distant verticillate brs in umbel-like clusters; glumes lacking; lemma and palea very loosely enveloping the ovary, the lemma 1-nerved, glumelike, narrowed to a slender awnlike tip; palea ca half as long as the lemma, 2–4-toothed at the tip; rachilla not prolonged; stamens 2; lods lacking; fr slender, > the lemma and palea; tiny ann with open sheaths, short membranous ligs, and very short, non-aur bls. (Gr *koleos*, sheath, and *anthos*, fl, in reference to the inflated sheath from which the infl projects).

C. subtilis (Tratt.) Seidel. Spreading, ± matted, barely 5 cm; sheaths inflated, ligs 1–1.5 mm; bls ca 1 (2) cm × 0.5–1.5 mm; pan 1–5 cm, the verticils of brs gen 3–6; lemma ca 1 mm, the awnlike tip ca = the ovate basal portion; stamens exserted, the anthers ca 0.3 mm; fr brownish, fusiform, barely 1 mm; European weed, intro and persistent along sand bars of the CR and on islands near Portland, Ore.

Corynephorus Beauv.

Spikelets pan, art above the glumes, 2-fld; glumes subequal, membranous, keeled, 1-nerved, exceeding the 2nd flt; rachilla hairy, not prolonged beyond the upper flt; lemmas hyaline-membranous, rounded on the back, 1-nerved, lightly bearded on the callus, awned from just above the base; awn bearded and jointed near midlength, the lower segm brownish or reddish, terete, the upper segm slenderly clavate, gen pale green; stamens 3; lods 2; ann or caespitose per (ours) with hollow culms, open sheaths, membranous ligs, and narrow, involute, non-aur bls. (Gr *corynephoros*, club-bearing, in reference to the clavate awns).

C. canescens (L.) Beauv. Gray hairgrass. Tufted, glaucous per 1–3 (4) dm; lvs finely and copiously scab, the ligs (1) 2–4 mm, acute, puberulent-scab; bls < 1 mm broad, filiform, involute; pan 2–10 cm, gen purplish-tinged, narrow, the brs erect-ascending; glumes 3–4 mm, scab on the keel; lemmas ca 1.5 mm; awn ca twice as long as the lemma, bent at the joint; anthers orange or purple, ca 1.2 mm; native to Europe, reported on waste ground or old ballast-dumping areas, in our region well estab on VI and recently collected in Seattle, Wn, in neither place near old ballast areas.

Cynodon Rich.   Cynodon

Spikelets sessile, borne in 2 rows on one side of the rachis, in several digitate terminal spikes, art above the glumes, gen 1-fld, the rachilla prolonged and bristle-like, very occ bearing a rudimentary 2nd flt; glumes subequal, narrow, 1-nerved; lemma ± indurate, strongly compressed, > the glumes, ciliate-pubescent on the dorsal and 2 submarginal nerves; stamens 3; lods 2, ± wedge-shaped; low per, stoloniferous and rhizomatous, with hollow or pith-filled culms, open sheaths, ligs mainly of long straight hairs, and flat to involute, non-aur bls rolled in the bud. (Gr *kuon*, dog, and *odous*, tooth, because of the hardened, sharp, toothlike scales of the rhizomes).

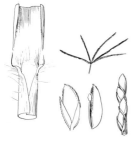

C. dactylon (L.) Pers. Bermuda grass. Per with elongate, scaly rhizomes, the aerial sts decumbent and freely rooting; upright culms gen 1–3 dm; collars and throats long-hairy; ligs with membranous base scarcely 0.4 mm, fringed with hairs 2–4 mm; bls flat, mostly 1.5–3 (5) mm broad; spikes mostly 4–5, up to 5 cm; glumes ca 1.3 mm; lemma ca 2 mm; Eurasian, widespread in most of the warmer areas of the world, often used as a lawn grass, but also a pernicious weed; in our area reported (prob as a waif) in Ida, and repeatedly collected in and around Portland, Ore, where it seems to persist along railroad tracks and dock yards (*Capriola d.*).

Cynosurus L.   Dogtail; Dog's-tail Grass

Spikelets subsessile in much-contracted, strongly secund, spikelike pan, dimorphic, borne mostly in 2's, on very short brs, one spikelet of the pair fertile, the other short-ped, much flattened and fanlike, consisting of the glumes and several similar empty lemmas; fertile spikelets (gen more numerous than the sterile) 2–3 (4)-fld, the upper flt often rudimentary; glumes narrow, acuminate, 1-nerved, the margin finely serrulate; lemmas ± lopsided, rounded on the back, awn-tipped, with 5 rather obscure convergent nerves; stamens 3; lods 2; ann or per with hollow culms, open sheaths, prominent membranous ligs, and mostly flat, non-aur bls. (Gr *kuon*, dog, and *oura*, tail, in reference to the infl).

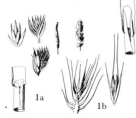

1a Pl per, 4–8 dm; ligs truncate, barely 1.5 mm, finely erose-ciliolate; pan < 1 cm thick; awns < 1 mm; native of Europe; sometimes cult, and intro in many places in US and Can, occurring in our area in Ida and occ w Cas in Wn and Ore to Cal; crested d.   1 **C. cristatus** L.
1b Pl ann, (1) 2–5 dm; ligs 2–5 (11) mm, erose to toothed; pan > 1 cm thick; awns 3–10 mm; European weed, intro in many parts of US, well estab w Cas, sw BC to WV, Ore; hedgehog d.   2 **C. echinatus** L.

Dactylis L.   Orchard-grass; Cock's-foot Grass

Spikelets compressed, 3 (–5)-fld, art above the glumes, subsessile in dense 1-sided clusters at the ends of comparatively few stiff pan brs; glumes keeled, the 1st lopsided, gen 2 (3)-nerved, both with short, soft, awnlike tips; lemmas keeled and ciliate on the keel above, gen awn-tipped, with 5 rather obscure, convergent nerves; stamens 3; lods 2; per with hollow culms, closed and ± compressed sheaths, long membranous ligs, and broad, non-aur bls folded in the bud. (Gr *daktulos*, finger, said to be in reference to the stiff pan brs).

D. glomerata L. Strongly tufted, up to 12 dm; sheaths glab to slightly scab; ligs mostly 3–9 mm, gen sparsely pubescent, obtuse to acute, the margin finely erose-ciliolate, but the upper half gen turned back and split in several places; bls flat, (2) 3–11 mm broad, mostly scab, gen the 2 margins freed from the sheath at different levels; spikelets compressed, 5–9 mm; glumes 4–6 mm; lemmas 5–8 mm, the awn tip ca 1 mm; waste places, roadsides, and meadows; intro in much of N Am, common throughout our area at lower elev; Eurasia; commonly cult for hay and pasture and occ weedy. The flts rarely are modified into bulblets.

Danthonia Lam. & DC.   Danthonia; Oatgrass

Spikelets at least 1 cm, several-fld, art above the glumes, borne in rather small, often much-reduced pan, occ even solitary; glumes gen > the flts, rounded to keeled, mostly 3- or 5-nerved; lemmas broadly rounded, pilose on the margins and over the back, strongly bifid, with a flattened, twisted and ± bent awn from just below the lobes; callus bearded but the rachilla itself glab; stamens 3; lods 2; caespitose pers, mostly 2–7 (1–10) dm, with hollow culms, open sheaths, ligs composed almost entirely of rather short hairs, and narrow, non-aur bls rolled in the bud. (For Etienne Danthione, a French botanist of the early 19th century). Of fair palatability but of 2ndary importance as range grasses, as seldom abundant.

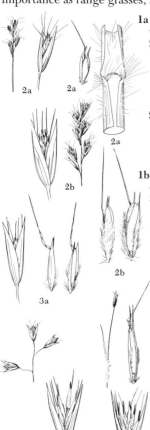

1a Lemmas ± pilose over the back as well as along the margins; pan narrow, with (3) 5–10 spikelets, the brs erect, at least after anthesis
　　2a Glumes 9–12 mm; lemmas 4–5 mm, the apical teeth acute to acuminate, 0.5–2 mm; sheaths often pubescent; anthers 2–2.5 mm; sandy to rocky soil, mostly in dry woods but also in semidry meadows; Alas to Wn and Ore, both sides Cas, e to Newf and s to most of c, s, and e US and in RMS to NM; poverty d., common wild o. *(D. pinetorum, D. thermale)*; ours the var. *pinetorum* Piper   **1 D. spicata** (L.) Beauv.
　　2b Glumes 16–22 mm; lemmas 10–13 mm, the apical teeth aristate, 2–3 mm; sheaths glab; anthers 5–6 mm; sagebr foothills to light mont for, in moist meadows or on open slopes and ridges up to near timberl; RMS, Alta to NM, but not in Ida; Parry's d. or o. *(D. p.* var. *longifolia)*
　　　　　　　　　　　　　　　　　　　　　　2 D. parryi Scribn.
1b Lemmas glab over the back, pilose only along the margins and on the callus; pan often open, with (1) 2–5 (10) spikelets
　　3a Pan narrow, the brs and pedicels erect, ± secund; spikelets mostly (4) 5–10; sheaths gen glab; anthers ca 4 mm; dry to moist, often rocky soil; prairies or grassl to timbered areas or to alp ridges above timber; Alas to Cal, from near the coast inl, also in mts of all the w states, e across Can to Newf and s to Mich; timber d. or o. *(D. cusickii, D. i.* var. *c.)*
　　　　　　　　　　　　　　　　　　　　3 D. intermedia Vasey
　　3b Pan open, the brs gen spreading; spikelets mostly 1–4 (5); sheaths gen pilose if the infl at all narrow and congested
　　　　4a Pls mostly > 5 (3–10) dm; spikelets gen > 1, borne on divergent brs; hairs of the collar and throat mostly 1–2 mm; terminal teeth of lemmas up to 4 mm; open grassy meadows to rocky ridges, from coastal prairies to mont for; BC s, both sides Cas, to s Cal, e to Mont and s in RMS to NM; Chile; Cal d., Cal o. *(D. americana, D. c.* var. *a., D. macounii, D. c.* var. *palousensis, D. c.* var. *piperi)*   **4 D. californica** Boland.
　　　　4b Pls mostly 1–3 dm; spikelets gen 1, if > 1 the lower one(s) short-ped and suberect; some of the hairs of the collar and throat 3–4 mm; terminal teeth of lemma 1–2 mm; dry to occ moist prairies, foothills, and open parks and ridges in mont for; BC s on both sides Cas to Cal, e to Alta, Mont, Wyo, Utah, and Colo; onespike d., few-fld wild o. *(D. californica* var. *u.)*   **5 D. unispicata** (Thurb.) Munro

Deschampsia Beauv.   Hairgrass

Spikelets in open to contracted pan, often glistening or purplish, gen 2 (3)-fld, art above the glumes, the rachilla prolonged beyond the uppermost flt; glumes acute, subequal to unequal, > at least the lower flt; lemmas with 5 gen obscure nerves, truncate and 2–4-toothed, bearded on the callus, awned from about midlength or below, the awn twisted to bent; stamens 3; lods 2; ann or caespitose per with hollow culms, open sheaths, rather long, membranous, pubescent ligs, and flat to narrow and involute, non-aur bls rolled or folded in the bud. (For L. A. Deschamps, 1774–1849, a French botanist).

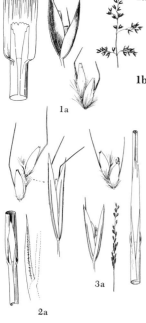

1a Bls flat, 3–6 mm wide, rolled in the bud; ligs obtuse to truncate, 1–3.5 mm; spikelets purplish; awns stout, scarcely 3 mm; pl per, 1.5–6 dm; glumes ca 5 mm; mt meadows, streambanks, and lake margins to alp ridges; Alas s through Wn and Ore to Cal, and in RMS to Ida, Mont, Wyo, and Colo, across Can to Newf and NH; Eurasia, S Am; mt h. (*Aira a.; A. latifolia; D. l.,* misapplied; *D. a.* var. *minor);* ours the var. *latifolia* (Hook.) Scribn.

1 **D. atropurpurea** (Wahl.) Scheele

1b Bls often folded or involute, gen < 3 mm broad, folded in the bud; ligs mostly acute or acuminate, at least some gen > 3.5 mm; spikelets mostly greenish or tawny; awns gen slender and > 3 mm

  2a Pl ann; bls filiform, rarely as much as 1.5 mm broad; glumes 5–8 mm; roadsides, dry banks, and drying vernal pools to dry mt meadows; Alas to Baja Cal, on both sides Cas, e to w Mont, Utah, and Ariz; Chile; ann h. (*Aira d.)*   2 **D. danthonioides** (Trin.) Munro

  2b Pl per; bls often > 1.5 mm broad; glumes often < 5 mm

    3a Pan narrow; glumes gen = or > the upper flt; anthers < 1 mm; bls filiform to as much as 1.5 mm broad; ligs acuminate, 3–9 mm; sandy or gravelly, gen moist banks and slopes or borders of streams or lakes, sometimes in woods; from near sea level to alp drainages; Alas to Cal, e to Alta and Mont and s to NM and n Mex; S Am; slender h. (*Aira e., D. e.* var. *ciliata, D. c., A. vaseyana*)   3 **D. elongata** (Hook.) Munro

    3b Pan gen open, or narrow but the glumes < the upper flt; anthers 1.2–2.2 mm; bls 1.5–2 (4) mm; ligs acute or obtuse, mostly ca 4 (to 8) mm; coastal marshes and prairies to alp ridges and talus slopes; Alas to Greenl, s to most of US and n Mex; Eurasia; tufted h.; 3 vars.

4 **D. cespitosa** (L.) Beauv.

    a1 Glumes mostly 5–7 mm; lemmas ave 4 mm; pan open during anthesis; Alas to VI and n coast of Wn (*D. beringensis, D. c.* ssp. *b., D. b.* var. *atkensis*)   var. **arctica** Vasey

    a2 Glumes mostly <5 mm; lemmas ave ca 3.5 mm, sometimes longer but then the pan closed during anthesis

      b1 Pan closed during anthesis, narrow, the brs erect; glumes up to 5.5 (6) mm; lemmas up to 4 mm; along coast from sw VI to c Cal (*D. c.* var. *maritima? D. holciformis, D. c.* var. *h.*)

var. **longiflora** Beal

      b2 Pan open during anthesis, the brs spreading or drooping; glumes < 5 mm; lemmas ca 3.5 mm; range of the sp. for N Am, except rare along coast (*Aira alpicola, D. c.* var. *alpina, D. pungens, D. c.* var. *p.*)

var. **cespitosa**

Digitaria Heister   Crabgrass

Spikelets art below the glumes, strongly obcompressed, mostly (ours) paired and in 2 rows in linear racemes on one side of a flattened and 3-angled rachis, 1 spikelet short-ped, the other subsessile, the racemes gen several, approximate to wholly digitate; 1st glume greatly reduced (ours) or even lacking; 2nd glume well developed; flts 2, lower one sterile; sterile lemma gen as long as the fertile, mostly 5-nerved; fertile lemma and palea ca =, much-hardened, smooth; stamens 3; lod 2; ann (ours) with hollow culms, open sheaths, membranous ligs, and flat bls rolled in the bud, tending to root freely at the nodes and to form large patches. (L *digitus,* finger, in reference to the fingerlike, or digitate, arrangement of the racemes). (*Syntherisma*). Our spp. noxious weeds.

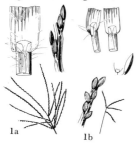

1a Sheaths and collars strongly pustulate-hirsute; 1st glume ovate-lanceolate, not transparent, ca 0.5 mm; 2nd glume ca 2/3 as long as the ca 3 mm fertile lemma; bls 4–6 (10) mm broad, hirsute toward the base at least; chiefly in lawns, and in waste places or gardens in most of the US; cosmopolitan weed from Europe; hairy c. (*Panicum s.*)   1 **D. sanguinalis** (L.) Scop.

1b Sheaths and collars gen nearly or quite glab; 1st glume semicircular, transparent, scarcely 0.3 mm; 2nd glume ca = the 2 (2.5) mm fertile lemma; bls 2–4 (6) mm, glab to slightly scab, or rarely with a few pustulose hairs near the base; lawns, roadsides, and wasteland, gen where moist; ubiquitous weed from Europe; smooth c. (*Panicum i.*)

2 **D. ischaemum** (Schreb.) Schreb.

Distichlis Raf.   Saltgrass

Spikelets in compact pan, few–∝-fld, art above the glumes; glumes < the 1st lemma, unequal, the 1st 3 (5)-nerved, the 2nd 5–7 (9)-nerved; lemmas hardened, unawned, with 9–11 convergent, rather obscure nerves; paleas subequal to the lemmas, the 2 keels ± winged; stamens 3; lod 2; ♂, ♀, rhizomatous per with solid culms, open sheaths, short membranous ligs that are ± erose-ciliolate and sometimes tipped with long stiff hairs near the front, and stiff, sharp-pointed, non-aur bls rolled in the bud. (Gr *distichos*, referring to the distichous [i.e., 2-ranked] lvs). Wiry and perhaps unpalatable grasses that are seldom grazed except during extreme drought.

1a Keel wing of the ♀ paleas 0.2–0.3 mm broad, ± serrulate; long hairs gen on the collar as well as at the top of the sheath margins; glumes ca 2.5 and 5 mm respectively; pan comparatively loose, the peds of the spikelets mostly discernible; alkali flats and des seeps to sandy lakeshores; s BC s, e Cas, to Baja Cal, e to Sask, Mo, Okla, and Tex; alkali s.; 2 vars.
                                                                        1 **D. stricta** (Torr.) Rydb.
   a1 Upper surface of the lvs commonly sparsely to copiously pilose-hirsute; keels of the ♀ paleas subentire to remotely dentate; RM primarily, Sask to Colo, w to e Wn and Ore *(Uniola stricta, D. maritima* var. *s., D. spicata* var. *s.)*   var. **stricta**
   a2 Upper surface of the lvs glab to very sparsely pilose-hirsute; keels of ♀ paleas often prominently dentate; common var. from BC to Cal and e to Ida and Utah *(D. d.)*   var. **dentata** (Rydb.) Hitchc.
1b Keel wing of the ♀ paleas ca 0.1 mm broad, entire; hairs gen lacking on the collar, present only near the top of the sheath margins; glumes ca 2–2.5 and 2.3–5 mm; pan compact, the peds mostly visible; coastal beaches and salt marshes, VI to Cal and on the Atl coast; seashore s. *(Brizopyrum boreale, D. spicata* var. *b.)*; ours the var. *borealis* (Presl) Beetle
                                                                        2 **D. spicata** (L.) Greene

Echinochloa Beauv.   Cockspur; Barnyard-grass

Spikelets subsessile to short-ped in rather stiff pans, borne singly or in clusters along one side of the pan brs, art below the glumes, 2–fld, the lower flt sterile, dorsiventrally compressed and slightly flattened on the side; glumes unequal, mostly 3 (5)-nerved, often hispid to echinulate, unawned; sterile lemma = the 2nd glume and similar in texture, often awned, including a membranous palea ca half its length; fertile lemma and palea hardened, unawned; stamens 3; lods 2; ours anns with pithy or hollow culms, open, ± compressed sheaths, and flat bls rolled in the bud, lacking both aurs and ligs. (Gr *echinos*, hedgehog, and *chloa*, grass, in reference to the echinate spikelets). Although some spp. are regarded as fair forage, ours are troublesome weeds, esp in irrigated fields e Cas.

1a Racemes (brs of the pan) simple, gen only 1–2 (3) cm; spikelets scarcely 3 mm, the sterile lemma unawned; nodes gen pubescent; known in our area only from collections near Portland, Ore, where not persistent; small b.
                                                                        1 **E. colonum** (L.) Link
1b Racemes in part gen br, mostly at least 3 cm; spikelets 3–4 mm, the sterile lemma often awned; nodes glab; moist areas, esp in irrigating ditches; widespread weed in temperate to tropical regions; large b., watergrass *(E. muricata* var. *occidentalis, E. o., E. pungens* var. *wiegandii)*
                                                                        2 **E. crusgalli** (L.) Beauv.

Eleusine Gaertn.   Eleusine

Spikelets several-fld, art above the glumes, sessile and spicate in 2 rows on one side of the rachis, the spikes sometimes only 1 or 2 but gen several and digitate or 1–2 additional below the terminal group; glumes unequal, < the 1st lemma, several-nerved but the keel greenish and apparently the only nerve; lemmas unawned, strongly keeled, several-nerved, but gen the keel and 2 lateral nerves greenish and most prominent; stamens 3; lods 2; hollow-culmed anns with open sheaths, short membranous ligs, gen long hairs on the throat, and folded (ours) non-aur bls rolled in the bud. (Named for *Eleusis*, an ancient Gr town).

E. indica (L.) Gaertn. Goose-grass. Basally br, glab ann up to 6 dm; throat with hairs up to 6 mm, the lig itself ca 0.5 mm, finely ciliolate; bls mostly folded, 3–8 mm broad; spikes mostly (2) 3–6, up to 10 cm; glumes ca 3 and 4 mm; lemmas ca 4 mm, the paleas not much > 1/2 as long; occ weed in w Ore, more common in Cal, widespread in the tropics (*Cynosurus i., E. gracilis*). *E. tristachya* Lam., an African sp. differing chiefly in having only 1–3 spikes scarcely 2.5 cm, was collected several times some 60 yrs ago on ballast at Albina (Portland), Ore.

Elymus L.   Wildrye; Ryegrass

Spikelets in a single terminal spike or spikelike infl, 2–6 (12)-fld, art above the glumes, borne flatwise to the rachis, gen sessile and 2 per node, the rachis continuous, but not rarely the spikelets either 1 or > 2 at some nodes, and frequently 1– more of the spikelets shortly to prominently pedicellate, the spike even occ compound; glumes = to slightly unequal, narrowly lanceolate to subulate, firm to rigid, 1–5 (6)-nerved, acute to slenderly awned, often ± parallel and borne in front of the rest of the spikelet; lemmas rounded on the back, faintly nerved, awnless to prominently and often divergently awned; paleas gen nearly = the body of the lemmas; lods 2, often very prominent; stamens 3; ann or gen caespitose to rhizomatous pers with hollow culms, open sheaths, gen well-developed aurs, mostly either very short or rather prominent membranous ligs, and stiff and involute to broad and flat bls rolled in the bud. (Gr *elumos*, the ancient name for some kind of a grain). Most spp. are fair to good forage, many being eaten during the winter, although some are too coarse to be used to any extent. The rhizomatous spp. are good soil binders and several are planted to help stabilize shifting sand along highways and seacoast.

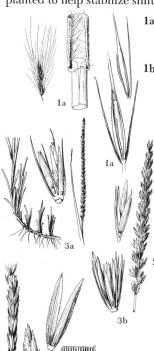

1a Pl ann; spikes 1–5 cm; spikelets with 1 ⚥ fl; glumes with awn up to 3 cm; lemma with flattened awn 3–10 cm; bls 1–1.5 mm broad; European, now widely intro in sagebr area of Wn, Ore, and Ida to Cal; medusahead w. (*Hordeum c., Taeniatherum c.*)   1 **E. caput-medusae** L.
1b Pl per; spikes gen > 5 cm
  2a Ligs mostly at least 2 mm; lf bls gen flat, (5) 8–20 mm broad; spikelets commonly > 2 at most nodes of the spike
    3a Pl strongly rhizomatous; culms rarely > 1 m; spikelets all sessile; local on sand dunes near Washtucna, Adams Co, Wn, where intro as a sandbinder; native to Russia; Siberian w.   2 **E. giganteus** Vahl
    3b Pl short-rhizomatous; culms 1–2 m; spikelets in part ped; along streams and gullies, on gravelly hills and on sand dunes in sagebr-covered to wooded areas; e Cas, BC to Cal, e to Alta, Minn, Colo, and NM; giant w.; 2 vars.   3 **E. cinereus** Scribn. & Merr.
      a1 Sheaths soft-hairy, the hairs spreading to appressed; bls pubescent; sagebr des, c Wn (Grant Co) to s Ore, occ to Wyo (*E. condensatus* ssp. *p.*)   var. **pubens** (Piper) Hitchc.
      a2 Sheaths glab to very finely puberulent; bls glab to scab, rarely puberulent; range of the sp. (*E. c.* f. *laevis; E. condensatus*, misapplied)   var. **cinereus**
  2b Ligs rarely if ever as much as 2 mm; lf bls often either involute or < 8 mm broad; spikelets rarely > 2 per node
    4a Rhizomes well developed
      5a Glumes lanceolate, prominently 3–6-nerved; anthers 5–9 mm; lemmas pubescent; bls 6–15 mm broad; culms stout, 5–15 dm, gen copiously hairy above (glab); coastal dunes; Greenl and e arctic Am and w N Am from Alas to Cal; nw Asia, rarely inl; dune w., Am dunegrass (*E. aleuticus; E. arenarius*, misapplied; *E. dives; E.* × *uclueletensis; E. vancouverensis; E. m.* ssp. *m.* var. *m.* f. *simulans*)   4 **E. mollis** Trin.
      5b Glumes narrow, often subulate, mostly 1–3-nerved; anthers gen (2) 3–5 (7) mm; either pls from e Cas or lemmas glab
        6a Lemmas gen glab (sparsely puberulent), gen with a 1–6 mm awn; bls 3–6 mm broad; culms 5–10 (12) dm; dry to moist, often saline meadows, river flats, and sand dunes from sagebr des to lower mts; c Wn and CRG to Baja Cal, e to Mont, Wyo, Colo, and NM; creeping or beardless w.; 2 vars.   5 **E. triticoides** Buckl.
          a1 Pl pubescent with mostly spreading hairs; esp common in Wn and Ore, s to c Nev   var. **pubescens** Hitchc.

a2 Pl glab or minutely strigillose throughout; range of sp., the more common phase in CRG, most of Cal, and the more e and s part of the range (*E. acicularis, E. condensatus* var. *t.*)

var. **triticoides**

6b Lemmas strongly pubescent, often awnless

   7a Spikes gen at least 10 cm; lemmas awnless; aurs lacking or poorly developed; bls 3–5 mm broad; culms 6–10 (12) dm; sand dunes, open flats, and ditch and roadbanks; c Wn and Ore to SR plains of Ida and to n Madison Co, Mont; yellow or sand w. (*E. arenicola, Leymus a.*)

6 **E. flavescens** Scribn. & Smith

   7b Spikes rarely as much as 10 cm; lemmas awned or awn-tipped; aurs gen prominent; bls stiff, 2–4 mm broad; culms 4–8 (10) dm; sandy meadows, streambanks, and rocky hillsides to open lodgepole or spruce for; Alas and BC to Alta, s to Mont, Wyo, and SD (*E. brownii*)   7 **E. innovatus** Beal

4b Rhizomes lacking

  8a Glumes narrowly lanceolate, gen broadest somewhat below mid-length, nearly parallel at the flattened, not strongly indurate base, concealing at least the 1st joint of the rachilla; lemmas not strongly scab-pubescent on the back; awns, if any, mostly straight; culms 5–12 dm

    9a Spike rather flexuous; lemmas strongly ciliate-margined, the cilia several times as long as any dorsal pubescence; bls flat, 4–10 mm broad; prairies and woods to open, dry to moist slopes; Alas to nw Ore, w slope Cas to coast in Wn; hairy w., n rye-grass (*E. borealis; E. ciliatus,* misapplied)   8 **E. hirsutus** Presl

    9b Spike stiff and erect; lemmas not ciliate or if so the hairs of the margins no > the pubescence on the back of the lemma; bls flat, (3) 5–10 (15) mm broad; prairies, open woods, and dry to moist hillsides, lowl to midmont; s Alas to s Cal, e to Ont, Mich, Ind, Ia, Colo, and NM; blue w., w rye-grass; 3 vars.

9 **E. glaucus** Buckl.

     a1 Lemmas awnless or awn-tipped, the awn rarely > 5 mm; s Alas to c Cal, mostly close to coast, but up to 6000 ft elev in Wn (*E. virescens, E. g.* var. *v.*)   var. **breviaristatus** Davy

     a2 Lemmas with awn 1–2 (3) cm

      b1 Sheaths and bls gen ± hirsute or pilose with spreading to slightly retrorse hairs; BC to Cal, from coast e to Mont, Ida, and Nev (*E. edentatus, E. g.* var. *tenuis, E. petersonii*)

var. **jepsonii** Davy

      b2 Sheaths and bls glab to scab; range of the sp. as a whole (*E. sibiricus* var. *g., E. nitidus, Clinelymus g.* ssp. *coloratus*)

var. **glaucus**

  8b Glumes linear and tapered from the base, or if broadest somewhat below midlength then their bases hardened and not parallel, thus exposing the 1st joint of the rachilla; lemmas not long-ciliate, but often very strongly scab-pubescent; awns, if any, mostly strongly divergent

   10a Glumes often 3–5-nerved, gen long-awned; lemmas mostly long-awned; bls flat, 5–20 mm broad, many of them cauline

    11a Lemmas mostly strongly scab-pubescent, with a gen strongly divergent awn 2–3 cm; glumes not strongly bowed out at base, often flattened; common; streambanks and sandy, dry to moist meadows and mt canyons; Alas s, e Cas in BC, Wn, and Ore, to n Cal, e to much of US and s Can, incl RMS; Can or nodding w. (*Hordeum c., Sitanion brodiei*)   10 **E. canadensis** L.

    11b Lemmas glab to scab, with a gen straight awn rarely as much as 2 cm; glumes semiterete and strongly bowed out at the base; a variable sp. of c and e US and s Can, of which a short-awned phase, var. *submuticus* Hook., is known in our area from ne Wn, Boundary Co, Ida, and in both w and e Mont; ter-rell-grass, Va w.   11 **E. virginicus** L.

   10b Glumes mostly subulate, unnerved or 1–2-nerved, awnless or awn-tipped; lemmas awnless or with an awn rarely > 5 mm; bls mostly involute or folded, rarely > 5 mm broad, mostly basal; in light and sandy to heavy clay soil, sagebr or short-grass prairie to

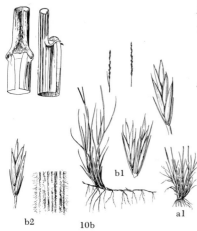

pinyon or juniper woodl; Alas s, mostly e RM, to Colo, w side RM from s Ida to ne Ariz; 3 vars.   12 **E. ambiguus** Vasey & Scribn.

a1 Lvs mostly in a basal tuft, bls rarely as much as 10 cm, glab to puberulent, not soft-hairy; glumes mostly < spikelet, often greatly reduced; pl non-rhizomatous; mostly in sagebr des or grassl, s Ida to Ariz, e to Wyo and Colo *(E. strigosus, E. a. var. s.)*
                                                              var. **ambiguus**

a2 Lvs largely cauline, the bls mostly 10–20 cm, sometimes soft-hairy; glumes well developed, often ca = the spikelet; lemmas mostly with awn tip 1.5 mm

   b1 Sheaths glab or scab, not soft-pubescent; pl gen rhizomatous; awn tip of lemma often 1.5–5 mm; mostly in pinyon or juniper woodl to mont at 6000–9000 ft; Alta to Colo and Utah, possibly not in Mont; salina w. *(E. s.)*
                                                   var. **salina** (Jones) C. L. Hitchc.

   b2 Sheaths copiously soft-pubescent; awn tip of the lemma mostly 1–1.5 mm; Salmon R, Ida, near Challis
                                                   var. **salmonis** C. L. Hitchc.

Eragrostis Beauv.   Lovegrass; Eragrostis

Spikelets in open to contracted pans, several—∝-fld, art above the glumes but the glumes and lemmas often deciduous before the rachilla shatters; glumes from subequal to strongly unequal, < the 1st flt, mostly 1-nerved but the 2nd one sometimes 3-nerved, often keeled; lemmas awnless, 3-nerved (the nerves faint to prominent, nonconvergent), rounded to keeled on the back, gen deciduous before the somewhat shorter paleas; lods 2; stamens 3 (ours); fr loose in the lemma; ann (ours) or per with hollow culms, open sheaths, ligs a fringe of straight hairs (throat often with longer hairs), and flat to folded or involute, non-aur lvs rolled in the bud. (Gr name for lovegrass, from *eros*, love, and *agrostis*, grass.)

1a Pl per; bls tough, involute, sharp pointed; collected on ballast at Linnton, near Portland, Ore, but not believed to exist in our flora presently
                                                     E. cyperoides (Thunb.) Beauv.

1b Pl ann; bls gen flat, not tough

   2a Depressed, saucerlike glands on the pan brs, lf bls, and often on the keel of the glumes and lemmas

      3a Pan narrow; spikelets 1.5 (2) mm broad; lemmas 1.5–2 mm, mostly with few, if any, glands; culms 0.5–2 dm; mainly along river banks, se Wn, e Ore, and sw Ida to Cal, Ariz, Colo, and n Mex; yellow l., viscid e.   1 **E. lutescens** Scribn.

      3b Pan broad at base and tapered upward; spikelets ca 3 mm broad; lemmas ca 2.5 mm, strongly glandular on the keel; culms 1–5 dm; intro from the Old World, now widely distributed on waste or disturbed areas and along streambanks and pond- or lake-margins; c Wn to Me, s to Cal and Mex, WI, and Argentina; stinkgrass, candy-grass *(E. megastachya, E. m. var. c., E. eragrostis var. m.)*
                                                       2 **E. cilianensis** (All.) Mosher

   2b Depressed glands lacking

      4a Lemmas scarcely 1.5 mm; spikelets 3–10-fld, ca 1 mm broad (unpressed); hairs at the top of sheath 2–4 mm, not pustular-based; glumes ca 0.5 and barely 1 mm respectively; European, intro and weedy in N Am, known to me only from near Indian Valley, Adams Co, Ida; India l., tufted e.   3 **E. pilosa** Beauv.

      4b Lemmas slightly to considerably > 1.5 mm; spikelets sometimes > 11-fld, gen either > 1 mm broad or with hairs of the sheath top either < 2 mm or pustular-based

         5a Sts prostrate, < 1 dm, often rooting at the nodes and forming mats; pan 1.5–6 cm; spikelets 9–25-fld, up to 1 cm; glumes scarcely 1 and 1.5 mm respectively; mud flats along streams, ponds, and lakes, se Can and most of US to S Am, in our area mostly w Cas and along CR and Snake R; creeping e.   4 **E. hypnoides** (Lam.) B.S.P.

         5b Sts ascending to erect, not nodally rooting or matforming

            6a Pls mostly 6–10 dm; spikelets ca 1 mm broad; margins of the sheaths sparingly pustular-bristly; pastures, ditch banks, stream- and pond-margins, and waste places; common in sw US, intro in

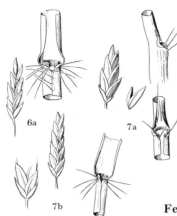

our area in Canyon Co, Ida, Yakima Valley, Wn, near Portland, Ore, and prob elsewhere; Orcutt's l.   5 E. orcuttiana Vasey

6b Pls gen < 6 dm; spikelets mostly ca 1.5 mm broad; margins of the sheaths gen glab

    7a Glumes ca 0.5 and 1 mm; ligs proper barely 0.3 mm; anthers scarcely 0.2 mm; spikelets 4–9-fld; native to Eurasia, occ in our area in waste places, collected near Portland, Ore, on ballast dumping ground, and on sand bar of CR opp Vancouver, Wn; many-std e.   6 E. multicaulis Steud.

    7b Glumes ca 1 and 1.5 mm; ligs 0.3–0.5 mm; anthers ca 0.4 mm; spikelets mostly 7–13-fld; sandy fields, waste ground, and banks of streams and ponds; e Wn to s Cal and Mex, e to prob all of the US; purple e. (*E. brizoides, E. nuttalliana, E. diffusa, E. purshii* var. *d., E. caroliniana* misapplied)

7 **E. pectinacea** (Michx.) Nees

Festuca L.   Fescue; Fescue-grass

Spikelets (1) 2–12-fld, borne in open to contracted pans sometimes reduced to a single raceme, art above the glumes; glumes gen unequal, < the lemmas, unawned, the 1st 1-nerved, the 2nd 3-nerved; lemmas rounded on the back, faintly to evidently 5-nerved, acute to obtuse, rarely minutely bifid, awnless or with a short to long, simple, persistent awn; stamens 1–3, the anthers from very short (fls cleistogamous) to elongate and often completely exserted; ann or per, caespitose to rhizomatous grasses with hollow culms, open sheaths, membranous ligs that are mostly very short and highest on the sides, and involute or folded to flat, aur or non-aur bls rolled in the bud; the glume-lemma-awn lengths given in the key refer to the 1st glume, 2nd glume, lemma, and awn, in that order. (Ancient L name for a grass). (*Bucetum, Gnomonia, Vulpia, Zerna*). Most of the native spp. are valuable forage grasses; the intro per spp. are used chiefly in meadow hay or in lawn mixtures; the anns, often referred to a separate genus, *Vulpia*, have little forage value.

1a Pl ann

  2a 1st glume 0.6–2.5 mm, < half as long as the ca 4 (2.5–5) mm 2nd glume

    3a Lemmas strongly ciliate; fields, waysides, and waste or overgrazed areas; common, BC to s Cal, e to Nev and Ariz and occ farther; prob in S Am and Europe; perhaps only a phase of the next; foxtail f., w six-weeks f.   1 **F. megalura** Nutt.

    3b Lemmas not ciliate; widespread weedy sp. of waste places and dry, mostly overgrazed areas; BC to Mex; S Am; prob native to Europe; rat-tail f.   2 **F. myuros** L.

  2b 1st glume rarely < 3 mm, always at least half as long as the 2nd

    4a Awn gen < the lemma; fls gen 7–12 (6–15) per spikelet; average relative length of glumes, lemmas, and awn ca 3, 4, 4.2, and 2.5 mm; slender f., six-weeks f.; 3 vars. in our area   3 **F. octoflora** Walt.

      a1 Glumes ca 3 and 3.5 mm; lemmas ca 3.5 mm, awns gen scarcely 2 mm; flts gen > 8 per spikelet; primarily C States and adj Can, w to Mont, occ elsewhere   var. **tenella** (Willd.) Fern.

      a2 Glumes mostly ave 3 and 4 mm; lemmas ca 4.2 mm; awns mostly 2–5 mm; flts often not > 8 per spikelet

        b1 Pls gen < 2 dm; lemmas puberulent or puberulent-scab; primarily in sw US and Baja Cal, but occ in our area

var. **hirtella** (Piper) Hitchc.

        b2 Pls often > 2 dm; lemmas mostly glab or scab; mostly Wn and BC e, although occ throughout range of sp. as a whole

var. **octoflora**

    4b Awn = or > the lemmas; flts gen < 6, although not rarely 6–7, but then the relative length of the glumes, lemma, and awn at least 4, 6, 6, and 7 mm

      5a Spikelets and brs of the infl erect; pls glab; lemmas scab but never pubescent; ocean beaches and salt marshes to sagebr des; intro from Europe, common from s BC to s Cal and Ariz, occ farther e; barren f., six-weeks f. (*F. dertonensis, F. sciuroides*)   4 **F. bromoides** L.

      5b Spikelets or brs of the infl spreading to reflexed; sheaths often hairy; lemmas sometimes hairy; coastal strand to inl valleys, des, and woodl or foothill for, often weedy; BC to Baja Cal, e to Mont, Utah, and NM; S Am; small f., Nuttall's f. (*F. arida, F. confusa, F.*

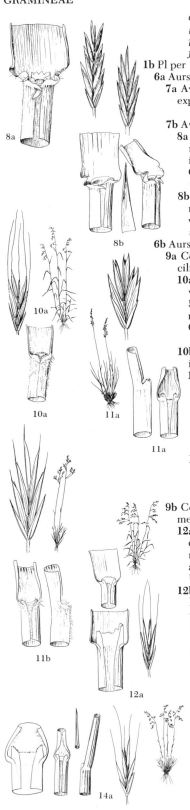

dives, F. eastwoodae, F. eriolepis of Piper, *F. grayi, F. microstachys g., F. m.* var. *ciliata, F. m.* var. *pauciflora, F. m.* var. *subappressa, F. pacifica, F. p.* var. *ciliata, F. p.* var. *simulans, F. reflexa, F. suksdorfii, F. subbiflora, F. tracyi)*   5 **F. microstachys** Nutt.

1b Pl per

  6a Aurs well developed; bls flat, rarely < 4 mm broad

    7a Awn 2–3 times > the lemma; European, intro and occ in US, to be expected but not known from our area; tall brome

                                                  F. gigantea (L.) Vill.

    7b Awn not > the lemma, often lacking

      8a Aurs ciliate-margined; bls 4–10 mm broad; glumes 3–5 and 5–7 mm; lemmas 7–9 mm; meadowl and fallow fields and ditch banks; intro from Europe and well estab, esp in Ida, but also in w Wn and Ore; tall or alta or reed f. *(F. elatior* var. *a.)*

                                          6 **F. arundinacea** Schreb.

      8b Aurs nonciliate; bls mostly 3–5 (7) mm broad; glumes ca 3 and 4 mm; lemmas 5–7 mm; valuable forage grass intro from Europe, widely estab; meadow f., English f. *(F. elatior, F. e.* var. *p., F. fluitans* var. *p.)*   7 **F. pratensis** Huds.

  6b Aurs lacking

    9a Collars or ligs (or both) strongly villous-hirsute; ligs margined with cilia nearly or quite as long as the membranous lower portion

      10a Awns mostly 10–15 (at least 5) mm; bls pubescent on the ventral (upper) surface, lax and gen flat, 3–8 mm broad; pl 5–10 dm; in light to heavy woods and on moist slopes and in meadows, from near sea level into lower mts; w Cas, BC to nw Cal; Coast Range f., crinkle awn f. *(F. ambigua, F. denticulata)*   8 **F. subuliflora** Scribn.

      10b Awn lacking or < 5 mm; bls scab but not pubescent, stiff and gen involute or folded

        11a Pan gen compressed, lower brs closely ascending; lemmas ca (7) 8 mm, acute, only rarely awn-tipped; collars not hairy; pl very strongly tufted, producing thick mats of persistent sheaths and culm bases, (4) 6–10 dm; dry areas e Cas; BC to se Ore, e to Newf and to ND, Mont, Wyo, and Colo; rough f., buffalo bunchgrass *(F. altaica* var. *s., F. s.* var. *major, F. altaica* var. *m., Daluca campestris, F. c.)*   9 **F. scabrella** Torr.

        11b Pan open, the lower brs ± spreading; lemmas ca 9 mm, mostly awn-tipped; collars often hairy; dry, open slopes to moist streambanks; w Cas, Clackamas Co, Ore, to s Cal; Cal f. *(F. aristulata)*   10 **F. californica** Vasey

    9b Collars and ligs not villous; ligs never with cilia as long as the basal membranous portion

      12a Bls lax and flat, mostly > 4 (3–10) mm broad; pan brs gen drooping; lemma awn up to 17 mm, gen at least = the lemma; ligs not markedly higher on the sides; moist to dry woods, streambanks, and meadows; both sides Cas, Alas to Cal, e to Mont, Wyo, and Utah; bearded f., nodding f.   11 **F. subulata** Trin.

      12b Bls mostly stiff and folded or involute, or pan brs stiff and erect, or lemmas with awns < body; ligs projecting much higher on the sides than at the back (i.e., opp the throat)

        13a Lemmas unawned, ca (7) 8 mm; bls involute (see lead 11a)

                                      9 **F. scabrella** Torr.

        13b Lemmas mostly either < 7 mm or awned; bls sometimes flat

          14a Pan gen open, often > 15 cm; 2nd glume rarely > 4 mm; lemmas < 6 mm; some of the awns = or > the lemmas; streambanks, lake margins, and moist woods to ponderosa pine and mont for; BC s, along coast and in the mts, to c Cal, e to Ida, Mont, Wyo, Mich, and Ont; western f. *(F. ovina* var. *polyphylla)*   12 **F. occidentalis** Hook.

          14b Pan often narrow to congested, mostly not > 15 cm; either 2nd glume > 4 mm and the lemma at least 6 mm or all awns much < the lemmas

            15a Glumes, lemmas, and awn ave ca 2.7, 3.5, 4.2, and 1.5 mm; anthers 0.3–1.7 mm; pan congested, rarely as much as 10 cm; bls filiform, the basal rarely as much as 10 cm

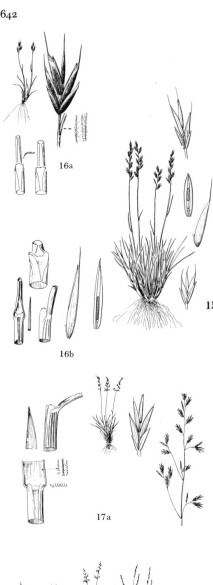

16a

16b

17a

18a

18b

16a Culms crisp-puberulent; anthers 0.3–0.5 mm; pan gen purplish; arctic-alp, Alas to Greenl, recently collected from Beartooth Plateau, Mont-Wyo border, also in RM in Can and in Colo; Baffin f.   13 **F. baffinensis** Polunin

16b Culms glab to slightly scab; anthers rarely < 0.7 mm; pan gen greenish; gravelly prairies to alp slopes; Alas to Newf, s to Cal, NM, and NY; S Am, Europe; sheep f., several vars., 2 common in our area   14 **F. ovina** L.

   a1 Lemmas unawned; intro as lawn element in e US and said to occur in Ore, occ escape in w Wn

                                                                        var. **capillata** (Lam.) Alef.

   a2 Lemmas awned

      b1 Pan open and spreading; Eurasian, occ intro in lawn mixtures, not seen from our area   var. **ovina**

      b2 Pan strongly congested; pls native

         c1 Culms mostly > 2 dm; anthers > 1 mm; BC to Newf, s to Ore, Utah, Colo, and Neb; common var. from non-alp areas *(F. saximontana, F. brachyphylla* ssp. *s.)*   var. **rydbergii** St.-Yves

         c2 Culms mostly 0.5–2 dm; anthers seldom > 1 mm; alp and subalp; Alas to Me, s to Cal, Colo, and Mich; Eurasia; alpine f. *(F. brachyphylla)*

                                                                var. **brevifolia** (R. Br.) Wats.

15b Glumes, lemma, and awn ave > 2.7, 3.5, 4.2, and 1.5 mm (lemma sometimes unawned); anthers 2–4 mm; pan often open or > 10 cm; bls often > 10 cm or not filiform

   17a Bls in part flat, 1.5–2.5 mm broad; lemmas 6–8.5 (ave 7) mm, awnless or with an awn tip rarely as much as 1.5 (gen ca 0.3) mm; subalp or alp slopes and rock slides and meadows to well above timberl; BC to Cal, not w Cas, e to Mont, Ida, and ne Wn; green f., mt bunchgrass *(F. howellii, F. v.* var. *h., F. v.* var. *vaseyana)*   15 **F. viridula** Vasey

   17b Bls all gen folded-involute and < 1.5 mm broad (as folded); lemmas often averaging < 7 mm, the awn rarely < 1 mm

      18a Pl truly caespitose, non-rhizomatous; basal sheaths firm, greenish, rarely at all red, persistent, not shredding into fibers; awns 2–5 (ave 3.3) mm; grassl and sagebr des to dry and rocky mt slopes and meadows, up to 12,000 ft in some areas; BC to Alta, s in OM and Cas through Ore to Sierran Cal, and to Nev, Mont, Wyo, Utah, and Colo; Ida f., blue bunchgrass; 2 vars.

                                                             16 **F. idahoensis** Elmer

         a1 Bls short, mostly < 7 (10) cm; pan gen spreading, the spikelets loosely fld, 10–15 mm; alp and subalp in w Wn and Ore, s to Cal *(F. ovina* var. *o.)*

                                                       var. **oregona** (Hack.) Hitchc.

         a2 Bls often much > 10 cm; pan often contracted, the spikelets rather closely fld, often < 10 mm; range of the sp. but rare or lacking in OM and at lower levels than var. *oregona* in Cas *(F. o.* var. *columbiana, F. o.* var. *ingrata, F. i.)*   var. **idahoensis**

      18b Pl gen either decumbent-based or short-rhizomatous; basal sheaths reddish, then brown, thin, strongly nerved, ultimately shredding into filiform fibers (the old veins); awns mostly < 3 (ave barely 1.5) mm; coastal marshes and sand dunes to mont for and meadows; Alas to Newf, s to s Cal, NM, Tex, and SC; Europe; red f.; 2 vars.   17 **F. rubra** L.

         a1 Pl mostly 3–6 dm; coastal BC to Cal, gen on dunes, the ave glume-lemma-awn ratio 3, 4.5, 5.5, < 1 mm

                                                            var. **littoralis** Vasey

         a2 Pls often much > 6 dm; gen, sometimes even coastal; glume, lemma, awn ratio more nearly 3.5, 5.4, 7, 1.5–2 *(F. oregona, Poa tenuifolia* var. *o., F. ovina* var. *rubra, F. r.* ssp. *densiuscula, F. richardsoni, F. vallicola)*   var. **rubra**

·Gastridium Beauv.

Spikelets 1-fld, art above the glumes, borne in dense, ± glistening spikelike pan, the ped enlarged at the tip; glumes 1-nerved, swollen and rounded at the base, then narrowed to an elongate, linear-lanceolate tip, the 1st the longer; lemma much < glumes, rounded, gen awned from just below the tip; rachilla prolonged as a tiny bristle; stamens 3; lods 2; ann with hollow culms, open sheaths, membranous ligs, and flat bls rolled in the bud. (Gr *gaster*, stomach or belly, referring to the pouch formed by the swollen base of the glumes).

G. ventricosum (Gouan) Schinz and Thell. Nitgrass. Culms 1–several, 1–3.5 dm; sheaths smooth; ligs 1–2.5 mm, slightly scab-puberulent, truncate to obtuse, erose-jagged; bls 1–2.5 mm broad, scab on the margins; pan 2–8 cm; glumes 3–5 mm, the 1st ca 1 mm > the 2nd; lemma ca 1 mm, pubescent; anthers ca 0.7 mm; lods ca 0.2 mm; native to Europe, this grass is intro and well estab from the s end of WV, Ore, to s Cal; it is sporadic in ne US.

Glyceria R. Br.   Mannagrass

Spikelets in narrow to open pan, art above the glumes, (3) 4–15-fld, linear and subterete to oblong or ovate in outline; glumes < the 1st lemma, gen papery, 1-nerved; lemmas awnless, broad, rounded on the back, firm, with 5–9 rather prominent, non-convergent nerves; lods ca 0.2 mm, truncate, fleshy; stamens 2–3; rhizomatous or stoloniferous, glab to scab pers with hollow culms, closed sheaths, membranous and gen prominent ligs, and flat (folded), non-aur bls rolled in the bud. (Gr *glukeros*, sweet, supposedly in reference to the grain). *(Panicularia)*. The spp. are all tender and readily eaten, but furnish comparatively little forage, occurring mostly in wet places, in limited quantity.

1a Spikelets terete, mostly much > 1 cm, linear or narrowly oblong in outline, several times as long as thick
  2a 1st lemma ca 5.5 mm; anthers 1.2–1.6 mm; culms up to 1.5 m; ligs 7–12 mm; bls 4–8 (12) mm broad; marshy areas and stream and pond margins; VI to c Cal, from w Cas e to Ida; western m.
            1 **G. occidentalis** (Piper) Nels.
  2b 1st lemma 3–4 mm; anthers < 1.2 mm
    3a Lemma slightly > 3 mm, scab over the back; bls scab (not papillate) on the ventral surface, 3–7 mm broad; spikelets 12–18 mm, 8–13 (15)-fld; 1st glume 1.2–1.5 mm; anthers slightly > 0.5 mm; swamps and margins of water; Alas and PT and OP s, w Cas, to c Cal; slender-spike m. *(P. davyi)*   2 **G. leptostachya** Buckl.
    3b Lemmas ca 3.5 (3–4) mm, scab only on the nerves, the internerves gen glab; bls minutely papillate on the ventral surface, mostly 3–5 (2–6) mm broad; spikelets 10–12 (15) mm, 6–11-fld; 1st glume 1.5–2.5 mm; anthers almost 1 mm; swamps and pond margins, often in 1–3 ft of water; Alas to c Cal, e to Newf, Me, and Pa, and throughout our area to Ariz and NM; northern m. *(G. fluitans var. angustata)*
            3 **G. borealis** (Nash) Batch.
1b Spikelets mostly flattened, often < 1 cm, ovate to broadly oblong in outline, rarely > 3 times as long as broad
  4a Sheaths smooth, often open above; ligs glab; 1st glume 1.3–2 (ave at least 1.5) mm, lanceolate; lemmas gen purplish; anthers mostly 3 (2 or only 1); lf bls mostly ave 10 (6–15) mm broad; culms 9–20 dm; sloughs, meadows, lake- and stream-borders, and damp ground gen; Alas to nw Ore and n Nev, only occ w Cas, e to e Can and ne US and s to Ariz, NM, Ill, and Va; Am m., reed m. *(G. flavescens, G. maxima ssp. g.)*
            4 **G. grandis** Wats.
  4b Sheaths retrorsely scab; ligs often pubescent-scab or partially closed (sheaths of course then completely closed); 1st glume often ave < 1 (1.2) mm, gen ovate; lemmas often greenish; anthers 2 (3); bls ave < 10 mm broad
    5a Ligs gen closed in front, the lower ones mostly 1.5–3 mm; bls 2–5 (6) mm broad; pls (2) 3–8 (11) dm; glumes ovate, rounded to obtuse, the 1st 0.7–0.8 (1) mm; lemmas barely 2 mm; wet places, bogs, and mt meadows; Alas to n Cal, e to Newf, Fla, Tex, and n Mex; fowl m. *(G. nervata, G. n.* var. *rigida, G. n.* var. *stricta);* ours the var. *stricta* (Scribn.) Fern.   5 **G. striata** (Lam.) Hitchc.

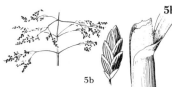

5b Ligs open in front, mostly 3–6 mm; bls 6–10 (12) mm broad; pls mostly 10–15 dm; glumes lanceolate-ovate, mostly acute, the 1st 0.7–1.2 mm; lemmas mostly 2–2.2 mm; tall m. *(G. latifolia, G. nervata* var. *e.)* wet places, lowl to Mont, widespread   **6 G. elata** (Nash) Jones

Heleochloa Host   Heleochloa

Spikelets 1-fld, borne in much-contracted pan, flattened, art above the glumes, but ultimately the glumes also deciduous; glumes awnless, slightly < the awnless lemma; palea subequal to lemma; pl ann; culms hollow; sheaths open; aurs lacking; ligs fringed; blades small, rolled in the bud. (Gr *helos,* marsh, and *chloa,* grass).

H. alopecuroides (Pill. & Mitterp.) Host. Spreading, many-std ann, gen 5–10 cm; sheaths glab except for the pilose throat; lig scarcely 1 mm; bls flat, 1.5–3 cm × 1–2 mm, scab-puberulent above; pan 10–20 × 3–4 mm, gen fully exserted from the sheath; 2nd glume ca 2 mm, slightly exceeding 1st glume and subequal to the lemma; European, intro mainly on ballast, but now apparently estab along the sand dunes and arid banks of the CR, in both Ore and Wn.

Helictotrichon Besser

Spikelets pan, 3–6-fld, large, art above the glumes; glumes slightly unequal, the 1st < the 1st lemma, 3-nerved, the 2nd > the 1st lemma, 5-nerved; lemmas coriaceous at base, hyaline above, 5–7-nerved, rounded on the back, the tip entire or shallowly bilobed, awned from ca midlength, the awn twisted and bent, exceeding the spikelet; paleas thin, < the lemmas; rachilla silky-bearded; stamens 3; lods 2; caespitose pers with hollow culms, open sheaths, short membranous ligs, and rather tough non-aur bls that are rolled in the bud and erect and narrow in the seedling stage. (Gr *helicos,* twisted, and *trichon,* hair, referring to the twisted awn).

H. hookeri (Scribn.) Henrard. Spike-oat. Pl tufted, culms several, mostly 3–4 dm; lvs somewhat scab, the sheaths compressed, the bls 2–4 mm broad, gen folded, the margins thickened (gen whitish); ligs erose-lacerate, 1–3 mm glab; pan tight, up to 10 cm, the brs erect and mostly with 1–2 spikelets; spikelets mostly 4–6-fld, the glumes ca 12 and 14 (15) mm, the 1st lemma 10–13 mm; awn up to 15 mm; prairies and foothills; Alta to Man, s to Mont, NM, and Minn, in our area in GNP and in Gallatin Co, Mont (*Avena pratensis* var. *americana, A. h., Avenochloa h.*).

Hesperochloa (Piper) Rydb.   Spike-fescue.

Spikelets compactly 3–5-fld, laterally compressed, art above the glumes, borne in rather congested pan; glumes broadly lanceolate, < the 1st lemma, the 1st 1-nerved, the 2nd 3-nerved; lemmas with 5 faint, converging nerves, very slightly keeled on the back, awnless; pls functionally ♂, ♀; stamens 3, the anthers exserted; per with hollow culms, open sheaths, membranous ligs, and non-aur bls that are rolled in the bud. (Gr *hesperos,* western, and *chloa,* grass).

H. kingii (Wats.) Rydb. Strongly caespitose per 3–6 dm, smooth to strongly scab-puberulent, forming large dense clumps of persistent old sts and sheaths; ligs 0.5–2 (3.5) mm, erose-ciliolate, finely puberulent; bls stiff, erect, glaucous, involute to flat, (2) 3–6 mm broad; pan mostly 1–2 dm, the brs short, erect, spikelet-bearing almost to the base; spikelets ca 1 cm; glumes 4–6.5 mm, subequal or the 1st the shorter; lemmas finely scab, ca 6–7 (8) mm; anthers ca 3 mm; moist to gen dry grassl or rolling hills and open ridges or talus slopes up to 11,000 ft elev; se Ore to s Cal, e through s and c Ida to Mont and Wyo, Neb, and Colo (*Festuca k., Leucopoa k., Poa k., Wasatchia k., F. watsoni*).

Hierochloe R. Br.   Sweetgrass; Vanillagrass

Spikelets pan, art above the glumes, 3-fld, the 1st 2 fls ♂, the top one ⚥; glumes broad, subequal, 3-nerved, glab; lemmas ca =, the 1st 2 gen more pubescent and less firm than the 3rd, acute to rounded, entire to deeply bifid, awnless or awned gen from a bifid apex; stamens 3; lods 2; rhizomatous, mostly sweet-smelling pers with hollow culms, open sheaths, membranous or membranous and fringed ligs, and narrow to rather broad, flat to involute, non-aur bls that are rolled in the bud, those of the innovations gen much > those of the culms. (Gr *hieros,* sacred, and *chloe,* grass; "holy grass," *H. borealis* and *H. odorata,* were strewn in front of churches in n Europe on saints' days). *(Savastana, Torresia).*

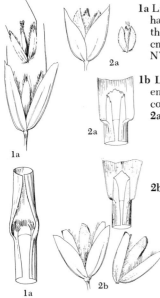

1a Ligs scarcely 1 mm, consisting ca half and half of membrane and a fringe of hairs; lower lemmas awned from a deeply bifid apex, the terminal lobes of the lemma acute; pl 2–4 dm; bls 1–2 mm broad; pan contracted, 1.5–4.5 cm; arctic or alp meadows; Alas to Newf, s in higher mts to BC, n Mont, NY, NH, and Vt; Greenl, Europe; alpine s. *(Holcus a.)*

1 **H. alpina** (Swartz) R. & S.

1b Ligs > 1 (mostly 3–6) mm, almost entirely membranous; lower lemmas entire and acute or rounded and shallowly lobed at the tip, awnless or inconspicuously awn-tipped

2a Lemmas acute, entire, rather uniformly pubescent over the back and margins; bls rarely > 5 mm broad, those of the culms rarely as much as 5 cm; pl 3–5 dm; moist soil of lower mont to subalp meadows and slopes; Alas to Lab, s to Ore, Nev, Ariz, NM, SD, Gr Lakes region, and Pa; Eurasia; s., v., holy grass, Seneca grass *(Holcus o., Avena o., H. arctica)*

2 **H. odorata** (L.) Beauv.

2b Lemmas rounded and shallowly bilobed, only puberulent over the back, but pilose-ciliate along the margins; bls mostly > 5 (up to 12) mm broad, those of the culms averaging at least 10 (to 40) cm; pl 6–9 dm; moist to rather dry, mostly for areas; Klickitat Co, Wn, s in w Ore and in coastal mts to Monterey Co, Cal; Cal s., Cal v. *(H. macrophylla)*

3 **H. occidentalis** Buckley

Holcus L.   Velvet-grass

Spikelets in rather congested pan, art below the glumes, 2–fld, the lower fl ☿, the upper ♂; glumes subequal, > flts, strongly keeled, membranous, the 1st 1-nerved, the 2nd 3-nerved; lemmas obscurely 5-nerved, shining, somewhat coriaceous, the lower one awnless, the upper with a short awn from just below the tip; stamens 3; lods 2; pers with hollow culms, open sheaths, short membranous ligs, and flat, non-aur bls that are rolled in the bud. (L name used for some grass of uncertain identity). *(Notholcus).*

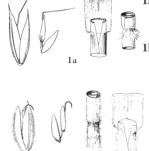

1a Pl rhizomatous; awn bent, exserted beyond the glumes; culms glab except at the nodes; ligs 2–3 mm; in lawns and damp areas; intro from Europe, estab in w Wn and Ore to Cal and scattered in e US; creeping softgrass or v.   1 **H. mollis** L.

1b Pl caespitose (or stoloniferous); awn hooked, included or exserted sidewise from the glumes; culms gen pubescent on the internodes as well as the nodes; ligs 1–2 mm; roadsides, railroad embankments, and waste ground to open woods in the lower mts; intro from Europe and widely estab in N Am fron Alas to Cal, esp w Cas, e to Newf and Ga, not common in RMS, but known in both Ida and Mont, prob intro as a meadow grass, although not particularly palatable; now a bad lawn weed, where it becomes stoloniferous; common v., Yorkshire-fog   2 **H. lanatus** L.

Hordeum L.   Barley

Spikelets in a terminal spike, gen 3 (2) per node, mostly 1 (2)-fld, the rachis gen disart at maturity (rarely continuous); c spikelet at each node sessile, the flt ☿, sessile or with a short to rather long stalk-like rachilla joint, the lemma gen long-awned, the rachilla prolonged and bristle-like, sometimes with the rudiments of a 2nd flt; lateral spikelets gen short-ped (sessile), the flt sessile or with a stalk-like rachilla joint, gen ♂ to rudimentary (rarely ☿); glumes very narrow, often awnlike the entire length, sometimes slightly broadened below, borne in front of (rather than lateral to) the enclosed flt; lemmas rounded on the back, 5-nerved, gen awned; ann or non-rhizomatous per with hollow culms, open sheaths, short membranous ligs, and gen flat, glab to soft-pubescent, strongly aur to non-aur bls, the spikes mostly deciduous into the nodal groups of 3 spikelets. (The classical L name for barley). Most of our spp. are weedy. *H. brachyantherum* and *H. jubatum* are of some importance as forage grasses, but the ann spp. are an annoyance to humans, because of the awned spikelets that work into the clothing, and to other animals because of the danger of penetration of the eyes and ears.

1a Aurs well developed, at least some of them much > 1 mm

  2a Bls 5–15 mm wide; lemmas gen with awns 10–15 cm, but sometimes unawned but prominently 3-lobed near the tip; rachis not disart; pl ann, 8–12 dm; ligs 0.5–1 mm; widely cult and often persistent for 1–2 years, but prob never truly estab; cult b. (*H. sativum* var. *v.*)   **1 H. vulgare** L.

  2b Bls mostly < 5 (7) mm wide; lemmas never 3-lobed near the tip, the awn < 5 cm; rachis ultimately disart

    3a Flt of central spikelet sessile or subsessile, the rachilla joint not nearly so long as the ped of the lateral spikelets; inner glume of the lateral spikelets narrower than either glume of the c spikelet; intro from Europe, fairly frequent in s BC and w Wn, weedy; mouse or wall b., foxtail   **2 H. murinum** L.

    3b Flt of c spikelet with a rachilla joint = or subequal to ped of the lateral spikelets, or the spikelet itself with a distinct ped; inner glume of the lateral spikelets as broad as those of the c spikelet

      4a Joints of the rachis rarely as much as 2 mm; anthers barely 0.5 mm; rachilla of the lateral spikelets prolonged beyond the flt for scarcely 2 mm; sheaths smooth; bls 3–5 (7) mm broad; intro from the Mediterranean region, now abundant in much of w US; not common in our area but collected from c Wn and Ore to Ida; seagreen b. (*H. stebbinsii*)   **3 H. glaucum** Steud.

      4b Joints of the rachis at least 2 (3) mm; anthers mostly ca 1 (0.8–1.3) mm; rachilla of the lateral spikelets gen prolonged beyond the flt for at least 2.5 (2–3.5) mm; sheaths glab to pilose; bls 2-4 mm broad; intro from Europe, common throughout our area except in the mts; charming b. (*H. murinum* var. *l.*)   **4 H. leporinum** Link

1b Aurs lacking or < 1 mm

  5a Glumes awnlike, 2–6 cm, the spike (awns included) gen nearly or quite as thick as long; aurs gen developed on some lvs but barely 0.5 mm; tufted per 2–6 dm, glab to densely soft-pubescent; bls 2–4 (5) mm broad; very common weedy sp. of dry to moist areas, from sagebr des to mt for, often in slightly disturbed areas; Alas s, both sides Cas, to Cal, e to Atl coast and s to Tenn, Tex, and Mex; squirrel-tail, foxtail b. (*H. caespitosum; H. j.* var. *c.; H. j.* ssp. × *intermedium; H. comosum*, misapplied)   **5 H. jubatum** L.

  5b Glumes often broadened at base, never so much as 2 cm, the spike (including awns) much longer than thick; aurs often lacking

    6a Pl per, mostly 4–8 (2–10) dm; aurs lacking; bls 2–6 (9) mm broad; spike 5–10 cm; anthers 1–1.5 mm; coastal to mont in sagebr des, gen where moist; Alas to Cal, e to RMS and occ in e Can and e US; meadow b. (*H. boreale*, misapplied; *H. jubatum* ssp. *breviaristatum; H. nodosum*, misapplied)   **6 H. brachyantherum** Nevski

    6b Pl ann, often < 3 dm; bls mostly 2–4 mm broad

      7a Glumes dissimilar, the inner 1 of the lateral spikelets widened above the base (those of the c spikelet often slightly widened also), gen 0.8–1.8 mm broad; ped of the lateral spikelets < 1 mm, gen strongly curved; grassl and des areas, esp on saline or waste land; e Wn to s Cal, e to Atl and s to Fla and Mex; S Am; little b. (*H. p.* var. *pubens*)   **7 H. pusillum** Nutt.

      7b Glumes all linear and not at all (or very slightly) widened above the base, gen < 0.7 mm broad; ped of the lateral spikelets ca 1 mm, gen straight

        8a Lateral spikelets with a rudimentary, awnless flt; glumes slender, ascending to suberect; aurs lacking; weedy and mostly in waste areas, esp where moist or saline; VI s, both sides Cas, to s Cal, e to Ida; meadow b., low b. (*H. nodosum* var. *d.*)

            8 H. depressum (Scribn. & Smith) Rydb.

        8b Lateral spikelets with gen awned flt; glumes rather stout, curved outward from the base and ± spreading; some lvs gen with tiny aurs barely 0.5 mm; intro from Europe, very common in our area on waste land, on dry to moist soil; s BC s, both sides Cas, to s Cal, less common e to Ida, (Mont?), Utah, and Ariz; Mediterranean b. (*H. gussonianum, H. hystrix*)   **9 H. geniculatum** All.

Koeleria Pers.   Koeleria

Spikelets in much-congested, gen glistening, spikelike pan, (1) 2–4-fld, art above the glumes; glumes keeled, the 1st 1-nerved, gen slightly < and much narrower than the 2nd, the 2nd 3 (5)-nerved, gen subequal to or slightly > the 1st flt; lemmas 5-nerved, awnless or very short-awned from a minutely bifid apex, keeled; stamens 3; lods 2; ann or caespitose per with hollow culms, open sheaths, membranous ligs, and flat to involute, non-aur bls folded in the bud. (For George Wilhelm Koeler, 1765–1807, German botanist).

1a Pl ann; lemmas strongly hirsute; collected once or twice in our area from near Portland, Ore, where apparently not persistent; native to the Old World; bristly k.   **K. phleoides** (Vill.) Pers.

1b Pl per; lemmas scab but never hirsute; strongly caespitose, 2–6 dm, glab to downy; margins of the sheaths gen not overlapping above, the collar or esp the margins of the bl often with long straight hairs 1–1.5 mm; ligs 0.5–2 mm, sometimes highest in front, gen strongly pubescent, erose to subentire, but ± ciliate; bls mostly 1–2 (4.5) mm broad, gen folded to involute (flat), the tips strongly prow-shaped; sagebr des, prairies, and open for to subalp ridges, mostly on sandy to rocky soil; BC to n Mex, e Cas, e to Ont and Me, Del, La, and Tex; Eurasia; prairie Junegrass; Koeler's grass *(K. c.* var. *nuttallii, K. c.* var. *major, K. nitida, K. n.* var. *munita, K. pseudocristata* var. *oregana, K. robinsoniana, K. r.* var. *australis)*; one of the better range grasses in much of w US   1 **K. cristata** Pers.

1b

Leersia Soland.

Spikelets strongly compressed, pan, 1-fld, falling entire, the glumes lacking (ours) or greatly reduced; lemmas leathery, 5-nerved, the lateral nerves marginal; palea = the lemma, 3-nerved; stamens in ours 3; lods 2; per of wet places with hollow culms, open sheaths, short membranous ligs, and flat to ± folded, non-aur bls that are rolled in the bud. (For Johann Daniel Leers, 1727–1774, a German apothecary).

L. oryzoides (L.) Swartz. Cutgrass, rice c. Rhizomatous per 6–15 (3–20) dm; sheaths gen retrorsely scab; ligs truncate, ca 1 mm; bls flat, mostly 6–10 mm broad, strongly scab on the margins, abruptly narrowed just above the base; main pan 1–2 dm, upper sheaths gen with reduced, axillary pans often with cleistogamous spikelets; lemmas ca 5 mm, bristly-pubescent; wet places, often in fairly deep (2–3 ft) water; BC to Que, s to most of the US, in all our area including Mont, and on both sides Cas; Europe *(Homalocenchrus o., Oryza clandestina, O. o., Phalaris o.)*.

Leptochloa Beauv.   Sprangletop

Spikelets (ours) several-fld, art above the glumes, sessile to short-ped on one side of the rachis, in numerous spikes or racemes arising from a c axis; glumes 1-nerved, considerably < the 1st lemma, gen unequal, keeled; lemmas rather broad, short-awned from a bifid apex, prominently 3-nerved, the nerves (ours) pubescent; stamens 3; lods 2; ann or per with hollow culms, open sheaths, membranous ligs, and flat to involute, non-aur bls that are rolled in the bud. (Gr *leptos*, slender, and *chloa*, grass, referring to the slender spikes).

L. fascicularis (Lam.) Gray. Loose-fld s., clustered salt-grass. Scab but otherwise glab ann with spreading to erect culms 2–8 dm; ligs 3–7 mm, gen lacerate; bls 1–2 (3) mm broad, often involute; pan gen partially enclosed by upper sheath, up to 2.5 dm; spikelets 7–12-fld, short-ped in spikelike racemes up to 10 cm; glumes ca 2 and 3 mm; lemmas 3.5–4 (5) mm; awn 1–3 mm; anthers mostly 0.2 mm, the fls apparently largely cleistogamous; coastal, in brackish water; native along Atl coast and inl in moist habitats from SD to Tex, also in S Am; intro and occ in Ida, Wn, and Ore, mostly along SR and CR; more common from Cal to Colo and NM *(Cynodon f., Festuca f.)*.

Lolium L.   Ryegrass

Spikelets ∝ , ∝ -fld, laterally flattened, sessile in a single terminal spike on opp sides of a continuous rachis to which they are turned edgewise, the 2nd glume of the spikelet (on the opp side of the spikelet from the lowest flt) well developed, but the 1st glume (which would be next to the rachis) lacking on all but the terminal spikelet; glumes 5–9-nerved, unawned; lemmas 5-nerved, awned or awnless; lods 2; stamens 3; ann to per with hollow culms, open sheaths, mostly well-developed aurs, short membranous ligs, and flat to folded bls either rolled or folded in the bud. (L name for darnel).

1a Glume gen > rest of the spikelet, 7–9-nerved; pl ann, 4–9 dm, ± scab but otherwise glab; aurs 1–2 mm; bls flat, 3–10 mm broad; intro from Europe, widely distributed in N Am in waste places mainly; ann r., darnel
                                                                      1 L. temulentum L.
1b Glume < rest of the spikelet; pl mostly bien or per
  2a Lemmas (at least the upper ones) awned; bls involute when young, mostly 3–8 mm broad; culms not compressed; pls often bien, gen without innovations; intro from Europe; well estab on waste ground and roadsides throughout our area; coarser than *L. perenne*, and less desirable as a lawn grass; Italian r., Australian r. *(L. italicum, L. m.* var. *i., L. perenne* var. *i., L. p.* var. *m., L. temulentum* var. *m.)*   **2 L. multiflorum** Lam.
  2b Lemmas unawned; bls folded when young, mostly 2–4 mm broad; culms mostly ± compressed; pl per, gen with basal innovations; intro from Europe and commonly cult as a forage pl in many parts of the world, often used as a cover crop or as a component of lawn mixtures; well estab in much of our area; per r., English r. *(L. p.* var. *cristatum)*
                                                                      3 L. perenne L.

1a   2b

Melica L.   Melic; Oniongrass

Spikelets in open to contracted pans, art above the glumes (ours), with (1) 2–several ☿ fls, the upper 2–4 flts sterile and represented by empty lemmas only, each of which is enfolded by the one below, the whole forming a spindle-like body (the "rudiment"); glumes < 1st lemma, rather papery or scarious, unequal to subequal, rounded on the back, the 1st 3 (5)-nerved, the 2nd gen 5 (7)-nerved; lemmas firmer than the glumes, rounded on the back, blunt to acuminate, awnless or awned (often from between the teeth of a bifid apex), with 7 (to 11) obscure to rather prominent, nonconvergent nerves; callus not bearded, the joint terete to (rarely) inflated or punky; stamens 3; lods fused and forming a collarlike, semifleshy structure ca 0.5 mm, extending 1/2–2/3 around the base of the ovary; per, caespitose to rhizomatous pls with hollow culms that often are enlarged and bulbous at the base; sheaths closed, sometimes to the top; ligs membranous, often closed in front, erose to deeply lacerate; bls flat to involute, non-aur, rolled in the bud. (Derivation uncertain; said to be a classical name of some grass). *(Bromelica).* The spp. are readily grazed but rarely abundant enough to constitute an appreciable percentage of the forage available in any area.

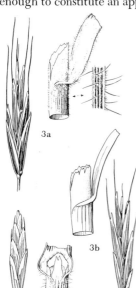

1a Culms not bulbous-based; lemmas often awned or awn-tipped; sheaths and often ligs closed in front unless mechanically ruptured
  2a Lemmas with an awn 4–10 mm, not long-ciliate-pilose at base; ligs not projecting upward in front as a prominent tooth
    3a Ligs gen pubescent to the top; bls with very closely spaced veins, strongly scab, stiff, 2.5–5 mm broad; pan brs gen paired; awn mostly ca 8 (6-10) mm; dry mont woods; Cas, sw Wn to Sierran Cal, rare in our area; bearded m.   **1 M. aristata** Thurb.
    3b Ligs gen glab except at base; bls with widely spaced veins, weakly scab, lax, 5–10 mm broad; pan brs gen solitary; awn mostly 4–7 (3-10) mm; moist woods, s BC to Ore, e to Alta, Mont, and Wyo and from n Mich to Ont; Smith's m. *(M. retrofracta, Schizachne s.)*
                                                              2 M. smithii (Porter) Vasey
  2b Lemmas merely awn-tipped (awn rarely up to 4 mm), strongly pilose-ciliate at base; lig projecting upward in front as a prominent tooth; culms sometimes solid, 4–12 dm; bls 1–4 mm broad; dry areas, lower mts; VI and OP and Cas of s Wn to coastal mts and Sierra Nevada, Cal; Harford's m. *(M. h.* var. *tenuis, M. h.* var. *viridifolia)*
                                                              3 M. harfordii Boland.
1b Culms bulbous-based; lemmas unawned; sheaths mostly open above (rarely closed their full length)
  4a Rachilla joints swollen, wrinkled or spongy and gen brownish or yellow when dry; culms mostly 1.5–4 (1–6) dm, clustered on short rhizomes; sheaths scab to strongly hirsute; ligs collar-like but nearly always split in front, 1–3 mm; bls 1.5–3 (4.5) mm broad, scab and often pubescent on the ventral surface; anthers 1.5–2 mm; dry woods to rocky valleys and

open mont for up to 7000 ft elev; e slope Cas, ne Wn to Cal, e to w Ida and Nev; little o. (*M. f.* ssp. *madophylla, M. f.* var. *inexpansa, M. macbridei, M. f.* var. m.)   **4 M. fugax** Boland.

4b Rachilla joints slender, terete, greenish or straw-colored; culms gen much > 4 (3–18) dm

   5a Lemmas acuminate, gen pilose-ciliate near the base, 9–13 mm; sheaths gen closed to the top and even the lig closed in front; rachilla joints 2–3 mm; pan narrow, the brs ascending; anthers ca 1.5 (2) mm; open slopes to thick, dry or moist woods, from near sea level to mid-mont; s Alas to Cal, e to Mont, Ida, and Wyo; Alas o.; 2 vars.

                                     5 M. subulata (Griseb.) Scribn.

     a1 Lemmas 10–13 mm, acuminate, strongly veined, gen prominently pilose-strigose (ciliate) along the marginal veins and over the back; range of the sp. as a whole (*M. acuminata*)   **var.** subulata

     a2 Lemmas 9–12 mm, acute to acuminate, often not strongly veined, sparsely pilose-strigose (ciliate), the cilia sometimes almost or quite lacking; occ from w Ida to YNP and Sweetgrass Co, Mont (*M. p.*)
                                var. pammelii (Scribn.) Hitch.

   5b Lemmas acute to obtuse, not acuminate, never pilose-ciliate near the base, often < 10 mm; sheaths mostly open above; rachilla joints often only 1.5–2 mm; pan often open, the lower brs spreading

     6a Pan open, the brs spreading; lower rachilla joints mostly at least 2 mm; lemmas obscurely 7-nerved, ave 7–10 mm; culm bases tightly clustered on a thick rhizome; culms 7–18 dm; fls 2–6 mm broad; anthers 3–4 mm; dry woods and open slopes at lower levels in the mts; nw Ore to the coastal mts and Sierra Nevada, Cal; Geyer's o. (*M. bromoides, Glyceria bulbosa, M. b.* var. *howellii*)
                                   6 M. geyeri Munro

     6b Pan narrow, the brs short, gen erect, lower rachilla joints mostly 1.5–2 mm; lemmas often prominently 7 (or obscurely 9–11)-nerved, 6–12 mm; culm bases sometimes spaced 1–3 dm apart on a slender rhizome

       7a Culms gen not tightly clustered, but attached from 1–3 cm apart on a rather slender rhizome; lemmas ca 7 (6–8) mm; anthers 2–2.5 mm; in wet to rather dry meadow to subalp ridges, but gen in moist loamy meadows or open parks in mont for; s BC s mostly E Cas, to n Cal, e to Alta, Mont, Wyo, and Colo; showy o. (*M. scabrata*)   **7 M. spectabilis** Scribn.

       7b Culms tightly clustered on a short thick rhizome; lemmas ave 9–12 mm; anthers (2.5) 3–4 mm; open sagebr hills and ponderosa pine for to subalp rocky knolls and talus slopes; BC to Cal, mostly e Cas, e to Mont, Wyo, and Colo; oniongrass; 2 vars.
                                 8 M. bulbosa Geyer

       a1 Infl dense, the spikelets of the upper half strongly overlapping; herbage often copiously pubescent; Mt Stuart and Wen Mts, Wn (*M. bella* ssp. *i.*)   **var. intonsa** (Piper) Peck

       a2 Infl more open, the spikelets often not overlapping; herbage gen glab; range of the sp. except rare in Wen Mts (*M. bella, M. inflata, M. b.* var. *i., M. b.* var. *caespitosa*)   **var.** bulbosa

Molinia Schrank

Spikelets subterete, 2–4-fld, art above the glumes, borne in long slender pans; glumes 1-nerved (ours), < the 1st flt; lemmas awnless, rather strongly 3-nerved, the nerves convergent; lods flabellate, membranous, ca 0.5 mm; stamens 3; tall caespitose pers with hollow culms, open sheaths, ligs mostly of short hairs, and rather tough, elongate, flat to involute, non-aur bls rolled in the bud. (For Juan Ignazio Molina, 1740–1829, a Jesuit missionary and botanist noted for his study of the flora of Chile).

M. caerulea (L.) Moench. Purple moorgrass. Culms tufted, erect, 6–12 dm; sheaths gen glab (sparsely pilose along the margin); hairs of the ligs scarcely 0.5 mm, those of the throat up to 2 mm; bls stiff, gen erect, flat to involute, 2–7 mm broad; pan 1–2 (3) dm, very slender, the brs short, appressed; glumes 1-nerved, subequal, 1/2–2/3 the length of the 1st flt; 1st lemma 3–4 mm, the upper ones smaller, the uppermost flt often sterile or rudimentary; native to Eurasia, intro in a few places in ne N Am, and believed to be estab in marshes near Newport, Lincoln Co, Ore (*Festuca c., Cynodon c.*).

Muhlenbergia Schreb.   Muhly; Muhlenbergia

Spikelets in open to much-contracted, spikelike pans, 1 (2)-fld, art above the glumes; glumes from much < to occ somewhat > the lemma, acuminate and sometimes awned to blunt, the 1st 1-nerved, the 2nd sometimes 3-nerved; lemma 3-nerved, often ± pilose at the base, the callus inconspicuous, gen glab but long-bearded in 1 sp.; rachilla not prolonged behind the palea; stamens 3; lods 2; ann or tufted to rhizomatous pers with solid or hollow culms, open sheaths, membranous ligs, and rather narrow, non-aur bls rolled in the bud. (For G. H. E. Muhlenberg, 1753–1815, a minister of Pennsylvania who was a student of the grasses). In our area the spp. are seldom common enough to be considered important forage grasses, although they are mostly fairly palatable.

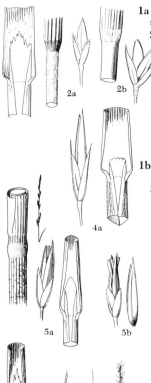

1a Spikelets on long slender peds in open, spreading pans not rarely 1/3–2/3 total height of the pl; culms solid
  2a Pl ann, 1–3 dm; culms pubescent below the nodes; glumes gen ciliate; ligs 1–2.5 mm, obtuse to acute, lacerate but not ciliate; sandy riverbanks, moist meadows, or open rocky and apparently dry slopes; c Mont and SD to Tex and Mex, w to Ida and sporadically to Yakima Co, Wn, and Union Co, Ore, and to Nev and Cal; ann or littleseed m. *(Sporobolus minutissimus; S. microspermus, missapplied)*   1 **M. minutissima** (Steud.) Swallen
  2b Pl per; culms glab below the nodes, 1–3 (4) dm; glumes not ciliate; ligs 0.5–1 mm, finely erose-ciliolate, truncate; dry to moist, often alkaline places e Cas in our area; BC to Cal and Mex, e to Ida and Mont and to Miss Valley and Tex; S Am; alkali m., rough-lvd dropseed *(Sporobolus a.)*   2 **M. asperifolia** (Nees & Meyen) Parodi
1b Spikelets subsessile or short-ped in contracted (often spikelike) pans gen < 1/3 the total height of the pl
  3a Lemmas neither slenderly awned (sometimes awn-tipped) nor pilose on the lower half (sometimes puberulent); culms solid
    4a Ligs 0.5–1 (1.5) mm, truncate, finely erose-ciliolate; tufted per at least 2 dm, without elongate rhizomes or stolons, the culms pubescent below the nodes but not nodulose-roughened; bls gen flat, 1–1.5 (2) mm broad; foothills and prairies; e base RM, Alta to Sask, to NM, O, and Ky; along our border in Mont; plains m., prairie rush-grass *(Sporobolus c., S. brevifolius, M. b.)*   3 **M. cuspidata** (Torr.) Rydb.
    4b Ligs gen 1–3 mm, not ciliolate, gen acute; ann or slenderly (sometimes almost filiformly) rhizomatous or stoloniferous per, often much < 2 dm; culms gen not pubescent below the nodes, but sometimes nodulose-roughened
      5a Sheaths and culms minutely nodulose-roughened; pl per, 0.5–3 (6) dm, strongly rhizomatous; anthers ca 1.5 mm; lvs 1–1.5 mm broad; moist to dry, lowl to mont prairies, meadows, and rocky slopes; E Cas, BC to Baja Cal, e to NB and Me, Daks, Neb, Colo, and NM; mat m., short-lvd m. *(M. squarrosa, M. brevifolia var. r., Sporobolus r., S. aspericaulis, S. depauperatus)*   4 **M. richardsonis** (Trin.) Rydb.
      5b Sheaths and culms not nodulose-roughened; pl ann or ± stoloniferous per; anthers ca 0.5 (0.8) mm; lvs 1–2.5 mm broad; gen near springs or seepage, sometimes in fairly hot water, or in moist meadows of mt valleys to subalp slopes; BC to Alta and SD, s to Cal, Ariz, and NM, not w Cas; pullup m., slender m. *(M. aristulata, M. idahoensis, Sporobolus aristatus, S. filiformis, S. depauperatus var. f., S. simplex var. thermale)*   5 **M. filiformis** (Thurb.) Rydb.
  3b Lemmas with slender awn at least 1.5 mm or pilose on lower half, or both; culms often hollow
    6a Ligs acute, not erose-ciliolate, (2) 3–5 mm; pl non-rhizomatous; culms solid, 2–4 (1–6) dm; bls involute, ca 1.5 (1–2) mm broad; awn slender to stout, (1.5) 2–12 mm; anthers ca 1.5 mm; prairies to foothills and mid-elev in the mts; Mont to Tex and Mex, w in the sw to Utah, Ariz, and Cal, possibly in our area in Mont; mt m.
                                6 **M. montana** (Nutt.) Hitchc.
    6b Ligs obtuse or truncate, erose-ciliolate, < 2 mm; pl rhizomatous; culms hollow
      7a Hairs of the callus almost or fully as long as the lemma; culms 4–10 dm; ligs truncate, ca 1 mm; bls 2–5 mm broad; glumes acuminate to awn tipped; awn of lemma 4–8 mm; anthers ca 0.6 mm; mont or submont where moist; e Wn to Cal, e to Mont and NM; foxtail m., hairy m. *(M. comata)*   7 **M. andina** (Nutt.) Hitchc.

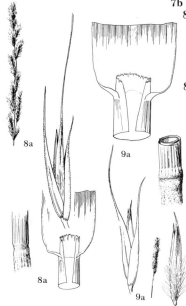

8a

9a

8a

9a

7b Hairs of the callus < half as long as the lemma

> **8a** Glumes, including the awn tips, < 4 mm; lemma ca = or > the body of the glumes, with an awn tip or slender awn up to 10 mm; anthers ca 0.5 mm; bls 2–4 (5) mm; dry to moist ground; BC s, e Cas, to Cal, e to Que and most of US except the se; lfy m., wire-stem m. (*M. ambigua, M. foliosa* spp. *a., M. setiglumis*)
> 8 **M. mexicana** (L.) Trin.

> **8b** Glumes, including the awn tips, (4) 5–6.5 mm, > the awnless or merely awn-tipped lemma; anthers 0.5–1 (1.5) mm; ligs ca 1 (to 3) mm; bls 2–7 mm

>> **9a** Pl gen in dry, often rocky areas; culms tending to br at mid-length, slightly compressed, internodes glab; sheaths slightly keeled; anthers 0.5–0.8 mm; lemmas pilose at base, only; mostly on plains e RM, Alta to NM and Ariz, w to w Mont, ne Ida, and perhaps occ in e Wn; not readily separable from no 10; satin-grass, wild timothy (*Agrostis r., Cinna r., Polypogon r.*)
>> 9 **M. racemosa** (Michx.) B.S.P.

>> **9b** Pl gen in moist areas, often even in bogs; culms br basally, semiterete, internodes often ± puberulent; sheaths scarcely keeled; anthers 0.8–1.5 mm; lemmas pilose at base and along the margins; n and e US and adj Can, w in s Can to BC, and occ in e Wn, ne Ore, and nw Mont; marsh m. (*Polypogon g.*)
>> 10 **M. glomerata** (Willd.) Trin.

Munroa Torr.

Spikelets (2) 3–5-fld, clustered and closely subtended by scarcely reduced lvs, 2–3 per node, mostly on one side of a short rachis of 3–5 nodes, unlike in size, the lower ones smallest, all disart above the glumes; true glumes small, 1-nerved, 1 or all the nerves protruding as very short awn tips; stamens 3; lods lacking; low, spreading ann with hollow terete culms, open sheaths, short ligs consisting chiefly of hairs, and short, strongly white-margined, non-aur, gen folded bls with stiff pungent tips. (For William Munro, 1818–1880, an English agrostologist).

M. squarrosa (Nutt.) Torr. False buffalograss. Matted ann 3–10 cm; lvs tufted at the nodes, the internodes mostly bare, strongly striate, scab-puberulent; throat with stiff hairs up to 2 mm; ligs scarcely 1 mm; bls mostly 10–15 × ca 2 mm, pungent; clusters of spikelets much < lf bls, the spikelets 6–9 mm; glumes subequal, < the 5–8 mm 1st lemma; lemmas pilose on the margins near midlength; anthers ca 1.5 mm; grassl of prairie and rocky foothills; Alta and Mont, where w as far as Missoula and Gallatin cos, to ND, s to Tex, NM, Ariz, and Cal (*Crypsis s.*).

Nardus L.   Wirebent; Mat-grass; Nard

Spikelets 1 per node, borne in 2 interlocking rows on 1 side of a secund terminal spike, sessile, partially sunken in concavities of the continuous rachis, 1-fld; 1st glume absent, the 2nd (inner) glume minute or sometimes lacking also; lemma hardened, 2-nerved, enveloping the palea, bearing a short erect awn; lods absent; stamens 3; style 1; tufted per with hollow culms, open sheaths, short membranous ligs, and non-aur bls. (Gr *nardos*, a name for an aromatic herb, the application to this grass not understood).

N. stricta L. Wiry, tufted per 1–3 (5) dm, glab; ligs 0.5–1 mm; bls ca 1 mm broad, folded, the lower ones spreading at nearly a right angle to the culm, scab; spike 3–10 dm; lemma gen purplish, 7–10 mm, tapered almost uniformly into a short awn; style gen > lemma; intro from Europe, uncommon, but reported from Ida, and possibly destined to become another weedy grass of dry areas.

Oryzopsis Michx.   Ricegrass; Mountain-rice

Spikelets 1-fld, art above the glumes, borne in open to congested and often reduced pans; glumes = to slightly unequal, obtuse to acuminate, membranous to firm, gen 3–5-nerved; lemma < to ca = the glumes, indurate, appressed-pubescent to hirsute (glab), awned, the callus short, blunt or very slightly pungent, the bearding, if any, mostly not noticeably > the gen pubescence; awn terminal or subterminal, deciduous, gen ± bent and often twisted; rachilla not prolonged; stamens 3; lods 2 or 3, conspicuous; caespitose pers with solid or hollow culms, open sheaths, membranous ligs, and non-aur, flat to strongly involute bls rolled in the bud. (Gr *oruza*, rice, and *opsis*, like). (*Eriocoma*, × *Stiporyzopsis*). The spp. are palatable and nutritious, but seldom abundant except for *O. hymenoides*. *Nassella chilensis* (Trin.) Desv., a Chilean pl, has been collected on ballast near Portland, Ore, but is not believed to persist there. It resembles *Oryzopsis* but has small lemmas scarcely 2 mm, glab and shining, and strongly lopsided at the tip.

1a Lemma smooth or < 3 mm, or both; ligs truncate, rarely > 1 mm; tips of the anthers not hairy
    2a Glumes obtuse; culms 1–3 (0.5–4) dm, semi-solid; lemma brownish, (3.8) 4–5 mm, the awn (3) 6–10 mm, soon deciduous; bls scarcely 1 mm broad; ligs often higher on sides than at the back, lower ones ca 0.5 mm, upper ones up to 1 mm; dry, rocky soil in sagebr or yellow pine for; e side Cas, Kittitas and Yakima cos, Wn, and Jefferson Co, Ore; Henderson r. (*O. exigua* var. *h.*)   **1 O. hendersonii** Vasey
    2b Glumes acute to acuminate; culms 3–7 dm, hollow; lemma greenish, 2–2.5 mm, the awn firmly attached, 6–8 mm; bls 0.5–2 mm broad; ligs ca 0.5 (1) mm; open woods and rocky ridges or slopes; BC to Alta and ND, s in RM and e to Neb, Okla, NM, Ariz, and Nev; not known from Wn, Ore, or Ida, and only from as far w in Mont as the Big Belt Mts; small-fld r.   **2 O. micrantha** (Trin. & Rupr.) Thurb.
1b Lemma pubescent, at least 3 mm; ligs often acute or > 1 mm; tips of the anthers often hairy
    3a Bls 3–7 mm broad, flat, tapered at each end, those of the culm lvs greatly reduced or sometimes lacking; ligs truncate, scarcely 1 mm; lemma 6–8 mm, appressed-pubescent, strongly crisp-hairy on the upper part of the callus; anthers ca 3.5 mm, the tips hirsute-tufted; lods ca 5 mm; mostly in open conif woods; BC to Newf and Me and W Va, s to Daks and in RMS to Utah and NM, reported also from Ida and ne Wn; roughlf r., white-grained m. (*O. mutica*)   **3 O. asperifolia** Michx.
    3b Bls < 3 mm broad or strongly involute, or the ligs acute and much > 1 mm; lemma either much < 6 mm or else pilose-hirsute
        4a Lemma pilose-hirsute; glumes 5–9 mm
            5a Pan 6–18 cm, open, brs divaricate and dichotomously forked; lemma 3–4 mm, the callus blunt; culms 3–6 (8) dm, thick walled but hollow; ligs acuminate, 5–7 mm; grassl and des plains and foothills, esp on rocky or sandy soil; e Cas, BC to s Cal and ne Mex, e to Alta, the Daks, and Tex; Indian r. (*O. caduca, Stipa c., Stipa bloomeri, O. b., Stipa membranacea* Pursh, *O. m., Stipa h., O. cuspidata*); the sp. tends to hybridize with various spp. of *Stipa* and has apparently produced several intermediate phases which have been named as separate spp., as the synonymy indicates
                4 O. hymenoides (R. & S.) Ricker
            5b Pan brs ascending to erect, not dichotomously forked; lemma 6–7 mm, the callus more-sharply pointed; culms solid, 1–3 dm; ligs scarcely 0.5 mm, gen higher on sides than in back; dry open flats and foothills to des ranges, mostly with sagebr, chiefly in Nev and Cal, but also collected in Crook Co, Ore, and in Clark Co, Ida; very possibly a *Stipa-Oryzopsis* hybrid, but fairly common, and prob self-perpetuating; Webber's r. (*Stipa w.*)
                5 O. webberi (Thurb.) Benth.
        4b Lemma strigose-pubescent; glumes mostly < 6 mm
            6a Ligs acute, (2) 3–4 mm; bls 0.5–1 mm broad; glumes 4–6 mm; lemma ca = glumes; mt meadows and open valleys to rocky slopes and ridges, in sandy to rocky soil; s BC, Wn, and Ore, e Cas, to Mont and s to Utah, Colo, and Nev; little r.   **6 O. exigua** Thurb.
            6b Ligs truncate or higher on the sides than in the center, ca 0.3 mm; glumes 5–6 mm; lemma ca 1.5 mm < glumes; sagebr valleys and foothills, on calcareous soil, se Custer and Lemhi, and w Clark cos, Ida; Swallen's r.   **7 O. swallenii** Hitchc. & Spell.

Panicum L.   Witchgrass; Panicgrass

Spikelets in open to contracted pans, art below the glumes, ± dorsiventrally flattened, 2-fld, the 1st fl gen represented by an empty lemma; 1st glume always much < the 2nd; 2nd glume mostly· = the 1st (sterile) lemma and with it gen completely enclosing the fertile flt; 1st lemma gen empty but not rarely with a reduced palea; fertile lemma much hardened and very faintly nerved in comparison to the glumes and sterile lemma, the margins overlapping the equally firm palea; stamens 3; lods 2; ann or per with hollow culms, open sheaths, ligs mainly of straight hairs, and mostly flat bls rolled in the bud. (L name for a millet, millet being a name that applies to several spp. of grass, at least one of which is a *Panicum*).

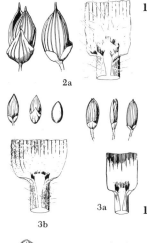

1a Pl ann; glumes and sterile lemma not short-pilose
    2a Spikelets 4–5 mm; 1st glume at least half as long as the spikelet; pl pi-lose-hirsute, up to 1 m; hairs in part pustulose, 2–5 mm; bls 7–15 mm broad; intro from Asia, and cult in various parts of US, where occ persistent in abandoned fields or along railroads and highways; broomcorn millet   **1 P. miliaceum** L.
    2b Spikelets < 3.5 mm; 1st glumes often < half as long as the spikelet
        3a Sheaths glab; 1st glume < half as long as the spikelet; culms up to 1 m; ligs 2–3 mm; native in c and e US and se Can, intro and estab in our area in Indian Valley, Adams Co, Ida, and in Malheur and Jackson cos, Ore; western w.   **2 P. dichotomiflorum** Michx.
        3b Sheaths pilose; 1st glume at least half as long as the spikelet; culms 2–7 dm; ligs 1.5–2 mm; mostly a weed in irrigated or moist areas, or along streams or around ponds; BC to Que, s to most of the US; common w., old-witch grass (*P. barbipulvinatum, P. b.* var. *hirsutipes, P. c.* var. *brevifolium, P. c.* var. *occidentale, P. elegantulum*)
                3 P. capillare L.
1b Pl per; glumes and sterile lemma often short-pilose
    4a Spikelets ca 3.2 mm; 1st glume ca 1.5 mm; dry prairies or rocky areas to sandy streambanks; BC to Me, s, mainly w Cas, to n Cal, also in e Wn and Ore and from Ida and Mont to Utah, Ariz, NM, and most of Miss Valley and ne US; Scribner w.   **4 P. scribnerianum** Nash
    4b Spikelets < 3 mm; 1st glume < 1 mm
        5a Spikelets 2.2–2.4 mm; rare, but collected at least once fairly recently near Portland, Ore; hairy w.; ours the var. *pseudopubescens* (Nash) Fern.   **5 P. villosissimum** Nash
        5b Spikelets 1.5–2.0 mm; 1st glume ca 0.5 mm; rocky or sandy riverbanks or lake margins to open woods, marshy areas to dry prairies, from sea level up to 7800 ft around hot springs; s BC to Banff, Alta, s along the coast and in the mts to s Cal, occ in WV, Ore, e Cas mainly along water courses or near springs in the mts, to w Mont and Wyo, occ s to Ariz; western w. (*P. brodiei, P. ferventicola, P. f.* var. *papillosum, P. f.* var. *sericeum, P. lassianum, P. pacificum, P. pubescens*)
                6 P. occidentale Scribn.

Parapholis C. E. Hubb.

Spikelets in a terminal spike, unawned, gen 1 per node, sunken in the tardily disarticulating rachis, 1-fld; glumes ca =, hardened, strongly 5-nerved, borne side by side in front of the slightly shorter, hyaline lemma and palea; stamens 3; lods 2; ann with narrow, non-aur lvs, membranous ligs, and open sheaths. (Gr *para*, beside, and *pholis*, scale, referring to the glumes which are side by side).

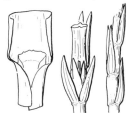

P. incurva (L.) C. E. Hubb. Sickle-grass. Culms ± decumbent, up to 4 dm; sheaths glab, blades flat to invol, 1–2 mm broad; ligs ca 1 mm; spikes cylindrical, 4–10 cm; glumes 6–8 mm, acute; lemma slightly < glumes, membranous; European weedy sp., rare in our area, but estab at Gold Beach, Ore, and coll at Portland long ago, so to be expected elsewhere along the coast (*Pholiurus i.*).

Paspalum L.

Spikelets dorsiventrally flattened, art below the glumes, borne (ours) in 2 rows on one side of the broadened rachis, ± sunken in the rachis, 2-fld, the upper fl ⚥, the lower one merely an empty lemma; spikes in ours gen in terminal pairs; 1st glume gen lacking, 2nd glume and sterile lemma = and slightly > the hardened lemma and palea; stamens 3; lods 2; ours per with solid culms, open sheaths, membranous ligs, and flat, non-aur bls rolled in the bud. (Gr name, *paspalos,* for some kind of millet). Several of the spp. are used as pasture or hay grasses.

P. distichum L. Knotgrass. Strongly stoloniferous per with upright culms 4–10 dm; nodes strongly pubescent; sheaths (esp the throat) gen pilose; ligs 1–1.5 mm, slightly erose; bls 3–8 (10) mm broad, the margins slightly thickened and antrorsely scab; spikelets mostly 3–3.5 mm; 1st glume often lacking, but sometimes lanceolate and up to 1.5 mm; 2nd glume appressed-pubescent, 3-nerved; sterile lemma gen glab; along ditches and streams, always where the soil is moist at least part of the season; common in much of e and s US to Mex and S Am; occ in sw and wc Ida, and in Wn and Ore where mostly w Cas and along CR *(Digitaria d.).*

Phalaris L.   Canarygrass

Spikelets in congested, often spikelike pans, art above the glumes (but spikelets sometimes deciduous as a whole), strongly compressed, (1–2) 3-fld, the uppermost flt gen ♀, the lower one(s) represented by sterile and often greatly reduced lemmas; glumes ± =, greatly compressed, often strongly keeled, gen 3-nerved; sterile lemmas from > half as long as the fertile to greatly reduced and one or even both sometimes lacking; fertile lemma gen hardened, mostly appressed-hairy, rounded to acute, much < the glumes; stamens 3; lods 2 (rarely lacking); ann or per with hollow culms, open sheaths, membranous ligs, and flat bls rolled in the bud. (Gr name, *Phalaris,* for some grass). Although several spp. have been reported for our area, some of these are not known to be estab and 2 or 3 have prob been collected only a very few times.

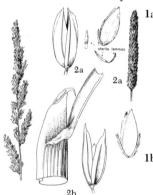

sterile lemmas

1a Pl rhizomatous per; glumes not prominently keeled, the keel, if any, < 0.5 mm wide
  2a Lvs mostly 3–8 mm broad; pan gen not lobed at anthesis; glumes ca 6 mm; 1st (sterile) lemma gen absent or much smaller than the 2nd, the 2nd ca 1.5 mm; culms up to 2 m; ligs 4–8 mm, blunt; fields and waste places, recently escaped from cult in Ore, also in Cal and occ elsewhere in US; Harding grass *(P. tuberosa, P. stenoptera)*   **1 P. aquatica** L.
  2b Lvs mostly 7–17 mm broad; pan plainly lobed or br at anthesis; glumes ca 4.5–5 mm; sterile lemmas 2, ca 1–1.5 mm; culms 7–14 dm; ligs 4–10 mm; wet places, esp along highways; fairly common from Alas across Can, and s to all but se US; Eurasia; reed c. (var. *picta*—the phase with variegated lvs, is sometimes grown as an orn)   **2 P. arundinacea** L.
1b Pl ann; glumes gen with a keel at least 0.5 mm wide
  3a Spikelets in groups of (5) 6–7, shed as a unit (glumes not remaining), often all but 1 of the spikelets sterile and with greatly modified (even club- or bulb-like) glumes, the keel gen projecting as an erect finlike tooth; bls 3–8 mm broad, margins gen joined with the sheath at different levels and 1 or both often free as a distinct aur; waste places, only very occ in our area in Wn and Ore, s to Cal and Ariz; intro from Europe; paradox c.   **3 P. paradoxa** L.
  3b Spikelets all ± alike, the glumes persistent after the flts are shed, the keel of ± uniform width
    4a Glumes 5–6 mm, acute, broadest near the middle, the keel with a wing scarcely 1 mm broad; fertile lemma ca 3 (to 4?) mm, the 2 sterile lemmas 1/3-1/2 as long; ligs 1.5–2.5 (3.5) mm; bls 3–6 (7) mm, margins freed at slightly different levels and 1 or both sometimes forming minute aurs scarcely 0.2 mm; moist or wet areas often drying by summer; Ore to Cal, e through the sw to NM and n Mex and most of se US; Carolina c.   **4 P. caroliniana** Walt.
    4b Glumes often > 6 mm, shortly acute, broadest well above midlength, the keel with a wing gen ca 1 mm broad; fertile lemma 4–5.5 (6) mm, or sterile lemmas only 1
      5a Fertile lemma ca 3 mm; sterile lemma 1; collected in our area only once or twice, on ballast near Portland, Ore; small c., Mediterranean c.   **P. minor** Retz
      5b Fertile lemma 4–6 mm; sterile lemmas 2
        6a Fertile lemma 4–5 mm, the sterile lemmas scarcely 1/4–1/3 as long; collected in our area only once or twice on ballast near Portland, Ore; shortspike c.   **P. brachystachys** Link

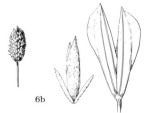

6b Fertile lemma 5–5.5 (6) mm, the sterile lemmas at least half as long; bls 3–10 mm broad, margins freed unequally and occ forming aurs scarcely 0.3 mm; ligs 4–8 mm; intro from the Mediterranean region and used in bird seed, frequently escaping and becoming estab, known throughout our area but nowhere common; c. *(P. aviculare)*   **5 P. canariensis** L.

Phleum L.   Timothy

Spikelets 1-fld, art above the glumes (but often shed as a unit late in the season), strongly flattened laterally, borne in cylindrical spikelike pan; glumes =, strongly keeled, often long-ciliate on the keel, 3-nerved, abruptly narrowed (acute to truncate) to a short to prominent awn; lemma thin, membranous, 5-nerved, truncate or obtuse, gen pubescent; palea 2-nerved, subequal to the lemma; rachilla lacking; stamens 3; lods 2; ann or tufted per with hollow culms, open sheaths, often tiny rounded aurs, membranous ligs, and flat bls rolled in the bud. (Gr *phleos*, the name for some reedy grass).

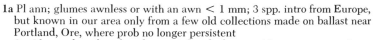

1a Pl ann; glumes awnless or with an awn < 1 mm; 3 spp. intro from Europe, but known in our area only from a few old collections made on ballast near Portland, Ore, where prob no longer persistent
  2a Glumes broadened and semi-inflated above, with a stout awn tip ca 0.3–0.4 mm, scab and ± stiff-ciliate on the lower half
                                                              P. paniculatum Huds.
  2b Glumes narrowed and not at all inflated above, unawned, either glab or stiff-ciliate the full length
    3a Glumes glab or scab, but not at all ciliate
                                                    P. subulatum (Savi) Asch. & Graebn.
    3b Glumes stiffly long-ciliate the full length of the keel   **P. arenarium** L.
1b Pl per; glumes with awns at least 1 mm
  4a Culms gen bulbous at base; pan gen > 4.5 (3–13) × < 1 cm when pressed; anthers 1.6–2.3 mm; culms up to 1 m; intro from Europe; commonly cult and freely escaped and estab, lowl to lower mt valleys, often in waste places; common throughout our area except in dry des; common t.   **1 P. pratense** L.
  4b Culms not bulbous at base; pan 1–4.5 × often > 1 cm when pressed; anthers 1–1.5 mm; culms 1.5–5 dm; streambanks and meadows in the mts; Alas to Newf, s to most of mont-subalp w US, to S Am; Europe; alpine t., mt t. *(P. haenkeanum, P. pratense* var. *a., P. a.* var. *americanum, P. a.* var. *scribnerianum)*   **2 P. alpinum** L.

Phragmites Trin.   Reed

Spikelets in large feathery pan, several-fld, art above the glumes; glumes rounded on the back, slender, the 2nd much > the 1st, but < 1st lemma; rachilla with ∞ long silky hairs; lemmas 3-nerved, rounded on the back; stamens 2 or 3; lods 2; per, rhizomatous reeds with hollow culms, open sheaths, short membranous ligs gen fringed with hairs = or > the membrane, and flat, non-aur bls rolled in the bud. (Ancient Gr name, said to mean "growing in hedges," because of the dense stands fringing streams and ponds).

P. communis (L.) Trin. Common r. Culms stout, erect, 2–3 m; sheaths glab, loose, twisting in the wind and aligning the bls on one side; ligs 1.5–3 mm, half membrane and half hairs but the innovations with mostly membrane, the fringe of hair late in developing; bls flat, mostly 20–40 × 1–4 cm, gen breaking from the culms by winter; pan 15–35 cm, often purplish but later straw-colored; spikelets 10–15 mm, 3–6-fld; glumes 4–6 and 6–9 mm; 1st lemma 9–12 mm, unawned, the upper ones gen smaller and with awns often as long as the body, glab but exceeded by the silky hairs of the rachilla; fairly common around ponds and marshes, springs, and rivers; throughout most of s Can and all but extreme se US, ± cosmopolitan *(Arundo phragmites)*; too coarse for forage, but sometimes used for thatching and matting.

Pleuropogon R. Br.   Semaphoregrass; Pleuropogon

Spikelets loosely racemose, many-fld, art above the glumes; glumes membranous-hyaline, rather broad, erose to lacerate at the tip, < the subtended flts, unequal, the 1st 1-nerved, the 2nd faintly 3-nerved; lemmas rounded on the back, with 7 prominent, nonconvergent nerves, the tip often lobed and bearing a short to prominent awn; palea ± winged on the lower part of the 2 prominent keels; lods 2; rhizomatous per of wet places, with at least partially closed sheaths, membranous ligs, hollow culms, and rather soft, flat, non-aur bls rolled in the bud. (Gr *pleura*, side, and *pogon*, beard, referring to the awns at the base of the palea in the first-described sp.).

1a Palea subequal to the lemma, awned from each keel somewhat below midlength, the awns 2–7 mm; pl glab, 4–9 dm; sheaths closed ca 3/4 their length; ligs 2–5 mm, acute, glab; racemes 1–1.5 dm; lemmas ca 6 mm; streambanks and wet meadows; known from only a few collections in Union and Lake cos, Ore; Ore s.   **1 P. oregonus** Chase

1b Palea slightly to considerably < lemma, the wing of each keel prominent and free at the tip for 0.5–1 mm, but unawned; pl glab to scab, 10–15 dm; sheaths closed at base but open above for 2–7 cm; ligs 1.5–3.5 mm, ciliolate; racemes 1–2.5 dm; lemmas 7–8 mm; wet places, such as bogs, streambanks, swampy meadows, and shaded woods, from near sea level up to perhaps 5000 ft elev; s BC s, in OM and Cas, to nw Cal; nodding s. (*Lophochlaena r.*)   **2 P. refractus** (Gray) Benth.

Poa L.   Bluegrass

Spikelets in compact to open and diffuse pans, 2–7 (10)-fld, art above the glumes; flts gen ⚥, but often imperfect and the pls mostly ♂, ♀, although ♂ pls rare or nonexistent in some spp., ♀ fls often with vestigial stamens; glumes subequal or unequal, gen ± keeled, mostly 1- or 3-nerved, gen < the 1st lemma, but sometimes at least the 2nd glume = or > it; lemmas unawned (rarely the midvein freed as a tiny apiculation), with gen 5 obscure to rather prominent, convergent nerves, mostly strongly keeled, but sometimes rounded on the back, often with a tuft of cobwebby, crumpled hairs at base, the surface gen scab to pubescent or pilose at least on the keel and marginal nerves; stamens 3; lods 2; ann or strongly caespitose to rhizomatous or stoloniferous pers with hollow culms, gen partially closed sheaths, membranous ligs, and flat to folded or involute, non-aur bls folded in the bud, the tips gen ± upcurved and prow-like. (The Gr name for grass). (*Atropis*). *Poa* is an important grass for hay, pasture, and lawns, and of great value as range forage; taxonomically the genus is notoriously difficult.

1a Pl ann or winter ann, the remains of old culms lacking

2a 2nd glume broadened above midlength; lemmas not webbed at base; culms mostly < 2 dm, freely rooting at the nodes and forming small mats; lawns, gardens, roadsides, and waste ground to open woods, ± cosmopolitan; gen in our area, but most common w Cas where a common weed in gardens and lawns, tending to start growth early in the spring and to die or brown during the middle of the summer; sometimes behaving, esp in lawns, as a per; ann b.   **1 P. annua** L.

2b 2nd glume as well as the 1st tapered from the base; lemmas webbed at base; culms mostly > 2 dm

3a Lemmas pubescent over the back, rather plainly 5-nerved; pl scab on the bls and often also on the sheaths; bls slender-pointed, 5–10 cm; pan 20–25 cm; prairies and lightly wooded areas, gen where moist, mostly near sea level in our area; VI and mainl sw BC s to Cal, w Cas, and in CRG; Howell's b. (*P. bolanderi* var. *h.*)

2 P. howellii Vasey & Scribn.

3b Lemmas glab or scab over the back, obscurely 5-nerved; pl glab on the sheaths, scab only on the lf margins and midvein; bls abruptly and bluntly pointed, mostly 2–5 cm; pan 10–15 cm; dry to moist areas in the mts, mostly at 5000 ft or above; Blue Mts, Wn and Ore, s to s Cal, e to wc Ida and w Nev; Bolander's b. (*P. horneri*) 3 **P. bolanderi** Vasey

1b Pl per, sometimes fl the 1st year, but gen bearing the remains of culms or lvs of previous year(s)

4a Rhizomes present, often extensive

5a Pls ♂, ♀, growing on the sand or dunes along the coast, forming tufts; bls involute; sheaths closed for < half their length

6a Lemmas < 5 (2.5–4) mm, obscurely 5-nerved; bls scarcely 2 mm broad; ligs acute, 0.8–2 mm; spikelets 3–4-fld; glumes 2.5–3 and 3–4 mm; VI to near San Francisco, along the coast; coastline b.

4 P. confinis Vasey

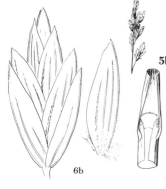

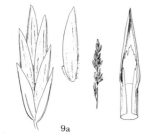

6b Lemmas > 5 (7–10) mm, rather plainly 7–11-nerved; bls 2–3 mm broad; ligs 1–1.5 mm; spikelets 4–6-fld; glumes 8–9 and 9–10 mm, the 2nd ca= the 1st lemma; VI and PS region to nw Cal, along the ocean or sound; seashore b. *(P. douglasii* ssp. *m.)*

5 **P. macrantha** Vasey

5b Pl either ♂ or not growing along the coast on sandy soil; bls often flat or folded

7a Pl partially to completely ♂, ♀, gen ♀

8a ♀ lemmas ± concealed by long silky hairs; Okla and Tex to Ga; intro in Ida and Ore, but not believed to be self-perpetuating anywhere in our area; Tex b.   **P. arachnifera** Torr.

8b ♀ lemmas scab to pubescent but not covered with long silky hairs; pls native to our area

9a Uppermost culm lf gen with bl much-reduced or lacking; lower bls stiff, gen strongly scab; sheaths closed < 1/3 their length; lemmas often with longest hairs along the marginal nerves and keel; 2nd glume 3.5–5 mm; ligs from truncate and 0.5–1 mm to elongate, acuminate, and up to 10–12 mm above; sagebr des to wooded areas at mostly middle elev in the mts; e Cas, BC to Cal, e to Man and RMS s to Tex and NM; mutton-grass *(P. longiligula)*   6 **P. fendleriana** (Steud.) Vasey

9b Uppermost culm lf gen with well-developed bl; lower bls gen rather lax, rarely strongly scab except along the margins; sheaths gen closed at least half their length; lemmas variously pubescent or scab but not longer-hairy on the marginal nerves and keel; 2nd glume ca 3.5 mm; ligs 1–2 (4) mm; exposed ridges and talus slopes to open woods, mont to alp; BC to Alta, s to Cal, Nev, Colo, and NM; Wheeler's b.; 2 vars. of

7 **P. nervosa** (Hook.) Vasey

a1 Collar and throat pubescent, the hairs > any on the sheath; pls mostly ♀, but sometimes ♂-fld; w Cas, Wn and Ore, limited to lower CR and adj tributaries *(Festuca n., P. columbiensis)*   var. **nervosa**

a2 Collar and throat glab or at least with hairs no > any on the adj sheath; pls all functionally ♀; range of the sp. except not W Cas in Wn and rarely so in Ore *(P. olneyae, P. vaseyana, P. w.)*   var. **wheeleri** (Vasey) Hitchc.

7b Pls mostly ♂-fld

10a Lemmas webbed at the base but otherwise glab; reported for our area from Corvallis, Ore, but the record believed to be erroneous, the sp. apparently confined to coastal Cal; Kellogg's b.

P. kelloggii Vasey

10b Lemmas not webbed, or webbed but also with other pubescence

11a Culms strongly flattened, 2-edged; pl strongly rhizomatous; lemmas sparsely if at all webbed; pl 2–4 (6) dm; sheaths open to near the base; bls flat to folded, (1) 2–4 mm broad; ligs truncate to rounded, 0.5–1.5 (2) mm; pan compact, 3–9 (12) cm; waste ground, roadsides, and gardens to open woods and meadows, mostly where fairly moist; Alas to Newf, s to much of US and throughout our region; Can b.; widely used as a constituent of pastures and ± as a substitute for *P. pratensis* in lawns   8 **P. compressa** L.

11b Culms terete or only slightly flattened, not 2-edged; pl often either weakly rhizomatous or with strongly webbed lemmas

12a Pan narrow, the brs ascending or appressed; lemmas scab to pubescent on all the nerves, but not webbed

13a Lemmas not strongly keeled, mostly rather evenly scab or scab-puberulent over the lower half, rarely pubescent on the keel and marginal nerves, but then the basal bls flat and < 3 cm

14a Bls thick, very prominently nerved and with narrow, thickened, whitish margins, gen flat, scarcely 3 cm × 1–3 mm; ligs prominent, 2–5 mm, strongly decurrent; Wen Mts, Wn; Mt Stuart b., little mt b.

9 **P. curtifolia** Scribn.

14b Bls mostly thin, involute, or > 3 cm; ligs various,

15a

15b

16a

17b

18a

19a

sometimes < 2 mm and not strongly decurrent, but sometimes much longer; widespread (the *P. sandbergii* complex)   **Group III** p. 663

13b Lemmas strongly keeled, villous on the keel and marginal nerves, but gen glab between; basal bls gen involute, much > 3 cm

15a Pl 2–4 (6) dm; bls mostly involute-folded, 1–3 mm broad, the uppermost often only 1–2 cm; spikelets 4–7 (9)-fld; 2nd glume 3–3.5 (4.2) mm, < the 1st lemma; pls of the prairies and lower mts, often in moist saline areas; Alta to Man, s to Utah and NM, e to Daks, Ia, and Tex, mostly e RM in Mont, where w to near Deer Lodge, Powell Co; dryland b., prairie spear-grass *(P. pratericola, P. fendleriana* var. *a.)*   **10 P. arida** Vasey

15b Pl 3.5–8 dm; bls mostly flat, 2–4 mm broad, the uppermost 5–10 cm; spikelets mostly 3–4-fld; 2nd glume (4) 4.5–5 mm, often > 1st lemma; meadows, creek bottoms, and ditches to open high plains or gravelly or rocky slopes; se BC and Alta s to Ida, Utah, and Ariz, e to Minn, Neb, and Colo; pale-lf b.

11 P. glaucifolia Scribn. & Will.

12b Pan open, the brs mostly spreading; lemmas often webbed at base, or glab on the internerves

16a Pl alp or subalp; lemmas pubescent over the back and silky on the 5 nerves, gen webbed at base; culms 2–6 dm; bls flat or folded, gen 2–4 mm broad; ligs 0.5–4 mm, mostly erose-jagged; anthers ca 2 mm; on screes, open ridges, meadowl, and streambanks; BC to sw Alta, s in RMS to Utah and NM; apparently also in Cas, in a modified form, in Wn and n Ore; Gray's b. *(P. alpicola, P. arctica, P. cenisia* misapplied, *P. longipila)*

12 P. grayana Vasey

16b Pl often of lower elev; lemmas gen glab on the internerves or not webbed

17a Lemmas not webbed at base, from glab overall or scab-puberulent on the keel and marginal nerves only, to finely scab-puberulent over the back; sheaths often retrorsely puberulent; closed > half their length

18a Bls flat, (2) 3–6 mm broad, those of the culms rarely as much as 10 cm, scab but not puberulent; sheaths not puberulent and gen not purplish; collar not pubescent; ligs mostly finely lacerate-erose; shaded, gen moist areas, mossy rocks and mt meadows, well below timberl; wc Ida to nw Wyo, s to Utah; short b.

13 P. curta Rydb.

18b Bls often folded, mostly > 10 cm × 2–3.5 (5) mm, often puberulent on the ventral surface; sheaths often purplish, mostly retrorsely puberulent; collar often pubescent; ligs mostly ciliate, but rarely lacerate-erose; pls from w Cas to Mont (see lead 9b)

7 P. nervosa (Hook.) Vasey

17b Lemmas often webbed at base or pilose on the keel and marginal nerves, at least   **Group I** p. 660

4b Rhizomes lacking, although pl sometimes stoloniferous

19a Culms gen enlarged and bulbous at base; flts mostly or all modified into small purplish bulblets without stamens or pistil; pl glab, 1.5–4 (6) dm; sheaths open almost to the base; ligs blunt to acuminate, 1.5–3 (1–5) mm; bls flat to involute, 1–2.5 mm broad; intro from Europe; widespread in much of N Am and now common in most of our area in other than mont habitats, but mostly e Cas; bulbous b.

14 P. bulbosa L.

19b Culms not bulbous at base; flts not modified into bulblets, but with stamens, a pistil, or both

20a Lemmas with long, tangled, cobwebby hairs at base

21a Lemmas glab except for the cobwebby base, or pubescent also but only on the keel

22a Lemmas barely 3 mm; fls ♀; culms (3) 4–10 dm, ± stolonous;

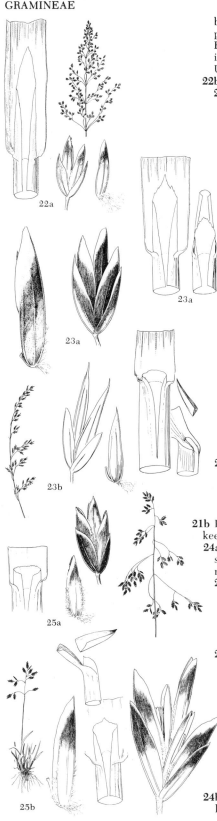

22a

23a

23a

23a

23b

25a

25b

bls flat, lax, up to 15–20 cm × 2–4 mm, the tips only slightly prow-like; ligs mostly 3–7 mm; anthers 1–2 mm; intro from Europe, widely estab, mostly in moist woods or in wet places; in our area mainly w Cas and in CRG, but also occ in Mont, Utah, and Colo; roughstalk b. *(P. callida)*   **15 P. trivialis** L.

22b Lemmas (4) 4.5–7 mm; fls often ♂, ♀

  23a Ligs in part 1–6 mm; sheaths closed ca half their length; fls ♂, ♀, mostly ♀, rarely ♂; culms 1.5–5 dm; anthers 2.2–3.2 mm; sagebrush plains to alp meadows and ridges; BC to Cal, in both OM and Cas, e to Alta and Sask, Nev, Utah, and Colo; 3 vars.   **16 P. cusickii** Vasey

    a1 Basal lvs filiform, involute, scarcely 1 mm broad, gen ± scab, not rarely up to 20 cm; culm bls narrow, seldom flat or > 1 mm broad (as rolled or folded); lemmas mostly not > 5 mm, glab to scab; pls chiefly of sagebr des and ponderosa pine for, ranging upward on dry, open rocky ridges sometimes to timberl; e side Cas, Wn and Ore to Cal, e to Alta and Sask and s in RM to Mont, Wyo, and n Colo; Cusick's b. *(P. cottoni, P. filifolia, P. idahoensis, P. scabrifolia, P. spillmani, P. subaristata)*   var. **cusickii**

    a2 Basal bls gen ave < 10 cm × at least 1 mm; culm bls gen flat (folded), (1) 1.5–3.5 mm broad; lemmas sometimes > 5 mm and pilose on the midnerve; pls mont to alp

      b1 Lemmas 5.5–7.2 mm, ± pilose on the keel and not rarely with a slight basal web; culm bls 2–3.5 mm broad, flat; BC s in Cas (and OM) to Crater Lake, Ore *(P. alpina* var. *purpurascens* Vasey, Descr. Cat. Grasses US 75. 1885, not Beal, Grasses N. Am. 2:543. 1896, as cited by Hitchc. V. P. P. N. W. 1:659. 1969)

        var. **PURPURASCENS** (Vasey) C. L. Hitchc. hoc loc.

      b2 Lemmas mostly < 6 mm, glab to scab but not at all pilose or webbed; culm bls rarely > 2.5 mm broad, often folded; BC to Alta, s and in both Cas and OM, to Sierran Cal and to Nev, Utah, and Colo, perhaps less common than var. *cusickii* in Ida and Mont; skyline b. *(P. epilis, P. purpurascens* var. *e.)*

        var. **epilis** (Scribn.) Hitchc.

  23b Ligs gen ca 1 mm; sheaths closed almost their full length; fls ♂; bls flat, soft, 1–2.5 mm broad; anthers 0.4–0.6 mm; spikelets mostly 2-fld; moist areas in the coastal mts, VI s to nw Ore, known from several localities, but rather rarely collected; withered b.   **17 P. marcida** Hitchc.

21b Lemmas pubescent on the marginal nerves as well as on the keel

  24a Pan loose, the brs slender, lower ones 1–3 per node, spreading to reflexed; spikelets gen purplish; ligs glab, gen > 1 mm; anthers not > 1 mm

    25a Lower pan brs reflexed, 1–3 per node; glumes subequal, ca 3 (2.5–3.5) mm; lemmas ca 3 mm; subalp to alp meadows, streambanks, and slopes, e BC and Mont to Ida and reportedly to e Ore, s to Nev, Ariz, and NM, very common in the RM; nodding b. *(P. acuminata, P. leptocoma* var. *r.)*

        18 P. reflexa Vasey & Scribn.

    25b Lower pan brs gen not reflexed, mostly 2 per node; tufted per 1–5 dm; glumes acute, the 2nd much broader and ca 1/4 > the 1st; lemmas ca 3.5 (3–4) mm; wet places along streams to open subalp or alp ridges and meadows; Alas s, in Cas and OM, to Cal, e to Alta and in RMS to Colo and NM; bog b.; 2 vars. of   **19 P. leptocoma** Trin.

      a1 Sheaths smooth; pl gen only 1–2 (3) dm; pan mostly 4–9 cm; Alas to Alta, s to Mt Rainier, Wn, and to c Mont *(P. p.)*   var. **paucispicula** (Scribn. & Merr.) Hitchc.

      a2 Sheaths minutely scab; pl mostly 3–5 dm; pan 7–15 cm; Alas and Yuk to Cal, e to RMS from Mont to NM *(P. stenantha* var. *l.)*   var. **leptocoma**

  24b Pan often narrow, the brs ascending or > 3 per node; spikelets gen greenish; ligs often puberulent-scab, sometimes all <

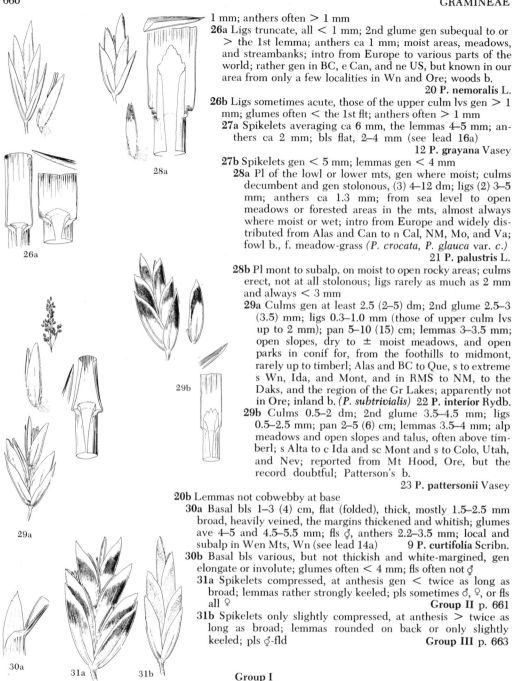

1 mm; anthers often > 1 mm

26a Ligs truncate, all < 1 mm; 2nd glume gen subequal to or > the 1st lemma; anthers ca 1 mm; moist areas, meadows, and streambanks; intro from Europe to various parts of the world; rather gen in BC, e Can, and ne US, but known in our area from only a few localities in Wn and Ore; woods b.
20 **P. nemoralis** L.

26b Ligs sometimes acute, those of the upper culm lvs gen > 1 mm; glumes often < the 1st flt; anthers often > 1 mm

27a Spikelets averaging ca 6 mm, the lemmas 4–5 mm; anthers ca 2 mm; bls flat, 2–4 mm (see lead 16a)
12 **P. grayana** Vasey

27b Spikelets gen < 5 mm; lemmas gen < 4 mm

28a Pl of the lowl or lower mts, gen where moist; culms decumbent and gen stolonous, (3) 4–12 dm; ligs (2) 3–5 mm; anthers ca 1.3 mm; from sea level to open meadows or forested areas in the mts, almost always where moist or wet; intro from Europe and widely distributed from Alas and Can to n Cal, NM, Mo, and Va; fowl b., f. meadow-grass *(P. crocata, P. glauca var. c.)*
21 **P. palustris** L.

28b Pl mont to subalp, on moist to open rocky areas; culms erect, not at all stolonous; ligs rarely as much as 2 mm and always < 3 mm

29a Culms gen at least 2.5 (2–5) dm; 2nd glume 2.5–3 (3.5) mm; ligs 0.3–1.0 mm (those of upper culm lvs up to 2 mm); pan 5–10 (15) cm; lemmas 3–3.5 mm; open slopes, dry to ± moist meadows, and open parks in conif for, from the foothills to midmont, rarely up to timberl; Alas and BC to Que, s to extreme s Wn, Ida, and Mont, and in RMS to NM, to the Daks, and the region of the Gr Lakes; apparently not in Ore; inland b. *(P. subtrivialis)* 22 **P. interior** Rydb.

29b Culms 0.5–2 dm; 2nd glume 3.5–4.5 mm; ligs 0.5–2.5 mm; pan 2–5 (6) cm; lemmas 3.5–4 mm; alp meadows and open slopes and talus, often above timberl; s Alta to c Ida and sc Mont and s to Colo, Utah, and Nev; reported from Mt Hood, Ore, but the record doubtful; Patterson's b.
23 **P. pattersonii** Vasey

20b Lemmas not cobwebby at base

30a Basal bls 1–3 (4) cm, flat (folded), thick, mostly 1.5–2.5 mm broad, heavily veined, the margins thickened and whitish; glumes ave 4–5 and 4.5–5.5 mm; fls ♂, anthers 2.2–3.5 mm; local and subalp in Wen Mts, Wn (see lead 14a) 9 **P. curtifolia** Scribn.

30b Basal bls various, but not thickish and white-margined, gen elongate or involute; glumes often < 4 mm; fls often not ♂

31a Spikelets compressed, at anthesis gen < twice as long as broad; lemmas rather strongly keeled; pls sometimes ♂, ♀, or fls all ♀
Group II p. 661

31b Spikelets only slightly compressed, at anthesis > twice as long as broad; lemmas rounded on back or only slightly keeled; pls ♂-fld
Group III p. 663

Group I

Pl rhizomatous; lemmas webbed at base, or pilose on keel and marginal nerves, or both

32a Culms and gen also the sheaths and bls retrorsely scab; ligs 1–2 (4) mm; lemmas webbed at base and pubescent on the keel and marginal nerves, the intermediate nerves obscure; pl 9–12 dm; sheaths closed to about midlength; anthers scarcely 1 mm; moist woods to rocky open slopes, from near sea level to lower elev in the mts; s Alas to the OP and to Multnomah, Clackamas, and Benton cos, Ore; rare, or at least rarely collected; loose-fld b. *(P. leptocoma* ssp. *elatior, P. remissa)* 24 **P. laxiflora** Buckley

32b Culms gen not scab or ligs in part > 3 mm, or lemmas not pubescent on the marginal nerves, the intermediate nerves often prominent

33a Anthers mostly 0.5–0.9 (never > 1) mm; lower pan brs mostly in 2's (see lead 25b)   19 **P. leptocoma** Trin.

33b Anthers at least 1 (1.8–2) mm; lower pan brs gen in 3's or 5's

  34a Lemmas not webbed at base (see lead 15b)   11 **P. glaucifolia** Scribn. & Will.

  34b Lemmas ± strongly webbed at base

    35a Pl truly rhizomatous, gen not growing in wet places; ligs mostly not > 1.5 (the uppermost up to 3) mm, truncate; lemmas ca 3.5 mm; bls flat or folded, 2–4 (5) mm broad, the tips strongly prow-like; anthers ca 1.7 mm; sheaths closed ca half their length; roadsides and waste areas to dry or moist prairies and woodl; widespread in most of temp Can and all but se US; Eurasia; Kentucky b., one of the best of lawn grasses (*P. agassizensis*, *P. peckii*)
                                   25 **P. pratensis** L.

    35b Pls stoloniferous but not truly rhizomatous, often growing in wet places; ligs mostly (2) 3–7 mm, often acute; lemmas 2.5-3 mm

      36a Internerves of lemma obscure; marginal nerves hairy; ligs mostly 3–5 mm (see lead 28a)   21 **P. palustris** L.

      36b Internerves of lemma prominent; marginal nerves not hairy; ligs 3–7 mm (see lead 22a)   15 **P. trivialis** L.

Group II

Pl non-rhizomatous; spikelets compressed, the lemmas keeled on back

37a Lemmas pubescent (gen villous or pilose) on the keel and marginal nerves or hairy over the entire lower portion, or sericeous, or pilose on the nerves and hairy between

  38a Pls almost completely ♂, ♀, gen functionally ♀

    39a Pls rarely as much as 3 (0.5–3) dm; bls lax, filiform, gen ca 1 mm broad; ligs ca 1 (the basal) to as much as 4 mm, acute; pan 2–8 cm; functional anthers 2.8–3.2 mm; mostly in shade of rocks, gen where moistened by seepage until early summer, often on cliffs, from sagebr plains and foothills to lightly wooded mts; CRG, e along CR and SR to Asotin Co, Wn, and scattered in n Ore from near Mt Hood to Malheur Co; Leiberg's b. (*P. pulchella*, *P. vaseyochloa*, *P. gracillima* var. *v.*)
                                   26 **P. leibergii** Scribn.

    39b Pls mostly > 3 dm; bls often stiff or much > 1 mm broad

      40a Lemmas long-villous on the keel and marginal nerves (see also *P. cusickii* which does not have greatly reduced bls on the upper culm lvs and which is pilose on the keel, only); bls stiff, gen strongly scab, the uppermost ones on the culms gen greatly reduced or even wanting; sheaths glab or scab, rarely closed as much as 1/3 their length (see lead 9a)   6 **P. fendleriana** (Steud.) Vasey

      40b Lemmas uniformly pubescent, not long-villous on the keel and marginal nerves; bls rather lax, mostly scab on the margins, only, the uppermost not greatly reduced; sheaths often puberulent, gen closed at least 1/2 their length (see lead 9b)
                                   7 **P. nervosa** (Hook.) Vasey

  38b Pls mostly ⚥-fld, both the stamens and pistil functional

    41a Pls strictly coastal; lemmas ca 4 (3.5–4.5) mm, pubescent only at the base; sheaths open almost to the base; ligs (1) 2–5 mm; bls mostly flat (folded-involute), 1–1.5 mm broad; pan congested, 2–6 cm; spikelets 3–8-fld; glumes (2) 3–3.5 and 3.5–4 mm; anthers 1.75–2.5 mm; known only from ocean cliffs, Ilwaco, Pacific Co, Wn; thickglumed b., seacliff b.   27 **P. pachypholis** Piper

    41b Pls never coastal; lemmas gen pubescent to above the middle

      42a Pan narrow, contracted, several times as long as broad, but gen not > 6 cm; pls rarely > 2 dm

        43a 2nd glume 3.5–4.2 (ave almost 4) mm, subequal (or =) to the 1st lemma; pan very compact, peds or pan brs scarcely visible, rarely as long as the spikelets; pls mostly 0.5–1.5 (2) dm (see lead 29b)   23 **P. pattersonii** Vasey

        43b 2nd glume 2.5–3.5 (ave ca 3.2) mm, < the 1st lemma; pan more open, the brs and peds mostly visible, some of them > the spikelets; pls mostly 1–2 (2.5) dm; lemmas strongly keeled, ave barely 3 mm, not cobwebby at base, but at least the keel and marginal

nerves ± silky, and often the entire lower half densely pubescent; meadows to dry, open, often rocky slopes, esp abundant on limestone, mostly near or above timberl; Yuk, BC, and Alta, s to n Wn and through RMS to Utah, Colo, and NM, e to SD, w to ne Ore, Nev, and Cal; timberl b.   28 **P. rupicola** Nash

42b Pan either open and with spreading or ascending to drooping brs, or > 6 cm, or pls > 2 dm

    44a Spikelets 2/3 as broad as long, subcordate; bls flat, 2–4 mm broad, the uppermost below midlength of the culm; ligs coarsely erose, truncate to obtuse, glab, (1) 1.5–3 (4) mm; pl 1–3 (4) dm; subalp to alp meadows, ridges, and talus slopes; Alas to Que, s in the Cas to near Mt Rainier, Wn, and to ne Ore, Utah, Colo, Ida, Mont, and n Mich; Eurasia; alpine b.   29 **P. alpina** L.

    44b Spikelets gen < 2/3 as broad as long, or oblong rather than subcordate; bls often involute or < 2 mm broad, the uppermost gen above midlength of the culm; ligs various

        45a Ligs gen truncate, 0.3–1.0 (occ the longest 1.5–2) mm; spikelets mostly 2–3 (4)-fld; 2nd glume 2.5–3 (3.5) mm; pl 2.5–4.5 (2–5) dm; open slopes, dry to rather moist meadows, and open parks in conif for from the foothills to middle elev in the mts, rarely almost to timberl (see lead 29a)   22 **P. interior** Rydb.

        45b Ligs gen obtuse to acute, mostly at least 1 (longest 2–4) mm; spikelets mostly 3–5-fld; 2nd glume rarely < 4 mm

            46a Lemmas pilose-sericeous along the 5 nerves; bls 2–4 mm broad, flat or folded; pan brs gen with only 1–3 spikelets at the tips (see lead 16a)   12 **P. grayana** Vasey

            46b Lemmas pilose-sericeous only on the keel and marginal nerves; bls 1–2.5 mm broad, often involute; pan brs gen with several spikelets; sheaths closed scarcely 1/4 their length; open flats in mont for to subalp meadows and talus slopes; Alas to OM and Cas, Wn, and at Crater Lake, and in Josephine Co, Ore, e to Ida and Mont, and in Colo; Trinius' b. *(P. englishii)*   30 **P. stenantha** Trin.

37b Lemmas glab or scab to scab-puberulent, not villous or pilose on the keel and marginal nerves and not truly pubescent elsewhere

    47a Pls gen < 1 dm, ♂-fld; anthers ca 0.5 mm; glumes subequal, ca 3 (2.5–3.5) mm, 3-nerved, glab, gen = or > 1st 2 lemmas; alp ledges and ridges; higher peaks of Cas in BC and Wn and in Cal and Nev, e to Mont, Utah, Wyo, and Colo; Letterman's b.   31 **P. lettermanii** Vasey

    47b Pls gen > 1 dm, often ♂, ♀; anthers gen at least 1 mm; glumes gen < 1st lemma

        48a Pls (0.5) 1–3 dm; bls soft, ca 1 mm broad; pan loose, ± open, peds slender, evident, gen at least as long as the spikelets; mostly on sagebr plains and foothills, CRG e (see lead 39a)   26 **P. leibergii** Scribn.

        48b Pls > 3 dm, or with bls > 1 mm broad, or mont, or coastal; pan often congested and peds mostly < spikelets and not evident

            49a Pls strictly coastal, known only from Ilwaco, Pacific Co, Wn; ligs puberulent (see lead 41a)   27 **P. pachypholis** Piper

            49b Pls other than coastal; ligs often pubescent

                50a Pans open, the peds evident; sheaths of basal culm lvs often puberulent or reddish-purple, those of the upper part of the culms closed to well above midlength; bls mostly 2–3.5 (5) mm broad; lower ligs truncate, strongly pubescent, ciliate, ave 0.5–1 mm; glumes ca 3 and 3.5 mm (see lead 9b) 7 **P. nervosa** (Hook.) Vasey

                50b Pan congested, the peds mostly concealed; sheaths of basal lvs rarely either puberulent or reddish-purple, those of the upper part of the culm rarely closed as much as half the length; bls often much < 2 mm broad; lower ligs often acute, or much > 1 mm, or not pubescent; glumes tending to ave > 3 and 3.5 mm

                    51a Pls in part ♂-fld, in part ♀; anthers 1–1.3 mm; pan narrow, mostly not > 1 cm thick, gen purplish, often not much > the basal lvs at anthesis; alp on a few peaks in w Wn and n Ore, and in c Sierran Cal; Suksdorf's b.

                                32 **P. suksdorfii** (Beal) Vasey

                    51b Pls almost exclusively ♀, the rare stamen-bearing pl with anthers 2–3 mm; pan broader, gen > 1 cm thick, often tawny or greenish, gen > basal lvs (see lead 23a)   16 **P. cusickii** Vasey

Group III

Pl non-rhizomatous; spikelets scarcely compressed, lemmas mostly rounded on back

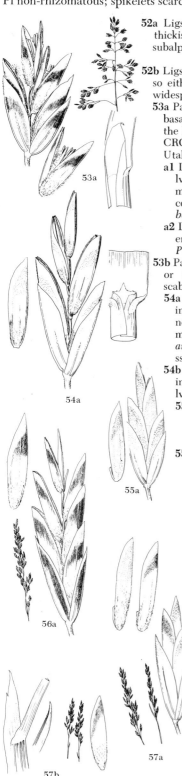

52a Ligs prominent, (1.5) 2–5 mm, strongly decurrent; basal bls flat but thickish and stiff, rarely > 3 cm × 1.5–3 mm, often whitish-margined; subalp in the Wen Mts, Kittitas Co n to Chelan Co, Wn (see lead 14a)
                                                                    9 **P. curtifolia** Scribn.

52b Ligs often neither prominent nor decurrent; basal bls gen not flat, but if so either > 3 cm or < 1.5 mm broad, rarely if ever whitish-margined; widespread

  53a Pan open, at one stage the lateral brs spreading, often at right angles; basal bls gen flat, lax, 1–2 mm broad; lemmas gen ± crisp-pubescent on the lower half; pls mostly subalp or alp, but down to near sea level in the CRG and near the Pac Ocean; BC to Alta, s to Cal, e to Mont, nw Wyo, Utah, and Colo; Pac b., slender b.; 2 vars.   33 **P. gracillima** Vasey

    a1 Ligs truncate and ciliolate, rarely > 1 mm, even on the upper culm lvs; bls mostly lax, the basal > half as long as the culms; s side CRG, mainly, in rocky, shaded cliffs mostly near waterfalls, much less common on n side of the Gorge, and along lower WV, Ore *(Sporobolus bolanderi, P. m.)*   var. **multnomae** (Piper) Hitchc.

    a2 Ligs acute, nonciliolate, (1) 1.5–5 mm; range of the sp. except apparently absent on s side CRG *(P. alcea, P. buckleyana* var. *stenophylla, P. invaginata, P. saxatilis, P. g.* var. *s.)*   var. **gracillima**

  53b Pan gen narrow, the brs mostly ascending to erect, or basal bls involute or folded or stiff and > 2 mm broad; lemmas often scab or scab-puberulent rather than pubescent; pls often other than subalp or alp

    54a Ligs thickish, strongly ciliolate, truncate to rounded, those of the innovations and basal culm lvs scarcely visible from the side, mostly not > 0.5 (to 1) mm, those of the upper culm lvs mostly 1–1.5 (2–2.5) mm; sagebr areas, often where alkaline, e Cas to e RMS; alkali b. *(P. ampla, P. brachyglossa, P. confusa, P. fendleriana* var. *juncifolia, P. j.* ssp. *porteri, P. laeviculmis, P. truncata)*   34 **P. juncifolia** Scribn.

    54b Ligs thin and membranous, rarely ciliolate, gen acute, those of the innovations and lower culm lvs gen > 1 mm, those of the upper culm lvs mostly 2–7 mm

      55a Lemmas glab, or scab (but not crisp-puberulent) on the lower half or fourth, at least, sometimes all over; moister areas of sagebr valleys to ponderosa pine or lower mont for e Cas; Yuk and BC to s Cal, e to Mont, Wyo, Colo, and Ariz; Nev b. 35 **P. nevadensis** Vasey

      55b Lemmas crisp-puberulent on the lower half or fourth or sometimes only at the base, never scab all over

        56a Pl greenish, gen > 4 dm; spikelets mostly tawny or light green, the lemmas sometimes purple-banded but gen not purplish all over; basal lf bls gen > 5 cm × 1–3 mm; sagebr des to mont for and ridges, from near sea level in Cal to 10,000 ft in the RM; e Cas in Wn, Yuk, and BC s to most of Cal, e to Alta, Mich and Minn, Wyo, and Colo; pine b., Malpais b. *(P. canbyi, Atropis laevis, P. l., P. laevigata, P. nevadensis* var. *laevigata, P. acutiglumis, P. helleri, P. limosa, P. leckenbyi)*
                                                                    36 **P. scabrella** (Thurb.) Benth.

        56b Pl often ± reddish-tinged, mostly < 3 (very rarely > 4) dm; spikelets mostly purplish-tinged; basal bls gen filiform and not > 1 mm broad, involute to folded, mostly < 5 (but very rarely up to 8) cm

          57a Pls gen of the des or of dry exposed areas in the lower mts, tending to fl between Apr and late June, gen strongly purplish-tinged; lvs mostly involute and scarcely 1 mm broad; wide-ranging, Yuk and BC s, mostly E Cas, to Cal, e to Sask, Daks, Neb, Colo, and NM; Sandberg's b. *(Atropis tenuifolia; Paneion s.; Poa buckleyana; P. b.* var. *s.; P. secunda,* misapplied)   37 **P. sandbergii** Vasey

          57b Pls mont, gen on talus, rocky ridges, or open slopes, rarely purplish, tending to fl from June to Aug; lvs mostly 1–1.5 mm broad, often flat; BC to Cal, in both OM and Cas, e to Alta, Mont, Wyo, and Nev; curly b.
                                                                    38 **P. incurva** Scribn. & Will.

Polypogon Desf.   Polypogon; Beard-grass

Spikelets 1-fld, borne in dense contracted pans, art slightly below the glumes and stipitate by the short, persistent segm of the ped; glumes keeled and flattened, equal, narrowly acuminate or slightly bilobed at the tip, the awn terminal or from between the slight lobes; lemma much < the glumes, gen short-toothed at the tip, awned, the awn terminal or subterminal, mostly straight and slightly exceeding the glumes; rachilla not prolonged behind the palea; lods 2; stamens 1–3; ann or per with hollow culms, open sheaths, prominent membranous ligs, and non-aur, gen flat bls rolled in the bud. (Gr *poly*, much, and *pogon*, beard, in reference to the long awn of the glumes). The pls are of little forage value, although in moist alkaline flats *P. monspeliensis* is often abundant enough to be grazed.

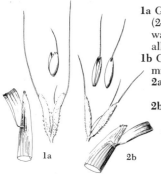

1a Glumes shortly bilobed, the awns 6–10 mm; pl ann, 0.5–7 dm; ligs 3–10 (2–12) mm; dry to wet waste areas, often in vernal pools esp where the water is brackish; intro from Europe; in w N Am from BC to Baja Cal, e to all RMS; rabbitfoot p., ann b. *(Alopecurus m.)* 1 **P. monspeliensis** (L.) Desf.

1b Glumes acute or rounded but not bilobed, the awns rarely as much as 6 mm; pl per

  2a Glumes 1.5–2 mm; S Am, intro and collected once, near Bingen, Klickitat Co, Wn; not believed to be estab   **P. australis** Brongn.

  2b Glumes 2.5–3 mm; pl per, 2–8 dm; ligs 3–6 mm; moist ground at lower elev; occ, BC to Cal, mostly w Cas, e across s US, Mex to S Am; Europe; ditch p. *(Alopecurus i., P. lutosus)*   2 **P. interruptus** H.B.K.

1a   2b

Puccinellia Parl.   Alkaligrass

Spikelets in open to contracted pans, art above the glumes, several-fld; glumes < the lowest lemma, the 1st gen 1-nerved, the 2nd gen 3-nerved; lemmas awnless, rounded on the back, obtuse to acute, with 5 parallel (nonconverging), indistinct to fairly prominent nerves that do not extend through the scarious margin above; lods 2, triangular-obovate, gen shallowly lobed; stamens 3; ann or (ours) caespitose to stoloniferous or rhizomatous pers with hollow culms, open or partially closed sheaths, membranous ligs, and narrow, flat to involute, non-aur bls that are rolled in the bud. (For Benedetto Puccinelli, 1808–1850, Italian botanist). Pls fairly palatable, but seldom abundant except in muddy areas where often severely damaged when grazed, in this respect resembling *Glyceria*.

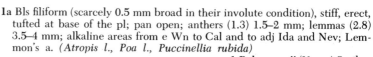

1a Bls filiform (scarcely 0.5 mm broad in their involute condition), stiff, erect, tufted at base of the pl; pan open; anthers (1.3) 1.5–2 mm; lemmas (2.8) 3.5–4 mm; alkaline areas from e Wn to Cal and to adj Ida and Nev; Lemmon's a. *(Atropis l., Poa l., Puccinellia rubida)*

                                              1 **P. lemmonii** (Vasey) Scribn.

1b Bls gen > 1 mm broad, not tufted at the base of the pl, often lax; pan sometimes narrow; anthers often < 1.3 mm; lemmas often < 2.8 mm

  2a Anthers mostly ca 2 mm; lemmas (3) 4 (5) mm; pan brs mostly smooth; pl glab, 2–7 dm; bls 1–2 mm broad; salt or brackish marshes and tidelands, PS area, Wn, more common on Atl coast, perhaps intro on ship-ballast from the Old World; coast a. *(Atropis m., A. distans* var. *m., Panicularia m., Poa m.)*   2 **P. maritima** (Huds.) Parl.

  2b Anthers < 2 mm, or lemmas < 3 mm, or pan brs scab; pls often not maritime

    3a Nerves of the lemma prominent; pls occurring in fresh water; bls flat, 3–12 (15) mm broad; ligs erose or lacerate, 3–9 mm; anthers 0.5–0.7 mm; Alas s along the coast to Cal, e to Alta, SD, Wyo, Colo, and NM, from near sea level to high mont; weak a.; 3 vars.

                                   3 **P. pauciflora** (Presl) Munz

    a1 Pl mostly dwarf, 1.5–6 (8) dm; bls mostly 3–6 (7) mm broad; ligs gen 5–9 mm; 1st lemma mostly 1.9–2.1 mm, gen 5-nerved, but sometimes with an additional nerve on 1 or both margins; common mont phase in RM, Alta to NM, w to Wallowa Mts, Ore, where intergradient with var. *microtheca (Panicularia h.)*

                                    var. **holmii** (Beal) Hitchc.

    a2 Pl gen > 6 dm or bls in part > 7 mm broad; ligs mostly < 7 mm; 1st lemma 2.2–3 mm, often with 7 prominent nerves

      b1 Lower pan brs mostly in 2's; 1st lemma 2.2–2.6 mm, gen with 7

1a   2a   2a   3a

5a

5b

6a

7a

8a

8b

well-developed nerves, the outer ones submarginal; w Cas and mostly coastal, Alas to PS area and OP (*Glyceria otisii, Torreyochloa o., Panicularia multiflora, Panicularia flaccida, Glyceria p., Panicularia p., Torreyochloa p.*)   var. **pauciflora**

b2 Lower pan brs more commonly in 3's; 1st lemma 2.2–3 mm, 5-nerved or with shorter and less-prominent submarginal nerves also; mostly in and on w slope Cas, BC to Cal, e to Ida and Nev and perhaps to Utah and Ariz (*Glyceria m.*)
var. **microtheca** (Buckl.) Hitchc.

3b Nerves of the lemma rather indistinct; pls gen of saline or alkaline soil; bls often involute, mostly < 3 and rarely > 4 mm broad; ligs entire or subentire, rarely > 3 mm

4a Pan brs smooth

5a Pls mostly 1–2 (3) dm; pan narrow, (3) 5–10 cm, the brs few, gen erect; lemmas entire-margined above; ligs 1–2 mm; tideflats and salt marshes; Alas to Mendocino Co, Cal, also on Atl coast; dwarf a. (*Glyceria pumila, P. p., P. alaskana, P. paupercula* var. *a., Atropis a., P. langeana* var. *a., P. angustata*)
4 **P. pumila** (Vasey) Hitchc.

5b Pls mostly much > 2 dm; pan often open or > 10 cm, the brs sometimes spreading; lemmas often erose on the upper margin; ligs mostly 1.5–3.5 mm; sea coast from n Alas to s BC (VI) and in PS to Seattle and Silverdale where prob no longer persistent; Alas a., Pac a. (*Poa n.*)   5 **P. nutkaensis** (Presl) Fern. & Weath.

4b Pan brs scab

6a Lemmas mostly 1.6–1.8 (1.4–2) mm; anthers 0.5–0.8 mm; lower pan brs gen reflexed, spikelet-bearing chiefly near the tip; moist soil, esp where alkaline; Yuk and nc BC to Cal, only rarely w Cas in Wn and Ore, e through most of Can and in US to e base of RM and Gr Lakes region and N Eng states; Eurasia; weeping a., European a. (*Glyceria d., Festuca d., Poa d., Panicularia d., Puccinellia suksdorfii*)   6 **P. distans** (L.) Parl.

6b Lemmas mostly > 2 mm; anthers often > 0.8 mm; lower pan brs often erect, gen spikelet-bearing most of their length

7a 2nd glume 3–3.5 mm; lemmas (3.2) 3.5–4.2 mm; anthers ca 1 mm; along the coast, often on tidel or around brackish swamps; Alas to San Mateo Co, Cal; Que; shining a. (*P. grandis*)
7 **P. lucida** Fern. & Weath.

7b 2nd glume < 3 mm; lemmas rarely as much as 3.2 mm

8a Anthers (1) 1.5–1.8 (2) mm; 2nd glume 2–2.8 mm; lemmas 2.6–3.2 mm; gen in fairly moist, ± alkaline areas; Klickitat Co, Wn, e Ore, Ida, and Mont, to ND and Wyo; Cusick's a.
8 **P. cusickii** Weath.

8b Anthers mostly 0.7–1 (1.2) mm; 2nd glume 1.5–2.2 (2.5) mm; lemmas (2) 2.5–3.2 mm; saline areas along the coast and moist, ± alkaline areas inl; BC to Cal, e to Alta, Sask, and Wisc to Kans, NM, and n Mex; mostly e Cas but common on VI and along shores of PS, Wn; Nuttall's a. (*Atropis airoides, A. nuttalliana, Panicularia distans airoides, Glyceria montana*)   9 **P. nuttalliana** (Schult.) Hitchc.

Schedonnardus Steud.   Tumblegrass; Schedonnardus

Spikelets 1-fld, unawned, art above the glumes, sessile in 2 rows on one side of the rachis in several racemose spikes, partially sunken in the rachis; stamens 3; lods 2; caespitose per with solid culms, open sheaths, membranous ligs, and white-margined, flat to folded, non-aur bls. (Gr *schedon*, near, and *Nardus*, a genus it was thought to resemble). Of little forage value.

S. **paniculatus** (Nutt.) Trel. Tufted, scab-puberulent per 1.5–3 dm; sheaths compressed, white-keeled; ligs acute, ca 3 mm; bls 1–2 mm broad, with obviously thickened and wavy, whitish margins; spikes several, divaricate, up to 10 cm; spikelets rather distant; glumes acuminate, 1-nerved, ca 2.5 and 3.5 mm; lemma ca 4 mm, short-pilose at base; prairies and plains; Alta and Sask to Tex and Ariz, e to Ill and La; Argentina (*Lepturus p., Spirochloe p.*).

Schizachne Hack.

Spikelets few in open, loose pan, art above the glumes, 4–5 (2–6)-fld, the upper 1–2 flts reduced to small sterile lemmas; glumes membranous, the 1st 3-nerved, the 2nd 5-nerved; fertile lemmas rounded on the back, strongly bearded on the callus, awned from slightly below the sinus of a rather deeply bifid apex; stamens 3; lods 2; tufted per with hollow culms, closed sheaths, membranous and often closed ligs, and flat to subinvolute, non-aur bls rolled in the bud. (Gr *schizo*, split, and *achne*, chaff, referring to the bifid apex of the lemma).

S. purpurascens (Torr.) Swallen. False melic. Pl glab, (4) 6–10 dm; ligs rarely > 1 mm, gen longest in front, subentire, glab; bls mostly 2–3 (5) mm broad, slightly scab; pan ave ca 10 cm, the few brs ± drooping; glumes gen purplish, ca 5 and 7 mm; lemmas ca 10 mm, bifid 1/4–1/5 their length, the bearding of the callus ca 3 mm, the awn gen = or slightly > the lemma, ultimately divergent; anthers 1.5–2 mm; open grassy or wooded, moist to dry and rocky areas; Alas to BC, across Can and s in RM to n Mex; e Asia (*Trisetum p., Melica p., Avena torreyi, Avena striata, Melica s., Bromelica s., S. s.*).

Sclerochloa Beauv.

Spikelets 3-fld (the 3rd fl sterile), subsessile and closely crowded on one side of the rachis in terminal spikelike racemes, falling entire, the rachis continuous; glumes rounded at the tip, compressed; lods 2; ann with hollow or semisolid st, compressed and partially closed sheaths, membranous ligs, and non-aur bls ± boat-shaped at tip and folded in the bud. (Gr *skleros*, hard, and *chloa*, grass, referring to the hardened spikelets).

S. dura (L.) Beauv. Hardgrass. Glab, low and spreading, mostly 3–9 (15) cm; ligs 0.5–1.5 mm, somewhat lacerate; racemes 1–5 cm, gen < upper lvs; spikelets 6–10 mm; 1st glume 2–3 mm, 3-nerved; 2nd glume 4–6 mm, 7-nerved; lemmas 4–6 mm, compressed, rounded at the tip, with 5 fairly prominent, nonconvergent nerves, unawned, hyaline-margined; palea much smaller and shorter than the lemma, sharply 2-keeled, the keels semiwinged; anthers barely 1 mm; intro from Europe; well estab as a weed in e Wn and Ore and sw Ida.

Scleropoa Griseb.   Scleropoa

Spikelets borne in stiff pans, several-fld, art above the glumes, the rachilla joints remaining attached below and forming stipelike bases to the flts; glumes < 1st flt, 1–3-nerved; lemmas rounded on the back, unawned, with 5 ± faint, convergent nerves; stamens 3; lods 2; glab ann with hollow culms, open sheaths, and bls folded in the bud. (Gr *skleros*, hard, and *poa*, grass, referring to the rigid infl).

S. rigida (L.) Griseb. Basally br, glab, 5–30 cm; ligs membranous, 1–3 mm; sheaths open; bls non-aur, 1–2 mm broad, flat to folded; pan narrow, stiff, rather dense, 3–8 cm, 1-sided, the brs short, spreading, spikelet-bearing to the base; spikelets 4–9-fld; glumes lanceolate, acute, keeled, the 1st ca 1.5 mm, 1-nerved, the 2nd only slightly longer, 3-nerved; lemmas ca 2.5 mm; intro from Europe, estab in many parts of the US, but known in our area only from Ore, where in Baker Co, and near Portland and Salem.

Scolochloa Link   Scolochloa

Spikelets in open pans, closely 3–4-fld, art above the glumes; glumes rounded on the back; lemmas unawned, rounded on the back, firm, with 7 fairly evident, converging nerves, strongly bearded-villous on the callus; lods 2; stamens 3; rhizomatous per with hollow culms, open sheaths, and non-aur bls rolled in the bud. (Gr *scolops*, prickle, and *chloa*, grass, the application not obvious).

S. festucacea (Willd.) Link. Fescue s. Culms up to 1.5 m; ligs membranous, 2–6 mm, lacerate; bls flat, 5–10 mm broad, firm, elongate and narrowed gradually to a slender tip; pan 15–25 cm, the brs ascending and mostly naked below midlength; 1st glume 4–6 mm, 3-nerved, < 1st flt; 2nd glume 5–7.5 mm, 5-nerved, ca = 1st flt; lemmas ca 6 mm; anthers 3–4 mm; marshes and at the edge of lakes or streams, gen in standing water; native to Eurasia, intro in much of n US and in Can, known in our area from se Ore and from Flathead Co, Mont.

Scribneria Hackel

Spikelets 1-fld, laterally compressed, art above the glumes, borne in terete, terminal, simple spikes, 1 (2) per node on opp sides of a continuous rachis and sunken in it; glumes narrow, rigid, slightly unequal, covering and concealing the flt; lemma membranous, < glumes, keeled, shortly awned from a bifid apex; lods 2; stamen 1; small ann with hollow culms, open sheaths, and narrow, non-aur bls. (For Frank Lamson-Scribner, 1851–1938, an Am agrostologist).

S. bolanderi (Thurb.) Hackel. Scribner-grass. Glab ann 5–30 cm; ligs membranous, 2–4 mm, slightly puberulent, entire; bls involute, 1–3 cm × ca 0.5 mm; spike gen ca half the height of the pl; spikelets (3) 4–7 mm; glumes gen reddish (purplish)-tinged, > the lemma, the 1st 2-nerved, the 2nd 4-nerved, strongly keeled on the outer nerve, the side next the rachis membranous and unnerved; callus short-bearded; lemma with a straight awn 2–4 mm; dry, sandy to rocky soil, often along roadsides, mostly from the foothills and lower mts; Klickitat Co, Wn, sw Ore, and Cal, where more gen; rarely collected, but perhaps not entirely because of rarity (*Lepturus b.*).

Secale L.   Rye

Spikelets 1 per node in a terminal spike, art above the glumes, borne flatwise to the continuous or ultimately disart rachis, sessile; lods 2; stamens 3; tall anns with hollow culms, open sheaths, fairly prominent aurs, short membranous ligs, and flat bls rolled in the bud. (Classical name for rye).

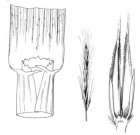

S. cereale L. Cult r. Pl mostly 6–15 dm; ligs erose, ca 1 mm; bls (3) 4–10 mm broad; spikes 8–15 cm; glumes stiff, slender, subulate, 1-nerved, < the lemmas; flts gen 2, more nearly alongside than above one another, the rachilla prolonged between the two; lemmas keeled asymmetrically, 5-nerved, curved, strongly ciliate on the keel and exposed margin, with an awn 4–7 cm; widely cult, esp in poor, dry soil where sometimes persistent for > a season on fallow ground, frequently along highways and railroads; intro from the Old World. *S. montanum* Guss., an Asiatic sp., which has a disart rachis and shorter awns (1–2 cm), has been reported as an escape from experimental plantings in e Wn.

Setaria Beauv.   Bristlegrass; Foxtail

Spikelets art below the glumes, unawned, subterete, 2-fld, borne in terminal, greatly congested pan, each subtended by 1–several persistent, scab or scab-barbellate bristles (sterile brs); glumes membranous, of the same texture as the lower lemma, the 1st much the shorter of the two, gen 3 (5)-nerved, the 2nd = or slightly < the 1st lemma, gen 5-nerved; 1st (lower) fl ♂ or sterile, if sterile the palea often greatly reduced or lacking; upper fl ♀, the lemma largely enclosing the palea, coriaceous, minutely cross-wrinkled or corrugate when mature; stamens 3; lods 2; ann or per with hollow culms, open and often keeled sheaths, ligs ca half membrane and half terminal fringe, and flat to folded, non-aur bls rolled in the bud. (L *seta*, bristle, referring to the bristle-like sterile brs of the infl). (*Chaetochloa*). Our spp. all intro and weedy.

1a Bristles retrorsely scab-barbellate; pan brs verticillate; pan 5–10 × ca 1 cm when pressed; waste or irrigated areas or in gardens; intro from Europe to most of US, but not common in our area; bur b., bristly f.

1 **S. verticillata** (L.) Beauv.

1b Bristles antrorsely scab-barbellate; pan brs not verticillate

2a Spikelets mostly subtended by (4) 6 or more bristles, the bristles very slender, rarely as much as 3 times as long as the spikelet; 2nd glume rarely > 2/3 as long as the 1st lemma; throat and collar glab; ligs scarcely 1 mm

3a Pl ann, non-rhizomatous; lower fl ♂; spikelets gen ca 3 mm; roadsides, waste areas, and irrigated land; intro from Europe and rather gen in much of US and Can, occ in our area on both sides Cas, e to Mont; yellow b. or f.   2 **S. lutescens** (Weigel) Hubb.

3b Pl per, short-rhizomatous; lower fl with lemma and palea only; spikelets ca 2.5 mm; collected as a weed in Portland, Ore, where not believed to be persistent; per f., knotroot b.   **S. geniculata** (Lam.) Beauv.

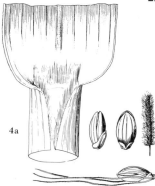

2b Spikelets subtended by 2–3 (1-4) fairly stiff bristles often > 3 times their length, gen some spikelets rudimentary or abortive, the others then apparently subtended by > 4 bristles; 2nd glume ca = the 1st lemma; throat and gen collar ± ciliate or pubescent; ligs gen > 1 mm

    4a Fertile lemma and palea closely invested by the sterile lemma and 2nd glume, not shed separately from them; pan gen greenish, cylindric, rarely > 10 cm × 15 mm when pressed; widespread weed in s Can and most of the US, intro from Eurasia; green b. (*S. italica* ssp. *v.*)
                                                               3 S. viridis (L.) Beauv.

    4b Fertile lemma and palea loosely surrounded by the sterile lemma and 2nd glume, often deciduous separately from them; pan often enlarged and lobed near the base or middle, often yellow or purple; collected as a weed near Portland, Ore, but apparently not estab there; foxtail millet, Hungarian grass
                                                              S. italica (L.) Beauv.

Sieglingia Bernh.   Sieglingia

Infl a much-reduced pan or raceme, the spikelets few, 3–5-fld, art above the glumes; glumes subequal, > the 1st flt (gen = or > the uppermost flt), rounded on the back, 1-nerved; lemmas unawned, the 1st often sterile and glumelike, the upper fertile ones coriaceous, with 7 obscure, convergent nerves, rounded on the back, silky-pilose at base and on the margin up to midlength; stamens 3; lods 2; per with hollow culms, open sheaths, short-fringed ligs, and tough, elongate, flat to involute, non-aur bls rolled in the bud. (For Prof. Siegling, a German botanist of the early 19th century).

S. decumbens (L.) Bernh. Heathgrass. Culms tufted, 2–6 dm; ligs barely 1 mm, the hairs of the collar ∝, up to 2.5 mm; sheaths and bls glab or sparsely pilose, the hairs of the sheath slightly pustulose-based; bls 2–3 mm broad; pan 4–10 cm, the brs erect, mostly with only 1 spikelet; spikelets 8–11 (12) mm; lemmas 5–6 mm, with 3 blunt apical teeth scarcely 0.5 mm; anthers barely 0.2 mm, the flts mostly cleistogamous; cleistogamous and much-modified (bulblike) spikelets also gen borne in the basal sheaths; European sp., rare in most of N Am, although possibly native in e Can; intro and apparently estab in open woods near Long Beach, Wn, and in somewhat marshy flats near Newport, Lincoln Co, Ore.

Sitanion Raf.   Squirreltail

Spikelets in a terminal spike, 2–6 (1–8)-fld, borne flatwise to the rachis, sessile and gen 2 (sometimes 1, 3, or even 4) per node, the rachis rather readily disart; glumes narrow, 1–2 (3)-nerved, simple or commonly bifid or deeply cleft and attenuate into 2–7 (9) elongate, awned segms; flts varying, sometimes all fertile, or the lowermost 1–2 sterile and the lemma glumelike, or all flts except that of the central spikelet sterile; fertile lemmas rounded on the back, gen faintly 5-nerved, the midnerve (and sometimes 1–2 lateral nerves) extending into slender awns; lods 2; stamens 3; caespitose pers with hollow culms, open sheaths, gen well-developed aurs, short membranous ligs, and rather narrow, flat to folded or involute bls rolled in the bud. (Gr *sitos*, grain). Although seldom abundant, the spp. have fair palatability early in the season, but the long awns of the spikelets are a nuisance to man and a minor hazard to grazing animals.

    1a Glumes entire or 2 (rarely 3)-cleft; aurs often lacking on many lvs, always < 1 mm; pl 1–6 dm; along coast and inl, esp on des plains and prairies but also in the mts to above tree line; BC to Alta, s to s Cal, Ariz, n Mex, Tex, and w SD, Neb, and Okla; bottlebrush s.; 3 vars.   **1 S. hystrix** (Nutt.) Smith

        a1 Spikelets gen 2 per node in the lower half of the spike, the 4 glumes simple; sterile, glumelike lemmas absent; pls mostly 3–6 dm; very common from Mex and Tex to s Cal, less frequent n, to SC, Mont, Ida, and Wn, where esp intergradient with var. *hystrix (Elymus b., S. b.)*
                                                            var. brevifolium (Smith) Hitchc.

        a2 Spikelets either 3 per node (the lateral gen sterile) or 2 per node, but with at least the lowest lemma of 1 spikelet modified and glumelike, the glumes apparently 5 or more; pls often < 3 dm

            b1 Spikelets mostly 3 per node, the lateral ± reduced, often with 3 or more glumes and pseudoglumes; pls mostly < 2 dm; des region into ponderosa pine for; c and e Wn and Ore to n Cal, nc Nev, and se Ida (*S. h.*)
                                                             var. hordeoides (Suksd.) Hitchc.

b2 Spikelets 2 (1) per node, but one or both with at least the lowest lemma glumelike, the glumes simple to bifid or trifid; pls often > 2 dm; range of the sp. as a whole *(Aegilops hystrix, Elymus h., E. elymoides, E. sitanion, Hordeum elymoides, S. albescens, S. basalticola, S. ciliatum, S. elymoides, S. latifolium, S. longifolium, S. rigidum, S. montanum, S. strigosum, S. velutinum)* var. **hystrix**

1b Glumes 3– ∝ -cleft; aurs mostly present, some gen at least 1 mm; dry prairies to rocky hillsides and open woods, often where soil disturbed; se and ec Wn to s Cal, e to w Ariz, Utah, and w Ida; big s. *(Elymus j., Elymus multisetus, S. m., S. strictum, S. villosum)* 2 **S. jubatum** Smith

1b

Sorghum Moench Sorghum

Infl a large pan, the brs ending in few-jointed racemes that ultimately disart; spikelets in 2's at each joint of the ultimate divisions of the pan, art below the glumes, ± dorsiventrally compressed, one spikelet sessile, ♀-fld, the other ped and ♂ (the ♂ spikelets 2 at the tip of each rachis); ped spikelet unawned; sessile spikelet gen with a bent, twisted, readily deciduous awn arising dorsally on the fertile lemma, the glumes hardened, nerveless, shiny and ± appressed-puberulent, the sterile lemma, fertile lemma, and palea smaller, papery and delicate; stamens 3; lods 2; ann or (ours) per with solid culms (occ the lowermost internodes hollow), open sheaths, short, membranous and fimbriate ligs, and large, flat to folded, non-aur bls rolled in the bud. (*Sorgho,* the Italian vernacular name for the pl). Besides the following sp., *S. vulgare* Pers., an ann, is freely grown in many areas as a source of stock feed, but it does not persist.

S. halepense (L.) Pers. Johnsongrass. Strongly rhizomatous per 1–1.5 m; ligs ca 2 mm, mostly membrane with a terminal fringe of fine hairs, hairy externally; bls gen 1–2 cm broad; pan up to 3 dm, the ultimate br tips raceme-like, consisting of 2–5 nodes, ending in 3 spikelets, the other nodes each with a sessile, ♀-fld spikelet and a ped ♂ spikelet, the 2 spikelets ca 6 (5–7) mm; awn of the sessile spikelet 9–15 mm, readily deciduous; anthers ca 3 mm; lods ca 0.5 mm, cuneate, finely ∝ -veined, delicately long-pilose from the upper corners; well estab in waste ground, esp where moist, but rarely if ever persistent in our area.

Spartina Schreb. Cordgrass

Spikelets 1-fld, art below the glumes, strongly compressed, borne in greatly congested spikes in 2 rows on one side of the rachis, the spikes 2–∝ and racemosely arranged, the rachis gen prolonged; glumes strongly keeled, blunt, apparently 1-nerved (lateral nerves obscure); lemmas keeled, apparently 1-nerved; stamens 3; lods lacking; rhizomatous pers with hollow culms, open sheaths, short ligs composed chiefly of a copious fringe of fine, straight hairs, and flat to strongly involute, non-aur lvs rolled in the bud. (Gr *spartine*, a cord made from *S. juncea*). Several of the spp. are used as sand binders, and spp. other than the following may be fleeting introductions).

1a Ligs gen ca 1 mm; bls rarely > 5 mm broad at base; glumes long-ciliate on the keel, not scab, the 2nd acute but not awned; culms gen 3–6 (10) dm; marshes and ditches to dry plains, gen where ± alkaline; e Cas, BC to Sask, s to Cal, Ariz, Mont, Kans, and NM; alkali c. 1 **S. gracilis** Trin.

1b Ligs gen at least 1.5 (3) mm; bls gen at least 5 mm broad at the base; glumes scab but not long-ciliate, the 2nd often awned; culms gen > 10 dm

2a Glumes strongly scab, both gen awn-tipped, the 1st nearly = the flt, the 2nd much > flt; ligs gen 2–3 mm; ditches, ponds, and fresh water marshes, not common in our area; e Wn and Ore e through Ida and Mont to Newf, s to Utah, NM, Tex, Tenn, and NC; prairie c. (*S. cynosuroides × gracilis*) 2 **S. pectinata** Link

2b Glumes not strongly scab, neither one awn-tipped, the 1st considerably the shorter, the 2nd ca = the flt; ligs mostly (1) 1.5–2 mm; coastal

3a 2nd glume 9–11 mm; spikes mostly at least 6, only 4–8 cm; intro from Europe, known in our range only from Pacific Co, Wn, where said to have been intro through the oyster-growing industry; smooth c. 3 **S. alterniflora** Loisel.

3b 2nd glume > 11 mm; spikes 2–5, mostly 10–15 cm; native to Europe, of hybrid origin, valued as a soil stabilizer and intro in 1961 on mudflats at Stanwood, Snohomish Co, Wn, where apparently becoming estab; Townsend's c. **S. townsendii** H. & G. Groves

Sphenopholis Scribn.   Wedgegrass; Prairie-grass

Spikelets in narrow pans, gen 2 (3)-fld, art below the glumes and deciduous as a unit; glumes strongly keeled and compressed, the 1st 1-nerved and very narrow, the 2nd much broader, ± oblanceolate (ours), rather plainly 3-nerved and gen obscurely nerved between (thus with 2 obscure and 3 strong nerves), scarious-margined, gen 1/5–1/4 < the 1st lemma; lemmas awnless, obscurely 3-nerved, keeled, often scab; per or winter ann with hollow culms, open sheaths, membranous ligs, and flat, non-aur bls rolled in the bud and with margins often freed from the sheaths at slightly different levels. (Gr *sphen*, wedge, and *pholis*, scale, referring to the wedge-shaped 2nd glume). *(Eatonia)*. Patalable to livestock, but seldom abundant.

S. obtusata (Michx.) Scribn. Caespitose per sometimes fl as a winter ann, 2–8 (11) dm; sheaths glab or scab to strongly pubescent; ligs 1.5–2 (1–2.5) mm, gen slightly scab externally, irreg erose-fimbriate and ± toothed; bls scab to pubescent, 3–5 (2–6) mm broad; pan (2) 5–15 cm, dense and spikelike to ± open and interrupted, the brs erect to spreading; glumes scab, the 1st very narrow, ca 2 (1.6–2.5) mm, the 2nd 3–4 times as broad, much flattened, slightly cucullate, ca 2–2.5 mm; lemmas oblong, 2–3 mm, unawned (rarely awn-tipped), scab; rachilla prolonged and with a rudiment; mostly on moist soil, from stream- and lake-margin to grassl; s Alas and BC s, e Cas and along CR, to Cal and Mex, e to Me and Fla; W Ind. *(Aira obtusata, Eatonia o., E. annua, S. a., E. intermedia, S. i.)*.

Sporobolus R. Br.   Dropseed; Rush-grass

Spikelets 1-fld, borne in open and diffuse to contracted and spikelike pans, art above the glumes, but glumes sometimes deciduous with or before the flt; glumes 1-nerved, subequal to unequal, slightly to considerably < the lemma; lemma awnless, gen acute, 1 (3)-nerved, gen glab; rachilla not prolonged behind the palea; anthers 3; lods 2; pericarp of the fr tending to separate readily from the seed, esp when wet, sometimes before the spikelet is shed; ann or per, gen caespitose (rhizomatous), with solid (occ hollow) culms, open sheaths often pubescent along the margins, notably short ligs with a membranous base and a hairy upper portion at least as long as the membrane, and non-aur bls rolled in the bud. (Gr *spora*, seed, and *ballein*, to throw, because of the seeds that often "drop" from the pericarp).

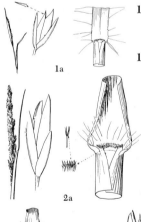

1a Pl ann, up to 5 dm; pan 1.5–3 cm, gen concealed by the swollen sheath; culms solid; bls ca 2 mm broad; dry places, often in sandy or rocky soil, Mont to Que and Me, s to Ariz, NM, and most of c and e US, also near Spokane, Wn, where prob intro; small r.   1 **S. neglectus** Nash
1b Pl per, strongly tufted; pan gen > 3 cm and projecting from the sheath
  2a Spikelets at least 3.5 mm, the 1st glume 2.5–3 mm, the 2nd 3.5–4 mm, the lemma 4–5 mm; anthers 1.5–2.5 mm; dry soil, often where sandy, prairies to foothills; e Wn and Ore across the US to Vt, s to Ariz, NM, and Tex; rough d., long-lvd r. *(S. pilosus, S. a.* var. *p.)*
                                                  2 **S. asper** (Michx.) Kunth
  2b Spikelets < 3 mm; anthers < 1.5 mm
    3a Pan spikelike, the brs short, strictly erect; glumes obtuse, gen not > half as long as the lemma; known from our area only as a waif on ballast near Portland, Ore, where non-persistent; smutgrass
                                        S. poiretii (R. & S.) Hitchc.
    3b Pan rather open, the brs elongate, spreading to ascending; glumes acute, > half the length of the lemma
      4a Culms solid, grooved lengthwise; throat of the lvs pilose-lanate, the collar with a dense ring of divergent hairs 1–3 mm; ligs ca 0.5 mm; pan only moderately open, the brs simple to again br, stiffly ascending, spikelet-bearing to near the base; sagebr des, esp where sandy, to grassl and into the foothills; e Cas, BC to s Cal and n Mex, e to Ont, Que, and Me, and most of the US except the extreme se; sand d. *(Agrostis c., Vilfa c., Vilfa triniana)* 3 **S. cryptandrus** (Torr.) Gray
      4b Culms terete, pithy, sometimes hollow; throat glab or ciliate only at the top; collar glab; ligs < 0.5 mm; pan open, the brs slender, spreading, spikelet-bearing chiefly near the tips; prairies, des, and mt foothills, gen where ± moist, and esp characteristic of ± alkaline soil; se BC and e Wn to SD, s (e Cas) to Cal, Mex, and Tex; alkali Sacaton, hairgrass d.   4 **S. airoides** (Torr.) Torr.

Stipa L.   Needlegrass; Needle-and-thread; Porcupine-grass; Speargrass

Spikelets 1-fld, art above the glumes, borne in open to much-contracted pans; glumes membranous, acute to long-acuminate; flt gen hardened, the rachilla persistent as part of the hardened, sharp, bearded callus; lemma lightly (gen 5-) nerved, convolute, prolonged into a slender, twisted and ± bent, persistent awn; palea enclosed by the lemma except sometimes at anthesis, slightly to considerably < the lemma; lods 2; stamens 3; caespitose pers with hollow culms, open sheaths, membranous ligs, and non-aur, gen ± involute bls. (Gr *stupe*, tow, because of the feathery appearance of the type sp., due to the plumose awns). Our spp. are fairly abundant, palatable much of the year, and readily grazed; those with a particularly hardened and sharp callus (esp S. *comata*) are sometimes injurious to livestock.

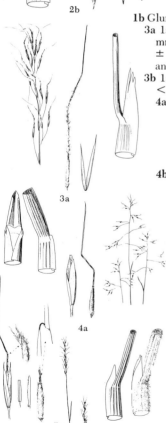

1a Glumes 15–40 mm; lemma 8–25 mm, the awn gen (6) 10–20 cm
  2a Lemma pubescent distally only along the margin, dark brown when
    mature; ligs mostly 0.5–2.5 (5) mm, rounded to emarginate; prairies and
    grassy benchlands; BC to Ont, s to nw and e Mont, Wyo, Colo, NM, and
    e through most of c US to Pa; ours mainly var. *curtiseta* Hitchc., the
    shorter-awned phase (awns 5–8 cm, lemmas 13–18 mm); p.
                                   1 S. spartea Trin.
  2b Lemma pubescent all over, although gen sparsely so distally, green to
    pale brownish when mature; ligs mostly 3–6 mm, rounded to acute or
    acuminate, gen lacerate; plains and prairies to mont for, gen on rocky
    soil; BC to Ont, e Cas in Wn and Ore, to Cal, Tex, Neb, and Ind;
    n.-and-t.; 2 vars.   **2 S. comata** Trin. & Rupr.
    a1 Terminal segm of the awn firm, nearly or quite straight, gen < 5.5
      cm; pls of e Wn and adj Ore to Ida, Mont, and Wyo, occ s to NM;
      Tweedy's n. (S. *tweedyi*, S. *spartea* var. *t.*)
                           var. intermedia Scribn. & Tweedy
    a2 Terminal segm of awn flexuous, curved, often > 5.5 cm; range of
      the sp. as a whole (S. *c.* ssp. *intonsa*, S. *c.* var. *suksdorfii*) **var. comata**
1b Glumes < 15 mm; lemma mostly < 8 mm, the awn rarely > 6 cm
  3a 1st 2 segms of the awn plumose, some of the hairs > 1 mm; ligs 2–5
    mm, those of the culm lvs rarely < 3 mm; culms 3–7 dm, nodes gen
    ± pubescent; sagebr des and ponderosa pine woodl; nc Wn to Cal
    and Nev, e to sw Mont and ne Wyo; Thurber's n. **3 S. thurberiana** Piper
  3b 1st 2 segms of the awn glab to semi-plumose, the hairs, if any, mostly
    < 1 mm; ligs rarely as much as 2.5 mm
    4a Pan open, the brs spreading, spikelet-bearing only near the tip; awn
      twice bent, 18–25 mm; culms 4–10 dm; ligs barely 0.5 mm; stamens
      ca 3 mm; flt 5–6 mm, callus barely 1 mm; grassl or sagebr des to
      open parks in for of lodgepole or ponderosa pine; sc BC to ne Wn, e
      to Sask, SD, Wyo, and Colo; Richardson's n. (Oryzopsis *r.*, S. *r.* var.
      major)   **4 S. richardsonii** Link
    4b Pan narrow, the brs ascending to erect, often spikelet-bearing to
      near the base; awn often > 25 mm
      5a Awn 18–29 mm, subterminal, the edges of the lemma free above
        the point of insertion of the awn and gen thickened; lemma in-
        durate, 5–7.5 × 1–1.5 mm, scarcely 6 times as long as thick,
        callus barely 0.5 mm; foothills into the mts, commonly in ponder-
        osa pine for; sc BC s, e Cas in Wn, but in CRG and s through
        WV, Ore, to Cal, e to sw Mont, n Ida, w Nev, and Ariz; Lem-
        mon's n.; 2 vars.   **5 S. lemmonii** (Vasey) Scribn.
      a1 Sheaths and bls puberulent to short-villous; longer ligs 1.5–2.2
        mm; Wen Mts, Wn, occ s to Klickitat Co, Wn, also in Wallowa
        Mts, Ore, and the Sierra Nevada, Cal var. **pubescens** Crampton
      a2 Sheaths and bls glab or subglab; longer ligs gen < 2 mm;
        range of the sp. as a whole (S. *pringlei* var. *l.*) var. **lemmonii**
      5b Awn terminal or edges of the lemma free but produced into
        slender hyaline teeth rather than thickened; lemma often not
        indurate, gen > 6 times as long as thick, often villous; upper ligs
        often < 1.5 mm (except in S. *viridula*)
        6a Margins of the lemma projecting upward past the base of the
          awn as 2 slender hyaline teeth; lemma villous; awn subglab;
          culms 1–3 (5) dm; ligs scarcely 0.5 mm; rocky benchland and
          lower slopes to mont ridges; se and Sierran Cal e through Nev

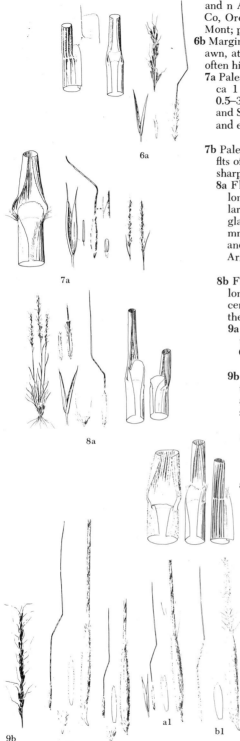

6a

7a

8a

9b

c2

c1

a1

b1

and n Ariz to Utah, Colo, and Wyo, n to Steens Mts, Harney
Co, Ore, and to our area in Clark Co, Ida, and reportedly in
Mont; pine n.   **6 S. pinetorum** Jones
6b Margins of the lemma gen not projecting above the base of the
awn, at least as 2 slender teeth; lemma often not villous; awn
often hirsute
   7a Palea glab, < half as long as the lemma; flts plump, 5–6 ×
ca 1 mm; callus ca 0.5 mm, blunt; collar ± villous; ligs
0.5–3 mm; bls 3–6 mm; grassl and sagebr des; s BC to Alta
and Sask, s mostly e Cont Div through Mont to NM and Ariz
and e to Wisc and Ill; green n. *(S. nuttalliana)*
                                                        7 S. viridula Trin.
   7b Palea either hairy or > half as long as the lemma, or both;
flts often > 6 times as long as thick; callus often slender and
sharp; collar sometimes glab
      8a Flt < 6 times as long as thick, the palea at least 2/3 as
long as the lemma; awn ca 15 (20) mm, glab or scab; collars glab; callus blunt, the inner curve without an oval
glab area; culms 2–6 dm, glab or subglab; ligs 0.3–0.8
mm; bls filiform-involute, 1–1.5 mm broad; sagebr hills
and benchland to subalp ridges; c Ida and s Mont to Nev,
Ariz, and NM; Letterman's n. *(S. viridula* var. *l.)*
                                                  8 S. lettermanii Vasey
      8b Flt either > 6 times as long as thick or palea < 2/3 as
long as the lemma, or both; awn often > 20 mm, pubescent to semiplumose; collars often villous; callus acute,
the inner curve often with an oval glab area
         9a Lemma 5–6.5 mm, ca twice (1.7–2.3 times) as long as
the palea, with hairs at the tip ca 1.5 mm; reported from
Craters of the Moon, Ida, but mainly well s of our area
                                              S. nevadensis Johnson
         9b Lemma 5–8 mm, > twice (gen > 2.3 times) as long as
the palea, the hairs at the tip often < 1.5 mm; grassl
and sagebr des to subalp for and ridges; Yuk and BC to
s Cal, e to Sask, Daks, NM, sw Tex, and n Mex; 4 vars.
                                              9 S. occidentalis Thurb.
            a1 Hairs at the tip of the lemma much > those of the
rest of lemma or on the awn; se Wn s, and on both
sides Cas in Ore, to Sierran Cal, e to Ida and Nev; Cal
n. *(S. c.)*   var. **californica** (Merr. & Davy) Hitchc.
            a2 Hairs at the tip of the lemma not much > those of
the rest of the lemma or on the awn
              b1 Awn semi-plumose over the 1st segm and gen
also over the 2nd segm; much the range of the sp.
as a whole; western n. *(S. elmeri, S. oregonensis,
S. viridula* var. *pubescens)*   var. **occidentalis**
              b2 Awn glab to scab, not semi-plumose, even on the
1st segm
                c1 Awn ave < 30 (18–35) mm; callus relatively
blunt, scarcely 1 mm, the inner curve gen with-
out a glab area above except for the short
naked tip; c BC s through Cas and OM to Cal,
e to w Alta, Mont, Wyo, and Colo; small n. *(S.
columbiana, S. m., S. viridula* var. *m., S.
williamsii)*   var. **minor** (Vasey) Hitchc.
                c2 Awn ave much > 30 (to 60) mm; callus
sharp, 1–1.7 mm, the inner curve with a glab
area extending from the naked tip; ec Wn to w
Cal, e to e Ida, sw Mont, and nw Utah; Nelson's
n. *(S. n., S. columbiana* ssp. *n.)*
                                        var. **nelsonii** (Scribn.) Hitchc.

Trisetum Pers.   Trisetum

Spikelets 2 (3–5)-fld, in open to spikelike pans, art above the glumes (ours), the rachilla gen prolonged and often with a terminal rudiment, the joints strongly hirsute; glumes subequal to very

unequal, keeled, the 1st 1-nerved, the 2nd 3-nerved, from < the 1st lemma to almost = the upper flt; lemmas obscurely 5-nerved, ± strongly keeled, slenderly bifid and gen awned from well above midlength, mostly from just below the terminal teeth, the awn bent, but in *T. wolfii* the lemmas awnless or with a straight awn-tip or short awn; palea subequal to the lemma, the 2 nerves ± excurrent as bristle tips; callus short-hirsute; stamens 3; lods 2; per (ours), mostly caespitose, with hollow culms, open sheaths, membranous and gen erose ligs, and non-aur, flat to ± involute bls rolled in the bud. (L *tres*, three, and *seta*, bristle, referring to the 3-awned lemmas of the type sp.). Fairly palatable spp., but seldom abundant enough to constitute much of the range forage, with the exception of *T. spicatum*.

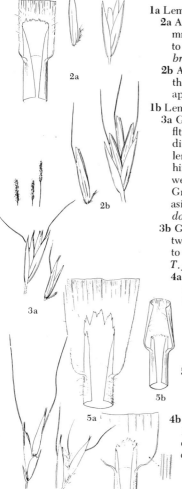

1a Lemmas awnless or with a short straight awn not > 6 mm
  2a Awn lacking or < the lemma; bls 2–4 (6) mm broad; lemmas 4.5–5.5 (6) mm; mostly where wet, as in meadows or along streams in the mts; e Wn to Mont, s to Cal and NM; Wolf's t., beardless t. (*Graphephorum w., G. brandegei, T. b.*)   1 **T. wolfii** Vasey
  2b Awn 5–6 mm, extending past the lemma 2–4 mm; known only through the type collection from Lolo Hot Springs, Missoula Co, Mont, where apparently no longer existent   2 **T. orthochaetum** Hitchc.
1b Lemmas with a curved or bent awn
  3a Glumes subequal, the 2nd < twice as broad as the 1st, but gen > 1st flt; infl spikelike, the spikelets subsessile; ovary glab; ligs < 2 mm; awns divaricate, 5–6 mm, arising barely 1.5 mm below the tip of the bifid lemma; pl 1–5 (7) dm, finely scab to puberulent or strongly pilose-hirsute; mont for to alp slopes in dry and rocky to moist areas, often weedy on waste areas; Alas to Greenl, s to most of the mts of w US, the Gr Lakes area, and the Appalachian Mts and through Mex to S Am; Eurasia; often treated as several taxa; spike t., downy oat-grass (*T. congdoni, T. alaskanum*)   3 **T. spicatum** (L.) Richter
  3b Glumes gen very unequal, the 2nd often twice as long and gen at least twice as broad as the 1st, but mostly not > the 1st lemma; infl an open to congested (but not spikelike) pan; ovary often pubescent; ligs (except *T. flavescens*) ave much > 2 mm
    4a 2nd glume (4) 4–7 mm, not erose, gen < twice as long as the 1st; pan rather compact, the brs ascending
      5a Ligs 1.5–4 mm; bls flat, ave at least 4 mm broad; spikelets green to straw colored; lods mostly slenderly fimbriate; ovary pubescent; pan loose, 1–2 (2.5) dm; from near sea level into lower mts, esp in ponderosa pine for, in moist to dry areas; s BC s, e and w Cas, to c Cal, e to Ida, w Mont, and ne Ore; tall t. (*T. elatum, T. cernuum* var. *c.*)   4 **T. canescens** Buckl.
      5b Ligs 0.5–1 mm; bls often involute or folded, gen < 4 mm broad; spikelets gen yellowish-brown; lods ± lobed but not fimbriate; ovary gen glab; pan 0.5–1 (1.5) dm; intro from Europe, reported for Wn; yellow oat (*Avena f.*)   5 **T. flavescens** (L.) Beauv.
    4b 2nd glume 3.5–4.5 mm, erose near the tip, gen > twice as long as the 1st; pan gen open, the brs spreading to drooping; moist woods to edge of water; Alas s along coast to n Cal, e to Alta, Ida, and w Mont; nodding t. (*Avena nutkaense, T. n., T. sandbergii*)   6 **T. cernuum** Trin.

Triticum L.   Wheat

Spikelets 2–5-fld, borne in a terminal spike flatwise to the continuous rachis, 1 per node, sessile, art above the glumes; stamens 3; lods 2; ann or bien with hollow culms, open sheaths, well-developed aurs, short, membranous ligs, and flat bls rolled in the bud. (The classical name for wheat).

T. aestivum L. Cult w. Pl ann or winter ann, up to 1.5 m; ligs scarcely 1 mm; bls 5–20 mm broad; spike 5–12 cm; glumes firm, keeled, strongly 3–several-nerved, acute to awned, the awns 1–several; lemmas broad, keeled, asymmetrical, with several non-convergent nerves, sharp-pointed to awned; along roadsides and in fallow fields, not persistent more than a few seasons at most; intro from the Old World. There are various types, including awned or bearded and beardless phases; some are considered to be vars. or strains of *T. aestivum*, others are treated as distinct spp.

Ventenata Koeler   Ventenata

Spikelets 2–3-fld, art above the glumes, borne in open pan; glumes firm, membranous-margined, prominently 6–9-nerved, slightly unequal, acuminate, ± keeled, < the 1st flt; lemmas obscurely 5-nerved, the 1st with a terminal, gen straight awn, the upper one(s) bifid-aristate, awned from near midlength, the awn twisted and bent; callus bearded; stamens 3; lods 2; hollow-culmed ann with open sheaths, prominent membranous ligs, and narrow, non-aur bls. (For Pierre Etienne Ventenat, 1757–1805, a professor of botany at Paris).

V. dubia (Leers) Coss. & Dur. Basally br, 3–7 dm, glab to puberulent; ligs 1–8 mm, glab, obtuse and gen lacerate; bls flat but becoming involute, 1–3 mm broad, gen glab on the lower surface but scab above; pan open, 1–4 dm, the brs spreading to drooping; spikelets gen 3-fld; glumes sometimes awn-tipped, the 1st 5–6.5 mm, 6–7-nerved, the 2nd 7–9 mm, 8–9-nerved; 1st lemma slightly > the glumes, awned from the acuminate tip, the awn 1–3 mm; upper 1 or 2 lemmas with bristlelike teeth 1.5–2 mm and with a dorsal twisted and bent awn 10–16 mm; intro from Europe recently, but well estab in Kootenai Co, Ida, and near Spokane and Yakima, and along n side CR in Klickitat Co, Wn.

Zizania L.   Indian Rice; Wild Rice; Water-oats

Spikelets in large pan, linear, often terete, falling entire, 1-fld, the fls ♂♀, the ♂ spikelets lowermost in the infl; glumes apparently lacking, but often at least 1 of them vestigial in the ♂ spikelet (up to 1 mm) and both reduced to a slightly raised rim surrounding the base of the ♀ spikelets; ♂ spikelets ± compressed, not hardened, the lemma 5-nerved, acuminate to awn-tipped; stamens 6; ♀ spikelets hardened, subterete but strongly angled, the lemma long-awned; lods 2; aquatic ann (ours) with hollow culms, open sheaths, prominent membranous ligs, and large, flat, non-aur bls rolled in the bud. (Gr name *Zizanion,* for a weed of grain fields, believed to be the "tares" of Biblical reference).

Z. aquatica L. Culms 1–3 m, nodes and collars gen pubescent; sheaths strongly cross-nerved; ligs firm, truncate to acute, entire or gen ± lacerate, 4–15 mm, glab externally, marked in front by the upturned margins of the bls; bls flat, mostly 6–40 mm broad; pan open, 2–5 dm; ♂ spikelets 6–8 mm, anthers 6–7 mm; ♀ spikelets subterete, 10–15 (in fr to 20) mm, pubescent toward the tip, tapered into an awn 6–40 mm; wet meadows, marshes, and lakes, often in muck at the edge of lakes in 1–3 dm of water; common in most of s Can and the US, e RM; said to be estab in Ida. Wild rice is still harvested to a limited extent in the region of the Gr Lakes, and is often intro into lakes and ponds in Wn, Ore, Ida, and Mont as a lure for wild fowl. Three vars. of the pl are sometimes recognized, based chiefly on the width of the bls, and 2 of them, var. *aquatica,* with bls 1–4 cm broad, and var. *interior* Fassett, with bls scarcely 1 cm broad, have been reported from our range.

SPARGANIACEAE   Bur-reed Family

Fls ♂♀, borne in several capitate-globose clusters toward the tip of the main st, the upper 1–several clusters ♂, the lower ones ♀; ♂ fls gen with 3–5 stamens and 3–5 minute chaffy bracts; ♀ fls gen with somewhat larger bracts, the pistil simple or sometimes 2-carpellary, the styles and lateral stigmas 1 (2), ovary 1 (2)-celled, each cell 1-ovuled; fr a hardened and nutletlike, 1 (2)-seeded, strongly beaked achene; per, rhizomatous herbs of marshy or aquatic habitats, with alt, linear, erect to floating, basally sheathing lvs.

Sparganium L.   Bur-reed

♂ heads gen more ∝ than the ♀; filaments much > the anthers and tiny bracts; ovary gen ca = by the bracts; frs gen broadly fusiform to truncate-pyriform. (Gr *sparganion,* a band, supposedly in reference to the narrow lvs).

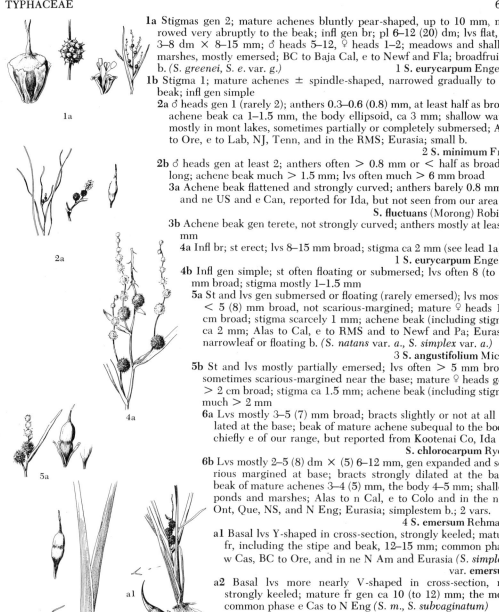

1a Stigmas gen 2; mature achenes bluntly pear-shaped, up to 10 mm, narrowed very abruptly to the beak; infl gen br; pl 6–12 (20) dm; lvs flat, (2) 3–8 dm × 8–15 mm; ♂ heads 5–12, ♀ heads 1–2; meadows and shallow marshes, mostly emersed; BC to Baja Cal, e to Newf and Fla; broadfruited b. (S. greenei, S. e. var. g.)   1 **S. eurycarpum** Engelm.

1b Stigma 1; mature achenes ± spindle-shaped, narrowed gradually to the beak; infl gen simple

2a ♂ heads gen 1 (rarely 2); anthers 0.3–0.6 (0.8) mm, at least half as broad; achene beak ca 1–1.5 mm, the body ellipsoid, ca 3 mm; shallow water, mostly in mont lakes, sometimes partially or completely submersed; Alas to Ore, e to Lab, NJ, Tenn, and in the RMS; Eurasia; small b.

2 **S. minimum** Fries

2b ♂ heads gen at least 2; anthers often > 0.8 mm or < half as broad as long; achene beak much > 1.5 mm; lvs often much > 6 mm broad

3a Achene beak flattened and strongly curved; anthers barely 0.8 mm; c and ne US and e Can, reported for Ida, but not seen from our area

S. fluctuans (Morong) Robins.

3b Achene beak gen terete, not strongly curved; anthers mostly at least 1 mm

4a Infl br; st erect; lvs 8–15 mm broad; stigma ca 2 mm (see lead 1a)

1 **S. eurycarpum** Engelm.

4b Infl gen simple; st often floating or submersed; lvs often 8 (to 12) mm broad; stigma mostly 1–1.5 mm

5a St and lvs gen submersed or floating (rarely emersed); lvs mostly < 5 (8) mm broad, not scarious-margined; mature ♀ heads 1–2 cm broad; stigma scarcely 1 mm; achene beak (including stigma) ca 2 mm; Alas to Cal, e to RMS and to Newf and Pa; Eurasia; narrowleaf or floating b. (S. natans var. a., S. simplex var. a.)

3 **S. angustifolium** Michx.

5b St and lvs mostly partially emersed; lvs often > 5 mm broad, sometimes scarious-margined near the base; mature ♀ heads gen > 2 cm broad; stigma ca 1.5 mm; achene beak (including stigma) much > 2 mm

6a Lvs mostly 3–5 (7) mm broad; bracts slightly or not at all dilated at the base; beak of mature achene subequal to the body; chiefly e of our range, but reported from Kootenai Co, Ida

S. chlorocarpum Rydb.

6b Lvs mostly 2–5 (8) dm × (5) 6–12 mm, gen expanded and scarious margined at base; bracts strongly dilated at the base; beak of mature achenes 3–4 (5) mm, the body 4–5 mm; ponds and marshes; Alas to n Cal, e to Colo and in the n to Ont, Que, NS, and N Eng; Eurasia; simplestem b.; 2 vars.

4 **S. emersum** Rehmann

a1 Basal lvs Y-shaped in cross-section, strongly keeled; mature fr, including the stipe and beak, 12–15 mm; common phase w Cas, BC to Ore, and in ne N Am and Eurasia (S. simplex)

var. **emersum**

a2 Basal lvs more nearly V-shaped in cross-section, not strongly keeled; mature fr gen ca 10 (to 12) mm; the more common phase e Cas to N Eng (S. m., S. subvaginatum)

var. **multipedunculatum** (Morong) Reveal

TYPHACEAE   Cat-tail Family

Fls ♂♀, naked, but subtended by capillary hairs, borne in densely crowded, terminal, cylindrical spikelike infl, the ♂ above the ♀; ♂ fls gen 2–5 stamens, the filaments joined below or distinct, borne directly on main axis, intermixed with ∝ slender hairs; ♀ fls with a single 1-carpellary pistil, the ovary short-stipitate, 1-celled and 1-ovuled, the stipe with slender, slightly clavate hairs, the basal one somewhat bractlike, gen broadened at the tip; fr dry, eventually dehiscent, buoyant by the elongate, slender hairs and by the persistent, slender style and expanded stigma; tall per herbs of wet places, with extensive rhizomes and erect, simple, cylindric, pithy sts; lvs alt, sheathing, linear, rather spongy.

Typha L.   Cat-tail; Bulrush; Reedmace

♂ portion of infl contiguous with or separated from the ♀ portion, the anthers much > the filaments; ♀ fls often in part sterile and then with the ovary terminal on a short stipe, the style and stigma obsolete; pls tending to form extensive, almost pure stands in marshy areas. (From *typhe*, the ancient Gr name for the pl.). (*Massula*). The cat-tail (often erroneously called *tule*) is an important refuge for many animals, esp water fowl.

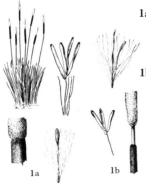

1a Lvs mostly 8–20 mm broad; pollen in tetrads; pls 1–3 m; ♂ and ♀ portions of the spike ± contiguous; ♀ fls gen lacking the subtending bract; stigmas oblanceolate-obovate; abortive ovaries pear-shaped; in shallow, standing or slow-moving water; Alas to Mex, e to most of s Can and throughout US; Eurasia, n Africa; common c.   **1 T. latifolia** L.
1b Lvs mostly ca 5 (3–10) mm broad; pollen grains single; pls 1–1.5 m; ♂ and ♀ portions of the spike not contiguous, ♀ fls gen with subtending bract; stigmas linear; abortive ovaries flattened and ± cuneate; shallow, quiet to slow-moving water; chiefly along Atl coast, w to Neb and Mo, occ to YNP, c Cal; much Mont material seems to be intermediate between this sp. and *T. latifolia*; lesser c.   **2 T. angustifolia** L.

ARACEAE   Arum or Calla-lily Family

Fls small and reduced, sessile and crowded, gen ± sunken into a fleshy axis (spadix) that is partially enclosed in a sometimes showy bract (spathe), either ♂ (ours) or ♂♀, or rarely ♂, ♀; perianth of 4 or 6 scalelike segms (sometimes lacking); stamens gen 2, 4, or 8; ovary 1–several-celled, style gen short or lacking; fr 1–∞-seeded, often baccate or leathery, indehiscent; ours per herbs gen with acrid or milky juice and entire to lobed, ± netted-veined lvs.

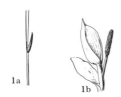

1a Lvs linear, < 3 cm broad; spathe much like the lvs, green, not enclosing the spadix   **Acorus**
1b Lvs lanceolate to oblong-ovate, much > 3 cm broad; spathe yellow or white, enclosing the spadix   **Lysichitum**

Acorus L.   Sweet Flag

Fls ♂, borne on a rather slender, elongate spadix; spathe lf-like, linear, seemingly a prolongation of the scape, the spadix apparently laterally inserted; tepals 6; stamens 6; ovary 2–3-celled; fr 1–3-seeded; rhizomatous per herbs with basal, linear lvs and naked scapes. (Gr name, *akoras*, perhaps for some other pl).

A. calamus L. Pl aromatic; lvs 5–20 dm × 1–2 (2.5) cm, the midvein gen much off-center; scapes 2–6 dm, the spathe lf-like and often = or > the scape; spadix 5–8 (10) × scarcely 1 cm at anthesis, ca twice as thick in fr; fls yellow or brownish; ponds and marshes; native to Eurasia, but widely intro and estab in e N Am, and in our area at Flathead Lake, Mont.

Lysichitum Schott   Skunk Cabbage

Fls all ♂, on a fleshy spadix 4–10 × 1–2 cm, the spathe large, white or yellowish; perianth 4-lobed; stamens 4; ovary 2-celled, 2–4-ovuled; stigma capitate, sessile; per herbs with short, fleshy, upright underground sts and large, simple lvs. (Gr *lysis*, a loosening, and *chiton*, tunic, in reference to the large spathe).

L. americanum Hultén & St. John. Glaucous, mephitic, acaulescent per; lvs lanceolate to oblong-ovate, mostly 4–10 (15) dm; spathe yellow, up to 2 dm; fls greenish-yellow; fr baccate, 1–2-seeded; swampy areas, Alas to Cal, e (but less common) to Mont and Ida. On the whole a choice garden pl for the boggy area or water's edge, easily transplanted or propagated by divisions of the underground st.

LEMNACEAE   Duckweed Family

Fls ♂♀, naked, gen (1) 2 ♂ and 1 ♀ borne together in a small pouch or sheath, the ♂ a single stamen with a 1–2-celled anther, the ♀ a single carpel with a 1-celled ovary and 1–several ovules; fr utricular; small to minute, ♂♀, free-floating to submersed, fleshy, colonial, thalloid pls without definite st or lvs, rootless or with 1–several simple roots, reproducing chiefly vegetatively.

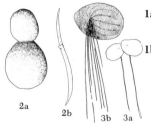

1a Individual thalli unnerved, rootless
  2a Thalli scarcely 1 mm, subglobose or ellipsoid   **Wolffia**
  2b Thalli several mm, flattened, much longer than broad   **Wolffiella**
1b Individual thalli 1–12-nerved, with 1 or more roots
  3a Pl 1–5-nerved, with 1 root per frond   **Lemna**
  3b Pl 4–12-nerved, with 2–several roots per frond   **Spirodela**

2a   2b   3b   3a

Lemna L.   Duckweed

Fls ca 3 (2 ♂ and 1 ♀) together in a marginal cleft of the thallus, the anther 2-celled, the ovary 1–7-ovuled; fr mostly 1-seeded; fronds flattened, gen colonial. (Gr name for a water pl, used by Theophrastus, possibly from *limne*, lake, in reference to the usual habitat). *(Hydrophace, Staurogeton)*.

1a Fronds oblong to lanceolate, 6–10 (12) mm, faintly 3-nerved, finely serrulate toward the tip, many with slender stalks = the main body, gen remaining attached together to form large mats, often submersed; new thalli arising on opp sides near the base of the parent thallus and at right angles to it; in quiet streams and standing water throughout most temp and tropical regions; star d., ivy d.   **1 L. trisulca** L.
1b Fronds more nearly oval, < 6 mm, not long-stalked, tending to separate into individual pls or small colonies, floating
  2a Fronds papillose, purplish-mottled, and bulged on the upper surface, 2–5 mm, 3–5-nerved; Eurasian sp. intro in much of US and Can, known from BC but not seen from our area; inflated d., wind bags   **L. gibba** L.
  2b Fronds either not papillose or not purplish-mottled, flat on the upper surface, mostly 2–3 (4) mm, often only 1-nerved
    3a Fronds 2–3 times as long as broad, nerveless, slightly falcate or otherwise asymmetrical; fairly common in c, e, and sw US, but not seen from our area although reported from Ida and from se Ore   **L. valdiviana** Phil.
    3b Fronds gen < twice as long as broad, ± symmetrical, mostly 3-nerved; cosmopolitan in quiet fresh water of temp and subtropical areas; water lentil   **2 L. minor** L.

Spirodela Schleid.

Fls gen 3 in each pouch, 2 ♂ and 1 ♀, the anthers 2-celled; ovary with 2 ovules; fronds suborbicular, (4) 7–12-nerved, gen in colonies, with triangular fl pouches along the margin; rootlets several, fascicled in the center of the frond. (Gr *speira*, cord, and *delos*, evident, referring to the cluster of rootlets).

S. polyrhiza (L.) Schleid. Great duckweed. Fronds round to obovate, (3) 4–8 mm, gen in small colonies, dark green above, purplish beneath, 5–11-nerved; rootlets (4) 5–15, the root-cap prominent, sharp-pointed; fls enclosed in a membranous sac (spathe) inside the marginal reproductive pouch; quiet water, mostly in marshes, ponds, lakes, or slow-moving streams; widely distributed in temp to tropical regions and known throughout the US and s Can; very rarely found in fl *(Lemna p.)*.

Wolffia Horkel   Water-meal

Fls 2 (1 ♂ and 1 ♀) on the upper surface of the frond, the anther 1-celled; utricle subglobose; fronds ellipsoid, flat to globular, very small, nerveless and rootless. (For Johann Friedrich Wolff, 1778–1806, a German botanist).

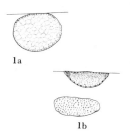

1a

1b

1a Pl floating below the surface of the water, the top of the thallus just touching the surface film; thallus ovoid to subglobose, 0.5–1.0 mm, the upper surface rounded, green, not puncticulate; stomata 1–6; widely distributed in N and S Am, but in our area rare, only collected in WV, Ore, where growing in mixture with *W. punctata*   1 **W. columbiana** Karst.

1b Pl floating on the surface of the water, the thallus oblong, 0.5–1.0 (1.2) mm and 1/2 as broad or less, the upper surface flattish, minutely white-puncticulate; stomata ∝ ; a free-floating pl often in association with *Lemna, Spirodela,* or *Azolla;* common in much of c and ne US; in our area in Thurston Co, Wn, and in WV, Ore   2 **W. punctata** Griseb.

Wolffiella Hegelm.   Wolffiella

Fronds flattened, nerveless and rootless, straplike or sickle-shaped; fls rarely developed, but said to be 2 in a pouch near the basal end of the thallus, 1 a solitary stamen, the other a single pistil. (Diminutive of *Wolffia).*

W. floridana (J. D. Smith) Thompson. Florida w. Pls aggregated into small to large masses, submerged except at the base; single fronds elongate and ± sickle-shaped, 5–9 (12) × ca 0.5 mm, ± acuminate, brownish-punctate on both surfaces; c and e US, very rare in our area, but collected at least once, in Thurston Co, Wn, where growing with *Lemna, Spirodela,* and *Wolffia punctata.*

PONTEDERIACEAE   Pickerel-weed Family

Fls racemose or pan, subtended by a spathe-like bract, complete, reg (ours), 3-merous; perianth biseriate, segms gen colored alike and mostly connate into a basal tube; stamens 3 or 6, adnate to base of the perianth; ovary superior, 1-celled with 3 parietal placentae or 3-celled; style 1, stigmas 3; fr a caps or indehiscent and achene-like; per, aquatic or palustrine herbs with sheathing and gen emersed or floating lvs.

Heteranthera   Ruiz & Pavon

Fls small, 1–several per spathe; perianth salverf, the tube slender, the tepals linear, ca equal; stamens 3, inserted in the tube of the perianth; ovary 1-celled or 3-celled by placental intrusion; fr a ∝ -seeded caps; per herbs with weakly ascending to floating or creeping sts. (Gr *hetero*, different, and *anthera*, anther, some spp. having unequal stamens).

H. dubia (Jacq.) MacMill. Water star-grass. Submersed to partially floating per with freely rooting, flaccid sts and linear, stip lvs 7–15 cm × (1) 2–5 mm; spathes 1-fld; fls pale yellow, gen borne at the water surface, perianth tube very slender, 2–4 cm, the tepals linear, ca 5 mm; stamens slightly < the tepals; caps 1-celled; quiet streams and ponds and lakes, fairly common in c, s, and e US and se Can; known in our area from a few collections in w Wn, from near Portland (Sauvie's I), and from WV, Ore, s to Cal.

LILIACEAE   Lily Family

Fls ⚥ to ⚥ ♂ ♀ or ♂, ♀, 3 (rarely 2) -merous; perianth mostly 2 rather similar sets of corolla-like tepals, or clearly differentiated into a mostly greenish calyx and an often highly colored corolla, the segms distinct to strongly connate; stamens mostly 6 (sometimes 3 of these sterile and staminoid), or occ only 3, hypog or adnate to the perianth tube; ovary superior to ± inferior, mostly 3-celled; styles distinct to fully connate and the stigma entire to 3-cleft; fr gen a membranous to leathery caps, or a berry; seeds few to ∝ ; ours per, herbaceous or rarely with a ± woody base *(Yucca)*, with scaly or tunicated bulbs, solid corms, or short to greatly elongated rhizomes, the sts simple to br, scapose to strongly lfy; lvs linear to ovate or obovate, sometimes aur or cordate, mostly alt but sometimes whorled.

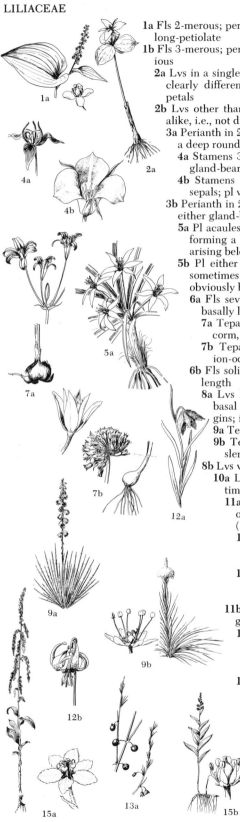

1a Fls 2-merous; perianth segms 4, alike, ca 2.5 mm; lvs 1 or 2, bl cordate and long-petiolate **Maianthemum**
1b Fls 3-merous; perianth segms 6, alike or dissimilar, gen > 2.5 mm; lvs various

  2a Lvs in a single whorl of 3 at the top of the otherwise naked st; perianth clearly differentiated into green sepals and purplish or white to red petals **Trillium**
  2b Lvs other than in a single whorl of 3; perianth segms often all much alike, i.e., not differentiated into a greenish calyx and colored corolla

    3a Perianth in 2 very dissimilar series, either the sepals or the petals with a deep round to oblong gland

      4a Stamens 3; petals smaller than the sepals, erect; sepals spreading, gland-bearing; pls rhizomatous; lvs mottled **Scoliopus**
      4b Stamens 6; petals gland-bearing, larger than the non-spreading sepals; pl with a bulb; lvs not mottled **Calochortus**

    3b Perianth in 2 series of essentially similar segms (tepals), all tepals gen either gland-bearing or glandless

      5a Pl acaulescent; lvs grasslike, tufted, linear; tepals strongly connate, forming a basal narrow tube 4–10 cm; fls apparently scapose, peds arising below ground level on the rhizome **Leucocrinum**
      5b Pl either caulescent or with non-grasslike lvs; tepals distinct or sometimes connate up to half their length, if connate the fls very obviously borne on an aerial st

        6a Fls several–∝ in a terminal umbel or head on a lfless or only basally lfy scape; lvs linear; pl with a tunicated bulb or corm

          7a Tepals 1.5–4 cm, connate gen up to half their length; pl with corm, not onion-odored **Brodiaea**
          7b Tepals gen < 1.5 cm, distinct; pl with bulb, mostly onion-odored **Allium**

        6b Fls solitary to racemose or pan; sts gen lfy or bracteate most of length

          8a Lvs linear, tough and stiff, persistent more than 1 year in a basal rounded clump, often sharp pointed or with fibrous margins; infl a large raceme or pan terminal on an ann st

            9a Tepals 4–5 cm; lvs pungent; style 1, thick **Yucca**
            9b Tepals scarcely 1 cm; lvs not pungent; styles 3, distinct, slender **Xerophyllum**

          8b Lvs various, but never persistent > 1 year; infl various

            10a Lvs mostly on the fl st and mostly above the base, sometimes all cauline, gen not strongly reduced upward on st

              11a Pl with a deep-seated scaly bulb; lvs linear to lanceolate, often whorled; fls showy, white or yellow to purple; tepals (1.2) 2–9 cm

                12a Perianth bell-shaped, tepals 1.2–3.5 cm, neither spreading nor recurved; anthers attached below the midpoint **Fritillaria**
                12b Perianth more nearly funnelf, tepals (3) 4–9 cm, often strongly spreading to recurved; anthers attached near the midpoint **Lilium**

              11b Pl rhizomatous; lvs scalelike to ovate, never whorled; fls gen not showy; tepals often < 1.2 cm

                13a Lvs scalelike, nongreenish, bearing in their axils several short, filiform, greenish brs; fls functionally ♂, ♀, borne singly or in pairs along the st on slender, jointed peds; sts freely br, wiry **Asparagus**
                13b Lvs large and greenish; fls ⚲, borne variously, but rarely on jointed peds; sts often simple

                  14a Fls several–∝ in a terminal raceme or pan on an unbr st

                    15a Styles 3; stamens perig; pollen sacs joined in their tip; tepals 6–17 mm; pl rarely < 10 dm; lvs very prominently veined **Veratrum**
                    15b Styles 1; stamens hypog; pollen sacs not joined at the tip; tepals 2–7 mm; pl 2–9 (10) dm; lvs not prominently veined **Smilacina**

                14b Fls 1–few, borne along st or in small clusters at tip of the gen br main st

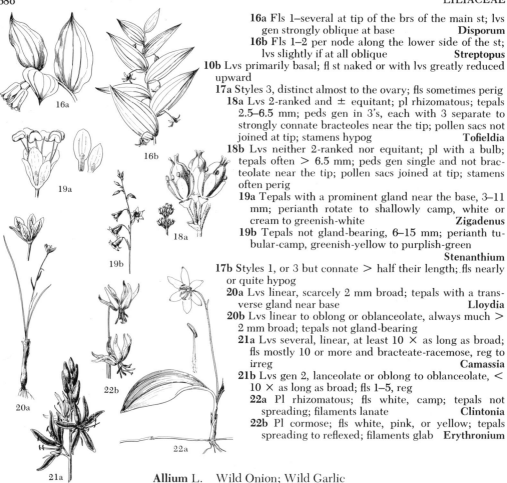

16a Fls 1–several at tip of the brs of the main st; lvs gen strongly oblique at base **Disporum**

16b Fls 1–2 per node along the lower side of the st; lvs slightly if at all oblique **Streptopus**

10b Lvs primarily basal; fl st naked or with lvs greatly reduced upward

17a Styles 3, distinct almost to the ovary; fls sometimes perig

18a Lvs 2-ranked and ± equitant; pl rhizomatous; tepals 2.5–6.5 mm; peds gen in 3's, each with 3 separate to strongly connate bracteoles near the tip; pollen sacs not joined at tip; stamens hypog **Tofieldia**

18b Lvs neither 2-ranked nor equitant; pl with a bulb; tepals often > 6.5 mm; peds gen single and not bracteolate near the tip; pollen sacs joined at tip; stamens often perig

19a Tepals with a prominent gland near the base, 3–11 mm; perianth rotate to shallowly camp, white or cream to greenish-white **Zigadenus**

19b Tepals not gland-bearing, 6–15 mm; perianth tubular-camp, greenish-yellow to purplish-green **Stenanthium**

17b Styles 1, or 3 but connate > half their length; fls nearly or quite hypog

20a Lvs linear, scarcely 2 mm broad; tepals with a transverse gland near base **Lloydia**

20b Lvs linear to oblong or oblanceolate, always much > 2 mm broad; tepals not gland-bearing

21a Lvs several, linear, at least 10 × as long as broad; fls mostly 10 or more and bracteate-racemose, reg to irreg **Camassia**

21b Lvs gen 2, lanceolate or oblong to oblanceolate, < 10 × as long as broad; fls 1–5, reg

22a Pl rhizomatous; fls white, camp; tepals not spreading; filaments lanate **Clintonia**

22b Pl cormose; fls white, pink, or yellow; tepals spreading to reflexed; filaments glab **Erythronium**

Allium L.   Wild Onion; Wild Garlic

Fls several–∝, subcapitate to openly umbellate, the infl invol; tepals 6, alike, distinct; stamens 6, filaments basally ± connate, gen adherent to base of the opposing tepals; ovary 3-celled, sessile, 3-lobed, gen with 2 (occ more) ovules per cell; caps often crested, 3–6 (rarely more) -seeded; per scapose herbs with tunicated bulb reformed each year, 1–several linear flat to terete lvs, and gen an onion or garlic odor. (L name for garlic). The cellular reticulations on the bulb coats of some spp. represent the persistent walls of the cells of the inner epidermis of the bulb coats. They become apparent only after the outer layers of the bulb scale have been removed by decay, and are frequently stripped from the bulb at the time of collection.

1a Cells of ovary each with 6–12 ovules; pl inodorous; lvs flat, up to 3 dm × 1–2.5 cm; scape terete, hollow, up to 1 m × 1 cm; umbel ∝ -fld, peds 3–5 cm; tepals ca 8 mm, white with median green nerve; only occ as an escape, mainly in WV, Ore; black garlic, Homer's g.   **1 A. nigrum** L.

1b Cells of ovary each with 2 ovules; pl gen onion or garlic scented; lvs, scape, and tepals various

2a Inner filaments laterally strongly flattened and broadened and with slender lateral appendages gen surpassing the anther; lvs terete, hollow, strongly sheathing; tepals 3–4 mm, greenish to purplish; pl strongly garlic-scented; crow g., field g.; occ escape, known at least from nw Ore and VI   **2 A. vineale** L.

2b Inner filaments not greatly flattened and not laterally appendaged; lvs and tepals various; pls not garlic-scented

3a Old bulb coats persistent on bulb as coarse-meshed fibrous nets; ovary with 6 low rounded crests

4a Lvs gen 2; tepals 5–7 mm, white or occ pinkish with a red to reddish-brown midrib, the tips of the inner ones ± spreading; plains

and low hills, sw Alta to SR, Ida, e to Man and Minn and s to NM; textile o. (*A. reticulatum, A. r.* var. *playanum*)

3 **A. textile** Nels. & Macbr.

4b Lvs gen at least 3; tepals (4) 6–8 (10) mm, gen pink, the tips of the inner ones erect; fls sometimes replaced by bulbils; Geyer's o.; widespread, with 2 vars. in our area   4 **A. geyeri** Wats.

    a1 Fls functional, not replaced by bulbils; low meadows and along streams, e base Cas, Wn, s to Ariz, e Ida, Nev, Utah, s Wyo, and w Tex, also in SD   var. **geyeri**

    a2 Fls mostly replaced by bulbils; VI and e Cas, Wn, ne Ore, and c Ida to s Mont (*A. fibrosum* Rydb., *A. rydbergii*)

                                         var. **tenerum** Jones

3b Old bulb coats non-persistent, or non-fibrous, or with parallel fibers that do not form a network, but inner epidermis of bulb coats often cellular-reticulate; ovary gen crestless

5a Lvs 1 or 2, terete, hollow or solid

    6a Lvs 1, semisolid, gen much > scape, tending to coil at the tip and to break off in age; infl umbellate, fl from the outside inward; scape 3–5 (10) cm; ovary and caps very strongly crested; bulbs ovoid; des, se Ore to s Ida, s to Ariz and w Colo, barely in our area in s Ida; Nev o. (*A. n.* var. *macropetalum*)

                                       5 **A. nevadense** Wats.

    6b Lvs gen 2, hollow, gen < scape, tip not tending to coil or to break off, ovary and caps crestless; infl subcapitate, fl from the inside outward; scape 2–5 dm; wet or moist places mostly; bulbs elongate, slender, often clustered; circumboreal, s in N Am to Cas and along CR, Wn, ne Ore, n Ida, in RM to Colo, and to Minn, NY, and Newf; chives   6 **A. schoenoprasum** L.

5b Lvs 1–several, strongly flattened to channeled, not hollow

    7a New bulbs formed outside the bulb coats of the parent bulb, all but the roots of the latter gen disappearing by anthesis, the new bulb present at that time gen ± oblique at base and without roots of its own, sometimes attached to a short rhizome; ovary and caps crestless; pls very local in Blue Mts, Wn, or in Yamhill Co, Ore

        8a Bulbs attached to a rather evident rhizome; lvs (1) 2–3; tepals pale pink or occ white, 10–17 mm, lanceolate, acute, entire, becoming papery and spreading in fr, but not involute; in our area in Yamhill Co, Ore, more common from se Ore to below San Francisco; one-lvd o.   7 **A. unifolium** Kell.

        8b Bulb often not attached to an evident rhizome; lvs 2; tepals bright pink, 10–15 mm, lanceolate, attenuate, finely serrulate-denticulate, becoming firm and keeled-involute in fr; known only from Columbia Co, Blue Mts, Wn; Blue Mt o.

                                    8 **A. dictuon** St. John

7b New bulbs formed inside the bulb coats of the parent bulb which gen persists until after anthesis, mostly symmetrical; ovary and caps crestless to strongly crested

    9a Scape terete full length or slightly flattened and even somewhat winged near the umbel; lvs mostly slightly channeled or ± V-shaped in cross-section, not curved or sickle-shaped, tending to persist with the scape until after seeds are shed or pl is dried; bulbs sometimes elongate and borne terminally on a short rhizome, the rhizome gen covered with remnants of previous bulbs   **Group I**, lead 10a

    9b Scape slightly to strongly flattened, even toward the base, and gen ± wing-margined; lvs mostly thick and flat, often curved or sickle-shaped, gen deciduous with the scape as soon as the seeds mature or the pl dries; bulbs mostly ovoid, not attached to a rhizome   **Group II**, lead 24a

Group I

10a Bulbs slender and elongate, not ovoid, borne terminally on a short rhizome which gen bears remains of older bulbs; outer bulb coats with prominent elongate cell walls in reg rows

11a Ovary strongly crested with 6 thin, flattened, entire or toothed pro-

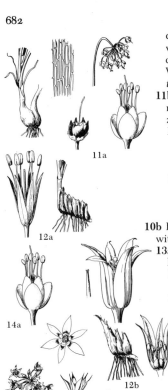

cesses; stamens > tepals; tepals 4–6 mm, elliptic-ovate, ± obtuse, pink or white; bulbs obscurely to evidently short-rhizomatous; umbel nodding; common in somewhat moist areas; transcont in Can, s to coastal and inl Wn and Ore, c Ida, w Mont, and to Mex and Ga; nodding o. *(A. recurvatum, A. c. f. alba)*   9 **A. cernuum** Roth

11b Ovary without crests; stamens often < tepals; tepals often > 6 mm, narrowly lanceolate and acuminate; bulbs evidently attached to a rhizome; umbel erect

    12a Stamens and style > tepals; stigma capitate; caps longer than broad; tepals 8–10 mm, pink; wet places, mont to alp; s BC s along e side Cas to sw Ore and Sierran Cal, e in BC and s to w Ida, e Ore, and ne Nev; Pac o., swamp o.   10 **A. validum** Wats.

    12b Stamens and style only ca half length of tepals; stigma 3-lobed; caps broader than long; tepals 10–13 mm, pink; wet places, mont to alp; c Mont and ne Ida to ne Utah and Colo; short-style o.

                                            11 **A. brevistylum** Wats.

10b Bulbs gen ovoid, never attached to a rhizome; outer bulb coats seldom with prominent elongate cell walls in reg rows

    13a Ovary and caps strongly crested with 6 thin and slender, often papillose-denticulate processes

        14a Tepals 4–6 mm, white or pink, obtusish; stamens > tepals (see lead 11a)   9 **A. cernuum** Roth

        14b Tepals gen (4) 7–10 mm, white to purplish, acuminate; stamens < to ca = tepals

            15a Lvs gen green at anthesis and persistent in fr; tepals 7–10 mm, lilac to pale pink or white, becoming papery in age, the tips neither keeled nor strongly involute-margined; scape gen 2–3 (1–4) dm; meadows to open slopes; s Cas, Ore, to sw Cal, e to s Ida, Nev, and Utah; disjunct and in our area in the Selway R region of n Ida; Palmer's o., Patis o.   12 **A. bisceptrum** Wats.

            15b Lvs gen withering by anthesis and often deciduous in fr; tepals 7–8 (4–10) mm, purplish (rarely white), gen with purple blotch near base, becoming rigid in age, the tips strongly keeled and with involute margins; scape gen < 1.5 (0.5–3) dm; dryish places, mont to alp; nc Ore to Cal and Nev; Sierra o.   13 **A. campanulatum** Wats.

13b Ovary crestless or only shallowly crested with rounded, low processes

    16a Lvs (2) 3–4, narrowly linear, flexuous, < the scape, gen persistent after seed maturity; scapes 1–4 dm; tepals lanceolate, acute to acuminate; bulb coats thick, with prominent cellular reticulations (under strong lens)

        17a Outer tepals lanceolate, gen 10–12 (8–17) mm, acuminate, becoming involute-margined and keeled, the tips spreading to recurved, commonly rose-purple, less commonly pink, occ white, the inner series gen rather obscurely serrulate-denticulate, much < outer series; ovary very slightly crested with rounded processes; outermost bulb coats becoming thickened, prominently marked with squarish reticulations; dry places; VI and nw Wn and e side Cas, Wn and Ore, s to Cal and Ariz, e to s Wyo and w Colo; tapertip o., Hooker o. *(A. a. var. cuspidatum)*

                              14 **A. acuminatum** Hook.

        17b Outer tepals elliptic-lanceolate, mostly 5–11 mm, acute, becoming papery and connivent over the caps, white or pink, both series entire, the inner series ca as broad and long as the outer; outermost bulb coats with cross-reticulations ± sinuate, forming a herringbone pattern; scattered in our area, VI and nw Wn, CRG, WV and se Ore to s Cal; slim-lf o.   15 **A. amplectens** Torr.

16b Lvs 1 or 2, gen broad and flat and often ± curved or sickle-shaped, often > scape; scapes gen < 1 (but occ up to 4) dm, gen breaking off at ground level, along with the lvs, at maturity of the seeds

    18a Stamens at least = the tepals; bulb never with a cluster of bulblets at base

        19a Scape 1–2 dm, gen much < the lvs; anthers reddish or purple, lightly corrugate; tepals 5–7 mm, white or pink with green or reddish midrib, lanceolate, acute or acuminate, becoming papery and the midrib prominent; dry gravelly soil, scattered and local; c and se Wn, to sc Ore; rock o. *(A. equicaeleste)*

                              16 **A. macrum** Wats.

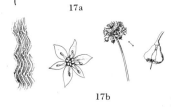

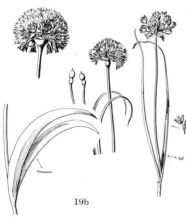

19b Scape 2–3 (1–4) dm, gen > the lvs; anthers yellow or pinkish, smooth; tepals 7–8 (6–10) mm, pink or occ white, narrowly lanceolate, acuminate, becoming rigid and strongly keeled in fr; Douglas' o.; 4 vars.   17 **A. douglasii** Hook.

  a1 Lvs broad and flat, often curved; scape not thickened below the umbel; s Whitman and e Walla Walla cos, Wn, and in Wheeler, Grant, and e Umatilla cos, Ore *(A. hendersoni)*
                        var. **douglasii**

  a2 Lvs rather narrow, channeled, often not curved; scape often thickened below the umbel

    b1 Scape not thickened below the umbel; e slope Cas, Kittitas Co, Wn, to n Ore *(A. n.)*
             var. **nevii** (Wats.) Ownbey & Mingrone

    b2 Scape thickened just below the umbel

      c1 Scape thickened below the umbel but not constricted; along Wn-Ida line from Kootenai to Idaho Co, Ida, and Spokane to n Whitman Co, Wn, also in Sander Co, Mont
          var. **columbianum** Ownbey & Mingrone

      c2 Scape constricted between the swollen upper portion and the umbel; Upper Grand Coulee, Grant Co, Wn
          var. **constrictum** Ownbey & Mingrone

18b Stamens < tepals, or bulb with a basal cluster of bulblets, or both

  20a Scape 1–2 dm, gen at least as long as lvs; larger bulbs with a basal cluster of bulblets; tepals 6–10 mm, lanceolate, obtuse to acuminate, white with green midribs, or pinkish, becoming papery and keeled in fr; areas wet early in spring; Blue Mts, e Ore, and Valley and Adams cos, Ida; swamp o.
                18 **A. madidum** Wats.

  20b Scape < 1 dm, except in *A. fibrillum*, where 0.3–2.5 dm; bulbs never with a basal cluster of bulblets; tepals various

    21a Lvs gen <twice length of scape; fl peds gen > tepals; tepals 5–8 mm, entire, white with greenish midrib to pinkish, becoming papery in fr; reticulations of outer bulb coats gen visible under high magnification; scape 0.3–2.5 dm

      22a Scape 3–15 (25) cm; ovary sometimes slightly crested; inner epidermis of outer bulb coats with very irreg reticulations that are strongly contorted and never in transverse rows; moist places, ec and se Wn and ne Ore through mts of w and n Ida to near GNP, Mont; fringed o. *(A. collinum)*.
           19 **A. fibrillum** Jones

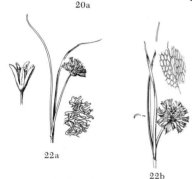

      22b Scape 3–10 cm; ovary rarely at all crested; inner epidermis of outer bulb coats with reg, gen hexagonal reticulations arranged either in a honeycomb pattern or transversely elongate in rather reg rows; meadows, ne Ore and Ida, s of Salmon R, to sw Mont, w Wyo, Utah, and w Colo; Brandegee o. *(A. minimum)*   20 **A. brandegei** Wats.

    21b Lvs > twice length of scape; fl peds rarely > tepals; tepals various but gen serrulate-denticulate, 5–9 mm; reticulations of outer bulb coats obscure, scarcely discernible even under high magnification; scape gen < 0.5 dm

      23a Tepals white with green midrib; ovary crested with low, rounded processes; mts of c and sw Ida, foothills to alp slopes; dwarf o.   21 **A. simillimum** Hend.

      23b Tepals deep pink, fading to white; ovary crestless; known only from near Boise, Ada Co, and from Gem Co, Ida; Aase o.   22 **A. aaseae** Ownbey

Group II

24a Ovary strongly crested with 6 elongate, thin processes; bulbs gen clustered, either elongate and slenderly ovoid or broadly ovoid but oblique at base

  25a Scape 1–5 dm, not greatly flattened, the umbel nodding; tepals 4–6 mm, elliptic-ovate, ± obtuse, white or pink; bulb elongate, not oblique; lvs 1–6 mm broad, channeled to nearly plane (see lead 11a)
                9 **A. cernuum** Roth

  25b Scape mostly < 1 dm, strongly flattened, the umbel erect; tepals 10–12

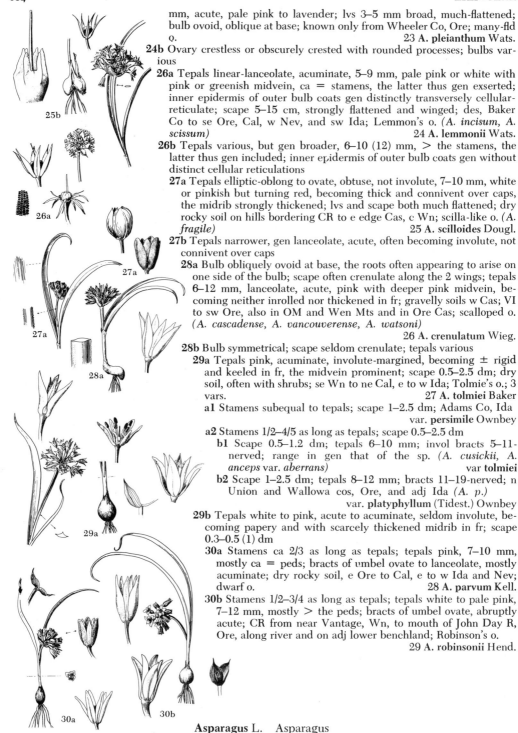

mm, acute, pale pink to lavender; lvs 3–5 mm broad, much-flattened; bulb ovoid, oblique at base; known only from Wheeler Co, Ore; many-fld ♀.
23 **A. pleianthum** Wats.

24b Ovary crestless or obscurely crested with rounded processes; bulbs various

 26a Tepals linear-lanceolate, acuminate, 5–9 mm, pale pink or white with pink or greenish midvein, ca = stamens, the latter thus gen exserted; inner epidermis of outer bulb coats gen distinctly transversely cellular-reticulate; scape 5–15 cm, strongly flattened and winged; des, Baker Co to se Ore, Cal, w Nev, and sw Ida; Lemmon's o. *(A. incisum, A. scissum)*
24 **A. lemmonii** Wats.

 26b Tepals various, but gen broader, 6–10 (12) mm, > the stamens, the latter thus gen included; inner epidermis of outer bulb coats gen without distinct cellular reticulations

  27a Tepals elliptic-oblong to ovate, obtuse, not involute, 7–10 mm, white or pinkish but turning red, becoming thick and connivent over caps, the midrib strongly thickened; lvs and scape both much flattened; dry rocky soil on hills bordering CR to e edge Cas, c Wn; scilla-like o. *(A. fragile)*
25 **A. scilloides** Dougl.

  27b Tepals narrower, gen lanceolate, acute, often becoming involute, not connivent over caps

   28a Bulb obliquely ovoid at base, the roots often appearing to arise on one side of the bulb; scape often crenulate along the 2 wings; tepals 6–12 mm, lanceolate, acute, pink with deeper pink midvein, becoming neither inrolled nor thickened in fr; gravelly soils w Cas; VI to sw Ore, also in OM and Wen Mts and in Ore Cas; scalloped o. *(A. cascadense, A. vancouverense, A. watsoni)*
26 **A. crenulatum** Wieg.

   28b Bulb symmetrical; scape seldom crenulate; tepals various

    29a Tepals pink, acuminate, involute-margined, becoming ± rigid and keeled in fr, the midvein prominent; scape 0.5–2.5 dm; dry soil, often with shrubs; se Wn to ne Cal, e to w Ida; Tolmie's o.; 3 vars.
27 **A. tolmiei** Baker

     a1 Stamens subequal to tepals; scape 1–2.5 dm; Adams Co, Ida
var. **persimile** Ownbey

     a2 Stamens 1/2–4/5 as long as tepals; scape 0.5–2.5 dm

      b1 Scape 0.5–1.2 dm; tepals 6–10 mm; invol bracts 5–11-nerved; range in gen that of the sp. *(A. cusickii, A. anceps var. aberrans)*
var **tolmiei**

      b2 Scape 1–2.5 dm; tepals 8–12 mm; bracts 11–19-nerved; n Union and Wallowa cos, Ore, and adj Ida *(A. p.)*
var. **platyphyllum** (Tidest.) Ownbey

    29b Tepals white to pink, acute to acuminate, seldom involute, becoming papery and with scarcely thickened midrib in fr; scape 0.3–0.5 (1) dm

     30a Stamens ca 2/3 as long as tepals; tepals pink, 7–10 mm, mostly ca = peds; bracts of umbel ovate to lanceolate, mostly acuminate; dry rocky soil, e Ore to Cal, e to w Ida and Nev; dwarf o.
28 **A. parvum** Kell.

     30b Stamens 1/2–3/4 as long as tepals; tepals white to pale pink, 7–12 mm, mostly > the peds; bracts of umbel ovate, abruptly acute; CR from near Vantage, Wn, to mouth of John Day R, Ore, along river and on adj lower benchland; Robinson's o.
29 **A. robinsonii** Hend.

Asparagus L. Asparagus

Fls ♂ or (ours) functionally ♂, ♀, borne in 2's or singly along the st on slender peds jointed near midlength; tepals 6, distinct nearly to base; stamens 6, inserted on the tepals near their base; ovary 3-celled; style short, slender; stigmas 3; fr a few-seeded berry; rhizomatous per herbs with freely-br, erect to ± climbing sts; lvs alt, scalelike, scarious, each subtending gen several short filiform (ours) to flattened brs that resemble and function as lvs. (The ancient Gr name).

A. officinalis L. New sts fleshy and simple, with scarious scale lvs up to 5 mm, ultimately br freely and mostly 1–1.5 (2) m, the clusters of axillary reduced brs filiform, mostly 10–12 (20) mm; fls drooping on filiform peds up to 2 cm; perianth greenish-white, bell-shaped, 3–7 mm, the ♂ fls slightly the larger; berry red, 6–8 mm thick; extensively grown commercially and commonly escaping and persisting along fence rows, in orchards, or along river banks.

Brodiaea Sm.   Brodiaea

Fls few (rarely only 1) to ∝ in a capitate to open invol umbel on peds jointed near tip; perianth gen connate at least half the length, the tube slenderly funnelf to camp, the 6 tepals erect to spreading, much alike; stamens 6 and all functional or 3 functional and the alt 3 modified into thin, broadened and bifid staminodia; ovary sessile or short-stipitate; style 1, stigmas 3; fr caps; per, scapose herbs with deep-seated, tunicated corms, the lvs 1–5, linear, greatly elongate, often appearing long before the scape and withering by anthesis. (Named for James J. Brodie, a Scotch botanist). *(Dichelostemma, Hookera, Triteleia)*. The spp. are mostly attractive-fld; they look and do best on well-drained slopes where seldom watered.

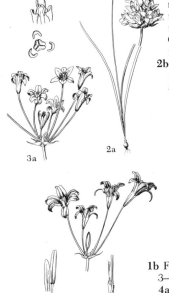

1a Fertile stamens 3, anthers mostly (4) 5–9 mm; caps sessile or subsessile

  2a Infl subcapitate, the peds all < 10 mm; staminodia adnate to the level of the anther tips, with deeply bifid free tips and broad free margins along the adnate portion; perianth lavender to bluish-purple, ca 2 cm, the tube narrowly camp, the lobes ca = tube, slightly spreading; grassy meadows to glacier outwash "prairies" or sagebr slopes; Island Co, Wn, s and occ through PT and n and e side of OP to CRG, e to Goldendale, Klickitat Co, Wn, and Wasco Co, Ore, and s through WV to sw Ore and much of n and c Cal; northern saitas   **1 B. congesta** Smith

  2b Infl open-umbellate, the peds nearly all > 10 mm; staminodia oblong, neither deeply bifid nor adnate above the base of the anthers

    3a Staminodia curved (involute), > the stamens, blunt to retuse, tending to lean toward the center of the fl and partially to enfold the anthers; anthers (4) 5–6 (7) mm; perianth violet-purple, mostly 2–3.5 cm, narrowly camp at base, the lobes 1.5–2 × as long as the tube; gravelly prairies to grassy slopes and rocky bluffs; s VI and is and mainl of PS, occ on OP, up CRG to Wasco Co, Ore, and Yakima Co, Wn, s in Ore w Cas to Cal; bluedicks b., harvest b. *(B. grandiflora)*   **2 B. coronaria** (Salisb.) Engl.

    3b Staminodia plane, subequal to the stamens, not leaning toward center of the fl or enfolding the anthers, obtuse to acute; anthers 7–9 (10) mm; perianth gen violet-purple (to almost pink), mostly 3–4 cm, the base more obconic or funnelf than camp, the lobes > the tube; dry plains and grassy, dry hills, Hood R and Clackamas cos, Ore, occ s in the WV to se Ore, more common in Cal; elegant b.   **3 B. elegans** Hoover

1b Fertile stamens 6, anthers mostly 1.5–3 (4) mm; caps with slender stalk 3–6 mm

  4a Perianth gen pale yellow, purple-lined, 15–20 (30) mm, the tube very slender toward the base, flaring above; anthers blue, ca 2.5 mm, filaments scarcely at all flared toward the base, all inserted at the same level and of approximately the same length; stipe slender, gen > caps; prairies and dry hills to oak or pine for; Lane to Josephine and Jackson cos, Ore; Henderson's b.   **4 B. hendersonii** Wats.

  4b Perianth white or blue to purple, the tube rather broadly camp, not flared toward the top; anthers yellow; filaments either greatly flared below and confluent at base, or inserted at different levels and the 2 sets of anthers at very different levels

    5a Perianth shallowly camp or camp-rotate, not concealing the ovary, gen white, sometimes bluish-tinged to clear light blue, 10–16 mm, the lobes 2–3 × > the tube; filaments broadly ovate-triangular, = inserted and of = length, confluent at their bases; grassy, open flats to midmont meadows; s BC s, w Cas in Wn and Ore, to Sierran Cal, and common in sagebr des e Cas, from Chelan Co, Wn, prob to n Nev, e to Ida; hyacinth b. *(B. grandiflora,* misapplied; *Hesperoscordum* h.; *Milla* h.; *Calliprora* h.; *H. lewisii)*   **5 B. hyacinthina** (Lindl.) Baker

    5b Perianth turbinate, concealing the ovary, from rather deep blue to

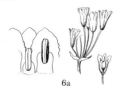

sometimes light blue or nearly white, mostly 18–30 mm, the lobes ca = the tube; filaments broadly ovate-oblong to linear, inserted at different levels or of unequal length, the alt stamens in 2 sets 2–3 mm apart

6a Filaments flat, several times as broad as thick, very unequal in length but inserted at nearly the same level, the adnate portion not strongly keeled; perianth nearly white to fairly deep blue, the 2 series similar, neither strongly ruffled, the inner ones arising at nearly the same level as the outer, all flared together; both sides Cas, coastal bluffs and prairies to sagebr des and lower hills; sw BC s through PT and WV to sw Ore, up CRG to e base of Cas from Chelan and Grant cos, Wn, to Gilliam Co, Ore; Howell's b. (*T. grandiflora* var. *h.*, *B. douglasii* var. *h.*, *B. bicolor*)

                                                               6 **B. howellii** Wats.

6b Filaments nearly as thick as broad, unequally inserted, the adnate portion strongly keeled; perianth pale to deep blue, the outer series not ruffled, the inner somewhat the broader, strongly ruffled, flared at a level ca 2–3 mm higher than the outer series; e Cas, grassl and sagebr des to ponderosa pine woodl; s BC s to se Ore, e to Ida and w Mont, w Wyo, and n Utah; Douglas' b. (*T. grandiflora*, *Milla* g.)

                                                               7 **B. douglasii** Wats.

Calochortus Pursh   Butterfly, Mariposa, or Star Tulip; Mariposa or Sego Lily; Mariposa; Cats-ear

Fls 1–several on simple or br st, white or yellow to reddish or purplish; sepals narrow, greenish, gen spotted near base; petals broad, gen spotted or blotched near base and with a round to elongate, depressed and often fringe-margined gland; ovary 3-celled, superior; caps linear to oval, 3-angled to strongly 3-winged, erect to drooping or nodding; per from a deep-seated bulb, the lvs few, the basal lf gen much the largest. (Gr *kalo*, beautiful, and *chortos*, grass). Most spp. propagate freely from seeds, but it takes 3–5 years for seedlings to grow into fl bulbs; on the other hand, collected bulbs are apt not to survive.

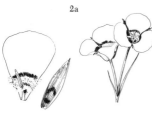

1a Basal lf narrow, channeled and ± V-shaped in cross-section, seldom much broader than the st lvs, often < half as long as the st; ovary and caps elongate, many times longer than thick, not winged

2a Sepals long-acuminate, mostly > the petals, (3) 4–7 cm; petals white to deep lavender, with a median greenish stripe, deltoid-cuneate, acuminate, 3–6 (7) cm; gland oblong-oval, ± sagittate at base, bordered with a ± continuous, fringed membrane, the surface covered with elongate, gen freely br, thick hairs; anthers > the filaments; fls 1–3; dry places e Cas, sc BC to ne Cal, e to Ida, sw Mont, and n Mex; sagebr m., green-banded star-tulip (*C. acuminatus*, *C. cyaneus*)

                                                               1 **C. macrocarpus** Dougl.

2b Sepals gen < the petals; petals various, often not acuminate; gland various, often circular or broader than long, seldom at all sagittate

3a Gland much broader than long, oblong to ± crescent-shaped; petals cuneate, rounded at the tip or only slightly acute

4a Hairs surrounding the gland thick, forked or much enlarged near the tip, glandular; gland oblong, slightly arched, not crescent-shaped; petals white or yellowish to purple, greenish inside, sometimes purple-banded or -spotted on either side of the gland, but spotting never conspicuous; anthers strongly acute or gen apiculate, > the filaments; caps linear-oblong, 4–6 × < 1 cm; meadows to light woods; RM, c Mont to NM, Utah, and e Ariz; Gunnison's m., sego-lily   2 **C. gunnisonii** Wats.

4b Hairs surrounding the gland very slender, unbr, not glandular; gland crescent-shaped, acute at each end; petals various, gen strongly spotted or blotched within; anthers blunt or rounded at tip, gen < the filaments; caps winged (this feature sometimes not evident in fl material); see leads 9a–9b for distinction between

                                                               7 **C. nitidus** Dougl.
                                                               8 **C. eurycarpus** Wats.

3b Gland at least as long as broad; petals often strongly acute to acuminate

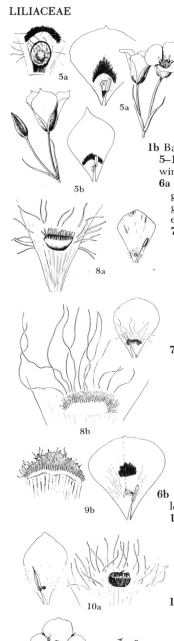

5a Petals bearded adj to gland with few–∝ slender hairs, white to lilac-tinged, gen yellow near base and with a brownish-purple spot above the gland, not greenish-striped; anthers gen ca = filaments; dry areas, e Mont to the Daks, s to NM, w to SRP, e Ida, Nev, and Ariz; sego-lily, Nuttall's s.   3 **C. nuttallii** T. & G.

5b Petals glab or with few short thickish hairs near the gland, white to lavender-tinged, gen with a median greenish stripe and a reddish-purple spot above the gland; anthers gen < the filaments; sagebr des from se Ore to e Cal, e to sw Mont, e Ida, and ne Utah; Bruneau m. *(C. nuttallii* var. *b.)*   4 **C. bruneaunis** Nels. & Macbr.

1b Basal lf broad and flat, often = fl st, seldom < half as long, mostly (3) 5–15 (20) mm broad, much broader than st lvs; ovary winged; caps broadly winged, gen < 2 (3) times as long as wide when pressed

  6a Petals neither fringed nor ciliate, although sometimes with a few marginal hairs near base, with bearding on inner surface mostly near the gland, (2.5) 3–4 (5) cm (often < 2.5 cm in *C. uniflorus);* caps gen erect (sometimes reflexed in *C. uniflorus)*

    7a Lowest bract very close to the basal lf, bearing a small bulblet in its axil; petals pinkish-lavender to lilac, drying purplish, (1.5) 2–3 (3.5) cm; st 1–3 dm

      8a Petals mostly 1.5–2.5 cm, bearded sparsely near the gland; st mostly 1–2 (2.5) dm; caps nodding to erect; w Cas, Lane Co, Ore, to wc Cal; Monterey m., large-fld s.   5 **C. uniflorus** H. & A.

      8b Petals mostly (2) 2.5–3 (3.5) cm, sparsely to densely long-bearded on lower half; pl mostly (1.5) 2–3 dm; caps erect; e slope Cas, Yakima Co, Wn, s to n Cal and to w Cas in sw Ore; long-bearded s. *(C. l.* var. *peckii)*   6 **C. longebarbatus** Wats.

    7b Lowest bract often distant from 1st lf, not bulblet-bearing; petals mostly white to pale lavender and deeply purple-spotted, or purplish and gen > 3.5 cm; st mostly 2–5 dm

      9a Petals (3.5) 4–4.5 cm, pale to deep lavender, with a dark purple, crescent-shaped blotch above the crescent-shaped gland; dried swales and prairies, se Wn (Whitman Co) to adj Ida; broad-fr m. *(C. pavonaceus)*   7 **C. nitidus** Dougl.

      9b Petals 2.5–5 cm, white to very pale lilac, with a large, ± circular, central purple spot; grassl and sagebr slopes to open conif for; se Wn (Asotin Co) to nc and ne Ore, e and common in c Ida to sw Mont and w Wyo; wide-fr m., big-pod m. *(C. parviflorus, C. euumbellatus, C. nitidus* var. *e.)*   8 **C. eurycarpus** Wats.

  6b Petals strongly fringed or ciliate, the inner face often densely long-bearded, often clawed at base, mostly not > 2.5 (3) cm

    10a Gland purplish-black, subrotund, scarcely 2 mm broad, bordered by a short, black-purple fringe; anthers long-apiculate, tipped by a setaceous process 1–2 mm; petals yellowish-white, 2–2.5 cm, deltoid-obovate, obtuse to acute, strongly bearded on the inner face; caps recurved-nodding; dryish meadowl to sparse or deep woods, se BC and ne Wn to sw Alta and nw Mont; pointed m., Baker's m.

                                   9 **C. apiculatus** Baker

    10b Gland transversely elongate, gen ± crescent-shaped, at least 2 mm broad, rarely blackish-bordered; anthers blunt to apiculate; petals various

      11a Petals (3.5) 4–4.5 cm, purplish, cuneate-obovate, sparingly ciliate and rather sparsely hairy on the inner face mainly just above the gland; pl 2–4 dm; caps erect; anthers obtuse, < the filaments (see lead 9a where this will gen key)   7 **C. nitidus** Dougl.

      11b Petals various, but gen < 3.5 cm or white or yellowish, strongly ciliate-fringed, the inner face gen strongly hairy over much of the length; pl often < 2 dm; caps erect to nodding; anthers often acute and ± apiculate

        12a Petals ovate-lanceolate to oblanceolate, strongly clawed, acute, gen ca twice as long as broad, white or greenish-gray to purplish tinged, gen with a purplish, crescent-shaped blotch above the crescent-shaped gland; caps erect to nodding

          13a Caps erect; basal lf gen < uppermost fl, the st 1–3 (5) dm; fls (1) 2–9, white or purplish-tinged; petals broadly lanceolate; sagebr slopes to open for; e slope Cas, BC to n Yakima Co, Wn, mostly w CR; Lyall m. *(C. ciliatus)*   10 **C. lyallii** Baker

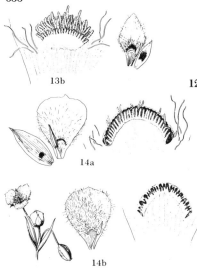

13b

14a

14b

13b Caps reflexed and nodding; basal lf gen much > uppermost fl, the st 0.5–1 (1.5) dm; fls mostly 1–2 (3), greenish-white; petals oblanceolate to narrowly rhombic; open grassy hills to conif for; se Wn to sw Ore and n Cal, e across Ida to w Mont; northwest m., elegant s. or c. *(C. selwayensis, C. e. var. s., C. s. vars. major and minor)*   11 **C. elegans** Pursh

12b Petals cuneate to obovate, narrowed gradually to base (not clawed), < twice as long as broad, yellowish-white to lavender; caps reflexed and nodding

14a Anthers strongly apiculate, the slender tip gen at least 1 mm; st gen unbr; petals yellowish-white, sometimes lavender-tinged; open subalp to alp slopes; Cas, s Wn to c Ore, w to Mt St Helens, Wn; not clearly separable from the next, esp in Wn, where pls often have br sts; mont m. *(C. lobbii, misapplied)*   12 **C. subalpinus** Piper

14b Anthers merely acute to very shortly apiculate, the slender tip < 0.5 mm; st gen br; petals white or yellowish-white to reddish- or purplish-tinged; lowl valleys to subalp, ± for areas; WV s, w Cas, to Cal; Tolmie's m. *(C. elegans var. lobbii, C. purdyi, C. maweanus var. roseus, C. galei)*

13 **C. tolmiei** H. & A.

Camassia Lindl.   Camas

Fls in bracteate racemes, rather showy, reg to irreg, the tepals alike, distinct, narrow, white to deep blue or violet; style slender, stigmas 3, ovary superior; caps ovoid to subglobose; seeds lustrous black, pear-shaped to ovoid-ellipsoid, 2–4 mm; scapose pers from deep-seated bulbs with several linear basal lvs. (From the Indian name, *camas* or *quamash*). *(Quamasia)*. The spp. are easily transplanted and grown in the garden where they make a distinct contribution to the bulb or rock garden. The bulbs are edible and are still harvested and eaten by the Indians.

1a Lvs rarely < 10, mostly 3–5 dm × 1–3.5 (5) cm; bulbs becoming clustered and remaining ± attached as the clumps enlarge, 5–10 × 3–4 (5) cm, ill-smelling, very mucilaginous and foul tasting; scape 4–6 dm, the raceme nearly as long, very densely fld, the fls overlapping; tepals 2–2.5 (3) cm, light blue, 1 declined from the others, all withering separately to the base of the developing caps; steep moist hillsides near the SR, along Pine Cr, Baker Co, and possibly elsewhere (as along upper Imnaha R, Wallowa Co) in Ore; Cusick's c.   1 **C. cusickii** Wats.

1b Lvs gen < 10, mostly 1.5–4 (6) dm × 0.5–2.5 (3) cm; bulbs single, not becoming clustered, 2–4 (5) × 1–2.5 (3) cm, not particularly ill-smelling or mucilaginous and not bad tasting; scape mostly 1–3 dm, the raceme somewhat shorter to much longer (to 6 dm), loosely to closely fld; tepals white to deep bluish-purple

2a Fls reg; tepals twisting together after anthesis and remaining over the ovary, (2) 2.5–3.5 (4) cm, white to deep blue-violet (ours); anthers 4–7 mm; meadows, prairies, and hillsides where moist, at least early in spring; s BC s, w Cas, to sw Ore and Sierran Cal; Leichtlin's c.; the var. *leichtlinii*, from s of our area, is white-fld; our plants are referable to var. *suksdorfii* (Greenm.) Hitchc. *(C. s., C. l. ssp. s.)*

2 **C. leichtlinii** (Baker) Wats.

2b Fls slightly irreg, the lowest tepal curved outward away from the st; tepals 1.5–3.5 (4) cm, light to deep blue, mostly withering separately after anthesis and not covering the ovary, at least as the sp. exists w Cas; scape gen 1.5–3.5 dm, raceme nearly as long, loosely to closely fld; moist areas, often where dry by late spring; BC s, e and w Cas, to Cal, e to sw Alta, Mont, Wyo, and Utah; common camas; several vars.

3 **C. quamash** (Pursh) Greene

a1 Tepals tending to twist together and cover the ovary as they wither, finally being ruptured by the young caps, 3 (5) -nerved; peds 7–20 mm, incurved and erect in fr; e Cas

b1 Tepals 2–3 (4) cm, 3- or 5-nerved; pl gen very robust, up to 7 dm; lvs 1–2.5 cm broad; peds 10–20 mm; anthers gen dark blue; from near SR, s Baker Co, Ore, to Custer Co, Ida, e to YNP and Utah *(C. q. ssp. u.)*   var. **utahensis** (Gould) Hitchc.

b2 Tepals 1.5–2 cm, mostly 3 (5) -nerved; pl gen < 5 dm; lvs 1–1.7

1a

2a

2b

d1

b2

c1

2b

cm broad; peds 7–15 mm; anthers gen yellow; sc and se Wn through e Ore to ne Cal *(C. q.* ssp. *b.)*

var. **breviflora** (Gould) Hitchc.

a2 Tepals mostly spreading and withering separately after anthesis, not covering the base of the ovary; mostly w Cas or n SR in Wn, through n Ida to w Mont

  c1 Peds spreading in fr, curved upward but not at all incurved, the caps held at ca a 45° angle from the rachis; tepals pale bluish-violet, 3–6 mm broad, 5–7 (9) -veined; prairies and grassy clearings, PT, from Pierce Co to Grays Harbor Co, and on both sides OP, Wn *(Q. a., C. q.* ssp. *a.)* var. **azurea** (Heller) Hitchc.

  c2 Peds erect-appressed, or tepals deep violet-blue or < 5-veined

    d1 Tepals 3 (5) -veined, mostly < 30 × 6 mm; peds gen tightly appressed in fr; common e Cas, from Wn and s BC to sw Alta, n Ida, and Mont *(Q. q., Phalangium q., C. esculenta, C. leichtlinii* var. *watsoni, C. q.* f. *pallida, C. teapeae, C. q.* var. *t.)*

var. **quamash**

    d2 Tepals mostly 5–9-veined, 20–35 × often > 6 mm; peds erect to ascending in fr; w Cas

      e1 Fls pale blue; peds erect and appressed in fr; Lane Co to se Ore *(C. q.* ssp. *i.)* var. **intermedia** (Gould) Hitchc.

      e2 Fls deep blue to violet; peds often spreading, esp in the more n part of the range; sw BC through w Wn to se end of WV, Ore *(C. q.* ssp. *m.)* var. **maxima** (Gould) Hitchc.

Clintonia Raf.   Beadlily; Clintonia

Fls solitary, ⚲, showy, white (ours); perianth bell-shaped, the segms alike, distinct, narrow; stamens 6, hypog, flat but slender; style slender, stigma shortly 3-lobed; fr a blue berry; rhizomatous per with 2–3 (ours) or more broad lvs and a scape-like peduncle with 1–several fls. (For DeWitt Clinton, 1769–1828, naturalist and governor of New York).

C. uniflora (Schult.) Kunth. Queen's cup, bride's bonnet, blue-bead. Pl widely rhizomatous; lvs oblong or elliptic to oblong-obovate, rounded to abruptly acute, mostly 7–25 cm and ca 1/3 as broad; peduncle 1 (2) -fld, 1/2–3/4 as long as the lvs; fls ca 2 cm, tepals 7–9 -nerved, nearly twice as long as the ± lanate filaments; anthers 4–5 mm; berry 6–10 mm; conif for, often where rather moist, from the foothills to midmont; Alas to Cal, from the coast inl to sw Alta, Mont, Ida, and e Ore *(Smilacina borealis* var. *u., S. u.).*

Disporum Salisb.   Fairy-bell

Fls solitary on a short ped to 2–several and ± umbellate at the br ends, pendent and often largely concealed by the lvs, tubular to ± narrowly camp, creamy-white (ours); tepals distinct to base, oblong-lanceolate to ± oblong-elliptic; stamens hypog, filaments slender; ovary glab to pubescent or papillate, 3-celled; style slender, stigma very shallowly to fairly deeply 3-lobed; fr a yellow to reddish berry; rhizomatous per herbs with simple or sparingly br, lfy sts and broad, prominently veined, mostly sessile, rounded or cordate, gen oblique-based lvs. (Gr *dis*, double, and *spora*, seed, the seeds often 2 per ovary cell).

  1a Perianth 15–28 mm; tepals oblong-lanceolate, much > the stamens and concealing them, erect, flared only at the tips, narrowed only slightly at the base and completely covering the ovary; ovary ellipsoidal; lvs glab or glabrate, more commonly rounded than cordate at base; stigma lobes 0.5–1 mm; w Cas, BC to Cal, in light to deep woods, mostly where rather moist; fairy lantern, Smith f., large-fld f. *(Uvularia s., D. menziesii)*

1a

1 **D. smithii** (Hook.) Piper

  1b Perianth 8–18 mm; tepals gen < the stamens or if > then not concealing them, oblanceolate or elliptic but narrowed toward the base and flaring, the ovary always partially exposed in pressed specimens; ovary often obovoid; berry sometimes papillate or hairy; style mostly sparsely hairy or glab; stigma sometimes lobed much < 0.5 mm; lvs often hairy, commonly cordate; both sides Cas

    2a Fr 6–15 -seeded, papillose, the papillae visible at anthesis as wrinkles on the upper third of the distinctly rounded, ovoid to obovoid, glab ovary;

2a

2b

style glab; stigma lobed 0.3–1.0 mm; lvs glab on upper surface, the marginal cilia spreading, not pointing forward; wooded slopes, often near streams; e Cas, c BC to Alta, s to Okanogan Co, Wn, and in e Wn to the Blue Mts, ne Ore, s in RM to much of Ida and Mont and to Colo and Ariz, e to Daks and Neb; wartberry f., Sierra f. (*Prosartes t., Lethea t., D. majus*)   2 **D. trachycarpum** (Wats.) Benth. & Hook.

2b Fr 4–6 -seeded, never papillose, often hairy, the ovary ellipsoid or ellipsoid-obovoid and distinctly pointed; style gen hairy; stigma subentire, lobed scarcely 0.3 mm; lvs gen hairy on upper surface, strongly ciliate with forward-pointing hairs; both sides Cas, in wooded, gen moist or shaded areas; BC to nw Ore, e to Alta, wc and n Ida and w Mont; Hooker f.; ours is the var. *oreganum* (Wats.) Jones

3 **D. hookeri** (Torr.) Nicholson

Erythronium L.   Adder's-tongue; Trout-lily; Fawn-lily; Glacier-lily; Dogtooth-violet

Fls solitary or 2–5 in a loose raceme on a naked peduncle, large and showy, nodding, ours white or yellow to deep pink; tepals distinct, spreading at anthesis but eventually reflexed, often closing with age, lanceolate, acute or acuminate, alike or gen the inner set somewhat the broader, both sets or only the inner with 2–4 inflated saclike appendages near the base abutting against the filaments; stamens 6; style slender, gen ca= the tips of the longest anthers; stigma (ours) shallowly to deeply 3-lobed; caps± obovoid to cylindric-clavate; per herbs from deep-seated, elongate corms; lvs gen a basal pair, petiolate to subsessile, often mottled. (Gr *erythro*, red, in reference to the pink or red fls of some spp.). Several spp., but esp *E. revolutum* and *E. oreganum*, thrive in cult.

2a

2b

3a

3b

1a Lvs mottled with irreg patches of pale green on a dark greenish background; tepals pink or white, never yellow in life; filaments much-expanded and (2) 3–4 mm broad somewhat above the base, concealing the ovary and surrounding the style rather closely; w Cas, rarely above 2500 ft elev

2a Tepals white, but ± purplish near the base without and often drying ± pinkish, mostly (3.5) 4–5 cm; anthers 10–12 mm before dehiscence, 5–7 (9) mm afterward; stigma lobes 4–7 mm; caps (3.5) 4–5 cm; moist woods to open gravelly prairies; BC s through PT, but not on w side OP, to CR and through WV to Josephine Co, Ore; giant f. or t. (*E. giganteum* ssp. *leucandrum*)   1 **E. oregonum** Applegate

2b Tepals deep pink, drying to pinkish-purple, mostly 3.5–4 (5) cm; anthers 7–8 mm before dehiscence, ca 4 mm afterward; stigma lobes 2–3 (4) mm; caps 3–4 cm; river banks and light to fairly thick woods; s BC to nw Cal, scattered in Wn, but known from Skagit Co and occ on w side OP, more frequent near the coast in Ore; pink or coast f. or t. (*E. johnsonii, E. r.* var. *j., E. grandiflorum* f. *r.*)   2 **E. revolutum** Smith

1b Lvs not mottled, but uniformly green; tepals yellow or white, but if white the pls either high mont (above 3000 ft elev) or from e Cas; filaments linear, only slightly broadened toward the base where scarcely 1 mm wide, not concealing the ovary and not closely surrounding the style

3a Lf bls gen rather abruptly narrowed and rounded to a distinct petiole mostly 2–4 cm; tepals white (often drying pinkish), broadly lanceolate, much-widened above the base where 7–17 mm broad; anthers yellow; subalp to alp for and meadows, rarely below 3500 ft elev; VI s to OM and to n Ore; alpine f., avalanche lily or f.   3 **E. montanum** Wats.

3b Lf bls narrowed gradually to broad petioles, or subsessile; tepals lanceolate, mostly 4–8 (rarely > 10) mm wide, gen yellow or if white the pl gen from e Cas; anthers white, red, yellow, or purple; sagebr slopes to mont for, sometimes upward to near timberl; rather gen in much of s BC, OM and Cas, Wn, and n Ore, e to Mont, Wyo, and Colo; pale f. or d., yellow f.; 3 vars.   4 **E. grandiflorum** Pursh

a1 Tepals white or cream with a yellow band at base; se Wn and adj Ida (*E. idahoense, E. g.* var. *i., E. i.* f. *tricolor, E. g.* ssp. *c.*)

var. **candidum** (Piper) Abrams

a2 Tepals pale to deep yellow

b1 Stigmas very short, ca 0.5 mm; lvs short-petiolate; s drainage of middle fork of Salmon R, extreme n Custer Co, Ida (*E. n.*)

var. **nudipetalum** (Applegate) Hitchc.

b2 Stigmas 1–2 (3) mm; gen in the range of the sp. as a whole (*E. giganteum, E. g.* f. *grandiflorum, E. nuttallianum, E. g.* var. *parviflorum, E. p.* var. *obtusatum, E. leptopetalum, E. pallidum, E. g.* var. *p., E. g.* ssp. *chrysandrum*)   var. **grandiflorum**

Fritillaria L. Fritillary

Fls solitary or 2–several and loosely racemose, camp, gen nodding, yellow or red to purplish, often mottled; tepals much alike, distinct, gland-bearing near the base; stamens 6, < the tepals; ovary sessile; styles from connate to the tip and the stigma scarcely lobed, to distinct nearly to the base and stigmatic only at the tip of the brs; caps ∝ -seeded, sometimes winged; seeds flat, brownish; per herbs from small bulbs consisting of few fleshy scales and ∝ rice-grainlike offset bulblets; st erect, glab, unbr, with several alt to ± whorled, linear to lanceolate lvs. (L *fritillus*, dice box, in reference to the appearance of the caps). The spp. are rather attractive in the native garden, but are fast disappearing from much of their range.

1a Fls 1 (2–3), yellow but fading to red or purplish; tepals oblong-lanceolate to oblanceolate, 12–26 × 4–10 mm, rounded; lvs gen 2 and subopp or > 2 and alt to semiwhorled, 3–16 × 3–12 mm; style 1, the stigma discoid, scarcely lobed; grassl and sagebr des to pine or mixed conif for; e Cas, BC to n Cal, e to Alta, Mont, Wyo, Utah, and Nev; yellow bell or f. (*Lilium p.*, *F. washingtonensis*, *F. oregonensis*, *F. oreodoxa*)

1 **F. pudica** (Pursh) Spreng.

1b Fls (1) 2–several, purplish, mostly ± spotted or mottled with white or yellow; styles 3, connate only near the base, the stigmas 3

2a Tepals dark greenish-brown to brownish-purple, sometimes streaked or spotted with yellow, but gen not mottled, 20–30 × 7–12 mm, the inner surface prominently ridged over the veins; filaments < twice as long as the anthers; caps not winged; pl ill-smelling; lvs mostly in 1–3 whorls of 5–11 each, narrowly to broadly lanceolate, 4–10 cm × 5–25 mm; moist areas w Cas, from near tideflats to mt meadows, Kodiak I and coastal Alas to Whidbey I and Snohomish Co, Wn; Indian rice, black lily (*Lilium c.*)

2 **F. camschatcensis** (L.) Ker-Gawl

2b Tepals spotted or mottled with white or yellow, 12–35 mm, not prominently ridged over the veins; filaments gen > twice as long as the anthers; caps winged; pl not ill-smelling; lvs scattered and ± imperfectly whorled; both sides Cas

3a Lvs rather narrowly linear, 2–7 mm broad, at least 15 × as long; tepals 12–20 (22) mm, ± rhombic, tapered almost equally from the center; caps 10–15 mm; grassy slopes to conif for or rocky mont ridges; se Ore to Sierran Cal, e across c and s Ida to the Daks and Neb, Colo, and NM; checker lily, chocolate 1.

3 **F. atropurpurea** Nutt.

3b Lvs linear to lanceolate or ovate-lanceolate, 3–25 mm broad, gen < 10× as long; tepals 20–30 (35) mm; prairies and grassy bluffs to woodl and conif for from near sea level to above 5000 ft elev; BC to Cal, both sides Cas in Wn, but w Cas in Ore, e in BC and Wn to n Ida; checker lily, rice-root f., mission bells (*F. lunellii*)

4 **F. lanceolata** Pursh

Leucocrinum Nutt.

Fls white, rather showy, borne in clusters; perianth persistent; stamens 6; style 1, persistent, stigmatic lobes short; caps subterranean, obovoid, cartilaginous and ± corrugated; seeds black, lustrous; stless per with short, deeply buried rhizome and ∝ lvs. (Gr *leuco*, white, and *krinon*, lily).

L. montanum Nutt. Sand lily, star lily, mt lily, wild tuberose, star-of-Bethlehem. Lvs tufted, linear, up to 20 cm × 2–8 mm, ± whitish-membranous margined and sheathing at base; peds not reaching ground level; fls fragrant, the tube mostly (4) 5–8 (10) cm, the free segms linear-lanceolate to narrowly oblong-elliptic, 20–25 × up to 7 mm; anthers 4–6 mm; caps 5–7 mm; sagebr des to open mont for, in sandy to rocky areas or in fairly heavy soil; Jefferson Co, Ore, to Cal, e across Ida to Mont and SD, and s to NM.

Lilium L. Lily

Fls large, showy, funnelf to somewhat camp, solitary or gen 2–several in a loose, terminal, often lfy-bracteate raceme; tepals 6, ours white or pink to orange or brick red and often spotted with red, brown or purple, distinct, spreading to strongly recurved; stamens 6, hypog; style 1, short-lobed; caps ∝ -seeded; glab per herbs from fleshy-scaled bulbs or short scaly rhizomes, with unbr sts and ∝ alt to whorled, gen lanceolate lvs. (L form of *leirion,* the classical Gr name). The lilies are a constant temptation to the avaricious gardener, but should be left strictly alone in the wild, as our spp. do not survive transplantation satisfactorily.

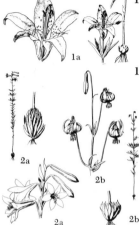

1a Fls 1–3, stiffly erect; tepals orange-red to brick red, 5–7 × 1.2–2.2 cm, narrowly clawed at base and not concealing the ovary, spreading but not reflexed; heavy, often ± alkaline meadowl to mont for, esp in aspen groves; BC s along e edge RM to NM, e to Sask, Mich, and Ohio; red l., wood l., Philadelphia l. *(L. umbellatum, L. andinum, L. p. f. or ssp. a., L. montanum);* ours is var. *andinum* (Nutt.) Ker-Gawl   1 L. **philadelphicum** L.

1b Fls often several and loosely racemose, spreading to pendent; tepals either white or strongly reflexed, not clawed, concealing the ovary at base; w RM

2a Tepals white or pinkish, fading to deep pink or purplish, not strongly reflexed, 6–9 cm; in chaparral or open for e side Mt Hood s through Cas to sw Ore and n and Sierran Cal, up to ca 5000 ft in our area; Washington l., Shasta l. *(L. w.* var. *purpureum, L. w.* var. *purpurascens)*
2 L. **washingtonianum** Kell.

2b Tepals yellow-orange to reddish-orange, strongly reflexed, 3–6 cm; prairies, thickets, and conif for; up to 6000 ft in drier mts; BC s, both sides Cas, to n Cal, e to n Ida and Nev; tiger l., Ore l., Columbia l. *(L. parviflorum, L. canadense* var. *walkeri, L. bakeri, L. purdyi)*
3 L. **columbianum** Hanson

Lloydia Salisb.   Lloydia

Fls solitary to few in a loose, terminal, lfy-bracteate raceme; tepals distinct, alike, white with greenish or purplish veins, transversely corrugate, with a small gland near the base; stamens 6, inserted on base of the tepals, very slightly perig; ovary 3-celled; style 1, short, persistent, short-lobed; caps globose-obovoid, ∝-seeded; per herbs with linear lvs. (For Edward Lloyd, 1660–1709).

L. serotina (L.) Sweet. Alpine lily. Pl with short thick rhizome, the erect sts and basal lvs of each year surrounded by brownish, sheathing, persistent bracts that wither away near ground level and simulate a tunicated bulb; basal lvs 4–10 (20) cm × ca 1 mm; fl sts 5–15 cm, the 2–4 lvs 1–2 mm broad; fls mostly 1 (2); perianth turbinate; tepals white with greenish or purplish veins or tinge, 9–12 mm; stamens ca = style; caps 6–8 mm; gravelly ridges, cliffs, and rock crevices; Alas to nw Ore, in both OM and Cas, e to Alta and s to NM and Nev; Eurasia *(Bulbocodium s., Anthericum s.).*

Maianthemum Web.

Fls small, white, racemose; tepals 4, alike, distinct; stamens 4, borne at base of the tepals; pistil 2-carpellary; style 1, bilobed; fr a 1–4-seeded reddish berry; rhizomatous per herbs with erect, lfy, unbr fl sts. (Gr *maios,* May, and *anthemon,* flower, referring to the time of blossoming).

M. dilatatum (Wood) Nels. & Macbr. Beadruby, deerberry, may-lily, false lily-of-the-valley. Sts 1–3.5 (4.5) dm, erect; lvs commonly 2 (1–3), the uppermost often much-reduced or bractlike, cordate to sagittate, 5–11 cm, ca as broad; racemes 3–6 cm; tepals ca 2.5 mm; berry red, globose, 5–6 mm; shaded or moist streambanks and open to dense woods where at all moist, from sea level up to ca 3500 ft elev; Alas to Cal, from the coast inl, e to c BC and n Ida *(M. bifolium* var. *d., Unifolium d., M. b.* var. *kamtschaticum);* attractive in the wild but apt to prove a nasty pest in the native garden.

Scoliopus Torr.   Fetid Adder's-tongue

Fls 1–few in a nearly sessile umbel on long slender peds, the peduncle entirely subterranean; perianth grayish-yellow, mottled and heavily lined with purple or chocolate, the segms distinct, the 2 series dissimilar, the sepals much the broader and spreading, the petals erect; stamens 3, opp sepals; styles connate ca half length; caps thin-walled; short-rhizomatous per herbs with gen 2 ± mottled basal lvs, the peduncle of the umbel subterranean and the peds thus pseudoscapose. (Gr *skolios,* crooked, and *pous,* foot, referring to the recurving peds).

S. hallii Wats. Slink lily, slink pod, Ore f. St subterranean, bracteate at base, 2–6 cm; lvs oblong-elliptic to oblong-oblanceolate, abruptly acute, 6–20 cm, ± purplish-mottled, the petioles partly subterranean; peds gen 2–4, up to 8 cm at anthesis, then soon lengthening and recurving; fls foetid; perianth scarcely 1 cm, grayish-yellow with purplish veins and mottling; sepals 3–4 mm broad, spreading and slightly recurved; petals ca 0.5 mm broad, erect and arched over the pistil; anthers purple; caps ellipsoid, ca 1.5 cm, gen brought in contact with the soil by the recurving of the peds and opening more by decay than dehiscence; damp woods, esp along streams; w slope Cas and coastal mts of Ore, from Tillamook Co s almost to Cal.

Smilacina Desf.   Solomon-plume; False Solomon's Seal

Fls racemose to pan, small, white; peds jointed just below the fls; tepals distinct, alike; stamens 6, filaments < to > and often broader than the tepals; style 1, short, obscurely 3-lobed; fr a few-seeded, globose, greenish to red berry; widely rhizomatous per herbs with erect, lfy, unbr fl sts, the lvs ∝. (Diminutive of *Smilax*). Our spp. have also been placed in *Convallaria, Unifolium,* and *Vagnera*.

1a Fls ∝, pan; tepals narrowly oblong, 1.5 (2) mm; peds 0.5–2 mm; filaments ovate-lanceolate and much broader and > the tepals; moist woods and streambanks to open for, from near sea level to medium elev in the mts; Alas to Cal, both sides Cas, e to NS and s to Ga, Miss, Mo, Colo, and Ariz; false spikenard, western s. *(S. amplexicaulis, S. r. var. a., S. r. var. brachystyla, V. brachypetala)*   1 **S. racemosa** (L.) Desf.
1b Fls 5–10 (20), gen in a simple raceme or rarely the raceme compound near the base; tepals narrowly oblong or lanceolate, (3) 4–7 mm; peds 3–20 mm; filaments slender, < the tepals; moist woods and streambanks to rocky, well-drained, often fully exposed sidehills; Alas to Cal, on both sides Cas, e to Atl coast, s in RM to Colo, Ariz, and Nev; starry s., star-fld s. *(S. sessilifolia, S. s. f. paniculata)*   2 **S. stellata** (L.) Desf.

Stenanthium (Gray) Kunth   Stenanthium

Fls racemose-pan, ⚥ (ours), perig; tepals alike, slightly connate at base; filaments free or very slightly connate basally; ovary hollow well above the ovuliferous portion, the styles almost distinct, persistent; fr a membranous, slenderly 3-beaked caps, in growth becoming almost superior (free of the short hypan); seeds 3–4 in each cell, the testa loose and forming an elongate wing at each end; per herbs from small tunicated bulbs, main lvs 2–3, basal, narrowly lanceolate; st gen with 2–3 much reduced bracts. (Gr *stenos*, narrow, and *anthos*, fl, in reference to the narrow tepals).

S. occidentale Gray. Western s. Bulb 2–4 cm; basal lvs 10–30 × 0.3–3.2 cm; pl 1–4 (5) dm, the st with gen 1–2 lanceolate bracts below; fls 3–25, racemose to pan, peds 1–3 cm; fls tubular-camp, 8–15 mm; tepals pale greenish-yellow to deep purplish-green, oblong-lanceolate with spreading acuminate tips; caps up to 2 cm; wet cliffs, rock crevices, and moist mont thickets, meadows, and scree; mostly alp or subalp, but down to near sea level on OP and in CRG; BC to n Cal, e to Alta, Mont, and Ida.

Streptopus Michx.   Twisted-stalk

Fls borne beneath the lvs, 1 (2) per peduncle, the peduncle slender, gen adnate to the st for 1 internode and apparently arising opp the lf axil above that in which it had its origin; perianth camp to saucer-shaped; tepals rather narrow, distinct, erect to spreading or recurved, outer series gen flat, the inner series often ± keeled, slightly the narrower; stamens 6, hypog, filaments broad and flat, anthers apiculate to setose-aristate; ovary 3-celled; fr a greenish to red berry; widely rhizomatous per herbs with simple to several-brd sts and ∝ ovate to oblong-lanceolate st lvs with mostly clasping bases and acute or acuminate tips. (Gr *streptos*, twisted, and *pous*, foot, in reference to the bent or twisted peduncles).

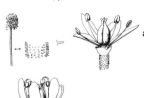

1a Perianth saucer-shaped, greenish, often purple-tinged; tepals spreading, 3–4 mm; anthers minutely apiculate; style bulbous, < 1 mm, unbr; st simple, gen < 2 dm; lvs 3–5.5 cm and scarcely half so broad, acute or acuminate; dense, conif midmont woods in our area; Alas to OM and Cas, Wn, e to e BC and n Ida; Kruhsea; ours the var. *brevipes* (Baker) Fassett, lacking the ciliate lvs of the Asiatic var. *streptopoides (Kruhsea s., Smilacina s., St. brevipes, St. s. var. b., St. s. ssp. b.)* 1 **S. streptopoides** (Ledeb.) Frye & Rigg
1b Perianth camp, tepals nearly erect, only the tips spreading, 6–15 mm; anthers long-setose; style stout, elongate, at least 1 mm, sometimes br; st often br, mostly > 2 dm
   2a Anthers ca 2 mm, subequal to the filaments; tepals only slightly spreading, 6–10 mm, white or greenish-yellow with reddish-purple streaks or spots to rose with white tips; style short-brd near tip; st simple or occ br, 1.5–3 (4) dm; peduncles curved but rarely geniculate, pubescent; streambanks and damp mont woods, mostly at 3000–6000 ft elev in our area; Alas s through Cas and OM to n Ore, e to se BC; rosy t., simple-std t., sessile-lvd t.; ours the var. *curvipes* (Vail) Fassett, differing only slightly from the more e var. *roseus (S. curvipes)* 2 **S. roseus** Michx.
   2b Anthers ca 3–3.5 mm, the outer series 3–4 × as long as the filaments; tepals spreading-reflexed, 9–15 mm, white but strongly greenish-tinged and ± yellow-green; styles connate to the tip, not brd; sts freely brd, mostly 5–12 dm; peduncles gland-bearing near midlength where sharply geniculate, glab; moist for, streambanks, and mt thickets; Alas to Cal, e through much of Can and US; clasping-lvd t., cucumber-root; 2 vars.
3 **S. amplexifolius** (L.) DC.
      a1 Sts coarsely pubescent below the 1st brs; lvs mostly with minute marginal teeth (these prob more pronounced in e than in w material); Alas to Cal, from Cas to coast in Wn and Ore, the only phase in Cal, e in n Can to GL region and most of e US *(S. a.* var. *americanus, S. a.* var. *grandiflorus)* var. **americanus** Schult.
      a2 Sts glab; lf margins entire; Alas to n Ore, mostly e Cas in Wn and Ore, e to sw Alta and s in RM through Ida and Mont to Colo and NM var. **chalazatus** Fassett

Tofieldia Huds.   Tofieldia

Fls small, in small terminal racemes; tepals white to greenish-white, distinct, erect or ± spreading; stamens 6, hypog; styles 3, distinct, short, subulate; fr a membranous caps apiculate by the persistent style; seeds ∝, often with loose spongy testa; per, short-rhizomatous herbs with ∝ tufted but ± 2-ranked linear basal lvs and gen 1–3 sheathing lvs on the lower 1/4–1/2 of the scape. (For Thomas Tofield, a British botanist, 1730–1799).

T. glutinosa (Michx.) Pers. Sticky t. St 1–5 (8) dm, glab or ± glandular-hairy below, but always copiously glandular-hairy above and in the infl; lvs mostly in a basal tuft, linear, 5–15 (20) cm × 3–8 mm; raceme subcapitate and 1–2 cm at anthesis, elongating to 3–7 cm in fr; peds mostly in 3's, 1–6 (10) mm, bearing at the tip 3 separate or gen almost completely connate invol bracts; tepals white to ± greenish, (2.5) 3–5.5 mm, oblong obovate, the inner set somewhat the narrower and longer; stamens = or slightly > the tepals, anthers scarcely 1 mm; caps 4–9 mm; testa spongy, grown to body of the seed except at the ends, or inflated and free of the seed; meadows, bogs, and streambanks to alp ridges; Alas to Cal, e to Atl coast and s to Wyo and NC; 3 vars. in our area
   a1 Hairs of upper st and infl relatively slender, 3–4 × as long as thick, nearly uniform in diameter; outer tepals ca 3.5 mm, the inner ca 4 mm; s Can RM to Mont, most of Ida, and Wyo (*T. g.* ssp. *m.*)   var. **montana** (Hitchc.) Davis
   a2 Hairs of upper st and infl relatively thick, mostly ca 2 (3) times as long as thick, tapered and semi-papillose; outer tepals ca 3.8 mm, the inner ca 4.2 mm
      b1 Invol bracts (just below the fls) only partially connate, often free to the base; appendages of seeds lacking or < half as long as the body of the seed; tepals mostly retuse; known only from near Priest Lake, Ida *(T. intermedia* ssp. *a., T. g.* ssp. *a.)* var. **absona** (Hitchc.) Davis
      b2 Invol bracts connate at least half their length; appendages of seeds gen at least half as long as the body; tepals rarely at all retuse; s Alas, VI and mainl BC, s in OM and Cas to s Ore, e to Selkirk Mts in se BC *(T. g.* ssp. *b.)* var. **brevistyla** (Hitchc.) Hitchc.

Trillium L.   Wake-robin; Trillium

Fls normally solitary, sessile or pedunculate, terminal and subtended by a whorl of gen 3 large lvs at the top of an otherwise naked st; sepals 3, persistent, distinct, greenish; petals 3 (4 or even 5), withering, white, pinkish, yellow, or purple, distinct; stamens 6; ovary 3-celled; fr a $\propto$-seeded, berry-like caps; glab per herbs with a short, thick, horizontal to erect rhizome and 1–more erect fl sts, the lvs mostly broadly ovate. (L name, supposedly from *tres*, three, in reference to the lvs). The trilliums are rather easily grown, and *T. ovatum* in particular is an excellent garden pl, but it is a shame to dig them in the wild, esp since they grow readily from seed.

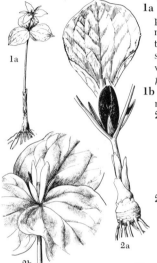

1a Fl borne on a peduncle (1) 2–8 cm; sepals 1.5–6 cm; petals white or pinkish (aging to pink or red), = or > the sepals, up to 3 cm broad; lvs mostly acute to acuminate, sessile, not mottled, 5–15 (20) cm; streambanks to open or thick woods, often where boggy in the spring, lowl to mont; BC s, from coast inl, to c Cal, e to sw Alta, Mont, Wyo, and Colo; white t., western w. or t. (*T. crassifolium, T. scouleri, T. venosum, T. o.* var. *stenosepalum*)   1 **T. ovatum** Pursh

1b Fl sessile; petals yellowish to purple; lvs mostly either long-petiolate or mottled, mostly rounded to obtuse

   2a Lvs long-petiolate, bl 8–13 cm, not mottled; st largely underground, 0.5–1.8 cm; sepals 2.5–6 (7) cm; petals gen purple (green or yellow to red or brownish), gen ca same width and length as the sepals; streambanks and moist thickets or meadows; e side Cas, Chelan Co, where below 2500 ft elev, to Spokane Co, Wn, and adj Ida, s to Blue Mts, Ore, and w in n Ore to Grant, Gilliam, and Morrow cos; purple t., petioled w.   2 **T. petiolatum** Pursh

   2b Lvs sessile, 6–12 (18) cm, mottled; st 1–3 dm; sepals 1.5–4 (5.5) cm; petals white to pink or rather deep purplish except at the whitish base, 2.5–7 cm, gen much broader and > the sepals; moist woods w Cas, Pierce Co, Wn, and WV, Ore, to Cal; giant t. or w. (*T. sessile* var. *c., T. s.* var. *californicum*)   3 **T. chloropetalum** (Torr.) Howell

Veratrum L.   False Hellebore; Skunk-cabbage

Fls in large pan or compound raceme, mostly polygamous, the upper ones ⚥ and the lower ♂; perianth white or yellowish to greenish, broadly bell- to saucer-shaped; tepals distinct, sometimes clawed; stamens 6, perig, a short hypan fused with the base of the ovary; ovary 3-celled; styles distinct, short; caps tipped by the persistent style, very slightly inferior; seeds $\propto$, flat, broadly winged by the loose testa; tall per herbs from thick rhizomes, the st simple, with $\propto$ broad, coarsely veined lvs. (L name for hellebore, said to be derived from *vere*, true, and *ater*, black, because of the black roots).

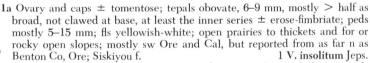

1a Ovary and caps ± tomentose; tepals obovate, 6–9 mm, mostly > half as broad, not clawed at base, at least the inner series ± erose-fimbriate; peds mostly 5–15 mm; fls yellowish-white; open prairies to thickets and for or rocky open slopes; mostly sw Ore and Cal, but reported from as far n as Benton Co, Ore; Siskiyou f.   1 **V. insolitum** Jeps.

1b Ovary and caps glab or only sparsely pubescent with straight hairs; tepals mostly lanceolate to narrowly oblanceolate, 6–17 mm, broadly clawed, mostly entire; peds rarely > 6 mm; fls sordid white to rather deep green

   2a Pan densely fld, the brs crowded, spreading to ascending, not drooping; tepals white or slightly greenish-tinged except at base where deeper greenish, (8) 10–17 mm, mostly entire; swamps, creek bottoms, meadows, and moist woodl, lowl to subalp; w Wn to s Cal, e to Ida and Mont and s to Mex; Cal f.; 2 vars.   2 **V. californicum** Durand

     a1 Pan compound nearly or quite to the top; tepals mostly broadly elliptic to ± obovate, mostly (8) 10–15 mm; range of the sp. except not w Cas in Wn or Ore (*V. speciosum, V. jonesii*)   var. **californicum**

     a2 Pan unbr above, the upper 1/3–1/2 an erect, densely-fld raceme; tepals mostly lanceolate, 12–17 mm; Cas and w from PT and OP s to sw Ore, only very occ e Cas (*V. c.*)   var. **caudatum** (Heller) Hitchc.

   2b Pan loosely fld, the brs rather distant, lower ones drooping; tepals yellow-green to green, 6–10 mm, at least the inner series ± finely serrulate; wet thickets, swamps, and lowl to mont meadows; Alas to n Cas of Ore, through Cas and OM, e to RM, BC and Alta to Mont and Ida, and in Can to Ont and Que and s to NC; Am f., Indian-poke, green f. (*V. eschscholtzii, V. v.* var. *e., V. v.* ssp. *e., V. v.* var. *escholtzianoides*)   3 **V. viride** Ait.

Xerophyllum Michx.   Beargrass

Fls long-ped in a terminal raceme, cream-colored; tepals distinct, spreading, glandless, persistent; stamens 6, hypog; ovary 3-celled; styles elongate, distinct; fr caps; per herbs with short thick rhizome, any particular offshoot pl ultimately fl once and then dying, the fl stalk erect, unbr, strongly bracteate. (Gr *xeros*, dry, and *phyllon*, leaf, the lvs being dry and tough).

X. tenax (Pursh) Nutt. Indian basket-grass, w turkey-beard. Rhizome 1–2 cm thick, basal lvs in large clumps, tough and wiry, strongly scab, 1.5–6 (8) dm × 1.5–4 (6) mm; fl st up to 15 cm, copiously lfy, the lvs reduced upward, the infl at first corymb-like, but ultimately elongating to as much as 5 dm; peds slender, 2.5–5 cm; perianth semirotate, the tepals oblong, mostly (5) 6–8 (10) mm; stamens = or > the perianth; styles proper scarcely 1 mm, the brs ca 3 (4) mm, stigmatic the full length; caps 5–7 mm; open woods and clearings; BC to Cal, e to RM from BC to Ida and Mont, from near sea level on OP up to over 7000 ft in the RM (*X. douglasii*).

Yucca L.   Yucca

Fls in large terminal pan or raceme, mostly drooping, large and showy; perianth camp, gen cream or white; tepals alike, distinct, withering-persistent; stamens 6, hypog; ovary 3-celled; stigmas short, 2-lobed; fr caps; coarse pls with stout, upright, short to rather large, simple or br, semi-woody caudex and ∞ persistent, linear lvs; infl large, terminal on an ann, bracteate peduncle. (Said to be an Indian or Haitian name for manihot, an entirely different pl).

Y. glauca Nutt. Soapwell, beargrass. St short, often prostrate; lvs stiff, 2–6 dm × 6–12 mm, sharp-pointed, the white margins shredding into stiff fibers; fl st 1.5–10 dm, the infl up to 1 m; fls greenish-white, 4–5 cm; caps woody, 5–7 cm; seeds black; prairies and light woodl into the lower mts, almost entirely e RM; Mont to Tex and Ariz, e to Daks, Ia, and Mo, barely in our area in c Mont (*Y. angustifolia*).

Zigadenus Michx.   Zigadenus; Zygadene; Death-camas

Fls racemose or pan, white to yellow-green, subrotate to shallowly camp, withering-persistent; tepals ovate to oblong-lanceolate, alike, or the outer series the shorter, each with a yellow or greenish gland near the often ± clawed base; stamens 6, hypog to perig, = or > the perianth; ovary superior to ca 1/3 inferior; styles 3, distinct; caps several–∞-seeded; per herbs with tunicated bulbs, the sts simple, lfy at base and with reduced lvs or membranous bracts above; lvs linear, mostly basal, gen glab and often glaucous. (Gr *zugon*, yoke, and *aden*, gland, in reference to the paired or yoked glands of the type sp.). (*Toxicoscordion*). The genus includes some of our most virulent stock-poisoning pls.

1a Tepals 8–11 mm, not clawed, the gland obcordate; stamens perig, the ovary 1/4–1/3 inferior; infl 3–several -fld, racemose to freely br and pan, sometimes > 3 dm, the peds gen 10–35 mm, ascending; for, meadows, and rocky slopes to open grassl, mostly mont to alp; Alas and BC to Cas and OM, Wn, through e Ore to Nev and Ariz, e to Alta, Mont, and Daks and s to Colo, Tex, NM, and n Mex; glaucous z. (*Anticlea e., Z. chloranthus, Z. longus, Z. alpinus*)   **1 Z. elegans** Pursh

1b Tepals < 7 mm, at least the inner series clawed, the gland broadly ovate; stamens hypog, the ovary superior

2a Infl nearly always pan; fls often polygamous, those of the main axis gen ⚲, those of the lower pan brs often functionally ♂; outer tepals scarcely clawed, rather broadly ovate-triangular, 3–4.5 mm, acute to acuminate; stamens mostly 1–2 mm > the tepals; e Cas, sagebr des to ponderosa pine or lodgepole pine for; Chelan Co, Wn, to Sierran Cal, e to c Wn, ne Ore, and through c and s Ida to sw and c Mont and to w Wyo, Utah, w Colo, n Ariz, and nw NM; panicled d. (*Helonias p.*)

2 Z. paniculatus (Nutt.) Wats.

2b Infl gen racemose, but occ br below; fls all functionally ⚲; tepals often with a short (0.3–1 mm) claw, mostly ovate-lanceolate, (4) 4.5–5 mm, obtuse to acute; stamens mostly ca = the tepals, rarely up to 1 mm >; both sides Cas, from coastal "prairies" and rocky bluffs to grassy hill-

sides, sagebr slopes, and mont for in exposed places; s BC to Baja Cal, e to Alta and sw Sask, Daks, Neb, and Colo; deadly z., meadow d.; 2 vars. in our area   3 **Z. venenosus** Wats.

a1 Upper st lvs, except the greatly reduced bracts of the infl, all sheathing; outer tepals almost clawless, the claw scarcely 0.5 mm; raceme sometimes compound; sc BC s, along the e base of the Cas, to nc Ore, e to Alta and Sask, and s in RM to Colo; the common var. in e Wn, n Ida, and Mont (*Z. g., Z. intermedium*)   **var. gramineus** (Rydb.) Walsh

a2 Upper st lvs not sheathing; outer tepals mostly with claw 0.5–1 mm; raceme rarely compound; sw BC to Baja Cal, in our area the only var. w of the Cas, where along the coast and s through PT to w Ore, also along e base of Cas in Wn and through e Ore to s Ida, Nev, and Utah (*Z. salinum*)   **var. venenosus**

IRIDACEAE   Iris Family

Fls umbellate to racemose or pan, bracteate, ☿, gen showy, reg (ours); perianth segms 6, connate and similar, or distinct and often dissimilar; stamens 3, gen connate at least basally; ovary inferior, 3-celled; style brs 3, often flattened, stigmas sometimes 2-cleft; fr a loculicidal caps; per herbs, gen with rhizomes but sometimes with bulbs, the lvs linear to ensiform, 2-ranked and ± equitant, the fl st lfless or lfy.

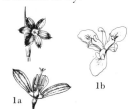

1a Perianth segms essentially similar, < 3 cm; style brs slender, not at all petaloid   **Sisyrinchium**
1b Perianth segms dissimilar, the outer more showy, gen spreading to reflexed, much > 3 cm; style brs petaloid   **Iris**

Iris L.   Iris; Flag; Fleur-de-lis

Fls 1–several; perianth segms fused at base, forming a short to much-elongate, slender to ± flaring tube; sepals spreading and gen reflexed, strongly pencilled with brown to purple; petals erect, gen narrower than the sepals; style brs often petaloid, opp the sepals and gen curved over them, with 2 terminal lobes (crests), the stigma gen a short triangular to broad flap projecting on the lower side at the juncture of the style br and the crests; stamens opp the sepals; caps fusiform to subglobose, gen coriaceous; per herbs with linear, chiefly basal lvs and lfy to naked fl sts. (Gr name for the genus, from *Iris*, the rainbow). Most spp. are valuable garden subjects.

1a Fls pale to bright yellow; pls gen ca 1 m, growing in wet places; European sp., intro and well estab around lakes and ponds and along streambanks; yellow f.   1 **I. pseudacorus** L.
1b Fls gen white or blue, occ yellow but then the pls much < 1 m and growing in dry places
2a Perianth tube at least 4 (to 10) cm; fls off-white to pale yellow or ± bluish-tinged; open woods, mostly in ponderosa pine and Douglas fir; Polk and Marion cos, Ore, to nw Cal; slender-tubed i. (*I. tenax* spp. c.)
                                                    2 **I. chrysophylla** Howell
2b Perianth tube < 1.5 cm; fls white to blue
3a Pls with thick, widespreading rhizomes; fl sts gen at least 4 dm, lfless or with a single lf; fls pale to fairly deep blue; moist meadows and streambanks, often where becoming very dry in midsummer; the common sp. e Cas, esp in sagebr des to ponderosa pine for; w Cas only on islands of PS, BC to s Cal, e to Daks and s to n Mex; w blue f., RM f.   3 **I. missouriensis** Nutt.
3b Pls with slender, but seldom extensive rhizomes; fl sts rarely if ever as much as 4 dm, gen with 2–3 lvs; pls exclusively w Cas
4a Lvs rarely > 5 mm broad; fl st unbr; caps 25–35 mm; fls mostly lavender-blue to purple, but also occ white to yellow, pink, or orchid; our common sp. in w Wn and nw Ore; Ore i. (*I. gormani, I. t.* ssp. *g.*)   4 **I. tenax** Dougl.
4b Lvs lax, all gen 7–18 mm broad; st br in the infl; fls white with purple venation and gen some yellow or purplish blotching; caps scarcely 15 mm; known only from moist shaded streambanks in Clackamas Co, Ore; Clackamas i.   5 **I. tenuis** Wats.

Sisyrinchium L.   Sisyrinchium

Fls 1–few, umbellate, subtended by a 2-lvd spathe; perianth rotate to ± camp, the segms much alike, slightly connate at base, yellow, blue, or pink to rose-purple, rarely white; filaments ± connate, sometimes to near the tip; style brs slender; caps with several subglobose seeds per cell; pl sometimes short-rhizomatous, lvs gen equitant and sheathing at base, bls flattened and grasslike to semi-terete; sts simple to rarely brd, flattened and often winged to semi-terete, lfy at least at the base. (Name used by Theophrastus for an *Iris*-like pl.)

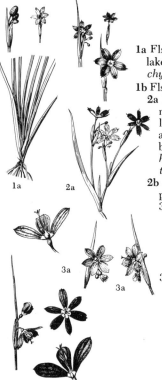

1a Fls yellow; pls blackening when dried; wet places, esp at edge of bogs or lakes; VI to s Cal, w Cas in our area; golden-eyed grass (*S. borealis, S. brachypus*)   1 S. **californicum** (Ker-Gawl.) Dryand.

1b Fls white, blue, or reddish to purple; pls not blackening when dried

  2a Filaments connate to near the tip; perianth more bluish than reddish; marshes and ditches, always where moist in the spring, from near sea level to mont; s Alas to Baja Cal, e across Can, and to e side RM in Mont and Wyo and s; extremely variable and often regarded as several spp.; blue-eyed grass, blue star, eye-bright (*S. bellum, S. birameum, S. idahoense* var. *b., S. i., S. sarmentosum, S. occidentale, S. macounii, S. segetum, S. septentrionale*)   2 S. **angustifolium** Mill.

  2b Filaments connate only 1/3–1/2 their length; perianth (white) reddish-purple

    3a Filament tube inflated just above the base; perianth mostly light purple, shallowly and broadly camp at base; grassy areas in sagebr-juniper des to open parks in woodl, gen where moist in early spring but dry during the summer, chiefly e Cas, se BC to Cal, e to Ida and n Utah; purple-eyed grass, grass-widows (*S. inalatum, S. i.* f. *alba*)   3 S. **inflatum** (Suksd.) St. John

    3b Filament tube only slightly enlarged above the base; perianth deep reddish-purple, broadly and shallowly conic at base; prairies to rocky slopes, in sagebr to oak or pine woodl, often apparently in very dry places, but where moist in early spring; from near sea level to above 6000 ft elev; VI s, w Cas, to Cal, up CRG and along e base Cas from Wasco Co, Ore, to n Kittitas Co, Wn; grass-widows (*S. grandiflorum*, misapplied)   4 S. **douglasii** A. Dietr.

ORCHIDACEAE   Orchid Family

Fls complete (rarely ♂♀), solitary to pan, gen showy, strongly irreg, epig; perianth segms 6 (sometimes 2 or more fused), the 2 series similar, or more often the outer greenish or otherwise dissimilar to the inner in size or coloration, in orientation (because the ovary and ped twist a half turn in development) 1 sepal, gen the largest, becomes uppermost and the other 2 lateral; corolla gen highly colored but sometimes dull white to brownish or pale to deep green, the 2 upper petals often connate with the top sepal and forming a slight to prominent hood, the 3rd and lowermost (the lip or labellum), gen larger and otherwise unlike the other perianth segms, often grooved or concave and forming an orifice leading to a basal nectary or into a sac or spur; stamens ± completely fused with the style and stigmas to form a short to prominent structure (the column); fertile anthers gen 1, but occ (*Cypripedium*) 2, attached on the back or at the tip of the column, pollen sacs 2, often separated by a broad connective, the pollen of each sac adhering in 1–4 masses (pollinia); pollinia often with a stalklike base (caudicle); stigmas 3, gen forming a sticky concave to convex surface on the lower side and near the tip of the column; ovary inferior, 1-celled with 3 parietal placentae (or 3-celled); fr gen a dry caps; seeds ∝, minute; per herbs (ours) to vines or semishrubs, ours all terrestrial, sometimes saprophytic and without green lvs, but mostly with 1–several linear to orbicular, often sheathing lvs, frequently rhizomatous or with fleshy, enlarged clusters of roots. Our spp. should be left strictly alone in their native habitat in the hope that they will somehow be preserved for others to see and enjoy. They do not do well in cultivation, and will persist in the garden a few seasons at most.

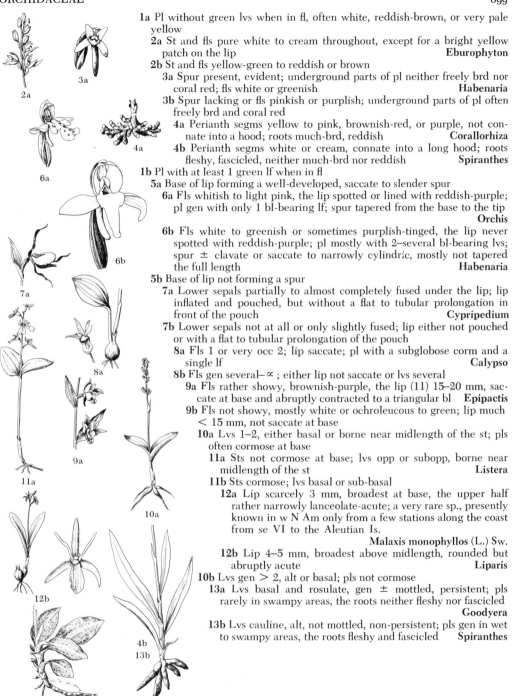

1a Pl without green lvs when in fl, often white, reddish-brown, or very pale yellow
  2a St and fls pure white to cream throughout, except for a bright yellow patch on the lip   **Eburophyton**
  2b St and fls yellow-green to reddish or brown
    3a Spur present, evident; underground parts of pl neither freely brd nor coral red; fls white or greenish   **Habenaria**
    3b Spur lacking or fls pinkish or purplish; underground parts of pl often freely brd and coral red
      4a Perianth segms yellow to pink, brownish-red, or purple, not connate into a hood; roots much-brd, reddish   **Corallorhiza**
      4b Perianth segms white or cream, connate into a long hood; roots fleshy, fascicled, neither much-brd nor reddish   **Spiranthes**
1b Pl with at least 1 green lf when in fl
  5a Base of lip forming a well-developed, saccate to slender spur
    6a Fls whitish to light pink, the lip spotted or lined with reddish-purple; pl gen with only 1 bl-bearing lf; spur tapered from the base to the tip   **Orchis**
    6b Fls white to greenish or sometimes purplish-tinged, the lip never spotted with reddish-purple; pl mostly with 2–several bl-bearing lvs; spur ± clavate or saccate to narrowly cylindric, mostly not tapered the full length   **Habenaria**
  5b Base of lip not forming a spur
    7a Lower sepals partially to almost completely fused under the lip; lip inflated and pouched, but without a flat to tubular prolongation in front of the pouch   **Cypripedium**
    7b Lower sepals not at all or only slightly fused; lip either not pouched or with a flat to tubular prolongation of the pouch
      8a Fls 1 or very occ 2; lip saccate; pl with a subglobose corm and a single lf   **Calypso**
      8b Fls gen several–∝ ; either lip not saccate or lvs several
        9a Fls rather showy, brownish-purple, the lip (11) 15–20 mm, saccate at base and abruptly contracted to a triangular bl   **Epipactis**
        9b Fls not showy, mostly white or ochroleucous to green; lip much < 15 mm, not saccate at base
          10a Lvs 1–2, either basal or borne near midlength of the st; pls often cormose at base
            11a Sts not cormose at base; lvs opp or subopp, borne near midlength of the st   **Listera**
            11b Sts cormose; lvs basal or sub-basal
              12a Lip scarcely 3 mm, broadest at base, the upper half rather narrowly lanceolate-acute; a very rare sp., presently known in w N Am only from a few stations along the coast from se VI to the Aleutian Is.   **Malaxis monophyllos** (L.) Sw.
              12b Lip 4–5 mm, broadest above midlength, rounded but abruptly acute   **Liparis**
          10b Lvs gen > 2, alt or basal; pls not cormose
            13a Lvs basal and rosulate, gen ± mottled, persistent; pls rarely in swampy areas, the roots neither fleshy nor fascicled   **Goodyera**
            13b Lvs cauline, alt, not mottled, non-persistent; pls gen in wet to swampy areas, the roots fleshy and fascicled   **Spiranthes**

Calypso Salisb.   Fairy-slipper; Venus-slipper

Fls showy; lip much larger than the other perianth segms, calceolate, thickened basally, the tip shortly bilobed-horned, the opening to the sac with a distal, suborbicular, flange-like appendage with 3 short rows of bearding leading into the orifice; column concave, winged and petaloid; caps erect; low herb with a subglobose corm, 1 green lf, and a short, 2–3-bracteate scape. (For the sea nymph, *Kalypso*, the word meaning covered or hidden from view). (*Cypripedilum, Cytherea*).

C. bulbosa (L.) Oakes. Scape 5–20 cm; corm ovoid, 1.5–2.5 cm; lf produced at tip of the corm in the fall and persistent through the winter, withering in the summer, bl > petiole, broadly ovate-elliptic to oblong-ovate, mostly 3–6 cm; sepals and 2 petals alike, erect to ascending, (10) 15–22 (25) mm, lanceolate, magenta with 3 darker veins; lip pendent, ca 10 mm wide, whitish (rarely) to yellowish-purple or reddish-purple, variously spotted on the flanged appendage and streaked inside with brownish-purple; deep shade of cool, moist for, from near sea level to midmont; Alas to Lab, s to Cal, Colo, and Ariz; Eurasia *(C. b. f. occidentalis)*.

Corallorhiza Chat.   Coral-root

Fls racemose, short-ped, yellow to reddish-brown or purple; sepals =, lower pair sometimes extending in front of the column and with it forming a gibbosity or very short spur near top of the ovary; lip short-clawed, simple or with lateral lobes near the base, the main lobe or bl gen obovate, entire to crenate, longitudinally ridged on either side of the midvein; column curved above the lip, compressed and concave on the lower side; caps pendent; scapose, gen yellowish to brownish-red pers with extensive coral-like rhizomes, the sts with several membranous bracts, but without geen lvs. (Gr *korallion*, coral, and *rhiza*, root).

1a Sepals and petals pinkish, strongly 3 (4–5) -striped with reddish-brown or purple, (7) 10–17 mm; lip not lobed; spur lacking; deep conif to deciduous for; BC to Sierran Cal, on both sides Cas, e to Que, Mich, Utah, and NM; hooded c., striped c. *(Neottia s.)*   1 **C. striata** Lindl.
1b Sepals and petals yellow or pink to wine-red, not striped although often lightly red-veined, often < 10 mm, lip often lobed; spur often present
  2a Sepals (4.5) 5–6 mm, yellow or greenish-yellow to off-white (occ purplish-tinged), 1-nerved; lip toothed-lobed near the base, almost white, 3.5–4.5 mm; moist shaded areas, mont-subalp; Alas to Lab, s to Wn, ne Ore, Ida, Colo, and NJ; Eurasia; early c., yellow c.   2 **C. trifida** Chat.
  2b Sepals 6–13 mm, often reddish, gen 3-nerved; lip sometimes not lobed
    3a Lip neither lobed nor toothed on the side, white and gen few-spotted, 6–7 (× 4–5) mm, the margin crenulate, upturned; sepals 6–8 mm; moist areas in deep woods; e side RM in c Mont (Lewis & Clark to Park cos), also in Fremont Co, Ida, e to Mo, Pa, Tex, and Fla; Wister c.   3 **C. wisteriana** Conrad
    3b Lip gen lobed or toothed on the side, (6) 7–10 mm, the margin mostly not upturned; sepals (6) 8–10 (13) mm
      4a Column slender, 6–8 mm; spur 1–3 mm; lip deep pink to red, unspotted but gen with 2 reddish-purple callosities near the midnerve; moist conif for; Alas to se BC, Mont, Ida, Wyo, and Cal; western c., Mertens' c. *(C. vancouveriana, C. maculata ssp. m.)*
                                       4 **C. mertensiana** Bong.
      4b Column stout, 3.5–4.5 (5) mm; spur lacking or a mere gibbosity scarcely 1 mm; lip white, gen freely wine-reddish spotted (nonspotted and albino pls not uncommon); moist to fairly dry woods; BC to NS, s to s Cal, NM, Ind, and NC; C Am; Pac c., spotted c. *(C. multiflora* var. *sulphurea, C. leimbachiana, C. hortensis, C. m.* var. *immaculata)*   5 **C. maculata** Raf.

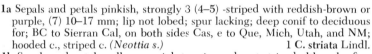

Cypripedium L.   Lady's-slipper

Fls 1–several, showy; sepals spreading, lateral pair fused under the lip, sometimes only the tips free; petals gen similar to the sepals but often slightly longer; lip inflated, pouched, somewhat puckered and inrolled around the orifice; column bent over the orifice; stigma slightly 3-lobed, the lateral lobes often ± covering the fertile anthers; pubescent and gen glandular, caulescent, somewhat rhizomatous herbs with large, flat, often sheathing lvs. (Gr *Kypris*, Venus, and *pes*, foot, in reference to the moccasin-like lip).

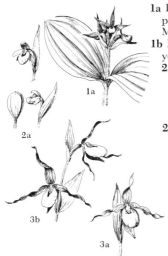

1a Lvs 2, subopp; fls (1) 2–4 in a terminal cluster; st 0.5–2 dm; lip strongly purplish-mottled or -tinged; open conif for; s BC to Cal, E Cas, e to Ida, Mont, and Colo; clustered l.　　　　　　　　　　**1 C. fasciculatum** Kell.

1b Lvs several, alt; fls 1–3, widely spaced; st mostly 1.5–6 dm; lip white or yellow, only lightly purple-tinged

　　2a Sepals green, rarely > 2 cm, obtuse, neither undulate nor twisted, the lower pair often nearly distinct; lip white, but with reddish-purple spots within, 12–15 (20) mm; moist places, often along streams or around hot springs; Alas to Que, s to se BC, n Mont (Flathead and Lewis & Clark cos), and Lake Superior; northern l.　　　　　　**2 C. passerinum** Richards.

　　2b Sepals greenish-yellow to deep purplish-brown, 2.5–6 cm, gen undulate-margined and often twisted, the lower pair fused to near the tip; lip yellow or white, not spotted within, 20–30 mm

　　　　3a Lip yellow, often purple-dotted around the orifice; fls gen 1 (very occ 2); bogs and damp woods; BC to Wn and Ore, e Cas, e to Ida, Wyo, Colo, and much of e Can and US; Eurasia; yellow l.; ours the var. *parviflorum* (Salisb.) Fern. (*C. p.*)　　　　　　**3 C. calceolus** L.

　　　　3b Lip white to purplish-tinged, not purple-dotted; fls gen 2 (1–3); dry to moist shrub- or for-covered valleys and mt sides; Alas to Cal, rarely w Cas, CRG, e to sw Alta, Mont, and Wyo; mt l.　　**4 C. montanum** Dougl.

Eburophyton Heller　　Phantom-orchid; Snow-orchid

Fls several, subsessile, but racemose; lateral sepals slightly spreading, the upper 1 and the 2 petals connivent although distinct; lip saccate to very slightly spurred at base, with erect margins that touch the column, bearing a yellowish gland within, constricted ca midlength, the terminal lobe-like portion broadly ovate-triangular, with 3 prominent callosities leading into the basal sac; column slender, terete, nearly = the lip and arching over it; saprophytic, the lvs reduced to membranous-scarious, sheathing bracts. (L *ebur*, ivory, and Gr *phyton*, plant, referring to the pl color). (*Cephalanthera*).

E. austiniae (Gray) Heller. Pl white throughout, aging to brown, 2–5 dm; fls 5–20; sepals and petals ca =, mostly elliptic-lanceolate, pure white to cream, 10–20 mm; lip somewhat < the sepals, terminal lobe ca 5 × 5–6 mm, ± deflexed, bright yellow on the upper margin leading into the ascending basal sac; column 7–10 mm; moist, gen deep, mostly conif woods; OP and Cas, Wn, to s Cal, also in wc Ida (*C. oregana*).

Epipactis Sw.　　Helleborine

Fls several, bracteate, racemose; sepals subequal, the lower 2 spreading, the upper 1 and the 2 petals subequal, free but ± connivent; lip saccate (not spurred) at the base and with the upper margins rounded distally, abruptly constricted to a triangular bl, declined at ca a 45° angle, flattened distally but trough-like leading into the basal sac; column short; sts 1–several, lfy. (Said to be from *epipaktis*, the ancient Gr name for hellebore). (*Serapias*).

1a Sepals gen < 12 (8–14) mm; lip 10–15 mm, greenish with purplish tinge, not lobed; intro from Europe and occ escaping, reported as estab on VI and in Lewis & Clark Co, Mont (*E. latifolia*)　　　**1 E. helleborine** (L.) Crantz

1b Sepals 12–16 mm; lip 15–20 mm, greenish and purplish, purple-lined, 3-lobed, the basal lobes prominent; streambanks, lake margins, and near springs and seepage, esp near thermal waters, often in des regions; BC s, on both sides Cas, to Baja Cal, w to RM and n Mex; giant h.　　　　　　　　　　　　　　　　　　　　**2 E. gigantea** Dougl.

Goodyera R. Br.　　Rattlesnake-plantain; Lattice-leaf

Fls small, gen in ± 1-sided, tight racemes, ours dull white to ± greenish, glandular-pubescent; upper sepal and lateral petals connivent, forming a hood covering the lip; lateral sepals free, not joined below the tip, a spur lacking; lip unlobed (ours), deeply saccate, the sides erect, the tip pointed and gen slightly reflexed; column rather short, but slenderly beaked; caps erect; scapose, short-rhizomatous, glandular-hairy herbs with few to many alt to basal and rosulate, gen ± mottled lvs. (For John Goodyer, 1592–1664, an English botanist). (*Peramium*).

1a Hood 3–3.7 mm; lip deeply saccate, 3–3.5 mm, the basal pouch 2–3 mm deep; lvs 1–3 cm, 5-nerved but neither white-mottled nor with white midrib; pls 1–2 (2.5) dm; mossy to dryish woods; Alas to Newf, s to BC and in RM to NM, possibly not within our area; northern r., dwarf r.

1 **G. repens** (L.) R. Br.

1b Hood 6–10 mm; lip scarcely saccate, 5–9 mm, the tip slightly recurved and semi-beaked; lvs 3.5–9 cm, gen ± mottled or striped with white, esp along midrib; pls (2) 2.5–4.5 dm; dry to moist woods or for; Alas to NS, s to most of w US and Mich to Me; western r., giant r. (*G. decipiens, G. o.* var. *reticulata*)

2 **G. oblongifolia** Raf.

Habenaria Willd.   Bog-orchid; Rein-orchid

Fls few–∝ in a loose to spikelike raceme, white to yellowish-green or green, often purplish-tinged; upper sepal gen erect or concave and ± hooded, often connivent with upper petals; lateral sepals spreading to ± reflexed; lip pendent to upcurved, entire to lobed or fringed; column rather short; spur well developed, from strongly saccate and didymous (scrotiform) to clavate or narrowly cylindric, straight to strongly curved; glab pers of dry to gen wet areas, often with fleshy or tuberous roots; lvs sometimes 1–2 and basal, but gen several and reduced upward, sometimes withering by anthesis. (L *habena*, reins or narrow strap, in reference to the narrow lip of some spp.). Our spp. with fairly close mutual relationship, but often treated in part under other generic names. (*Coeloglossum, Lysias, Lysiella, Limnorchis, Orchis, Piperia*).

1a Lvs all on the lower 1/3 of the st, often strictly basal; pls ± scapose, the fl st with 1–several reduced bracts on the lower half; fls gen greenish

  2a Spur saccate, scarcely 1 mm; lip 1.5–2 mm, nearly as broad, < the sepals; pls of wet places, 0.5–1.5 (2) dm; lvs 2.5–6 cm; Alas to VI and at one time at Lake Serene, Snohomish Co, Wn; Choriso b.

1 **H. chorisiana** Cham.

  2b Spur at least 2.5 mm; lip at least 2.5 mm, 1/3–3/5 as broad; pls often in dry places, gen > 2 dm

    3a Sepals 1-nerved; lvs 1–5, always at least 5 times as long as broad, borne close together on the lower 1/3 of the st, but not all basal, tending to wither by anthesis; fls subsessile; lip < 6 mm; pls mostly of dry areas

      4a Spur 8–11 mm

        5a Fls greenish, with only the spur whitish; sepals 3.5–4.5 mm; raceme mostly > 10 cm × < 15 mm when pressed; orifice narrowly rectangular or dumb-bell shaped; mostly in dry woods, but occ where moist; sw BC to Cal, e and w Cas, e to nw Mont and Ida; elegant or hillside r. (*P. elongata, P. transversa*)

2 **H. elegans** (Lindl.) Boland.

        5b Fls white, with only the upper sepal and petals at all greenish, the spur green-tipped; sepals 5–5.5 mm; raceme < 10 cm × at least 15 mm when pressed; orifice square to broadly rectangular; cliffs and dry hills where moist early in the spring; islands of PS and along Ore coast to Cal, reported from BC; Greene's b. (*H. maritima*, misapplied)

3 **H. greenei** Jeps.

      4b Spur 3–5 mm; sepals 2–3.5 mm; lip 2.5–4.5 mm; gen in dry woods, to gravelly streambanks and open mt sides; Alas to Que, s and gen to Baja Cal, Nev, Colo, SD, and Ont; Alas r.

4 **H. unalascensis** (Spreng.) Wats.

    3b Sepals 3–several-nerved; lvs 1–2 (3), < 5 times as long as broad, basal, not withering by anthesis; peds (1) 2–7 mm; lip (4) 5–20 mm (if < 6 mm the basal lvs only 1 or rarely 2); pls gen in moist areas

      6a Spur 15–25 mm, clavate, strongly curved; lvs gen 2 (3), ± flat on the ground, gen 6–16 × 3–14 cm, rarely as much as twice as long as broad; lip 10–20 × 2–4 mm; moist mossy for; Alas to Newf, s on both sides Cas to Ore and to Ida, nw Mont, Ind, and Ga; large round-lvd r. (*H. menziesii*)   5 **H. orbiculata** (Pursh) Torr.

      6b Spur 5–8 mm, tapered from the base, slightly curved; lvs gen 1 (2), semi-erect, 3–10 × 1–3 cm, gen > twice as long as broad; lip (4) 5–9 × scarcely 1 mm; damp to wet for, mont; Alas to Newf, s to BC, ne Ore, Ida, Mont, Colo, Wisc, and NY; Europe; blunt-lf r., one-lf r., small n b.   6 **H. obtusata** (Banks) Richards.

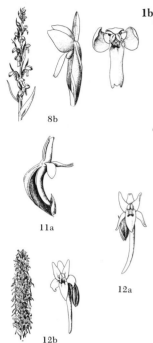

8b

11a

12a

12b

13a

b1

a1

b2

14a

15a

1b Lvs several–∝ , all cauline, the sts lfy most of their length; fls often white

7a Lip shallowly 2–3-lobed or -toothed at the tip; fls mostly greenish

8a Spur > the lip, clavate; lip shallowly and equally 3-lobed; reported for Mont from a single collection believed to have been a cult pl, otherwise well to the n of our area   **H. clavellata** (Michx.) Spreng.

8b Spur 1/4–1/2 as long as the lip, scrotiform; lip unequally 3-lobed, the middle lobe the shortest (sometimes lacking); dryish to moist woods and open mt slopes; Alas to Newf, s to BC and perhaps Wn, and in RM to Colo, e to Ill, W Va, and NC; Eurasia; frog orchis; the N Am pls are known as var. *bracteata* (Muhl.) Gray   **7 H. viridis** (L.) R. Br.

7b Lip neither lobed nor toothed at the tip, although sometimes shallowly lobed behind the tip; fls white to greenish

9a Sepals 1-nerved; sts lfless and not lfy-bracteate above the middle

10a Spur < 1.5 mm; lip rarely as much as 2 mm (see lead 2a)

1 H. chorisiana Cham.

10b Spur at least 2.5 mm; lip at least 2.5 mm

11a Spur 3–5 mm (see lead 4b)   **4 H. unalascensis** (Spreng.) Wats.

11b Spur 8–18 mm

12a Fls greenish, only the spur at all white; sepals 3.5–4.5 mm; raceme mostly > 10 cm × < 15 mm when pressed; orifice narrowly rectangular or dumb-bell shaped (see lead 5a)

2 H. elegans (Lindl.) Boland.

12b Fls white, only the upper sepals and petals greenish centered, the spur green-tipped; sepals 5–5.5 mm; raceme 5–8 cm × at least 15 mm when pressed; orifice square to broadly rectangular (see lead 5b)   **3 H. greenei** Jeps.

9b Sepals 3-nerved, the lateral nerves sometimes obscure; st lfy or lfy-bracteate to above the middle

13a Spur scrotiform, much enlarged near the tip where ± didymous, scarcely at all curved, gen < 2/3 as long (rarely almost as long) as the lip; fls greenish; lip narrowly oblong to elliptic, not much-widened near the base; lowermost st lvs oblong to oblong-elliptic, rounded to obtuse; raceme gen open; wet to boggy places; Alas to sw Alta, s on both sides Cas to n Cal and nw Nev and through RM, Ida and Mont, to NM; slender b. *(H. stricta)*

8 H. saccata Greene

13b Spur linear to clavate, rarely at all scrotiform, gen slightly to strongly curved, < to > the lip, if < 2/3 as long as the lip then fls white and lip gen much-widened near the base; lowermost st lvs often acute; raceme mostly ± congested

14a Fls white or pale greenish; lip rhombic-lanceolate or -oblong, much broadened just above the base and gen abruptly narrowed to a lanceolate or narrowly oblong distal half, often cellular-serrulate along the margin; spur from < the lip to almost twice as long, only slightly if at all clavate; wet to boggy ground; Alas to Greenl, s and gen to Cal, Nev, NM, SD, Mich, Pa, and NY; white b., boreal b., lfy white orchis, bog-candle, scent-bottle; 3 vars. in our area   **9 H. dilatata** (Pursh) Hook.

a1 Spur much < lip; pls mostly < 6 dm; largely RM, esp e Cont Div   var. **albiflora** (Cham.) Correll

a2 Spur slightly < to much > lip; pls 2–10 (13) dm

b1 Spur very slender, strongly curved, gen at least half again as long as the lip; mainly far w, the common phase near the coast, Alas to Cal, occ e to Mont, Ida, and Utah *(H. graminifolium, H. l.)*   var. **leucostachys** (Lindl.) Ames

b2 Spur thicker, not so strongly curved, mostly ca = lip; range of the sp., but only occ in range of other vars. *(H. gracilis, H. leucostachys* var. *robusta)*   var. **dilatata**

14b Fls greenish; lip lanceolate to linear, sometimes broadened at the base but not rhombic, the margin not cellular-serrulate; spur rarely > lip, if so gen strongly clavate

15a Infl almost always open and lax, the lower fls often rather distant; fls relatively large, lip linear to linear-lanceolate, pendent, mostly (6) 7–10 (14) mm; lateral sepals 6–10 mm; column at least 2.5 mm, gen > half as long as the upper sepal; wet to boggy areas; mostly sw Ore to Baja Cal and e to NM, but reported from Skamania Co, Wn, although possibly never there,

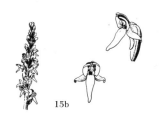

15b

since pls seen from that area are somewhat aberrant but referable to *H. saccata*; canyon b.   10 **H. sparsiflora** Wats.
15b Infl gen dense, the fls mostly strongly overlapping; fls relatively small, the lip gen linear-lanceolate to lanceolate or even
± ovate-lanceolate, mostly not pendent, (3) 4–7 (9) mm; lateral sepals 3–8 mm; column scarcely half as long as the upper
sepal; moist to boggy areas, often on soil derived from limestone; Alas to Newf, s to Cal, Ariz, NM, Neb, and Pa, in our
area almost entirely e Cas; n green b. (*H. borealis* var. *viridiflora*)   11 **H. hyperborea** (L.) R. Br.

Liparis L. C. Rich.   Liparis; Twayblade

Fls several in a minutely bracteate raceme on a naked scape; sepals narrowly lanceolate,
spreading; petals much narrower and somewhat < the sepals, spreading; lip entire (ours) to
lobed, gen declined; spur lacking; small scapose pers with gen 2 basal lvs and an enlarged, cormose base. (Gr *liparos*, fat, supposedly in reference to the rather fleshy, shining lvs).

L. loeselii (L.) L. C. Rich. Pl glab; lvs gen erect, oblong-elliptic, 5–15 cm,
narrowed to winged petioles; scape 7–20 cm, including the rather loosely-fld
raceme; fls white to yellowish-green; sepals 5–7 mm, 3-nerved; petals
1-nerved, 4–5 mm; lip 4–5 mm, oblong-oval, abruptly acute and with a narrower base, curved downward, 5–7-veined; column ca (2) 2.5 mm, very broad
at base; around springs and in bogs; once collected in Klickitat Co, Wn, and
possibly still there, but otherwise only known from well e RM; Europe.

Listera R. Br.   Listera; Twayblade

Fls rather small, greenish to somewhat purplish-green, racemose; sepals and petals similar,
1-nerved, spreading to reflexed; lip somewhat > the sepals, pointing forward and almost horizontal to somewhat declined, rounded to shallowly or deeply bilobed, mostly aur or toothed near
the base and sometimes clawed, spurless, but gen with a small, pocket-like nectary directly under
the stigma; column subterete; pls small, rhizomatous, with 2 broad, sessile, opp or subopp lvs ca
midlength of the st, mostly glandular-pubescent above the lvs but glab below. (For Dr. Martin
Lister, 1638–1711, an English naturalist).

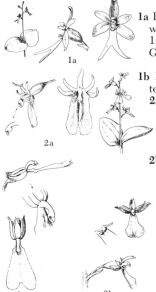

1a Lip cleft ca half the length into linear-lanceolate, ± divergent lobes and
with 2 erect teeth near base; lvs subcordate; column and anther scarcely
1.5 mm; damp to boggy places or sometimes where rather dry; Alas to
Greenl, s to Cal, NM, and NC; Eurasia; heart-lf l. or t. (*L. nephrophylla*)
1 **L. cordata** (L.) R. Br.
1b Lip rounded to retuse or shallowly bilobed, often without basal erect
teeth; lvs rarely at all cordate; column and anther at least 2 mm
2a Lip puberulent, at least along the margins, oblong in outline but narrowed toward the gen aur base, without erect teeth at base; column and
anther at least 3.5 mm; floral bracts scarcely 3 mm; light to deep woods,
where moist or wet, gen mont; Alas to e Can, s to nc Wn (Okanogan Co),
Ida, Mont, Wyo, and Utah; n t.   2 **L. borealis** Morong
2b Lip often not puberulent, gen narrowed considerably and sometimes
clawed, but not aur (sometimes slenderly toothed) near the base; column
and anther gen < 3.5 mm; floral bracts mostly much > 3 mm
3a Column and anther ca 3 (3.5) mm; lip almost horizontal, abruptly contracted to a short claw, the tip prominently retuse, the margin ciliolate; moist woods and streambanks; Alas to Newf, s to Cal, Colo,
Mich, and NY; broad-lipped t.   3 **L. convallarioides** (Sw.) Nutt.
3b Column and anther ca 2 (2.5) mm; lip declined up to a 45° angle,
gradually narrowed to the base, not clawed, the tip rounded to only
very slightly retuse, the margin glab; moist places, lowl to subalp; Alas
to w Alta, s to nw Cal, Ida, and Mont; w or nw t. (*L. retusa*)
4 **L. caurina** Piper

Orchis L.   Orchis

Fls racemose, rather showy; sepals subequal; petals smaller than the sepals, connivent with the upper sepal and forming a hood above the column, lip simple or oblong and ± 3-lobed, slightly connate with the base of the column and produced into a prominent spur; small herbs with lfy sts, or scapose and with 1–few basal lvs. (Gr *orchis,* testicle, the name used by Theophrastus because of the swollen tubers of some spp.).

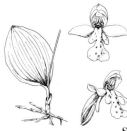

O. rotundifolia Banks. Small round-lvd o. St with 1–3 sheathing bracts below the single, subbasal, sessile to short-petiolate lf; lf bl broadly elliptic to suborbicular, 3–9 cm; sepals (3) 5-nerved, whitish to light pink, 6–11 mm; petals pink, gen 3-nerved; lip white to pinkish, magenta-purple spotted (or streaked), 6–9 mm, oblong in outline, gen prominently lobed on either side ca midlength, then considerably narrowed before flaring to a broad, undulate-crenate and retuse terminal lobe; spur 5–6 mm; column 3–4 mm; wet woods and streambanks where drainage is good, often on limestone; Alas to Greenl, s to s BC, n and c Mont, nc Wyo, and Mich and NY.

Spiranthes L. C. Rich.   Ladies-tresses; Pearl-twist

Fls closely spicate, gen in spiral longitudinal rows, whitish but often tinged with yellow, pink, red, or brownish-purple; sepals free or (ours) connivent with the petals and forming a hood enclosing the column and much of the lip; lip short-clawed or sessile, strongly grooved-concave near the base and ± enclosing the column, spreading to recurved at tip; pls small to rather large, with fleshy, fascicled roots and narrow, basal or cauline lvs reduced above to sheathing bracts. (Gr *speira,* coil, and *anthos,* flower, in reference to the twisted infl). (*Gyrostachys, Ibidium, Orchiastrum*).

S. romanzoffiana Cham. Pl glab, 1–6 dm; lvs gen several, near-basal, linear to narrowly oblong, (5) 8–25 cm × 5–10 (13) mm, abruptly transitional upward to short, sheathing, lanceolate bracts; spike 3–12 (17) cm; sepals viscid-pubescent, the upper one and the lateral petals subequal, 7–12 mm; lip sharply deflexed, ca = the sepals; column 2–4 mm, with a slender, forked, terminal beak; moist to swampy areas; Alas to Newf, s to Cal, Ariz, NM, Ia, and NY; 2 vars. in our area

a1 Lip constricted below the tip, without prominent basal callosities or terminal puberulence; fls white to cream; range of the sp.; hooded l.

var. **romanzoffiana**

a2 Lip nearly triangular, only slightly if at all constricted below the slightly puberulent lip, the base with rather prominent callosities; fls cream to greenish-white; Utah to Cal, w to Marion Co, Ore, and Klickitat Co, Wn; w l. (*S. p.*)

var. **porrifolia** (Lindl.) Ames & Correll

a1

a2

Index to Scientific (Latin) and to Common Names

Common names and accepted generic and specific names are printed in Roman type, names of families in CAPITALS, nonaccepted scientific names (synonyms) in *italics;* new combinations published in this work are shown in **BOLD FACE CAPITALS.**

In general, neither common nor scientific names are listed for the individual species in genera containing fewer than 13 species in our area. Infraspecific names (varieties and subspecies) are indicated by indentation under the appropriate specific names, without further designation of rank.